TABLE 2 Areas and centroids of areas

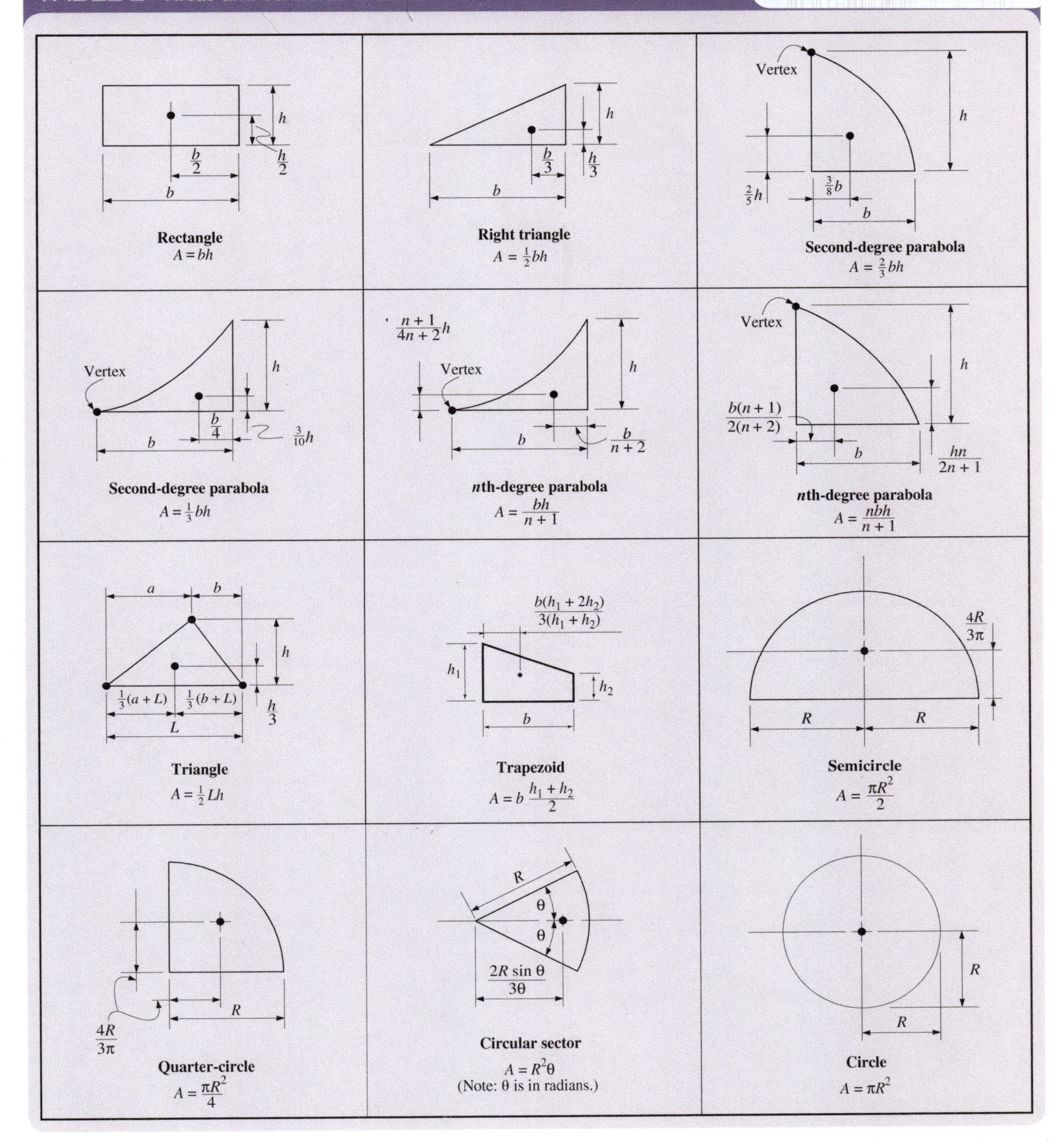

APPLIED STATICS AND STRENGTH OF MATERIALS

APPLIED STATICS AND STRENGTH OF MATERIALS

Sixth Edition

George F. Limbrunner, P.E.
Hudson Valley Community College (Emeritus)

Craig T. D'Allaird, P.E.
Hudson Valley Community College

Boston Columbus Hoboken Indianapolis New York San Francisco
Amsterdam Cape Town Dubai London Madrid Milan Munich Paris Montréal Toronto
Delhi Mexico City São Paulo Sydney Hong Kong Seoul Singapore Taipei Tokyo

Editorial Director: Andrew Gilfillin
Editorial Assistant: Nancy Kesterson
Director of Marketing: David Gesell
Marketing Manager: Darcy Betts
Program Manager: Holly Shufeldt
Project Manager: Janet Portisch
Procurement Specialist: Deidra M. Skahill
Senior Art Director: Diane Ernsberger
Cover Designer: Cenveo Publisher Services
Cover Image: Shutterstock
Full-Service Project Management and Composition: Integra Software Services, Ltd.
Printer/Binder: LSC Communications
Cover Printer: LSC Communications
Text Font: 10/12 Minion Pro Regular

Library of Congress Cataloging-in-Publication Data

Limbrunner, George F.
Applied statics and strength of materials/George F. Limbrunner, P.E., Hudson Valley Community College (Emeritus), Craig T. D'Allaird, P.E., Hudson Valley Community College.—Sixth edition.
pages cm
ISBN 978-0-13-384054-4 (alk. paper)—ISBN 0-13-384054-9 (alk. paper)
1. Statics. 2. Strength of materials. I. D'Allaird, Craig T. II. Title.
TA351.S64 2015
620.1'123—dc23

2014030424

ISBN 10: 0-13-384054-9
ISBN-13: 978-0-13-384054-4

25 2020

BRIEF CONTENTS

CONTENTS

PREFACE

The sixth edition of *Applied Statics and Strength of Materials* presents an elementary, analytical, and practical approach to the principles and physical concepts of statics and strength of materials. It is written at an appropriate mathematics level for engineering technology students, using algebra, trigonometry, and analytic geometry. An in-depth knowledge of calculus is not required for understanding the text or solving the problems.

This book is intended primarily for use in two-year or four-year technology programs in engineering, construction, or architecture. Much of the material has been classroom-tested in our Accreditation Board for Engineering and Technology (ABET) accredited engineering technology programs. The text can also serve as a concise reference guide for undergraduates in a first engineering mechanics (statics) and/or strength of materials course in an engineering program. Although written primarily for technology students, this book can be a valuable reference also for those preparing for state licensing exams in engineering, architecture, or construction.

The book emphasizes mastery of basic principles, since it is this mastery that leads to successful solutions of real-world problems. This emphasis is achieved through numerous, step-by-step example problems, a logical and methodical presentation of material, and selection of topics geared toward student needs. This step-by-step approach to solving problems provides consistent and comprehensive solutions to problems that can be used as references. The principles and applications presented are applicable to many fields of engineering technology including civil, mechanical, construction, architectural, industrial, and manufacturing.

This sixth edition updates the content where necessary and rearranges and revises some of the material to enhance teaching aspects of the text. Some of the changes made to this text include:

- Addition of units to all example problems to assist in student understanding of work.
- Symbols and notations have been changed to reflect commonly used notations, or to provide consistency to applicable design codes. Most notable is the discontinuance of "s" to represent stress. This version of the book will use σ to represent direct stress, τ to represent direct shear stress, and f and F to represent calculated and known bending stresses, respectively.
- Minor typographical errors as well as errors in the selected answers were corrected.
- All shape properties in Appendices A through D, Appendix I, and Appendix J have been updated to match the American Institute of Steel Construction (AISC) *Manual of Steel Construction,* 14th ed.

The book includes the following features:

- Each chapter is prefaced with learning objectives to emphasize the important concepts in the chapter.
- Problems at the end of each chapter are grouped and referenced to a specific section within the chapter. These are followed by a group of supplemental problems. Problems are generally arranged in order of increasing difficulty.
- A summary at the end of each chapter provides a concise reference of the important concepts presented in that chapter.
- Tables of properties of areas and conversion factors for U.S. Customary/SI conversions are printed inside the covers for easy access.
- Most chapters contain computer problems following the section problems. These problems require students to develop computer programs to solve problems pertinent to the topics of the chapter. Any appropriate computer software may be used. The computer problems are another tool with which to reinforce students' understanding of concepts.
- Answers to selected problems are included at the back of the text.
- The primary unit system in this book remains the U.S. Customary System. SI, however, is fully integrated in both the text and the problems. Although full conversion to the metric system in the United States is not likely to happen soon, engineers and technicians must be fluent in both systems to participate in a global market.
- Design and analysis aids are furnished in the appendices, providing data in both U.S. Customary and SI units, to allow users to work though problems without additional references.
- Calculus-based proofs are presented in the appendices.
- The Instructor's Manual includes complete solutions for all the end-of-chapter problems in the text.

There is sufficient material in this book for two semesters of work in statics and strength of materials. In addition, by selecting certain chapters, topics, and problems, the instructor can adapt the book to other situations such as separate courses in statics (or mechanics) and strength of materials.

ACKNOWLEDGMENTS

First and foremost I would like to thank George F. Limbrunner for his trust and confidence in allowing me to contribute to the sixth edition of this book. I learned statics and strength of materials using an earlier version of this textbook. I used this textbook as a reference throughout college, as a practicing engineer, and when preparing for both the Fundamentals of Engineering and the Professional Engineering Licensing exams. Having used this textbook for over 20 years, I am excited to have the opportunity to be a part of future editions of the book.

In addition, I want to thank my fellow faculty at Hudson Valley Community College for their input and feedback in preparing this sixth edition of the book.

Special thanks to my wife and son, Elizabeth and Tommy, for their patience and understanding in working on this project. They are my whole world and I could not have completed this project without their support.

Craig T. D'Allaird

CHAPTER
ONE

INTRODUCTION

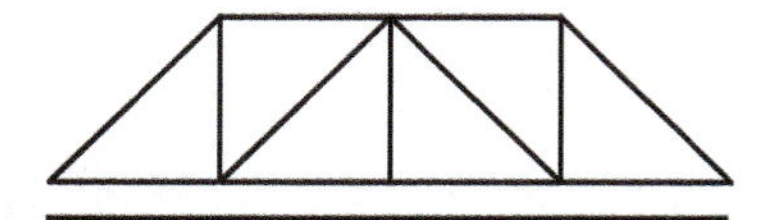

LEARNING OBJECTIVES

Upon completion of this chapter, readers will be able to:

- Develop an understanding of engineering mechanics as it applies to statics.
- Define methods for solving right and oblique triangles mathematically.
- Discuss and use significant figures for engineering purposes.
- Discuss the importance of units in calculations and discuss both U.S. Customary units and the SI unit system. Furthermore, readers should be able to convert between different units and perform dimensional analysis of equations.

1.1 MECHANICS OVERVIEW

Mechanics is the oldest and most fundamental of the physical sciences. Its laws and principles underlie all branches of engineering. In fact, so universal are the applications of mechanics that they often seem nonspectacular, of no compelling interest. Like mathematics, mechanics is sometimes thought of as a necessary evil, a means to perhaps more interesting ends, such as design and analysis or research and development of space vehicles, buildings, bridges, automobiles, aircraft, and the like. Although mechanics is essential to understand and participate in these endeavors, it also deserves attention in its own right.

Mechanics deals essentially with the study of forces and their effects on bodies that are at rest or in motion. Figure 1.1 illustrates the broad categories encompassed by the field of mechanics.

The first eight chapters of this text are concerned solely with solid mechanics and with only that specialized area designated as *statics*. In statics, we consider forces and force systems acting on rigid bodies that are, and that remain, at rest. In the study of statics, it is assumed that all solid bodies (parts of the structure or machine being considered) are perfectly rigid and do not deform, even under large forces. Statics is basic to the understanding of how structural components and complex systems of buildings, bridges, machines, and equipment perform their function. The spectrum of applications ranges from the very simple (e.g., a person standing on a plank that spans a creek) to the highly complex (e.g., aircraft/spacecraft structural systems).

Statics also provides a basic foundation for the study of strength of materials, which may be described as a study of the relationships between external forces acting on solid bodies and the internal responses generated by these forces. Here, solid bodies are assumed to be deformable, not rigid.

To describe the scope and use of statics and strength of materials, consider an example: the design of a structure or machine to serve some definite purpose. Such design almost always involves consideration of the following questions: (a) What are the loads that come upon the structure and its parts? (b) How large, in what form, and of what material should these parts be made so that they may sustain these loads safely and economically?

Some of the loads to be supported may be known at the outset; they may be prescribed by codes or specifications. Some loads may have to be assumed based on engineering experience and judgment. Whatever the case, the principles of statics are used to determine and describe the system of forces that acts on the structure as a whole and on each individual part of the structure.

When all the forces that act on a given part are known, their effect with respect to the physical integrity of the part (that is, their effect in elongating, bending, twisting, compressing, or breaking it) still must be determined. The study of the relations between the forces that act on a body and the changes they produce in its size and form, or the tendency they have to break it, is the province of strength of materials.

Thus, the sequence of statics and strength of materials allows for a beginning understanding of the basic laws and principles involved in both the design and investigation of machine and structural elements.

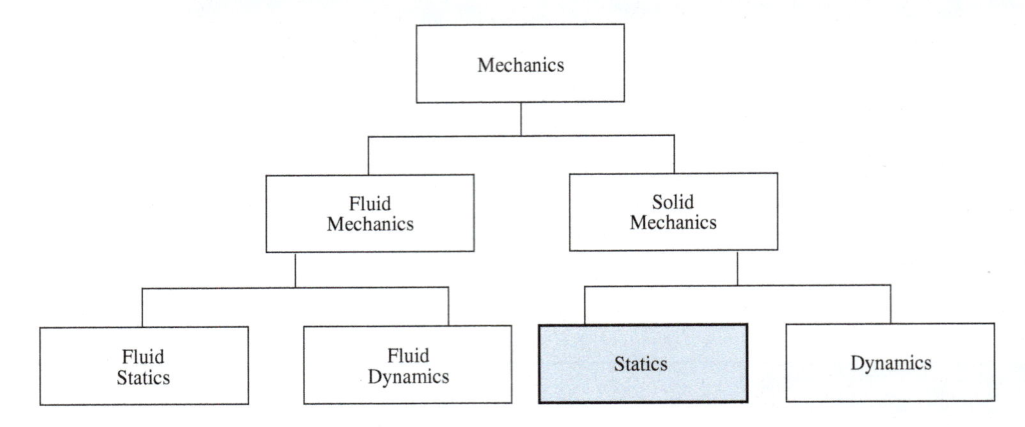

FIGURE 1.1 The field of mechanics.

1.2 APPLICATIONS OF STATICS

Perhaps some of us remember the mistake of teeter-tottering with a bigger kid who laughingly held us aloft at the high end of the plank (see Figure 1.2). Was there any thought, at the time, that we were at the mercy of the principles of statics (rather than at the mercy of the bigger kid)? Probably not. We quickly learned, however, to move the plank so that the pivot point would be farther away and more of the plank would be on our side. We couldn't explain why it worked, but it did, and we added that experience to our accumulated knowledge. What we were actually doing was applying one of the principles of statics.

Many everyday examples may be cited in which the understanding of an application is made possible through the science of what we call statics. A few of these are illustrated in Figure 1.3. All involve the analysis of forces and force systems. The basic understanding of forces in many structures and machines is intuitive, or perhaps based on experience, but a detailed analysis can be made only through the rigorous application of the principles of statics.

One such application is a simple floor system in residential construction that appears straightforward enough at first glance. The principles of statics will make possible a detailed analysis of the magnitude of forces involved and how the forces pass from floor deck to supporting joists, to bearing beams, to posts, and eventually into the building's foundation. Another application of statics is found in the analysis of a truss, a type of structure sometimes used to support roofs or bridges. Trusses are composed of individual members so connected to form triangles (see Figure 1.3). The principles of statics allow one to determine the forces induced in each of the individual members.

FIGURE 1.2 Teeter-totter example.

1.3 THE MATHEMATICS OF STATICS

Statics is an analytical subject that usually requires the physical conceptualization, as well as the mathematical modeling, of a problem. Complicated mathematics is not required in our treatment of the subject. A knowledge of basic arithmetic, algebra, geometry, and trigonometry is sufficient. Because of the importance of directions of forces and the geometric layout of typical structures, familiarity with trigonometry is necessary. A brief review of essential trigonometric relationships follows.

1.3.1 Right Triangles

In Figure 1.4a right triangle ABC is shown. The right angle (90°) at C is indicated. Angles A and B are acute (less than 90°) angles. The sum of the three interior angles is 180°. The sides opposite angles A, B, and C are denoted a, b, and c, respectively. Side c is the hypotenuse of the right triangle, and the other two sides are the legs (or simply, the sides).

The ratios formed between various sides of the right triangle are termed *trigonometric functions* (or *trig functions*) of the acute angles. The functions of importance to us are the sine, cosine, and tangent. These are abbreviated as sin, cos, and tan, and are defined as follows:

$$\sin = \frac{\text{opposite side}}{\text{hypotenuse}}$$

$$\cos = \frac{\text{adjacent side}}{\text{hypotenuse}}$$

$$\tan = \frac{\text{opposite side}}{\text{adjacent side}}$$

From the preceding definitions, and with reference to the right triangle of Figure 1.4a, the following may be written:

$$\sin B = \frac{b}{c} \qquad \cos B = \frac{a}{c} \qquad \tan B = \frac{b}{a}$$

$$\sin A = \frac{a}{c} \qquad \cos A = \frac{b}{c} \qquad \tan A = \frac{a}{b}$$

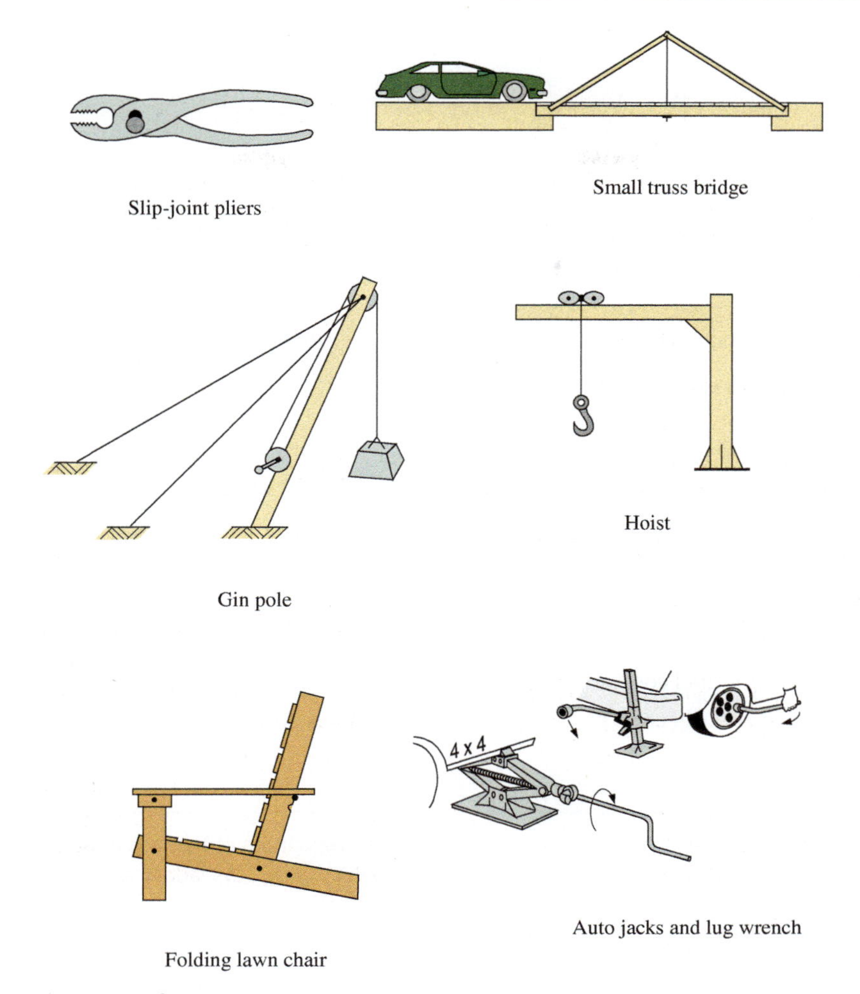

FIGURE 1.3 Everyday applications of statics.

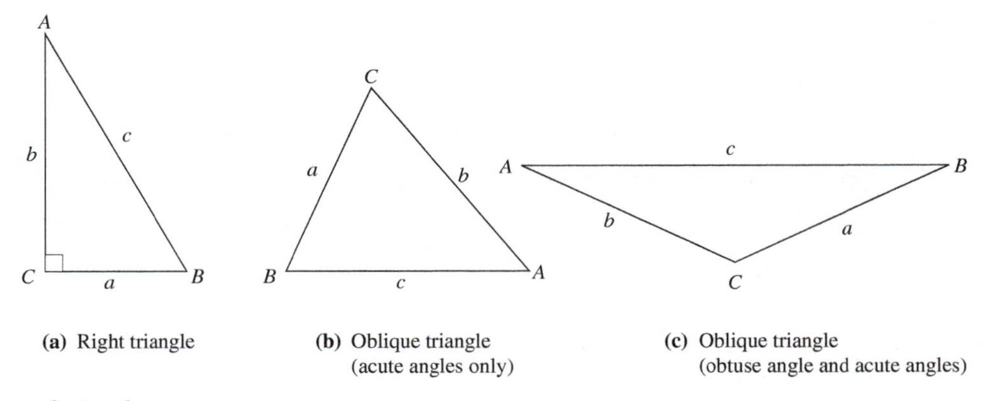

FIGURE 1.4 Types of triangles.

These values are constant for a given angle, regardless of the size of the triangle; they may be obtained from reference books and scientific handheld calculators.

A relationship formulated by Pythagoras, a Greek philosopher and mathematician, gives us another tool for use with right triangles. The Pythagorean theorem states that in a right triangle, the square of the hypotenuse equals the sum of the squares of the other two sides. With reference to Figure 1.4a,

$$c^2 = a^2 + b^2$$

Knowing two sides of a right triangle, or one side and one of the acute angles, the unknown sides and angles can be computed using the Pythagorean theorem and/or the trig functions.

1.3.2 Oblique Triangles

An *oblique triangle* is one in which no interior angle is equal to 90°. It may have three acute (less than 90°) angles, or two acute angles and one obtuse (greater than 90°) angle, as shown in Figure 1.4b and c. As with the right triangle, the sum of the three interior angles is 180°.

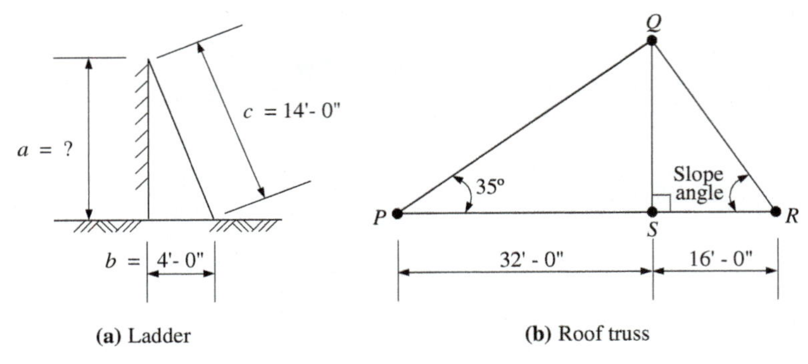

FIGURE 1.5 Mathematics of statics examples.

Knowing three sides, or two sides and the included angle, or two angles and the included side, the unknown sides and angles can be computed using the following laws:

1. The law of cosines:

$$a^2 = b^2 + c^2 - 2bc(\cos A)$$
$$b^2 = a^2 + c^2 - 2ac(\cos B)$$
$$c^2 = a^2 + b^2 - 2ab(\cos C)$$

2. The law of sines:

$$\frac{a}{\sin A} = \frac{b}{\sin B} = \frac{c}{\sin C}$$

The letter designations are shown in Figure 1.4b and c.

The following examples illustrate solutions of both the right triangle and oblique triangle. (Refer to Figure 1.5 for Examples 1.1 and 1.2.)

EXAMPLE 1.1 A 14-ft-long ladder leans against a wall with the bottom of the ladder placed 4 ft from the base of the wall, as shown in Figure 1.5a. How high on the wall will the ladder reach?

Solution The Pythagorean theorem is used:

$$c^2 = a^2 + b^2$$

Rewrite, substitute, and solve for a:

$$a^2 = c^2 - b^2 = (14 \text{ ft})^2 - (4 \text{ ft})^2 = 180 \text{ ft}^2$$

from which

$$a = \sqrt{180 \text{ ft}^2} = 13.42 \text{ ft}$$

Conversion of this result from decimal feet to feet and fractional inch units is discussed in Section 1.5.

EXAMPLE 1.2 For the roof truss shown in Figure 1.5b, determine the height QS, the length of the steep slope QR, and the slope angle at R.

Solution To determine QS, use the triangle PQS:

$$\tan 35° = \frac{\text{opposite}}{\text{adjacent}} = \frac{QS}{32}$$
$$QS = 32 \text{ ft}(\tan 35°) = 22.4 \text{ ft}$$

To determine QR, use the Pythagorean theorem for triangle QRS:

$$(QR)^2 = (QS)^2 + (SR)^2$$
$$= (22.4 \text{ ft})^2 + (16 \text{ ft})^2 = 758 \text{ ft}^2$$
$$QR = 27.5 \text{ ft}$$

Now find the angle at R using any of the three trig functions (since all three sides of the triangle QRS are known):

$$\tan R = \frac{\text{opposite}}{\text{adjacent}} = \frac{22.4 \text{ ft}}{16 \text{ ft}} = 1.40$$

Determine the angle that has a tangent of 1.40. This is called the *arc tangent* of 1.40 and is written as

$$R = \tan^{-1}(1.40)$$
$$R = 54.5°$$

or

$$R = \sin^{-1}\left(\frac{\text{opposite}}{\text{hypotenuse}}\right) = \sin^{-1}\left(\frac{22.4 \text{ ft}}{27.5 \text{ ft}}\right) = 54.5°$$

or

$$R = \cos^{-1}\left(\frac{\text{adjacent}}{\text{hypotenuse}}\right) = \cos^{-1}\left(\frac{16 \text{ ft}}{27.5 \text{ ft}}\right) = 54.4°$$

The slight difference in the solution is due to rounding and can be neglected.

EXAMPLE 1.3 Compute the angle B between cables AB and BC if a force (load) is applied as shown in Figure 1.6.

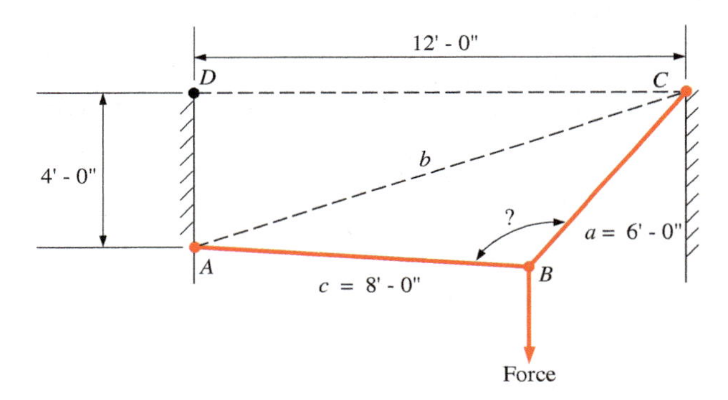

FIGURE 1.6 Cable structure.

Solution The sides of triangle ABC (which is *not* a right triangle) have been designated a, b, and c, as shown. Compute distance AC using right triangle ACD and the Pythagorean theorem:

$$(AC)^2 = (12 \text{ ft})^2 + (4 \text{ ft})^2$$
$$AC = 12.65 \text{ ft}$$

Compute angle B using oblique triangle ABC and the law of cosines:

$$b^2 = a^2 + c^2 - 2ac(\cos B)$$
$$12.65^2 = 6^2 + 8^2 - 2(6)(8)\cos B$$
$$\cos B = -0.625$$

Therefore,

$$B = \cos^{-1}(-0.625) = 128.7°$$

Note that since $\cos B$ is negative, the angle B must lie in the second or third quadrant (where the cosine is negative) and must have a value between 90° and 270°. We select 128.7° because it is apparent that angle B cannot exceed 180°.

EXAMPLE 1.4 A rigging boom is supported by means of a boom cable *BC* as shown in Figure 1.7. Compute the length of the cable and the angle it makes with the boom (angle *C* in oblique triangle *ABC*).

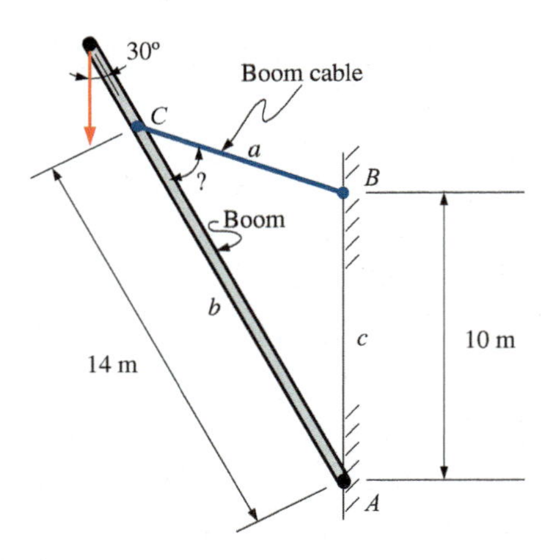

FIGURE 1.7 Cable-supported boom.

Solution The sides of triangle *ABC* are designated *a*, *b*, and *c*, as shown. Compute the length of the boom cable *a* using the law of cosines. Note that the data needed for the law of cosines are two sides and the included angle and that the side to be found is opposite the known angle. Also note that angle $A = 30°$ by alternate interior angles.

$$\begin{aligned} a^2 &= b^2 + c^2 - 2bc(\cos A) \\ &= (14\text{ m})^2 + (10\text{ m})^2 - 2(14\text{ m})(10\text{ m})\cos 30° \\ &= 53.51\text{ m}^2 \end{aligned}$$

from which

$$a = 7.32\text{ m}$$

Then compute the angle that the cable makes with the boom (angle *C*) using the law of sines:

$$\frac{a}{\sin A} = \frac{c}{\sin C}$$

$$\frac{7.32\text{ m}}{\sin 30°} = \frac{10.0\text{ m}}{\sin C}$$

$$\sin C = \frac{10.0\text{ m}(\sin 30°)}{7.32\text{ m}} = 0.683$$

from which

$$C = \sin^{-1}(0.683) = 43.1°$$

In addition to a required familiarity with trigonometry, one must also be familiar with various algebraic manipulations and equations. One type of problem that is often encountered involves the need to solve for two or more unknown quantities that are related by linear equations. Such equations are called *simultaneous equations*.

The following examples will illustrate two solution methods for this type of problem.

EXAMPLE 1.5 An engineer lives 5 mi from his office. In an attempt to get some regular exercise, he decides to ride his bicycle for part of the distance and jog the rest. He knows that he can average 18 mph on the bike and 6.0 mph jogging.

He must get to work in one-half hour. How long should he ride and how long should he jog?

Solution Let

$$x = \text{the length of time to ride (hr)}$$
$$y = \text{the length of time to jog (hr)}$$

The two equations may be expressed as follows:

$$x + y = 0.5 \text{ hr} \quad \textbf{(1)}$$

and

$$18 \text{ mph}(x) + 6 \text{ mph}(y) = 5.0 \text{ mi}$$

from which

$$18x + 6y = 5.0 \text{ mi} \quad \textbf{(2)}$$

Algebraic solution method: To eliminate one unknown, multiply Equation 1 by −6 and then add the two equations:

$$-6x - 6y = -3.0 \quad \textbf{(1)}$$

$$\begin{array}{l} +18x + 6y = +5.0 \\ \hline +12x \quad\quad = +2.0 \end{array} \quad \textbf{(2)}$$

from which

$$x = 0.1667 \text{ hr}$$

Since x + y = 0.5, substitute for *x*:

$$0.1667 + y = 0.5$$
$$y = 0.5 - 0.1667$$
$$y = 0.333 \text{ hr}$$

Substitution method: Solve one of the equations for one of the variables and substitute this expression into the other equation.

Solve Equation 1 for *y*:

$$x + y = 0.5 \text{ hr} \quad \textbf{(1)}$$
$$y = 0.5 - x$$

Substitute this expression into Equation 2:

$$18x + 6y = 5.0 \text{ mi} \quad \textbf{(2)}$$
$$18x + 6(0.5 - x) = 5.0$$
$$18x + 3.0 - 6x = 5.0$$
$$+12x = 2.0$$
$$x = 0.1667 \text{ hr}$$

Then

$$y = 0.5 - x$$
$$y = 0.5 - 0.1667 = 0.333 \text{ hr}$$

1.4 CALCULATIONS AND NUMERICAL ACCURACY

Solutions to problems cannot be more accurate than the engineering data that are used. When dealing with statics problems, we must consider that the dimensions of structural and machine parts and the loads used in the analysis are accurate only to a certain degree. While calculation methods and tools are capable of handling numbers having many digits, this degree of accuracy is usually not warranted.

A *significant digit* is a meaningful digit, one that reflects a quantity that has been measured (or counted) and is thought to be accurate. The accuracy of a number is implied by the significant digits shown. The number 58 has two significant digits, while the number 7 has only one significant digit.

An ordinary electronic calculator will yield the following:

$$\frac{58}{7} = 8.285714286$$

The 10 digits of the result shown, however, indicate an accuracy far greater than that of the numbers going into the calculation (one significant digit in the denominator). Logically the result of a calculation should not reflect an accuracy greater than that of the data from which that result is obtained. Therefore, one should guard against implying unwarranted accuracy in this manner and simply round the result of the preceding calculation to 8 (not 8.0 or 8.00, since these contain two and three significant digits, respectively).

To determine the number of significant digits in a number, begin at the left with the first nonzero digit and count left to right across the number. Stop counting at the last nonzero digit unless any trailing zeros are to the right of the decimal point, in which case they are considered significant. Note that the location of the decimal point does not establish the number of significant digits. For example, each of the following numbers has three significant digits:

4.78 47.8 0.478 0.00478 4.78×10^6 0.470

A predicament may exist with a number such as 47,800. The usual assumption is that the two zeros exist to place the decimal point only and are not significant. Yet the possibility exists that the two zeros were measured or counted and that there are actually five significant digits. One way around this problem is to use exponential notation:

$$478 \times 10^2 \text{(three significant digits)}$$
$$478.00 \times 10^2 \text{(five significant digits)}$$

Although it may be difficult to attest to the accuracy of the data available, it is generally agreed that engineering data are rarely known to an accuracy of greater than 0.2%, which is equivalent to a possible error of 100 lb in a 50,000-lb load:

$$\frac{100}{50{,}000}(100) = 0.2\%$$

Rather than tediously determining 0.2% of each numerical solution, a general rule of thumb for engineering calculations has evolved: *Represent solutions numerically to an accuracy of three significant digits. If the number begins with 1, then use four significant digits.* This rule keeps us true to the spirit (if not within the letter) of the 0.2% guideline.

In rounding numbers, the following method is used:

1. If the digit to be dropped is 5 or greater, the digit to the left is increased by 1. Example: 47.68 becomes 47.7.
2. If the digit to be dropped is less than 5, the digit to the left remains unchanged. Example: 47.62 becomes 47.6.

Adhering to the preceding method, we would round as follows:

18,435.35 rounds to 18,440

2.2321 rounds to 2.23

0.096831 rounds to 0.0968

We have attempted to maintain consistency in our problem solutions by rounding intermediate and final numerical solutions in accordance with this method. For the text presentation, we have used the rounded intermediate solution in the subsequent calculations. When working on a calculator, however, one would normally maintain all digits and round only the final answer. For this reason, the reader may frequently obtain numerical results that differ slightly from those given in the text. Such a difference should not cause undue concern.

1.5 CALCULATIONS AND DIMENSIONAL ANALYSIS

An integral part of the calculation process in mechanics deals with the proper handling of units. In most cases (not all), a unit must be included with a numerical result to accurately describe the quantity in question. If the result is to be a calculated distance, for example, the associated unit must be a length unit (ft, in., m, etc.).

Thoroughness in the calculation process, particularly for the beginner in the mechanics field, should incorporate inclusion of all units in the calculation. For instance, for the simple conversion of 87.3 ft to miles (mi), the calculation is

$$87.3\ \cancel{\text{ft}}\left(\frac{1\text{ mi}}{5280\ \cancel{\text{ft}}}\right) = 0.01653\text{ mi}$$

Notice that the foot units of the original quantity will be canceled by the foot units in the denominator of the conversion factor (within the parentheses), since ft/ft = 1.

Therefore, the resulting unit will be the mile. The canceling process is indicated by strike marks through the units.

In a similar example, when converting 185 yards to miles, using familiar conversions,

$$185\ \cancel{\text{yd}}\left(\frac{3\ \cancel{\text{ft}}}{1\ \cancel{\text{yd}}}\right)\left(\frac{1\text{ mi}}{5280\ \cancel{\text{ft}}}\right) = 0.1051\text{ mi}$$

The inclusion of units is also important (and not only for the beginner) when formulas and calculations become long and complex. When all units are included, they point out any necessary conversions and occasionally also provide clues as to substitution errors. Here are some conversions and equivalencies that the reader may find useful:

1 mile (mi)	=	5280 ft
1 mi^2	=	640 acres
1 acre	=	43,560 ft^2
1 yard (yd)	=	3 ft
1 rod	=	5.50 yd
1 kip (k)	=	1000 lb
1 ton (short ton)	=	2000 lb
1 ft^3 fresh water	=	62.4 lb
1 ft^3	=	7.481 gallons (gal)

EXAMPLE 1.6 Determine the weight (in kips) of water contained in a cylindrical tank having a radius r of 20 in. and a height h of 5 ft. (*Note:* The kip [short for kilopound] is a unit of force and equals 1000 lb.)

Solution The weight of the water is calculated from

$$\text{weight} = \text{unit weight} \times \text{volume}$$

Note that the unit weight of freshwater is 62.4 lb/ft^3. Therefore, the volume must be found in units of cubic feet (ft^3), and any length units of inches must be converted to units of feet. The volume of the cylinder is determined from

$$V = \pi r^2 h$$

Therefore,

$$\text{weight} = \left(62.4\frac{\text{lb}}{\text{ft}^3}\right)(\pi)(20\text{ in.})^2\left(\frac{1\text{ ft}}{12\text{ in.}}\right)^2(5\text{ ft})\left(\frac{1\text{ kip}}{1000\text{ lb}}\right)$$

$$= 2.72\text{ kips}$$

Note in the final equation of Example 1.6 that the fourth term on the right side converts the square of the radius (in.^{2}) to square feet (ft^2).

When using the U.S. Customary (inch-pound) System of units for construction applications, it is common to express dimensions in feet and fractional inches. Calculations, however, are carried out using decimal feet (or decimal inches).

EXAMPLE 1.7 In Example 1.1, the solution for the vertical leg of the right triangle was 13.42 ft. This was a rounded value. Convert this decimal feet dimension to feet and fractional inches.

Solution

1. The 13.42 ft dimension contains 13 full feet and this will not change.
2. The remaining 0.42 ft must be converted to decimal inches.

$$0.42\text{ ft} \times 12\text{ in./ft} = 5.04\text{ in.}$$

3. Thus far, then, we have 13 ft and 5 (full) inches, and we need now to convert the remaining 0.04 in. to fractional inches. The accuracy of the fraction (the number used for the denominator of the fraction) depends on the application. For instance, in structural steel fabrication, dimensions to $\frac{1}{16}$ in. are common. To convert 0.04 in. to sixteenths of an inch, multiply by 16 (there are 16 sixteenths of an inch per inch):

$$0.04\text{ in.} \times 16 = 0.64\,(\text{sixteenths of an inch})$$

and round 0.64 (sixteenths) to 1 (sixteenth).

4. Therefore, summing the results of steps 1, 2, and 3:

$$13.42\text{ ft} = 13\text{ ft} + 5\text{ in.} + \tfrac{1}{16}\text{ in.} = 13\text{ ft} - 5\tfrac{1}{16}\text{ in. or } 13' - 5\tfrac{1}{16}''$$

(*Note*: The dash is common usage and should not be interpreted as a minus sign.)

If we want to show 13.42 ft to an accuracy of $\frac{1}{64}$ in., step 3 would change to:

$$0.04 \times 64 = 2.56\,(\text{sixty-fourths})$$

Therefore,

$$13.42\text{ ft} = 13\text{ ft} - 5\tfrac{3}{64}\text{ in.}$$

EXAMPLE 1.8 Convert 13 ft − $5\frac{3}{64}$ in. to decimal feet.

Solution

1. The 13 full feet dimension does not require conversion.
2. Convert 5 inches to decimal feet:

$$5 \text{ in.} \times \frac{1 \text{ ft}}{12 \text{ in.}} = 0.4167 \text{ ft}$$

3. Convert the fractional inch part of the dimension to decimal feet:

$$\frac{3}{64} \text{ in.} \times \frac{1 \text{ ft}}{12 \text{ in.}} = 0.0039 \text{ in.}$$

4. Sum steps 1, 2, and 3 for the result:

$$13 \text{ ft} - 5\tfrac{3}{64} \text{ in.} = 13 \text{ ft} + 0.4167 \text{ ft} + 0.0039 \text{ ft} = 13.42 \text{ ft}$$

Sometimes it is helpful to check the units (or dimensions) in an equation by substituting only the units for each term and then carrying out the algebraic operations. The dimensional results on the left and right side of the equation must be the same:

$$\text{weight (kips)} = \left(\frac{\text{lb}}{\text{ft}^3}\right)(\text{in.}^2)\left(\frac{\text{ft}^2}{\text{in.}^2}\right)(\text{ft})\left(\frac{\text{kips}}{\text{lb}}\right) = \text{kips}$$

(O.K.)

Therefore, the units are consistent. This procedure is sometimes called *dimensional analysis*. That the units cancel out, leaving only kips, is a requirement (but does not guarantee a correct numerical result).

The preceding discussion applies when one is dealing with fundamental physical equations that derive from physical laws. Any system of units can be used. Normally, calculations will be performed using the unit system indicated in the given (or known) data. With proper conversions, however, one may utilize any valid system of units.

1.6 SI UNITS FOR STATICS AND STRENGTH OF MATERIALS

The U.S. Customary System of weights and measures is used as the primary unit system in this text. The replacement of the U.S. Customary System with the metric system in the United States will not occur in the near future, although the global economy is providing increased pressure to do so. The change to the metric system has progressed at different rates within the various industrial fields. In the U.S. construction industry, the conversion to the metric system has been faltering. Nevertheless, some construction projects and many products and machines are designed and produced in metric units. Therefore, engineers and technologists must be familiar with both unit systems.

The metric system, in various forms, has been in use for centuries. In 1960 the International General Conference on Weights and Measures adopted an extensive revision and simplification of the system. The name *Le Système International d'Unités* (International System of Units), with the international abbreviation SI, was adopted for this modern metric system.

The SI consists of a limited number of base units that correspond to fundamental quantities and a large number of derived units (derived from the base units) to describe other quantities. SI base units and derived units pertinent to statics and strength of materials are listed in Tables 1.1 and 1.2.

Because the orders of magnitude of many quantities in the SI cover wide ranges of numerical values, prefixes are used to deal with decimal point placement. SI prefixes representing steps of 1000 are recommended to indicate orders of magnitude. Those recommended for use in statics and strength of materials are listed in Table 1.3. Referring to Table 1.3, it is clear that one has a choice of ways to represent numbers. It is preferable to use numbers between 0.1 and 1000 by selecting the appropriate prefix. For example, 18 m (meters) is preferred to 0.018 km (kilometers) or 18,000 mm (millimeters).

In the presentation of numbers, the recommended international practice is to set off groups of three digits with a gap as shown in Table 1.3. Note that this method is used on both the left and right sides of the decimal marker for any string of five or more digits. A group of four digits on either side of the decimal marker need not be separated.

TABLE 1.1 SI base units and symbols

Quantity	Unit	SI Symbol
Length	meter	m
Mass	kilogram	kg
Time	second	s
Angle[a]	radian	rad
Temperature[b]	kelvin	K

[a]It is also permissible to use the arc degree and its decimal submultiples when the radian is not convenient.

[b]It is also permissible to use degrees Celsius (°C).

TABLE 1.2 SI derived units

Quantity	Unit	SI Symbol	Formula
Acceleration	meter per second squared	—	m/s^2
Area	square meter	—	m^2
Density (mass per unit volume)	kilogram per cubic meter	—	kg/m^3
Force	newton	N	$\frac{kg \cdot m}{s^2}$
Pressure or stress	pascal	Pa	N/m^2
Volume	cubic meter	—	m^3
Section modulus	meter to third power	—	m^3
Moment of inertia	meter to fourth power	—	m^4
Moment of force (torque)	newton meter	—	$N \cdot m$
Force per unit length	newton per meter	—	N/m
Mass per unit length	kilogram per meter	—	kg/m
Mass per unit area	kilogram per square meter	—	kg/m^2
Energy or work	joule	J	$N \cdot m$
Power	watt	W	J/s or $\frac{N \cdot m}{s}$
Rotational speed	radian per second	—	rad/s

TABLE 1.3 SI prefixes

Prefix	SI Symbol	Factor	Example
giga	G	$1\,000\,000\,000 = 10^9$	Gm = 1 000 000 000 meters
mega	M	$1\,000\,000 = 10^6$	Mm = 1 000 000 meters
kilo	k	$1000 = 10^3$	km = 1000 meters
milli	m	$0.001 = 10^{-3}$	mm = 0.001 meter
micro	μ	$0.000\,001 = 10^{-6}$	μm = 0.000 001 meter

It should be noted that in the U.S. Customary System, the terms *force* and *weight* are used interchangeably and are expressed in the same units of measure (e.g., pounds, kips, tons). In the SI, force is derived from the base unit of mass. Mass represents a quantity of matter in an object and is expressed in kilograms, whereas force is expressed in newtons. When the term *weight* is used, it should not be confused with mass. Weight is defined as the force of gravity on a body. Since gravity may vary, the weight of a body may vary. Weight may be computed using Newton's second law of motion, which may be stated mathematically as

$$F = ma$$

where

F = force
m = mass
a = acceleration

For the determination of weight, which has been defined as the force of gravity on a body,

$$\begin{aligned} \text{weight} &= \text{force of gravity on a body} \\ &= (\text{mass of the body}) \times (\text{acceleration of gravity}) \end{aligned}$$

or

$$\text{weight} = mg$$

where g is the acceleration of gravity.

For the determination of the weight of a given mass for purposes of analysis, we will use a value of 9.81 m/s^2 for the acceleration of gravity. This approximate value is sufficiently accurate for our calculations. The acceleration of gravity varies slightly over the surface of the earth. For the types of problems in this text, the variation can be neglected.

As an example, for a truck carrying 500 kg (kilograms) of stone, the weight of the stone can be obtained from

$$\begin{aligned} \text{weight} &= mg = (500\text{ kg})(9.81\text{ m/s}^2) \\ &= 4910\text{ kg}\cdot\text{m/s}^2 \\ &= 4910\text{ N} \\ &= 4.91\text{ kN} \end{aligned}$$

Note that the preceding reflects that 1 $\text{kg}\cdot\text{m/s}^2$ is defined as one newton (N).

Table 1, inside the front cover of this book, is provided to enable rapid conversion from the U.S. Customary System to SI units. This table includes those quantities frequently used in statics and strength of materials.

EXAMPLE 1.9 A steel beam, shown in Figure 1.8a, weighs 48 lb/ft. Determine the weight per unit length in SI units.

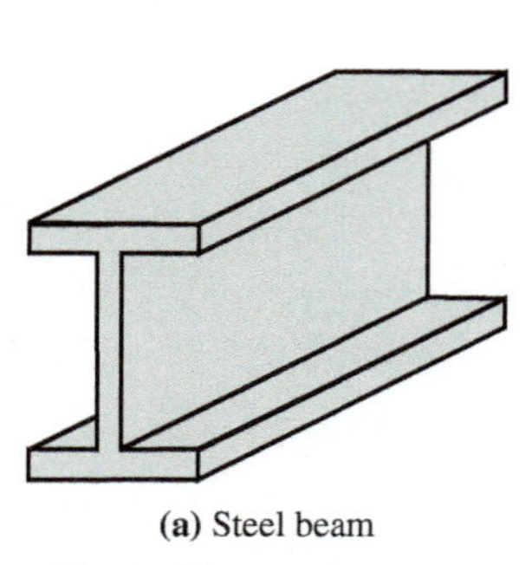

(a) Steel beam

(b) Reinforced concrete beam

FIGURE 1.8 Typical beams

Solution The weight of lb/ft in the U.S. Customary System will be converted to N/m (newtons per meter) in the SI. In the following expression, the second and third terms will convert pounds to kilograms and feet to meters, which will yield mass per unit length. The fourth term will convert the mass to force (weight) using $F = ma$. All conversion factors are from Table 1 inside the front cover.

$$48\frac{\text{lb}}{\text{ft}}\left(\frac{0.4536\text{ kg}}{1\text{ lb}}\right)\left(\frac{1\text{ ft}}{0.3048\text{ m}}\right)\left(9.81\frac{\text{m}}{\text{s}^2}\right) = 701\frac{\text{kg}\cdot\text{m}}{\text{m}\cdot\text{s}^2} = 701\text{ N/m}$$

or, using the direct conversion factor from Table 1,

$$48\text{ lb/ft} \times 14.594 = 701\text{ N/m}$$

EXAMPLE 1.10 In the design of a reinforced concrete beam such as that shown in Figure 1.8b, the weight of the beam itself is always considered as a load, or force, per unit length of span. In the U.S. Customary System, reinforced concrete is usually assumed to weigh 150 lb/ft^3 (pcf).

Calculate the load per unit length of beam in the SI if the width of the beam is 500 mm and the depth of the beam is 1000 mm.

Solution Use the conversion factors from Table 1 to determine mass per unit volume:

$$150\frac{\text{lb}}{\text{ft}^3}\left(\frac{1\text{ ft}}{0.3048\text{ m}}\right)^3\left(\frac{0.4536\text{ kg}}{1\text{ lb}}\right) = 2403\text{ kg/m}^3$$

The value of 2403 kg/m^3 represents a mass per unit volume where the unit volume is one cubic meter. To obtain a force or weight per cubic meter, Newton's law must be applied:

$$\begin{aligned} \text{weight} = mg &= \left(2403\frac{\text{kg}}{\text{m}^3}\right)\left(9.81\frac{\text{m}}{\text{s}^2}\right) \\ &= 23\,573\frac{\text{kg}\cdot\text{m}}{\text{m}^3\cdot\text{s}^2} = 23.6\frac{\text{kN}}{\text{m}^3} \end{aligned}$$

Since the beam dimensions are 500 mm by 1000 mm (or 0.500 m by 1.000 m), the weight or load per unit length can be calculated from

$$(0.500\text{ m})(1.000\text{ m})23.6\frac{\text{kN}}{\text{m}^3} = 11.80\text{ kN/m}$$

Thus, the load per unit length of one meter, due to the weight of the beam, is 11.80 kN.

SUMMARY BY SECTION NUMBER

1.1 Statics deals with forces and force systems acting on rigid bodies at rest.

1.3 A right triangle may be solved using the Pythagorean theorem and/or trigonometry functions (sin, cos, tan). An oblique triangle may be solved using the law of cosines or the law of sines.

1.4 Numerical accuracy of calculations will be considered sufficient, for our purposes, if solutions are rounded to three significant digits. If the number begins with 1, then use four significant digits. When working on a calculator, maintain all digits and round only the final answer.

1.5 Dimensional analysis is a process used in computations to establish that units on each side of an equation are of the same dimensional form. Inclusion of all units in calculations is recommended.

1.6 Engineers and technologists must be familiar with both the U.S. Customary and the metric system of units. Both unit systems are used in this text. Conversion factors from U.S. Customary to SI units are inside the front cover of this book.

PROBLEMS

Section 1.3 The Mathematics of Statics

1.1 In the right triangle shown,
(a) $a = 10$ ft, $b = 7$ ft, find c.
(b) $c = 20$ m, $a = 16$ m, find b.

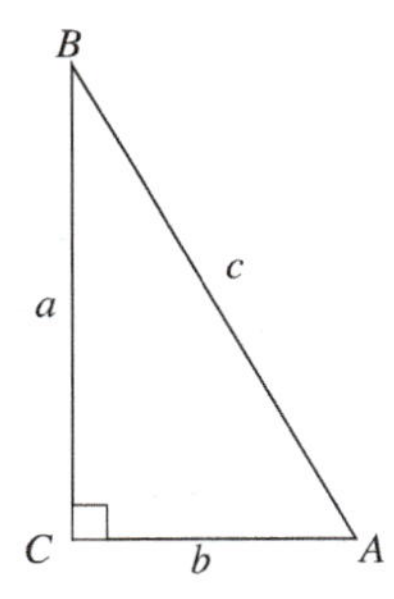

PROBLEM 1.1

1.2 In the right triangle ABC shown, $c = 25$ ft and angle $A = 48°$.
(a) Determine side a.
(b) Determine side b.
(c) Determine height h.

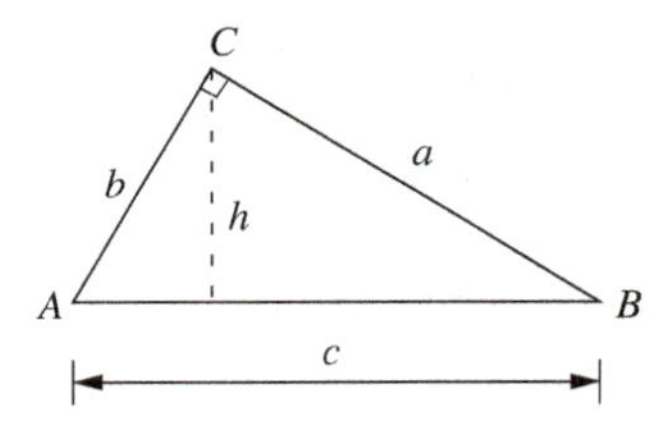

PROBLEM 1.2

1.3 Determine the length of side a and the measure of the two acute angles.

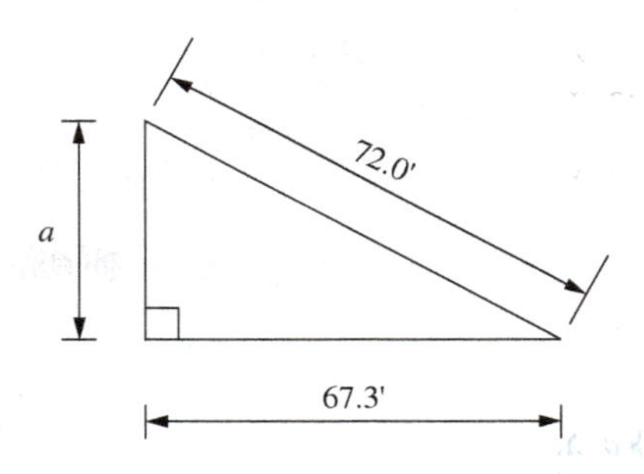

PROBLEM 1.3

1.4 A 28-ft-long ladder leaning against a wall makes an angle of 70° with the ground. Determine the vertical height to which the ladder will reach.

1.5 Two legs of a right triangle are 6 ft and 10 ft. Determine the length of the hypotenuse and the angle opposite the short side.

1.6 In the roof truss shown, determine the lengths of the top chords AB and BC and find the angles at A and C.

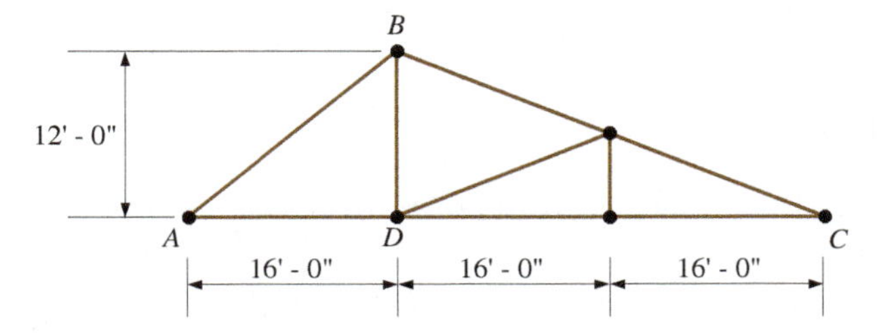

PROBLEM 1.6

1.7 A level scaffold is to be supported at the center of a 12-ft ladder, as shown. Determine the required length x and the angle θ.

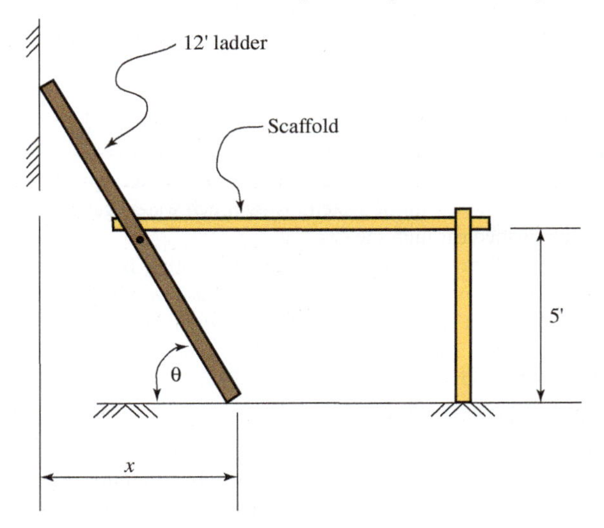

PROBLEM 1.7

1.8 A wanderer hikes due east for 2 mi, then southeast for 3 mi, then due north for 6 mi, then southwest for 6 mi. She then notices a storm gathering. How far is she from her starting point?

1.9 Freehand sketch the following triangles and solve completely. (Find the unknown sides and/or angles.)
(a) $a = 11$ ft, $b = 13$ ft, $C = 80°$
(b) $b = 78$ ft, $c = 85$ ft, $A = 72°$
(c) $a = 7$ ft, $b = 24$ ft, $c = 25$ ft

1.10 One side of a triangular lot is 100 ft and the angle opposite this side is 55°. Another angle is 63°. Sketch the shape of the lot and determine how much fencing is needed to enclose it.

1.11 A box containing 8 shock absorbers and 10 brake pad sets weighs 101.6 lb. Another box containing 10 shock absorbers and 6 brake pad sets weighs 106.2 lb. Determine the weight of one shock absorber and one set of brake pads. Neglect the weight of the boxes.

1.12 Find the cable lengths *AB* and *BC*, the angle *C*, and the angle at *B* (angle *ABC*).

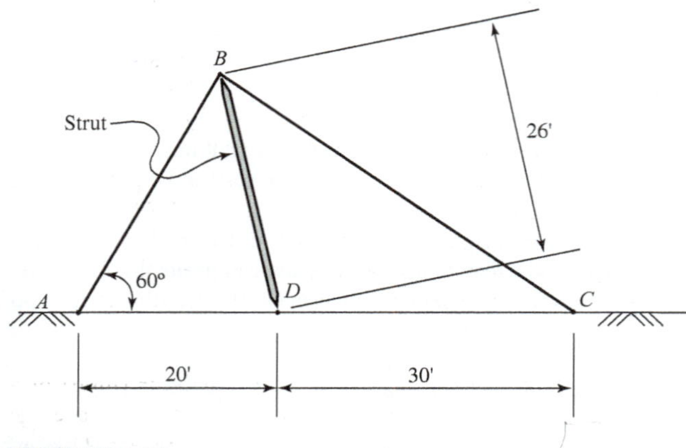

PROBLEM 1.12

Section 1.5 Calculations and Dimensional Analysis

1.13 Express 0.015 ton **(a)** in pounds (lb) and **(b)** in ounces (oz).

1.14 Express 5 mi **(a)** in yards (yd) and **(b)** in feet (ft).

1.15 Express 60 miles per hour (mph) in units of feet per second (fps).

1.16 Find the number of square rods in one acre.

1.17 A reservoir contains an estimated 125 billion gallons of water.
- **(a)** Express this volume in acre-ft.
- **(b)** Estimate the weight of this volume of water in tons.

1.18 Convert the following:
- **(a)** 27 ft – $7\frac{3}{4}$ in. to decimal feet (nearest 0.01 ft)
- **(b)** 1.815 ft to feet and fractional inches (nearest $\frac{1}{32}$ in.)

1.19 A water system of 2 in. (inside diameter) pipe, having a length of $\frac{1}{4}$ mile, is to be flushed with a volume of water equal to twice that contained in the system. How much water (gallons) must be flushed through the system?

1.20 The soil pressure under a concrete footing is calculated to be 2 tons per square foot. Express this value in pounds per square foot (psf) and pounds per square inch (psi).

Section 1.6 SI Units for Statics and Strength of Materials

1.21 Using Table 1, establish your height in millimeters (mm), your weight in newtons (N), and your mass in kilograms (kg).

1.22 The dimensions on a baseball field are as follows: pitcher's mound to home plate, 60.5 ft; first base to second base, 90.0 ft; and home plate to center field fence, 413.0 ft. Calculate the SI equivalents in meters.

1.23 Convert the following quantities from U.S. Customary System units to the designated SI units:
- **(a)** 2212 lb to kg
- **(b)** 87 ft^2 to mm^2 and m^2
- **(c)** 18,460 psi to MPa and kPa

1.24 Convert the following quantities to SI units of millimeters and meters:
- **(a)** 18.6 ft
- **(b)** 8 ft – 10 in.
- **(c)** $27\frac{1}{2}$ in.

1.25 Convert the following quantities to SI units of newtons and kilonewtons:
- **(a)** 248 lb
- **(b)** 3.65 kips
- **(c)** 8.7 tons

1.26 Convert the following quantities to the designated SI units:
- **(a)** 627 in^2 to mm^2 and m^2
- **(b)** 14 yd^2 to mm^2 and m^2
- **(c)** 3.5 kips/ft to kN/m
- **(d)** 8470 psi to Mpa and kPa
- **(e)** 1740 psf to Pa and kPa
- **(f)** 2.8 kips/ft^2 to kPa
- **(g)** 247 pcf to kN/m^3

1.27 Rework Problem 1.19 using an inside pipe diameter of 50 mm and a system length of 400 m. Find the volume of flushing water in liters.

1.28 A spherical tank has a radius of 27 in. Find the volume of the tank in **(a)** cubic inches (in^3), **(b)** cubic millimeters (mm^3), **(c)** cubic meters (m^3), and **(d)** liters.

Supplemental Problems

1.29 A ladder rests against a vertical wall at a point 12 ft from the floor. The angle formed by the ladder and the floor is 63°. Calculate **(a)** the length of the ladder, **(b)** the distance from the base of the wall to the foot of the ladder, and **(c)** the angle formed by the ladder and the wall.

1.30 A surveyor measures a zenith angle of 70° 03′20″ to the top of a flagpole and a horizontal distance to the flagpole of 93.7 ft as shown. The height of the transit is 6.00 ft. Calculate the height of the flagpole.

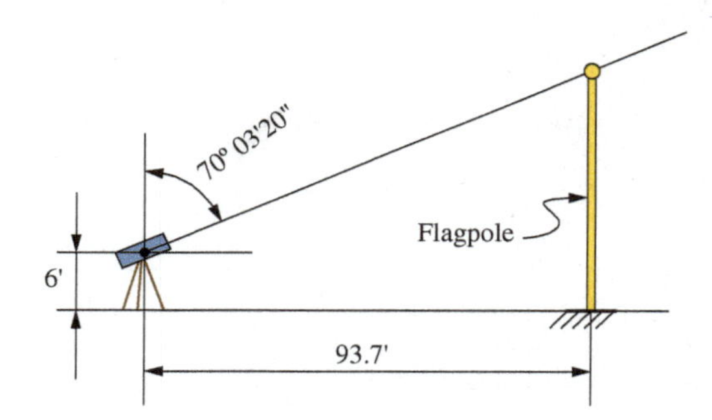

PROBLEM 1.30

1.31 The tower of a tower crane casts a shadow (on level ground) of 32 m when the sun is 55° above the horizon. What is the height of the tower?

1.32 From the top of a 30-ft building, the angle of depression to the foot of a building across the street is 17° and the angle of elevation to the top of the same building is 60°. How tall is the building across the street? (Assume level ground.)

1.33 Two planes, starting from airport *X*, fly for 2 hr, one at 800 km/hr and the other at 1050 km/hr. The slower plane flies in a direction of 60° east of north and the faster plane flies in a direction of 10° south of east, with reference to *X*. How far apart are the planes at the end of the 2 hr period?

1.34 Rework Problem 1.31 assuming that the ground slopes upward at 1:5 from the base of the crane to the end of the shadow, as shown.

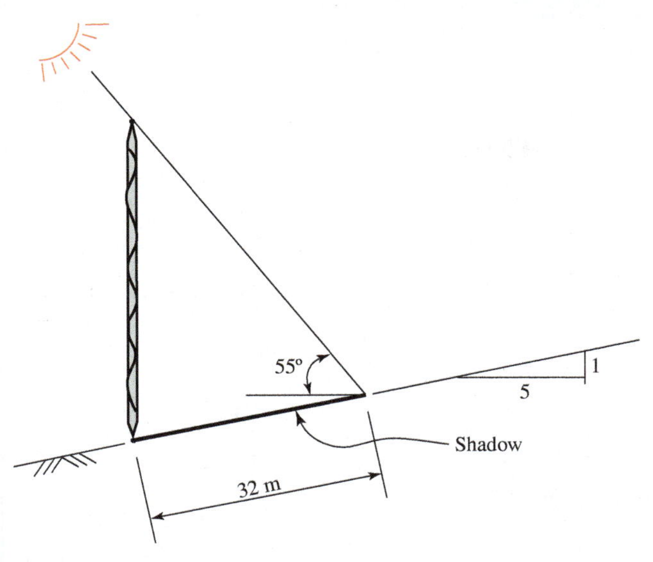

PROBLEM 1.34

1.35 An Egyptian pyramid has a square base and symmetrical sloping faces. The inclination of each sloping face is 53°07′. At a distance of 600 ft from the base, on level ground, the angle of inclination to the apex is 27°09′. Find the vertical height and the slant height of the pyramid and the width of the pyramid at its base.

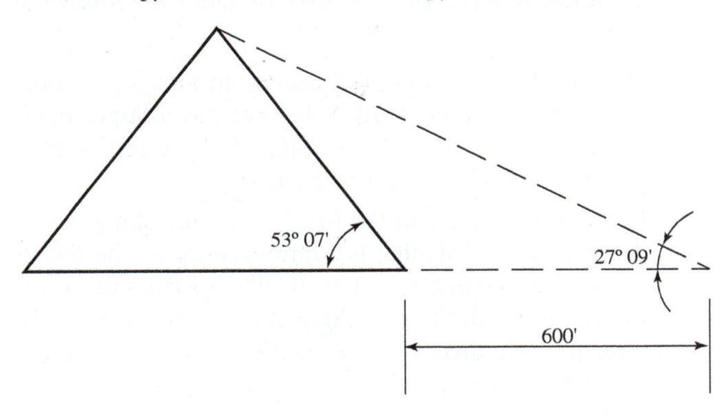

PROBLEM 1.35

1.36 Two observation towers *A* and *B* are located 300 ft apart, as shown. An object on the ground between the towers is observed at point *C*, and the observer on tower *A* notes that the angle *CAB* is 68° 10′. At the same time, the observer on tower *B* notes that the angle *CBA* is 72° 30′. Assuming a horizontal surface and towers of equal height, how far is the object from each tower and how high are the towers?

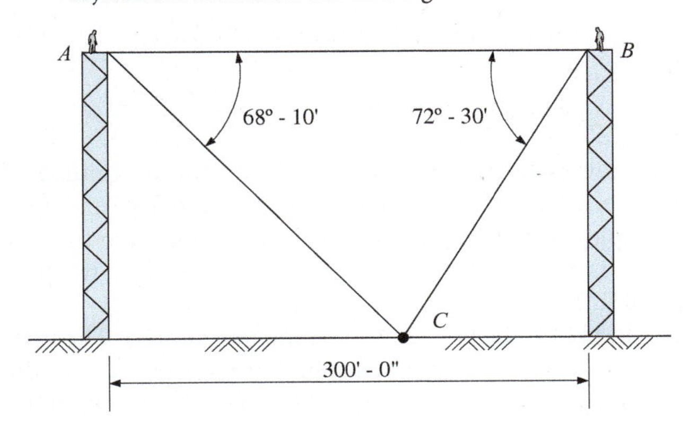

PROBLEM 1.36

1.37 Two people fishing from rowboats on a lake are 100 m apart and have hooked their lines into the same sunken log. The first has 85 m of line out, and the second has 68 m of line out. Assuming that the lines both lie in the same vertical plane, meet at a point (on the log), and are taut, how deep is the sunken log?

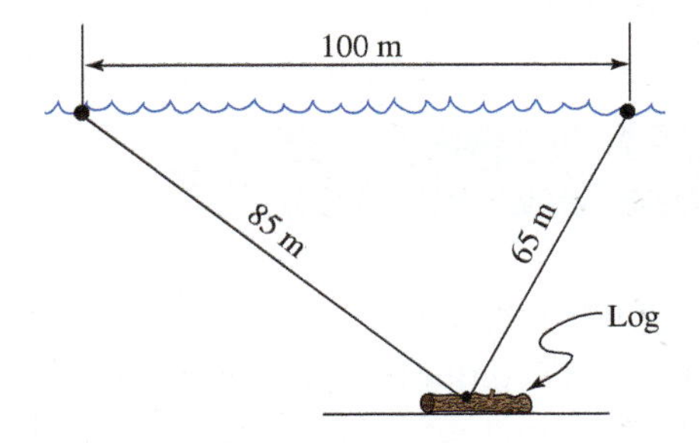

PROBLEM 1.37

1.38 What is the average speed in miles per hour (mph) of a runner who runs a mile in 4 min? What is the average speed (mph) of a runner who runs a 26-mi race in 2 hr and 47 min?

1.39 The volume flow of a river is expressed as 5000 cubic feet per second (cfs). Express the flow in gallons per minute and cubic miles per year.

1.40 A 55-gal drum filled with sand weighs 816 lb. The empty drum weighs 38 lb. Find the unit weight of the sand in pounds per cubic foot (pcf).

1.41 The total area to be occupied by a building and a parking lot is 90,000 ft^2. If the area occupied by the parking lot is to be $\frac{1}{2}$ acre plus three times the area occupied by the building, what is the area occupied by each?

1.42 An electronic distance measurement device (EDM) is to be used to lay out a canoe race course on a lake. Buoys *B* and *C* have been placed. The EDM is set up at point *A* as shown. *AB* is 175.83 m and *AC* is 287.21 m. The angle at *A* is 42°37′20″.

(a) Find the length *CB*.

(b) Keeping the angle at *A* and the line *AB* unchanged, find the required length of *AC* (to 0.01 m) so that *BC* is 200.00 m.

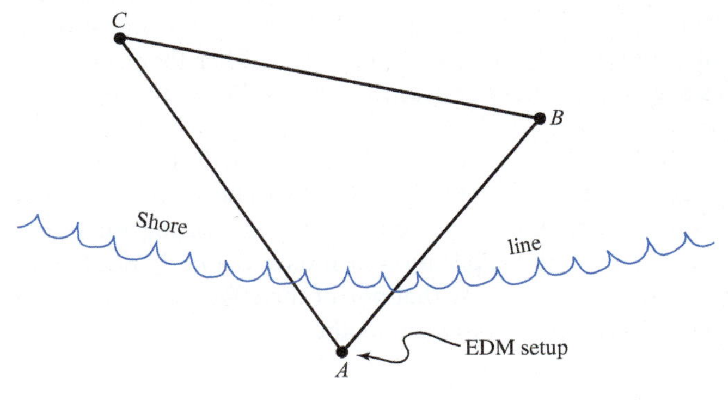

PROBLEM 1.42

CHAPTER TWO

PRINCIPLES OF STATICS

LEARNING OBJECTIVES

Upon completion of this chapter, readers will be able to:

- Define and describe the four qualities of a force.
- Discuss the units of force and multiples of those units in both U.S. and SI systems.
- Identify the types of forces (loads) that act on a structure.
- Discuss the principles of forces and how they affect a body.
- Identify and analyze various force systems to determine properties of those forces.

2.1 FORCES AND THE EFFECTS OF FORCES

Statics has been previously described as the science that treats the relationships between forces acting on rigid bodies at rest. With this description in mind, it is sufficient to define a force as a push or a pull exerted on one body by another. A more precise definition of a force (although outside the realm of statics) would be a push or pull tending to produce a change in the motion of the body acted upon. This definition applies to the external effects of a force.

In general, forces have two effects on a body: (1) to cause it to move if it is at rest or to change its motion if it is already moving and (2) to deform it. In statics we are not concerned with the deformation of a body. The deformation and the internal behavior of a body when subjected to a force lie within the province of strength of materials. Also, the study of the relationships between forces and motion lies within the province of dynamics, rather than statics. In statics our primary concern is with bodies at rest (or moving with zero acceleration), which we will hereafter refer to as bodies in equilibrium. In statics, then, the effect of a force may be described as a tendency to preserve or to upset equilibrium. Also, since supporting or reacting forces are produced by the application of forces to a body not free to move, we may also say that one of the effects of a force is to produce, or bring into action, other forces.

2.2 CHARACTERISTICS OF A FORCE

For practical applications, a force must be completely described. The graphical representation of a force in a diagram is an arrow, a line ending with an arrowhead. The complete description of a force includes the following information:

1. **Magnitude:** Refers to the size or amount of the force in acceptable units. A 1000-lb force has a larger magnitude than a 500-lb force. Magnitude is represented graphically by the length of the arrow.
2. **Direction:** Refers to the path of the line along which the force acts. This line is commonly called the *line of action*. The force may act vertically, horizontally, or at some angle with the vertical or horizontal. Graphically, direction of a force is represented by the direction of the shank of the arrow.
3. **Sense:** Refers to the way in which a force acts along its line of action. The direction of a force may be vertical, but the sense could be up or down. Similarly, the direction may be horizontal, but the sense could be to the left or to the right. Graphically, the arrowhead of the arrow denotes sense.
4. **Point of application:** Refers to the point on, or in, the object at which the force is applied. Graphically, this is the point at which the arrowhead contacts the body.

Figure 2.1 portrays the characteristics of a force.

2.3 UNITS OF A FORCE

The unit generally used for expressing the magnitude of a force in the U.S. Customary System is the pound (lb). Multiples of the pound, also commonly used, are the kip, which is equal to 1000 lb, and the ton (short ton), which is equal to 2000 lb.

In the SI, the unit of force is the newton (N). The various prefixes discussed in Chapter 1 are also used, with the kilonewton (kN), which is equal to 1000 N, being common.

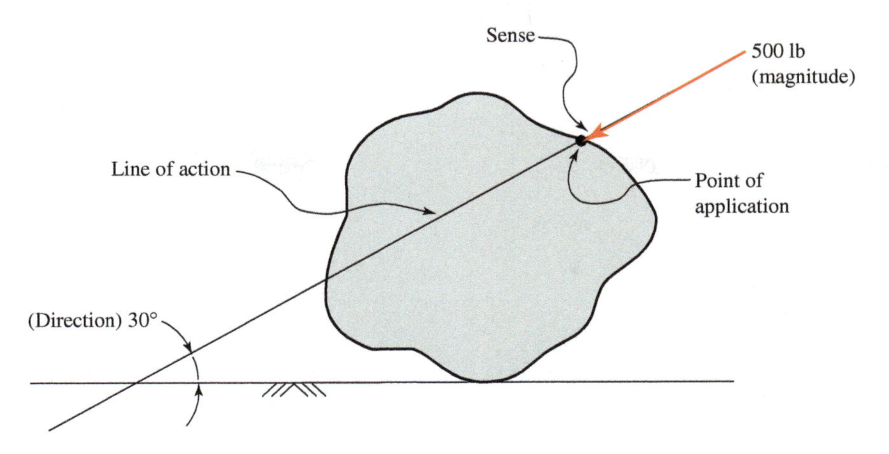

FIGURE 2.1 Characteristics of a force.

2.4 TYPES AND OCCURRENCE OF FORCES

Forces may be broadly classified according to the way in which they are exerted. Forces may be exerted by contact of one solid body on another solid body (e.g., through a push or pull). Forces may also be exerted directly against a solid body by wind or by water. In addition, forces may be exerted without contact, as in the gravitational attraction of one body for another. Forces in engineering applications that may be analyzed using the principles of statics encompass any of the foregoing forces acting singly or in combination. When a part of a structure or a machine resists these forces, the element is said to be loaded, and the applied forces are often referred to as *applied loads*. In this text, we use the terms *force* and *load* almost interchangeably because most applications discussed here involve structures, machines, or their elements, with forces acting on them as loads.

As shown in Figure 2.1, if a force (load) is assumed to act at a point, it is frequently referred to as a *concentrated force* (*concentrated load*). This idealization is commonly used when the area of application is very small in comparison to the total surface area of the body on which the force is acting and is satisfactory for most engineering solutions. When the area of application of the force (load) is large, it is referred to as a *distributed force*, or a *distributed load* and has units such as pounds per square foot (psf or lb/ft^2). Examples of distributed loads are snow on a roof, water pressure against the side of a tank or dam, and wind pressure against the side of a building. If the force is spread out along a line or distributed along a narrow strip, it is referred to as a *distributed load* and has units such as kips per foot (kips/ft) or newtons per meter (N/m). An example of a distributed load is the weight of a beam or the weight of a lane of bumper-to-bumper traffic on a long bridge. Distributed loads are discussed in detail in Section 3.5.

Forces may also be categorized as external or internal. A force is said to be external to a body if it is exerted on that body by some other body. It is said to be internal if it is exerted on a part of that body by some other part of the same body. For example, the external forces acting on the truss of Figure 2.2 produce internal forces in the various truss members. The externally applied forces are P_1, P_2, P_3, and P_4. These forces are, in turn, transmitted from their points of application through the various members of the truss to the external supports at A and B. Internal forces are developed in the truss members as a result of the members pushing or pulling on other members at points of intersection.

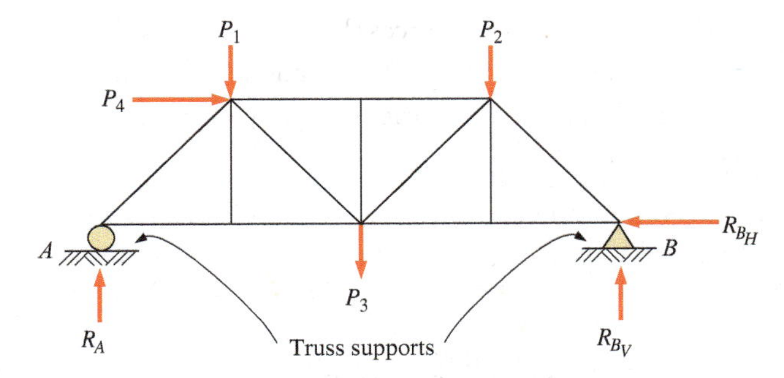

FIGURE 2.2 Forces acting on a truss.

Note that the externally applied forces are all concentrated forces and are exerted by other bodies. They are said to be acting on the truss. External resisting forces R_A, R_{B_V}, and R_{B_H} are shown at the external supports. These are commonly called *reacting forces*, or *reactions*.

2.5 SCALAR AND VECTOR QUANTITIES

A *scalar quantity* is a quantity that has magnitude only. For example, the population of a city, a volume of water, a duration of time, and an amount of money are all scalar quantities. Scalar quantities may be added algebraically, with the result still having magnitude only.

A *vector quantity* is any quantity that has magnitude, direction, and sense. These attributes can be represented by a *vector*, which is a segment of a straight line, with the sense expressed by an arrowhead at one end. If drawn to scale, the vector will be of a definite length, representing its magnitude. A force is a vector quantity because a push or pull is exerted in a certain direction and sense as well as with a certain magnitude. Other quantities commonly represented by vectors are displacements, velocities, and accelerations.

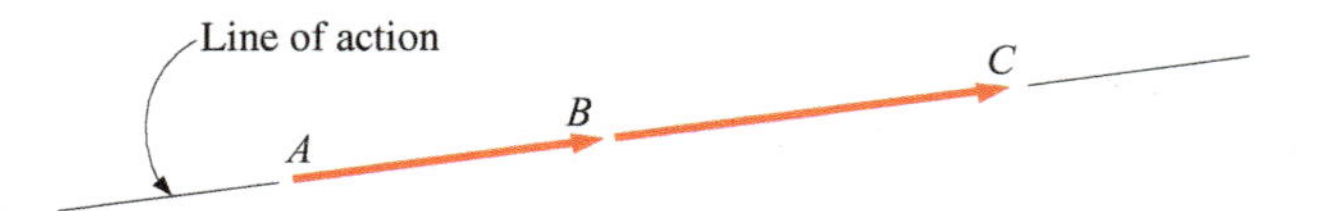

FIGURE 2.3 Vectors with same line of action.

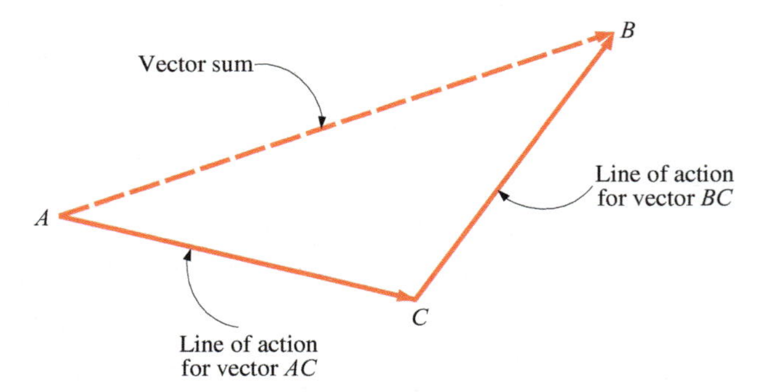

FIGURE 2.4 Vectors with different lines of action.

Vector quantities cannot be added algebraically, since the direction of the vector is as significant as the magnitude. One exception is illustrated in Figure 2.3 where two vectors (forces) have the same line of action. In such a case, the vectors' magnitudes may be added algebraically, in the manner of scalar quantities.

With forces having different lines of action, the vectors must be added vectorially (geometrically). For example, in Figure 2.4, vector AB, shown as a dashed line, represents the vector sum of vectors AC and BC; it could replace them and produce the same effect.

2.6 THE PRINCIPLE OF TRANSMISSIBILITY

In Figure 2.5, a block is being pulled along a horizontal surface by a horizontal force P. The motion of the block is independent of whether it is pulled or pushed by the force, as long as P has the same magnitude, direction, and sense and acts along the same line of action. This illustrates the *principle of transmissibility*, that the external effect of a force on a body is the same for all points of application along its line of action. The effect is independent of the point of application.

When applying the principle of transmissibility, one must be aware that only external effects behave this way. The internal effect, that is, within a body, of a force is dependent on its point of application. Consider the forces within the block of Figure 2.5. If the force P were applied at A, the block would tend to be crushed; on the other hand, if the force were applied at B, the block would tend to be elongated. Therefore, we see that two completely different types of internal behavior would result.

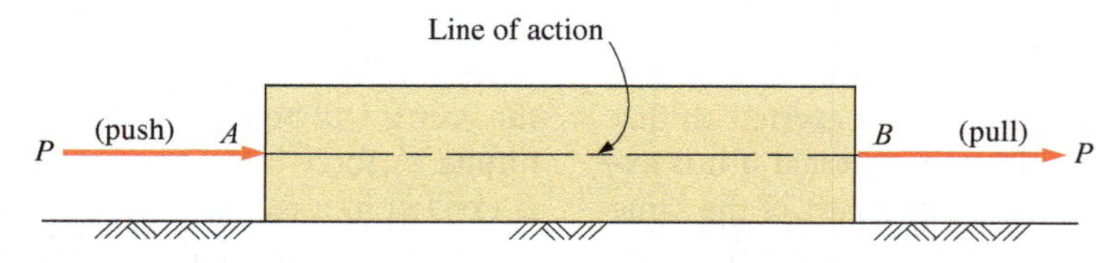

FIGURE 2.5 Principle of transmissibility.

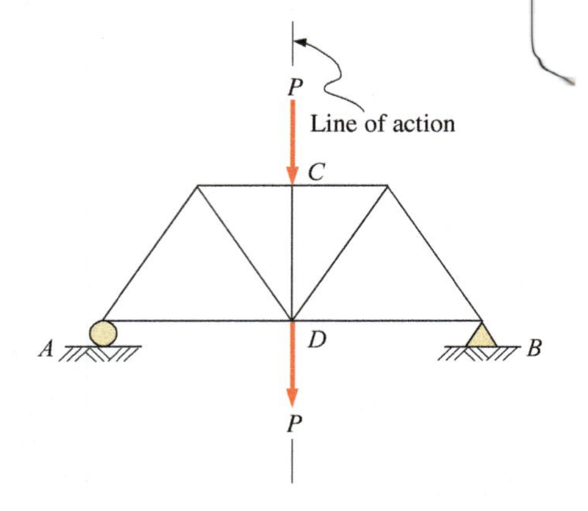

FIGURE 2.6 Principle of transmissibility.

Similarly, the external support forces at A and B of the truss shown in Figure 2.6 are independent of whether P is applied at C or at D. The internal forces in some of the truss members, however, will be appreciably different depending on whether the point of application is at C or D.

2.7 TYPES OF FORCE SYSTEMS

A *force system* may be defined as any number of forces that are collectively considered. There are two broad classifications of force systems: coplanar and noncoplanar. A coplanar force system is one in which the lines of action of all the forces lie in the same plane. A noncoplanar system of forces is one in which the lines of action of all the forces do not lie in the same plane.

The coplanar and noncoplanar force systems may be further classified. When the lines of action of all the forces intersect at a common point, the system is said to be concurrent. If the lines of action of all the forces are parallel, the system is said to be a parallel force system. If the action lines do not intersect at a common point and are not parallel, the system is said to be a nonconcurrent system of forces.

If all the forces in a parallel system act along a single line of action, the system is said to be collinear. Actually, this situation is considered a special case of the coplanar concurrent force system.

The classification of force systems may be summarized as follows:

1. Coplanar concurrent
2. Coplanar nonconcurrent
3. Coplanar parallel
4. Noncoplanar concurrent
5. Noncoplanar nonconcurrent
6. Noncoplanar parallel

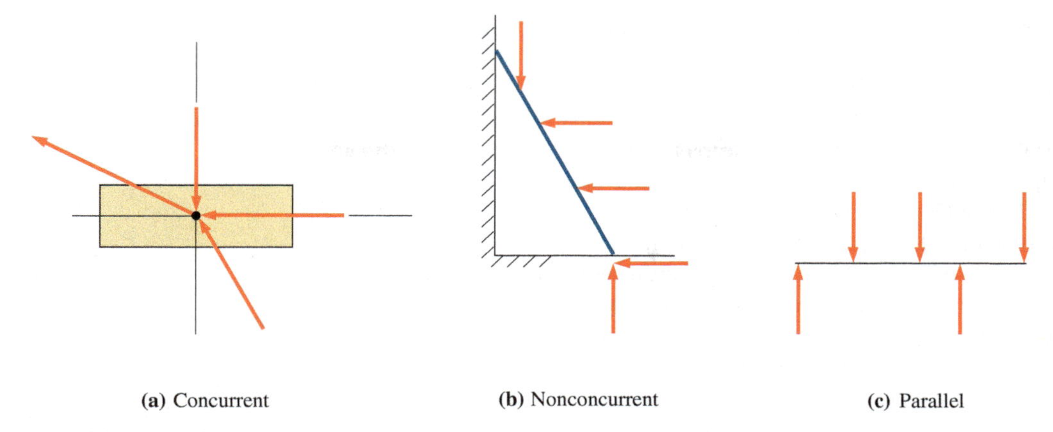

(a) Concurrent (b) Nonconcurrent (c) Parallel

FIGURE 2.7 Types of coplanar force systems.

Since most structural systems can be reduced to coplanar force systems, from a design and analysis standpoint, only coplanar force systems are considered in this text. These systems are summarized graphically in Figure 2.7.

2.8 ORTHOGONAL CONCURRENT FORCES: RESULTANTS AND COMPONENTS

In structural analysis, it is sometimes convenient to join the effects of two (or more) forces into a single force. We first discuss the case where the two original forces act at a right angle (orthogonal) with respect to one another. An example is shown in Figure 2.8a. A barge is being maneuvered by two tugboats.

Tugboat *A* is pushing along the *Y* axis with a force of P_y, and tugboat *B* is pushing along the *X* axis with a force of P_x. Imagine in Figure 2.8b that the barge is reduced to a point at *O*, the point of concurrence of the lines of action of the forces, and using the principle of transmissibility, the forces are moved to that point. The combined effect of the two forces can be determined by adding the forces vectorially, as shown in Figure 2.8c, by completing the rectangle of which the forces are a part. The combined effect, called the *resultant vector*, the *resultant force*, or just the *resultant*, designated here as *P*, is the diagonal of the rectangle. The two forces P_x and P_y are called the *component vectors, component forces*, or simply the *components* of the resultant. (Resultants are discussed in detail in Chapter 3.) A slightly different method, using a *force*

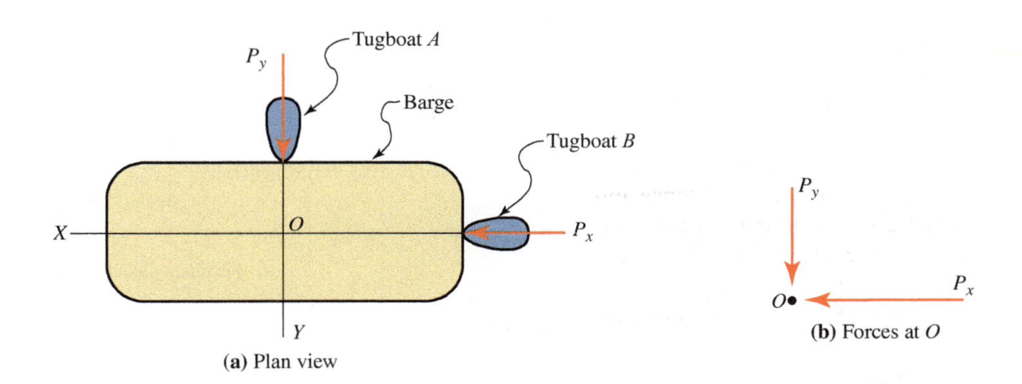

(a) Plan view

(b) Forces at *O*

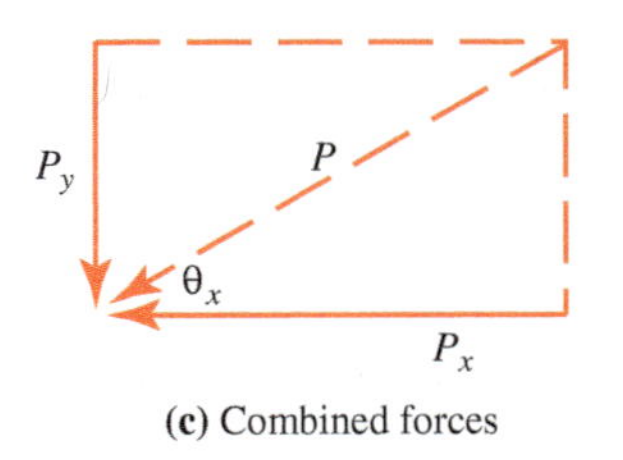

(c) Combined forces

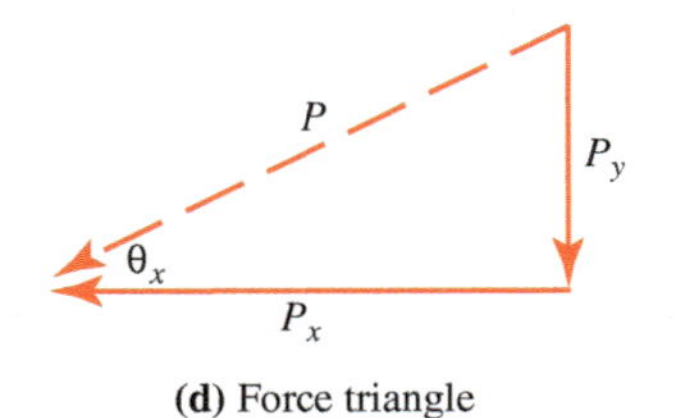

(d) Force triangle

FIGURE 2.8 Resultant determination.

triangle, is shown in Figure 2.8d, where the two vectors are added by positioning the arrowhead (or tip) of the P_y vector in contact with the end (or tail) of the P_x vector. Note that the forces still act through point O as shown in Figure 2.8a. The force triangle is merely a convenient means for showing the relationship between concurrent force vectors and their resultant. In each case (Figure 2.8c and d), the magnitude of the resultant can be found using the Pythagorean theorem.

The force triangle of Figure 2.8d also shows how a convenient graphical solution can be accomplished. Assuming that the magnitudes of the forces are known, draw the component forces accurately to scale and arranged as in Figure 2.8d. The resultant force can be drawn from the tail of P_y to the tip of P_x. This process is a graphical vector addition. The magnitude of the resultant can be scaled from the diagram and the direction determined with a protractor, which can serve as a quick check for the trigonometric calculations. The magnitude of the resultant in Figure 2.8c (and d) is determined analytically using the Pythagorean theorem:

$$P^2 = P_x^2 + P_y^2$$

from which

$$P = \sqrt{P_x^2 + P_y^2} \quad \textbf{(2.1)}$$

In addition, the direction (inclination with the horizontal X axis) of the vector P is

$$\theta_x = \tan^{-1}\frac{P_y}{P_x} \quad \textbf{(2.2)}$$

The sense of P is determined by the signs of the component vectors, as previously discussed, in combination with a diagram such as that of Figure 2.8d. In this case, the sense would be downward and to the left.

It should be noted at this point that the preceding discussion has been developed using the notation of P for vector, or force. This is a general notation. Forces are frequently, but not always, denoted P; resultant forces are frequently, but not always, denoted R. Other designations are equally valid and a sketch will usually help clarify the issue. Most important is that once the notation scheme for a particular problem has been chosen, it should be adhered to carefully.

EXAMPLE 2.1 The rectangular components of a force are +100 lb in the Y direction (P_y) and −200 lb in the X direction (P_x). Both forces act through point O, as shown in Figure 2.9a. Calculate the magnitude, inclination with the X axis, and the sense of the resultant force P.

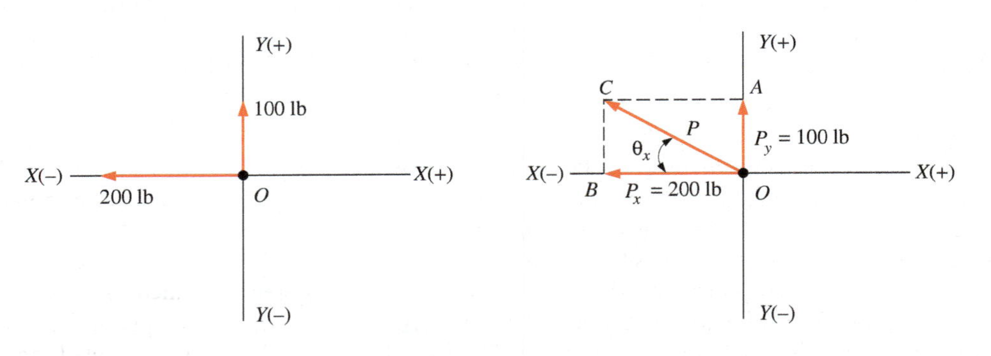

FIGURE 2.9 Force relationships.

Solution Using an X–Y coordinate axes system (see Figure 2.9b), construct a rectangle by drawing lines AC parallel and equal to P_x and through the tip of P_y and BC parallel and equal to P_y and through the tip of P_x. The diagonal of the rectangle represents the unknown force P.

Use the right triangle OBC where $BC = P_y$ along with Equation (2.1):

$$P = \sqrt{P_x^2 + P_y^2} = \sqrt{(-200\text{ lb})^2 + (100\text{ lb})^2} = 224\text{ lb}$$

The inclination of the resultant with the X axis can be obtained using Equation (2.2):

$$\theta_x = \tan^{-1}\frac{P_y}{P_x} = \tan^{-1}\frac{100\text{ lb}}{200\text{ lb}} = 26.6°$$

The sense of force P is upward and to the left and is shown in Figure 2.9b.

EXAMPLE 2.2 Two forces, F_x and 4000 lb, intersect at O on a heavy object that is to be pulled along a line defined by a 40° angle with the Y axis, as shown in Figure 2.10. Determine the required force F_x and find the magnitude of the resultant of the two forces.

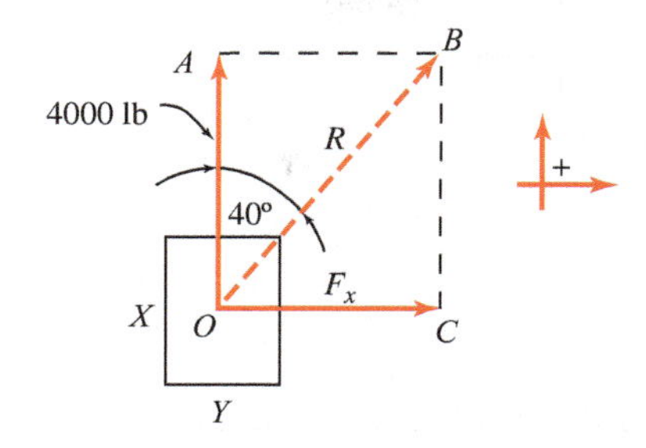

FIGURE 2.10 Plan view for Example 2.2.

Solution In Figure 2.10, the rectangle $OABC$ with sides parallel and perpendicular to the X–Y coordinate axis system establishes the relationship between the forces. The diagonal OB represents the resultant R. Use right triangle OAB, where side AB equals force F_x:

$$\tan 40° = \frac{AB}{4000 \text{ lb}}$$

Therefore,

$$F_x = AB = 4000 \text{ lb} (\tan 40°) = 3360 \text{ lb}$$

From Equation (2.1),

$$R = \sqrt{(3360 \text{ lb})^2 + (4000 \text{ lb})^2} = 5220 \text{ lb}$$

It is frequently convenient to replace a given force to simplify an analysis. Any given force may be replaced by an alternate set of forces provided that the combined replacement forces have the same effect as that of the original force.

As an example, consider the force due to wind against the roof of a building. The direction of the wind force is perpendicular to the sloping surface, as shown in Figure 2.11a. To simplify certain aspects of this problem, we may choose to replace the inclined force with the vertical and horizontal forces shown in Figure 2.11b. The vector sum of these two forces would be equivalent to the inclined force. In other words, the inclined force would be replaced by an equivalent rectangular force system, with the two forces acting at right angles to each other and having the same effect as the applied inclined force.

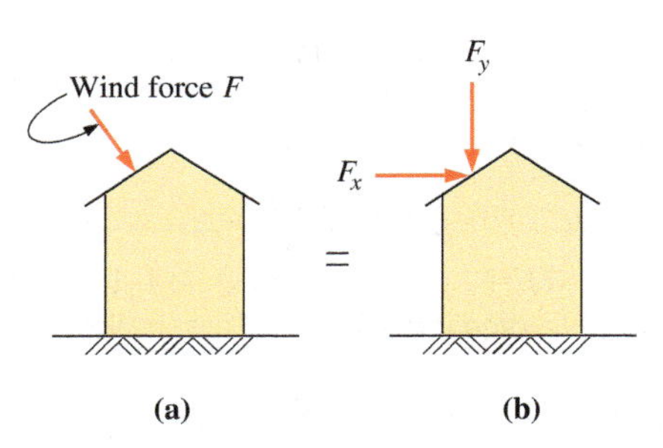

FIGURE 2.11 Inclined wind force.

The process of replacing a force with a rectangular system in which the two forces are at right angles to each other, or the special case in which the forces are vertical and horizontal, is called *force resolution*, or resolving a force into its components. When the two components are at right angles to each other, they are called *rectangular components* or *orthogonal components* and are usually subscripted x and y when the conventional X–Y coordinate axes are used.

There is an infinite number of possible sets of components for a given force, but usually only one or two specific sets are of interest. In most engineering problems, the components of interest are the rectangular components with forces at right angles to each other.

Rectangular components may be computed analytically using the conventional X–Y (horizontal–vertical) coordinate axes system, as shown in Figure 2.12. An arbitrary vector representing a force F is shown applied to a hypothetical body at point O. The force acts at angle θ_x with the horizontal X axis. As an arbitrary sign convention, horizontal forces acting to the right and vertical forces acting upward will be taken as positive, whereas horizontal forces to the left and

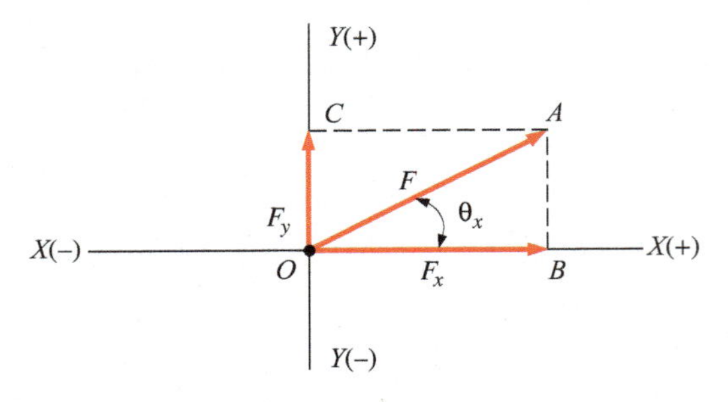

FIGURE 2.12 Rectangular components of a force.

vertical forces acting downward will be taken as negative. The sign convention is indicated in the figure.

If the vector F is projected upon the X and Y axes as shown, the rectangular components F_x and F_y are formed and can then be determined. Note that the components form a concurrent force system.

Since triangle OAB is a right triangle, the relationship between the components and the force F can be determined by the sine and cosine functions of the angle θ_x.

Since $BA = F_y$ and $CA = F_x$,

$$\cos \theta_x = \frac{F_x}{F} \quad \text{and} \quad \sin \theta_x = \frac{F_y}{F}$$

which can be rewritten as

$$F_x = F \cos \theta_x \tag{2.3}$$

$$F_y = F \sin \theta_x \tag{2.4}$$

There are some cautions to note. The choice of the orientation of the X and Y axes is arbitrary. Any orientation may be chosen, not necessarily vertical and horizontal (although this is usually the most convenient). Also, Equations (2.3) and (2.4) are appropriate for the situation of Figure 2.12 where the direction of the force is referenced to the X axis. The equations will change if the direction is referenced to the Y axis.

EXAMPLE 2.3 Compute the rectangular components (vertical and horizontal) for the 400-lb force shown in Figure 2.13.

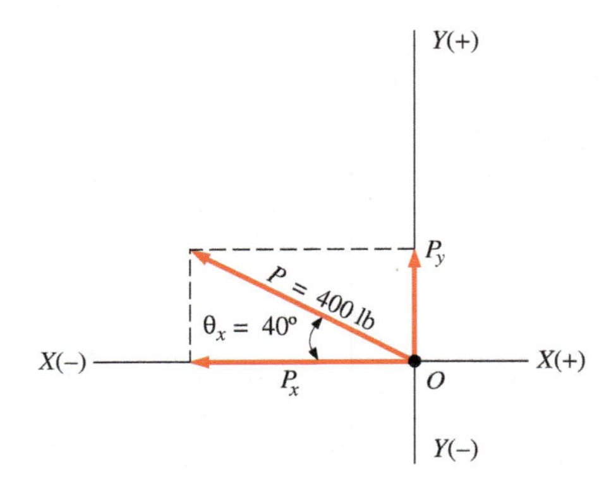

FIGURE 2.13 Rectangular components of a force.

Solution The force (labeled F) is shown acting through the origin of X–Y coordinate axes. Its inclination with the horizontal X axis is 40°. Note that the diagram does not have to be drawn to scale; to the extent possible, however, the various parts should be drawn in proportion to each other.

Projecting force P upon the X and Y axes reveals that the sign of P_y is positive (acting upward) and that the sign of P_x is negative (acting to the left). Therefore, assign a negative sign to the P_x component. Using Equations (2.3) and (2.4),

$$P_x = -P \cos \theta_x = -400 \text{ lb}(\cos 40°) = -306 \text{ lb}$$

$$P_y = +P \sin \theta_x = +400 \text{ lb}(\sin 40°) = +257 \text{ lb}$$

EXAMPLE 2.4 The roof of a building is subjected to a wind force F of 600 N, as shown in Figure 2.14. The wind force acts perpendicular to the surface of the roof. Resolve this force into its vertical and horizontal rectangular components.

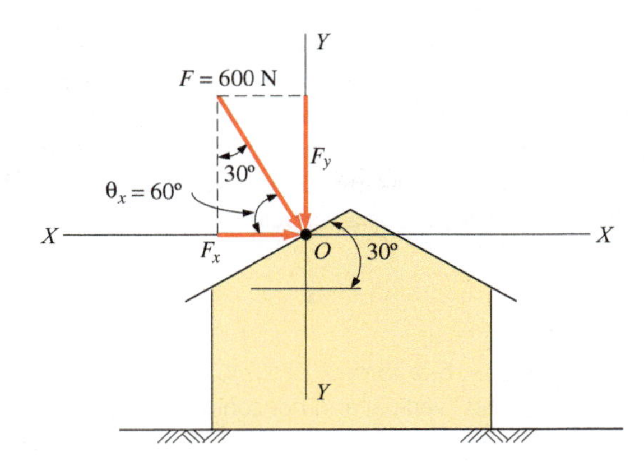

FIGURE 2.14 Wind force on inclined surface.

Solution The wind force F is shown acting through the origin of the X–Y coordinate axes. Due to the roof slope, the inclination of the force with the horizontal X axis is 60°. Projecting force F upon the X and Y axes reveals that the sign of F_y is negative (acting downward) and the sign of F_x is positive (acting to the right). Therefore, use Equations (2.3) and (2.4) to obtain

$$F_x = +F\cos\theta_x = +600\text{ N}(\cos 60°) = +300\text{ N}$$
$$F_y = -F\sin\theta_x = -600\text{ N}(\sin 60°) = -520\text{ N}$$

EXAMPLE 2.5

Figure 2.15 shows a body on a 20° inclined plane. The weight of the body represented by the force W is 50 lb. Resolve this force at point O into components parallel and perpendicular to the inclined plane.

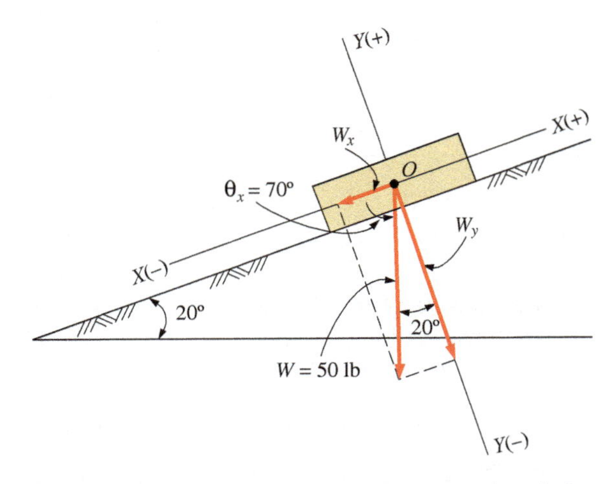

FIGURE 2.15 Rectangular components of a force acting on an inclined plane.

Solution The X–Y coordinate axes are sketched with the X axis parallel to the incline and the Y axis perpendicular to the incline. The 50-lb force is shown acting through the origin. Note that since the force W represents the weight of the body, it is a gravity force and acts vertically downward.

Projecting the force W upon the X and Y axes reveals that the signs of both components are negative. W_x acts to the left and W_y acts downward. Use Equations (2.3) and (2.4):

$$W_x = -W\cos\theta_x = -50\text{ lb}(\cos 70°) = -17.1\text{ lb}$$
$$W_y = -W\sin\theta_x = -50\text{ lb}(\sin 70°) = -47.0\text{ lb}$$

Directions of forces have, so far, been defined with an angle between the force and a reference line. In Figure 2.16a, note that the direction of the force is defined with a *slope triangle*, a right triangle having its hypotenuse coincident with the force P. For clarity, the slope triangle is shown in Figure 2.16b. This slope triangle has a vertical leg of 1 and a horizontal leg of 2. This is sometimes referred to as a 1:2 slope (one vertical unit on two horizontal units). The direction of the force could be described equally well with a slope angle α between the force P and a horizontal line (where $\alpha = \tan^{-1}\left(\frac{1}{2}\right) = 26.6°$).

EXAMPLE 2.6 In Figure 2.16, a force P of 25 lb is acting at a slope as shown. Resolve the force into its rectangular components (vertical and horizontal).

Solution Note that the direction of the force is defined by a slope triangle, as shown in Figure 2.16b, rather than by an angular value. Projecting force P upon the X and Y axes reveals that the signs of both components are positive. P_y acts upward and P_x acts to the right. Note that $P_y = AB$.

The force triangle is shown in Figure 2.16c. Note the similarity between the force triangle and the given slope triangle. Since the triangles are similar, the corresponding sides are proportional to each other. Hence, the components P_x and P_y can be computed as follows:

$$\frac{P_y}{1} = \frac{P_x}{2} = \frac{P}{\sqrt{5}}$$

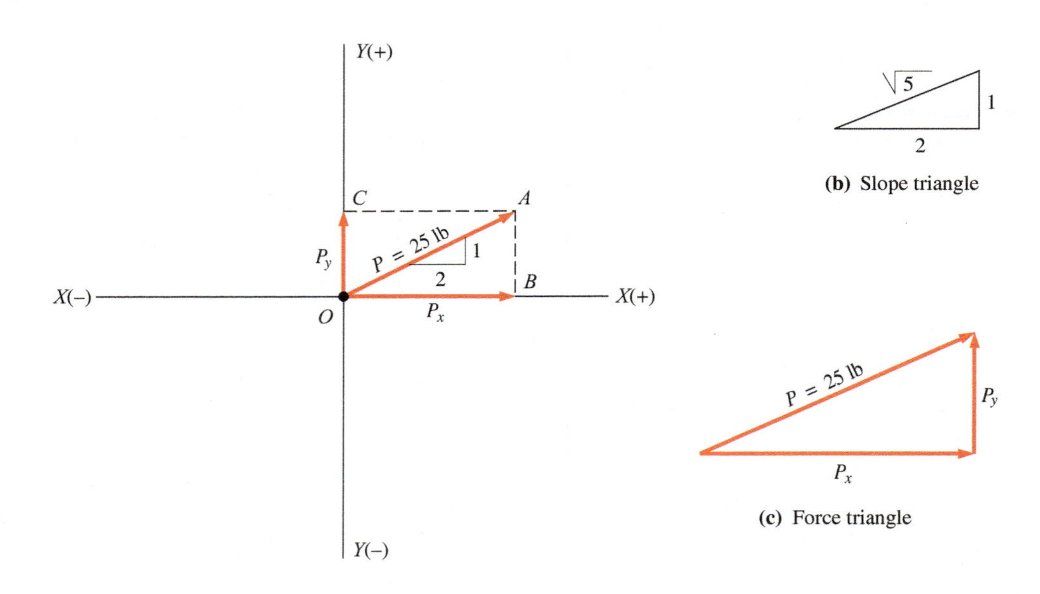

(a) X-Y axes with applied force

FIGURE 2.16 Rectangular components of a force.

from which

$$P_x = +\frac{2P}{\sqrt{5}} = +\frac{2(25\text{ lb})}{\sqrt{5}} = +22.4\text{ lb}$$

$$P_y = +\frac{1P}{\sqrt{5}} = +\frac{1(25\text{ lb})}{\sqrt{5}} = +11.2\text{ lb}$$

EXAMPLE 2.7 A force P of 950 N is applied to a body at a slope, as shown in Figure 2.17. Resolve the force into its rectangular components (vertical and horizontal).

Solution The direction of the force is defined by a slope triangle, rather than by an angle. Projecting force P upon the X and Y axes reveals that the signs of both components are positive. P_y acts upward and P_x acts to the right.

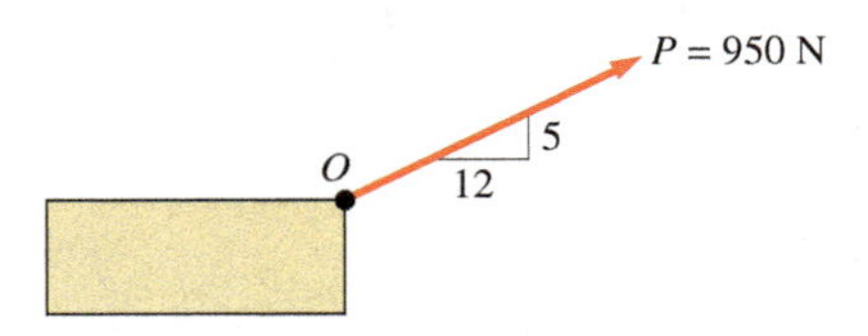

FIGURE 2.17 Inclined force acting on a body.

In Figure 2.18a, P is shown as diagonal OA of rectangle $OCAB$. Note that in $OCAB$, AB is equal to P_y. Triangle OAB may be considered to be a force triangle, as shown in Figure 2.18c. Note that the force triangle and the slope triangle are similar, with the corresponding sides proportional to each other. Express this mathematically as

$$\frac{P_x}{12} = \frac{P}{13} = \frac{P_y}{5}$$

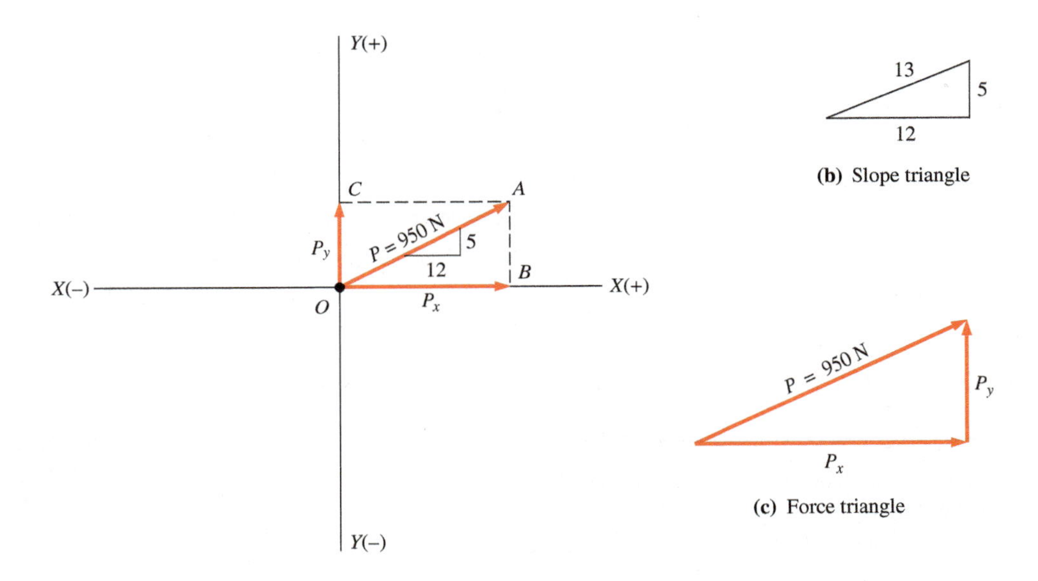

FIGURE 2.18 Rectangular components of a force.

from which

$$P_x = \frac{12P}{13} = \frac{12(950\text{ N})}{13} = 877\text{ N}$$

and

$$P_y = \frac{5P}{13} = \frac{5(950\text{ N})}{13} = 365\text{ N}$$

EXAMPLE 2.8 A block having a mass of 400 kg rests on an inclined surface, as shown in Figure 2.19. Determine the components of gravitational force exerted on the block perpendicular and parallel to the inclined surface.

Solution For convenience, reference X–Y axes are established with the X axis parallel to the incline. The weight W (which is the force of gravity exerted on the block), shown in Figure 2.19a, can be calculated from

$$W = mg = 400\text{ kg}(9.81\text{ m/sec}^2) = 3924\text{ N} = 3.924\text{ kN}$$

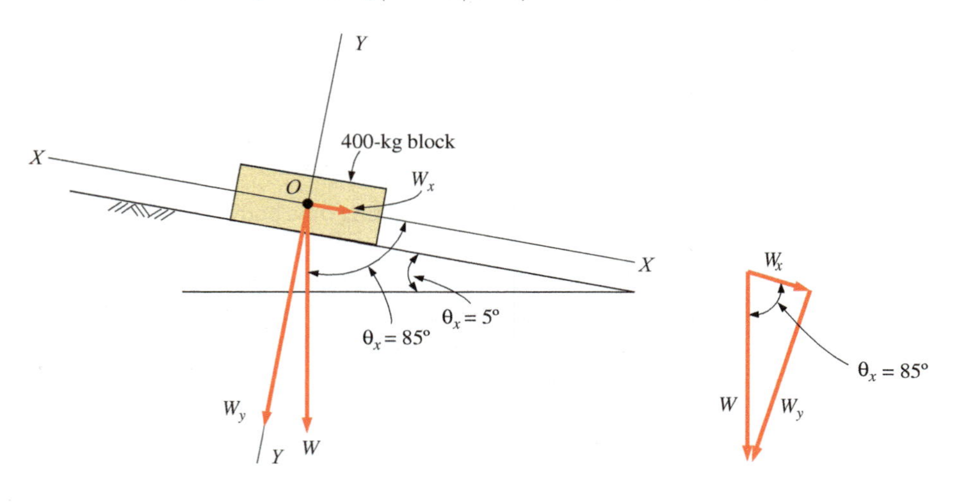

FIGURE 2.19 Rectangular components of a force.

Noting θ_x and the force triangle of Figure 2.19b, calculate the rectangular components from Equations (2.3) and (2.4):

$$W_x = W\cos\theta_x = (3.924\text{ kN})(\cos 85°) = 0.342\text{ kN}$$
$$W_y = -W\sin\theta_x = (-3.924\text{ kN})(\sin 85°) = -3.91\text{ kN}$$

SUMMARY BY SECTION NUMBER

2.1 A force is a push or a pull that tends to change the state of motion of a body.

2.2 A force is completely described by four quantities: (a) magnitude, (b) direction, (c) sense, and (d) point of application.

2.3 The unit of force is the pound (lb) in the U.S. Customary System and the newton (N) in the SI. Also used are kip, ton, and kilonewton.

2.4 A distributed load has a large area of application with respect to the total surface area. A concentrated load has a small area of application with respect to the total surface area. An external resisting force is called a *reaction*.

2.5 A force is a vector quantity. A vector is a graphical representation of a vector quantity. A force is usually denoted with an uppercase letter and sometimes subscripted.

2.6 The principle of transmissibility states that the external effect of a force on a body is the same for all points of application along its line of action.

2.7 Forces can be classified as coplanar concurrent, coplanar nonconcurrent, coplanar parallel, noncoplanar concurrent, noncoplanar nonconcurrent, or noncoplanar parallel.

2.8 Coplanar forces lie in the same plane. Noncoplanar forces do not lie in the same plane. The lines of action of concurrent forces intersect at a common point; those of nonconcurrent forces do not. Force systems (coplanar and noncoplanar) may be categorized as (a) concurrent, (b) nonconcurrent, or (c) parallel. A force system is collinear if all forces act along a single line of action.

If two orthogonal forces are known, their vector sum, called the resultant, can be obtained using the Pythagorean theorem:

$$P = \sqrt{P_x^2 + P_y^2} \quad \textbf{(2.1)}$$

and the inclination (θ_x) of the force P with the X axis is expressed as

$$\theta_x = \tan^{-1}\frac{P_y}{P_x} \quad \textbf{(2.2)}$$

Conversely, a single force may be replaced by two or more forces called *components* that will produce the same effect as the single force they replace. The rectangular components of a single force F, in terms of the angle θ_x at which it is inclined with the X axis, are expressed by

$$F_x = F\cos\theta_x \quad \textbf{(2.3)}$$
$$F_y = F\sin\theta_x \quad \textbf{(2.4)}$$

PROBLEMS

Section 2.8 Orthogonal Concurrent Forces: Resultants and Components

2.1 Find the resultant force for each system of forces shown.

2.2 Rework Problem 2.1 solving for the resultants using a graphical solution.

2.3 Compute the vertical and horizontal components for each of the forces shown.

2.4 Compute the vertical and horizontal components for each of the forces shown.

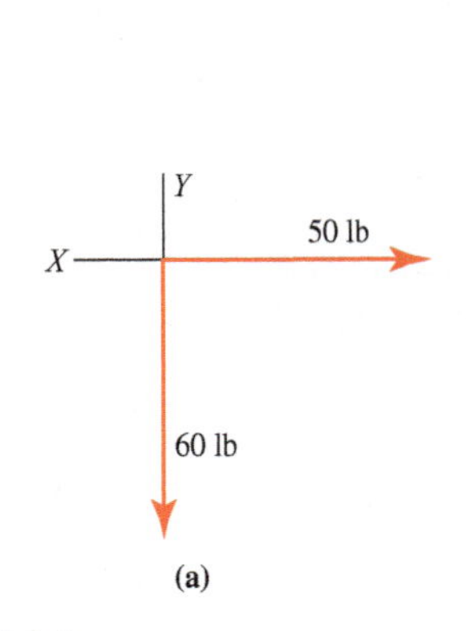

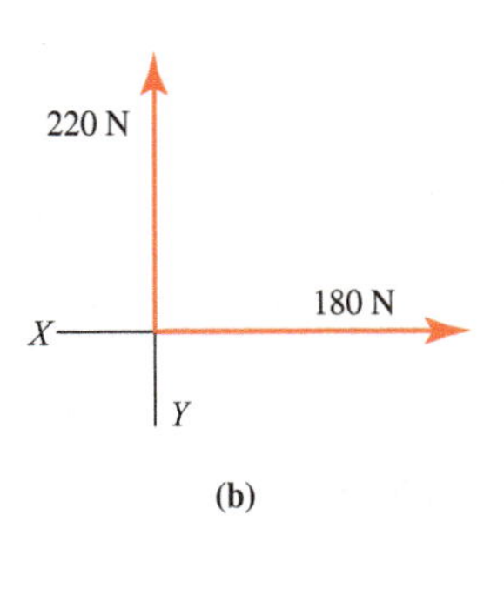

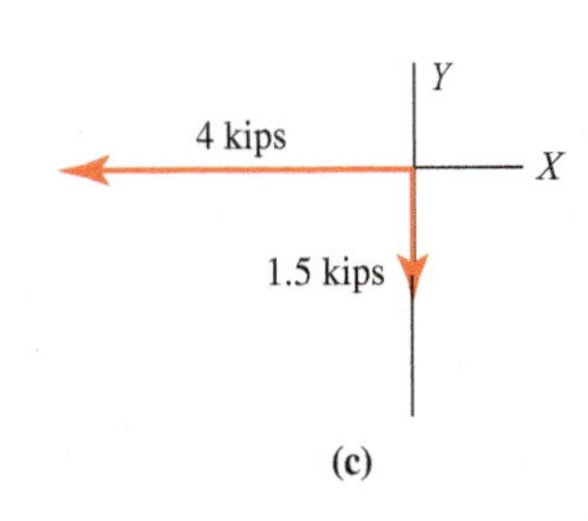

PROBLEM 2.1

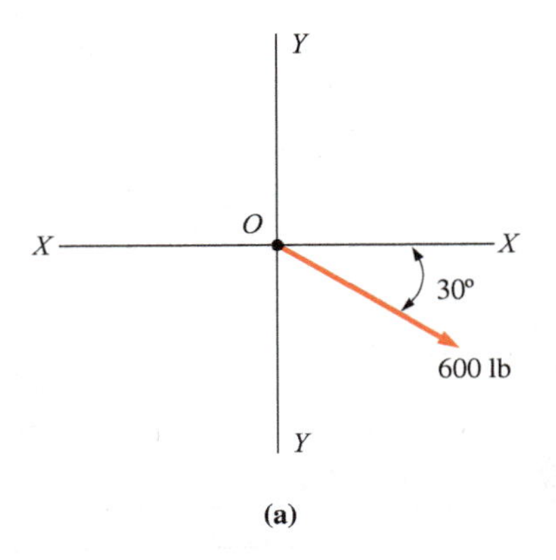

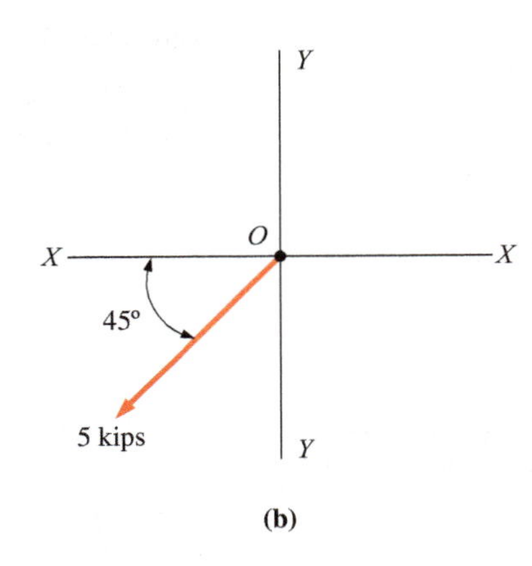

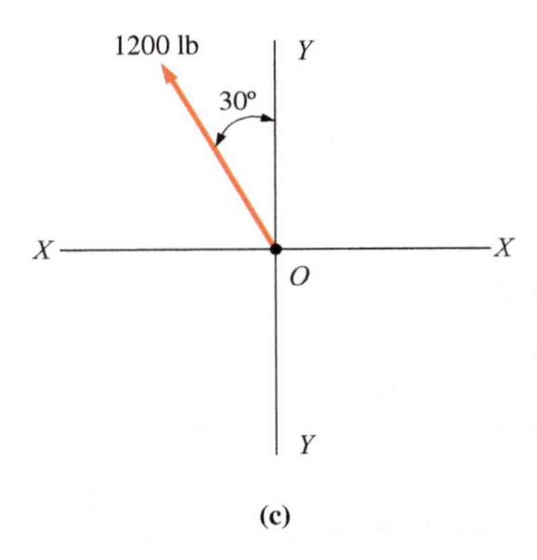

PROBLEM 2.3

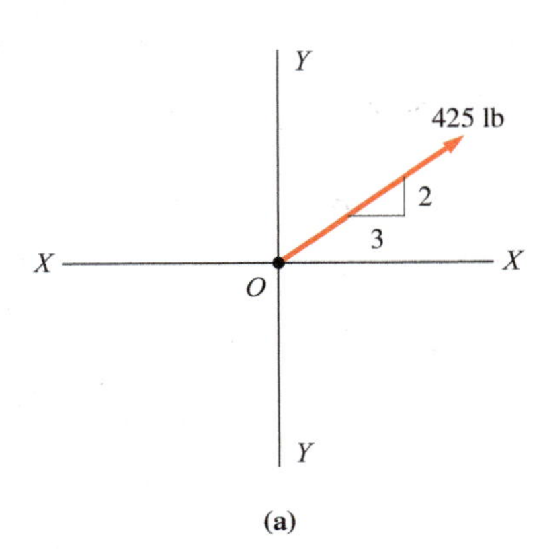

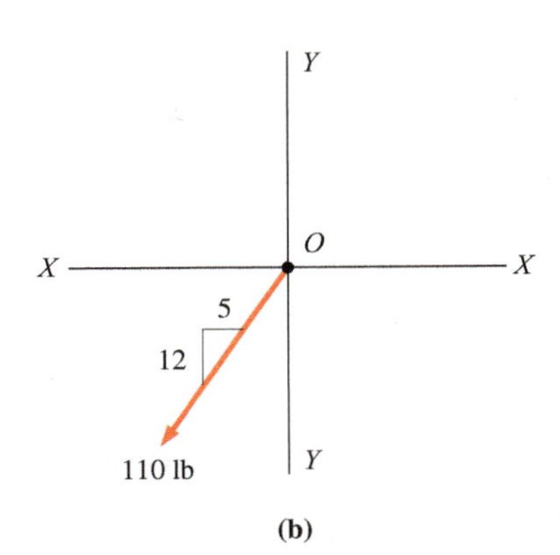

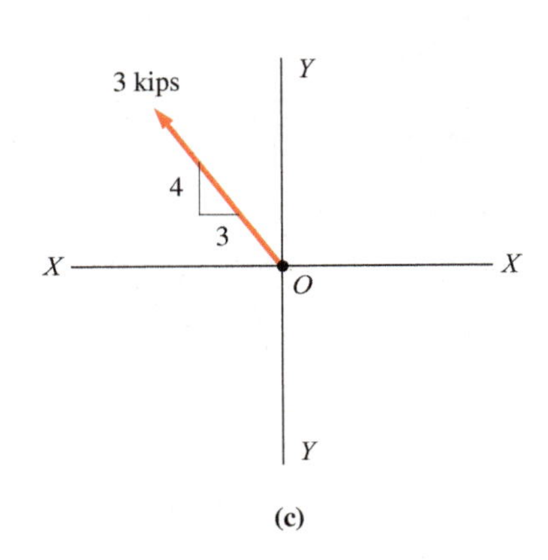

PROBLEM 2.4

2.5 Compute the vertical and horizontal components for the given values of P and θ. The angle is measured clockwise from the positive X axis to the force P, as shown.

(a) 320 lb, 20°
(b) 640 lb, 30°
(c) 320 lb, 40°
(d) 320 lb, 88°

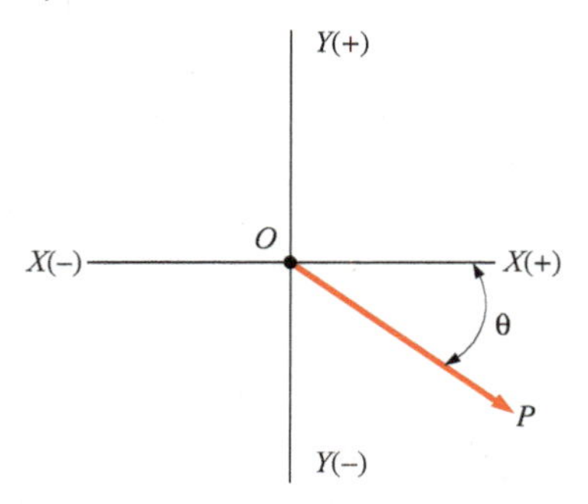

PROBLEM 2.5

2.6 The rectangular components of a force are $F_y = +1.3$ kN and $F_x = +103$ N. Calculate the magnitude, inclination with the horizontal axis, and the sense of the resultant force F.

2.7 Find the resultant force on the screw eye. One rope is horizontal, the other rope is vertical. The force in the rope is 175 lb.

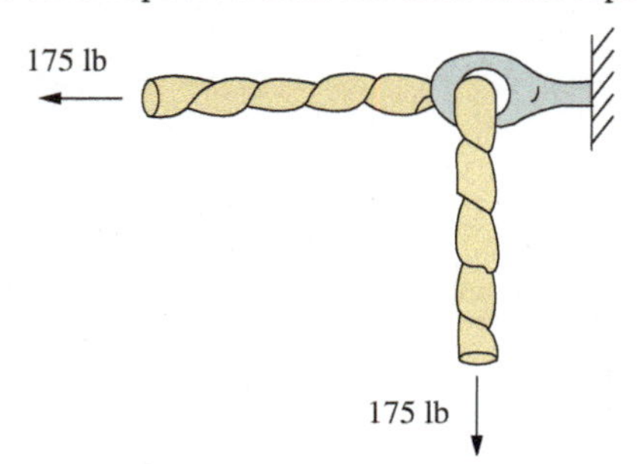

PROBLEM 2.7

Computer Problems

For the following computer problems, any appropriate software may be used. Input prompts should fully explain what is required of the user (the program should be user-friendly). The resulting output should be well labeled and self-explanatory.

2.8 Write a program that will calculate the rectangular components of a force. User input is to be the magnitude of the force (lb) and its inclination with the positive X axis, as defined in Figure 2.12.

2.9 Write a program that will calculate the magnitude of a resultant force, given the X and Y rectangular components of the force. Also include the calculation of the inclination of the resultant force with the positive X axis.

2.10 Calculate and display the vertical and horizontal components for a given force (user input). The angle of the force with the positive X axis is to vary from 0° to 90° in steps of 2°.

Supplemental Problems

2.11 A plastic barrel containing water is used as an anchor for a large tent. The rope tie anchoring the tent exerts a 120-lb pull on the barrel as shown. Determine the vertical and horizontal components of the force in the tie.

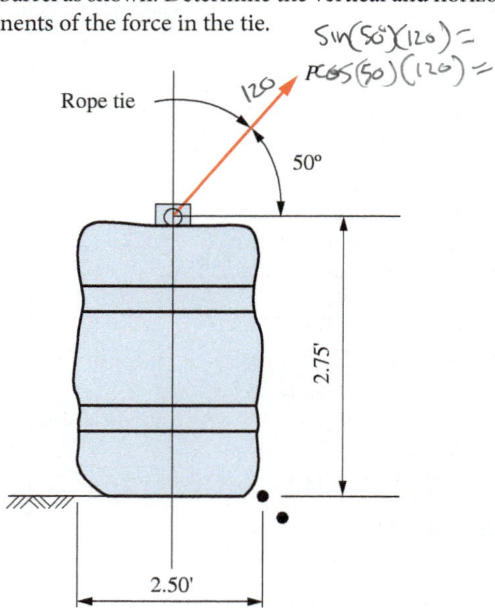

PROBLEM 2.11

2.12 A telephone pole is braced by a guy wire that exerts a 300-lb pull on the top of the pole, as shown. The guy wire forms an angle of 42° with the pole. Compute the vertical and horizontal components of the pull on the pole

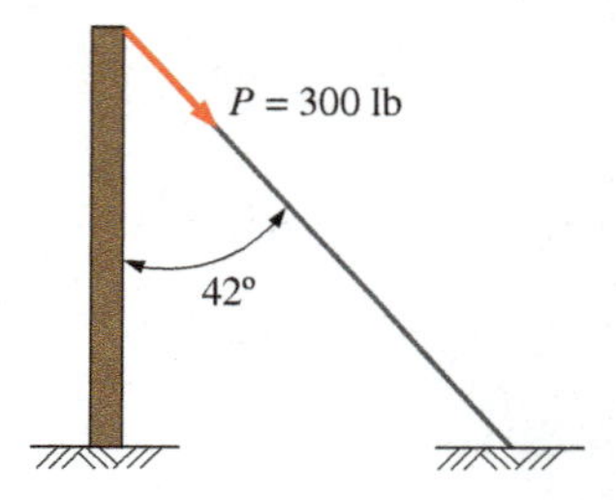

PROBLEM 2.12

2.13 Compute the vertical and horizontal components for each of the forces shown. In each case the force is perpendicular to the incline.

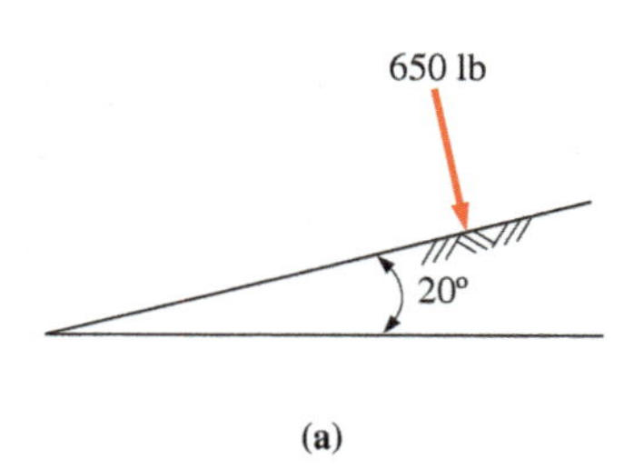

(a)

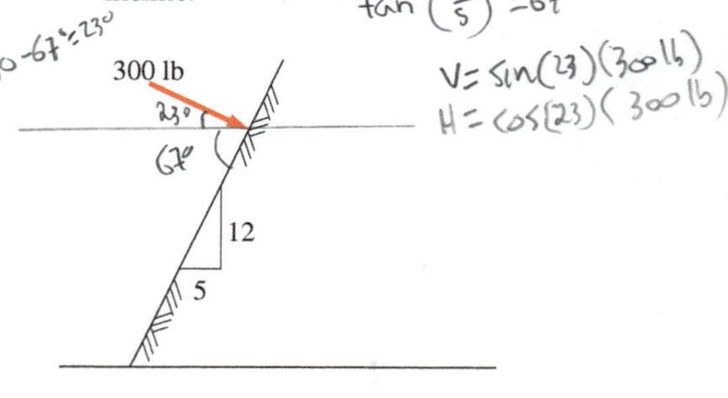

(b)

PROBLEM 2.13

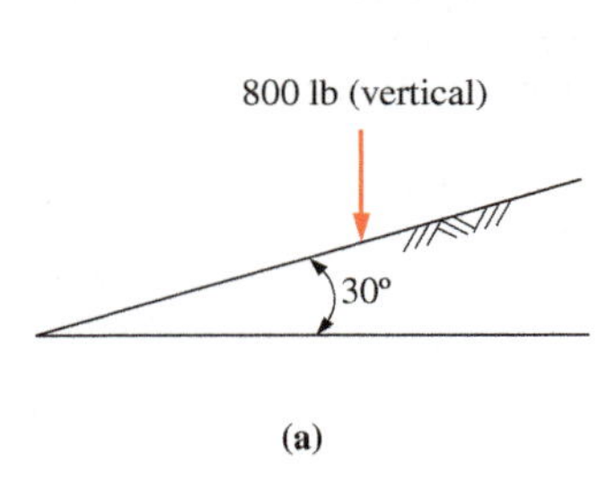

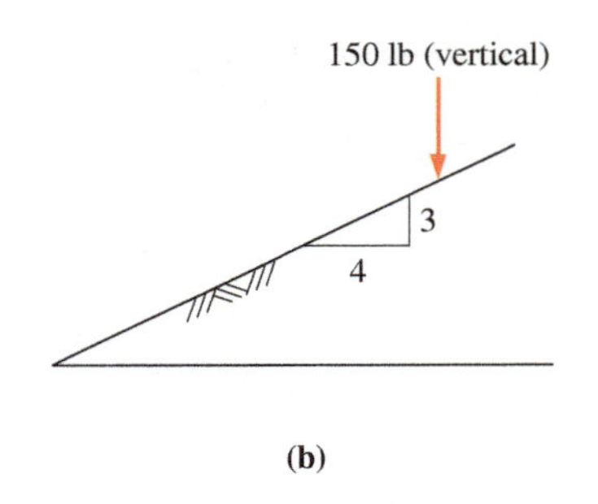

PROBLEM 2.15

2.14 Compute the vertical and horizontal components of P for the given values of P and θ. The angle is measured clockwise from the positive X axis to the force P, as shown in Problem 2.5.

(a) 120 kN, 30°

(b) 120 kN, 75°

(c) 120 kN, 275°

2.15 Compute the rectangular components parallel and perpendicular to the inclined planes shown.

2.16 Determine the vertical and horizontal components of the 1000 lb force that acts at the tip of the boom shown. (The force acts perpendicular to the boom.)

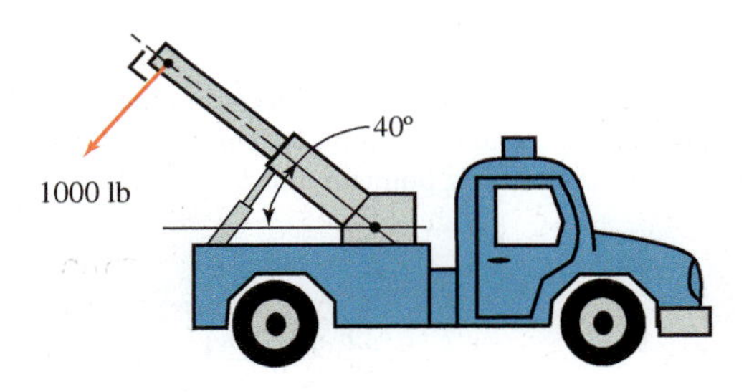

PROBLEM 2.16

2.17 Find the vertical and horizontal components of each of the given forces.

(a) A force of 80 kips acting downward and to the left at an angle of 20° with the vertical.

(b) A force of 60 kN acting upward and to the left at an angle of 30° with the vertical.

2.18 An inclined cable is used to pull on a log as shown. Determine the magnitude of the force tending to lift the log and the magnitude of the force tending to slide the log.

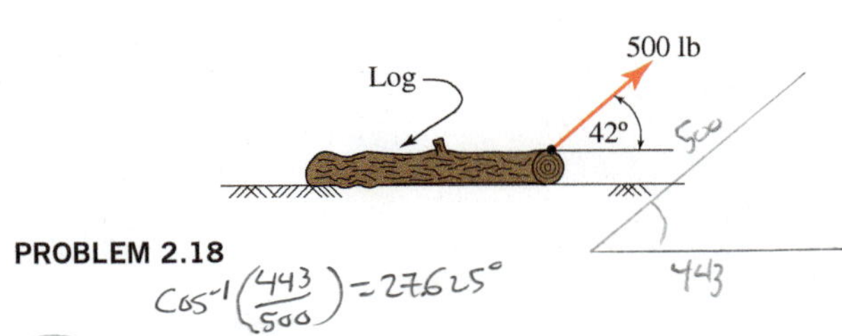

PROBLEM 2.18

2.19 In Problem 2.18, if the log will slide when a force of 443 lb is applied horizontally, what is the maximum angle at which the 500-lb force can be applied if the log is to slide?

2.20 A rope is tied to the top of a 10-m flagpole. The rope is tensioned to 120 N and then tied to a stake in the ground 20 m from the bottom of the pole. The ground is level. Find the vertical and horizontal components of force applied at the top of the flagpole.

2.21 A rope tow is used to pull skiers up a snow-covered slope that is 1000 ft long and has a slope of 17° measured from the horizontal. On a recent good skiing day it was noted that the maximum number of people on the tow never exceeded 38. Calculate the maximum tension in the rope under these conditions. Assume that the total weight of each skier is 175 lb. Assume that the slope is frictionless.

2.22 Determine the slope triangle dimension s so that the horizontal components of the two forces shown are equal in magnitude and opposite in sense. Then find the resultant of the force system.

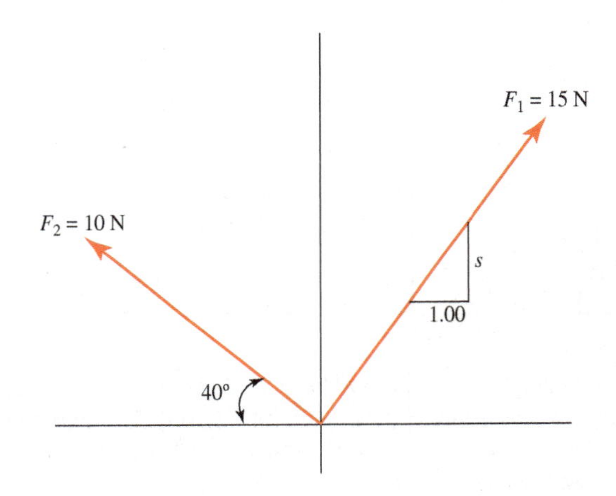

PROBLEM 2.22

2.23 The crane wheel applies a 9500-lb force to the rail as shown. Find the vertical and horizontal components.

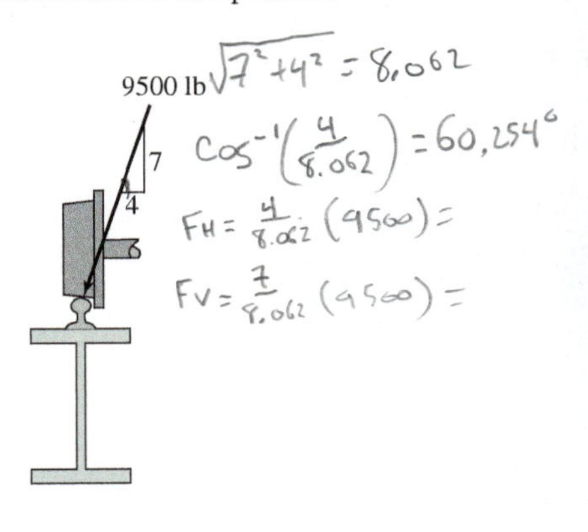

PROBLEM 2.23

2.24 The horizontal and vertical components, respectively, of a force P are given. Compute the magnitude, inclination with the X axis, and the sense of the force P.

(a) 300 lb, −200 lb
(b) −500 lb, −300 lb
(c) −240 lb, 360 lb
(d) 250 lb, 460 lb

2.25 Calculate the X and Y components of each force shown.

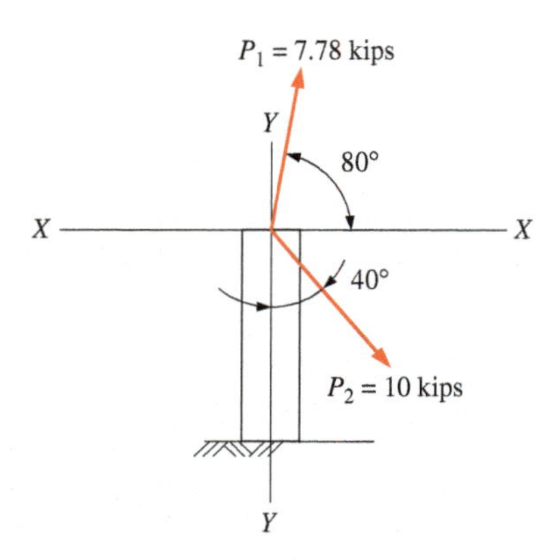

PROBLEM 2.25

2.26 Find the resultant of the two forces acting on the dam. Locate where the line of action of the resultant passes through the bottom of the dam.

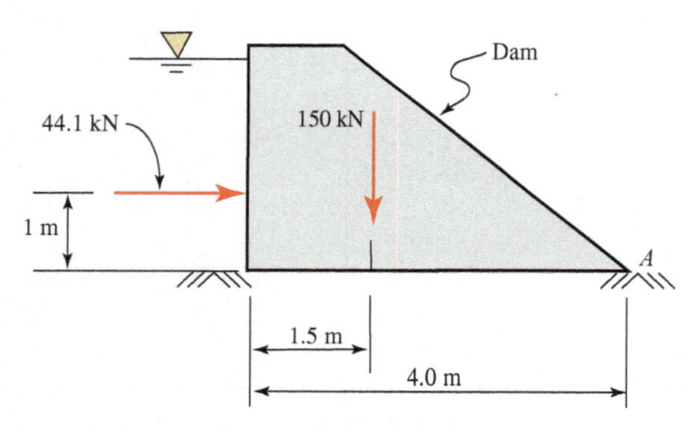

PROBLEM 2.26

2.27 A drawbar support assembly is shown in the figure. The force in the lifting-link is 2000 lb. Determine the vertical and horizontal components of the force.

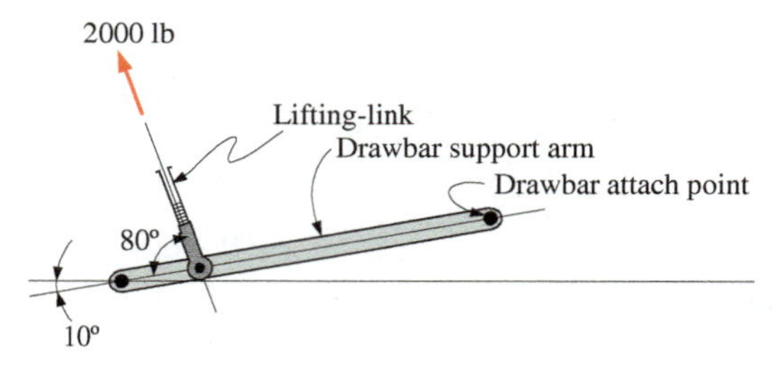

PROBLEM 2.27

2.28 The Howe roof truss shown is subjected to wind loads as shown. The wind loads are perpendicular to the inclined top chord. Resolve each force into vertical and horizontal components.

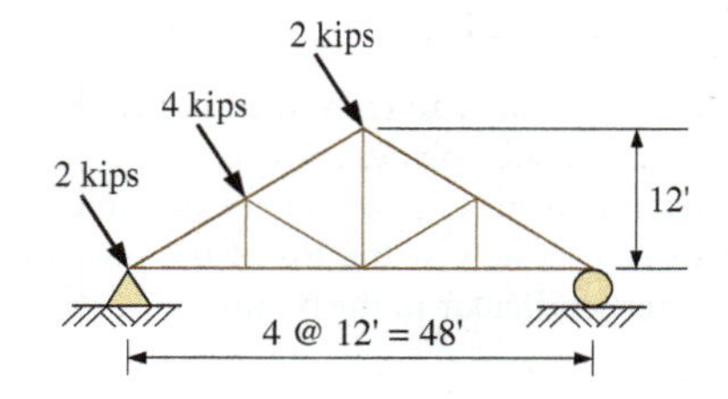

PROBLEM 2.28

2.29 The release cam mechanism is subjected to the forces shown. Resolve each force into components parallel and perpendicular to a line connecting points A and B.

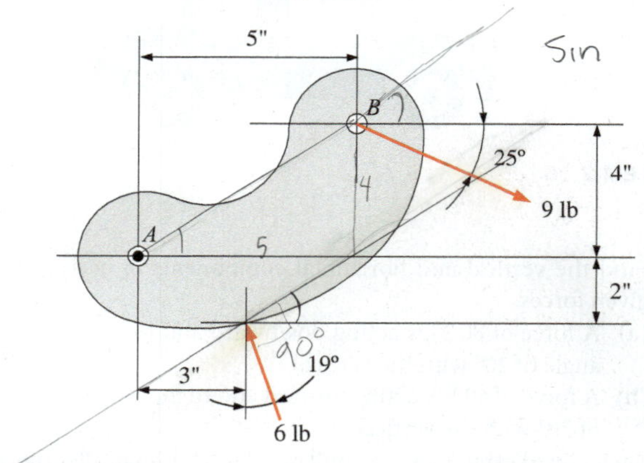

PROBLEM 2.29

RESULTANTS OF COPLANAR FORCE SYSTEMS

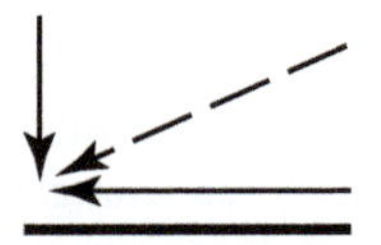

LEARNING OBJECTIVES

Upon completion of this chapter, readers will be able to:

- Determine an equivalent point load to represent a distributed load.
- Find the resultants of concurrent, parallel, and nonconcurrent coplanar force systems using parallelogram theorem and method of components.
- Calculate magnitude of moment of force for parallel and nonconcurrent force systems.
- Analyze a force couple.

3.1 RESULTANT OF TWO CONCURRENT FORCES

In Chapter 2 we established a rationale and method for determining a resultant of two orthogonal forces. It was shown that the resultant force was, in effect, equivalent to the sum of the two original forces. The reverse process in which two orthogonal components were derived from a single force was also described. The effect of the two components, acting together, was the same as the effect of the original force.

In both those situations, the two forces that acted in conjunction with each other were orthogonal, at right angles to one another, usually collinear with an X–Y axis system. This was a special case.

A more general case is one in which the coplanar concurrent forces are not acting at right angles to each other, as shown in Figure 3.1a where a barge is being towed by two tugboats. The forces on the barge are represented as P_1 and P_2. These forces are concurrent at point O.

The resultant of these two nonrectangular forces can be computed using the parallelogram law, as shown in Figure 3.1b. The principle of the parallelogram law is that two concurrent forces can be replaced by their resultant, which is represented by the diagonal of a parallelogram, the sides of which are equal and parallel to the two forces.

Note that the parallelogram $OABC$ is constructed by drawing AB and CB parallel to OC and OA (which are, respectively, forces P_1 and P_2). The parallelogram composed of forces (vectors) and angles may be drawn to some selected scale and the resultant determined using a graphical method of solution. Using an electronic calculator, however, makes it simpler and more accurate to use a freehand sketch of the parallelogram along with geometric and trigonometric relationships to obtain the resultant. Our discussion will be

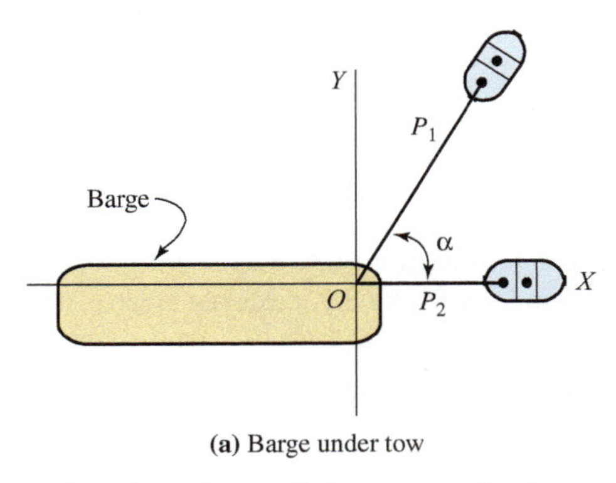

(a) Barge under tow

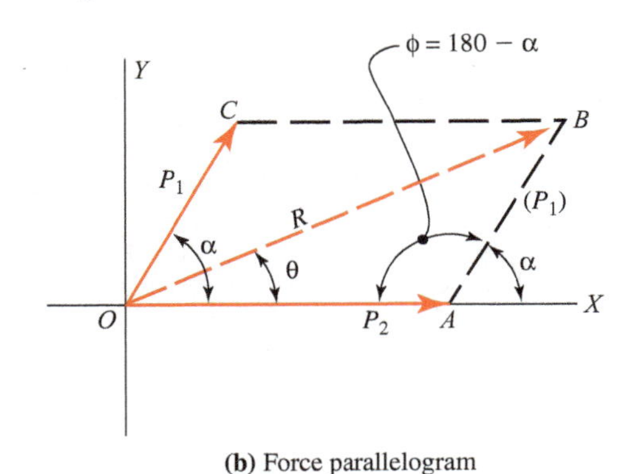

(b) Force parallelogram

FIGURE 3.1 Resultant by parallelogram method.

limited to mathematical solutions rather than to graphical solutions. You may wish to verify the solutions graphically.

To compute the magnitude and direction of the resultant, the magnitudes and directions of the two forces P_1 and P_2 must be known. It may be recognized in the parallelogram of Figure 3.1b that triangle OAB has been created, in which two sides and the included angle are known. The resultant R can then be calculated using the law of cosines, which, with reference to Figure 3.1b, can be written as

$$R^2 = P_1^2 + P_2^2 - 2P_1P_2 \cos \phi$$

The direction of the resultant can be obtained using the law of sines, which, again with reference to Figure 3.1b, can be written

$$\frac{R}{\sin \phi} = \frac{P_1}{\sin \theta}$$

from which

$$\sin \theta = \frac{P_1 \sin \phi}{R}$$

Note that in a concurrent force system the line of action of the resultant will always pass through the point of concurrence (point O). Therefore, the location of the resultant is always known, and it is necessary to determine only its magnitude, direction, and sense.

An examination of Figure 3.1b reveals that it is necessary to sketch and make computations for only one of the triangles of the parallelogram. Both triangles will yield the same result. The triangle is simply a force triangle, similar to that shown in Figure 2.16, except that this force triangle is not a right triangle. The relationship shown by the force triangle is sometimes set forth as a principle itself and is given the name the *triangle law*. The basis for this principle is that if the tail end of either force vector is placed at the arrow end of the other, the resultant force vector is the third side of the triangle, and it has a direction from the tail end of the first vector to the arrow end of the other. Note that the computations to determine the resultant are identical, based on trigonometric relationships, whether a parallelogram or a triangle is considered. A graphical solution may be used in either case.

EXAMPLE 3.1 Determine the magnitude, direction, and sense of the resultant of the two concurrent forces P_1 and P_2 shown in Figure 3.2. The forces have magnitudes of 100 lb and 140 lb, respectively, and are concurrent at point O. P_1 acts at an angle of 30° above the X axis and P_2 acts at an angle of 45° below the X axis.

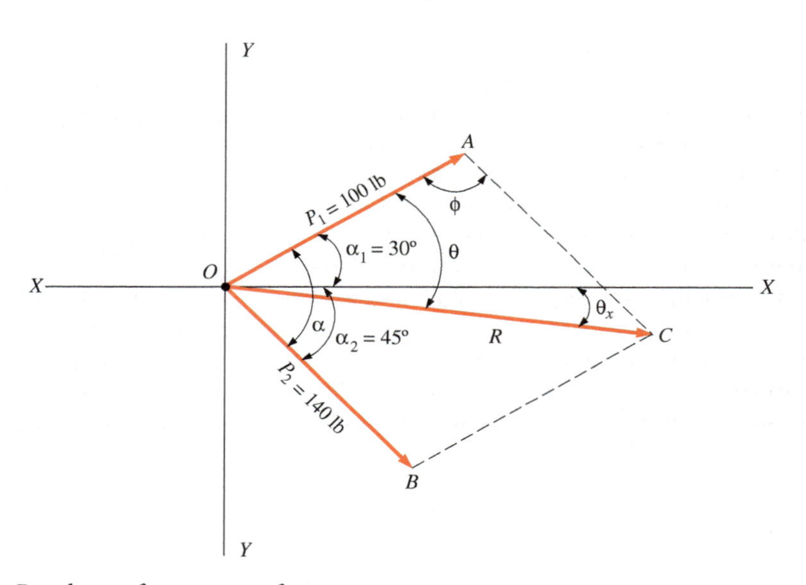

FIGURE 3.2 Resultant of concurrent force system.

Solution The parallelogram law will be used. First, construct parallelogram $OACB$ as shown, with AC equal and parallel to P_2 and BC equal and parallel to P_1. Then, from Figure 3.2, calculate angle α (the angle at O between line segments OA and OB, which represent the forces):

$$\alpha = \alpha_1 + \alpha_2 = 30° + 45° = 75°$$

Since OB and AC are parallel, angle ϕ can be calculated:

$$\phi = 180° - \alpha = 180° - 75° = 105°$$

Using triangle OAC, now calculate the magnitude of R, recognizing that side AC is equal to P_2 (or 140 lb):

$$R^2 = P_1^2 + P_2^2 - 2P_1P_2\cos\phi$$
$$= (100\text{ lb})^2 + (140\text{ lb})^2 - 2(100\text{ lb})(140\text{ lb})(-0.259)$$

from which

$$R = 192\text{ lb}$$

Note that the cosine of 105° is a negative value. Recall that the cosine of an angle between 0° and 90° is positive and that it is negative from 90° to 180°.

The direction θ of the resultant with respect to force P_1 can be calculated using the law of sines. Again, using triangle OAC,

$$\frac{R}{\sin\phi} = \frac{AC}{\sin\theta}$$

$$\theta = \sin^{-1}\left(\frac{AC\sin\phi}{R}\right)$$

Substituting, noting that the length AC is equal to the length of P_2 and equal to 140 lb:

$$\theta = \sin^{-1}\left(\frac{140\text{ lb.}(\sin 105°)}{192\text{ lb}}\right) = \sin^{-1}(0.7043)$$
$$= 44.8°$$

The direction of the resultant with respect to the horizontal X axis, designated θ_x, can be determined from Figure 3.2:

$$\theta_x = \theta - \alpha_1 = 44.8° - 30° = 14.8°$$

This angle is measured clockwise from the X axis. Therefore, the sense of the resultant is downward and to the right.

Note that in Example 3.1 the geometric diagram of the forces was based on the parallelogram law. A parallelogram was sketched and the triangular portion OAC was used to calculate the unknown resultant. Instead of the parallelogram, a force triangle could have been sketched based on the triangle law as previously defined and as shown in Figure 3.3. Note that this triangle is the same as triangle OAC in Figure 3.2. The computations using the law of cosines and the law of sines would be identical.

Another method of determining the resultant of two coplanar concurrent forces is known as the *method of components*. This method is probably the most general method, and therefore the most common method used, with applications in other types of force problems as we will see in subsequent sections. The method is shown in Figure 3.4 graphically superimposed on a parallelogram method solution.

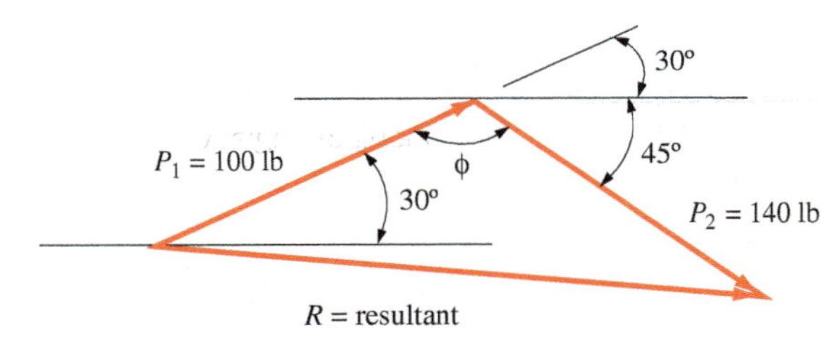

FIGURE 3.3 Force triangle.

The method of components again makes use of a selected X–Y rectangular coordinate axis system. The sequence of steps in the application of the method is as follows:

1. Calculate and algebraically sum the X components for each force (ΣF_x).
2. Calculate and algebraically sum the Y components for each force (ΣF_y).
3. The results from steps 1 and 2 represent the X and Y rectangular components of the resultant force. Designating ΣF_x as R_x and ΣF_y as R_y, the resultant R can be calculated using Equation (2.1):

$$R = \sqrt{R_x^2 + R_y^2}$$

This equation represents the magnitude of the resultant. To determine the sense of the resultant, a sketch should be drawn indicating the two rectangular components, with the resultant superimposed on an X–Y coordinate set of axes.

4. The angle of inclination between the resultant and the X axis can be observed in the sketch of step 3. It can be calculated using Equation (2.2):

$$\tan\theta_x = \frac{R_y}{R_x} = \frac{\Sigma F_y}{\Sigma F_x}$$

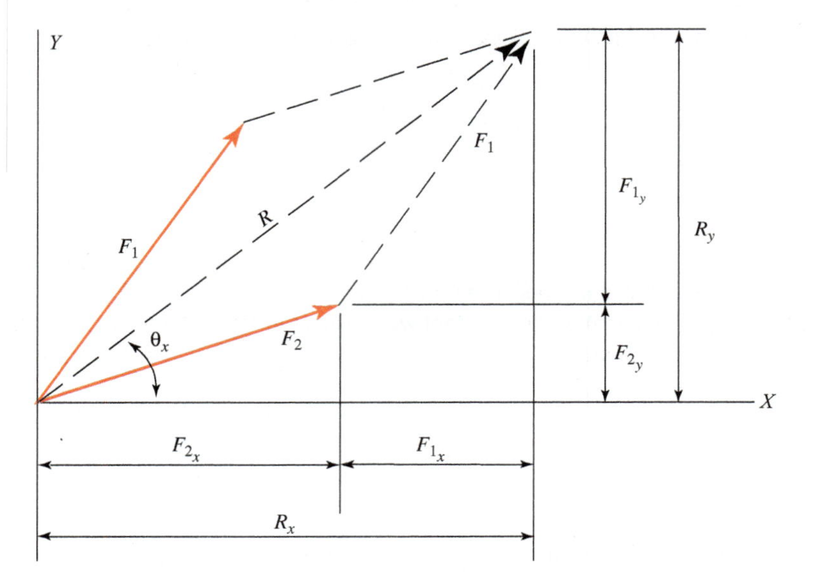

FIGURE 3.4 Resultant by method of components.

EXAMPLE 3.2 Determine the resultant R of the two-force coplanar system of Figure 3.5. Compute its magnitude, sense, and angle of inclination with the horizontal X axis.

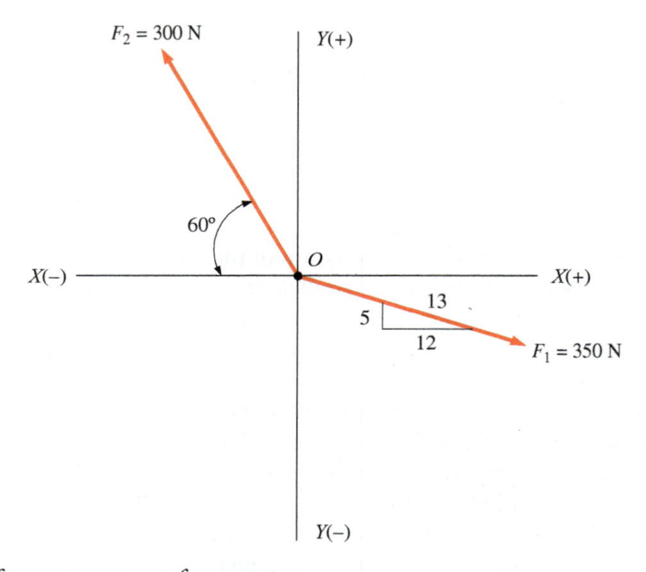

FIGURE 3.5 Two-force concurrent force system.

Solution Use the method of components. The X–Y coordinate axes with forces F_1 and F_2 are shown in Figure 3.5. Positive directions are upward and to the right.

1. The X component R_x of the resultant force is equal to the algebraic summation of the X components of F_1 and F_2:

$$\begin{aligned} R_x &= \Sigma F_x = F_{1_x} + F_{2_x} \\ &= 350\text{ N}\left(\frac{12}{13}\right) - 300\text{ N }(\cos 60^\circ) \\ &= +173.1\text{ N}\rightarrow \end{aligned}$$

The positive sign (and the arrow) indicates that the sum of the force components acts to the right.

2. The Y component R_y of the resultant force is equal to the algebraic summation of the Y components of F_1 and F_2:

$$R_y = \Sigma F_y = F_{1_y} + F_{2_y}$$
$$= -350\text{ N}\left(\frac{5}{13}\right) + 300\text{ N }(\sin 60°)$$
$$= +125.2\text{ N}\uparrow$$

The positive sign (and the arrow) indicates that the sum of the force components acts upward.

3. The results obtained in steps 1 and 2 are shown in Figure 3.6. Knowing R_x and R_y, the parallelogram (actually a rectangle) $OCBA$ is sketched along with the resultant shown as diagonal R. Using triangle OAB (in which $AB = OC$),

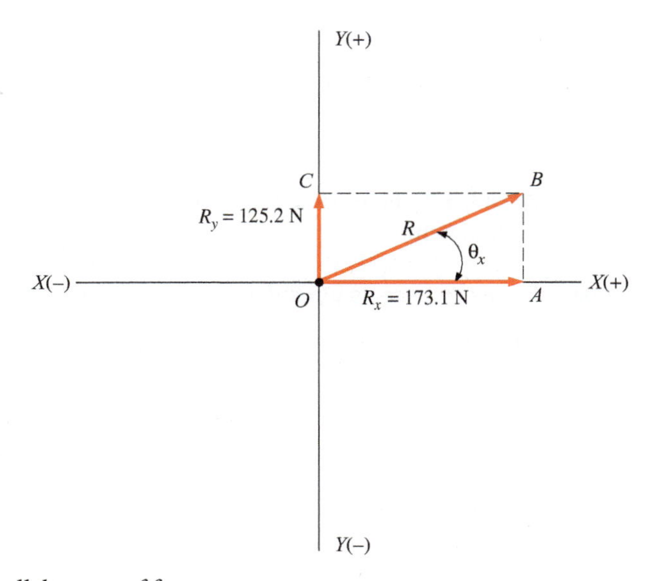

FIGURE 3.6 Parallelogram of forces.

$$R = \sqrt{R_x^2 + R_y^2}$$
$$= \sqrt{(173.1\text{ N})^2 + (125.2\text{ N})^2} = 214\text{ N}$$

The sense of the resultant is upward and to the right.

4. Using triangle OAB, compute the angle of inclination θ_x:

$$\theta_x = \tan^{-1}\frac{R_y}{R_x} = \tan^{-1}\frac{125.2\text{ N}}{173.1\text{ N}} = 35.9°$$

3.2 RESULTANT OF THREE OR MORE CONCURRENT FORCES

The method of components, as used in Section 3.1, may not appear to offer much advantage when only two forces are involved. For three or more concurrent coplanar forces, however, it is the recommended method.

The sequence of steps described in Section 3.1 is followed, with the only difference being the number of forces involved. The following example is typical of cases involving three or more concurrent forces.

EXAMPLE 3.3 Determine the resultant R of the four-force coplanar force system of Figure 3.7. Compute its magnitude, sense, and angle of inclination with the horizontal X axis. Use the method of components.

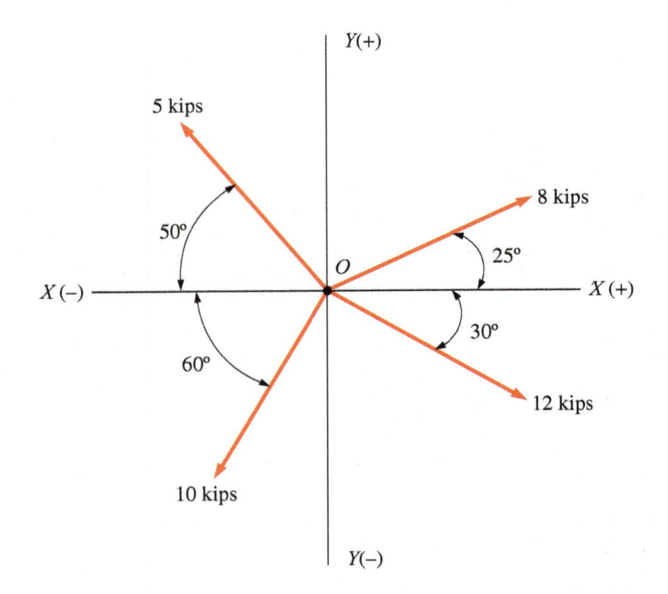

FIGURE 3.7 Four-force concurrent force system.

Solution Force components upward and to the right are considered positive. Use the following steps:

1. Calculate the X component of the resultant:

$$\begin{aligned} R_x &= \Sigma F_x \\ &= -5\text{ k}(\cos 50°) - 10\text{ k}(\cos 60°) + 8\text{ k}(\cos 25°) + 12\text{ k}(\cos 30°) \\ &= +9.43\text{ k}\rightarrow \end{aligned}$$

2. Calculate the Y component of the resultant:

$$\begin{aligned} R_y &= \Sigma F_y \\ &= +5\text{ k}(\sin 50°) - 10\text{ k}(\sin 60°) + 8\text{ k}(\sin 25°) - 12\text{ k}(\sin 30°) \\ &= -7.45\text{ k}\downarrow \end{aligned}$$

3. The results of steps 1 and 2 represent the X and Y components of the resultant force and are shown in Figure 3.8. Parallelogram $OABC$ is drawn, with diagonal R indicating the resultant. Using triangle OAB (where $AB = OC$ by construction),

$$R = \sqrt{R_x^2 + R_y^2} = \sqrt{(9.43\text{ k})^2 + (-7.45\text{ k})^2} = 12.02\text{ k}$$

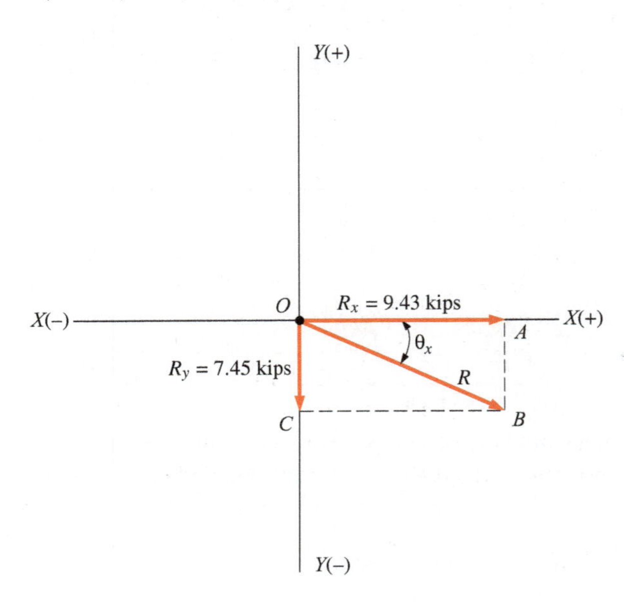

FIGURE 3.8 Parallelogram of forces.

4. Again using triangle *OAB* compute the angle of inclination of the resultant with the *X* axis:

$$\theta_x = \tan^{-1}\frac{R_y}{R_x} = \tan^{-1}\frac{7.45\ \text{k}}{9.43\ \text{k}} = 38.3°$$

Note that the equations of steps 1 and 2 could also be shown in the tabular format shown in Table 3.1.

TABLE 3.1 Tabular format for Example 3.3

Force (kips)	F_x (kips)	F_y (kips)
5	−5 cos 50° = −3.21	+5 sin 50° = +3.83
10	−10 cos 60° = −5.00	−10 sin 60° = −8.66
8	+8 cos 25° = +7.25	+8 sin 25° = +3.38
12	+12 cos 30° = +10.39	−12 sin 30° = −6.00
Σ	+9.43	−7.45

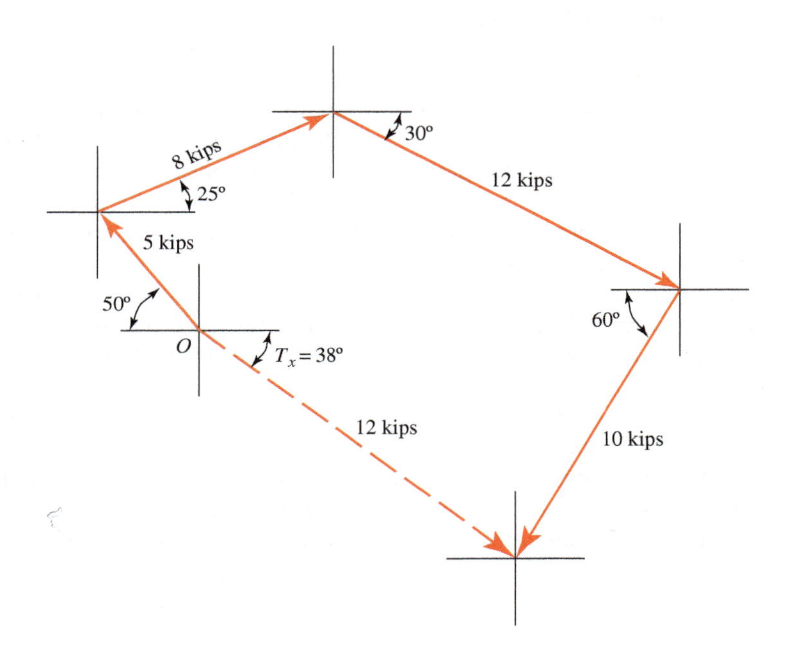

FIGURE 3.9 Graphical solution for Example 3.3.

In Section 2.8, a graphical solution for the resultant of two orthogonal forces was discussed. A graphical solution can also be employed to determine the resultant of three or more concurrent coplanar forces. This solution results from the drawing of a force polygon. For the four-force system of Example 3.3, the polygon is constructed using the magnitudes, directions, and senses from Figure 3.7 beginning at point O with the 5-kip force and placing the tail of the 8-kip force at the tip of the 5-kip force, and so on. The result is shown in Figure 3.9. The resultant closes the polygon. The magnitude and direction of the resultant shown here have been scaled from the drawing, and they compare favorably with the analytical results.

3.3 MOMENT OF A FORCE

If a force is applied to a body "at rest," the body can be disturbed in two different ways from the standpoint of planar motion. Either it can be moved as a whole up or down, to the right, or to the left (translation), or it can be turned about some fixed point (rotation). For example, as shown in Figure 3.10a, a force *P* applied at the midpoint of a free, rigid, uniform object will slide the object such that every point moves an equal distance. The object is said to *translate*. If the same force is applied at some other point as in Figure 3.10b, then the object will both translate and rotate. The amount of rotation depends on the point of application of the force. If a point on the

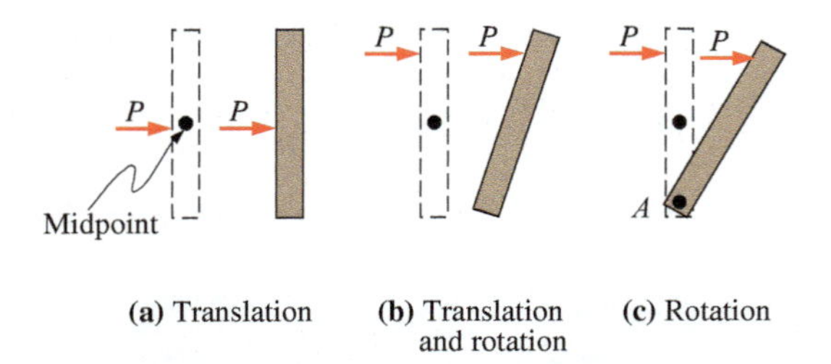

(a) Translation (b) Translation and rotation (c) Rotation

FIGURE 3.10 Types of motion.

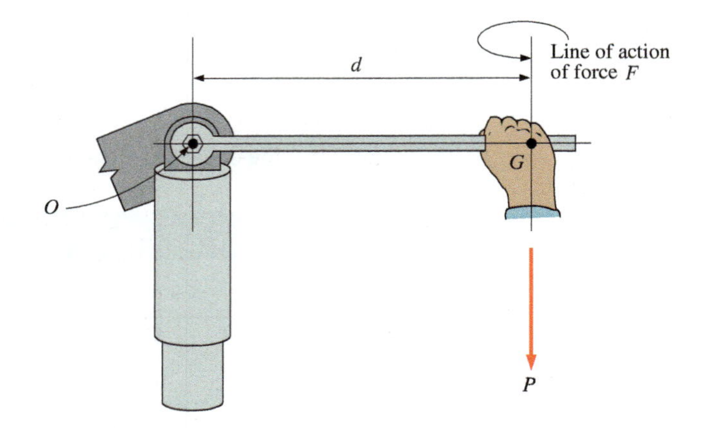

FIGURE 3.11 Moment of a force.

object is fixed against translation, as at point A in Figure 3.10c, then the applied force P causes the object to rotate only. Such rotation occurs when one pushes on a door or pulls on the handle of a wrench to turn a nut. This tendency of a force to produce rotation about some point is called the *moment* of a force. The magnitude of this tendency is directly proportional to (a) the magnitude of the force and (b) the perpendicular distance between the point of rotation and the line of action of the force.

The perpendicular distance from the line of action of the force to the point about which rotation is assumed to take place is called the *moment arm* (or lever arm). Since the tendency to cause rotation depends on both the magnitude of the force and the length of the moment arm, the moment of a force (or simply, the moment) is defined as the product of the force and its moment arm.

With reference to Figure 3.11, the moment about point O may be expressed as the product Pd. P represents a force. Point O, called the *moment center*, is a point on an axis of rotation that is perpendicular to the plane of the page. OG, of length d, is the moment arm and extends from the moment center to the point of application of force P. The moment arm OG is perpendicular to the force P and perpendicular to the axis of rotation. Therefore, as previously stated, the moment of P with respect to point O is equal to the product Pd.

Since moment is the product of force and distance, the units of moment are the product of the applicable units generally used for force and distance. In the U.S. Customary System, units for moment are lb-in., lb-ft, k-in., and k-ft. In the SI, the units are newton-meter (N • m).

Since a moment tends to produce rotation about an axis or point, a sign convention is generally used to identify the *sense* of the moment. An acceptable convention, and the one used throughout this text, is to identify a clockwise rotation as negative and a counterclockwise rotation as positive. For example, in Figure 3.12a, the force F acts along the line of action shown. Its moment arm has a length of d, and the moment of the force about the point (or axis) O is Fd. The moment has a tendency to produce rotation about point O, the moment center. It is seen that the rotation is in a clockwise direction. This moment, according to our sign convention, would have a negative (−) sense.

In Figure 3.12b, the force P, acting tangent to the wheel, creates a moment about O, which is equal to Pr, where r is the radius of the wheel. The moment Pr tends to turn the wheel in a counterclockwise direction. This moment would be considered to have a positive (+) sense.

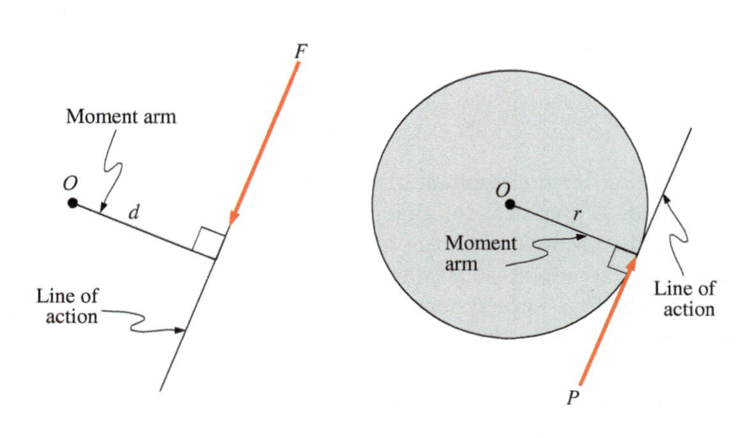

(a) Clockwise rotation of a moment—negative (−) (b) Counterclockwise rotation of a moment—positive (+)

FIGURE 3.12 Moment examples.

EXAMPLE 3.4 Three coplanar concurrent forces act on a body at point *A* as shown in Figure 3.13. (a) Calculate the moment of each of these forces about point *O*. (b) Calculate the algebraic sum of the three moments about point *O* and determine the direction of the rotation.

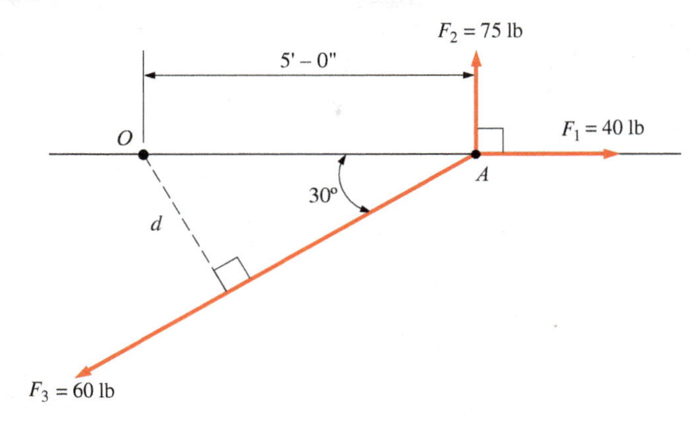

FIGURE 3.13 Moment of concurrent forces.

Solution (a) Note that point *O* lies on the line of action of the 40-lb force. Therefore, the moment of this force (about point *O*) is zero.

The moment of the 75-lb force is calculated from

$$M = Fd = 75\text{ lb}(5\text{ ft}) = +375\text{ lb-ft}$$

The positive sign is assigned because the moment is counterclockwise.

The moment of the 60-lb force is calculated from

$$M = Fd = 60\text{ lb}[5.0\text{ ft}(\sin 30^\circ)] = -150\text{ lb-ft}$$

Since the rotation is clockwise, the moment is assigned a negative sign.

(b) The algebraic summation of the three moments is

$$\Sigma M = 0 + (375\text{ lb-ft}) - (150\text{ lb-ft}) = +225\text{ lb-ft}$$

which indicates that the effect of the moments of the three forces about point *O* is to produce a counterclockwise rotation.

3.4 THE PRINCIPLE OF MOMENTS: VARIGNON'S THEOREM

In Chapter 2 it was shown that the components of a force will produce the same effect on a body as the original force. Therefore, with respect to any point, the algebraic summation of the moments of the components of a force must equal the moment of the original force. This principle is known as *Varignon's theorem*. A numerical example will illustrate the principle.

EXAMPLE 3.5 Calculate the moment about point *O* of the 200-lb force that lies in the *X–Y* plane of Figure 3.14. Use the following techniques: (a) Solve directly using the perpendicular distance from the line of action to point *O*. (b) Resolve the force into rectangular components at point *M*. (c) Resolve the force into rectangular components at point *N*.

Solution (a) Compute distance *d* shown in Figure 3.14:

$$a = 2.0\text{ ft}(\tan 30^\circ) = 1.15\text{ ft}$$

$$b = 3.0\text{ ft} - a = 3.0\text{ ft} - 1.15\text{ ft} = 1.85\text{ ft}$$

$$d = b\cos 30^\circ = (1.85\text{ ft})(0.866) = 1.60\text{ ft}$$

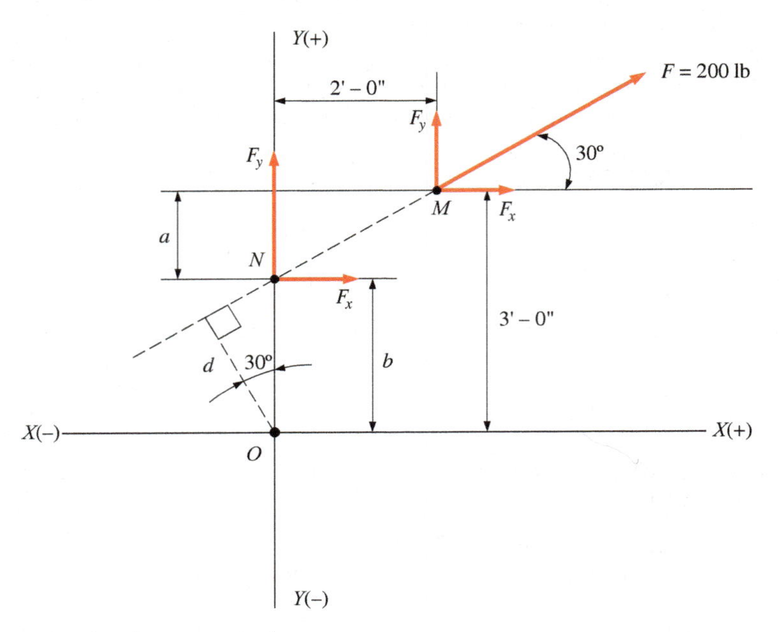

FIGURE 3.14 Example of Varignon's theorem.

Therefore, using $M = Fd$ and assigning a negative sign since the rotation is clockwise,

$$M = -200 \text{ lb}(1.60 \text{ ft}) = -320 \text{ lb-ft}$$

(b) Resolving the force F into X and Y components at point M yields

$$F_x = (200 \text{ lb}) \cos 30° = 173.2 \text{ lb}$$
$$F_y = (200 \text{ lb}) \sin 30° = 100.0 \text{ lb}$$

The algebraic summation of moments about point O is

$$\begin{aligned} \Sigma M_O &= -F_x(3.0 \text{ ft}) + F_y(2.0 \text{ ft}) \\ &= -173.2 \text{ lb}(3.0 \text{ ft}) + 100 \text{ lb}(2.0 \text{ ft}) \\ &= -320 \text{ lb-ft} \end{aligned}$$

where the negative sign indicates a clockwise rotation.

(c) Using the principle of transmissibility, the force F can be resolved into X and Y components at point N instead of point M, and the moment about point O should be the same. Note that the moment of F_y about the moment center O is equal to zero since the line of action of this component passes through point O, making the moment arm equal to zero.

$$\Sigma M_O = -F_x b = -173.2 \text{ lb}(1.85 \text{ ft}) = -320 \text{ lb-ft}$$

Again, the negative sign indicates a clockwise rotation.

In Example 3.5, the moment about point O was the same in all three cases. It is apparent, then, that the algebraic summation of moments of the components of a force is equal to the moment of the force itself.

3.5 RESULTANTS OF PARALLEL FORCE SYSTEMS

A *parallel force system* is one in which the lines of action of all the forces are parallel. The resultant of such a system will be parallel to the lines of action of the forces in the system. The magnitude, direction, and sense of the resultant can be determined by an algebraic summation of the forces. The position (location) of the line of action of the resultant may be determined by the principle of moments (Varignon's theorem); namely, that the moment of the resultant force is equal to the algebraic sum of the moments of the given forces.

As an example, consider the system of vertical parallel forces F_1, F_2, and F_3 applied to a horizontal member, as shown in Figure 3.15. Assume that the member is weightless. The magnitude and direction of the resultant R is

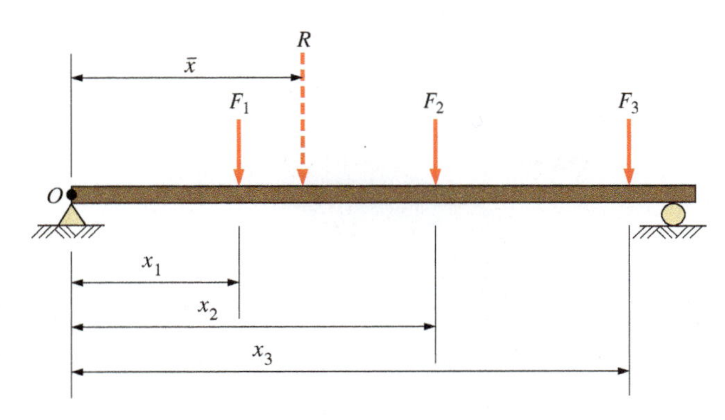

FIGURE 3.15 Parallel force system.

computed from an algebraic summation of the vertical forces, which can be written as

$$R = \Sigma F_y = F_1 + F_2 + F_3$$

The resultant is shown as a dashed arrow. The position at which the resultant is indicated is an assumed position and is dimensioned as distance $\bar{x}$ ("x-bar" or "bar x") from a moment center at O. Applying Varignon's theorem to solve for the distance $\bar{x}$,

$$\Sigma M_O = R\bar{x} = F_1x_1 + F_2x_2 + F_3x_3$$

from which

$$\bar{x} = \frac{F_1x_1 + F_2x_2 + F_3x_3}{R}$$

In more general terms, for n forces,

$$R = \Sigma F_y = F_1 + F_2 + \cdots + F_n \tag{3.1}$$

$$\bar{x} = \frac{\Sigma M}{R} \tag{3.2}$$

EXAMPLE 3.6 Determine the resultant of the parallel force system of Figure 3.16 acting on the horizontal beam *AB*. All forces are vertical. Neglect the weight of the beam.

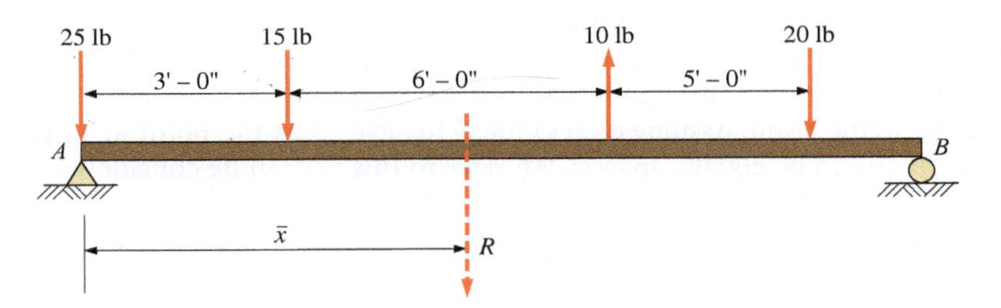

FIGURE 3.16 Parallel force system.

Solution The resultant force is equal in magnitude to the algebraic summation of the vertical forces it replaces. Assuming upward acting forces to be positive,

$$R = \Sigma F_y = -25 \text{ lb} - 15 \text{ lb} + 10 \text{ lb} - 20 \text{ lb} = -50 \text{ lb} \downarrow$$

The resultant acts vertically downward. (Recall that it must be parallel to the forces in the parallel force system.) The resultant (dashed arrow) is shown in the diagram at an arbitrary location $\bar{x}$ from point *A*. The location of the resultant can be determined by applying Equation (3.2). Using point *A* as the moment center and taking clockwise moments as negative,

$$\Sigma M_A = -15 \text{ lb}(3 \text{ ft}) + 10 \text{ lb}(9 \text{ ft}) - 20 \text{ lb}(14 \text{ ft}) = -235 \text{ lb-ft}$$

This summation of moments yields a clockwise moment about point *A*. Further, note that since the resultant force must also produce a clockwise moment about point *A*, the relative position of *R* with respect to point *A* has been correctly assumed. Equation (3.2) then yields

$$\bar{x} = \frac{\Sigma M_A}{R} = \frac{235 \text{ lb-ft}}{50 \text{ lb}} = 4.70 \text{ ft}$$

Again, note that the downward-acting resultant produces a clockwise moment, as did the original forces. Always verify the result, at this point, to ensure that it is logical. A force acting with a negative sense does not necessarily produce a negative moment. The sign conventions for forces and moments should be considered as separate entities.

(a) Snow load on a roof

Each sandbag weighs 60 lb

(b) Sandbags on a beam

FIGURE 3.17 Distributed loads.

Up to this point, problems and discussions have been limited to concentrated forces. The principles and methods used in determining the resultant of a parallel force system, however, are also applicable when dealing with distributed loads or with combinations of distributed and concentrated loads as applied on structures and structural elements.

Distributed loads were defined and briefly discussed in Section 2.4. Consider a flat roof on which a generally uniform depth of snow has fallen as shown in Figure 3.17a. Unlike a concentrated load, the snow is spread out over the surface of the roof. This is referred to as a *uniformly distributed load*. Figure 3.17b depicts a beam that supports a line of sandbags. The load (the sandbags) is distributed along the length of the beam. Assume each sandbag to have a similar weight and to be equally spaced. We refer to this load as a *uniformly distributed line load*. In this text, distributed loads are predominantly distributed line loads.

We will discuss distributed loads that cover surfaces later in this section (hydrostatic forces) and in Chapter 15 (floors and roofs).

3.5.1 Uniformly Distributed Line Loads

The load of sandbags shown on the beam in Figure 3.17b is an example of a uniformly distributed line load. If each sandbag were to weigh 60 lb and occupy 1 lineal foot along the length of the beam, we would say that the intensity of this distributed line load was 60 lb/ft. The graphical representation of a uniformly distributed line load is a series of arrows within a rectangle, as shown in Figure 3.18. This representation defines both the intensity of the uniformly distributed line load and the length of the member over which it acts. The notation commonly used for the intensity of the uniformly distributed line load is w. Units for the load intensity are usually lb/ft or k/ft in the U.S. Customary System and N/m in the SI. As an example, if we consider a beam with a uniform cross section throughout its length, the weight of the beam (lb/ft) would be a uniformly distributed line load.

We can visualize the distributed line load as being equivalent to a series of concentrated loads. In the example of the sandbags in Figure 3.17b, the sandbag load is equivalent to a series of 60-lb concentrated loads placed along the beam at 1-ft intervals. If the beam were 8 ft long (8 sandbags), the total load would be $60\ \text{lb/ft} \times 8\ \text{ft} = 480\ \text{lb}$. This load is a parallel force system, and the location of the resultant is easily found using procedures already discussed.

Consider the distributed line load of Figure 3.18 to be a rectangular area with the length being the length of the beam (8 ft) and the height being the load intensity (60 lb/ft). The magnitude of the resultant of the force system is equal to the area (length $\times$ height) of the rectangle. The resultant is located at the center of the rectangle, which is called the *centroid* of the rectangle; it is the point at which all of the area within the rectangle can be considered to be concentrated. The centroid of an area is similar to the center of gravity of a body, the center of gravity being that point at which all the weight of the body is considered concentrated.

Another way to visualize centroid and center of gravity is as a balance point. If a rectangular area is balanced on the tip of a finger as shown in Figure 3.19a, the finger is positioned directly below the centroid. The center of gravity of a baseball bat can be located using trial and error by balancing the bat on a fulcrum as shown in Figure 3.19b. Centroids and centers of gravity are discussed in detail in Chapter 7.

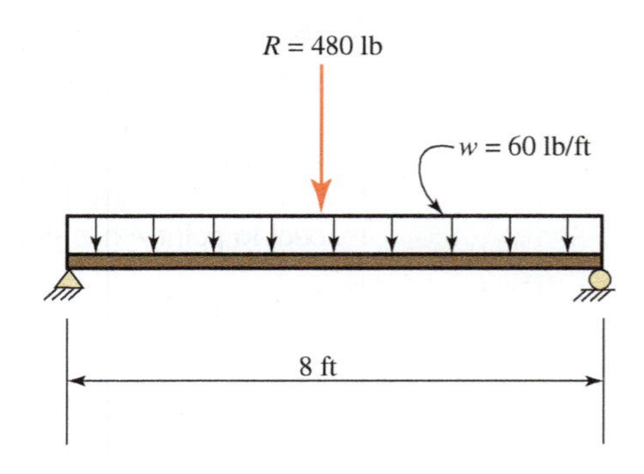

FIGURE 3.18 Uniformly distributed line load.

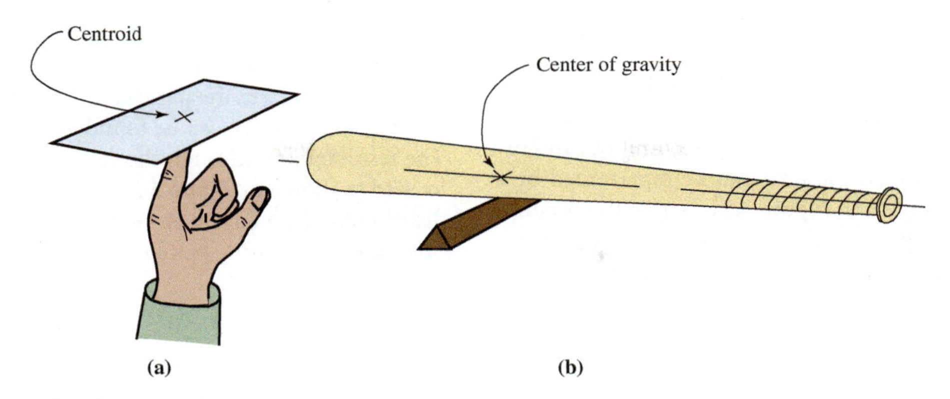

FIGURE 3.19 Centroid and center of gravity.

EXAMPLE 3.7

Determine the magnitude and location of the resultant R of the parallel force system of Figure 3.20. The force system acts on a horizontal beam AB. All forces are vertical. Neglect the weight of the beam.

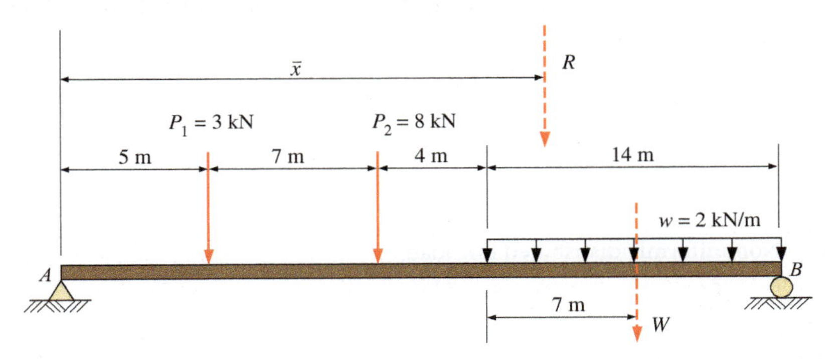

FIGURE 3.20 Parallel force system.

Solution Note that this is a combined force system consisting of concentrated loads and a uniformly distributed load. The uniformly distributed load may be replaced by its equivalent concentrated resultant force. This equivalent concentrated force is often denoted as W. Therefore,

$$W = 2\ \text{kN/m} \times 14\ \text{m} = 28\ \text{kN}$$

Note also that the resultant W of the uniformly distributed load is indicated by a dashed arrow to distinguish it from the given concentrated loads and that the location of the dashed arrow is shown at the center of the distributed-load portion. (The equivalent concentrated resultant force of the uniformly distributed load always acts through the centroid of the distributed-load portion.)

The magnitude of the resultant force of the total force system is equal to the algebraic summation of the vertical forces it replaces. Assume upward acting forces are positive:

$$R = \Sigma F_y = -3\ \text{kN} - 8\ \text{kN} - 28\ \text{kN} = -39\ \text{kN} \downarrow$$

The resultant force acts vertically downward. In Figure 3.20, it is indicated by a dashed arrow R. The location of the resultant is unknown at this time, but it is shown acting a distance $\overline{x}$ from a moment center at point A. Determine $\overline{x}$ by applying Equation (3.2). Use point A as the moment center and assume clockwise moments negative:

$$\Sigma M_A = -3\ \text{kN}(5.0\ \text{m}) - 8\ \text{kN}(12\ \text{m}) - 28\ \text{kN}(23\ \text{m}) = -755\ \text{kN}\cdot\text{m}$$

The negative sign indicates that the moment about point A is clockwise. Resultant R must also produce an equal clockwise moment about point A. Equation (3.2) yields

$$\overline{x} = \frac{\Sigma M_A}{R} = \frac{755\ \text{kN}\cdot\text{m}}{39\ \text{kN}} = 19.4\ \text{m}$$

Since the moment of the given forces was clockwise, the downward acting R must be located to the right of point A as shown, to also produce a clockwise moment.

3.5.2 Nonuniformly Distributed Line Loads

Another common type of distributed line load is one in which the load intensity varies along the length of the load. Although the variation can be complex, we limit our discussion to the case in which the load can be represented by a triangular area, as shown in Figure 3.21, where the nonuniformly distributed line load is fully described if the maximum intensity and the length of the line load are known. The intensity decreases at a steady rate along the length of the load until it reaches zero. Once again, a triangular area can be used to represent the parallel force system. The magnitude of the resultant force can be found as the area of the triangle. The resultant acts through the centroid of the triangle and is located as shown in Figure 3.21. Therefore, the nonuniformly distributed line load can be replaced by its resultant force for further use in treating parallel force systems.

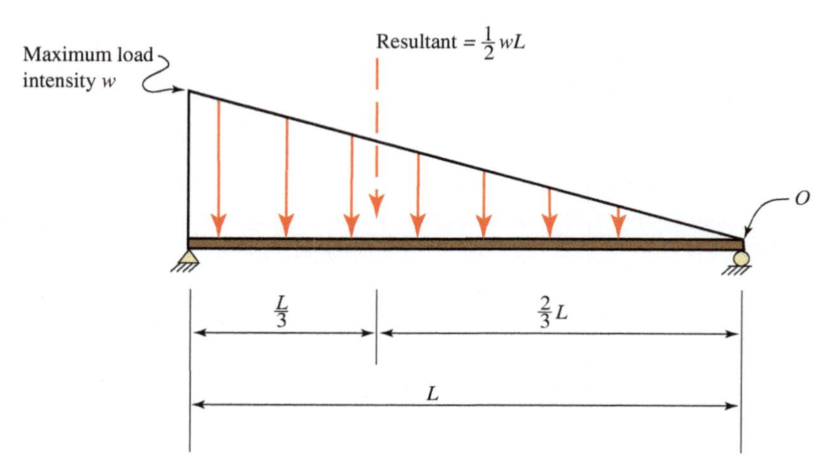

FIGURE 3.21 Nonuniformly distributed line load.

EXAMPLE 3.8 Determine the magnitude and location of the resultant of the parallel force system of Figure 3.22a. The force system acts on a horizontal beam AB. All forces are vertical. Neglect the weight of the beam.

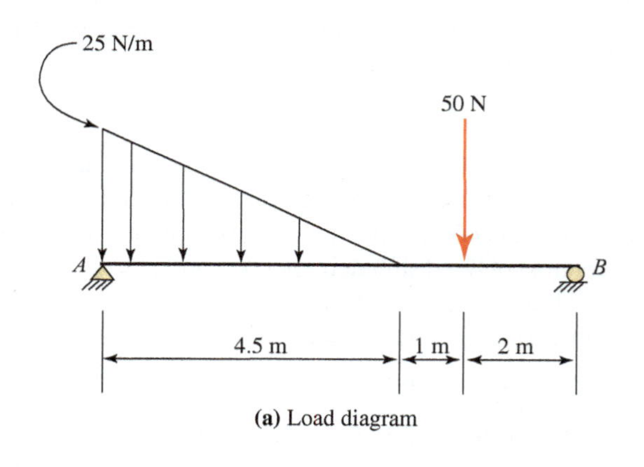

(a) Load diagram

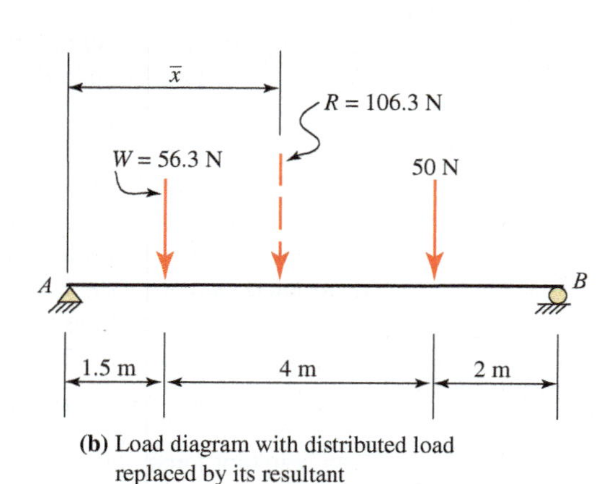

(b) Load diagram with distributed load replaced by its resultant

FIGURE 3.22 Diagram for Example 3.8.

Solution The magnitude of the resultant of the nonuniformly distributed line load is calculated from the area of the triangle:

$$W = \frac{4.5\ \text{m}(25\ \text{N/m})}{2} = 56.3\ \text{N}$$

The location of the resultant of this distributed load is one-third of the base $\left(\frac{1}{3} \times 4.5\ \text{m} = 1.5\ \text{m}\right)$ from A and is shown in Figure 3.22b. The magnitude of the resultant of the entire force system (assuming upward acting forces are positive) is calculated from

$$R = \Sigma F_y = -56.3\ \text{N} - 50\ \text{N} = -106.3\ \text{N} \downarrow$$

and is shown in Figure 3.22b. Take moments of the applied forces about A (assume clockwise moments negative):

$$\Sigma M_A = -56.3\ \text{N}(1.5\ \text{m}) - 50\ \text{N}(5.5\ \text{m}) = -359\ \text{N}\cdot\text{m}$$

From Equation (3.2) solve for $\overline{x}$:

$$\overline{x} = \frac{\Sigma M_A}{R} = \frac{359\ \text{N}\cdot\text{m}}{106.3\ \text{N}} = 3.38\ \text{m}$$

3.5.3 Hydrostatic Forces

A practical example of a distributed load is the force imparted by water on the side of a pool, tank, or other container or on any submerged object. A hydrostatic analysis reduces a physical system to equivalent loads or forces that can then be handled using familiar techniques. Although hydrostatic concepts can be generalized to other liquids, we limit our discussion to water.

Water is nearly incompressible, and its unit weight (or specific weight) is usually assumed constant for a particular temperature regardless of the pressure on the water. For ordinary temperatures, the unit weight of water is assumed to be 62.4 lb/ft^3 in U.S. Customary units or 9800 N/m^3 in SI units.

Water in an open tank conforms to the shape of the tank and fills it to a certain depth. The surface of the water in contact with the atmosphere is called the *free surface*. Most hydrostatic problems involving open tanks are analyzed using relative or gage pressure, which assumes the pressure at the free surface to be zero. Actually, the pressure at the free surface is that of the atmosphere, and absolute pressure p_{abs} is the sum of the gage pressure p_g and atmospheric pressure p_{atm}. This sum is expressed as

$$p_{abs} = p_g + p_{atm}$$

If you have gone diving for coins in a pool of water, you have felt the pressure increase in your ears as you swim deeper. Pressure in a pool or tank (river, ocean, etc.) increases with depth and can be found using the relation

$$p = \gamma h$$

where p = gage pressure
γ = unit weight of water
h = the depth below the free surface

The pressure, and therefore the hydrostatic force, will always act only perpendicular (or normal) to any submerged surface or object.

EXAMPLE 3.9 A 4 ft × 4 ft square plate is held at the bottom of a tank with 8 ft of water in it, as shown in Figure 3.23. Find the equivalent hydrostatic force on the plate.

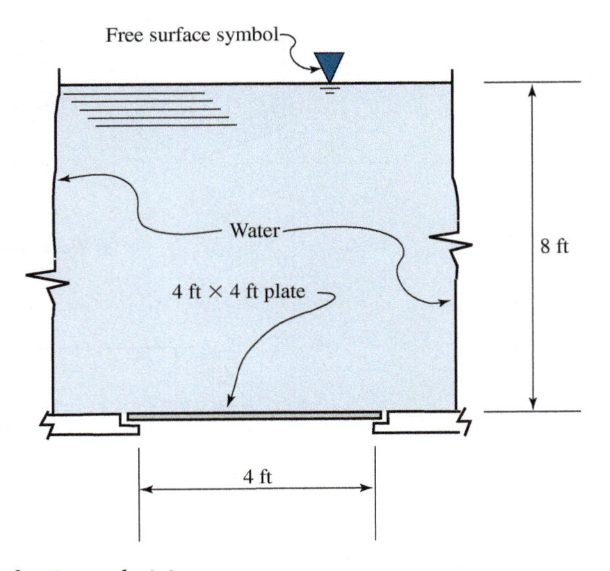

FIGURE 3.23 Sketch for Example 3.9.

Solution We first calculate the pressure on the plate. Note that this will be a uniform pressure since the height of water is constant.

$$P = \gamma h = (62.4\ \text{lb/ft}^3)(8\ \text{ft}) = 499\ \text{lb/ft}^2$$

Pressure is defined as force per unit area ($p = F/A$); therefore, we can find force F from

$$
\begin{aligned}
F &= pA \\
&= (499\ \text{lb/ft}^2)(4\ \text{ft})^2 \\
&= 7990\ \text{lb}
\end{aligned}
$$

The pressure on the plate can be represented as a uniformly distributed load as shown in Figure 3.24. The resultant force of 7990 lb acts through the centroid of the distributed load and through the centroid (center) of the plate.

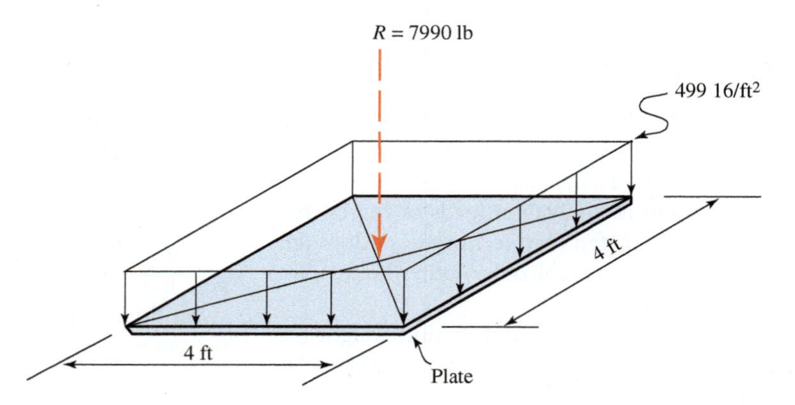

FIGURE 3.24 Force on a 4 ft × 4 ft plate.

EXAMPLE 3.10 Water completely fills the tank shown in Figure 3.25a. Find the equivalent hydrostatic resultant force on a typical 1-m width of wall *ABCD*.

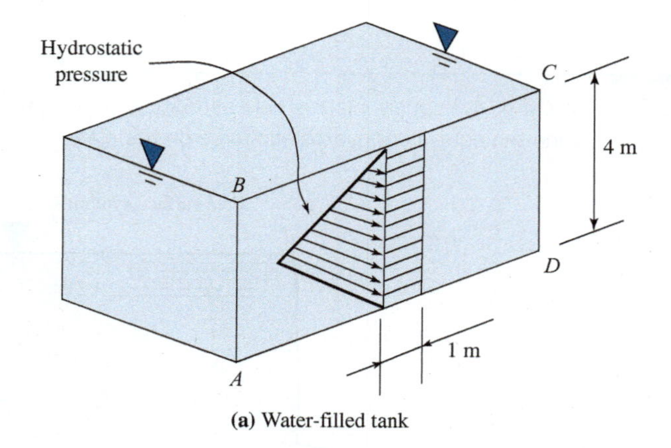

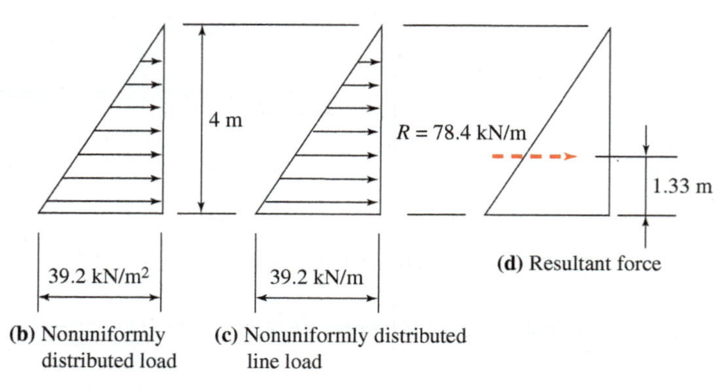

FIGURE 3.25 Sketches for Example 3.10.

Solution Since $p = \gamma h$, the pressure increases with depth, and this creates a nonuniformly distributed load on the vertical wall of the tank. The maximum pressure is at the bottom of the tank and has a magnitude of

$$p = \gamma h = (9800\ \mathrm{N/m^3})(4\ \mathrm{m}) = 39.2\ \mathrm{kN/m^2}$$

as shown in Figure 3.25b. Recognizing that we are analyzing a 1-m-wide strip of wall, we can consider this to be a load intensity of 39.2 kN/m (per meter). The nonuniformly distributed line load then appears as a triangle, as shown in Figure 3.25c.

We next find the magnitude of the resultant force by calculating the area of the triangular load:

$$R = \frac{(39.2\ \mathrm{kN/m})\ (4\ \mathrm{m})}{2} = 78.4\ \mathrm{kN}\ (\text{per meter})$$

The location of the resultant force is at the centroid of the triangle, one-third of the height above the base:

$$y = \frac{1}{3}(4\ \mathrm{m}) = 1.33\ \mathrm{m}$$

The resultant is shown in Figure 3.25d.

3.6 COUPLES

A coplanar parallel force system composed of two parallel forces having different lines of action, equal in magnitude but opposite in sense, constitutes a special case. This force system is called a *couple* and is illustrated in Figure 3.26.

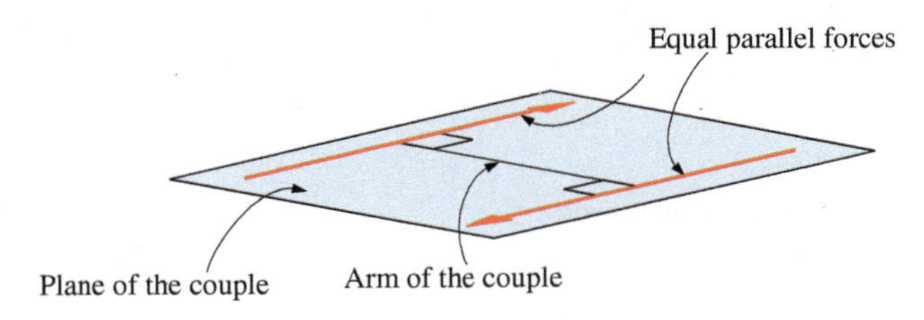

FIGURE 3.26 Couple.

The perpendicular distance between the lines of action of the two parallel forces is called the *arm of the couple*, and the plane in which the lines of action lie is called the *plane of the couple*. A couple either causes or tends to cause rotation about an axis perpendicular to its plane. When the driver of an automobile grasps opposite sides of the steering wheel and turns it, a couple is being applied to the wheel.

The following characteristics of couples should be noted:

1. The moment of a couple is the product of one of the forces and the arm of the couple.
2. The moment of a couple is independent of the choice of the axis of moments (moment center). The moment of a couple is the same with respect to any axis perpendicular to the plane of the couple (or any point in the plane of the couple).
3. The magnitude of a resultant force of a couple is zero. Therefore, a couple cannot be replaced with a single equivalent resultant force.
4. If a given force system is composed entirely of couples in the same plane, the resultant moment will consist of another couple equal to the algebraic summation of the original couples.
5. A couple can be balanced only by an equal and opposite couple in the same plane.
6. A couple is fully defined by its magnitude and sense of rotation. If the sense of rotation is counterclockwise, it will be considered positive.
7. A couple may be transferred to any location in its plane and still have the same effect.

EXAMPLE 3.11 Calculate the magnitude of the moment of the two parallel forces shown in Figure 3.27. (a) Use point O as the moment center. (b) Use point A as the moment center.

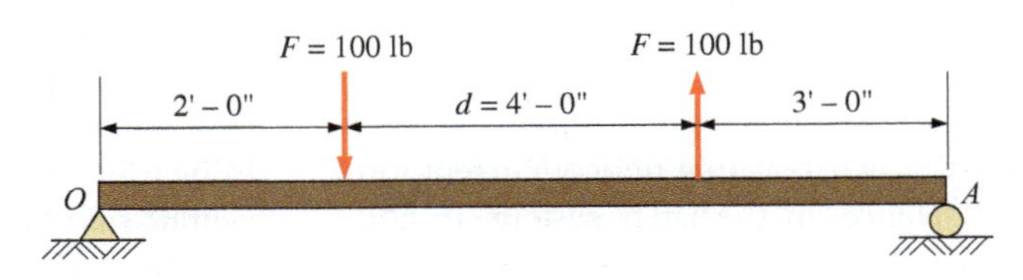

FIGURE 3.27 Special parallel force system.

Solution Since the two forces are parallel with different lines of action, equal in magnitude, and opposite in sense, they constitute a couple. The magnitude of a couple is equal to the product of one of the forces and the perpendicular distance between the forces and may be expressed as

$$M = Fd = 100 \text{ lb}(4.0 \text{ ft}) = +400 \text{ lb-ft}$$

Note that the result is a positive value since the couple is acting counterclockwise.

One of the characteristics of a couple is that its magnitude is the same with respect to any point in the plane of the couple. Next compute the moment of the couple with respect to points *O* and *A*:

$$\Sigma M_O = -100 \text{ lb}(2 \text{ ft}) + 100 \text{ lb}(6 \text{ ft}) = +400 \text{ lb-ft}$$
$$\Sigma M_A = -100 \text{ lb}(3\text{ft}) + 100 \text{ lb}(7 \text{ ft}) = +400 \text{ lb-ft}$$

Note that the couple has the same counterclockwise moment effect of 400 lb-ft, irrespective of where the moment center is chosen.

EXAMPLE 3.12 Four vertical forces with parallel lines of action acting on a horizontal beam are shown in Figure 3.28. Calculate the resultant moment with respect to (a) point *O*, (b) point *B*, and (c) point *C*.

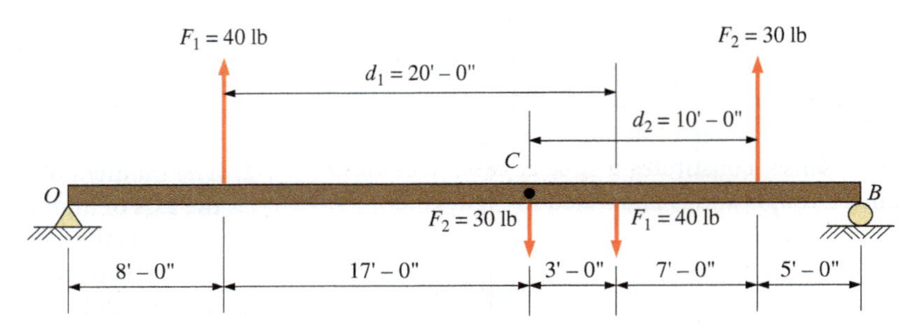

FIGURE 3.28 Parallel force systems.

Solution Moment summations at the individual points are taken as follows:

$$\Sigma M_O = +(40 \text{ lb})(8 \text{ ft}) - (30 \text{ lb})(25 \text{ ft}) - (40 \text{ lb})(28 \text{ ft}) + (30 \text{ lb})(35 \text{ ft}) = -500 \text{ lb-ft}$$
$$\Sigma M_B = -(40 \text{ lb})(32 \text{ ft}) + (30 \text{ lb})(15 \text{ ft}) + (40 \text{ lb})(12 \text{ ft}) - (30 \text{ lb})(5 \text{ ft}) = -500 \text{ lb-ft}$$
$$\Sigma M_C = -(40 \text{ lb})(17 \text{ ft}) + (30 \text{ lb})(10 \text{ ft}) - (40 \text{ lb})(3 \text{ ft}) = -500 \text{ lb-ft}$$

Note that the moment effect is the same in each case and is irrespective of the moment center. Also note that the member is subjected to two couples. You can determine a resultant couple equal to the algebraic summation of the two couples acting on the member as follows:

$$M = M_1 + M_2 = F_1 d_1 + F_2 d_2$$
$$= -(40 \text{ lb})(20 \text{ ft}) + (30 \text{ lb})(10 \text{ ft}) = -500 \text{ lb-ft}$$

The result—a clockwise moment of 500 lb-ft—is identical to the moment summations at the individual points.

3.7 RESULTANTS OF NONCONCURRENT FORCE SYSTEMS

In a concurrent force system, the lines of action of the forces meet at a common point, whereas in a nonconcurrent force system, they do not. Hence, in the latter system, in addition to the unknown magnitude, sense, and direction, the location of the line of the action of the resultant is also unknown. As with every force, the resultant of a coplanar nonconcurrent force system is defined by its magnitude, direction, sense, and the location of its line of action.

The magnitude, direction, and sense can be calculated in a way similar to that for the concurrent force system, that is, by using an *X–Y* coordinate axis system and taking an algebraic summation of the *X* and *Y* components of each force. The location of the line of action of the resultant can then be determined by computing its moment arm with respect to some convenient moment center. This latter step is similar to the procedure for locating the resultant of a coplanar parallel force system and is based on Varignon's theorem.

A common error in the analysis of this type of force system is to place the resultant on the wrong side of the moment center. As with the parallel force system, however, the location of the resultant with respect to a moment center is based on the fact that the resultant must produce the same effect as the original force system.

The following example will illustrate a method of determining the resultant of a coplanar nonconcurrent force system.

EXAMPLE 3.13 Determine the magnitude, direction, and sense of the resultant of the coplanar nonconcurrent force system of Figure 3.29. Locate the resultant with respect to point O.

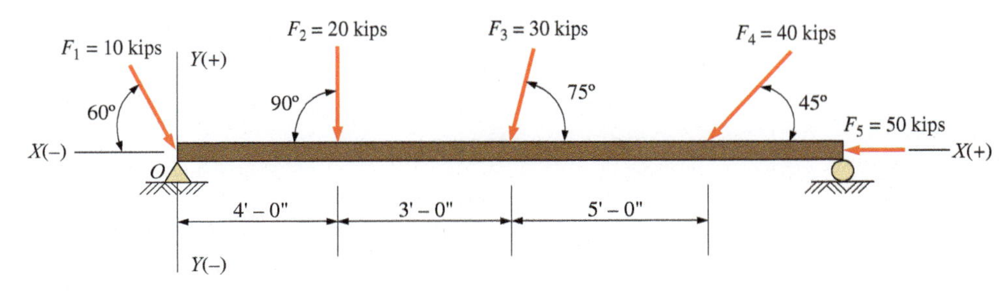

FIGURE 3.29 Nonconcurrent force system.

Solution Forces upward and to the right are positive.

1. First compute the X component of the resultant force:

$$\begin{aligned} R_x &= \Sigma F_x \\ &= +10 \text{ k} \cos 60° - 30 \text{ k} \cos 75° - 40 \text{ k} \cos 45° - 50 \text{ k} \\ &= -81.0 \text{ k} \leftarrow \end{aligned}$$

2. Next compute the Y component of the resultant force:

$$\begin{aligned} R_y &= \Sigma F_y \\ &= -10 \text{ k} \sin 60° - 20 \text{ k} - 30 \text{ k} \sin 75° - 40 \text{ k} \sin 45° \\ &= -85.9 \text{ k} \downarrow \end{aligned}$$

3. Using R_x and R_y, sketch a parallelogram with the resultant shown as the diagonal (see Figure 3.30). Using triangle OAB (with AB equal to OC, by construction), compute the magnitude of the resultant from Equation (2.1):

$$R = \sqrt{R_x^2 + R_y^2} = \sqrt{(-81.0 \text{ k})^2 + (-85.9 \text{ k})^2} = 118.1 \text{ k}$$

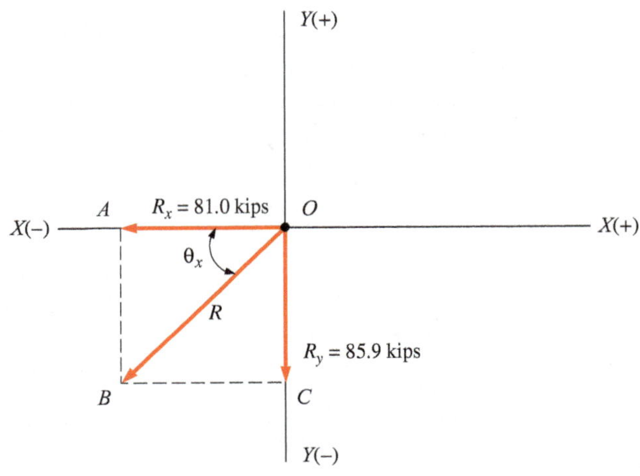

FIGURE 3.30 Parallelogram of forces.

Note that the sense of the resultant is downward and to the left.

4. Again using triangle OAB, compute the angle of inclination θ_x:

$$\theta_x = \tan^{-1}\frac{R_y}{R_x} = \tan^{-1}\frac{85.9\text{ k}}{81.0\text{ k}} = 46.7°$$

5. Note that the magnitude, direction, and sense of the resultant have been determined. You must, however, determine its location relative to some moment center. The point O in Figure 3.31 will be used as the moment center, and the location of the resultant ($\bar{x}$ from point O) will be determined.

   Use Varignon's theorem, taking moments about point O in Figure 3.29. The most convenient solution is to consider the moments due to the Y components of the applied forces. The moment arms for these components are measured along the X axis. Since the lines of action of the X components pass through the moment center at O, their effects disappear.

The summation of moments of the Y components of the applied forces with respect to point O (clockwise negative) is

$$\begin{aligned}\Sigma M_O &= -(20\text{ k})(4\text{ ft}) - (30\text{ k})(\sin 75°)(7\text{ ft}) - (40\text{ k})(\sin 45°)(12\text{ ft})\\ &= -622\text{ k-ft}\end{aligned}$$

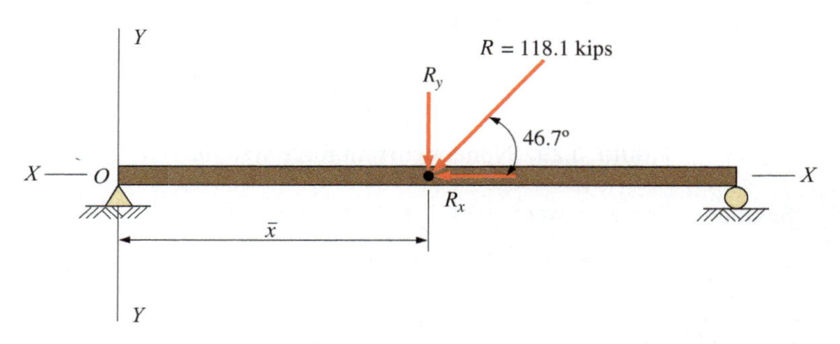

FIGURE 3.31 Location of resultant.

Since the resultant must have the same moment effect about point O as all the other given forces, it is evident from Figure 3.31 that the location of the resultant must be to the right of point O, as shown. This location will provide a clockwise moment, which is in agreement with the moment effect of the given forces.

Resolve the resultant R into its components (where it intersects the X axis) and take moments about point O:

$$\Sigma M_O = R_y\bar{x}$$

from which

$$\bar{x} = \frac{\Sigma M_O}{R_y} = \frac{622\text{ k-ft}}{85.9\text{ k}} = 7.24\text{ ft}$$

Note that the signs of the moment and the force in the preceding expression for $\bar{x}$ are neglected. The signs for moment and force are separate entities. Using this approach, as discussed, the $\bar{x}$ distance will always be positive, and you establish the necessary correlation of the senses of the force and moment in the final calculation by inspection of Figures 3.29 through 3.31.

SUMMARY BY SECTION NUMBER

3.1 and 3.2 In these sections, the resultant is defined. Resultants of concurrent coplanar force systems may be found using the parallelogram law, the triangle law, or the method of components.

3.3 The magnitude of the moment of a force equals the product of the force and moment arm. The moment arm is the perpendicular distance from the line of action of the force to the moment center. In the sign convention for this text, counterclockwise is positive.

3.4 Varignon's theorem: the algebraic sum of the moments of the components of a force system is equal to the moment of the resultant force of the system.

3.5 Resultant of a parallel force system:

$$R = \Sigma F_y = F_1 + F_2 + \cdots + F_n$$

$$\bar{x} = \frac{\Sigma M}{R}$$

Resultants of distributed loads act at the centroids of the load areas.

3.6 A couple is a special case of a coplanar parallel force system. It is composed of two equal, opposite, and parallel forces that are noncollinear.

3.7 Resultants of nonconcurrent coplanar force systems are defined by magnitude, direction, sense, and position of the line of action. The resultant must produce the same effect as the original force system.

PROBLEMS

Section 3.1 Resultant of Two Concurrent Forces

3.1 through 3.3 Determine the magnitude, direction, and sense of the resultant for the coplanar concurrent force systems shown. Use the parallelogram law. Also sketch the force triangle.

3.4 through 3.6 Solve Problems 3.1 through 3.3 using the method of components.

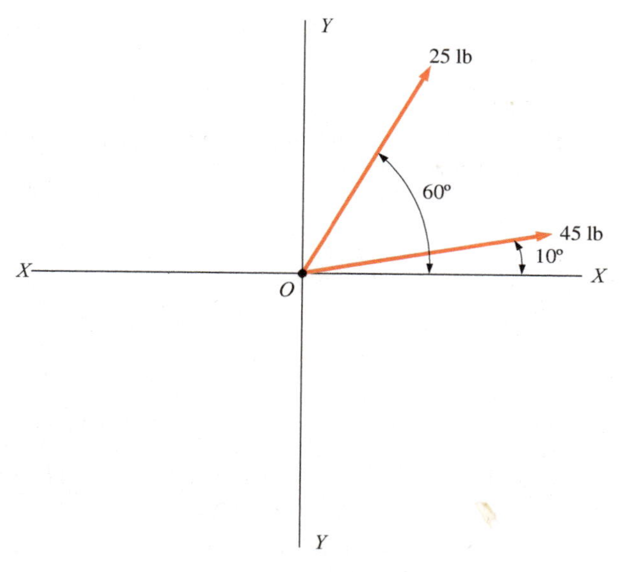

PROBLEM 3.1

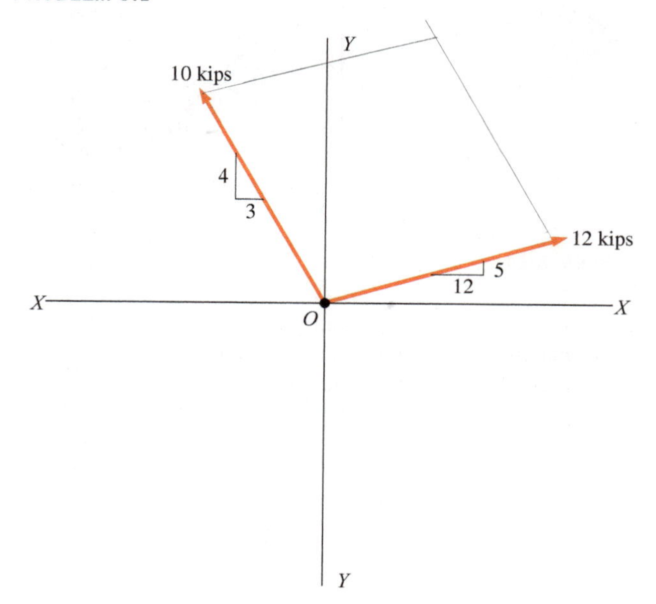

PROBLEM 3.2

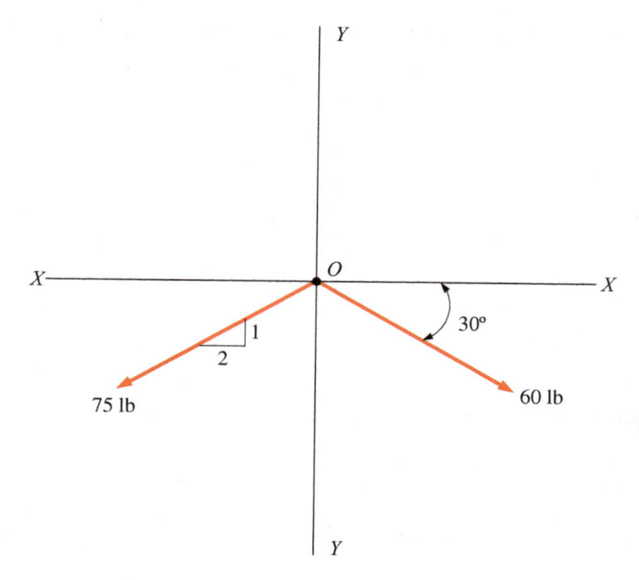

PROBLEM 3.3

3.7 The 150-lb force shown is the resultant of two forces, one of which is shown. Determine the other force.

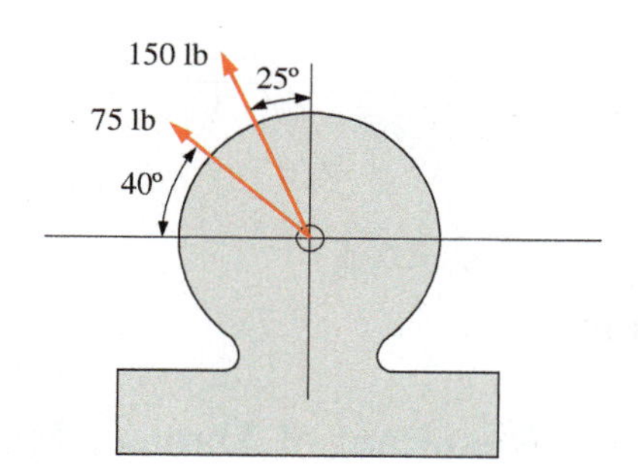

PROBLEM 3.7

3.8 Find the resultant force P exerted on the tree.

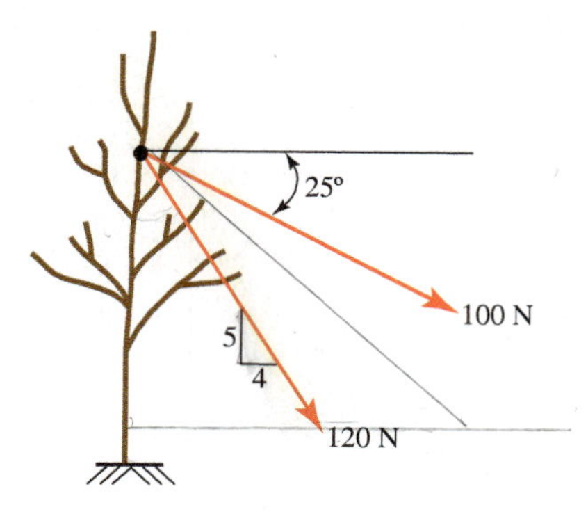

PROBLEM 3.8

3.9 Find the resultant force R exerted on the pole.

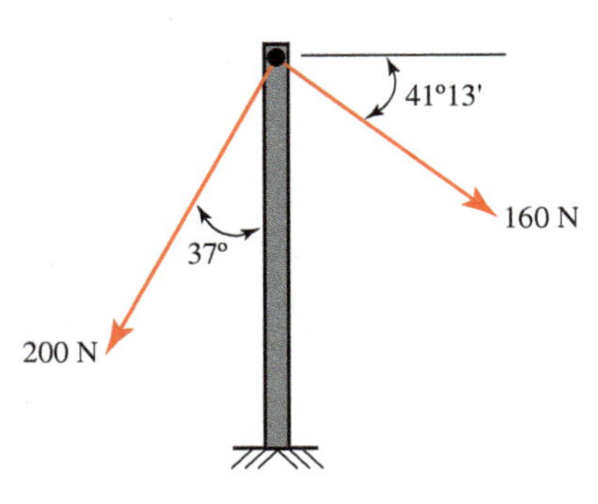

PROBLEM 3.9

3.10 Calculate the resultant force on the screw eye. One part of the rope is horizontal, and the other part is at the indicated slope.

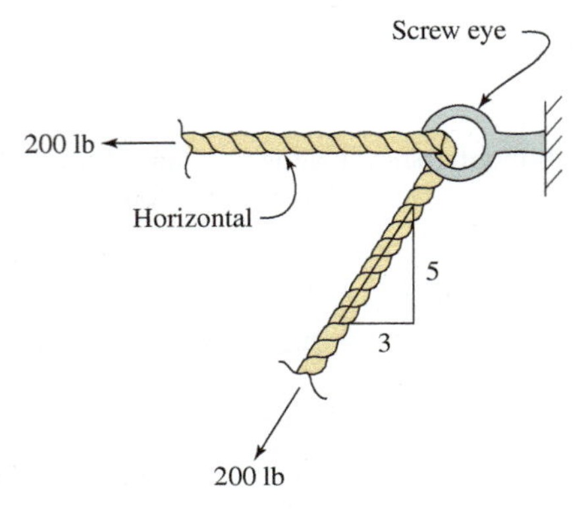

PROBLEM 3.10

Section 3.2 Resultant of Three or More Concurrent Forces

3.11 Determine the resultant of the coplanar concurrent force system shown. Determine the magnitude, sense, and angle of inclination with the X axis. Use the method of components.

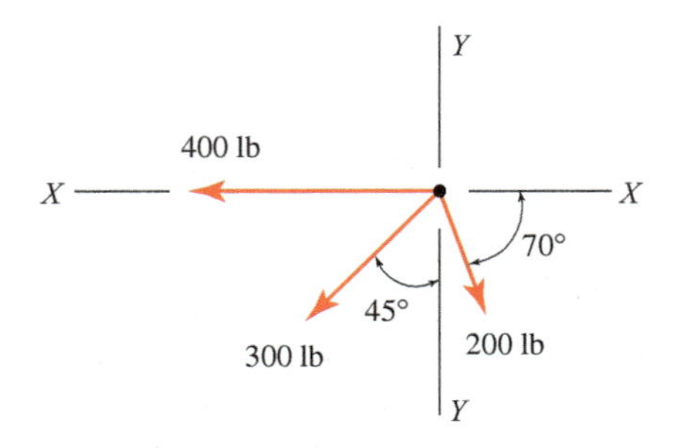

PROBLEM 3.11

3.12 Use the parallelogram law to find the following resultants. Specify direction with an angle referenced to the positive X axis.

(a) The resultant $R_{1,2}$ of forces F_1 and F_2

(b) The resultant $R_{3,4}$ of forces F_3 and F_4

(c) Use the results of parts (a) and (b) to find the resultant of the four-force system.

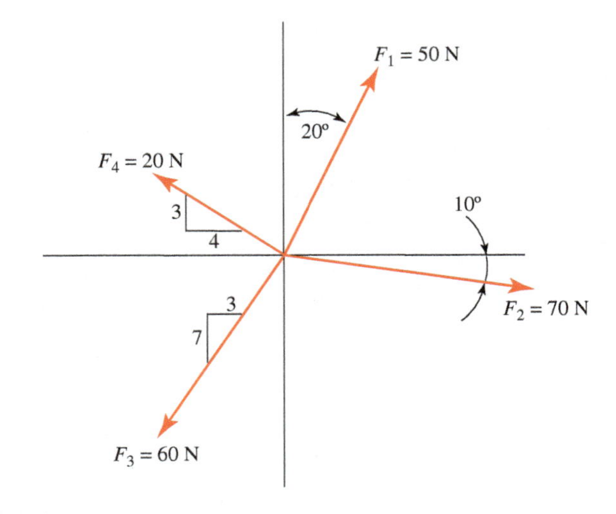

PROBLEM 3.12

3.13 Verify the results of Problem 3.12c by solving for the resultant of the four-force system

(a) using the method of components.

(b) using a graphical method.

3.14 Determine the resultant of the coplanar concurrent force system shown. Compute the magnitude, sense, and angle of inclination with the X axis. Use the method of components.

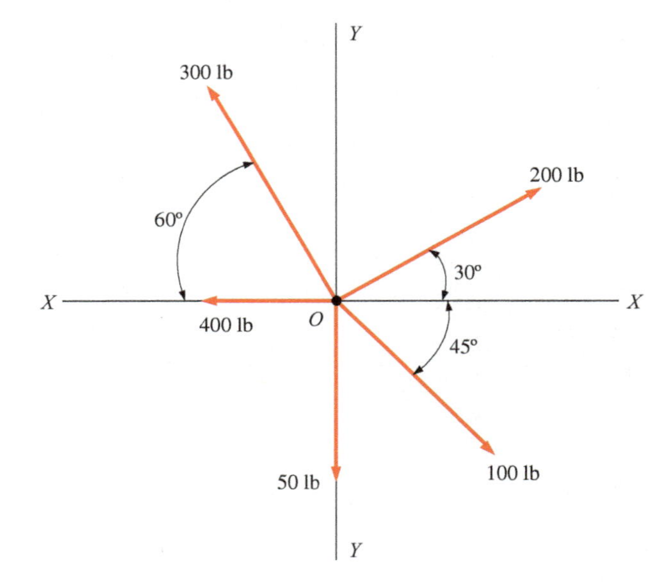

PROBLEM 3.14

3.15 The resultant of the concurrent force system shown has a magnitude of 300 lb acting vertically upward along the Y axis. Compute the magnitude of the force F_1 and its required angle of inclination θ for the resultant to act as described.

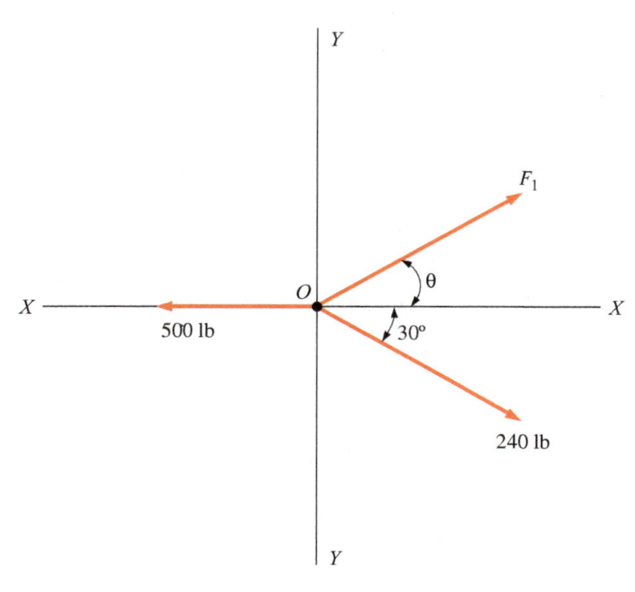

PROBLEM 3.15

3.16 Three forces of 900 lb, 1000 lb, and 600 lb are acting on a boat. The first force acts due north, the second acts due east, and the third acts 30° east of south. Find the magnitude and direction of the resultant force on the boat.

Section 3.3 Moment of a Force

3.17 The four forces shown have parallel lines of action. Member *AB* is perpendicular to the lines of action of these forces.

(a) Determine the moment of each force with respect to point *A*.

(b) Compute the algebraic summation of the moments with respect to point *A*. Be sure to note the sense of the resultant moment.

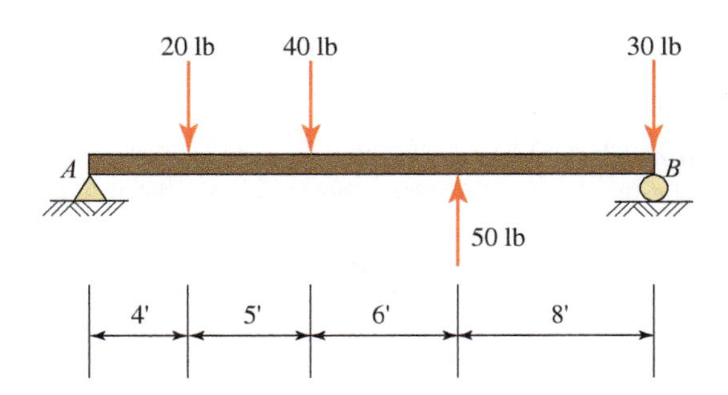

PROBLEM 3.17

3.18 Three coplanar concurrent forces act as shown.

(a) Calculate the moment of each force about point *O* that lies in the line of action of the F_1 force.

(b) Calculate the algebraic summation of the three moments about point *O* and determine the sense of the moment sum.

(c) Calculate the magnitude of the resultant and angle of inclination with the *X* axis.

(d) Compute the moment of the resultant about point *O* and compare with the result of part (b).

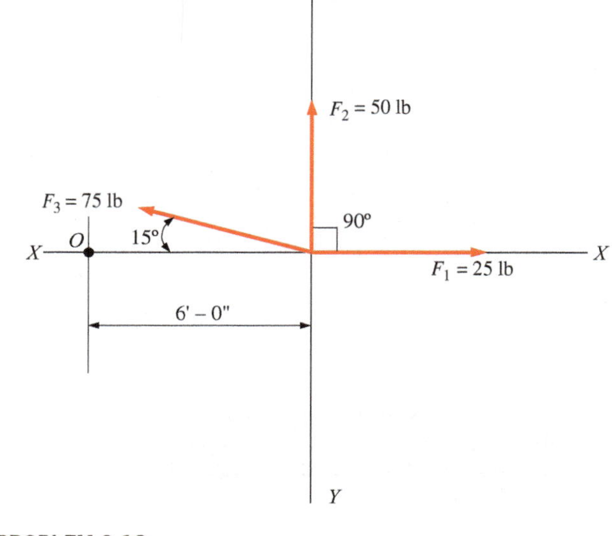

PROBLEM 3.18

3.19 Four coplanar concurrent forces act as shown.

(a) Calculate the moment of each force about point *O*.

(b) Calculate the algebraic summation of the four moments about point *O* and determine the sense of rotation.

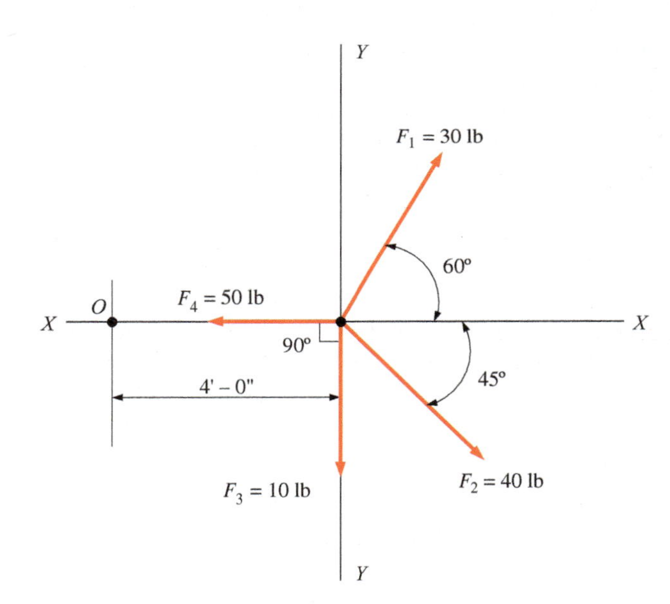

PROBLEM 3.19

3.20 Determine the resultant of the four forces of Problem 3.19 (magnitude and angle of inclination with respect to the *X* axis). Compute the moment of the resultant with respect to point *O* and compare with the results of Problem 3.19.

3.21 For the concrete wall and footing shown:

(a) Calculate the algebraic summation of the moments of the forces shown about point *A*.

(b) Calculate the algebraic summation of the moments of the forces about point *B*.

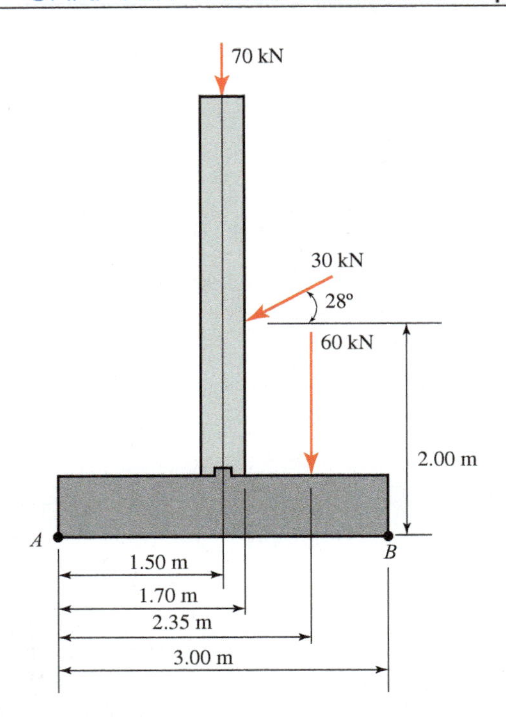

PROBLEM 3.21

Section 3.4 The Principle of Moments: Varignon's Theorem

3.22 Calculate the moment of the 550-lb force about point O shown without using Varignon's theorem. Make a similar calculation using the theorem, resolving the force into its X and Y components at point A.

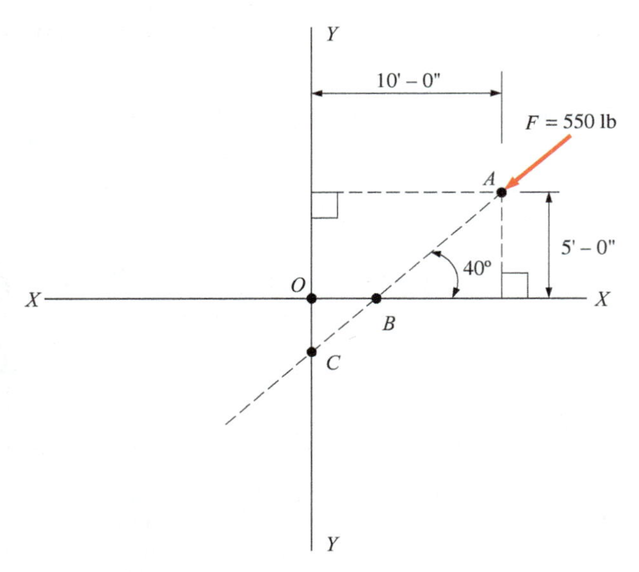

PROBLEM 3.22

3.23 In Problem 3.22, calculate the moment about point O, using Varignon's theorem, resolving the force into its X and Y components at point B and at point C.

3.24 Compute the moment about point A for the linkage shown.

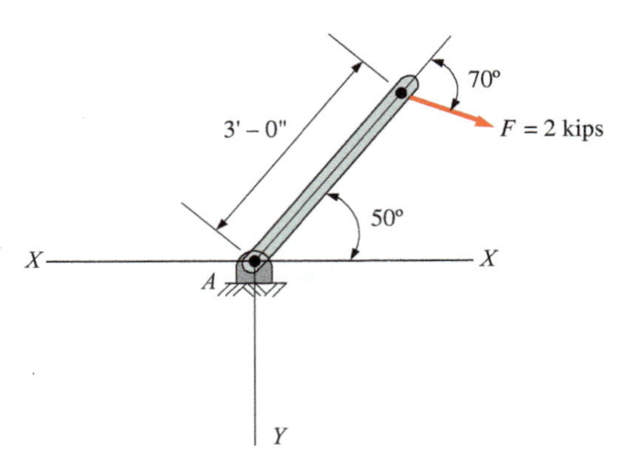

PROBLEM 3.24

3.25 Compute the moment of the force F about point A for the conditions shown.

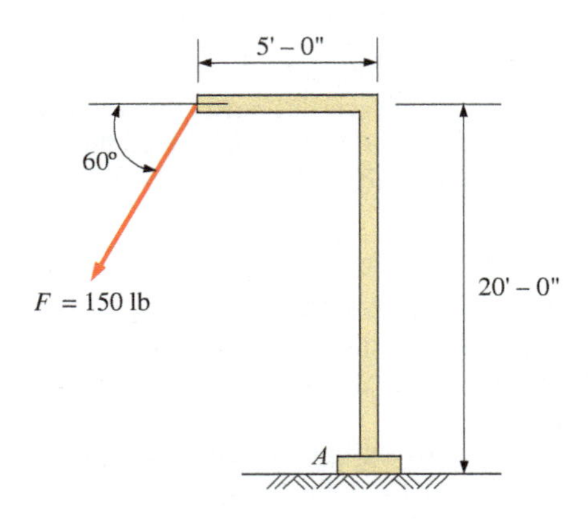

PROBLEM 3.25

Section 3.5 Resultants of Parallel Force Systems

3.26 through 3.29 Determine the magnitude of the resultant of the parallel force systems shown. Locate the resultant with respect to point A. Assume that all forces are vertical and that the members are horizontal.

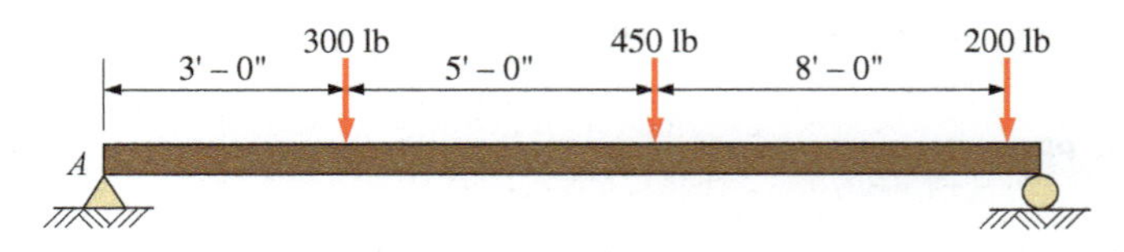

PROBLEM 3.26

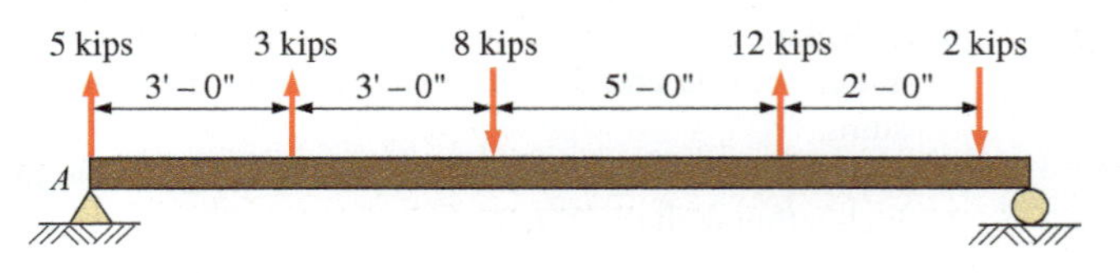

PROBLEM 3.27

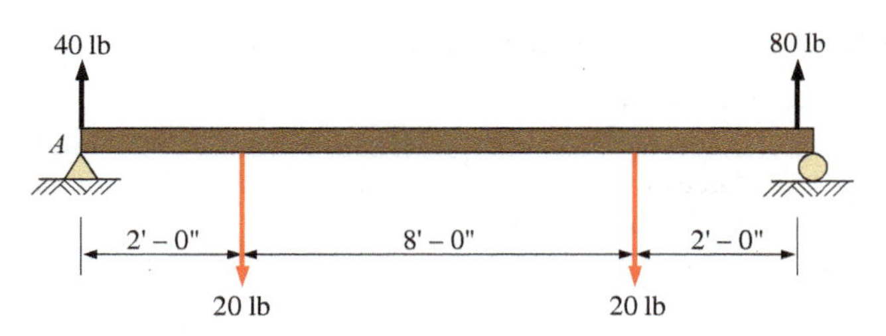

PROBLEM 3.28

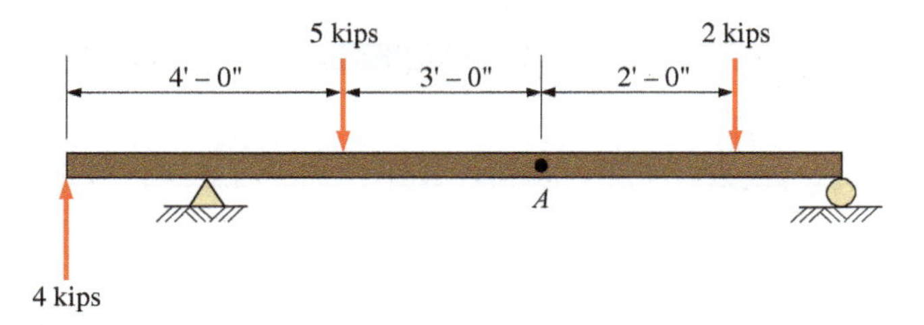

PROBLEM 3.29

3.30 Determine the resultant and its location for the standard truck loading used in highway bridge design. The loads shown represent axle loads.

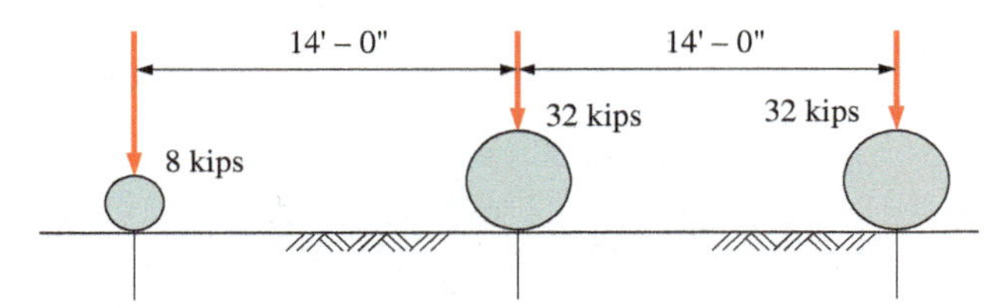

PROBLEM 3.30

3.31 and 3.32 Compute the magnitude, sense, and location of the required vertical force F_1 (not shown) if the other forces of the parallel force system and the resultant are as shown.

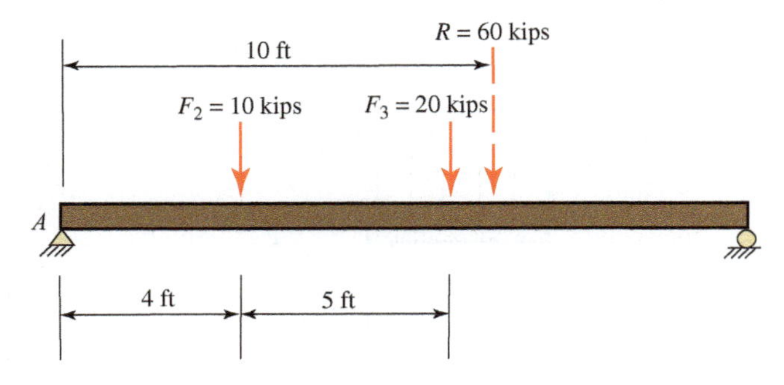

PROBLEM 3.31

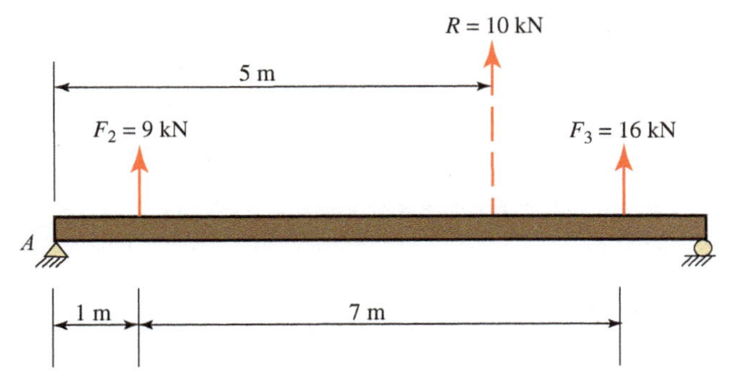

PROBLEM 3.32

3.33 Compute the magnitude and location of the resultant for the load system shown.

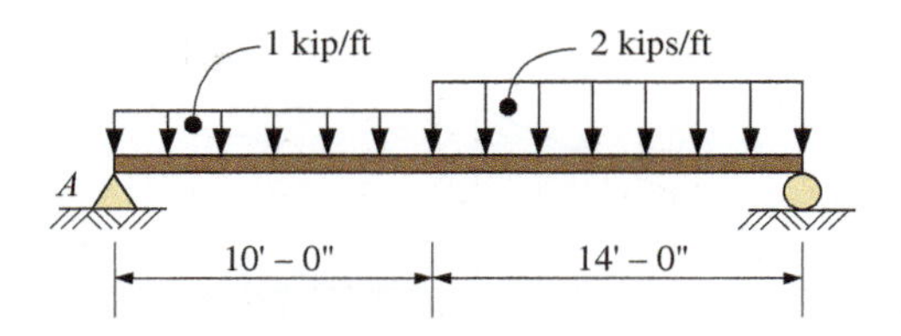

PROBLEM 3.33

3.34 Determine the magnitude and location of the resultant force for the load system shown.

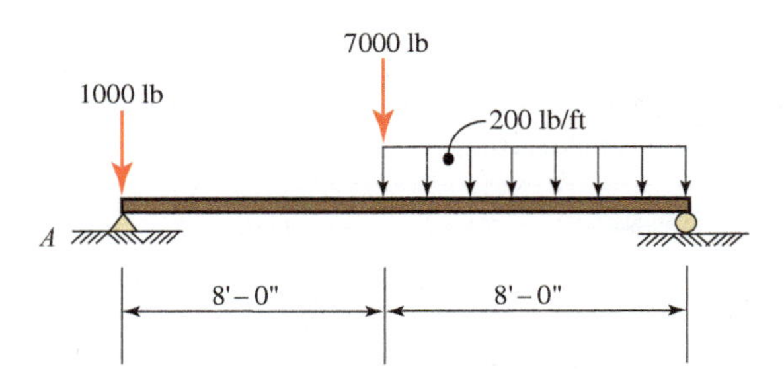

PROBLEM 3.34

3.35 Fresh water is impounded behind a dam to a height h of 18 ft, as shown.

(a) Determine the maximum pressure behind the dam.

(b) Determine the resultant force on a typical 1-ft-wide (18-ft-high) segment of the dam.

(c) Locate the position of the resultant force with respect to the bottom of the dam.

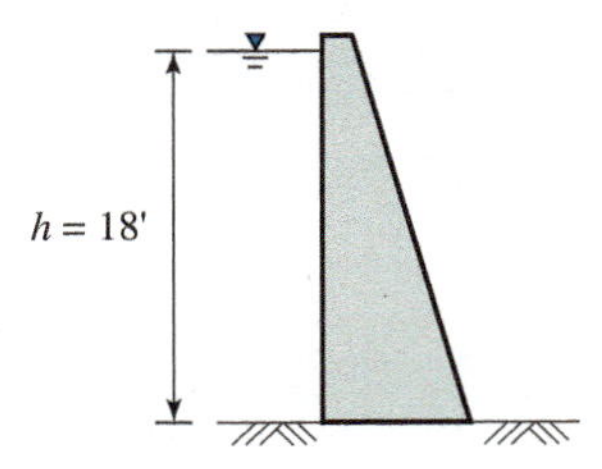

PROBLEM 3.35

3.36 Determine the magnitude and location of the resultant for the load system shown.

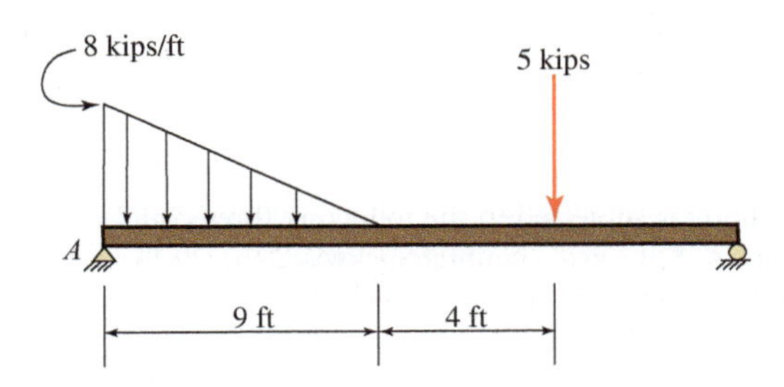

PROBLEM 3.36

3.37 Determine the magnitude and location of the resultant for the load system shown.

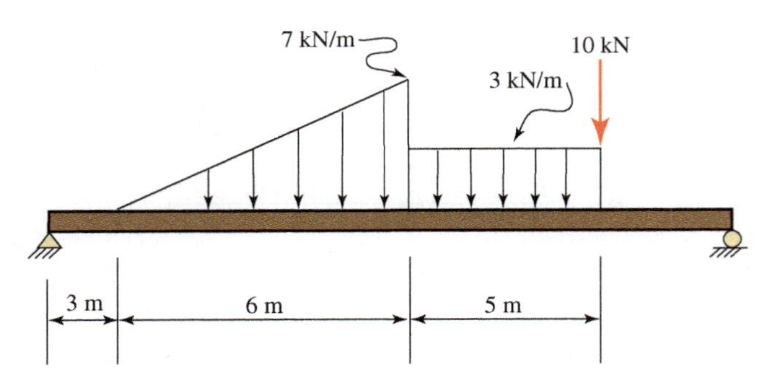

PROBLEM 3.37

Section 3.6 Couples

3.38 through 3.40 Compute the magnitude and direction of the resultant couples acting on the bodies shown.

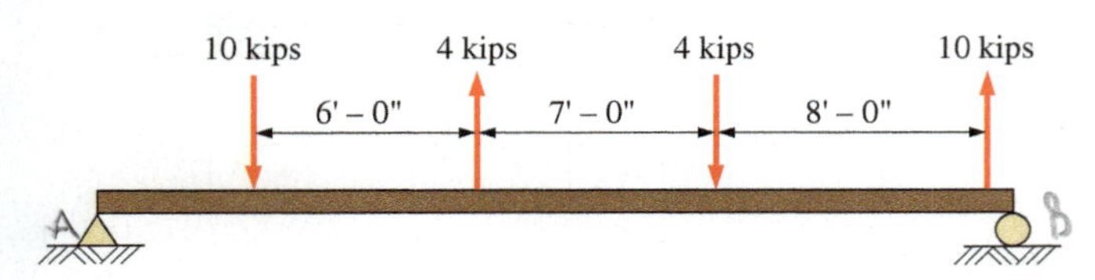

PROBLEM 3.38

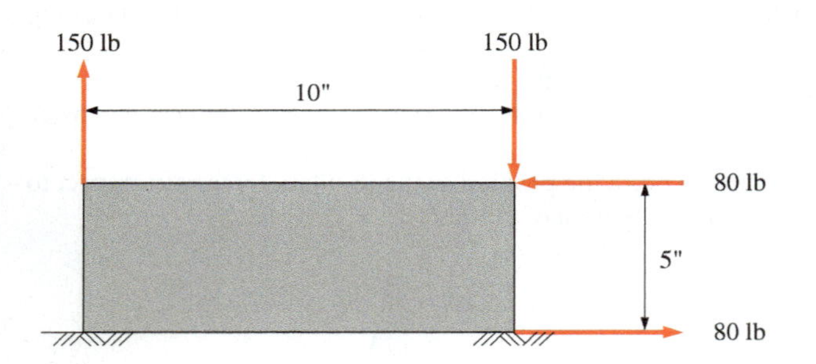

PROBLEM 3.39

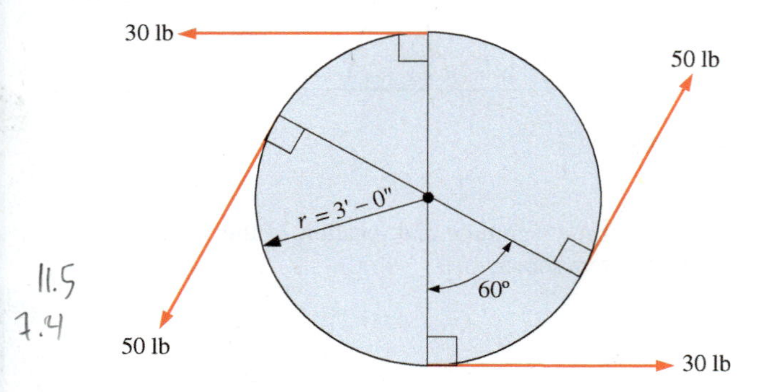

PROBLEM 3.40

3.41 A body is subjected to the following three couples: (a) 30-lb forces, 3-in. arm, counterclockwise; (b) 20-lb forces, 6-in. arm, counterclockwise; (c) 10-lb forces, 5-in. arm, clockwise. Determine the required magnitude of the forces of a single resultant couple, equivalent to the three given couples, and having a 2.5-in. arm.

Section 3.7 Resultants of Nonconcurrent Force Systems

3.42 Determine the magnitude, direction, and sense of the resultant force of the nonconcurrent force system shown. Locate the resultant with respect to point O.

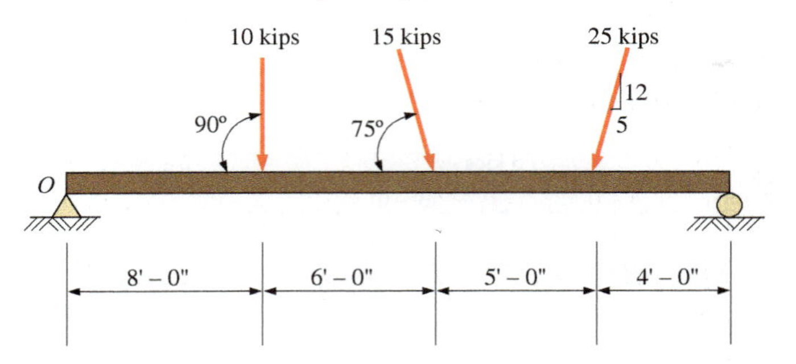

PROBLEM 3.42

3.43 Determine the magnitude, direction, and sense of the resultant of the forces shown. Determine where the resultant intersects the bottom of the body with respect to point O.

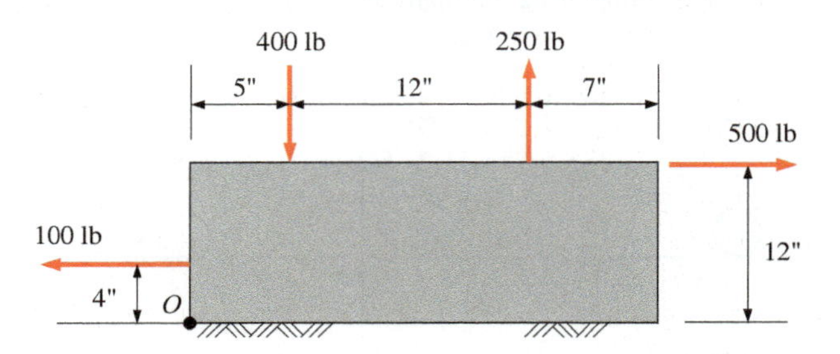

PROBLEM 3.43

3.44 Determine the resultant of the load system shown. Locate where the resultant intersects grade with respect to point A at the base of the structure.

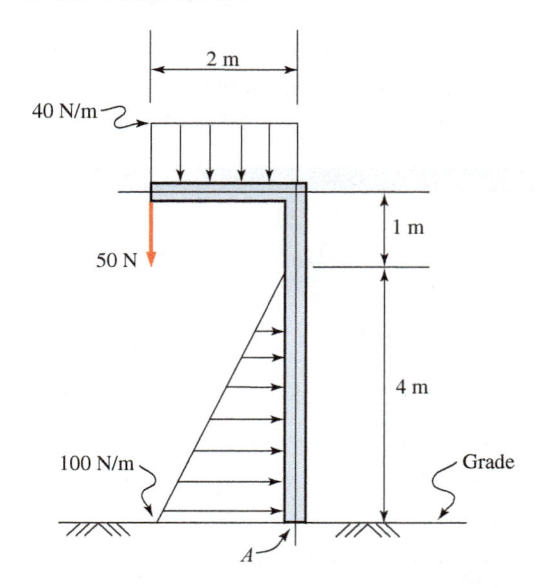

PROBLEM 3.44

3.45 For the concrete structure shown, determine the magnitude, direction, and sense of the resultant force. Determine where the resultant intersects the bottom of the footing with respect to point A.

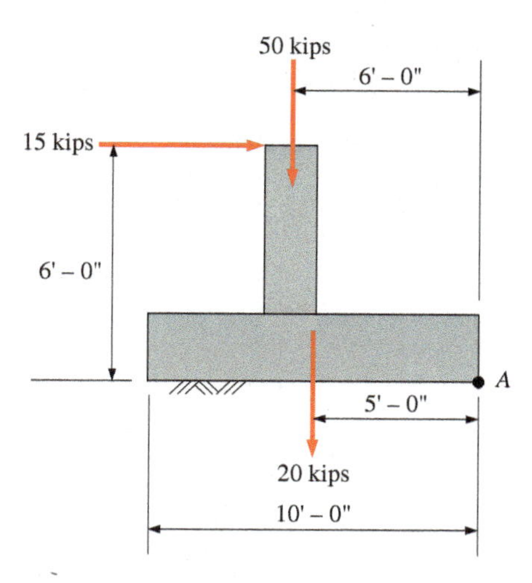

PROBLEM 3.45

Computer Problems

For the following computer problems, any appropriate software may be used. Input prompts should fully explain what is required of the user (the program should be "user friendly"). The resulting output should be well labeled and self-explanatory.

3.46 Write a program that will calculate the magnitude and direction of the resultant of two concurrent forces that lie in the first quadrant, as shown in Problem 3.1. User input is to be the magnitude of each force and its angle of inclination with the positive X axis.

3.47 Write a program that will determine the resultant of n concurrent forces. User input is to be the number of forces (n) and the magnitude and direction of each. For convenience, assume that the layout is similar to that shown in Figure 3.7. (*Hint*: You may wish to consider azimuth angles from the positive X axis, or designation of quadrants.)

3.48 Write a program to solve Problem 3.30. The distance between the two 32-kip loads may range from 14 ft to 30 ft. The user should be allowed to input this axle spacing.

Supplemental Problems

3.49 Determine the magnitude, direction, and sense of the resultant force of the coplanar concurrent force system shown.

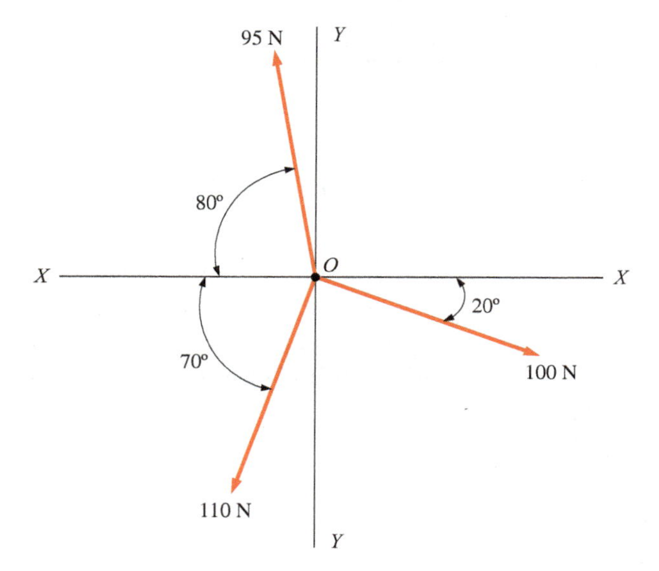

PROBLEM 3.49

3.50 The resultant and one-component force of a two-force concurrent force system are shown. Compute the other component force F_2 (not shown) that would be required. Determine magnitude, direction, and sense.

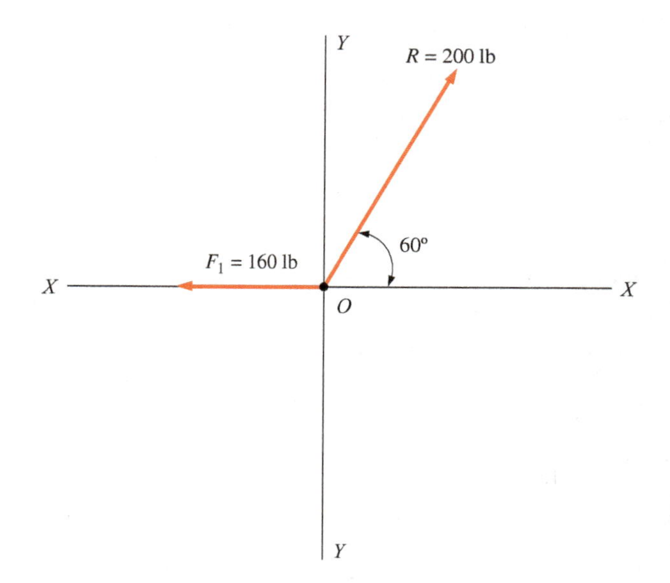

PROBLEM 3.50

3.51 The resultant force of a concurrent force system is shown. Determine the magnitude of the two concurrent components if their angles of inclination are known.

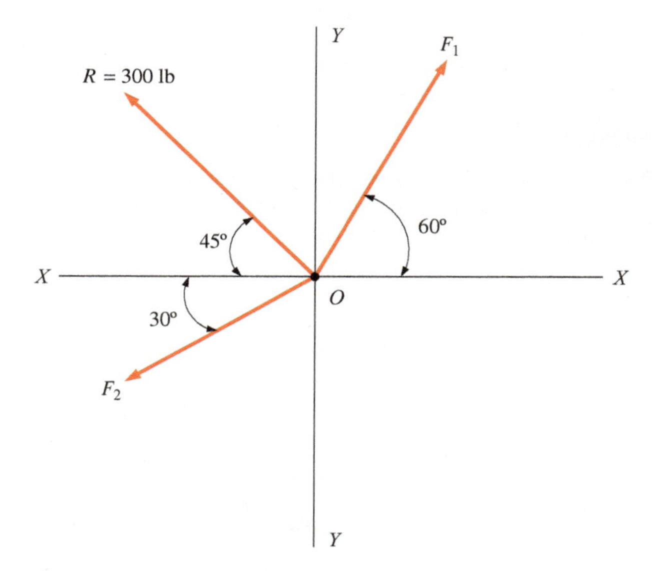

PROBLEM 3.51

3.52 Determine the magnitudes of forces P_1 and P_2 such that the resultant of the three-force system is zero.

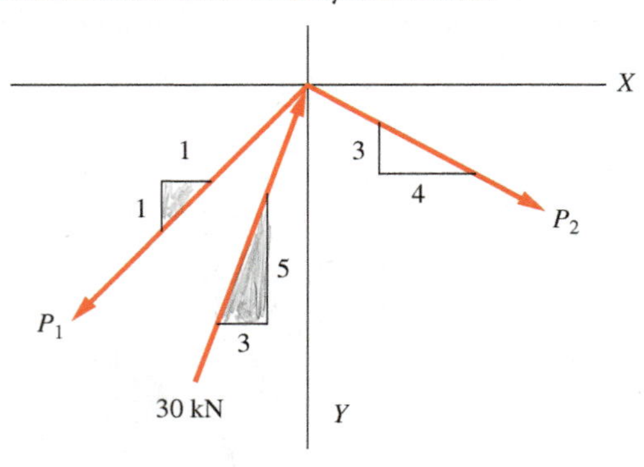

PROBLEM 3.52

3.53 The resultant force of a concurrent force system is shown. Determine the magnitudes of the components F_1 and F_2 if their directions and senses are as shown.

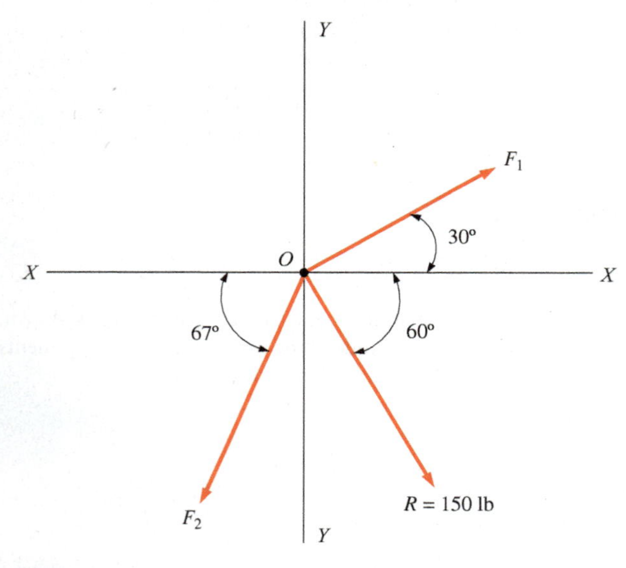

PROBLEM 3.53

3.54 A hockey puck is acted on simultaneously by two sticks as shown. Determine the resultant force (magnitude and direction) on the puck.

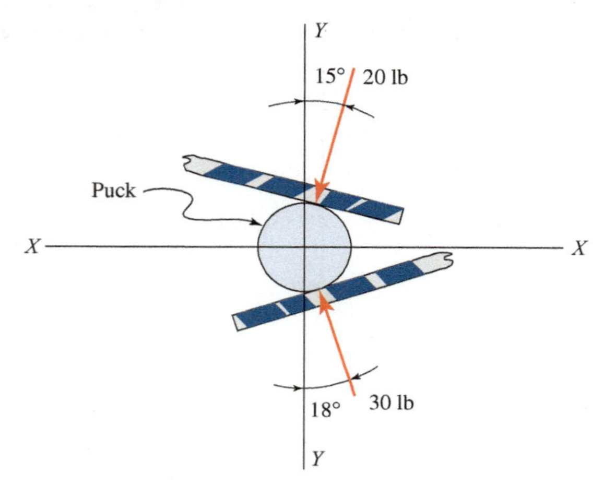

PROBLEM 3.54

3.55 and 3.56 Determine the resultant force for each of the coplanar concurrent force systems shown. Compute the magnitude, sense, and angle of inclination with the horizontal X axis. Use the method of components.

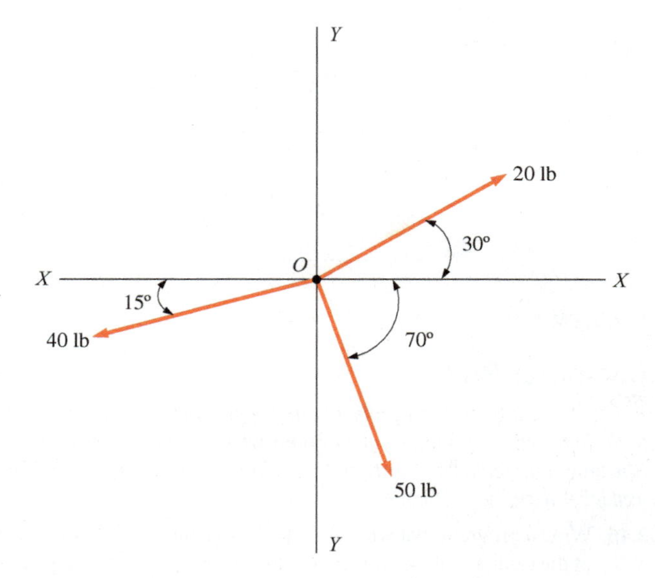

PROBLEM 3.55

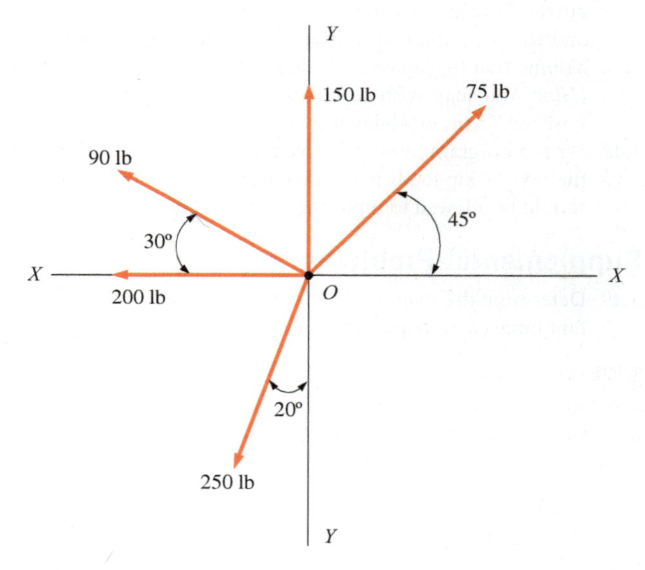

PROBLEM 3.56

3.57 The resultant of the three concurrent forces shown is 100 lb and has an angle of inclination of 20° to the X axis. Determine the magnitude of F_1 and F_2.

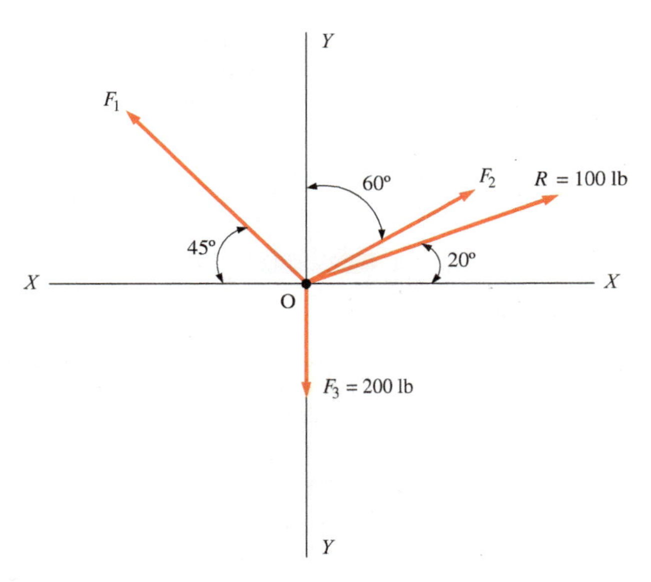

PROBLEM 3.57

3.58 The transmission tower shown is subjected to a horizontal wind force of 400 lb acting at point *B*. Supported cables transmit a force of 900 lb at two different levels acting as shown. Calculate the moment about point *A* at the base of the tower.

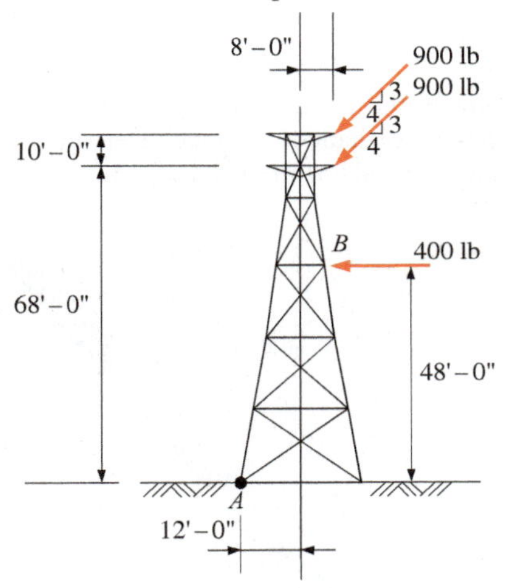

PROBLEM 3.58

3.59 A gravity-type masonry dam, as shown, depends on its own weight for stability. There is a horizontal force due to the pressure of the water acting on a section of the dam having a width of 1 ft (into the page). The vertical force represents the weight of a 1-ft width of section. Find the resultant of these two forces and locate the point where its line of action intersects the base *AB* of the dam. This point should fall within the middle third of the length *AB*. Does it?

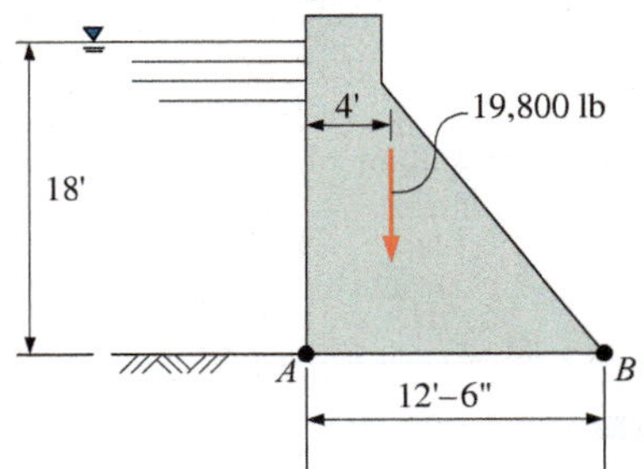

PROBLEM 3.59

3.60 The transformer (as shown) must be lifted vertically out of the pit. Angle $\theta = 18°$. Determine the required magnitude of force *F* so that the resultant of the two forces will be vertical.

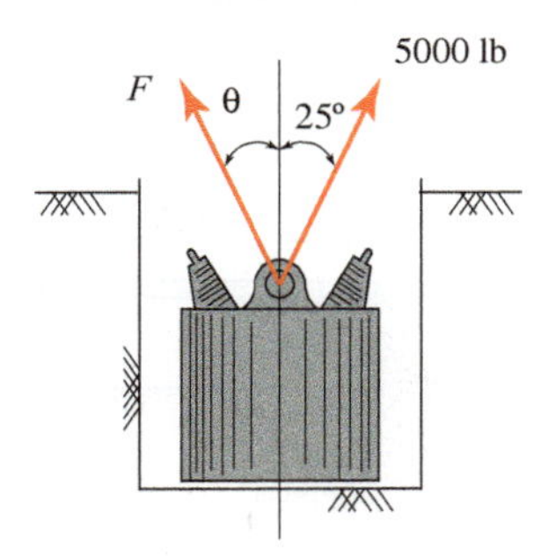

PROBLEM 3.60

3.61 Refer to the diagram for Problem 3.60. Assume that the transformer weighs 6000 lb. Determine θ and the magnitude of the force *F* so that the resultant of the lifting force (6000 lb) will act vertically upward.

3.62 The plastic barrel tent anchor of Problem 2.11 contains 100 gallons of water. Determine the magnitude and sense of the resultant moment about point *O*. Be sure to include the weight of the barrel (60 lb) as well as the weight of its contents.

3.63 Calculate the moment of the forces shown with respect to point *B*. The forces are vertical and the member is horizontal.

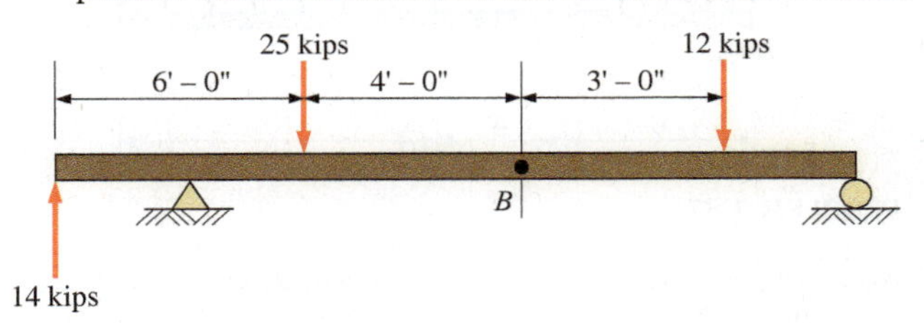

PROBLEM 3.63

3.64 Determine the magnitude and location of the resultant of the parallel force system acting on the horizontal member *AB* shown. Use point *A* as the reference point. Neglect the weight of the member.

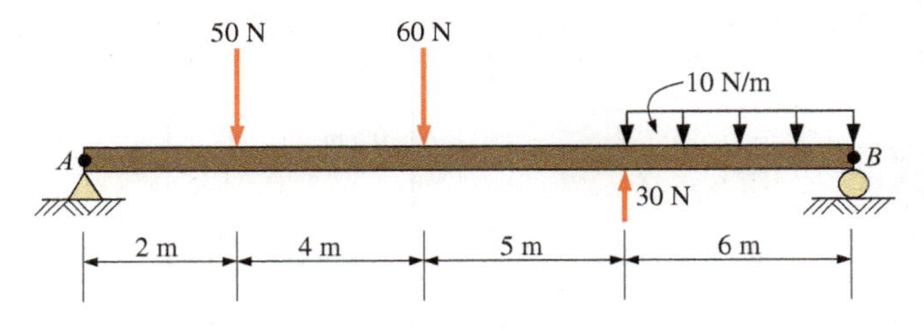

PROBLEM 3.64

3.65 Determine the moment (about point *A*) of the applied loads shown.

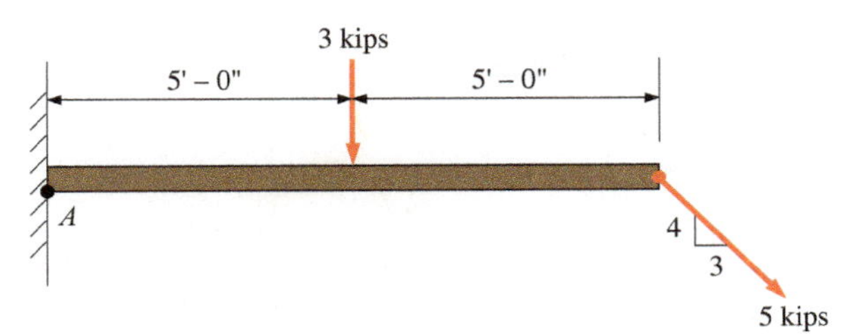

PROBLEM 3.65

3.66 The lift force on the wing of an aircraft is approximately represented as shown. Calculate the magnitude and location of the resultant of the distributed forces.

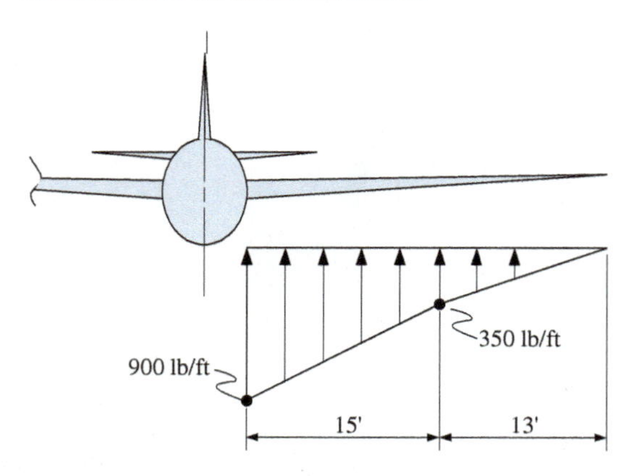

PROBLEM 3.66

3.67 A beam is subjected to distributed loads as shown. Determine the magnitude and location of the resultant of the distributed forces.

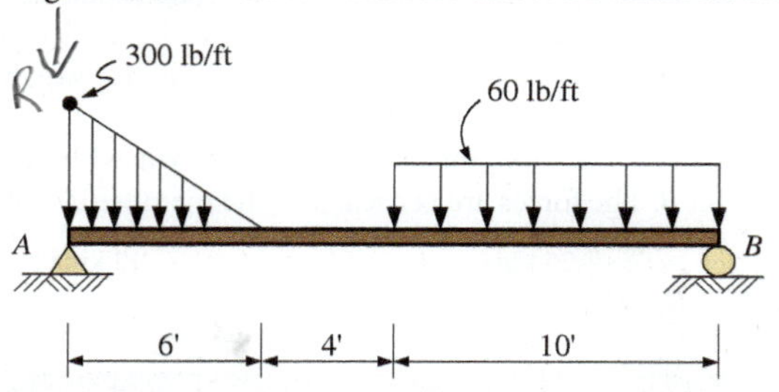

PROBLEM 3.67

3.68 For the concrete gravity wall shown, determine the magnitude, direction, and sense of the resultant force. Determine where the resultant intersects the base of the wall with respect to point *A*.

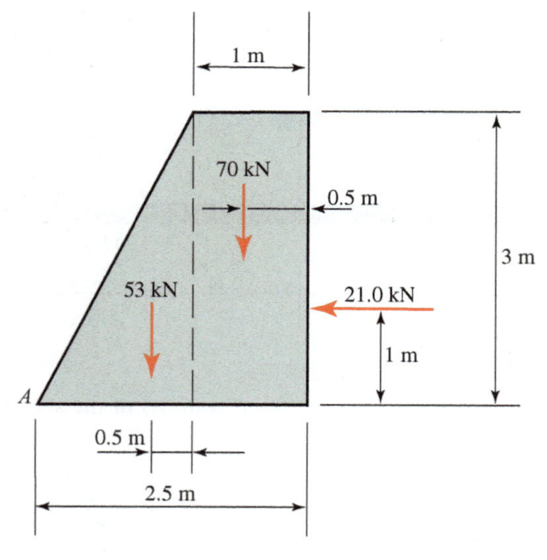

PROBLEM 3.68

3.69 Fresh water is impounded to a height of 8 ft behind a dam as shown. Determine the magnitude and location of the resultant force on the 3-ft by 3-ft gate.

(a) There is no water on the left side of the gate.

(b) Water on the left side of the gate has risen to a height of 2 ft.

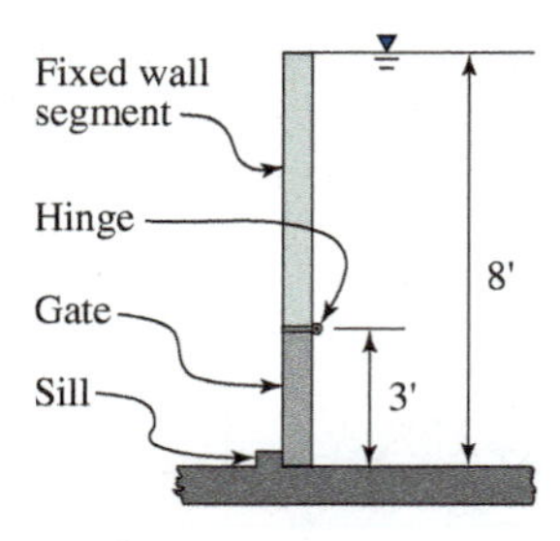

PROBLEM 3.69

3.70 Planks, 2 in. by 10 in. in cross section and 5 ft long, are used at the top of a dam to control the level of the impounded water, as shown. Find the resultant force (magnitude and location) on each plank.

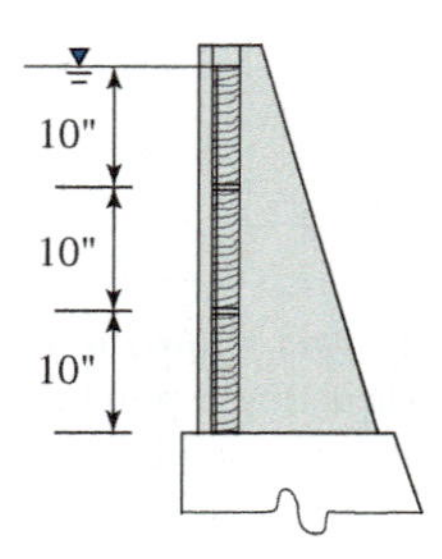

PROBLEM 3.70

3.71 **(a)** Compute the moment (about point *A*) of the forces shown.

(b) Find the resultant of the forces. Determine where it intersects a vertical line through point *A*.

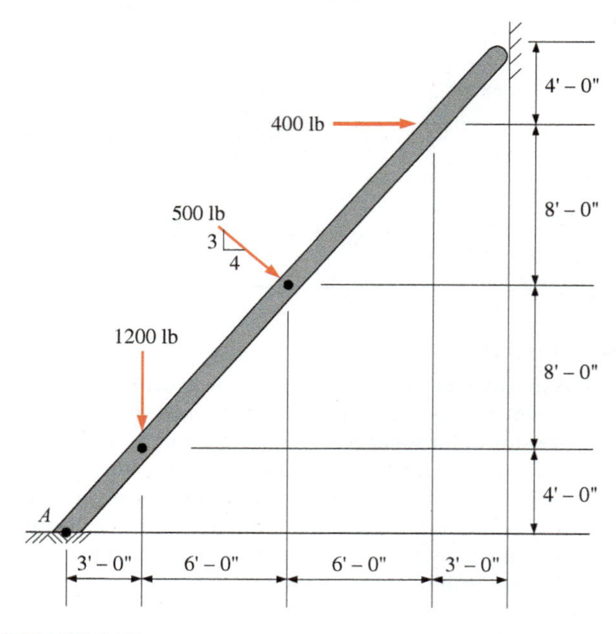

PROBLEM 3.71

3.72 Determine the resultant of the three forces acting on the horizontal beam shown.

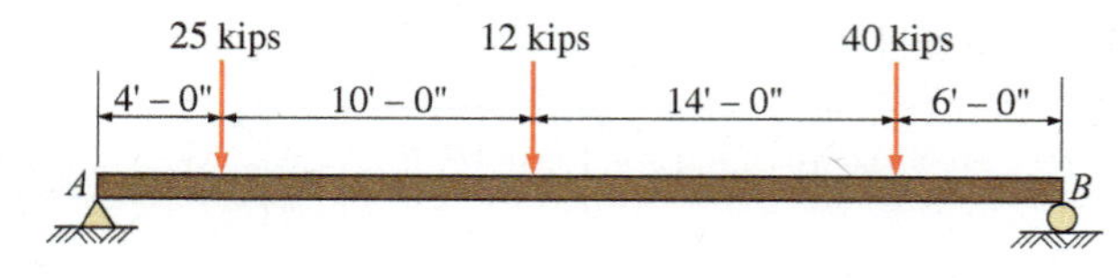

PROBLEM 3.72

3.73 **(a)** Calculate the moments about points A and B due to the nonconcurrent force system shown.
(b) Determine the location and direction of the resultant force.

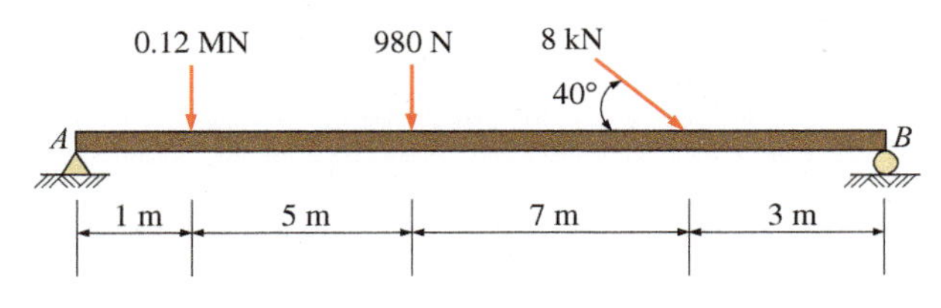

PROBLEM 3.73

3.74 Determine the magnitude of F_1 and F_2 shown such that the resultant will be a counterclockwise couple with a moment of 220 lb-ft.

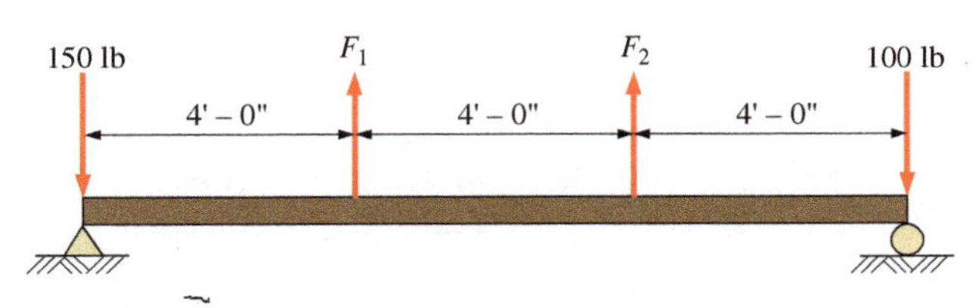

PROBLEM 3.74

3.75 Calculate the magnitude, direction, and sense of the resultant force of the noncurrent force system shown. Determine where the resultant intersects the bottom of the shape with respect to point A.

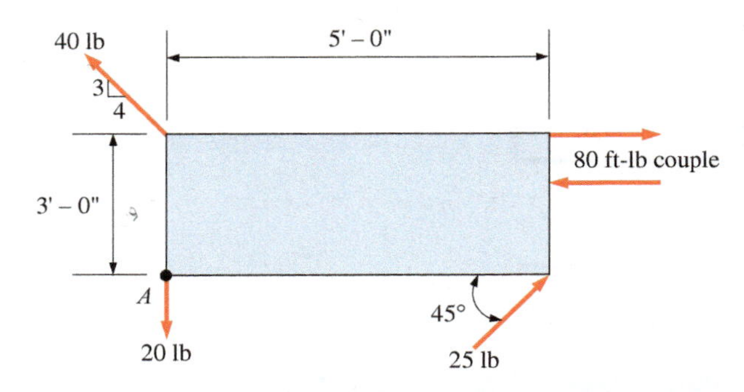

PROBLEM 3.75

CHAPTER
FOUR

EQUILIBRIUM OF COPLANAR FORCE SYSTEMS

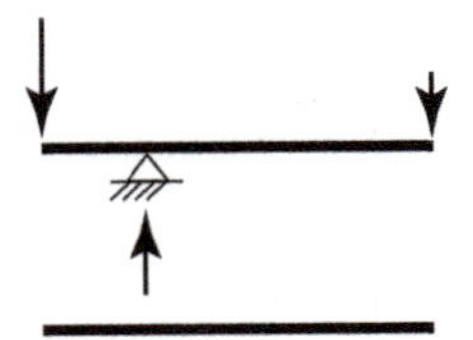

LEARNING OBJECTIVES

Upon completion of this chapter, readers will be able to:

- Discuss the concept of equilibrium with respect to structural systems.
- Draw a free-body diagram for any body or structural system being acted upon by a force.
- Identify and calculate reaction forces for concurrent, parallel, and nonconcurrent coplanar force systems using the concepts of equilibrium.

4.1 INTRODUCTION

As we have learned, statics deals essentially with the action of forces on rigid bodies at rest, a state also defined as one of equilibrium, or of zero motion. We may therefore conclude that, with zero motion, both the body and the entire system of external forces and moments acting on the body, no matter how complex, are in equilibrium.

If a system of forces is in equilibrium, the resultant of the force system is equal to zero. When a force system that is in equilibrium acts upon a body, the body is also said to be in equilibrium.

In this chapter we establish the conditions a force system must satisfy for it to be in equilibrium.

4.2 CONDITIONS OF EQUILIBRIUM

The resultant of a force system must be zero if the force system is to be in equilibrium. Two general conditions, based on fundamental laws, must be satisfied if the resultant is to be zero:

1. For a force system to be in equilibrium, the algebraic sum of all forces (or components of forces) along any axis, in any direction, must be equal to zero.
2. For a force system to be in equilibrium, the algebraic sum of the moments of the forces about any axis or point must be equal to zero.

These two conditions for equilibrium can be expressed mathematically as

$$(1)\ \Sigma F = 0 \quad \text{and} \quad (2)\ \Sigma M = 0$$

Considering only the two-dimensional case and the usual X–Y coordinate axes system, the algebraic summation of forces must be zero in both the X and the Y directions. The $\Sigma F = 0$ condition can then be rewritten as

$$\Sigma F_x = 0 \quad \text{and} \quad \Sigma F_y = 0$$

which states that, for equilibrium, the algebraic sums, respectively, of the X and Y components of the force system must equal zero. Frequently, the X–Y coordinate axes system is oriented with a vertical Y axis (rather than inclined), as defined by the direction of gravity. In this situation, the horizontal and vertical directions are, for convenience, denoted as such. Therefore, it is common to use $\Sigma F_H = 0$ and $\Sigma F_V = 0$ in place of the preceding expressions.

The moment condition for equilibrium can still be expressed as

$$\Sigma M = 0$$

which states that, for equilibrium, the algebraic sum of the moments of the forces of the system about any point in the plane must equal zero. This, in essence, states that the sum of the clockwise moments must equal the sum of the counterclockwise moments.

It should be noted that $\Sigma F_x = 0$, $\Sigma F_y = 0$, and $\Sigma M = 0$ represent what are generally termed the *three laws of equilibrium*. They are fundamental laws for bodies at rest and cannot be proven mathematically. These laws, based on the results of observations, were first advanced by Sir Isaac Newton (1642–1727) in his statement on the laws of motion. They stem from Newton's first law that states, in part, that when a body is at rest, the resultant of all the forces acting on the body is zero.

4.3 THE FREE-BODY DIAGRAM

The following sections of this chapter deal with the applications of the conditions of equilibrium. Our primary purpose is to determine desired information about certain forces that result from the effects of other forces acting on bodies.

Most problems encountered in statics result from the interaction of bodies. To solve these problems generally requires that a body be isolated and that the given force system acting on the body be identified and analyzed so that unknown forces can be determined. The method most useful in meeting these requirements for solution is known as the *free-body diagram method.*

The free-body diagram is a sketch or pictorial representation (not necessarily to scale) indicating (a) the body in question, by itself, entirely isolated from the other bodies and (b) all the external forces exerted on that body as a result of the interaction between the free body and other bodies. The external interactive forces acting on the free body may be direct forces due to contact between the free body and other bodies external to it (which could be solid, liquid, or gaseous) or indirect forces, such as gravitational or magnetic forces, that act without bodily contact. The free-body diagram is one of the most important and useful analytical tools in engineering mechanics.

Figure 4.1a shows a weight suspended by a rope. Free-body diagrams of the rope and the weight are shown in Figure 4.1b and c. The rope and the weight are isolated in their free-body diagrams and the external forces acting on them are shown. In each case, the line of action of the force must coincide with the vertical centerline of the rope.

Figure 4.2a shows a weight suspended from a rope sling. In situations such as this, the point of concurrency of the forces (the knot in the sling) is the key to the solution. A free-body diagram of the knot is shown in Figure 4.2b. Note how the knot is isolated and the ropes are replaced with the forces expected from them.

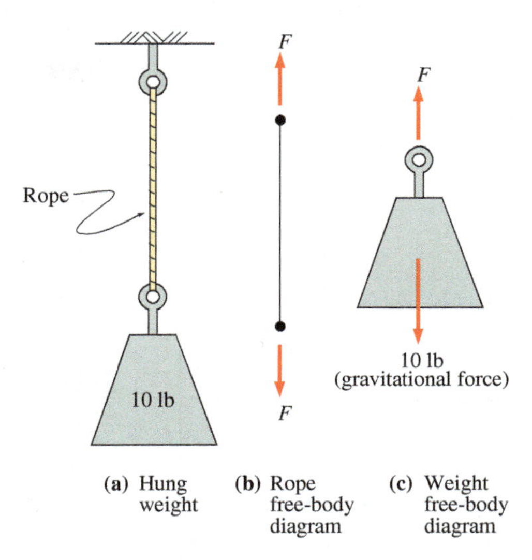

FIGURE 4.1 The free-body diagram.

A summary of the procedure for sketching a free body is as follows:

1. Sketch the subject body isolated from all other bodies. This body may be an entire structure consisting of many component parts but treated as a unit, or it may be any individual part of the structure. Free bodies of linear structures or elements (such as beams and truss members) are commonly represented by their centerlines (a single line). Dimensions on the free-body diagram are the same as the dimensions on the actual body.
2. Show all the known forces acting on the body, indicating a magnitude, line of action, and sense. These forces will represent the effect of the contact bodies removed and the noncontact gravitational or magnetic effects.
3. Indicate all desired unknown forces with a symbol and include as much known information as possible. Point of application is usually known; direction *may* be known. Sense, components, or both may have to be assumed.

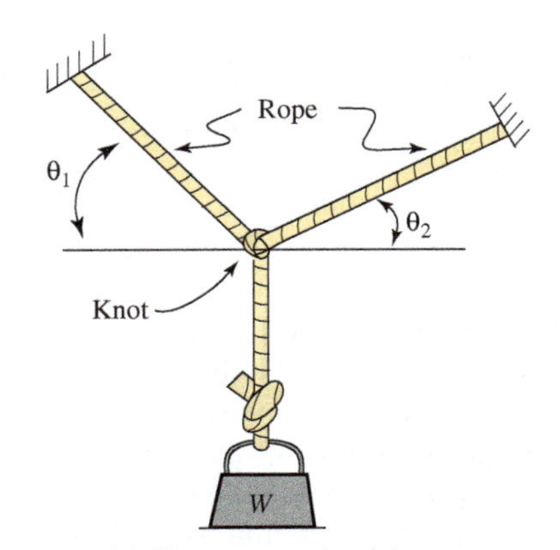

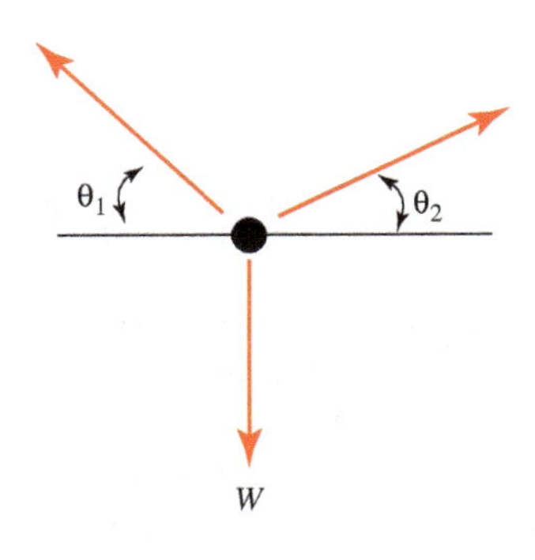

FIGURE 4.2 The free-body diagram.

	Description of Support	Sketch of Idealized Support	Effect on Free-Body (How Represented)	Number of Reactions
(a)	Flexible cable, rope, chain, or wire	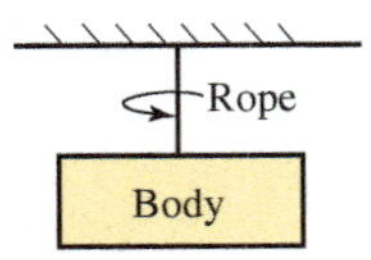 Rope; Body	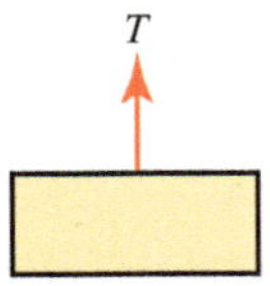 T	1
(b)	Roller (zero friction)	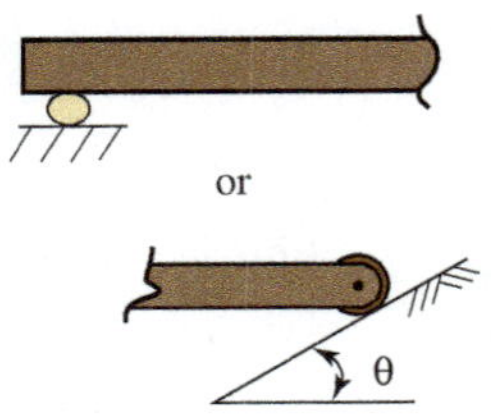 or; θ	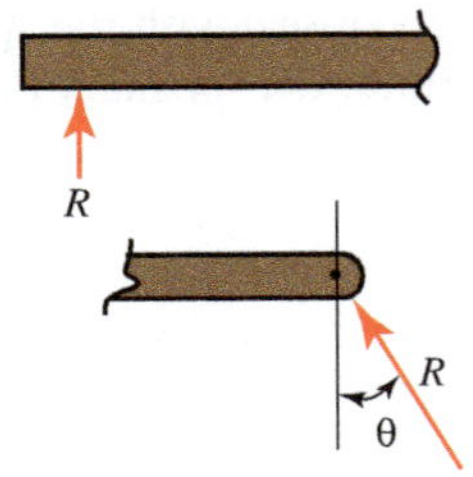 R; R; θ	1
(c)	Smooth surface (zero friction)	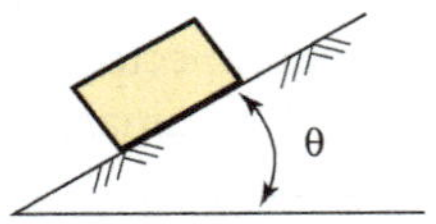 θ	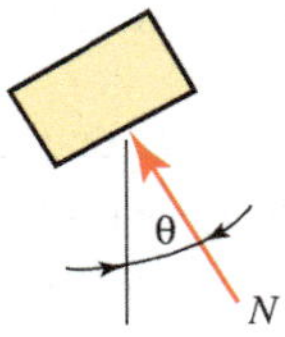 θ; N	1
(d)	Pinned, hinged, or knife-edged (rough surface)	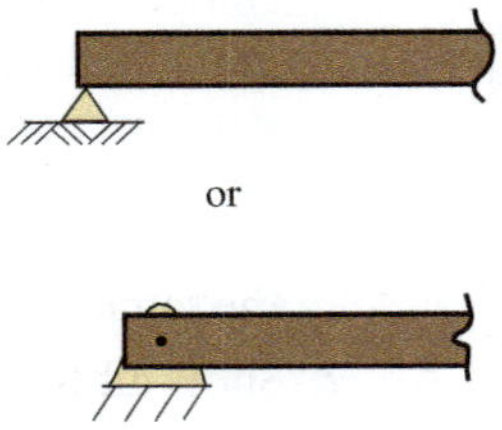 or	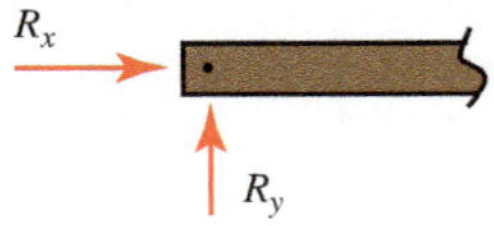 R_x; R_y	2
(e)	Fixed	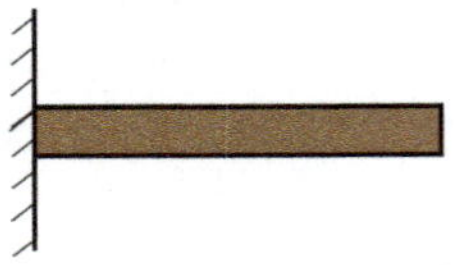	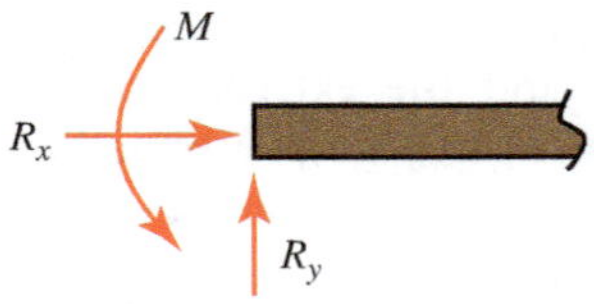 M; R_x; R_y	3

FIGURE 4.3 Supports and their representation in free-body diagrams.

Since the free-body diagram is a sketch showing known and unknown external forces acting on the free body, conventional symbols should be used to represent the action of the forces. Figure 4.3 shows some types of bodily contact or support and illustrates how the specific conditions should be represented in a free-body diagram. If the supporting member of a body is flexible, as in the case of a cable or rope (Figure 4.3a), the force, in the nature of a pull (tension), is directed along the centerline of the member. A roller-type contact or support (Figure 4.3b) permits motion parallel to the supporting surface. Therefore, the roller provides a reaction (an external resisting force; see Section 2.4) that is perpendicular to the supporting surface. The action of the smooth surface (Figure 4.3c) on the free body is similar to the roller-type support in that it provides a reaction perpendicular to the supporting surface. A pinned-type contact or support—also called a hinged, or knife-edged, support—(Figure 4.3d) does not permit any linear motion, but does permit rotational motion. The direction of the reaction at a pinned support is unknown; therefore, the reaction is generally indicated as two independent components. This situation is similar to the case in which a body is supported by a rough surface. A fixed-type contact or support (Figure 4.3e) allows no linear or rotational motion. Therefore, the reaction is indicated as two independent components and a resisting couple (moment).

EXAMPLE 4.1 The beam shown in Figure 4.4a is supported by a pinned support at A and a roller on a horizontal surface at B. Sketch the free-body diagram for the beam.

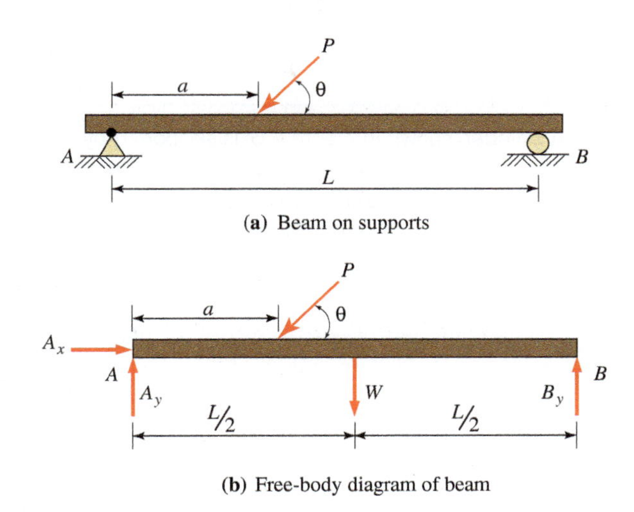

FIGURE 4.4 Beam for Example 4.1.

Solution The free-body diagram for the beam is shown in Figure 4.4b. The weight of the beam itself, which acts vertically downward through the center of the beam, is represented by the force W. The inclined load P is indicated on the free-body diagram exactly as on the beam diagram. Since the beam is supported by a pinned connection at A, the direction of the reaction at this point is unknown. The reaction is represented by its rectangular components A_x and A_y with assumed senses as shown. The reaction supplied at B is represented by a single vertical force B_y, which is perpendicular to the horizontal surface on which the roller is supported.

EXAMPLE 4.2 The horizontal beam shown in Figure 4.5a is supported by a pinned support at A and a roller on an inclined plane at B. Sketch the free-body diagram of the beam.

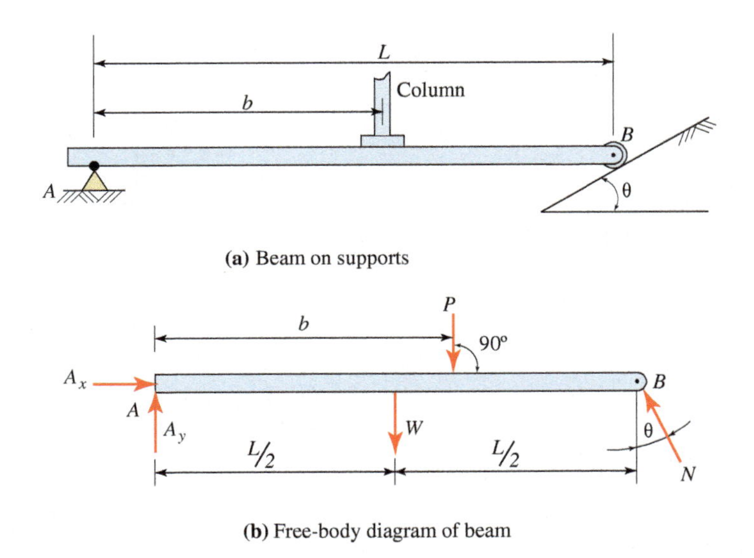

FIGURE 4.5 Beam for Example 4.2.

Solution The free-body diagram for the beam is shown in Figure 4.5b. The weight of the beam and the reaction at A are similar to that in Example 4.1 and are indicated in the same manner. The external force from the column, which is supported by the beam, is represented by P and is acting vertically downward.

The beam is supported by a roller at point *B*. The reaction *N* acting on the beam must be perpendicular to the supporting surface (since the roller permits motion parallel to the inclined plane). Therefore, *N* is known to be acting at an angle of inclination of θ with the vertical, as shown.

EXAMPLE 4.3 A cylinder, shown in Figure 4.6, is supported by a vertical wall, a beam that is pin-connected to the wall at point *A*, and a flexible cable. The two contact surfaces with the cylinder are smooth. Sketch the free-body diagram for (a) the entire system, considering the cylinder, beam, and cable as a single body; (b) the cylinder alone; and (c) the beam alone. (To simplify this diagram, no dimensions are shown.)

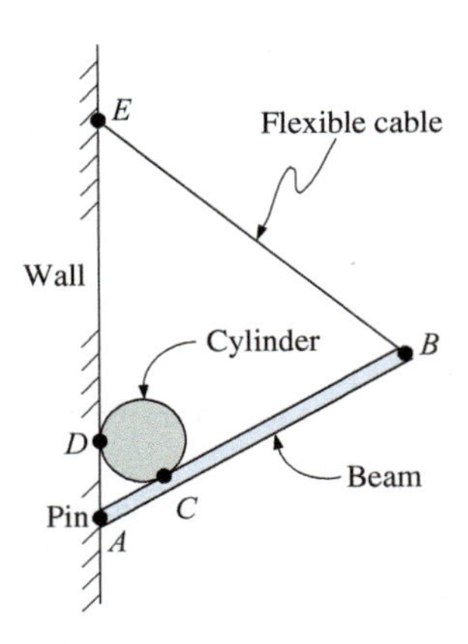

FIGURE 4.6 Support system for cylinder.

Solution (a) The free-body diagram for the entire system is shown in Figure 4.7a. The external forces are the weight of the cylinder W_C, the weight of the beam W_B, the pull *T* of the wall on the cable, the push *P* of the wall at point *D*, and the force exerted by the pin at point *A*. Note that *T* must act along the centerline (or axis) of the cable, since the cable is flexible. The sense of *T* is upward and to the left since the flexible cable can support only tension and not compression. Also note that *P* must be horizontal, acting perpendicular to the vertical wall, since the smooth contact surface of the wall permits vertical motion.

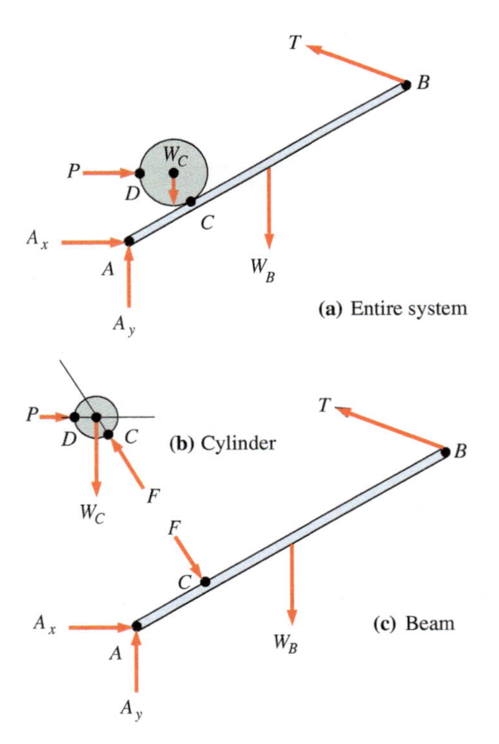

FIGURE 4.7 Free-body diagrams.

Also note that the direction of the force exerted at point A is unknown. Therefore, this force is represented by its rectangular components A_y and A_x with assumed senses as shown.

(b) The free-body diagram for the cylinder is shown in Figure 4.7b. The external forces are the weight of the cylinder W_C acting vertically downward, the push P of the wall acting perpendicularly to the surface of the wall, and the reaction of the beam on the cylinder (known to be perpendicular to the beam).

(c) The free-body diagram for the beam is shown in Figure 4.7c. The external forces are the weight of the beam W_B acting vertically downward, the pull T of the cable acting along its axis, the force F of the cylinder, and the rectangular components A_y and A_x of the unknown force exerted by the pin-connection at point A. In particular, note the equal and opposite relationship of force F acting on the beam and the cylinder.

4.4 EQUILIBRIUM OF CONCURRENT FORCE SYSTEMS

When a coplanar concurrent force system is in equilibrium, the algebraic sum of the vertical and horizontal components of all the forces must, respectively, equal zero. This case has been expressed in Section 4.2 as

$$\Sigma F_y = 0 \quad \text{and} \quad \Sigma F_x = 0$$

Conversely, if it can be demonstrated that if $\Sigma F_y = 0$ and $\Sigma F_x = 0$ in a concurrent force system, then we can say that the system is in equilibrium and that the resultant is equal to zero.

Recall from previous chapters that a concurrent force system is one in which the action lines of all the forces intersect at a common point. This force system cannot cause rotation of the body on which it acts, thereby implying that the two equations of force equilibrium are sufficient for analyzing this type of system.

Problems involving concurrent force systems frequently involve special members that are in equilibrium under the action of two equal and opposite forces, called *two-force members*. The forces (which may be resultants of concurrent force systems) are normally applied at the ends, as shown in Figure 4.8. Note that the bar MN (assumed to be weightless) is in equilibrium under the effects of two forces and that the points of application lie on the longitudinal axis of the bar. For the resultant of the two forces to be zero, the forces must be equal, collinear, and opposite in sense. Should one of the forces not have a line of action coincident with the line joining the points of application of the forces, there will be a component of force that will upset equilibrium. Therefore, for a two-force member, the lines of action of the applied forces must be coincident with the line joining the points of application. For the straight two-force members that we consider in this text, this line coincides with the longitudinal axis of the member.

The recognition of two-force members will often be a very important key in the solution of a problem. Note that we are assuming the members to be weightless. A more practical assumption is that the effect of the weights of the members is extremely small when compared with the effects of the applied forces. With this assumption in mind, we note that the cable RQ and the boom OQ in Example 4.4 are two-force members. The beam AB in Figure 4.6 (and Figure 4.7) is a multiple-force member. These members are discussed in Sections 4.6 and 5.6.

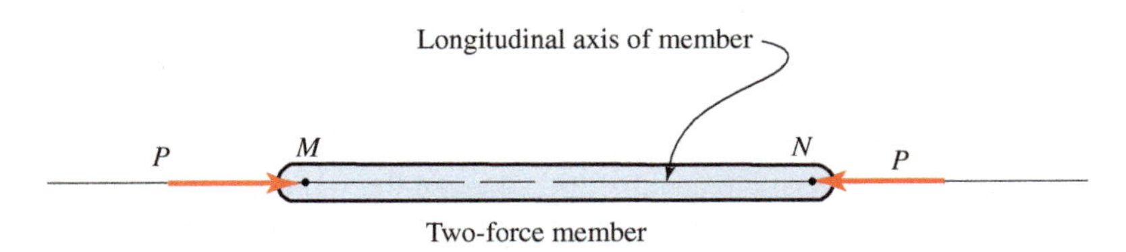

FIGURE 4.8 Two-force member.

EXAMPLE 4.4 A 100-lb weight is supported by a tied boom, as shown in Figure 4.9a. Determine the magnitude of the force C in the boom and the force T in the flexible cable.

Solution The free-body diagram of the joint at Q is shown in Figure 4.9b. Note that the vertical and horizontal force components are designated with V and H subscripts. The boom and the cable are two-force members. Therefore, the lines of action of unknown forces C and T are known. The two forces are

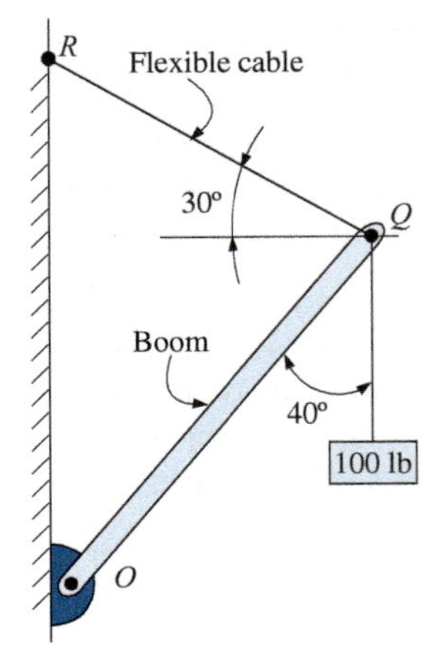

(a) Tied boom-supported weight

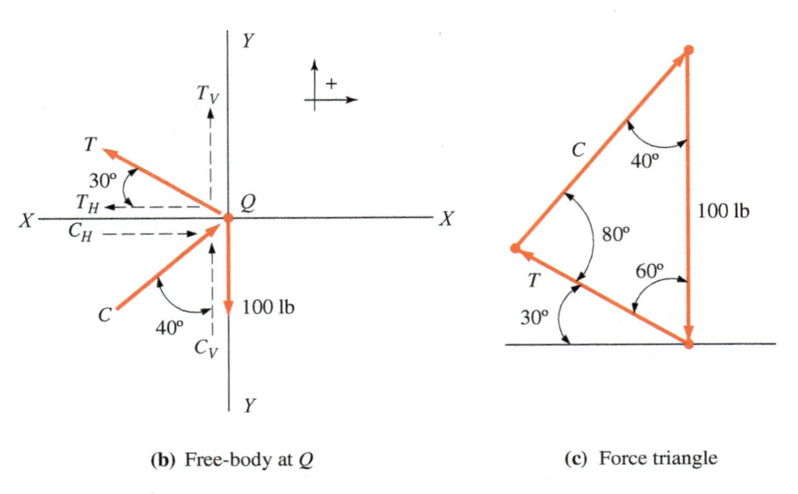

(b) Free-body at Q (c) Force triangle

FIGURE 4.9 Concurrent force system.

shown acting with assumed senses, which are readily apparent in this case. At times, however, the situation will not be as clear. In those cases, a sense should be assumed for each unknown force and verified by calculation.

This force system is categorized as a coplanar concurrent force system. The two unknown forces may be found either by the force triangle method or by the method of components, applying the two laws of equilibrium. For illustrative purposes, we show both methods.

The Force Triangle Method

A concurrent coplanar force system in equilibrium must have a zero resultant. If such a system is composed of three forces in equilibrium and the forces are arranged in a force triangle with the tip (arrowhead) of each force touching the tail of one of the other forces, a zero resultant is indicated by the closed triangle, as shown in Figure 4.9c.

It is not necessary to assume the sense of the unknown force in either the cable or the boom; they can be determined from the force triangle. As shown in Figure 4.9c, begin the sketch of the force triangle with the 100 lb force, which is known in magnitude, direction, and sense. Then draw lines parallel to the lines of action of the two unknown forces, one through the tail and one through the tip of the known 100 lb force vector. The lines must intersect to form a closed triangle because equilibrium exists. Since the force vectors must lie tip to tail, the senses of the unknown forces are established in the force triangle. These senses are then related to the forces acting at the point of concurrency Q in Figure 9b. Force C acts upward and to the right; force T acts upward and to the left.

Use the law of sines to solve for the unknown forces:

$$\frac{100 \text{ lb}}{\sin 80^\circ} = \frac{T}{\sin 40^\circ} = \frac{C}{\sin 60^\circ}$$

from which

$$T = \frac{\sin 40^\circ}{\sin 80^\circ}(100 \text{ lb}) = 65.3 \text{ lb}$$

$$C = \frac{\sin 60^\circ}{\sin 80^\circ}(100 \text{ lb}) = 87.9 \text{ lb}$$

The Method of Components Applying the two laws of force equilibrium ($\Sigma F_H = 0$ and $\Sigma F_V = 0$), and with reference to Figure 4.9b, the two unknown forces may be determined. Assume positive senses to be upward and to the right. This sign convention is indicated by a symbol such as that shown in the upper right portion of Figure 4.9b. The senses shown for the forces T (upward and to the left) and C (upward and to the right) are, at this point, assumed senses.

Sum forces in the horizontal direction:

$$\Sigma F_H = -T_H + C_H = 0$$
$$= -T\cos 30^\circ + C\sin 40^\circ = 0$$

from which

$$T = \frac{\sin 40^\circ}{\cos 30^\circ}C = 0.7422C \quad (1)$$

Sum forces in the vertical direction:

$$\Sigma F_V = +T_V + C_V - 100 = 0$$
$$= +T\sin 30^\circ + C\cos 40^\circ - 100 \text{ lb} = 0$$

Substituting from Equation 1,

$$0.7422C(\sin 30^\circ) + C\cos 40^\circ - 100 \text{ lb} = 0$$
$$0.7422C(0.500) + C(0.7660) - 100 \text{ lb} = 0$$

from which

$$C = +87.9 \text{ lb}$$

From Equation 1,

$$T = 0.7422C = +65.2 \text{ lb}$$

The positive signs indicate that the senses of the forces are, in fact, as they were assumed. This concept is important. A positive sign resulting in the final calculation does not necessarily mean that the force is acting with a positive sense. It means that the sense of the force is as it was assumed. Except for an insignificant rounding error in force T, this solution checks with the force triangle solution.

Note that a problem involving three coplanar concurrent forces is very conveniently solved using the force triangle method. This method is limited in applicability to the case of three coplanar, concurrent forces. Nevertheless, it is recommended for this type of problem, particularly where the two unknown forces are sloping forces (not perpendicular or parallel to the reference axes). For problems involving more than three concurrent forces, the method of components is applicable.

EXAMPLE 4.5 A block of 200-kg mass is supported by two cables, as shown in Figure 4.10a. Find the magnitude of the forces in the cables.

Solution The load due to the block is calculated from

$$W = mg = 200 \text{ kg}(9.81 \text{ m/s}^2) = 1962 \text{ N} = 1.962 \text{ kN}$$

The force system is coplanar and concurrent. Both cables must be in tension. (Cables can never be in compression.) The force system may be solved by using the method of components and applying

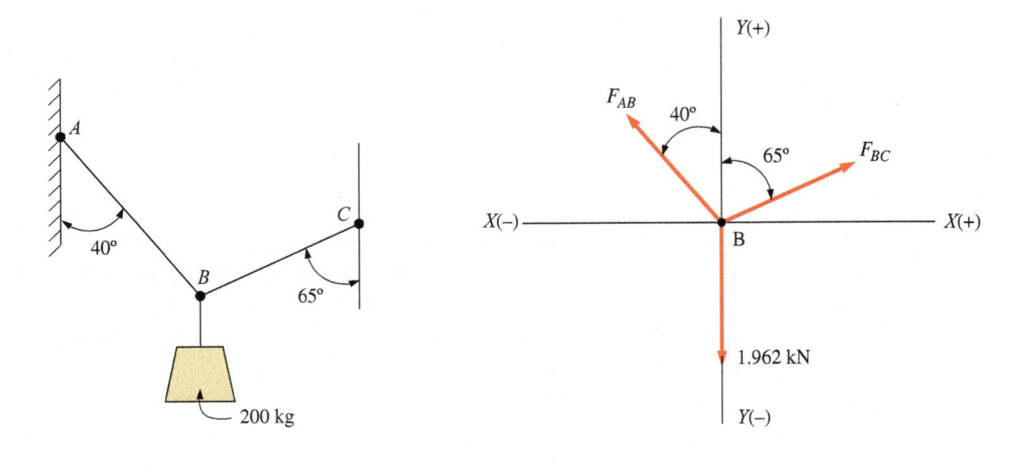

(a) Suspended weight

(b) Free-body diagram of joint *B*

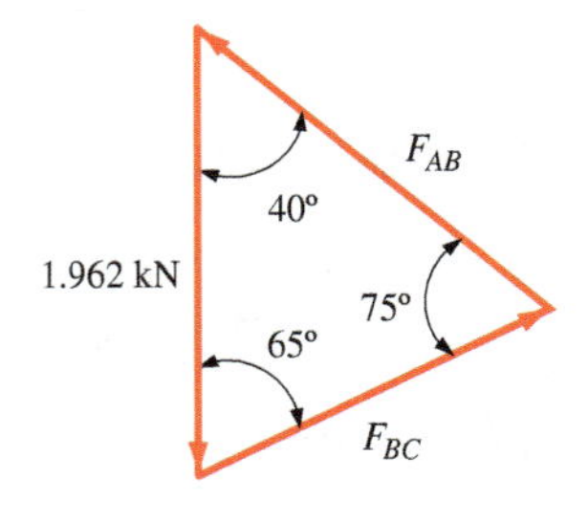

(c) Force triangle

FIGURE 4.10 Equilibrium of concurrent forces.

the two equations of equilibrium to the free body of joint *B*, shown in Figure 4.10b, or by solution of the force triangle, shown in Figure 4.10c. The latter is the more direct solution. Thus,

$$\frac{1.962\text{ kN}}{\sin 75^\circ} = \frac{F_{BC}}{\sin 40^\circ} = \frac{F_{AB}}{\sin 65^\circ}$$

$$F_{BC} = \frac{\sin 40^\circ}{\sin 75^\circ}(1.962\text{ kN}) = 1.306\text{ kN}$$

$$F_{AB} = \frac{\sin 65^\circ}{\sin 75^\circ}(1.962\text{ kN}) = 1.841\text{ kN}$$

EXAMPLE 4.6 Two straight, rigid bars, *AB* and *BC*, are pin-connected to a horizontal supporting floor at their lower ends and to each other at their upper ends, as shown in Figure 4.11a. Applied loads at point *B* are 2000 lb vertically and 1800 lb horizontally. Assume that the weights of the bars are negligible and that the system is coplanar. Compute the magnitude and sense of the forces in the two bars using the method of components.

Solution The free-body diagram for the pin at point *B* is shown in Figure 4.11b. The forces exerted on the pin are the horizontal pull of 1800 lb, the vertically downward pull of 2000 lb, the force exerted by *AB*, and the force exerted by *BC*. Note that *AB* and *BC* are two-force members. Therefore, the lines of action of forces F_{AB} and F_{BC} will lie, respectively, along the axes of members *AB* and *BC*. That is, the directions of the forces are known. Only the magnitudes and senses must be determined.

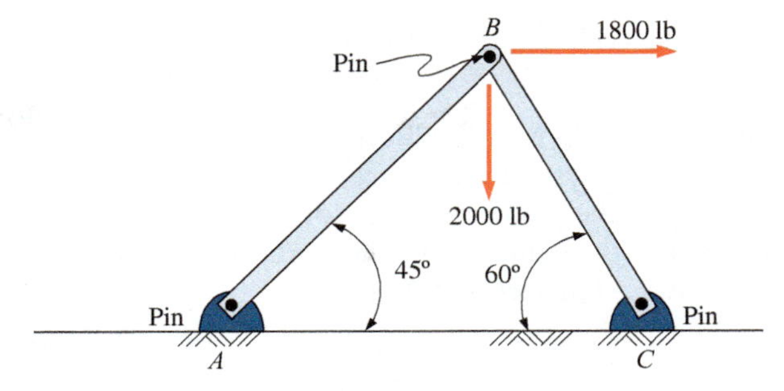

(a) Rigid pin-connected structure

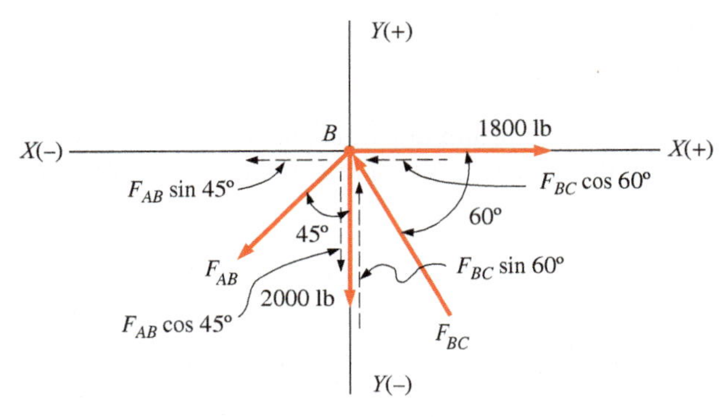

(b) Free-body diagram of pin at point B

FIGURE 4.11 Structure for Example 4.6.

The senses of the forces are assumed, as shown in Figure 4.11b. The sense will be determined by the sign of the computed magnitude of the force: a positive sign will indicate that the sense of the force is as assumed; a negative sign will indicate that the sense of the force is opposite to that assumed.

Since the force system is coplanar and concurrent, the two equations of equilibrium may be applied using the conventional *X–Y* coordinate axes. Forces upward and to the right will be assumed positive. The assumed senses of F_{AB} and F_{BC} are as shown. We next apply the two equations of force equilibrium:

$$\Sigma F_x = +1800 \text{ lb} - F_{BC} \cos 60° - F_{AB} \sin 45° = 0$$

$$= +1800 \text{ lb} - 0.5\, F_{BC} - 0.707\, F_{AB} = 0 \quad \textbf{(1)}$$

$$\Sigma F_y = -2000 \text{ lb} + F_{BC} \sin 60° - F_{AB} \cos 45° = 0$$

$$= -2000 \text{ lb} + 0.866\, F_{BC} - 0.707\, F_{AB} = 0 \quad \textbf{(2)}$$

Solve Equation 2 for F_{AB} in terms of F_{BC}:

$$0.707\, F_{AB} = 0.866\, F_{BC} - 2000 \text{ lb}$$

Substituting the preceding in Equation 1 yields

$$+1800 \text{ lb} - 0.5\, F_{BC} - (0.866\, F_{BC} - 2000 \text{ lb}) = 0$$

$$+1800 \text{ lb} - 1.366\, F_{BC} + 2000 \text{ lb} = 0$$

from which

$$F_{BC} = +2782 \text{ lb}$$

Substituting this in Equation 2 and solving yields

$$-2000 \text{ lb} + 0.866(2782 \text{ lb}) - 0.707\, F_{AB} = 0$$

$$F_{AB} = +579 \text{ lb}$$

The positive signs indicate that the senses of the forces are as assumed. Member *BC* acts toward the pin (at *B*) and member *AB* pulls away from the pin.

4.5 EQUILIBRIUM OF PARALLEL FORCE SYSTEMS

When a coplanar parallel force system is in equilibrium, the algebraic sum of the forces of the system must equal zero. In addition, the algebraic sum of the moments of the forces of the system about any point in the plane must equal zero. These requirements were expressed in Section 4.2 as

$$\Sigma F = 0 \quad \text{and} \quad \Sigma M = 0$$

Conversely, if it can be demonstrated that the two preceding requirements are satisfied (where the moment M may be about any point in the plane), then we can say that the parallel force system is in equilibrium and that the force and moment resultants are equal to zero.

It is important to note that equilibrium of parallel force systems cannot be verified through the use of the force summation equations only. In all cases, at least one moment summation equation must be considered.

A common type of problem associated with parallel force systems is determining two unknown support reactions for a beam or member. These are external reactions and, as shown in later chapters, must be calculated prior to the internal behavior investigation of the member.

In computing reactions of parallel force systems, care must be taken to adhere to a sign convention. Whether a clockwise moment (rotation) is taken as positive or negative is inconsequential, but the adherence to one or the other convention throughout the solution of any particular problem is of utmost importance. Here we assume a counterclockwise moment (rotation) about a moment center to be positive and a clockwise moment to be negative. Further discussion on computing reactions with examples using more complex members and loadings is provided in Chapter 13.

EXAMPLE 4.7 A beam carries vertical concentrated loads as shown in Figure 4.12a. The beam is pin-supported at A and supported by a roller on a horizontal surface at B. A beam of this type with the indicated supports is called a *simple beam*. The supports are called *simple supports*. The reactions at the supports are assumed to be parallel to the loads. Calculate the reactions at each support. Neglect the weight of the beam.

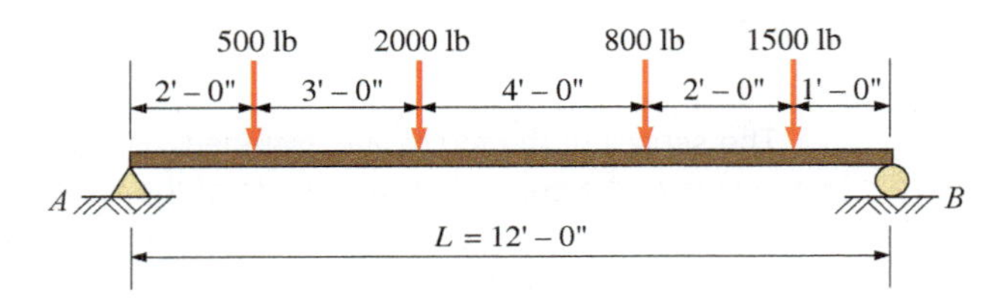

(a) Parallel force system on beam

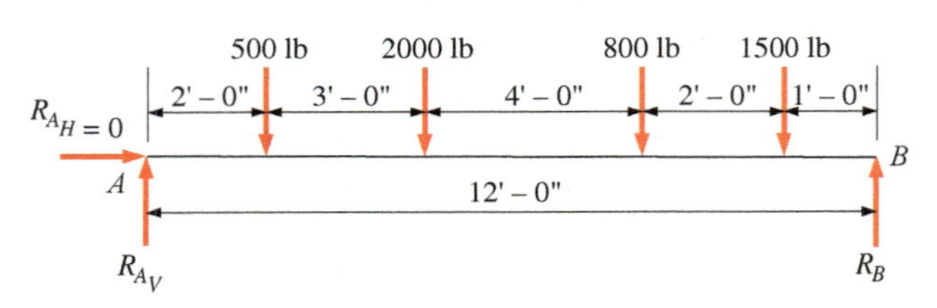

(b) Free-body diagram

FIGURE 4.12 Beam for Example 4.7.

Solution The free-body diagram is shown in Figure 4.12b. Note that the supports at A and B have been replaced with forces (reactions) with a known line of action and assumed sense. The pin support at A could provide a horizontal reaction, but there are no horizontally applied forces or components. Therefore, this reaction would be zero and may be neglected. As will be discussed further in Chapter 13, the free-body diagram of the beam is commonly called a *load diagram*. The load diagram has the same geometry as the original structure (beam), the same loads are shown, and the supports have been replaced with the reactions expected from those supports.

Assuming counterclockwise rotation is positive, the reaction at B may be computed by taking an algebraic summation of the moments of the forces about point A. Units are lb and ft.

$$\Sigma M_A = +R_B(12.0) - 500(2.0) - 2000(5.0) - 800(9.0) - 1500(11.0) = 0$$

from which

$$R_B = +2892 \text{ lb}$$

The positive sign indicates that the sense (upward) is as assumed for the reaction at point *B*. Had this result turned out to be negative, it would have meant that the sense of the reaction was opposite to that assumed, that is, that the reaction was actually downward.

The reaction at point *A* may be computed by taking an algebraic summation of the moments of the forces about point *B*. Again, units are lb and ft.

$$\Sigma M_B = -R_{A_V}(12.0) + 1500(1.0) + 800(3.0) + 2000(7.0) + 500(10.0) = 0$$

from which

$$R_{A_V} = +1908 \text{ lb}$$

As an independent check of the computations, an algebraic summation of all the vertical forces should be made. For the beam to be in equilibrium, it is required that the sum be equal to zero. Assuming upward acting forces to be positive and downward to be negative,

$$\Sigma F_V = +1908 + 2892 - 500 - 2000 - 800 - 1500 = 0 \qquad \textbf{(O.K.)}$$

EXAMPLE 4.8

A simple beam is subjected to vertical concentrated and uniformly distributed loads as shown in Figure 4.13a. The beam is pin-supported at *A* and roller-supported (on a horizontal surface) at *B*. Calculate the reactions at each support. Neglect the weight of the beam.

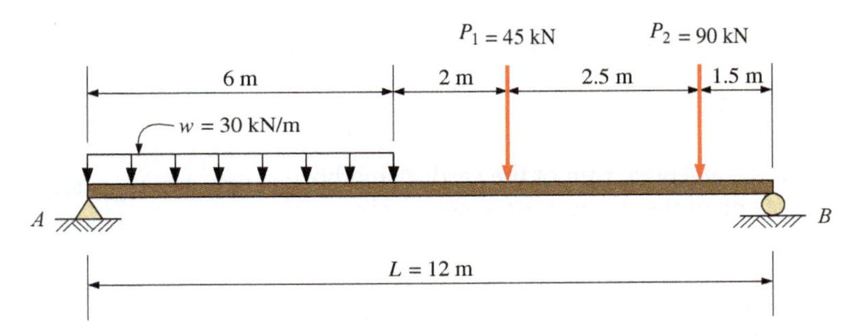

(a) Parallel force system on simple beam

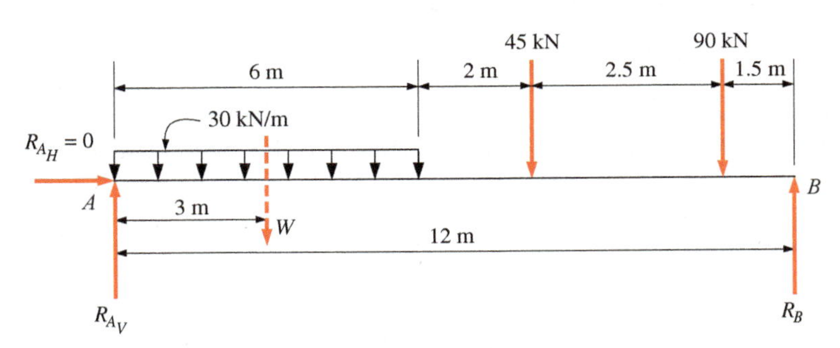

(b) Free-body diagram

FIGURE 4.13 Beam for Example 4.8.

Solution

The free-body diagram (the load diagram) is shown in Figure 4.13b. The reactions are assumed to be parallel to the loads. The supports at *A* and *B* have been replaced with forces (reactions) with known lines of action and assumed senses. The pin support at *A* could provide a horizontal reaction, but there are no horizontally applied loads or components. Therefore, this reaction would be zero and may be neglected.

The equivalent concentrated resultant force for the uniformly distributed load is

$$W = 30 \text{ kN/m} \times 6 \text{ m} = 180 \text{ kN}$$

This force is indicated by a dashed arrow acting at the center of the distributed load.

Next compute the reaction at B by taking an algebraic summation of the moments of the forces about point A. Assume counterclockwise rotation is positive and also that both reactions are acting upward. Units are kN and m.

$$\Sigma M_A = +R_B(12) - 90(10.5) - 45(8) - 180(3) = 0$$

from which

$$R_B = +153.8 \text{ kN}$$

Since the result is positive, the sense of the reaction is as assumed (upward).

The reaction at A may be computed by summing the moments of the forces about point B. Again, units are kN and m.

$$\Sigma M_B = -R_{A_V}(12) + 180(9) + 45(4) + 90(1.5) = 0$$

from which

$$R_{A_V} = +161.3 \text{ kN}$$

Finally, check the calculations with a vertical summation of forces (upward is positive):

$$\Sigma F_V = +153.8 + 161.3 - 180 - 45 - 90 = +0.10 \text{ kN} \approx 0 \qquad \textbf{(O.K.)}$$

4.6 EQUILIBRIUM OF NONCONCURRENT FORCE SYSTEMS

When a coplanar, nonconcurrent, nonparallel force system is in equilibrium, the algebraic sum of the vertical and horizontal components of all the forces must, respectively, equal zero. In addition, the algebraic sum of the moments of the forces about any point in the plane must equal zero.

Conversely, if $\Sigma F_V = 0$, $\Sigma F_H = 0$, and $\Sigma M = 0$, then we can say that the force system is in equilibrium and that the force and moment resultants are equal to zero.

Equilibrium of this system cannot be verified through the use of the force summation equations only. In all cases, at least one moment summation equation must be considered. In choosing the moment center about which to sum moments, we must remember that forces having lines of action passing through the moment center have zero moment about that moment center. Therefore, a moment center should be selected that will eliminate as many forces as possible (or selected forces) from the moment equation.

EXAMPLE 4.9 Compute the reactions at A and B on the truss shown in Figure 4.14a. There is a roller support at A and a pin support at B.

Solution The free-body diagram for the truss is shown in Figure 4.14b. The external forces are as follows:

1. On the bottom chord: one vertical concentrated load.
2. On the top chord: one inclined load acting at the upper right corner. (For convenience, the inclined load is resolved into its components.)
3. The reaction at A, known to be vertical since the support is furnished by a roller. The sense is assumed upward.
4. The reaction at B. Since the support is pinned, the direction of the reaction is unknown; therefore, the reaction is represented by its vertical and horizontal components. The senses are assumed upward and to the right, respectively.

This external force system is coplanar, nonconcurrent, and nonparallel. Therefore, use the three laws of equilibrium: $\Sigma F_V = 0$, $\Sigma F_H = 0$, and $\Sigma M = 0$. Assume forces upward and to the right and counterclockwise moments are positive. So

$$\Sigma F_H = +R_{B_H} - 423 \text{ lb} = 0$$

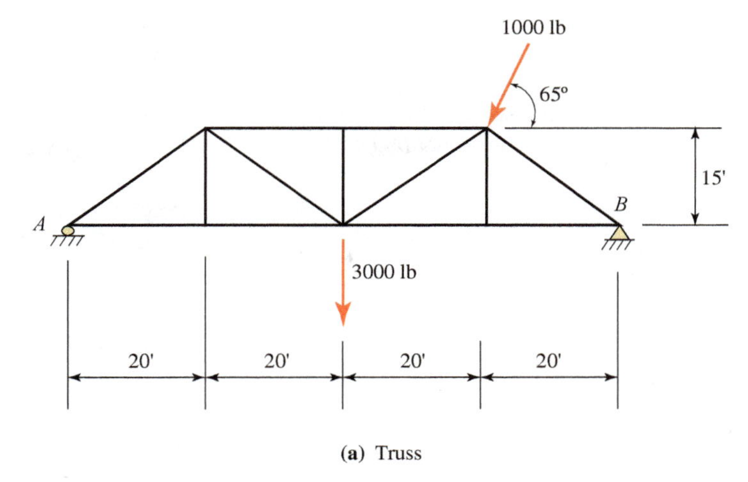

(a) Truss

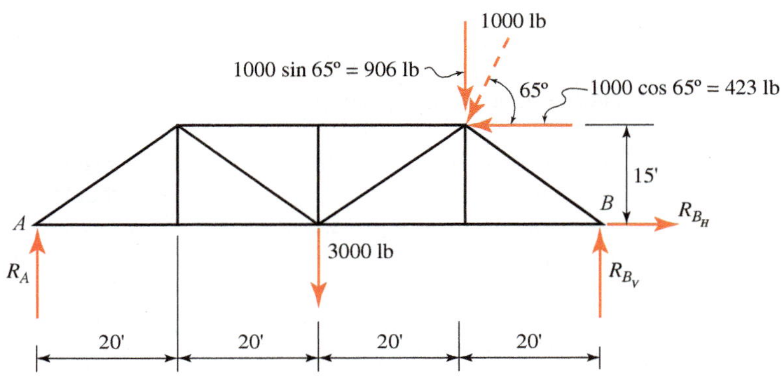

(b) Free-body diagram

FIGURE 4.14 Truss for Example 4.9.

from which

$$R_{B_H} = +423 \text{ lb} \rightarrow$$

The positive sign indicates that the sense was assumed correctly and R_{B_H} does, in fact, act to the right. Next determine R_A by summing moments about point B. The selection of this point as the moment center will eliminate R_{B_V} and R_{B_H} from the moment equation, leaving only one unknown, R_A. Units are lb and ft.

$$\Sigma M_B = -R_A(80) + 3000(40) + 906(20) + 423(15) = 0$$

from which

$$R_A = +1806 \text{ lb} \uparrow$$

Use point A as the moment center to calculate R_{B_V}, after which an algebraic summation of the vertical forces can be done as a check. Again, units are lb and ft.

$$\Sigma M_A = -3000(40) - 906(60) + 423(15) + R_{B_V}(80) = 0$$

from which

$$R_{B_V} = 2100 \text{ lb} \uparrow$$

Finally, use an algebraic summation of vertical forces as a check:

$$\Sigma F_V = +1806 - 3000 - 906 + 2100 = 0 \qquad \textbf{(O.K.)}$$

EXAMPLE 4.10

A cylinder is supported by an inclined wall and a vertical bar, as shown in Figure 4.15. The bar is pin-connected to the wall at point *A* and connected to a flexible cable at point *B*. All surfaces are smooth. The cylinder weight *W* is 2500 lb. Determine the reaction at point *A* and the pull in the cable at point *B*. Neglect the weight of the bar.

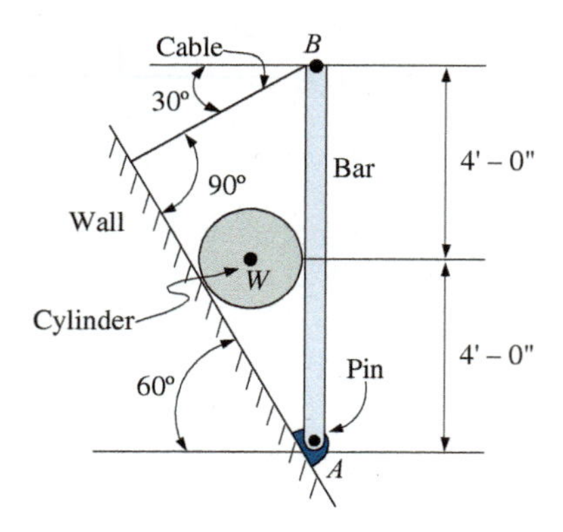

FIGURE 4.15 Cylinder support system.

Solution The free-body diagram for the bar is shown in Figure 4.16b. The force system acting on the bar is coplanar, nonconcurrent, and nonparallel. Note that four unknown forces are acting on the bar. Since there are only three equations of equilibrium, you will not be able to determine all these forces using this free body alone. One of the forces must be determined elsewhere.

The cylinder is in contact with the bar and creates the force *N* on the bar. Investigate the cylinder as a free body and, using the laws of equilibrium, determine the contact force *N*. As shown in Figure 4.16a, the force system acting on the cylinder is coplanar and concurrent. For the cylinder, assuming forces upward and to the right to be positive, apply $F_V = 0$ and $\Sigma F_H = 0$ as follows:

$$\Sigma F_V = -W + F \sin 30° = -2500 \text{ lb} + F(0.5) = 0$$

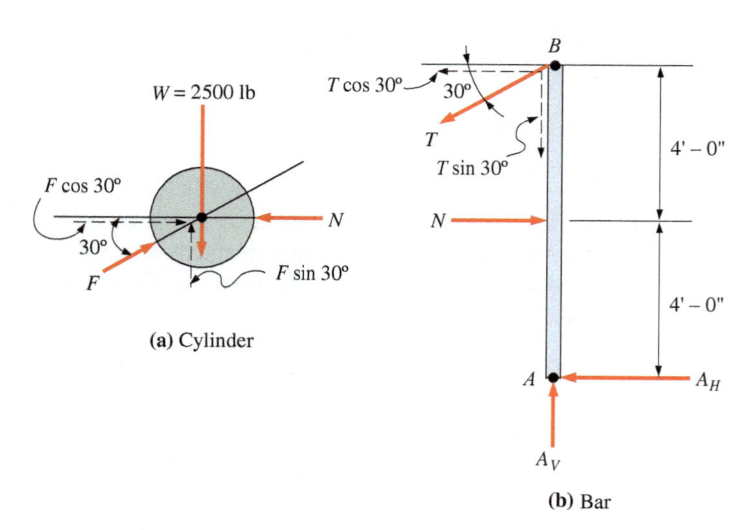

FIGURE 4.16 Free-body diagrams.

from which

$$F = +5000 \text{ lb}$$

and

$$\Sigma F_H = -N + F \cos 30° = -N + 5000 \text{ lb } (0.866) = 0$$

from which

$$N = +4330 \text{ lb}$$

The positive signs for F and N indicate that the senses for these forces are as assumed.

With force N known, the laws of equilibrium may now be applied to the free body of the bar shown in Figure 4.16b. Using point A as a moment center will eliminate A_V and A_H from the summation. Assuming counterclockwise moment to be positive and assuming T to be acting as shown (the vertical component of T has a line of action passing through A; therefore, its moment about A is zero), the moment summation yields

$$\Sigma M_A = -N(4.0) + T\cos 30°(8.0) = -4330(4.0) + T(0.866)(8.0) = 0$$

from which

$$T = +2500 \text{ lb}$$

Since the result is positive, the sense of T is as assumed. Next, with T and N known, the two components of the reaction at point A may be determined. Assuming the forces at A to be acting as shown,

$$\Sigma F_H = +N - T\cos 30° - A_H = +4330 \text{ lb} - (2500 \text{ lb})(0.866) - A_H = 0$$

from which

$$A_H = +2165 \text{ lb}$$

For the vertical forces,

$$\Sigma F_V = +A_V - T\sin 30° = +A_V - (2500 \text{ lb})(0.5) = 0$$

from which

$$A_V = +1250 \text{ lb}$$

The positive signs of the results indicate that the senses for A_H and A_V are as assumed.

Previously we discussed two-force members. Bar AB in Example 4.10 is acted on by forces at three points along its length. Members such as this one, acted on by forces at three or more points along their lengths, are sometimes called *multiple-force members*. They differ radically from two-force members in that some of, or all, the forces are applied transversely to the axis of the member, causing it to bend. Recall that in the two-force member the forces, of necessity, were collinear with (acted along) the axis of the member. The direction of a force applied to a multiple-force member (such as at A on bar AB) will frequently be unknown. Multiple-force members in frames are discussed in Section 5.6.

SUMMARY BY SECTION NUMBER

4.1 Equilibrium is an "at rest" state, or a state of zero motion.

4.2 The resultant of any force system acting on a body in equilibrium must be zero. For the two-dimensional case, the three laws of equilibrium are

$$\Sigma F_x = 0 \qquad \Sigma F_y = 0 \qquad \Sigma M = 0$$

4.3 The free-body diagram is a sketch of an isolated body indicating all forces that act on the body.

4.4 Coplanar concurrent force systems are in equilibrium if $\Sigma F_x = 0$ and $\Sigma F_y = 0$. Generally, such systems involve two-force members. In a two-force member, forces are collinear with the axis of the member and act at two points only, usually at the ends.

4.5 Coplanar parallel force systems are in equilibrium if $\Sigma F = 0$ and $\Sigma M = 0$. A common parallel force system problem involves determining two unknown support reactions for a beam.

4.6 Coplanar nonconcurrent force systems are in equilibrium if $\Sigma F_x = 0$, $\Sigma F_y = 0$, and $\Sigma M = 0$. Such systems may involve multiple-force members, which are members acted on by forces at three or more points along their length. The reactions for multiple-force members are not collinear with the member.

PROBLEMS

Section 4.3 The Free-Body Diagram

4.1 and 4.2 Sketch free-body diagrams for the members shown.

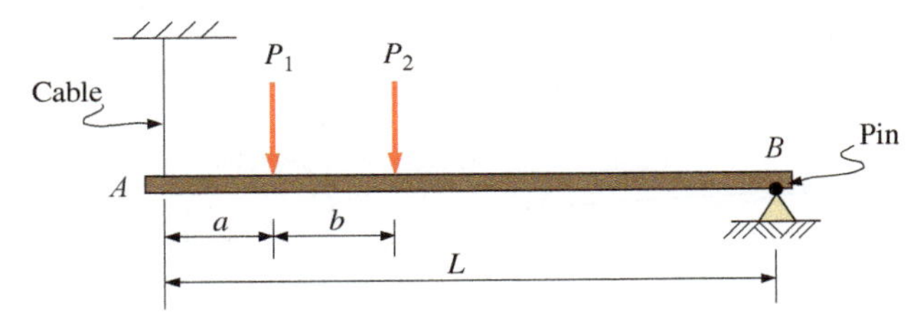

PROBLEM 4.1

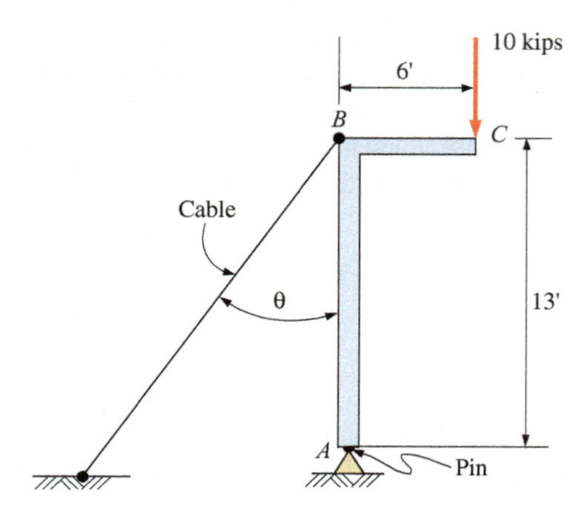

PROBLEM 4.2

4.3 A steel cylinder having a mass of 120 kg is supported by two inclined planes, as shown. Sketch the free-body diagram for the cylinder.

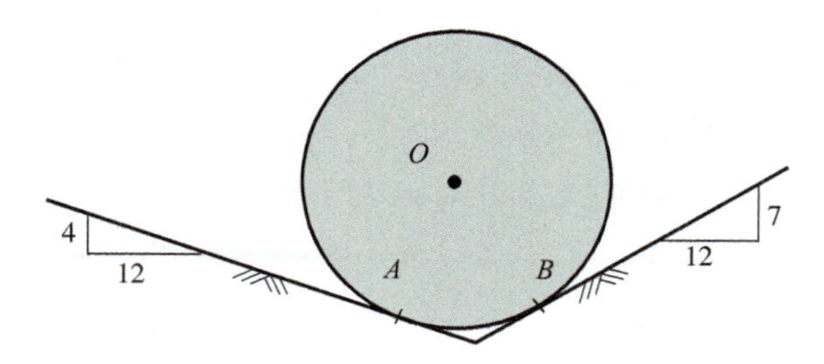

PROBLEM 4.3

4.4 A 50-lb block is supported by a pin support and a flexible cable, as shown. Draw the free-body diagram for the block.

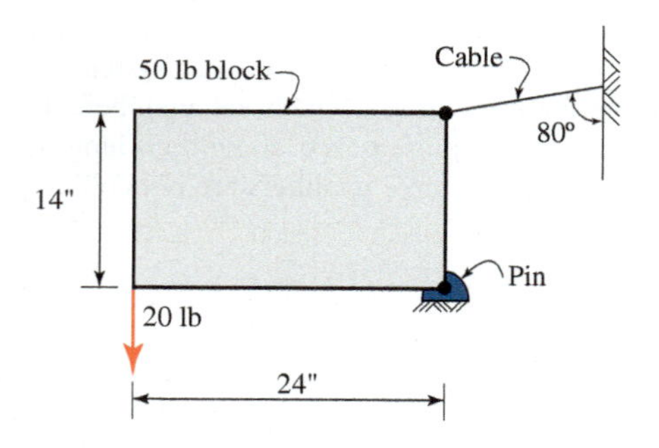

PROBLEM 4.4

4.5 A cylinder weighing 200 lb is supported on an inclined plane by a cable, as shown. Sketch the free-body diagram for the cylinder.

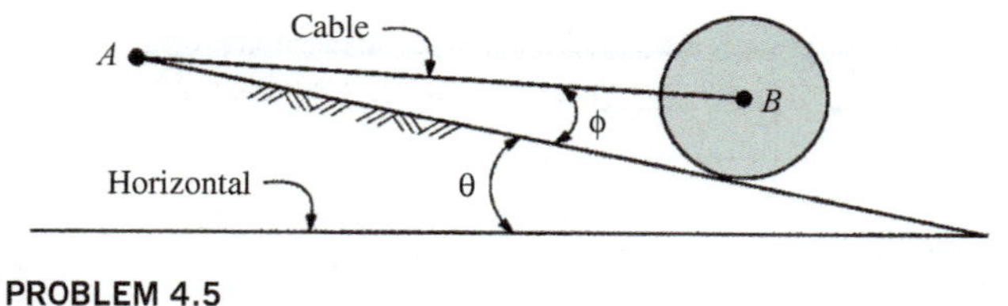

PROBLEM 4.5

4.6 A weight W is supported by a flexible cable and an inclined bar, as shown. The bar is pin-connected at a vertical wall. Sketch the free-body diagram for the bar.

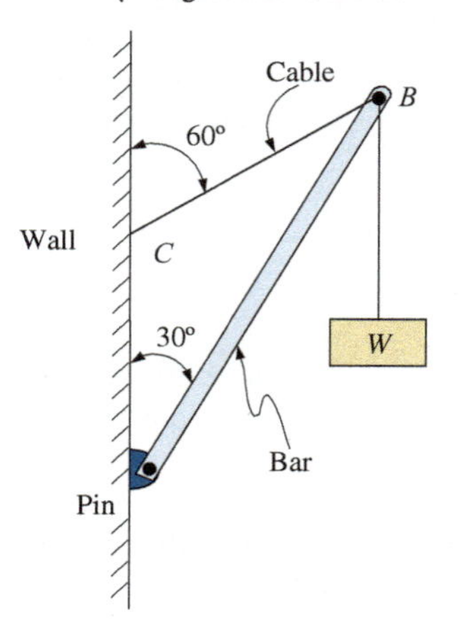

PROBLEM 4.6

4.7 The ladder shown is supported by a smooth frictionless vertical wall and is pin-connected at point A. The ladder has a length of 16 ft and supports a weight W at point C. Assuming the weight of the ladder is acting at its center of gravity, sketch the free-body diagram for the ladder.

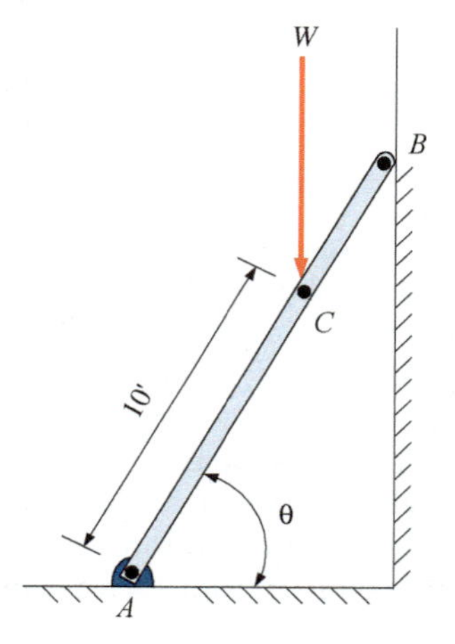

PROBLEM 4.7

Section 4.4 Equilibrium of Concurrent Force Systems

4.8 What horizontal force F applied at the center of the cylinder shown is required to start the cylinder to roll over the 5-in. curb? What is the reaction of the curb? The cylinder weighs 250 lb.

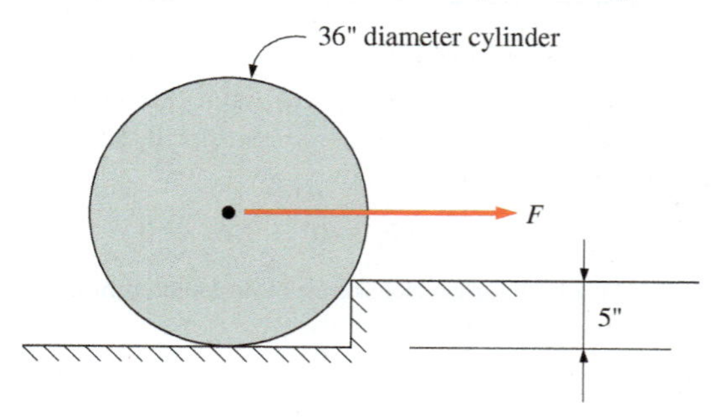

PROBLEM 4.8

4.9 Calculate the force in cable AB and the angle θ for the support system shown.

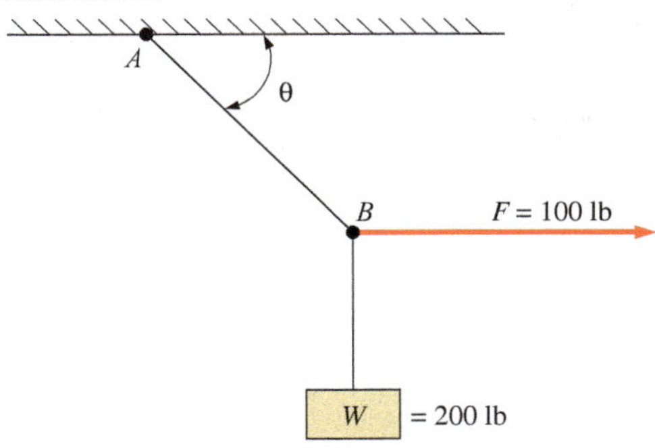

PROBLEM 4.9

4.10 Calculate the horizontal force F that should be applied to the 100 lb weight shown so that the cable AB is inclined at an angle of 30° with the vertical.

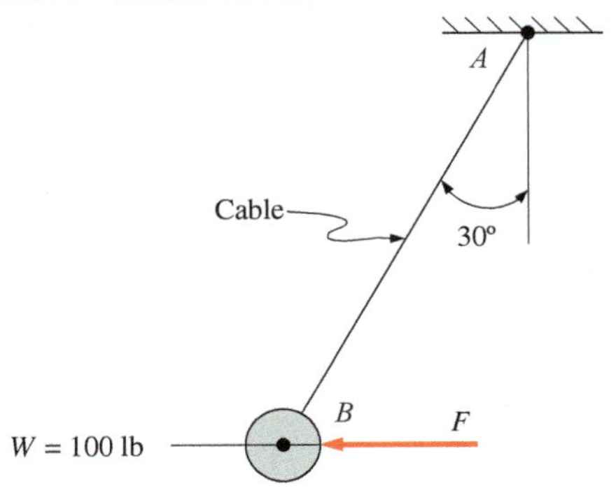

PROBLEM 4.10

4.11 Calculate the reactions of the two smooth inclined planes against the cylinder shown. The cylinder weighs 100 lb.

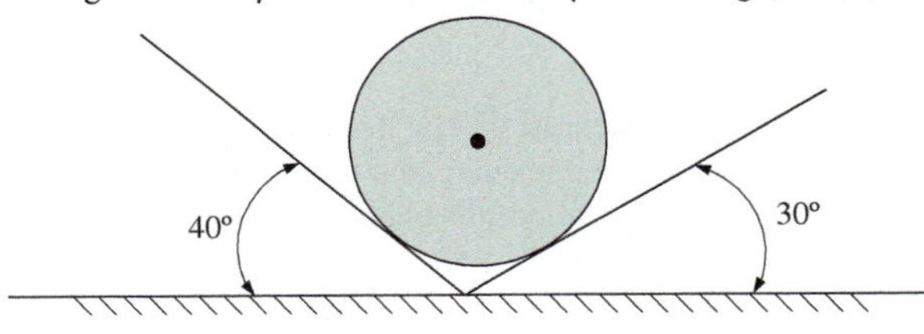

PROBLEM 4.11

4.12 Calculate the force in each cable for the suspended weight shown.

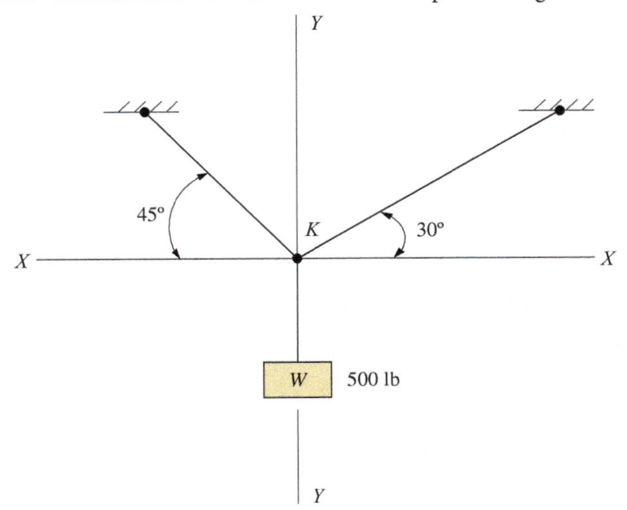

PROBLEM 4.12

4.13 Three members of a truss intersect at joint B as shown. The forces in the members are concurrent at the joint. Force F_1 is horizontal. Determine forces F_1 and F_2.

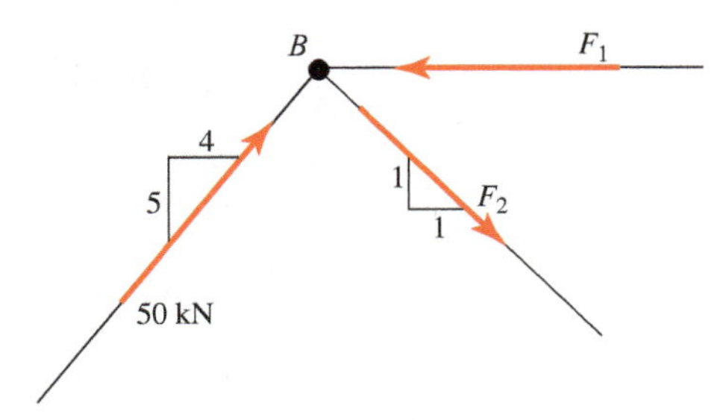

PROBLEM 4.13

4.14 Four concurrent forces in equilibrium act at point C, a truss joint, as shown. The 15 kN force acts vertically. Determine forces F_3 and F_4.

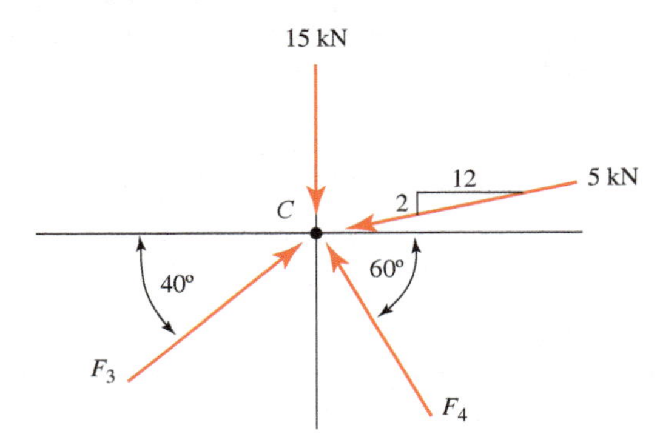

PROBLEM 4.14

Section 4.5 Equilibrium of Parallel Force Systems

4.15 The beam shown carries vertical concentrated loads. Calculate the reaction at each support. Neglect the weight of the beam.

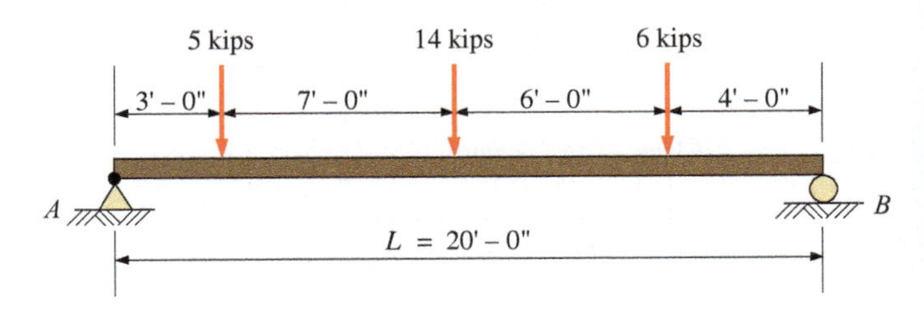

PROBLEM 4.15

4.16 Find the reactions at A and B for the beam shown. The uniformly distributed load of 1 kip/ft includes the weight of the beam.

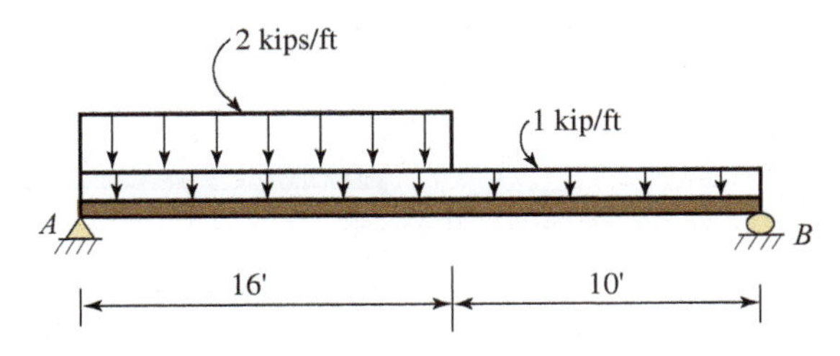

PROBLEM 4.16

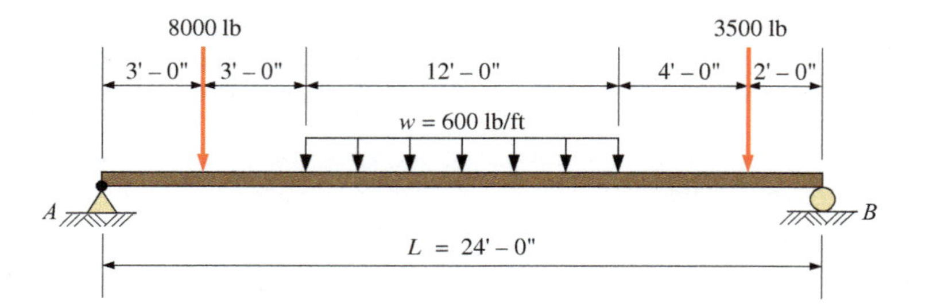

PROBLEM 4.18

4.17 A simply supported beam spans 10 m. The beam supports a single concentrated load of 4 kN. How far from the left support should the 4-kN load be placed so that the left reaction is 2.70 kN? Neglect the weight of the beam.

4.18 The beam shown carries vertical loads. Calculate the reaction at each support. Neglect the weight of the beam.

4.19 Calculate the reaction at each support for the truss shown. Neglect the weight of the truss.

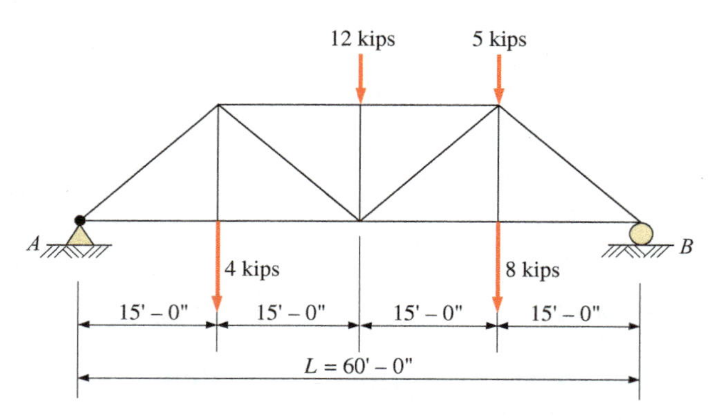

PROBLEM 4.19

4.20 Calculate the reactions at *A* and *B* for the beam shown. Neglect the weight of the beam.

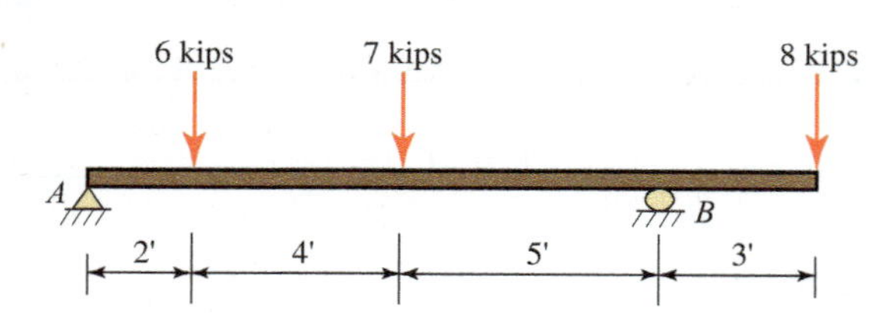

PROBLEM 4.20

4.21 Calculate the reactions at *A* and *B* for the beam shown. The weight of the beam is 600 N/m.

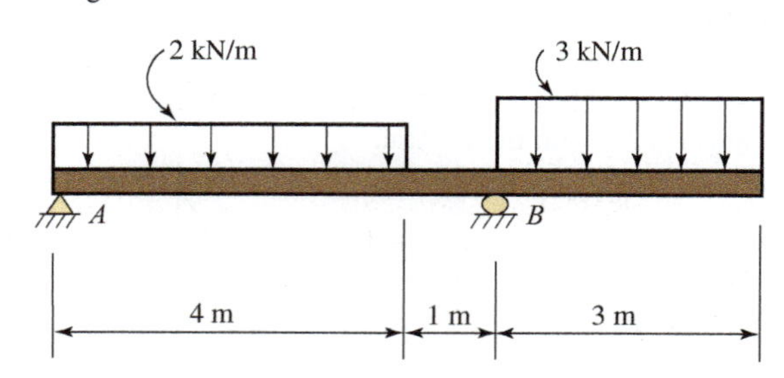

PROBLEM 4.21

4.22 A 12-ft simple beam is supported at each end. It supports a concentrated load of 800 lb at 3 ft from the left support. Where should a second concentrated load of 1500 lb be placed so that the beam reactions will be equal? Neglect the weight of the beam.

4.23 The beam shown carries vertical loads as indicated. Calculate the reaction at each support. Neglect the weight of the beam.

Section 4.6 Equilibrium of Nonconcurrent Force Systems

4.24 Determine the reactions for the beam shown. The beam has a pinned support at one end and a roller support at the other end. Neglect the weight of the beam.

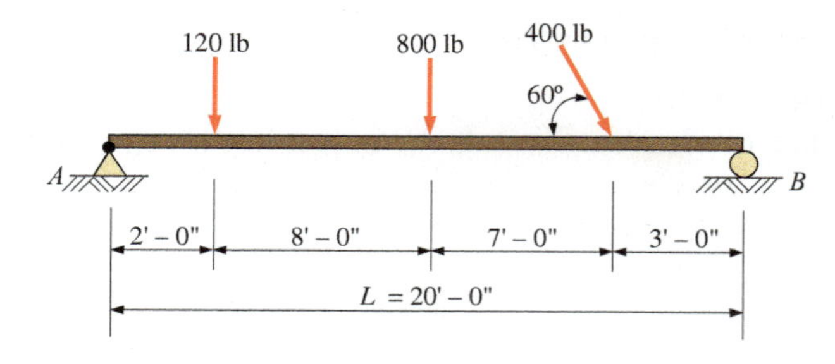

PROBLEM 4.24

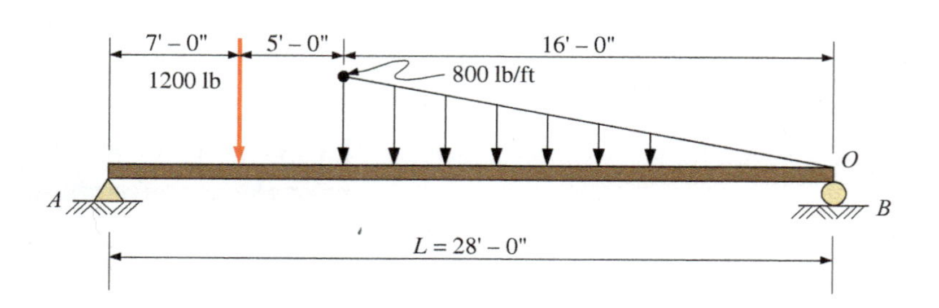

PROBLEM 4.23

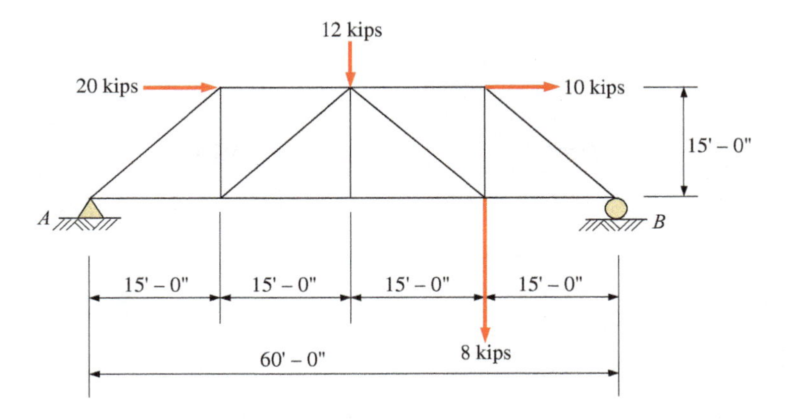

PROBLEM 4.25

4.25 Calculate the reaction at each support for the truss shown. Neglect the weight of the truss.

4.26 Calculate the wall reactions for the cantilever truss shown. The upper support is pinned and the lower support is a roller. Neglect the weight of the truss.

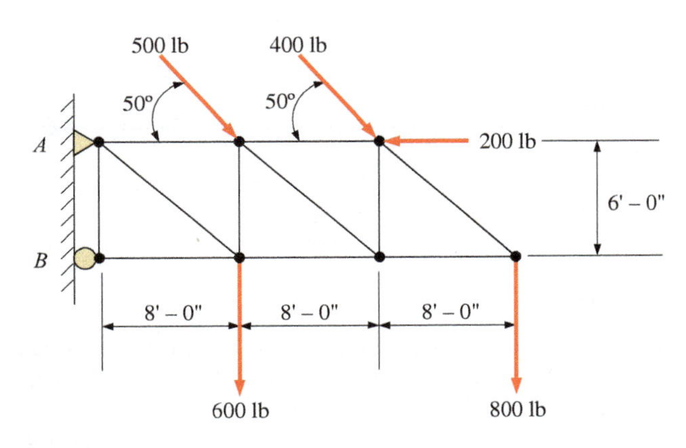

PROBLEM 4.26

4.27 Determine the reactions at supports A and B of the beam shown. Neglect the weight of the beam.

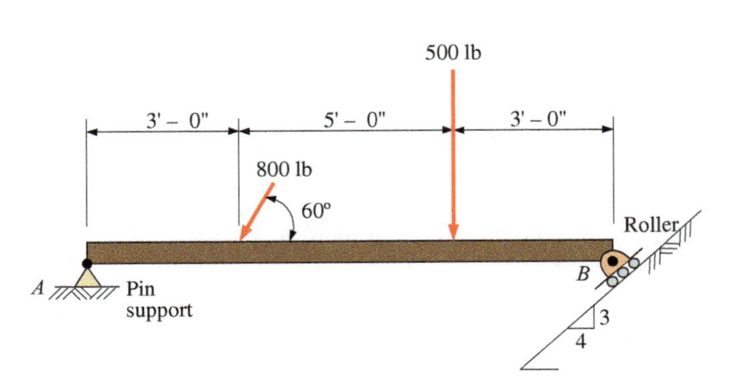

PROBLEM 4.27

4.28 A mass M of 300 kg is supported by a boom, as shown. Determine the tensile force in the cable and the reaction at A. The cable is horizontal.

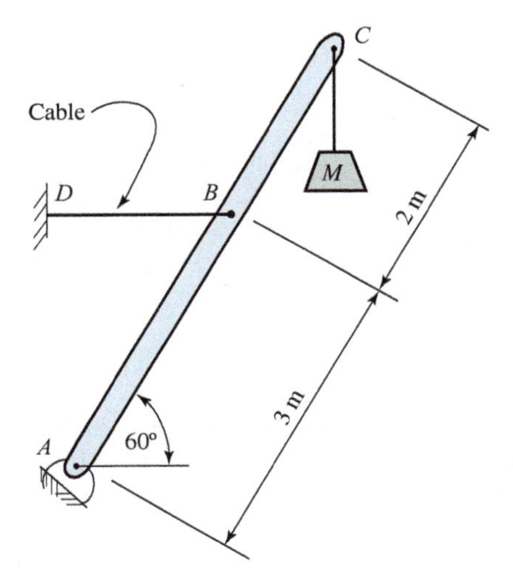

PROBLEM 4.28

4.29 Rework Problem 4.28 assuming that point D has been elevated so that cable DB makes an angle of 15° with the horizontal.

4.30 Calculate the force in the tie rod BC and the reaction at the pinned support at point A for the rigid frame shown.

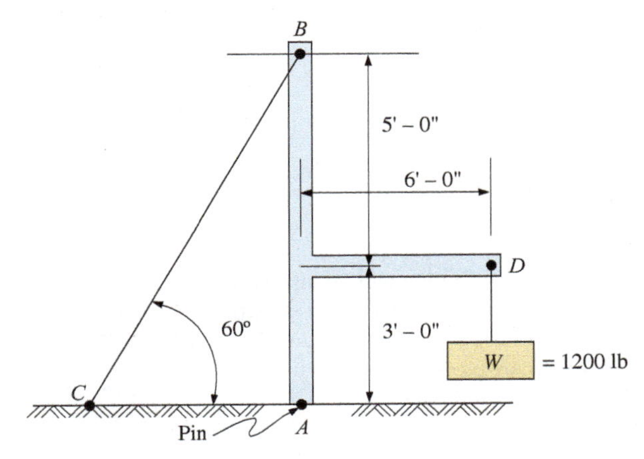

PROBLEM 4.30

4.31 The davit shown is used in pairs for supporting lifeboats on ships. The load of 3500 lb represents that portion of the weight of the boat and its occupants supported by one davit. *CG* represents the center of gravity of the davit. The weight of the davit itself is 900 lb. The ship has a "list" of 12°. Calculate the reactions at supports *B* and *C*. Assume that the reaction at *B* is at 90° to the axis of the davit and that the support at *C* is a pocket.

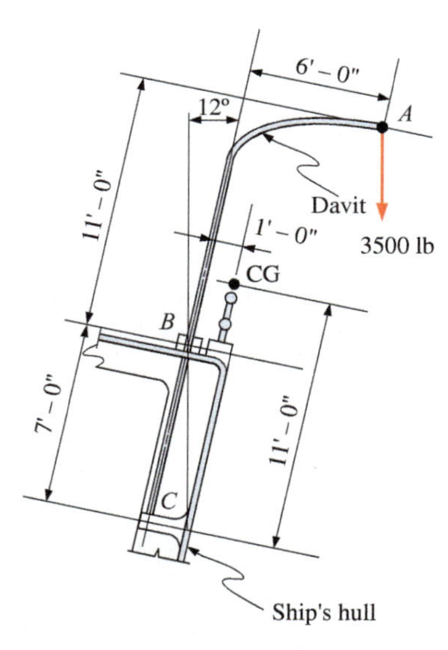

PROBLEM 4.31

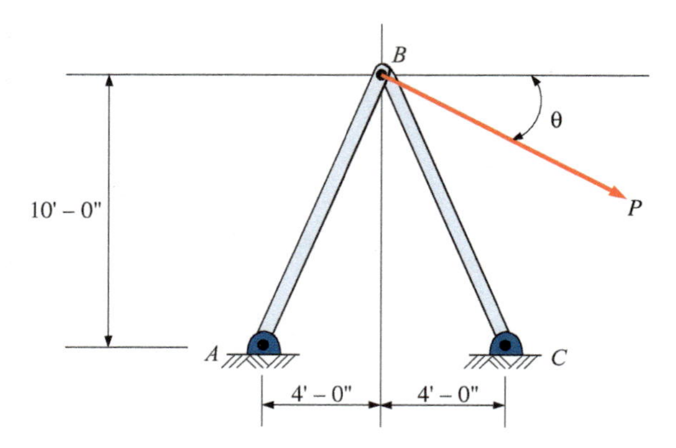

PROBLEM 4.33

4.34 With reference to Problem 4.25, write a program that will calculate the force in the cable *DB*. User input is to be the mass *M* as well as lengths *AB* and *BC*. In addition, for each combination of input, the angle at *A* is to vary from 30° to 60° in steps of 5°.

Computer Problems

For the following computer problems, any appropriate software may be used. Input prompts should fully explain what is required of the user (the program should be user-friendly). The resulting output should be well labeled and self-explanatory.

4.32 Write a program that will calculate the reactions for the beam shown. User input is to be L, L_1, L_2, P, and w. Neglect the weight of the beam.

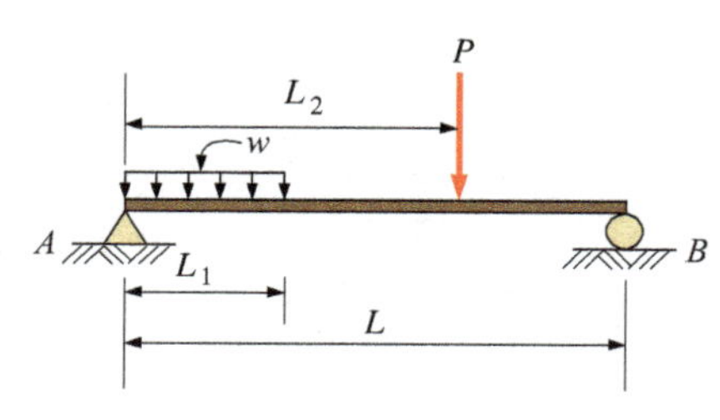

PROBLEM 4.32

4.33 Write a program that will calculate the forces in members *AB* and *CB* for the support frame shown. User input is to be θ and *P* (where $0° \leq \theta \leq 90°$ and *P* is not limited). All connections are pinned.

Supplemental Problems

4.35 For the structure shown, draw free-body diagrams for both the beam *ABC* and the link *BD*. The members are weightless.

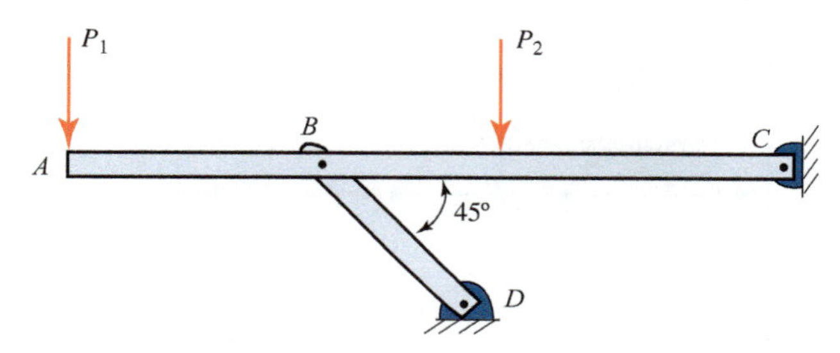

PROBLEM 4.35

4.36 A 1200-lb load is supported by a cable that runs over a small pulley at *E* and is anchored to a bar *DA*, as shown. Sketch free-body diagrams of bars *EB* and *DA* and of the pulley. The bars are pin-connected at each end. Neglect the weights of the members and the pulley.

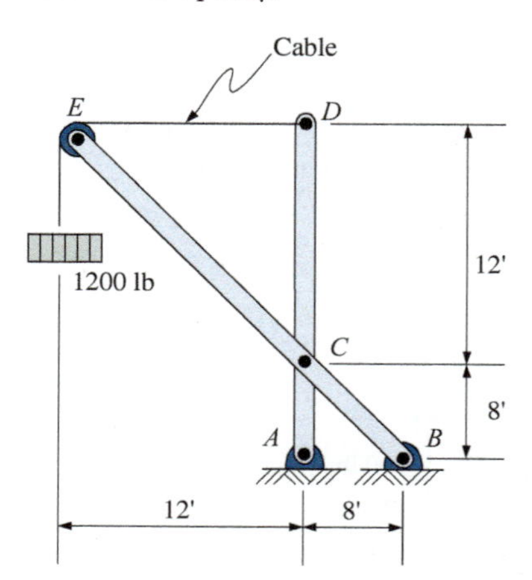

PROBLEM 4.36

4.37 For the pin-connected frame shown, sketch a free-body diagram of (**a**) the entire frame, (**b**) the member *AC*, and (**c**) the member *DF*. Neglect the weight of the members.

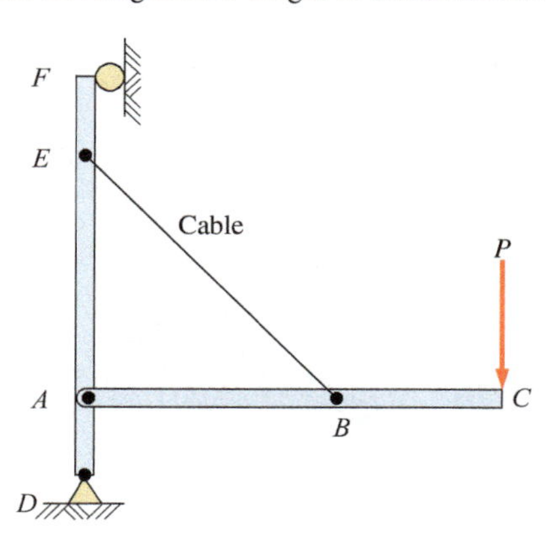

PROBLEM 4.37

4.38 For the concurrent force system shown, calculate the maximum load *W* that could be supported if the maximum allowable force in each cable were 1200 lb.

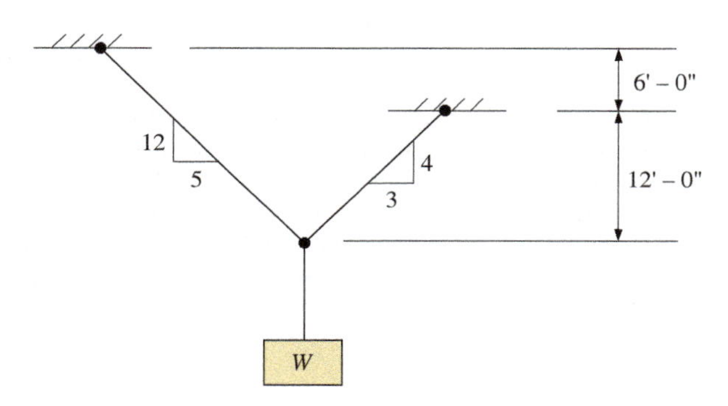

PROBLEM 4.38

4.39 A strut having a mass of 40 kg/m is supported by a cable, as shown. The structure supports a block having a mass of 500 kg. Find the force in the cable and the horizontal and vertical reactions at the pin-connection at *B*. Neglect the mass of the cable.

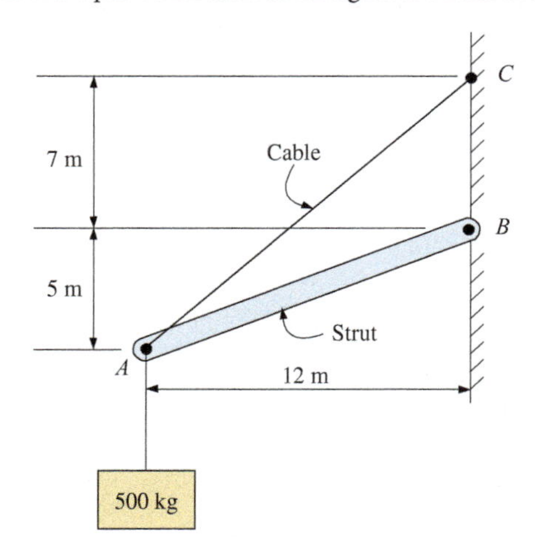

PROBLEM 4.39

4.40 Calculate the reaction at each support for the beam shown. Neglect the weight of the beam.

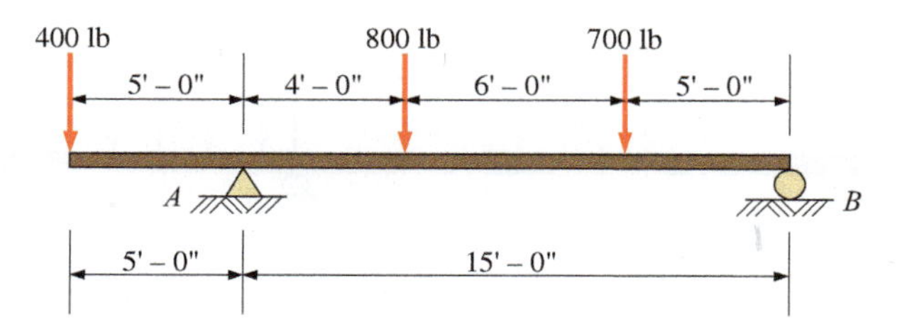

PROBLEM 4.40

4.41 Calculate the reaction at each support for the beam shown. The beam weight is 40 lb/ft.

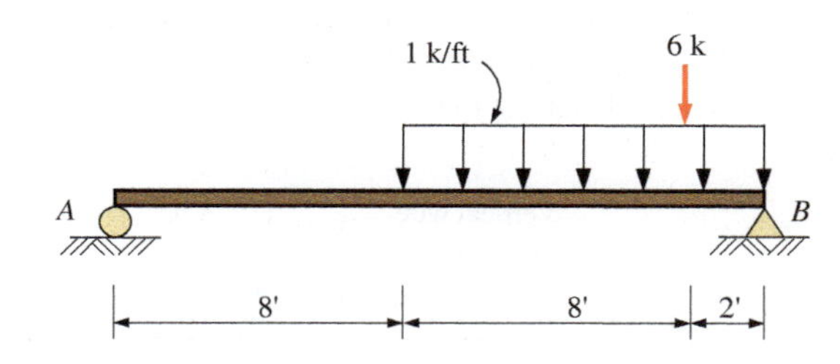

PROBLEM 4.41

4.42 A beam supports a nonuniformly distributed load as shown. Calculate the reaction at each support. Neglect the weight of the beam.

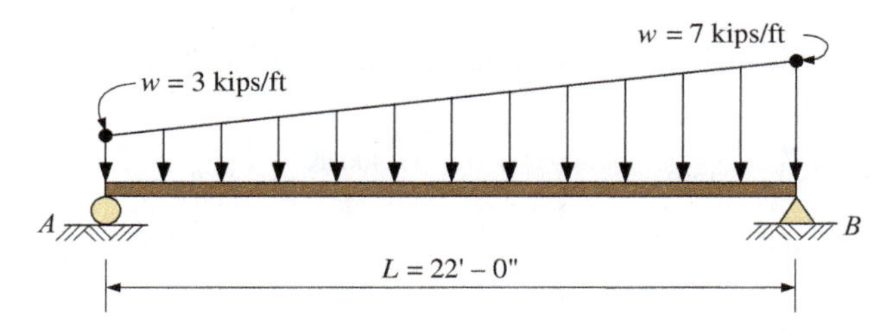

PROBLEM 4.42

4.43 Calculate the reactions at each support for the beam shown. Neglect the weight of the beam.

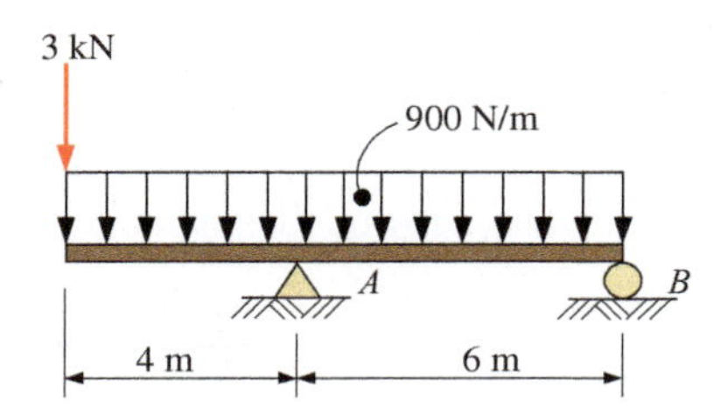

PROBLEM 4.43

4.44 Compute reactions at each support for the beam shown. The beam weight is 50 lb/ft.

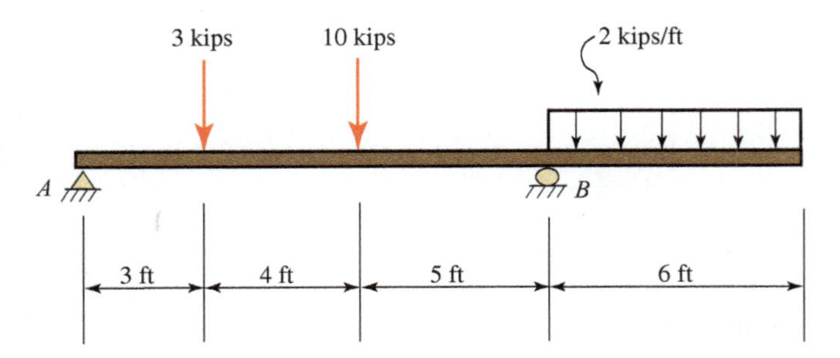

PROBLEM 4.44

4.45 A rod of uniform cross section weighs 4 lb/ft and is pin-connected at point *A*, as shown. The rod supports a load of 48 lb at point *B* and is held horizontal by a vertical wire attached 2 ft from point *B*. With a force of 85 lb in the wire, determine the length of the rod.

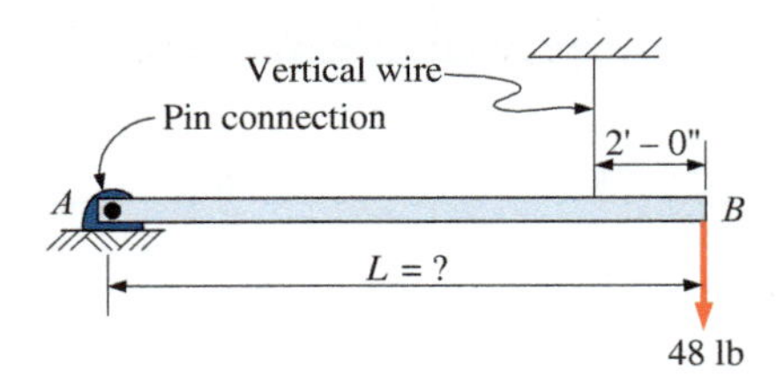

PROBLEM 4.45

4.46 A 12-ft-long weightiness member supports two loads, as shown. The member is held horizontal by two flexible cables attached to its ends. Determine the force in each cable and the angle θ required for the system to be in equilibrium.

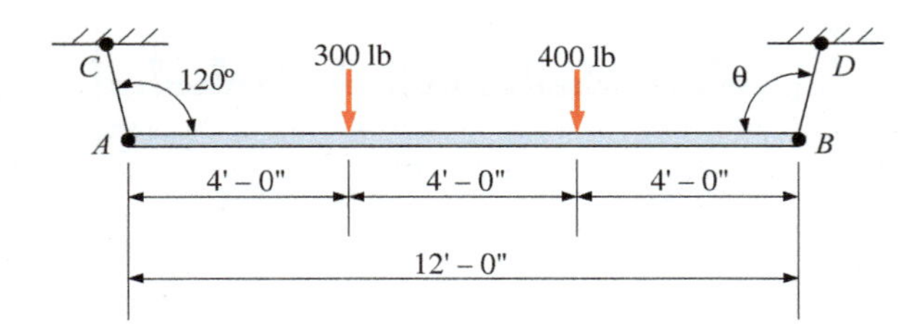

PROBLEM 4.46

4.47 A uniform rod *AB*, having a weight of 5.00 lb and a length of 20.0 in., is free to slide within a vertical slot at *A*. A rope supports the rod at point *B*. For the situation shown, determine the angle θ and the tension in the rope.

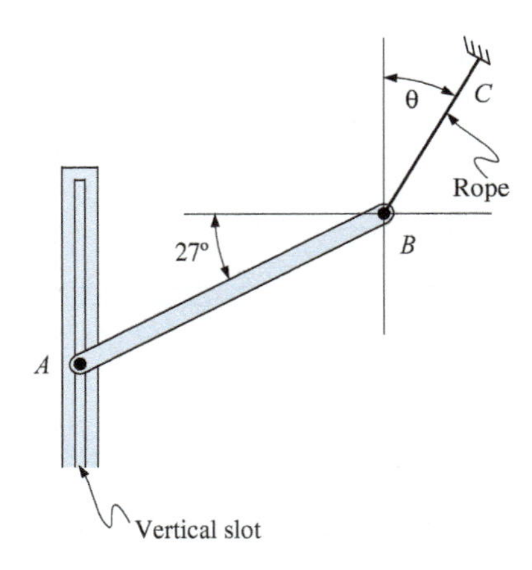

PROBLEM 4.47

4.48 The plastic barrel tent anchor of Problem 2.11 contains 100 gallons of water. The weight of the empty barrel is 60 lb. Determine the force in the rope required to cause the barrel to begin to tip about point *O*.

4.49 Compute the reactions at *A* and *B* for the bracket shown.

4.50 The truss shown is supported by a pin at *A* and a roller at *B*. Determine the reactions at these points.

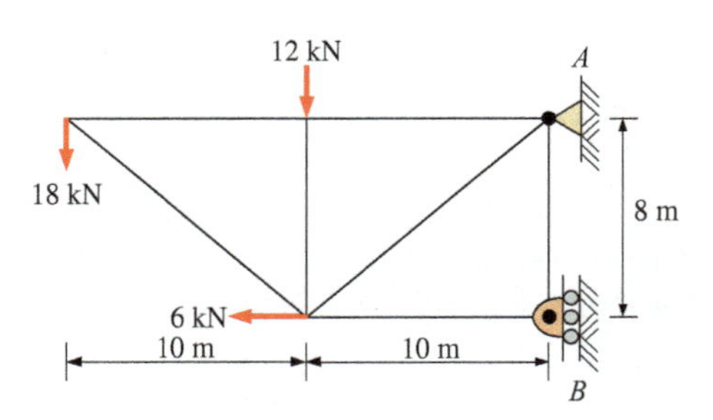

PROBLEM 4.50

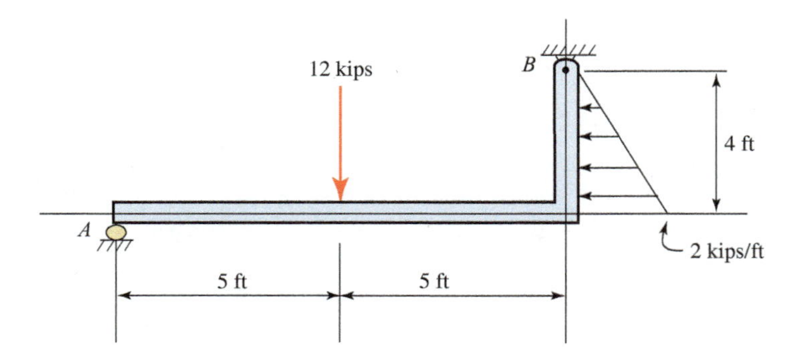

PROBLEM 4.49

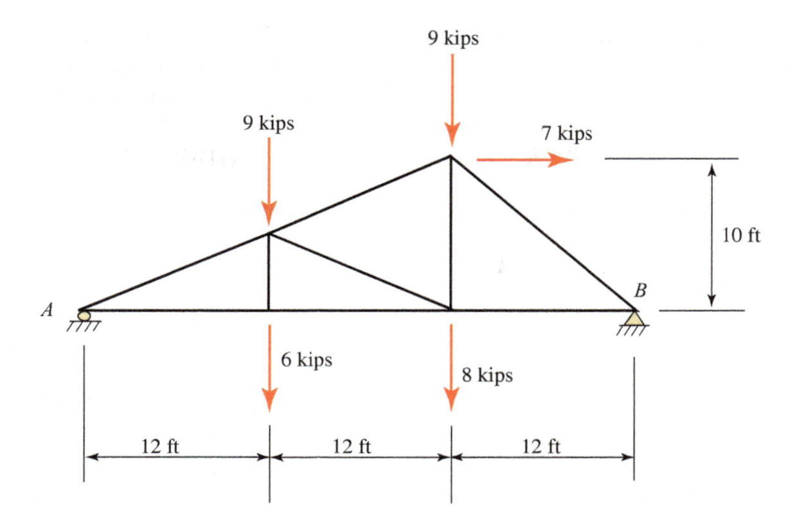

PROBLEM 4.51

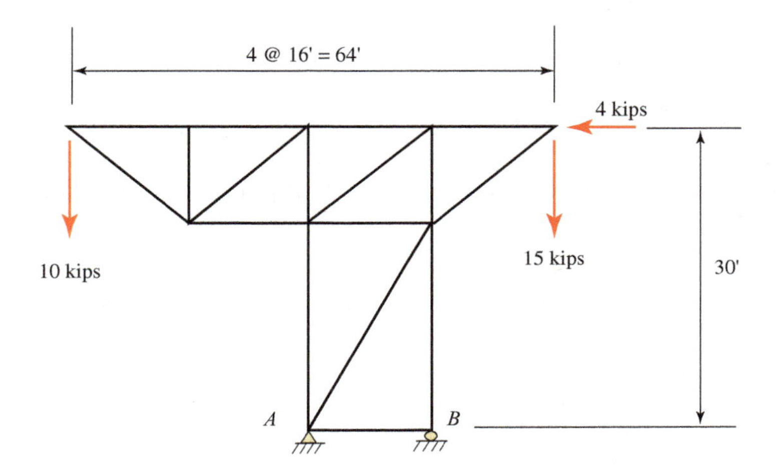

PROBLEM 4.52

4.51 Find the reactions at supports *A* and *B* for the truss shown.

4.52 Find the reactions at supports *A* and *B* for the tower truss shown.

4.53 Determine the reactions at *A* and *B* for the truss shown. The two 2.6-kip loads are perpendicular to the upper member. There is a roller at *A* and a pin at *B*.

4.54 A 40-ft ladder weighing 130 lb is pin-connected to the floor at point *A* and rests against a smooth, frictionless wall at point *B*, as shown. The ladder forms an angle of 60° with the horizontal floor and supports a load of 200 lb located 5 ft from the top end. Calculate the reactions at the top and bottom of the ladder.

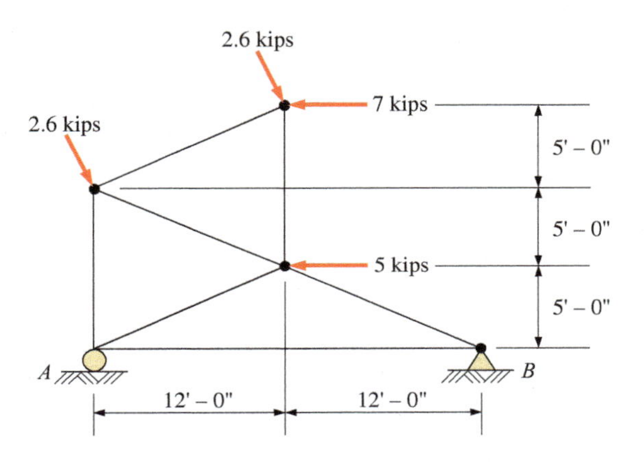

PROBLEM 4.53

PROBLEM 4.54

4.55 The frame shown is pin-connected at point A and held horizontal by the cable CD. The frame is subjected to a vertical load of 16 kips applied at point B. Calculate the force in the cable and the vertical and horizontal components of the reaction at point A.

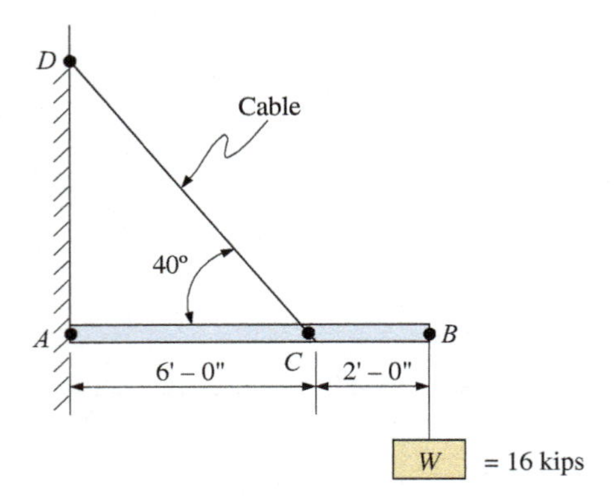

PROBLEM 4.55

4.56 A crane consists of a post AB, a boom CD, and a brace EF, weighing, respectively, 400 lb, 600 lb, and 300 lb. The crane supports a load of 3000 lb, as shown. The post is pin-connected at point A and laterally supported at point B. Thus, the reaction at point B will be horizontal. Determine the reactions at points A and B.

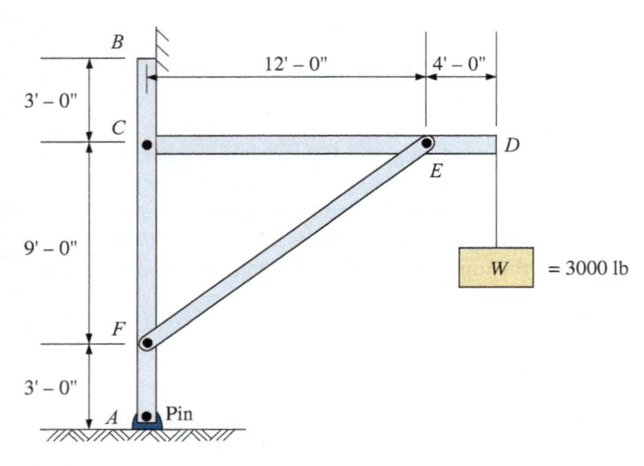

PROBLEM 4.56

4.57 A horizontal beam is pin-connected to a wall at one end and braced diagonally at point D, as shown. The beam carries a uniformly distributed load of 800 lb/ft and a concentrated load at its free end of 500 lb. Determine the horizontal and vertical components of the reaction at point A and the force in member BD. Neglect the weight of the members.

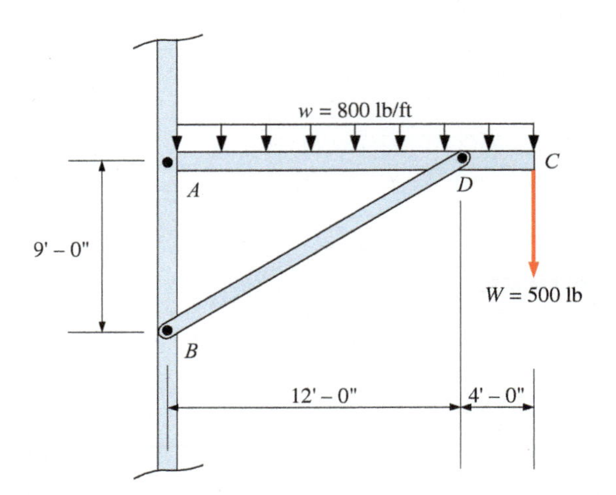

PROBLEM 4.57

4.58 Calculate the force in the cable for the structure shown.

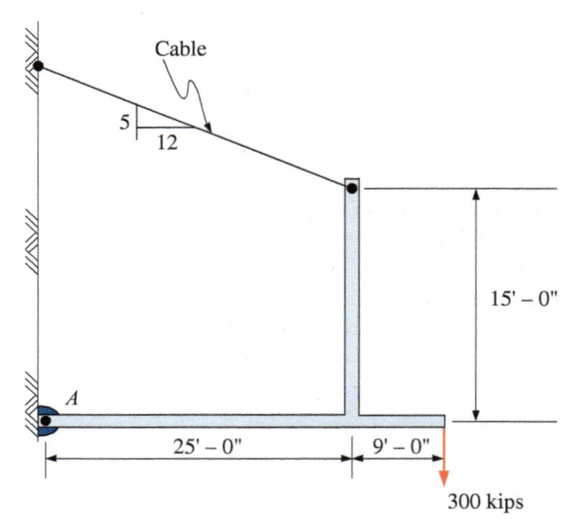

PROBLEM 4.58

4.59 The Thénard shutter dam shown was originally developed and used in 1831. Assume frictionless hinges at *B* and *C* and neglect the weight of the dam. Calculate the reactions at *C* and *D*, assuming a 1-ft width of dam (perpendicular to the page). Note that the force of the water acts perpendicular to the surface of the dam.

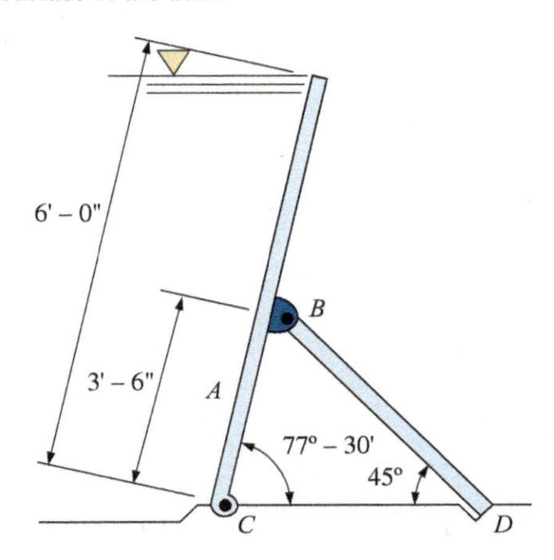

PROBLEM 4.59

4.60 An inclined railway can be used to lift heavy loads up steep inclines, as shown. Determine the tension in the tow cable and the reactions at each wheel. Neglect the weight of the car.

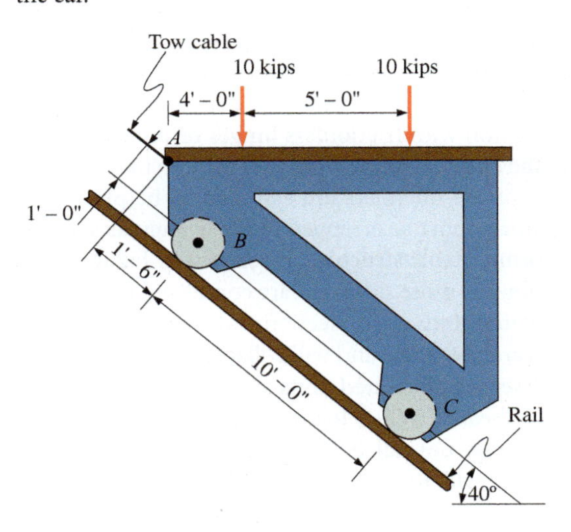

PROBLEM 4.60

4.61 Two cylinders are supported in a box, as shown. Cylinder *A* has a mass of 75 kg and a diameter of 400 mm. Cylinder *B* has a mass of 50 kg and a diameter of 200 mm. All surfaces are smooth and frictionless. Show a free-body diagram for each cylinder and find all forces acting on each cylinder.

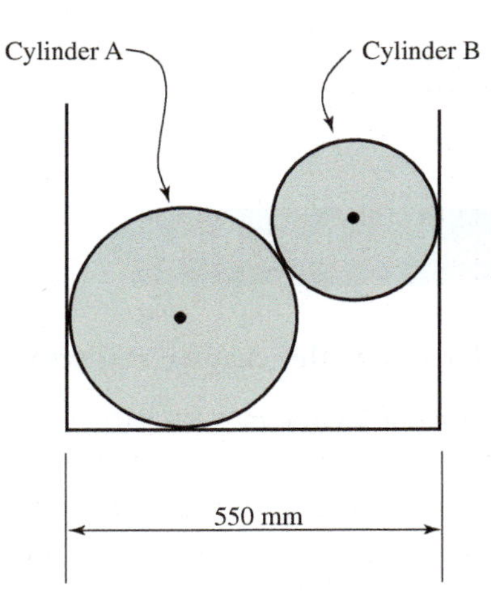

PROBLEM 4.61

ANALYSIS OF STRUCTURES

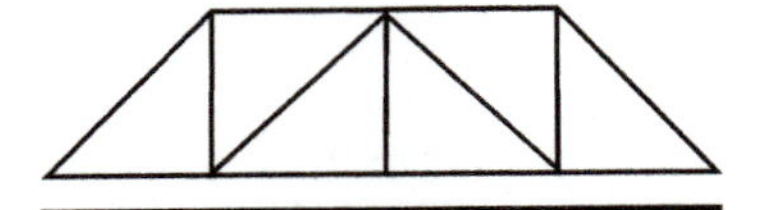

LEARNING OBJECTIVES

Upon completion of this chapter, readers will be able to:

- Discuss the difference between a truss and a frame and be able to identify each.
- Analyze a truss using the method of joints and equilibrium equations introduced in Chapter 4.
- Use the method of sections and the equilibrium equations to analyze portions of a truss.
- Analyze a frame structure using the three equations of equilibrium.

5.1 INTRODUCTION

The function of a structure is to transmit applied loads through the structure to its external supports. The goal of an analysis of a structure is the determination of the force or forces that each member of the structure must resist due to application of a load system on the structure.

The two types of structures considered in this chapter are pin-connected trusses and pin-connected frames. The difference between them is as follows: In trusses, all the members are two-force members, meaning that the member is subjected to equal, opposite, and collinear forces (either a push or a pull), with lines of action coinciding with the longitudinal axis of the member. In frames, all or some of the members are multiple-force members, in which there is a bending of the member in combination with a longitudinal push or pull.

The analysis process, as described in this chapter, involves the analytical application of the conditions and laws of equilibrium of coplanar force systems, which were presented in Chapters 3 and 4. The object of our analyses is to determine the forces that are developed in, or on, the various members of the structures.

5.2 TRUSSES

A *truss* may be described as a structural framework consisting of straight individual members, all lying in the same plane and connected to form a triangle or a series of triangles. The triangle is the basic stable element of the truss. It may be readily observed that practically all trusses are composed of members placed in triangular arrangements, although some specialized trusses do not have this configuration. The trusses considered in this text are planar trusses; that is, all the truss elements and all the applied loads or forces lie in the same plane. Further, all loads are applied at points of intersection of the members.

In the common types of trusses shown in Figure 5.1, the truss members are assumed to be connected at their points of intersection with frictionless hinges or pins; in effect, they permit the ends of the member the freedom to rotate. Because the ends of the members are assumed to be pin-connected, the members must be arranged in the triangular shape if they are to form a stable structure. As shown in Figure 5.2, structures of four or more sides that are connected with frictionless pins at their points of intersection and are not subdivided into triangles are not stable and will collapse under load.

Trusses are fabricated units and may be considered to be very large beams. When loads are extremely heavy or spans are very long, normal beam sections will not be adequate and trusses will be used. Most trusses are constructed of

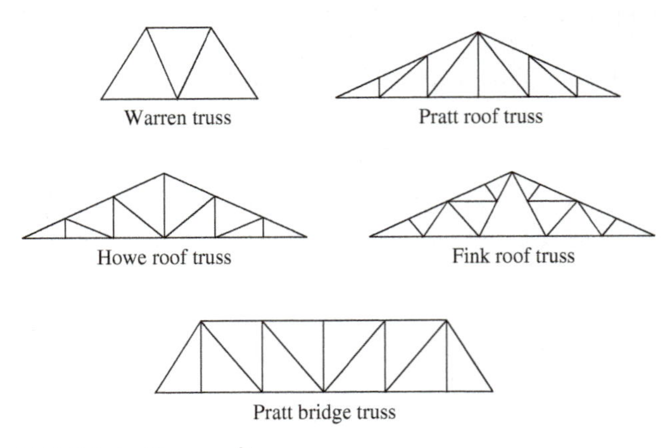

FIGURE 5.1 Types of trusses.

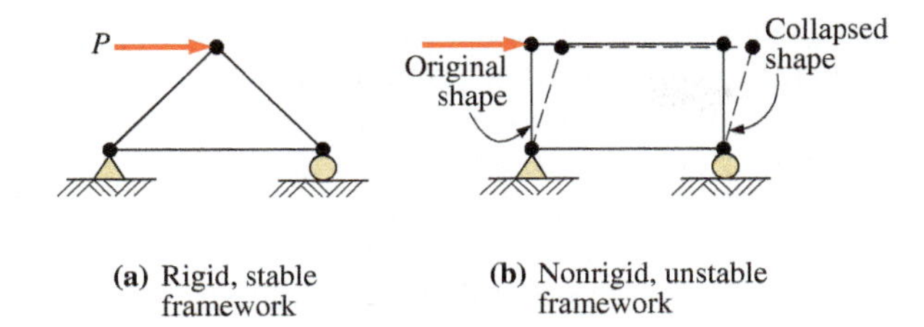

FIGURE 5.2 Member relationships.

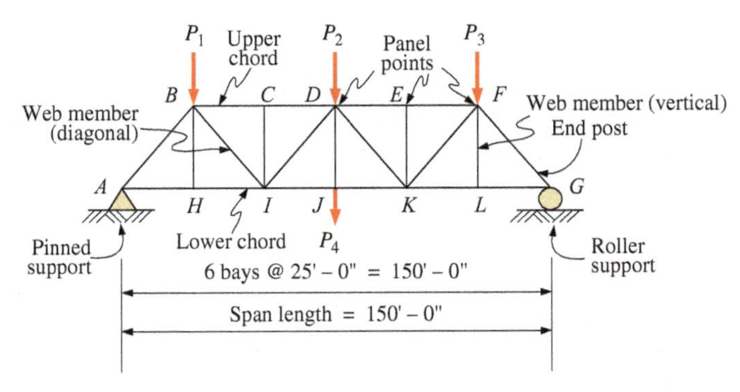

FIGURE 5.3 Truss terminology.

either metal or wood. Compared to a solid bending member, trusses are generally economical with respect to material; fabrication costs, however, are high.

Trusses, being somewhat specialized, have an associated terminology descriptive of the various component parts. As indicated in Figure 5.3, members that form the upper and lower outline of the truss are generally termed the *upper chord* and the *lower chord*, respectively.

Points *A*, *B*, *C*, *D*, and so forth are called *joints* or *panel points*. They are points of intersection of the longitudinal axes of the truss members. The interior members connecting the joints of the chords are called *web members* (either *vertical* or *diagonal*, depending on their direction in the web system). The distance between panel points (e.g., between *B* and *C*, or *D* and *E*) represents a *panel length*. The general area between panel points is commonly called a *bay*. Assuming equal panel lengths for all bays, the total span length of a truss is the number of bays multiplied by the panel length.

In the ideal case, trusses are loaded at their panel points (joints), and each member plays a role in transmitting the applied loads to the supports. The members are two-force members, and the applied forces are collinear with the longitudinal axes of the members. The applied forces tend to either stretch or shorten the members. They are called *axial forces* or *direct forces*, as distinguished from forces that produce bending (see Figure 5.4). A member that is stretched is said to be in *tension*, and a member that is shortened is said to be in *compression*. The analysis of a truss involves the determination of the magnitude of force in the members as well as the determination of whether the member is in tension or compression.

5.3 FORCES IN MEMBERS OF TRUSSES

To simplify the analysis of a truss, the following assumptions are made:

1. All members of the truss lie in the same plane.
2. Loads and reactions are applied only at the panel points (joints) of the truss.
3. The truss members are connected with frictionless pins.
4. All members are straight and are two-force members; therefore, the forces at each end of the member are equal, opposite, and collinear.
5. The line of action of the internal force within each member is axial.
6. The change in length of any member due to tension or compression is not of sufficient magnitude to cause an appreciable change in the overall geometry of the truss.
7. The weight of each member is very small in comparison with the loads supported and is therefore neglected. (If the weight is to be considered, it may be assumed to be a concentrated load acting partially at each end of the member.)

Based on these assumptions, and using the principles and laws of static equilibrium as discussed in Chapter 4, the force in each member (tension or compression) may be determined by means of either of two analytical techniques. One is called the *method of joints*, and the other is called the *method of sections*.

At this point, it should be noted that prior to determining the internal force in each truss member, the external equilibrium of the truss must be considered. This means that the truss reactions at the external supports should be determined as was described in Chapter 4 for beams and frames.

5.4 THE METHOD OF JOINTS

The method of joints consists of removing each joint in a truss and considering it as if it were isolated from the remainder of the truss. A free-body diagram of the pin is the basis of the approach. The free body is prepared by cutting through all the members framing into the joint being considered. Since all members of a truss are two-force members

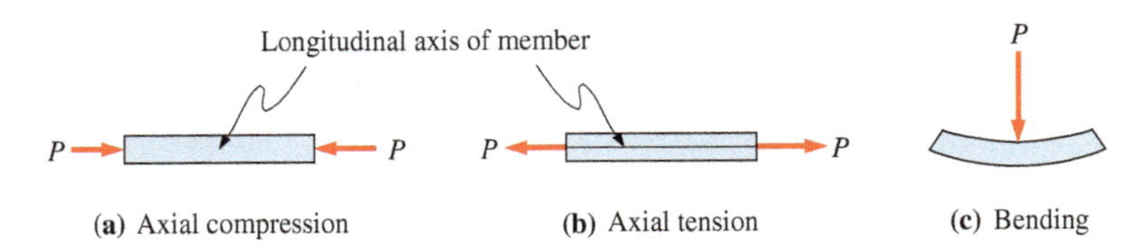

FIGURE 5.4 Member behavior.

carrying axial loads, the free-body diagram of each joint will represent a coplanar concurrent force system.

Because the truss as a whole is in external equilibrium, any isolated portion of it must likewise be in equilibrium. Therefore, each joint must be in equilibrium under the action of the external loads and the internal forces of the cut members that frame into the joint.

As shown in Chapter 4, only the two equations of force equilibrium ($\Sigma F_x = 0$ and $\Sigma F_y = 0$) are necessary to determine the unknown forces in a coplanar concurrent force system. Again, for convenience, we use the subscripts H and V to indicate horizontal and vertical directions, since, in our applications here, the Y axis is assumed to be vertical (coincident with the direction of gravity). Therefore, our two equations of force equilibrium are written as $\Sigma F_H = 0$ and $\Sigma F_V = 0$. The third equation of equilibrium ($\Sigma M = 0$) is not applicable, since all the forces intersect at a common point.

The use of force components and the two equations of force equilibrium is usually the method of choice at each joint. In situations where only three members are connected at a joint, however, and particularly when two unknown force members are sloped at that joint, analysis of the force triangle will provide a quick solution, saving time and effort.

When using the method of joints, no more than two unknown member forces can be determined at any one joint. Once these unknown forces have been calculated for one joint, their effects on adjacent joints are known. Successive joints may be then considered until the unknown forces in all the members have been determined. This procedure is demonstrated in Example 5.1.

EXAMPLE 5.1 A simply supported truss is shown in Figure 5.5. There is a pin support at A and a roller support at C (the roller support shown is sometimes called a *roller nest*). Determine the support reactions and the internal force in each member. Use the method of joints.

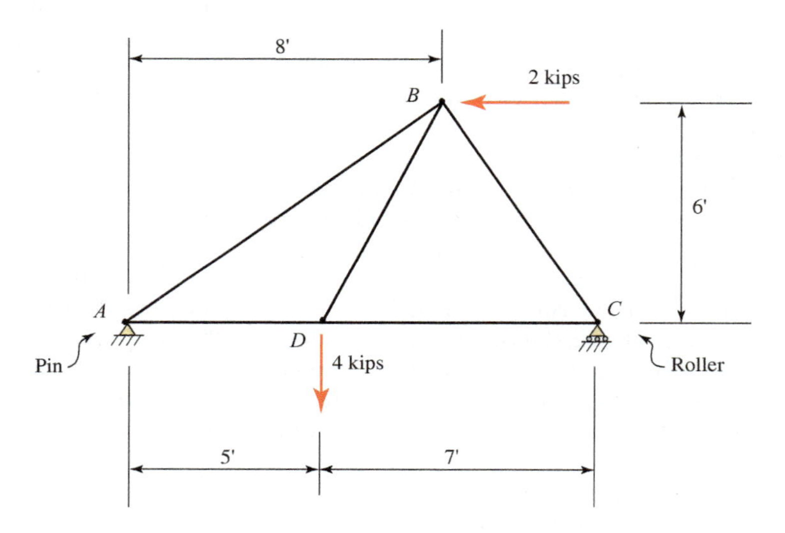

FIGURE 5.5 Truss for Example 5.1.

Solution The reactions at A and C will be determined first. A free-body diagram of the entire truss is shown in Figure 5.6. The unknown reactions are shown. Forces upward, to the right, and counterclockwise moments are considered positive. Take moments about point A:

$$\Sigma M_A = -(4\text{ k})(5\text{ ft}) + (2\text{ k})(6\text{ ft}) + R_{C_V}(12\text{ ft}) = 0$$

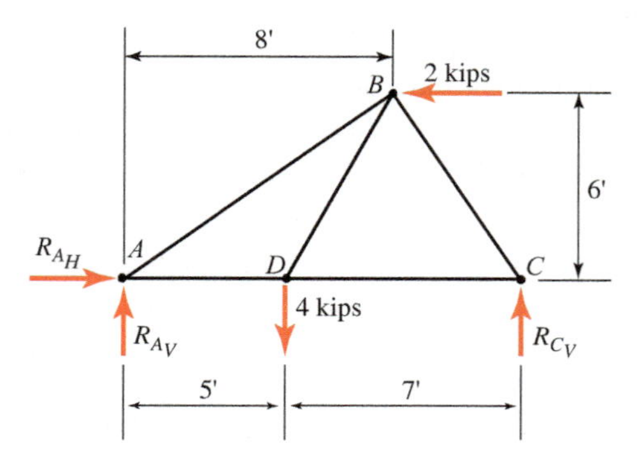

FIGURE 5.6 Truss free-body diagram.

from which

$$R_{C_V} = +0.667 \text{ k}$$

Take moments about point C:

$$\Sigma M_C = -R_{A_V}(12 \text{ ft}) + (4 \text{ k})(7 \text{ ft}) + (2 \text{ k})(6 \text{ ft}) = 0$$

from which

$$R_{A_V} = +3.33 \text{ k}$$

Check the calculations by a summation of vertical forces:

$$\Sigma F_V = +3.33 \text{ k} - 4 \text{ k} + 0.667 \text{ k} \approx 0 \qquad \textbf{(O.K)}$$

Then sum the horizontal forces:

$$\Sigma F_H = +R_{A_H} - 2 \text{ k} = 0$$

from which

$$R_{A_H} = +2 \text{ k}$$

Now, with the truss reactions known, the internal forces in the truss members can be calculated. Each joint will be isolated in sequence as a free body. Since a joint with more than two unknown forces cannot be solved completely, we must begin at either joint C or joint A.

A free-body diagram for joint C is shown in Figure 5.7a. Note that the force system is a coplanar, concurrent system consisting of one known force R_{C_V} and two unknown forces. Start by drawing the joint (the pin) and labeling it. Then show the known force R_{C_V} and carefully indicate its direction, sense, and magnitude. The unknown forces are designated CD and BC and should be interpreted as the forces in members CD and BC. Since all the truss members are straight two-force members, the directions of the two unknown forces are known to coincide with the axes of the members (CD is horizontal, and the slope of BC is indicated by a slope triangle). The senses and magnitudes, however, are unknown. If the sense of an unknown force is not obvious, it must be assumed. If calculations show the resulting force to have a negative sign, the sense of the force is opposite to that assumed.

In Figure 5.7a force BC is shown acting toward the joint, which means that member is assumed to be in compression and is pushing into the joint. Force CD is shown acting away from the joint. Member CD is therefore assumed to be in tension, pulling away from the joint.

A force triangle solution is very convenient for this situation and is shown in Figure 5.7b. The hypotenuse of the slope triangle is calculated from

$$\sqrt{6^2 + 4^2} = 7.21$$

Using similar triangles, we have

$$\frac{0.667 \text{ k}}{6} = \frac{CD}{4} = \frac{BC}{7.21}$$

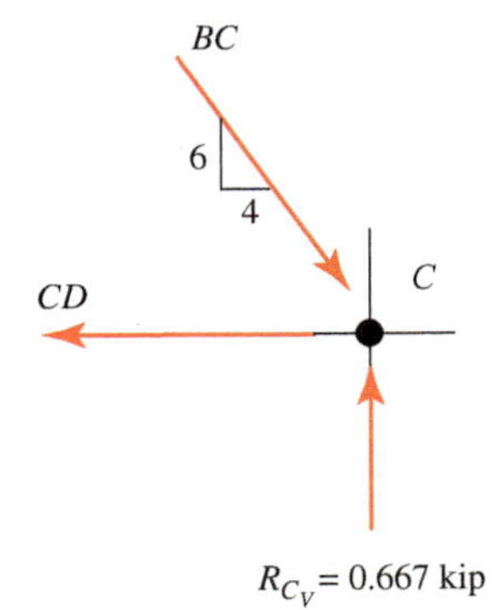

(a) Free-body diagram

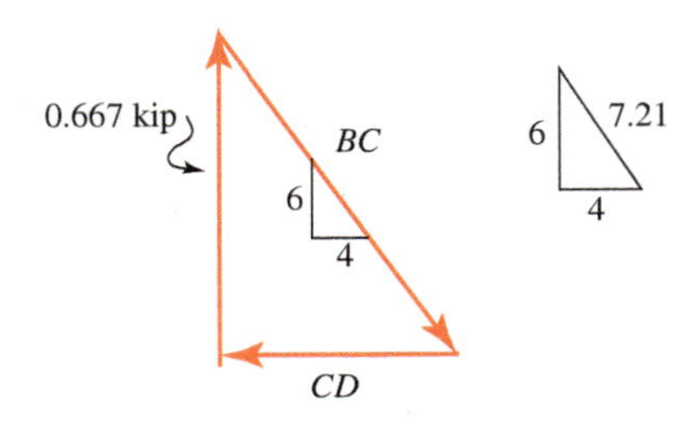

(b) Force triangle

FIGURE 5.7 Analysis of joint C.

Therefore,

$$CD = \frac{4}{6}(0.667\text{ k}) = 0.445\text{ k (tension)}$$

$$BC = \frac{7.21}{6}(0.667\text{ k}) = 0.802\text{ k (compression)}$$

Note how the senses of the forces correlate in the free-body diagram and the force triangle. If the senses of either or both of the unknown forces are assumed incorrectly, the force triangle will not close.

With the forces in *BC* and *CD* determined, any of the remaining three joints can be solved next, since two unknown forces remain at each joint. At joint *B* the two unknown forces are in sloping members and a set of simultaneous equations with two unknowns will result. Alternatively, joint *B* could be solved using a force triangle or joints *A* and *D* could be solved in either order. For illustrative purposes, joint *B* will be analyzed as the next step, using simultaneous equations.

Figure 5.8 shows a free-body diagram of joint *B*. The known forces are the 2-kip horizontally applied force and the force *BC*. Note that force *BC* was found to be in compression at joint *C*. Therefore, it must also be in compression at joint *B* and is appropriately shown in Figure 5.8 acting into the joint. The components of *BC* are needed and can be obtained from the analysis of joint *C* and the force diagram of Figure 5.7b. The unknown forces *AB* and *BD*, their components, and slope triangles are shown. As before, the senses of the unknown forces must be assumed, and the calculations will verify whether the assumptions were correct or not.

The relationships between the unknown forces and their respective components, using similar triangles, are

$$\frac{AB}{10} = \frac{AB_H}{8} = \frac{AB_V}{6}$$

$$\frac{BD}{6.71} = \frac{BD_H}{3} = \frac{BD_V}{6}$$

from which

$$AB_H = \frac{8}{10}AB = 0.800AB$$

$$AB_V = \frac{6}{10}AB = 0.600AB$$

$$BD_H = \frac{3}{6.71}BD = 0.447BD$$

$$BD_V = \frac{6}{6.71}BD = 0.894BD$$

Joint *B* is in equilibrium and ΣF_V and ΣF_H must be equal to zero. Therefore, summing vertical forces, we have

$$\Sigma F_V = +AB_V - BD_V + 0.667\text{ k} = 0$$

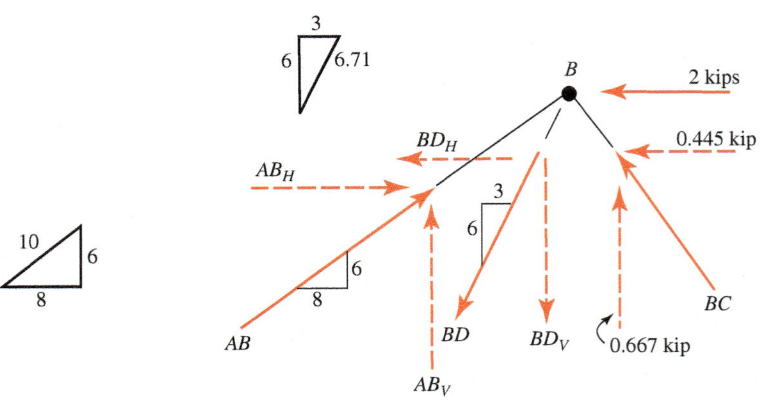

FIGURE 5.8 Analysis of joint *B*.

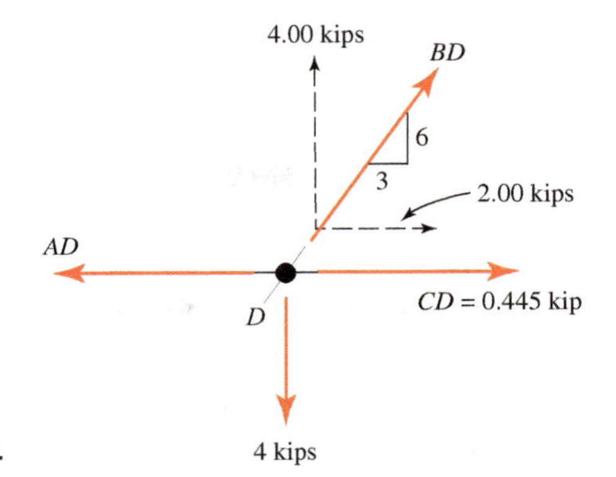

FIGURE 5.9 Analysis of joint D.

Substituting for AB_V and BD_V results in

$$0.600AB - 0.894BD + 0.667\text{ k} = 0 \quad (1)$$

Similarly, for horizontal forces:

$$\begin{aligned}\Sigma F_H &= +AB_H - BD_H - 0.445\text{ k} - 2\text{ k} = 0 \\ &= 0.800AB - 0.447BD - 2.445\text{ k} = 0 \quad (2)\end{aligned}$$

Equations 1 and 2 are simultaneous equations that may be solved to yield $AB = 5.55$ k and $BD = 4.47$ k. Both are positive; therefore, the senses were assumed correctly: AB is in compression and BD is in tension.

Joint D will be analyzed next. The free-body is shown in Figure 5.9. From the analysis of joint B, the components of BD are

$$\begin{aligned}BD_H &= 0.447BD = 0.447(4.47\text{ k}) = 2.00\text{ k} \\ BD_V &= 0.894BD = 0.894(4.47\text{ k}) = 4.00\text{ k}\end{aligned}$$

These components are shown on the free-body diagram.

Although force BD has been replaced by its components, it is also shown in Figure 5.9. Force CD of 0.445-kip tension (from the analysis at joint C) is shown acting away from joint D, as is the applied tensile load of 4 kips. Force AD is the only unknown and is assumed to be tensile. Note that the sum of the vertical forces is zero. Summing horizontal forces at joint D gives

$$\Sigma F_H = -AD + 2.00\text{ k} + 0.445\text{ k} = 0$$

from which

$$AD = 2.45\text{ k (tension)}$$

All the member forces have now been found. Joint A has not been used in the analysis, and it can therefore be used to make a check on the calculations. The forces acting at joint A must be in equilibrium. The free-body diagram of joint A is shown in Figure 5.10. The components of AB are calculated from

$$\begin{aligned}AB_H &= 0.800AB = 0.800(5.55\text{ k}) = 4.44\text{ k} \\ AB_V &= 0.600AB = 0.600(5.55\text{ k}) = 3.33\text{ k}\end{aligned}$$

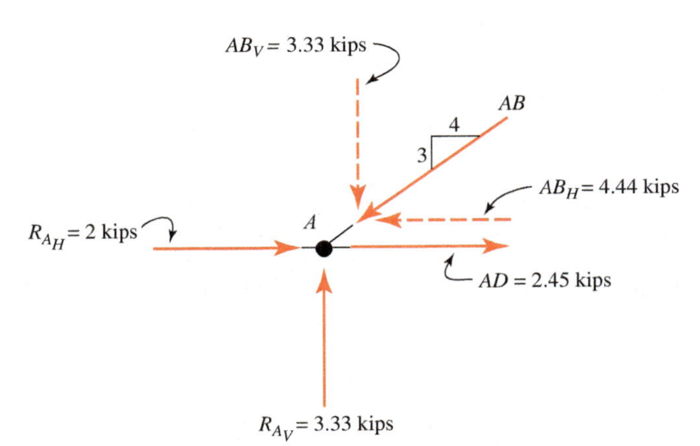

FIGURE 5.10 Check at joint A.

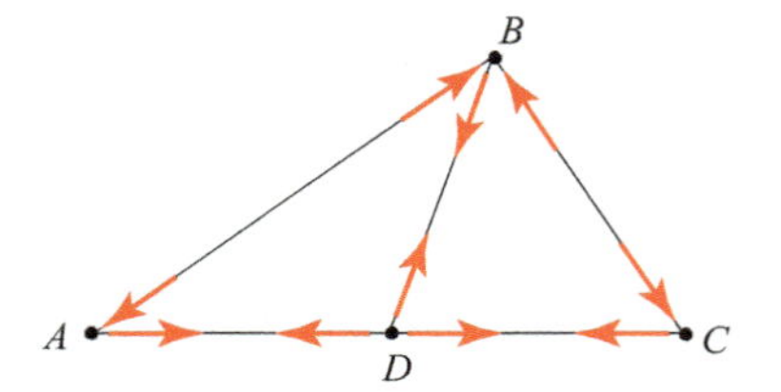

FIGURE 5.11 Member force sense summary.

These and all other forces acting at joint *A* are shown with senses carefully indicated. Summing vertical and horizontal forces gives

$$\Sigma F_V = +3.33 \text{ k} - 3.33 \text{ k} = 0 \qquad \textbf{(O.K)}$$

$$\Sigma F_H = +2 \text{ k} + 2.45 \text{ k} - 4.44 \text{ k} \approx 0 \qquad \textbf{(O.K)}$$

Figure 5.11 shows a view of the truss that displays the senses of the member forces. It is a composite of the joints that have been isolated in the analysis process. The senses of the member forces are shown, but dimensions, external loads, and reactions are omitted. Note that compressive forces in members act into the joints and tensile forces in members act away from the joints.

As a means of summarizing the internal forces in the truss members, a force summary diagram, as shown in Figure 5.12, should be sketched. Note that the member forces are designated (T) or (C) for tension or compression, respectively, and that no dimensions, external loads, or reactions are shown.

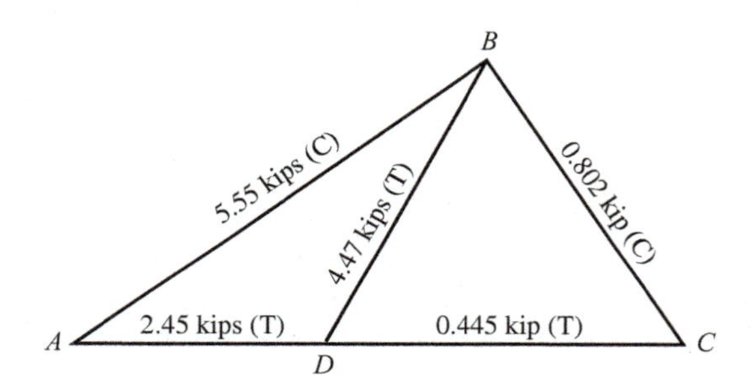

FIGURE 5.12 Force summary diagram.

5.4.1 Summary of Procedure: Method of Joints

1. Replace sloping forces with their vertical and horizontal components. Calculate the reactions for the truss using a free-body diagram for the entire truss.
2. Isolate each joint in sequence as a free body and sketch the free-body diagram showing all forces acting on the joint. Each joint represents a coplanar concurrent force system and can involve no more than two members in which the forces are unknown.
3. By applying the two equations of force equilibrium ($\Sigma F_V = 0$ and $\Sigma F_H = 0$), the two unknown member forces for each joint may be determined. These forces are then carried over to successive adjacent joints. In some cases, an alternative force triangle solution may be convenient.
4. It is good practice to check the calculations. Select a joint not used in the calculations and verify that $\Sigma F_V = 0$ and $\Sigma F_H = 0$.
5. After all member forces have been determined, draw a force summary diagram.

5.5 THE METHOD OF SECTIONS

The method of sections is another technique for determining the forces acting in the various members of a truss. This method has some advantages when the analysis of only a few members is desired, even when those members are situated far from the supports of the truss. Unlike the method of joints, the analysis need not proceed from joint to joint, analyzing the entire truss. With the proper approach and a suitable truss layout, a rapid analysis of selected members may be accomplished.

In the method of sections, the truss is divided into two parts by a cutting plane. One of the two parts is isolated as a free body. The procedure involves cutting through a number of members in which the unknown forces are acting. Not more than three members with unknown forces may be cut along any cutting plane, except where the lines of action of all but one of the members intersect at a common point.

Since the truss as a whole is in external equilibrium, any isolated portion of it must likewise be in equilibrium. Therefore, the portion of the truss isolated as a free body must be in equilibrium under the action of the externally applied loads and the internal forces of the members intersected by the cutting plane. The forces (known and unknown) acting on the free body constitute a coplanar nonconcurrent force system. Therefore, the three equations of equilibrium ($\Sigma F_V = 0$, $\Sigma F_H = 0$, and $\Sigma M = 0$) are applicable.

In using this method, the truss may be cut at any location, and the force in any member may be determined independently of all others. In general, when selecting a cutting plane, the plane should cut the least number of members. The member in which the unknown force is being sought must be one of the cut members.

Depending on the forces to be found, more than one cutting plane may be required. Moreover, in some problems, it may be necessary to use a combined analysis approach using both the method of joints and the method of sections. The method of sections, using a single cutting plane, is illustrated in Example 5.2.

EXAMPLE 5.2 Determine the forces acting in members *BD*, *CD*, and *CE* of the Howe truss shown in Figure 5.13a. Use the method of sections. The truss is supported by a pin at *A* and a roller at *J*.

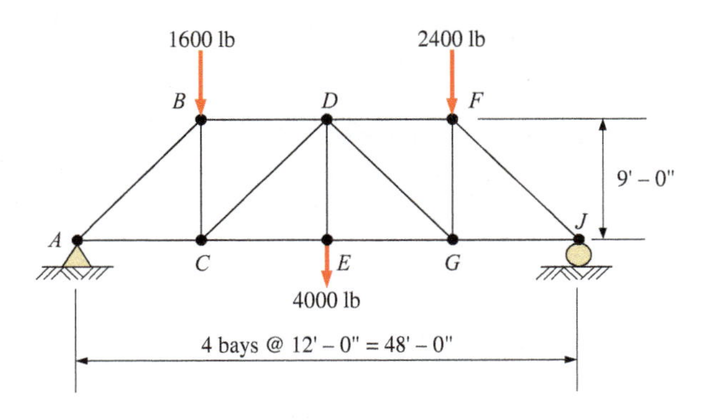

(a) Simply supported Howe truss

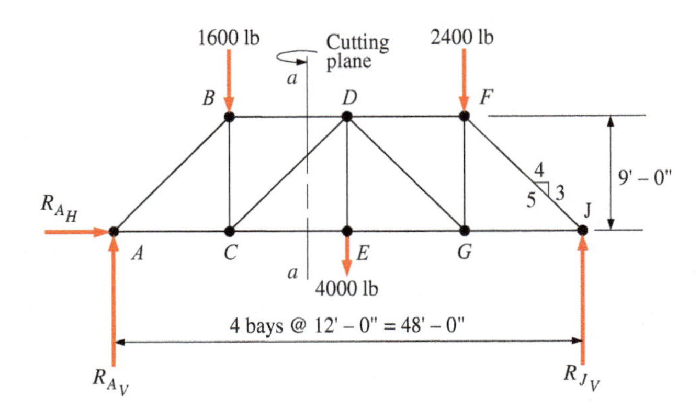

(b) Free-body diagram

FIGURE 5.13 Truss for Example 5.2.

Solution The reactions at *A* and *J* will be computed first. Figure 5.13b is a free-body diagram of the entire truss, which will be used for the calculation of the reactions. Forces upward and to the right and counterclockwise moments are considered positive. For the vertical reaction at *A*,

$$\Sigma M_J = -R_{A_V}(48\text{ ft}) + 2400\text{ lb}(12\text{ ft}) + 4000\text{ lb}(24\text{ ft}) + 1600\text{ lb}(36\text{ ft}) = 0$$

from which

$$R_{A_V} = +3800\text{ lb}$$

For the reaction at J,

$$\Sigma M_A = R_{J_V}(48\text{ ft}) - 1600\text{ lb}(12\text{ ft}) - 4000\text{ lb}(24\text{ ft}) - 2400\text{ lb}(36\text{ ft}) = 0$$

from which

$$R_{J_V} = +4200\text{ lb}$$

Checking the calculations using a summation of vertical forces,

$$\Sigma F_V = +3800\text{ lb} + 4200\text{ lb} - 2400\text{ lb} - 4000\text{ lb} - 1600\text{ lb} = 0 \qquad \textbf{(O.K)}$$

There are no horizontal forces or horizontal components of diagonal forces; therefore, $R_{A_H} = 0$.

Now, with the truss reactions known, the internal forces in members *BD*, *CD*, and *CE* can be calculated. This process is accomplished by passing a cutting plane *a–a* through these three members, as shown in Figure 5.13b, and isolating the left portion of the truss as a free body.

The free-body diagram for the left portion of the truss is shown in Figure 5.14. Note that the applicable external forces, as well as the internal forces of the cut members, are shown applied on the free body. The vertical and horizontal components of *CD* are shown as dashed arrows. The internal forces of the cut members are unknown in magnitude and sense. Since all the members are two-force members, the lines of action of the forces are known to be collinear with the members themselves and are established by the geometry of the truss. Assume member *CE* to be in tension and members *BD* and *CD* to be in compression. Note that an arrow pointing away from the free body (such as *CE*) means that member *CE* is assumed to pull on the body and therefore to be in tension.

Any of the three equations of equilibrium can be applied to the free-body diagram, since the force system is a coplanar nonconcurrent force system. The relationships between the vertical and horizontal components of *CD*, using the slope triangle, can be written as

$$\frac{CD}{5} = \frac{CD_H}{4} = \frac{CD_V}{3}$$

from which

$$CD_H = \frac{4}{5}CD \qquad \text{and} \qquad CD_V = \frac{3}{5}CD$$

Forces *BD* and *CE* are horizontal forces; therefore, they have no vertical components.

Next write a summation of vertical forces for the free body, from which you can determine the force *CD*:

$$\Sigma F_V = +3800\text{ lb} - 1600\text{ lb} - CD_V = 0$$

$$= +3800\text{ lb} - 1600\text{ lb} - \left(\frac{3}{5}CD\right) = 0$$

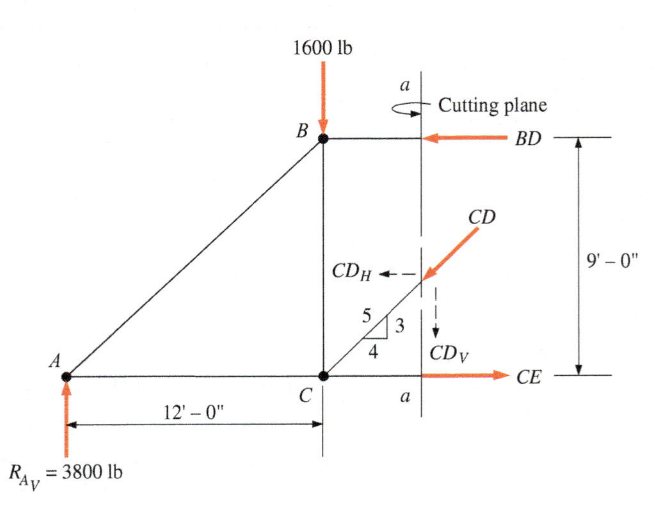

FIGURE 5.14 Free-body diagram.

from which

$$CD = +3667 \text{ lb (compression)}$$

Since the result is a positive value, the sense for force *CD* is as assumed. (Member *CD* is in compression.)

The force in member *BD* can be calculated by summing moments of all the known and unknown forces about point *C*. In effect, forces *CD*, *CE*, and the 1600 lb load are eliminated from the computation, since their lines of action pass through point *C*. Thus,

$$\Sigma M_C = -3800 \text{ lb}(12 \text{ ft}) + BD(9 \text{ ft}) = 0$$

from which

$$BD = +5067 \text{ lb (compression)}$$

The result is again a positive value, indicating that the sense for force *BD* is as assumed. (Member *BD* is in compression.)

The force in member *CE* can be calculated by summing the horizontal forces. Assuming forces acting to the right are positive,

$$\Sigma F_H = +CE - BD - CD_H = 0$$

$$= +CE - 5067 \text{ lb} - \frac{4}{5}(3667 \text{ lb}) = 0$$

from which

$$CE = +8000 \text{ lb (tension)}$$

The result is positive; therefore, the sense for force *CE* is as assumed. (Member *CE* is in tension.)

A check on the calculations can be accomplished by summing moments about some point not used thus far in the calculations. Point *B* is such a point. Move the components of *CD* to point *C* (principle of transmissibility):

$$\Sigma M_B = -3800 \text{ lb}(12 \text{ ft}) + CE(9 \text{ ft}) - CD_H(9 \text{ ft}) = 0$$

$$= -3800 \text{ lb}(12 \text{ ft}) + 8000 \text{ lb}(9 \text{ ft}) - \frac{4}{5}(3667 \text{ lb})(9 \text{ ft}) = 0 \text{ (very close)} \qquad \textbf{(O.K)}$$

5.5.1 Summary of Procedure: Method of Sections

1. Calculate the reactions for the truss using a free-body diagram for the entire truss.
2. Isolate a portion of the truss by passing a cutting plane through the truss, cutting no more than three members in which the forces are unknown (unless all but one of the lines of action of the cut members intersect at a point).
3. Apply the three equations of equilibrium to the isolated portion of the truss and solve for the unknown forces.
4. It is good practice to check the calculations by verifying $\Sigma F_H = 0$, $\Sigma F_V = 0$, and $\Sigma M = 0$. Use a moment center not previously used.

5.6 ANALYSIS OF FRAMES

Up to this point in the chapter, we have discussed trusses composed of two-force members. When forces are applied at more than two points along the length of a member, and when the forces are not collinear with the axis of the member, the member will be subject to bending. We now turn our attention to members of this type, sometimes called *multiple-force members*. Straight two-force and multiple-force members are shown in Figure 5.15.

Two-force members and their equilibrium under the action of two applied forces (or two resultants of concurrent force systems) were discussed in Section 4.4.

For the multiple-force member of Figure 5.15b, the resultant force at each end does not act along the axis of the member. A member that supports a transverse load, or for which the weight is not negligible, is always a multiple-force member.

Since the resultant forces at the ends of multiple-force members are not directed along the axis of the member but are usually unknown in direction and sense, it is customary to work with the rectangular components of each resultant force.

Structures composed partially or totally of pin-connected multiple-force members are called *frames*. As with trusses, the pins are assumed to be frictionless. The frames we will consider are coplanar; that is, all the members lie in a common plane.

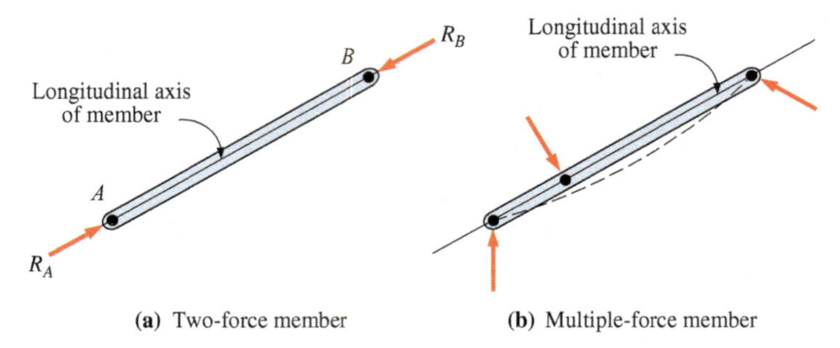

FIGURE 5.15 Types of members.

Our analysis of frames (with multiple-force members) differs significantly from the analysis of trusses. We make no attempt to determine the internal forces in the members of these frames; rather, we limit our analysis to a determination of the forces that act *on* the member. The forces resulting at the pins are called *pin reactions*. The pin reactions are normally determined as rectangular component forces. They may be combined into a resultant force using Equation (2.1).

The usual procedure to determine the pin reactions is as follows:

1. Compute the support reactions, or their components, considering the frame as a whole. If all the support reactions (or reaction components) cannot be determined, determine as many as possible and then proceed with the next steps. The unknown reactions will be determined at a later time.
2. Isolate each member as a free body, showing all known forces and unknown pin reactions (or components of the pin reaction).
3. Consider each multiple-force member individually. Apply the three equations of equilibrium. Compute the unknown pin reactions or components of the pin reaction. A maximum of three unknown forces (or components) can be found using any one free body. Any unknown forces remaining must be identified and carried forward.
4. Proceed to an adjacent member, carrying forward the newly determined forces, carefully accounting for the appropriate senses. Then repeat step 3.
5. It is good practice to check the calculations. Select a member not used in the analysis (or a member supporting many of the forces that have been determined) for the check. Verify that $\Sigma F_V = 0$, $\Sigma F_H = 0$, and $\Sigma M = 0$ for that member.

Step 4 requires some additional explanation. Once a force has been determined in magnitude and sense and is carried forward to other members, the magnitude will not change, but the sense will. If we consider the hung weight shown in Figure 4.1, the force F acts upward on the weight (Figure 4.1c). If this force were carried forward to the next member, the rope (Figure 4.1b), it would have to be shown acting downward. The rope pulls upward on the weight. The weight pulls downward on the rope, illustrating that forces occur in pairs. No single force can exist as an isolated force. Newton's third law of motion—to every action there is an equal and opposite reaction—describes this condition.

Since the procedure we have described considers each member individually, it is generally called the *method of members*.

EXAMPLE 5.3 A crane consisting of a vertical post, a horizontal boom, and an inclined brace supports a vertical load of 10,000 lb, as shown in Figure 5.16a. The crane is supported by a pin connection at *A*. The support at *B* permits only a horizontal reaction (smooth vertical surface). The members of the crane are pin-connected at points *C*, *D*, and *E*. The members have weights as follows: post, 1400 lb; boom, 1500 lb; and brace, 900 lb. The member weights may be considered to be acting at their respective midpoints. Calculate all the forces acting on each of the three members.

Solution First, consider the entire frame as a free body, as shown in Figure 5.16b. Assume the external reactions at *A* and *B* to act as shown. Apply the three equations of equilibrium, considering forces upward and to the right and counterclockwise moments positive. First,

$$\Sigma M_A = +B_H(14\text{ ft}) - 1500\text{ lb}(7\text{ ft}) - 900\text{ lb}(5\text{ ft}) - 10{,}000\text{ lb}(14\text{ ft}) = 0$$

from which

$$B_H = +11{,}070\text{ lb} \leftarrow$$

Next,

$$\Sigma F_H = +A_H - B_H = +A_H - 11{,}070\text{ lb} = 0$$

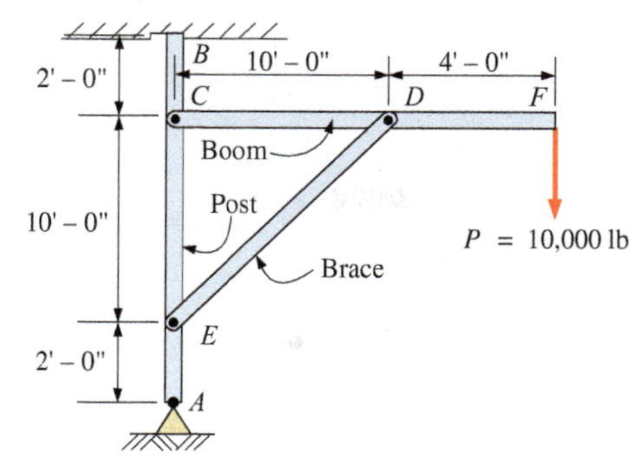

(a) Crane framework

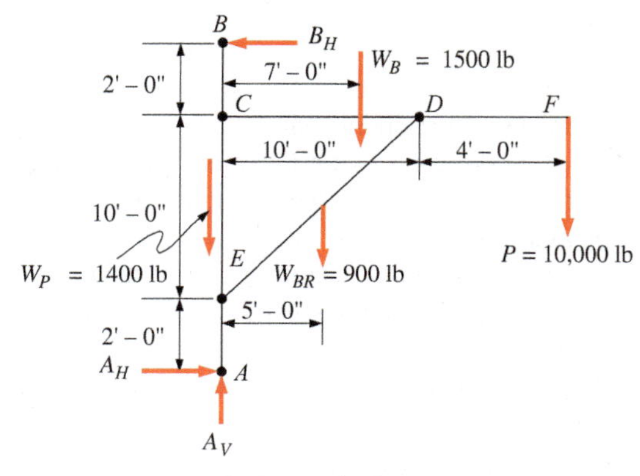

(b) Free-body diagram of entire frame

FIGURE 5.16 Framework for Example 5.3.

from which

$$A_H = +11{,}070 \text{ lb} \rightarrow$$

Finally,

$$\Sigma F_V = +A_V - 1500 \text{ lb} - 1400 \text{ lb} - 900 \text{ lb} - 10{,}000 \text{ lb} = 0$$

from which

$$A_V = +13{,}800 \text{ lb} \uparrow$$

The results are all positive values; therefore, the senses of the reactions are as assumed. Arrows indicate the sense in each case.

Next consider member *CF* as a free body, as shown in Figure 5.17. Note that member *CF* is a multiple-force member with pin connections at points *C* and *D*. The effects of the brace and the post on the boom are unknown; therefore, vertical and horizontal component forces are indicated at *C* and *D*. Solve for these component forces. The senses of the components are assumed in Figure 5.17. If the senses are as assumed, positive signs will result. Apply the three equations of equilibrium. First,

$$\Sigma M_C = +D_V(10 \text{ ft}) - 10{,}000 \text{ lb}(14 \text{ ft}) - 1500 \text{ lb}(7 \text{ ft}) = 0$$

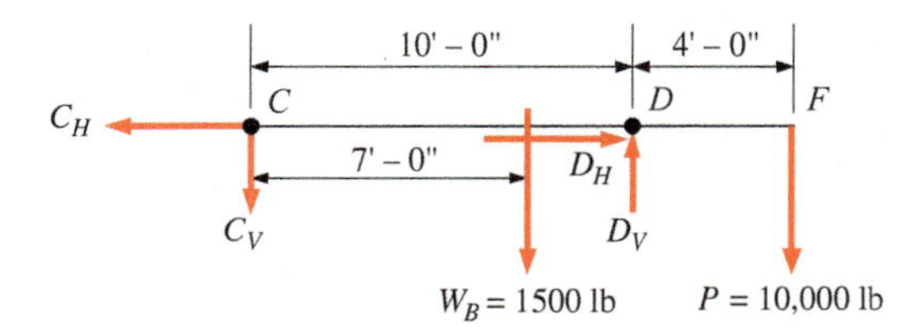

FIGURE 5.17 Free-body diagram of boom.

from which

$$D_V = +15{,}050 \text{ lb} \uparrow$$

Next,

$$\Sigma F_V = -C_V + D_V - 1500 \text{ lb} - 10{,}000 \text{ lb} = 0$$

from which

$$C_V = 15{,}050 \text{ lb} - 1500 \text{ lb} - 10{,}000 \text{ lb} = +3550 \text{ lb} \downarrow$$

Finally,

$$\Sigma F_H = +D_H - C_H = 0$$

from which

$$D_H = C_H$$

These two components cannot be determined from this free body (there were four unknown forces and only three equations of equilibrium). Therefore, proceed to the free body for member *DE*. Note that the results for the preceding calculations are all positive, indicating that the senses for the components are as assumed.

The free-body diagram for member *DE* is shown in Figure 5.18. Member *DE* is a multiple-force member with pin connections at points *D* and *E*. Note the senses of D_V and D_H that have been carried forward from the free body of member *CDF*. For example, D_V acted upward on member *CDF*; therefore, it must, of necessity, act downward on member *DE*. The senses of the components acting at point *E* are assumed as shown.

Apply the three equations of equilibrium. First,

$$\Sigma M_E = +D_H(10 \text{ ft}) - 15{,}050 \text{ lb}(10 \text{ ft}) - 900 \text{ lb}(5 \text{ ft}) = 0$$

from which

$$D_H = +15{,}500 \text{ lb} \leftarrow$$

Next,

$$\Sigma F_H = +E_H - D_H = 0$$

from which

$$E_H = +15{,}500 \text{ lb} \rightarrow$$

Finally,

$$\Sigma F_V = +E_V - 900 \text{ lb} - 15{,}050 \text{ lb} = 0$$

from which

$$E_V = +15{,}950 \text{ lb} \uparrow$$

Furthermore, since $C_H = D_H$, then $C_H = 15{,}500$ lb, which acts to the left on member *CDF*.

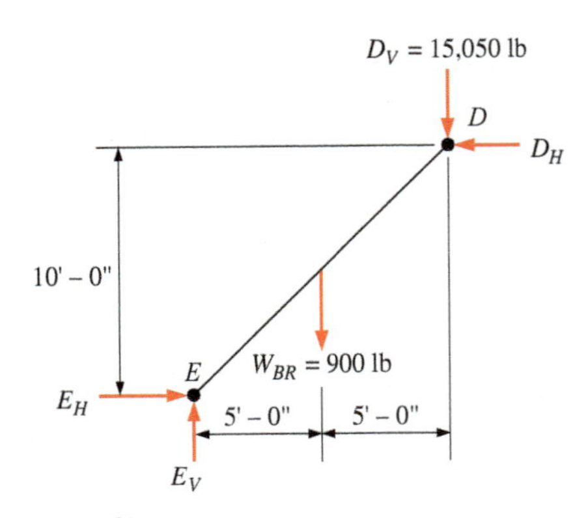

FIGURE 5.18 Free-body diagram of brace.

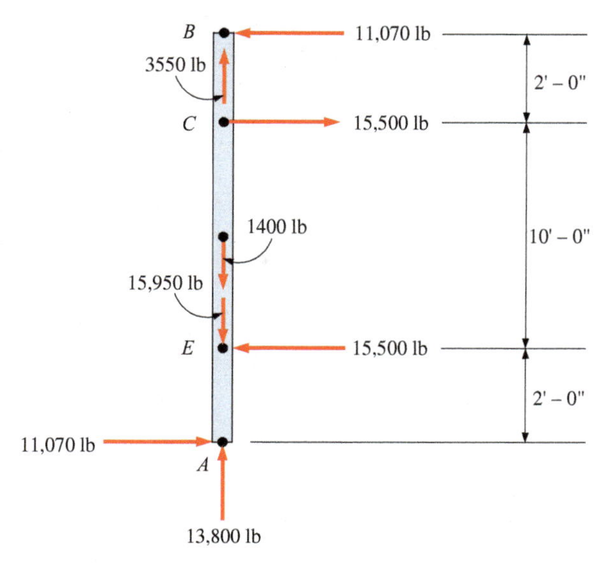

FIGURE 5.19 Free-body diagram of post.

Member *BCEA* may now be used as a final check of the calculations. The free-body diagram with all of the forces that have been determined is shown in Figure 5.19. You can verify that $\Sigma F_V = 0$, $\Sigma F_H = 0$, and $\Sigma M = 0$.

Table 5.1 lists the vertical and horizontal forces that were calculated in this example. The pin reactions—refer to Equation (2.1)—are also indicated. The results are also shown in Figure 5.20.

TABLE 5.1 Force summary

Pin	Vertical Component (lb)	Horizontal Component (lb)	Pin Reaction (lb)
A	13,800	11,070	17,690
C	3550	15,500	15,900
D	15,050	15,500	21,600
E	15,950	15,500	22,200

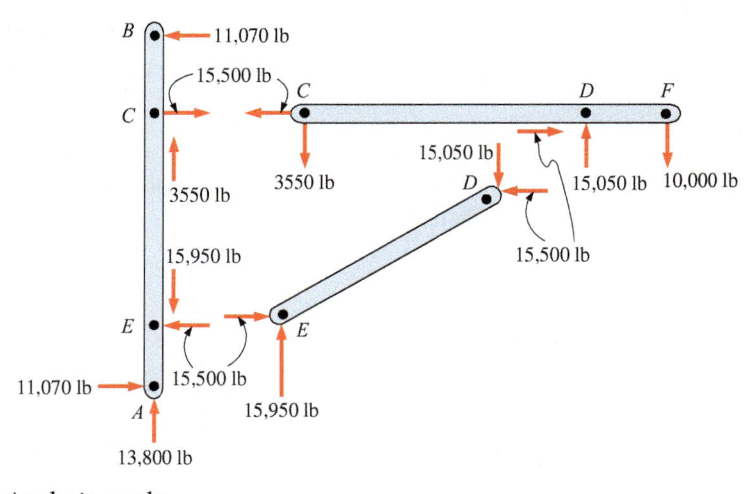

FIGURE 5.20 Analysis results.

EXAMPLE 5.4 The frame shown in Figure 5.21a is pin-connected at *A*, *B*, *D*, and *E*. Find the forces in members *AD*, *DB*, and *ED*. Check the calculations by an equilibrium check for member *ABC*.

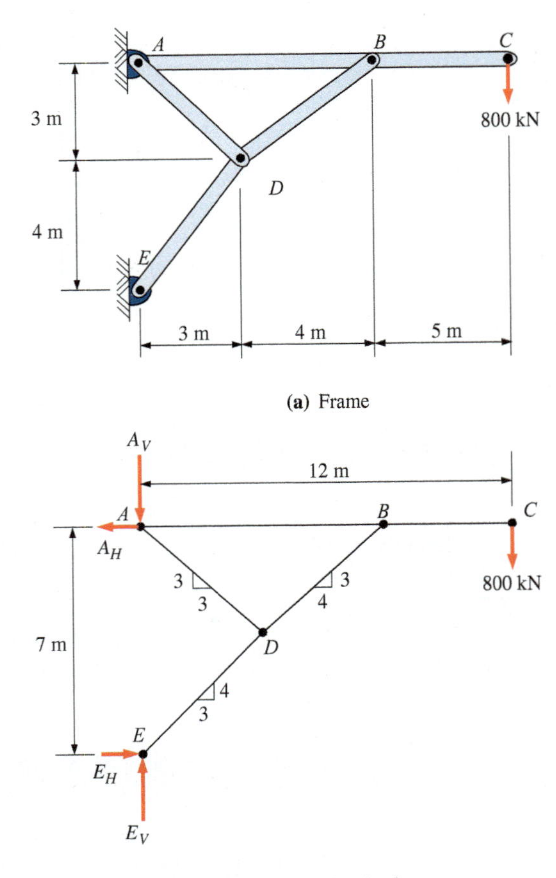

FIGURE 5.21 Framework for Example 5.4.

Solution The free-body diagram for the entire frame is shown in Figure 5.21b. Note that the pin supports at *A* and *E* have been replaced with vertical and horizontal reactions. Senses have been assumed. Forces acting upward or to the right and counterclockwise moments are considered positive. First, solve for E_H by summing moments about *A*:

$$\Sigma M_A = +E_H(7\text{ m}) - 800\text{ kN}(12\text{ m}) = 0$$

$$E_H = \frac{+800\text{ kN}(12\text{ m})}{7\text{ m}} = +1371\text{ kN} \rightarrow$$

The positive sign indicates that the sense is as assumed. (Member *ED* is in compression.)

Since member *ED* is a two-force member, the reaction at *E* must be directed along the longitudinal axis of *ED*. Therefore,

$$\frac{E_V}{4} = \frac{E_H}{3}$$

from which

$$E_V = \frac{4E_H}{3} = \frac{4(1371\text{ kN})}{3} = 1828\text{ kN} \uparrow$$

The force in member *ED* is then

$$ED = \sqrt{E_V^2 + E_H^2} = \sqrt{1828^2 + 1371^2} = 2285\text{ kN}$$

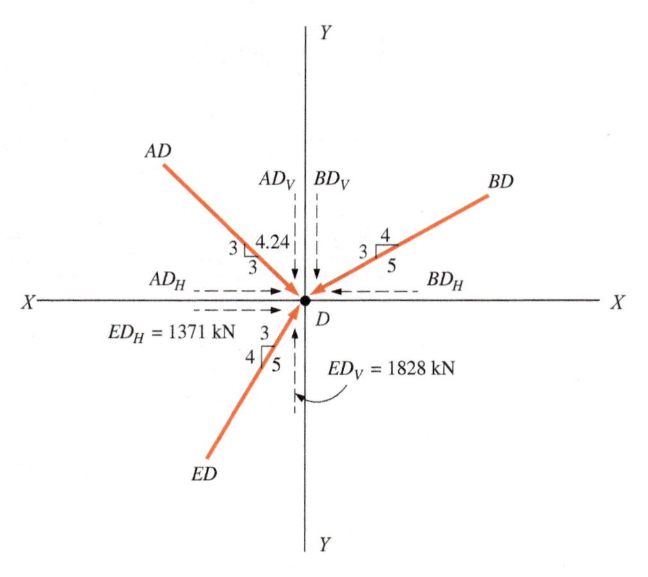

FIGURE 5.22 Free-body diagram of joint D.

Calculating A_V and A_H,

$$\Sigma F_V = 0 = 1828 \text{ kN} - A_V - 800 \text{ kN} = 0$$
$$A_V = +1028 \text{ kN} \downarrow$$
$$\Sigma F_H = 0 = -A_H + 1371 \text{ kN} = 0$$
$$A_H = +1371 \text{ kN} \leftarrow$$

Next, isolate the pin at joint D (Figure 5.22) and solve for AD and BD. Assume that both are in compression as shown. Writing the relationships between the vertical and horizontal components gives

$$\frac{AD}{4.24} = \frac{AD_H}{3} = \frac{AD_V}{3}$$
$$\frac{BD}{5} = \frac{BD_H}{4} = \frac{BD_V}{3}$$

from which

$$AD_V = AD_H = \frac{3(AD)}{4.24} = 0.7076AD$$
$$BD_V = \frac{3(BD)}{5} = 0.60BD$$
$$BD_H = \frac{4(BD)}{5} = 0.80BD$$

Summing vertical and horizontal forces at D yields

$$\Sigma F_H = +1371 \text{ kN} + AD_H - BD_H = 0$$
$$= +1371 \text{ kN} + 0.7076AD - 0.80BD = 0 \quad \textbf{(1)}$$
$$\Sigma F_V = +1828 \text{ kN} - AD_V - BD_V = 0$$
$$= +1828 \text{ kN} - 0.7076AD - 0.60BD = 0 \quad \textbf{(2)}$$

Equations 1 and 2 are simultaneous equations that may be solved to yield

$$AD = +646 \text{ kN}$$
$$BD = +2285 \text{ kN}$$

The components of AD and BD may be summarized as in Table 5.2 (where all forces and components are in units of kilonewtons).

TABLE 5.2 Force summary

Member	Compressive Force (kN)	Vertical Components (kN)	Horizontal Components (kN)
ED	2285	1828	1371
AD	646	457	457
BD	2285	1371	1828

Finally, check equilibrium of member *ABC* (see Figure 5.23):

$$\Sigma F_H = -1371 \text{ kN} - 457 \text{ kN} + 1828 \text{ kN} = 0 \quad \textbf{(O.K)}$$

$$\Sigma F_V = -1028 \text{ kN} + 457 \text{ kN} + 1371 \text{ kN} - 800 \text{ kN} = 0 \quad \textbf{(O.K)}$$

$$\Sigma M_A = +1371 \text{ kN}(7 \text{ m}) - 800 \text{ kN}(12 \text{ m}) = -3 \text{ kN}\cdot\text{m} \quad \textbf{(O.K)}$$

The final moment check reflects some slight rounding error.

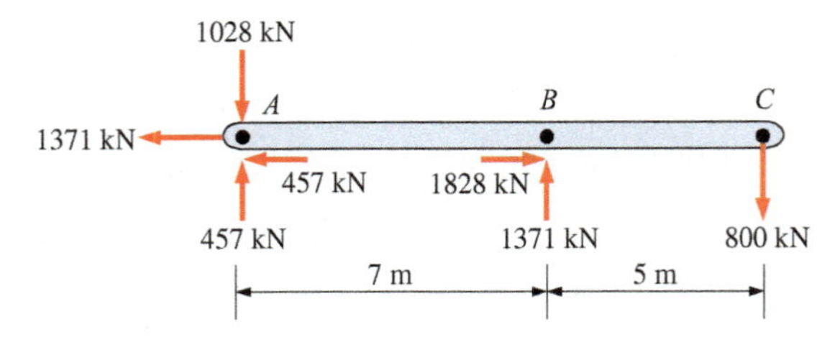

FIGURE 5.23 Free-body diagram of member *ABC*.

SUMMARY BY SECTION NUMBER

5.1 In trusses all members are two-force members. In frames all members, or some of them, are multiple-force members.

5.2 and 5.3 A truss is a structural framework consisting of straight individual members connected to form a triangle or series of triangles. Truss members are connected at joints (or panel points) with pins assumed to be frictionless. Loads are assumed to be applied at the joints. In the context of our presentation, all members are straight and are assumed to lie in the same plane.

5.4 The method of joints is used to determine the internal force in truss members. Each joint is isolated sequentially, beginning with one in which only two member forces are unknown, then proceeding to adjacent joints. No more than two unknown member forces can be determined at any one joint. Unknown forces are found by the method of components, applying the two equations of force equilibrium, or by solution of the force triangle.

5.5 The method of sections is also used to determine the internal force in a truss member. A portion of the truss is isolated as a free body by passing a cutting plane through the truss, cutting no more than three members in which the forces are unknown. Unknown forces are found by applying the three equations of equilibrium to the isolated portion of the truss. The method of sections usually has the advantage of allowing for the analysis of particular members without having to analyze the entire truss.

5.6 A frame is a structure composed partially, or totally, of pin-connected, multiple-force members. The analysis involves the determination of the forces acting at the pins joining the members. The procedure is to isolate each member as a free body, showing all known and unknown forces (or components of forces), and then to apply the three equations of equilibrium.

PROBLEMS

Section 5.4 The Method of Joints

5.1 through 5.7 Calculate the forces in all members of the trusses shown using the method of joints.

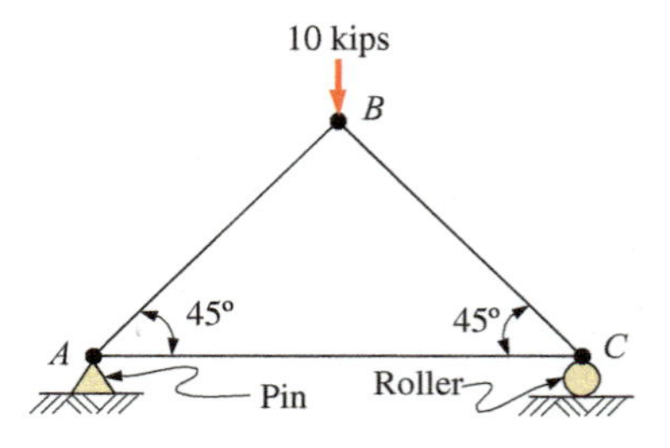

PROBLEM 5.1

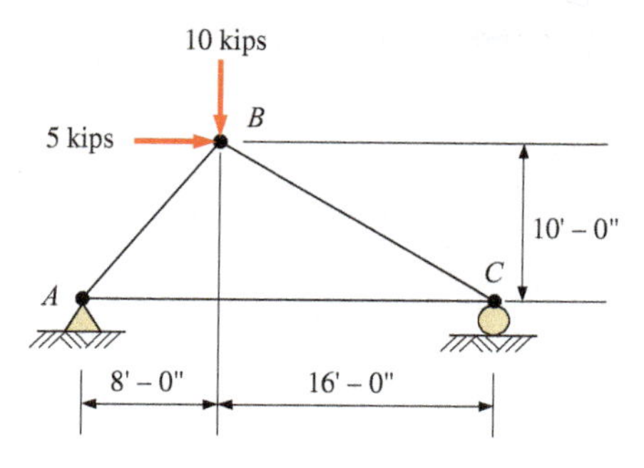

PROBLEM 5.2

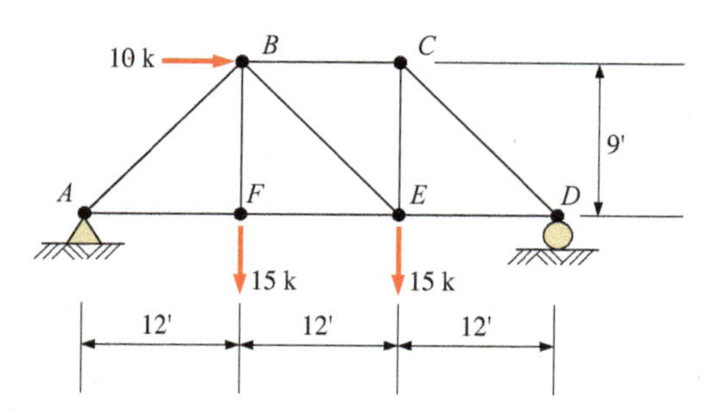

PROBLEM 5.3

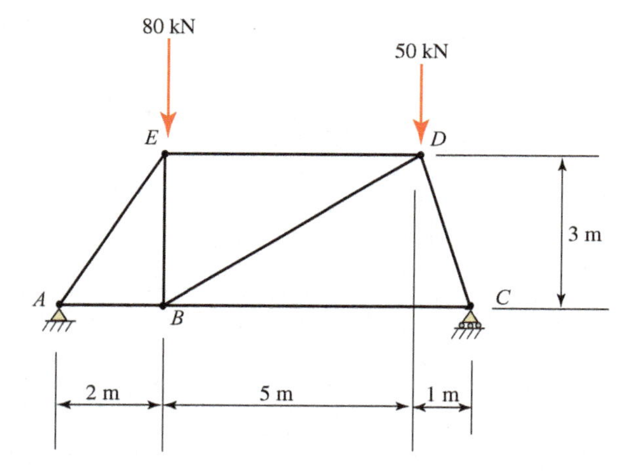

PROBLEM 5.4

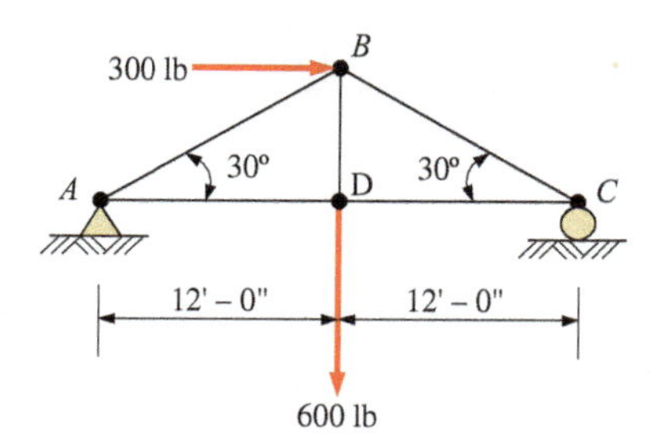

PROBLEM 5.5

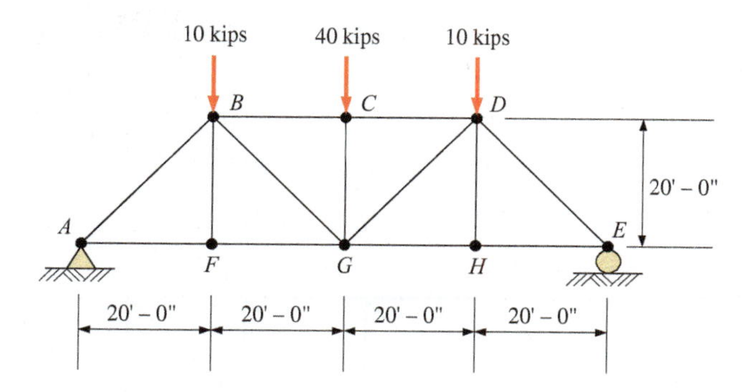

PROBLEM 5.6

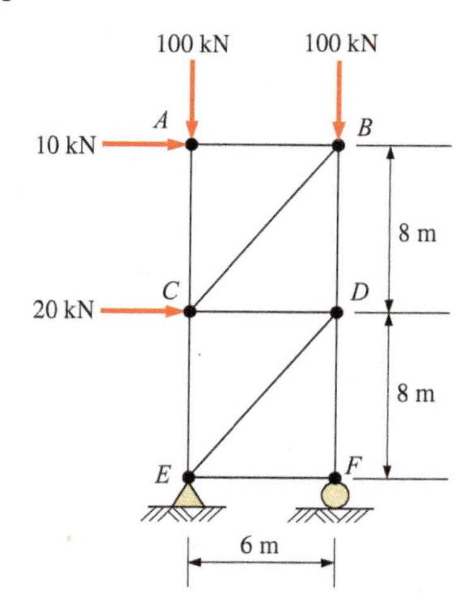

PROBLEM 5.7

Section 5.5 The Method of Sections

For Problems 5.8 through 5.14, use the method of sections.

5.8 Determine the forces in members *CD*, *DH*, and *HI* for the truss shown.

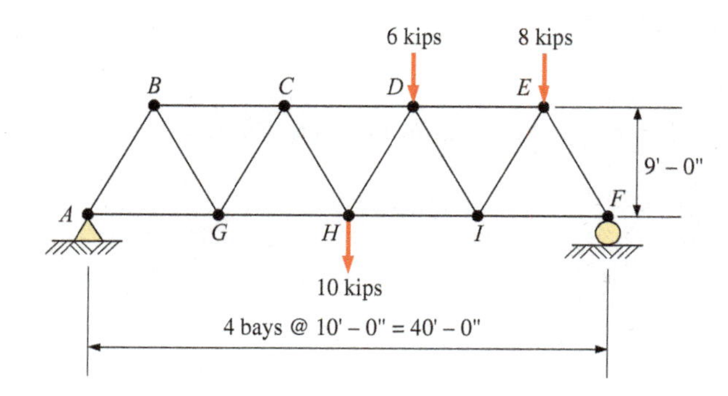

PROBLEM 5.8

5.9 Determine the forces in members *BC*, *BE*, and *FE* for the truss shown.

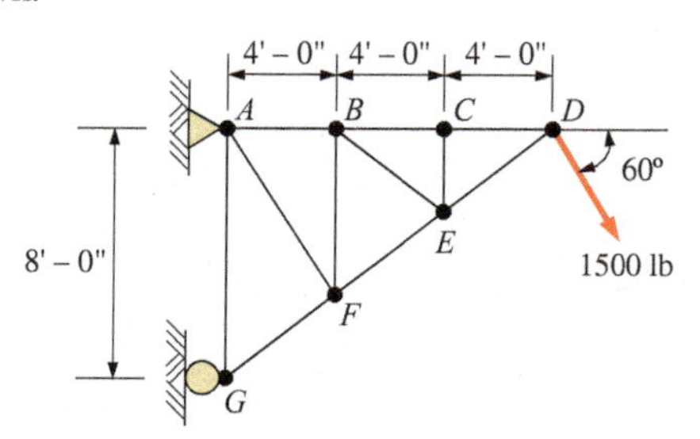

PROBLEM 5.9

5.10 Determine the forces in members *BC*, *CH*, and *CG* in the truss shown.

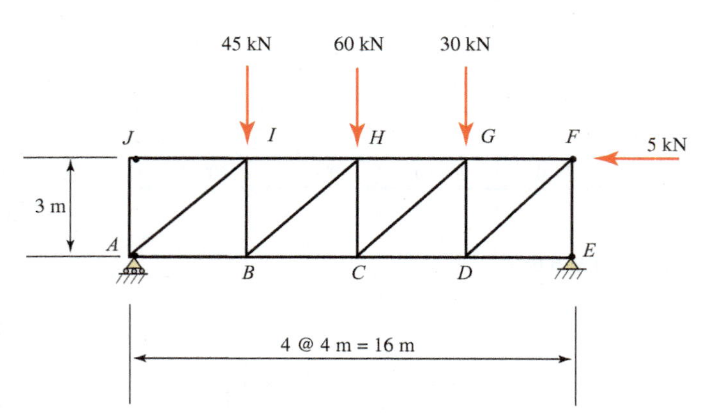

PROBLEM 5.10

5.11 For the Howe roof truss shown, determine the forces in members *BC*, *CI*, and *IJ*.

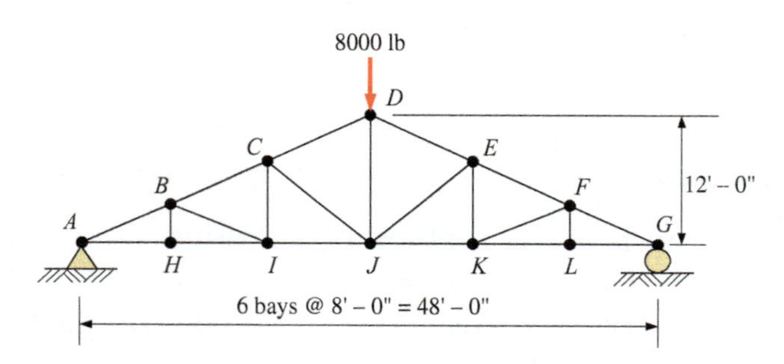

PROBLEM 5.11

5.12 Determine the forces in members *DE*, *CE*, and *BC* in the truss shown.

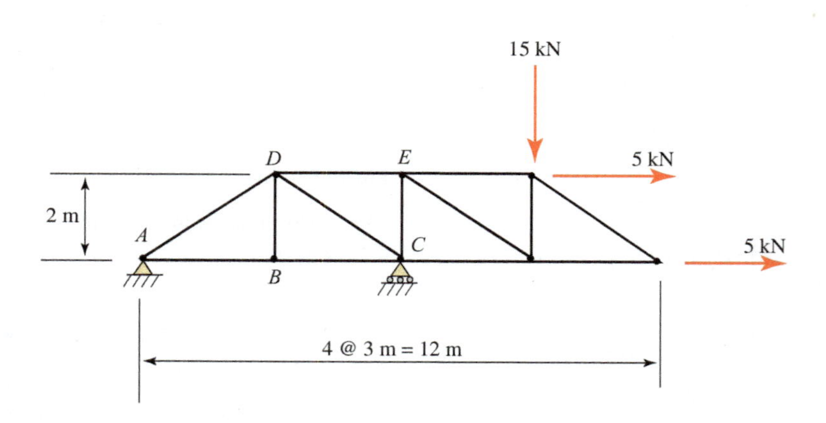

PROBLEM 5.12

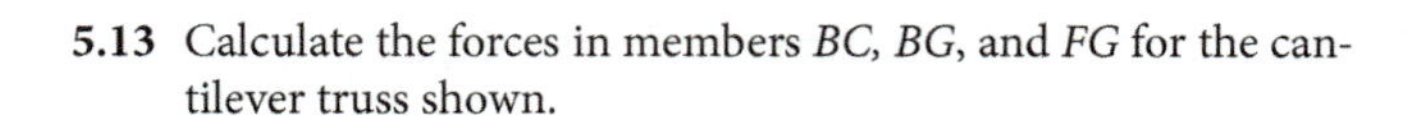

5.13 Calculate the forces in members *BC*, *BG*, and *FG* for the cantilever truss shown.

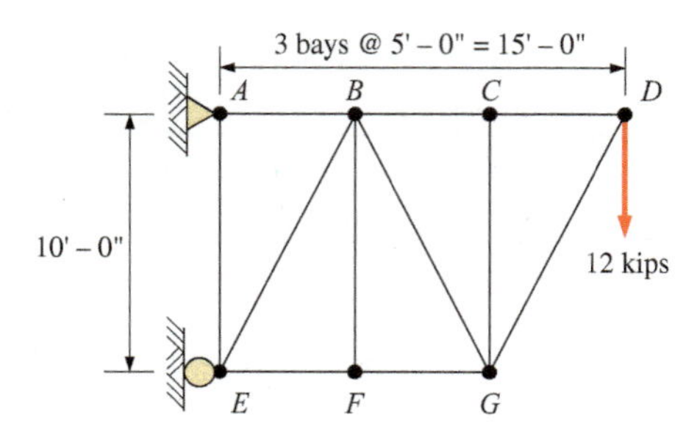

PROBLEM 5.13

5.14 Determine the forces in members *CD*, *BD*, *BE*, and *CB* in the yoke truss shown.

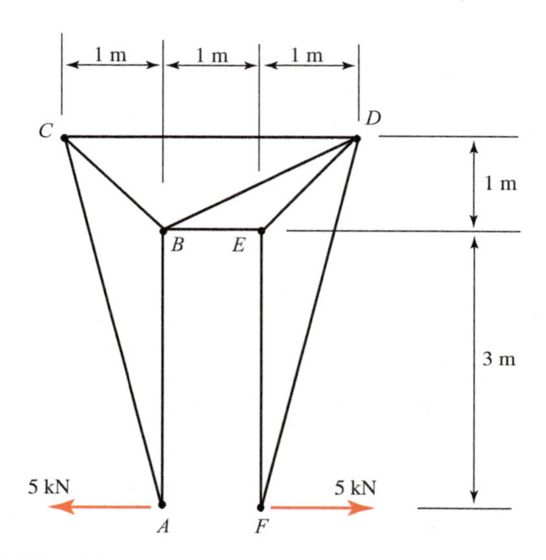

PROBLEM 5.14

Section 5.6 Analysis of Frames

5.15 A pin-connected A-frame supports a load, as shown. Compute the pin reactions at all of the pins. Neglect the weights of the members.

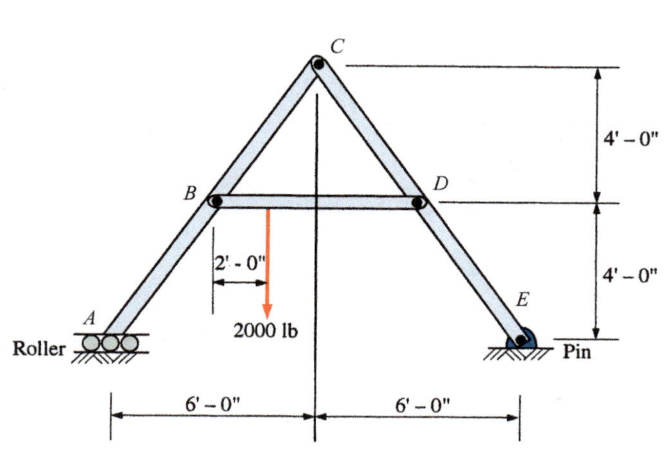

PROBLEM 5.15

5.16 Determine the pin reactions at pins *A*, *B*, and *C* in the frame shown. Neglect the weights of the members.

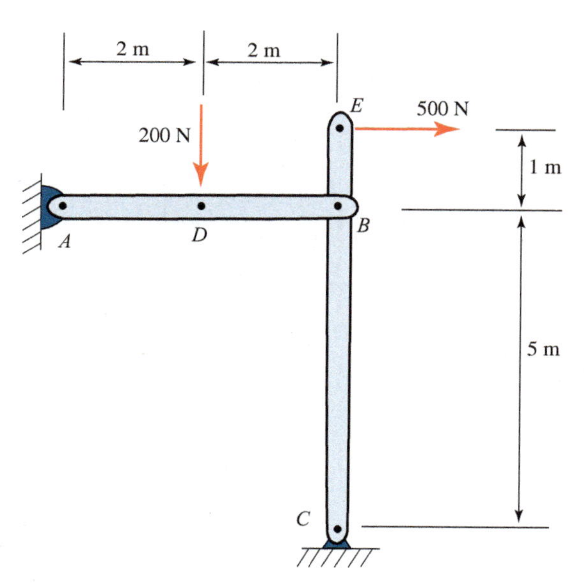

PROBLEM 5.16

5.17 Calculate the pin reactions at each of the pins in the frame shown.

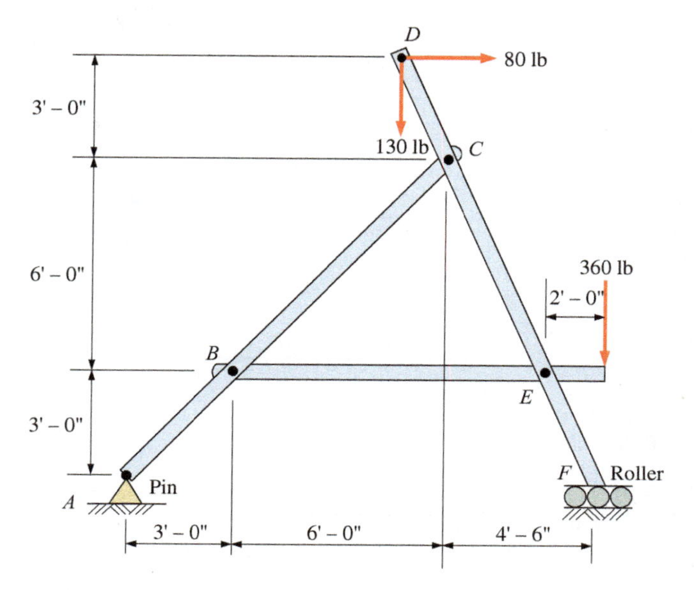

PROBLEM 5.17

5.18 A bracket is pin connected at points *A*, *B*, and *D* and is subjected to loads, as shown. Calculate the pin reactions. Neglect the weights of the members.

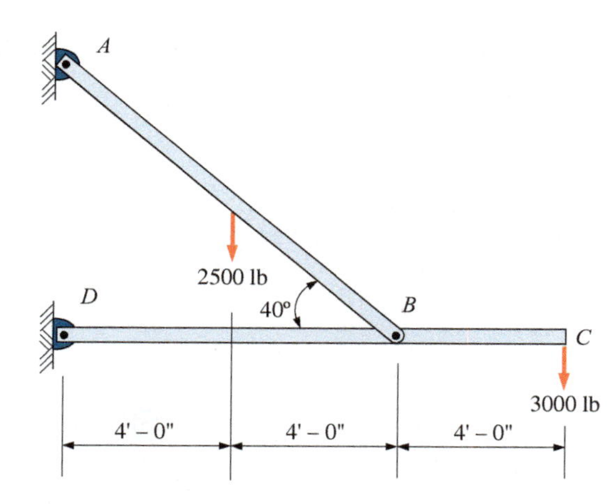

PROBLEM 5.18

5.19 A pin-connected frame is loaded, as shown. Calculate the pin reactions at *A*, *B*, and *C*. Neglect the weights of the members.

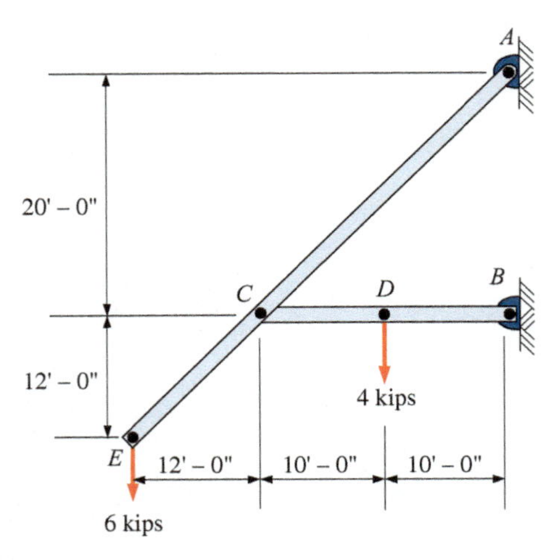

PROBLEM 5.19

5.20 The cylinder shown has a mass of 500 kg. Determine the force in member *AB* and the pin reaction at *C*. All surfaces are smooth. *A*, *B*, and *C* are frictionless pins.

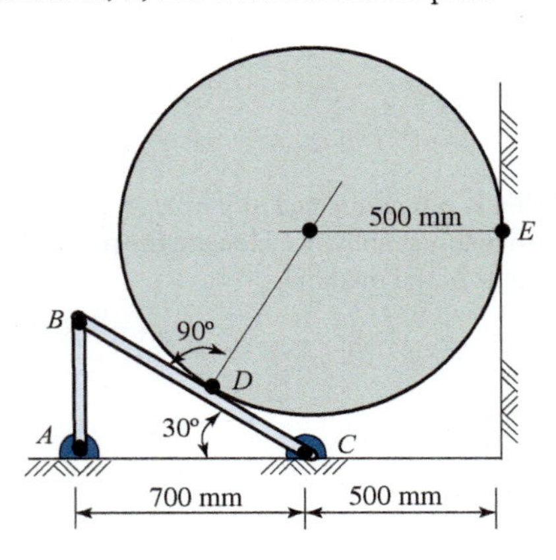

PROBLEM 5.20

5.21 A simple frame is pin connected at points *A*, *B*, and *C* and is subjected to loads as shown. Compute the pin reactions at *A*, *B*, and *C*. Neglect the weights of the members.

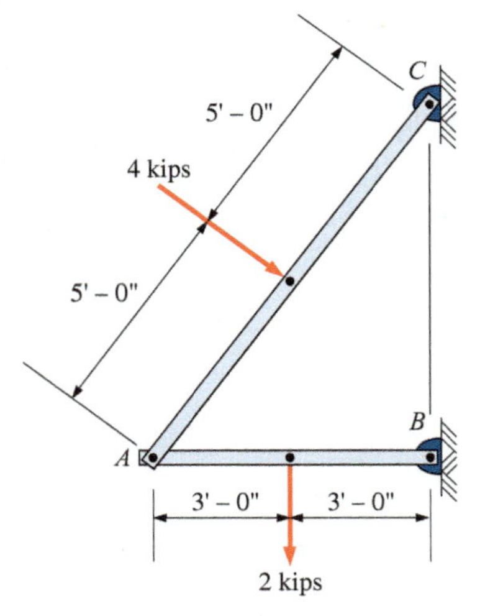

PROBLEM 5.21

5.22 Using the method of sections, determine the forces in *CE* and *ED* in the vertical truss of Problem 5.7.

5.23 Using the method of sections, determine the forces in members *BD* and *BE* of the truss shown.

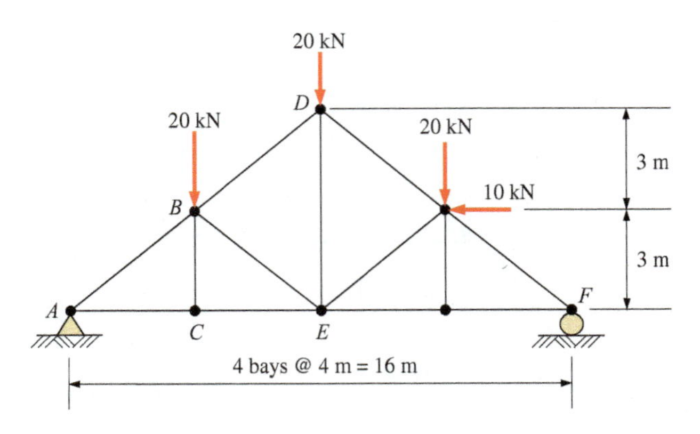

PROBLEM 5.23

Supplemental Problems

5.24 through 5.31 Calculate the forces in all members of the trusses shown, using the method of joints.

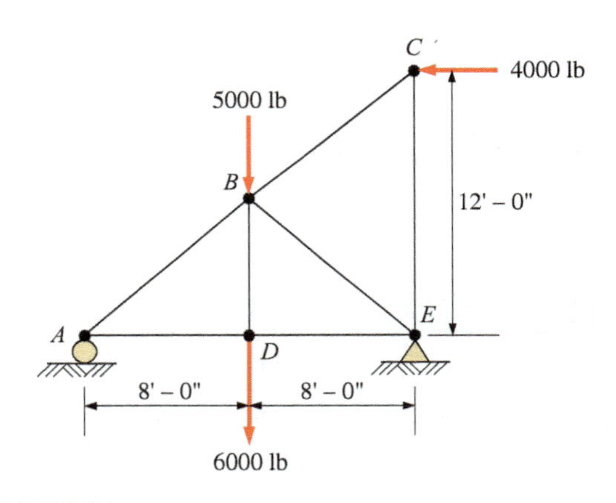

PROBLEM 5.24

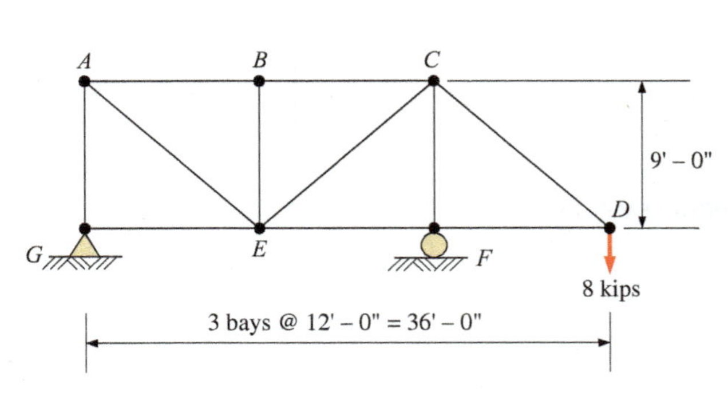

PROBLEM 5.25

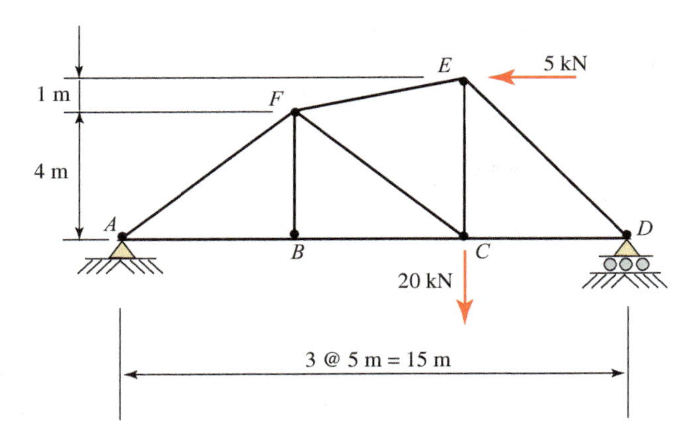

PROBLEM 5.26

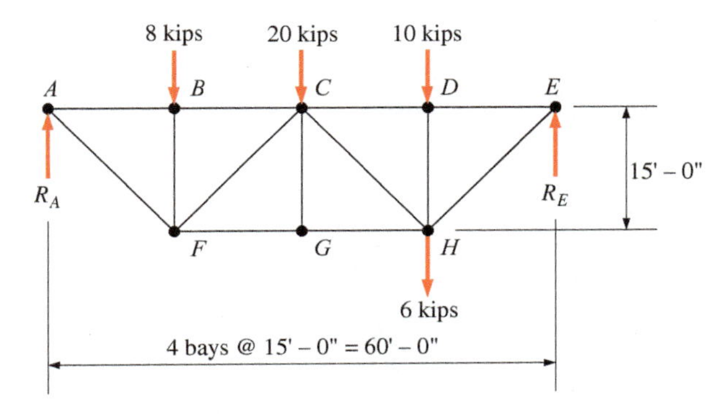

PROBLEM 5.27

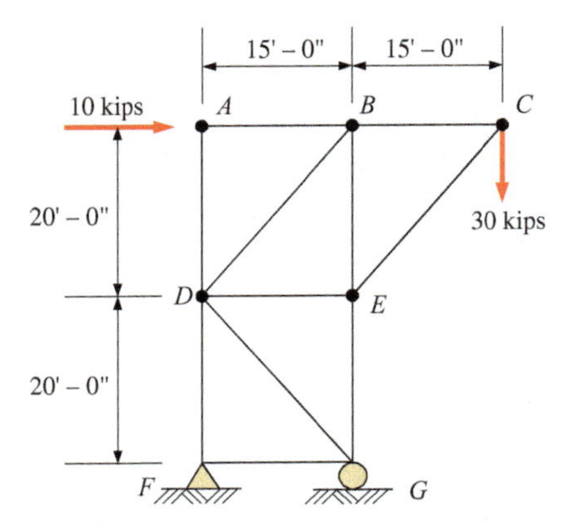

PROBLEM 5.28

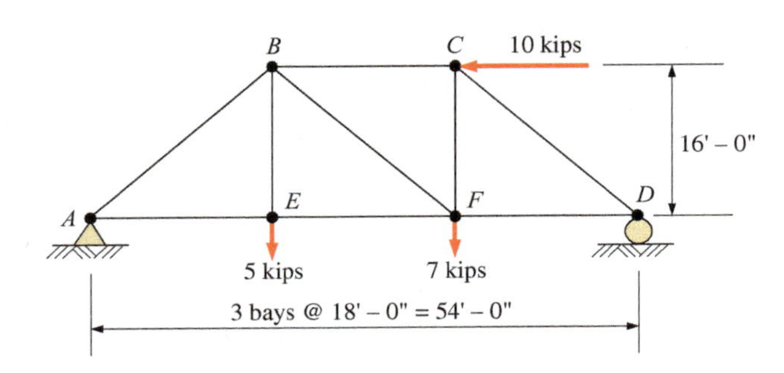

PROBLEM 5.29

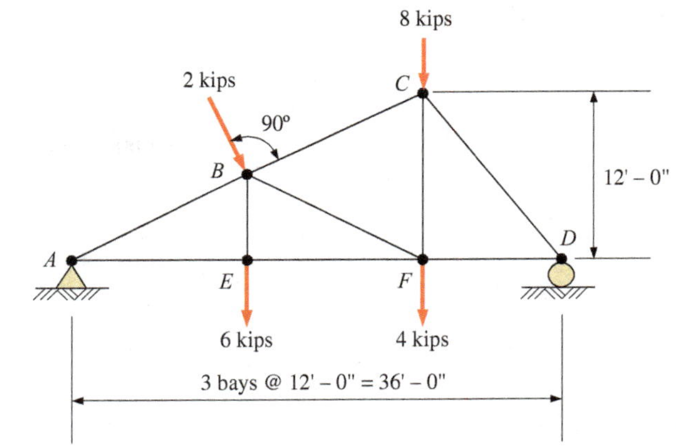

8 kips
2 kips
C
90°
B
12' – 0"
A
D
E
F
6 kips
4 kips
3 bays @ 12' – 0" = 36' – 0"

PROBLEM 5.30

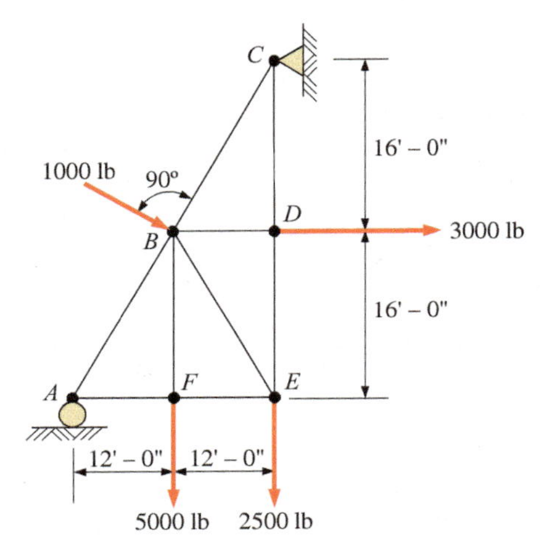

PROBLEM 5.31

For Problems 5.32 through 5.38, calculate the forces in the indicated members of the trusses shown. Use the method of sections.

5.32 Members *CD*, *CK*, and *EM*.

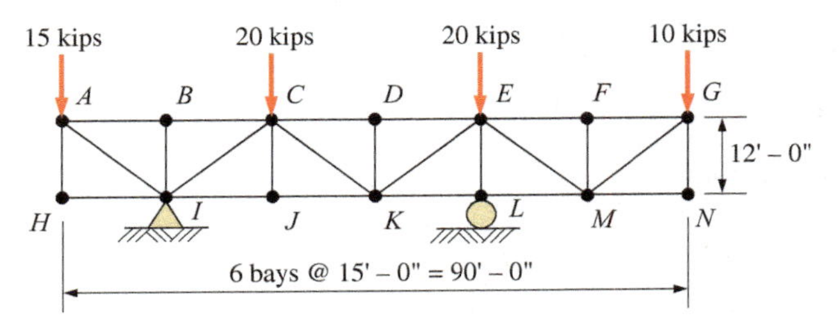

PROBLEM 5.32

5.33 Members *BF*, *BC*, and *EF*.

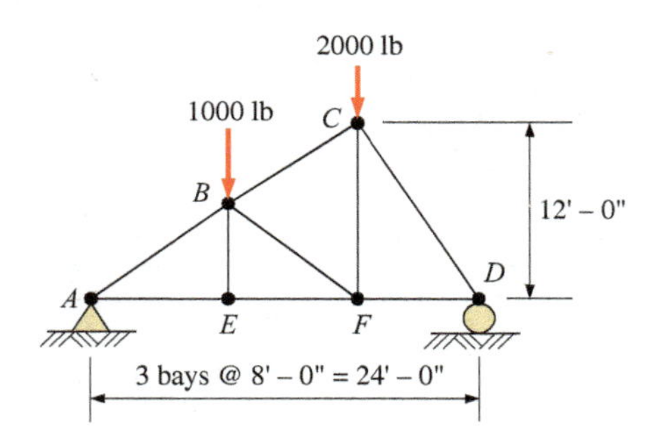

PROBLEM 5.33

5.34 Members *AF*, *BG*, and *FG*.

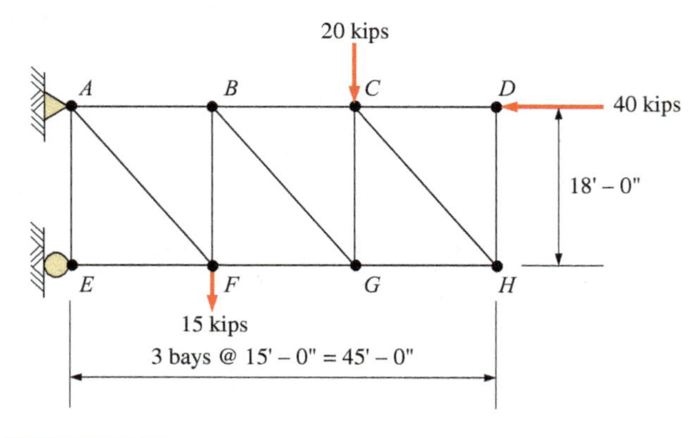

PROBLEM 5.34

5.35 Members *CD*, *BH*, and *HG*.

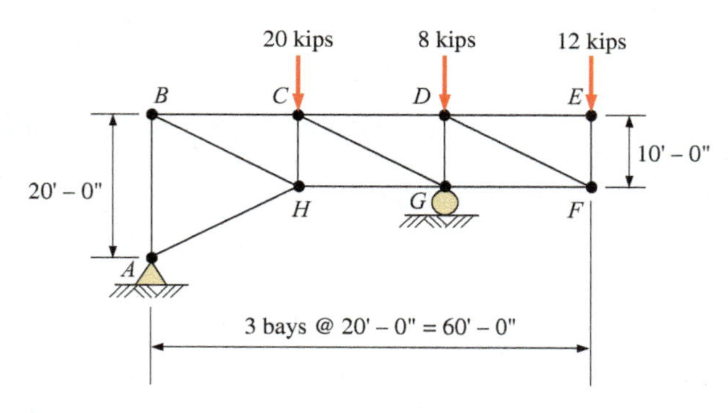

PROBLEM 5.35

5.36 Members *BE*, *CE*, and *CF*.

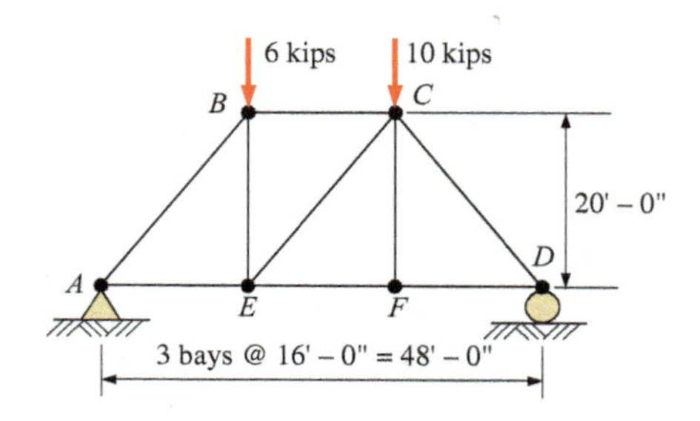

PROBLEM 5.36

5.37 Members *EH*, *CE*, *CB*, and *AB*.

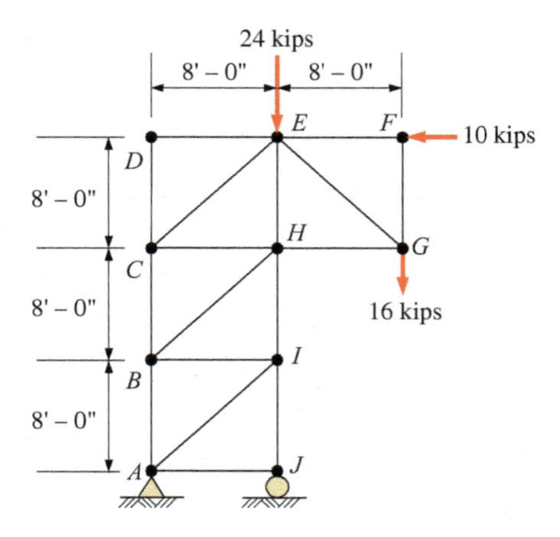

PROBLEM 5.37

5.38 Members *BC*, *CD*, and *AD*.

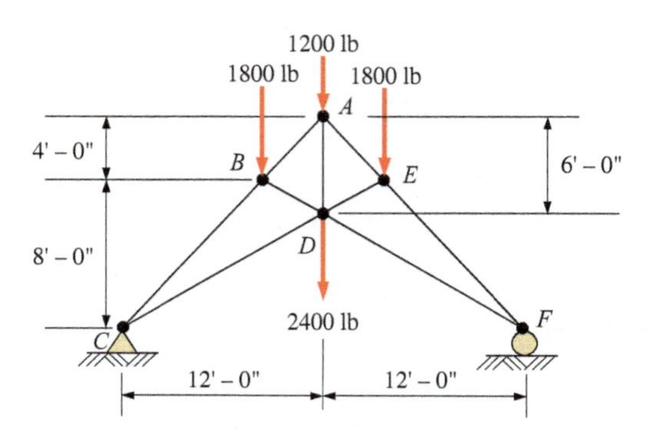

PROBLEM 5.38

5.39 A pin-connected crane framework is loaded and supported, as shown. The member weights are post, 700 lb; boom, 800 lb; and brace, 600 lb. These weights may be considered to be acting at the midpoint of the respective members. Calculate the pin reactions at pins *A*, *B*, *C*, *D*, and *E*.

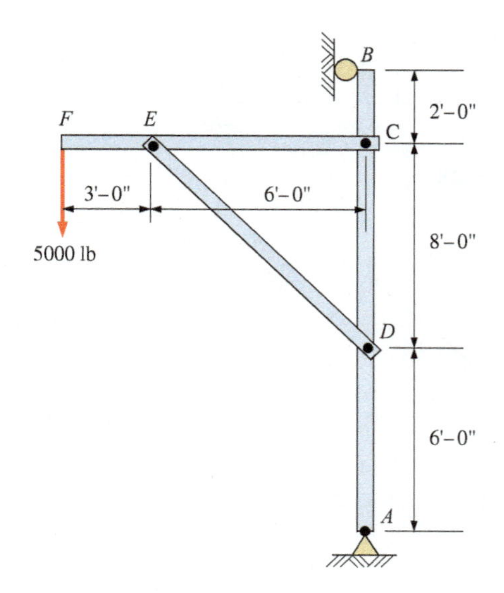

PROBLEM 5.39

5.40 Calculate the pin reactions at pins *A*, *B*, and *D* in the frame shown. Neglect the weights of the members.

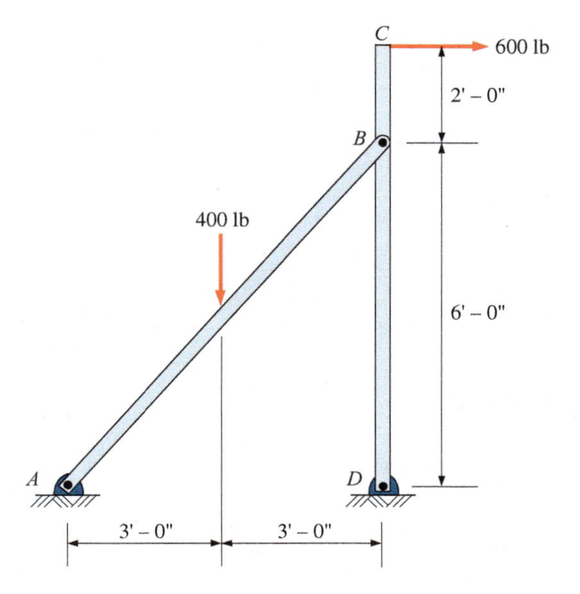

PROBLEM 5.40

5.41 Determine the pin reactions at pins *A*, *B*, and *C* in the frame shown. Neglect the weights of the members.

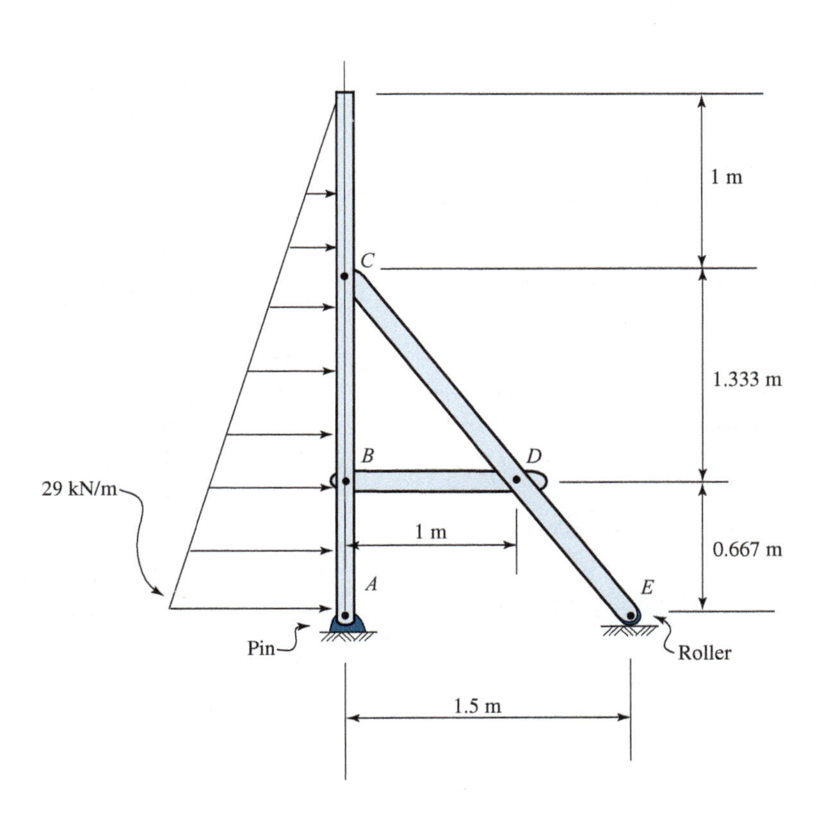

PROBLEM 5.41

5.42 The wall bracket shown is pin-connected at points *A*, *B*, and *C*. Calculate the pin reactions at these points. Neglect the weights of the members.

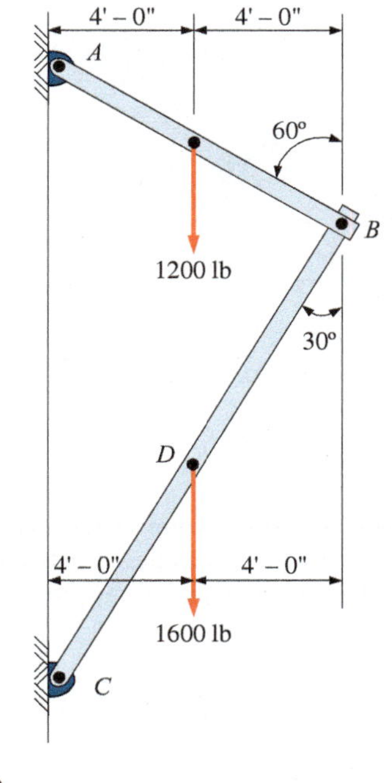

PROBLEM 5.42

5.43 Calculate the pin reactions at each of the pins in the frame shown.

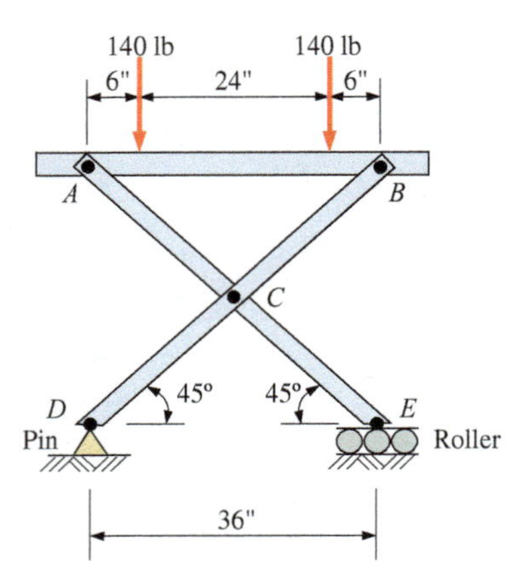

PROBLEM 5.43

5.44 The *A*-frame shown is pin-connected at *A, B, C,* and *D*. The surface at *E* is level and frictionless. Calculate the reaction at *E* and the reaction components at the pins.

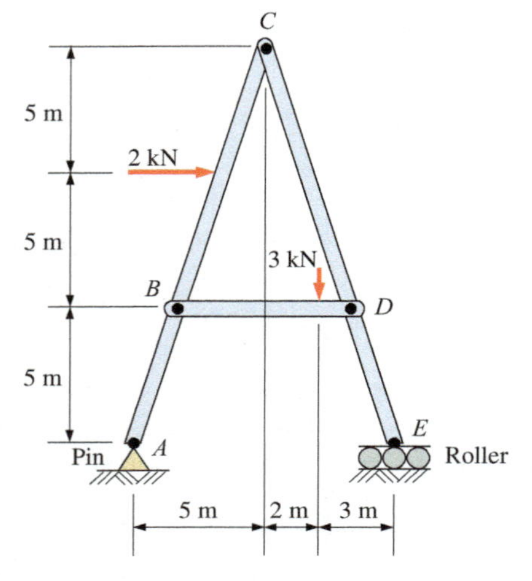

PROBLEM 5.44

5.45 The tongs shown are used to grip an object. For an input force of 12 lb on each handle, determine the forces exerted on the object and the forces exerted on the pin at *A*.

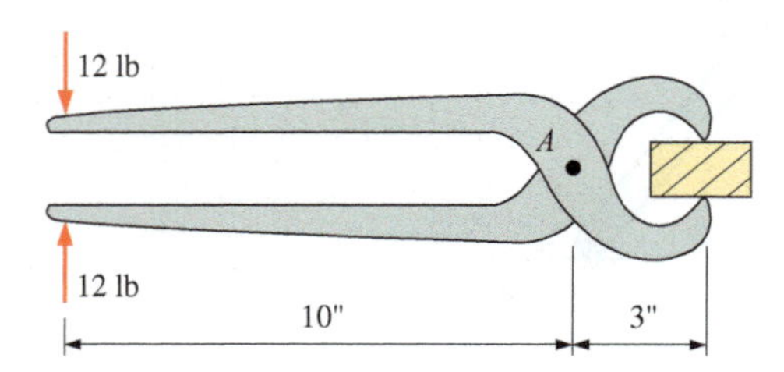

PROBLEM 5.45

5.46 A toggle joint is a mechanism by which a comparatively small force *P* produces or balances a force *F* that increases rapidly as the angle θ approaches 180°. For a given angle of θ, the toggle joint becomes equivalent to a simple truss. Force $P = 10$ lb, $a = 2$ ft, $b = 5$ ft, and $c = 1$ ft, as shown. Find force *F*. Neglect friction and the weights of the members.

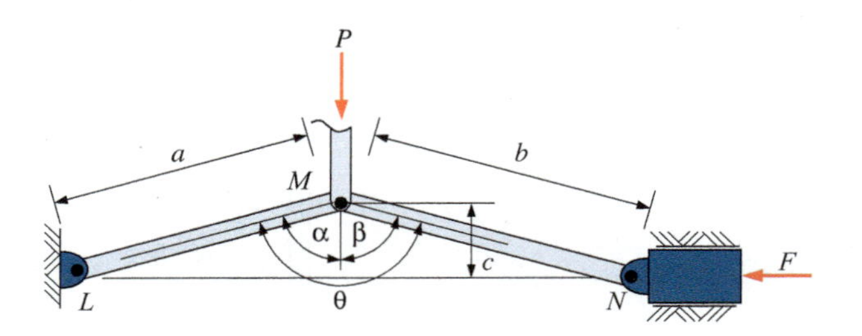

PROBLEM 5.46

5.47 In the toggle joint of Problem 5.46, assume that *a* and *b* are equal and $\theta = 160°$. What force *P* is required to balance a force of 100 lb at *F*?

5.48 A beverage can crusher can aid in the reduction of volume of recyclable items. For a constant force of 30 lb applied at a right angle to the handle, as shown, determine the force applied to the beverage can for values of θ of **(a)** 90° and **(b)** 135°. Neglect friction. Link *AB* is 9 in. long.

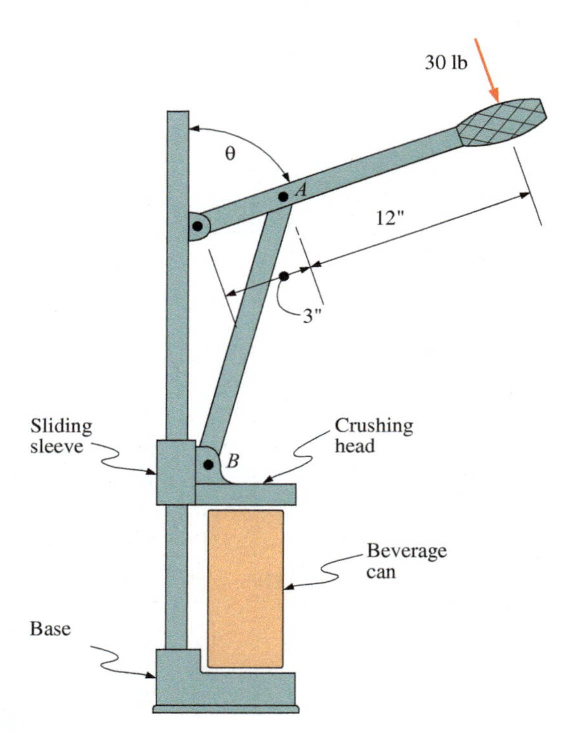

PROBLEM 5.48

CHAPTER SIX

FRICTION

LEARNING OBJECTIVES

Upon completion of this chapter, readers will be able to:

- Define friction and the forces associated with friction.
- Calculate the coefficient of static friction between objects.
- Calculate the angle of static friction.
- Solve motion-impending friction problems using the equilibrium equations from Chapter 4.
- Discuss the use of wedges and be able to calculate forces in a wedge system.
- Calculate belt tension in belts wrapped around cylindrical surfaces.
- Calculate the forces in a square-threaded screw.

6.1 INTRODUCTION

Over the centuries, many renowned scientists, among them Leonardo da Vinci (1452–1519) and Charles Augustin de Coulomb (1736–1806), have studied friction. Friction is one of the most familiar concepts in the field of engineering mechanics, yet it remains one of the least understood. Friction is a complicated phenomenon to which considerable research is still being devoted within the scientific community.

Friction is defined as a retarding force that resists the relative movement of two bodies in contact with each other. Its action is always opposite in direction to the actual or impending motion between the two bodies. In addition, its action is always parallel or tangent to the surfaces in contact. This retarding or resisting force is commonly called *friction force* or *frictional resistance*. Its magnitude is primarily dependent on the condition of the contact surface between the two bodies. Every surface, as shown in Figure 6.1, no matter how polished or carefully finished, will reveal projections and depressions when microscopically examined. It is the interlocking of these irregularities on the two surfaces that resists the sliding of one surface with respect to the other.

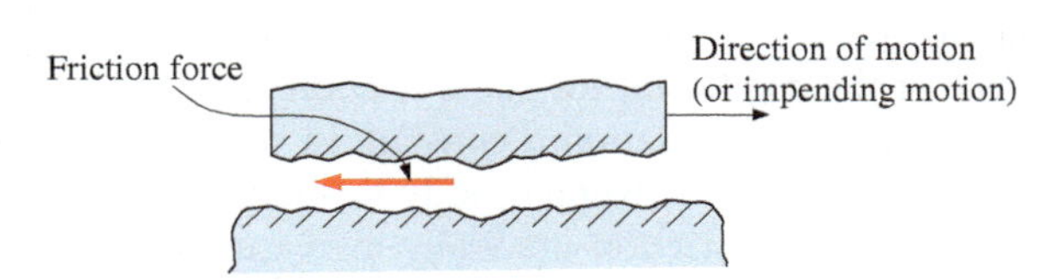

FIGURE 6.1 Contact surfaces magnified.

Research has indicated that other factors such as temperature, molecular adhesion, electrostatic attraction, lubrication, and relative velocities, in addition to surface irregularities, all contribute to frictional resistance in some measure. Thus, the friction phenomenon depends on, and is a function of, a combination of complex factors. In this chapter, we discuss friction only between dry, unlubricated surfaces. Dry friction is sometimes called *Coulomb friction;* Coulomb conducted friction research in the eighteenth century and developed some of its so-called empirical laws.

Dry friction may be categorized as either static or kinetic (dynamic). When the surfaces in contact are at rest (not moving) with respect to each other, the resistance to motion is called *static friction*. When the surfaces in contact are moving with respect to each other, a resistance to motion still exists; this resistance is called *kinetic friction* or *dynamic friction*.

Detailed discussions of other distinct types of friction are beyond the scope of this text. One is fluid friction, which develops between layers of a fluid moving at different velocities. It is generally termed *viscosity*. Another is *rolling friction*, which results when the supporting surface under a rolling load deforms, such as with a steel wheel or a bowling ball rolling on a wood floor.

6.2 FRICTION THEORY

Despite the complexity of the friction phenomenon, its study can be approached conceptually in a very simple manner when the discussion applies only to dry, unlubricated contact surfaces.

Consider a block of weight W supported on a horizontal surface. There is no motion, and the block is said to be "at rest." As the block rests on the supporting surface, the only forces acting are (a) the weight W of the block acting vertically downward and (b) a force perpendicular to the contact surface, which is the reaction of the supporting surface, equal and opposite to W. This normal force is designated N. There is no frictional resistance at this time, since frictional resistance is essentially a reaction, and will develop only if there is a force producing motion, or tending to produce motion.

Next, assume that a horizontal force P is applied to the block, as shown in Figure 6.2. To maintain equilibrium (assuming the block does not move), a frictional resistance F must develop at the horizontal contact surface. As P increases from zero, the frictional resistance F that is developed also increases. This process continues up to the point of motion impending (motion on the verge of occurring) and may be observed graphically in Figure 6.3. Note that up to the point of impending motion, the frictional resistance F that is developed is numerically equal to the horizontally applied force P. The frictional resistance that is developed prior to motion is static friction.

After motion occurs, the frictional resistance F decreases rapidly to a value that then remains relatively constant, for all practical purposes. The frictional resistance that develops while the block is in motion is kinetic or dynamic friction. As may be observed in Figure 6.3, kinetic friction is always

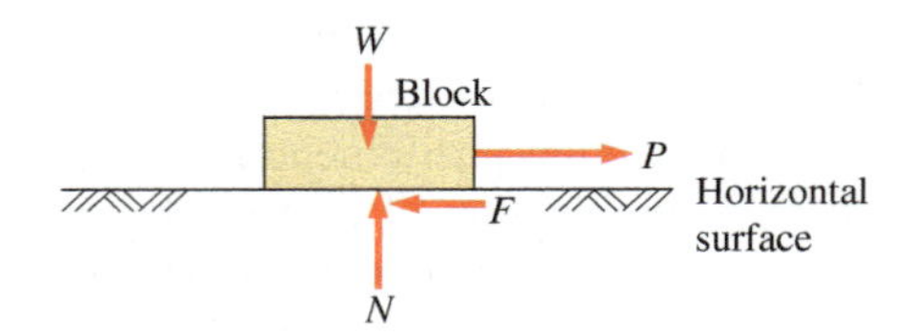

FIGURE 6.2 Free-body diagram.

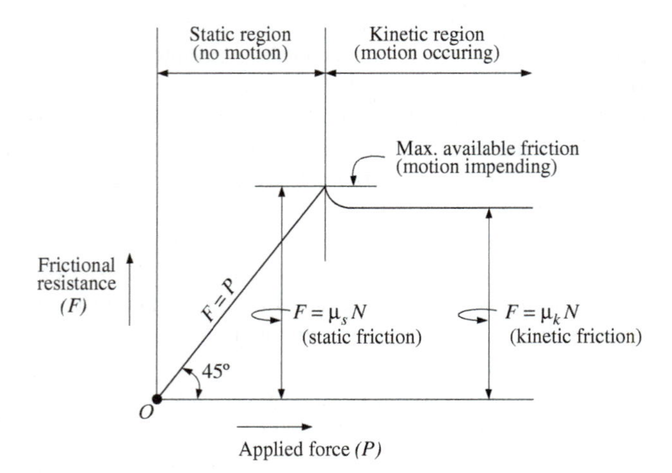

FIGURE 6.3 Frictional resistance.

less than the maximum available friction that occurs when motion is impending.

While the block remains at rest, the frictional resistance must equal the resultant force tending to cause motion. This is valid up to the instant at which the frictional resistance can no longer balance the resultant of the applied forces and motion occurs. Once motion takes place, the frictional resistance drops to a value below that which was in effect when the motion started. Thus, frictional resistance increases only up to a certain point and no further; that is, a limiting value of friction is developed beyond which it cannot increase. The block will not move until the applied horizontal force P is greater than the limiting value of F. In other words, in the "at rest" state of equilibrium, $W = N$ and $F = P$. But when P is greater than the available frictional resistance, the block will move, implying that the maximum value of F has been reached.

The ratio of the maximum frictional resistance F to the normal force N is called the *coefficient of static friction* and is designated μ_s (μ is the Greek lowercase letter mu). This is expressed mathematically as

$$\mu_s = \frac{F}{N} \tag{6.1}$$

where μ_s = the coefficient of static friction (unitless)
F = the maximum frictional resistance (lb, kips) (N)
N = the normal force perpendicular to the contact surface (lb, kips) (N)

Equation (6.1) can be rewritten as

$$F = \mu_s N \tag{6.2}$$

where F is the available maximum frictional resistance (lb, kips) (N) and μ_s and N are as previously defined.

Research indicates that the coefficient of static friction (μ_s) depends primarily on the nature of the contacting surfaces and is independent of the normal force. In other words, in a particular application, as N increases, F will increase proportionally, the factor of proportionality being μ_s. In Equations (6.1) and (6.2), note that the surface contact area does not enter into the problem, which is true for a wide range of values and conditions. In some applications, such as in brake and tire design, however, the contact area will affect the coefficient of friction. In these cases, the small contact area will not permit the heat developed during operation to be adequately dissipated. The increase in temperature will result in a lower coefficient of friction and frictional resistance.

Kinetic frictional resistance may also be determined by Equation (6.2), using the coefficient of kinetic friction μ_k in place of the coefficient for static friction μ_s. It is usually assumed that kinetic friction is constant in value. As shown in Figure 6.3, however, when a body in motion is brought to rest, the kinetic friction gradually increases at very low speeds up to the maximum value (static friction) as the body finally comes to rest.

Table 6.1 indicates a range of approximate values of coefficients of static and kinetic friction for combinations of

TABLE 6.1 Typical values of coefficients of friction for dry surfaces

Materials	Static μ_s	Kinetic μ_k
Steel on steel	0.4–0.7	0.3–0.5
Steel on brass	0.3–0.6	0.2–0.4
Aluminum on steel	0.4–0.6	0.3–0.5
Wood on steel	0.3–0.5	0.2–0.4
Teflon on steel	0.03–0.05	0.03–0.05
Wood on wood	0.3–0.6	0.2–0.5

some common materials. These values may vary appreciably due to contact surface conditions or extremes of temperature or velocity.

6.3 ANGLE OF FRICTION

We have discussed the general nature of friction problems with the illustration of a block resting on a supporting surface. At times it is convenient to represent the reaction of the supporting surface on the block as a single force, R, instead of two forces, F and N. It is apparent in Figure 6.4 that F and N are really components of the total reaction R. The angle between R and N depends on the value of the frictional resistance F. If F is zero, the angle will be zero. As F increases, so does the angle. When the maximum frictional resistance is acting, the value of the angle is termed the *angle of static friction* and is denoted ϕ_s (Greek lowercase phi). The maximum value of ϕ_s develops *only* when motion is impending.

Recalling that $\mu_s = F/N$, the angle of friction (in Figure 6.4) can be defined by

$$\tan \phi_s = \frac{F}{N} = \mu_s \qquad \textbf{(6.3)}$$

Therefore, the coefficient of friction can be obtained by determining the angle of friction. The angle of friction is easily obtained as follows. Place a block (of weight W) on a plane inclined to the horizontal at an angle θ, as shown in Figure 6.5. Gradually increase the angle of inclination θ from zero to some maximum value at which the block will be on the verge of sliding down the incline. At this point, the angle of inclination θ will be equal to the angle of friction ϕ_s. The angle of inclination θ when motion impends is called the

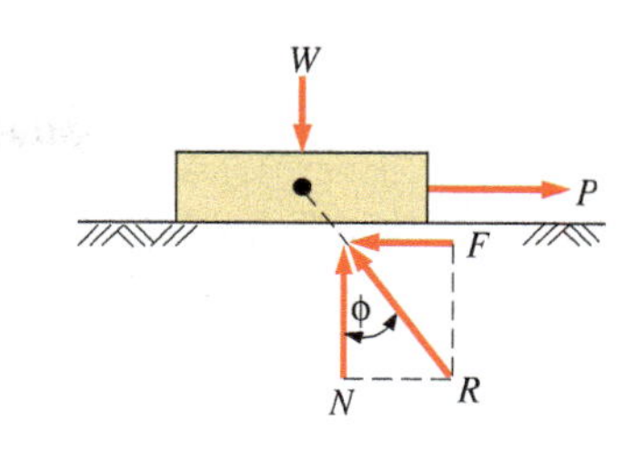

FIGURE 6.4 Angle of friction.

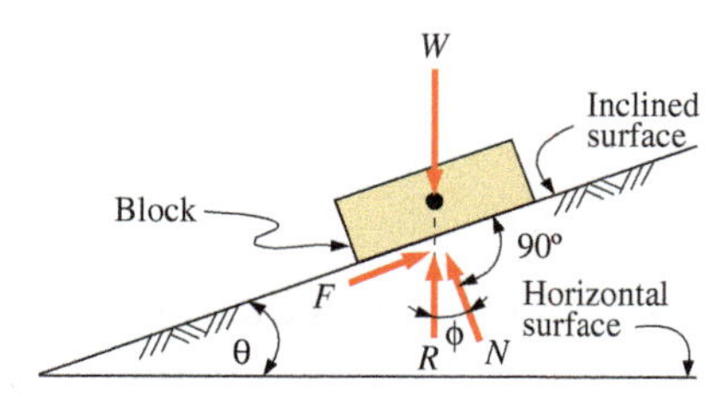

FIGURE 6.5 Block on incline.

angle of repose. Note that for equilibrium to exist, it is necessary that the weight W and the reaction R be collinear.

6.4 FRICTION APPLICATIONS

In this section, we use various introductory examples to provide a methodology that may be applied to friction problems. The general approach we use for "motion impending" problems is as follows:

1. Draw the free-body diagram(s). Show all applicable forces. Decide which way motion is impending and show frictional forces opposing the impending motion.
2. When forces are concurrent, apply the equations of force equilibrium and Equation (6.2):

$$\Sigma F_y = 0 \qquad \Sigma F_x = 0 \qquad F = \mu_s N$$

3. If forces are *not* concurrent, in addition to the three preceding equations, apply the equation of moment equilibrium:

$$\Sigma M = 0$$

Note that if motion is not impending, the preceding steps are still valid, except that the frictional force F will be less than $\mu_s N$. F will be equal and opposite to the resultant of the forces tending to cause motion.

EXAMPLE 6.1 A 400 lb block is supported by a horizontal floor. The coefficient of static friction μ_s between the block and the floor is assumed to be 0.40. Calculate the force P required to cause motion to impend. The force is applied to the block (a) horizontally and (b) downward at an angle of 30° with the horizontal.

Solution (a) The block is assumed to be in static equilibrium, and the free-body diagram is shown in Figure 6.6. The forces acting on the block are its own weight W of 400 lb acting vertically downward, unknown force P acting horizontally, and the reaction of the floor that consists of the normal component N acting vertically upward and the frictional resistance component F acting to the left to oppose sliding of the block. Note that since the forces W, P, and the reaction of the floor R must be concurrent, the normal component N of the floor reaction will not be collinear with the W force. Actually, the exact position of N is immaterial, since the solution is affected only by the horizontal and vertical force equilibrium conditions. This situation applies to subsequent problems as well in which the body under consideration is a block or similar object, the dimensions of which are not specified.

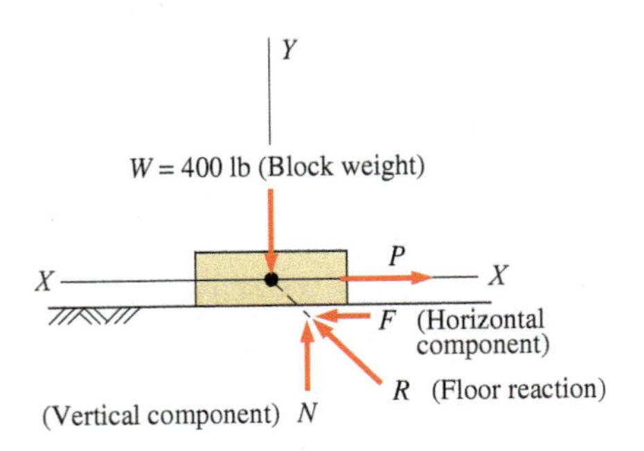

FIGURE 6.6 Free-body diagram.

There are three unknowns in this problem: N, F, and P. Since μ_s provides a relationship between F and N, the three unknowns can be determined using the two equations of force equilibrium in addition to Equation (6.2). Using the free-body diagram of Figure 6.6, which shows all the forces and reference axes, first compute N by summing vertical forces:

$$\Sigma F_y = +N - 400 \text{ lb} = 0$$

from which

$$N = 400 \text{ lb}$$

The available maximum frictional force is calculated from Equation (6.2):

$$F = \mu_s N = 0.40(400 \text{ lb}) = 160 \text{ lb}$$

Last, compute the horizontal force P that will cause motion to impend by summing horizontal forces. A positive sign $(+)$ will be used for the direction of impending motion. Thus,

$$\Sigma F_x = +P - F = 0$$

from which

$$P = 160 \text{ lb}$$

(b) This part is similar to part (a), except that the force P is applied downward at an angle. The free-body diagram is shown in Figure 6.7, and the block is assumed to be in static equilibrium.

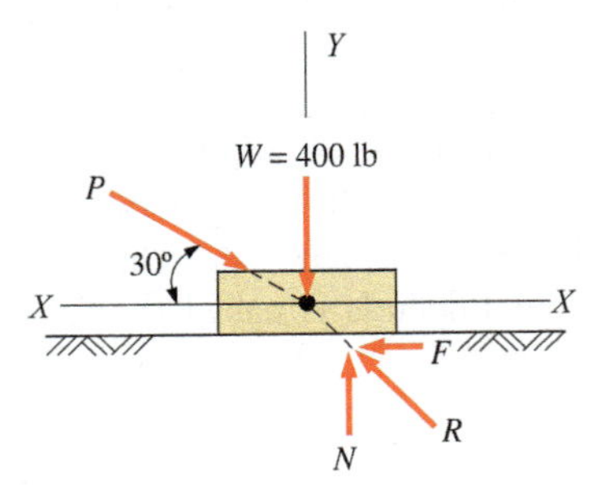

FIGURE 6.7 Free-body diagram.

As in part (a), the three unknown forces N, F, and P can be calculated using the two equations of force equilibrium and Equation (6.2). Note, however, that P now has a vertical component. First, sum vertical forces (upward-acting forces are positive):

$$\begin{aligned}\Sigma F_y &= +N - 400 \text{ lb} - P\sin 30^\circ = 0\\ &= +N - 400 \text{ lb} - 0.5P = 0\end{aligned}$$

from which

$$N = 0.5P + 400 \text{ lb}$$

Similarly, sum horizontal forces:

$$\Sigma F_x = +P\cos 30° - F = 0$$

from which

$$F = 0.866P$$

Next, compute P using Equation (6.2), substituting for F and N:

$$\begin{aligned} F &= \mu_s N \\ 0.866P &= 0.40(0.5P + 400 \text{ lb}) \\ 0.866P &= 0.20P + 160 \text{ lb} \end{aligned}$$

from which

$$P = 240 \text{ lb}$$

EXAMPLE 6.2 A body having a mass of 140 kg is supported on an inclined plane that forms an angle of 30° with the horizontal. As shown in Figure 6.8, a force P parallel to and acting up the plane is applied to the body. The coefficient of static friction is 0.40. (a) Find the value of P to cause motion to impend up the plane. (b) Find the value of P to cause motion to impend down the plane. (c) If $P = 450$ N, determine the magnitude and direction of the resisting friction force.

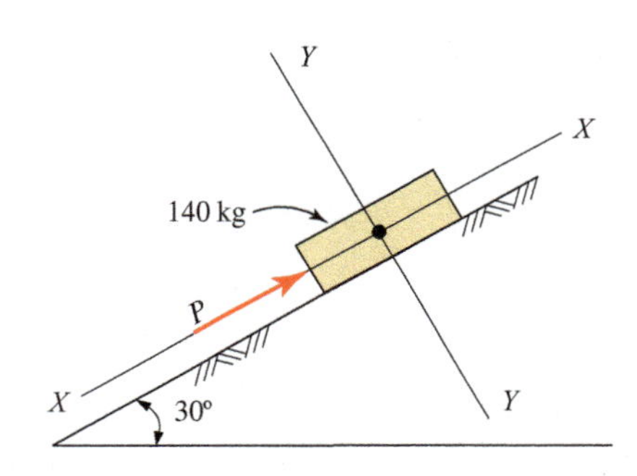

FIGURE 6.8 Body on inclined plane.

Solution First calculate the weight of the block (force due to gravity):

$$W = mg = 140 \text{ kg } (9.81 \text{ m/s}^2) = 1373 \text{ N}$$

(a) The body is assumed to be in static equilibrium, and the free-body diagram is shown in Figure 6.9a. Note that since the body has motion impending up the plane, the maximum friction force F will act down the plane to resist motion. Reference axes are selected so that the X axis is parallel to the inclined plane and forces are assumed positive in the direction of impending motion.

The three unknown forces N, F, and P can be calculated using Equation (6.2) along with the two equations of equilibrium for concurrent forces. Use the free-body diagram of Figure 6.9a and sum forces perpendicular to the plane:

$$\begin{aligned} \Sigma F_y &= +N - W\cos 30° = 0 \\ &= +N - (1373 \text{ N})(0.866) = 0 \end{aligned}$$

from which

$$N = 1189 \text{ N}$$

The available maximum friction force is calculated using Equation (6.2):

$$F = \mu_s N = 0.40(1189 \text{ N}) = 476 \text{ N}$$

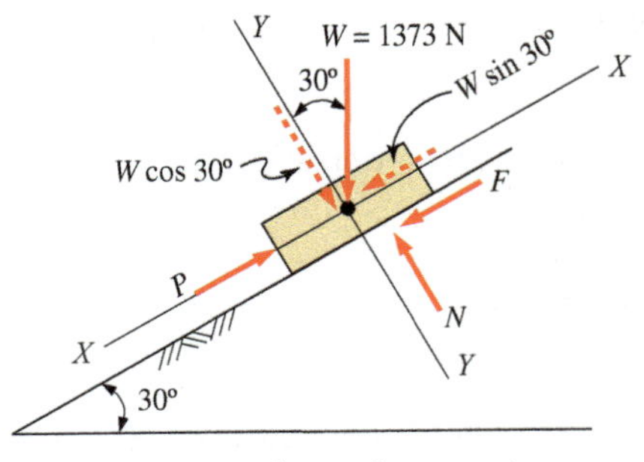

(a) Impending motion, upward

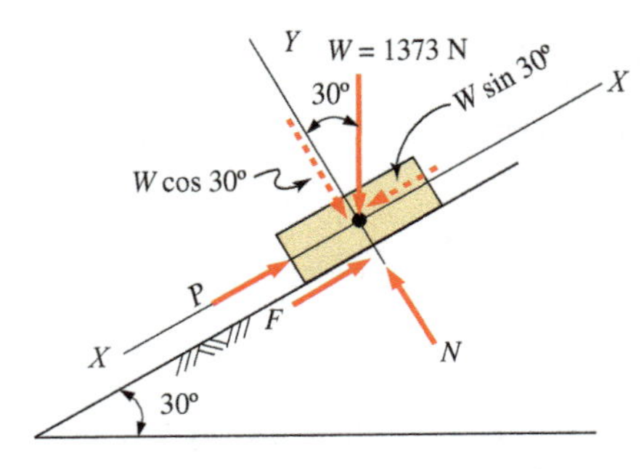

(b) Impending motion, downward

FIGURE 6.9 Free-body diagrams.

Last, the force P is calculated by summing forces parallel to the plane:

$$\Sigma F_x = +P - F - W \sin 30° = 0$$

from which

$$P = 476 + (1373 \text{ N})(0.5) = 1163 \text{ N}$$

(b) In this part, the body rests on the plane in static equilibrium, but motion is impending down the plane. The free-body diagram is shown in Figure 6.9b. Note that the maximum friction force F is acting up the plane to oppose the impending motion. The magnitudes of N and F are calculated in exactly the same way as in part (a) and have the same values as in part (a). You can then compute the force P that will prevent motion down the plane by summing forces parallel to the plane. (*Note*: The positive direction is taken in the direction of impending motion.) Thus,

$$\Sigma F_x = -P - F + W \sin 30° = 0$$
$$= -P - 476 \text{ N} + (1373 \text{ N})(0.5) = 0$$

from which

$$P = 211 \text{ N}$$

(c) Based on the results of parts (a) and (b), the body will remain at rest for all values of P between 211 N and 1163 N.

When P is equal to 450 N, acting up the plane, the body will tend to slide down the plane since $450 \text{ N} < 1373 \sin 30° \text{ N}$. Therefore, the friction force F will act up the plane, opposing the tendency of the block to move. The friction force will be less than the maximum available friction, as will be shown by summing the forces in the X direction. Use the free-body diagram of Figure 6.9b:

$$\Sigma F_x = -P - F + (1373 \text{ N}) \sin 30° = 0$$
$$= -450 \text{ N} - F + (1373 \text{ N})(0.5) = 0$$

from which

$$F = 237\text{ N}$$

This value represents the frictional resistance necessary for the body to be in static equilibrium when subjected to a force P of 450 N acting upward along the inclined plane.

Note that if $P > (1373\text{ N})\sin 30°$, the body would tend to slide up the plane and the friction force F would then act down the plane.

EXAMPLE 6.3 A body weighing 200 lb is supported on an inclined plane forming an angle of 40° with the horizontal. A horizontal force P is applied to the body, as shown in Figure 6.10. If the coefficient of static friction between the body and the plane is 0.25, (a) calculate the value of P required to prevent the body from sliding down the plane and (b) calculate the value of P required to cause motion to impend up the plane.

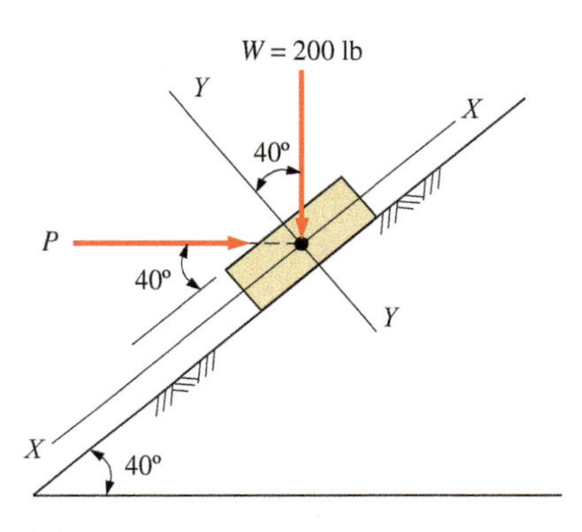

FIGURE 6.10 Body on inclined plane.

Solution (a) The value of P required to keep the body from sliding down the plane will be the smallest value required to maintain an "at rest" condition. Using the free-body diagram of Figure 6.11a, first calculate N in terms of P by summing forces perpendicular to the plane:

$$\begin{aligned}\Sigma F_y &= +N - W\cos 40° - P\sin 40° = 0\\ &= +N - (200\text{ lb})(0.766) - P(0.643) = 0\end{aligned}$$

from which

$$N = 153.2\text{ lb} + 0.643P$$

Next, compute F in terms of P by summing forces parallel to the plane. Recall that forces are positive (+) when acting in the direction of impending motion that, in this case, is down the plane. The friction force F is acting up the plane, opposing the motion. Thus,

$$\begin{aligned}\Sigma F_x &= -F - P\cos 40° + W\sin 40° = 0\\ &= -F - 0.766P + (200\text{ lb})(0.643) = 0\end{aligned}$$

from which

$$F = 128.6\text{ lb} - 0.766P$$

Last, compute the horizontal force P using Equation (6.2) and substituting for F and N:

$$\begin{aligned}F &= \mu_s N\\ 128.6\text{ lb} - 0.766P &= 0.25(153.2\text{ lb} + 0.643P)\\ 128.6\text{ lb} - 0.766P &= 38.3\text{ lb} + 0.161P\\ 0.927P &= 90.3\text{ lb}\end{aligned}$$

from which

$$P = 97.4\text{ lb}$$

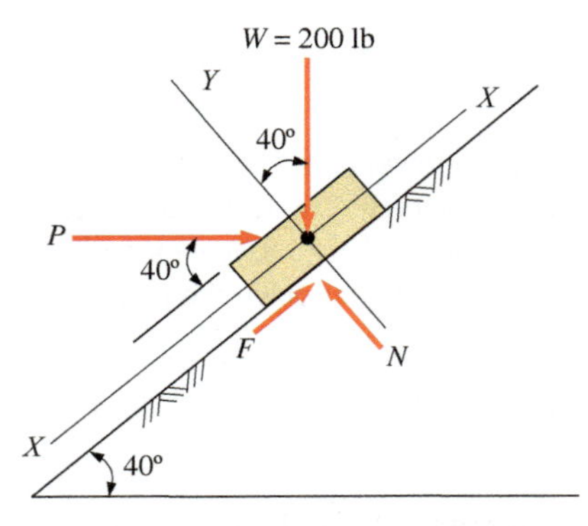

(a) Motion impending—downward

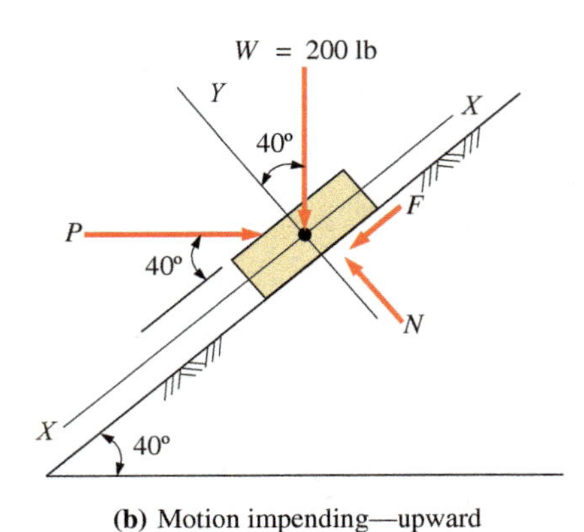

(b) Motion impending—upward

FIGURE 6.11 Free-body diagrams.

(b) Using the free-body diagram of Figure 6.11b, calculate the force P required to cause motion to impend up the plane. N is calculated exactly as in part (a). F is calculated by summing forces parallel to the plane. Note that since impending motion is up the plane, the friction force will act down the plane, opposing the motion. Thus,

$$\Sigma F_x = -F + P\cos 40^\circ - W\sin 40^\circ = 0$$
$$= -F + P(0.766) - (200\ \text{lb})(0.643) = 0$$

from which

$$F = -128.6\ \text{lb} + 0.766P$$

Last, compute the horizontal force P using Equation (6.2) and substituting for F and N:

$$F = \mu_s N$$
$$-128.6\ \text{lb} + 0.766P = 0.25(153.2\ \text{lb} + 0.643P)$$
$$-128.6\ \text{lb} + 0.766P = 38.3\ \text{lb} + 0.161P$$
$$0.605P = 166.9\ \text{lb}$$

from which

$$P = 276\ \text{lb}$$

EXAMPLE 6.4 A 250 lb block rests on a horizontal surface and is acted on by an inclined force P, as shown in Figure 6.12. Determine the magnitude of the force P that will cause the block to move. The coefficient of friction between the two contact surfaces is 0.25.

Solution Assuming that the force is gradually applied, motion may occur in two ways: (a) The block may tend to slide to the left or (b) it may tend to tip about point O.

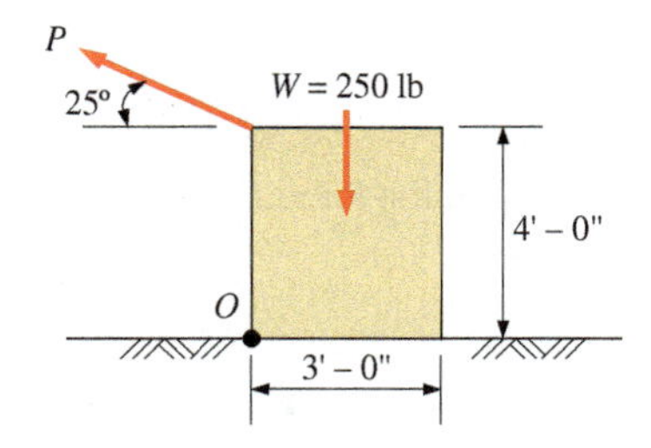

FIGURE 6.12 Block on horizontal plane.

(a) Initially compute the force P required to cause sliding (this neglects tipping). Figure 6.13a shows the free-body diagram for the sliding consideration.

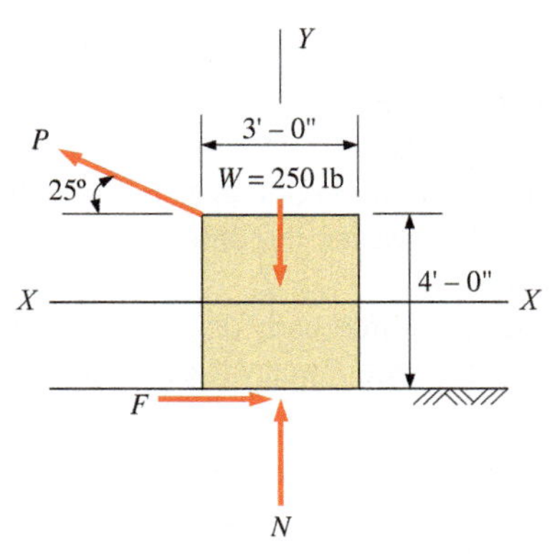

(a) Free-body diagram—sliding

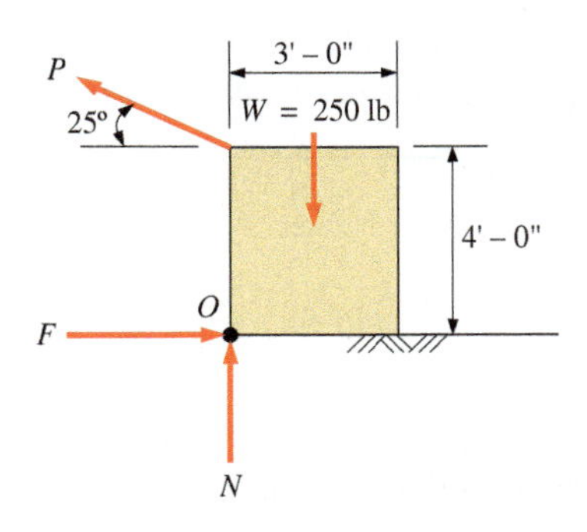

(b) Free-body diagram—tipping

FIGURE 6.13 Free-body diagrams.

The three unknown forces N, F, and P can be determined using Equation (6.2) and the two equations of force equilibrium. First, solve for N in terms of P by summing forces in the vertical direction:

$$\Sigma F_y = +N - W + P\sin 25° = 0$$
$$= +N - 250\text{ lb} + 0.423P = 0$$

from which

$$N = 250\text{ lb} - 0.423P$$

Next, compute F in terms of P by summing forces in the horizontal direction. Note that impending motion is to the left; therefore, the friction force F acts to the right, opposing motion. Then,

$$\Sigma F_x = -F + P\cos 25° = 0$$

from which

$$F = 0.906P$$

Finally, compute the force P using Equation (6.2) and substituting for F and N:

$$F = \mu_s N$$
$$0.906P = 0.25(250 \text{ lb} - 0.423P)$$
$$0.906P = 62.5 \text{ lb} - 0.106P$$
$$1.012P = 62.5 \text{ lb}$$

from which

$$P = 61.8 \text{ lb}$$

(b) Now compute the force P that will cause the block to tip about point O (this neglects sliding). Figure 6.13b shows the free-body diagram for the tipping consideration.

Since the block is assumed to tip about point O, the last contact point prior to motion will be point O, and the reaction of the supporting surface will pass through point O. Therefore, the normal force N and the friction force F (components of the supporting surface reaction) will pass through point O, as shown in Figure 6.13b.

There are three unknown forces in this problem. If a summation of moments is taken with respect to point O, however, the effects of forces F and N will be eliminated and the unknown force P can be calculated.

Using the free-body diagram of Figure 6.13b and summing moments about point O, calculate the value of the force P that will cause the block to tip:

$$\Sigma M_o = +P\cos 25°(4 \text{ ft}) - (250 \text{ lb})\left(\frac{3 \text{ ft}}{2}\right) = 0$$

from which

$$P = 103.5 \text{ lb}$$

Since a force of 61.8 lb will cause the block to slide, it is evident that sliding will occur first as P increases from zero to a maximum value.

EXAMPLE 6.5 The ladder shown in Figure 6.14a is supported by a horizontal floor and a vertical wall. It is 16 ft long and weighs 48 lb. The weight may be considered to be concentrated at its midlength. The ladder must support a person weighing 200 lb at point D. The coefficient of static friction between the ladder and the wall is 0.25; between the ladder and the floor it is 0.40. The base of the ladder has been moved to the left and the ladder is on the verge of slipping. Calculate (a) the horizontal and vertical reactions at points A and B and (b) the angle θ.

Solution (a) The horizontal and vertical reactions at points A and B can be computed using the free-body diagram of Figure 6.14b along with Equation (6.2) and the two force equations of equilibrium. The horizontal reaction at A designated F_A and the vertical reaction at B designated F_B represent the frictional resistance at the respective points opposing any movement of the ladder.

Using Equation (6.2), F_A and F_B can be expressed as follows:

$$F_A = \mu_s N_A = 0.40 N_A$$
$$F_B = \mu_s N_B = 0.25 N_B$$

Using the X–Y coordinate system shown and applying the two force equations of equilibrium, the reactions at A and B can be calculated. Note that motion impends to the left at point A and down at point B. As shown in Figure 6.14b, the direction of the friction forces F_A and F_B oppose the direction of motion. Forces will be assumed positive if acting in the direction of impending motion.

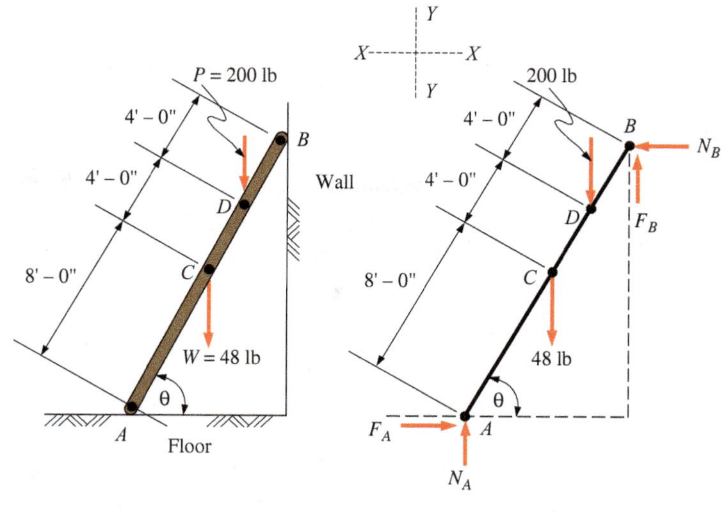

FIGURE 6.14 Ladder supported by wall and floor.

Sum forces in the X direction and substitute for F_A:

$$\Sigma F_x = +N_B - F_A = +N_B - 0.40N_A = 0$$

from which

$$N_B = 0.40N_A$$

Next, sum forces in the Y direction and substitute for F_B:

$$\begin{aligned}\Sigma F_y &= +P + W - N_A - F_B = 0\\ &= +200\text{ lb} + 48\text{ lb} - N_A - 0.25N_B = 0\end{aligned}$$

from which

$$N_A = 248\text{ lb} - 0.25N_B$$

Substitute for N_B in the preceding expression:

$$N_A = 248\text{ lb} - 0.25(0.40N_A)$$

from which

$$N_A = 225\text{ lb}$$

Having determined N_A, all other reactions can now be calculated:

$$\begin{aligned}N_B &= 0.40N_A = 0.40(225\text{ lb}) = 90.0\text{ lb}\\ F_A &= 0.40N_A = 90.0\text{ lb}\\ F_B &= 0.25N_B = 0.25(90.0\text{ lb}) = 22.5\text{ lb}\end{aligned}$$

(b) The angle θ at which motion is impending to the left can be computed using the third equation of equilibrium ($\Sigma M = 0$) with respect to point A. Use a sign convention of positive for counterclockwise moments and negative for clockwise moments:

$$\begin{aligned}\Sigma M_A &= -W(8\cos\theta) - P(12\cos\theta) + F_B(16\cos\theta) + N_B(16\sin\theta) = 0\\ &= -(48\text{ lb})(8\text{ ft})\cos\theta - (200\text{ lb})(12\text{ ft})\cos\theta + (22.5\text{ lb})(16\text{ ft})\cos\theta\\ &\quad + (90.0\text{ lb})(16\text{ ft})\sin\theta = 0\\ &= (384\text{ lb-ft})\cos\theta - (2400\text{ lb-ft})\cos\theta + (360\text{ lb-ft})\cos\theta\\ &\quad + (1440\text{ lb-ft})\sin\theta = 0\end{aligned}$$

$$(1440 \text{ lb-ft})\sin \theta = (1656 \text{ lb-ft})\cos \theta$$

$$\frac{\sin \theta}{\cos \theta} = \tan \theta = \frac{1656 \text{ lb-ft}}{1440 \text{ lb-ft}} = 1.150$$

from which

$$\theta = 49.0°$$

The ladder is on the verge of slipping at 49.0°. If the ladder is placed at an angle greater than 49.0°, it will not slip. The horizontal reaction N_B will be smaller, the vertical reaction N_A will be larger, and F_A will be less than its limiting value of $\mu_s N_A$.

EXAMPLE 6.6 A sliding block system is shown in Figure 6.15. Compute the horizontal force P necessary to cause motion of the 100 lb block to impend to the left. The coefficient of friction for the contact surfaces is 0.25; the pulley is assumed to be frictionless.

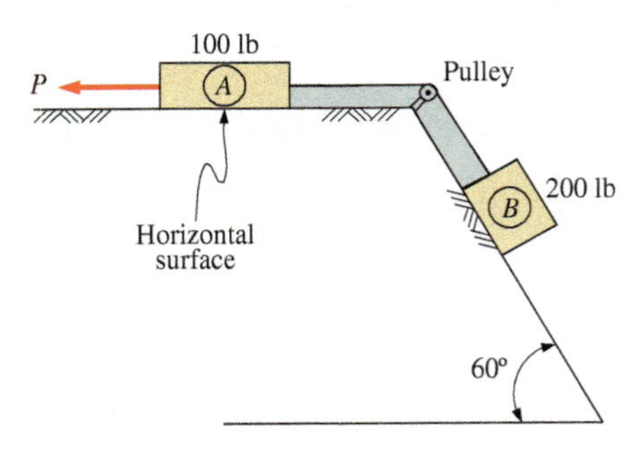

FIGURE 6.15 Blocks on two planes.

Solution Free-body diagrams of both blocks are shown in Figure 6.16. The free-body diagram of block A (Figure 6.16a) shows the unknown force P, the weight W, the tension T in the rope, the reaction N normal to the supporting surface, and the frictional resistance F that acts to the right (since the block tends to move to the left). Similarly, the forces acting on block B are shown in the free-body diagram of Figure 6.16b. Since block B tends to move up the plane, the frictional resistance F acts down the plane. The other forces shown are the tension T in the rope, the weight W acting vertically, and the reaction N, normal to the supporting surface. Reference axes X and Y are selected as shown.

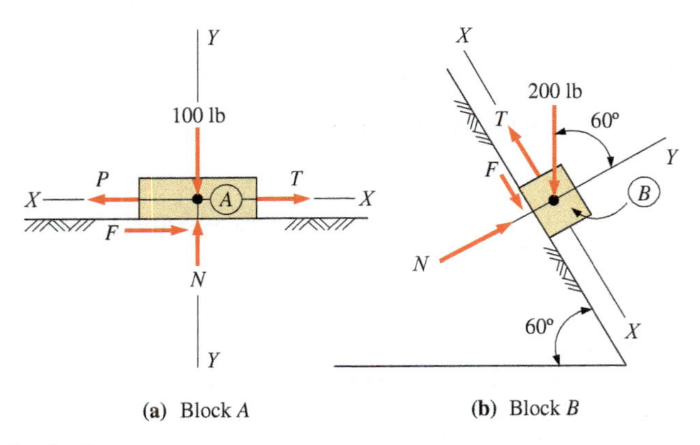

FIGURE 6.16 Free-body diagrams.

Considering block B and applying Equation (6.2) along with the two force equations of equilibrium, the forces N, F, and T can be calculated:

$$\Sigma F_y = +N - W\cos 60° = +N - (200 \text{ lb})(0.5) = 0$$

from which

$$N = 100 \text{ lb}$$

From Equation (6.2),

$$F = \mu_s N = 0.25(100 \text{ lb}) = 25 \text{ lb}$$

Next, compute T by summing forces parallel to the plane. Assume forces positive in the direction of impending motion, which is up the plane. Thus,

$$\Sigma F_x = +T - F - W\sin 60° = 0$$
$$= +T - 25 \text{ lb} - (200 \text{ lb})(0.866) = 0$$

from which

$$T = 198.2 \text{ lb}$$

The free-body diagram of block A (Figure 6.16a) may now be considered. Equation (6.2) and the two equations of force equilibrium will provide the solutions for forces N, F, and P. Then,

$$\Sigma F_y = +N - 100 \text{ lb} = 0$$

from which

$$N = 100 \text{ lb}$$

Next, compute the maximum friction force F acting on the horizontal surface. Equation (6.2) yields

$$F = \mu_s N = 0.25(100 \text{ lb}) = 25 \text{ lb}$$

Finally, compute P by summing forces in the X direction. Assume forces positive in the direction of impending motion, which is to the left:

$$\Sigma F_x = +P - T - F = 0$$
$$= +P - 198.2 \text{ lb} - 25 \text{ lb} = 0$$

from which

$$P = 223 \text{ lb} \leftarrow$$

6.5 WEDGES

A typical application of the wedge is shown in Figure 6.17. Its use makes possible the overcoming (lifting, in this case) of a large load W by means of a relatively small applied force P.

In order that P may start the wedge inward and thus move the load upward, the frictional resistance along planes A–A, B–B, and C–C must be overcome. To determine the necessary value of P, the block and the wedge must be considered separately as bodies in equilibrium with motion impending. The forces can then be determined using a method similar to that used previously in this chapter.

Free-body diagrams for the block and wedge are shown in Figure 6.18. Note that the N_2 and F_2 forces are common to both free bodies.

Example 6.7 demonstrates an approach for this type of problem.

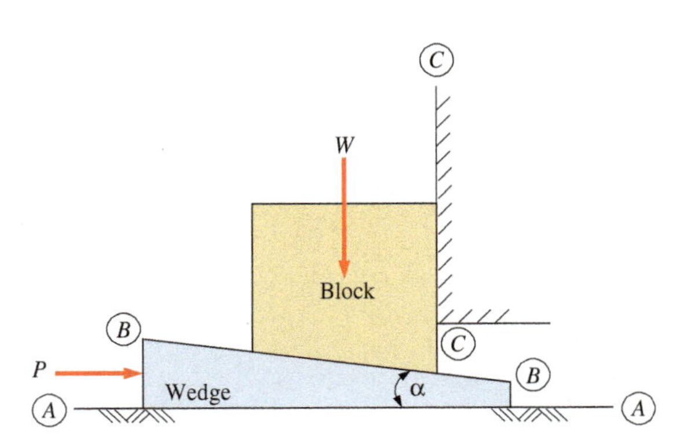

FIGURE 6.17 The wedge.

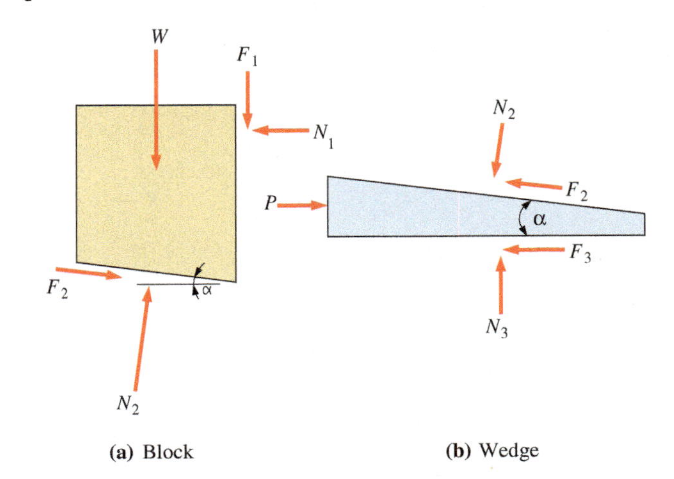

FIGURE 6.18 Free-body diagrams.

EXAMPLE 6.7 The block shown in Figure 6.19 supports a load W of 700 lb and is to be raised by forcing a wedge under it. The coefficient of friction on each of the three contact surfaces is 0.25. Calculate the force P required to start the wedge in motion under the block. Neglect the weights of the block and the wedge.

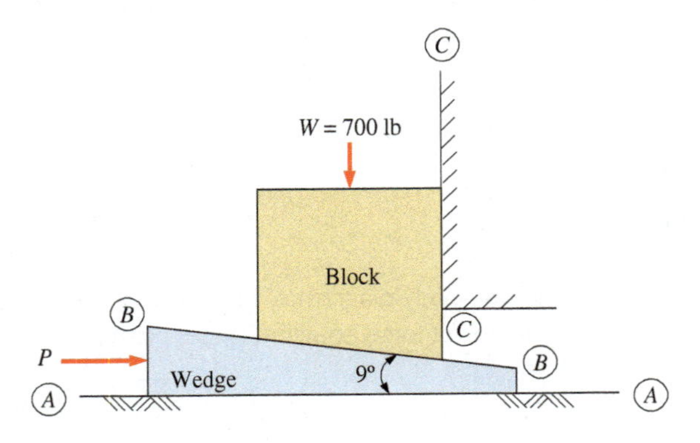

FIGURE 6.19 Block-and-wedge arrangement.

Solution The free-body diagrams for the block and the wedge are shown in Figure 6.20. For convenience, the X and Y components of the sloping forces are shown in each case.

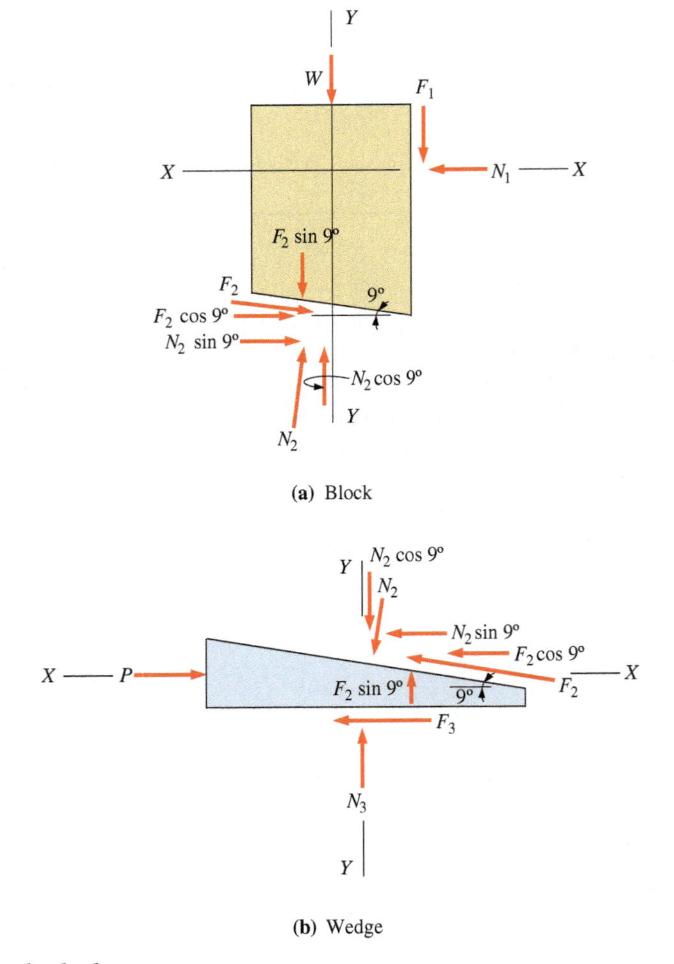

FIGURE 6.20 Free-body diagrams.

(a) Initially, consider the block of Figure 6.20a and calculate F_2 and N_2 using Equation (6.2) and the two equations of force equilibrium. Sum the forces in the Y direction, assuming forces to be positive when acting in the direction of impending motion (which will be upward):

$$\Sigma F_y = -W - F_1 - F_2 \sin 9° + N_2 \cos 9° = 0$$
$$= -700 \text{ lb} - F_1 - F_2(0.156) + N_2(0.988) = 0 \quad \textbf{(1)}$$

Sum forces in the X direction:

$$\Sigma F_x = -N_1 + F_2 \cos 9° + N_2 \sin 9° = 0$$
$$= -N_1 + F_2(0.988) + N_2(0.156) = 0 \quad \textbf{(2)}$$

From Equation (6.2), obtain F_1 and F_2:

$$F = \mu_s N$$
$$F_1 = \mu_s N_1 = 0.25N_1 \quad \textbf{(3)}$$
$$F_2 = \mu_s N_2 = 0.25N_2 \quad \textbf{(4)}$$

Substitute Equations 3 and 4 into Equation 1:

$$-700 \text{ lb} - 0.25N_1 - 0.25N_2(0.156) + N_2(0.988) = 0$$
$$-700 \text{ lb} - 0.25N_1 - 0.0390N_2 + 0.988N_2 = 0$$
$$-700 \text{ lb} - 0.25N_1 + 0.949N_2 = 0 \quad \textbf{(5)}$$

Substitute Equations 3 and 4 into Equation 2:

$$-N_1 + 0.25N_2(0.988) + N_2(0.156) = 0$$
$$-N_1 + 0.247N_2 + 0.156N_2 = 0$$
$$-N_1 + 0.403N_2 = 0$$
$$N_1 = 0.403N_2 \quad \textbf{(6)}$$

Substitute Equation 6 into Equation 5 and solve for N_2:

$$-700 \text{ lb} - 0.25(0.403N_2) + 0.949N_2 = 0$$
$$-700 \text{ lb} - 0.1008N_2 + 0.949N_2 = 0$$
$$-700 \text{ lb} + 0.848N_2 = 0$$
$$N_2 = 825 \text{ lb}$$

Finally, Equation 4 yields

$$F_2 = \mu_s N_2 = 0.25(825 \text{ lb}) = 206 \text{ lb}$$

(b) Having computed F_2 and N_2, now consider the free body of the wedge, as shown in Figure 6.20b. Again apply Equation (6.2) and the two equations of force equilibrium. Sum forces in the Y direction:

$$\Sigma F_y = +N_3 - N_2 \cos 9° + F_2 \sin 9° = 0$$
$$= +N_3 - (825 \text{ lb})(0.988) + (206 \text{ lb})(0.156) = 0$$

from which

$$N_3 = 783 \text{ lb}$$

Equation (6.2) yields

$$F_3 = \mu_s N_3 = 0.25(783 \text{ lb}) = 196 \text{ lb}$$

Sum forces in the X direction:

$$\Sigma F_x = +P - F_3 - N_2 \sin 9° - F_2 \cos 9° = 0$$
$$= +P - 196 \text{ lb} - (825 \text{ lb})(0.156) - (206 \text{ lb})(0.988) = 0$$

from which

$$P = 528 \text{ lb}$$

6.6 BELT FRICTION

When a flexible belt, rope, or band is wrapped around a circular pulley or a cylindrically shaped drum and then subjected to a tensile force, a frictional resistance is developed between the contact surfaces. This friction can be used either to transmit power from one shaft to another (for example, in belt-driven machinery) or to retard motion (for example, in band brakes or the tying up of a ship at a pier).

Figure 6.21a shows a flexible belt wrapped around a fixed drum with an arc length of contact subtending an angle β (Greek lowercase beta). This angle is commonly called the *angle of wrap*. The coefficient of friction between the belt and the drum is designated μ.

A free-body diagram of the segment of belt in contact with the drum (segment *AB*) is shown in Figure 6.21b. The drum is stationary (fixed against rotation), and the belt is on the verge of slipping (motion is impending). As shown, the forces acting on this piece of belt are the belt tensions T_L and T_S (with the subscripts *L* and *S* indicating large and small belt tensions, respectively); the distributed normal forces *N* (which are really radial, normal-acting to a tangent line at any given point); and tangential friction forces *F*, which oppose the direction of the impending motion. Since this free body must be in equilibrium, the sum of the moments of the previously mentioned forces about the center of the drum *O* must equal zero. The moment of T_L must equal the moment of T_S plus the moment of the friction forces *F* when motion is impending. Note that the friction forces are acting to retard motion of the belt around the fixed drum.

To establish a relationship between the belt tensions and the friction between the belt and the drum, the previously mentioned forces are resolved into *X* and *Y* components and equations of equilibrium applied. The derivation of the resulting mathematical expression is beyond the scope of this text. The expression itself, however, is not complex:

$$T_L = T_S e^{\mu\beta} \tag{6.4}$$

where T_L = the larger belt tension (lb) (N)
T_S = the smaller belt tension (lb) (N)
μ = the coefficient of friction between the belt and the drum (since motion impends, this would be the coefficient of static friction μ_s)
β = the angle of contact between the belt and the drum (radians) (recall that one radian $= 180°/\pi = 57.3°$)
e = the base of natural logarithms (2.718). The term $e^{\mu\beta}$ represents an exponential power of the number 2.718.

Equation (6.4) may also be expressed in alternative forms:

$$\ln \frac{T_L}{T_S} = \mu\beta \tag{6.5}$$

or

$$\ln T_L - \ln T_S = \mu\beta \tag{6.6}$$

where ln indicates the natural logarithm (base 2.718) and the other terms are as previously defined.

Belts are used extensively in transmitting power from one shaft to another. In Figure 6.22 we see a belt that passes around two pulleys that are, in turn, mounted on shafts. Pulley B is the driving pulley; therefore, the lower belt has more tension in it than the upper belt. If friction did not exist between the belt and the pulleys, the driving pulley could not drive the belt, nor could the driven pulley be turned by the belt. Note that both sides of the belt are in tension. The tension T_L on the tight side is greater than the tension T_S on the loose or slack side, thus resulting in a net driving force on the pulleys equal to

$$T_{\text{net}} = T_L - T_S$$

For pulley-and-shaft applications, the term *torque* is commonly used. Torque is a twisting action applied in a plane perpendicular to the longitudinal axis of a shaft. The torque that a belt can transmit can be computed by a moment

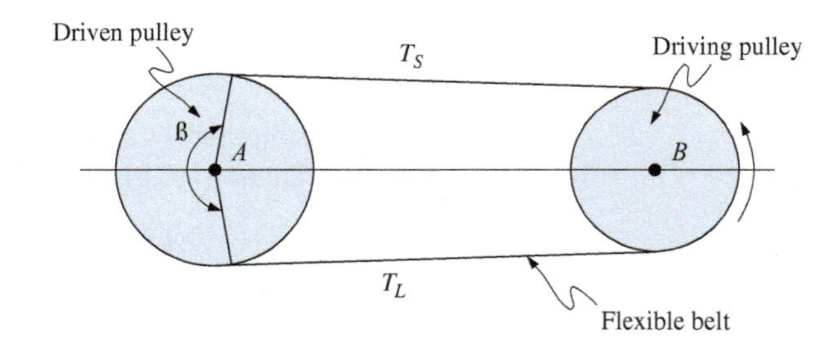

FIGURE 6.22 Open belt drive.

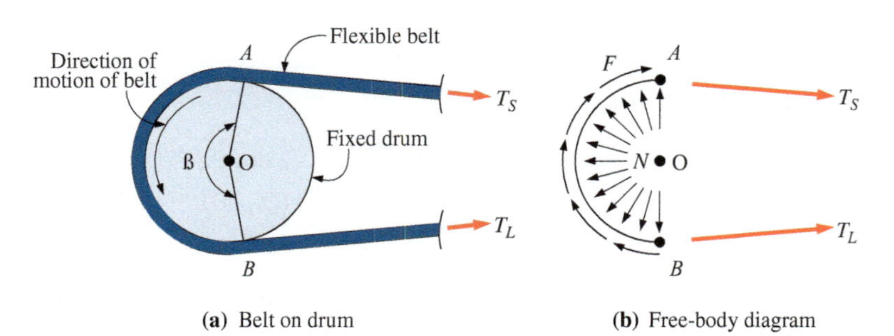

FIGURE 6.21 Flexible belt on fixed drum.

summation about the center of the pulley of the two forces T_L and T_S:

$$\text{torque} = T_L r - T_S r = (T_L - T_S) r = T_{\text{net}} r$$

where r represents the radius of the pulley.

Note that a free-body diagram of pulley A in Figure 6.22 would be identical to the free-body diagram of Figure 6.21b. Therefore, it can be concluded that the same type of analysis with the same resulting equations as that given for the fixed drum may be applied.

Equations (6.4), (6.5), and (6.6) are not applicable to all the many types of belts available. Our discussion to this point included flat belts or ropes; the most widely used type of belt drive, however, is the V-belt drive. This type makes contact on two sloping sides of a grooved pulley, as shown in Figure 6.23. The belt acts like a wedge and the normal forces on the contact surfaces are increased, thereby increasing the tractive force developed by the belt.

The relationship between the belt tensions and the friction forces is established using an analysis similar to that used for the flat belt and rope. The only additional parameter included in the analysis is the groove angle (2ϕ). The resulting equation, applicable to only the flexible V-belt where motion is impending, is

$$T_L = T_S e^{\mu\beta/\sin\phi} \qquad \textbf{(6.7)}$$

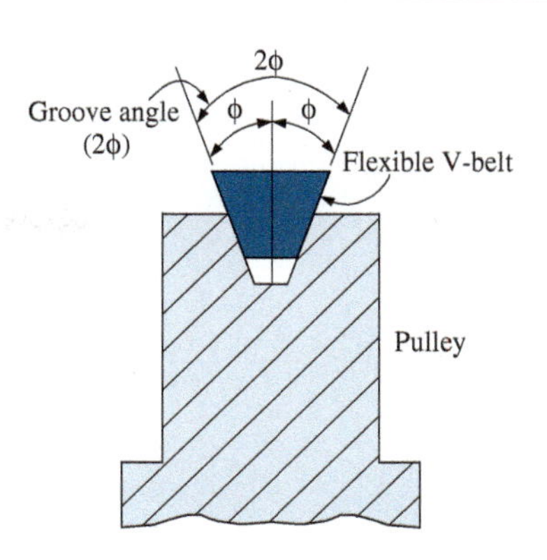

FIGURE 6.23 V-belt on pulley.

where ϕ = one-half the groove angle (see Figure 6.23) and the other terms are as previously defined.

Equation (6.7) may also be expressed in alternative forms:

$$\ln\frac{T_L}{T_S} = \frac{\mu\beta}{\sin\phi} \qquad \textbf{(6.8)}$$

or

$$\ln T_L - \ln T_S = \frac{\mu\beta}{\sin\phi} \qquad \textbf{(6.9)}$$

EXAMPLE 6.8 An object having a mass of 100 kg is suspended from a rope that passes over a rough cylindrical stationary drum, as shown in Figure 6.24. The coefficient of static friction between the rope and the drum is 0.30. Assuming that motion is impending, calculate the force P that must be applied to the free end of the rope to keep the weight from slipping downward. (a) The rope is in contact with half of the cylindrical surface (angle of wrap is 180°). (b) The rope is wrapped around the cylindrical drum $1\frac{1}{2}$ times (angle of wrap is 540°).

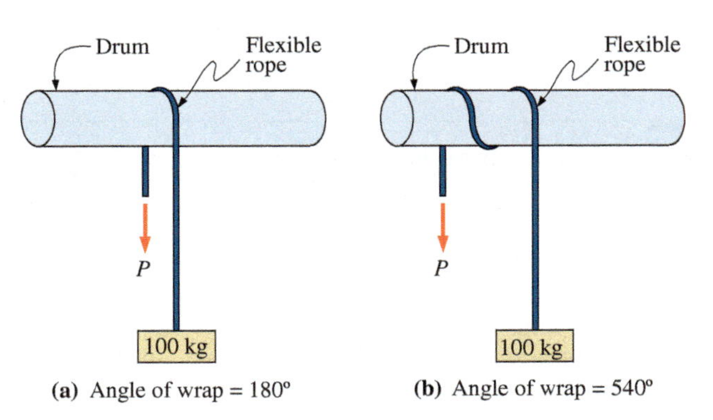

FIGURE 6.24 Rope-and-drum arrangement.

Solution First, calculate the weight of the object (force due to gravity):

$$W = mg = 100\text{ kg}(9.81\text{ m/s}^2) = 981\text{ N}$$

To determine the value of P to keep the object just from falling, consider that with a slight decrease in P, the object will fall. Therefore, the frictional resistance between the drum and the rope will act in a counterclockwise direction, as shown in the free-body diagram of Figure 6.25. This resistance will assist P in holding the object aloft. It is evident, then, that P will be numerically less than 981 N.

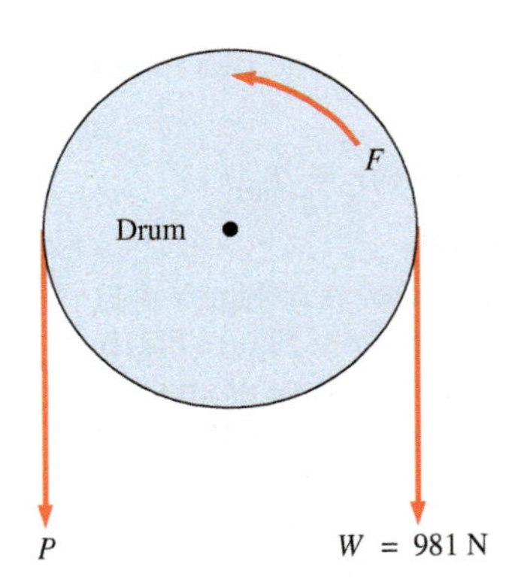

FIGURE 6.25 Free-body diagram.

(a) From Equation (6.4),

$$T_L = T_S e^{\mu\beta}$$
$$981 \text{ N} = Pe^{0.30(\pi)}$$
$$981 \text{ N} = 2.57P$$

from which

$$P = 382 \text{ N}$$

(b) From Equation (6.4),

$$T_L = T_S e^{\mu\beta}$$
$$981 \text{ N} = Pe^{0.30(3\pi)}$$
$$981 \text{ N} = 16.90P$$

from which

$$P = 58.0 \text{ N}$$

EXAMPLE 6.9 Using the basic information given in Example 6.8, calculate the force P required to raise the 100-kg object. Note that the drum is stationary.

Solution To solve for the force P to raise the 100-kg object, it is evident that the impending motion of the object would be upward. Therefore, the frictional resistance between the drum and the rope will act in a clockwise direction to oppose the motion. This frictional resistance would have to be overcome. It is logical, then, that P should be numerically larger than 981 N.

(a) From Equation (6.4),

$$T_L = T_S e^{\mu\beta}$$
$$P = (981 \text{ N})e^{0.30(\pi)} = (981 \text{ N})(2.57) = 2520 \text{ N} = 2.52 \text{ kN}$$

(b) From Equation (6.4),

$$T_L = T_S e^{\mu\beta}$$
$$P = (981 N)e^{0.30(3\pi)} = 16.58 \text{ kN}$$

EXAMPLE 6.10 A flat belt makes contact with a 24-in.-diameter pulley through half its circumference. If the coefficient of static friction is 0.25 and a torque of 4000 lb-in. is to be transmitted, calculate the belt tensions.

Solution Since there are two unknowns, T_L and T_S, two equations must be developed to establish a relationship between the unknowns. From Equation (6.4), substitute π radians for β:

$$T_L = T_S e^{\mu\beta} = T_S e^{0.25(\pi)} = 2.19 T_S$$

The torque to be transmitted is 4000 lb-in. and can be expressed as

$$\text{torque} = (T_L - T_S)(r)$$
$$4000 \text{ lb-in.} = (T_L - T_S)(12 \text{ in.})$$

from which

$$T_L - T_S = 333 \text{ lb}$$

Substitute the expression for T_L into the preceding equation:

$$2.19 T_S - T_S = 333 \text{ lb}$$

from which

$$T_S = 280 \text{ lb}$$

Last, solve for T_L:

$$T_L = 2.19 T_S = 2.19(280 \text{ lb}) = 613 \text{ lb}$$

EXAMPLE 6.11 Using the basic information given in Example 6.10, calculate the belt tensions if the belt used is a V-belt with a groove angle (2ϕ) of 60°.

Solution From Equation (6.7),

$$T_L = T_S e^{\mu\beta/\sin\phi} = T_S e^{0.25(\pi)/\sin 30°} = 4.81 T_S$$

As in Example 6.10,

$$\text{torque} = (T_L - T_S)(r)$$
$$4000 \text{ lb-in.} = (T_L - T_S)(12 \text{ in.})$$

from which

$$T_L - T_S = 333 \text{ lb}$$

Substitute for T_L:

$$4.81 T_S - T_S = 333 \text{ lb}$$

from which

$$T_S = 87.4 \text{ lb}$$

Finally, solve for T_L:

$$T_L = 4.81 T_S = 4.81(87.4 \text{ lb}) = 420 \text{ lb}$$

6.7 SQUARE-THREADED SCREWS

Screws are generally used for fastening purposes. In many types of machines and equipment, however, screws called *power screws* are used to translate rotary motion into uniform longitudinal motion, thereby playing an important part in transmitting force or power from one part of the machine to another. Square-threaded screws, one type of a broad category of power screws, are generally used where large loads (forces) are to be transmitted. The square thread is the most efficient of the power screw threads; it is relatively costly to manufacture, however. Common applications of square-threaded screws include jacks, valves, machine tools, and presses.

A square-threaded screw may be regarded as an inclined plane wrapped around a cylinder. The usual proportions for a square-threaded screw, as well as the concept of the inclined plane, are shown in Figure 6.26. A block moving up (or down) the plane is then equivalent to turning the screw through a set of fixed threads (a nut) or turning a nut on a fixed screw.

A useful mechanical device illustrative of the square-threaded screw is the ordinary jackscrew, shown in Figure 6.27. The jackscrew generally has a square-threaded screw that turns in a threaded hole of a stationary frame or base. As a force Q is applied to the handle (lever) of the jack at a distance a from the point of rotation, the screw turns in its fixed base and moves upward, lifting the axial load W. The thread in the fixed base represents the inclined plane and the applied load W, supported by the screw, is transmitted to this base.

Figure 6.28 shows a block on an inclined plane. The block represents the thread of the screw that may slide up or down the threads of the fixed base, represented by the inclined plane. The horizontal force P represents the minimum force that, if applied at the mean radius r of the screw, would slide the block up the plane, thus raising the load W.

The slope of the inclined plane θ depends on the mean radius r and the lead L of the screw. As shown in Figure 6.26c and Figure 6.27, by theoretically unwinding one turn of the screw thread onto a flat surface, a triangle is created with the base equal to the circumference of the mean radius of the thread $(2\pi r)$ and the rise equal to the pitch (or *lead*) of the thread L. The mean radius is equal to one-half the sum of the outer radius and the root radius of the thread. The lead of a screw is the distance that a nut will advance along the screw in one revolution. In the usual case of a single-threaded screw, the lead is the same as the pitch p, which is the distance between similar points on adjacent threads. All square-threaded screws considered in this text are single threaded.

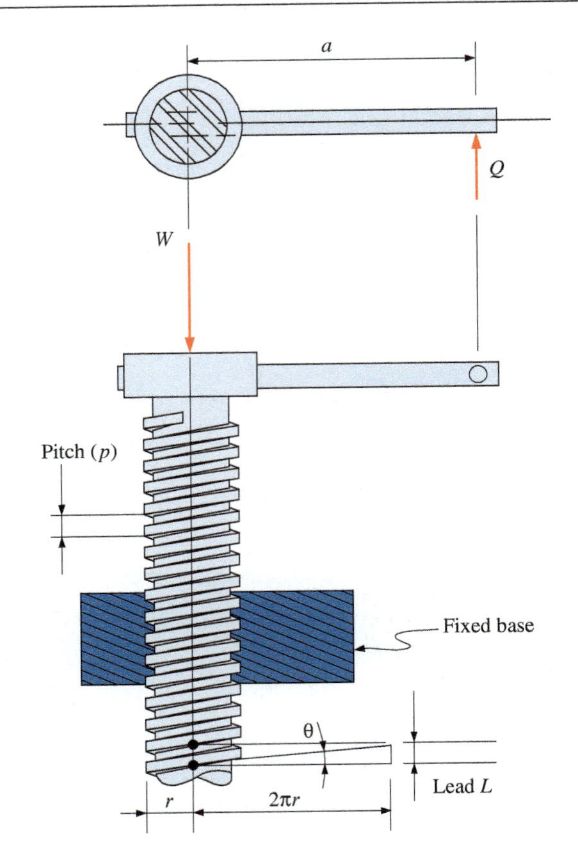

FIGURE 6.27 Jackscrew.

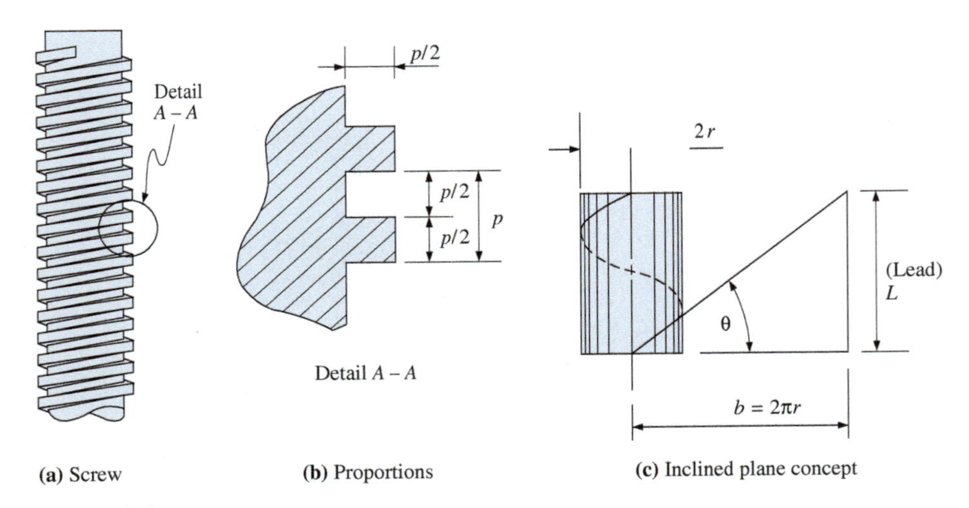

FIGURE 6.26 Square-threaded screw.

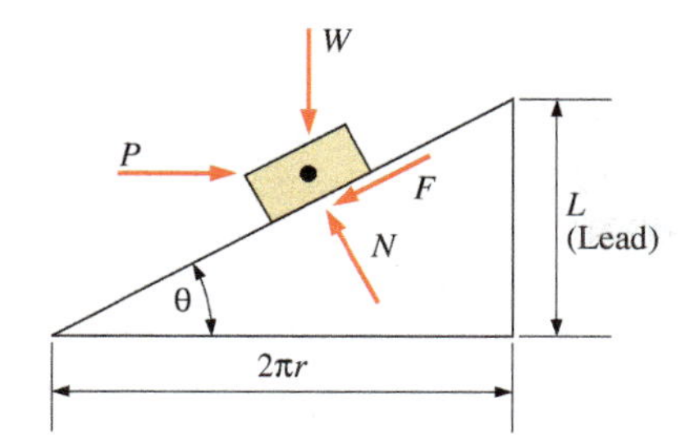

FIGURE 6.28 Force acting up the plane to raise load W.

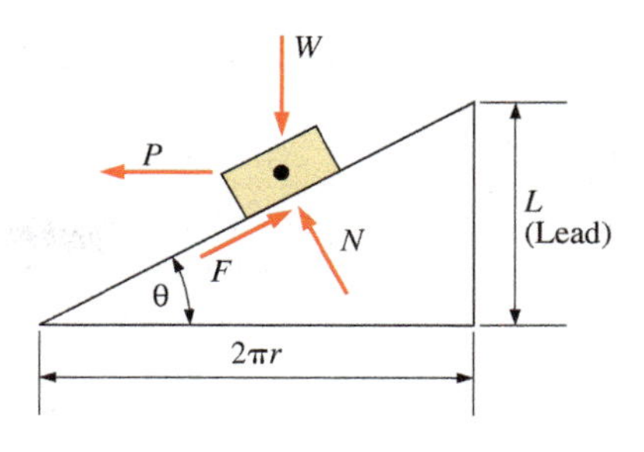

FIGURE 6.29 Force acting down the plane to lower the load W.

Therefore, the slope of the inclined plane, also called the *lead angle* θ, may be computed from

$$\theta = \tan^{-1}\frac{L}{2\pi r} = \tan^{-1}\frac{p}{2\pi r} \tag{6.10}$$

As discussed in Section 6.2, F represents the frictional resistance on the inclined plane ($F = \mu_s N$), and N represents a normal force perpendicular to the inclined plane.

After computing the angle θ, the horizontal force P, as shown in Figure 6.28, can be calculated using the equations of equilibrium and friction theory as discussed in Section 6.4. Once force P is computed, the required applied force Q at the end of the jack handle can be calculated based on the principle that the moment effect of Q, with respect to the longitudinal axis of the screw, must be equal to the moment effect of P with respect to the same axis. This is expressed as

$$Qa = Pr \tag{6.11}$$

If the lead angle θ is very steep, the frictional resistance may not be sufficient to overcome the tendency for the screw thread (or block) to slide under the action of the load, which would result in the load being lowered without any applied force Q. However, θ is usually small and the frictional resistance is large enough to prevent this unaided movement. Such a jackscrew is called *self-locking*, and this is a desirable characteristic for jacks and similar devices. For a screw to be self-locking, the coefficient of static friction must be greater than the tangent of the lead angle ($\mu_s > \tan\theta$).

To determine the force P required to lower the applied load W (which is equivalent to moving the block down the inclined plane), the free-body diagram as shown in Figure 6.29 is applicable.

EXAMPLE 6.12 The mean diameter of the jackscrew shown in Figure 6.27 is 3.50 in. and the pitch of the square-threaded screw is 0.50 in. The coefficient of static friction is 0.15. (a) Calculate the force Q required at the end of a 24-in. jack handle ($a = 24$ in.) to raise a weight of 4000 lb. (b) Calculate the force Q required (if any) at the end of the jack handle to lower a weight of 4000 lb.

Solution (a) The applicable free-body diagram is shown in Figure 6.30.

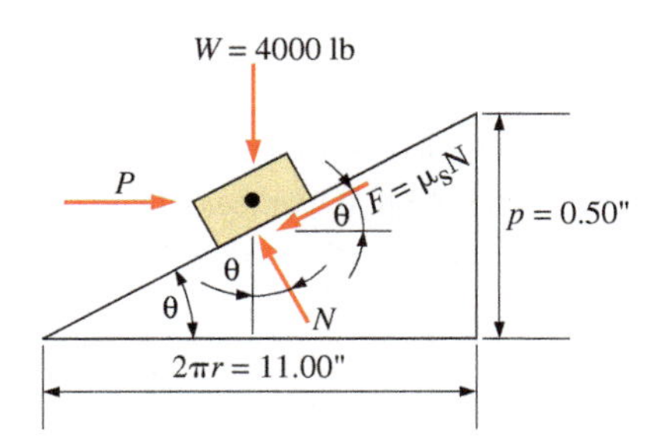

FIGURE 6.30 Free-body diagram.

1. Using Equation (6.10), compute the lead angle:

$$\theta = \tan^{-1}\left(\frac{0.50}{11.00}\right) = 2.60°$$

Therefore,

$$\sin \theta = 0.0454$$
$$\cos \theta = 0.999$$

2. Compute the horizontal force P. Take a summation of vertical forces:

$$\begin{aligned}\Sigma F_y &= -W + N\cos\theta - F\sin\theta = 0 \\ &= -4000 \text{ lb} + N(0.999) - 0.15(N)(0.0454) = 0\end{aligned}$$

from which

$$N = 4031 \text{ lb}$$

Take a summation of horizontal forces:

$$\begin{aligned}\Sigma F_x &= +P - N\sin\theta - F\cos\theta = 0 \\ &= +P - (4031 \text{ lb})(0.0454) - 0.15(4031 \text{ lb})(0.999) = 0\end{aligned}$$

from which

$$P = 787 \text{ lb}$$

3. Using Equation (6.11), compute force Q on the jack handle:

$$Qa = Pr$$

$$Q = \frac{(787 \text{ lb})(1.75 \text{ in.})}{24 \text{ in.}} = 57.4 \text{ lb}$$

(b) The applicable free-body diagram is shown in Figure 6.31.

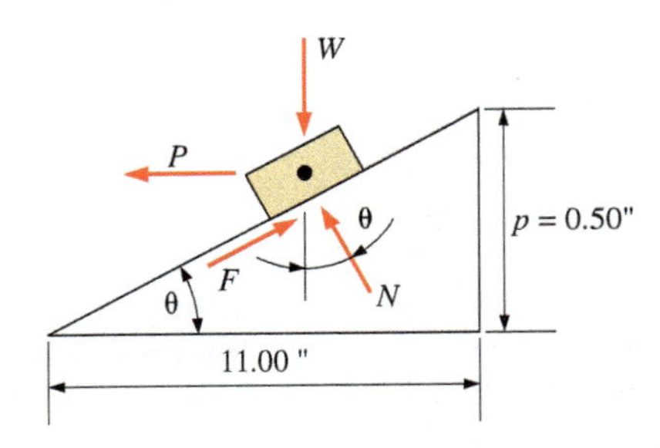

FIGURE 6.31 Free-body diagram.

1. Check whether the jackscrew is self-locking:

$$\mu_s = 0.15 > \tan\theta = 0.0454$$

Therefore, it is self-locking.

2. The lead angle θ was calculated in part (a) to be 2.60°. Compute the horizontal force P. Take a summation of vertical forces:

$$\begin{aligned}\Sigma F_y &= -W + N\cos\theta + F\sin\theta = 0 \\ &= -4000 \text{ lb} + N(0.999) + 0.15(N)(0.0454) = 0\end{aligned}$$

from which

$$N = 3977 \text{ lb}$$

Take a summation of horizontal forces:

$$\begin{aligned}\Sigma F_x &= -P - N\sin\theta + F\cos\theta = 0 \\ &= -P - (3977 \text{ lb})(0.0454) + 0.15(3977 \text{ lb})(0.999) = 0\end{aligned}$$

from which

$$P = 415 \text{ lb}$$

3. Compute force Q on the jack handle:

$$Qa = Pr$$

$$Q = \frac{(415 \text{ lb})(1.75 \text{ in.})}{24 \text{ in.}} = 30.3 \text{ lb}$$

SUMMARY BY SECTION NUMBER

6.1 Friction is defined as a retarding force that resists the relative movement of two bodies in contact. It acts to oppose motion and always acts parallel to the surfaces in contact.

6.2 The coefficient of static friction is the ratio of the maximum available frictional resistance F to the normal force N between the contacting surfaces:

$$\mu_s = \frac{F}{N} \tag{6.1}$$

The maximum available frictional force is developed when motion is impending. It is dependent on the type of materials and the nature of the contact surfaces. It is independent of the contact area and proportional to the normal force N:

$$F = \mu_s N \tag{6.2}$$

After motion begins, $F = \mu_k N$, where μ_k is the coefficient of kinetic friction.

6.3 The reaction of the supporting surface may be expressed as a single force R, which is the resultant of F and N. The angle ϕ_s between R and N is called the angle of static friction and is a maximum when motion is impending and F is a maximum:

$$\tan \phi_s = \frac{F}{N} = \mu_s \tag{6.3}$$

6.4 The general procedure for the solution of motion-impending friction problems in this text involves drawing the free-body diagram and deciding which way motion is impending. This is followed by the application of Equation (6.2) and the equations of force equilibrium for concurrent force systems, or the equations of force equilibrium and moment equilibrium for noncurrent force systems.

6.5 Wedges are devices used to overcome (lift or move) large loads by the means of relatively small applied forces. Wedge problems are solved using free bodies, friction considerations, and the equations of equilibrium.

6.6 The difference in belt tensions T_L and T_S caused by wrapping a belt around a cylindrical surface depends on the angle of contact (wrap) β and the coefficient of friction μ between the contact surfaces. When slip is impending, the following equations apply: for flat belts,

$$T_L = T_S e^{\mu\beta} \tag{6.4}$$

or

$$\ln \frac{T_L}{T_S} = \mu\beta \tag{6.5}$$

or

$$\ln T_L - \ln T_S = \mu\beta \tag{6.6}$$

and for V-belts,

$$T_L = T_S e^{\mu\beta/\sin\phi} \tag{6.7}$$

or

$$\ln \frac{T_L}{T_S} = \frac{\mu\beta}{\sin\phi} \tag{6.8}$$

or

$$\ln T_L - \ln T_S = \frac{\mu\beta}{\sin\phi} \tag{6.9}$$

6.7 Square-threaded screws are an application of the inclined plane and can be used to move heavy loads or transmit power. Only single-threaded screws are considered in this text. Performance of the screw is a function of the coefficient of friction between the parts, the lead, and mean radius of the thread. The lead angle is calculated from

$$\theta = \tan^{-1}\frac{L}{2\pi r} = \tan^{-1}\frac{p}{2\pi r} \tag{6.10}$$

6.8 The equations of equilibrium and frictional considerations can be used to determine the force P required to move a screw within a threaded block. The force Q required at the end of a handle to turn the screw can be calculated from

$$Qa = Pr \tag{6.11}$$

PROBLEMS

Section 6.4 Friction Applications

6.1 A 150 lb block rests on a horizontal floor. The coefficient of friction between the block and the floor is 0.30. A pull of 40 lb, acting upward at an angle of 30° to the horizontal, is applied to the block. Determine whether or not the block will slide.

6.2 A 200 lb block rests on a horizontal surface. The coefficient of static friction between the block and the supporting surface is 0.50. Calculate the force P required to cause motion to impend if the force applied to the block is (a) horizontal and (b) upward at an angle of 20° with the horizontal.

6.3 A tool locker having a mass of 140 kg rests on a wooden pallet. Assuming a coefficient of static friction of 0.38, determine the horizontal force that would cause sliding motion to impend.

6.4 A body weighing 100 lb rests on an inclined plane, as shown. The coefficient of static friction between the body and the plane is 0.40. Compute the force P that will cause impending motion

(a) up the inclined plane.
(b) down the inclined plane.

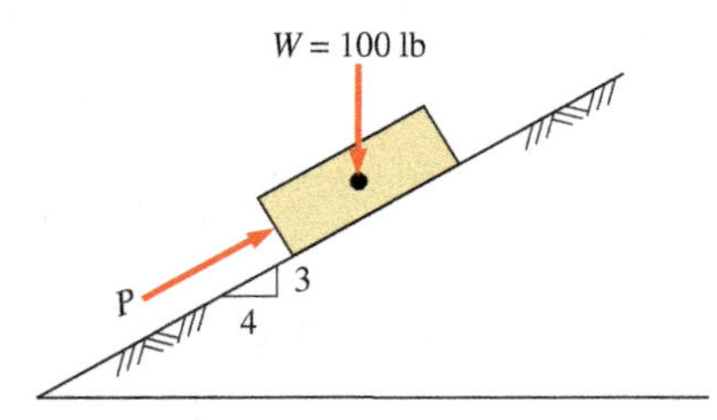

PROBLEM 6.4

6.5 For Problem 6.4, compute the friction force F when the force P acting up the plane is equal to

(a) 40 lb **(b)** 60 lb **(c)** 70 lb

6.6 Block A in the figure shown has a mass of 10 kg and block B has a mass of 20 kg. The coefficients of static friction are 0.5 between the two blocks and 0.15 between the lower block and the supporting surface. Calculate the force P that will cause motion to impend.

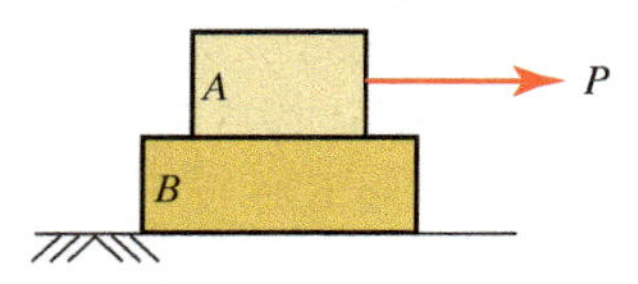

PROBLEM 6.6

6.7 Compute the horizontal force P required to cause motion to impend up the plane for the 100 lb block shown. The coefficient of static friction between the block and the plane is 0.25.

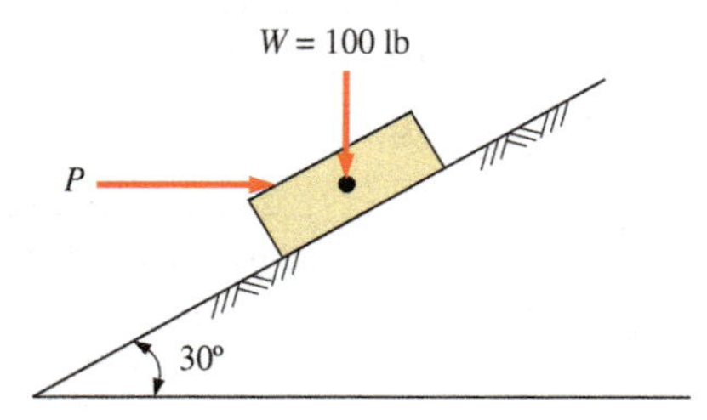

PROBLEM 6.7

6.8 Compute the horizontal force P required to prevent the block from sliding down the plane for the 175 lb block shown. Assume the coefficient of static friction to be 0.65.

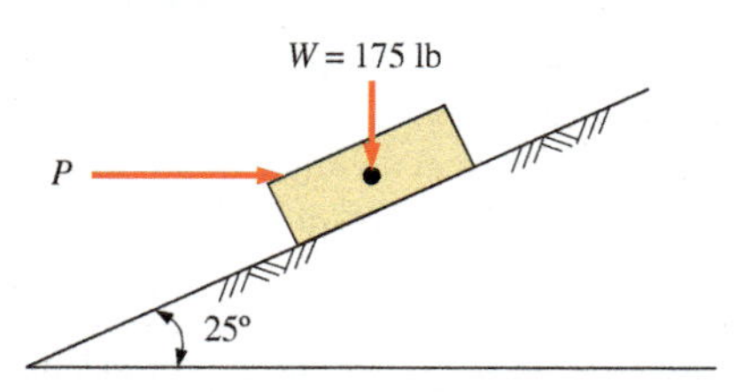

PROBLEM 6.8

6.9 Calculate the magnitude of the force P, acting as shown, that will cause the 200 lb crate to move, either sliding to the right or tipping. Assume the coefficient of static friction to be 0.30.

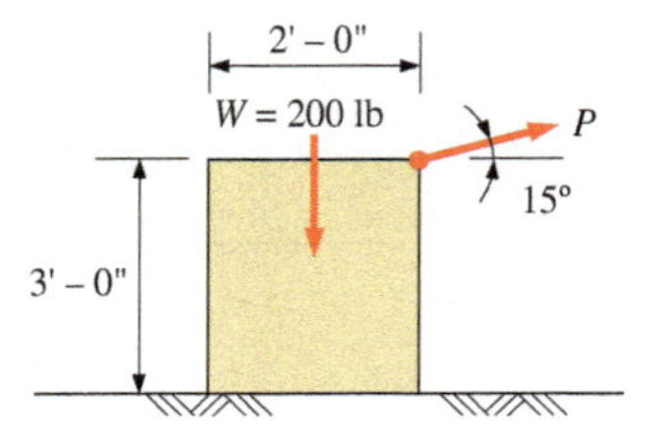

PROBLEM 6.9

Section 6.5 Wedges

6.10 For the block-and-wedge system shown in Figure 6.17, calculate the force P required to initiate upward motion of the block. The block supports a load of 700 lb, the slope angle of the wedge is 9°, and the coefficient of static friction on the two surfaces of the wedge is 0.25. The vertical surface (C–C) is frictionless. Compare the result with Example 6.7.

6.11 Calculate the value of the horizontal force P required to start the V-shaped wedge in motion to the right, raising the 200 lb block, as shown. The coefficient of static friction for the contact surfaces of the wedge is 0.364. The vertical surface is frictionless.

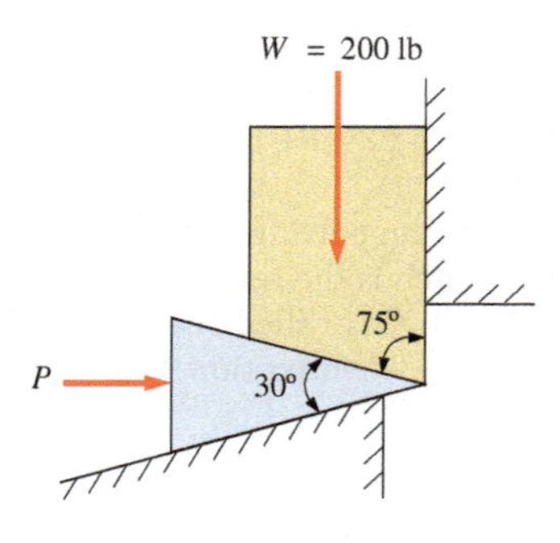

PROBLEM 6.11

6.12 Two blocks, each having a mass of 100 kg and resting on a horizontal surface, are to be pushed apart using a 30° wedge, as shown. The coefficient of static friction for all contact surfaces is 0.25. Calculate the vertical force P required to start the wedge and the blocks in motion.

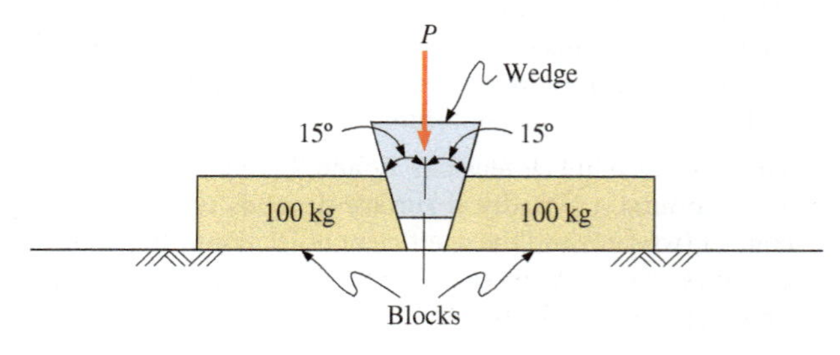

PROBLEM 6.12

6.13 Calculate the force P required to move the wedges and raise the 1000 lb block shown. The coefficient of static friction for all contact surfaces is 0.18.

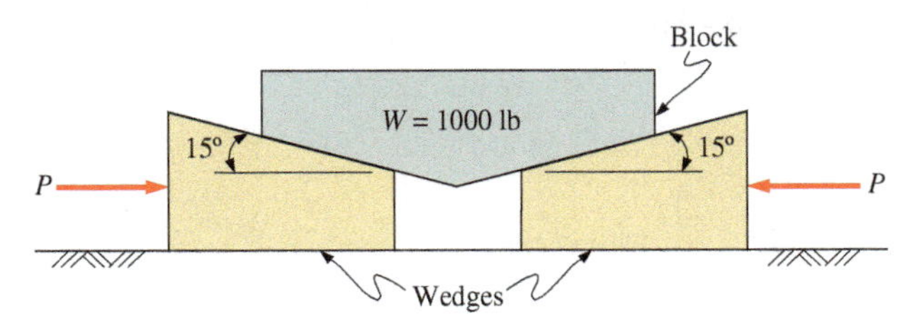

PROBLEM 6.13

Section 6.6 Belt Friction

6.14 A heavy machine is lowered into a pit by means of a rope wrapped around an 8-in.-diameter stationary pole placed horizontally across the top of the pit. The coefficient of static friction for the rope on the pole is 0.35. The rope makes $1\frac{1}{2}$ turns around the pole. Calculate the maximum weight that can be sustained if a person exerts a force of 50 lb on the end of the rope.

6.15 Calculate the maximum weight that the person in Problem 6.14 can sustain if $2\frac{1}{2}$ turns of rope are taken around the pole.

6.16 A flat belt passes halfway around a 1.80-m-diameter pulley. The maximum permissible tension in the belt is 2.0 kN. The coefficient of static friction for the belt on the pulley is 0.3. Calculate the maximum torque the belt can transmit to the pulley under these conditions.

6.17 A belt-and-pulley arrangement has a maximum belt tension of 150 lb on the tight side and 75 lb on the loose side. If the coefficient of static friction between the flat belt and the pulley is 0.30, calculate the minimum angle of contact required between the pulley and the belt.

6.18 Rework Problem 6.17 for a V-belt with a 40° groove angle, rather than a flat belt.

6.19 A mass of 320 kg is prevented from falling by a rope wrapped around a horizontal stationary circular pole. A force of 275 N is exerted on the free end of the rope. The coefficient of static friction between the pole and the rope is 0.25. How many turns of rope around the pole are necessary to keep the weight aloft?

6.20 A belt is wrapped around a pulley for 180°. The loose-side belt tension is 50 lb and the coefficient of static friction is 0.30. Calculate the maximum belt tension if motion is impending. Assume (a) a flat belt and (b) a V-belt with a groove angle of 60°.

Section 6.7 Square-Threaded Screws

6.21 A jackscrew has a square thread with a pitch of 0.50 in. The mean diameter of the thread is 2 in. The coefficient of static friction is 0.30. Calculate the force required at the end of a 15-in. jack handle to raise 2000 lb.

6.22 The mean diameter of a square-threaded jackscrew is 1.80 in. The pitch of the thread is 0.40 in. and the coefficient of static friction is 0.20.

(a) Calculate the force that must be applied at the end of an 18-in.-long jack handle to raise a load of 5000 lb.

(b) Calculate the force required (if any) on the handle to start the load down.

6.23 The woodworking vise shown is designed for a maximum applied force of 225 N at each end of the handle. The square-threaded screw has a pitch of 6 mm and a mean diameter of 22 mm. The coefficient of static friction is estimated to be 0.14. Determine the maximum clamping force.

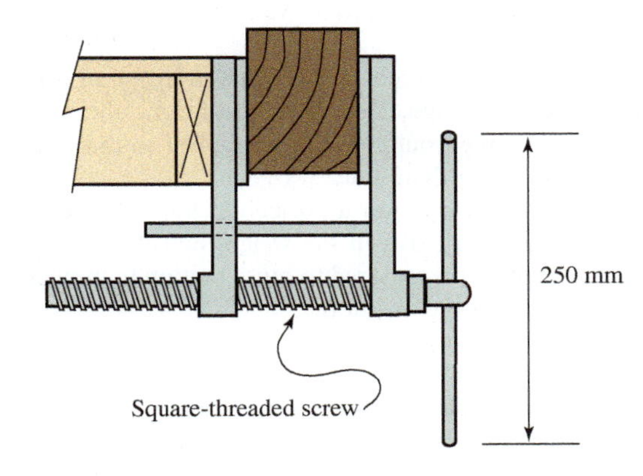

PROBLEM 6.23

6.24 A square-threaded screw is used in a press, shown, to exert a pressure of 4000 lb. The screw has a mean diameter of 3 in. and a pitch of 0.25 in. The coefficient of static friction is 0.15. Calculate the force that must be applied at the end of the press handle.

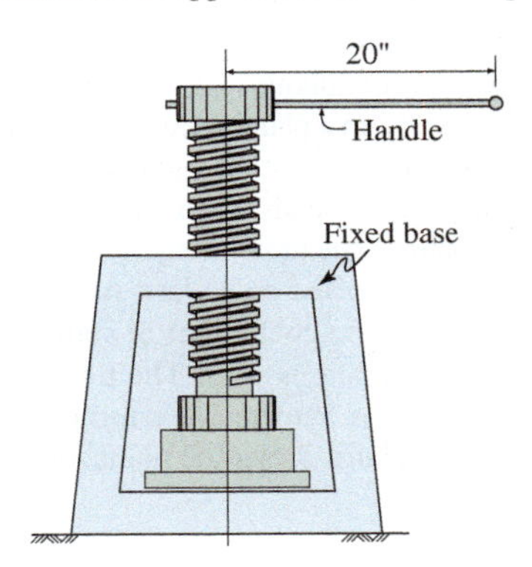

PROBLEM 6.24

Computer Problems

For the following computer problems, any appropriate software may be used. Input prompts should fully explain what is required of the user (the program should be user-friendly). The resulting output should be well labeled and self-explanatory.

6.25 Write a program that will solve Problem 6.36. User input is to be the angle of the inclined plane with the horizontal (not to exceed 90°), the weight of the block, and the coefficient of static friction.

6.26 A weight W is to be held aloft by wrapping a rope around a horizontal stationary pole. The coefficient of static friction is 0.33. A force of 20 lb will be applied to the free end of the rope. Write a program to generate a tabulation of weights that could be so supported as a function of the number of wraps around the pole, ranging from 1 wrap to 3 wraps, in $\frac{1}{4}$-wrap steps.

6.27 Write a program that will solve Problem 6.38. User input is to be width and weight of the crate. The output should state whether sliding or tipping governs the solution.

Supplemental Problems

6.28 A horizontal force of 18 lb is required to just start a 45-lb block in motion on a horizontal surface. Determine the coefficient of static friction between the block and the supporting surface.

6.29 A 90-lb block lying on a rough horizontal surface is subjected to a horizontal force P. The coefficient of static friction is 0.33. Calculate the maximum value that P can have before motion impends.

6.30 The tool locker of Problem 6.3 is 0.8 m by 0.8 m square (in plan) and 2 m high. Determine the height above which a horizontal force would have to be applied so that the locker would tip, rather than slide.

6.31 A 1-ton weight rests on a horizontal floor. The coefficient of static friction between the weight and the floor is 0.35. Calculate the force required to cause motion to impend if the force applied to the weight is

(a) horizontal

(b) downward at an angle of 15° with the horizontal

6.32 A 50-lb block rests on a rough inclined plane. If the coefficient of friction between the block and the plane is 0.50, calculate the inclination of the plane with the horizontal when motion impends.

6.33 If in Problem 6.32 the plane has an inclination with the horizontal of 20°, calculate

(a) the frictional resistance when the block is at rest

(b) the force parallel to the plane necessary for motion to impend up the plane

(c) the force parallel to the plane necessary to prevent motion down the plane

6.34 A 325-lb block rests on a plane inclined 25° with the horizontal. Calculate the coefficient of static friction between the block and the plane if motion impends when a force of 225 lb, parallel to the plane, is applied. The impending motion is up the plane.

6.35 A 47 lb body is supported on a plane inclined 33° to the horizontal, as shown. The coefficient of static friction between the body and the plane is 0.15. The body is subjected to a 29-lb force, acting as shown. Determine whether the body will move up the plane, down the plane, or remain at rest.

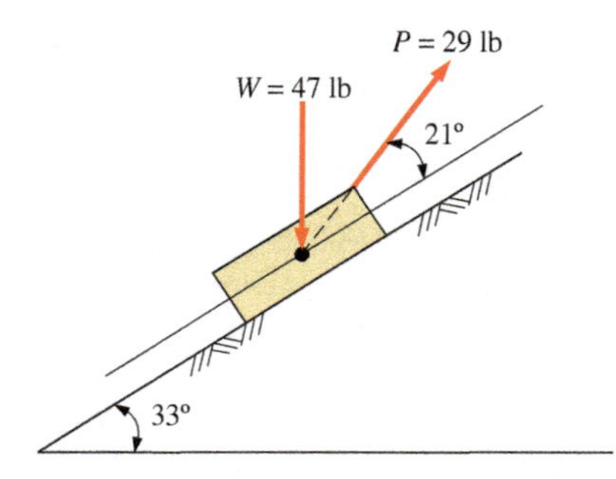

PROBLEM 6.35

6.36 Compute the horizontal force P required to prevent the 900-lb block, shown, from sliding down the plane. The coefficient of friction between the block and the plane is 0.20.

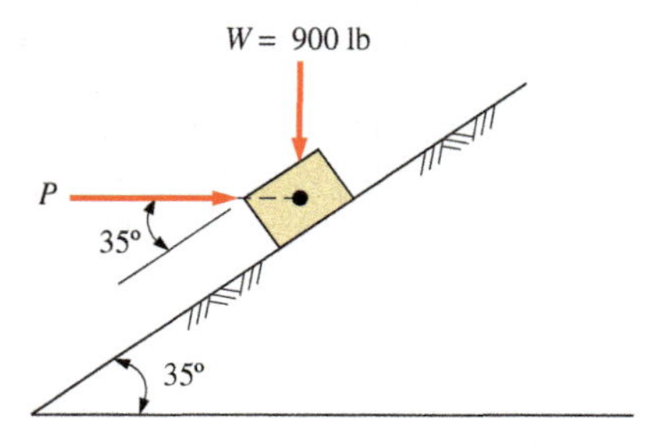

PROBLEM 6.36

6.37 In the figure shown, block A weighs 200 lb and block B weighs 400 lb. The coefficient of static friction for all contact surfaces is 0.35. Block A is anchored to the wall with a flexible cable. Calculate the force P required for motion to impend for block B.

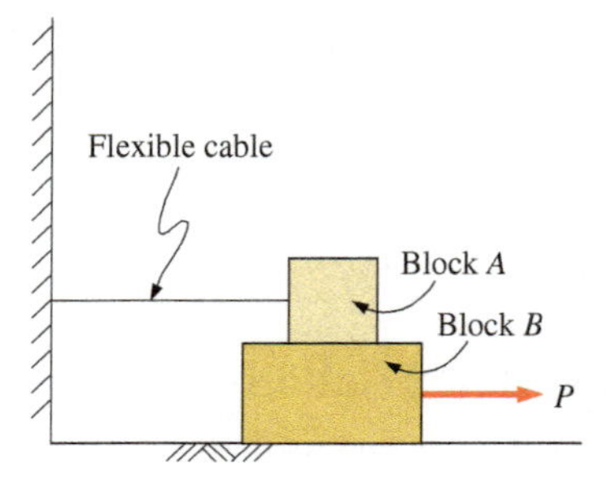

PROBLEM 6.37

6.38 Calculate the magnitude of the horizontal force P, acting as shown, that will cause motion to impend for the 450-lb crate. (The motion may be either sliding or tipping.) Assume the coefficient of static friction to be 0.40.

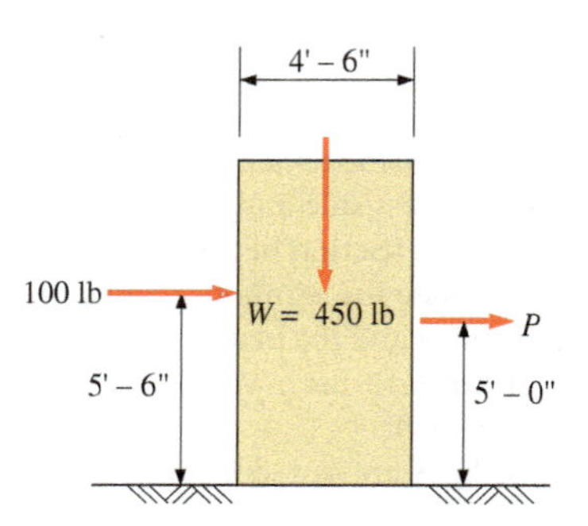

PROBLEM 6.38

6.39 In the figure shown, the coefficient of static friction between the 160-lb crate and the floor is 0.35. Calculate the value of the force P and the distance h that will cause the crate to tip and slide simultaneously.

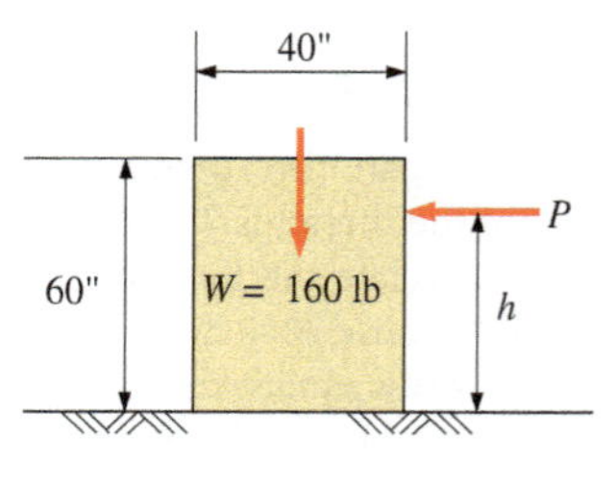

PROBLEM 6.39

6.40 A 500-lb block rests on a horizontal surface, as shown. The coefficient of static friction is 0.25. Calculate the maximum value of the horizontal force P so that neither sliding nor tipping will occur. Assume that P is gradually applied.

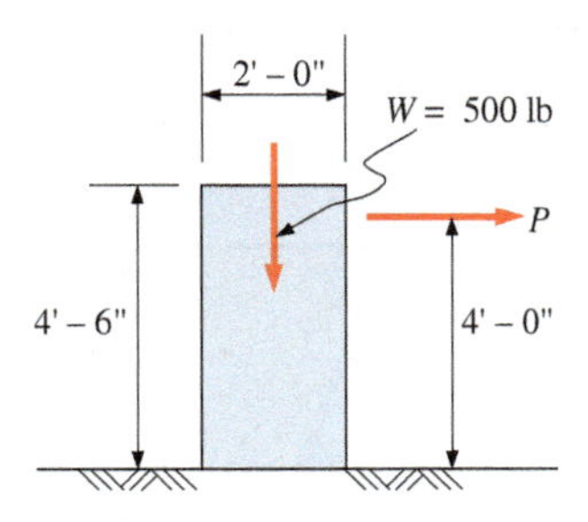

PROBLEM 6.40

6.41 A ladder, 8 m long and having a mass of 25 kg, rests on a horizontal floor and is supported by a vertical wall. The ladder is inclined 62°, as shown, and starts to slip when a person having a mass of 73 kg has climbed halfway up it. The coefficient of friction at the wall is 0.20. Assume the weight of the ladder to be concentrated at its midpoint. Calculate the coefficient of friction at the floor.

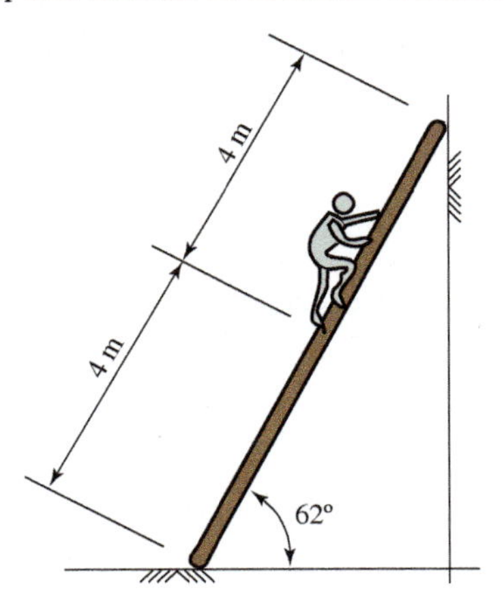

PROBLEM 6.41

6.42 The ladder shown is supported by a horizontal floor and a vertical wall. It is 12 ft long, weighs 30 lb (assumed to be concentrated at its midlength), and supports a person weighing 175 lb at point D. The vertical wall is smooth and offers no frictional resistance. The coefficient of static friction at point A is 0.30. Calculate the minimum angle θ at which the ladder will stand without slipping to the left.

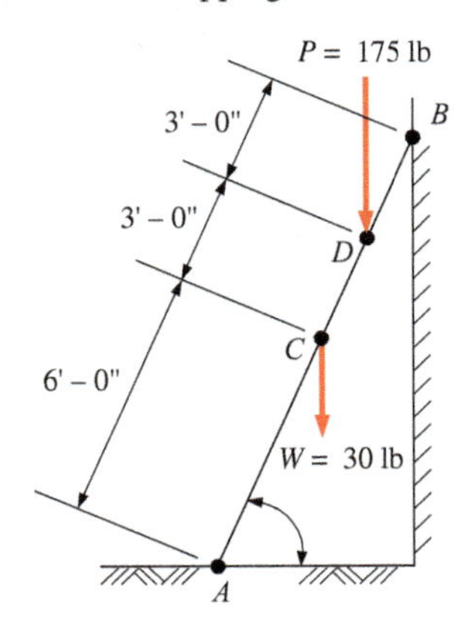

PROBLEM 6.42

6.43 A 16-ft ladder weighing 62 lb (assumed concentrated at its midlength) rests on a horizontal floor and is supported by a vertical wall, as shown. The lower end is prevented from slipping by friction and by a rope attached to the base of the wall. The ladder supports a 200 lb person at the top. The coefficient of static friction at the wall is 0.25; at the floor it is 0.30. Calculate the tension P in the rope necessary to prevent slipping.

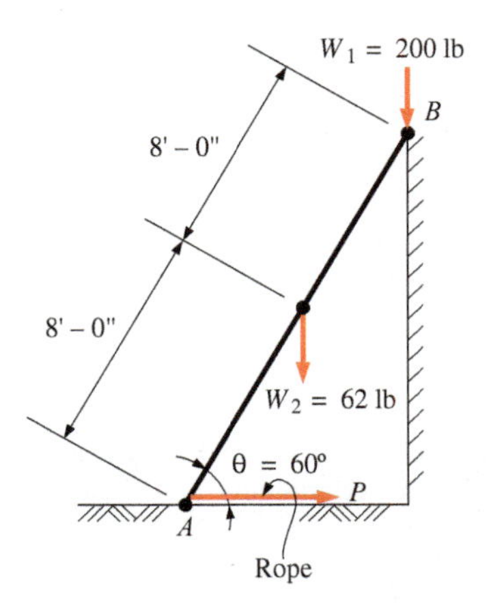

PROBLEM 6.43

6.44 Assume that the rope is removed from the base of the ladder in Problem 6.43. Calculate the minimum angle θ at which the ladder will stand without slipping to the left.

6.45 Compute the minimum weight of block B that will prevent the two-block system shown from sliding. Block A weighs 100 lb and the coefficient of static friction for the contact surfaces is 0.20. Assume the pulley to be frictionless.

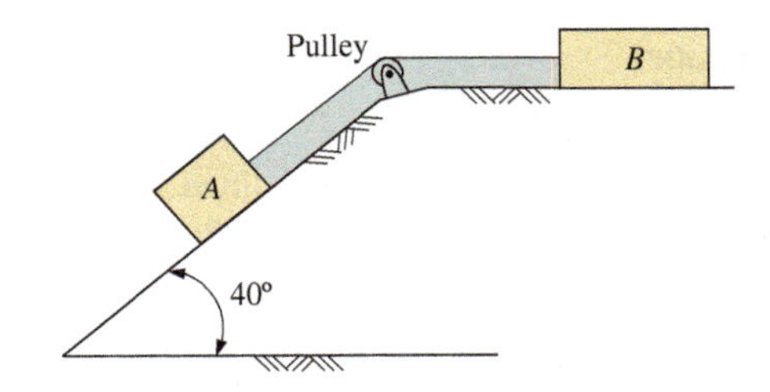

PROBLEM 6.45

6.46 In the figure shown, block R has a mass of 90 kg and block Q has a mass of 45 kg. They are connected by a rope parallel to the inclined plane. The coefficient of static friction between R and the plane is 0.10, and between Q and the plane is 0.40. Calculate the value of the angle θ at which motion impends down the plane. Calculate the magnitude of the tension in the rope.

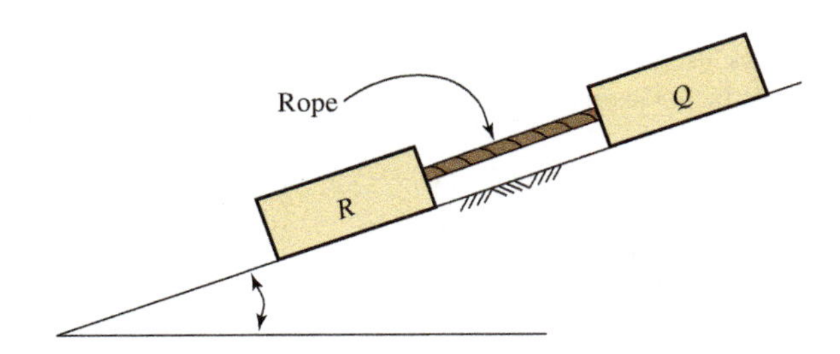

PROBLEM 6.46

6.47 Compute the minimum weight of block A shown for motion to be impending down the plane. Assume the pulley to be frictionless. Block B weighs 500 lb. The coefficient of static friction for all contact surfaces is 0.20.

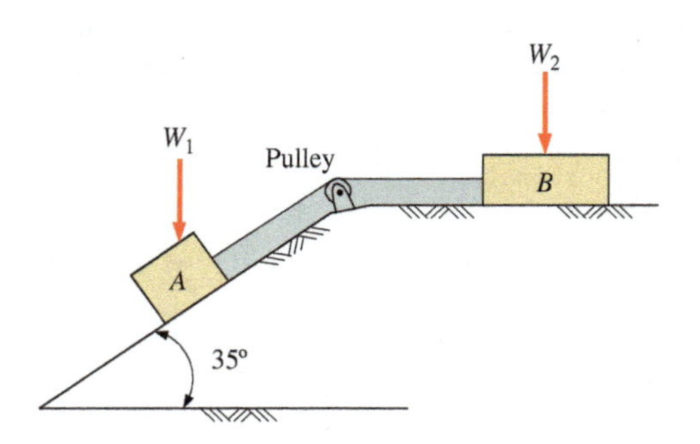

PROBLEM 6.47

6.48 The blocks shown are separated by a solid strut attached to the blocks with frictionless pins. The coefficient of static friction for all surfaces is 0.325. Determine the value of the horizontal force P that will cause motion to impend to the right. Neglect the weight of the strut.

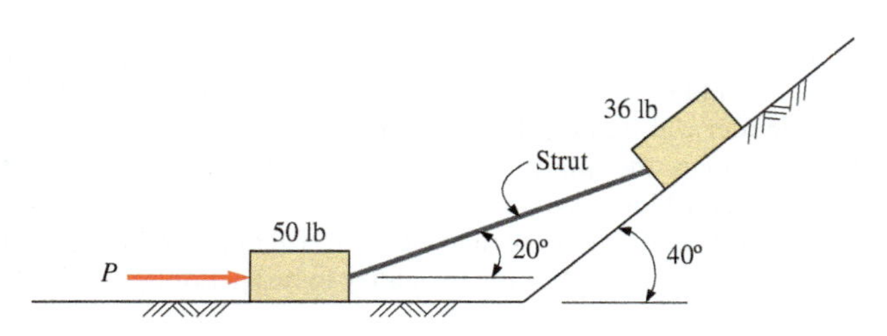

PROBLEM 6.48

6.49 The end of a timber beam is to be lifted using a steel wedge at B, as shown. Determine the force P that must be applied to the wedge to cause it to move to the left. The coefficient of static friction for the contact surfaces is 0.30.

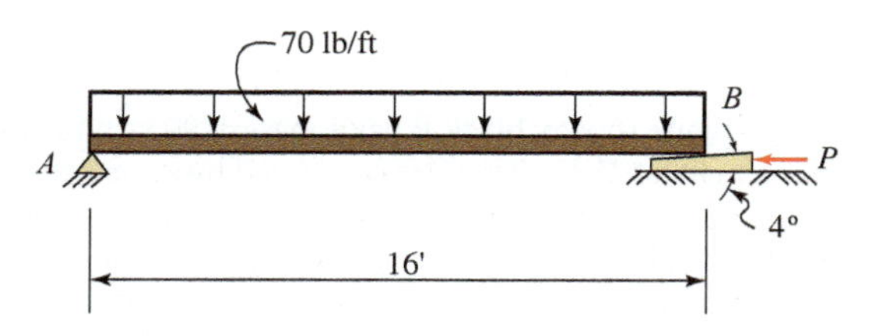

PROBLEM 6.49

6.50 Rework Problem 6.11 assuming the coefficient of static friction to be 0.364 on all contact surfaces.

6.51 Calculate the force P necessary to start the block C sliding upward, as shown. The coefficient of static friction for all contact surfaces is 0.25. Neglect the weights of blocks A and C.

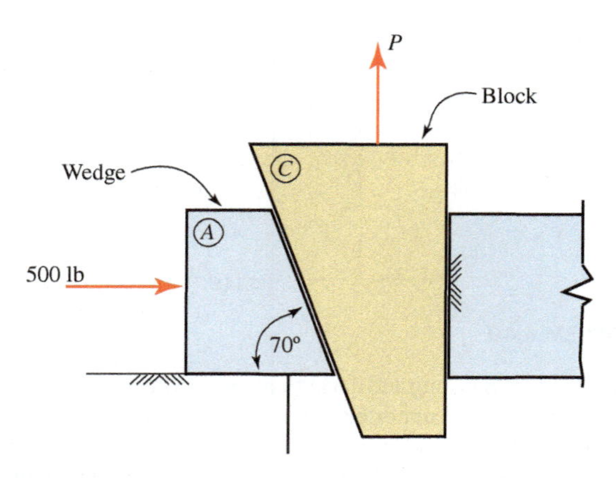

PROBLEM 6.51

6.52 A machine having a mass of 500 kg is to be raised using a wedge, as shown. The coefficient of friction on all surfaces is 0.25. Determine the force P required to raise the machine.

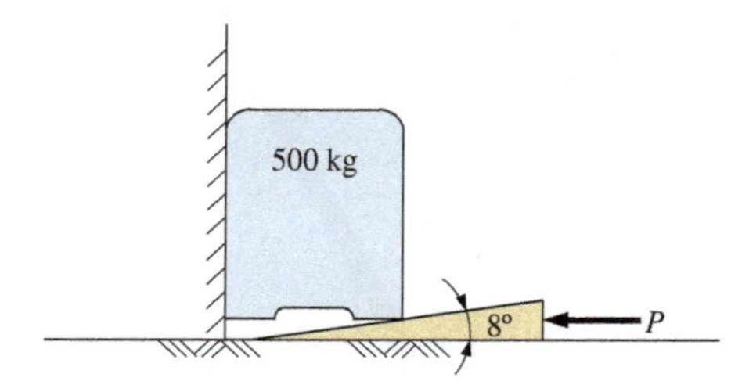

PROBLEM 6.52

6.53 A ship may exert an estimated pull of 8000 lb on its hawser (a heavy rope), which is wrapped around a mooring post on the dock. Determine the number of turns required of the hawser around the post so that the pull at the free end will not exceed 40 lb. Assume a coefficient of static friction of 0.35.

6.54 A sailor wraps a heavy rope around a bollard (a vertical post) to secure a drifting ship to a pier. The sailor can apply a force of 400 N to the free end of the rope. With what force can the ship pull on the rope before another sailor will have to be called? Consider **(a)** one wrap, **(b)** two wraps, **(c)** three wraps. Assume a coefficient of friction of 0.35.

6.55 When a large rope is wrapped twice around a post, a pull of 100 lb at the free end will withstand a pull of 2 tons at the other end. What pull would the 100 lb withstand if the rope were wrapped only once around the post?

6.56 A band brake is in contact with drum C through an angle of 180° and is connected to the horizontal lever at *A* and *B*, as shown. The drum diameter is 16 in. The coefficient of kinetic friction between the brake band and the drum is 0.40. Force *P* is 90 lb. Determine the tensions at *A* and *B* in the band brake if the drum is rotating
(a) clockwise
(b) counterclockwise

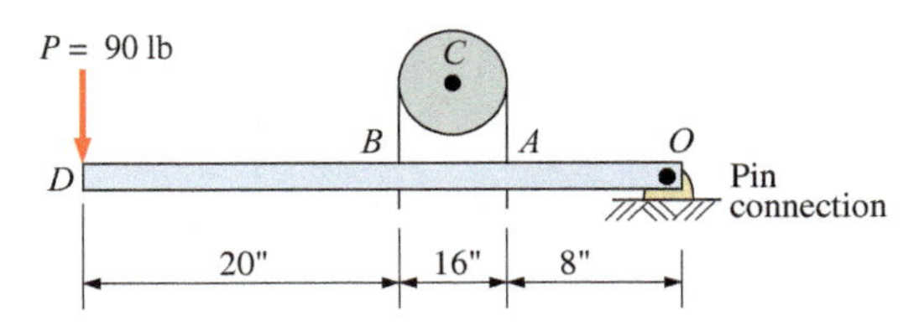

PROBLEM 6.56

6.57 A V-belt with a groove angle of 60° drives a 0.37-m-diameter pulley on a small concrete mixer. The contact angle is 210° and the coefficient of friction is 0.33. The mixer requires a torque of 50 N·m. Determine the required belt tensions.

6.58 The mean diameter of a square-threaded jackscrew is 3.00 in., the pitch is 0.40 in., and the coefficient of static friction is 0.15. Calculate the maximum load that can be raised by a force of 75 lb applied at the end of a 14-in. jack handle.

6.59 The manually operated apple cider press shown has a 36-in. handle. The square-threaded screw has a mean diameter of 1.700 in. The screw must be turned through $2\frac{1}{2}$ turns to advance the head of the press 1.00 in. The coefficient of static friction is estimated to be 0.13.
(a) Find the force on the stack if a force of 60 lb is applied at each end of the handle.
(b) How much is the resulting force decreased if the coefficient of static friction goes to 0.18 due to lack of lubrication on the screw?

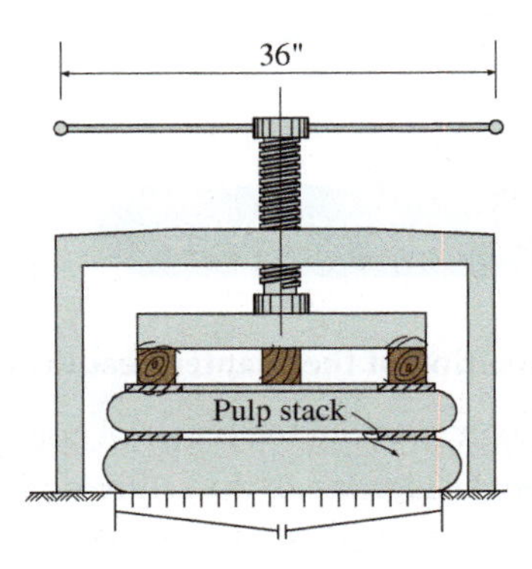

PROBLEM 6.59

CHAPTER SEVEN

CENTROIDS AND CENTERS OF GRAVITY

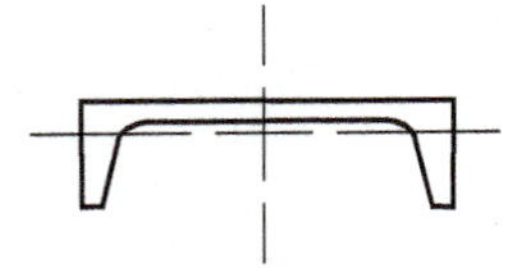

LEARNING OBJECTIVES

Upon completion of the chapter, readers will be able to:

- Differentiate between the center of gravity of an object and the centroid of an object.
- Calculate the location of the centroid of a shape using either the weight method or the area method.
- Calculate the location of the centroid of a composite shape.

7.1 INTRODUCTION

All bodies may be considered to be composed of a multitude of small particles, each being acted upon by a gravitational force. When algebraically added, these forces exerted on the particles of a body represent the weight of the body. For all practical purposes, these forces are assumed to be parallel and to act vertically downward. Hence, the force system may be categorized as a parallel force system, and the algebraic sum (which is the resultant of the system) is called the *weight* of the body. The resultant of these individual gravity forces will always act through a definite point, regardless of how the body is oriented. This point is called the *center of gravity*.

Weight is a force, and it may be treated and represented as a vector. Therefore, it must have magnitude, direction, sense, and a point of application, all of which describe a force vector. Since the direction and sense of the force of gravity are always known, only the magnitude and the point of application must be determined. The magnitude and location of this resultant can be determined experimentally; for the purpose of analysis and design, however, we limit our discussion to an analytical determination of both magnitude and location. In effect, the problem of locating the center of gravity of a body becomes one of determining the point through which the resultant weight of the body acts.

7.2 CENTER OF GRAVITY

The procedure for determining the magnitude and location of the center of gravity is the same as that for determining the magnitude and location of the resultant force of a parallel force system, as described in Chapter 3. As an example, consider the irregularly shaped flat plate of uniform thickness and homogeneous material shown in Figure 7.1. The plate is divided into infinitesimal elements, the typical element being located a distance x from the reference Y–Y axis and a distance y from the reference X–X axis. The weight w of each element may be thought of as being concentrated at its center. The weights of the elements form a parallel force system, the resultant of which is the total weight W of the plate. The magnitude of the total weight can be written mathematically as

$$W = \Sigma w$$

The weight W (the weight of the plate), by definition, acts through the center of gravity of the plate. The coordinates

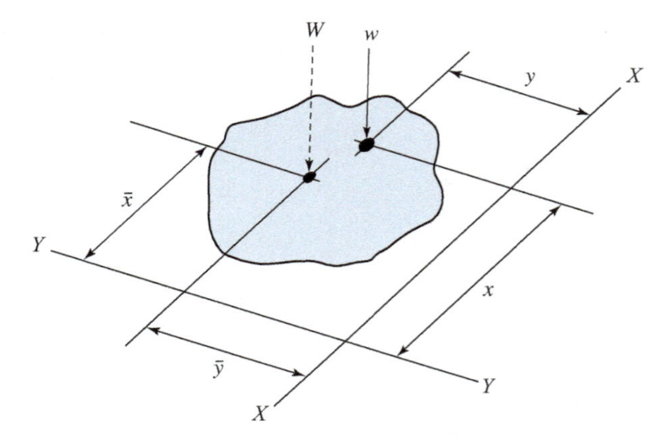

FIGURE 7.1 Center of gravity.

of the center of gravity are designated $\bar{x}$ and $\bar{y}$. To determine the location of W, and thus to determine the location of the center of gravity, moments of the weights of the individual elements are taken with respect to each of the axes shown. Using Varignon's theorem that the moment of the resultant about any point or axis must equal the algebraic sum of the moments of the individual weights about the same point or axis, the following expressions can be established to locate the resultant:

$$W\bar{x} = \Sigma wx$$
$$W\bar{y} = \Sigma wy$$

Solving for the center of gravity locations gives

$$\bar{x} = \frac{\Sigma wx}{W} \quad \text{or} \quad \frac{\Sigma wx}{\Sigma w} \tag{7.1}$$

$$\bar{y} = \frac{\Sigma wy}{W} \quad \text{or} \quad \frac{\Sigma wy}{\Sigma w} \tag{7.2}$$

It may be readily apparent, and can easily be shown, that if the plate has an axis of symmetry, then the center of gravity will lie somewhere on that axis of symmetry. If the plate has two mutually perpendicular axes of symmetry (in the case, for example, of a rectangular or a circular plate), then the center of gravity will lie at the intersection of the axes of symmetry.

Note that our discussion in this section applies strictly to bodies that have mass and, therefore, weight. Frequently, however, the center of gravity of an area is desired. (This may be thought of as the plate in Figure 7.1 having a zero thickness.) Since an area does not have mass, it cannot have weight or, theoretically, a center of gravity. The point in the area that would be analogous to the center of gravity of a body having mass, however, is commonly called the *centroid* of the area. (It is not uncommon in the applied engineering fields to use *centroid* and *center of gravity* interchangeably.)

EXAMPLE 7.1 A 10-in.-diameter steel sphere is anchored firmly to the top of a 12-in.-square concrete pedestal. The pedestal is 18 in. high. The two bodies are considered to be a single unit. The resulting member, shown in Figure 7.2, may be called a *built-up member.* Locate the center of gravity of the member.

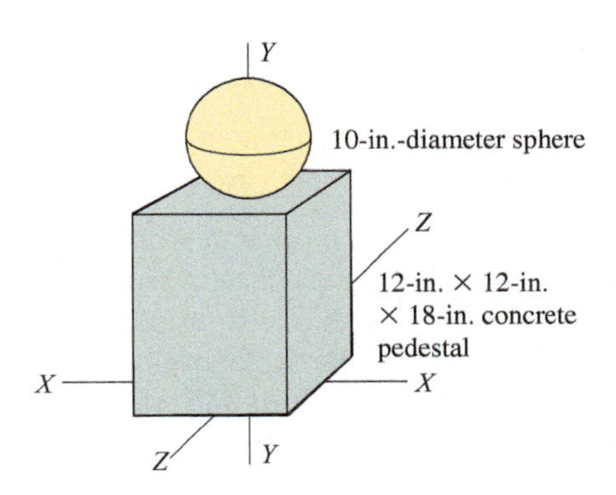

FIGURE 7.2 Two-body unit.

Solution The unit is symmetrical with respect to the Y–Y axis. Therefore, the center of gravity will lie on the Y–Y axis, which is the vertical axis of the unit. It is necessary to compute only $\bar{y}$, which in this problem represents the distance from the bottom of the concrete pedestal to the center of gravity of the member.

The unit weights of the materials can be obtained from Appendix G. Denote the concrete pedestal as w_1 and the steel sphere as w_2. The weight of each component part is calculated as the product of the volume (in cubic feet) and the unit weight (in pounds per cubic foot) as follows:

$$w_1 = (12\text{ in.})(12\text{ in.})(18\text{ in.})\left(\frac{1\text{ ft}^3}{1728\text{ in.}^3}\right)\left(150\frac{\text{lb}}{\text{ft}^3}\right) = 225\text{ lb}$$

$$w_2 = \left(\frac{4}{3}\pi R^3\right)\left(490\frac{\text{lb}}{\text{ft}^3}\right) = \frac{4}{3}\pi(5\text{ in.})^3\left(\frac{1\text{ ft}^3}{1728\text{ in.}^3}\right)\left(490\frac{\text{lb}}{\text{ft}^3}\right) = 148.5\text{ lb}$$

The total weight is then

$$W = w_1 + w_2 = 225\text{ lb} + 148.5\text{ lb} = 374\text{ lb}$$

Using the bottom of the concrete pedestal as the reference axis (shown as axis X–X in Figure 7.3), $\bar{y}$ is calculated from Equation (7.2). Note that the Z–Z axis shown in Figure 7.2 could also serve as a

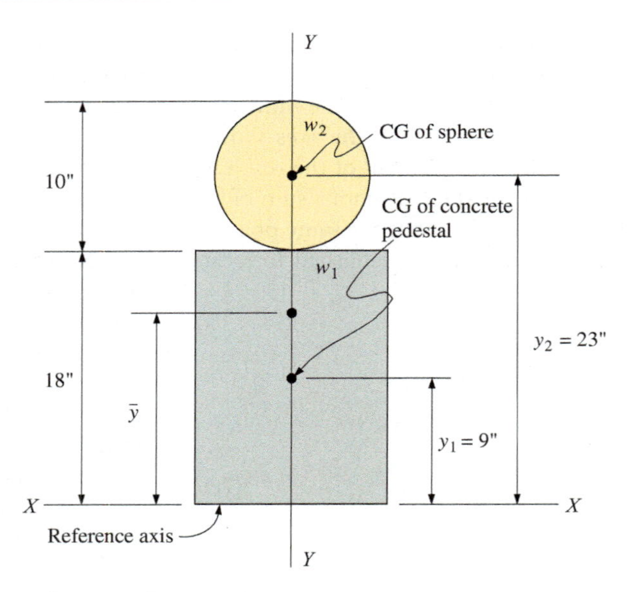

FIGURE 7.3 Location of center of gravity.

reference axis. The y distances, from the reference axis to the centers of gravity of each component, are shown in Figure 7.3. So

$$y = \frac{\Sigma wy}{W} = \frac{w_1 y_1 + w_2 y_2}{W}$$

$$= \frac{(225\text{ lb})(9\text{ in.}) + 148.5\text{ lb}(23\text{ in.})}{374\text{ lb}} = 14.55\text{ in.}$$

EXAMPLE 7.2 The performance of an aircraft is affected by the positioning of the cargo load it carries. The location of the center of gravity is critical. Assume that the load shown in Figure 7.4a is composed of five crates (numbered 1 through 5) with masses, in kilograms, as follows: 500, 350, 200, 800, and 700, respectively. Assume that the center of gravity of each individual crate is at the center of that crate. Locate the center of gravity of the load with reference to point A.

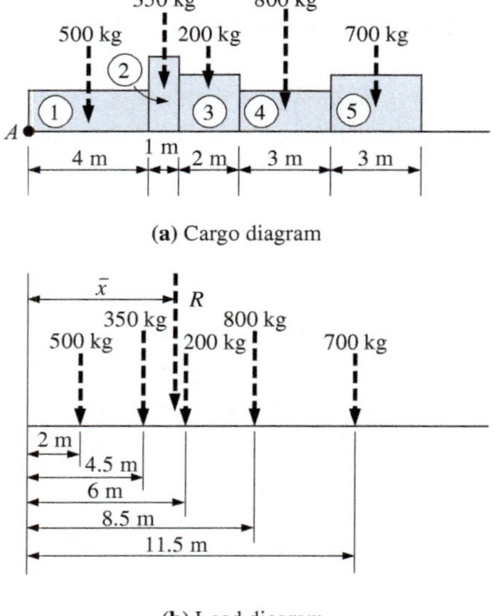

FIGURE 7.4 Center of gravity.

Solution The weight of each crate could be determined, but this is unnecessary since weights will be directly proportional to the known masses. The given masses will be used. The resultant mass R is

$$R = \Sigma m = 500\text{ kg} + 350\text{ kg} + 200\text{ kg} + 800\text{ kg} + 700\text{ kg} = 2550\text{ kg}$$

R is shown in Figure 7.4b. The location of the resultant (which passes through the center of gravity) is calculated from Varignon's theorem,

$$R\bar{x} = \Sigma mx$$

where $\bar{x}$ is the distance from point A to the resultant (or to the center of gravity).

$$\bar{x} = \frac{\Sigma mx}{R} = \frac{500\text{ kg}(2\text{ m}) + 350\text{ kg}(4.5\text{ m}) + 200\text{ kg}(6\text{ m}) + 800\text{ kg}(8.5\text{ m}) + 700\text{ kg}(11.5\text{ m})}{2550\text{ kg}}$$

$$= 7.30\text{ m}$$

7.2.1 Summary of Procedure: Center of Gravity of Built-Up Members

1. Sketch the member showing all dimensions. Note any axes of symmetry.
2. Divide the member into component parts. Each part must be proportioned so that its weight can be determined and its center of gravity can be located.
3. Select a reference axis.
4. Apply Equation (7.1) and/or Equation (7.2) to determine $\bar{x}$ and/or $\bar{y}$. (A similar equation may be written for $\bar{z}$ if necessary.)

7.3 CENTROIDS AND CENTROIDAL AXES

If we assume that the irregularly shaped flat plate shown in Figure 7.1 is homogeneous and has a uniform thickness, the weight of the plate would be directly proportional to the area. Thus, we can use areas instead of weights (forces) in the equations of Section 7.2 to determine the location of the centroid of the area. This strategy is equivalent to allowing the thickness of the plate to approach zero and finding its center of gravity.

The procedure for finding the centroid of an area is exactly the same as the procedure described in Section 7.2 for finding the center of gravity, except for the following substitutions: a replaces w and A replaces W. The term a represents an infinitesimal component of area, and A represents Σa (or, the total area).

Applying Varignon's theorem, the moment of the total area about any axis will be equal to the algebraic sum of the moments of the component areas about the same axis. Note that the moment of an area is analogous to the moment of a force, except that the moment of a force has physical meaning, whereas the moment of an area is a mathematical concept. Moment of area has units of length cubed (in.^{3}, mm^3). Referring to Figure 7.1, and making the substitutions just mentioned, equations for $\bar{x}$ and $\bar{y}$ may be written as follows. With respect to the Y–Y axis,

$$A\bar{x} = \Sigma ax$$

from which

$$\bar{x} = \frac{\Sigma ax}{A} \quad \text{or} \quad \frac{\Sigma ax}{\Sigma a} \qquad (7.3)$$

With respect to the X–X axis,

$$A\bar{y} = \Sigma ay$$

from which

$$\bar{y} = \frac{\Sigma ay}{A} \quad \text{or} \quad \frac{\Sigma ay}{\Sigma a} \qquad (7.4)$$

The terms $\bar{x}$ and $\bar{y}$ represent coordinates of the centroid of an area. An axis that passes through the centroid is generally termed a *centroidal axis*. A centroidal axis, then, of great significance in statics and strength of materials, is an axis on which the centroid of the area lies.

Areas and centroids for some frequently encountered geometric shapes have been determined mathematically and are shown in Table 2, inside the front cover of this book. Examples for locating the centroids of a triangle and a semicircle using integration (calculus) are provided in Appendix K.

7.4 CENTROIDS AND CENTROIDAL AXES OF COMPOSITE AREAS

A composite, or built-up, area may be described as one made up of a number of simple geometric areas or standardized shapes. Many of the members used in structures and machines are combinations of various shapes so connected (usually by welding) as to act as a single unit. To determine the location of the centroid of a composite area, the area is generally divided into two or more component areas, each with a known area and a known centroid location. The centroid location for the composite area can then be determined by applying Equations (7.3) and (7.4). The centroidal axes are two axes, at right angles to each other, that intersect at the centroid. These axes are used in many applications. The centroidal axes of importance normally coincide with axes of symmetry or are parallel/perpendicular to major elements of the composite area.

In determining the location ($\bar{x}$ and $\bar{y}$) of the centroid or centroidal axes of a composite area, reference axes must be established. It is customary that a conventional reference X–Y coordinate axes system be used. One technique is to establish the axes in such a manner that the entire composite area lies in the upper right quadrant (the first quadrant) of the coordinate axes system. The lowest edge of the area should lie on the X axis and the left edge of the area should lie on the Y axis (see Figure 7.6 in Example 7.3). This technique avoids any need for a sign convention, since x distances and y distances measured from the reference axes are upward and to the right and are therefore considered positive values.

After establishing the reference axes, the composite area should be divided into simple geometric areas, such as rectangles, triangles, or standard shapes. Moments of these component areas are then taken with respect to each reference axis.

If a hole is cut out of the area, it must be treated as a *negative* area. The negative sign will then remove the effect of that area (which is absent) in the summation of Σax or Σay and Σa. Similarly, if an area is added to the composite area, a positive sign in the summations will include the effect of that area.

EXAMPLE 7.3 Determine the location of the centroid of the area shown in Figure 7.5.

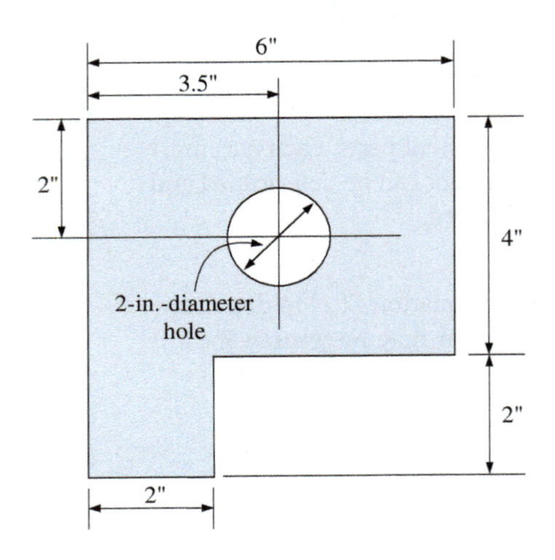

FIGURE 7.5 Composite area.

Solution First establish a reference X–Y coordinate axes system as shown in Figure 7.6. The entire composite area is placed in the first quadrant and then divided into three component areas as shown. The areas are then determined:

$$a_1(\text{rectangle}) = (6\text{ in.})(4\text{ in.}) = 24\text{ in.}^2$$
$$a_2(\text{rectangle}) = (2\text{ in.})(2\text{ in.}) = 4\text{ in.}^2$$
$$a_3(\text{circle}) = \frac{\pi d^2}{4} = 0.7854d^2 = 0.7854(2\text{ in.})^2 = -3.14\text{ in.}^2$$

Note that the area of the circle carries a negative sign since it represents a cutout from the composite area and, in effect, reduces the total area. The total area is calculated:

$$A = a_1 + a_2 + a_3$$
$$= 24\text{ in.}^2 + 4\text{ in.}^2 + (-3.14\text{ in.}^2) = 24.86\text{ in.}^2$$

The component areas are all simple geometric shapes with the location of the centroid of each area known. These locations are shown in Figure 7.6 with respect to the reference axes.

Next, Equation (7.3) is used to determine $\bar{x}$. It may be helpful to visualize this process as one of pivoting the component areas about reference axis Y–Y. Thus,

$$\bar{x} = \frac{\Sigma ax}{A} = \frac{a_1x_1 + a_2x_2 + a_3x_3}{A}$$
$$= \frac{(24\text{ in.}^2)(3\text{ in.}) + (4\text{ in.}^2)(1\text{ in.}) + (-3.14\text{ in.}^2)(3.5\text{ in.})}{24.86\text{ in.}^2} = 2.62\text{ in.}$$

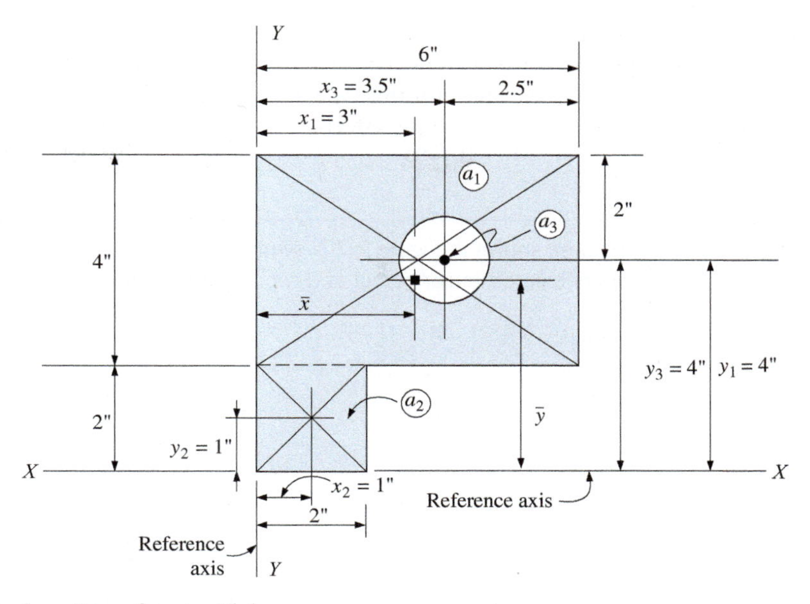

FIGURE 7.6 Location of centroidal axes.

Equation (7.4) yields

$$\bar{y} = \frac{\Sigma ay}{A} = \frac{a_1y_1 + a_2y_2 + a_3y_3}{A}$$

$$= \frac{(24 \text{ in.}^2)(4 \text{ in.}) + (4 \text{ in.}^2)(1 \text{ in.}) + (-3.14 \text{ in.}^2)(4 \text{ in.})}{24.86 \text{ in.}^2} = 3.52 \text{ in.}$$

Table 7.1 shows a tabular format that can be used. Therefore, substitute values from the table:

TABLE 7.1 Tabular format for Example 7.3

Component Area	a (in.^{2})	x (in.)	ax (in.^{3})	y (in.)	ay (in.^{3})
a_1	24.0	3.0	72.0	4.0	96.0
a_2	4.0	1.0	4.0	1.0	4.0
a_3	−3.14	3.5	−10.99	4.0	−12.56
Σ	24.86		65.01		87.44

$$\bar{x} = \frac{\Sigma ax}{\Sigma a} = \frac{65.01 \text{ in.}^3}{24.86 \text{ in.}^2} = 2.62 \text{ in.}$$

and

$$\bar{y} = \frac{\Sigma ay}{\Sigma a} = \frac{87.44 \text{ in.}^3}{24.86 \text{ in.}^2} = 3.52 \text{ in.}$$

Thus, the centroid of the area has been located. Note that as shown in Figure 7.6 the centroid actually falls within the cutout circle. Other common cross sections in which this would occur are such standard shapes as pipes, tubes, angles, and channels.

If we were to superimpose an X–Y coordinate axes system on the area so that the origin coincided with the centroid of the area, we would establish the X–Y centroidal axes. (Recall that the centroidal axes intersect at the centroid of a cross section.) These axes would normally be parallel to their respective X–X and Y–Y reference axes.

The computer spreadsheet is a tool that can be readily adapted to these calculations, following the concept of the tabular format. The power of the spreadsheet for this type of application is in the ability to see very quickly the effects of changes made to the original cross section. Several packages are available that would serve nicely, such as Excel, a product of the Microsoft Corporation. The tabular approach is further extended for the calculations of Chapter 8.

EXAMPLE 7.4 The area shown in Figure 7.7 is symmetric about the Y–Y axis and is composed of a rectangular area with one semicircular cutout and two triangular cutouts. Determine the location of the centroid.

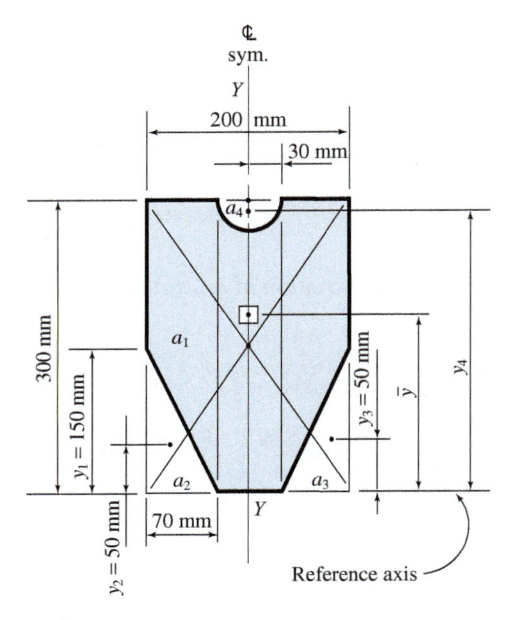

FIGURE 7.7 Sketch for Example 7.4.

Solution Since the area is symmetric about the Y–Y axis, the centroid must lie on the Y–Y axis and only the vertical location $\bar{y}$ of the centroid must be calculated. A horizontal reference axis is established at the bottom of the area as shown. The area is then divided into simple geometric component areas as follows: a 200-mm-by-300-mm rectangular area a_1, two 150-mm-by-70-mm right triangles a_2 and a_3, and one 30-mm-radius semicircular area a_4. The magnitudes of these component areas are calculated from

$$a_1 = bh = 200\text{ mm}(300\text{ mm}) = 60{,}000\text{ mm}^2$$

$$a_2 \text{ and } a_3 = \frac{1}{2}bh = \frac{1}{2}(70\text{ mm})(150\text{ mm}) = 5250\text{ mm}^2$$

$$a_4 = \frac{\pi R^2}{2} = \frac{\pi(30\text{ mm})^2}{2} = 1414\text{ mm}^2$$

The centroid locations of the component areas with respect to the reference axis are noted as y_1, y_2, y_3, and y_4. The location of the centroid of the semicircle y_4 is determined using data from Table 2 (inside the front cover):

$$y_4 = 300\text{ mm} - \frac{4R}{3\pi} = 300\text{ mm} - \frac{4(30\text{ mm})}{3\pi} = 287\text{ mm}$$

Applying Equation (7.4) and noting that a_2, a_3, and a_4 are taken as negative gives

$$\bar{y} = \frac{\Sigma ay}{A} = \frac{a_1y_1 + a_2y_2 + a_3y_3 + a_4y_4}{a_1 + a_2 + a_3 + a_4}$$

$$= \frac{60{,}000\text{ mm}^2(150\text{ mm}) + (-5250\text{ mm}^2)(50\text{ mm}) + (-5250\text{ mm}^2)(50\text{ mm}) + (-1414\text{ mm}^2)(287\text{ mm})}{60{,}000\text{ mm}^2 + (-5250\text{ mm}^2) + (-5250\text{ mm}^2) + (-1414\text{ mm}^2)}$$

$$= 167.8\text{ mm}$$

Table 7.2 shows a tabular format that could be used in place of the long equation for $\bar{y}$. From Equation (7.4),

TABLE 7.2 Tabular format for Example 7.4

Component Area	a (mm^2)	y (mm)	ay (mm^3)
a_1	60,000	150	9,000,000
a_2	−5250	50	−262,500
a_3	−5250	50	−262,500
a_4	−1414	287	−405,800
Σ	48,100		8,070,000

$$\bar{y} = \frac{\Sigma ay}{A} = \frac{8{,}070{,}000 \text{ mm}^3}{48{,}000 \text{ mm}^2} = 167.8 \text{ mm}$$

In the analysis and design of structural members and machine parts, commercially available structural shapes of various materials are commonly used. Appendices A through E provide sizes and properties for some of four common structural steel shapes and for some timber members of square and rectangular cross sections. The profiles of these shapes are shown in Figure 7.8. Both U.S. Customary System and SI information is given in the appendices.

Appendix A contains sizes and properties of wide-flange (W) shapes, the most commonly used hot-rolled structural steel shapes. A typical designation is W8 × 31, which indicates the basic shape (W), the approximate depth (8 in.), and the weight in pounds per foot (31). The pipe in Appendix B is designated by diameter and weight (standard or extra strong). The channels of Appendix C are designated by depth and weight (a C15 × 50 is 15 in. deep and weighs 50 lb/ft). Angles of Appendix D are designated by leg sizes and thickness (an L4 × 3 × $\frac{1}{4}$ is an angle having one 4 in. leg and one 3 in. leg, with both legs $\frac{1}{4}$ in. thick). SI designations follow a similar pattern, with units of mass per meter provided in place of weight per foot.

Appendix E contains sizes and properties of solid, rectangular timber members. The sizes by which the sections are designated are *nominal* ("in name only") sizes such as 2 in. × 4 in. or 75 mm × 200 mm. The *dressed* sizes are the actual dimensions of the finished product after surfacing (planing). Most structural timbers are surfaced on all four sides, a treatment designated as S4S (surfaced four sides).

In the calculations of center of gravity of members and centroids of areas, we use the common steel and timber shapes from the appendices. For these calculations, only areas, weights, and dimensions are required. Other tabulated properties are discussed in Chapter 8.

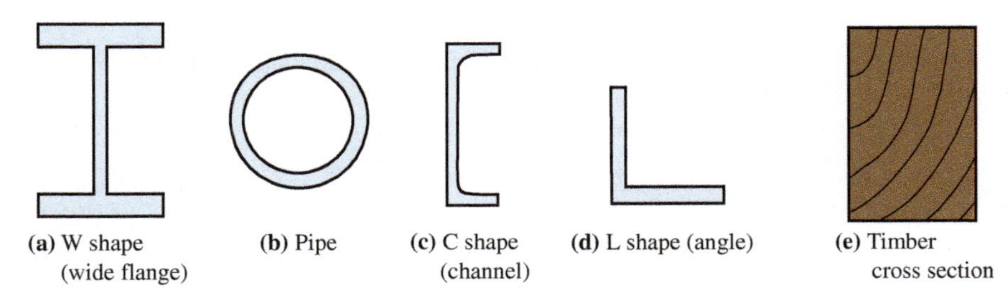

FIGURE 7.8 Some typical shapes for structural members.

EXAMPLE 7.5 A built-up steel member, shown in Figure 7.9, is fabricated from two C15 × 40 American Standard channels, a 16 in. by 1 in. top plate, and a 14 in. by $\frac{1}{2}$ in. bottom plate. All of the components are welded together securely so as to act as a single unit. Locate the X–X centroidal axis.

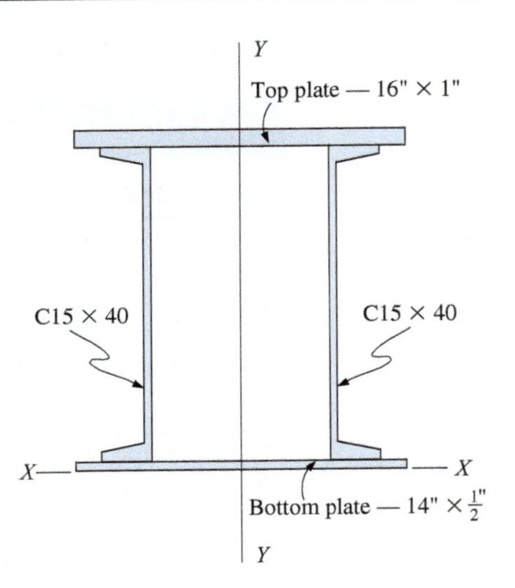

FIGURE 7.9 Built-up steel member.

Solution In this problem, the member is symmetrical with respect to the vertical *Y–Y* axis shown. Recall that an axis of symmetry is always a centroidal axis. Only $\bar{y}$, locating the *X–X* centroidal axis, must be determined. The lower edge of the member (the bottom of the bottom plate) is chosen as the reference *X–X* axis.

The component areas are shown in Figure 7.10. Areas are determined as follows:

$$a_1 = 16(1) = 16 \text{ in.}^2$$
$$\left.\begin{aligned} a_2 &= 11.8 \text{ in.}^2 \\ a_3 &= 11.8 \text{ in.}^2 \end{aligned}\right\} \text{(See Appendix C)}$$
$$a_4 = 14(0.5) = 7 \text{ in.}^2$$
$$A = \Sigma a = 46.6 \text{ in.}^2$$

Note that the components include two rectangular plates and two standard channels. All necessary properties and dimensions of the channels may be found in Appendix C. The locations of the centroids of the components (the *y* dimensions) are calculated as follows and are shown in Figure 7.10:

$$y_1 = 0.5 \text{ in.} + 15 \text{ in.} + 0.5 \text{ in.} = 16 \text{ in.}$$
$$y_2 = y_3 = 7.5 \text{ in.} + 0.5 \text{ in.} = 8 \text{ in.}$$
$$y_4 = 0.25 \text{ in.}$$

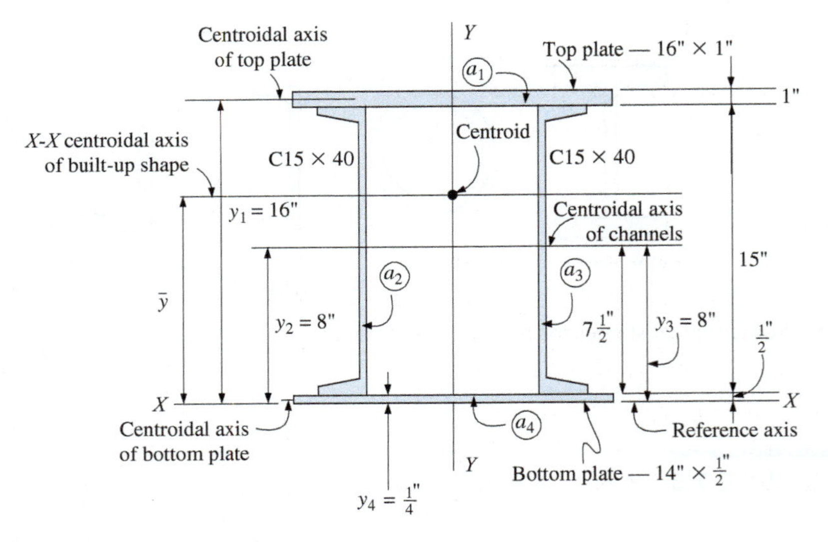

FIGURE 7.10 Location of centroidal axis.

Applying Equation (7.4) gives

$$\bar{y} = \frac{\Sigma ay}{A} = \frac{a_1y_1 + a_2y_2 + a_3y_3 + a_4y_4}{A}$$

$$= \frac{(16\text{ in.}^2)(16\text{ in.}) + (11.8\text{ in.}^2)(8\text{ in.}) + (11.8\text{ in.}^2)(8\text{ in.}) + (7\text{ in.}^2)(0.25\text{ in.})}{46.6\text{ in.}^2} = 9.58\text{ in.}$$

Therefore, the *X–X* centroidal axis lies 9.58 in. above the reference axis. Table 7.3 shows a tabular format that can be used as well.

TABLE 7.3 Tabular format for Example 7.5

Component Area	a (in.^{2})	y (in.)	ay (in.^{3})	Notes
a_1	16.0	16.0	256.0	Top plate
a_2	11.8	8.0	94.4	Channel
a_3	11.8	8.0	94.4	Channel
a_4	7.0	0.25	1.75	Bottom plate
Σ	46.6		447	

After the summations Σa and Σay have been determined, the final calculation for $\bar{y}$ can be made:

$$\bar{y} = \frac{\Sigma ay}{\Sigma a} = \frac{447\text{ in.}^3}{46.6\text{ in.}^2} = 9.59\text{ in.}$$

7.4.1 Summary of Procedure: Centroids of Areas

1. Sketch the area showing all known dimensions.
2. Note any axis of symmetry. Establish a reference *X–Y* coordinate axes system. It is usually most convenient for calculation purposes if the reference axes are placed at the bottom edge and/or the left edge of the area.
3. Divide the area into component areas. Each component area must be proportioned so that its area and the location of its centroid can be determined.
4. Apply Equation (7.3) and/or Equation (7.4).

SUMMARY BY SECTION NUMBER

7.1 The center of gravity of a body is the point through which the line of action of its total weight passes.

7.2 The center of gravity of a body is located in the same way the resultant force of a parallel force system is located. The location may be determined using

$$\bar{x} = \frac{\Sigma wx}{W} \quad \text{or} \quad \frac{\Sigma wx}{\Sigma w} \qquad (7.1)$$

$$\bar{y} = \frac{\Sigma wy}{W} \quad \text{or} \quad \frac{\Sigma wy}{\Sigma w} \qquad (7.2)$$

The term *centroid* is used when referring to the center of gravity of an area, which may be thought of as a plate having zero thickness.

7.3 The location of a centroid may be determined using

$$\bar{x} = \frac{\Sigma ax}{A} \quad \text{or} \quad \frac{\Sigma ax}{\Sigma a} \qquad (7.3)$$

$$\bar{y} = \frac{\Sigma ay}{A} \quad \text{or} \quad \frac{\Sigma ay}{\Sigma a} \qquad (7.4)$$

7.4 A composite area is made up of a number of simple geometric areas or standardized shapes. The location of the centroid of the composite area may be determined by using Equations (7.3) and (7.4).

PROBLEMS

Section 7.2 Center of Gravity

For the following problems, refer to Appendix G for information on unit weights.

7.1 A cylindrical cast-iron casting has an axial hole extending partway through the casting, as shown. Locate the center of gravity of the casting.

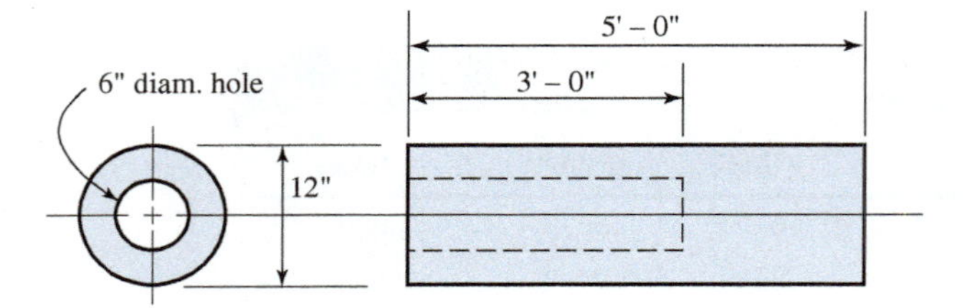

PROBLEM 7.1

7.2 Locate the center of gravity of the cast-iron casting of Problem 7.1, assuming the hole is filled with a magnesium alloy.

7.3 A 5-in.-diameter steel sphere is rigidly attached to the end of a 24-in.-long, $1\frac{1}{2}$-in.-diameter aluminum rod. Locate the center of gravity of the composite member.

7.4 A solid steel shaft is fabricated as shown. Locate the center of gravity with respect to the left end.

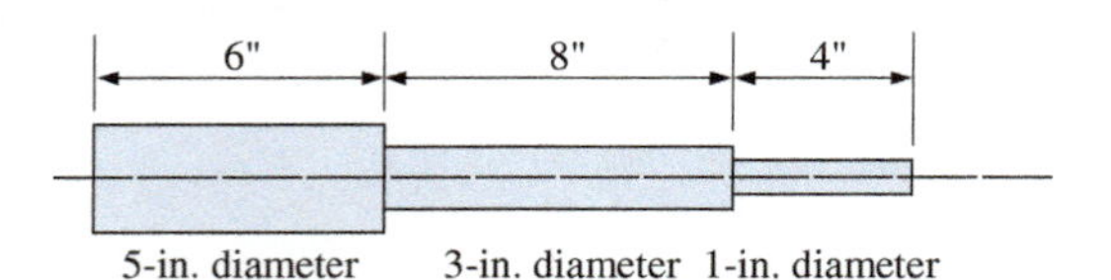

PROBLEM 7.4

7.5 A wood mallet has a cylindrical head 6 in. long and 3 in. in diameter. If the cylindrical handle is 1 in. in diameter, how long must it be for the mallet to balance at a point 6 in. from where the handle enters the mallet head? (Assume the unit weight of the wood to be 40 lb/ft^3.)

7.6 The built-up member shown is composed of two W460 × 60 wide-flange shapes. There is a 300-mm-diameter hole located on the centerline in the lower shape, as shown. Locate the center of gravity vertically and horizontally from point A.

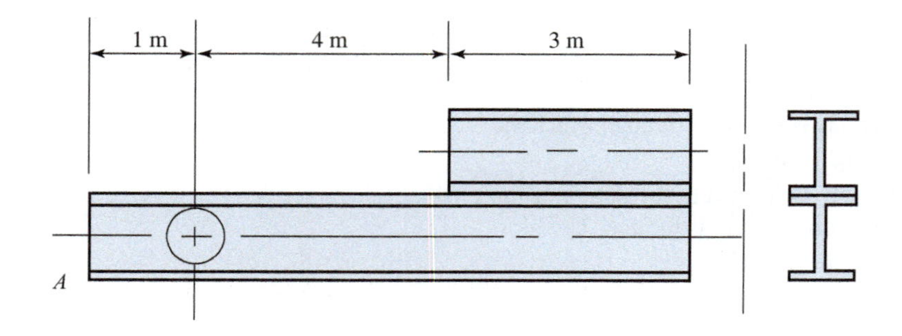

PROBLEM 7.6

Section 7.4 Centroids and Centroidal Axes of Composite Areas

7.7 A concrete member has a cross section as shown. Locate the centroid.

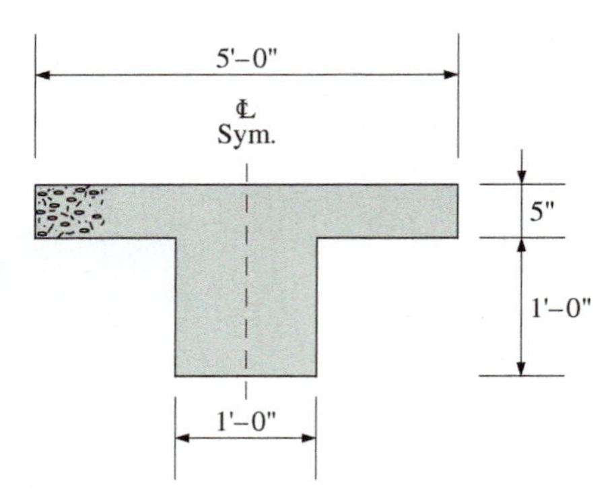

PROBLEM 7.7

7.8 A thin steel plate, having the dimensions shown, contains three 2-in.-diameter holes. Note the axis of symmetry. Locate the centroid of the area.

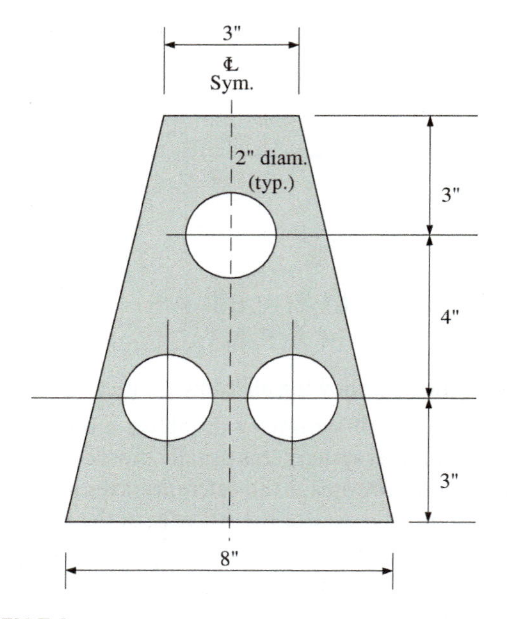

PROBLEM 7.8

7.9 The U-shaped built-up section shown is composed of dressed (S4S) timber sections. (See Appendix E for dressed sizes.) Locate the centroid of the area.

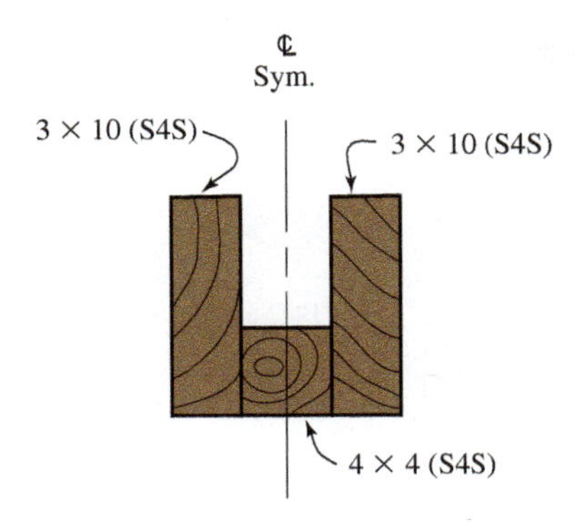

PROBLEM 7.9

7.10 Locate the centroid of the area shown. In part (b), note the symbol signifying that the area is symmetric about the centerline.

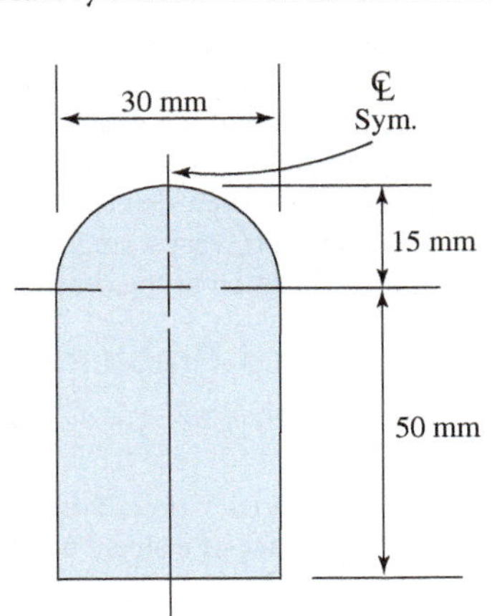

(a)

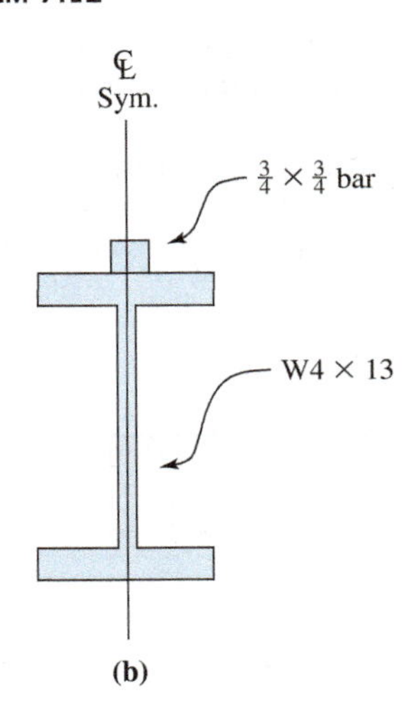

(b)

PROBLEM 7.10

7.11 Locate the X–X and Y–Y centroidal axes for the areas shown.

7.12 A built-up steel member is composed of a W18 × 71 wide-flange section with a 12-in.-by-$\frac{1}{2}$-in. plate welded to its top flange, as shown. Locate the X–X centroidal axis.

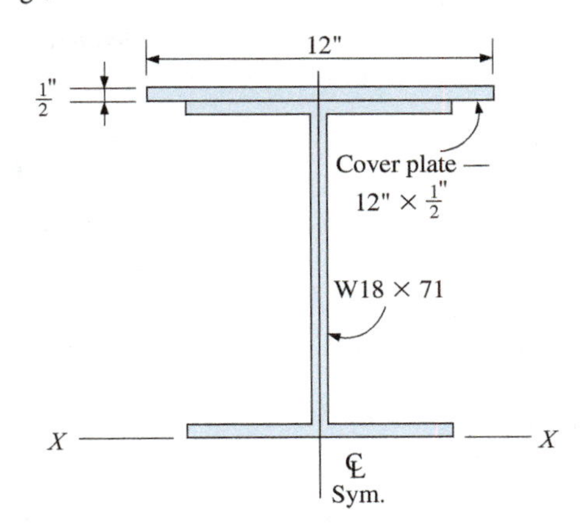

PROBLEM 7.12

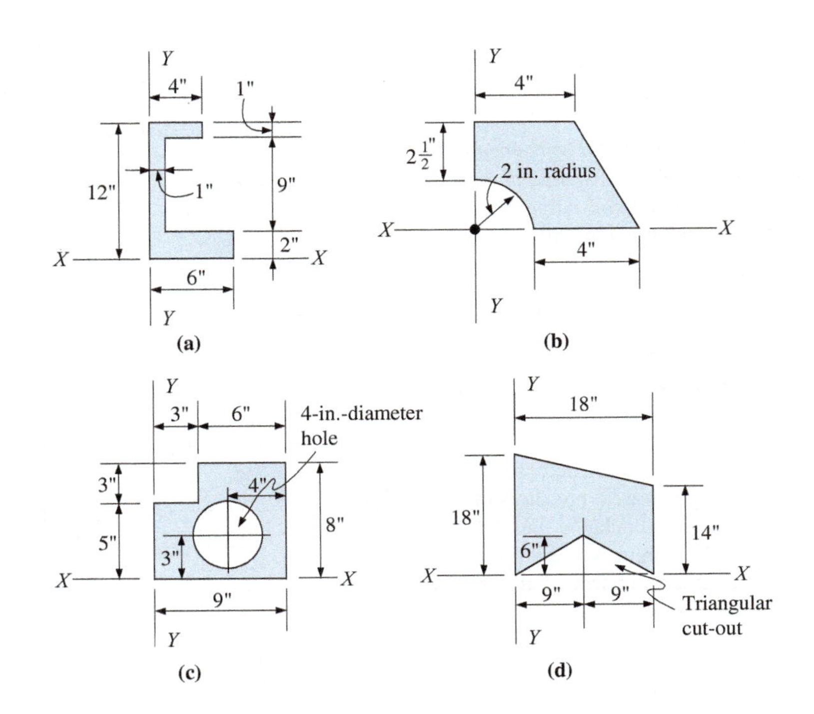

PROBLEM 7.11

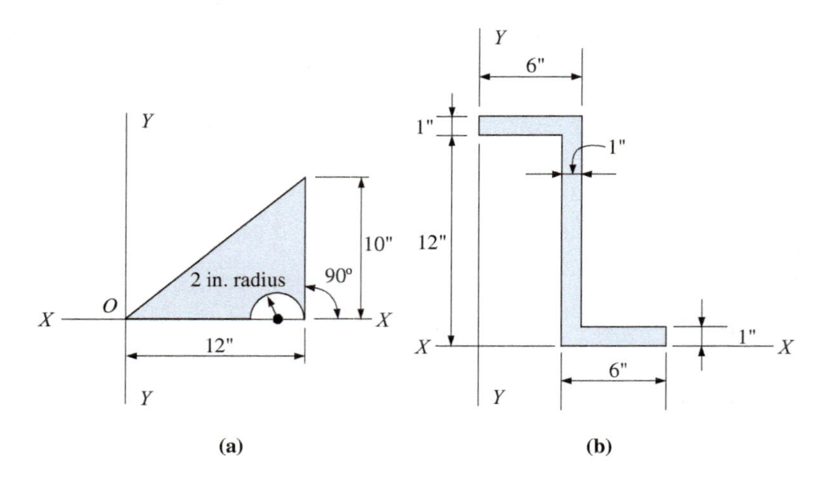

PROBLEM 7.13

7.13 Locate the X–X and Y–Y centroidal axes for the areas shown.

7.14 Find the center of gravity for the three-axle truck shown. The front axle supports a mass of 3.5 Mg, the middle axle supports a mass of 12.0 Mg, and the rear axle supports a mass of 14.0 Mg. The middle axle is 4.5 m from the front axle and 10 m from the rear axle. Locate the center of gravity with respect to the front axle.

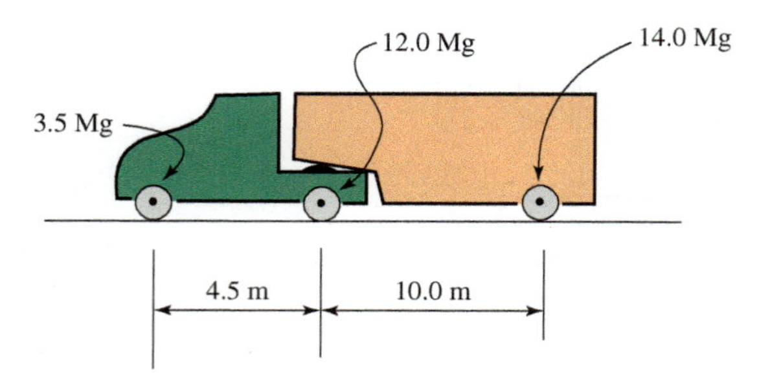

PROBLEM 7.14

Computer Problems

For the following computer problems, any appropriate software may be used. Input prompts should fully explain what is required of the user (the program should be user-friendly). The resulting output should be well labeled and self-explanatory. For spreadsheet problems, any appropriate software may be used.

7.15 Write a program that will calculate the location of the centroid for a tee shape similar to that of Problem 7.7. User input is to be the width and the depth of the two rectangles. (The upper element is called the *flange*; the lower element is called the *web*.)

7.16 Write a program that will calculate the location of the centroid of a semicircle (measured from the diameter line). This may be accomplished by dividing the area into slices of uniform width, either parallel or perpendicular to the diameter, and applying Equation (7.4). User input is to be the radius and width of the slice. Check the result with that obtained from Table 2. Note how the accuracy varies with the choice of the slice width.

7.17 Use a spreadsheet program to solve for the centroid location of built-up areas such as shown in Problems 7.12 and 7.28. The user should be able to input the appropriate geometric properties for any shapes of the types shown, and the program should then determine the centroid location and display the result. Vary a single shape or dimension and observe how the location of the centroid is affected.

Supplemental Problems

For the following problems, refer to Appendix G if unit weights are not stated.

7.18 A 2-in.-diameter hole, 5 in. long, is drilled into the center of the top face of a steel cube that is 6 in. on each side. The hole is filled with lead, which has a unit weight of 710 pcf. Locate the center of gravity of the cube with respect to the bottom.

7.19 Locate the center of gravity of the cube in Problem 7.18, assuming the hole is left empty.

7.20 The head of a maul is made of steel and is 3 in. in diameter and 5 in. long. It has a wood handle 1 in. in diameter and 3 ft long. The unit weight of the wood is 80 pcf. Calculate the distance from the end of the handle to the center of gravity.

7.21 Locate the X–X centroidal axis for the cross section shown.

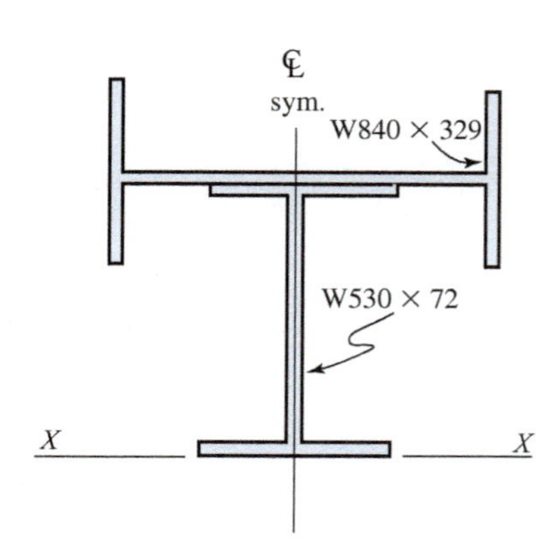

PROBLEM 7.21

7.22 A built-up steel member is composed of a W21 × 62 wide-flange section with a C12 × 25 American Standard channel welded to its top flange, as shown. Locate the *X–X* centroidal axis.

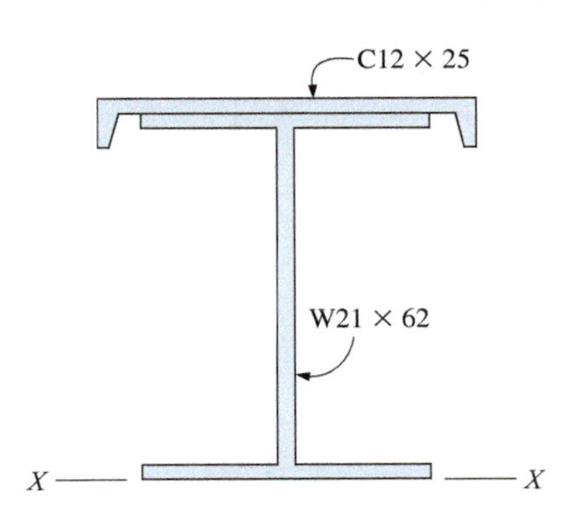

PROBLEM 7.22

7.23 Locate the *X–X* centroidal axis for the cross section shown.

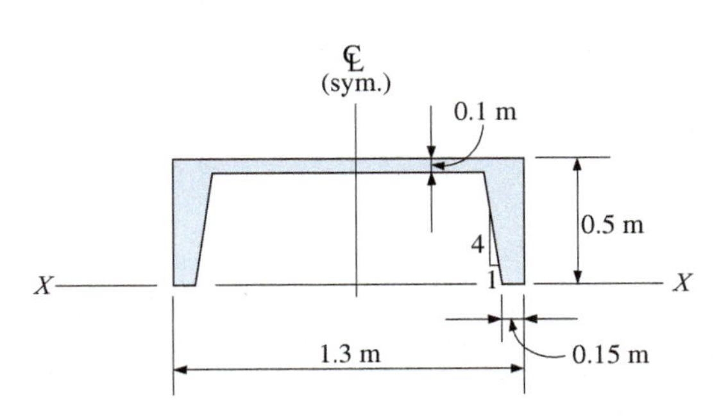

PROBLEM 7.23

7.24 Locate the *X–X* and *Y–Y* centroidal axes for the cross section shown.

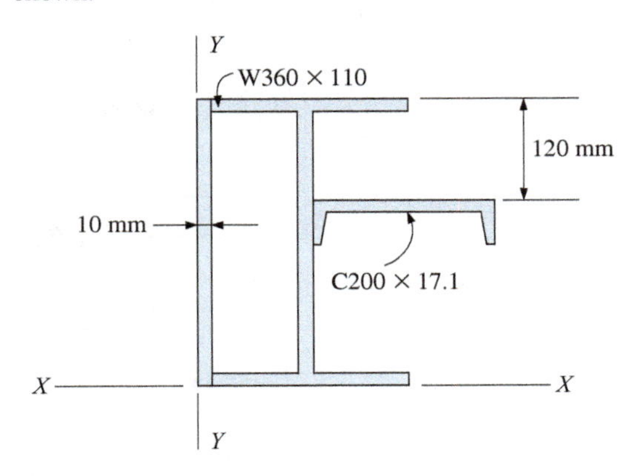

PROBLEM 7.24

7.25 Locate the *X–X* and *Y–Y* centroidal axes for the areas shown. Assume standard weight pipe.

7.26 For the area shown, the *X–X* centroidal axis is required to be located at 10 in. above the reference axis. Calculate the required dimension *h*.

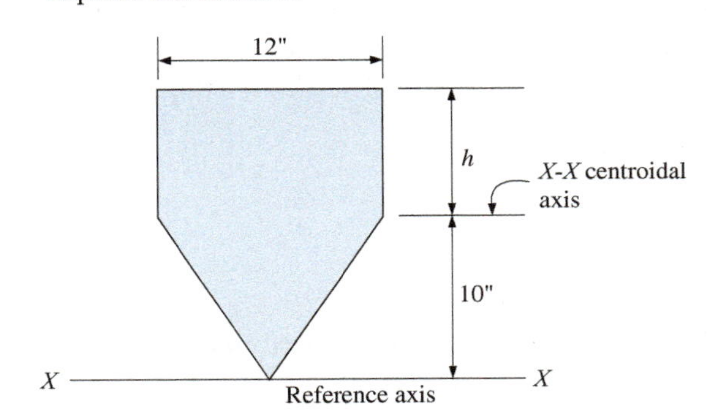

PROBLEM 7.26

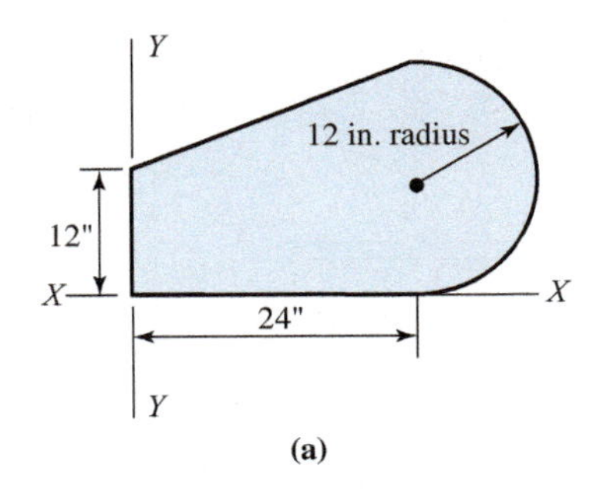

(a)

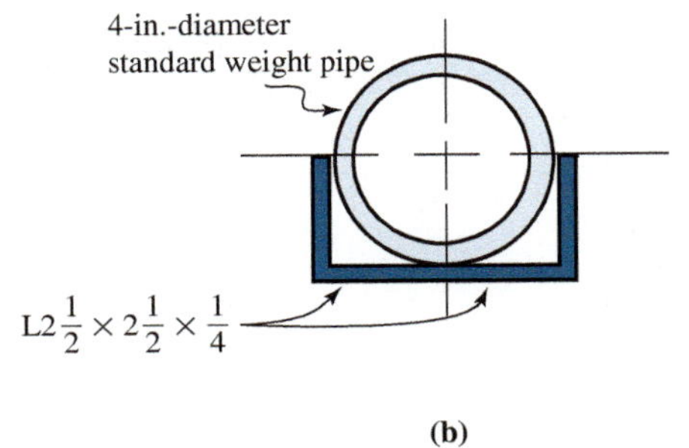

(b)

PROBLEM 7.25

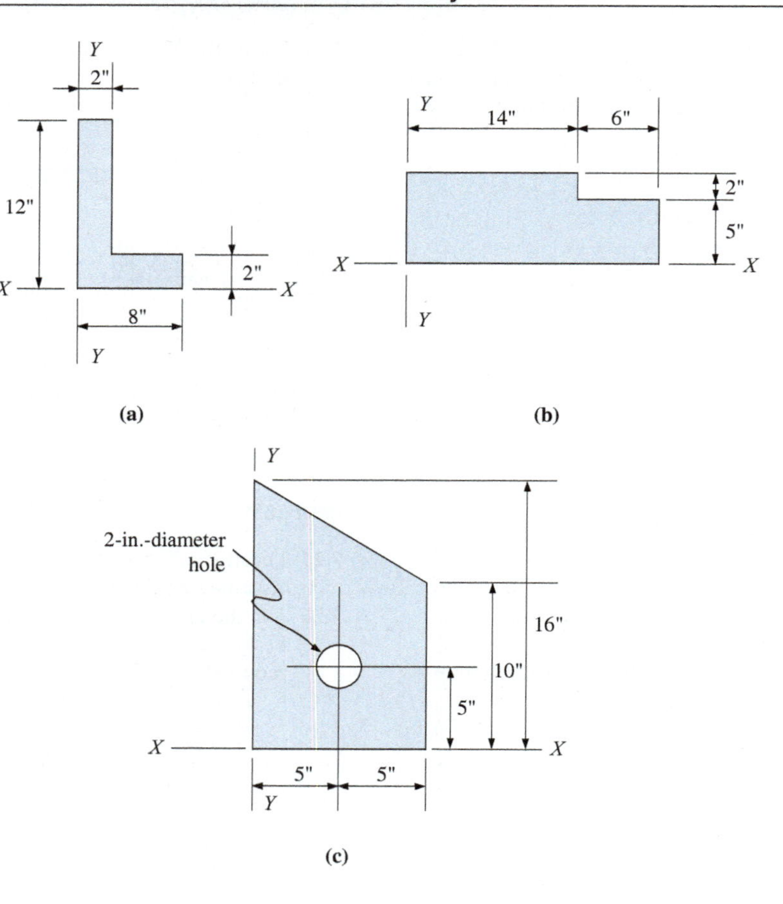

PROBLEM 7.27

7.27 Locate the X–X and Y–Y centroidal axes for the areas shown.

7.28 Locate the X–X and Y–Y centroidal axes for the built-up structural steel areas shown.

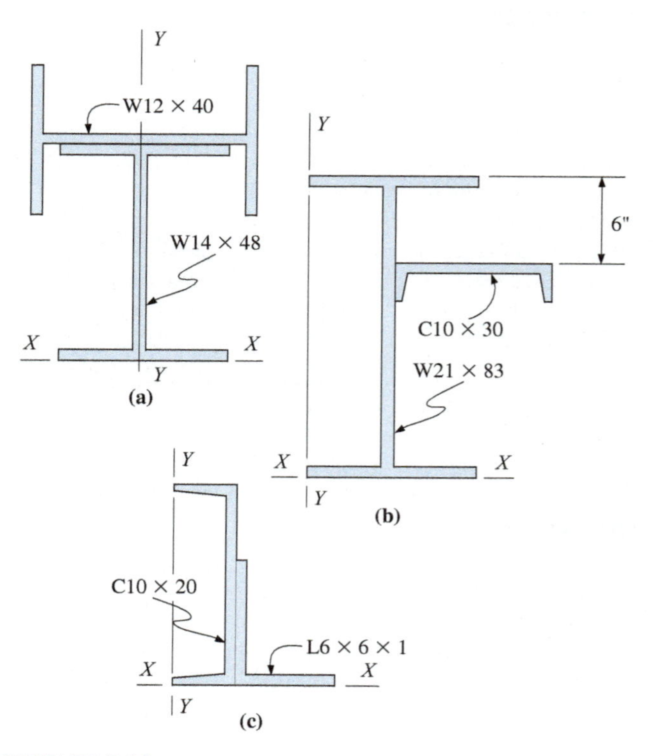

PROBLEM 7.28

CHAPTER EIGHT

AREA MOMENTS OF INERTIA

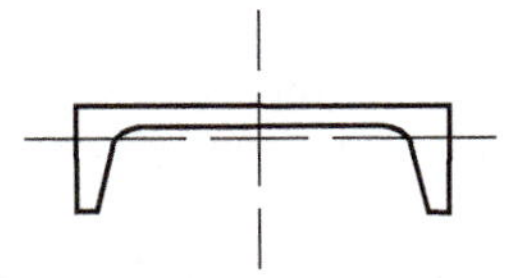

LEARNING OBJECTIVES

Upon completion of this chapter, readers will be able to:

- Discuss the theory of the second area of moment and how it applies to general shapes.
- Determine the moment of inertia of an object about an axis that is parallel to the centroidal axis using the parallel axis theorem.
- Calculate the moment of inertia of a composite, or built up, shape about its centroidal axis using the transfer theorem.
- Calculate the radius of gyration for a shape using the moment of inertia.

8.1 INTRODUCTION AND DEFINITIONS

In Chapters 9 through 21 of this text, we consider aspects of the strength of machine parts and structural members. In the development of the equations dealing with the bending of beams, the buckling of columns, and the twisting of circular shafts, we make use of several quantities that are dependent on the size and shape of the cross sections of the members. These quantities are used so widely that they have been given special names and are tabulated in numerous sources as common properties. Foremost among them is the property called the *moment of inertia*, which is denoted I.

The moment of inertia of an area represents one of the more abstract concepts in engineering mechanics. It is not a readily observable property of the area; rather, it is a purely mathematical quantity, albeit a very important one.

The moment of inertia of an area may be defined by considering a plane area designated A, as shown in Figure 8.1. Let X–X and Y–Y be any set of rectangular axes in the same plane as the area. Area A is divided into small areas (represented by a), and each small area is located with respect to the axes. The coordinates of a are distances x and y. A moment of inertia must always be computed with respect to a specific axis. Therefore, in Figure 8.1, we may have a moment of inertia with respect to axis X–X, denoted I_x, or with respect to axis Y–Y, denoted I_y. The moment of inertia is defined as the sum of all the small areas, each multiplied by the square of its distance (moment arm) from the axis being considered.

Thus, as shown in Figure 8.1, the moment of inertia about the X–X axis is the summation of the products of each area a and the square of its moment arm y, which gives

$$I_x = \Sigma a y^2 \tag{8.1}$$

Similarly, the moment of inertia about the Y–Y axis is given by

$$I_y = \Sigma a x^2 \tag{8.2}$$

These mathematical expressions give what are sometimes referred to as the *second moments* of the area, since each small area, when multiplied by its moment arm, gives the moment of the area (or first moment of the area). When multiplied a second time by the moment arm, the result is the second moment of the area (or moment of inertia). These moments of inertia are also referred to as *rectangular moments of inertia*.

The expression "moment of inertia of an area" is actually a misnomer, since plane areas have no thickness and,

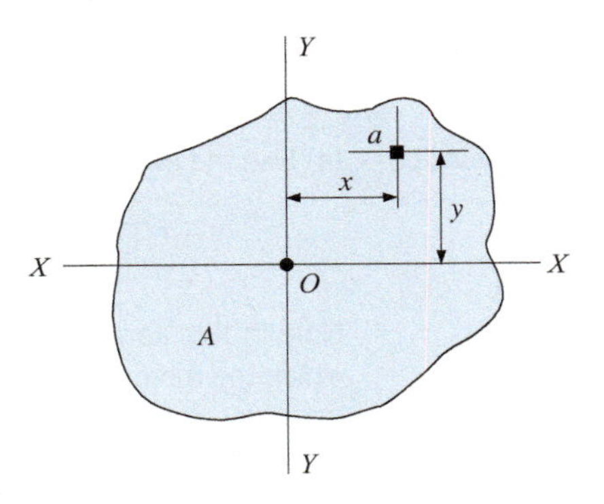

FIGURE 8.1 Moment of inertia of an area.

therefore, no mass or inertia. It is a traditional expression, however, and we use it throughout this text.

Since the moment of inertia is an area multiplied by the square of a distance, the resulting units will be length to the fourth power. The moment of inertia is generally expressed as in.^4 in the U.S. Customary System. In the SI, the recommended units are mm^4 or m^4. Note that the moment of inertia is always a positive quantity.

The magnitude of the moment of inertia is a measure of the ability of a cross-sectional area to resist bending or buckling. If we consider two beams of the same material but of different cross sections, the beam having the cross-sectional area with the greater moment of inertia would have the greater resistance to bending. The beam with the greater moment of inertia, however, does not necessarily have the greater cross-sectional area. It is the distribution of the area relative to the reference axis that determines the magnitude of the moment of inertia.

Generally, the moment of inertia desired is with respect to axes that pass through the centroid of a section. We are usually interested in either the maximum or the minimum moment of inertia, which, for symmetrical sections, will be with respect to the centroidal axis (or axes), coinciding with the axis (or axes) of symmetry. A comprehensive treatment of moment of inertia about inclined axes and of maximum/minimum moments of inertia for shapes with no axis of symmetry is beyond the scope of this text.

8.2 MOMENT OF INERTIA

Using the calculus form of Equations (8.1) and (8.2) and assuming a total area divided into infinitesimal component areas, exact theoretical formulas have been mathematically derived for determining the moment of inertia of simple geometric shapes. Derivations of the moment-of-inertia formulas for rectangular and triangular areas are given in Appendix L. Moment-of-inertia formulas for the most commonly used geometric areas with respect to the designated axes are given in Table 3, inside the back cover of this book.

An approximate determination of the moment of inertia of an area can be obtained by dividing the total area into finite component areas. The moment of inertia of each component area can then be calculated using Σay^2 or Σax^2. The moment of inertia of the total area about any particular axis is then equal to the sum of the moments of inertia of the component areas about the same axis.

This sum will result in an approximate moment of inertia, with the degree of accuracy a function of the size selected for the component areas. The smaller the size of the component areas, the greater the accuracy.

This approximate method is used chiefly with irregular areas for which exact derived formulas are either not applicable or excessively cumbersome to use. Example 8.1 illustrates the technique of the approximate method and compares the result with the theoretically exact method.

EXAMPLE 8.1 Calculate the moment of inertia with respect to the *X–X* centroidal axis for the area shown in Figure 8.2. (a) Use the exact formula. (b) Use the approximate method and divide the area into four similar horizontal strips parallel to the *X–X* axis. (c) Use the approximate method, but use eight similar horizontal strips. For parts (b) and (c), compare the result with part (a) and calculate the percent error.

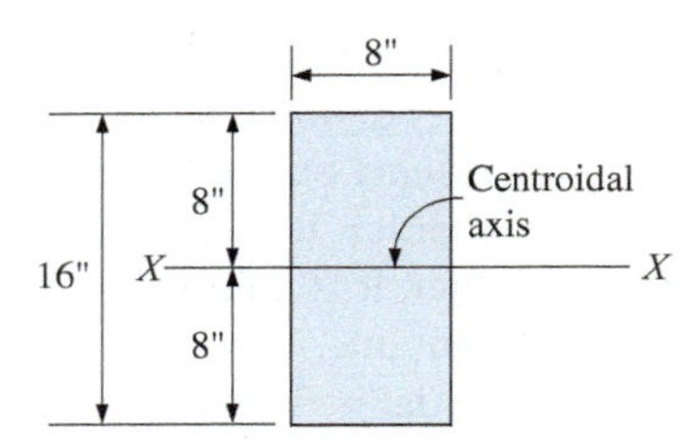

FIGURE 8.2 Rectangular area.

Solution (a) Use the exact formula from Table 3 (inside the back cover):

$$I_x = \frac{bh^3}{12} = \frac{(8\text{ in.})(16\text{ in.})^3}{12} = 2731\text{ in.}^4$$

Note in this application that b is the dimension of the side of the rectangle parallel to the axis about which the moment of inertia is being calculated and that h is the dimension of the side perpendicular to that axis. Note that the dimension parallel to the axis is the linear term and the dimension perpendicular to the axis is cubed. Carefully study the expressions for I_{x_o} and I_{y_o} in Table 3.

(b) Divide the area into four 4-in. × 8-in. horizontal-strip elements, as shown in Figure 8.3. Each strip has an area of 32 in^2. The perpendicular distance from the centroid of each strip (component areas a_1 and a_2) to the X–X centroidal axis is shown in Figure 8.3; these distances are designated y_1 and y_2. So,

$$y_1 = 6 \text{ in.} \quad \text{and} \quad y_2 = 2 \text{ in.}$$

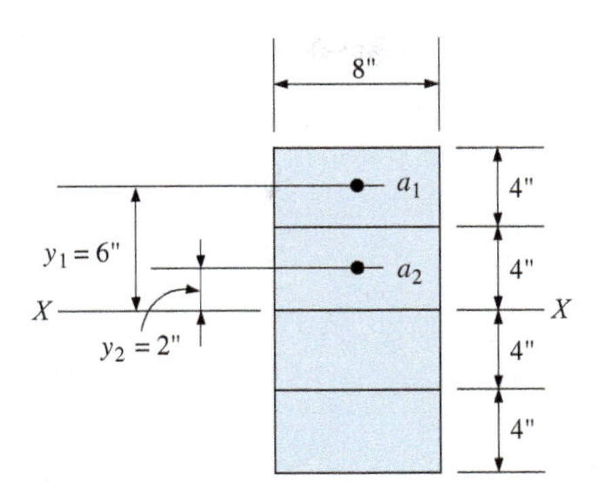

FIGURE 8.3 Approximate moment of inertia.

Note that due to symmetry with respect to axis X–X, the moment of inertia of the upper half of the area equals that of the lower half. Therefore, it is necessary to compute only the moment of inertia for either the upper or lower half and then multiply by two to obtain the moment of inertia for the total area. Using Equation (8.1),

$$\begin{aligned} I_x &= \Sigma a y^2 \\ &= 2(a_1 y_1^2 + a_2 y_2^2) \\ &= 2[(32 \text{ in.}^2)(6 \text{ in.})^2 + (32 \text{ in.}^2)(2 \text{ in.})^2] = 2560 \text{ in.}^4 \end{aligned}$$

Comparing with the exact moment of inertia, the percentage error is

$$\frac{2560 \text{ in.}^4 - 2731 \text{ in.}^4}{2731 \text{ in.}^4} \times 100 = -6.3\%$$

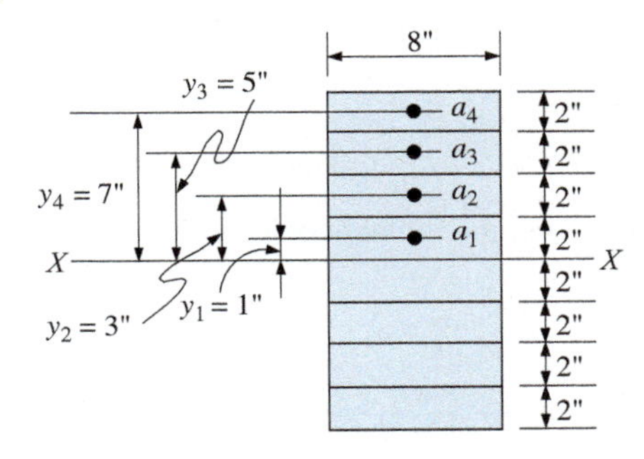

FIGURE 8.4 Approximate moment of inertia.

(c) Divide the area into eight 2-in. × 8-in. horizontal-strip elements, as shown in Figure 8.4. Each strip has an area of 16 in^2. The perpendicular distance from the centroid of each strip to the X–X centroidal axis is shown in Figure 8.4. The y dimensions are noted. Equation (8.1) yields

$$\begin{aligned} I_x &= \Sigma a y^2 \\ &= 2(a_1 y_1^2 + a_2 y_2^2 + a_3 y_3^2 + a_4 y_4^2) \\ &= 2[16 \text{ in.}^2(1 \text{ in.})^2 + 16 \text{ in.}^2(3 \text{ in.})^2 + 16 \text{ in.}^2(5 \text{ in.})^2 + 16 \text{ in.}^2(7 \text{ in.})^2] = 2688 \text{ in.}^4 \end{aligned}$$

Comparing with the exact moment of inertia, the percent error is

$$\frac{2688 \text{ in.}^4 - 2731 \text{ in.}^4}{2731 \text{ in.}^4} \times 100\% = -1.6\%$$

As the component area decreases in size, the moment of inertia approaches the exact theoretical value obtained by using the exact formula.

Note that in the following example the fact that the member is concrete plays no role. The result would be the same if the material were steel or wood. Moment of inertia is a geometric property of a shape; it is not dependent on material.

EXAMPLE 8.2 Compute the moment of inertia with respect to the X–X centroidal axis of the hollow-core precast concrete member shown in Figure 8.5.

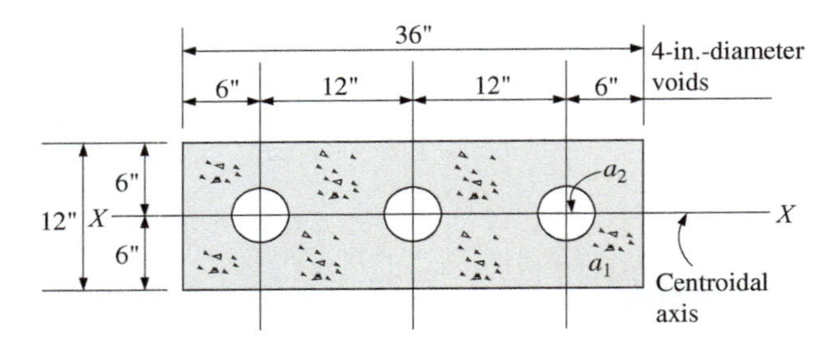

FIGURE 8.5 Hollow-core precast concrete member.

Solution Let us define a_1 and I_1 as the area and moment of inertia of a solid rectangular area 12 in. by 36 in. in cross section and a_2 and I_2 as the area and moment of inertia of one of the voids. Note that the centroids of the circular voids lie on the centroidal X–X axis of the member and that the centroidal X–X axis of area a_1 and the member coincide. The moment of inertia of the member is expressed as

$$I_X = I_{X_1} - 3I_{X_2}$$

where all moments of inertia are with respect to the X–X axis. Also note that since the circular areas represent voids, effectively reducing the rectangular area, the moments of inertia of the circles must be subtracted from the moment of inertia of the rectangle.

Compute the moments of inertia using formulas of Table 3:

$$I_{X_1} = \frac{bh^3}{12} = \frac{(36\text{ in.})(12\text{ in.})^3}{12} = 5184\text{ in.}^4$$

$$I_{X_2} = \frac{\pi d^4}{64} = \frac{\pi(4\text{ in.})^4}{64} = 12.57\text{ in.}^4$$

Therefore, the moment of inertia of the total unit is

$$I_X = I_{X_1} - 3I_{X_2}$$
$$= 5184\text{ in.}^4 - 3(12.57\text{ in.}^4) = 5146\text{ in.}^4$$

8.3 THE TRANSFER FORMULA

It is frequently necessary to determine the moment of inertia of an area about a noncentroidal axis that is parallel to a centroidal axis. This is accomplished using an expression called the *transfer formula*. With reference to Figure 8.6, the moment of inertia of an area about any axis (X'–X' in this case) that is parallel to a centroidal axis is equal to the moment of inertia about the centroidal axis plus the product of the area times the distance between the two axes squared. Mathematically, this is written as

$$I = I_o + ad^2 \tag{8.3}$$

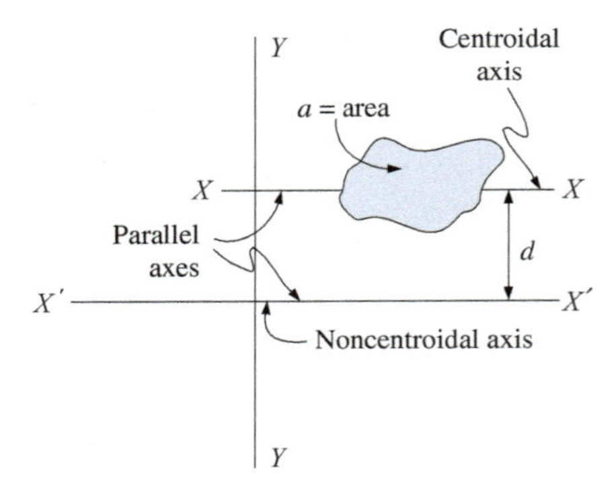

FIGURE 8.6 Moment of inertia of an area with respect to a noncentroidal axis.

where, as illustrated in Figure 8.6, the terms may be defined as follows:

I = the moment of inertia of an area with respect to any axis (in.^{4}) (mm^4)

I_o = the moment of inertia of an area with respect to its own centroidal axis (in.^{4}) (mm^4)

a = the area under consideration (in.^{2}) (mm^2)

d = the perpendicular distance between parallel axes, referred to as the *transfer distance* (in.) (mm)

Note that the axes between which the transfer is made must be parallel. Because the axes involved are parallel, Equation (8.3) is also called the *parallel axis theorem*.

8.4 MOMENT OF INERTIA OF COMPOSITE AREAS

As discussed previously in Section 7.4, a composite area is one that is made up of a number of simple geometric component areas or standardized shapes. Each of the component areas may have a centroidal axis different from that for the total composite area. If an area is composed of n component areas, where the areas are designated a_1, a_2, and so on, the transfer formula, given in Equation (8.3), is applied to each component area. The moment of inertia of the entire area is then the sum of the moments of inertia of all the component areas. Mathematically,

$$I = (I_{o_1} + a_1 d_1^2) + (I_{o_2} + a_2 d_2^2) + \cdots + (I_{o_n} + a_n d_n^2)$$

$$I = \Sigma(I_o + ad^2) \qquad \textbf{(8.4)}$$

To determine the moment of inertia of the composite area with respect to its centroidal axes, the following sequence of steps is recommended:

1. Divide the composite area into common simple geometric component areas or shapes and designate them a_1, a_2, and so on.
2. Determine the location of the centroidal axes for the composite area (as described in Section 7.4).
3. Determine the transfer distances from the centroidal axis of the composite area to the centroidal axis of each of the component areas and designate them d_1, d_2, and so on. Note that the axes must be parallel.
4. Compute the moment of inertia of each component area with respect to its own centroidal axis and designate these moments of inertia I_{o_1}, I_{o_2}, and so on. Use Table 3 for formulas and standardized tables for the moments of inertia.
5. Compute the moment of inertia of the composite area about its centroidal axis using Equation (8.4).

EXAMPLE 8.3 Compute the moments of inertia with respect to the *X–X* and *Y–Y* centroidal axes for the composite area shown in Figure 8.7.

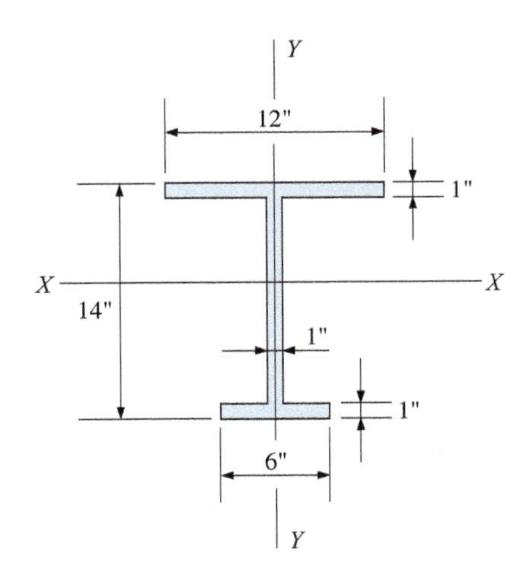

FIGURE 8.7 Composite area.

Solution The vertical *Y–Y* axis is a centroidal axis because it is an axis of symmetry. To determine the location of the *X–X* centroidal axis, a reference axis is selected at the bottom of the composite area that has been divided into three rectangular component areas, as shown in Figure 8.8. Table 8.1 shows a tabular format that can be used to organize the information. Therefore, from Table 8.1,

$$\bar{y} = \frac{\Sigma ay}{\Sigma a} = \frac{249 \text{ in.}^3}{30 \text{ in.}^2} = 8.30 \text{ in.}$$

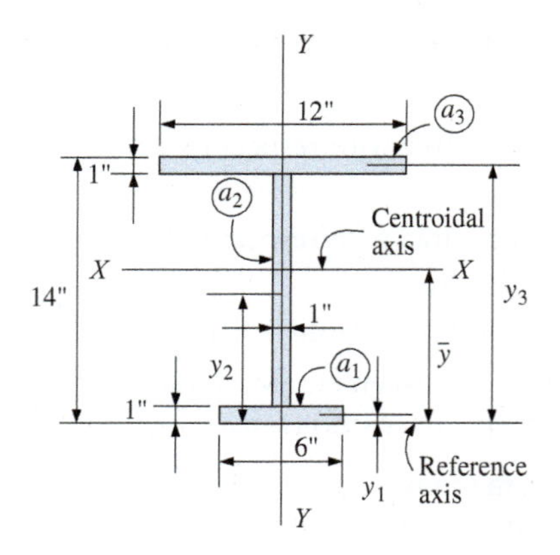

FIGURE 8.8 Location of centroidal axis.

TABLE 8.1 Tabular format for Example 8.3

Component Area	a (in.^{2})	y (in.)	ay (in.^{3})
a_1	6	0.5	3.0
a_2	12	7.0	84.0
a_3	12	13.5	162.0
Σ	30		249

Now compute the moment of inertia of the composite area with respect to the X–X centroidal axis. With reference to Figure 8.9, the transfer distances are

$$d_1 = 8.3 - 0.5 = 7.8 \text{ in.}$$
$$d_2 = 8.3 - 7 = 1.3 \text{ in.}$$
$$d_3 = 14 - 8.3 - 0.5 = 5.2 \text{ in.}$$

The moment of inertia of each component area (a rectangle) with respect to its own centroid is determined with reference to Table 3 (inside the back cover):

$$I = \frac{bh^3}{12}$$

from which

$$I_{o_1} = \frac{(6 \text{ in.})(1 \text{ in.})^3}{12} = 0.5 \text{ in.}^4$$
$$I_{o_2} = \frac{(1 \text{ in.})(12 \text{ in.})^3}{12} = 144 \text{ in.}^4$$
$$I_{o_3} = \frac{(12 \text{ in.})(1 \text{ in.})^3}{12} = 1.0 \text{ in.}^4$$

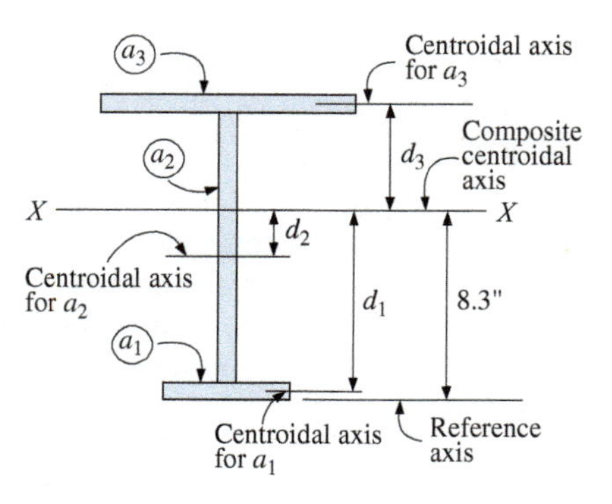

FIGURE 8.9 Determination of transfer distances.

Finally, calculate the moment of inertia of the composite area about the centroidal *X–X* axis using Equation (8.4):

$$\begin{aligned} I_x &= \Sigma(I_o + ad^2) \\ &= [0.5 + 6(7.8)^2] + [144 + 12(1.3)^2] + [1.0 + 12(5.2)^2] \\ &= 855 \text{ in.}^4 \end{aligned}$$

TABLE 8.2 Tabular format for Example 8.3

Component Area	a (in.^{2})	y (in.)	ay (in.^{3})	d (in.)	ad^2 (in.^{4})	I_o (in.^{4})
a_1	6	0.5	3.0	7.8	365.0	0.5
a_2	12	7	84.0	1.3	20.3	144.0
a_3	12	13.5	162.0	5.2	324.5	1.0
Σ	30		249		709.8	145.5

Table 8.2 shows how the entire solution can be accomplished through the use of a tabular format. Therefore from Table 8.2,

$$\bar{y} = \frac{\Sigma ay}{\Sigma a} = \frac{249 \text{ in.}^3}{30 \text{ in.}^2} = 8.30 \text{ in.}$$

and

$$\begin{aligned} I_x &= \Sigma(I_o + ad^2) \\ &= 145.5 \text{ in.}^4 + 709.8 \text{ in.}^4 = 855 \text{ in.}^4 \end{aligned}$$

The moment of inertia with respect to the *Y–Y* centroidal axis is somewhat easier to calculate since the centroidal axis for each component area coincides with the composite *Y–Y* centroidal axis. Therefore, the ad^2 term for each component area is zero. The transfer formula shows, then, that the moment of inertia of the composite area is the sum of the moments of inertia of the component areas about their own centroidal axes that are coincident with and parallel to the composite *Y–Y* centroidal axis. The moment of inertia about the *Y–Y* centroidal axis is

$$\begin{aligned} I_y &= \Sigma I_o = \Sigma \frac{bh^3}{12} \\ &= \frac{1 \text{ in.}(6 \text{ in.})^3}{12} + \frac{12 \text{ in.}(1 \text{ in.})^3}{12} + \frac{1 \text{ in.}(12 \text{ in.})^3}{12} \\ &= 163 \text{ in.}^4 \end{aligned}$$

Note that for the determination of I_y a tabular format offers no real advantage; hence, it is not shown.

You may again recognize, as was discussed briefly in Chapters 7, that computer spreadsheeting has application for this type of problem.

EXAMPLE 8.4 Calculate the moment of inertia with respect to the *X–X* centroidal axis for the built-up (composite) structural steel member shown in Figure 8.10.

Solution This member is symmetrical with respect to both the *X–X* and *Y–Y* axes. Therefore, no centroidal axis computations are necessary. From Appendix A, the cross-sectional area of the W18 × 71 is 20.8 in.^{2} and the moment of inertia with respect to the *X–X* centroidal axis is 1170 in.^{4} The latter is the I_o term for the wide-flange shape.

The member has additional area and additional moment of inertia due to the cover plates welded to the top and bottom flanges of the wide-flange shape. The area of each cover plate is 15 in^2. The vertical dimensions shown in Figure 8.10 (parallel to the web) can be readily verified.

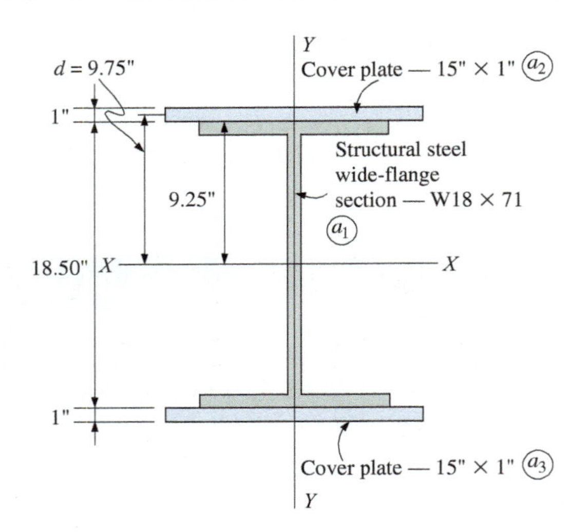

FIGURE 8.10 Built-up steel member.

For the application of Equation (8.4), the built-up area may be considered to be composed of three component areas: one standard structural steel shape and two rectangular steel plates. The summation is simplified because the plates are identical in their effect, and the transfer distance d for the standard shape is zero. For each plate, the moment of inertia about its own centroidal axis (parallel to the X–X axis) is

$$I_o = \frac{bh^3}{12} = \frac{15\text{ in.}(1\text{ in.})^3}{12} = 1.25\text{ in.}^4$$

Equation (8.4) then yields

$$\begin{aligned} I_x &= \Sigma(I_o + ad^2) \\ &= 1170\text{ in.}^4 + 2\left[1.25\text{ in.}^4 + 15\text{ in.}^2(9.75\text{ in.})^2\right] = 4024\text{ in.}^4 \end{aligned}$$

Notice how little the centroidal moment of inertia I_o for each cover plate contributes to the overall moment of inertia. Since I_o for a cover plate with respect to the axis parallel to its long side is so small, it is common practice to neglect it.

The calculation of moment of inertia involves length raised to the fourth power. When using SI units of measure, the preferred unit for moment of inertia is mm^4.

EXAMPLE 8.5 The cross section of the timber beam shown in Figure 8.11 is made up of planks 50 mm thick. Determine the moment of inertia with respect to the X–X centroidal axis.

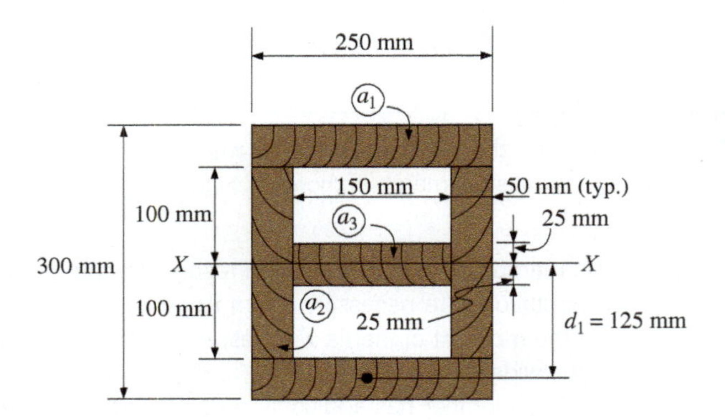

FIGURE 8.11 Timber box beam.

Solution The X–X centroidal axis is shown. It is an axis of symmetry. Component areas have been designated a_1, a_2 (two of each), and a_3. Area calculations yield

$$a_1 = (250\text{ mm})(50\text{ mm}) = 12{,}500\text{ mm}^2$$
$$a_2 = (200\text{ mm})(50\text{ mm}) = 10{,}000\text{ mm}^2$$
$$a_3 = (150\text{ mm})(50\text{ mm}) = 7500\text{ mm}^2$$

Only one transfer distance is required ($d_1 = 125$ mm). The centroids of a_2 and a_3 lie on the centroidal X–X axis (the transfer distances are zero).

The moment of inertia of each component area with respect to its own centroidal axis, parallel to the X–X centroidal axis, may be calculated using the appropriate formula from Table 3:

$$I_{o_1} = \frac{bh^3}{12} = \frac{(250\text{ mm})(50\text{ mm})^3}{12} = 2.60 \times 10^6\text{ mm}^4$$

$$I_{o_2} = \frac{bh^3}{12} = \frac{(50\text{ mm})(200\text{ mm})^3}{12} = 33.3 \times 10^6\text{ mm}^4$$

$$I_{o_3} = \frac{bh^3}{12} = \frac{(150\text{ mm})(50\text{ mm})^3}{12} = 1.56 \times 10^6\text{ mm}^4$$

Calculating the moment of inertia of the composite area about the centroidal X–X axis using Equation (8.4) gives

$$\begin{aligned} I_x &= \Sigma(I_o + ad^2) \\ &= 2\left[(2.60 \times 10^6\text{ mm}^4) + (12{,}500\text{ mm}^2)(125\text{ mm})^2\right] + 2(33.3 \times 10^6\text{ mm}^4) + (1.56 \times 10^6\text{ mm}^4) \\ &= 464 \times 10^6\text{ mm}^4 \end{aligned}$$

EXAMPLE 8.6 A rectangular area with a semicircular portion removed is shown in Figure 8.12. Calculate the moment of inertia with respect to (a) the X–X centroidal axis, (b) the Y–Y centroidal axis, and (c) the X'–X' reference axis.

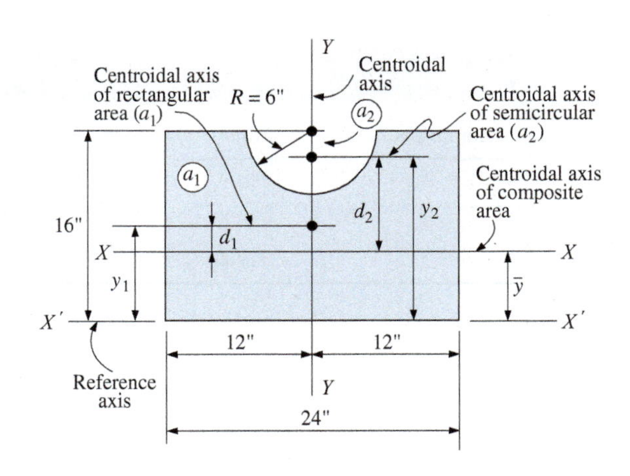

FIGURE 8.12 Composite area.

Solution The centroid of the composite area must first be located. Consider two component areas, a_1 being a rectangular area 16 in. by 24 in. and a_2 being the semicircular cutout area having a radius of 6 in. The latter is considered a negative area. The area calculations are

$$a_1 = (16\text{ in.})(24\text{ in.}) = 384\text{ in.}^2$$
$$a_2 = \frac{1}{2}(\pi R^2) = \frac{\pi(6\text{ in.})^2}{2} = -56.5\text{ in.}^2$$
$$A = \Sigma a = 384\text{ in.}^2 - 56.5\text{ in.}^2 = 327.5\text{ in.}^2$$

Compute the distance from the reference axis (axis $X'-X'$) to the centroid of each component area:

$$y_1 = 8 \text{ in.}$$

$$y_2 = 16 \text{ in.} - \frac{4R}{3\pi} = 16 \text{ in.} - \frac{4(6 \text{ in.})}{3\pi} = 13.45 \text{ in.}$$

Then apply Equation (7.4):

$$\bar{y} = \frac{\Sigma ay}{A} = \frac{a_1y_1 + a_2y_2}{A}$$

$$= \frac{(384 \text{ in.}^2)(8 \text{ in.}) + (-56.5 \text{ in.}^2)(13.45 \text{ in.})}{327.5 \text{ in.}^2} = 7.06 \text{ in.}$$

(a) The moment of inertia of the composite area will be determined by calculating the moment of inertia of the full rectangle (16 in. by 24 in.) and then subtracting the moment of inertia of the semicircular area (both are with respect to the *X–X* centroidal axis of the composite area).

The moments of inertia of a_1 and a_2 about their respective centroidal axes parallel to axis *X–X* (refer to Table 3) are

$$I_{o_1} = \frac{bh^3}{12} = \frac{(24 \text{ in.})(16 \text{ in.})^3}{12} = 8192 \text{ in.}^4$$

$$I_{o_2} = 0.1098R^4 = 0.1098(6 \text{ in.})^4 = -142.3 \text{ in.}^4$$

The transfer distances (*d* dimensions) are

$$d_1 = y_1 - \bar{y} = 8 \text{ in.} - 7.06 \text{ in.} = 0.94 \text{ in.}$$

$$d_2 = y_2 - \bar{y} = 13.45 \text{ in.} - 7.06 \text{ in.} = 6.39 \text{ in.}$$

From Equation (8.4), the moment of inertia for the composite area with respect to its *X–X* centroidal axis is

$$I_x = \Sigma(I_o + ad^2)$$

$$= \left[(8192 \text{ in.}^4) + (384 \text{ in.}^2)(0.94 \text{ in.})^2\right] + \left[(-142.3 \text{ in.}^4) + (-56.5 \text{ in.}^2)(6.39 \text{ in.})^2\right]$$

$$= 6082 \text{ in.}^4$$

TABLE 8.3 Tabular format for Example 8.6

Component Area	a (in.^{2})	y (in.)	ay (in.^{3})	d (in.)	ad^2 (in.^{4})	I_o (in.^{4})
a_1	384	8	3072	0.94	339	8192
a_2	−56.5	13.45	−760	6.39	−2307	−142.3
Σ	327.5		2312		−1968	8050

Table 8.3 shows a tabular format of the preceding computations. Therefore, from Table 8.3,

$$\bar{y} = \frac{\Sigma ay}{\Sigma a} = \frac{2312 \text{ in.}^3}{327.5 \text{ in.}^2} = 7.06 \text{ in.}$$

and

$$I_x = \Sigma(I_o + ad^2)$$

$$= (8050 \text{ in.}^4) + (-1968 \text{ in.}^4) = 6082 \text{ in.}^4$$

(b) The *Y–Y* centroidal axis coincides with the vertical axis of symmetry and is shown in Figure 8.12. The ad^2 term for each component area will be zero. Calculating the centroidal moments of inertia of the component areas about the *Y–Y* centroidal axis for the rectangle yields

$$I_{o_1} = \frac{bh^3}{12} = \frac{(16 \text{ in.})(24 \text{ in.})^3}{12} = 18{,}430 \text{ in.}^4$$

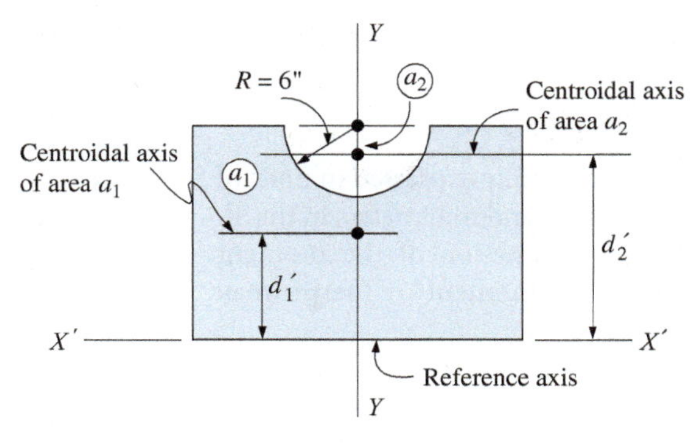

FIGURE 8.13 Moment of inertia about the base.

And for the semicircle, with reference to Table 3,

$$I_{o_2} = 0.393R^4 = (0.393)(6 \text{ in.})^4 = 509 \text{ in.}^4$$

The total moment of inertia is then

$$I_y = 18{,}430 \text{ in.}^4 - 509 \text{ in.}^4 = 17{,}920 \text{ in.}^4$$

(c) The moment of inertia of the composite area with respect to the reference axis X'–X' is calculated using Equation (8.4). In Figure 8.13, d' is the distance to the particular component area centroid from the X'–X' axis:

$$d_1' = 8 \text{ in. and } d_2' = 13.45 \text{ in.}$$

Equation (8.4) then yields

$$\begin{aligned} I_{X'} &= \Sigma(I_o + ad^2) \\ &= \left[(8192 \text{ in.}^4) + (384 \text{ in.}^2)(8 \text{ in.})^2\right] + \left[(-142.3 \text{ in.}^4) + (-56.5 \text{ in.}^2)(13.45 \text{ in.})^2\right] \\ &= 22{,}400 \text{ in.}^4 \end{aligned}$$

TABLE 8.4 Tabular format for Example 8.6, Part (c)

Component Area	a (in.^{2})	d (in.)	ad^2 (in.^{4})	I_o (in.^{4})
a_1	384	8	24,576	8192
a_2	−56.5	13.45	−10,221	−142.3
Σ	327.5		14,355	8050

Table 8.4 shows a tabular format for part (c). Therefore from Table 8.4,

$$\begin{aligned} I_{X'} &= \Sigma(I_o + ad^2) \\ &= 8050 \text{ in.}^4 + 14{,}355 \text{ in.}^4 = 22{,}400 \text{ in.}^4 \end{aligned}$$

8.5 RADIUS OF GYRATION

The *radius of gyration* of an area is difficult to describe in a physical sense. It is often described as the distance from a reference axis at which the entire area may be assumed to be located without changing its moment of inertia. This description is of questionable significance. A more practical interpretation of the radius of gyration of an area with respect to a given axis is that it is a convenient concept devised to replace a mathematical relationship between the moment of inertia and the area as encountered in the analysis and design of columns and other structural buckling calculations. It is usually denoted by the symbol r and expressed as

$$r = \sqrt{\frac{I}{A}} \tag{8.5}$$

where

r = the radius of gyration with respect to a given axis (in.) (mm)

I = the moment of inertia with respect to the same given axis (in.^{4}) (mm^4)

A = the cross-sectional area (in.^{2}) (mm^2)

Note that the radius of gyration is usually expressed in units of inches in the U.S. Customary System and millimeters in the SI.

The radius of gyration is a function of the moment of inertia. The magnitude of the moment of inertia may be different with respect to different axes of a member. Therefore, the magnitude of the radius of gyration may also be different with respect to different axes of a member. Formulas for the radius of gyration of commonly encountered simple geometric areas are given in Table 3 (inside the back cover). Chapter 18, which deals with columns, includes some practical applications of the radius of gyration.

EXAMPLE 8.7 Calculate the radius of gyration with respect to both centroidal axes for the cover-plated W shape shown in Figure 8.14

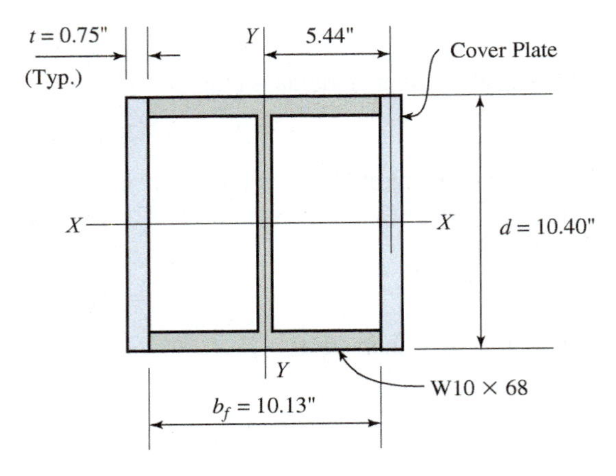

FIGURE 8.14 Cover-plated W shape.

Solution For the W10 × 68, from Appendix A, $A = 20.0$ in.^{2}, $d = 10.40$ in., $b_f = 10.13$ in., $I_x = 394$ in.^{4}, and $I_y = 134$ in^4. For each cover plate, the centroidal moments of inertia and area are:

$$I_x = \frac{1}{12}bh^3 = \frac{1}{12}(0.75\text{ in.})(10.40\text{ in.})^3 = 70.3\text{ in.}^4$$

$$I_y = \frac{1}{12}bh^3 = \frac{1}{12}(10.40\text{ in.})(0.75\text{ in.})^3 = 0.366\text{ in.}^4$$

$$a = bh = (0.75\text{ in.})(10.40\text{ in.}) = 7.80\text{ in.}^2$$

The built-up section is doubly symmetric so the location of the centroid is known. The transfer distance from the Y–Y axis to the centroid of the plate (shown in Figure 8.14) is calculated from

$$\frac{b_f}{2} + \frac{t}{2} = \frac{10.13\text{ in.}}{2} + \frac{0.75\text{ in.}}{2} = 5.44\text{ in.}$$

We next calculate the centroidal moments of inertia:

$$I_x = \Sigma(I_o + ad^2) = 394\text{ in.}^4 + 2(70.3\text{ in.}^4) = 535\text{ in.}^4$$
$$I_y = \Sigma(I_o + ad^2) = 134\text{ in.}^4 + 2(0.366\text{ in.}^4) + 2(7.80\text{ in.}^2)(5.44\text{ in.})^2 = 596\text{ in.}^4$$

Finally, using a total area $A = 20.0 + 2(7.80) = 35.6$ in.^{2}, we calculate the centroidal radii of gyration:

$$r_x = \sqrt{\frac{I_x}{A}} = \sqrt{\frac{535\text{ in.}^4}{35.6\text{ in.}^2}} = 3.88\text{ in.}$$

$$r_y = \sqrt{\frac{I_y}{A}} = \sqrt{\frac{596\text{ in.}^4}{35.6\text{ in.}^2}} = 4.09\text{ in.}$$

8.6 POLAR MOMENT OF INERTIA

We have previously discussed the calculation of rectangular moments of inertia, which are second moments of an area taken about axes that lie in the plane of the area. The moment of inertia of an area calculated with respect to an axis perpendicular to the plane of the area is called the *polar moment of inertia.*

In Figure 8.15, the *Z–Z* axis represents an axis perpendicular to the plane of the given area. Thus, the moment of inertia about the *Z–Z* axis is the summation of the product of each area a and the square of its moment arm r. The polar moment of inertia is normally denoted as J. Therefore,

$$J = \Sigma ar^2 \qquad \textbf{(8.6)}$$

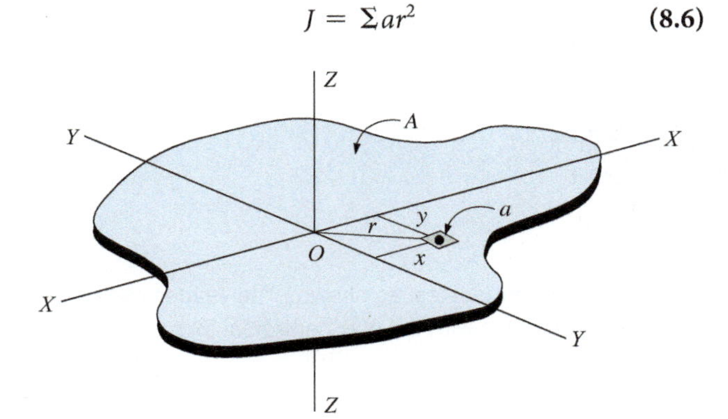

FIGURE 8.15 Polar moment of inertia.

Since for any right triangle

$$r^2 = x^2 + y^2$$

substituting in Equation (8.6) gives

$$\begin{aligned} J &= \Sigma a(x^2 + y^2) \\ &= \Sigma ax^2 + \Sigma ay^2 \end{aligned}$$

With reference to Equations (8.1) and (8.2), this equation may be written as

$$J = I_x + I_y \qquad \textbf{(8.7)}$$

Therefore, we see that the polar moment of inertia of a given area with respect to an axis perpendicular to its plane is equal to the sum of the moments of inertia about any two mutually perpendicular axes in its plane that intersect the polar axis. Formulas for the polar moment of inertia of solid and hollow circular areas are given in Table 3. The polar moment of inertia of circular members is a property required for the solution of problems involving shafts subjected to torsional loading. For further discussion on the applications of the polar moment of inertia, refer to Chapter 12.

EXAMPLE 8.8 Calculate the polar moment of inertia for a hollow circular shaft with an outside diameter of 4 in. and an inside diameter of 3 in., as shown in Figure 8.16.

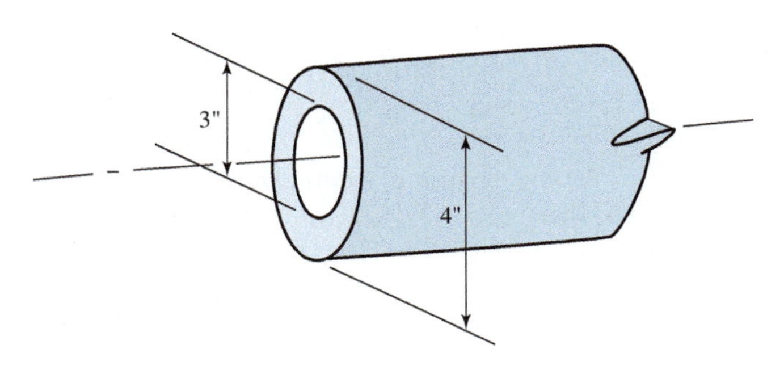

FIGURE 8.16 Hollow shaft.

Solution From Table 3, the expression for the polar moment of inertia taken about the center of gravity is

$$J_{CG} = \frac{\pi}{32}(d^4 - d_1^4)$$

Substitute the given data:

$$J_{CG} = \frac{\pi}{32}\left[(4\text{ in.})^4 - (3\text{ in.})^4\right] = 17.18\text{ in.}^4$$

EXAMPLE 8.9 For the T-shaped area shown in Figure 8.17, calculate the following: (a) the centroidal (rectangular) moments of inertia, (b) the radii of gyration with respect to the centroidal axes, and (c) the polar moment of inertia with respect to an axis perpendicular to the plane of the area through its centroid.

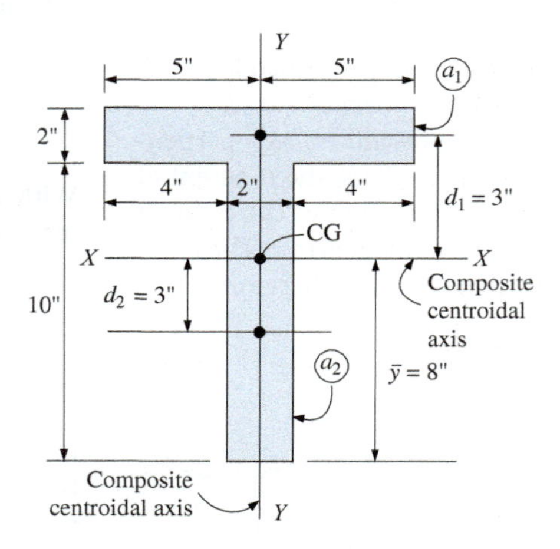

FIGURE 8.17 Composite area.

Solution The *X–X* centroidal axis of the composite area has been located as shown. The reader may wish to verify that $\bar{y} = 8$ in. The vertical axis of symmetry coincides with the *Y–Y* centroidal axis.

(a) First calculate I_x. The moments of inertia of a_1 and a_2 about their own centroidal axes, which are parallel to the *X–X* centroidal axis of the composite area, are

$$I_{o_1} = \frac{bh^3}{12} = \frac{(10 \text{ in.})(2 \text{ in.})^3}{12} = 6.67 \text{ in.}^4$$

$$I_{o_2} = \frac{bh^3}{12} = \frac{(2 \text{ in.})(10 \text{ in.})^3}{12} = 166.7 \text{ in.}^4$$

The transfer distances (*d* dimensions) are shown in Figure 8.17. Equation (8.4) yields

$$\begin{aligned} I_x &= \Sigma(I_o + ad^2) \\ &= \left[(6.67 \text{ in.}^4) + (20 \text{ in.}^2)(3 \text{ in.})^2\right] + \left[(166.7 \text{ in.}^4) + (20 \text{ in.}^2)(3 \text{ in.})^2\right] \\ &= 533 \text{ in.}^4 \end{aligned}$$

For the moment of inertia about the *Y–Y* axis, Equation (8.4) is again applied. The ad^2 terms are zero. Then

$$\begin{aligned} I_y &= \Sigma(I_o + ad^2) = \Sigma I_o = \Sigma\frac{bh^3}{12} \\ &= \frac{(2 \text{ in.})(10 \text{ in.})^3}{12} + \frac{(10 \text{ in.})(2 \text{ in.})^3}{12} = 173.3 \text{ in.}^4 \end{aligned}$$

(b) The total area of the T-shaped member is

$$A = a_1 + a_2 = 20 \text{ in.}^2 + 20 \text{ in.}^2 = 40 \text{ in.}^2$$

The radii of gyration with respect to the centroidal axes are then calculated from Equation (8.5):

$$r_x = \sqrt{\frac{I_x}{A}} = \sqrt{\frac{533 \text{ in.}^4}{40 \text{ in.}^2}} = 3.65 \text{ in.}$$

$$r_y = \sqrt{\frac{I_y}{A}} = \sqrt{\frac{173.3 \text{ in.}^4}{40 \text{ in.}^2}} = 2.08 \text{ in.}$$

(c) The polar moment of inertia with respect to the *Z–Z* axis through point *CG* is calculated from Equation (8.7):

$$J_{CG} = I_x + I_y$$
$$= 533 \text{ in.}^4 + 173.3 \text{ in.}^4 = 706 \text{ in.}^4$$

EXAMPLE 8.10 Two C310 × 31 structural steel channel shapes are welded together at their flange tips as shown in Figure 8.18. Determine (a) the least radius of gyration and (b) the polar moment of inertia.

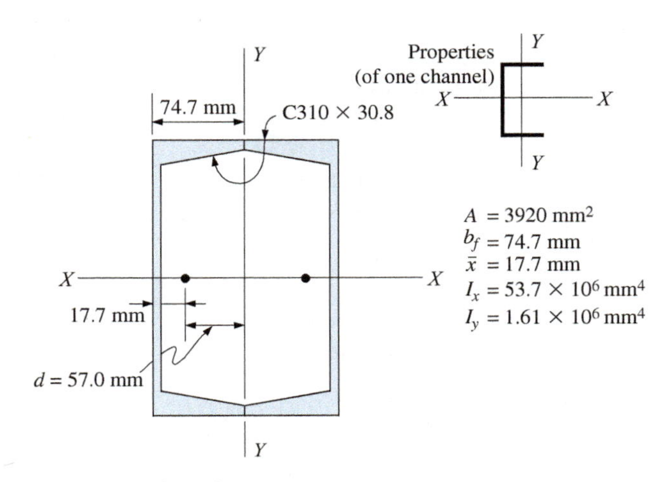

FIGURE 8.18 Built-up structural steel cross section.

Solution The necessary properties (from Appendix C) of a single C310 × 31 channel are given in Figure 8.18. To calculate the least radius of gyration, it is first necessary to determine the least moment of inertia; therefore, both I_x and I_y will be found. Both are required for the determination of the polar moment of inertia.

(a) The transfer distance *d* for the determination of I_y is calculated from $(b_f - \bar{x})$ and is shown in Figure 8.18. From Equation (8.4),

$$I_y = \Sigma(I_o + ad^2)$$
$$= 2(1.61 \times 10^6 \text{ mm}^4) + 2(3920 \text{ mm}^2)(57.0 \text{ mm})^2$$
$$= 28.7 \times 10^6 \text{ mm}^4$$

and

$$I_x = \Sigma(I_o + ad^2)$$
$$= 2(53.7 \times 10^6 \text{ mm}^4) = 107.4 \times 10^6 \text{ mm}^4$$

Therefore, the *Y–Y* axis is the weak axis and the radius of gyration *r* will be determined for this axis:

$$r_y = \sqrt{\frac{I_y}{A}} = \sqrt{\frac{28.7 \times 10^6 \text{ mm}^4}{2(3920 \text{ mm}^2)}} = 60.5 \text{ mm}$$

(b) The polar moment of inertia is calculated from Equation (8.7), where the I_x and I_y values are for the built-up cross section:

$$J = I_x + I_y$$
$$= 107.4 \times 10^6 \text{ mm}^4 + 28.7 \times 10^6 \text{ mm}^4$$
$$= 136.1 \times 10^6 \text{ mm}^4$$

SUMMARY BY SECTION NUMBER

8.1 The moment of inertia of an area (also called the *second moment of an area*) is a mathematical quantity that affects the load-carrying capacity of bending members and compression members. Units of moment of inertia are in.^{4}, mm^4, or m^4. With respect to an X–Y rectangular axes system, moment of inertia is expressed as

$$I_x = \Sigma a y^2 \quad \textbf{(8.1)}$$

$$I_y = \Sigma a x^2 \quad \textbf{(8.2)}$$

8.2 Equations (8.1) and (8.2) yield an approximate moment of inertia, the accuracy of which is a function of the size of the small component areas, represented by a's. An exact moment of inertia may be obtained using the formulas in Table 3 (inside the back cover) or by using integral calculus (see Appendix L).

8.3 The moment of inertia of an area with respect to a noncentroidal axis that is parallel to a centroidal axis is determined using the transfer formula

$$I = I_o + ad^2 \quad \textbf{(8.3)}$$

8.4 The moment of inertia of a composite area is found by summing the moments of inertia of the component parts as determined by the transfer formula:

$$I = \Sigma(I_o + ad^2) \quad \textbf{(8.4)}$$

8.5 The radius of gyration (r) of an area is a mathematical quantity expressed as

$$r = \sqrt{\frac{I}{A}} \quad \textbf{(8.5)}$$

Its most significant application is in the design and analysis of compression members (see Chapter 18).

8.6 The polar moment of inertia (J) of an area with respect to a polar axis perpendicular to its plane is defined as the sum of the moments of inertia about any two mutually perpendicular axes in its plane that intersect the polar axis:

$$J = I_x + I_y \quad \textbf{(8.7)}$$

PROBLEMS

Section 8.2 Moment of Inertia

8.1 Calculate the moment of inertia with respect to the X–X centroidal axes for the areas shown.

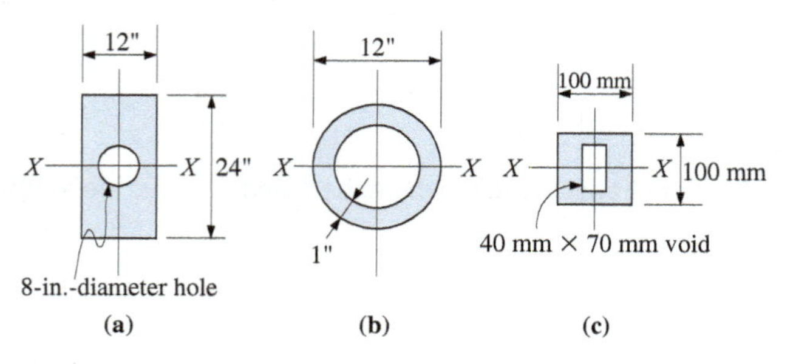

PROBLEM 8.1

8.2 Calculate the moment of inertia of the triangular area shown with respect to the X–X centroidal axis and with respect to the base of the triangle.

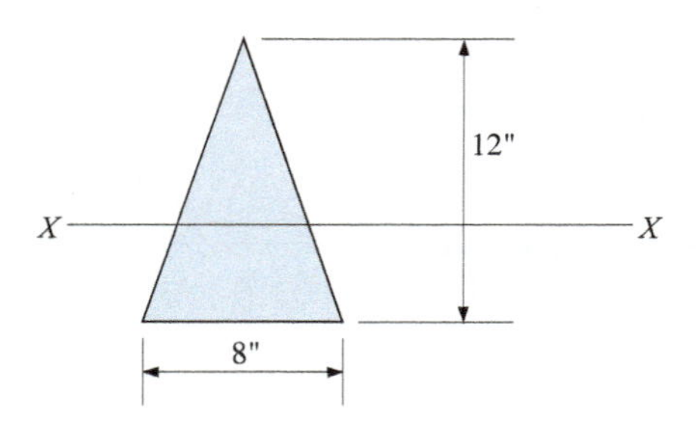

PROBLEM 8.2

8.3 A structural steel wide-flange section is reinforced with two steel plates attached to the web of the member, as shown. Calculate the moment of inertia of the built-up member with respect to the X–X centroidal axis.

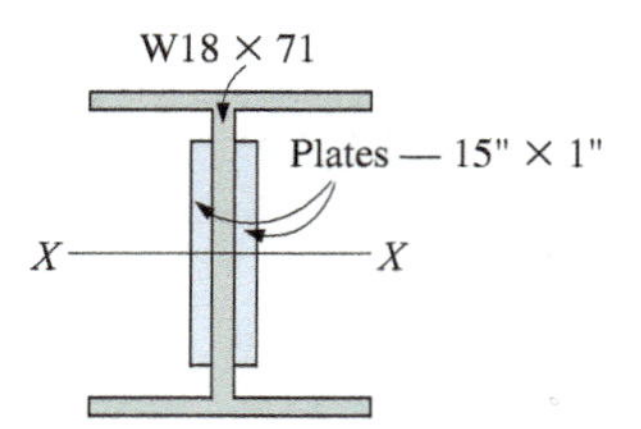

PROBLEM 8.3

8.4 The concrete block shown has wall thicknesses of 40 mm, voids that are 95 mm × 110 mm, and is 190 mm × 390 mm overall. Calculate the moment of inertia about the centroidal X–X axis.

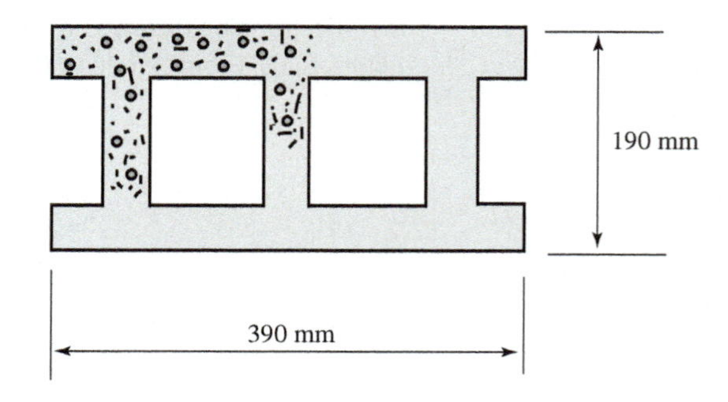

PROBLEM 8.4

8.5 A rectangle has a base of 6 in. and a height of 12 in. Using the approximate method, determine the moment of inertia with respect to the base.

(a) Divide the area into four equal horizontal strips.

(b) Divide the area into six equal horizontal strips.

8.6 For the area of Problem 8.5, calculate the exact moment of inertia and then calculate the percent error relative to the results obtained using the approximate method.

8.7 Check the tabulated moment of inertia for a 300 × 610 nominal-size timber cross section.

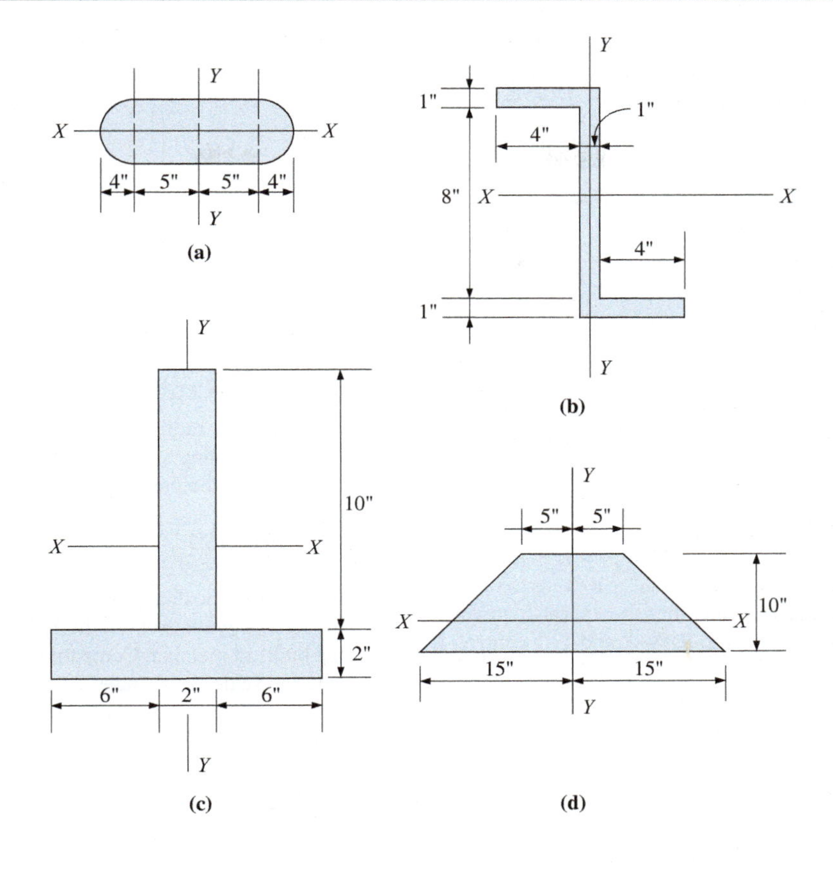

PROBLEM 8.9

Section 8.4 Moment of Inertia of Composite Areas

8.8 For the cross section of Problem 8.3, calculate the moment of inertia with respect to the Y–Y centroidal axis.

8.9 Calculate the moments of inertia with respect to both centroidal axes for the areas shown.

8.10 Calculate the moments of inertia with respect to both centroidal axes for the areas shown.

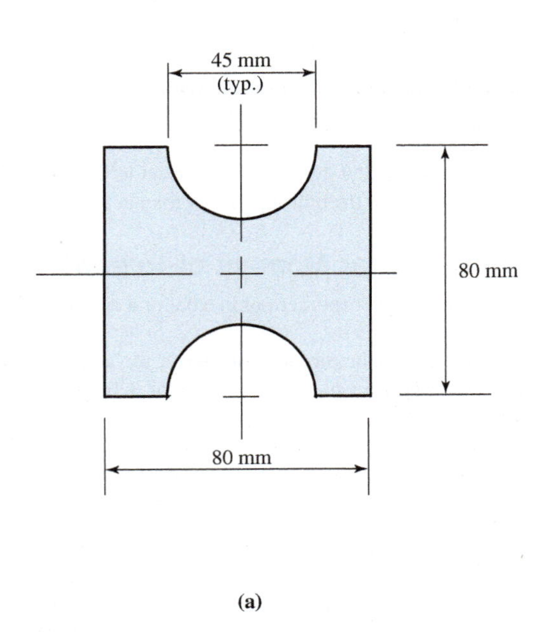

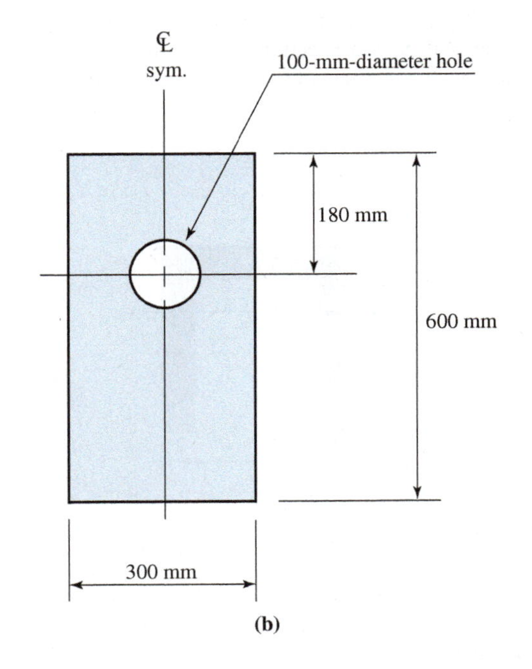

PROBLEM 8.10

8.11 The rectangular area shown has a square hole cut from it. Calculate the moments of inertia of the area with respect to its X–X centroidal axis and its base (X'–X').

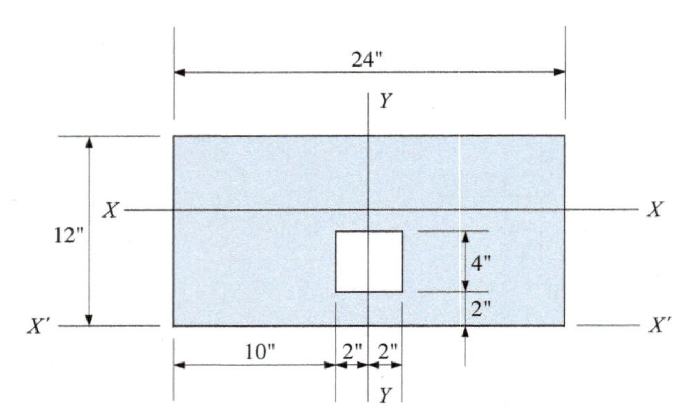

PROBLEM 8.11

8.12 For the built-up structural steel member shown, calculate the moments of inertia with respect to its X–X and Y–Y centroidal axes.

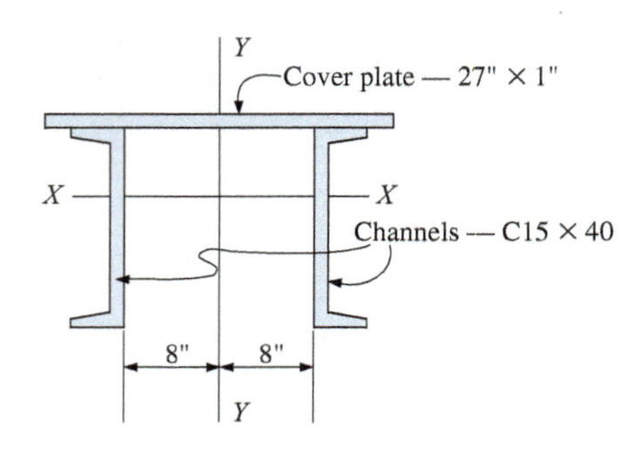

PROBLEM 8.12

8.13 Calculate the moments of inertia about both centroidal axes for the built-up cross section of Problem 7.21.

8.14 Calculate the moment of inertia with respect to the X–X centroidal axis of the built-up timber member shown. Use the rough dimensions given.

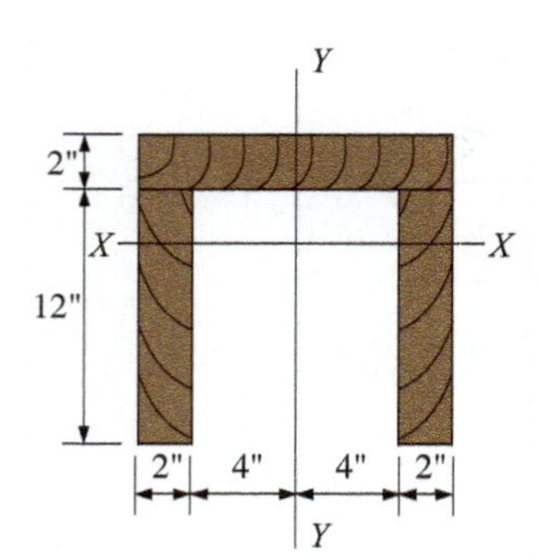

PROBLEM 8.14

8.15 For the two channels shown, calculate the spacing s required for the moment of inertia with respect to the X–X centroidal axis to be equal to the moment of inertia with respect to the Y–Y centroidal axis.

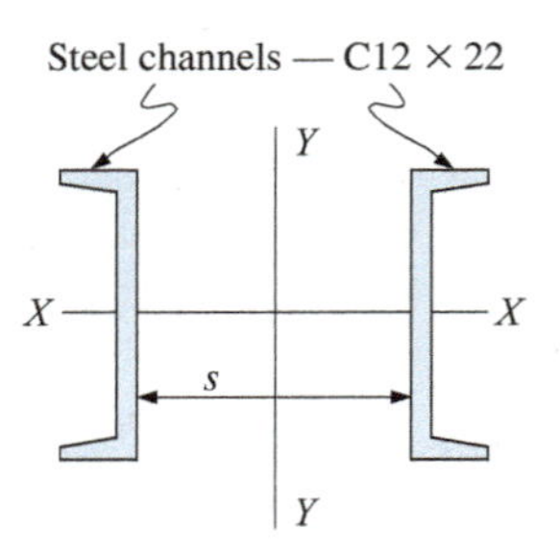

PROBLEM 8.15

Section 8.5 Radius of Gyration

8.16 Compute the radii of gyration about both centroidal axes for the following structural steel shapes and compare your results with tabulated values:
(a) W10 × 54
(b) C10 × 15.3

8.17 Two C10 × 15.3 channels are welded together at their flange tips to form a boxlike section as shown in Figure 8.18. Compute the radii of gyration about the X–X and Y–Y centroidal axes for the built-up member. Compare your results with the radius of gyration values for a single C10 × 15.3.

8.18 Compute the radii of gyration with respect to the X–X and Y–Y centroidal axes for the built-up steel member of Problem 8.12.

8.19 Compute the radii of gyration with respect to the X–X and Y–Y centroidal axes for the aluminum extruded shape shown.

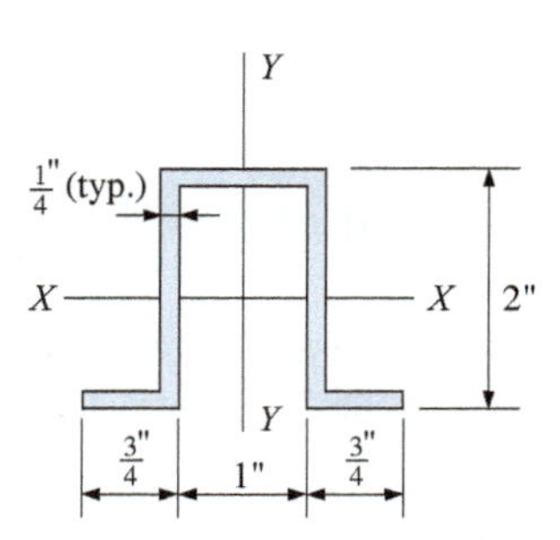

PROBLEM 8.19

8.20 Compute the radii of gyration with respect to the X–X and Y–Y centroidal axes for the built-up steel member of Problem 8.3.

8.21 Compute the radii of gyration with respect to the X–X and Y–Y centroidal axes for the built-up timber member of Problem 8.14.

Section 8.6 Polar Moment of Inertia

8.22 Calculate the polar moment of inertia for a circular solid steel shaft 3 in. in diameter.

8.23 Calculate the polar moment of inertia for a circular hollow steel shaft with an outside diameter of 3 in. and an inside diameter of $2\frac{1}{2}$ in.

8.24 For the areas (a) and (b) of Problem 8.9, calculate the polar moment of inertia with respect to an axis perpendicular to the plane of the area through its centroid.

Computer Problems

For the following computer problems, any appropriate software may be used. Input prompts should fully explain what is required of the user (the program should be user-friendly). The resulting output should be well labeled and self-explanatory. For spreadsheet problems, any appropriate software may be used.

8.25 Write a program that will calculate rectangular moments of inertia, radii of gyration, and polar moments of inertia for a rectangular area. Both the X–X (horizontal) and Y–Y (vertical) axes should be considered. User input is to be width and height of the rectangular area.

8.26 Write a program that will calculate the moments of inertia about the X–X and Y–Y centroidal axes for the semicircle of Problem 7.16. The calculation is to be based on Equation (8.4) and a summation of the effects of component areas. User input is to be as stated in Problem 7.16.

8.27 Use a spreadsheet program to solve one or more of the following: Example 8.4, Problem 8.31, 8.32, or 8.35. The user should be able to input the appropriate geometric properties for any shapes of the types shown and the program should then determine the appropriate quantities and display the result(s). Vary a single shape or dimension and observe the effect on the properties of the built-up cross section.

Supplemental Problems

8.28 For the cross section shown, calculate the moments of inertia with respect to both centroidal axes.

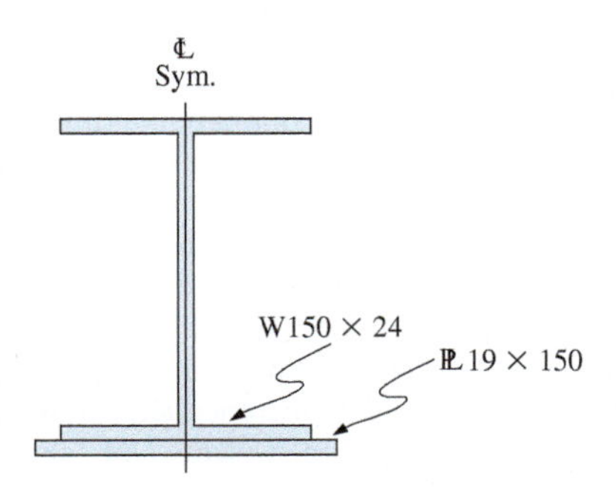

PROBLEM 8.28

8.29 Calculate the moments of inertia of the area shown with respect to both centroidal axes.

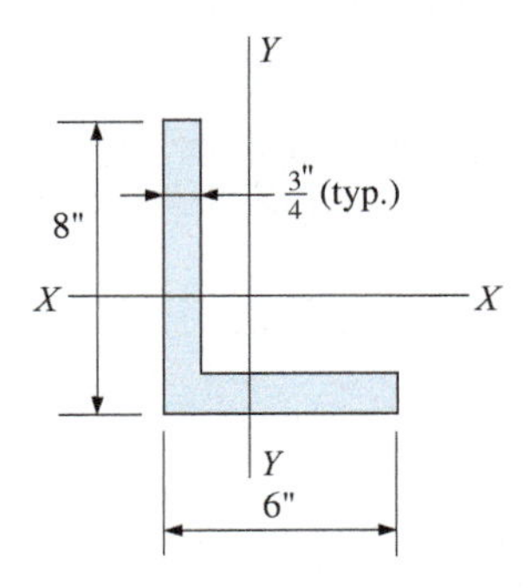

PROBLEM 8.29

8.30 and 8.31 For the cross-sectional areas shown, calculate the moment of inertia with respect to the horizontal X–X centroidal axis.

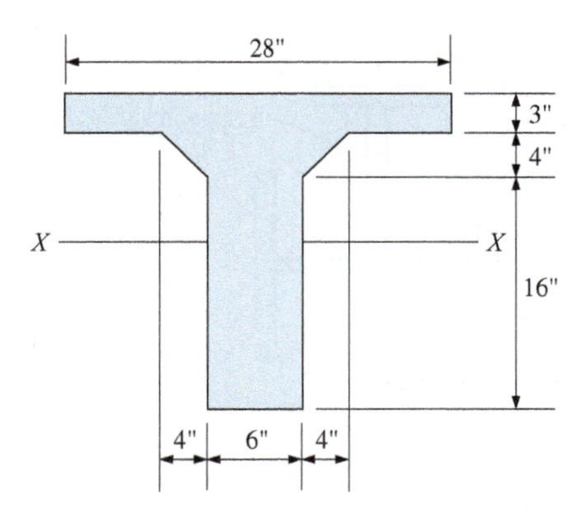

PROBLEM 8.30

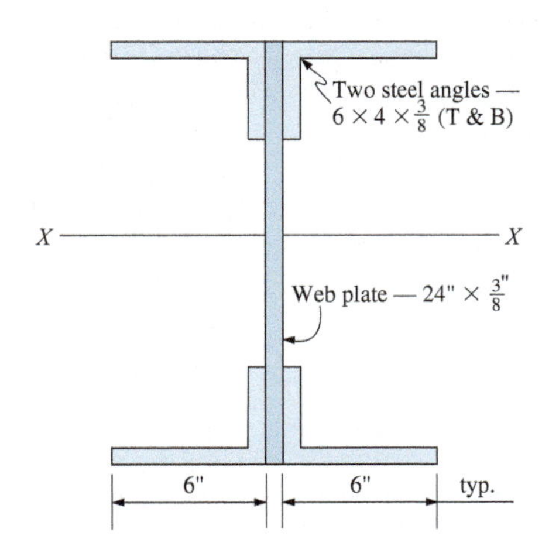

PROBLEM 8.31

8.32 Calculate the moments of inertia of the built-up steel member shown with respect to both centroidal axes.

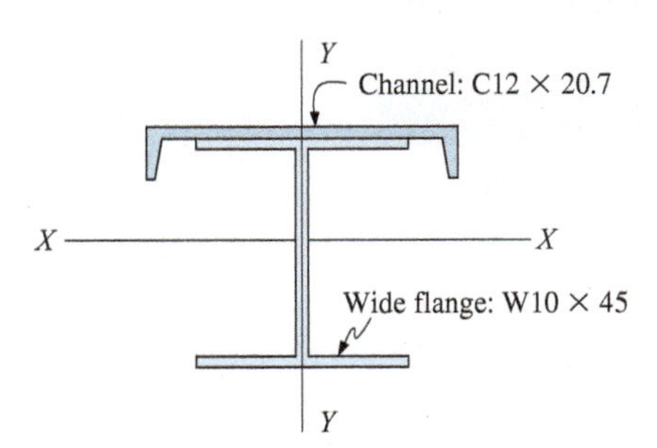

PROBLEM 8.32

8.33 Calculate the moments of inertia about both centroidal axes for the built-up steel member shown.

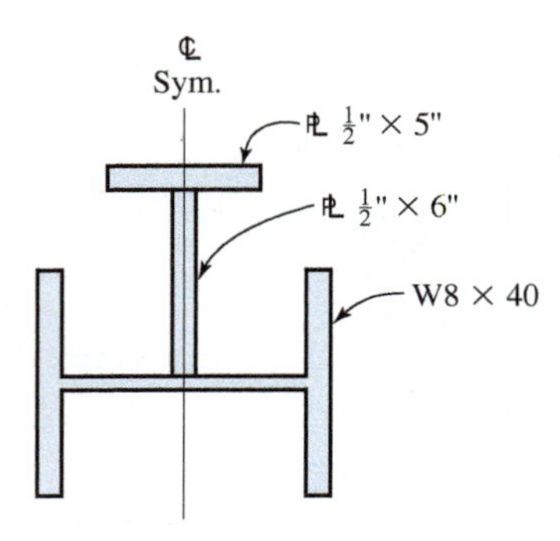

PROBLEM 8.33

8.34 Calculate I_x and I_y of the built-up steel members shown. In part (b), assume that the pipe is extra strong.

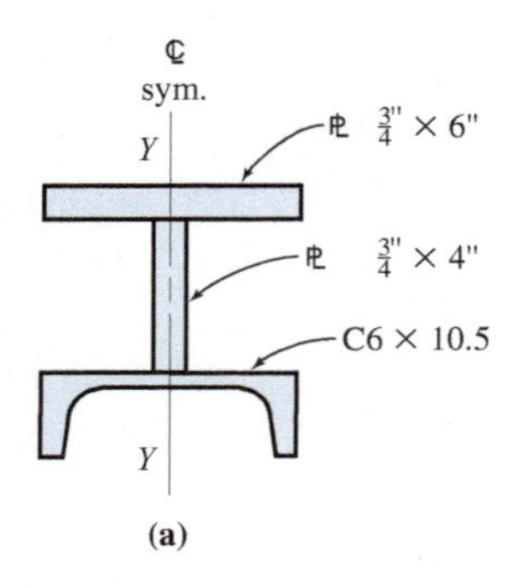

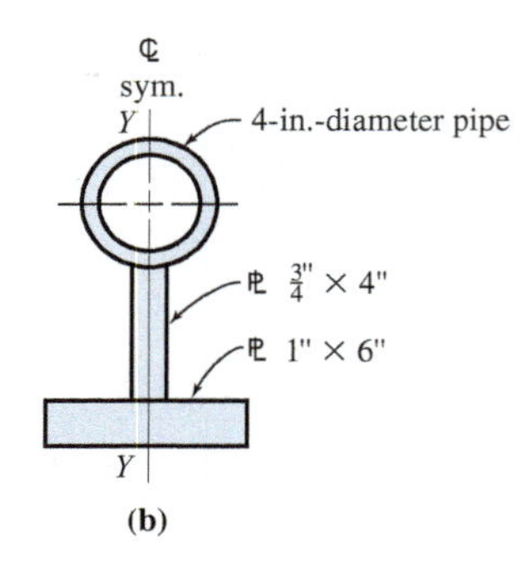

PROBLEM 8.34

8.35 Calculate the least radius of gyration for the areas shown. Use nominal timber sizes.

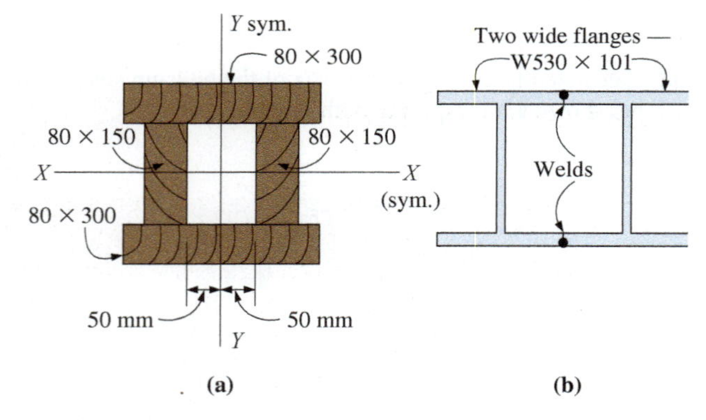

PROBLEM 8.35

8.36 A structural steel built-up section is fabricated as shown. Calculate **(a)** the moments of inertia and **(b)** the radii of gyration with respect to the X–X and Y–Y centroidal axes.

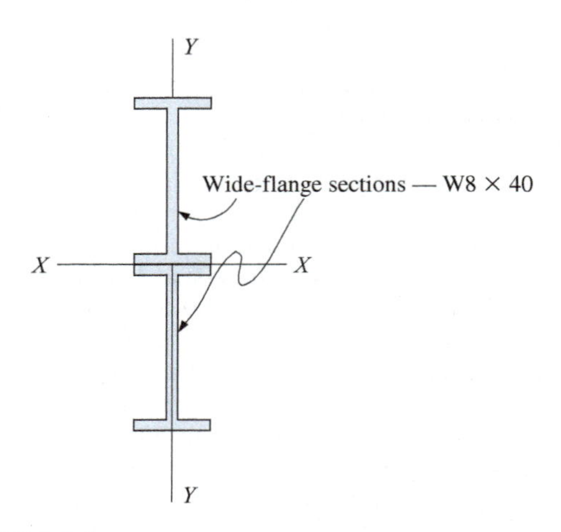

PROBLEM 8.36

8.37 Calculate the polar moment of inertia for the following shapes:

(a) a standard-weight steel pipe with a 152-mm nominal diameter

(b) a W530 × 101 wide-flange steel shape

(c) a 200 × 200 (nominal dimensions) timber cross section

8.38 Determine the polar moment of inertia for the areas of Problem 8.35.

8.39 Compute the radii of gyration with respect to the X–X and Y–Y centroidal axes for the areas indicated in Problem 8.32.

8.40 Calculate the polar moment of inertia about the centroid of an area that is hexagonal in shape and 6 in. on each side.

8.41 The area of the welded member shown is composed of a 4-in.-diameter standard weight pipe and four angles L$2\frac{1}{2} \times 2\frac{1}{2} \times \frac{1}{4}$. Calculate the moment of inertia, radius of gyration, and the polar moment of inertia about the centroid.

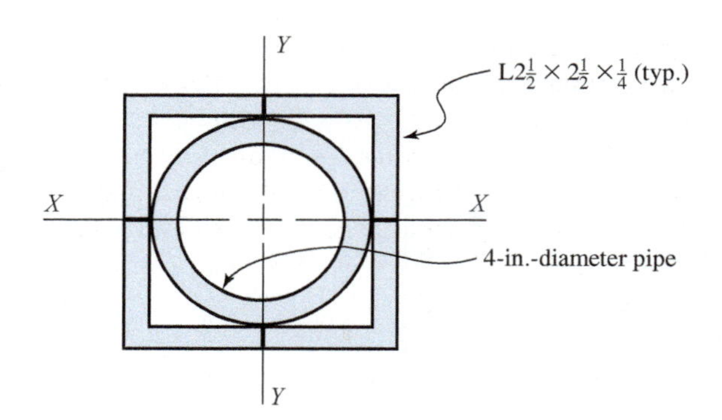

PROBLEM 8.41

CHAPTER NINE

STRESSES AND STRAINS

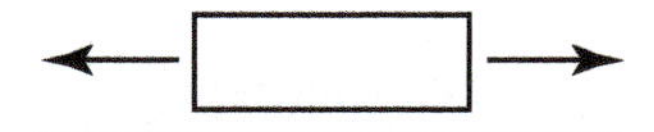

LEARNING OBJECTIVES

Upon completion of this chapter, readers will be able to:

- Discuss the forms of direct stress that can affect an object or structure as well as the concept of allowable stress.
- Discuss shear stresses and the concept of ultimate shear stress.
- Calculate direct and shear stresses on objects as well as determine acceptable design limits based on the allowable and ultimate stresses.
- Discuss the concept of strain and deformation of objects when subjected to loads and stress.
- Discuss Hooke's law, which states that stress is proportional to strain.
- Learn to calculate the modulus of elasticity, E, and the modulus of rigidity, G, of materials subject to normal and shear forces, respectively.

9.1 INTRODUCTION

As previously described, statics is a study of forces and force systems acting on rigid bodies at rest. *Strength of materials* may be described as a study of the relationships between external forces acting on elastic bodies and the internal stresses and strains generated by these forces. Based on the principles of strength of materials, we can establish the internal conditions that exist in an elastic body when it is subjected to various loading conditions.

In our study of statics, we neglected any dimensional changes (bodies were assumed rigid). For our study of strength of materials, however, the bodies will no longer be assumed to be rigid. Deformation and dimensional changes will be important considerations. We consider machine and structural elements that have applications in various fields of engineering technology with respect to both analysis (investigation) and design (selection) of these elements. Our approach is rational and analytical, and it is based on the principles of strength of materials.

9.2 TENSILE AND COMPRESSIVE STRESSES

Figure 9.1a represents an initially straight metal bar *BC* of constant cross section throughout its length. A bar of constant cross section such as this is termed a *prismatic bar*. The bar is acted on (loaded) at its ends by two equal and opposite axial forces (loads) *P*. An axial force, as shown in Figure 9.1a, is one that coincides with the longitudinal axis of the bar and acts through the centroid of the cross section.

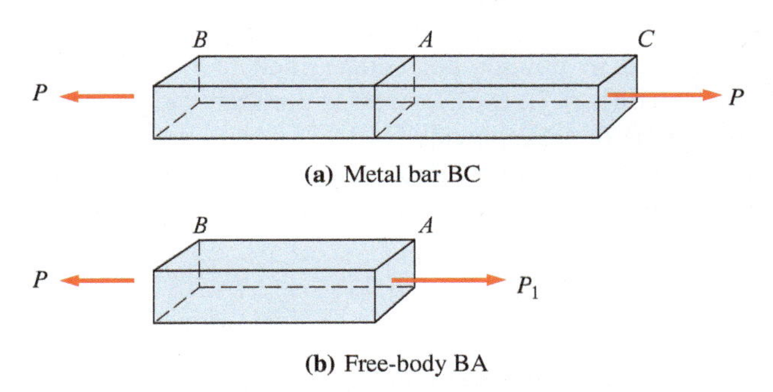

FIGURE 9.1 Prismatic metal bar in tension.

These forces, called *tensile forces*, tend to stretch or elongate the bar. The bar is said to be in tension.

Similarly, Figure 9.2a represents a straight prismatic bar acted on by two equal and opposite forces *P* that are directed toward each other. These forces, called *compressive forces*,

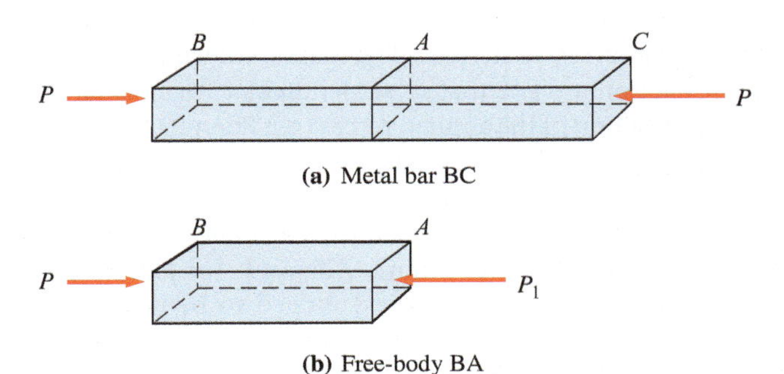

FIGURE 9.2 Prismatic metal bar in compression.

tend to shorten or compress the bar. The bar is said to be in compression.

Under the action of the two forces (whether tensile or compressive), internal resisting forces are developed within the bar and may be determined by imagining that a transverse plane is passed through bar BC (that is, perpendicular to its longitudinal axis), cutting it into two parts at point A. We consider the segment of the bar on the left of point A as a free body, as shown in Figures 9.1b and 9.2b. For the segment to be in equilibrium, a force P_1, equal and opposite to force P, must exist. The force P_1, which shows as an external force acting on the segment, is in reality an internal force in the original bar. It is considered to be an internal resisting force because it resists the action of external force P. In addition, the internal resisting force is assumed to be uniformly distributed over the cross section of the bar. This distribution, in effect, makes P_1 an axial force and also the resultant of the system of uniformly distributed forces that compose the internal resistance of the section.

To permit comparisons with standard and acceptable values (i.e., varying material breaking strengths), the total internal resistance P_1 acting on cross section A is converted to a unit basis and is expressed in terms of force per unit area or intensity of force. This is termed *unit stress*. Throughout the remainder of this book, the word *stress* is used to signify unit stress. The stress represents the internal resistance developed by a unit area of the cross section of the member. In the U.S. Customary System of units, stress generally is expressed in pounds per square inch (psi) or kips per square inch (ksi). In the SI, stress is preferably expressed in pascals (Pa) or megapascals (MPa), where $1\text{ Pa} = 1\text{ N/m}^2$ and $1\text{ MPa} = 1\text{ N/mm}^2$. Stress may be computed from the expression

$$\sigma = \frac{P}{A} \tag{9.1}$$

where σ = the average computed stress (lb/in.^{2}, kips/in.^{2}) (Pa, MPa)

P = the external applied load or force (lb, kips) (N)

A = the cross-sectional area over which stress develops (in.^{2}) (m^2, mm^2)

The preceding equation for stress is commonly termed the *direct stress formula*.

The tensile forces of Figure 9.1 produce internal tensile stresses, and the compressive forces of Figure 9.2 produce internal compressive stresses. The maximum values of these stresses act on a plane that is perpendicular (or normal) to the line of action of the applied forces (see Figure 9.8a and b); hence, they are sometimes called *normal stresses*. They are also referred to as *direct stresses*. It is usually necessary to limit stresses so as not to exceed the capability of materials to support them without failure. *Allowable* (or *permissible*) *stress*, $\sigma_{(\text{all})}$, is that level of stress judged to be acceptable. It is an upper limit that should not be exceeded. (For a more comprehensive discussion of allowable stress, refer to Section 10.6.)

The direct stress formula may be rewritten in several ways for use in various applications. For analysis problems in which the capacity of a member is to be determined,

$$P = \sigma_{(\text{all})}A \tag{9.2}$$

where P = the axial load capacity (maximum allowable axial load)

$\sigma_{(\text{all})}$ = the permissible or allowable axial stress

A = the cross-sectional area of the axially loaded member

In design-type problems where it is necessary to support a given load without exceeding an allowable stress,

$$A = \frac{P}{\sigma_{(\text{all})}} \tag{9.3}$$

where A = the required cross-sectional area of the axially loaded member being designed

P = the external applied axial load or force

$\sigma_{(\text{all})}$ = the permissible or allowable axial stress

Table 9.1 summarizes the three forms of the direct stress formula.

It should be noted that Equations (9.1), (9.2), and (9.3) may be used directly with tension members. Yet in the analysis and design of compression members, other than short, stocky ones, the length of a member plays an important role. For instance, a typical wooden yardstick, loaded in compression, will buckle before it fractures. Therefore, for the present, when we deal with compression members, we will limit our discussion to short compression members, unaffected by length considerations. Compression members are further discussed in Chapter 18.

So far, we have discussed tensile and compressive stresses, which imply an internal condition. Another type of stress is a *bearing stress*, which we designate σ_p. A bearing stress is basically a compressive stress exerted on an external surface of a body. It may be considered a contact pressure between separate bodies, rather than a stress. We usually speak of pressure with respect to gases or fluids, such as air pressure in an automobile tire or water pressure on a submerged submarine, but the terms *bearing pressure* and *bearing stress* are used almost interchangeably when they refer to situations in which a body bears on soil. When the footing of a concrete foundation bears on supporting soil,

TABLE 9.1 Direct stress formula summary

Formula	Usage
$\sigma = \frac{P}{A}$	Analysis for stress
$P = \sigma_{(\text{all})}A$	Analysis for load capacity
$A = \frac{P}{\sigma_{(\text{all})}}$	Design for required area

the bearing stress (pressure) is obtained by dividing the applied load by the contact area between the footing and the soil. The previous expressions for stress are also applicable for bearing stress (pressure) problems.

EXAMPLE 9.1 (a) Compute the tensile stress developed in the square steel bar 2 in. by 2 in. in cross section shown in Figure 9.3. The bar is subjected to an axial tensile load of 95 kips. (b) Determine the tensile stress σ_t if the member is a structural steel W8 × 31. (The load is the same, 95 kips.)

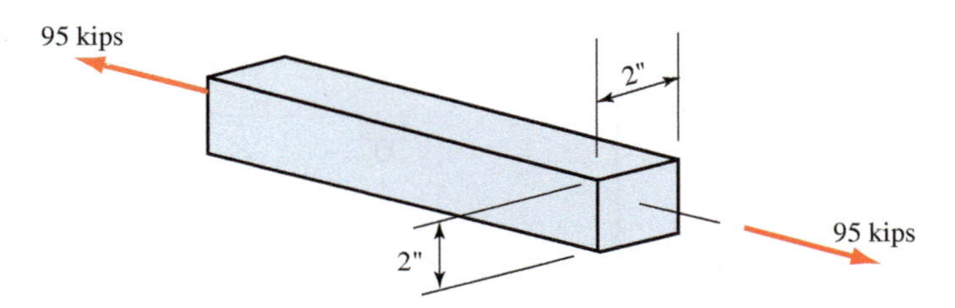

FIGURE 9.3 Steel bar in tension.

Solution (a) The cross-sectional area of the square steel bar is 2 in. × 2 in. = 4 in.^{2}. Using the direct stress formula,

$$\sigma_t = \frac{P}{A} = \frac{95\text{ k}}{4\text{ in.}^2} = 23.8\frac{\text{k}}{\text{in.}^2}\text{ (or 23.8 ksi)}$$

(b) From Appendix A, the cross-sectional area of a W8 × 31 is 9.13 in.^{2}. Therefore,

$$\sigma_t = \frac{P}{A} = \frac{95\text{ k}}{9.13\text{ in.}^2} = 10.41\text{ ksi}$$

EXAMPLE 9.2 Rework Example 9.1 assuming the steel bar to be 50 mm by 50 mm in cross section and loaded with an axial tensile load of 400 kilonewtons (kN).

Solution Using the direct stress formula, Equation (9.1),

$$P = 400\text{ kN} = 400 \times 10^3\text{ N}$$

$$A = (50\text{ mm})(50\text{ mm}) = 2500\text{ mm}^2 = 2.5 \times 10^{-3}\text{ m}^2$$

$$\sigma_t = \frac{P}{A} = \frac{400 \times 10^3\text{ N}}{2.5 \times 10^{-3}\text{ m}^2} = 160 \times 10^6\text{ N/m}^2 = 160\text{ MPa}$$

Since 1 N/mm^2 = 1 MPa, we may alternatively (and more simply) write:

$$\sigma_t = \frac{P}{A} = \frac{400 \times 10^3\text{ N}}{(50\text{ mm})^2} = 160\text{ N/mm}^2 = 160\text{ MPa}$$

EXAMPLE 9.3 Steel rod suspenders are to support pipes in a power plant, as shown in Figure 9.4. Each rod is $\frac{3}{8}$ in. in diameter and has an allowable axial tensile stress of 24,000 psi in the body of the rod. Calculate the allowable axial tensile load in each rod.

Solution The cross-sectional area of the $\frac{3}{8}$-in.-diameter rod is calculated from

$$A = \frac{\pi d^2}{4} = \frac{\pi(0.375\text{ in.})^2}{4} = 0.1104\text{ in.}^2$$

Therefore, the allowable axial tensile load is

$$P = \sigma_{t(\text{all})}A = (24{,}000\text{ lb/in.}^2)(0.1104\text{ in.}^2) = 2650\text{ lb}$$

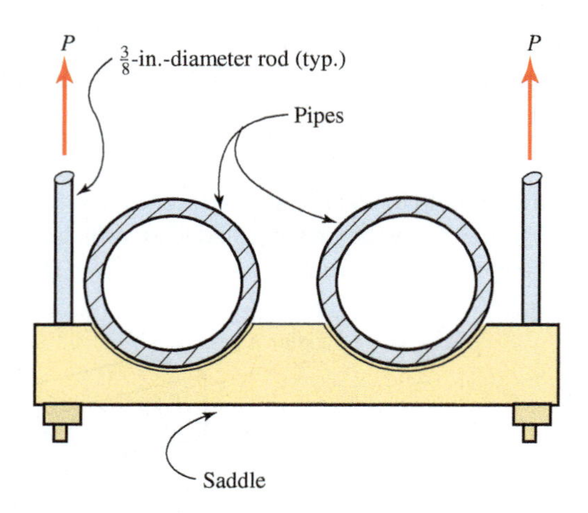

FIGURE 9.4 Pipe support system.

EXAMPLE 9.4 Compute the required size of a short, square, dressed (S4S) timber post shown in Figure 9.5. The post is subjected to a compressive load of 100 kN. The allowable axial compressive stress is 5.5 MPa.

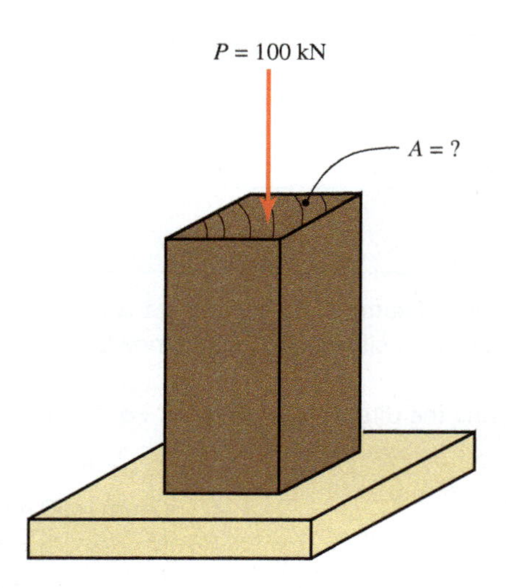

FIGURE 9.5 Timber post.

Solution This is a design-type problem, since the size of a member is to be found. The required cross-sectional area is found using Equation (9.2). The load is shown in newtons. So,

$$A = \frac{P}{\sigma_{c(\text{all})}} = \frac{100 \times 10^3 \text{ N}}{5.50 \text{ MPa}} = 18.18 \times 10^3 \text{ mm}^2$$

From Appendix E, Table E.2, a nominal 150-mm-by-150-mm square timber member has a cross-sectional area of 19.6×10^3 mm^2, based on dressed dimensions of 140 mm by 140 mm. Therefore, select a 150-mm-by-150-mm timber post.

EXAMPLE 9.5 An S4S timber column (nominal 6 in. by 6 in.) is subjected to a compressive load of 22,000 lb, as shown in Figure 9.6. The column is supported by a 2-ft-square concrete footing that is, in turn, supported by the soil. Compute (a) the bearing stress on the contact surface between the column and the footing and (b) the bearing stress at the base of the footing. Neglect the weight of the footing and the column.

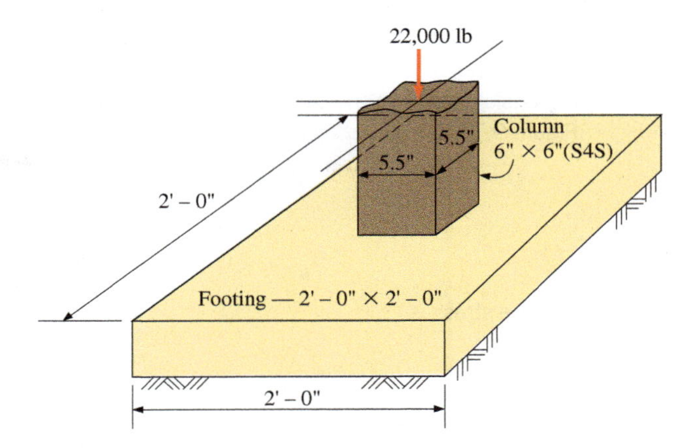

FIGURE 9.6 Column on footing.

Solution (a) The dressed dimensions of the column are $5\frac{1}{2}$ in. by $5\frac{1}{2}$ in. and the dressed area is 30.25 in^2. The bearing stress on the column-footing contact surface is

$$\sigma_p = \frac{P}{A} = \frac{22{,}000 \text{ lb}}{30.25 \text{ in.}^2} = 727 \text{ psi}$$

(b) The bearing stress at the base of the footing is

$$\sigma_p = \frac{P}{A} = \frac{22{,}000 \text{ lb}}{(24 \text{ in.})^2} = 38.2 \text{ psi}$$

Bearing pressure (stress) on soil is generally expressed in pounds per square foot (psf) rather than psi. Therefore,

$$\sigma_p = \left(38.2 \frac{\text{lb}}{\text{in.}^2}\right)\left(144 \frac{\text{in.}^2}{\text{ft}^2}\right) = 5500 \text{ psf}$$

When calculating stress, as in the preceding examples, the area in question must be determined carefully. If there is a decrease in the cross-sectional area, such as occurs at section B–B in the bar shown in Figure 9.7, there will naturally be a higher stress at that section. We neglect, for now, the stress concentrations produced in this situation. Stress concentrations are covered later in Section 11.4.

EXAMPLE 9.6

A flat steel bar, $\frac{1}{2}$ in. thick and 4 in. wide, is subjected to a 20-kip tensile load. Two $\frac{3}{4}$-in.-diameter holes are located as shown in Figure 9.7a. Determine the average tensile stress at sections A–A and B–B. Neglect stress concentrations adjacent to each hole.

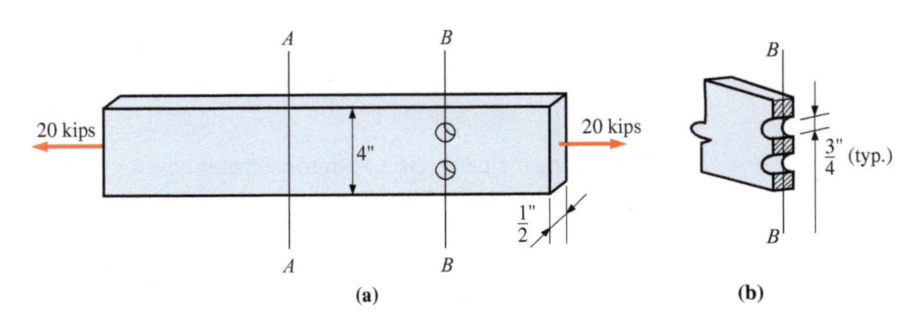

FIGURE 9.7 Flat bar with holes.

Solution At section A–A the cross-sectional area, called the *gross area*, is 2.0 in^2. The stress is calculated from

$$\sigma_t = \frac{P}{A} = \frac{20 \text{ k}}{2.0 \text{ in.}^2} = 10.0 \text{ ksi}$$

At section *B–B* the cross-sectional area is reduced by the two holes. The area reduction caused by each hole is a rectangle, with one dimension equal to the diameter d_H of the hole and the other dimension equal to the thickness *t* of the bar. The remaining area, or the *net area*, shown shaded in Figure 9.7b, is calculated from

$$\begin{aligned} A &= A_{\text{gross}} - A_{\text{holes}} \\ &= A_{\text{gross}} - 2d_H t \\ &= 2.0 \text{ in.}^2 - 2(0.75 \text{ in.})(0.50 \text{ in.}) = 1.25 \text{ in.}^2 \end{aligned}$$

and the stress is

$$\sigma_t = \frac{P}{A} = \frac{20 \text{ k}}{1.25 \text{ in.}^2} = 16.0 \text{ ksi}$$

EXAMPLE 9.7 A 25-mm-diameter bolt extends vertically through a timber beam, as shown in Figure 9.8. The maximum tensile stress developed in the bolt is 83 MPa. Determine the minimum required diameter (d) for a circular steel plate that is to be placed under the head of the bolt. The bearing stress on the timber is not to exceed 3.4 MPa.

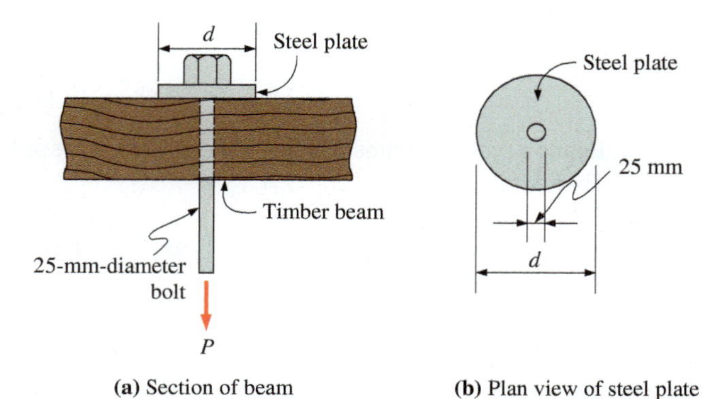

FIGURE 9.8 Bolt through timber beam.

Solution The cross-sectional area of the bolt is 491 mm². The applied tensile load *P* that induces the tensile stress of 83 MPa in the bolt is obtained from Equation (9.2). Substitute in terms of newtons and meters:

$$P = \sigma_t A = (83 \times 10^6 \text{ N/m}^2)(491 \times 10^{-6} \text{ m}^2) = 40{,}753 \text{ N}$$

The required bearing area between the steel plate and the timber beam is then computed from Equation (9.3):

$$\begin{aligned} \text{required } A = \frac{P}{\sigma_{p(\text{all})}} = \frac{40{,}753 \text{ N}}{3.4 \times 10^6 \text{ N/m}^2} &= 11{,}986 \times 10^{-6} \text{ m}^2 \\ &= 11{,}986 \text{ mm}^2 \end{aligned}$$

Assuming the plate has a 25-mm-diameter hole for the bolt, the net contact-bearing area between the plate and the timber beam is

$$A = 0.7854d^2 - 0.7854(25 \text{ mm})^2$$

Equating this to the required area and solving for *d*,

$$11{,}986 \text{ mm}^2 = 0.7854d^2 - 0.7854(25 \text{ mm})^2$$

from which

$$\text{required } d = 126 \text{ mm}$$

If a vertical bar rests on a horizontal base and supports no load except its own weight, the stress at any cross section may be calculated by dividing the weight of the part of the bar *above* the cross section by the cross-sectional area. If a bar is suspended from its upper end and supports no load except its own weight, the method of solution is similar, but the weight of the bar *below* the cross section must be considered.

EXAMPLE 9.8 A 100-ft-long steel bar is suspended vertically from its upper end. The cross-sectional area of the bar is 4.0 in.^{2} (2 in. by 2 in.). The unit weight of the steel is 490 pounds per cubic foot (pcf). Compute (a) the maximum tensile stress due to the bar's own weight and (b) the maximum load P (the allowable load) that can be safely supported at the lower end of the bar if the allowable tensile stress is 24,000 psi.

Solution Calculate the weight per foot for this bar as follows:

$$\text{weight per foot} = (\text{volume})(\text{unit weight})$$

$$\frac{(2\text{ in.})(2\text{ in.})(12\text{ in.})}{(12\text{ in./ft})^3}(490\text{ lb/ft}^3) = 13.61\text{ lb/ft}$$

Note that the volume calculated is for a length of one foot and must be determined in cubic feet, hence the use of the conversion factor of (12 in./ft)^3 or 1728 in.^{3}/ft^3.

(a) The maximum tensile stress in the bar due to its own weight will occur at its upper end:

$$\text{bar weight} = (100\text{ ft})(13.61\text{ plf}) = 1361\text{ lb}$$

The induced tensile stress is

$$\sigma_t = \frac{P}{A} = \frac{1361\text{ lb}}{4.0\text{ in.}^2} = 340\text{ psi}$$

(b) Since the allowable tensile stress is 24,000 psi and the induced stress due to the bar's own weight is 340 psi, the difference between the two represents a stress that can be induced by P. Hence, the allowable load P the bar can safely support at its lower end is calculated as follows:

$$\begin{aligned} P &= \sigma_t A \\ &= (24{,}000\text{ psi} - 340\text{ psi})(4.0\text{ in.}^2) \\ &= 94{,}640\text{ lb} \end{aligned}$$

9.3 SHEAR STRESSES

In Section 9.2 we discussed how tensile and compressive stresses are developed in a direction perpendicular (or normal) to the surfaces on which they act. They are sometimes called *normal stresses*. Another type of stress, called *shear stress* (also called *tangential stress*) is developed in a direction parallel to the surface on which it acts. Normal stresses and shear stresses are illustrated in Figure 9.9.

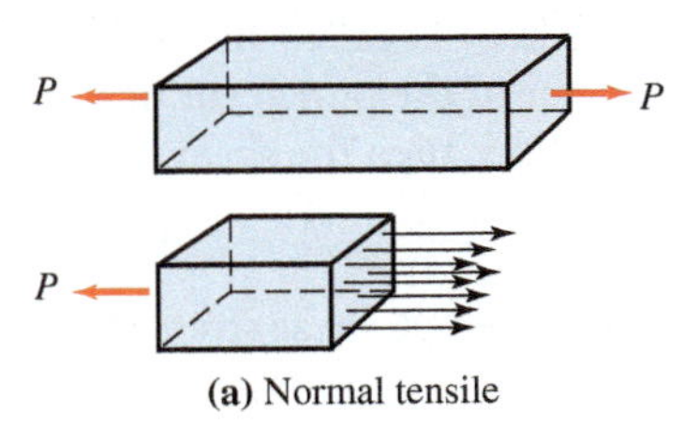

(a) Normal tensile

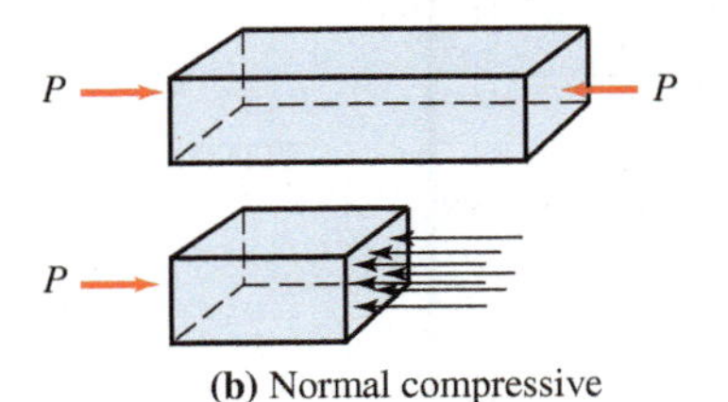

(b) Normal compressive

(c) Shear

FIGURE 9.9 Types of stress.

Shear stress is also illustrated in Figure 9.10a, where equal and opposite forces P are applied to two flat plates bonded together by an adhesive. The contact surface, shown shaded, is subjected to a shearing action. In the absence of the adhesive, the two surfaces would tend to slide past one another. The shear force is assumed to be uniformly distributed across the contact area. The result is the development of a shear stress, the magnitude of which is computed from the expression

$$\tau = \frac{P}{A} \tag{9.4}$$

where τ = the average computed shear stress (psi, ksi) (Pa, MPa)

P = the external applied shear force (lb, kips) (N)

A = the area over which shear stress develops (in.^{2}) (m^2, mm^2)

The uniformly distributed shear stress, as assumed here, is less likely to occur than is uniformly distributed tensile or compressive stress. This being the case, the shear stress computed from $\tau = P/A$ should be interpreted as an *average* value. In this particular application, along with other applications in this chapter, any nonuniform stress distribution is neglected.

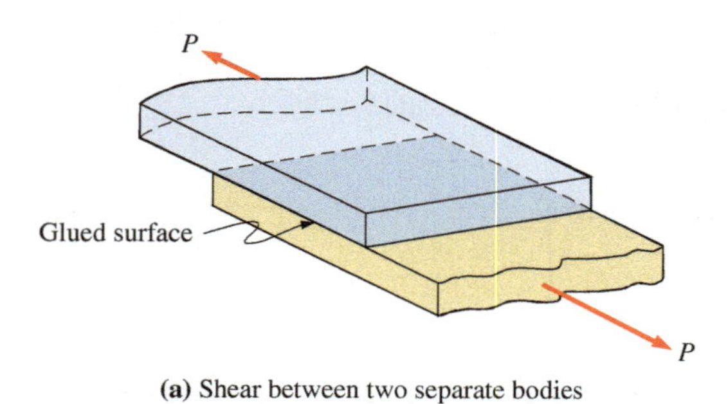

(a) Shear between two separate bodies

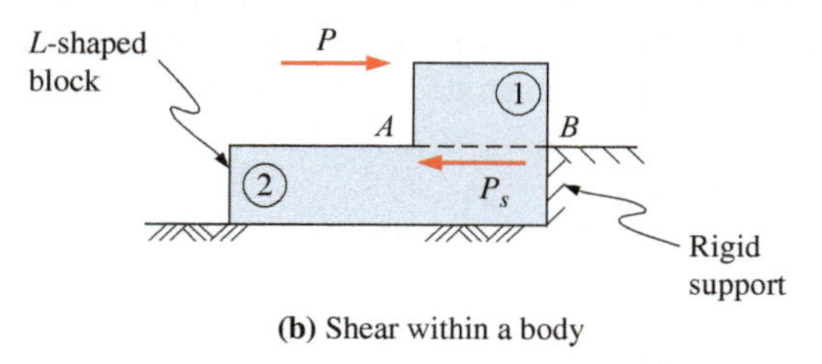

(b) Shear within a body

FIGURE 9.10 Shear stress examples.

Shear stress can also develop within a body when various layers of the material tend to slide with respect to each other. This action is illustrated in Figure 9.10b. A force P is applied as shown. A resisting force P_s acts in plane AB to prevent a sliding action between component parts 1 and 2. This resisting force constitutes an internal shear force. Assuming that the resistance is uniformly distributed, the resistance per unit area, or shear stress, can be computed using Equation (9.4).

Since the applied force P and the resisting force P_s are equal and parallel, all horizontal planes located between them have the same tendency to slide with respect to one another and each plane will develop the same intensity of shear stress.

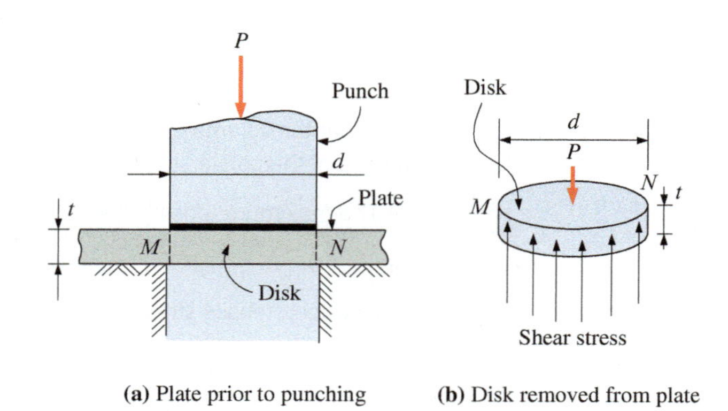

(a) Plate prior to punching (b) Disk removed from plate

FIGURE 9.11 The punching process.

Another shear stress illustration is shown in Figure 9.11 where a force P is applied to a punch so as to punch a hole through a metal plate. Since there is no support for the plate directly under the punch and since the summation of the vertical forces must equal zero, shear stresses are developed over a resisting area equal to the circumference of the punch multiplied by the plate thickness. The material removed by the punch will have a disklike shape and a thickness equal to the plate thickness. The metal around the circumference of the disk must fail in shear as the disk is separated from the rest of the plate. The magnitude of the shear stress achieved in the material when it fails in shear (when the hole is punched) is called the *ultimate shear strength.* If d is the diameter of the punch (or of the disk) and t is the thickness of the plate, then the sheared area is

$$A = \pi dt$$

and the average shear stress is

$$\tau = \frac{P}{A} = \frac{P}{\pi dt}$$

EXAMPLE 9.9 The lap joint in Figure 9.12 consists of two steel plates connected by two $\frac{3}{4}$-in.-diameter bolts. For a tensile load P of 18,000 lb, compute the average shear stress in the bolts.

Solution Assume that the load is resisted equally by each bolt and that the shear stress developed is uniformly distributed across the cross section of each bolt. Since there is only one plane of shear per bolt, the

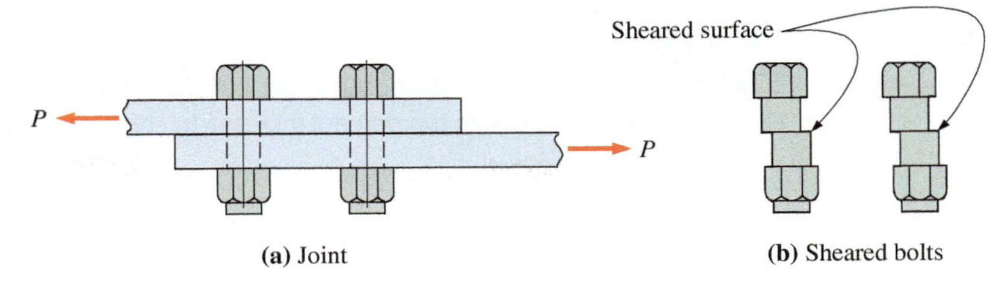

FIGURE 9.12 Bolted joint.

resisting shear (circular) area of each $\frac{3}{4}$-in.-diameter bolt is 0.442 in^2. Each bolt resists 9000 lb (one-half of the total load). The average shear stress is therefore

$$\tau = \frac{P}{A} = \frac{9000 \text{ lb}}{0.442 \text{ in.}^2} = 20{,}400 \text{ psi}$$

EXAMPLE 9.10 Three pieces of wood are glued together to form an assembly (shown in Figure 9.13) that can be used to test the shear strength of a glued joint. A load P of 50 kN is applied. Compute the average shear stress in each joint.

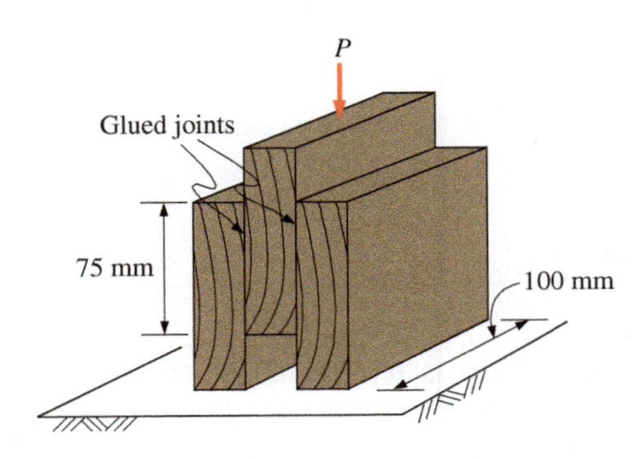

FIGURE 9.13 Glued wood joints.

Solution The resisting shear area in each joint is 7500 mm^2. Assume that the load P is equally resisted by each joint and that the shear stress developed is uniformly distributed throughout the joint. Therefore, each joint resists 25 kN (one-half the total load). The average shear stress is calculated as

$$\tau = \frac{P}{A} = \frac{25 \times 10^3 \text{ N}}{7.50 \times 10^3 \text{ mm}^2} = 3.33 \text{ N/mm}^2 = 3.33 \text{ MPa}$$

EXAMPLE 9.11 A punching operation is being planned in which a $\frac{3}{4}$-in.-diameter hole is to be punched in an aluminum plate $\frac{1}{4}$ in. thick (refer to Figure 9.11). For this aluminum alloy, the ultimate shear strength is 27,000 psi. Compute the force P that must be applied to the punch. Assume that the shear stress is uniformly distributed.

Solution The resisting shear area is the circumference of the punch multiplied by the thickness of the plate:

$$A = \pi d t = \pi(0.75 \text{ in.})(0.25 \text{ in.}) = 0.589 \text{ in.}^2$$

Next, rewrite Equation (9.4) and, by substituting the ultimate shear stress for τ, solve for P, the required applied force that will induce an ultimate shear stress of 27,000 psi:

$$P = A\tau_{(\text{ult})} = (0.589 \text{ in.}^2)(27{,}000 \text{ psi}) = 15{,}900 \text{ lb}$$

EXAMPLE 9.12 A tractor drawbar is connected to an implement as shown in Figure 9.14. The tractor pulls with a force of 50 kN that must be transmitted by the shear bolt. (a) Calculate the shear stress in the bolt if the bolt diameter is 19 mm. (b) Calculate the percent increase in shear stress if the bolt diameter is reduced to 16 mm.

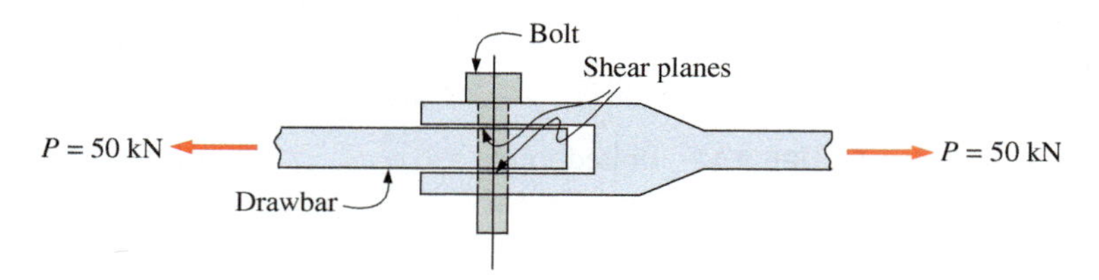

FIGURE 9.14 Implement hitch.

Solution In this connection there are two planes of shear resisting the load. Each shear plane resists 25 kN, one-half of the 50-kN total load.

(a) The cross-sectional area of the 19-mm bolt, per shear plane, is 284 mm^2. The average shear stress is

$$\tau = \frac{P}{A} = \frac{25 \times 10^3 \text{ N}}{284 \text{ mm}^2} = 88.0 \text{ N/mm}^2 = 88.0 \text{ MPa}$$

(b) The cross-sectional area of the 16-mm bolt, per shear plane, is 201 mm^2. The average shear stress is

$$\tau = \frac{P}{A} = \frac{25 \times 10^3 \text{ N}}{201 \text{ mm}^2} = 124.4 \text{ N/mm}^2 = 124.4 \text{ MPa}$$

The percent increase in shear stress is

$$\frac{124.4 \text{ MPa} - 88.0 \text{ MPa}}{88.0 \text{ MPa}}(100) = 41\%$$

The previous group of problems is representative of analysis-type problems. Design-type problems are those in which the size and/or shape of a member must be determined. The designed member must support expected loads without exceeding an allowable stress. In this case, Equation (9.4) must be rewritten to provide the required shear area. This is the same procedure that was presented initially in Section 9.2; see Equations (9.1) and (9.3). For shear considerations, the definitions of two of the three terms are slightly different:

$$A = \frac{P}{\tau(\text{all})} \tag{9.5}$$

where A = the required area that will limit the shear stress to the allowable shear stress

P = the applied (or expected) load or force

$\tau_{(\text{all})}$ = the allowable shear stress

The allowable shear stress depends on the application and on material properties, which are discussed in Chapter 10.

EXAMPLE 9.13 The rod shown in Figure 9.15 is to support a load P of 20,000 lb. The material of which the rod is to be made will be a steel designated by the American Iron and Steel Institute as AISI 1020. The steel has an allowable shear stress of 7500 psi. Determine the required diameter and select a diameter to use, assuming bar diameters are available in $\frac{1}{8}$ in. increments.

Solution Since there are two planes of shear, each shear plane will resist 20,000/2 or 10,000 lb. The required cross-sectional area per plane is

$$A = \frac{P}{\tau_{(\text{all})}} = \frac{10{,}000 \text{ lb}}{7500 \text{ psi}} = 1.33 \text{ in.}^2$$

Since $A = \pi d^2/4$, or $0.7854d^2$, the required diameter is calculated from

$$d = \sqrt{\frac{A}{0.7854}} = \sqrt{\frac{1.33 \text{ in.}^2}{0.7854}} = 1.30 \text{ in.}$$

Use a $1\frac{3}{8}$-in.-diameter rod.

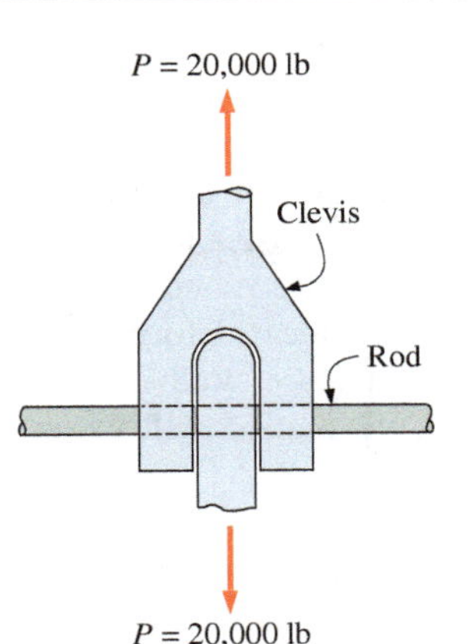

FIGURE 9.15 Clevis connection.

9.4 TENSILE AND COMPRESSIVE STRAIN AND DEFORMATION

The terms *deformation* and *strain* both represent dimensional change. A body subjected to either a tensile force or a compressive force will undergo a change in length. It will be elongated (lengthened) by the tensile force and compressed (shortened) by the compressive force.

With some materials (rubber, for example), small loads produce relatively large deformations. Other engineering materials respond in a similar way, although the amount of the deformation may be extremely small. Even such very rigid materials as steel, when subjected to load, will exhibit a small deformation. The total deformation, or change in length, of a member is generally designated δ (Greek lowercase delta).

To permit comparisons with standard values, the total deformation is converted to a unit basis and is expressed in terms of deformation per unit length. It is generally termed *unit strain*. For the remainder of this book, the word *strain* is used to signify unit strain.

In the determination of tensile or compressive strain, the assumption is made that each unit of length of a member will elongate or shorten the same amount. Strain, represented in this text by ϵ (Greek lowercase epsilon), can be computed by dividing the total deformation by the original length of the member. Mathematically, strain is written as

$$\epsilon = \frac{\text{total deformation}}{\text{original length}} = \frac{\delta}{L} \tag{9.6}$$

Since the strain is a ratio of two lengths, it is dimensionless and considered a pure number. It is common practice, however, to express strain in terms of inches per inch in the U.S. Customary System, since both numerator and denominator are expressed in inches. In the SI, strain may be expressed in terms of millimeters per millimeter or meters per meter. The units in the numerator and denominator must be consistent. In a perfectly straight bar (member) of homogeneous material, of constant cross section and under constant load, ϵ represents the theoretical strain in the bar. With any variation in these parameters, the quantity δ/L represents only the average strain along the length L.

If a member is suspended from its upper end and supports only its own weight, the strain varies uniformly from zero at the lower end to a maximum value at the upper end. The average strain is given by the expression δ/L, where δ and L are as previously defined. But the *maximum* strain occurs at the upper end, where the force in the member is the greatest. Since the strain variation is linear, the strain at the top of the member will be twice the average, or $2\delta/L$.

EXAMPLE 9.14 Compute the total elongation (deformation) for a steel wire 60 ft in length if the strain is 0.00067 in./in.

Solution The length L and the (unit) strain ϵ are given. The definition of strain is

$$\epsilon = \frac{\delta}{L}$$

Rewriting this expression for the total elongation δ gives

$$\delta = \epsilon L = 0.00067\,\frac{\text{in.}}{\text{in.}}\,(60\text{ ft})\left(12\frac{\text{in.}}{\text{ft}}\right) = 0.482\text{ in.}$$

9.5 SHEAR STRAIN

When an axial tensile load is applied to a body, it will cause a longitudinal tensile deformation: an elongation. Similarly, an axial compressive load will cause a longitudinal compressive deformation: a shortening. When a shear force is applied to a body, it will cause a shear deformation in the same direction as the applied force. Rather than an elongation or shortening, this deformation will be an angular distortion.

A motor mount is shown in Figure 9.16a. The motor mount is composed of a block of elastic material with attachments to allow for connection to the base of the motor and the supporting structure. A force P is applied at the top of the block. This action subjects the block (of height L) to a pair of shear forces as shown in Figure 9.16b. If we imagine that the block is composed of many thin layers and that each layer will slide slightly with respect to its neighbor, we can visualize how the angular distortion will develop. As may be observed, the total shear deformation in the length L is δ_s and the shear strain ϵ_s, similar to tensile and compressive strains, is the total shear deformation divided by the length L:

$$\epsilon_s = \frac{\delta_s}{L} \tag{9.7}$$

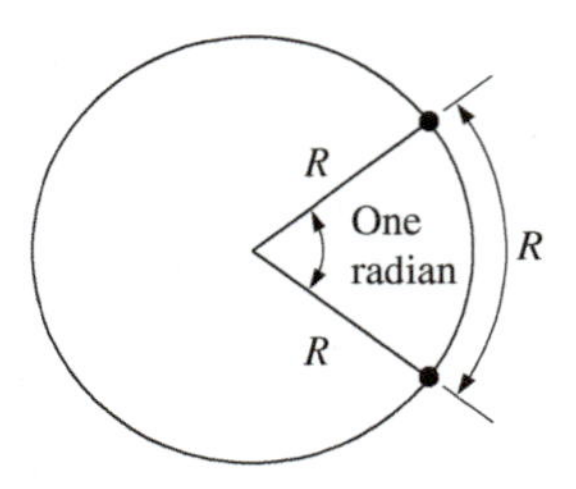

FIGURE 9.17 Radian definition.

Note also that in Figure 9.16 an angular relationship exists in which

$$\tan \phi = \frac{\delta_s}{L} = \epsilon_s$$

For small angles, which is generally the case, the tangent of the angle is approximately equal to the angle expressed in radians. A radian is defined as the central angle subtended by an arc length equal to the radius of the circle (see Figure 9.17). From this definition, we see that the angle, in radians, is the arc length divided by the radius of the circle. For extremely short arc lengths, the arc approaches a straight line and the line becomes almost perpendicular to the radius. Therefore, the tangent of the angle (opposite side/adjacent side) is approximately equal to the angle in radians (arc length/radius). Therefore, the angle ϕ in radians is very nearly equal to the shear strain.

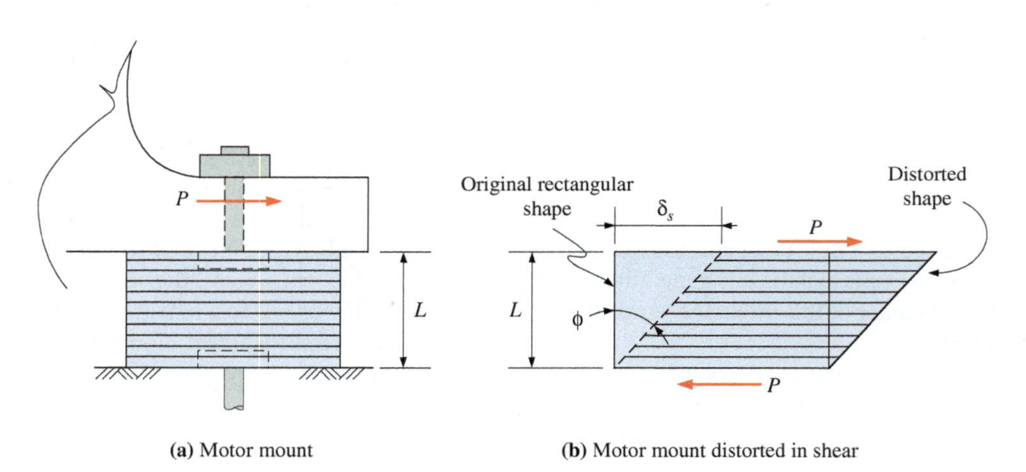

FIGURE 9.16 Shear strain.

EXAMPLE 9.15 In Figure 9.16, assume that load P is applied at the top of the block and that it displaces the top a horizontal distance of 0.0024 in. with respect to the bottom of the block. Assume the height L of the block to be 1.4 in. Compute the shear strain.

Solution

$$\epsilon_s = \frac{\delta_s}{L} = \frac{0.0024 \text{ in.}}{1.4 \text{ in.}} = 0.0017 \text{ in./in.}$$

9.6 THE RELATION BETWEEN STRESS AND STRAIN (HOOKE'S LAW)

For most engineering materials, a relationship exists between stress and strain. For each increment in stress there is a closely proportional increase in strain, provided that a certain limit of stress is not exceeded. If the induced stress exceeds this limiting value, the corresponding strain will no longer be proportional to the stress. This limiting value is called the *proportional limit* and may be concisely defined as that value of stress up to which strain is proportional to stress.

The proportional relationship between stress and strain was originally stated by Robert Hooke in 1678 and became known as *Hooke's law*. As an example, assume that a bar of material that obeys Hooke's law is subjected to a tensile load of P_A that produces a stress σ_A and a strain ϵ_A. Then subject the bar to a tensile load P_B that produces a stress σ_B and a strain ϵ_B. (Both stresses are less than the proportional limit.) The stresses and strains can be plotted on a stress–strain diagram, as shown in Figure 9.18. The ratio of stress to strain, which is also the slope of the line drawn from the origin and joining the plotted points, is constant. Mathematically,

$$\frac{\sigma_A}{\epsilon_A} = \frac{\sigma_B}{\epsilon_B} = \text{constant}$$

This constant is now known as the *modulus of elasticity* or *Young's modulus* (after Thomas Young, who is credited with having defined it in 1807). The modulus of elasticity for members in tension or compression is generally represented by the symbol E and is expressed by the equation

$$E = \frac{\text{stress}}{\text{strain}} = \frac{\sigma}{\epsilon} \quad \textbf{(9.8)}$$

Since strain is a pure number, it is evident that E has the same units as does stress (σ), which, in the U.S. Customary System, is usually pounds per square inch (psi) and, in the SI, pascals (Pa) or megapascals (MPa). For most of the common engineering materials, the modulus of elasticity in compression is equal to that found in tension for all practical purposes. In the case of steel and other ductile metals, tension tests are more commonly performed than compression tests. Therefore, it is generally the tensile modulus of elasticity that is determined and used.

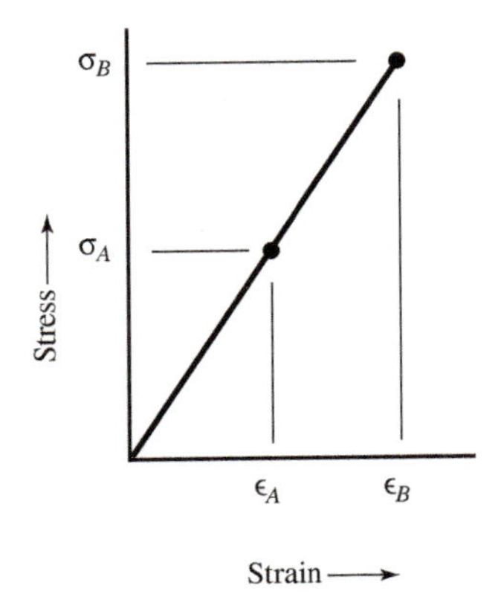

FIGURE 9.18 Stress-strain diagram.

The standard tension test used to determine the modulus of elasticity is discussed in Chapter 10, along with the stress limitations within which modulus of elasticity is valid. We briefly discuss modulus of elasticity in this section because of the importance of the constant ratio between stress and strain in the design process for many engineering materials.

The modulus of elasticity of steel (in tension and compression) is commonly assumed to be 29,000,000 psi or 30,000,000 psi in the U.S. Customary System. In the SI, these values convert to 200,000 MPa or 207,000 MPa. The particular values depend on the type of steel. For the purposes of this text, values of the modulus of elasticity for steel are taken as 30,000,000 psi (207,000 MPa). (Note that when values are shown in this form, the first number will be the value in the U.S. Customary System and the parenthetical value will be its approximate equivalent in the SI.)

For other materials, the values of the modulus of elasticity may be as low as 1,000,000 psi (7000 MPa) or less. Average values of modulus of elasticity for some common engineering materials are given in Appendix G (see also Appendices Notes).

Physically, the modulus of elasticity is a measure of the stiffness of a material in its response to an applied load and represents a definite property of that material. Material stiffness may be defined as the property that enables a material to withstand high stress without great strain. In Chapter 8 we discussed geometric properties of cross-sectional areas and introduced the concept of stiffness as a direct result of those geometric properties. We now see that the total stiffness of a member can be attributed to a combination of both material properties and geometric properties.

As with axially loaded bodies in tension and compression, the shear stress is proportional to the shear strain, as long as the proportional limit in shear has not been exceeded. This constant of proportionality is known as the *modulus of elasticity in shear* or the *modulus of rigidity*. It is denoted G and is expressed as

$$G = \frac{\text{shear stress}}{\text{shear strain}} = \frac{\tau}{\epsilon_s} \quad \textbf{(9.9)}$$

Average values of the modulus of rigidity for some common engineering materials are given in Appendix G. Note that these values and the tabulated values of modulus of elasticity are significantly different. The modulus of rigidity and its relationship to other material properties is further discussed in Chapter 11.

EXAMPLE 9.16 A 24-in.-long bar with a cross-sectional area of 1.0 in.^{2} is subjected to an axial tensile load of 10,000 lb as shown in Figure 9.19. Compute the stress, strain, and the total elongation if the bar material is (a) steel with $E_{ST} = 30{,}000{,}000$ psi; (b) aluminum with $E_{AL} = 10{,}000{,}000$ psi; and (c) titanium with $E_{TI} = 16{,}500{,}000$ psi. Assume the proportional limit for each material to be as follows: steel = 34,000 psi, aluminum = 34,000 psi, and titanium = 125,000 psi.

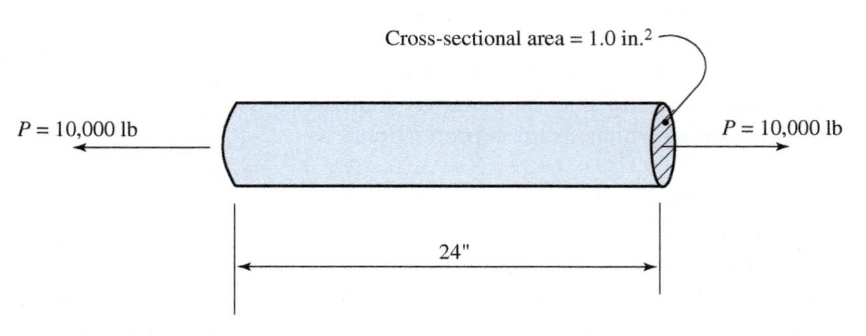

FIGURE 9.19 Tension bar.

Solution

1. The stress for each of the materials is

$$\sigma_t = \frac{P}{A} = \frac{10{,}000 \text{ lb}}{1.0 \text{ in.}^2} = 10{,}000 \text{ psi}$$

This value is less than the proportional limit for any of the materials; therefore, Hooke's law applies.

2. Compute the strains for each of the materials:

(a) Steel:

$$\epsilon_{ST} = \frac{\sigma_t}{E_{ST}} = \frac{10{,}000 \text{ psi}}{30{,}000{,}000 \text{ psi}} = 0.000333$$

(b) Aluminum:

$$\epsilon_{AL} = \frac{\sigma_t}{E_{AL}} = \frac{10{,}000 \text{ psi}}{10{,}000{,}000 \text{ psi}} = 0.001$$

(c) Titanium:

$$\epsilon_{TI} = \frac{\sigma_t}{E_{TI}} = \frac{10{,}000 \text{ psi}}{16{,}500{,}000 \text{ psi}} = 0.000606$$

3. Compute the total elongation for each material:

(a) Steel:

$$\delta_{ST} = \epsilon_{ST}L = 0.000333(24 \text{ in.}) = 0.00799 \text{ in.}$$

(b) Aluminum:

$$\delta_{AL} = \epsilon_{AL}L = 0.0010(24 \text{ in.}) = 0.0240 \text{ in.}$$

(c) Titanium:

$$\delta_{TI} = \epsilon_{TI}L = 0.000606(24 \text{ in.}) = 0.01454 \text{ in.}$$

The strain and total deformation in the aluminum bar are three times larger than those of the steel bar. The strain in the titanium bar is approximately twice that of the steel bar. Thus, the stiffness of the steel is significantly greater than that of either aluminum or titanium.

We have established expressions for stress, strain, and modulus of elasticity E. These may now be combined and a convenient expression developed to determine directly the total deformation δ for a homogeneous axially loaded prismatic member. We begin with the definition of modulus of elasticity and substitute for stress and strain:

$$E = \frac{\sigma}{\epsilon} = \frac{P/A}{\delta/L} = \frac{PL}{A\delta}$$

Solving for δ,

$$\delta = \frac{PL}{AE} \tag{9.10}$$

where δ = the total axial deformation (in.) (m, mm)
P = the total applied external axial load (lb, kips) (N)
L = the length of the member (in.) (m, mm)
A = the cross-sectional area of the member (in.^{2}) (m^2, mm^2)
E = the modulus of elasticity (psi, ksi) (Pa, MPa)

This expression is valid only when the stress in the member does not exceed the proportional limit.

EXAMPLE 9.17 A tensile member in a machine is subjected to an axial load of 5000 lb as shown in Figure 9.20. It has a length of 30 in. and is made from a steel tube having an O.D. of $\frac{3}{4}$ in. and an I.D. of $\frac{1}{2}$ in. Compute the tensile stress in the tube and the total axial deformation. Assume E = 30,000,000 psi and a proportional limit of 34,000 psi.

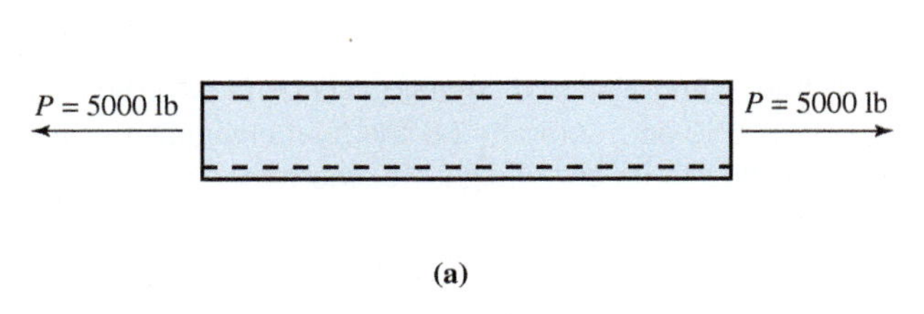

O.D. = $\frac{3}{4}$"
I.D. = $\frac{1}{2}$"
(b)

FIGURE 9.20 Tensile member for machine part.

Solution

1. The cross-sectional area of the hollow tube is calculated from

$$A = \frac{\pi}{4}(d^2 - d_i^2) = 0.7854[(0.75\text{ in.})^2 - (0.50\text{ in.})^2] = 0.245\text{ in.}^2$$

2. Calculate the axial deformation (elongation):

$$\delta = \frac{PL}{AE} = \frac{(5000\text{ lb})(30\text{ in.})}{(0.245\text{ in.}^2)(30{,}000{,}000\text{ psi})} = 0.0204\text{ in.}$$

3. Calculate the tensile stress developed and verify that it is not in excess of the proportional limit:

$$\sigma_t = \frac{P}{A} = \frac{5000\text{ lb}}{0.245\text{ in.}^2} = 20{,}400\text{ psi} < 34{,}000\text{ psi} \quad \textbf{(O.K.)}$$

EXAMPLE 9.18 A force of 6700 N is applied to a steel wire 7.5 m in length. The tensile stress in the wire must not exceed 138 MPa and the total elongation (deformation) must not exceed 4.5 mm. Compute the required diameter of the wire. Neglect the weight of the wire. Use E = 207,000 MPa and a proportional limit of 234 MPa.

Solution The required cross-sectional area is calculated based on the allowable tensile stress of 138 MPa. Make the substitutions in terms of newtons and meters:

$$\text{required } A = \frac{P}{\sigma_{t(\text{all})}} = \frac{6700\text{ N}}{138 \times 10^6\text{ N/m}^2} = 48.6 \times 10^{-6}\text{m}^2 = 48.6\text{ mm}^2$$

The required cross-sectional area is calculated based on allowable total deformation using Equation (9.10) rewritten for the required area:

$$\text{required } A = \frac{PL}{\delta E} = \frac{(6700\text{ N})(7.5\text{ m})}{(4.5 \times 10^{-3})(207{,}000 \times 10^{6}\text{ N/m}^2)}$$
$$= 0.0539 \times 10^{-3}\text{ m}^2 = 53.9\text{ mm}^2$$

The larger of the two areas (53.9 mm^2) must be provided. Therefore, the required diameter is calculated from

$$d = \sqrt{\frac{A}{0.7854}} = \sqrt{\frac{53.9}{0.7854}} = 8.28\text{ mm}$$

Check to ensure that the tensile stress developed is less than the proportional limit:

$$\sigma_t = \frac{P}{A} = \frac{6700\text{ N}}{53.9 \times 10^{-6}\text{ m}^2}$$
$$= 124 \times 10^{6}\text{ Pa} = 124\text{ MPa} < 234\text{ MPa} \qquad \textbf{(O.K.)}$$

EXAMPLE 9.19 A copper wire 150 ft long and $\frac{1}{8}$ in. in diameter is suspended vertically from its upper end. The unit weight of the copper is 550 pcf. Compute (a) the total elongation of the wire due to its own weight, (b) the average strain due to its own weight, (c) the total elongation of the wire δ (which includes the elongation due to its own weight) if a weight of 70 lb is attached to its lower end, and (d) the maximum load P that this wire can safely support at its lower end if the allowable tensile stress is 12,000 psi.

Solution For a wire suspended vertically from its upper end, the total elongation produced by the weight of the wire is equal to that produced by a load of half its weight applied at the end (this is the same as the average load being applied throughout the length of the wire).

(a) The weight of the wire is equal to the total volume times the unit weight of the copper:

$$\text{total weight} = \frac{\pi d^2}{4}(L)(550\text{ lb/ft}^3)$$
$$= \frac{\pi(0.125\text{ in.})^2}{4(144\text{ in.}^2/\text{ft}^2)}(150\text{ ft})(550\text{ lb/ft}^3) = 7.03\text{ lb}$$

The area of $\frac{1}{8}$-in.-diameter wire is 0.0123 in^2. The total elongation is calculated using E from Appendix G:

$$\delta = \frac{(P/2)L}{AE} = \frac{[(7.03\text{ lb}/2)](150\text{ ft})(12\text{ in./ft})}{(0.0123\text{ in.}^2)(15{,}000{,}000\text{ psi})} = 0.0343\text{ in.}$$

(Note that the length is converted to inches.)

(b) The average strain is

$$\epsilon = \frac{\delta}{L} = \frac{0.0343\text{ in.}}{(150\text{ ft})(12\text{ in./ft})} = 0.000019$$

(c) The additional elongation due to a 70-lb weight is

$$\delta = \frac{PL}{AE} = \frac{(70\text{ lb})(150\text{ ft})(12\text{ in./ft})}{(0.0123\text{ in.}^2)(15{,}000{,}000\text{ psi})} = 0.683\text{ in.}$$

from which

$$\text{total } \delta = (0.683\text{ in.}) + (0.0343\text{ in.}) = 0.717\text{ in.}$$

(d) To find the maximum load P that can be safely supported (see Example 9.8), calculate the maximum stress due to the weight of the wire:

$$\sigma_t = \frac{P}{A} = \frac{7.03 \text{ lb}}{0.0123 \text{ in.}^2} = 572 \text{ psi}$$

Therefore, the portion of the allowable stress that remains to resist the load P is

$$12{,}000 \text{ psi} - 572 \text{ psi} = 11{,}430 \text{ psi}$$

and the maximum (or allowable) load is

$$P = A\sigma_t = (0.0123 \text{ in.}^2)(11{,}430 \text{ psi}) = 140.6 \text{ lb}$$

EXAMPLE 9.20 A steel bar having a cross-sectional area of 1.0 in.^{2} is suspended vertically and subjected to loads as shown in Figure 9.21. The total length of the bar is 7 ft. Compute the total elongation of the bar. Use $E = 30{,}000{,}000$ psi. Neglect the weight of the bar.

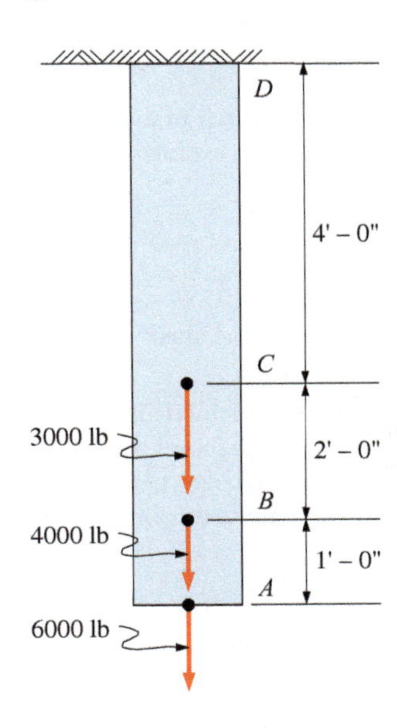

FIGURE 9.21 Suspended steel bar.

Solution The total elongation of the bar will be the sum of the elongations of the three segments AB, BC, and CD. The elongation of AB is

$$\delta = \frac{PL}{AE} = \frac{(6000 \text{ lb})(1 \text{ ft})(12 \text{ in./ft})}{(1.0 \text{ in.}^2)(30{,}000{,}000 \text{ psi})} = 0.0024 \text{ in.}$$

The elongation of BC is

$$\delta = \frac{PL}{AE} = \frac{(10{,}000 \text{ lb})(2 \text{ ft})(12 \text{ in./ft})}{(1.0 \text{ in.}^2)(30{,}000{,}000 \text{ psi})} = 0.0080 \text{ in.}$$

The elongation of CD is

$$\delta = \frac{PL}{AE} = \frac{(13{,}000 \text{ lb})(4 \text{ ft})(12 \text{ in./ft})}{(1.0 \text{ in.}^2)(30{,}000{,}000 \text{ psi})} = 0.0208 \text{ in.}$$

Thus,

$$\text{total } \delta = (0.0024 \text{ in.}) + (0.0080 \text{ in.}) + (0.0208 \text{ in.}) = 0.0312 \text{ in.}$$

SUMMARY BY SECTION NUMBER

9.2 *Stress*, as used herein, represents an internal resistance developed by a unit area of a member. An axial force is collinear with the longitudinal axis of a member, acts through its centroid, and develops a uniform axial stress distribution. An axial tensile force tends to elongate a member and develops uniform internal tensile stress over the cross section of the member. Similarly, an axial compressive force tends to shorten the member and develops uniform internal compressive stress. A bearing stress is a compressive stress exerted on external surfaces of separate bodies in contact with each other, and which transmits force. All these stresses are called *normal stresses* (also *direct stresses*), since the resisting area is normal to the direction of the force. They may be computed using the direct stress formula,

$$\sigma = \frac{P}{A} \tag{9.1}$$

Allowable stress $\sigma_{(\text{all})}$ is a level of stress that is judged to be acceptable.

9.3 A shear force develops a shear stress over an area in a direction parallel to the surface on which it acts. An average shear stress may be computed from

$$\tau = \frac{P}{A} \tag{9.4}$$

Ultimate shear strength is the magnitude of the shear stress that will cause a material to fail in shear.

9.4 *Deformation* (δ), as used herein, represents a total dimensional change of a stressed member. *Strain* (ϵ), as used herein, represents a unit strain, or a dimensional change per unit length of a stressed member, and may be computed from

$$\epsilon = \frac{\text{total deformation}}{\text{original length}} = \frac{\delta}{L} \tag{9.6}$$

9.5 Axial tensile or compressive forces applied to a body cause longitudinal tensile or compressive deformations. A shear force causes a shear deformation (δ_s), which is an angular distortion in the same direction as the applied force. Shear strain (ϵ_s) may be computed from

$$\epsilon_s = \frac{\delta_s}{L} \tag{9.7}$$

9.6 Hooke's law refers to the relationship in which stress is proportional to strain. This law applies to most engineering materials and is valid provided that the stress does not exceed the proportional limit of the material. This constant ratio of stress to strain is called the *modulus of elasticity (E)* and is defined for members in tension or compression, as

$$E = \frac{\text{stress}}{\text{strain}} = \frac{\sigma}{\epsilon} \tag{9.8}$$

The modulus of elasticity is a measure of a material's resistance to deformation when loaded.

The shear modulus of elasticity (G) is defined as

$$G = \frac{\text{shear stress}}{\text{shear strain}} = \frac{\tau}{\epsilon_s} \tag{9.9}$$

Equations (9.1), (9.6), and (9.8) may be combined to yield an expression for total deformation:

$$\delta = \frac{PL}{AE} \tag{9.10}$$

PROBLEMS

For the following problems, assume the modulus of elasticity for steel to be 30,000,000 psi (207,000 MPa) and the proportional limit to be 34,000 psi (234 MPa). Refer to Appendices F or G for other materials.

Section 9.2 Tensile and Compressive Stresses

9.1 Write the direct stress formula in its three forms and explain the meaning of each one.

9.2 A 6-in.-diameter concrete test cylinder is loaded with a compressive force of 113.1 kips, as shown. Determine the stress in the cylinder.

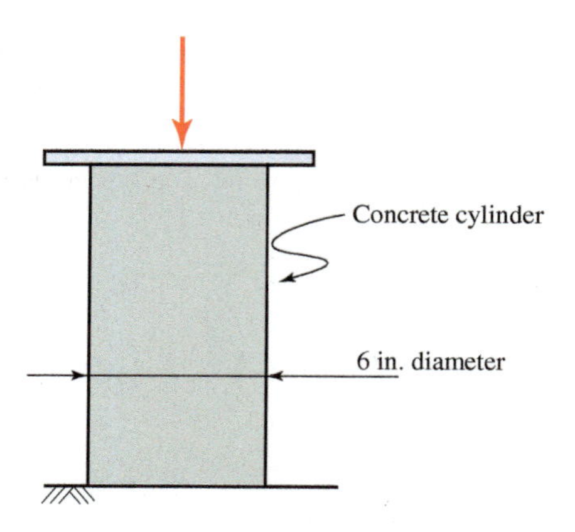

PROBLEM 9.2

9.3 Determine the tensile stress in each segment of the bar shown. Segment *AB* has a square cross section 2 in. by 2 in. Segment *BC* has a circular cross section with a diameter of 1.75 in. The tensile load $P = 75$ kips.

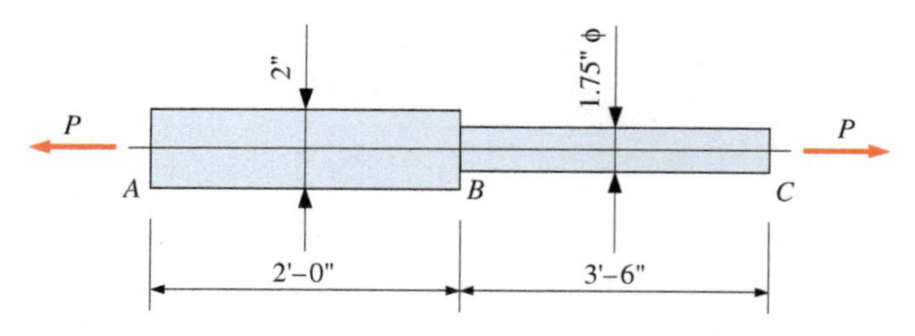

PROBLEM 9.3

9.4 Calculate the stress developed in the following members that are each subjected to an axial tensile load of 70 kN:

(a) steel bar, 25 mm by 50 mm

(b) 150-mm-diameter wood post (use the given diameter)

(c) 25-mm-diameter steel tie rod

9.5 The following members are subjected to axial tensile loads of 16,000 lb. Determine the stress induced.

(a) steel bar 2 in. by 2 in.

(b) W4 × 13

9.6 Determine the stresses in the two segments of the bar shown. Segment A has a diameter of $1\frac{1}{4}$ in. and segment B has a cross section 2 in. by 2 in.

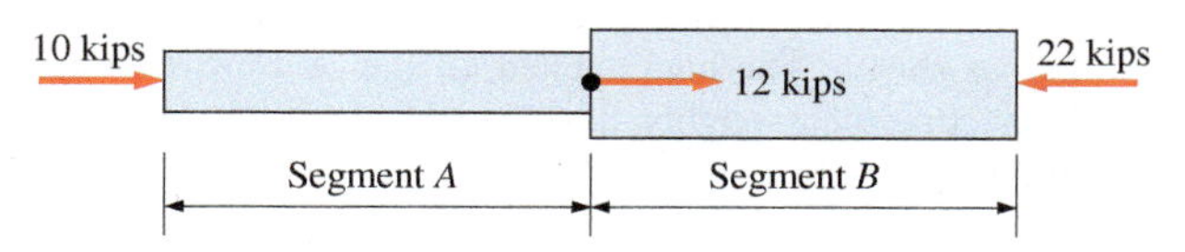

PROBLEM 9.6

9.7 A bin weighing 8 tons is supported by three steel tension rods. Each rod is $\frac{5}{8}$ in. in diameter. Assume that the rods support equal loads. Find the stress in the rods.

9.8 Diameters of small commercially available steel rods vary by sixteenths of an inch. Select the required commercial size of the rod required to support a tensile load of 35,000 lb if the tensile stress cannot exceed 20,000 psi.

Section 9.3 Shear Stresses

9.9 A No. 32 (metric designation) reinforcing bar for concrete has a diameter of 32 mm. The bar is cut to length in a shear apparatus as shown. The required force P for the shearing operation is 300 kN. Determine the ultimate shear strength $\tau_{(ult)}$ for this bar.

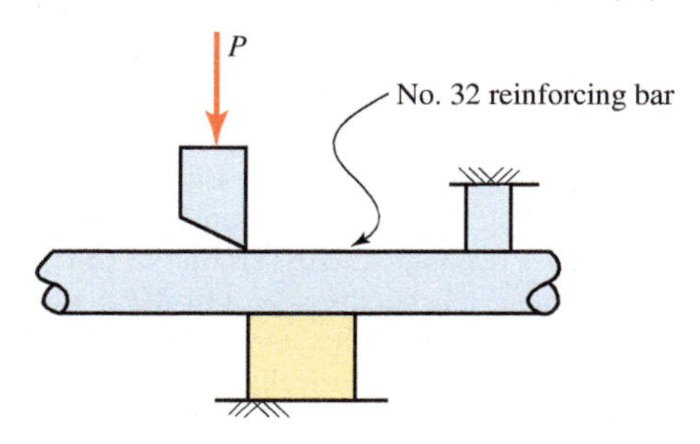

PROBLEM 9.9

9.10 In the bolted connection of Figure 9.12, assume that the bolts are $\frac{7}{8}$ in. in diameter and that the allowable shear stress for the bolts is 14,500 psi. Based on bolt shear only, compute the safe allowable load P that may be applied.

9.11 Compute the force required to punch a 1-in.-diameter hole through a $\frac{1}{2}$-in.-thick boiler plate. The ultimate shear strength for the material is 42,000 psi.

9.12 The $\frac{3}{4}$-in.-diameter bolt shown is subjected to a 6000-lb tensile load. The bolt passes through a $\frac{3}{4}$-in. thick plate. Compute
(a) the tensile stress in the bolt
(b) the shear stress in the head of the bolt

Section 9.4 Tensile and Compressive Strain and Deformation

9.13 Why is strain unitless (or dimensionless)?

9.14 (a) Given $\delta = 1.2$ in. and $L = 100$ ft, calculate ϵ.
(b) Given $\delta = 0.5$ ft and $\epsilon = 0.00030$, calculate L.
(c) Given $L = 1000$ in. and $\epsilon = 0.00042$, calculate δ.

9.15 A short compression member 2 in. by 2 in. in cross section and 8 in. long is subjected to a 40,000 lb axial compressive load. As a result, the length of the member is shortened to 7.85 in. Compute the strain in the member.

9.16 A 100-ft-long rod is suspended from one end. The diameter of the rod is 1 in. and the total elongation is 0.80 in. Calculate the maximum strain.

Section 9.6 The Relation Between Stress and Strain (Hooke's Law)

9.17 Compute the total elongation of a steel bar, originally 25 in. long, if the induced tensile stress is 15,000 psi.

9.18 A steel rod $\frac{3}{4}$ in. in diameter and 25 ft long is subjected to an axial tensile load of 15,000 lb. Compute
(a) stress
(b) strain
(c) total elongation

9.19 An aluminum rod 25 mm in diameter and 4 m long is subjected to an axial tensile load of 65 kN. Compute
(a) stress
(b) strain
(c) total elongation

9.20 A steel rod 10 ft long is made up of two 5 ft lengths, one $\frac{3}{4}$ in. in diameter and the other $\frac{1}{2}$ in. in diameter. How much will the bar elongate when subjected to an axial tensile load of 5000 lb?

9.21 A titanium alloy bar elongates 0.500 in. when subjected to an axial tensile load of 100 k. The bar is 16 ft long and has a cross-sectional area of 2.25 in^2. Compute the modulus of elasticity E. Use a proportional limit of 125 ksi.

Computer Problems

For the following computer problems, any appropriate software may be used. Input prompts should fully explain what is required of the user (the program should be user-friendly). The resulting output should be well labeled and self-explanatory.

9.22 Write a program that will calculate the allowable axial tensile load that may be applied on a round rod. User input is to be the rod diameter and the allowable axial tensile stress.

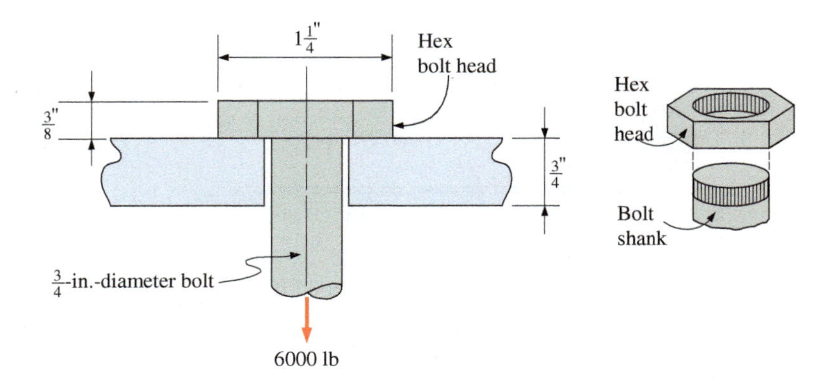

PROBLEM 9.12

9.23 Write a program that will compute stress, strain, and total deformation for an aluminum bar. Use $E = 10{,}000{,}000$ psi. User input is to be cross-sectional area, length, and axial tensile load. An error message is to be displayed if the stress exceeds the proportional limit (35,000 psi).

9.24 The Viking Bin Company manufactures suspended bins weighing from 4 tons to 20 tons (in 2-ton increments). Each bin is suspended by three steel rods. Two grades of steel are available. For one, the allowable tensile stress is 22,000 psi; for the other, 29,000 psi. Write a program that will generate a table of required rod diameters (to the next $\frac{1}{8}$ in.) as a function of bin weight and allowable tensile stress.

9.25 Write a computer program that will calculate the allowable load for a single-shear bolted lap connection similar to that shown in Figure 9.12. The computed allowable load is to be based on bolt shear only. User input is to be number of bolts, bolt diameter, and allowable shear stress for the bolts.

Supplemental Problems

9.26 The joint between a diagonal and a chord in a timber truss is shown. The compressive force P in the diagonal is 20 kN. Determine the compressive stress in the diagonal, the bearing stress on plane ab, and the shear stress on plane ac. Both the chord and the diagonal are 150 mm by 150 mm (full nominal size) in cross section.

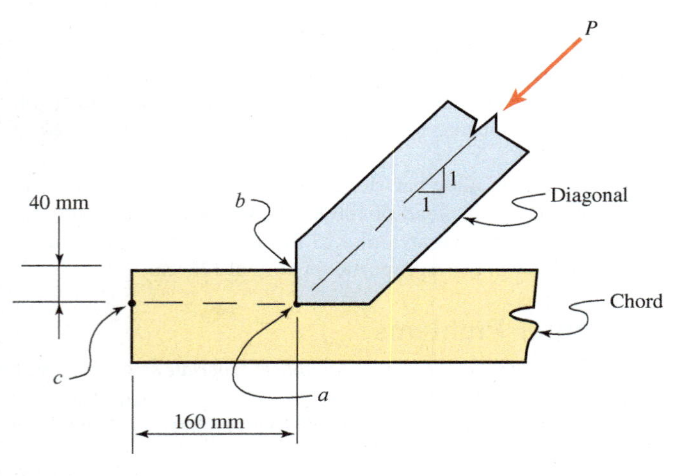

PROBLEM 9.26

9.27 In Problem 9.26, find the required length of the shear plane ac assuming the timber to be hem-fir. (Refer to Table F-2 in Appendix F.)

9.28 A column is supported by a base plate, pedestal, and footing, as shown. The load applied on the column is 60 kips. Neglecting their weights, find the bearing stress under each of the four elements.

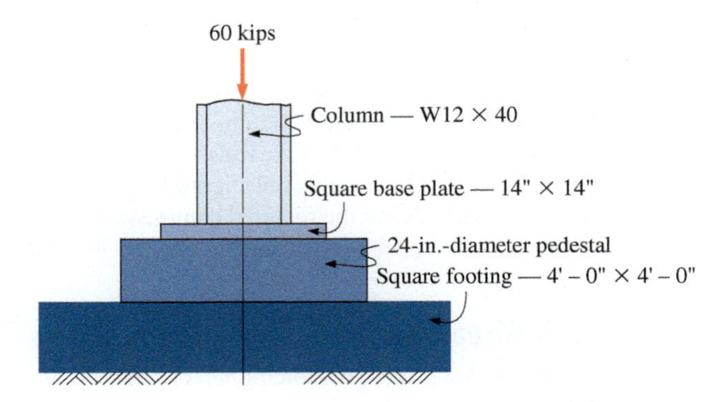

PROBLEM 9.28

9.29 If the soil pressure under the footing of Problem 9.28 were limited to 2.5 ksf, what would be the new required dimensions for the square footing? (Use 2-in. increments.)

9.30 A hopper weighing 75 kN is supported by three 19-mm-diameter steel rods. Calculate the tensile stress developed in the rods, assuming that the rods support equal loads.

9.31 A steel bar has a rectangular cross section 25 mm by 100 mm and an allowable tensile stress of 140 MPa. Calculate the allowable axial tensile load that may be applied.

9.32 A steel wire is suspended vertically from its upper end. The wire is 400 ft long and has a diameter of $\frac{3}{16}$ in. The unit weight of steel is 490 pcf. Compute **(a)** the maximum tensile stress due to the weight of the wire and **(b)** the maximum load P that could be supported at the lower end of the wire. Allowable tensile stress is 24,000 psi.

9.33 A W12 × 40 shape is subjected to a tensile load of 190 kips, as shown. Find the stress at a transverse section where a total of six holes have been drilled. The holes are $\frac{3}{4}$ in. in diameter. There are two holes in each flange and two in the web. (Neglect stress concentration.)

9.34 Calculate the required diameter of steel tie rods necessary to support a balcony. Each rod, suspended from a roof truss,

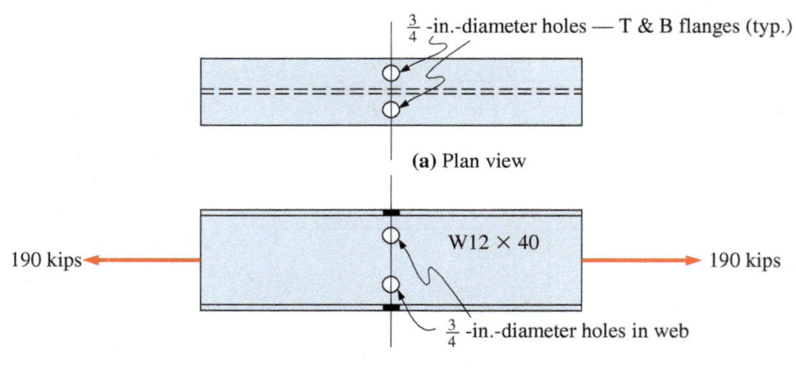

PROBLEM 9.33

supports a floor area of 7.0 m^2. The combined balcony floor load is 10.5 kPa, including dead load and live load. The allowable tensile stress in the steel tie rods is 140 MPa.

9.35 A 30-ft-long steel rod of circular cross section is to support a balcony. The tensile load in the rod will be 20 kips. The maximum elongation allowed is to be $\frac{1}{4}$ in. Allowable tensile stress is 24 ksi. Select the rod diameter (to $\frac{1}{16}$ in.) Neglect the weight of the rod.

9.36 Consider the bolted lap joint shown in Figure 9.12. Assume that the joint consists of two steel plates connected by two 25-mm-diameter bolts and that the load is equally resisted by the two bolts. Calculate the average shear stress in the bolts when the joint is subjected to a tensile load of 120 kN.

9.37 An inclined member is braced with a glued block, as shown. The ultimate shear stress in the glued joint is 1050 psi. Determine the dimension x at which the glued joint will fail.

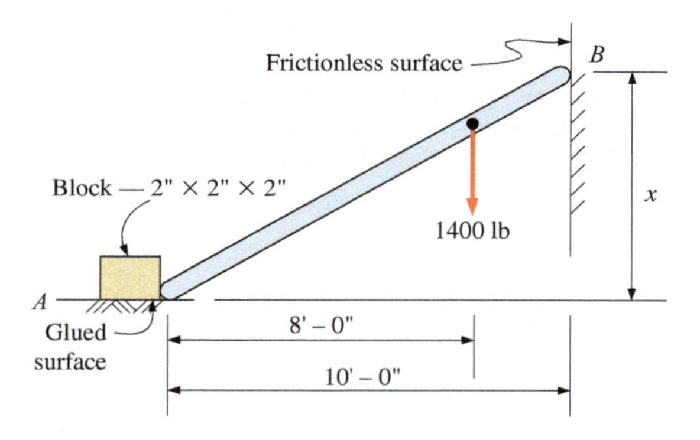

PROBLEM 9.37

9.38 Calculate the force a punch press must exert if it is to punch out a large washer from a strip of 5-mm-thick sheet steel with an ultimate shear strength of 400 MPa. The washer dimensions are as follows: hole diameter = 27 mm, outside diameter = 64 mm.

9.39 A $\frac{3}{4}$-in.-diameter punch is used to punch a hole through a steel plate $\frac{1}{2}$ in. thick. The force necessary to drive the punch through the plate is 60,000 lb. Compute the shear stress developed in the plate.

9.40 A control arm is keyed to a 1-in.-diameter shaft and a link transmits a 40 lb force to the end of the arm, as shown. The key is $\frac{1}{4}$ in. by $\frac{1}{4}$ in. and has a length of $\frac{7}{8}$ in. Find the shear stress in the key.

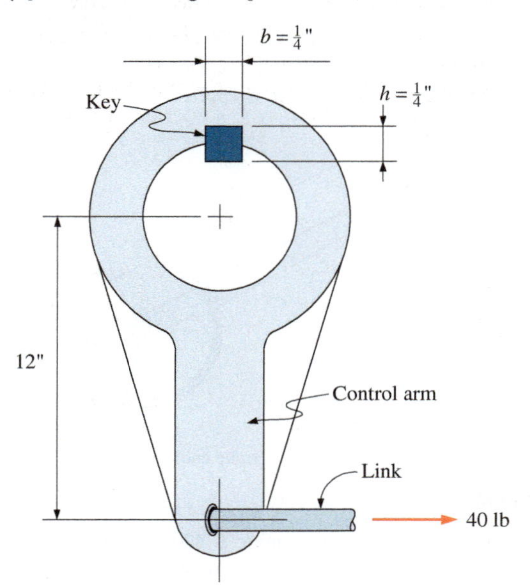

PROBLEM 9.40

9.41 Calculate the required width b for the key of Problem 9.40 if the link transmits a 60-lb force and the length of the key is $1\frac{1}{2}$ in. The allowable shear stress for the key is 4000 psi.

9.42 A 25-mm-diameter aluminum rod, 3 m long, is subjected to an axial tensile load of 67 kN. Assume a proportional limit of 207 MPa. Compute

(a) stress
(b) strain
(c) total elongation

9.43 A short timber post of Douglas fir is subjected to an axial compressive load of 40,000 lb. The nominal dimensions of the post area are 8 in. by 8 in. Its length is 8 ft–6 in. Assume a proportional limit of 6000 psi. Using *dressed* dimensions, compute

(a) stress
(b) strain
(c) total deformation

9.44 A 100-ft surveyor's steel tape with a cross-sectional area of 0.005 in.^{2} is subjected to a 15-lb pull when in use. Compute

(a) the total elongation of the tape
(b) the tensile stress developed in the tape

9.45 An 18-in.-long steel rod is subjected to a tensile load of 12,000 lb. The rod has a diameter of $1\frac{1}{4}$ in. Compute

(a) the tensile stress
(b) the strain
(c) the total elongation

9.46 Compute the magnitude of the tensile load that would produce a strain of 0.0007 in a 26-mm-diameter steel rod.

9.47 A 5-mm-diameter steel wire, 18 m in length, is used in the manufacture of prestressed concrete beams. Calculate the tensile load P required to stretch the wire 45 mm, and calculate the stress developed in the wire.

9.48 A structural steel rod $1\frac{1}{2}$ in. in diameter and 20 ft long supports a balcony and is subjected to an axial tensile load of 30,000 lb. Compute

(a) the total elongation
(b) the diameter of rod required if the total elongation must not exceed 0.10 in.

9.49 A rectangular structural steel eyebar $\frac{3}{4}$ in. thick and 24 ft long between centers of end pins is subjected to an axial tensile load of 50,000 lb. If this bar must not elongate more than 0.20 in., compute

(a) the required width of bar
(b) the tensile stress developed at maximum elongation, assuming the required width is provided

9.50 A $\frac{3}{4}$-in.-diameter steel rod, 100 ft long, is hanging vertically. Compute the total elongation of the rod due to its own weight and a suspended 8000 lb load applied at its free end. Use a unit weight of 490 pcf.

9.51 For the truss shown, compute the total deformation of member CD due to the applied load of 100 kips. Member CD is made of steel and has a cross-sectional area of 2.5 in^2. Indicate whether the member is in tension or compression.

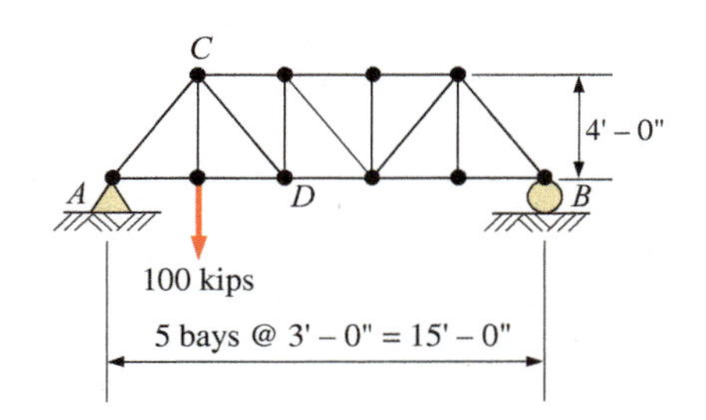

PROBLEM 9.51

9.52 A steel bar with a cross section of $\frac{1}{2}$ in. by $\frac{1}{2}$ in. is subjected to the axial loads shown. Compute the total elongation of the bar.

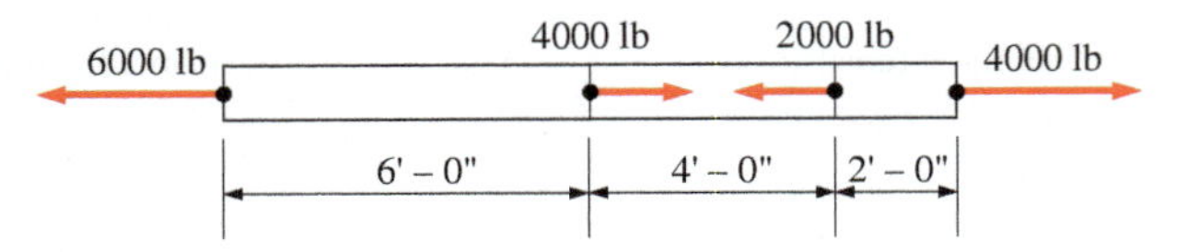

PROBLEM 9.52

9.53 Rework Problem 9.52, changing the second load from the left to 14,000 lb and the third load from the left to 12,000 lb.

9.54 In the structure shown, the tie-back *BC* is a round steel rod having a diameter of 17 mm. The steel strut *AC* is a tubular section having a cross-sectional area of 1350 mm². Find the total deformation in each member.

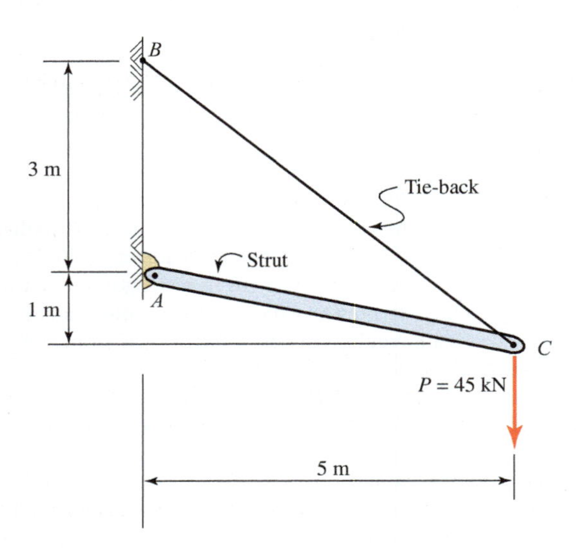

PROBLEM 9.54

9.55 A hook is suspended by two steel wires, as shown. The wires are $\frac{1}{4}$ in. in diameter. A load of 1 ton is hung on the hook. Compute the distance that the hook will drop due to the stretch in the wires.

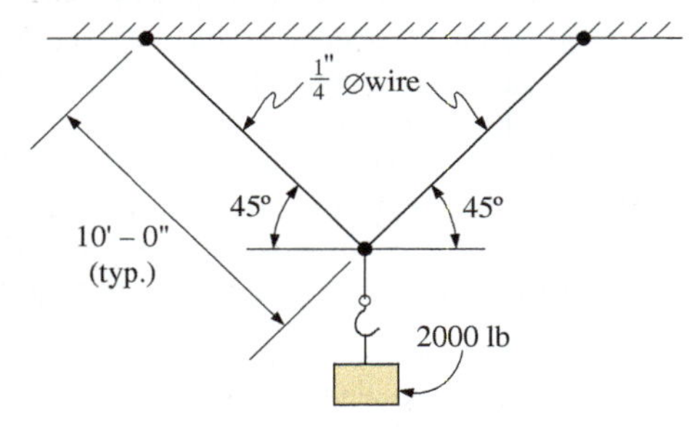

PROBLEM 9.55

9.56 The trolley of a small hoist is supported on a horizontal beam, as shown. The maximum mass to be lifted is 6000 kg. The tie-back is a round rod, 33 mm in diameter. The tensile stress in the tie-back is not to exceed 170 MPa. The maximum elongation allowed for the tie-back is 2.50 mm. Determine the maximum dimension *L* permitted.

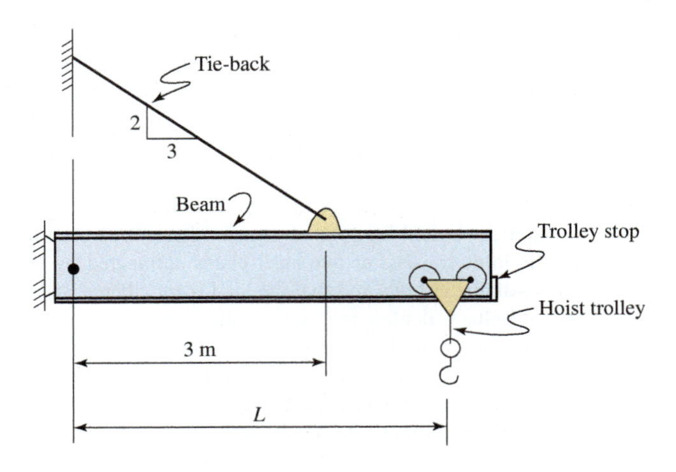

PROBLEM 9.56

9.57 The steel piston rod to the master cylinder has a diameter of $\frac{5}{16}$ in. Determine the stress in the piston rod when a force of 29 lb is applied to the brake pedal, as shown.

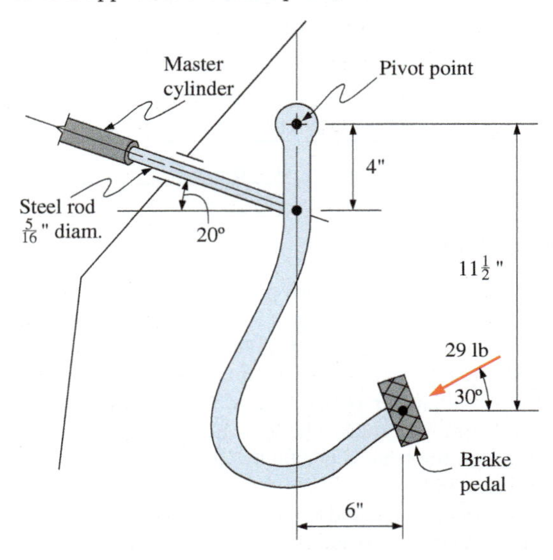

PROBLEM 9.57

9.58 A stranded steel brake cable is composed of 7 wires, each having a diameter of 0.047 in. Determine the stress in the cable when a force of 17 lb is applied to the lever, as shown.

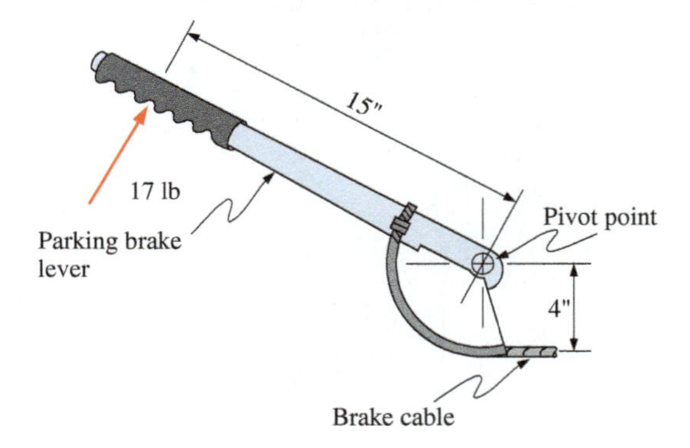

PROBLEM 9.58

9.59 The piston of a steam engine is 400 mm in diameter and its stroke is 600 mm. The maximum steam pressure is 1.72 MPa. Calculate the required diameter of the piston rod if the allowable stress is 69.0 MPa.

PROPERTIES OF MATERIALS

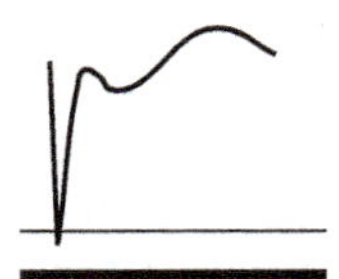

LEARNING OBJECTIVES

Upon completion of this chapter, readers will be able to:

- The tensile test for determining mechanical properties of engineering materials.
- The relationship between stress and strain and the use of the stress–strain diagram in determining the mechanical properties of engineering materials.
- Different types and classifications of various engineering materials.
- The concepts of allowable stress and factors of safety.
- The difference between elastic and inelastic behavior of engineering materials.

10.1 THE TENSION TEST

In the area of strength of materials, problems are primarily of two types: analysis and design. Problems of analysis could involve finding the greatest load that may be applied to a given body without exceeding specified limiting values of stress and strain. They could also involve determining load-induced stresses and strains for comparison with limiting values. Problems of design involve determining the required size and shape of a member to support given loads without exceeding specified limiting stress and/or strain.

Various properties of materials must be considered at this point since both analysis and design are essentially problems of materials. In particular, the mechanical properties are important to our study since they are the properties that affect the limiting values of stress and strain. Appendix G may be used as a reference source for the mechanical properties of the materials discussed in this text.

Probably the simplest and most informative test for establishing the mechanical properties of a large number of engineering materials is the tension test. Metals and plastics are normally subjected to a tension test, whereas a material such as concrete is tested in compression due to its low tensile strength and brittleness. The tension test is categorized as a static-loaded destructive-type test and has been standardized by the American Society for Testing and Materials (ASTM). It involves the axial tensile loading of a material specimen, one type of which is shown in Figure 10.1. The specimen is prepared for the test by placing two small punch

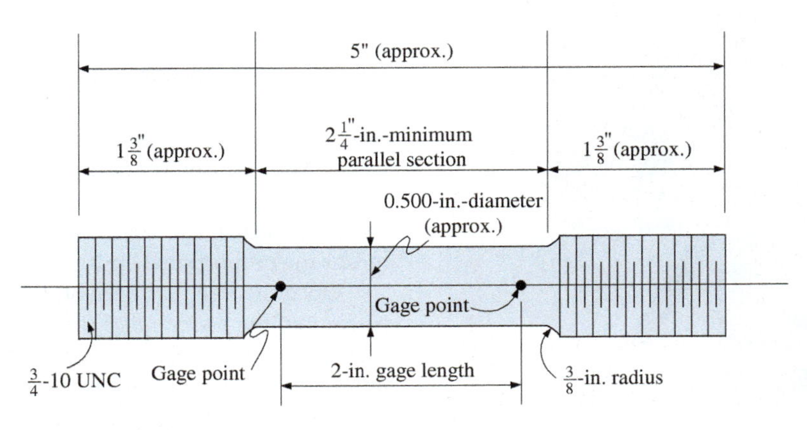

FIGURE 10.1 Standard tension test specimen.

marks (gage points) on it, which are a known distance apart (the gage length). Frequent simultaneous readings of the increasing applied load P and total change in gage length δ are made during the test. These are recorded in a tabular format. The specimen is loaded to failure. Additional information is obtained by taking the specimen from the machine, carefully fitting the ends together, and measuring the length between the gage points. The diameter of the specimen is also measured at the cross section at which failure occurs. Stress and strain values are computed and then plotted to form a stress–strain curve, which is discussed in Section 10.2.

Stress and strain values are computed based on the original cross-sectional area and original gage length of the specimen. The resulting stress is called the *nominal stress* (or the *engineering stress*), and the resulting strain is called the *nominal strain*. These are not the same as the true stress and the true strain that are discussed shortly. A plot is then drawn showing nominal values of stress versus strain. This plot is discussed in detail in Section 10.2.

The tension test is conducted in a laboratory using any one of several types of testing machines. Loads are read from dials or digital displays. Elongations are determined using measuring devices such as the extensometer shown in Figure 10.2. If elongations are large, handheld dividers and scales may be used. Some testing machines have the capability of automatically reading and recording the data and will also produce plotted results.

FIGURE 10.2 Extensometer on typical 0.505 tensile specimen.

10.2 THE STRESS–STRAIN DIAGRAM

After the tension test has been conducted, the data are converted to stress (based on original area) and strain (based on original length). These values are then plotted to form a stress–strain curve, or diagram, with the ordinate corresponding to stress and the abscissa corresponding to strain.

A smooth average curve is then drawn (rather than one that passes through every plotted point). The deviations of the smooth curve from the plotted points are mostly due to errors of instrumentation and observation.

A typical stress–strain diagram for a low-carbon, mild ductile structural steel is shown in Figure 10.3. Ductile materials are defined as materials that can undergo considerable plastic deformation under tensile load before actual rupture. (Brittleness is the opposite of ductility.) Figure 10.3a shows the full stress–strain curve from the beginning of the test through failure at point F. Figure 10.3b shows the left-most portion of the curve from a strain of zero to a strain of about 0.02. In Figure 10.3b, from the origin 0 to point A, the curve is a straight line, indicating that the steel conforms to Hooke's law (stress is proportional to strain). In that portion of the curve, the steel is in the elastic range (its behavior is said to be *elastic*). The slope of the straight-line portion of the curve is called the *modulus of elasticity, E*, defined in Chapter 9 as $E =$ stress/strain. The stress corresponding to the upper end of the straight-line portion of the curve (point A) is the *proportional limit* of the steel. The proportional limit may be defined as the greatest stress a material is capable of developing without deviation from straight-line proportionality between stress and strain.

For all practical purposes, the proportional limit is also generally accepted as the limit of elasticity and is termed the *elastic limit*. The elastic limit (point B) may be defined as the greatest stress a material is capable of developing without a permanent elongation remaining upon complete unloading of the specimen. Strains associated with stresses up to the elastic limit (therefore, within the elastic range) are small and reversible.

Determining the elastic limit is a tedious, trial-and-error procedure; therefore, it is not done very often. For most materials, the elastic limit and the proportional limit are almost identical in numerical value, and the terms are often used synonymously. Where a difference does exist, the elastic limit is generally higher than the proportional limit. Unless otherwise stated, the term *elastic limit*, as used in this text, means the proportional limit.

Once the steel has been stressed past the elastic limit, it passes into the *plastic range* (its behavior is said to be *plastic*). The strain is no longer reversible. That is, should the steel be unloaded, it will not return to its original length; rather, it will retain a permanent elongation (sometimes called a *permanent set*). For stresses above the elastic limit, the strain increases at a faster rate until, at point C on the curve shown in Figure 10.3b,

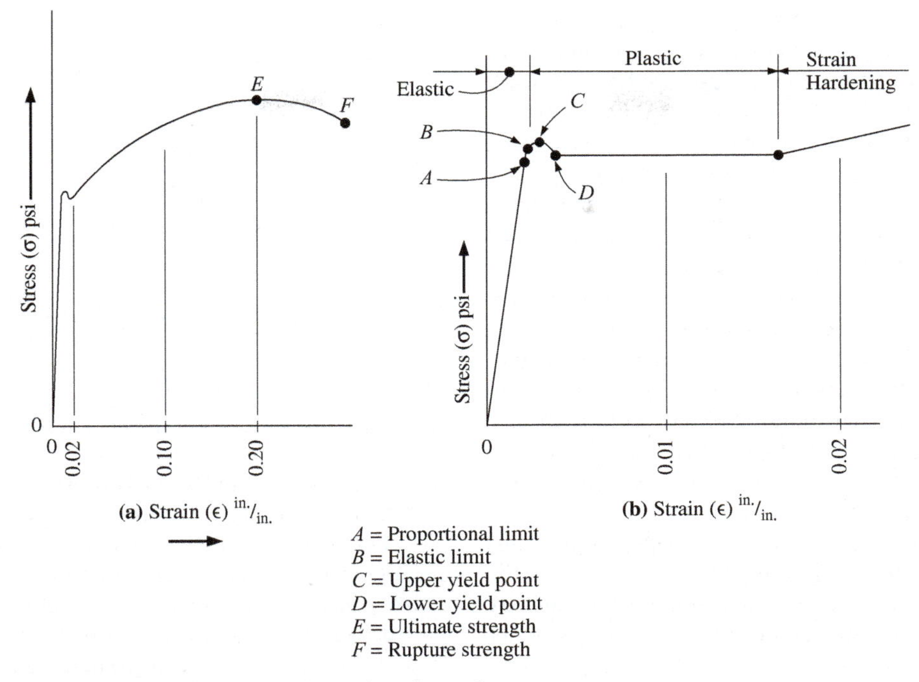

FIGURE 10.3 Stress–strain diagram for low-carbon, ductile steel.

the strain continues with little or no increase in applied load or stress. The point at which this occurs is called the *yield point* and the stress at this point is called the *yield stress*, which we will denote F_y. The yield stress may be defined as the stress at which there occurs a marked increase in strain without an increase in stress.

After the yield point has been reached (the steel is said to have yielded), the curve may dip downward to point *D*, also shown in Figure 10.3b, representing a condition in which the steel transmits less load as it elongates. This phenomenon occurs only with a low-carbon steel; it is not typical of all steels. This abrupt decrease in load (stress) results in upper and lower yield points. The upper yield point is influenced considerably by the shape of the test specimen and by the testing machine itself and is sometimes completely suppressed; hence, it is relatively unstable. The lower yield point, point *D*, is much less sensitive and is considered to be more representative of the true characteristics of the steel, characteristics that can serve as a basis for design criteria.

After the yield point is passed, the steel stretches still more at essentially constant stress up to a strain of about 0.015 (1.5%) at which time it begins to recover some of its strength and becomes capable of resisting additional load. The steel has passed from the plastic range into the strain-hardening range. With stress and strain increasing at different rates, the curve rises continuously, climbing until the maximum ordinate is reached at point *E* in Figure 10.3a, where the tangent to the curve is horizontal. This point represents the *ultimate strength* or *tensile strength*, which we will denote as F_u. The ultimate strength may be defined as the maximum stress a material is capable of developing. Tensile strength is similarly defined but it, of course, refers only to tension.

At point *E*, shown in Figure 10.3a, the steel continues to stretch accompanied by a decreasing ability to transmit load. Also, after passing point *E*, the steel visibly begins to decrease in diameter and to increase in length over a localized segment of the specimen. The localized decrease in the diameter is called *necking* and is depicted in Figure 10.4. Necking progresses rapidly until the steel suddenly ruptures in the plane of the reduced cross section. Point *F* on the curve in Figure 10.3a represents the breaking point or *rupture strength* of the steel. The rupture strength may be defined as the stress at which the specimen actually breaks.

As previously mentioned, the stresses (nominal values) plotted on the stress–strain diagram are calculated on the

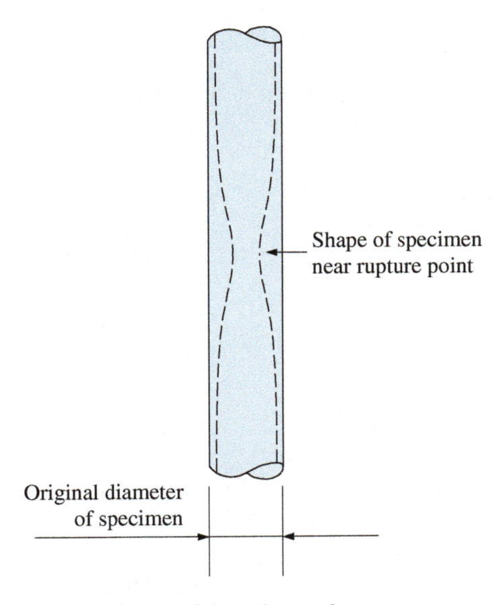

FIGURE 10.4 Necking of ductile steel specimen.

basis of the original cross-sectional area of the specimen, despite that the area is generally reduced appreciably by the time the ultimate and rupture points are reached. Also, the calculated unit strains (nominal values) are based on the original length of the specimen (which traditionally, using the U.S. Customary System, is either 2 in. or 8 in.).

The reduction in cross-sectional area is not significant until the specimen is strained well into the plastic range. The stress that is calculated using the reduced area, called the *true stress*, will, of course, be higher than the nominal stress. Similarly, as the tension test progresses, the gage length increases, and if this length is used in the strain calculation, the resulting strain is called the *true strain*. True strain will always be smaller than nominal strain. Although true stress and true strain are certainly more accurate depictions, the more easily determined nominal values are sufficiently accurate for most engineering applications.

Variations occur in the shape of stress–strain diagrams for different steels. The straight-line (elastic) portion of the curve, however, which represents the modulus of elasticity E, is approximately the same for all structural steels, whether low-carbon steels or high-strength steels. Some high-carbon steels and certain alloy steels exhibit no yield point when tested in tension and will rupture after much less elongation than the more ductile steels.

Materials other than steel exhibit different shapes of the stress–strain diagram. Many ductile metals, such as brass, copper, and aluminum, have gradually curving stress–strain diagrams beyond the proportional limit, rather than a definite yield point. Brittle materials, such as concrete and cast iron, are incapable of much plastic deformation. They do not exhibit a necking process and will rupture without warning. A tensile stress–strain curve for a brittle material

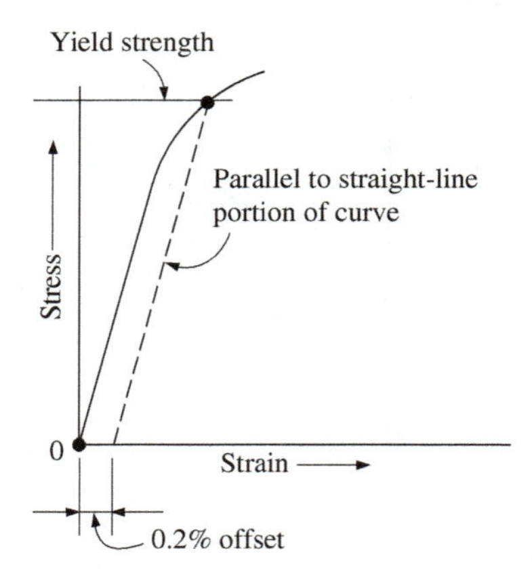

FIGURE 10.5 Offset method of determining yield strength.

ends before it becomes horizontal; consequently, the ultimate strength and the rupture strength are the same.

For materials without a well-defined yield point, such as higher strength steels and nonferrous metals, a property (stress value) analogous to the yield point is established by the offset method. It is common to define this stress as that which would cause a permanent set of 0.2%. This is shown in Figure 10.5, in which a line is drawn parallel to the straight-line portion of the stress–strain curve, but offset from it by 0.2% (0.002 in./in.). The stress read at the intersection of this line with the stress–strain curve is called the *yield strength*. It is also called the *yield point*, or, simply, the *yield stress;* it is not a true yield point.

The stress–strain diagram, in the context of our discussion, represents properties at normal temperatures. When materials are to be used at temperatures much above or below the normal range, it is necessary to establish stress–strain curves for the anticipated temperature range.

EXAMPLE 10.1 During a stress–strain test of a steel specimen, the strain at a stress of 5000 psi was calculated to be 0.000167 in./in. (point *A*, Figure 10.6) and at a stress of 20,000 psi to be 0.000667 in./in. (point *B*). If the proportional limit is 30,000 psi, calculate the modulus of elasticity *E* using the slope between the two points. Also, compute the stress corresponding to a strain of 0.0002 in./in.

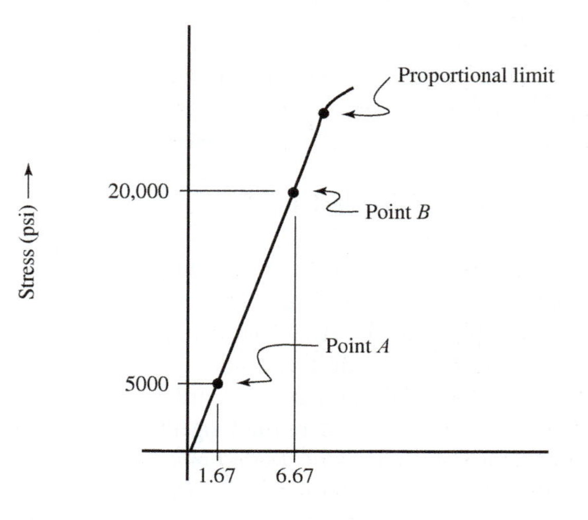

FIGURE 10.6 Stress–strain diagram.

Solution

$$\sigma_A = 5000 \text{ psi} \qquad \epsilon_A = 0.000167 \text{ in./in.}$$
$$\sigma_B = 20{,}000 \text{ psi} \qquad \epsilon_B = 0.000667 \text{ in./in.}$$

Since both points lie on the straight-line portion of the stress–strain curve, the modulus of elasticity will be equal to the slope of the line between the points (or between either of the two points and zero):

$$E = \frac{\text{change in stress}}{\text{change in strain}} = \frac{20{,}000 \text{ psi} - 5000 \text{ psi}}{0.000667 \text{ in./in.} - 0.000167 \text{ in./in.}} = 30{,}000{,}000 \text{ psi}$$

Calculate the stress corresponding to a strain of 0.0002 in./in. Since

$$E = \frac{\sigma}{\epsilon}$$

therefore,

$$\sigma = \epsilon E = (0.0002 \text{ in./in.})(30{,}000{,}000 \text{ psi}) = 6000 \text{ psi}$$

$$6000 \text{ psi} < 30{,}000 \text{ psi} \qquad \textbf{(O.K.)}$$

10.3 MECHANICAL PROPERTIES OF MATERIALS

As previously stated, the numerical stress values that may be obtained from a tension test are the proportional limit, elastic limit, yield stress, ultimate stress, and rupture stress. In addition, the modulus of elasticity, percent elongation, and percent reduction in cross-sectional area are also obtained. These values define those mechanical properties or qualities of a material that are significant in the applications of strength of materials.

In addition to the mechanical properties defined by numerical stress values, some other mechanical properties describe how a material responds to load and deformation (see Appendix G). These properties may be defined and summarized as follows:

1. *Stiffness* is the property that enables a material to withstand high stress without great strain. It is a resistance to any sort of deformation. Stiffness of a material is a function of the modulus of elasticity E. A material having a high value of E such as steel, for which $E = 30{,}000{,}000$ psi (207,000 MPa), will deform less under load (thereby exhibiting a greater stiffness) than a material with a lower value of E such as wood, where E may be equal to 1,000,000 psi (7000 MPa) or less.
2. *Strength* is the property determined by the greatest stress that the material can withstand prior to failure. It may be defined by the proportional limit, yield point, or ultimate strength. No single value is adequate to define strength, since the behavior under load differs with the kind of stress and the nature of the loading.
3. *Elasticity* is that property of a material enabling it to regain its original dimensions after removal of a deforming load. There is no known material that is completely elastic in all ranges of stress. Most engineering materials, however, are elastic, or very nearly elastic, over large ranges of stress. Steel is an elastic material only up to the elastic limit. Hence, the determination of the elastic limit establishes the elastic range or the limit of elasticity.
4. *Ductility* is that property of a material enabling it to undergo considerable plastic deformation under tensile load before actual rupture. A ductile material is one that can be drawn into a long thin wire by a tensile force without failure. Ductility is characterized by the percent elongation of the gage length of the specimen during the tensile test and by the percent reduction in area of the cross section at the plane of fracture. These quantities are defined by the following expressions:

$$\text{percent elongation} = \frac{\text{increase in gage length}}{\text{original gage length}}(100) \qquad \textbf{(10.1)}$$

$$\text{percent reduction in area} = \frac{\text{orig. area} - \text{final area}}{\text{original area}}(100) \qquad \textbf{(10.2)}$$

A high percent elongation indicates a highly ductile material. Most ASTM material specifications for metals have a required minimum percent elongation for either of the two standard specimen gage lengths (2 in. and 8 in.). A lower limit value commonly used to determine ductility is about 5% elongation. That is, a metal is considered to be ductile if its percent elongation is greater than about 5%.

5. *Brittleness* implies the absence of any plastic deformation prior to failure. A brittle material is neither ductile nor malleable and will fail suddenly without warning. A brittle material exhibits no yield point or necking-down process and has a rupture strength approximately equal to its ultimate strength. Brittle materials, such as cast iron, concrete, and stone, are comparatively weak in

tension and are generally not subjected to a tension test. They are usually tested in compression.

6. *Malleability* is that property of a material enabling it to undergo considerable plastic deformation under compressive load before actual rupture. Most materials that are very ductile are also quite malleable. A hammering or rolling operation would require a malleable material due to the extensive compressive deformation that accompanies the process.
7. *Toughness* is that property of a material enabling it to endure high-impact loads or shock loads. When a body is subjected to an impact load, some of the energy of the blow is transmitted and absorbed by the body. In absorbing the energy, work is done on the body. This work is the product of the strain and the average stress up to the rupture point. Hence, the measure of toughness is equal to the area under the stress–strain curve from the origin through the rupture point, as shown in Figure 10.7. A body that can be both highly stressed and greatly deformed without rupture is capable of withstanding a heavy blow and is said to be tough.
8. *Resilience* is that property of a material enabling it to endure high-impact loads without inducing a stress in excess of the elastic limit. It implies that the energy absorbed during the blow is stored and recovered when the body is unloaded. The measure of resilience is furnished by the area under the elastic portion of the stress–strain curve from the origin through the elastic limit, as shown in Figure 10.7.

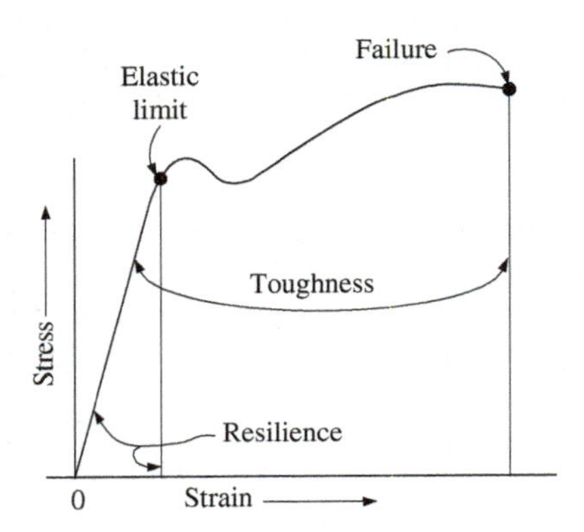

FIGURE 10.7 Resilience and toughness.

Many materials have been developed to satisfy the requirements of innumerable engineering applications. The variations in mechanical properties result, in part, from differing chemical compositions as well as from differing manufacturing processes. Various classification systems enable clear distinctions to be made between grades and/or types of a given material.

The various types of steels are classified by two systems, either ASTM or AISI (American Iron and Steel Institute). In the ASTM system, each structural steel has a number designation referring to the standard that defines the required minimum properties (both mechanical and chemical). The designation has the prefix letter *A* followed by one, two, or three numerals (e.g., ASTM A36, ASTM A588). Structural steel specified and manufactured according to the designated standard is expected to meet both the mechanical and the chemical property requirements.

The second way of designating steel is by means of the AISI number. The AISI uses a four-digit code to define each steel. The last two digits indicate the average percentage of carbon in the steel. For example, if the last two digits are 35, the steel has a carbon content of about 0.35%. The first two digits denote the major alloying elements in the steel other than carbon. For example, AISI 1020 is a plain carbon steel with approximately 0.20% carbon; AISI 2340 is a nickel steel with approximately 3% nickel and 0.40% carbon. The first digit (2) signifies that the steel is a nickel steel.

Other materials are classified according to similar coding systems, such as that developed for aluminum alloys by the Aluminum Association. The aluminum coding system was developed to identify the major alloying element and to indicate the temper imparted to the alloy.

EXAMPLE 10.2 A $\frac{1}{2}$-in.-diameter aluminum specimen (similar to that shown in Figure 10.1) is subjected to a tension test. After rupture, the two pieces are fitted back together. The distance between gage points, originally 2 in., is measured to be 2.83 in., and the final diameter of the specimen is measured to be 0.38 in. Calculate the percent elongation and the percent reduction in area.

Solution

$$\text{percent elongation} = \frac{\text{increase in gage length}}{\text{original gage length}}(100)$$

$$= \frac{2.83\text{ in.} - 2.0\text{ in.}}{2.0\text{ in.}}(100)$$

$$= 41.5\%$$

and

$$\text{percent reduction in area} = \frac{\text{orig. area} - \text{final area}}{\text{original area}}(100)$$

$$\text{original area} = 0.7854(0.5\text{ in.})^2 = 0.196\text{ in.}^2$$

$$\text{final area} = 0.7854(0.38\text{ in.})^2 = 0.113\text{ in.}^2$$

$$\text{percent reduction in area} = \frac{0.196\text{ in.} - 0.113\text{ in.}}{0.196\text{ in.}^2}(100) = 42.3\%$$

10.4 ENGINEERING MATERIALS: METALS

In this section, we consider some of the more common of the metals used in machines and structures. These metals are listed in Figure 10.8. The list is not all-inclusive and we do not consider any of the many exotic and specialized metals that have been developed for specific and peculiar purposes.

Metals are traditionally classified as ferrous and nonferrous. The term *ferrous*, derived from the Latin word *ferrum*, meaning iron, refers to those metals that contain a large percentage of iron. Nonferrous metals have special properties that make their use advantageous in some circumstances. The ferrous metals are by far the primary metals used in engineered structures.

10.4.1 Ferrous Metals

The basic and predominant component of the ferrous metals is iron, one of the most common chemical substances found in the earth's crust. Because of its great affinity for combination with other elements, iron is never found in its pure form in nature. The iron must be extracted from iron ores, which are mineral or rock deposits containing concentrations of iron. Most of the important iron ores found in the United States are in the form of iron oxides. These ores usually contain small amounts of impurities (phosphorus and silica, among others), most of which must be removed during the production of iron.

Iron is extracted from iron ore in a blast furnace. The production process requires a combination of iron ore, fuel, and a flux of crushed limestone to remove the impurities. The fuel used is coke, a coal product. The high temperature (in excess of 3000°F) achieved in the furnace results in the iron portion of the ore melting and the limestone flux being liquified. The end result is a molten iron combined with some carbon obtained from the coke and a molten slag formed by the limestone combining with the various impurities in the ore. The molten slag floats above the molten iron and is drawn off as a waste by-product. The molten iron is then also drawn off and cast into shallow molds

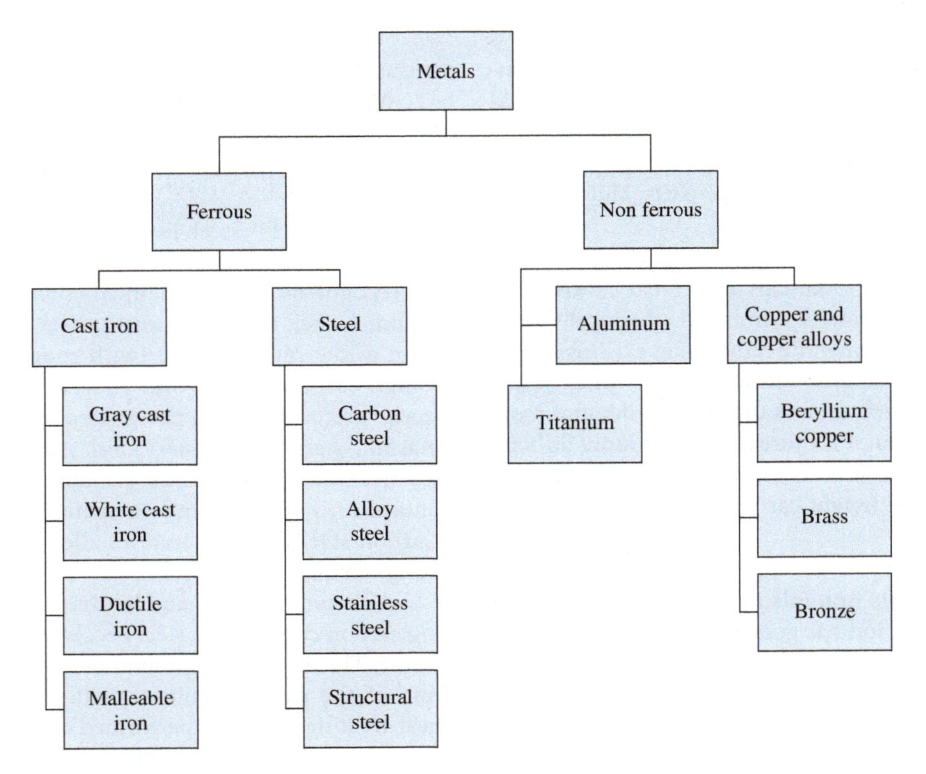

FIGURE 10.8 Some important metals for machines and structures.

called *pigs*. The cooled iron is called *pig iron*. Since pig iron is relatively brittle, additional processing and refining must be performed before it becomes a useful ferrous metal.

Most of the iron produced in the blast furnace is further processed to make steel. That portion not used to make steel is used primarily to make cast iron. Cast iron and steel are the most common forms of ferrous metals. They are principally iron–carbon alloys (mixtures) containing small amounts of sulfur, phosphorus, silicon, and manganese. In addition, other elements such as nickel and chromium may be added to alter the physical and mechanical properties. (An alloy is defined as a substance with metallic properties composed of two or more elements of which at least one is a metal.)

10.4.2 Cast Iron

Cast iron is the generic name for a group of metals that are alloys of carbon and silicon with iron. Included are gray cast iron, white cast iron, ductile iron, and malleable iron. Cast irons usually contain between 2% and 4.5% (by weight) of carbon. The manufacture of cast iron involves the mixing of pig iron, scrap cast iron, and a limestone flux in a small furnace fired with coke and an air blast. The properties of the resulting cast iron are established by differences in cooling rates, compositions, and subsequent heat treatment of the product. The suitability of a particular metal for an intended use is best determined by laboratory or service tests.

Gray Cast Iron Gray cast iron (also called *gray iron*) is the most commonly used type of cast iron. It is characterized by high percentages of carbon (up to 4%) and silicon (1% to 3%). It is cheap, easily cast, and relatively easy to machine. It is available in grades having a tensile strength ranging from 20 ksi to 60 ksi. Its ultimate compressive strength is much higher, ranging three to five times greater than its tensile strength. Gray iron is brittle and does not exhibit a yield point. It has excellent wear and corrosion resistance and has good vibration damping abilities. Gray iron is used in automotive engine blocks, gears, brake parts, clutch plates, machinery bases, and pipe installations.

White Cast Iron In white cast iron (also called *white iron*), most of the carbon is combined chemically with the iron in the form of cementite rather than existing as free carbon (graphite). A fractured surface appears white. White iron's advantage over gray iron is that it is harder and more resistant to abrasion, but it is more brittle and more difficult to machine and to cast. It is also less resistant to corrosion. White iron is used for freight car wheels, rolling-mill rolls, and plowshares.

Ductile Iron Ductile iron (also called *nodular cast iron*) is similar to gray iron and has good wear and corrosion resistance. It is also easily cast and relatively easy to machine, but it is far more ductile than gray iron and has a greater tensile strength and percent elongation. It thus has good toughness (shock resistance). Uses of ductile iron include crankshafts, casings, gears, hubs, and rolls.

Malleable Iron Malleable iron is white iron that has been heat-treated. It has moderate to high strength with a tensile strength comparable to ductile iron and an ultimate compressive strength greater than that of ductile iron. It also has good machinability and good wear resistance. Malleable iron is used for pipe fittings, guard rail fittings, construction machinery, and automotive and truck parts.

10.4.3 Steel

Steel is an alloy consisting almost entirely of iron in combination with small quantities of various elements. A wide range of properties may be achieved by varying the chemical composition of the steel. Carbon, generally contained in amounts between 0.05% and 2.0% (by weight), is the element that has the greatest effect on the properties of the steel. Up to a point, increasing the carbon content increases the hardness, strength, and abrasion resistance of steel. Ductility, toughness, impact properties, and machinability will be decreased, however.

The making of steel from pig iron is essentially a refining (purifying) process. Carbon, silicon, phosphorus, and sulfur levels of the pig iron are reduced to levels permissible by the steel specification. Other alloying elements may then be added to the mix to produce a steel having the desired properties. Steel is a recyclable product, and scrap steel is an important component in the making of new steel. The two processes commonly used to manufacture steel are the electric arc furnace process and the basic oxygen furnace process. The basic oxygen furnace process uses some scrap steel along with a charge of molten iron to produce a melt of new steel. The electric arc furnace uses all scrap steel. Most of the steel currently used in the production of structural steel for the construction industry is produced by the electric arc furnace process.

Many types of steels are available. We briefly discuss four general categories of steel: carbon steel, alloy steel, stainless steel, and structural steel.

Carbon Steel Steel is considered to be carbon steel when no minimum content is specified or required for the recognized alloying elements such as aluminum, chromium, nickel, and numerous others. Carbon steels range from a low-carbon grade (with roughly 0.1% carbon) to a high-carbon grade (with roughly 1.0% carbon). These steels are sometimes designated *plain carbon steel* or *machine steel* or *machinery steel*. All carbon steels contain small amounts of various elements such as manganese, phosphorus, silicon, and in many cases other elements. Carbon steel is considered an alloy even though it is not designated an alloy steel.

Although strength and hardness increase with increasing carbon content, the steel becomes less ductile and more brittle. The mechanical properties of carbon steel are a function of the carbon content, method of manufacture, and heat-treating processes performed on the steel. For example, carbon steel may have an ultimate tensile strength ranging from roughly 43,000 psi to 122,000 psi.

TABLE 10.1 AISI numbering system for steel

Steel	AISI No.	Steel	AISI No.
Plain carbon	10XX	Molybdenum-nickel 1.75% NI	46XX
Plain carbon[a]	11XX	Molybdenum-nickel-chromium	47XX
Manganese	13XX	Molybdenum-nickel 3.5% NI	48XX
Boron	14XX	Chromium	5XXX
Nickel	2XXX	Chromium-vanadium	6XXX
Nickel-chromium	3XXX	Nickel-chromium-molybdenum	8XXX
Molybdenum-chromium	41XX	Silicon-manganese	92XX
Molybdenum-chromium-nickel	43XX	Nickel-chromium-molybdenum (except 92XX)	9XXX

[a] With greater sulfur content for free-cutting.

Alloy Steel In addition to carbon, alloy steel contains significant quantities of recognized alloying metals. The most common alloying metals are aluminum, chromium, copper, manganese, molybdenum, nickel, phosphorus, silicon, titanium, and vanadium. Alloys are used to improve the hardenability of steel; to increase toughness, ductility, and tensile strength; and to improve low- and high-temperature properties.

Alloy steels are practically always heat-treated to develop specified properties. Heat treatment involves raising the temperature of the steel to some prescribed level and then cooling it rapidly by quenching. This procedure alters the crystalline structure of the steel and therefore alters its physical properties.

The method of designating carbon and alloy steels is by means of an ASTM number designation or an AISI number designation as discussed in Section 10.3. Table 10.1 furnishes the AISI two-digit code for identifying various types of alloy steels.

Stainless Steel *Stainless steel* is the designated name for a widely used grouping of iron–chromium alloys known for their corrosion resistance (notably their nonrusting quality). This ability to resist corrosion is attributable to a surface chromium oxide film that forms in the presence of oxygen. The film is essentially insoluble, self-healing, and nonporous. A minimum chromium content of 12% is required for the formation of the film, and 18% is sufficient to resist the most severe atmospheric corrosive conditions. For other reasons, the chromium content may go as high as 30%. Other elements, such as nickel, aluminum, silicon, and molybdenum may also be present.

Some applications for the stainless steels are chemical processing and oil processing equipment, cutlery, and automotive trim.

Structural Steel The term *structural steel* applies to hot-rolled steel of various shapes and forms and of varying alloy elements utilized to resist assorted types of loads and forces to which a structure may be subjected. The member may be a tension, compression, bending, or torsional member or a combination of these. The structure may be a building, bridge, transmission tower, or some other specialized type of structure. The steel shapes, as standardized by the AISC, are W shapes (wide-flange members), HP shapes (bearing pile members), S shapes (American standard beams; formerly called *I-beams*), M shapes (miscellaneous), C shapes (American standard channels), MC shapes (miscellaneous channels), and L shapes (angles). In addition, structural steel also includes plates, bars, steel pipe, and structural tubing.

Structural steel includes several types of steel. The five categories, as designated by the AISC, are carbon steel, high-strength low-alloy steel, corrosion-resistant high-strength low-alloy, quenched and tempered low-alloy, and quenched and tempered alloy. The common method of specifying structural steel is based on a standardized ASTM designation described in Section 10.3.

Carbon steel is considered the basic structural steel. The widely used carbon steel, designated A36, has a minimum yield stress of 36 ksi, except for plates in excess of 8 in. thick, which have a minimum yield stress of 32 ksi. A36 steel is ductile, weldable, and can be used in all types of structures.

A572 steel is a high-strength, low-alloy steel. It is produced in various grades and has a corrosion resistance roughly equal to twice that of A36. The yield stress of A572 steel varies from 40 ksi to 65 ksi for individual shapes, plates, and bars, depending on the grade of the steel and on the thickness of the elements of the cross section, trending downward for the thicker-element cross sections. A992, the preferred material specification for steel W shapes, is essentially equivalent to A572 grade 50. A992 properties are included in Appendix G.

A588 steel is a corrosion-resistant, high-strength, low-alloy steel. This steel has a corrosion resistance ranging from four to eight times that of A36 steel. The yield stress is 50 ksi for all shapes and for plates up to 4 in. thick. A588 steel can be used as exposed unpainted steel in structures and is suitable for welded or bolted structures.

Both categories of quenched and tempered steel (low-alloy and alloy) designated A852 and A514 are applicable for only high-strength plates. They are strong, tough steels with yield strengths in the 70 ksi to 100 ksi range and are primarily used in welded structures.

Various other ASTM grades of structural steel are approved for use in the construction of buildings and bridges.

10.4.4 Nonferrous Metals

Nonferrous metals and their alloys represent an important classification of engineering materials. Some have a high strength-to-weight ratio, whereas others have excellent resistance to corrosion. The mechanical properties of the primary nonferrous metals are a function of the principal element and the quantity and type of alloying element(s), as well as the method of manufacturing and the heat-treating process. A few of the more important nonferrous metals are briefly discussed next.

Aluminum Aluminum is one of the most abundant metals found in the crust of the earth, occurring as an oxide in most clay. The basic raw material from which aluminum is produced is bauxite, an ore containing a high percentage of aluminum oxide (also termed *alumina*). The total extraction process of obtaining aluminum from the ore results in a metallic aluminum of better than 99% purity. This high-purity aluminum is soft, weak, and ductile with an ultimate tensile strength of approximately 10,000 psi. Aluminum is light in weight and has a high resistance to corrosion under most service conditions. It also has good thermal conductivity and high electrical conductivity. Aluminum of high purity is used principally for electric-conductor purposes. Even for this application, however, small percentages of certain alloying elements usually are added to improve the mechanical properties and other characteristics with only a slight reduction in the electrical conductivity.

Most aluminum is used in the alloyed state. Mechanical properties can be considerably improved by alloying aluminum with small amounts of other metals. Aluminum alloys are generally harder and stronger than the high-purity aluminum. The alloys are widely used for structural and mechanical applications. Among aluminum's attractive properties are light weight (approximately one-third that of steel), good corrosion resistance, relative ease of machining, and a pleasing silver-white appearance. It has two (perhaps significant) disadvantages: a high coefficient of thermal expansion (approximately twice that of steel) and a modulus of elasticity of 10,000,000 psi (approximately one-third that of steel). The latter property indicates a less-stiff material. Similar to steel, a multitude of aluminum alloys are available that provide a wide range of properties. Aluminum can be alloyed to provide strengths exceeding that of some carbon steels. These strong alloys, however, have a lesser ductility and a lesser resistance to corrosion. Ultimate tensile strength levels up to approximately 80,000 psi may be obtained with suitable alloys, resulting in a high strength-to-weight ratio.

Aluminum and its alloys are divided into two classes according to how they are formed, whether wrought or cast. Roughly 75% of the aluminum produced in the United States is fabricated into wrought products, which include sheet and plate, tube, pipe, rolled structural and other shapes, extruded shapes, rod, bar, and wire. These products have applications within specific fields. Examples are the aerospace industry, architectural buildings, highway structures, tanks, pressure vessels, piping, and transportation structures.

Titanium Titanium production in commercial quantities began in the immediate post–World War II years. Titanium and its alloys have attractive engineering properties. They are about 45% lighter than steel and 70% heavier than aluminum. They also possess a very high strength, which may reach an ultimate tensile strength of 200,000 psi, depending on the alloying element. The combination of moderate weight and high strength gives titanium alloys the highest strength-to-weight ratio of any structural metal, roughly 30% greater than aluminum or steel. This high strength-to-weight ratio is roughly maintained within a temperature range of from −400°F to +1000°F.

Other notable properties for titanium and its alloys are excellent corrosion resistance to atmospheric and sea environments as well as to a wide range of chemicals, low thermal conductivity, low coefficient of thermal expansion, high melting point (higher than iron), and high electrical resistivity. In addition, the fatigue resistance of titanium and its alloys is good. The modulus of elasticity (stiffness) is 16,000,000 psi, which is 1.6 times the value used for aluminum. The high cost of the metal, however, limits its usefulness and range of applications.

The majority of the present applications for titanium and its alloys are in the aerospace industry. Examples are the structures of aircraft and spacecraft, sections of jet engines and allied components, landing gears, fuselage parts, and skins. Other applications have been pressure vessels, roofing, and various architectural building items such as fascias, flashing, and gravel stops.

Copper and Copper Alloys The term *copper*, in the United States, means copper containing less than 0.5% of impurities or alloying element. The most significant properties of copper are its high electrical conductivity (second only to that of silver), high thermal conductivity, good resistance to corrosion, and good malleability, formability, and strength.

Unalloyed copper is used widely as a structural material, for roofing and sheathing, in heat-exchange equipment, for large vessels and kettles, and in various kinds of equipment used in the production of chemicals, foods, and beverages. The greatest single use for unalloyed copper, however, is in the electrical field, where its electrical conductivity, corrosion resistance, and formability make it ideal for use in electrical devices and equipment of all kinds.

A few of the more important alloys of copper are beryllium copper, brass, and bronze. Each consists of copper in combination with different elements. All have different properties and all have a variety of specific applications. All these alloys generally possess good strength and corrosion resistance.

10.5 ENGINEERING MATERIALS: NONMETALS

There is no one simple classification for nonmetallic engineering materials. Some of the more common and significant materials such as concrete, wood, and plastics are briefly introduced here. These three materials constitute a diverse grouping with respect to structure, sources, and characteristics. Wood and concrete have broad application in the civil-architectural construction field, whereas plastics have application in virtually all fields of engineering and architecture.

10.5.1 Concrete

Concrete consists principally of a mixture of cement, fine and coarse aggregates (sand, gravel, crushed rock, and/or other materials), and water. The water is added as a necessary ingredient for the hardening of the mixture. The bulk of the mixture consists of the fine and coarse aggregates. The resulting concrete strength and durability are a function of the proportions of the mix as well as other factors such as the concrete placing, finishing, and curing history.

The compressive strength of concrete is relatively high, but it is a relatively brittle material with little tensile strength compared with its compressive strength. Steel reinforcing rods (which have high tensile and compressive strength) are used in combination with the concrete; the steel will resist the tension and the concrete will resist the compression. The result of this combination of steel and concrete is called *reinforced concrete*. In some instances, steel and concrete are positioned in members so that they both resist compression.

Cement, the binding material of the concrete mix, has as its basic ingredients limestone and clay. The ingredients are heated to a high temperature in a kiln and a greyish-black mass, called *clinker*, is formed. Upon cooling, a small amount of gypsum is added, followed by grinding to form the hydraulic cement product called *portland cement*. Chemically, this cement consists chiefly of calcium and aluminum silicates. Varying the composition of the ingredients results in various types of cement with specific properties such as rapid strength gain (type 3) and sulfate-resistant cement (type 5).

The cement requires water for the chemical reaction of hydration. In this process, the calcium silicate compounds produce new compounds that impart the stonelike quality to the mixture. In fresh concrete, the ratio of the amount of water to the amount of cement, by weight, is termed the *water/cement ratio*. For complete hydration, a water/cement ratio of 0.35 to 0.40 is required. Higher water/cement ratios, although leading to lower strengths, are generally used to expedite mixing, handling, and placing of the concrete.

In ordinary structural concrete, the aggregates occupy approximately 70% to 75% of the hardened mass. Gradation of aggregate size to produce close packing is desirable because it will generally result in better strength and durability. Aggregates are classified as fine or coarse. Fine aggregate is generally sand consisting of particles that will pass a No. 4 sieve (0.187 in. nominal opening). Coarse aggregate consists of particles that would be retained on a No. 4 sieve, with a maximum size governed by code requirements.

Concrete is designated according to its compressive strength, which may range from 2500 psi to 9000 psi depending on, among many factors, the mix proportions and the admixtures used. Higher strengths are possible with good quality control. Under favorable conditions the strength increases with age, the specified strength usually being that which occurs 28 days after the placing of the concrete.

Concrete applications in construction range from building foundations, highway pavements, and slabs-on-grade to structural framing for multistory buildings, bridges of all types, dams, and shell roofs for stadiums. Along with steel, it is a principal building material for structures of all sizes.

10.5.2 Wood

Wood is one of the oldest natural construction materials. It is a cellular organic material composed principally of cellulose (about 60%), which constitutes the longitudinal structural cells, and lignin (about 28%), which cements the structural cells together. The structural cells are hollow, very small in diameter, and oriented vertically in the growing tree.

Wood is divided into two classes, softwood and hardwood. The terms are misleading in that there is no direct relationship between these designations and the hardness or softness of the wood. Softwood comes from conifers (trees with needlelike or scalelike leaves), and hardwoods come from deciduous trees having broad leaves. Hardwoods shed their leaves at the end of each growing season, whereas most softwoods are evergreens. Most of the wood used in the United States for structural purposes is softwood; Douglas fir and southern pine are the most common.

Wood contains natural growth characteristics such as knots and slope of grain that may, depending on their peculiarities, adversely affect the strength properties of that member. Structural grading rules, in establishing allowable design stress values, take into account the effects of these growth characteristics on the strength of wood. Structural lumber is graded according to its intended use. Each piece is assigned a stress grade, designed to meet exacting requirements and strength values. Various lumber associations establish grade requirements and grade markings procedures for species of wood produced in their regions. A stamp, usually placed on the lumber at the mill, identifies the grade of each piece of lumber. Structural lumber is usually graded on the basis of visual inspection. Lumber that has been individually pretested by nondestructive means which supplements the visual grading is classified as *machine-graded* lumber. Allowable stresses for lumber must take into account species and grade (quality), as well as special conditions under which the lumber is used, such as load duration and moisture conditions.

10.5.3 Plastics

Plastics, as used herein, refers to a group of synthetic organic materials derived by a process called *polymerization*. There are many grades and formulations for each plastic, and each of these has its unique combination of characteristics and properties. Therefore, our discussion is very general.

All plastics fall into two broad classifications: thermoplastics and thermosetting plastics. A thermoplastic material can be repeatedly softened and made to flow by heating. The thermoplastics may be formulated in such a manner as to be capable of large plastic deformation or to be flexible. Some thermoplastics such as polyvinyl chloride (PVC) and polystyrene are rigid. The thermosetting plastics have no melting or softening point, although they may be damaged by heat. All thermosetting plastics are rigid, such as phenol-formaldehyde (Bakelite). The thermosetting plastics are brittle, hard, and strong, whereas the thermoplastics are generally ductile, low in strength, and resistant to impact. The thermal expansion of most thermoplastics is approximately 10 times that of steel. Most plastics have high creep characteristics that cause them to deform gradually under constant load.

Most thermoplastics will degrade due to weathering and exposure to the sun. This results in brittleness, hardness, cracking, and yellowing. Special formulations, when added to the plastic, will aid in resisting this degradation.

Some additional thermoplastics that are common are polyethylene, Teflon, nylon, Plexiglas, Lucite, and Delrin. Thermosetting plastics include epoxies, polyesters, silicones, urethanes, and urea-formaldehyde.

10.6 ALLOWABLE STRESSES AND CALCULATED STRESSES

The design and analysis of machines and structural elements are based on limiting values of materials with respect to stress and strain. The limiting values are in turn based partly on the mechanical properties of materials. The tension test and resulting stress–strain diagram is the most common test that provides information on the mechanical properties. After the various values of the stress–strain curve of a material are obtained, it is possible to establish the magnitude of the stress that may be considered as a safe or limiting stress for a given condition or problem. This stress is generally called the *allowable stress*. The allowable stress for a material may be defined as the maximum stress considered to be safe when a member made of that material is subjected to a particular loading condition.

Appropriate values for allowable stress depend on several factors, including

1. Material properties defined by numerical stress values, such as proportional limit, yield stress, and ultimate strength
2. Ductility of the material
3. Confidence in the prediction of loads
4. Type of loading: static, cyclic, or impact
5. Confidence in the analysis and design methods
6. Possible deterioration during the design life of the structure due to such factors as corrosion or chemical attack
7. Possible danger to life and property as a result of a failure
8. Design life of the structure, whether permanent or temporary

Allowable stresses and/or design approaches for materials used as structural components in buildings, bridges, and other civil engineering structures are established by the appropriate authoritative specification-writing agencies such as the AISC (American Institute of Steel Construction)[1] for structural steel buildings, AASHTO (American Association of State Highway and Transportation Officials)[2] for highway bridges of various materials, ACI (American Concrete Institute)[3] for reinforced concrete structures, AF&PA (American Forest & Paper Association)[4] for timber structures, and the Aluminum Association[5] for aluminum structures. The established specifications, including allowable stresses, are generally incorporated into the building codes of states or municipalities; therefore, they become part of the legality governing the various types of construction in a particular area.

In machine design, as well as in the aircraft and spacecraft industries, many different materials with widely varying mechanical properties are used under extreme and varied loading and environmental conditions. For these situations, allowable stresses may or may not be designated by standard specifications. In the absence of a specified allowable stress, a predetermined factor of safety may be used, as determined by the engineering department of the responsible company or manufacturing organization. Factor of safety is discussed further in Section 10.7.

The term *calculated stress* refers to the stress that is determined by calculation to be induced in a member as a result of the applied loads. The calculated stress may vary, depending on the magnitude of the loads. Although at times it may be only a small fraction of the allowable stress, the calculated stress must never exceed the allowable stress.

---

[1]American Institute of Steel Construction, Inc., *Specification for Structural Steel Buildings* (Chicago: AISC, 2005).

[2]American Association of State Highway and Transportation Officials, *Standard Specifications for Highway Bridges*, 17th ed. (Washington, DC: AASHTO, 2002).

[3]American Concrete Institute, *Building Code Requirements for Reinforced Concrete* (Detroit: ACI, 2005), 318–05.

[4]American Forest & Paper Association, *National Design Specification for Wood Construction* (Washington, DC: AF&PA, 2005).

[5]Aluminum Association, *Specifications for Aluminum Structures*, contained in *Aluminum Design Manual*, 2002. (Washington, DC: Aluminum Association, 2002).

EXAMPLE 10.3 An ASTM A36 steel rod 6 m long must be capable of supporting a tensile load of 11.0 kN without elongating more than 7 mm or exceeding an allowable tensile stress of 150 MPa. Calculate the required rod diameter to 1 mm. The proportional limit is 234 MPa. Refer to Appendix G for necessary mechanical properties.

Solution First calculate the required cross-sectional area of the rod based on stress:

$$\text{required } A = \frac{P}{\sigma_{t(\text{all})}} = \frac{11{,}000 \text{ N}}{150 \text{ MPa}} = 73.3 \text{ mm}^2$$

Equation (9.10), from Section 9.6, is then used to calculate the required cross-sectional area based on allowable elongation. Recall that Equation (9.10) is valid only if the stress does not exceed the proportional limit. In this situation, the condition is satisfied, since the allowable stress to be used in the calculation is less than the proportional limit:

$$\delta = \frac{PL}{AE}$$

from which

$$\text{required } A = \frac{PL}{\delta E} = \frac{11{,}000 \text{ N}(6000 \text{ mm})}{7 \text{ mm}(207 \times 10^3) \text{ MPa}} = 45.5 \text{ mm}^2$$

The largest required area is based on the allowable tensile stress. Since

$$A = \frac{\pi d^2}{4} = 0.7854\, d^2$$

the required diameter is calculated as

$$\text{required } d = \sqrt{\frac{73.3 \text{ mm}^2}{0.7854}} = 9.66 \text{ mm}$$

Use a 10-mm rod.

EXAMPLE 10.4 A tracked military vehicle, shown in Figure 10.9, is to operate on terrain where the bearing pressure under the tracks is not to exceed 10 psi. The vehicle weighs a maximum of 30 tons, and each of the two tracks is 20 in. wide. Determine the minimum required contact length L of the tracks.

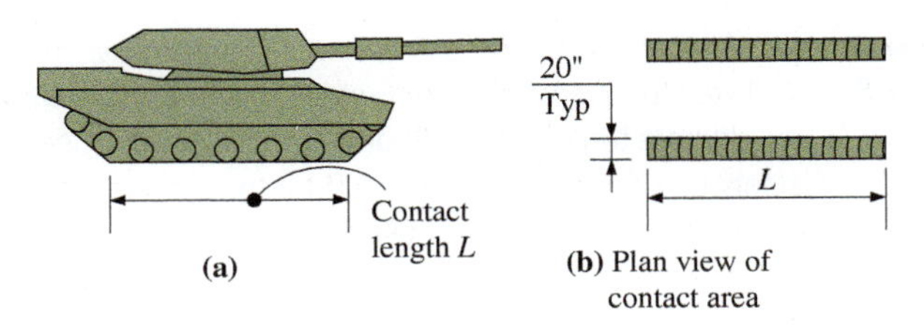

FIGURE 10.9 Tracked vehicle.

Solution The maximum bearing pressure under the tracks is to be 10 psi. Consider this to be an allowable bearing pressure $\sigma_{p(\text{all})}$. The bearing length L required will be based on this allowable pressure.

The required contact area for the two tracks is calculated from

$$\text{required area } A = \frac{P}{\sigma_{p(\text{all})}} = \frac{30 \text{ tons}\left(2000 \dfrac{\text{lb}}{\text{ton}}\right)}{10 \dfrac{\text{lb}}{\text{in.}^2}} = 6000 \text{ in.}^2$$

The calculated total contact area for the two tracks is

$$A = \text{width} \times \text{length} = 2(20 \text{ in.})L = 40\, L \text{ in.}$$

Note that L will have units of inches. Equating the required area and the calculated area:

$$40\ L \text{ in.} = 6000 \text{ in.}^2$$

$$\text{minimum } L = \frac{6000 \text{ in.}^2}{40 \text{ in.}} = 150 \text{ in.} = 12.50 \text{ ft}$$

The tracks must have a 12.5-ft contact length.

10.7 FACTOR OF SAFETY

The use of an allowable stress as an upper limit for induced stress enables a designer to create a margin of safety in a structure when subjected to a particular loading condition. The allowable stress will always be less than the failure stress. The ratio of the failure stress to the allowable stress is called the *factor of safety*. Mathematically, factor of safety (*F.S.*) is defined as

$$F.S. = \frac{\text{failure stress}}{\text{allowable stress}} \qquad \textbf{(10.3)}$$

In an elastic design approach, such as that used for the structural steels and aluminum, the attainment of yield stress in a member is considered to be analogous to failure. Although the steel or aluminum will not actually fail (rupture) at yield, significant and unacceptable deformations are on the verge of occurring. These deformations may render the structural element unusable. The factor of safety, then, is a factor of safety against yielding.

As an example, assume a structural steel with a yield stress of 36,000 psi and an allowable tensile stress of 24,000 psi. The factor of safety against yielding would then be

$$F.S. = \frac{\text{yield stress}}{\text{allowable stress}}$$

$$= \frac{36{,}000 \text{ psi}}{24{,}000 \text{ psi}} = 1.5$$

Another way to consider the case would be to think of the member as having a 50% reserve of strength against yielding in this particular application.

Since the factor of safety and allowable stress are interrelated, the recommended factor of safety values set forth in various specifications and codes depend on the same factors as discussed in Section 10.6 for allowable stress values. Recommended factors of safety and allowable stresses are the result of cumulative pooled experiences and history. They are limiting values that have been traditionally accepted as good practice.

Factor of safety may be based on a material yield strength (for ductile materials) as previously discussed or on a material ultimate strength (for brittle materials), and values may range from 1.5 to 20. For example, for such ductile metals as steel, which are subjected to static loads, a factor of safety of 1.5 based on the yield strength is often suggested. For brittle metals, such as cast iron or wood, which are subjected to shock or impact loads, a factor of safety of 20 based on the ultimate strength of the material may be suggested.

EXAMPLE 10.5 Test results of an ASTM A36 steel specimen (similar to the specimen shown in Figure 10.1) indicated an ultimate tensile strength of 60,000 psi and a yield stress of 36,000 psi. Assume a minimum factor of safety of 1.6 for yielding and 2.0 for tensile strength (or tensile rupture). Determine the allowable stress for (a) yielding and (b) tensile rupture.

Solution **(a)**

$$F.S. = \frac{\text{yield stress}}{\text{allowable stress}}$$

$$\text{allowable stress} = \frac{\text{yield stress}}{1.6} = \frac{36{,}000 \text{ psi}}{1.6} = 22{,}500 \text{ psi}$$

(b)

$$F.S. = \frac{\text{tensile strength}}{\text{allowable stress}}$$

$$\text{allowable stress} = \frac{\text{tensile strength}}{2.0} = \frac{60{,}000 \text{ psi}}{2.0} = 30{,}000 \text{ psi}$$

Allowable stresses such as these will be used to check different conditions in the design of a tension member.

EXAMPLE 10.6 Calculate the required diameter of a 3-m-long steel rod subjected to an axial tensile load of 67 kN. The yield stress is 345 MPa. Use a factor of safety of 2.5 based on the yield stress.

Solution Using Equation (10.4), solve for the allowable stress $\sigma_{t(\text{all})}$:

$$\sigma_{t(\text{all})} = \frac{\text{yield stress}}{F.S.} = \frac{345\text{ MPa}}{2.5} = 138\text{ MPa}$$

Next, find the required cross-sectional area:

$$\text{required } A = \frac{P}{\sigma_{t(\text{all})}} = \frac{67 \times 10^3\text{ N}}{138\text{ MPa}} = 486\text{ mm}^2$$

Circular area $A = \pi d^2/4 = 0.7854d^2$, from which

$$\text{required } d = \sqrt{\frac{A}{0.7854}} = \sqrt{\frac{486\text{ mm}^2}{0.7854}} = 24.9\text{ mm}$$

10.8 ELASTIC–INELASTIC BEHAVIOR

In Section 9.2 we discussed the analysis and design of axially loaded tension members. Design involves determining the required cross-sectional area of the member and then selecting the actual cross section to be used, subject to all the constraints such as the shape of the member. Note that the design process as previously discussed was based on an allowable axial stress that in turn was based on a margin of safety against failure. It is evident then, that the basis for design must hinge on a definition of failure.

The term *failure*, as used herein, means the condition that renders the load-resisting member unfit for resisting further increase in loads. In general, a member fails by inelastic deformation (yielding) if it is made of ductile material or by fracture (rupture or breaking) if the material is brittle. For ductile materials, the yield point stress has commonly been established as the stress at which inelastic deformation starts. From this stress, one can obtain the upper limit of the load that may be applied to a member without causing it to fail. That is, if the load is increased and the yield stress is reached, failure is said to be imminent. The rationale for this approach is that undesirable deformations will occur rendering the member unable to fulfill its intended function. When allowable stresses are used in the proportioning of members of a structural system, the approach may be called *allowable stress design* or *elastic design*. This approach implies that stresses and the accompanying strains will all lie within the elastic range.

As an example, consider a three-bar structure of ductile steel, as shown in Figure 10.10a. When one bar is stressed to its yield point, failure is considered to be imminent (by the elastic design approach) even though the other bars may not yet be stressed to their yield points. Note that in this structure the horizontal member can only translate (no rotation) in the vertical plane and that all three supporting bars must elongate the same amount. Since the strain ϵ in the center bar will be larger, the center bar will reach the yield point first. When it does, the structure is considered to have failed (or to be on the verge of failure).

An alternate design approach holds that failure is not considered to have occurred until all bars of the structure have been stressed to the yield point. From our study of the stress–strain curve for ductile materials, we recognize that the material will not be on the verge of rupture at this point, although all bars may have elongated significantly. When at least part of the structure is strained beyond the yield strain, the resulting behavior is said to be elastic–inelastic, meaning that the material strain of one bar has passed through the elastic range and into the inelastic range.

This alternate design approach, where all members are stressed to the yield point, will determine the maximum possible load, called the *ultimate load*, that can be applied to the structure. The assumptions are made that the stress–strain curve is of the type shown in Figure 10.10b and that the material has infinite ductility. This approach is sometimes referred to as *ultimate strength design* or *limit design*.

To compare the two approaches, we analyze the three-bar structure shown in Figure 10.10a.

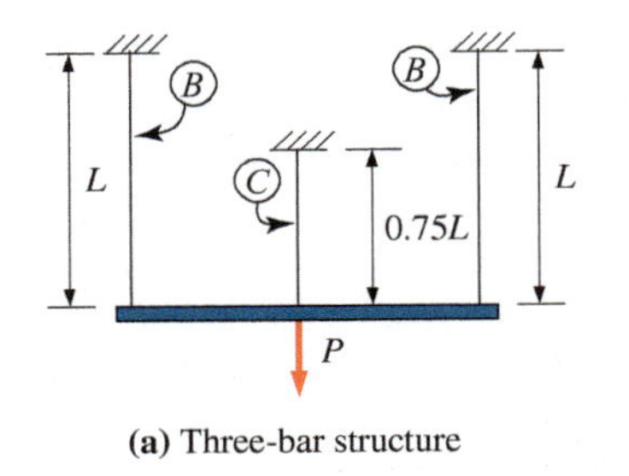

(a) Three-bar structure

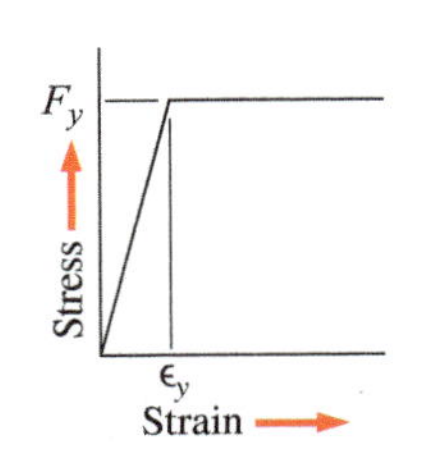

(b) Idealized stress-strain diagram

FIGURE 10.10 Elastic–inelastic behavior.

EXAMPLE 10.7 Determine the maximum load P (failure load) that can be applied to the three-bar structure shown in Figure 10.10a. All bars are steel and are vertical. The symmetrical arrangement allows the rigid horizontal member to deflect vertically without rotation while the three bars elongate by the same amount. The cross-sectional area of each bar is 1 in^2. Assume the steel to be ductile with a stress–strain diagram as shown in Figure 10.10b and a yield stress F_y of 40.0 ksi. The length L is 10 ft. (a) Assume the maximum load to be limited by attainment of first yield (elastic design approach). (b) Assume the maximum load to be limited by all three bars yielding (ultimate strength design approach).

Solution **(a)** For the elastic design approach, failure is defined as first yield in the structure. Since the elongation in each bar is the same, the (unit) strain ϵ will be largest in the center bar and it will reach yield point first. The other two bars will still be strained in the elastic range. Refer to the free-body diagram of Figure 10.11. For equilibrium,

$$P = 2P_B + P_C$$

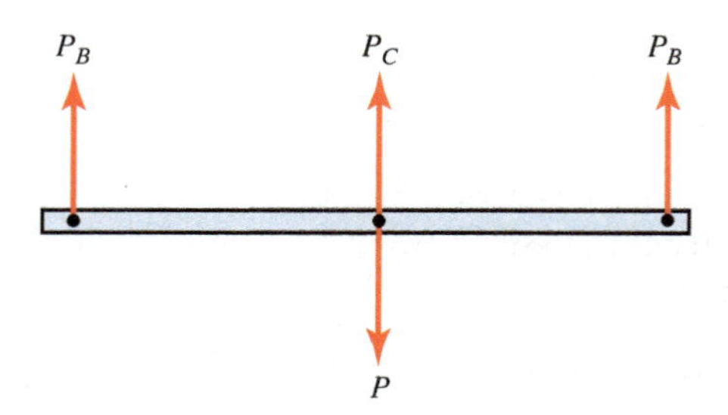

FIGURE 10.11 Free-body diagram.

where P_B and P_C are the forces in bars B and C, respectively. The force P_C in the center bar results from that bar being stressed to F_y (and strained to ϵ_y). By definition,

$$\epsilon_y = \frac{F_y}{E} = \frac{40.0 \text{ ksi}}{30{,}000 \text{ ksi}} = 0.001333$$

The total elongation in bar C is

$$\delta = \epsilon L = 0.001333(0.75)(10 \text{ ft})(12 \text{ in./ft}) = 0.1200 \text{ in.}$$

The elongation in each of the three bars is equal; therefore, the strain in bar B is

$$\epsilon = \frac{\delta}{L} = \frac{0.1200 \text{ in.}}{120 \text{ in.}} = 0.00100$$

from which the stress in bar B can be calculated:

$$\sigma_{t_B} = E\epsilon = (30{,}000 \text{ ksi})(0.00100) = 30.0 \text{ ksi}$$

The stresses in all three bars have now been determined and the maximum load (failure load based on an elastic design approach) is

$$P = 2P_B + P_C = 2\sigma_{t_B}A + F_yA$$
$$= 2(30.0 \text{ ksi})(1.0 \text{ in.}^2) + (40 \text{ ksi})(1.0 \text{ in.}^2) = 100 \text{ k}$$

(b) For the ultimate strength design approach to the determination of failure load, note carefully the illustration of the stress–strain relationship of Figure 10.10b. For strains less than (or equal to) the yield strain ϵ_y, the stress is proportional to strain. For strains greater than the yield strain, stress is constant and equal to the yield stress F_y.

Now assume that load P is further increased. The strain in the center bar will increase beyond the yield strain (at constant stress F_y) while the two outer bars will still be straining within the elastic range. The horizontal member will be supported by the elastically acting outer bars together with a constant resisting force (F_yA) furnished by the center bar. The value of P will increase until yielding begins in each of the outer bars, that is, when $P_B = F_yA$ and failure is assumed to have occurred.

Therefore, the maximum load (ultimate or failure load) that can be applied to the exist when all three bars have reached their yield point stress. This load is calculated from

$$P = 2P_B + P_C = 2F_yA + F_yA = 3F_yA = 3(40\text{ ksi})(1.0\text{ in.}^2) = 120\text{ k}$$

Note that this ultimate load is 20% greater than the maximum load that was based on elastic behavior.

In Example 10.7, allowable loads could have been found instead of failure loads. For part (a), an allowable stress would replace the yield stress in bar *C*. The rest of the load determination would follow the same procedure as illustrated. All three bars would be strained to less than ϵ_y. For part (b), a factor of safety would be applied to the failure load so that

$$P_{(\text{all})} = \frac{P_{(\text{failure})}}{F.S.}$$

In actual practice, no matter which design approach is used, elastic or ultimate strength (elastic–inelastic), appropriate factors of safety are introduced to provide adequate margins of safety and reasonable assurance against failure under expected loading conditions.

SUMMARY BY SECTION NUMBER

10.1 The tension test is a static-loaded destructive-type laboratory test used to establish the mechanical properties of many engineering materials. During the test, applied axial tensile loads and elongations are measured simultaneously.

10.2 A stress–strain diagram is a graphic representation of the results of the tension test. Some of the values obtained from the stress–strain diagram are modulus of elasticity, proportional limit, elastic limit, yield stress, ultimate stress, and rupture stress.

10.3 Values obtained from the stress–strain diagram establish those mechanical properties of a material describing how a material responds to an applied load. Significant mechanical properties are stiffness, strength, elasticity, ductility, brittleness, malleability, toughness, and resilience.

10.4 and 10.5 There are many different types of engineering materials, all of which have application in particular situations. They may be broadly classified as ferrous and nonferrous metals and as nonmetals. Iron and steel of various types comprise the ferrous metals. Aluminum is one of the principal nonferrous metals, being the most abundant metal found in the crust of the earth. Concrete, wood, and plastics are among the principal nonmetal engineering materials.

10.6 Allowable stress represents a safe (or limiting) stress for purposes of design and analysis. Calculated stress represents the stress developed in a member as a result of applied loads. The calculated stress must not exceed the allowable stress.

10.7 Factor of safety is the ratio of a failure stress to an allowable stress. The failure stress could be based on the yield stress or the ultimate stress of the material.

10.8 Elastic design considers that a structure has been loaded to capacity when it attains initial yielding, on the theory that inelastic deformation would terminate the utility of the structure. Ultimate strength design (inelastic design), on the other hand, recognizes that a structure may be loaded beyond initial yielding of some part of the structure provided that other parts that remain in the elastic stress range are capable of resisting the additional load.

PROBLEMS

For the following problems, unless noted otherwise refer to Appendices F and G for necessary mechanical properties of materials.

Section 10.2 The Stress–Strain Diagram

10.1 A $\frac{9}{16}$-in.-diameter steel rod is tested in tension and elongates 0.00715 in. in a length of 8 in. under a tensile load of 6500 lb. Compute **(a)** the stress, **(b)** the strain, and **(c)** the modulus of elasticity, *E*, based on this one reading. The proportional limit of this steel is 34,000 psi.

10.2 A concrete cylinder 150 mm in diameter was tested in compression and found to be shortened an amount of 0.074 mm over a length of 300 mm. The load at the time of the reading was 89 kN. Compute the modulus of elasticity.

10.3 A mild steel, known to have a proportional limit of 34,000 psi, was subjected to a tension test. Stresses and strains were calculated at three points as follows: (stress) 10,000 psi, 25,000 psi, and 40,000 psi; (associated strains) 0.000343, 0.000858, and 0.035 in./in. Calculate the modulus of elasticity, *E*, and comment on the data.

10.4 The data from the tension test of a steel specimen are given in Table 10.2. The gage length was 2.000 in. and the original diameter was 0.505 in. The final diameter was 0.397 in. Calculate stress and strain at each point and draw the stress–strain diagram. Estimate the value of the modulus of elasticity, upper and lower yield points, ultimate strength, rupture strength, and percent reduction in area. (*Hint*: Draw a second curve exaggerating the strain scale for the first eight points. See Figure 10.3.)

Section 10.6 Allowable Stresses and Calculated Stresses

10.5 An 18-in.-long titanium alloy rod is subjected to a tensile load of 24,000 lb. If the allowable tensile stress is 60 ksi and the allowable total elongation is not to exceed 0.05 in., compute the required rod diameter. Assume the proportional limit to be 120 ksi.

10.6 ASTM A36 steel rods are used to support a balcony. Each rod is suspended from a roof beam and supports a balcony floor area of 75 ft². If the combined dead load and live load to be carried by the floor is 250 psf, compute the rod diameter required. Use an allowable tensile stress of 22,000 psi.

TABLE 10.2 Data for Problem 10.4

Load (k)	Deformation (in.)	Load (k)	Deformation (in.)
0	0.0000	8.0	0.0361
2.0	0.0006	10.0	0.1100
4.0	0.0013	13.0	0.2320
6.0	0.0019	13.9	0.3200
8.0	0.0025	14.0	0.3600
7.83	0.0043	12.9	0.4000
7.80	0.0120	10.8	0.4200

10.7 A 450-mm-long AISI 1020 steel rod is subjected to a tensile load of 55 kN. The allowable tensile stress is 140 MPa and the allowable total elongation is not to exceed 0.2 mm. Calculate the required rod diameter. The proportional limit is 175 MPa.

10.8 A tension member in a roof truss is composed of two ASTM A36 structural steel angles that together have a net cross-sectional area of 8.62 in^2.

(a) Compute the allowable total load if the allowable tensile stress is 22,000 psi.

(b) If the total tensile load in the member is 150,000 lb, compute the actual tensile stress.

Section 10.7 Factor of Safety

10.9 A short, solid, compression member of circular cross section is to support a load of 100 kips. Determine the minimum required diameter for this machine member based on direct, compressive stress. The member is to be made of gray cast iron. Use a factor of safety of 6.0 against compression failure.

10.10 A main cable in a large bridge is designed for a tensile force of 2,600,000 lb. The cable consists of 1470 parallel wires, each 0.16 in. in diameter. The wires are cold-drawn steel with an average ultimate strength of 230,000 psi. What factor of safety was used in the design of the cable?

10.11 Test results of a steel specimen indicated an ultimate tensile strength of 827 MPa and a yield stress of 350 MPa. If the tensile allowable stress per design specifications is 210 MPa, compute the factor of safety based on

(a) yield stress

(b) tensile strength

10.12 A concrete canoe in storage is supported by two rope slings, as shown. Each sling supports 48 lb. The rope has a tensile breaking strength of 252 lb. Determine the maximum value of θ if there is to be a factor of safety of 4.0 against breaking.

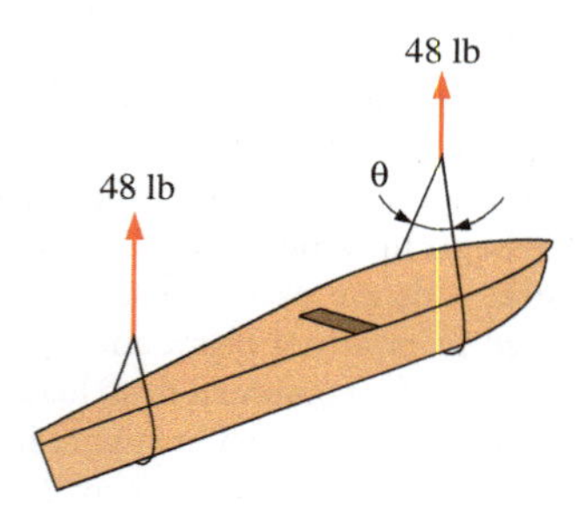

PROBLEM 10.12

Section 10.8 Elastic–Inelastic Behavior

10.13 A load is applied to a rigid bar that is symmetrically supported by three steel rods, as shown. The cross-sectional area of each of the rods is 1.2 in^2. Calculate the maximum load P that may be applied **(a)** using an elastic approach with an allowable stress of 22,000 psi and **(b)** using an ultimate strength approach (inelastic) with a factor of safety of 1.85. Assume a ductile material with a yield stress of 36,000 psi.

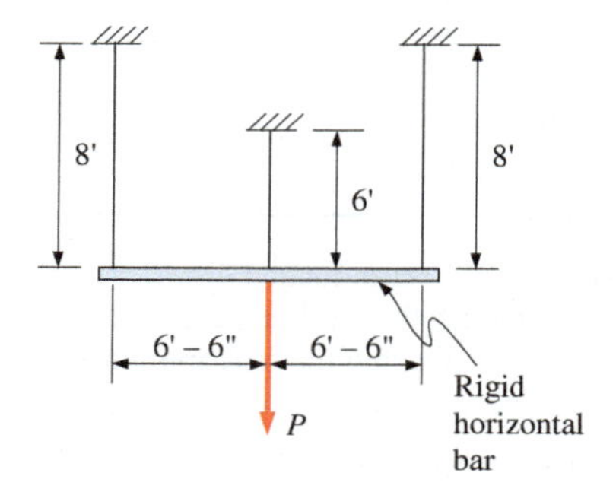

PROBLEM 10.13

Computer Problems

For the following computer problems, any appropriate software may be used. Input prompts should fully explain what is required of the user (the program should be user-friendly). The resulting output should be well labeled and self-explanatory.

10.14 Write a program that will calculate stress, strain, and modulus of elasticity for a rod of circular cross section that is loaded in tension. User input is to be rod diameter, load, original gage length, elongation, and the proportional limit for the material. Use the program to solve Problem 10.1.

10.15 Write a program that will allow a user to input the initial and final diameters and gage length for a tension test specimen along with a specified number of load-elongation combinations. The program should then calculate the stress and strain for each data set as well as the percent reduction in area. Use the program to check the calculations of Problem 10.4.

Supplemental Problems

10.16 A $\frac{1}{2}$-in.-diameter structural nickel steel specimen was subjected to a tension test. After rupture it was determined that the 2-in. standard-gage length had stretched to 2.42 in. The minimum diameter at the fracture was measured to be 0.422 in. Compute the percent elongation and percent reduction in area.

10.17 Compute the modulus of elasticity of a copper alloy wire that stretches 14.0 mm when subjected to a load of 320 N. The wire is 4.00 m long and has a diameter of 1.00 mm.

10.18 A concrete cylinder 6 in. in diameter was tested in compression and observed to have shortened by an amount of 0.0029 in. over a gage length of 12 in. Total load at the time of the reading was 20,000 lb. Compute the modulus of elasticity, E.

10.19 An aluminum bar 2 in. by $\frac{1}{2}$ in. in cross section is subjected to a tensile load of 16,000 lb. At this load, the axial strain is 1650×10^{-6} in./in. Assuming a proportional limit of 21,000 psi, calculate the modulus of elasticity for the material.

10.20 During a tensile test of a steel specimen, the strain at a stress of 35 MPa was calculated to be 0.000 170 (point A). The strain at a stress of 134 MPa was calculated to be 0.000 630 (point B). Determine the modulus of elasticity for this material using the slope between these two points. Calculate the expected stress that would correspond to a strain of 0.000 250. The proportional limit is 200 MPa.

10.21 A 12.5-mm-diameter steel rod was subjected to a tension test. After rupture, it was determined that the original 50-mm gage length had elongated to 60.7 mm. The minimum diameter at the fracture point was measured to be 10.7 mm. Compute the percent elongation and percent reduction in area.

10.22 In a tension test of steel, the ultimate load was 13,100 lb and the elongation was 0.52 in. The original diameter of the specimen was 0.50 in. and the gage length was 2.00 in. Calculate
(a) the ultimate tensile stress
(b) the ductility of the material in terms of percent elongation

10.23 A standard steel specimen having a diameter of 0.505 in. and a 2.00-in. gage length is used in a tension test. At what load P will the extensometer read 0.002 in. deformation? Assume a proportional limit of 34,000 psi.

10.24 A tension member in a structure is composed of stacked, parallel bars, each bar having a cross-sectional area of 225 mm by 32 mm. The allowable tensile stress for the bars is 165 MPa. How many bars will be required to carry a load of 6800 kN? With this number of bars, compute the tensile stress in each, assuming they are all stressed equally.

10.25 A pair of wire cutters is designed to operate under a maximum 35 lb force applied, as shown. Determine the required diameter of the pin (to the next higher $\frac{1}{32}$ in.). The allowable shear stress in the pin is 12,000 psi.

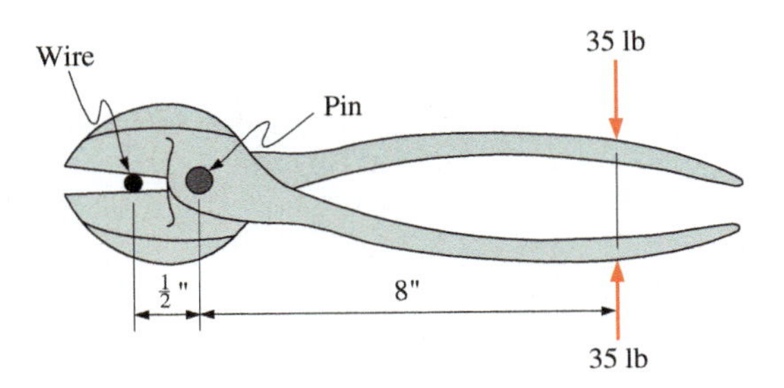

PROBLEM 10.25

10.26 Calculate the end bearing length required for a 10-in.-by-16-in. timber beam (dressed) that is supported on a reinforced concrete wall as shown. The beam reaction is 15,000 lb and the allowable compressive stress perpendicular to the grain for the timber member is 300 psi.

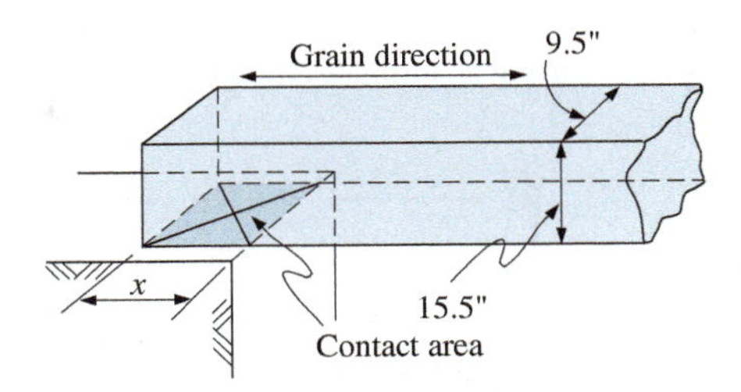

PROBLEM 10.26

10.27 Design a 3-m-long rod subjected to a tensile load of 67.5 kN. Using a factor of safety of 2.5 based on the yield stress, calculate the required rod diameter if it is to be made of
(a) steel with a yield stress of 310 MPa.
(b) aluminum alloy with a yield stress of 240 MPa.

10.28 The collar bearing shown is subjected to a compressive load P of 60,000 lb. Calculate the required diameter D_c of the collar. The diameter D_s of the shaft is 4 in. and the allowable bearing stress for the collar on its support is 300 psi. Assume a $\frac{1}{32}$-in. clearance all around between the shaft and the support.

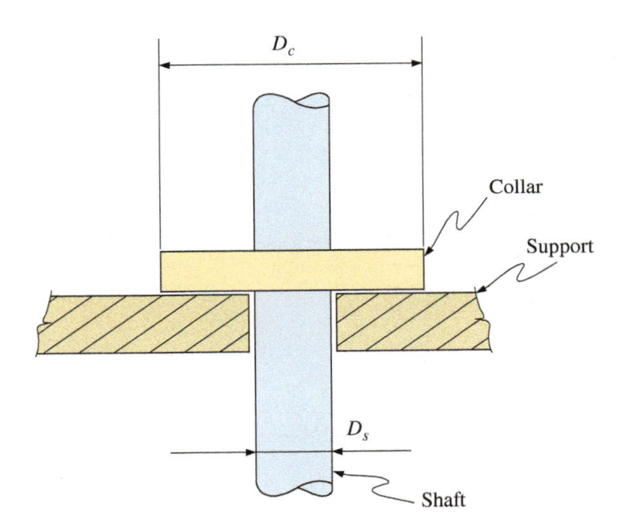

PROBLEM 10.28

10.29 A 10-ft-long steel member is subjected to a tensile load of 200,000 lb. The steel has an ultimate tensile stress of 95,000 psi.
(a) Calculate the cross-sectional area required using a factor of safety of 5.0.
(b) Calculate the cross-sectional area required assuming the maximum elongation to be 0.10 in.

10.30 Two steel bars A and B support a load P, as shown. Bar A has an area of 580 mm^2 and bar B has an area of 700 mm^2. The yield stress is 275 MPa. Both bars elongate equally at point O. Neglect the slope of the bars.

(a) Assume an allowable stress of 185 MPa and determine the allowable load using an elastic design approach.

(b) Assume a factor of safety of 1.80 and determine the allowable load using an ultimate strength design approach.

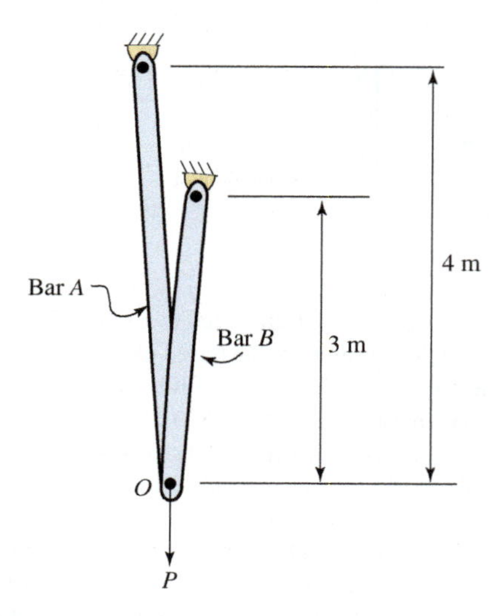

PROBLEM 10.30

10.31 A concrete slab of uniform thickness weighs 20,000 lb and is supported by two steel rods, as shown. The initial length of rod AB is 2 ft. The initial length of rod CD is 3 ft. Rod AB has an area of 1 $in.^2$ and rod CD has an area of 2 in^2. Calculate

(a) the elongation of each rod.

(b) the required ratio of the areas of AB and CD so that the elongation of each bar is the same.

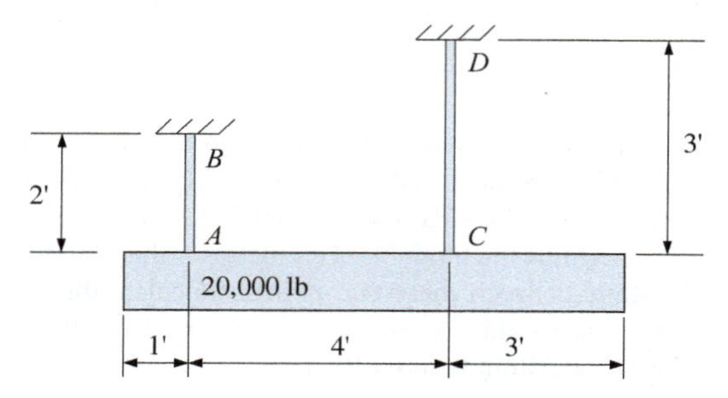

PROBLEM 10.31

CHAPTER ELEVEN

STRESS CONSIDERATIONS

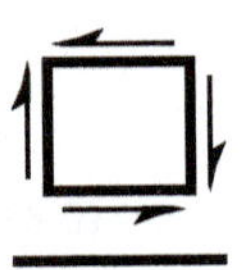

LEARNING OBJECTIVES

Upon completion of this chapter, readers will be introduced to:

- The relationship between lateral strain and axial strain through Poisson's ratio.
- How Poisson's ratio is related to modulus of elasticity and modulus of rigidity.
- How to determine the expansion and contraction of engineering materials caused by temperature variations.
- How to calculate the stresses induced in structural members by expansion or contraction.
- How to determine the effects of expansion and contraction on nonhomogenous members.
- How to calculate the normal and shear stresses in an object on planes inclined to the cross section.

11.1 POISSON'S RATIO

Tests have shown that if a body of elastic material is subjected to a tensile load, its transverse (or lateral) dimensions decrease at the same time that its axial dimensions in the direction of the load increase. This situation is shown pictorially in Figure 11.1. Similarly, if a body of elastic material is subjected to a compressive load, its transverse dimensions increase at the same time that its axial dimensions in the direction of the load decrease.

At stresses below the proportional limit, the transverse strain is proportional to the axial stress; similarly, the axial (longitudinal) strain is proportional to the axial stress. Since both lateral strain and axial strain are proportional to the axial stress, their ratio must be constant (and positive) for a given material. This ratio of lateral strain to axial strain is called *Poisson's ratio* and is represented by μ (Greek lowercase mu) and expressed by the equation

$$\mu = \frac{\text{transverse strain}}{\text{axial strain}} \quad (11.1)$$

Commonly used values for Poisson's ratio are given in Appendix G.

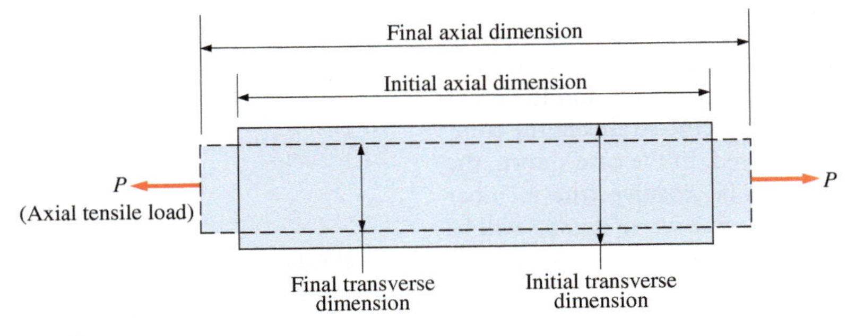

FIGURE 11.1 Dimensional changes due to axial tensile load.

EXAMPLE 11.1 A 10-ft-long rectangular ASTM A572 Grade 50 steel plate 1 in. by 12 in. in cross section is used as a hanger and subjected to a tensile load of 240,000 lb. The proportional limit of the steel is assumed to be 45,000 psi. Compute (a) axial stress, (b) axial strain, (c) transverse strain, (d) total axial dimensional change, and (e) total transverse (12-in. side) dimensional change. Refer to Appendix G for necessary mechanical properties.

Solution (a) The axial tensile stress (σ_t) is

$$\sigma_t = \frac{P}{A} = \frac{240{,}000 \text{ lb}}{(12 \text{ in.})(1 \text{ in.})} = 20{,}000 \text{ psi}$$

$$20{,}000 \text{ psi} < 45{,}000 \text{ psi} \qquad \textbf{(O.K)}$$

(b) For the axial strain (ϵ), since

$$E = \frac{\sigma}{\epsilon} = \frac{\text{stress}}{\text{strain}}$$

then

$$\epsilon = \frac{\sigma_t}{E} = \frac{20{,}000 \text{ psi}}{30{,}000{,}000 \text{ psi}} = 0.000667$$

(c) The transverse strain (ϵ) is

$$\text{transverse } \epsilon = \mu(\text{axial } \epsilon)$$
$$= 0.25(0.000667) = 0.000167$$

(d) For the total axial elongation (δ), since

$$\epsilon = \frac{\delta}{L}$$

therefore,

$$\delta = \epsilon L = 0.000667(10 \text{ ft})(12 \text{ in./ft}) = 0.080 \text{ in.}$$

(e) The total transverse change in the 12-in. dimension is

$$\delta = \epsilon L = 0.000167(12 \text{ in.}) = 0.0020 \text{ in.}$$

The transverse strain that accompanies the axial stress does not result from a transverse stress and does not cause a transverse stress. If the transverse strain is prevented in some way, however, a transverse stress will develop.

If an elastic body is loaded in two directions as shown in Figure 11.2, stresses are induced in both the x and y directions. Each total strain (ϵ_x and ϵ_y) will be the sum of two component parts, one due to axial stress and the other due to transverse stress. The signs of the strains in the summation must be carefully considered. In the case shown, the strains due to axial stresses will be positive (the member lengthens) while the strains due to transverse stresses will be negative (the member shortens).

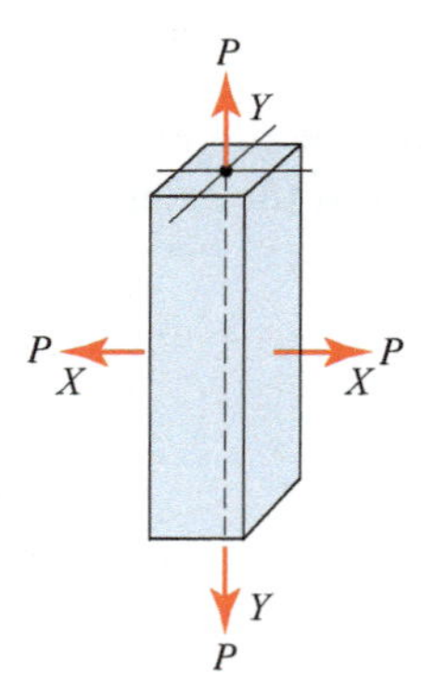

FIGURE 11.2 Tension member axially loaded in two directions.

EXAMPLE 11.2 A 12-in.-long 1-in.-by-3-in. ASTM A36 steel bar is loaded as shown in Figure 11.3. The proportional limit is 34,000 psi. Compute (a) the strains in the x and y directions and (b) the total dimensional changes in the x and y directions. Refer to Appendix G for necessary mechanical properties.

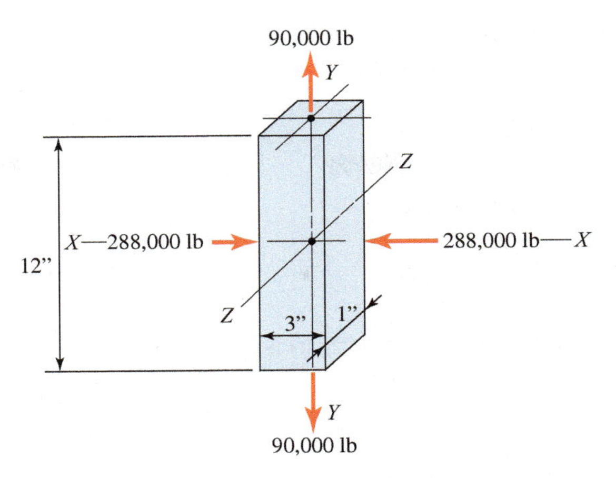

FIGURE 11.3 Load diagram.

Solution First, the axial stresses (σ_x and σ_y) are computed:

$$\sigma_y = \frac{P}{A} = \frac{90{,}000 \text{ lb}}{(3 \text{ in.})(1 \text{ in.})} = 30{,}000 \text{ psi}$$

$$30{,}000 \text{ psi} < 34{,}000 \text{ psi (tension)} \qquad \textbf{(O.K)}$$

and

$$\sigma_x = \frac{P}{A} = \frac{288{,}000 \text{ lb}}{(12 \text{ in.})(1 \text{ in.})} = 24{,}000 \text{ psi}$$

$$24{,}000 \text{ psi} < 34{,}000 \text{ psi (compression)} \qquad \textbf{(O.K)}$$

(a) Calculate the strains in the *x* and *y* directions. Denote strain in the *x* direction due to axial stress as ϵ_{xa} and due to transverse stress as ϵ_{xt}, and so on:

$$\epsilon_{xa} = \frac{\sigma_x}{E} = -\frac{24{,}000 \text{ psi}}{30{,}000{,}000 \text{ psi}} = -0.00080 \text{ (decrease)}$$

$$\epsilon_{ya} = \frac{\sigma_y}{E} = +\frac{30{,}000 \text{ psi}}{30{,}000{,}000 \text{ psi}} = +0.00100 \text{ (increase)}$$

$$\epsilon_{xt} = \frac{\mu\sigma_y}{E} = -\frac{0.25(30{,}000 \text{ psi})}{30{,}000{,}000 \text{ psi}} = -0.00025 \text{ (decrease)}$$

$$\epsilon_{yt} = \frac{\mu\sigma_x}{E} = +\frac{0.25(24{,}000 \text{ psi})}{30{,}000{,}000 \text{ psi}} = +0.00020 \text{ (increase)}$$

Summing the strains in each direction, carefully noting the signs:

$$\epsilon_x = \epsilon_{xa} + \epsilon_{xt} = (-0.00080) + (-0.00025) = -0.00105 \text{ (decrease)}$$

$$\epsilon_y = \epsilon_{ya} + \epsilon_{yt} = +0.0010 + 0.0020 = +0.00120 \text{ (increase)}$$

(b) Calculate the total dimensional change in the *x* and *y* directions:

$$\delta_x = \epsilon_x L = -0.00105(3 \text{ in.}) = -0.00315 \text{ in. (decrease)}$$

$$\delta_y = \epsilon_y L = +0.00120(12 \text{ in.}) = +0.0144 \text{ in. (increase)}$$

EXAMPLE 11.3 A 38-mm-diameter ASTM A36 steel bar is subjected to a tension test. Under a tensile load of 260 kN it was observed that the original gage length of 50 mm increased in length by 0.055 mm and the diameter decreased by 0.0105 mm.

The proportional limit is 234 MPa. Compute the modulus of elasticity *E* and Poisson's ratio μ.

Solution The axial stress is

$$\sigma_t = \frac{P}{A} = \frac{260{,}000\ \text{N}}{0.7854(38\ \text{mm})^2} = 229\ \text{MPa}$$

$$229\ \text{MPa} < 234\ \text{MPa} \qquad \textbf{(O.K)}$$

The axial strain is

$$\epsilon = \frac{\delta}{L} = \frac{0.055\ \text{mm}}{50\ \text{mm}} = 0.0011$$

The modulus of elasticity is

$$E = \frac{\sigma_t}{\epsilon} = \frac{229\ \text{MPa}}{0.0011} = 208 \times 10^3\ \text{MPa}$$

Poisson's ratio is

$$\text{transverse strain } \epsilon = \frac{\delta}{L} = \frac{0.0105\ \text{mm}}{38\ \text{mm}} = 0.000\ 276$$

$$\mu = \frac{\text{transverse } \epsilon}{\text{axial } \epsilon} = \frac{0.000\ 276}{0.0011} = 0.251$$

There is a relationship between the modulus of elasticity, modulus of rigidity, and Poisson's ratio.[1] In Section 9.6 we discussed the modulus of rigidity G as the ratio of shear stress to shear strain. For homogeneous elastic materials, the modulus of rigidity can be determined by means of a tensile test. Both the longitudinal (axial) strain and the transverse strain must be measured, however. The modulus of rigidity can then be calculated from

$$G = \frac{E}{2(1 + \mu)} \qquad \textbf{(11.2)}$$

where G = the modulus of rigidity (psi, ksi) (Pa, MPa)
E = the modulus of elasticity (tensile or compressive) (psi, ksi) (Pa, MPa)
μ = Poisson's ratio

Note that the modulus of rigidity G will always be less than E, since Poisson's ratio is always positive.

EXAMPLE 11.4 A 2-in.-diameter metal specimen is subjected to an axial compressive load of 40,000 lb. Transverse and longitudinal dimensional changes are measured with the use of electronic strain gages and the strains are determined to be 0.0012 longitudinally and 0.0004 transversely. Compute (a) Poisson's ratio μ, (b) the modulus of elasticity E, and (c) the modulus of rigidity G.

Solution (a) For Poisson's ratio,

$$\mu = \frac{\text{transverse strain}}{\text{longitudinal strain}} = \frac{0.0004}{0.0012} = 0.333$$

(b) For the modulus of elasticity, first calculate the compressive stress:

$$\sigma_c = \frac{P}{A} = \frac{40{,}000\ \text{lb}}{0.7854(2\ \text{in.})^2} = 12{,}730\ \text{psi}$$

Then calculate the modulus of elasticity:

$$E = \frac{\sigma_c}{\epsilon} = \frac{12{,}730\ \text{psi}}{0.0012} = 10{,}610{,}000\ \text{psi}$$

(c) For the modulus of rigidity, Equation (11.2) yields

$$G = \frac{E}{2(1 + \mu)} = \frac{10{,}610{,}000\ \text{psi}}{2(1 + 0.333)} = 3{,}980{,}000\ \text{psi}$$

---

[1] S. Timoshenko, *Strength of Materials*, Part I (New York: Van Nostrand, 1950).

EXAMPLE 11.5 Compute the modulus of rigidity for steel using the normal values of $E = 30{,}000{,}000$ psi and $\mu = 0.25$.

Solution Use Equation (11.2):

$$G = \frac{E}{2(1+\mu)} = \frac{30{,}000{,}000 \text{ psi}}{2(1+0.25)} = 12{,}000{,}000 \text{ psi}$$

11.2 THERMAL EFFECTS

Materials commonly used in engineering exhibit dimensional changes when subjected to temperature changes. For any particular material, the amount of dimensional change per unit temperature change is constant over moderate temperature ranges. Most materials expand as their temperatures rise and contract as their temperatures fall.

For most materials, standard values for dimensional change per degree of temperature change have been established by test. Such a value is called the *linear coefficient of thermal expansion* and is denoted by the Greek letter α. It is a measure of the change in length per unit of length per degree of temperature change. It will have the same numerical value for any particular material, no matter what unit of length is used. The coefficient is frequently given units such as in./in./°F in the U.S. Customary System and mm/mm/°C (where C denotes Celsius) in the SI. Since *any* convenient unit of length may be used, however, it is not uncommon for the coefficient to be expressed in units of 1/°F or 1/°C. For typical values, see Appendix G.

If a body is free to expand or contract due to temperature variations, there will be no stress induced in the member. The magnitude of the dimensional change can be expressed by

$$\delta = \alpha L(\Delta T) \tag{11.3}$$

where δ = the total change in length (in.) (mm)
α = the linear coefficient of thermal expansion; also, strain per degree change in temperature (1/°F) (1/°C)
L = the original length of the member (in.) (mm)
ΔT = the change in temperature (°F)(°C)

If a body is somehow partially or fully restrained so as to prevent a dimensional change due to temperature variations, internal stresses will develop. These are generally termed *temperature stresses* or *thermal stresses*. An expression for these stresses can be developed as follows.

1. Assume that a total dimensional change of δ is allowed to occur due to a temperature change:

$$\delta = \alpha L(\Delta T)$$

2. However, the member is restrained. Therefore, apply an axial force P to the member to restore it to its original length. This dimensional change is written as

$$\delta = \frac{PL}{AE} = \sigma\left(\frac{L}{E}\right)$$

3. Equating the two values of δ,

$$\sigma\left(\frac{L}{E}\right) = \alpha L(\Delta T)$$

from which

$$\sigma = \alpha E(\Delta T) \tag{11.4}$$

where σ = the temperature stress developed in a restrained member due to a temperature variation (invalid only if σ does not exceed the proportional limit) (psi) (MPa)
α = the coefficient of thermal expansion (1/°F) (1/°C)
E = the modulus of elasticity (psi) (MPa)
ΔT = the change in temperature (°F) (°C)

If a member is fully restrained and then cooled, the stress induced is tension. If the member is fully restrained and then heated, the stress induced is compression.

Table 11.1 summarizes the two preceding thermal effects equations.

TABLE 11.1 Thermal effects equations

	Equation	Use
Equation (11.3)	$\delta = \alpha L(\Delta T)$	Find dimensional change in unrestrained element caused by temperature change.
Equation (11.4)	$\sigma = \alpha E(\Delta T)$	Find stress in restrained element caused by temperature change.

EXAMPLE 11.6 A 100-in.-long ASTM A36 steel rod with a cross-sectional area of 2.0 in.^{2} is secured between rigid supports. If there is no stress in the rod at a temperature of 70°F, compute the stress when the temperature drops to 0°F. The proportional limit of the steel is 34,000 psi. Refer to Appendix G for necessary mechanical properties.

Solution Since no movement of the supports occurs,

$$\sigma = \alpha E(\Delta T)$$
$$= (0.0000065\ 1/°\text{F})(30{,}000{,}000\ \text{lb/in.}^2)(70°\text{F})$$
$$= 13{,}650\ \text{psi (tension)}$$
$$13{,}650\ \text{psi} < 34{,}000\ \text{psi} \qquad \textbf{(O.K)}$$

EXAMPLE 11.7 Compute the stress in the rod of the previous example if the supports yield and move together a distance of 0.02 in. as the temperature drops. Refer to Figure 11.4.

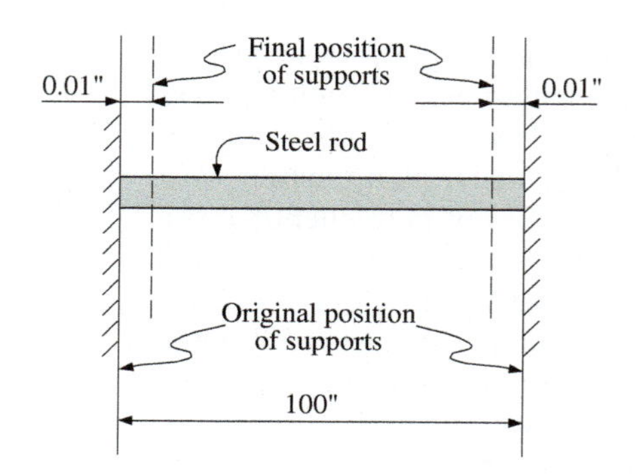

FIGURE 11.4 Support for steel rod.

Solution Note that in Example 11.6 the stress developed is independent of the length of the member. When yielding of the supports occurs, however, the length of the member does affect the stress developed. If the rod were free to contract, the amount that it would shorten could be calculated from

$$\delta = \alpha L(\Delta T)$$
$$= (0.0000065\ 1/°\text{F})(100\ \text{in.})(70°\text{F})$$
$$= 0.0455\ \text{in.}$$

If the supports were rigid, δ would be the restrained change in length. Since the supports yield 0.02 in., the restrained change in length is

$$\delta = 0.0455\ \text{in.} - 0.02\ \text{in.} = 0.0255\ \text{in.}$$

Therefore, the strain is

$$\epsilon = \frac{\delta}{L} = \frac{0.0255\ \text{in.}}{100\ \text{in.}} = 0.000255\ \text{in./in.}$$

from which

$$\sigma = E\epsilon$$
$$= (30{,}000{,}000\ \text{psi})(0.000255) = 7650\ \text{psi (tension)}$$
$$7650\ \text{psi} < 34{,}000\ \text{psi} \qquad \textbf{(O.K)}$$

EXAMPLE 11.8 An AISI 1040 steel fence wire 0.148 in. in diameter is stretched between rigid end posts with a tension of 300 lb when the temperature is 90°F. The proportional limit of the wire is 40,000 psi. Calculate the temperature drop that could occur without causing a permanent set in the wire. Refer to Appendix G for necessary properties.

Solution The cross-sectional area of the wire is

$$A = 0.7854(0.148\text{ in.})^2 = 0.0172\text{ in.}^2$$

The stress in the wire due to the tensile load of 300 lb is

$$\sigma = \frac{P}{A} = \frac{300\text{ lb}}{0.0172\text{ in.}^2} = 17{,}440\text{ psi}$$

The additional temperature stress to reach the proportional limit is

$$\sigma = 40{,}000\text{ psi} - 17{,}440\text{ psi} = 22{,}560\text{ psi}$$

The temperature change that would induce this stress in the wire is calculated from

$$\sigma = \alpha E(\Delta T)$$

Substitute the numerical values:

$$22{,}560\text{ psi} = (0.0000065\ 1/°\text{F})(30{,}000{,}000\text{ psi})(\Delta T)$$

from which

$$\Delta T = 115.7°\text{F (decrease)}$$

Therefore, the temperature would be

$$90°\text{F} - 115.7°\text{F} = -25.7°\text{F}$$

EXAMPLE 11.9 Steel crane rails are laid with their adjacent ends 3.20 mm apart when the temperature is 15°C. The length of each rail is 18 m, as shown in Figure 11.5. Refer to Appendix G for mechanical properties. (a) Calculate the temperature at which the rails will touch end to end. (b) Calculate the gap between adjacent ends when the temperature drops to −10°C. (c) Calculate the compressive stress in the rails when the temperature reaches 45°C.

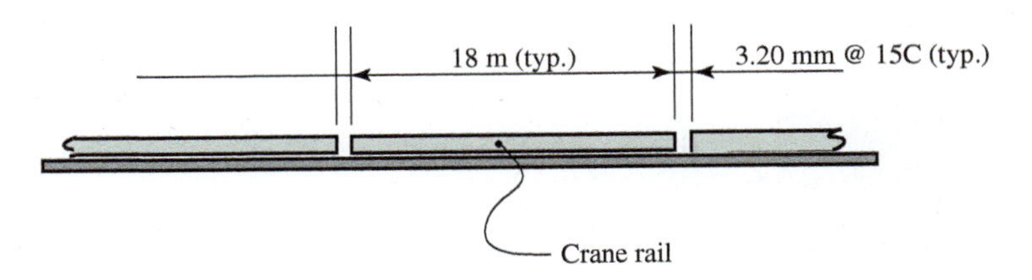

FIGURE 11.5 Sketch for Example 11.9.

Solution (a) The rails will touch end to end when the 3.2-mm gap is closed by a temperature rise. Use Equation (11.3) and solve for ΔT:

$$\Delta T = \frac{\delta}{\alpha L} = \frac{3.2 \times 10^{-3}\text{ m}}{\left(0.000\,011\,7\frac{1}{°\text{C}}\right)(18\text{ m})} = 15.2°\text{C}$$

This value represents the temperature change (increase) that will make the rail ends touch. The actual temperature will be

$$15°\text{C} + 15.2°\text{C} = 30.2°\text{C}$$

(b) The gap between the adjacent ends when the temperature drops to $-10°C$, which is a temperature change of $-25°C$, is also calculated from Equation (11.3):

$$\delta = \alpha L(\Delta T)$$
$$= \left(0.000\,011\,7\frac{1}{°C}\right)(18 \times 10^3\ \text{mm})(25°C)$$
$$= 5.27\ \text{mm}$$

The gap would then become

$$3.2\ \text{mm} + 5.27\ \text{mm} = 8.47\ \text{mm}$$

(c) A compressive stress in the rails will develop when the temperature rises above 30.2°C [from part (a)]. Therefore, if the temperature rose to 45°C, the temperature change would be

$$\Delta T = 45°C - 30.2°C = +14.8°C$$

and the stress, calculated from Equation (11.4), is

$$\sigma_c = \alpha E(\Delta T)$$
$$= \left(0.000\,011\,7\frac{1}{°C}\right)(207{,}000\ \text{MPa})(14.8°C)$$
$$= 35.8\ \text{MPa}$$

11.3 MEMBERS COMPOSED OF TWO OR MORE COMPONENTS

In some situations, structural members may be composed of two (or more) components. The simpler situation is represented by the post–pedestal combination, shown in Figure 11.6, where it is seen that the forces in the post and the pedestal are equal to P and the deformation of the composite member is cumulative—that is, the deformation of the member is equal to the sum of the deformations of the two component parts. This situation can be termed as two components "in series."

Consider a different case where the deformation is restrained. A member composted of two components, shown in Figure 11.7, is placed between two unyielding supports and is subjected to a temperature increase. There will be no net deformation (change in length) of the member and the resulting forces in the components must be equal.

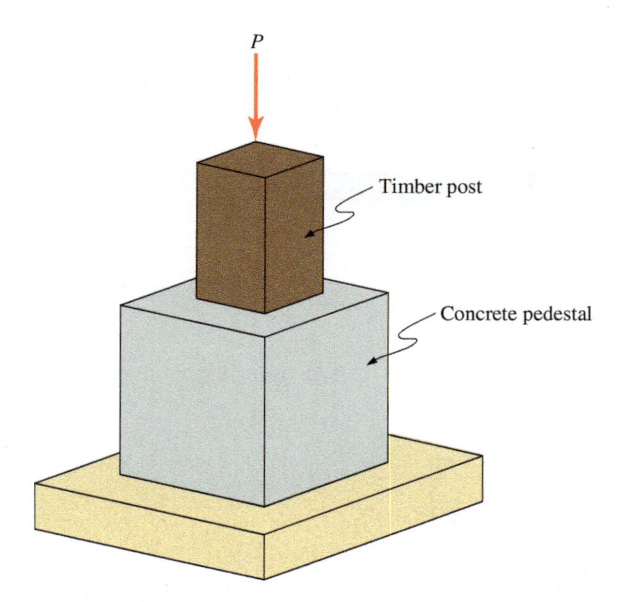

FIGURE 11.6

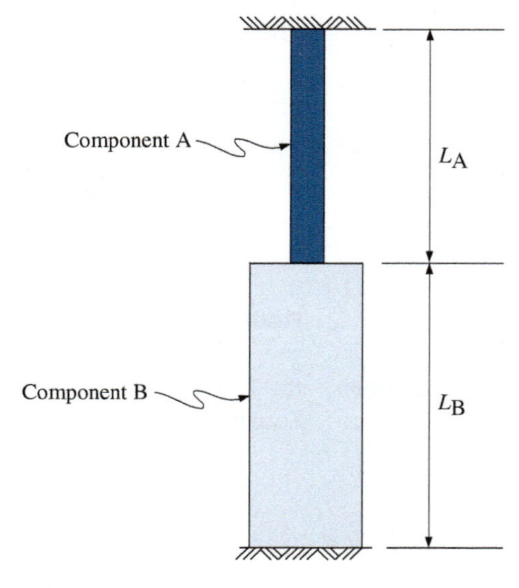

FIGURE 11.7

EXAMPLE 11.10 In Figure 11.7, assume components A and B are as described in Table 11.2. Material properties are from Appendix G. The supports are unyielding. There is no initial force in the two-component member. Determine the stress in each component of the member when the temperature rises 20°F.

TABLE 11.2 **Data for Example 11.10**

Component	Material	Area (in.^{2})	Length (in.)	E(ksi)	$\alpha(1/°F) \times 10^{-6}$
A	Aluminum alloy	6.00	60.0	10,000	13.1
B	Gray cast iron	8.00	18.0	15,000	5.9

Solution Consider two steps: unrestrained expansion followed by the application of sufficient force P to restore the member to its original length.

(a) Unrestrained expansion:

$$\begin{aligned}\delta &= \delta_{AL} + \delta_{CI}\\ &= [\alpha L(\Delta T)]_{AL} + [\alpha L(\Delta T)]_{CI}\\ &= \left(13.1 \times 10^{-6}\frac{1}{°F}\right)(60\text{ in.})(20°F) + \left(5.9 \times 10^{-6}\frac{1}{°F}\right)(18\text{ in.})(20°F)\\ &= 0.01784\text{ in.}\end{aligned}$$

(b) Apply force P to cause deformation equal to δ from part (a):

$$\delta = \left(\frac{PL}{AE}\right)_{AL} + \left(\frac{PL}{AE}\right)_{CI}$$

From which

$$P = \frac{\delta}{\left(\frac{L}{AE}\right)_{AL} + \left(\frac{L}{AE}\right)_{CI}} = \frac{0.01784\text{ in.}}{\left(\frac{60\text{ in.}}{(6\text{ in.}^2)\left(10{,}000\frac{\text{k}}{\text{in.}^2}\right)}\right) + \left(\frac{18\text{ in.}}{(8\text{ in.}^2)\left(15{,}000\frac{\text{k}}{\text{in.}^2}\right)}\right)}$$

$$P = 15.51\text{ k}$$

Calculate the stresses:

$$s_{AL} = \frac{P}{A} = \frac{15.51\text{ k}}{6\text{ in.}^2} = 2.59\text{ ksi}$$

$$s_{CI} = \frac{P}{A} = \frac{15.51\text{ k}}{8\text{ in.}^2} = 1.939\text{ ksi}$$

Another type of member composed of two (or more) different materials consists of components so arranged that deformations are equal but the induced stresses are different. This can be thought of as components "in parallel." One example is a building column composed of a steel shape encased in concrete. Another is a wood post strengthened with steel plates or channels, as shown in Figure 11.8. We assume that the materials in the reinforced post have different modulus of elasticity values and are so connected as to act as a single unit, each deforming equally under load. In this case, the stresses developed in the two materials by the applied load will be proportional to their moduli of elasticity.

For the same deformation, the stress developed in the material with the higher modulus of elasticity (material A) will be greater than the stress developed in the material with the lower modulus of elasticity (material B). Assuming the two materials are of the same length and deform equally,

$$\delta_A = \delta_B \quad \text{and} \quad \epsilon_A = \epsilon_B$$

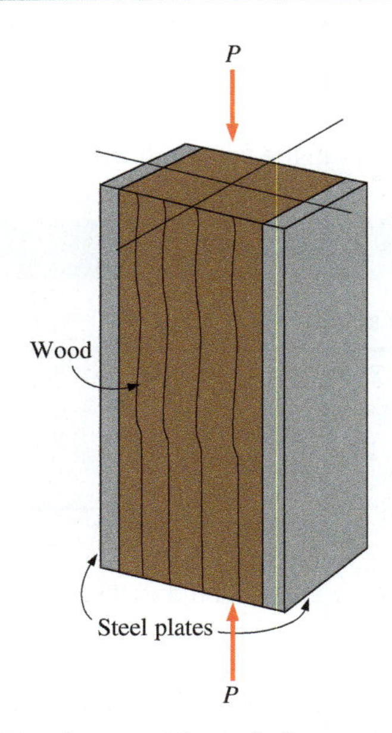

FIGURE 11.8 Wood post with steel plates.

Since

$$\epsilon = \frac{\sigma}{E}$$

it follows that

$$\frac{\sigma_A}{E_A} = \frac{\sigma_B}{E_B}$$

The preceding expression can be rearranged to yield the stress in material A (the higher stress):

$$\sigma_A = \frac{E_A}{E_B}\sigma_B$$

The ratio of the modulus of elasticity values is generally called the *modular ratio* and is denoted as n. Therefore,

$$\sigma_A = n\sigma_B \quad \textbf{(11.5)}$$

This concept may be carried one step further. Based on a condition of static equilibrium, the total load P must be resisted (carried) in part by each material. Therefore,

$$P = P_A + P_B$$

Substituting gives

$$P = A_A\sigma_A + A_B\sigma_B \quad \textbf{(11.6)}$$

Substituting for s_A from Equation (11.5) leads to

$$P = A_A(n\sigma_B) + A_B\sigma_B$$

which may be rearranged as

$$P = \sigma_B(nA_A + A_B) \quad \textbf{(11.7)}$$

The quantity nA_A is sometimes called an *equivalent area*. It is a hypothetical area that may be considered a replacement for area A_A. The resulting hypothetical cross section is then composed of a single material having the lower modulus of elasticity value. This concept sometimes facilitates the analysis process.

EXAMPLE 11.11 A short post (shown in Figure 11.9) consisting of a 6-in.-diameter standard weight steel pipe (see Appendix B) is filled with concrete, which has an ultimate compressive strength (s_c') of 3000 psi. The pipe is made of ASTM A501 steel. The post is subjected to an axial compressive load of 100,000 lb. Assuming both materials deform equally, compute the stress developed in the steel and the concrete. Refer to Appendix G for necessary material properties.

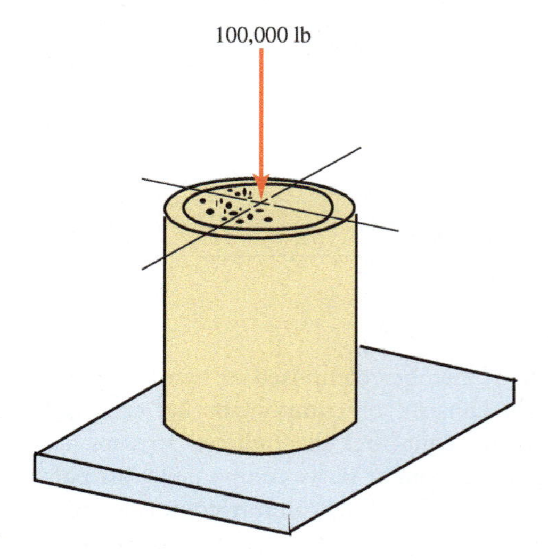

FIGURE 11.9 Steel–concrete post.

Solution The pipe has a cross-sectional area of 5.58 in.^{2} and an inside diameter of 6.065 in. The cross-sectional area of the concrete is calculated from

$$A_{CON} = 0.7854(6.065 \text{ in.})^2 = 28.89 \text{ in.}^2$$

The modular ratio is

$$n = \frac{E_{ST}}{E_{CON}} = \frac{30{,}000{,}000 \text{ psi}}{3{,}120{,}000 \text{ psi}} = 9.62$$

Use Equation (11.7),

$$P = \sigma_{CON}(nA_{ST} + A_{CON})$$

from which

$$\sigma_{CON} = \frac{P}{nA_{ST} + A_{CON}} = \frac{100{,}000 \text{ lb}}{9.62(5.58 \text{ in.}^2) + 28.89 \text{ in.}^2} = 1211 \text{ psi}$$

Then, from Equation (11.5),

$$\sigma_{ST} = n\sigma_{CON} = 9.62(1211 \text{ psi}) = 11{,}650 \text{ psi}$$

EXAMPLE 11.12 A 4-by-4 (S4S) Douglas fir wood truss tension member is strengthened by the addition of two ASTM A36 steel plates, as shown in Figure 11.10. Compute the allowable load for the composite member. In addition to mechanical properties from Appendices F and G, assume the allowable tensile stress for the steel to be 22,000 psi. Assume that the materials are of equal lengths and are connected to act as a single unit and deform equally.

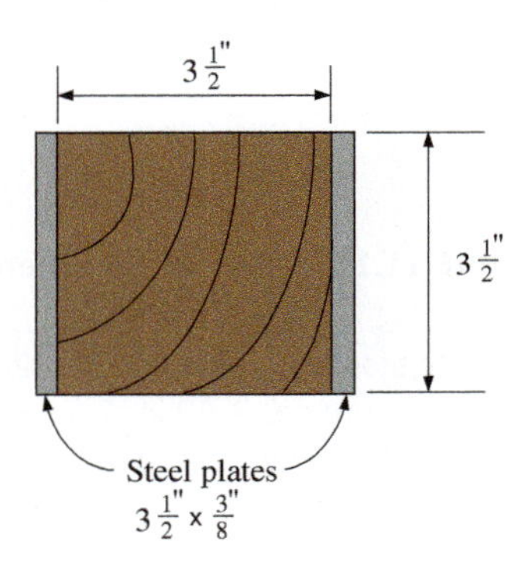

FIGURE 11.10 Composite member.

Solution Assume that the stress in the wood reaches its allowable stress of 625 psi (Appendix F) before the steel reaches its allowable stress. The stress in the steel is computed using Equation (11.5):

$$\sigma_{ST} = n\sigma_W = \frac{30{,}000{,}000 \text{ psi}}{1{,}700{,}000 \text{ psi}}(625 \text{ psi}) = 11{,}030 \text{ psi}$$

Therefore, when the stress in the wood is 625 psi, the stress in the steel is 11,030 psi. This value is acceptable because the steel stress of 11,030 psi is less than the allowable stress of 22,000 psi. The stresses may not increase beyond this point; if they did, the stress in the wood would exceed the allowable stress of the wood. This calculation shows, then, that the allowable stress of the wood limits the load-carrying capacity of the member.

The area of the wood is calculated from

$$A_W = (3.5 \text{ in.})^2 = 12.25 \text{ in.}^2$$

Use Equation (11.6) to calculate the allowable load:

$$\begin{aligned} P &= P_{ST} + P_W \\ &= A_{ST}\sigma_{ST} + A_W\sigma_W \\ &= (3.5 \text{ in.})(0.375 \text{ in.})(2)(11{,}030 \text{ psi}) + (12.25 \text{ in.}^2)(625 \text{ psi}) \\ &= 36{,}600 \text{ lb} \end{aligned}$$

We can also consider a type of system in which an axial load is simultaneously applied to two or more members of different materials and of different lengths. The analysis methodology is similar to the case of one member composed of two or more materials.

Assuming the total deformation of the members to be the same, but the lengths of the members *not* the same, Equation (11.5) cannot be used. We can express the equality of the total deformation of each member as follows (the two different members are denoted by subscripts A and B):

$$\left(\frac{PL}{AE}\right)_A = \left(\frac{PL}{AE}\right)_B \qquad \textbf{(11.8)}$$

The use of this equation is similar to the use of Equation (11.5).

EXAMPLE 11.13 The structural system of Figure 11.11 consists of a horizontal plate suspended by three rods. A load of 50 kips is applied to the plate. The plate is perfectly level prior to the application of the load and remains level after the load has been applied.

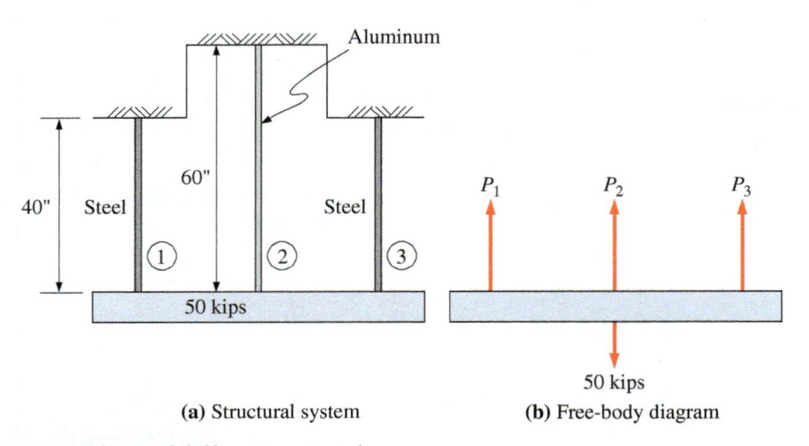

FIGURE 11.11 Members of different materials.

The steel rods are of AISI 1020 steel. Each is 40 in. long and has a cross-sectional area of 1.0 in^2. The aluminum rod is 60 in. long and has a cross-sectional area of 1.5 in.^{2}. Calculate the load that each rod will carry. Refer to Appendix G for necessary mechanical properties.

Solution Designate the rods as 1, 2, and 3 and assume that all three rods elongate by the same amount:

$$\delta_1 = \delta_2 = \delta_3$$

or

$$\left(\frac{PL}{AE}\right)_1 = \left(\frac{PL}{AE}\right)_2 = \left(\frac{PL}{AE}\right)_3$$

Substitute (where E is in units of ksi, A is in units of in.^{2}, and L is in in.):

$$\frac{P_1(40)}{(1.0)(30{,}000)} = \frac{P_2(60)}{(1.5)(10{,}000)} = \frac{P_3(40)}{(1.0)(30{,}000)} \qquad \textbf{(1)}$$

Due to symmetry, $P_1 = P_3$. Therefore, work with and solve for P_1 and P_2. Cross-multiplying the first two parts of Equation 1 gives

$$P_1(40)(1.5)(10{,}000) = P_2(60)(1.0)(30{,}000)$$

or

$$600{,}000\, P_1 - 1{,}800{,}000\, P_2 = 0 \qquad \textbf{(2)}$$

With reference to Figure 11.11b, a summation of vertical forces ($\Sigma F_V = 0$) yields

$$P_1 + P_2 + P_3 = 50 \text{ k}$$

or since $P_1 = P_3$,

$$2P_1 + P_2 = 50 \text{ k} \qquad \textbf{(3)}$$

Equations 2 and 3 are simultaneous equations that, when solved for P_1 and P_2, will yield

$$P_1 = 21.43 \text{ k} \qquad P_2 = 7.14 \text{ k}$$

Therefore, the steel rods each carry 21.43 kips and the aluminum rod carries 7.14 kips.

EXAMPLE 11.14 A 50-mm-diameter stainless steel rod, 254 mm in length, is placed inside a rolled brass tube having an inside diameter of 50 mm and an outside diameter of 75 mm, as shown in Figure 11.12. The member is subjected to an axial tensile load of 300 kN. The two materials are so connected to the rigid plates as to act as a unit, and both will elongate the same amount. Calculate (a) the stress developed in the steel and in the brass, (b) the magnitude of load supported by each material, and (c) the elongation of the system.

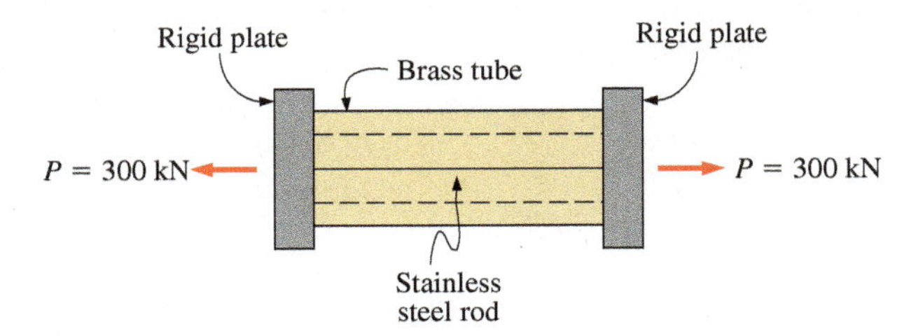

FIGURE 11.12 Load diagram.

Solution The cross-sectional area of the steel rod is calculated from

$$A_{ST} = 0.7854(50 \text{ mm})^2 = 1964 \text{ mm}^2$$

And, for the brass tube,

$$\begin{aligned} A_{BR} &= 0.7854(d_o^2 - d_i^2) \\ &= 0.7854\left[(75 \text{ mm})^2 - (50 \text{ mm})^2\right] = 2454 \text{ mm}^2 \end{aligned}$$

The modular ratio is

$$n = \frac{E_{ST}}{E_{BR}} = \frac{200{,}000 \text{ MPa}}{97{,}000 \text{ MPa}} = 2.06$$

(a) From Equation (11.7),

$$P = \sigma_{BR}(nA_{ST} + A_{BR})$$

from which

$$\begin{aligned} \sigma_{BR} &= \frac{P}{nA_{ST} + A_{BR}} = \frac{300 \times 10^3 \text{ N}}{[2.06(1964) + 2454] \text{ mm}^2} \\ &= 46.2 \text{ MPa (tension)} \end{aligned}$$

And, from Equation (11.5),

$$\sigma_{ST} = n\sigma_{BR} = 2.06(46.2 \text{ MPa}) = 95.2 \text{ MPa (tension)}$$

(b) The load supported by each material can be calculated from

$$\begin{aligned} P_{BR} &= \sigma_{BR}A_{BR} = (46.2 \text{ N/mm}^2)(2454 \text{ mm}^2) \\ &= 113{,}400 \text{ N} \\ &= 113.4 \text{ kN} \end{aligned}$$

$$\begin{aligned} P_{ST} &= \sigma_{ST}A_{ST} = (95.2 \text{ N/mm}^2)(1964 \text{ mm}^2) \\ &= 187{,}000 \text{ N} \\ &= 187.0 \text{ kN} \end{aligned}$$

(c) Elongation of the system can be obtained using the previous results for either the brass or the steel. Using the steel values, the strain is calculated from

$$\epsilon = \frac{s_{ST}}{E_{ST}} = \frac{95.2 \text{ MPa}}{200{,}000 \text{ MPa}} = 0.000476$$

The total deformation (elongation) is then calculated from

$$\delta = \epsilon L = (0.000476)(254 \text{ mm}) = 0.121 \text{ mm}$$

11.4 STRESS CONCENTRATION

As previously shown, when an axially loaded prismatic member is subjected to a tensile load, a tensile stress (P/A) will develop. This stress is assumed to be uniformly distributed over the cross-sectional area perpendicular to the direction of the load. This uniform stress distribution will occur on all planes except those in the vicinity of any load application points. Should abrupt changes (sometimes called *stress-raisers*) exist in the cross section of the member, however, large irregularities in the uniform stress distribution will develop. Examples of stress-raisers in flat axially loaded members and the resulting stress distributions are shown in Figure 11.13.

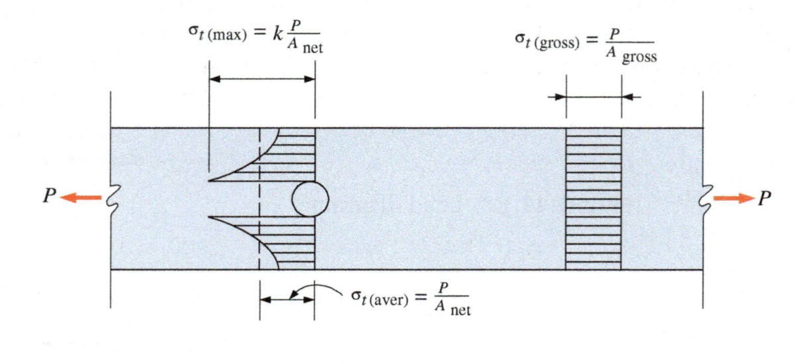

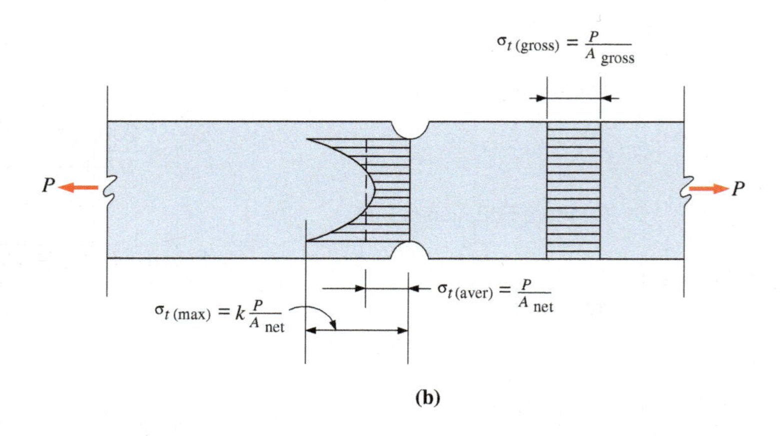

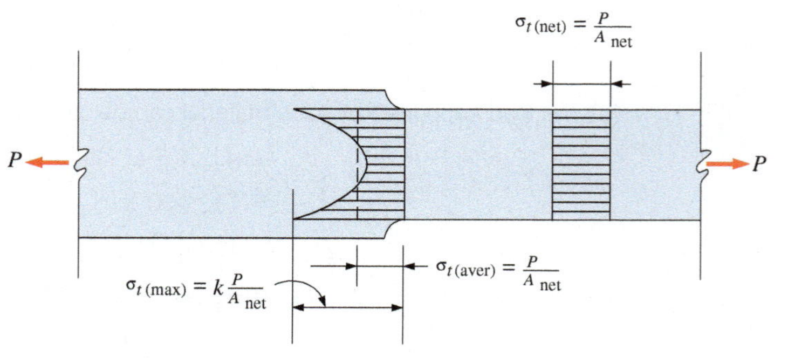

FIGURE 11.13 Tensile stress distributions.

Figure 11.13a indicates the presence of a circular hole centrally located in the member. Figure 11.13b indicates the presence of symmetrically placed semicircular side notches in the member. Figure 11.13c shows a member composed of two segments of different lateral dimensions joined with fillets. It is generally accepted that the tensile stress distribution as shown at the reduced cross section will return to a uniform stress distribution a short distance away.

In each case, as may be observed, a localized stress concentration develops immediately adjacent to the stress-raiser. This stress concentration represents a maximum tensile stress, the magnitude of which may be considerably in excess of an average tensile stress acting on the net cross-sectional area in the plane of the reduced cross section.

The calculation of this maximum tensile stress has been simplified with an experimentally determined *stress concentration factor* denoted by the letter k. The value of this factor depends on the geometric proportions of the member as well as on the type and size of the stress-raiser. Information on stress concentration factors is available in the technical literature in the form of tables and curves.

Figure 11.14 shows curves indicating stress concentration factors for flat axially loaded members with three types of change in cross section. With a stress concentration factor from the curves, the maximum tensile stress immediately adjacent to the stress-raiser can be calculated from

$$\sigma_{t(\max)} = k\left(\frac{P}{A_{\text{net}}}\right) \tag{11.9}$$

where $\sigma_{t(\max)}$ = the maximum tensile stress immediately adjacent to the discontinuity (psi, ksi) (Pa, MPa)

k = the stress concentration factor

P = the applied axial tensile load (lb, kips) (N)

A_{net} = the net cross-sectional area in the plane of the reduced cross section (in.^{2}) (m^2, mm^2)

This high stress concentration is not necessarily dangerous for ductile metals because plastic yielding and subsequent stress redistribution will occur. In the case of brittle materials, however, stress concentrations are much more serious. Cracks may occur in the material in areas of high localized stress due to the inability of the brittle material to deform plastically. Ductile materials, when subjected to repetitive-type loads, fare only slightly better. Should stress concentrations be unavoidable in a member, a reduction in allowable stresses should be considered.

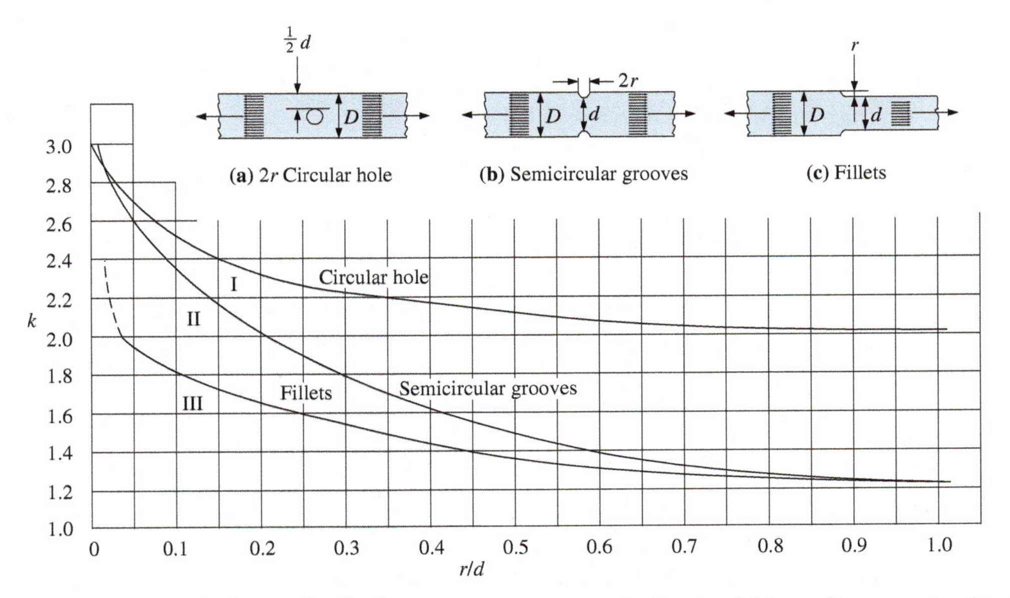

FIGURE 11.14 Stress concentration factors for flat bars (Source: M. M. Frocht, "Factors of Stress Concentration Photoelastically Determined," *ASME Journal of Applied Mechanics* 2:A67–A68 (1935).

EXAMPLE 11.15 A $\frac{3}{4}$-in.-diameter hole is drilled on the centerline of a flat steel bar as shown in Figure 11.15. The bar is subjected to a tensile load of 4000 lb. Calculate the average stress in the plane of the reduced cross section and the maximum tensile stress immediately adjacent to the hole.

Solution The radius of the hole is 0.375 in. The net width d (at the hole) is 2.5 in. − 0.75 in. = 1.75 in. Therefore,

$$\frac{r}{d} = \frac{0.375 \text{ in.}}{1.75 \text{ in.}} = 0.214$$

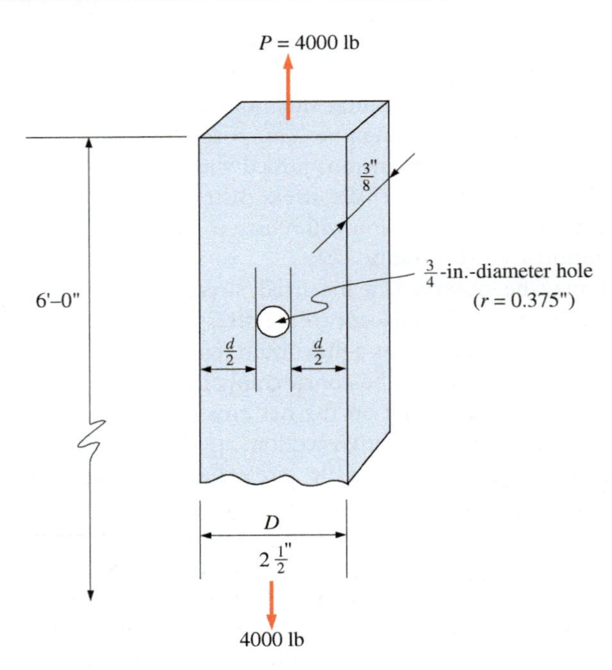

FIGURE 11.15 Long bar with centrally located hole.

From Figure 11.14, the value of k is approximately 2.3. The net area in the plane of the reduced cross section is then calculated:

$$A_{\text{net}} = (1.75 \text{ in.})(0.375 \text{ in.}) = 0.656 \text{ in.}^2$$

The average stress is then calculated as

$$\sigma_t = \frac{P}{A_{\text{net}}} = \frac{4000 \text{ lb}}{0.656 \text{ in.}^2} = 6100 \text{ psi}$$

and the maximum tensile stress is

$$\sigma_{t(\max)} = k\left(\frac{P}{A_{\text{net}}}\right) = 2.3(6100 \text{ psi}) = 14{,}030 \text{ psi}$$

EXAMPLE 11.16 A 1-m-long flat bar 75 mm wide by 6 mm thick has a circular hole 25 mm in diameter, centrally located. With an allowable tensile stress of 165 MPa, calculate the axial tensile load that may be applied to the bar.

Solution

$$\frac{r}{d} = \frac{12.5 \text{ mm}}{75 \text{ mm} - 25 \text{ mm}} = 0.25$$

From Figure 11.14, the value of k is approximately 2.3. Then

$$\sigma_{t(\max)} = k\left(\frac{P}{A_{\text{net}}}\right)$$

Since the maximum stress $\sigma_{t(\max)}$ is not to exceed 165 MPa, the upper limit of the average stress at the hole (P/A_{net}) can be calculated from Equation (11.9):

$$\frac{P}{A_{\text{net}}} = \frac{\sigma_{t(\max)}}{k} = \frac{165 \text{ MPa}}{2.3} = 71.7 \text{ MPa}$$

from which

$$P = A_{net}(71.7\ \text{MPa}) = (75\ \text{mm} - 25\ \text{mm})(6\ \text{mm})(71.7\ \text{MPa})$$
$$= 21{,}500\ \text{N}$$
$$= 21.5\ \text{kN}$$

11.5 STRESSES ON INCLINED PLANES

When a prismatic member is subjected to a uniaxial tensile or compressive force, maximum tensile and compressive stresses are developed on a plane perpendicular (normal) to the longitudinal axis of the member. This situation was shown in Figure 9.9a and b. Additionally, tensile and compressive stresses of lesser intensity, along with shear stresses, will also be developed on planes inclined to the cross section. If a member is composed of a material that does not exhibit the same strength in all directions, this consideration may be important.

In Figure 11.16a, a prismatic member is subjected to an axial tensile force P. The member is cut into two parts by plane CD in such a manner that the normal to the plane makes an angle θ with the longitudinal axis of the member. The bottom part is shown as a free body in Figure 11.16b. At section C–D the force P is resolved into two components, one parallel to plane C–D and the other perpendicular (normal) to plane C–D. The components have values of $P \sin\theta$ and $P \cos\theta$, respectively.

If the cross-sectional area of the member is denoted as A, then the area of the inclined plane C–D is equal to $A/\cos\theta$. The parallel component of the applied force acting on the inclined plane causes a shear stress τ' of

$$\tau' = \frac{P \sin\theta}{(A/\cos\theta)} = \frac{P}{A}\sin\theta\cos\theta = \frac{P}{2A}\sin 2\theta \qquad \textbf{(11.10)}$$

This expression becomes a maximum when $\sin 2\theta$ is a maximum, which occurs when $\sin 2\theta$ is 1. This maximum, in turn, occurs when $2\theta = 90°$ and $\theta = 45°$. When $\sin 2\theta = 1$, the previous expression becomes

$$\tau'_{(\max)} = \frac{P}{2A} \qquad \textbf{(11.11)}$$

and represents a maximum shear stress developed on a plane inclined at 45° to the cross section of the member.

The perpendicular component of the applied force acting on the inclined plane C–D causes a tensile stress of

$$\sigma_n = \frac{P\cos\theta}{(A/\cos\theta)} = \frac{P}{A}\cos^2\theta \qquad \textbf{(11.12)}$$

This expression becomes a maximum when $\cos^2\theta$ is a maximum. The maximum value of $\cos^2\theta$ is 1 and occurs when $\theta = 0°$. Note in Equation (11.11) that the maximum shear stress on the 45° inclined plane equals one-half the maximum normal tensile stress on the 0° plane.

Hence, we can conclude that at any point in a body subjected to load, there are different stresses or combinations of stresses developed on planes of different inclinations. If the applied force P were compressive, the same expressions would be obtained.

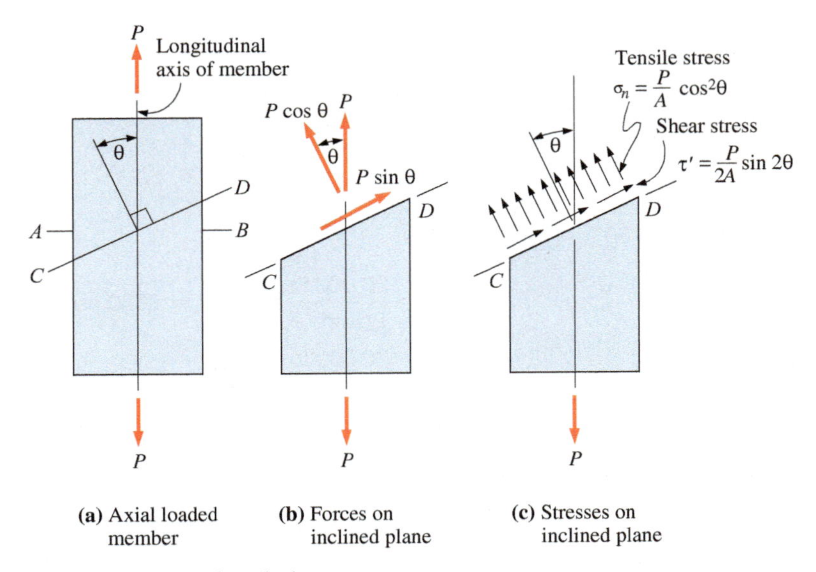

FIGURE 11.16 Forces and stresses on an inclined plane.

EXAMPLE 11.17 A square steel bar, 1 in. by 1 in. in cross section, is subjected to a tensile load of 20,000 lb, as shown in Figure 11.17. (a) Compute the shear stress and the tensile stress on inclined planes whose normals make angles of 30°, 45°, and 60° with the longitudinal axis of the bar. (b) Compute the maximum tensile stress and indicate on which plane this would be developed. (Refer to Figure 11.16.)

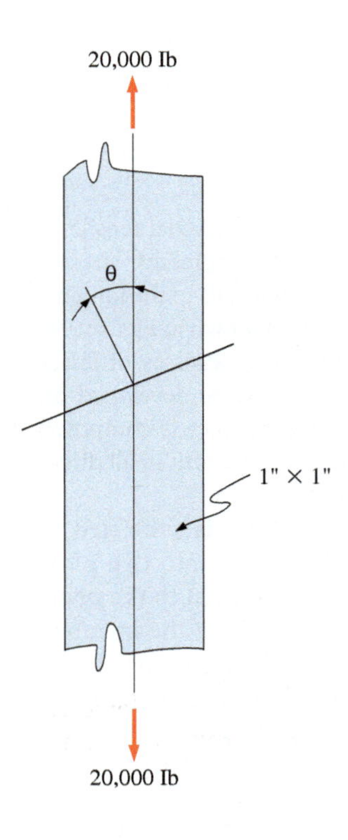

FIGURE 11.17

Solution **(a)** Normal line at 30°:

$$\tau' = \frac{P}{2A}\sin 2\theta = \frac{20{,}000\ \text{lb}}{2(1\ \text{in.}^2)}\sin 60^\circ = 8660\ \text{psi}$$

$$\sigma_n = \frac{P}{A}\cos^2\theta = \frac{20{,}000\ \text{lb}}{1\ \text{in.}^2}\cos^2 30^\circ = 15{,}000\ \text{psi}$$

Normal line at 45°:

$$\tau' = \frac{20{,}000\ \text{lb}}{2(1\ \text{in.}^2)}\sin 90^\circ = 10{,}000\ \text{psi}$$

$$\sigma_n = \frac{20{,}000\ \text{lb}}{1\ \text{in.}^2}\cos^2 45^\circ = 10{,}000\ \text{psi}$$

Normal line at 60°:

$$\tau' = \frac{20{,}000\ \text{lb}}{2(1\ \text{in.}^2)}\sin 120^\circ = 8660\ \text{psi}$$

$$\sigma_n = \frac{20{,}000\ \text{lb}}{1\ \text{in.}^2}\cos^2 60^\circ = 5000\ \text{psi}$$

(b) The maximum tensile stress occurs when $\theta = 0^\circ$, since $\cos 0^\circ = 1$. (This is a transverse section.) The expression can then be written as:

$$\sigma_n = \frac{P}{A}\cos^2\theta = \frac{P}{A}(1) = \frac{20{,}000\ \text{lb}}{1\ \text{in.}^2} = 20{,}000\ \text{psi}$$

EXAMPLE 11.18 A steel specimen having a diameter of 12.83 mm is subjected to a 45-kN axial tensile load. (a) Compute the maximum shear stress. (b) Compute the tensile stress acting on the same plane at which the shear stress is a maximum. (Refer to Figure 11.16.)

Solution **(a)** The maximum shear stress will occur on a plane whose normal is inclined at 45° to the longitudinal axis of the member:

$$\tau'_{(max)} = \frac{P}{2A}\sin 2\theta = \frac{45{,}000\text{ N}}{2(0.7854)(12.83\text{ mm})^2}\sin 90^\circ = 174.0\text{ MPa}$$

(b) The tensile stress on the plane whose normal is inclined at 45° to the longitudinal axis is calculated from

$$\sigma_n = \frac{P}{A}\cos^2\theta = \frac{45{,}000\text{ N}}{0.7854(12.83\text{ mm})^2}\cos^2 45^\circ = 174.0\text{ MPa}$$

11.6 SHEAR STRESSES ON MUTUALLY PERPENDICULAR PLANES

In this section we show that at any point in a stressed member where shear stresses exist on a plane there must simultaneously exist shear stresses of equal intensity on a perpendicular plane.

Let us consider a member subjected to a shear force as shown in Figure 11.18a. The shear force results in a shear stress τ_1 acting on the right-hand face of the member.

An infinitesimal element ABCD at the face of the member is then removed as a free body in equilibrium (see Figure 11.18b). Element ABCD is assumed to have a thickness of unity (1). If there is a shear stress of τ_1 acting on the right-hand face of the element, the shear force on this face is $\tau_1(h)(1)$. Furthermore, there must be an equal and oppositely directed shear force on the left face, since the summation of vertical forces must equal zero.

The two vertical forces described constitute a couple. To prevent rotation of the element, there must exist another couple made up of shear forces $\tau_2(w)(1)$ acting on the top and bottom faces of element ABCD. These two couples must be numerically equal and acting in opposite directions, as shown in Figure 11.18b.

Taking moments of the forces with respect to point A and equating the two couples gives

$$\tau_1(h)(1)(w) = \tau_2(w)(1)(h)$$

from which

$$\tau_1 = \tau_2$$

Therefore, we see that a shear stress on a plane cannot exist alone but induces an equal shear stress on a perpendicular plane.

In this discussion, no stresses other than shear stresses were considered. This condition is identified as a state of *pure shear*. Pure shear can exist in an element of a member that is being sheared (such as in a punching operation shown in Figure 9.11) or in circular shafts and in some beams. These latter two are discussed in later chapters.

11.7 TENSION AND COMPRESSION CAUSED BY SHEAR

Figure 11.19a shows a stressed element subjected to pure shear. In Section 11.6, it was proved that shear stresses on mutually perpendicular planes are equal. The shear stresses

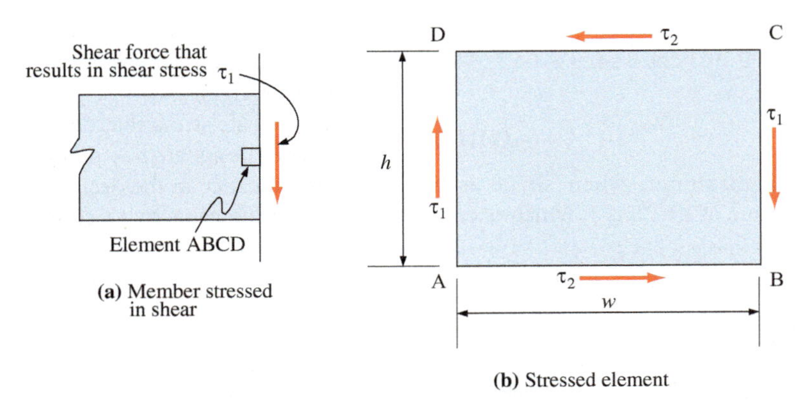

FIGURE 11.18 Shear stresses on perpendicular planes.

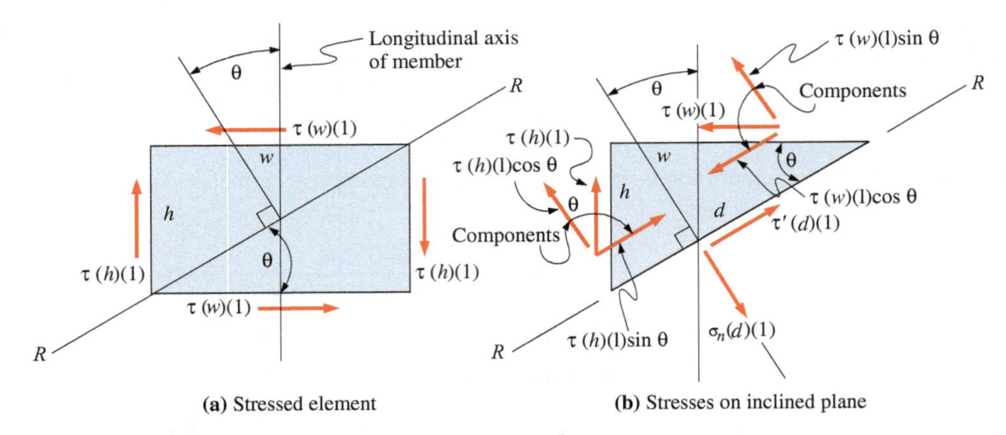

FIGURE 11.19 Stress at a point.

τ are shown on each of the four faces of the element. The arrows shown in Figure 11.19a, however, represent shear forces, which are obtained by multiplying the shear stress by the area on which the shear stress acts in each case.

A section *R–R* is cut through the element from corner to corner, and the upper left half is shown as a free body in Figure 11.19b. Angle θ is defined by the dimensions w and h. Angle θ is also the angle between the perpendicular to the diagonal plane and the longitudinal axis of the member. If d is the length of the diagonal, the forces acting on the diagonal surface are the shear force $\tau'(d)(1)$ and the tensile force $\sigma_n(d)(1)$. In these expressions, τ' is the shear stress acting on the diagonal and σ_n is the tensile stress acting normal to the diagonal.

Since the free body showing the cut element must be in equilibrium, the summation of forces perpendicular to the diagonal surface must equal zero. (Recall that the cutting of small elements into smaller free bodies is justified on the basis of equilibrium. If a body is in equilibrium, every portion of that body, no matter how small, must also be in equilibrium.) Taking the algebraic summation of forces perpendicular to the diagonal surface gives

$$\sigma_n(d)(1) = \tau(h)(1)\cos\theta + \tau(w)(1)\sin\theta$$

Dividing throughout by $(d)(1)$ yields

$$\sigma_n = \tau\left(\frac{h}{d}\right)\cos\theta + \tau\left(\frac{w}{d}\right)\sin\theta$$

Since $\sin\theta = h/d$ and $\cos\theta = w/d$,

$$\begin{aligned}\sigma_n &= \tau\sin\theta\cos\theta + \tau\cos\theta\sin\theta \\ &= 2\tau\sin\theta\cos\theta \\ \sigma_n &= \tau\sin 2\theta \qquad \textbf{(11.13)}\end{aligned}$$

This expression becomes a maximum when $\sin 2\theta$ is a maximum. The maximum value of $\sin 2\theta$ is 1, which occurs when $2\theta = 90^\circ$ and $\theta = 45^\circ$.

If it is desired to compute the shear stress on the diagonal surface, an algebraic summation of forces may be taken parallel to that surface. The signs of the forces in this summation are obtained by observation from Figure 11.19b:

$$\tau(d)(1) = \tau(w)(1)\cos\theta - \tau(h)(1)\sin\theta$$

Dividing throughout by $(d)(1)$ gives

$$\tau = \tau\left(\frac{w}{d}\right)\cos\theta - \tau\left(\frac{h}{d}\right)\sin\theta$$

Since $\sin\theta = h/d$ and $\cos\theta = w/d$,

$$\begin{aligned}\tau' &= \tau\cos^2\theta - \tau\sin^2\theta \\ &= \tau(\cos^2\theta - \sin^2\theta) \\ \tau' &= \tau\cos 2\theta \qquad \textbf{(11.14)}\end{aligned}$$

Note that when $\theta = 45^\circ$, the shear stress on the diagonal plane is zero.

A similar situation occurs on the other diagonal plane of the element subjected to pure shear. The stress normal to this diagonal plane, however, will be compressive rather than tensile. An equilibrium analysis will show that Equation (11.13) will yield the magnitude of this compressive stress. It will also reach a maximum value when $\theta = 45^\circ$.

Thus, for a stressed element subjected to pure shear, tensile and compressive stresses are developed on diagonal planes as a result of shear stresses. The maximum tensile and compressive stresses will develop on planes at angles of 45° with the applied shear stress and will be of an intensity equal to the shear stress.

The tensile stress that develops on the diagonal surface is generally designated as *diagonal tension*. This stress is of great significance in the design of reinforced concrete members, since the capacity of concrete to resist tension is very limited.

EXAMPLE 11.19 A shear stress (pure shear) of 8000 psi exists in a member. Compute the tensile normal stress developed on diagonal planes whose normals are oriented at angles of 15°, 30°, and 45° with the longitudinal axis as shown in Figure 11.20.

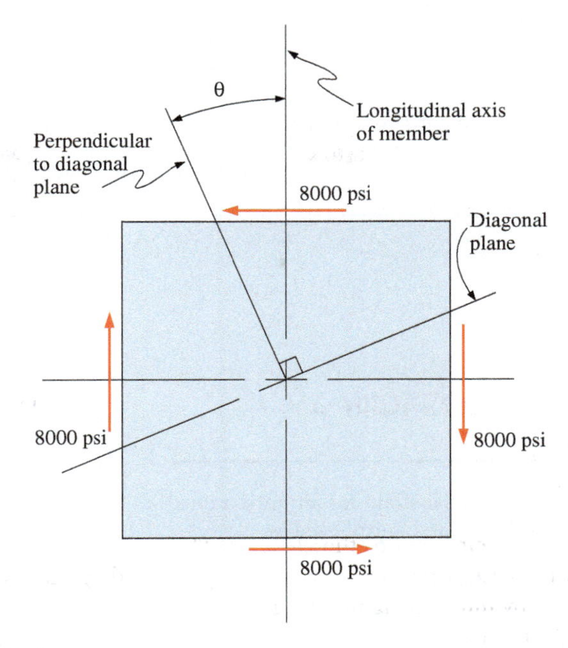

FIGURE 11.20 Stressed element.

Solution

$$\sigma_n = \tau \sin 2\theta$$

For $\theta = 15°$,

$$\sigma_n = (8000 \text{ psi}) \sin 30° = 4000 \text{ psi}$$

For $\theta = 30°$,

$$\sigma_n = (8000 \text{ psi}) \sin 60° = 6930 \text{ psi}$$

For $\theta = 45°$,

$$\sigma_n = (8000 \text{ psi}) \sin 90° = 8000 \text{ psi}$$

EXAMPLE 11.20 An element taken from a wood block as shown in Figure 11.21a is subjected to shear stresses on horizontal and vertical planes as shown in Figure 11.21b. The wood grain is at an angle of 20° with the axis of the member. Compute the shear stress and compressive normal stress developed on plane *A–A*, which is parallel to the grain.

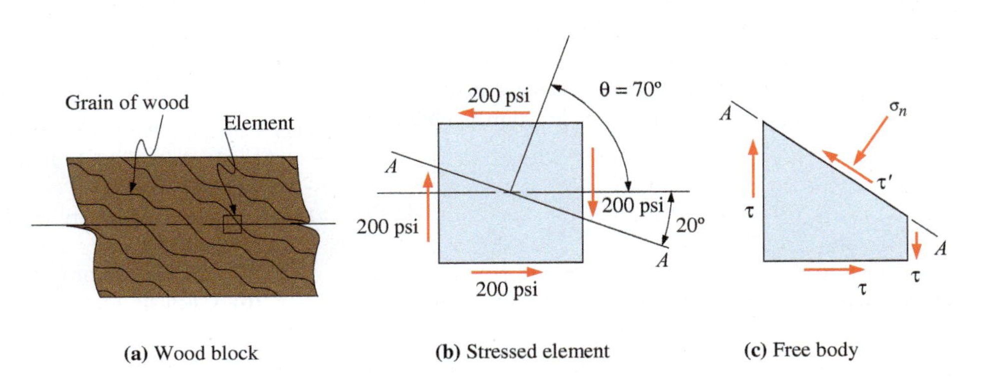

FIGURE 11.21 Wood block subjected to pure shear.

Solution The compressive normal stress on plane A–A is

$$\sigma_n = \tau \sin 2\theta = (200 \text{ psi}) \sin 140° = 128.6 \text{ psi}$$

The shear stress on plane A–A is

$$\tau' = \tau \cos 2\theta = (200 \text{ psi}) \cos 140° = 153.2 \text{ psi}$$

SUMMARY BY SECTION NUMBER

11.1 The ratio of lateral strain to axial strain for an unrestrained member is called *Poisson's ratio* and is expressed as

$$\mu = \frac{\text{transverse strain}}{\text{axial strain}} \qquad \textbf{(11.1)}$$

The relationship between modulus of rigidity G, modulus of elasticity E, and Poisson's ratio μ is expressed by

$$G = \frac{E}{2(1 + \mu)} \qquad \textbf{(11.2)}$$

11.2 Materials commonly used in engineering tend to change their dimensions (either expanding or contracting) when subjected to temperature changes. The linear coefficient of thermal expansion (α) is a measure of the change in length per unit of length per degree temperature change. For a body free to expand or contract, the magnitude of the dimensional change is

$$\delta = \alpha L(\Delta T) \qquad \textbf{(11.3)}$$

If a body is restrained against dimensional change due to temperature change, an internal stress (temperature stress) will develop. This value is calculated from

$$\sigma = \alpha E(\Delta T) \qquad \textbf{(11.4)}$$

11.3 In members made of two (or more) materials so connected that each deforms equally under load, the stresses developed in the two materials are proportional to their moduli of elasticity. The total load carried can be expressed as

$$P = A_A(n\sigma_B) + A_B\sigma_B = \sigma_B(nA_A + A_B) \qquad \textbf{(11.7)}$$

where $n = E_A/E_B$, E_A is the larger of the two moduli of elasticity, and σ_B is the lower of the two stresses.

11.4 Localized stress concentrations occur at abrupt changes in the cross section of axially loaded prismatic members in tension. The maximum stress at this location is obtained from

$$\sigma_{t(\max)} = k\left(\frac{P}{A_{\text{net}}}\right) \qquad \textbf{(11.9)}$$

11.5 In axially loaded members, shear stresses and normal stresses are developed on planes inclined to the cross section of the member. The shear stress (developed parallel to the inclined plane) is

$$\tau' = \frac{P}{2A}\sin 2\theta \qquad \textbf{(11.10)}$$

The normal stress (developed normal to the inclined plane) is

$$\sigma_n = \frac{P}{A}\cos^2\theta \qquad \textbf{(11.12)}$$

11.6 At any point in a stressed member where a shear stress exists on a plane, there must also exist a shear stress of equal intensity on a perpendicular plane.

11.7 At any point in a stressed member that is subjected to a pure shear condition, tensile and compressive stresses are developed on diagonal planes as a result of the shear stresses. The stress normal to the diagonal surface is

$$\sigma_n = \tau \sin 2\theta \qquad \textbf{(11.13)}$$

The stress parallel to the diagonal surface is

$$\tau' = \tau \cos 2\theta \qquad \textbf{(11.14)}$$

PROBLEMS

For the following problems, unless noted otherwise, refer to Appendices F and/or G for necessary mechanical properties. Unless noted otherwise, assume gray cast iron, as appropriate.

Section 11.1 Poisson's Ratio

11.1 A 2-in.-diameter AISI 1020 steel rod is 10 ft long. Under an applied load, the rod elongates by 0.48 in.
(a) Compute the axial (longitudinal) strain in the bar.
(b) If the transverse deformation of the rod is 0.0024 in., compute Poisson's ratio for the material.

11.2 A rectangular ASTM A36 steel bar 2 in. by 6 in. in cross section is subjected to an axial tensile load of 300,000 lb. The proportional limit of the steel is 34,000 psi. Compute the change in the transverse 6-in. dimension.

11.3 Calculate Poisson's ratio for a cast iron that has a modulus of elasticity E of 110 GPa and a modulus of rigidity G of 44 GPa.

11.4 Modulus of elasticity, modulus of rigidity, and Poisson's ratio are interrelated.
(a) Calculate G for $E = 30{,}000{,}000$ psi and $\mu = 0.30$.
(b) Calculate μ for $E = 16{,}000{,}000$ psi and $G = 5{,}800{,}000$ psi.

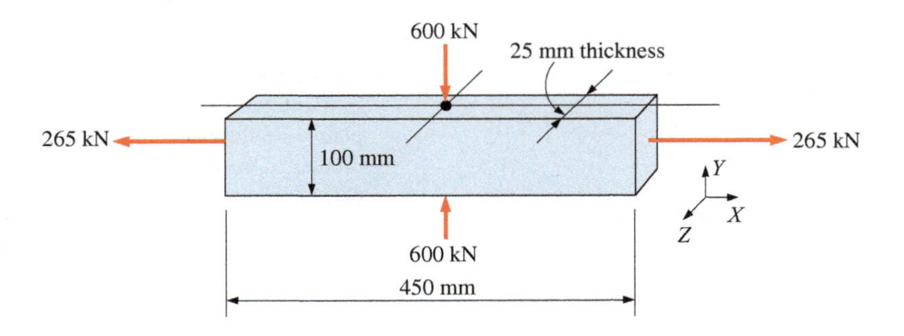

PROBLEM 11.5

11.5 Compute all the dimensional changes for the steel bar when subjected to the loads shown. The proportional limit of the steel is 230 MPa.

Section 11.2 Thermal Effects

11.6 A surveyor's steel tape is exactly 100 ft long between end markings at 70°F. What error is made in measuring a distance of 1000 ft when the temperature of the tape is 32°F?

11.7 An aluminum wire is stretched between two rigid supports. If the tensile stress in the wire is 5000 psi at 68°F, what is it at 32°F and 90°F? Assume that the proportional limit is 35,000 psi.

11.8 A 1-m-long copper bar is placed between two rigid, unyielding walls, as shown, with only one end attached. Calculate the stress developed in the bar due to a temperature rise of 50°C.

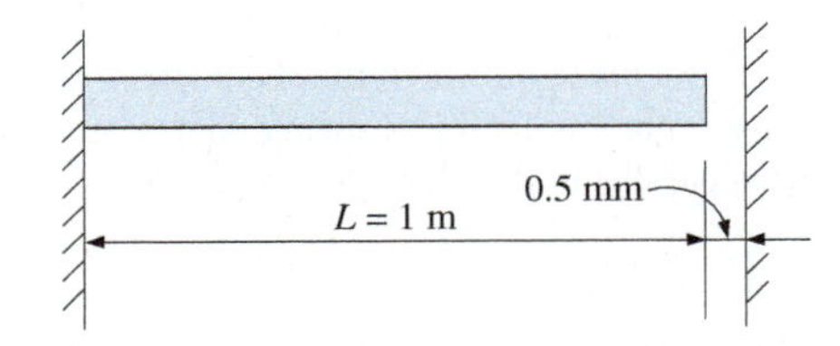

PROBLEM 11.8

11.9 A concrete roadway pavement is placed in 60-ft-long sections. An expansion joint between the ends of the sections is established as $\frac{1}{2}$ in. at 70°F. Calculate the width of the joint at temperatures of 30°F and 110°F.

11.10 An ASTM A36 structural steel beam 4.00 m long is placed between two rigid supports when the temperature is 16°C. Compute the stress developed in the beam if the temperature rises to 50°C. Assume that the proportional limit is 235 MPa.

Section 11.3 Members Composed of Two or More Components

11.11 A 4-in.-by-8-in. short wood post is reinforced on all four sides by ASTM A36 steel plates. Two plates are 4 in. by $\frac{1}{8}$ in. thick and two plates are 8 in. by $\frac{1}{4}$ in. thick. Using nominal dimensions, calculate the maximum axial compressive load that the member can safely carry. The wood is southern pine. The allowable stress is 20,000 psi for the steel.

11.12 A short post, 150 mm by 150 mm, of Douglas fir, is reinforced by the addition of two ASTM A36 steel plates. The cross section is shown. Assume that the components are of equal lengths and are so connected that they act as a single unit and deform equally. Use the nominal dimensions of the wood, as shown, and compute the allowable compressive load. Allowable compressive stress for the steel is 147 MPa.

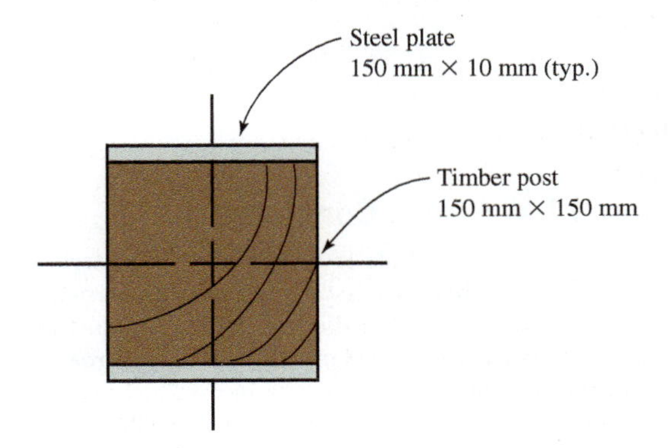

PROBLEM 11.12

11.13 The cables of a power line are copper-coated steel wire. The overall diameter of the wire is $\frac{5}{8}$ in. The steel core has a diameter of $\frac{1}{2}$ in. If the maximum tension in a wire is 8000 lb, what are the stresses in the steel and the copper?

11.14 A $5\frac{1}{2}$-in.-by-$11\frac{1}{2}$-in. Douglas fir column and a hem-fir column of the same size are bolted together to form a short composite column, as shown. What portion of a total load of 70,000 lb will each material carry?

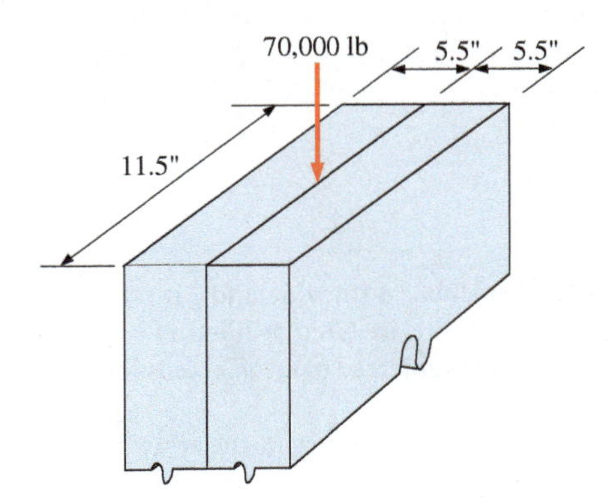

PROBLEM 11.14

11.15 For the short column shown, assuming that lateral buckling is prevented:

(a) Calculate the magnitude of the axial load P that will cause the total length of the member to decrease by 0.01 in.

(b) Calculate the compressive stress in the steel.

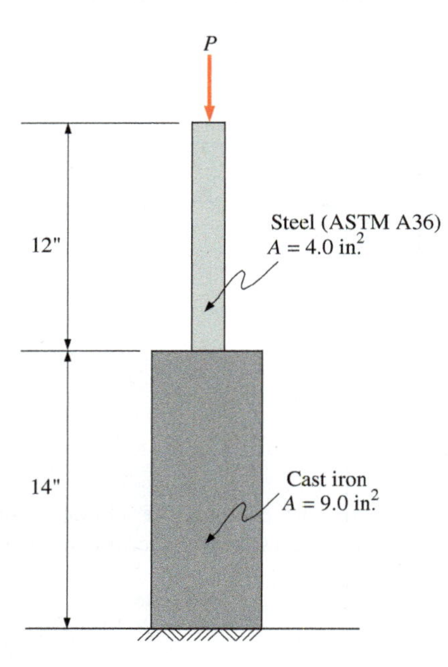

PROBLEM 11.15

Section 11.4 Stress Concentration

11.16 A 1.0-in.-diameter hole is drilled on the centerline of a long, flat steel bar that is $\frac{1}{2}$ in. thick and 4 in. wide. The bar is subjected to a tensile load of 30,000 lb. Calculate the average stress in the plane of the reduced cross section and the maximum tensile stress immediately adjacent to the hole.

11.17 A long, flat steel bar 13 mm thick and 120 mm wide has semicircular grooves as shown and carries a tensile load of 50 kN. Determine the maximum stress in the plate if
(a) $r = 8$ mm.
(b) $r = 21$ mm.
(c) $r = 38$ mm.

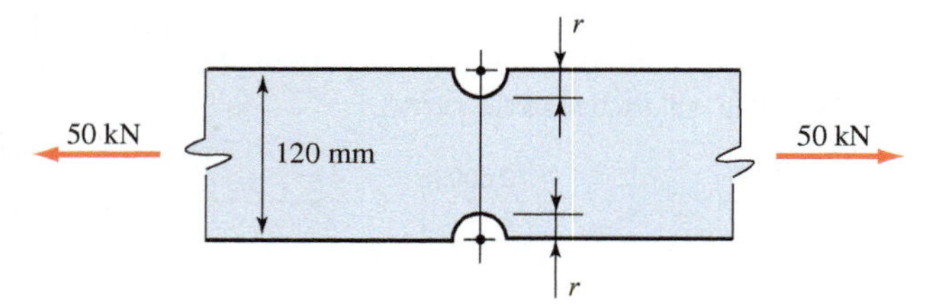

PROBLEM 11.17

11.18 A long, flat steel bar 4 in. wide and $\frac{3}{8}$ in. thick is reduced in width to 3 in. There are circular fillets of $\frac{1}{2}$ in. radius on each side. If the bar is subjected to an axial tensile load of 12,000 lb, calculate
(a) the average tensile stress in the wide portion of the bar some distance from the change in section.
(b) the average tensile stress in the narrow part of the bar.
(c) the maximum tensile stress adjacent to the circular fillet.

11.19 A long, flat steel bar 5 in. wide and $\frac{3}{8}$ in. thick has a circular hole 2 in. in diameter, centrally located. The allowable tensile stress for the steel is 22,000 psi. Calculate the axial tensile load that may be applied to the bar.

Section 11.5 Stresses on Inclined Planes

11.20 An aluminum specimen of circular cross section, 0.50 in. in diameter, ruptured under a tensile load of 12,000 lb. The plane of failure was found to be at 48° with a plane perpendicular to the longitudinal axis of the specimen.
(a) Compute the shear stress on the failure plane.
(b) Compute the maximum tensile stress.
(c) Compute the tensile stress on the failure plane.

11.21 A prismatic bar, 50 mm by 75 mm in cross section, is subjected to an axial tensile load of 50 kN.
(a) Compute the maximum shear stress developed in the bar.
(b) Compute the shear stress and tensile stress on a plane whose normal is inclined at 70° to the line of action of the axial load.

11.22 A short, square steel bar, 1 in. by 1 in. in cross section, is subjected to an axial tensile load of 8000 lb. Compute the shear stress and tensile stress acting on a plane whose normal is inclined at 60° to the longitudinal axis of the member.

11.23 A concrete cylinder, 6-in. in diameter, was loaded in axial compression and failed on a plane inclined at approximately 45° with the axis of the cylinder. The load on the cylinder at failure was 67 kips. Calculate the shear stress on the failure plane.

Section 11.7 Tension and Compression Caused by Shear

11.24 An element in a member is subjected to a pure shear of 10,000 psi.
(a) Sketch the element and show the shear forces.
(b) Determine the tensile normal stress on a plane at 35° with the horizontal. Show this plane on the sketch.

11.25 For the element of Problem 11.24:
(a) Locate the plane on which the shear stress is 6000 psi.
(b) Verify your answer by showing that a summation of forces parallel to the plane does equal zero.

Computer Problems

For the following computer problems, any appropriate software may be used. Input prompts should fully explain what is required of the user (the program should be user-friendly). The resulting output should be well labeled and self-explanatory.

11.26 Write a program that will compute the allowable load for a short composite member composed of two materials. User input is to be area, modulus of elasticity, and allowable stress for each material. The output should include the allowable load, the final stresses in each material, and an indication as to which of the two materials controls has reached its allowable stress.

11.27 Write a computer program that will solve for the shear stress and tensile stress developed in the plate of Problem 11.68. Have the program generate a table of these stresses for angles ranging from 0° to 75° in increments of 5°.

11.28 A metal bar is subjected to a temperature change. Write a program that will calculate
(a) the total change in the length of the bar and
(b) the stress developed in the bar if the ends were rigidly fixed and the bar were short enough to prevent compression buckling. The user should be prompted to input the original length of the bar, the temperature change, and the material of which the bar is made (steel, aluminum, or copper).

11.29 A surveyor's steel tape is exactly 100.000 ft long at 68°F. Write a program that will generate a table of errors (to 0.001 ft) that will occur in measurements from 70 ft to 100 ft (in 5-ft increments) and in the temperature range of 40°F to 80°F (in increments of 5°F).

Supplemental Problems

11.30 A 50-mm-diameter ASTM A36 steel rod is subjected to an axial tensile load of 270 kN. The proportional limit of this steel is 234 MPa.

(a) Compute the axial (longitudinal) strain and the transverse strain.

(b) If the rod is 1 m long, compute the total elongation.

11.31 A 4-ft-long square ASTM A36 steel bar, 2 in. by 2 in., is subjected to an axial tensile load of 64,000 lb parallel to its length. Calculate the dimensional changes in the bar's lateral and longitudinal dimensions.

11.32 A concrete test cylinder is 150 mm in diameter and 300 mm in length. During an axial compression test, the diameter increased by 0.0125 mm and the length decreased by 0.275 mm. Compute the value of Poisson's ratio and the magnitude of the compressive load. The concrete has an ultimate compressive strength of 21.0 MPa.

11.33 A 14-in.-long steel rod, $1\frac{1}{2}$ in. in diameter, was subjected to an axial tensile load of 60,000 lb in a universal testing machine. It was observed that a gage length of 2 in. near the midpoint of the rod increased in length by 0.0023 in. and the rod diameter decreased by 0.00043 in. Assume the steel proportional limit to be 34,000 psi. Calculate the modulus of elasticity and Poisson's ratio for the steel.

11.34 Determine the change in the diameter of an ASTM A36 steel rod subjected to an axial compressive load that results in a compressive stress of 30,000 psi. The original diameter is 3 in. and the proportional limit is 34,000 psi.

11.35 The steel bar shown in Figure 11.3 (see Example 11.2) is subjected to a compressive load of 648,000 lb in the z direction in addition to the forces shown. Calculate the new dimensions of the bar.

11.36 Compute the change in the thickness of the ASTM A36 steel bar when subjected to the loads shown. Assume a proportional limit of 34,000 psi.

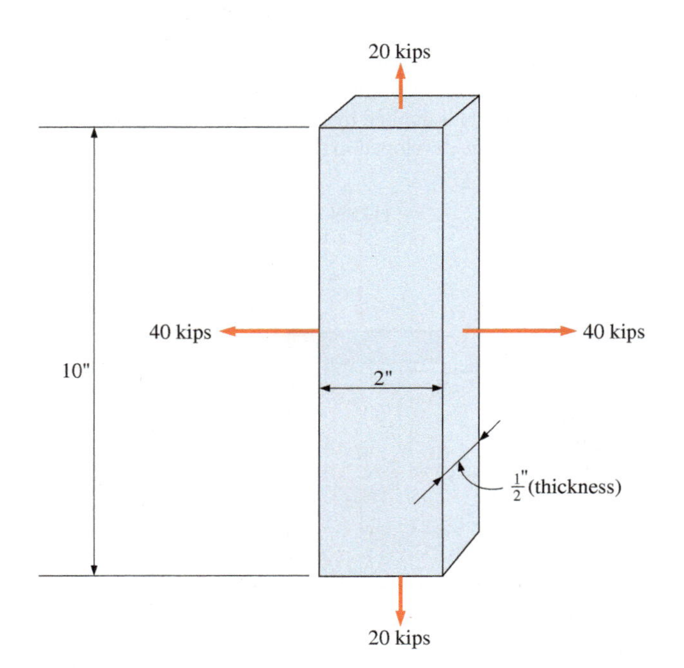

PROBLEM 11.36

11.37 The steel rails of a railroad track are laid in the winter at a temperature of 15°F with gaps of 0.01 ft between the ends of the rails. Each rail is 33 ft long. At what temperature will the rails touch end to end? What stress will result in the rails if the temperature rises to 110°F?

11.38 A rolled brass rod and a steel rod are secured to unyielding supports, as shown. Determine the temperature change required to close the 0.03-in. gap.

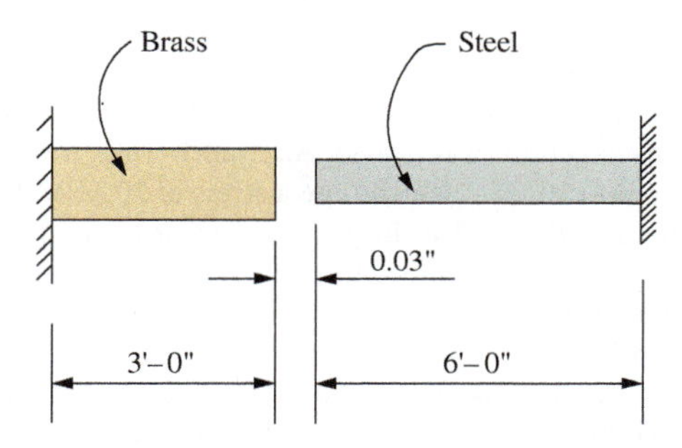

PROBLEM 11.38

11.39 A 3-m-long steel member is set snugly between two walls and then heated so that the temperature rise is 27°C. If each wall yields 0.38 mm, what is the compressive stress developed in the member?

11.40 The distance between two fixed points on a missile range in the desert was measured at a temperature of 112°F with a steel tape that was 100.00 ft long at 68°F. The distance was measured as 2208.56 ft. What is the distance corrected for temperature? What distance would have been recorded if the temperature at the time of measurement had been 20°F?

11.41 A steel truss is loaded as shown. The cross-sectional area of member CD is 2500 mm^2. Find the total change in length for member CD due to the applied loads shown and a temperature drop of 28°C.

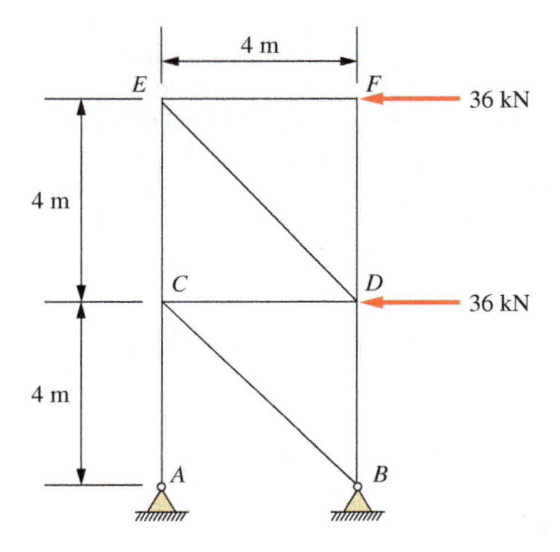

PROBLEM 11.41

11.42 For the truss of Problem 5.25, the members are of steel. Member CD has an area of 1 in.^{2} and member FD has an area of 3 in.^{2} Determine the total change in length for these members due to the applied loads and a temperature increase of 40°F.

11.43 A surveyor's steel tape has a cross-sectional area of 0.016 in.^{2} and is exactly 100.00 ft long at 70°F under a tensile pull of 15.0 lb. Compute the pull required to make it 100.00 ft long at 25°F.

11.44 A 1-in.-diameter ASTM A36 steel tie rod, 30 ft long, is used to tie together two walls of a building. The rod is tensioned to 10,000 lb. Compute the tensile stress in the rod if the temperature rises by 30 degrees and if it falls by 50 degrees. Assume that the walls do not move.

11.45 A copper wire is held taut between two supports 20 ft apart. How much may the temperature drop before a stress of 15,000 psi is reached

(a) if the supports are immovable?

(b) if one of the supports yields 0.05 in. while the temperature is dropping?

11.46 A horizontal steel member is anchored at each end. Just prior to installation it was heated. After anchoring, it was allowed to cool to 20°C, at which time a stress of 70 MPa developed. If the member is 3 m in length, to what temperature was it heated to have developed this stress upon cooling?

11.47 Three vertical steel wires are loaded as shown. At a temperature of 68°F, the bottom ends of the wires were at the same elevation with no load on the bar. A downward load of 500 lb was then centrally applied on the bar and the temperature was increased by 100 degrees. What load is carried by each wire assuming the load bar remains horizontal?

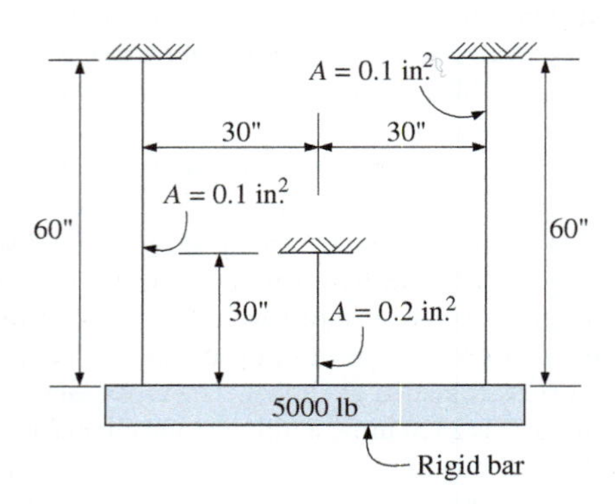

PROBLEM 11.47

11.48 Assume for Problem 11.47 that the same load is applied on the bar. To what temperature must the wires be heated before the load is entirely carried by the middle wire?

11.49 The rod shown is firmly attached to rigid supports. If there is initially no stress in the rod, compute the stress in each material if the temperature drops by 100 degrees and there is no movement of the supports.

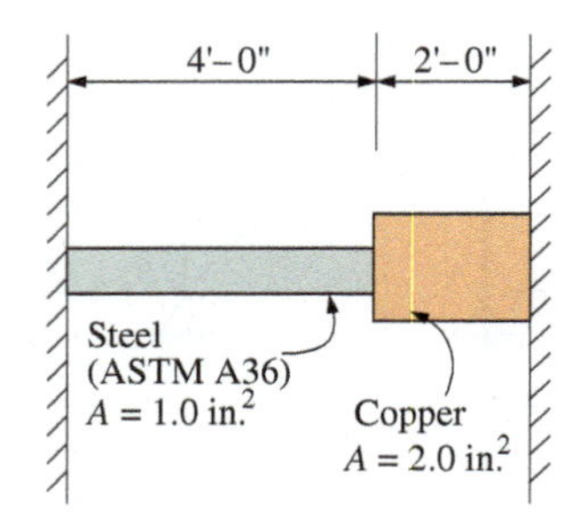

PROBLEM 11.49

11.50 A structural steel bar 100 mm in width and 12.7 mm in thickness has a copper bar 100 mm wide and 3.2 mm thick securely attached on each side. The assembly acts as a single unit. Calculate the stress and strain in each material caused by an axial tensile load of 225 kN.

11.51 A redwood timber member having a 16-in.-square cross section is reinforced with an ASTM A36 structural steel angle (L5 × 5 × $\frac{1}{2}$) at each corner, as shown. The materials are so attached as to act as a single unit. Compute the total allowable axial compressive load that may be applied through a rigid cap plate that covers the entire cross-sectional area. Use dressed dimensions and assume the materials have the same length. The allowable compressive stress for the steel is 20,000 psi.

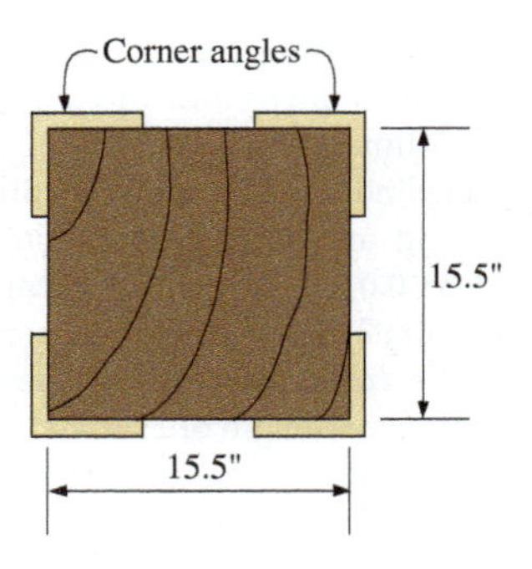

PROBLEM 11.51

11.52 A steel pipe has an outside diameter of 12 in. and an inside diameter of 11 in. A 3-in.-diameter solid cast-iron cylinder is placed in the pipe and is concentric with the pipe. The space between the two is filled with concrete (ultimate compressive strength = 3000 psi). Compute the stress in each material when the assembly is subjected to an axial load of 400,000 lb.

11.53 A short 14-in.-square concrete pier is reinforced with four longitudinal #7 bars ($\frac{7}{8}$ in. diameter). The ultimate compressive strength of the concrete is 4000 psi and the modulus of elasticity for the steel bars is 30,00,000 psi. The pier supports a load of 100 kips. Compute the portion of the load carried by the concrete and the portion carried by the steel. Compute the stress in each material.

11.54 An ASTM A36 steel rod, 375.06 mm long and having a cross-sectional area of 1250 mm^2, is loosely inserted into a copper tube, as shown. The copper tube has a cross-sectional area of 1800 mm^2 and is 375 mm long. If an axial load of 112 kN is applied by means of a rigid cap plate, what stresses will be developed in the two materials?

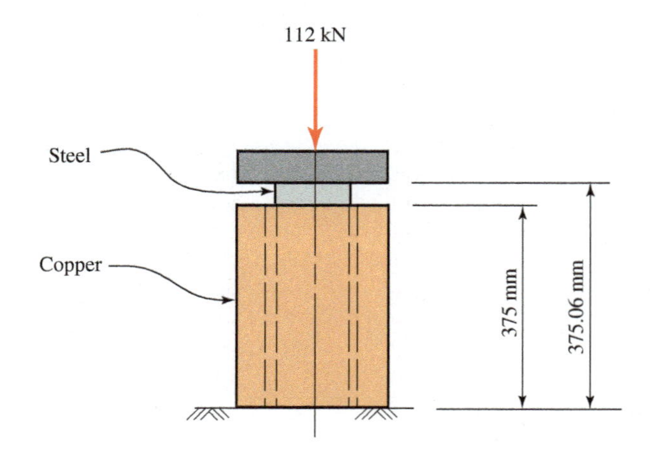

PROBLEM 11.54

11.55 An aluminum rod with an area of 1.5 in.^{2} and an AISI 1020 steel rod with an area of 1.0 in.^{2} support a rigid bar as shown. The bar is horizontal prior to the application of a 24,000 lb load. Find the distance a and the stresses in the two rods if the bar remains horizontal after the load has been applied.

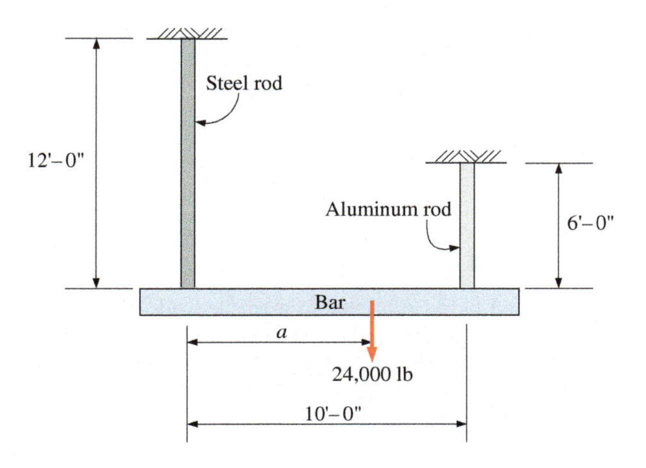

PROBLEM 11.55

11.56 A solid brass cylinder with a cross-sectional area of 3226 mm^2 is placed inside a steel tube having a cross-sectional area of 5162 mm^2. Both the cylinder and the tube are 216 mm long. They stand on end on a flat rigid surface and they support a rigid block. A load of 580 kN is applied to the block. Calculate the stress in each material.

11.57 Three $\frac{1}{4}$-in.-diameter wires are symmetrically spaced and are 20 ft long, as shown. The outer wires are steel and the middle wire is bronze. The wires support a rigid horizontal steel box weighing 2000 lb. Calculate

(a) the stresses in the wires.

(b) the total elongation of the wires.

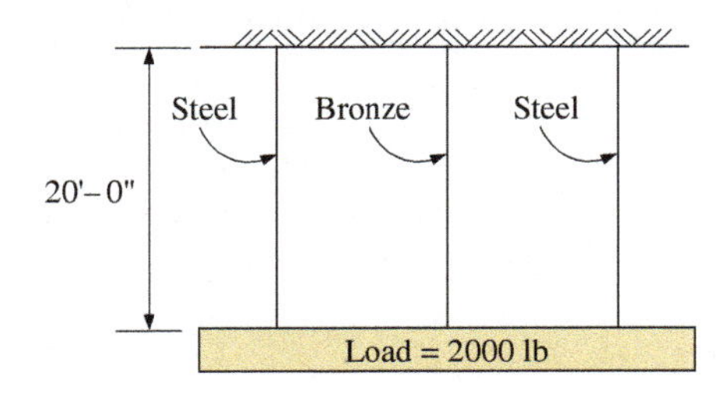

PROBLEM 11.57

11.58 Rework Problem 11.57 with outer wires of aluminum, a middle wire of steel, and a 1500 lb box.

11.59 Three rods support a weight, as shown. The supports are rigid and the load is applied uniformly through a rigid block. If the nuts on the three threaded rods were in the same horizontal plane before and after the applied loading, calculate the load carried by each rod.

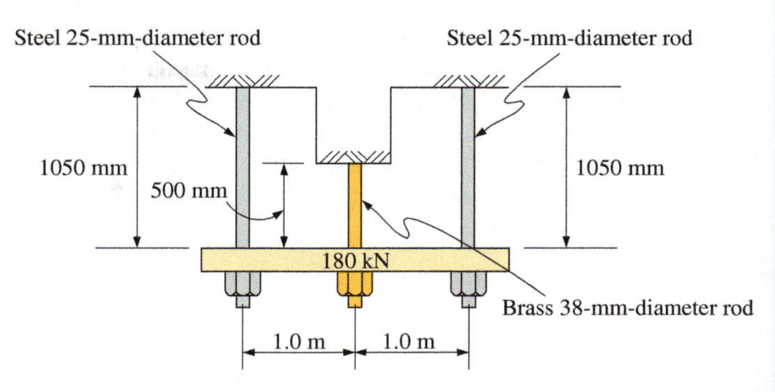

PROBLEM 11.59

11.60 A flat steel bar 4 in. wide and $\frac{1}{2}$ in. thick must be reduced to a $2\frac{1}{2}$-in. width, as shown. The allowable tensile stress is 15,000 psi. If the applied axial tensile load is 12,000 lb, compute the minimum size fillet radius r that may be used.

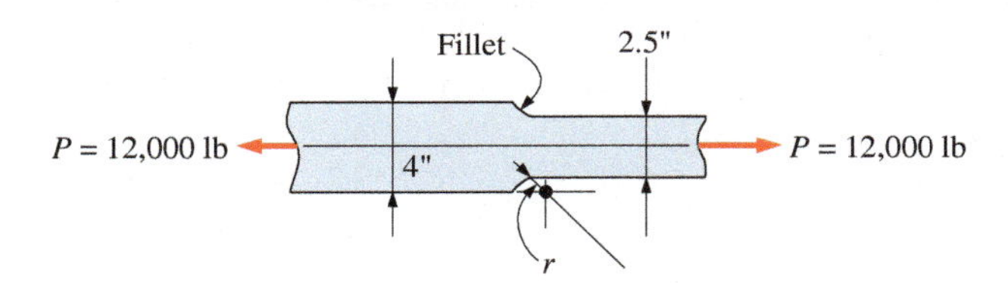

PROBLEM 11.60

11.61 A 19-mm-diameter hole is drilled on the centerline of a long, flat steel bar. The bar is 10 mm by 75 mm in cross section and is subjected to a tensile load of 18 kN. Calculate the average tensile stress in the plane of the reduced cross section and the maximum tensile stress immediately adjacent to the hole.

11.62 A flat bar is $\frac{3}{8}$ in. thick and has a centrally located drilled hole of diameter D. The bar is subjected to a 4000 lb tensile load. Compute the maximum tensile stress adjacent to the hole

(a) if the bar is 2 in. wide and $D = 0.25$ in.

(b) if the bar is 2.5 in. wide and $D = 0.50$ in.

(c) if the bar is 3 in. wide and $D = 1.00$ in.

11.63 A long, flat steel bar 125 mm wide and 10 mm thick has a circular hole of 38 mm diameter centrally located. The allowable tensile stress for the steel is 150 MPa. Calculate the axial tensile load that may be applied to the bar.

11.64 A short 6-in.-diameter compression member is made of a material with an ultimate shear strength of 5000 psi and a compressive strength of 12,000 psi. Compute the axial compressive load that may be applied before failure occurs.

11.65 A short timber compression block of rectangular cross section, 150 mm by 200 mm, has an allowable compressive stress of 6 MPa and an allowable shear stress of 1.6 MPa. Calculate the allowable load that may be applied to the member.

11.66 A rectangular block of wood, 2 in. by 2 in. in cross section, has an ultimate shear strength of 1200 psi. The block is subjected to an increasing tensile load until failure by shear occurs along the grain, which is at an angle of 10° with the longitudinal axis of the member, as shown. Compute the magnitude of the load P at which failure occurs. Use nominal dimensions.

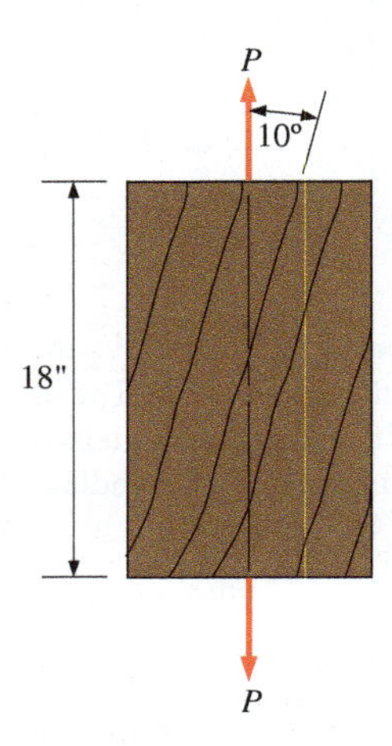

PROBLEM 11.66

11.67 A 25-mm-diameter rod is subjected to an axial tensile load of 80 kN. Compute

- **(a)** the normal and shear stresses developed on an inclined plane at an angle of 30° with the cross section of the rod.
- **(b)** the maximum normal and shear stresses developed in the rod.

11.68 The rectangular plate shown is subjected to a uniaxial tensile stress of 2000 psi. Compute the shear stress and the tensile stress developed on a plane forming an angle of 30° with the longitudinal axis of the member. (*Hint*: Assume a cross-sectional area of unity.)

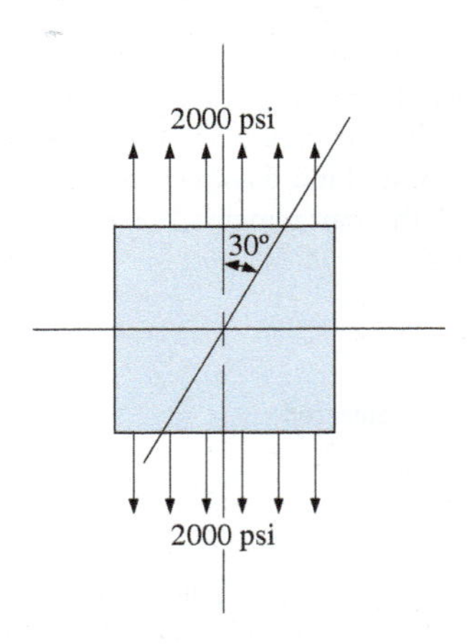

PROBLEM 11.68

11.69 A wood block, subjected to a tensile load, fails on a plane whose perpendicular is at an angle of 30° with the longitudinal axis, as shown in Figure 11.16a. The shear stress on the failure plane is 35 MPa. Compute the diagonal tension (normal stress) on this plane.

11.70 A shear stress (pure shear) of 5000 psi exists on an element.

- **(a)** Determine the maximum tensile and compressive stresses caused in the element due to this shear.
- **(b)** Sketch the element showing the planes on which the maximum tensile and compressive stresses act.

11.71 A shaft in a speed-reduction mechanism is loaded in torque. An element at the surface of the shaft is stressed to 9200 psi (pure shear).

- **(a)** Determine the maximum tensile and compressive stresses caused in the element due to this shear.
- **(b)** Sketch the element showing the planes on which the maximum tensile and compressive stresses act.

11.72 An axially loaded 50-mm-by-75-mm steel bar has a shear stress of 138 MPa developed on a plane at an angle of 50° with the cross section of the bar. Calculate

- **(a)** the applied axial load on the bar.
- **(b)** the normal stress developed on this plane.

CHAPTER TWELVE

TORSION IN CIRCULAR SECTIONS

LEARNING OBJECTIVES

Upon completion of this chapter, readers will be able to:

- Discuss the application of torque on an object or member.
- Calculate torsional shear stresses due to applied torque.
- Calculate the required side of a solid circular shaft to resist torsional forces.
- Discuss torsional rigidity and its relation to the allowable angle of twist, or flexibility.
- Calculate the power generated by a rotating shaft of circular cross section in both U.S. Customary and SI units.

12.1 INTRODUCTION

In previous chapters, our discussion centered on the analysis and design of members subjected to axial (concentric) loads or loads that caused direct shear stresses. In this chapter, we turn our attention to members subjected to a twisting action caused by a couple or a twisting moment. The twisting action, applied in a plane perpendicular to the longitudinal axis of the member, is commonly called *torque*, which is the terminology we use in this chapter. An example of this type of action may be observed in Figure 12.1, where the jaws of the bench vise are tightened by applying forces to the handle. A torque is applied to the threaded screw of the vise, turning it, which causes the jaws to tighten. An applied torque such as this is called an *external torque*.

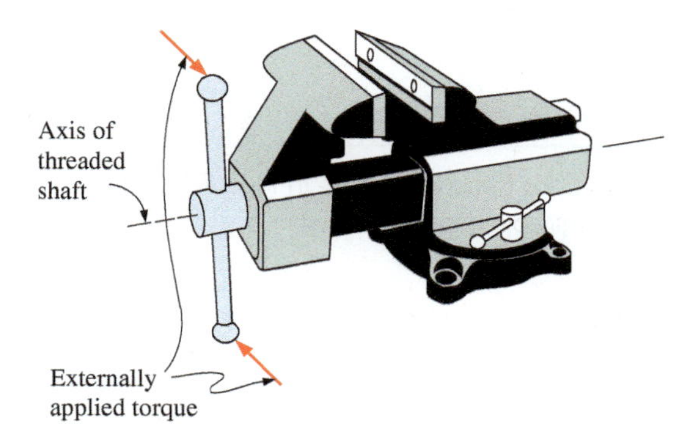

FIGURE 12.1 Bench vise.

12.2 MEMBERS IN TORSION

First, we consider members in static equilibrium when subjected to a twisting action caused either by a pair of externally applied equal and oppositely directed couples acting in parallel planes or by a single external couple applied to a member that has one end fixed against rotation. The fixed end, in effect, furnishes an internal resisting torque. The portion of the member between the two externally applied couples, or between the single externally applied couple and the fixed end, is subjected to a torque and is said to be in torsion, or under torsional load. This situation would occur in the screw of the bench vise when the jaws were fully tightened and the forces still applied to the handle. Next, we consider constant-speed rotating shaft-and-pulley systems that are in "dynamic" or "steady-state" equilibrium.

The case of two equal externally applied couples acting in opposite directions on parallel planes perpendicular to the longitudinal axis of the member is shown in Figure 12.2a. The magnitude of the torque in the bar is simply equal to that of one of the two couples (Fd). In Figure 12.2b the bar has been cut and the right-hand portion shown as a free body. It is apparent that for equilibrium there must be an internal resisting torque equal to the external torque.

In Figure 12.2c the bar is rigidly fixed against rotation at one end, and only one external couple is applied. In this case, equilibrium exists due to an equal and opposite internal

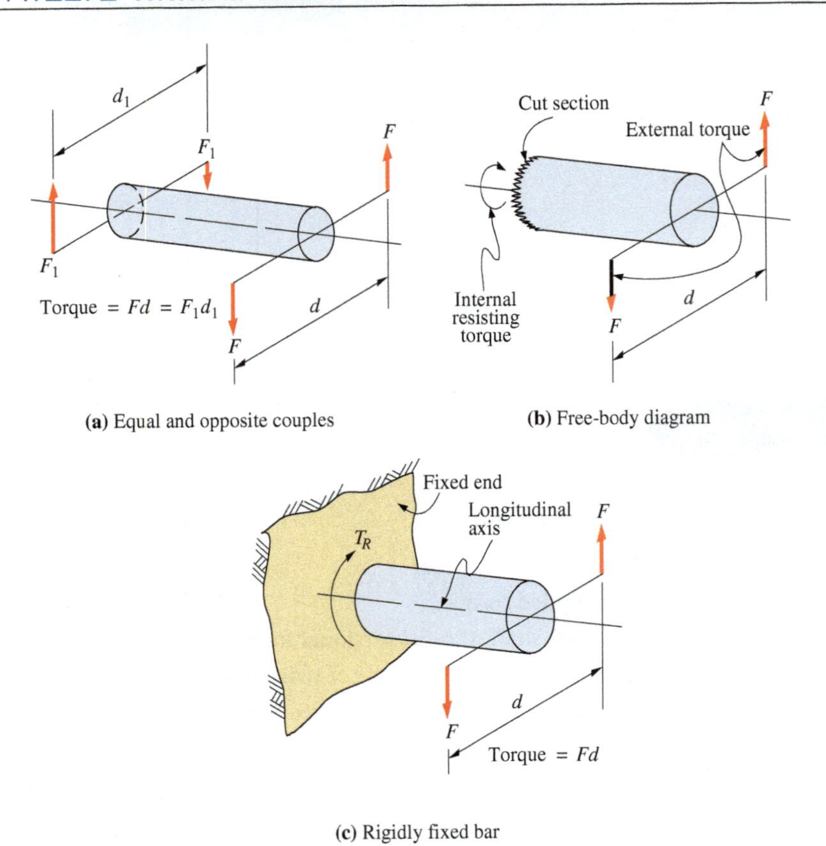

FIGURE 12.2 Members in torsion.

resisting torque at the fixed end. The magnitudes of the external and internal torques are, again, Fd. In fact, the free-body diagram of Figure 12.2b would apply for either of the two members shown.

The couple, or torque, is generally expressed in units of lb-in. in the U.S. Customary System and N · m in the SI. In the field of machine design, members subjected to torques or couples are generally shafts used for the transmission of power. Shafts are usually circular in cross section and may be either solid or hollow. The torque to which a shaft is subjected is generally applied through the use of pulleys or gears.

A shaft will commonly have several pulleys mounted on it. One of the pulleys (sometimes called the *driver pulley*) provides the torque to drive the shaft. The shaft then transmits torque to the other pulleys (sometimes called *power take-off pulleys*) that, in turn, provide the necessary torque to drive machines or equipment. In such applications the torque along the shaft length varies, depending on location and the magnitudes of the torques associated with the various types of pulleys. The torque at any cross section may be determined with the use of a free-body diagram. The internal torque must be equal to the algebraic sum of external torques on either side of the cross section in question.

One objective of this chapter is to determine the relationship between the torque and the resulting stresses and strains in shafts. The shafts will be assumed weightless, thereby making the effect of any bending negligible.

EXAMPLE 12.1 Calculate the internal torque at sections *R–R* and *S–S* for the shaft shown in Figure 12.3. The shaft is acted upon by the four torques indicated. Assume negligible bearing friction.

Solution Note that pulley B is the driver pulley. The other pulleys are power take-off pulleys. The torque of 2400 lb-in. at *B* is balanced by the three torques of 600 lb-in., 1000 lb-in., and 800 lb-in. at *A*, *C*, and *D*, respectively, which are opposite in direction. Therefore, the entire system may be thought of as being in steady-state equilibrium, that is, neither gaining nor losing speed. If the system as a whole is in equilibrium, every segment of the system must also be in equilibrium.

To determine the torque at section *R–R*, cut section *R–R* perpendicular to the longitudinal axis of the shaft anywhere between pulleys A and B and consider the left portion a free body, as shown in

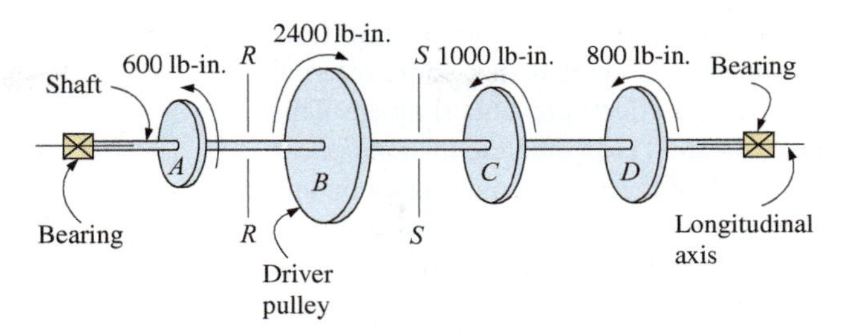

FIGURE 12.3 Shaft and pulleys.

Figure 12.4. For equilibrium, the summation of the torques must equal zero ($\Sigma T = 0$). This is the same as saying that the externally applied torque must be balanced by an internal resisting torque that is numerically equal but opposite in direction.

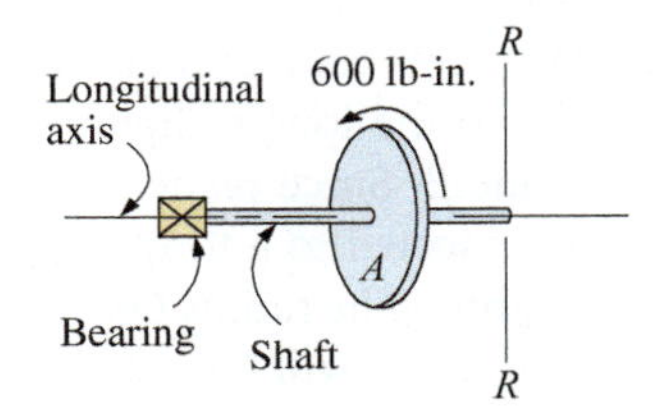

FIGURE 12.4 Free body.

Since the externally applied torque for this free body is 600 lb-in. counterclockwise, when observing from the left end of the shaft, the internal resisting torque between pulleys A and B must also be 600 lb-in. (but clockwise). A sign convention could be defined to describe the actual direction of the internal resisting torque, but for our purposes the use of the terms *clockwise* and *counterclockwise* are adequately descriptive.

Using the same approach to calculate the torque at section *S–S*, between pulleys B and C, cut section *S–S* and consider the left portion of the shaft a free body. This situation is shown in Figure 12.5.

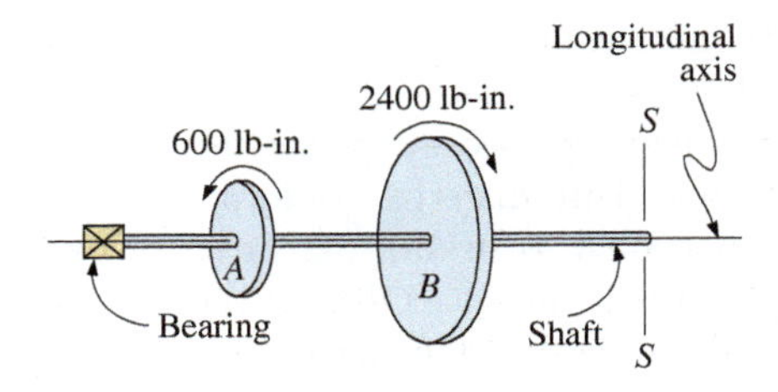

FIGURE 12.5 Free body.

Applying $\Sigma T = 0$, the internal resisting torque between pulleys B and C must be equal to the externally applied torque, which is clockwise. Therefore,

$$T_{\text{int}} = T_{\text{ext}} = (2400 \text{ lb-in.}) - (600 \text{ lb-in.}) = 1800 \text{ lb-in.}$$

which is a counterclockwise torque when observing from the left end of the shaft.

A similar approach results in an internal resisting torque of 800 lb-in. between pulleys C and D and zero torque between pulley D and the frictionless end bearing.

12.3 TORSIONAL SHEAR STRESS

Let us consider a torsionally loaded member of circular cross section fixed against rotation at one end and subjected to a torque at the other end, as shown in Figure 12.2c. Since couples cause neither bending nor direct tension or compression, this condition of loading develops pure shear stresses on each cross-sectional plane that lies between the couple and the fixed end.

If the torsionally loaded member is assumed to be made up of a series of thin plates bonded together, each thin plate tends to slip by, or shear, across the contact surface of the adjacent plate. Since the member is in equilibrium (and does not fracture), however, it is evident that some internal resistance is developed that, in effect, prevents slippage. This internal resistance (per unit area) is termed the *torsional shear stress*. The resultant, or total, of these resisting stresses on any cross-sectional plane constitutes an internal resisting torque.

Since all materials have limited shear strength, it is necessary for design purposes to develop a mathematical relationship between the torsional shear stress, the torque, and the physical properties of the member. Prior to this development, we evaluate a torsionally loaded circular member to establish a cross-sectional shear stress distribution based on stress–strain relationships.

Figure 12.6 shows a segment of a circular shaft that lies between two parallel planes A and B perpendicular to the longitudinal axis of the shaft. *CD* represents a straight line on the surface of the shaft, parallel to the longitudinal axis and extending from plane A to plane B. Since plane A is fixed against rotation, if the shaft is subjected to a torque applied at plane B, plane B will rotate slightly. The radius *OD* will then assume the position *OD′* and line *CD* will become *CD′*, part of a helix. Hence, the shear distortion of line *CD* is equal to *DD′* and the shear strain is $(DD')/L$.

Experiments have found that a plane cross section of a circular shaft will remain a plane after the shaft has twisted and also that a straight-line radius such as *OD* will remain a straight line as the shaft is twisted. Both of the foregoing conditions will hold, provided that the maximum stress developed in the shaft does not exceed the proportional limit of the material. Therefore, if we consider a line *EF* parallel to line *CD*, but halfway between the longitudinal axis and the outer surface, the distortion of line *EF*, equal to *FF′*, will be one-half that of line *CD*. The shear strain, $(FF')/L$, will also be one-half the shear strain at the outer surface.

Assuming that the developed stresses are below the proportional limit and that the shaft material conforms to Hooke's law (stress is proportional to strain), the shear stress at the location of line *EF* (radial distance *OF* from the longitudinal axis) will be equal to one-half the shear stress developed at the outer surface, line *CD*. Therefore, it may be concluded that the shear stress developed in a torsionally loaded circular shaft is proportional to the distance from the longitudinal axis of the member. Further, the stress is equal to zero at the longitudinal axis and varies linearly to a maximum at the outer surface of the shaft.

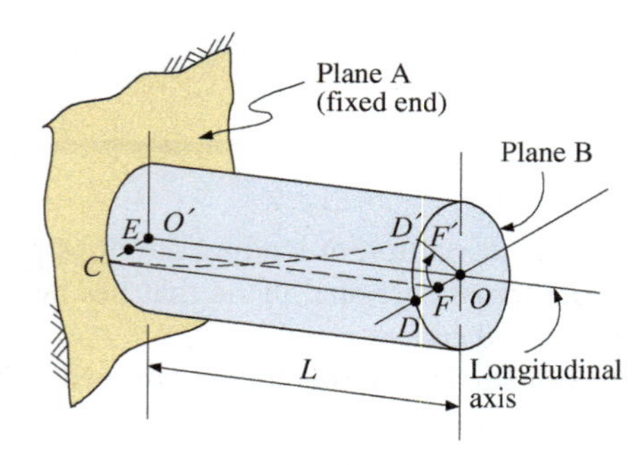

FIGURE 12.6 Circular shaft.

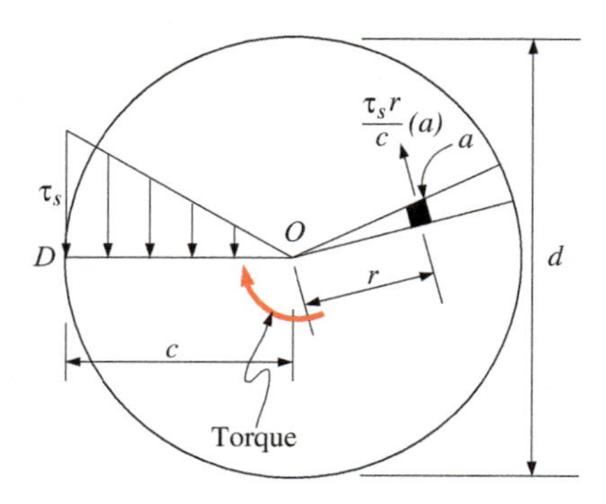

FIGURE 12.7 Cross-sectional stress distribution.

Figure 12.7 shows an enlarged view of the cross section of the shaft of Figure 12.6. The cross section is taken somewhere between plane A and plane B. Point *O* represents the centroidal longitudinal axis of the shaft. The variation of the shear stress on the cross section, which develops from an externally applied torque, is drawn using radius *OD* as a reference line. The radial distance from point *O* to the outer surface is denoted *c*. An infinitesimal area located a radial distance *r* from point *O* is denoted *a*. The shaft has a diameter *d*.

As a result of a torsional loading, a shear stress τ is developed on the cross section of the shaft at its outer surface (*c* distance from *O*). The shear stress developed at a radial distance *r* from point *O* can be calculated by proportion from

$$\frac{\tau(r)}{c}$$

Recalling that a force is equal to the product of stress and area, the resisting shear force developed on area *a* can be calculated from

$$\frac{\tau(r)}{c}(a)$$

Considering only area *a*, an internal resisting torque is developed with respect to point *O* by the resisting shear force acting on this area. This resisting torque can be written as

$$\frac{\tau(r^2)}{c}(a)$$

Finally, the total internal resisting torque about point *O*, from all the forces acting on the infinitesimal areas, can be calculated from

$$\frac{\tau}{c}\Sigma ar^2$$

The mathematical quantity of Σar^2 represents the moment of inertia of the circular shaft with respect to the centroidal longitudinal axis, which is perpendicular to the plane of the area. Recall from Chapter 8 that this is called the polar moment of inertia and is represented by the symbol *J*.

Therefore, the expression for the total internal resisting torque with respect to point *O* can be written as

$$\text{resisting torque} = \frac{\tau J}{c}$$

This total resisting torque must be equal and opposite in direction to the externally applied torque T. Therefore,

$$T = \frac{\tau J}{c} \qquad \textbf{(12.1)}$$

Solving for the shear stress gives

$$\tau = \frac{Tc}{J} \qquad \textbf{(12.2)}$$

where τ = the computed shear stress (psi, ksi) (Pa, MPa)
T = the externally applied torque (lb-in., k-in.) (N · m)
c = the radial distance from the centroidal longitudinal axis to the outer surface (in.) (mm, m)
J = the polar moment of inertia (in.^{4}) (mm^4, m^4) (see Table 3 inside the back cover of this book)

We could also rewrite Equation (12.1) to find the maximum resisting torque or allowable torque for a member. To use this formula, the allowable shear stress must be known:

$$T_R = \frac{\tau_{(\text{all})} J}{c} \qquad \textbf{(12.3)}$$

where T_R = the allowable torque (lb-in., k-in.) (N·m)
$\tau_{(\text{all})}$ = allowable shear stress (psi, ksi) (Pa, MPa)

Note that T_R (the allowable torque) is the maximum torque that can exist without exceeding the allowable shear stress. T_R is an upper limit of the torque that should be applied to the member. The maximum torque depends on the geometry and material of the member itself, not on the actual torque applied. Equations (12.1), (12.2), and (12.3) may be used for both solid and hollow circular shafts. The appropriate value for J can be obtained from Table 3.

For purposes of design of solid circular shafts, Equation (12.3) can be rewritten so as to enable direct calculation of the required diameter of a circular shaft to resist a given applied torque. We must provide a shaft diameter d such that T_R will be equal to or greater than the applied torque. Naturally, the allowable shear stress must be known.

Rewriting Equation (12.3) for the required centroidal polar moment of inertia J and substituting the applied torque T for T_R gives

$$\text{required } J = \frac{Tc}{\tau_{(\text{all})}}$$

For a solid circular shaft, $c = d/2$. Also, from Table 3,

$$J = \frac{\pi d^4}{32}$$

Substituting the two quantities gives

$$\frac{\pi d^4}{32} = \frac{T(d/2)}{\tau_{(\text{all})}}$$

Solving for the required d results in

$$\text{required } d = \sqrt[3]{\frac{16T}{\pi \tau_{(\text{all})}}} \qquad \textbf{(12.4)}$$

where all the terms have been previously defined. Note that this design expression is valid only for solid circular shafts. For hollow shafts, J depends on both the inner and the outer diameters (see Table 3). Therefore, two unknown diameters are to be determined. One solution to this problem is to establish a ratio between the two diameters, reducing the problem to one unknown diameter, which then can be determined. This approach is demonstrated in Example 12.5.

Design problems normally include the determination of a required size of the member. For circular sections this would be a diameter (or diameters, in the case of a hollow section). The natural conclusion is to select, or specify, the actual diameters to use. Shafts and bars are available in various sizes and can be machined to any size, within a reasonable tolerance, although this adds cost to the product. For our purposes, assuming machinery shafting, we adopt the practice of selecting diameters based on the increments given in Table 12.1.

TABLE 12.1 Shaft size increments

Diameter Range in Inches (mm)	Increment in Inches (mm)
$\frac{1}{2}$–$2\frac{1}{2}$ (12–63)	$\frac{1}{16}$ (1)
$2\frac{5}{8}$–4 (64–100)	$\frac{1}{8}$ (2)
$4\frac{1}{4}$–6 (101–150)	$\frac{1}{4}$ (5)

EXAMPLE 12.2 Calculate the allowable torque that can be applied to a circular shaft. The allowable shear stress is 83.0 MPa. (a) Assume that the shaft is solid and has a 150-mm diameter. (b) Assume that the shaft is hollow and has an outside diameter of 150 mm and an inside diameter of 125 mm.

Solution (a) For the solid shaft, the centroidal polar moment of inertia is calculated from

$$J = \frac{\pi d^4}{32} = \frac{\pi(150\text{ mm})^4}{32} = 49.7 \times 10^6 \text{ mm}^4$$

Calculate the allowable torque, using Equation (12.3):

$$T_R = \frac{\tau_{(\text{all})}J}{c} = \frac{(83.0\ \text{N/mm}^2)(49.7 \times 10^6\ \text{mm}^4)}{75\ \text{mm}} = 55.0 \times 10^6\ \text{N}\cdot\text{mm}$$
$$= 55.0\ \text{kN}\cdot\text{m}$$

(b) For the hollow shaft, the centroidal polar moment of inertia is calculated with an outer diameter d of 150 mm and an inner diameter d_1 of 125 mm:

$$J = \frac{\pi(d^4 - d_1^4)}{32} = \frac{\pi\left[(150\ \text{mm})^4 - (125\ \text{mm})^4\right]}{32} = 25.7 \times 10^6\ \text{mm}^4$$

Then, the allowable torque, from Equation (12.3) is

$$T_R = \frac{\tau_{(\text{all})}J}{c} = \frac{(83.0\ \text{N/mm}^2)(25.7 \times 10^6\ \text{mm}^4)}{75\ \text{mm}} = 28.4 \times 10^6\ \text{N}\cdot\text{mm}$$
$$= 28.4\ \text{kN}\cdot\text{m}$$

EXAMPLE 12.3 Calculate the maximum shear stress developed in a circular steel shaft when subjected to a torque T of 95,000 lb-in. Assume the shaft is (a) solid, with a diameter of 4 in., and (b) hollow, with an outside diameter of 4 in. and an inside diameter of 2 in.

Solution (a) For the solid shaft, the centroidal polar moment of inertia is calculated from

$$J = \frac{\pi d^4}{32} = \frac{\pi(4\ \text{in.})^4}{32} = 25.1\ \text{in.}^4$$

The maximum shear stress is calculated using Equation (12.2):

$$\tau = \frac{Tc}{J} = \frac{(95{,}000\ \text{lb-in.})(2\ \text{in.})}{25.1\ \text{in.}^4} = 7570\ \text{psi}$$

(b) For the hollow shaft, the centroidal polar moment of inertia is calculated from

$$J = \frac{\pi(d^4 - d_1^4)}{32} = \frac{\pi\left[(4\ \text{in.})^4 - (2\ \text{in.})^4\right]}{32} = 23.6\ \text{in.}^4$$

The maximum shear stress is calculated using Equation (12.2):

$$\tau = \frac{Tc}{J} = \frac{(95{,}000\ \text{lb-in.})(2\ \text{in.})}{23.6\ \text{in.}^4} = 8050\ \text{psi}$$

The shear stress varies linearly from 0 at the longitudinal center of the shaft to 8050 psi at the outer surface. The shear stress at the inner surface can be calculated by proportion, where

$$\tau = (8050\ \text{psi})\left(\frac{1}{2}\right) = 4025\ \text{psi}$$

EXAMPLE 12.4 Calculate the required diameter of a solid circular steel shaft that must resist a torque T of 20 kN · m. The allowable shear stress in the shaft is 83.0 MPa. Select a shaft diameter to use.

Solution We use Equation (12.4) for this solution, with all quantities in terms of newtons and meters:

$$T = 20\ \text{kN}\cdot\text{m} = 20 \times 10^3\ \text{N}\cdot\text{m}$$
$$\tau_{(\text{all})} = 83.0\ \text{MPa} = 83.0 \times 10^6\ \text{N/m}^2$$

Therefore,

$$\text{required } d = \sqrt[3]{\frac{16T}{\pi\tau_{(all)}}}$$

$$= \sqrt[3]{\frac{16(20 \times 10^3 \text{ N}\cdot\text{m})}{\pi(83.0 \times 10^6 \text{ N/m}^2)}} = 0.1071 \text{ m} = 107.1 \text{ mm}$$

Therefore, use a 110-mm shaft.

EXAMPLE 12.5 A circular AISI 1040 hot-rolled hollow steel shaft transmits a torque of 300,000 lb-in. under varying load conditions. Using a factor of safety of 3.0, calculate the required size of the hollow shaft if the inside diameter d_1 is to be approximately three-quarters times the outside diameter d. Select the diameters to use.

Solution First, determine the allowable shear stress. Use a shear yield strength of one-half the tensile yield strength:

$$0.5(42{,}000 \text{ psi}) = 21{,}000 \text{ psi}$$

The allowable shear stress is then calculated from

$$\tau_{(all)} = \frac{\text{shear yield strength}}{\text{F.S.}} = \frac{21{,}000 \text{ psi}}{3.0} = 7000 \text{ psi}$$

Next, calculate the centroidal polar moment of inertia in terms of the outside diameter d:

$$J = \frac{\pi(d^4 - d_1^4)}{32} = \frac{\pi(d^4 - (0.75d)^4)}{32} = 0.0671\, d^4$$

The required outside diameter is then calculated using Equation (12.3) and equating the applied torque to the allowable torque:

$$T_R = \frac{\tau_{(all)}J}{c}$$

$$300{,}000 \text{ lb-in.} = \frac{(7000 \text{ psi})(0.0671\, d^4)}{(d/2)}$$

Therefore,

$$\text{required } d^3 = \frac{300{,}000 \text{ lb-in.}}{(7000 \text{ psi})(0.0671)(2)}$$

from which the required outside diameter is obtained:

$$d = 6.84 \text{ in.}$$

The required inside diameter is then calculated as

$$d_1 = 0.75(6.84 \text{ in.}) = 5.13 \text{ in.}$$

Therefore, use a hollow steel shaft with an outside diameter of 7 in. and an inside diameter of 5 in. (Note that the inside diameter selected must be the required inside diameter or less.)

EXAMPLE 12.6 In Figure 12.8, pulleys B, C, and D are attached to the solid shaft supported on bearings at A and E. The shaft is driven at a uniform speed by pulley C. The shaft, in turn, drives pulleys B and D. The diameter of pulleys B, C, and D are 10 in., 12 in., and 14 in., respectively. Belt tensions are shown. The diameter of the shaft is $1\frac{1}{2}$ in. (a) Calculate the belt tension F_3. (b) Calculate the torque in the shaft between pulleys C and D. (c) Calculate the maximum shear stress developed from the torque of part (b).

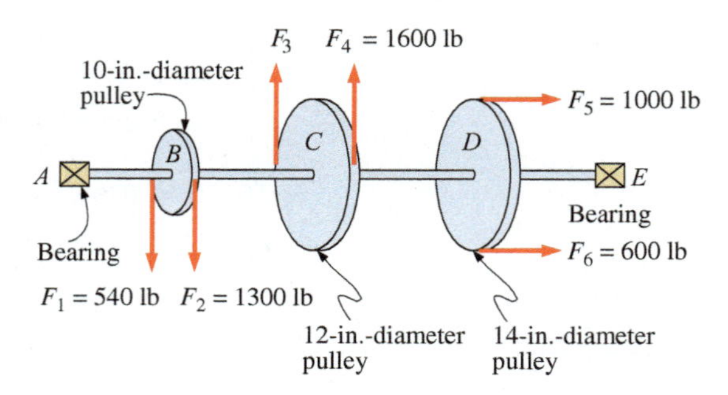

FIGURE 12.8 Shaft and pulley system.

Solution Note that all of the belt tensions are known except F_3. An imbalance between the belt tensions on either side of a pulley may be used to determine how much torque is transmitted between the pulley and the shaft. In each case, the torque equals the difference between the belt tensions times the radius of the pulley.

(a) Calculate belt tension F_3. Since the entire system is in equilibrium, the torque at C must be balanced by the torque at B and D. The torque from pulley B is

$$(F_2 - F_1)(5 \text{ in.}) = (1300 \text{ lb} - 540 \text{ lb})(5 \text{ in.}) = 3800 \text{ lb-in. (clockwise)}$$

The torque from pulley D is

$$(F_5 - F_6)(7 \text{ in.}) = (1000 \text{ lb} - 600 \text{ lb})(7 \text{ in.}) = 2800 \text{ lb-in. (clockwise)}$$

Pulley C drives the shaft. It must deliver a torque opposite in direction to the sum of the torques from pulleys B and D:

$$(2800 \text{ lb-in.}) + (3800 \text{ lb-in.}) = 6600 \text{ lb-in.}$$

The torque from pulley C is

$$(F_4 - F_3)(6 \text{ in.}) = (1600 \text{ lb} - F_3)(6 \text{ in.})$$

Equating gives

$$(1600 \text{ lb} - F_3)(6 \text{ in.}) = 6600 \text{ lb-in.}$$

from which

$$F_3 = 500 \text{ lb (belt tension)}$$

(b) Calculate the torque between pulleys C and D. A section S–S is cut between the two pulleys and the left portion taken as a free body, as shown in Figure 12.9.

Applying $\Sigma T = 0$, the external applied torque must equal the internal resisting torque:

$$(6600 \text{ lb-in.}) - (3800 \text{ lb-in.}) = T_{int} = 2800 \text{ lb-in.}$$

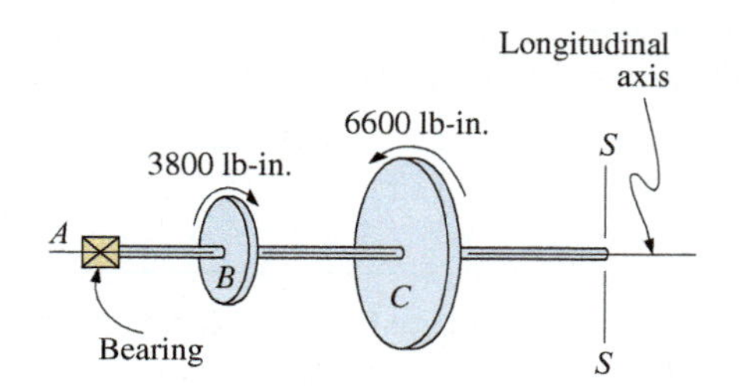

FIGURE 12.9 Free body.

(c) Calculate the maximum shear stress developed in the shaft between pulleys C and D. The shaft diameter is $1\frac{1}{2}$ in.; therefore, $c = 0.75$ in. Calculate the centroidal polar moment of inertia and the shear stress:

$$J = \frac{\pi d^4}{32} = \frac{\pi (1.5 \text{ in.})^4}{32} = 0.497 \text{ in.}^4$$

$$\tau = \frac{Tc}{J} = \frac{(2800 \text{ lb-in.})(0.75 \text{ in.})}{0.497 \text{ in.}^4} = 4255 \text{ psi}$$

12.4 ANGLE OF TWIST

If a circular shaft of length L is subjected to a torque T throughout its length, one end of the shaft will twist about its longitudinal axis relative to the other end. In Figure 12.10, part of a circular shaft is shown. AB represents a straight line on the surface of the untwisted shaft parallel to the longitudinal axis of the shaft. AB' represents a curve (part of a helix) that line AB assumes after the torque is applied. As a result of the applied torque, radius OB, shown on the end of the shaft, rotates and assumes a position OB'. The angle BOB' is called the *angle of twist;* it is generally expressed in radians (see Section 9.5) and is designated θ.

The deformation of a line on the surface of a shaft subjected to a torque is a shear deformation. Hence, the total shear deformation (δ_s) of line AB in the length L is BB'. The shear strain may then be expressed as

$$\epsilon_s = \frac{\delta_s}{L} = \frac{BB'}{L}$$

Since the shaft is circular in cross section, the magnitude of BB′ is equal to $c\theta$, where c is the radius of the shaft and θ is the angle of twist expressed in radians. Therefore,

$$\epsilon_s = \frac{BB'}{L} = \frac{c\theta}{L}$$

Assuming that Hooke's law is applicable, and using the shear stress–strain relationships, the modulus of rigidity G (see Chapter 9) is expressed as

$$G = \frac{\tau}{\epsilon_s}$$

Substituting for ϵ_s, the expression becomes

$$G = \frac{\tau}{(c\theta/L)} = \frac{\tau L}{c\theta}$$

Solving for θ yields

$$\theta = \frac{\tau L}{Gc} \qquad \textbf{(12.5)}$$

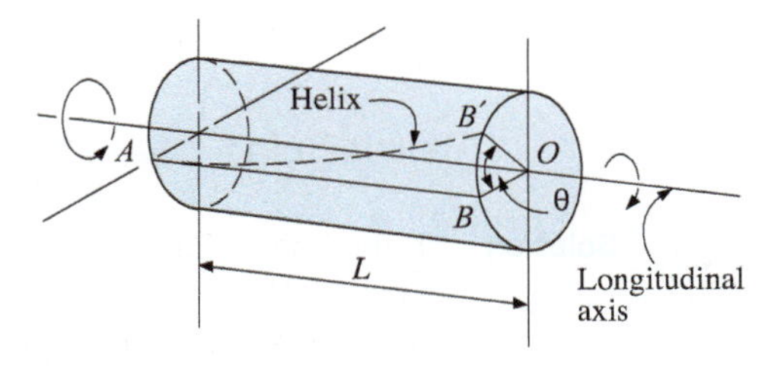

FIGURE 12.10 Angle of twist of a shaft.

which gives the angle of twist in terms of the maximum shear stress occurring at the outer surface.

Since

$$\tau = \frac{Tc}{J}$$

the angle of twist may also be expressed in terms of the torque T. Substituting for τ in Equation (12.5) gives

$$\theta = \frac{TcL}{JGc} = \frac{TL}{JG} \qquad \textbf{(12.6)}$$

where θ = the angle of twist (radians, where one radian = 57.3°)
T = the torque (lb-in., k-in.) (N · m)
L = the length of the shaft subjected to the torque (in.) (mm, m)
J = the polar moment of inertia (in.^{4}) (mm^4, m^4)
G = the modulus of rigidity (or modulus of elasticity in shear) (psi, ksi) (MPa, Pa)

Equations (12.5) and (12.6) are applicable to both solid and hollow circular shafts.

In some design cases, the size of the shaft necessary to transmit a given torque may be governed by the allowable angle of twist rather than by the allowable shear stress. A shaft may be strong enough to function properly but be entirely too flexible.

EXAMPLE 12.7 A 38-mm-diameter solid steel shaft, 2 m long, is subjected to a torque of 600 N · m, as shown in Figure 12.11. The steel is AISI 1020 hot-rolled. Calculate (a) the maximum shear stress and (b) the total angle of twist.

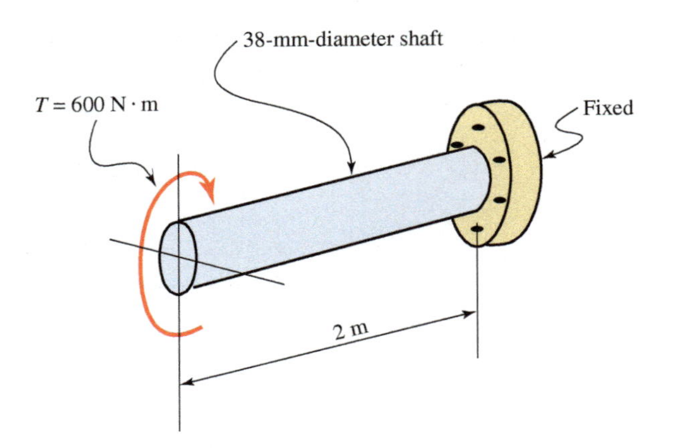

FIGURE 12.11 Sketch for Example 12.7.

Solution In the use of Equations (12.2) and (12.6), all quantities will be substituted in units of newtons and meters. (*Note*: 1 N/m² = 1 Pa.)

The centroidal polar moment of inertia J is calculated from

$$J = \frac{\pi d^4}{32} = \frac{\pi(38 \text{ mm})^4}{32} = 205 \times 10^3 \text{ mm}^4 = 205 \times 10^{-9} \text{ m}^4$$

$$c = \frac{d}{2} = 19 \text{ mm} = 19 \times 10^{-3} \text{ m}$$

(a) Calculate the maximum shear stress from Equation (12.2):

$$\tau = \frac{Tc}{J} = \frac{(600 \text{ N}\cdot\text{m})(19 \times 10^{-3} \text{ m})}{205 \times 10^{-9} \text{ m}^4} = 55.6 \times 10^6 \text{ Pa}$$

$$= 55.6 \text{ MPa}$$

(b) Use Equation (12.6) and G from Table G.2 in the appendices (note that 1 Pa = 1 N/m²):

$$G = 79.3 \times 10^3 \text{ MPa} = 79.3 \times 10^9 \text{ N/m}^2$$

$$\theta = \frac{TL}{JG} = \frac{(600 \text{ N}\cdot\text{m})(2 \text{ m})}{(205 \times 10^{-9} \text{ m}^4)(79.3 \times 10^9 \text{ N/m}^2)} = 0.0738 \text{ rad}$$

Since one radian = 57.3°,

$$\theta = (0.0738 \text{ rad})(57.3°/\text{rad}) = 4.23°$$

EXAMPLE 12.8

A solid steel shaft is to resist a torque of 300,000 lb-in. The angle of twist is not to exceed 1° in 5 ft, and the maximum shear stress is not to exceed 12,000 psi. Calculate the required shaft diameter and select a diameter to use. Assume G = 12,000,000 psi.

Solution To calculate the required diameter based on the allowable angle of twist, rewrite Equation (12.6) and solve for J:

$$J = \frac{TL}{\theta G}$$

For a solid circular shaft,

$$J = \frac{\pi d^4}{32}$$

Substituting this into the preceding expression and solving for the required d will yield

$$\text{required } d^4 = \frac{32TL}{\pi \theta G}$$

The maximum angle of twist is to be 1°. This value must be converted to radians:

$$\theta = \frac{1°}{57.3°/\text{rad}} = 0.01745 \text{ rad}$$

Then the required d can be calculated. Note that the radian, by definition, is a ratio and is therefore unitless.

$$\text{required } d = \sqrt[4]{\frac{32(300{,}000 \text{ lb-in.})(5 \text{ ft})(12 \text{ in./ft})}{(12{,}000{,}000 \text{ lb/in.}^2)(\pi)(0.01745)}} = 5.44 \text{ in.}$$

Next, use Equation (12.4) to calculate the required diameter based on the allowable shear stress:

$$\text{required } d = \sqrt[3]{\frac{16T}{\pi\tau_{(\text{all})}}} = \sqrt[3]{\frac{16(300{,}000 \text{ lb-in.})}{\pi(12{,}000 \text{ lb/in.}^2)}} = 5.03 \text{ in.}$$

The limitation on the angle of twist controls the design. Use a $5\frac{1}{2}$-in.-diameter shaft.

EXAMPLE 12.9 An 0.80-in.-diameter rod of an aluminum alloy was tested in a torsion testing machine. When the applied torque was 1135 lb-in., the angle of twist in a length of 8 in. was 3.21°. Calculate the modulus of rigidity G.

Solution The centroidal polar moment of inertia is calculated from

$$J = \frac{\pi d^4}{32} = \frac{\pi(0.80 \text{ in.}^4)}{32} = 0.0402 \text{ in.}^4$$

The angle of twist, in radians, is

$$\theta = \frac{3.21°}{57.3°/\text{rad}} = 0.0560 \text{ rad}$$

Then use Equation (12.6) and solve for G:

$$G = \frac{TL}{J\theta} = \frac{(1135 \text{ lb-in.})(8 \text{ in.})}{(0.0402 \text{ in.}^4)(0.0560)} = 4.03 \times 10^6 \text{ psi}$$

12.5 TRANSMISSION OF POWER BY A SHAFT

Rotating shafts are commonly used for transmitting power. If an applied torque turns a shaft, work is done by the torque. Recall from physics that work is defined as the energy developed by a force acting through a distance against a resistance. When the distance is linear, work can be expressed as

$$\text{work} = \text{force} \times \text{distance}$$

This definition must be changed somewhat with regard to a rotating shaft, in which case an applied torque turns the shaft through a circular distance. Here, work can be expressed as

$$\text{work} = \text{torque} \times (\text{angular distance}) = T\theta$$

To verify this expression, consider the bar shown in Figure 12.12. The bar is pivoted at O and is acted on by two equal and opposite forces separated by a distance d. The two forces constitute a couple (or a torque) T having a magnitude Fd. If the bar moves through an angle θ (in radians), then the distance through which each force moves will be $\theta(d/2)$. The work done by the two forces will be

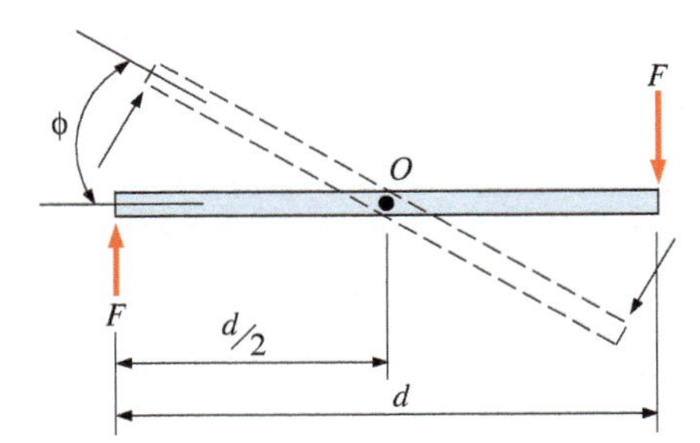

FIGURE 12.12 Work done by a couple.

$$\text{work} = 2F\theta\left(\frac{d}{2}\right) = F(d)(\theta) = T\theta$$

where T = the torque (lb-in. or k-in.)
θ = the angle through which the rotating body turns (radians)

and the units of work will be in lb-in. or k-in.

If a shaft is being rotated at a constant speed against a resistance, then the work done in one revolution will be $2\pi T$, since θ equals 2π radians per revolution (360°).

If T is in units of lb-in. and if n_r is defined as the number of revolutions per minute (rpm), then the work done in 1 minute is calculated from

$$\text{work per minute} = 2\pi T n_r \text{ (lb-in. per minute)}$$

Power is defined as work done per unit time:

$$\text{power} = \frac{\text{work}}{\text{time}}$$

The common unit of power in the U.S. Customary System is the horsepower (hp), the value of which is 33,000 lb-ft per minute, or 396,000 lb-in. per minute. Therefore, for a given combination of torque and shaft speed, the number of horsepower may be determined from

$$\text{hp} = \frac{\text{work per minute}}{396{,}000} = \frac{2\pi T n_r}{396{,}000}$$

Therefore,

$$\text{hp} = \frac{T n_r}{63{,}025} \tag{12.7}$$

where hp = the horsepower developed
T = the torque (lb-in.)
n_r = the number of revolutions per minute

The recommended unit for power in the SI is the watt (W), where 1 watt is defined as 1 newton meter per second:

$$1\text{ W} = 1\frac{\text{N}\cdot\text{m}}{\text{s}}$$

Power has been defined as work per unit time and is expressed further as work per minute, where

$$\text{power} = \text{work per minute} = 2\pi T n_r$$

with n_r expressed in units of revolutions per minute (rpm) and T expressed in units of lb-in. This relationship is still valid in the SI. Since power is expressed in watts and defined as N · m/s, however, the term n_r must be divided by 60 (to convert it to revolutions per second). In addition, T must be expressed in units of newton meters. Equation (12.7) may then be written as

$$\text{power} = W = \frac{2\pi T n_r}{60} \tag{12.7 [SI]}$$

where W = power in watts
T = torque (N · m)
n_r = revolutions per minute (r/min)

Note that the symbol for revolutions per minute is r/min in the SI, whereas in the U.S. Customary System it is rpm. Also note that the symbol for revolutions per second in the SI is r/s. Speed of rotation may also be expressed in terms of radians per second (rad/s). For use in this book, we prefer the more familiar r/min (or r/s).

Using the relationships of Equations (12.7) and (12.7 [SI]), the horsepower delivered by a shaft can be computed knowing the speed of rotation and the torque applied to the shaft, or the torque that a shaft delivers may be determined by using the speed of rotation and the horsepower of the driving motor or engine. As is usual with equations of this kind in which special constants have been introduced, one must be very careful to substitute data with prescribed units.

In the three examples that follow, friction considerations have been neglected, which is convenient for the purpose of discussion but unrealistic. In all applications, some power will be lost in overcoming frictional forces. The treatment of frictional considerations in these applications belongs to the realm of machine design and is beyond the scope of this text.

EXAMPLE 12.10 Calculate the maximum horsepower that can be transmitted by a $2\frac{1}{2}$-in.-diameter solid steel shaft operating at 300 rpm. The allowable shear stress in the shaft is 9000 psi.

Solution The centroidal polar moment of inertia is calculated from

$$J = \frac{\pi d^4}{32} = \frac{\pi(2.5\text{ in.})^4}{32} = 3.83\text{ in.}^4$$

Using Equation (12.3), the allowable torque can be calculated. This value is the maximum torque that could be transmitted without exceeding the allowable shear stress. Thus,

$$T_R = \frac{\tau_{(\text{all})} J}{c} = \frac{(9000\text{ psi})(3.83\text{ in.}^4)}{1.25\text{ in.}} = 27{,}600\text{ lb-in.}$$

Using Equation (12.7), the maximum horsepower is

$$\frac{T n_r}{63{,}025} = \frac{(27{,}600\text{ lb-in.})(300\text{ rpm})}{63{,}025} = 131.4\text{ hp}$$

EXAMPLE 12.11 A hollow steel shaft with an outside diameter of $3\frac{1}{4}$ in. and an inside diameter of 3 in. transmits 280 hp at 900 rpm. Calculate the maximum torsional shear stress developed in the shaft.

Solution The centroidal polar moment of inertia is calculated from

$$J = \frac{\pi(d^4 - d_1^4)}{32} = \frac{\pi[(3.25 \text{ in.})^4 - (3.0 \text{ in.})^4]}{32} = 3.00 \text{ in.}^4$$

Using Equation (12.7), the torque developed can then be calculated:

$$\text{hp} = \frac{Tn_r}{63{,}025}$$

from which

$$T = \frac{63{,}025(\text{hp})}{n_r} = \frac{63{,}025(280 \text{ hp})}{900 \text{ rpm}} = 19{,}610 \text{ lb-in.}$$

Using Equation (12.2), the torsional shear stress is

$$\tau = \frac{Tc}{J} = \frac{(19{,}610 \text{ lb-in.})[(3.25 \text{ in.})/2]}{3.00 \text{ in.}^4} = 10{,}620 \text{ psi}$$

EXAMPLE 12.12 A solid AISI 1020 steel shaft is required to transmit power of 50 kW. The speed of the shaft will be 6 r/s. The allowable shear stress is 67 MPa, and the allowable angle of twist (per meter of shaft length) is not to exceed 0.065 radian. Determine the required diameter of the shaft.

Solution Determine the required diameter based on shear stress and on twist of the shaft. The final required diameter will be the larger of the two. First, the magnitude of the torque to be transmitted is calculated. From Equation (12.7 [SI]), rewriting for torque:

$$T = \frac{60W}{2\pi n_r}$$

where n_r is defined in units of r/min. Therefore,

$$n_r = 6 \text{ r/s} \times 60 \text{ s/min} = 360 \text{ r/min}$$

and

$$T = \frac{60W}{2\pi n_r} = \frac{60(50 \times 10^3 \text{ watts})}{2\pi(360 \text{ r/min})} = 1.326 \times 10^3 \text{ N}\cdot\text{m}$$

Based on allowable shear stress, using Equation (12.4),

$$\begin{aligned}\text{required } d &= \sqrt[3]{\frac{16T}{\pi\tau_{(\text{all})}}} = \sqrt[3]{\frac{16(1.326 \times 10^3 \text{ N}\cdot\text{m})}{\pi(67 \times 10^6 \text{ N/m}^2)}}\\ &= 0.0465 \times 10^{-1} \text{ m}\\ &= 46.5 \text{ mm}\end{aligned}$$

Based on the allowable angle of twist, from Equation (12.6),

$$\theta = \frac{TL}{JG} = \frac{TL}{(\pi d^4/32)G}$$

from which

$$\begin{aligned}\text{required } d &= \sqrt[4]{\frac{32TL}{\pi\theta G}} = \sqrt[4]{\frac{32(1.326 \times 10^3 \text{ N}\cdot\text{m})(1 \text{ m})}{\pi(0.065)(77{,}000 \times 10^6 \text{ N/m}^2)}}\\ &= 0.4053 \times 10^{-1} \text{ m}\\ &= 40.5 \text{ mm}\end{aligned}$$

Therefore, the required diameter is 46.5 mm.

SUMMARY BY SECTION NUMBER

12.1 and 12.2 Torque is a twisting action (moment) applied in a plane perpendicular to the longitudinal axis of a member. A member subjected to a torque is said to be in *torsion*. Examples of such members are shafts used for the transmission of power.

12.3 Torsional shear stresses are developed on cross-sectional planes of a shaft as a result of an applied torque. In a shaft of circular cross section, the shear stress varies linearly from zero at the centroidal longitudinal axis to a maximum at the outer surface of the shaft and can be obtained from

$$\tau = \frac{Tc}{J} \quad \textbf{(12.2)}$$

The maximum resisting (or allowable) torque can be computed from

$$T_R = \frac{\tau_{(\text{all})}J}{c} \quad \textbf{(12.3)}$$

Solid shafts of circular cross section can be designed using

$$\text{required } d = \sqrt[3]{\frac{16T}{\pi\tau_{(\text{all})}}} \quad \textbf{(12.4)}$$

12.4 The design of a solid or hollow shaft of circular cross section may be governed by an allowable angle of twist rather than by an allowable shear stress. The angle of twist is computed from

$$\theta = \frac{TL}{JG} \quad \textbf{(12.6)}$$

and is indicative of the flexibility of a shaft. The torsional rigidity or stiffness of a shaft is the product JG.

12.5 Rotating shafts of circular cross section are commonly used for transmitting power. In the U.S. Customary System, the common unit of power is the *horse-power*. The power developed by a shaft rotating at n_r (revolutions per minute) as a result of an applied torque T (lb-in.) is computed from

$$\text{hp} = \frac{Tn_r}{63{,}025} \quad \textbf{(12.7)}$$

In the SI, the recommended unit of power is the *watt*. The power developed by a rotating shaft is computed from

$$W = \frac{2\pi Tn_r}{60} \quad \textbf{(12.7 [SI])}$$

where torque T is in units of N · m and n_r is r/min.

PROBLEMS

For the following problems, unless noted otherwise use a modulus of rigidity G of 12,000,000 psi (83,000 MPa) for steel and 4,000,000 psi (28,000 MPa) for aluminum.

Section 12.2 Members in Torsion

12.1 Determine the internal resisting torque in the shaft shown at *A*, *B*, and *C*. Show the free-body diagrams.

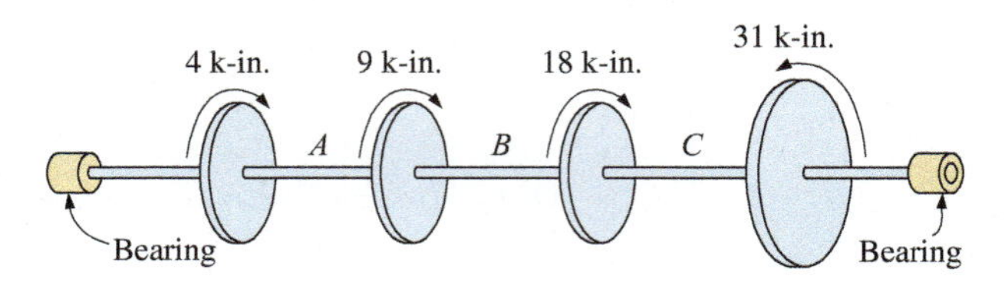

PROBLEM 12.1

12.2 Determine the internal resisting torque in the shaft shown at *A* and *B*. Show the free-body diagrams. Assume that the shaft is fixed against rotation at the fixed support.

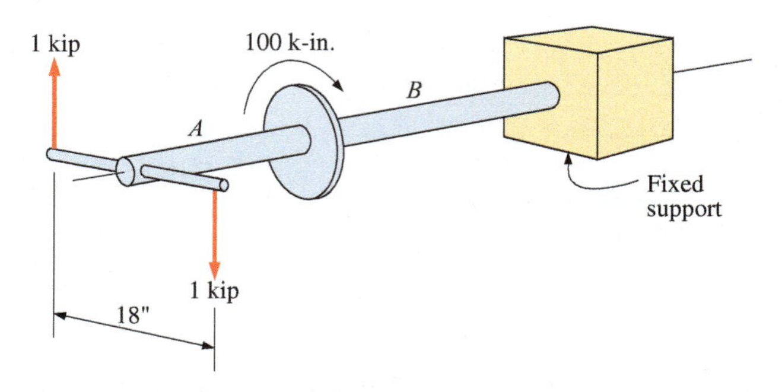

PROBLEM 12.2

Section 12.3 Torsional Shear Stress

12.3 Calculate the maximum shear stress developed in a $3\frac{1}{2}$-in.-diameter circular solid steel shaft subjected to an applied torque of 5000 lb-ft.

12.4 Calculate the allowable torque for a hollow steel shaft. The inside diameter is 40 mm and the outside diameter is 85 mm. The allowable shear stress is 68 MPa.

12.5 Calculate the allowable torque that may be applied to a 5-in.-diameter solid circular steel shaft if the allowable shear stress is 10,000 psi.

12.6 A hollow circular steel shaft has a 100-mm outside diameter and a 75-mm inside diameter. Calculate the allowable torque that can be transmitted if the allowable shear stress is 62 MPa. When the allowable torque is applied, calculate the shear stress at the inner surface of the shaft.

12.7 Design a solid circular steel shaft to transmit an applied torque of 25,000 lb-ft. The allowable shear stress is 10,000 psi.

12.8 Calculate the shear stresses at the outer and inner surfaces of a hollow circular steel shaft subjected to a torque of 350,000 lb-in. The outside diameter of the shaft is 6 in. and the inside diameter is 3 in.

12.9 A hollow shaft is produced by boring a 150-mm-diameter concentric core in a 225-mm-diameter solid circular shaft. Compute the percentage of the torsional strength lost.

12.10 Pulleys C and D are attached to shaft *AB*, as shown. The shaft is supported on bearings at each end. The shaft rotates

at a uniform speed. Pulley D is the driver and pulley C is the power take-off. The diameter of the shaft is $2\frac{1}{2}$ in. Calculate
(a) the belt tension P_3.
(b) the maximum shear stress in the shaft.

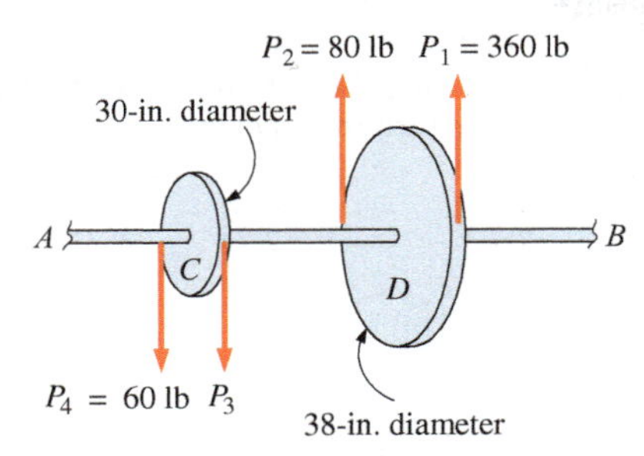

PROBLEM 12.10

Section 12.4 Angle of Twist

12.11 Calculate the angle of twist θ in a 3-in.-diameter solid steel shaft 4 ft long. The shaft is stressed to its allowable shear stress of 9000 psi.

12.12 Calculate the angle of twist θ in a 65-mm-diameter solid steel shaft 2 m long. The shaft is stressed to its allowable shear stress of 60 MPa.

12.13 Calculate the angle of twist θ in a 4-in.-diameter solid steel shaft 20 ft long. The shaft is subjected to a torque of 40,000 lb-in.

12.14 A 1-in.-diameter solid steel shaft, 60 in. long, was tested in a large torsion machine. An angle of twist of 5° was measured when the shaft was subjected to a torque of 1700 lb-in. Calculate the modulus of rigidity *G*.

12.15 A hollow aluminum tube has an outside diameter of 1 in. and an inside diameter of 0.50 in. The tube is subjected to a maximum shear stress of 7000 psi. Calculate the angle of twist per foot of length.

12.16 A solid steel shaft is to resist a torque of 9000 lb-in. The angle of twist is not to exceed 0.12° per foot of length, and the allowable shear stress is 8000 psi. Calculate the required diameter of the shaft and select an appropriate diameter to use.

12.17 A hollow steel shaft has a 50-mm outside diameter and a 40-mm inside diameter. The shaft is 2 m long. Compute the angle of twist when the maximum shear stress is 52 MPa.

12.18 If the shaft of Problem 12.17 were solid, with the same outside diameter and stressed to the same maximum shear stress, what would be the angle of twist? Compute the percentage increase in torque that the solid shaft could transmit. Compute the percentage increase in weight (use a unit weight of steel of 77 kN/m^3).

Section 12.5 Transmission of Power by a Shaft

12.19 An automobile engine develops 90 hp at 3500 rpm. What torque is developed?

12.20 Calculate the speed (rpm) at which a 1-in.-diameter solid steel shaft must operate so as to transmit 10 hp without exceeding an allowable shear stress of 8000 psi.

12.21 Select the diameter of a solid circular steel shaft for a motor that produces 7.5 kW operating at 1800 r/min. The allowable shear stress in the shaft is 50 MPa.

12.22 Select the diameter for a hollow steel shaft that is to transmit 36 hp at 1200 rpm. The allowable shear stress is 9000 psi, and the inside diameter of the shaft is to be three-fourths of the outside diameter.

12.23 A 6-ft-long solid steel shaft with a diameter of 4 in. transmits 250 hp at a speed of 250 rpm. Determine whether the following two requirements are satisfied:
(a) The maximum shear stress is not to exceed 10,000 psi.
(b) The angle of twist is not to exceed 1°.

12.24 The outside and inside diameters of a hollow steel shaft are 150 mm and 100 mm, respectively. Determine the maximum power that can be transmitted if the shaft rotates at 1200 r/min and has an allowable shear stress of 62 MPa.

12.25 Calculate the maximum shear stress developed in a $1\frac{1}{2}$-in.-diameter solid steel shaft that transmits 20 hp at a speed of 400 rpm.

Computer Problems

For the following computer problems, any appropriate software may be used. Input prompts should fully explain what is required of the user (the program should be user-friendly). The resulting output should be well labeled and self-explanatory.

12.26 Write a program that will calculate the allowable torque that may be applied to either a solid or a hollow circular shaft. User input is to be the allowable shear stress, the type of shaft, and the diameter (or diameters).

12.27 Write a program that will generate a table of required diameters (to $\frac{1}{8}$ in.) of solid circular shafts to resist torques ranging from 50 to 200 in.-kips (in increments of 25 in.-kips). Allowable shear stresses range from 10,000 psi to 12,000 psi (in increments of 500 psi). Note that there is no user input.

12.28 Rework the program of Problem 12.27 using SI units. Check your results with Example 12.4.

12.29 Write a program that will generate a table of power (hp) transmitted by a shaft as a function of shaft speed and torque. The shaft speed is to vary from 100 to 500 rpm (in increments of 50 rpm) and the torque is to vary from 1000 in.-kips to 3000 in.-kips (in increments of 500 in.-kips).

Supplemental Problems

12.30 Compute the maximum shear stress in the hollow steel shaft shown. The shaft has a 4-in. outside diameter and a 2-in. inside diameter.

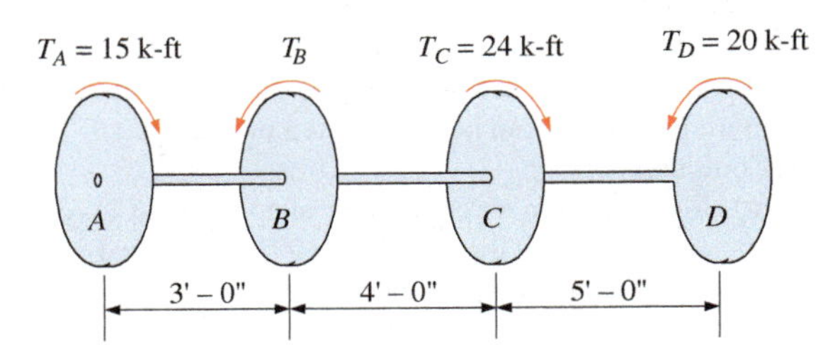

PROBLEM 12.30

12.31 Calculate the allowable torque that may be applied to a hollow circular shaft with an outside diameter of 150 mm and an inside diameter of 120 mm. The allowable shear stress is 75 MPa.

12.32 Design a hollow steel shaft to transmit a torque of 13,000 lb-in. The allowable shear stress in the shaft is 9000 psi. The outside diameter is to be twice the inside diameter.

12.33 A 32-in.-long solid steel circular shaft, 3 in. in diameter, is twisted through an angle of 0.012 radian. Calculate the maximum shear stress developed in the shaft.

12.34 The 65-mm-diameter solid shaft shown is subjected to torques of 600 N · m and 1400 N · m at points B and C, respectively. Determine the maximum shear stress in the shaft.

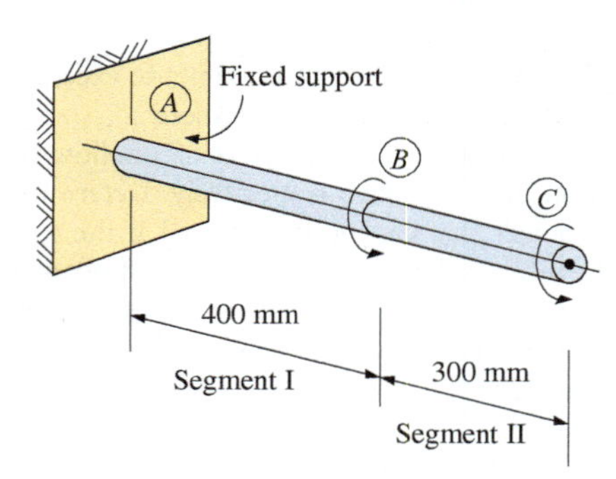

PROBLEM 12.34

12.35 Rework Problem 12.34, changing the diameter of segment II of the shaft to 55 mm.

12.36 Compute the maximum shear stress in the circular steel shaft shown if the shaft is subjected to the torques indicated. The shaft is solid and 2 in. in diameter for 21 in. of its length and is hollow with an outside diameter of 2 in. and an inside diameter of 1 in. for 21 in. of its length, as shown.

12.37 Determine the allowable torque a hollow steel shaft can transmit if its outside diameter is 4 in. and its inside diameter is 3 in. The allowable shear stress is 10,000 psi. Compute the angle of twist per foot of length.

12.38 A 1.00-m-long steel wire, 4 mm in diameter, is twisted through an angle of 61° by a torque of 2.25 N · m. Determine the modulus of rigidity of the wire.

12.39 Select the outside and inside diameters for a hollow steel shaft subjected to a torque of 1,000,000 lb-in. The maximum shear stress developed in the shaft is not to exceed 8000 psi. The inside diameter is to be approximately two-thirds the outside diameter. Calculate the angle of twist, in degrees, for a 14-ft length of this shaft.

12.40 A solid aluminum shaft, 6 ft in length, is to transmit a torque of 350 lb-ft. The allowable shear stress is 5000 psi and the angle of twist must not exceed 4° in a 6-ft length. Select the required diameter.

12.41 A 25-mm-diameter solid shaft with an allowable shear stress of 60 MPa rotates at a speed of 15 r/s. Determine the maximum power that can be transmitted by this shaft.

12.42 Compute

(a) the maximum shear stress developed in a 6-in.-diameter solid steel shaft that transmits 600 hp at a speed of 80 rpm.

(b) the shear stress if the speed were increased to 300 rpm.

12.43 What horsepower can a solid steel shaft 6 in. in diameter transmit at 160 rpm if the allowable shear stress is 5000 psi?

12.44 Calculate the maximum power that may be transmitted by a 60-mm-diameter solid steel shaft operating at 300 r/min. The allowable shear stress in the shaft is 65 MPa.

12.45 A small ski lift has a main cable driving wheel 11 ft in diameter. The cable speed is to be 500 ft per minute. The cable tension on one side of the driving wheel is 25,000 lb and on the other side is 28,800 lb, as shown. Calculate the horsepower required to turn the main cable driving wheel.

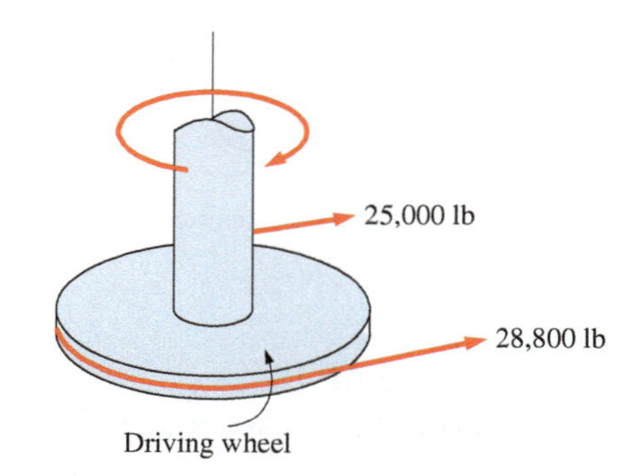

PROBLEM 12.45

12.46 A 32-mm-diameter solid shaft transmits 100 kW of power at a speed of 28 r/s. Determine the maximum shear stress in the shaft.

12.47 A solid steel shaft is to transmit power of 58 kW at a speed of 60 r/s. The allowable shear stress is 90 MPa. Determine the minimum permissible diameter for this shaft.

12.48 Select the diameter for a solid steel shaft that is to transmit 60 hp at 850 rpm without exceeding an allowable shear stress of 7000 psi.

12.49 A solid steel shaft is to transmit 100 hp at a speed of 1000 rpm. The allowable shear stress is 9000 psi. What is the required diameter of the shaft?

12.50 Two shafts—one a hollow steel shaft with an outside diameter of 100 mm and an inside diameter of 40 mm, the other a solid steel shaft with a diameter of 100 mm—are to transmit 90 kW each. Compare the shear stresses in the two shafts if both operate at 150 r/min.

12.51 A $1\frac{1}{2}$-in.-diameter solid steel shaft is 40 ft in length. If the torque necessary to operate a piece of equipment is 1500 lb-in., what horsepower must be delivered to the shaft to maintain a shaft speed of 1000 rpm? Calculate the maximum shear stress developed in the shaft.

12.52 A solid steel shaft is to transmit 120 hp. The allowable shear stress is 8000 psi.

(a) Select the diameter if the shaft speed is 3000 rpm.

(b) Select the diameter if the shaft speed is 300 rpm.

(c) Calculate the angle of twist of each shaft in a length of 10 ft. Use $G = 12{,}000{,}000$ psi.

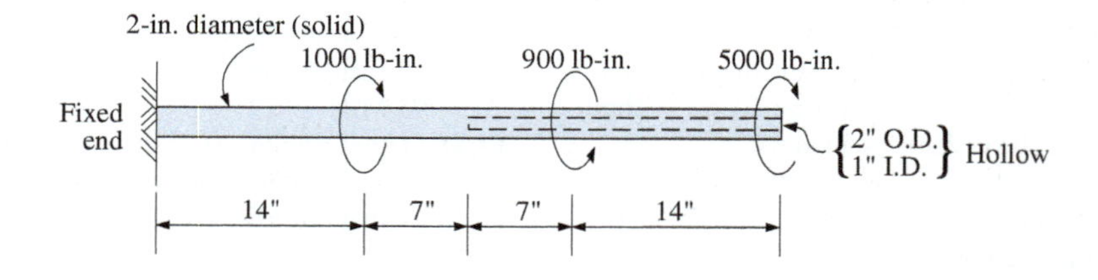

PROBLEM 12.36

SHEAR AND BENDING MOMENT IN BEAMS

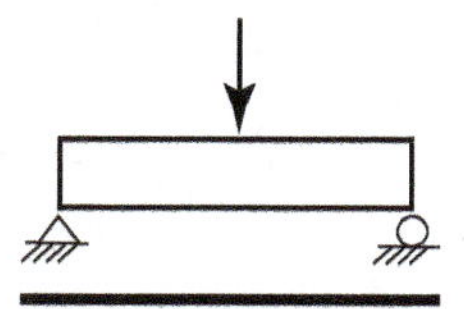

LEARNING OBJECTIVES

Upon completion of this chapter, readers will be able to:

- Identify statically determinate beams, including simply supported, cantilevered, and overhanging beams.
- Revisit types of loads and solve for reactions as explained in Chapters 2 and 4, respectively.
- Discuss the relationship between the externally applied loads and reactions, and the internal shear forces and bending moment in a beam.
- Calculate the internal shear and bending at any point along the length of a beam.
- Learn to graphically represent the shear forces within a beam using a shear diagram.
- Determine the relationship between the shear diagram and bending moments.
- Graphically represent the bending moments along the length of a beam using a moment diagram.
- Identify the critical points for shear and moment within the beam from the shear and moment diagrams.

13.1 TYPES OF BEAMS AND SUPPORTS

Beams are among the most common structural members. They carry loads applied at right angles to the longitudinal axis of the member, which causes the member to bend. A plank placed across a trench supporting a person passing over it is an example of a beam subjected to a load applied at a right angle.

Beams are generally oriented in a horizontal or near horizontal position. They may, on occasion, also exist in a vertical or sloping orientation and be subjected to loads that will produce bending.

In this chapter, we concern ourselves only with straight horizontal beams subjected to loads that will cause bending. Beams are sometimes called by other names, indicative of some specialized function. They may be called girders, stringers, floor beams, joists, lintels, spandrels, purlins, or girts. Many machine members with a specialized function, such as rotating shafts, are also subjected to bending.

A beam's type, as well as its behavior when subjected to load, is a function of the type and number of its supports. In Chapter 4, we discussed supports for beams and trusses. Recall that a roller support provides a reaction perpendicular to the contact surface. (For the beams considered here, a horizontal supporting surface for the roller is assumed; therefore, the reaction is in a vertical direction.) A pin support provides two reactions. (Again, for the beams considered here, one reaction is vertical and one is horizontal.) The pin and roller supports are sometimes called *simple supports*. These two supports are shown in Figure 13.1, which summarizes the real supports, the idealized supports, and the associated reactions. A fixed support is also indicated in Figure 13.1. This type of support may be constructed in a variety of ways. One easily visualized example would be a beam sufficiently anchored into a solid mass of concrete so as not to allow movement, as shown in Figure 13.1c. Note that the fixed support provides a vertical reaction, a horizontal reaction, and a moment reaction. The fixed support will theoretically allow no movement whatsoever: no horizontal, no vertical, and no rotational movement. The roller support, on the other hand, is assumed to permit horizontal and rotational movement, but no vertical movement. The pin support is assumed to permit rotational movement, but no horizontal or vertical movement.

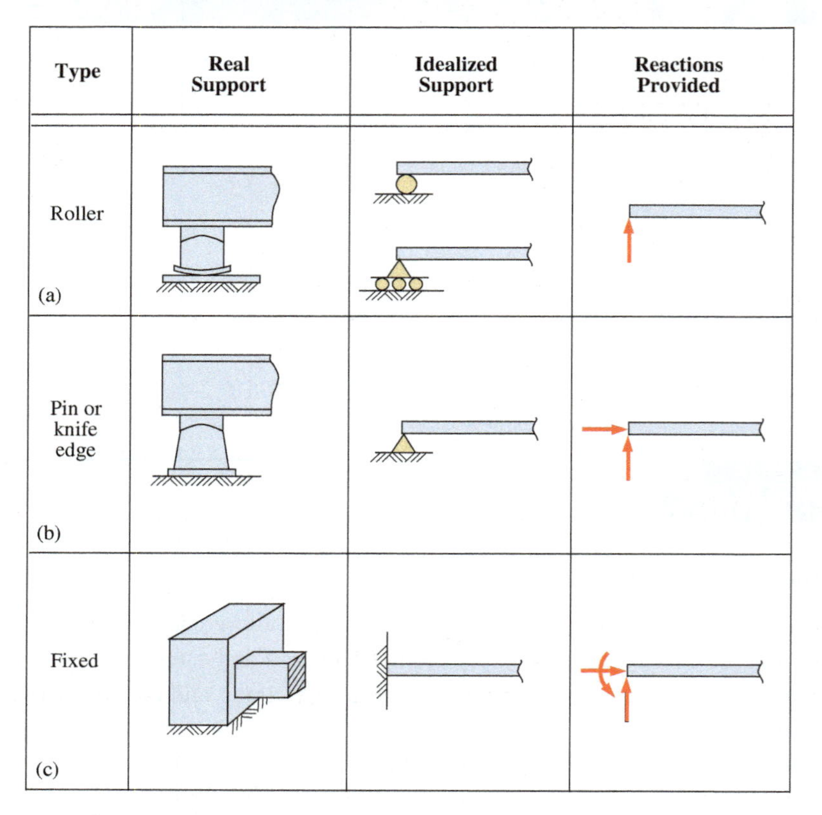

FIGURE 13.1 Beam supports and reactions.

The various types of beams in common use are shown in Figure 13.2, along with the deflected shape of the loaded beam in each case. It is convenient to represent a beam with a single line. We follow this convention for representing beams in load diagrams (which are also free-body diagrams). A diagram in which the beam depth is shown is used in connection with our discussion of internal stresses in beams.

Beams are categorized according to type and/or number of supports. A beam supported at the ends by simple supports and used to carry any system of loads between the supports is called a *simple beam*. A *fixed beam* (or totally restrained beam) is supported by two fixed supports that do not permit any end rotation or translation as the beam is loaded. A beam with a fixed support at one end, with no other support along its length, is called a *cantilever beam*. Again, the fixed support does not permit any end rotation or translation as the beam deflects under load.

Any beam with one or two simple supports that are not located at the ends of the beam is called an *overhanging beam*. A beam that is fixed at one end, similar to a cantilever beam, with a simple support at the other end, is called a *propped cantilever*. A beam supported on three or more supports is called a *continuous beam*.

The simple, cantilever, and overhanging beams are categorized as *statically determinate* because their reactions can be determined using the three basic laws of equilibrium discussed in Chapter 4: $\Sigma F_y = 0$, $\Sigma F_x = 0$, and $\Sigma M = 0$. Since the beams considered here are horizontal beams, for convenience we alter the notation slightly. Instead of subscripts y and x for the force summation equations, we use subscripts V and H to indicate vertical and horizontal: $\Sigma F_V = 0$ and $\Sigma F_H = 0$.

The fixed beam, the propped cantilever beam, and the continuous beam are categorized as *statically indeterminate* because their reactions cannot be determined by the three laws of equilibrium alone. Additional relationships based on the deflection of the beam must be introduced, as discussed in Chapter 21. Here we deal only with statically determinate beams.

13.2 TYPES OF LOADS ON BEAMS

Loads on beams are classified as *concentrated loads* and *distributed line loads*. In addition, distributed line loads may be categorized as *uniformly* or *nonuniformly* distributed.

Concentrated loads were dealt with in Chapters 2 and 3. Recall that they are assumed to act at a definite point. Although such loads are actually distributed over a small area (or a short length) of the beam, the distance is usually small in comparison with the length of the beam so that the load can be considered to be concentrated at a single point without affecting the accuracy of the calculations excessively. A single arrow is used to indicate the location, direction, and sense of the concentrated load, as shown in Figure 13.3. The unit for the concentrated load is generally pounds (lb) or kips in the U.S. Customary System and newtons (N) in SI.

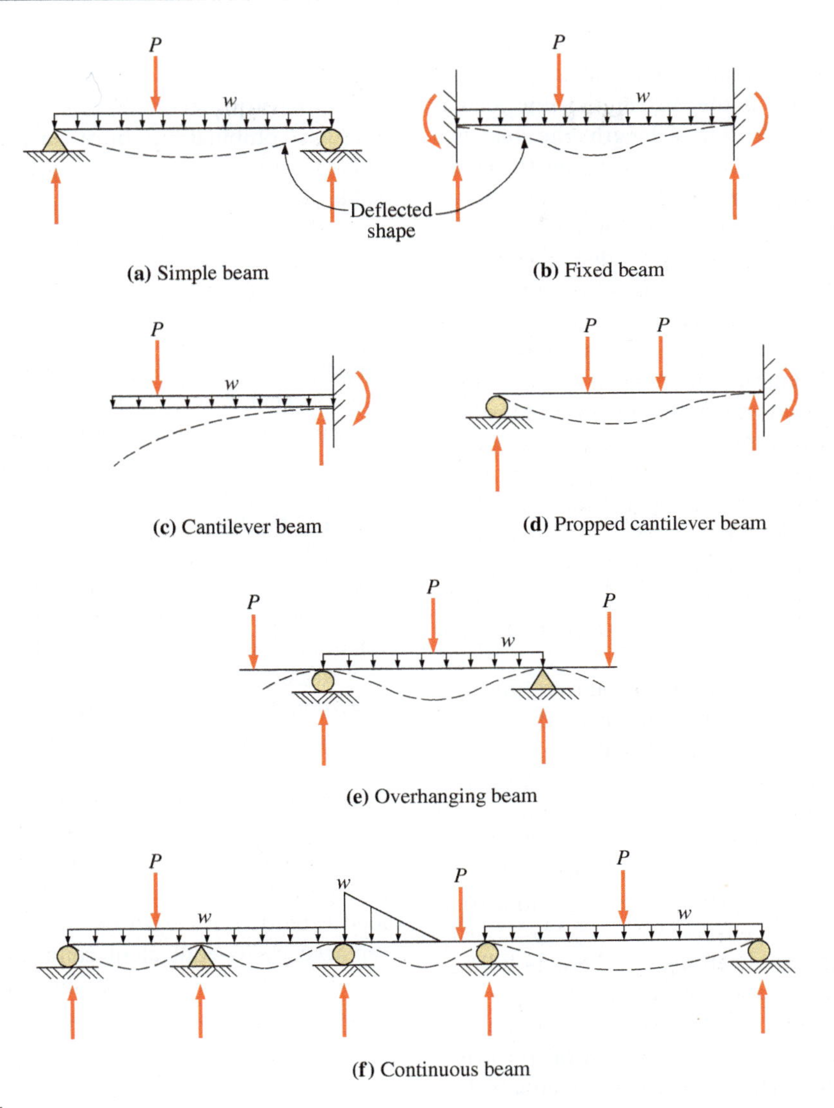

FIGURE 13.2 Types of beams.

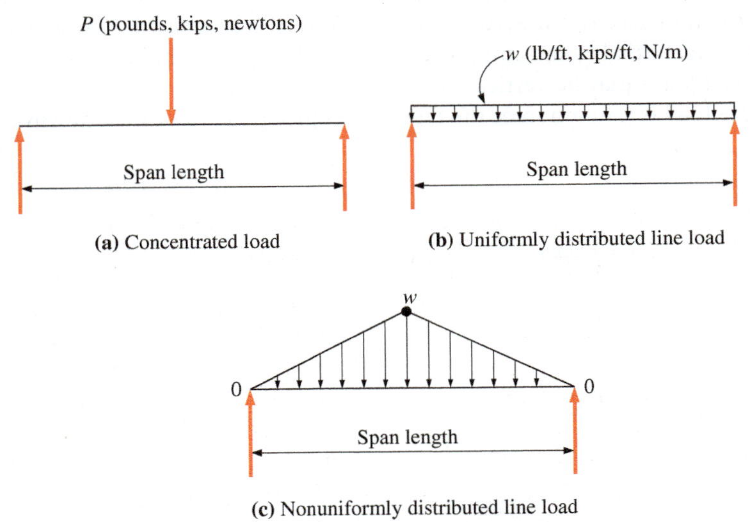

FIGURE 13.3 Types of loads.

Distributed line loads were discussed in Chapter 3. Recall that a distributed line load is one that is spread out over a length of the beam. If the distributed line load is of equal magnitude for each unit of length, the load is a uniformly distributed line load. It may exist over the entire length of the beam or over a portion of it. The load diagram usually used to indicate a uniformly distributed line load is a rectangular block. The block can be shaded or can show a system of arrows to indicate load direction and sense, as shown in Figure 13.3b. This diagram defines both the intensity of the distributed load and the length of the beam over which it exists. The length (whether full span or a portion of the span) is indicated with a dimension, and the load intensity is indicated with a notation. For the intensity of the uniformly distributed line load, w is commonly used. The units for the intensity are pounds per linear foot (lb/ft) or kips per linear foot (kips/ft) in the U.S. Customary System and newtons per meter (N/m) in SI. Assuming that a beam has a uniform cross section throughout its length, the weight of the beam would be a uniformly distributed line load.

The distributed line load may have a varying intensity. In this case, it is referred to as a nonuniformly distributed line load. Generally, a nonuniformly distributed line load increases or decreases at a known rate along the length of the beam. The diagram for this type of distributed load results in a triangular or trapezoidal block, as shown in Figure 13.3c. Note that, in this case, the maximum distributed load intensity is w at midspan and decreases to zero at the supports.

In structural design applications (bridges, buildings, etc.), loads are usually categorized as *dead loads* or *live loads*. Dead loads are static loads that produce vertical forces due to gravity; they include the weight of the structural framework and all materials permanently attached to it and supported by it. Reasonable estimates of the weight of the structure (or its component parts) can usually be made based on some preliminary calculations. Live loads may be defined as all loads that are not dead loads. They are loads that may or may not be present. Examples of live loads are vehicles, people, stored materials, snow, ice, wind, fluid pressure, earth pressure, earthquake, and impact due to moving loads. Some of these loads may be vertical, some may be lateral. The applicable load is a function of the type of structure and its intended use as well as its geographic location.

Loads can occur in combinations, and the probability of such combinations must be considered. Live loads on standard-type structures are generally specified by applicable building codes.

13.3 BEAM REACTIONS

To determine internal stresses at various points along the length of a beam, it is generally necessary to first compute the external reactions for the beam. The determination of reactions was discussed in Chapter 4. At this point, we briefly review the procedure. The determination of external reactions is accomplished for statically determinate beams by using the three laws of equilibrium: $\Sigma F_H = 0$, $\Sigma F_V = 0$, and $\Sigma M = 0$. The algebraic sum of all the external applied loads and reactions must equal zero. Also, the algebraic sum of moments about any point due to the externally applied loads and reactions must equal zero.

Recall that for the purpose of determining reactions, distributed loads may be replaced by an equivalent concentrated resultant load. As pointed out in Section 3.5, this equivalent concentrated resultant load may be represented as a dashed arrow to distinguish it from the given concentrated loads and is assumed to act through the centroid of the distributed load. This replacement is shown graphically in Figure 13.4, where a 1 kip/ft uniformly distributed line load extends full length on a 40-ft-long beam. The load is replaced by its equivalent concentrated resultant load, denoted W and shown as a dashed arrow at the center of the uniformly distributed line load. Therefore,

$$W = 1 \text{ kip/ft} \times 40 \text{ ft} = 40 \text{ kips}$$

This replacement is only for purposes of finding the external reactions. The internal effect of the uniformly distributed line load of 40 kips is very different from that of the concentrated resultant load of 40 kips.

The reactions of a simply supported beam will always be vertically upward if the applied loads are vertically downward. This does not necessarily hold for overhanging beams, in which case one of the reactions may be downward, even if all the applied loads are downward.

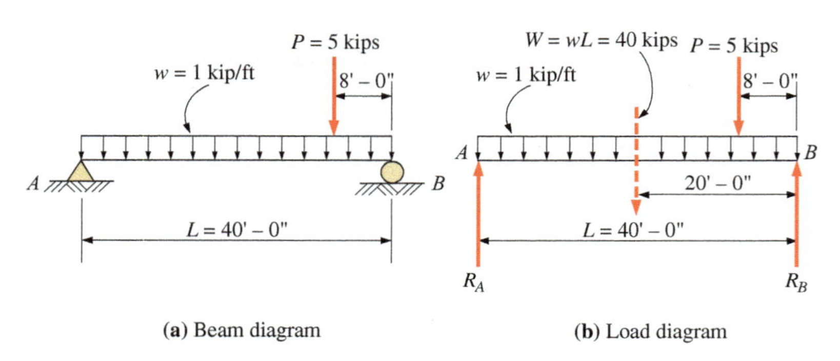

(a) Beam diagram

(b) Load diagram

FIGURE 13.4 Determination of beam reactions.

EXAMPLE 13.1 Compute the reactions for the simply supported beam *AB* in Figure 13.4a.

Solution Figure 13.4a, called a *beam diagram*, defines the type of beam; the types of supports; the span, and the types, magnitudes, and locations of the loads.

The *load diagram* is shown in Figure 13.4b. Note that the supports at points *A* and *B* are replaced with the expected reactions from those supports. The pin at *A* could provide a horizontal reaction, but since there is no horizontally applied load, this reaction would be zero and can be neglected. In reality, the load diagram is a free-body diagram of the beam.

Assuming counterclockwise rotation to be positive, the reaction at point *B* can be computed by summing the moments about point *A*:

$$\Sigma M_A = +R_B(40\text{ ft}) - (5\text{ k})(32\text{ ft}) - (1\text{ k/ft})(40\text{ ft})(20\text{ ft}) = 0$$

from which

$$R_B = +24\text{ k}\uparrow$$

Recall that the positive sign indicates that the sense was assumed correctly for the reaction at point *B*. No inference as to clockwise or counterclockwise moment should be made. Had this result turned out to be negative, it would mean only that the sense had been assumed incorrectly and that the reaction was actually downward.

The reaction at point *A* can be computed by taking moments about *B*:

$$\Sigma M_B = -R_A(40\text{ ft}) + (5\text{ k})(8\text{ ft}) + (1\text{ k/ft})(40\text{ ft})(20\text{ ft}) = 0$$

from which

$$R_A = +21\text{ k}\uparrow$$

An algebraic summation of all the vertical forces serves as a check. For the beam to be in equilibrium, it is required that this sum equal zero. Assuming upward to be positive,

$$+24\text{ k} + 21\text{ k} - 5\text{ k} - (1\text{ k/ft})(40\text{ ft}) = 0 \qquad \textbf{(O.K.)}$$

An alternate solution would be to compute the reaction at point *A* by taking moments about point *B* and then compute the reaction at *B* using $\Sigma F_V = 0$. If an error is made in computing the first reaction, however, the second computation will also be incorrect. At this point, it is recommended that each reaction be computed by means of a moment equation and then the calculations checked using a summation of vertical forces.

EXAMPLE 13.2 Compute the reactions for the overhanging beam in Figure 13.5a.

Solution The load diagram is shown in Figure 13.5b. The supports at *A* and *B* are replaced with the reactions that we expect from those supports. The pinned support at *A* could provide a horizontal reaction, but since there is no horizontally applied load, this reaction is zero and may be neglected.

Calculate the equivalent concentrated resultant load for the distributed load:

$$W = 12\text{ kN/m}(5\text{ m}) = 60\text{ kN}$$

The reaction at *B* can be computed by summing moments about point *A*. Assuming counterclockwise moments to be positive gives

$$\Sigma M_A = +R_B(8\text{ m}) - 30\text{ kN}(5\text{ m}) - 20\text{ kN}(3\text{ m}) - 60\text{ kN}(0.5\text{ m}) = 0$$

from which

$$R_B = +30\text{ kN}\uparrow$$

The reaction at *A* can be computed by summing moments about point *B*. Assuming counterclockwise moments positive,

$$\Sigma M_B = -R_A(8\text{ m}) + 30\text{ kN}(3\text{ m}) + 20\text{ kN}(5\text{ m}) + 60\text{ kN}(7.5\text{ m}) = 0$$

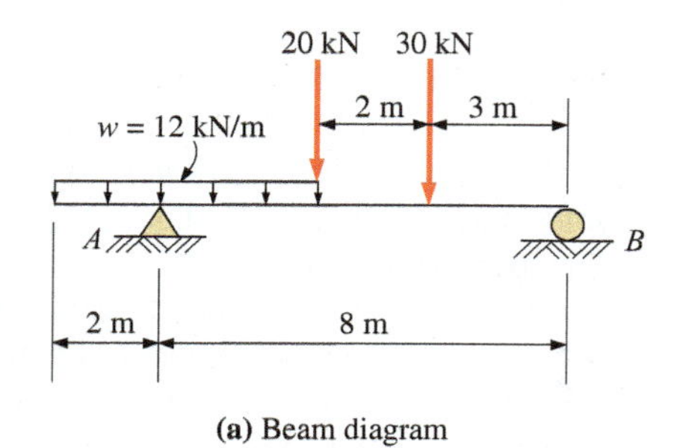

(a) Beam diagram

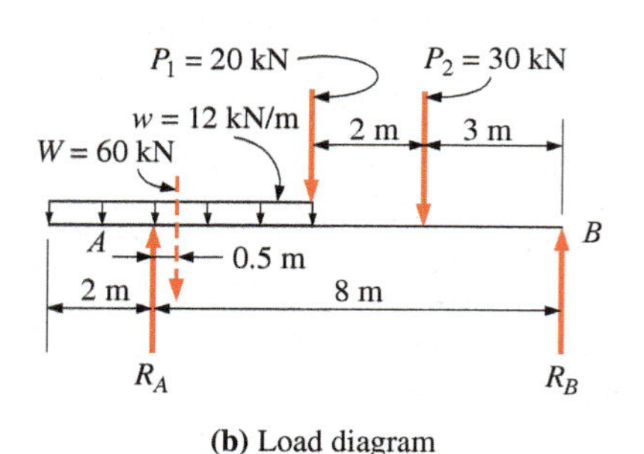

(b) Load diagram

FIGURE 13.5 Overhanging beam.

from which

$$R_A = +80 \text{ kN} \uparrow$$

Check the calculations by summing the vertical forces and assume upward positive,

$$\Sigma F_V = +80 + 30 - 60 - 20 - 30 = 0 \qquad \textbf{(O.K.)}$$

13.4 SHEAR FORCE AND BENDING MOMENT

We now investigate the internal effects of the externally applied loads and forces. When a beam is subjected to applied loads and the resulting reactions, bending of the beam will occur and internal stresses will develop. These stresses, termed *shear stresses* and *bending stresses*, are evaluated in Chapter 14 to establish their effect in the design and analysis of the bending members.

Our immediate objective concerning bending members is to evaluate the effects of the loading. These effects take the form of internal shear force and internal bending moment (usually called simply "shear" and "moment") that are developed by the externally applied loads and reactions. The magnitude of the shear and moment will directly affect the magnitude of the shear stress and the bending stress.

The method used to determine the shears and moments involves equilibrium considerations. External reactions must first be calculated. These will put the beam in external equilibrium. Recall that if the entire beam is in equilibrium, every segment must be in equilibrium. If we cut a section through the beam and isolate a small segment for analysis purposes, equilibrium considerations enable us to determine the internal shear and moment that must exist on the cut face of the segment. Depending on the loading and the conditions of support, shear and/or moment may vary throughout the length of the beam. We develop methods with which we can obtain a complete picture of shear and moment variations along the length of the beam.

EXAMPLE 13.3 Compute the shear force and bending moment (a) at 2 m, (b) at 5m, and (c) at 9 m from the left end of the beam having the load diagram shown in Figure 13.6. Neglect the weight of the beam.

Solution First, compute the beam reactions. Assuming counterclockwise to be positive,

$$\Sigma M_A = R_B(12 \text{ m}) - (30 \text{ kN})(3 \text{ m}) - 70 \text{ kN}(8.5 \text{ m}) = 0$$

from which

$$R_B = 57.1 \text{ kN} \uparrow$$

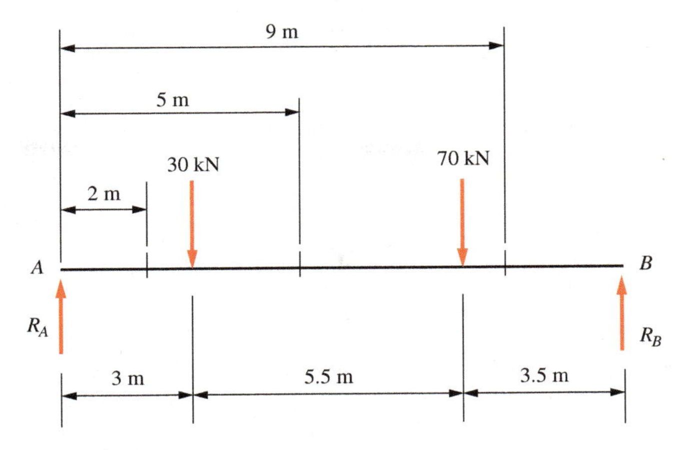

FIGURE 13.6 Load diagram for Example 13.3.

and

$$\Sigma M_B = -R_A(12\text{ m}) + 30\text{ kN}(9\text{ m}) + 70\text{ kN}(3.5\text{ m}) = 0$$

from which

$$R_A = 42.9\text{ kN}\uparrow$$

Check by summing vertical forces ($\Sigma F_V = 0$):

$$+42.9\text{ kN} - 30\text{ kN} - 70\text{ kN} + 57.1\text{ kN} = 0 \qquad \textbf{(O.K.)}$$

(a) *At 2 m from the left end:*

To compute the shear force and bending moment at a location 2 m from the left end, cut the beam at that location (designated point x in Figure 13.7) and consider the part on the left of the cutting plane as a free body.

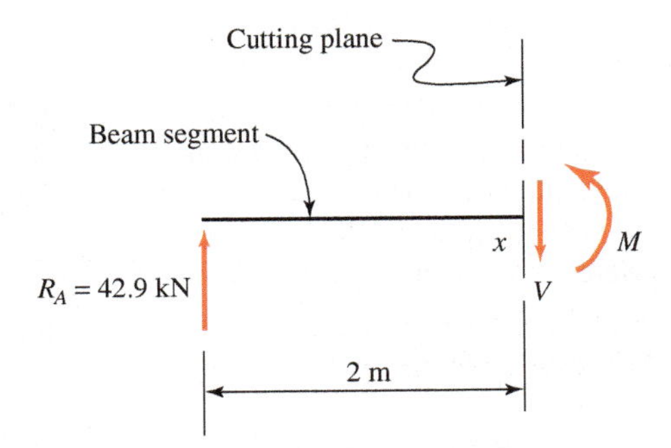

FIGURE 13.7 Free-body diagram for shear and moment determination.

Actually, the free body may be taken from either end of the beam. To be consistent with sign conventions (to be discussed later), however, it is recommended that the free body always be taken between the left end of the beam and the cutting plane.

The free body must be in equilibrium. In other words, the algebraic sum of the vertical forces must equal zero and the algebraic sum of the moments about any point must equal zero. Since there are no external horizontal forces or horizontal components of forces, the horizontal force consideration does not enter the problem.

Considering the free body of Figure 13.7, the only external vertical force is the 42.9-kN reaction at point A. Therefore, there is an unbalanced upward force (42.9 kN) and there must exist internally, at section x, a vertical force of 42.9 kN that will put the beam segment in vertical equilibrium. This force is the *internal shear*, which is equal in magnitude to the sum of the external forces acting on the segment to the left of the cutting plane. This is how we will define internal shear in a beam. We could achieve the same numerical result using the beam segment to the right of the section. The sense of the resulting 42.9-kN shear, however, would be opposite to that determined using the segment to the left of the section. To eliminate this ambiguity, we define a sign convention commonly used for shear. With reference to Figure 13.8,

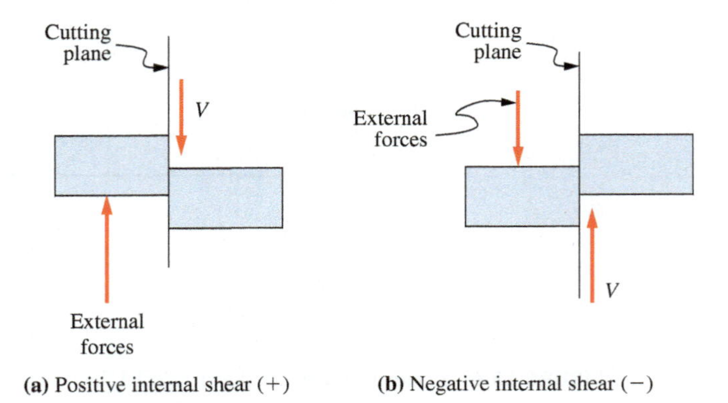

(a) Positive internal shear (+) **(b)** Negative internal shear (−)

FIGURE 13.8 Internal shear sign convention.

> *Internal shear is considered positive* (+) *in a beam if the segment of the beam to the left of a cutting plane tends to move upward with respect to the segment to the right. Negative internal* shear (−) *is indicative of the left segment moving downward with respect to the right segment.*

In taking a summation of vertical external forces on a free body that has been taken from the left end of the beam, the correct sign for the shear at the cutting plane (in accordance with the preceding convention) will always result if upward acting forces are taken to be positive and downward acting forces negative. We use this procedure whenever calculating shear.

Calculating the shear (see Figure 13.7) due to the external load (using upward acting forces positive) gives

$$V = +42.9 \text{ kN}$$

The shear at this location (2 m from the left end) is a positive 42.9 kN, indicating that the part of the beam to the left of the cutting plane tends to move upward with respect to the right part of the beam. The V force shown in Figure 13.7 represents the *internal shear*, which must be equal and opposite to the sum of the external loads. Shear of other magnitudes will occur at other locations along the length of the beam, as shown in parts (b) and (c) of the solution.

For the free body to be in equilibrium, it is necessary that all laws of equilibrium be satisfied. With reference to Figure 13.7, this requires that the sum of the moments taken at the right end of the beam section, at the cutting plane, must equal zero. It can then be seen that the internal moment M must be equal and opposite to the moment due to the external force of 42.9 kN. We will calculate the value of M shortly.

With reference to Figure 13.9, a sign convention for bending moment is defined as follows:

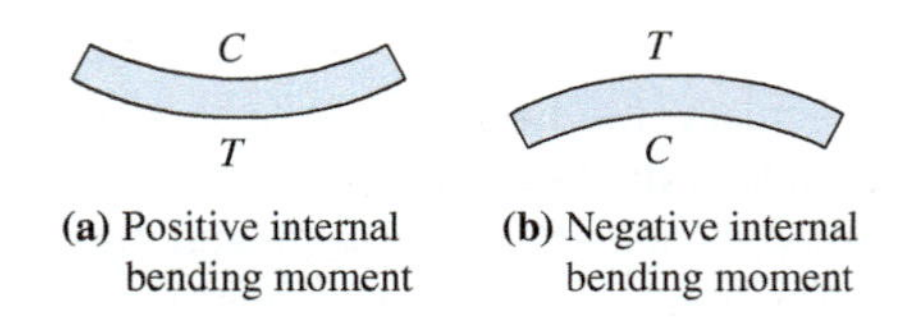

(a) Positive internal bending moment **(b)** Negative internal bending moment

FIGURE 13.9 Moment sign convention.

> *Internal bending moment is positive* (+) *in a beam when the bottom fibers are in tension and the top fibers are in compression. The internal bending moment is negative* (−) *when the top fibers are in tension and the bottom fibers are in compression.*

When determining the bending moment due to the external loads using segments as described in this example, the correct sign for the moment will always result if the moments due to the external upward acting forces are taken to be positive and moments due to the external downward acting forces are taken to be negative. We use this procedure when calculating bending moments using cut sections along a beam. Note that there is no consideration here as to whether the moment is clockwise or counterclockwise as was the case when determining reactions. The same sign for bending moment will result whether a beam segment to the left or to the right of the cutting plane is considered.

Now compute the bending moment about point x in Figure 13.7 due to the external loads (assuming upward acting forces producing positive moments and downward acting forces producing negative moments):

$$M = +(42.9 \text{ kN})(2 \text{ m}) = +85.8 \text{ kN} \cdot \text{m}$$

Therefore, the bending moment at a point 2 m from the left end of the beam is a positive 85.8 kN · m. The positive sign indicates that the bottom fibers are in tension and the top fibers are in compression.

The bending moment M shown in Figure 13.7 represents the *internal moment*, which must be equal and opposite to the computed moment found from the external loads. Bending moments of other magnitudes will occur at other locations along the length of the beam, as shown later in this problem.

(b) *At 5 m from the left end:*

To calculate the internal shear and bending moment at a location 5 m from the left end, cut the original beam again (but this time at the 5-m location) and consider the left part as a free body, as shown in Figure 13.10.

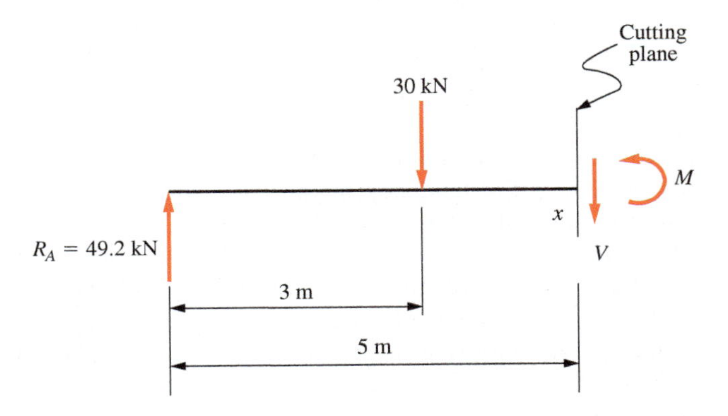

FIGURE 13.10 Free-body diagram (5 m).

Sum the external loads (assume upward acting forces are positive):

$$+42.9 \text{ kN} - 30 \text{ kN} = +12.9 \text{ kN}$$

For equilibrium, the internal shear V must act downward, as shown, and its magnitude must be 12.9 kN. Note that the part of the beam to the left of the cutting plane tends to move upward with respect to the part to the right of the cutting plane. Therefore, this is positive shear according to our sign convention: Internal shear equals +12.9 kN at this location.

Compute the internal bending moment at the right end of the segment shown in Figure 13.10. Upward forces produce positive moment; downward forces produce negative moment:

$$M = +(42.9 \text{ kN})(5 \text{ m}) - (30 \text{ kN})(2 \text{ m}) = +154.5 \text{ kN} \cdot \text{m}$$

The moment M shown at the right end of the beam section in Figure 13.10 represents the internal bending moment of +124.5 kN · m and is equal and opposite to the moment due to the external loads. The positive sign indicates that the bottom fibers are in tension and the top fibers are in compression.

(c) *At 9 m from the left end:*

As done previously, cut the beam at 9 m from the left end and show a free body of the section to the left of the cutting plane as shown in Figure 13.11. The internal shear is calculated by summing the external forces to the left of the cutting plane (upward acting forces are positive):

$$V = +42.9 \text{ kN} - 30 \text{ kN} - 70 \text{ kN} = -57.1 \text{ kN}$$

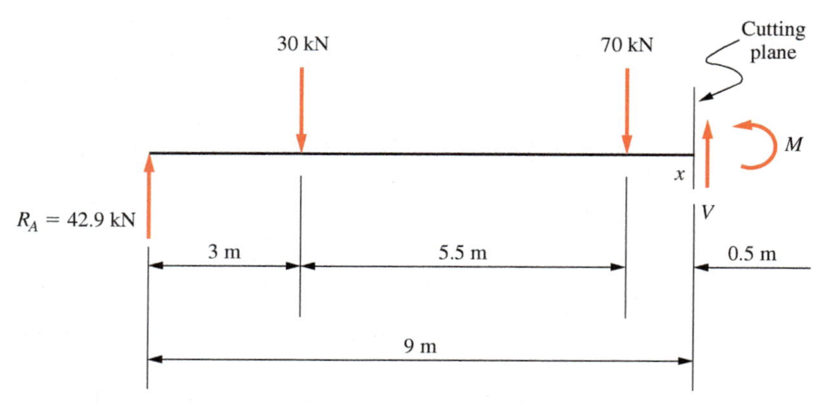

FIGURE 13.11 Free-body diagram (9 m).

The internal shear V is shown in Figure 13.11. Note that the sum of the external forces acts downward, hence the beam segment to the left of the cutting plane tends to move down with respect to the right and this is negative internal shear.

Compute the internal bending moment at the right end of the segment shown in Figure 13.11. Moment M is equal to the sum of the external moments taken at the cutting plane:

$$M = +(42.9\text{ kN})(9\text{ m}) - 30\text{ kN}(6\text{ m}) - 70\text{ kN}(0.5\text{ m}) = +171.1\text{ kN}\cdot\text{m}$$

In Example 13.3, the calculated bending moment due to the external loads represents the tendency of the left segment of the beam to rotate about x. Since the free body must be in equilibrium, an internal resisting moment must exist at point x that, in effect, resists the tendency to rotate. This internal resisting moment M, as shown in Figure 13.11, must be equal and opposite to the moment computed from the external loads. It is actually a couple (acting on the cut face) developed by the bending action of the beam.

As shown in Figure 13.12, a loaded simply supported beam will deflect between the end supports. As a result, the bottom fibers of the beam are lengthened and placed in tension, whereas the top fibers are shortened and placed in compression. Somewhere between the top and bottom fibers, a surface or plane must exist where there is no tension or compression and which remains at its original unloaded length. This plane is called the *neutral plane* of the beam.

Below the neutral plane, the beam is in tension. On the free body (Figure 13.12b), the total tension acting on the cutting plane may be denoted by T. Above the neutral plane, the beam is in compression and the total compression acting on the cutting plane may be denoted by C. Since the free body shown must be in equilibrium, the algebraic summation of all the horizontal forces must equal zero $(\Sigma F_H = 0)$. Therefore, C must equal T. Equal and opposite forces that are parallel constitute a *couple*. This particular couple is equal to (identical with) the internal resisting moment and is referred to as the *internal couple*.

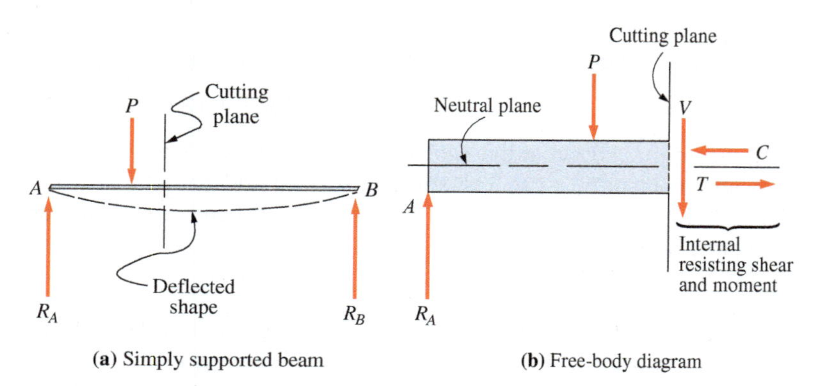

FIGURE 13.12 Shear and bending moment in beams.

EXAMPLE 13.4 Compute the internal shear and moment at 7 ft from the left end of the beam shown in Figure 13.13. The weight of the beam is included in the uniformly distributed load of 2 k/ft. The load indicated with a dashed arrow is the resultant of the distributed load.

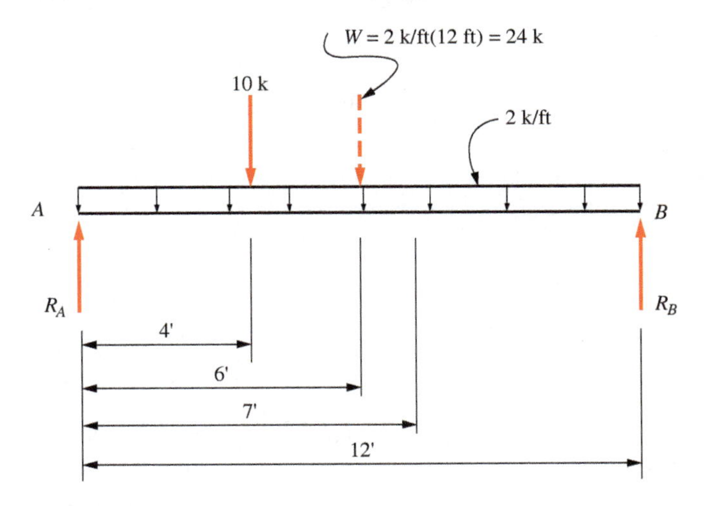

FIGURE 13.13 Load diagram.

Solution Compute the beam reactions (counterclockwise assumed positive):

$$\Sigma M_A = -(10\text{ k})(4\text{ ft}) - (24\text{ k})(6\text{ ft}) + R_B(12\text{ ft}) = 0$$

from which $R_B = 15.33$ k ↑
and

$$\Sigma M_B = -R_A(12\text{ ft}) + (10\text{ k})(8\text{ ft}) + (24\text{ k})(6\text{ ft}) = 0$$

form which $R_A = 18.67$ k ↑

Next, cut a section at *x*, 7 ft from the left end and draw a free-body diagram of the segment to the left of the cutting plane (see Figure 13.14). Note that there is a uniformly distributed load over the 7-ft segment. The 14-k resultant of this distributed load is shown. For purposes of determining internal shear *V* and internal moment *M*, the effect of the 14-kip resultant is the same as the effect of the 7-ft-long distributed load.

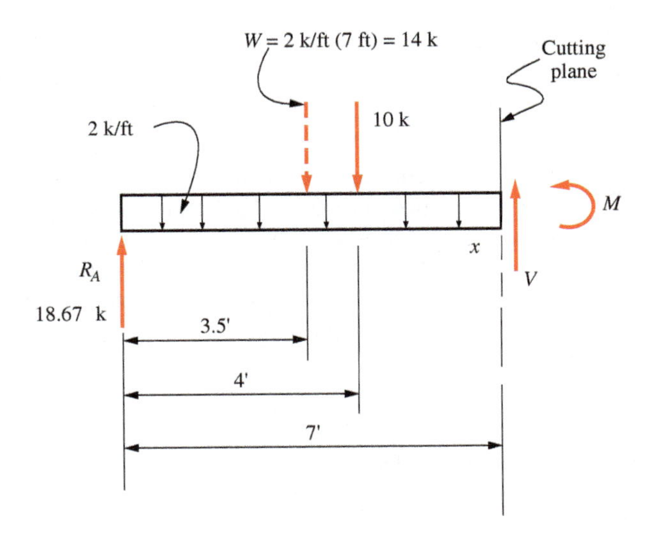

FIGURE 13.14 Free-body diagram (7′).

Calculate the internal shear V. Upward-acting external forces are taken as positive.

$$V = +18.67\text{ k} - 14\text{ k} - 10\text{ k} = -5.33\text{ k}$$

Calculated the internal moment M by summing moments at x:

$$M = +(18.67\text{ k})(7\text{ ft}) - (14\text{ k})(3.5\text{ ft}) - (10\text{ k})(3\text{ ft}) = +51.7\text{ k-ft}$$

EXAMPLE 13.5 Compute the shear and bending moment at (a) 5 ft and (b) 20 ft from the left end of the beam having the load diagram shown in Figure 13.15.

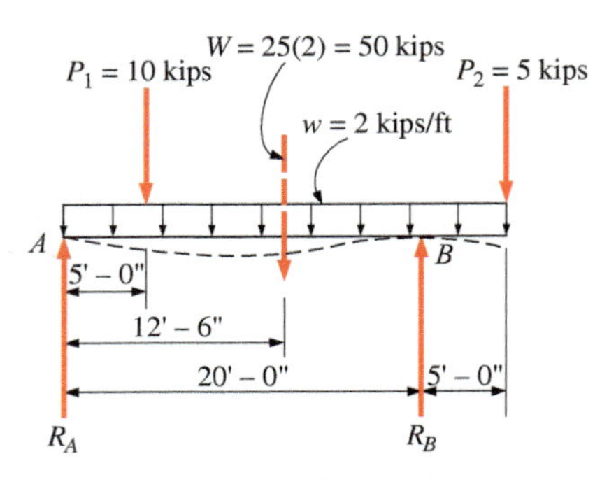

FIGURE 13.15 Load diagram.

Solution The location and magnitude of the resultant of the uniformly distributed load is shown on the load diagram. Compute the beam reactions:

$$\Sigma M_A = R_B(20\text{ ft}) - (10\text{ k})(5\text{ ft}) - (50\text{ k})(12.5\text{ ft}) - (5\text{ k})(25\text{ ft}) = 0$$

from which

$$R_B = +40.0\text{ k}\uparrow$$

and

$$\Sigma M_B = -R_A(20\text{ ft}) + (10\text{ k})(15\text{ ft}) + (50\text{ k})(7.5\text{ ft}) - (5\text{ k})(5\text{ ft}) = 0$$

from which

$$R_A = 25\text{ k}\uparrow$$

Use $\Sigma F_V = 0$ to check:

$$+40\text{ k} + 25\text{ k} - 10\text{ k} - 50\text{ k} - 5\text{ k} = 0 \qquad \textbf{(O.K.)}$$

(a) *At 5 ft from the left end:*

As may be observed, a concentrated load exists at 5 ft from the left end. Therefore, the shear at this point will change abruptly and will have one value at an infinitesimal distance to the left of the load and a different value at an infinitesimal distance to the right of the load. Two free bodies are shown in Figure 13.16 for purposes of computing the two shear values. Either free body may be used to compute the bending moment. In each case, note that the internal shear V and the internal moment M are shown acting on the cutting plane.

Calculate the shear force due to the external loads from Figure 13.16a:

$$V = +25\text{ k} - 10\text{ k} = +15\text{ k}$$

Calculate the shear force due to the external loads from Figure 13.16b:

$$V = +25\text{ k} - 10\text{ k} - 10\text{ k} = +5\text{ k}$$

Calculate the bending moment due to the external loads at the cutting plane (point x):

$$M = +(25\text{ k})(5\text{ ft}) - (10\text{ k})(2.5\text{ ft}) = +100\text{ k-ft}$$

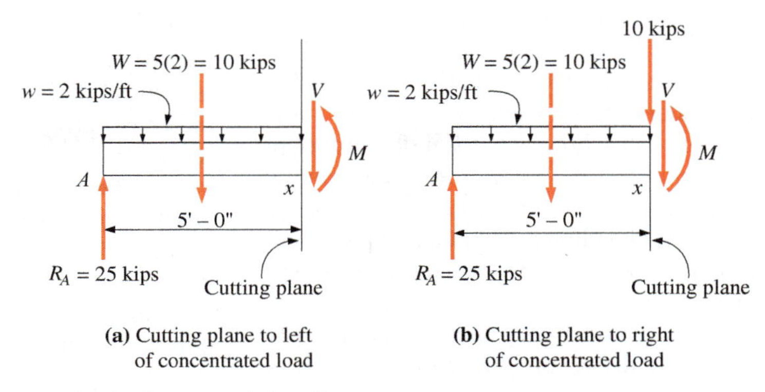

FIGURE 13.16 Free-body diagrams (5′ – 0″).

(b) *At 20 ft from the left end:*

A sudden change in the shear will exist because of the reaction at the support. Calculate the shear just to the left (Figure 13.17a) and just to the right (Figure 13.17b) of the support. From Figure 13.17a,

$$V = +25\text{ k} - 10\text{ k} - 40\text{ k} = -25\text{ k}$$

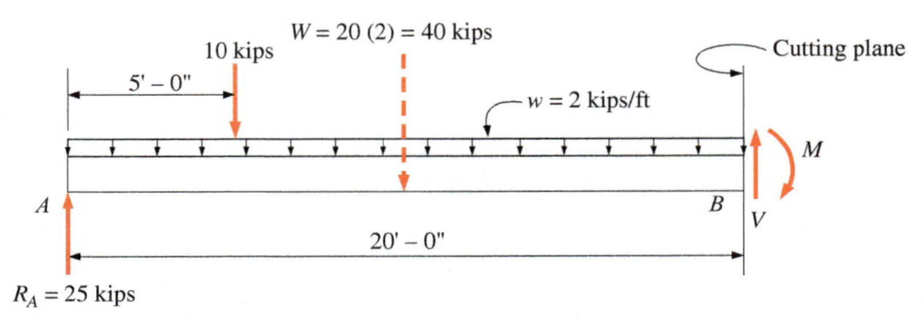

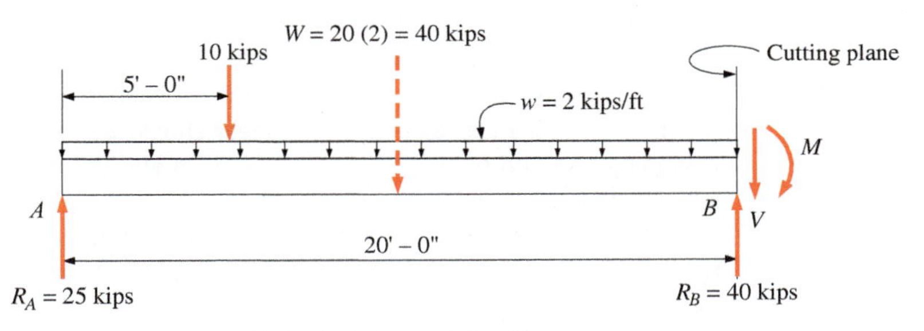

FIGURE 13.17 Free-body diagrams (20′ – 0″).

From Figure 13.17b,

$$V = +25\text{ k} - 10\text{ k} - 40\text{ k} + 40\text{ k} = +15\text{ k}$$

The moment can be calculated from either diagram:

$$M = +(25\text{ k})(20\text{ ft}) - (10\text{ k})(15\text{ ft}) - (40\text{ k})(10\text{ ft}) = -50\text{ k-ft}$$

The moment at the 20-ft point is negative, indicating that the top fibers are in tension and the bottom fibers are in compression. This situation may be verified by observing Figure 13.15; the approximate deflected shape shows that the cantilevered end at the right side of the beam will deflect downward.

13.5 SHEAR DIAGRAMS

In Section 13.4, shear and bending moments were computed at arbitrary locations along the span length of a beam. In the design and analysis of beams, however, it is more important to compute the maximum values of the shear and bending moment. In addition, it is also important to determine the variation of the shear and bending moment along the length of a beam, which may be accomplished using graphical representations known as *shear diagrams* and *moment diagrams*.

In this section, we consider only shear diagrams. The shear diagram is usually drawn directly below a sketch of the load diagram. (The beam diagram and the load diagram may be combined by superimposing the supports and the reactions.) A shear baseline, which represents zero shear, is drawn parallel to the beam. The abscissa along the baseline represents the locations of successive cross sections of the beam. The ordinate of the diagram represents the value of the shear at that particular cross section.

The drawing and construction of the shear diagram is best illustrated through a variety of examples. Generally, shear diagrams are not drawn to scale. Approximate proportions are satisfactory.

EXAMPLE 13.6 Draw the shear diagram for the simply supported beam having the load diagram shown in Figure 13.18. The reactions have been computed and are indicated. Neglect the weight of the beam.

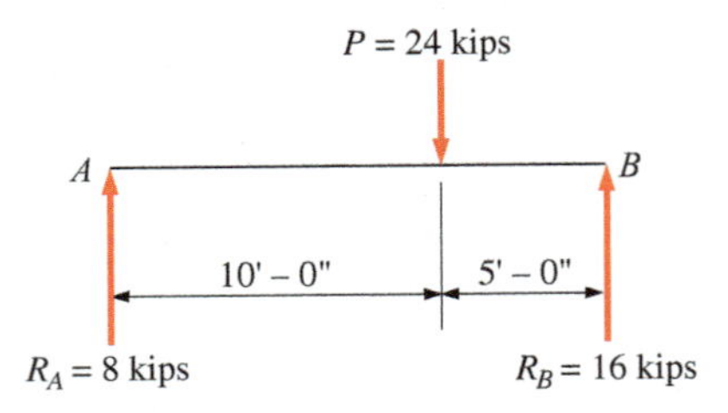

FIGURE 13.18 Load diagram.

Solution Initially, draw shear diagrams by computing the shear at various points along the beam. As demonstrated in Section 13.4, the shear at any point along the length of the beam may be computed utilizing free-body diagrams and performing an algebraic summation of vertical forces (ΣF_V). This computation may be done at 1-ft intervals if so desired and the computed values plotted as ordinates either above or below the zero shear line. When connected, these points will result in a graphic representation portraying the shear variation along the length of the beam. This method, however, is a tedious procedure. Instead, if the shear is computed at specific locations, sufficient shear values may be obtained from which to draw a shear diagram. The shear should be computed at the following locations: (a) at the beginning and end of all distributed-type loads and (b) at an infinitesimal distance to the right and/or left of each concentrated load and reaction.

Working from the left end of the beam, first compute the shear at an infinitesimal distance to the right of the left reaction. After drawing a free-body diagram of infinitesimal length and summing vertical forces as shown in Figure 13.19, the shear is computed to be

$$V = +8 \text{ k}$$

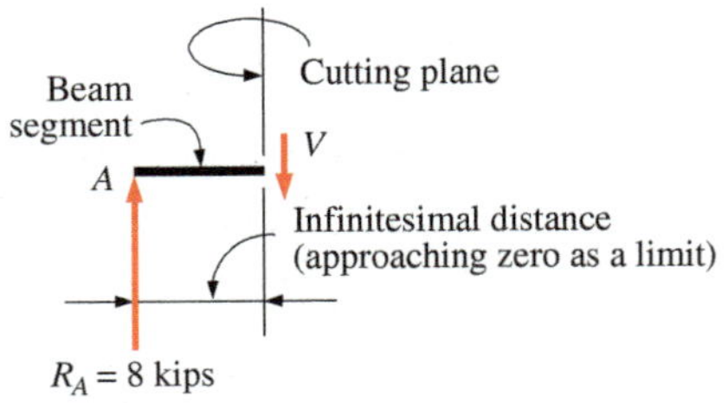

FIGURE 13.19 Free body.

This positive shear value is plotted above the zero shear line (refer to Figure 13.22) directly under the left reaction.

Now compute the shear an infinitesimal distance to the left of the concentrated load. Using a 10-ft-long free-body diagram as shown in Figure 13.20, the shear is computed to be

$$V = +8 \text{ k}$$

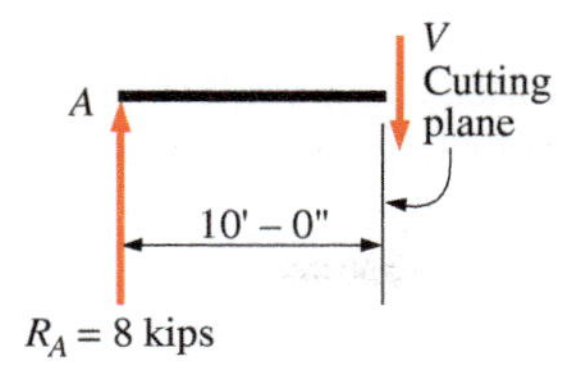

FIGURE 13.20 Free body.

This value is plotted directly under the concentrated load (refer to Figure 13.22) and, since it is positive shear, above the zero shear line. As may be observed, the shear at the left reaction and the shear just to the left of the concentrated load are both equal to +8 kips because there is no change of load between the two points. Therefore, for the length of beam from the left reaction to the concentrated load, there is a positive shear equal to 8 kips. This shear is shown on the shear diagram (see Figure 13.22) as a horizontal straight line between the two points.

Now compute the shear an infinitesimal distance to the right of the concentrated load. Drawing a free-body diagram as shown in Figure 13.21 and summing the vertical forces, the shear is computed to be

$$V = +8\text{ k} - 24\text{ k} = -16\text{ k}$$

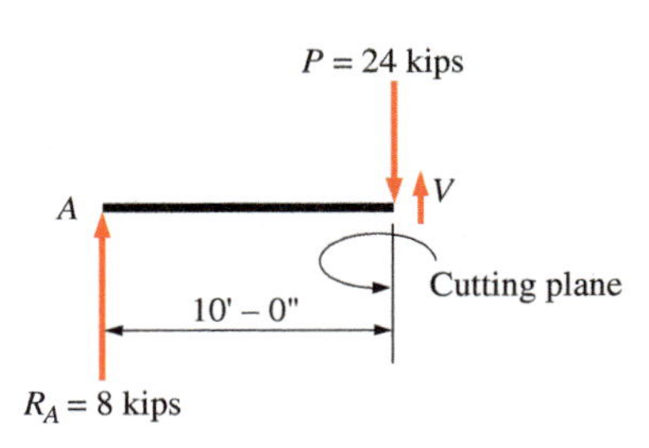

FIGURE 13.21 Free body.

This value is plotted directly under the concentrated load and, since it is negative, below the zero shear line (Figure 13.22). A positive shear of 8 kips exists just to the left of the concentrated load, and a negative shear of 16 kips exists just to the right. The shear diagram shows a sudden change in shear at this location. The shear is said to "go through zero."

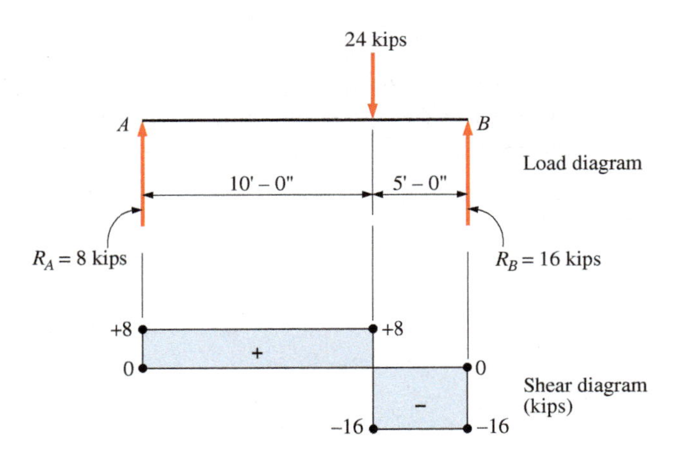

FIGURE 13.22 Load and shear diagrams.

As may be observed, there is no load change between the concentrated load and the right reaction. Therefore, the shear will not change and will remain at negative 16 kips over that length of the beam. This situation is shown on the shear diagram as a horizontal straight line between the two points. The completed shear diagram is shown in Figure 13.22, portraying the variation of shear along the length of the beam. Note that the absolute values of the positive and negative shear areas are equal.

EXAMPLE 13.7

Draw the shear diagram for the simply supported beam having a load diagram shown in Figure 13.23. The reactions have been calculated and are shown in the load diagram. Neglect the weight of the beam.

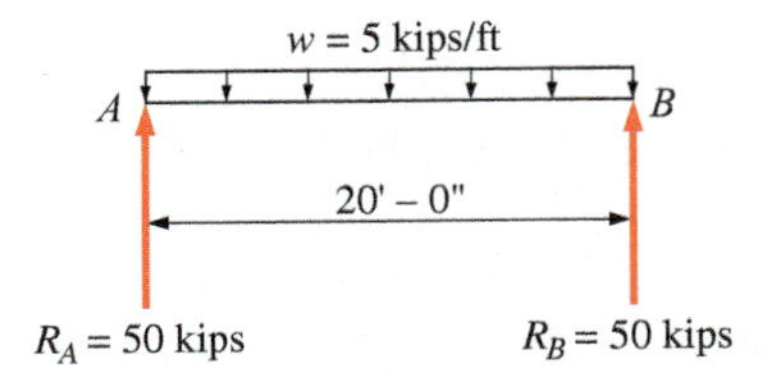

FIGURE 13.23 Load diagram.

Solution Working from the left end, as in Example 13.6, the shear just to the right of the left reaction is equal to +50 kips. The magnitude is equal to the left reaction. The sign (+) is in accordance with our sign convention for shear from Section 13.4.

If free-body diagrams were drawn at every foot along the length of the beam, we would note a change in the shear values of 5 kips per linear foot (kips/ft). A 10-ft-long free body is shown in Figure 13.24. The shear at the 10-ft point can be calculated as

$$V = +50 \text{ k} - 50 \text{ k} = 0$$

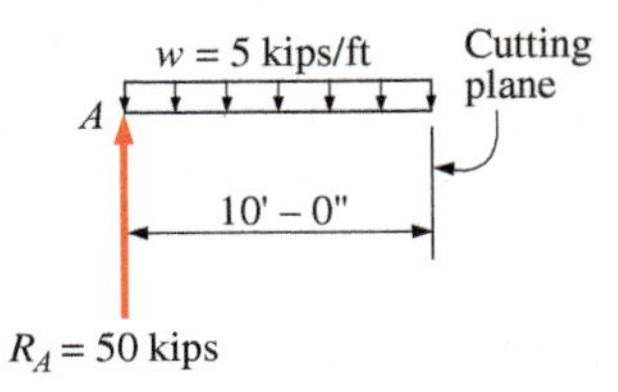

FIGURE 13.24 Free body.

The rate of change of shear is 5 kips/ft (equal to the uniformly distributed load). If an 11-ft-long free-body diagram is drawn and a summation of vertical forces performed, the shear will be found to be −5 kips. The negative shears will then continue to increase (become more negative) until, at the right reaction, the shear will be −50 kips. Note that the right reaction is 50 kips.

Plotting the positive shear values above the zero shear line and the negative shear values below the zero shear line results in the graphic representation of Figure 13.25. This diagram shows the shear variation along the length of the beam. Notice that the shape of the shear diagram is that of a sloping straight line. Also note that the absolute values of the positive and negative shear areas are equal.

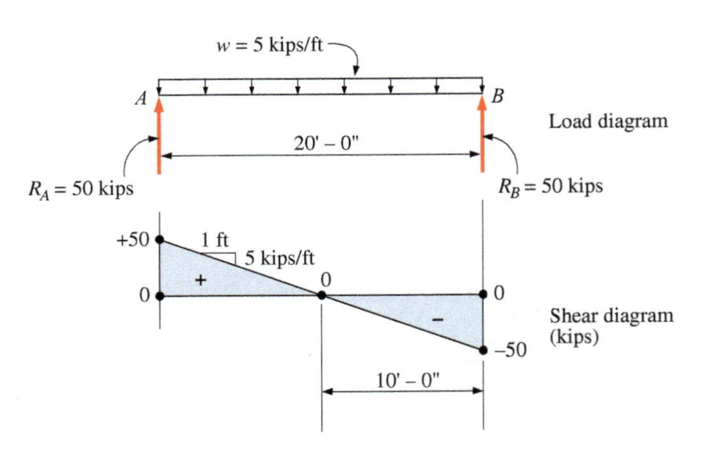

FIGURE 13.25 Load and shear diagrams.

Based on the two previous examples, the following general rules applicable to subsequent construction of shear diagrams can now be stated:

1. For any part of a beam where there are no loads, the shape of the shear diagram will be that of a horizontal straight line.
2. The shear diagram at the point of application of a concentrated load will be a vertical line; that is, there will be a sudden change in the shear.
3. For any part of a beam where there is a uniformly distributed line load, the shape of the shear diagram will be a straight sloping line (always upper left to lower right for gravity loads) with a slope equal to the intensity of the load.
4. For a simply supported, single span beam subjected to vertical loads, the absolute values of the magnitudes of the positive area and the negative area of the shear diagram will be equal.

EXAMPLE 13.8 Draw the shear diagram for the beam having a load diagram shown in Figure 13.26. The reactions have been computed and are shown. Neglect the weight of the beam.

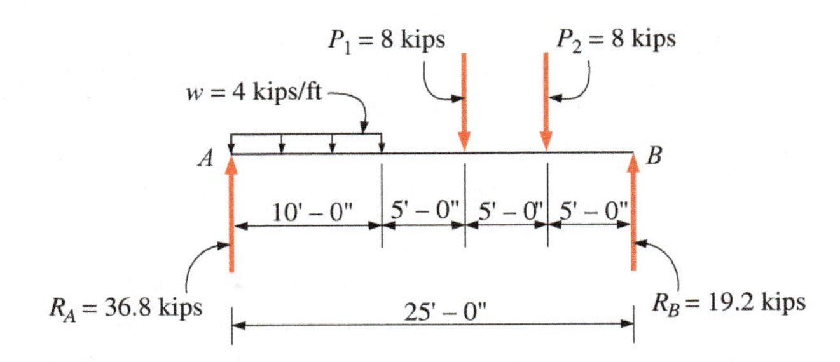

FIGURE 13.26 Load diagram.

Solution Working from the left reaction, compute the shear at the following points where changes will occur in the shear diagram:

(a) Just to the right of the left reaction
(b) At the right end of the uniformly distributed load
(c) Just to the left *and* right of each concentrated load
(d) Just to the left of the right reaction

The shears are then determined as follows:

(a) If we were to draw a free-body diagram and sum the vertical forces, the shear at an infinitesimal distance to the right of the left reaction would be found to be a positive 36.8 kips, which is equal to the reaction.
(b) To compute the shear at the end of the uniformly distributed load, a 10-ft-long free-body diagram is drawn, as shown in Figure 13.27. A summation of vertical forces will result in

$$V = +36.8\text{ k} - (4\text{ k/ft})(10\text{ ft}) = -3.2\text{ k}$$

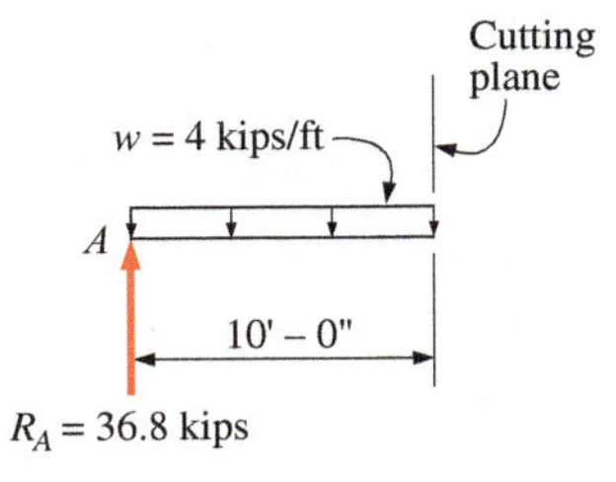

FIGURE 13.27 Free body.

(c) To compute the shear just to the left and right of the concentrated loads, free-body diagrams can be drawn and vertical forces summed as usual. The results of these calculations yield

$$\text{to the left of } P_1\text{: } V = -3.2\text{ k}$$
$$\text{to the right of } P_1\text{: } V = -11.2\text{ k}$$
$$\text{to the left of } P_2\text{: } V = -11.2\text{ k}$$
$$\text{to the right of } P_2\text{: } V = -19.2\text{ k}$$

(d) Similarly, the shear can be calculated an infinitesimal distance to the left of the right reaction. This calculation results in a shear of −19.2 kips, which is numerically equal to the right reaction. The completed shear diagram, obtained by connecting the shears at the various points, is shown in Figure 13.28 along with the load diagram.

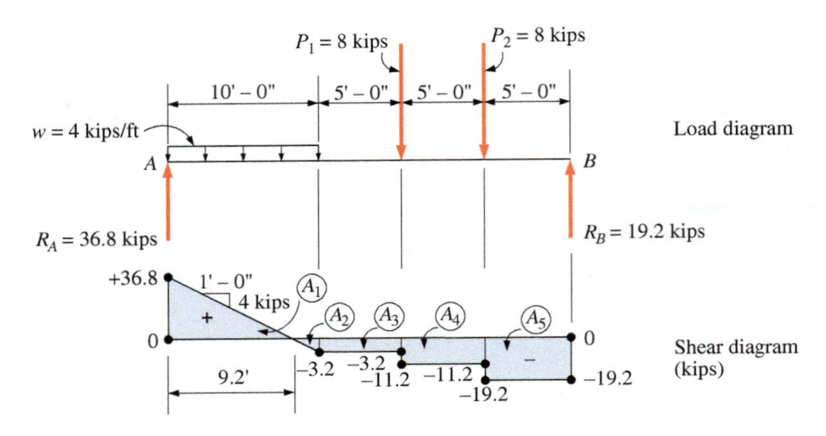

FIGURE 13.28 Load and shear diagrams.

A check on the shear diagram and the calculations is made by comparing the positive shear area with the negative shear area. Their absolute values should be equal. The areas are calculated by breaking the shear diagram into simple geometric shapes. First, determine the location of the zero shear point. Since the slope of the shear diagram in the uniformly distributed load portion of the beam is equal to 4 kips/ft, the distance to zero shear from the left reaction must be

$$\frac{36.8\text{ k}}{4\text{ k/ft}} = 9.2\text{ ft}$$

Therefore, for positive shear area,

$$A_1 = \left(\frac{1}{2}\right)(36.8\text{ k})(9.2\text{ ft}) = 169.3\text{ k-ft}$$

and for negative shear areas,

$$A_2 = \left(\frac{1}{2}\right)(3.2\text{ k})(0.8\text{ ft}) = 1.3\text{ k-ft}$$

$$A_3 = 3.2\text{ k}(5.0\text{ ft}) = 16.0\text{ k-ft}$$

$$A_4 = 11.2\text{ k}(5.0\text{ ft}) = 56.0\text{ k-ft}$$

$$A_5 = 19.2\text{ k}(5.0\text{ ft}) = 96.0\text{ k-ft}$$

The total negative shear area is 169.3 k-ft, which is equal in absolute value to the positive shear area. Therefore, the shear diagram and calculations check.

EXAMPLE 13.9 Draw the shear diagram for the cantilever beam in Figure 13.29. Neglect the weight of the beam.

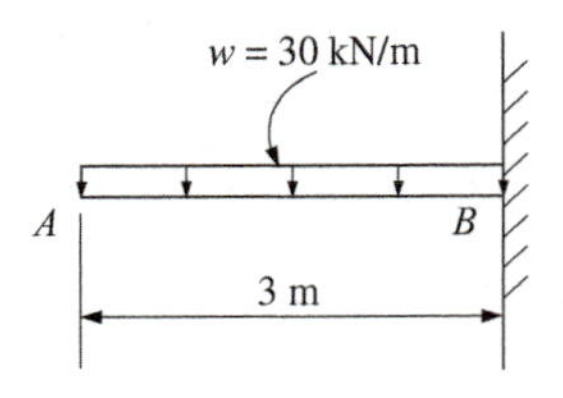

FIGURE 13.29 Beam diagram.

Solution Reactions for a cantilever beam include both vertical and moment reactions. Although only the vertical reaction will be required for the construction of the shear diagram, calculate both reactions. Figure 13.30 shows the load diagram for the given beam with the two reactions to be found. From a summation of vertical forces,

$$R_B = wL = (30\ \text{kN/m})(3\ \text{m}) = 90\ \text{kN}$$

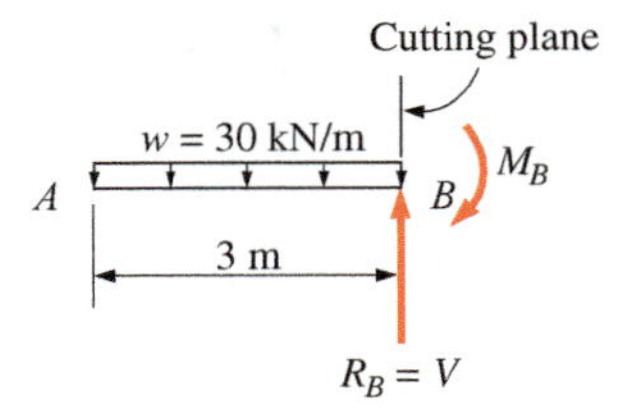

FIGURE 13.30 Load diagram.

From a summation of moments about the right support,

$$M_B = -\frac{wL^2}{2} = -\frac{(30\ \text{kN/m})(3.0\ \text{m})^2}{2} = -135\ \text{kN}\cdot\text{m}$$

The negative sign indicates tension in the top. Free-body diagrams may next be drawn from the free left end to any location along the beam length and a summation of vertical forces performed, as previously done. Recall that the shear should be computed at the beginning and end of all distributed loads. The beginning of the uniformly distributed load is at the free end, and, since no load (or beam) exists to the left of this point or at this point, the shear at the free end is zero. The shear at the right end of the load can be computed by using the load diagram (which is the free-body diagram of the whole beam), as shown in Figure 13.30, and summing vertical forces:

$$V = -(30\ \text{kN/m})(3\ \text{m}) = -90\ \text{kN}$$

Note that this is equal in magnitude to the vertical reaction at the support. The negative sign for shear indicates that the left side of the section moves down with respect to the right.

After computing the two shear values, the shear diagram may be drawn as shown in Figure 13.31. The slope is equal to the magnitude of the uniformly distributed load (30 kN/m). Note that the shear for a cantilever beam will always be negative assuming that the loads are vertically downward and that the beam is drawn with the free end on the left. Note too that the maximum shear in a cantilever beam will always occur at the fixed end.

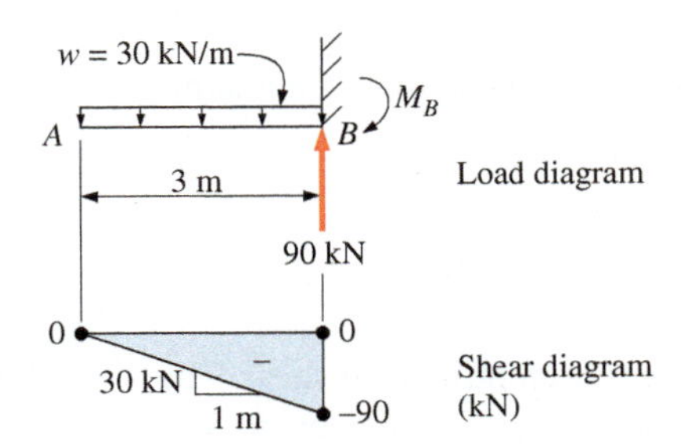

FIGURE 13.31 Load and shear diagrams.

By now it may be evident that the drawing of free bodies may really be neglected in the process of constructing shear diagrams. Once reactions have been determined, the shear diagram may be drawn by successive summations of forces (and reactions, as shown on a load diagram) beginning at the left side. If upward acting forces are considered positive and downward acting forces negative, the correct signs for shear will result. The method is illustrated in Example 13.10.

EXAMPLE 13.10 Draw the shear diagram for the beam having the load diagram shown in Figure 13.32.

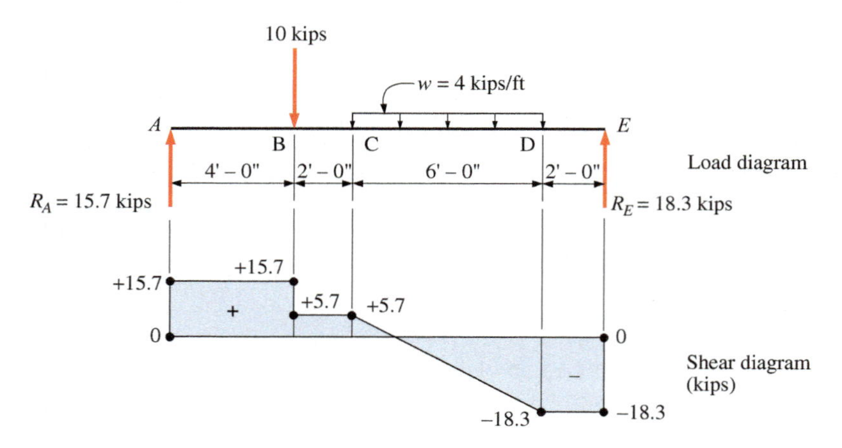

FIGURE 13.32 Load and shear diagrams.

Solution For convenience, the load diagram has been labeled with points *A* through *E*. Since the reactions are given, the first step is to draw the horizontal baseline for the shear diagram.

Starting at the left, point *A*, the shear diagram goes up to a value of +15.7 kips due to the reaction. Proceeding to the right, there is no change in the shear until the load at point *B* is encountered. At that point, the sum of the vertical forces decreases (the load is downward) by 10 kips and the shear diagram drops to +5.7 kips. Again proceeding to the right, there is no change until the beginning of the uniform load at *C* is encountered. The total change in the sum of the vertical forces between *C* and *D* is −24 kips, resulting in shear of −18.3 kips at *D*. The shear diagram between *C* and *D* must be a sloping straight line. Note that the line slopes downward to the right. Again proceeding to the right, there is no change in shear between *D* and *E*, resulting in a shear of −18.3 kips at *E*. Note that the right reaction is 18.3 kips. Such details as locating the point of zero shear and checking negative and positive shear areas may then be accomplished in the usual way.

13.5.1 Summary of Procedure: Shear Diagram Construction

The procedure for construction of the shear diagram is summarized as follows:

1. Sketch the load diagram.
2. Compute the reactions and note these on the load diagram.
3. Draw a horizontal baseline representing zero shear. The baseline should be the same length as the beam and directly below the load diagram.
4. Draw projection lines vertically downward from the following key points on the load diagram:
   (a) Each reaction and concentrated load.
   (b) The beginning and end of each distributed load.
5. Sketch the shear diagram using a continuous algebraic summation of vertical forces and reactions from the load diagram starting from the left end of the beam. Alternatively, compute the shear values at the key points from step 4 using free-body diagrams (calculate shear on each side of concentrated load locations), plot the values, and sketch the shear diagram.
6. Locate the point(s) of zero shear.

13.6 MOMENT DIAGRAMS

Just as a shear diagram shows the amount of shear at any point along the length of a beam, a moment diagram shows the amount of bending moment at any location. The moment diagram is drawn directly under the shear diagram. A moment baseline representing zero bending moment is drawn parallel to the shear baseline. The abscissa along the baseline represents the locations of successive cross sections of the beam. The ordinate of the diagram represents the value of the bending moment at that particular cross section.

As with shear diagrams, moment diagrams need not be drawn to scale. Moment diagrams are most conveniently drawn using moments that have been determined at certain key points where moment changes occur. The points to consider are as follows:

1. At each concentrated load and reaction
2. At points of zero shear and where the shear diagram goes through zero
3. At the beginning and end of all distributed loads

The following important general rules concerning moment diagrams should be noted:

1. The bending moment at the ends of a simply supported, single-span beam will always be equal to zero.
2. With loads acting vertically downward, the bending moment at the free end of a cantilever beam (fixed at one end and free at the other end) will always be equal to zero and the maximum moment will always occur at the fixed end. The shear will also be maximum at the fixed end.
3. The bending moment will always be positive for a simply supported, single-span beam and negative for a cantilever beam, assuming that all loads act vertically downward.
4. Except for cantilever beams, the maximum bending moment will always occur at a point of zero shear or where the shear diagram goes through zero.

The construction of the moment diagram is best illustrated through a variety of examples.

EXAMPLE 13.11 Draw the bending moment diagram for the beam in Figure 13.33. (The shear diagram for this beam was drawn in Example 13.6.)

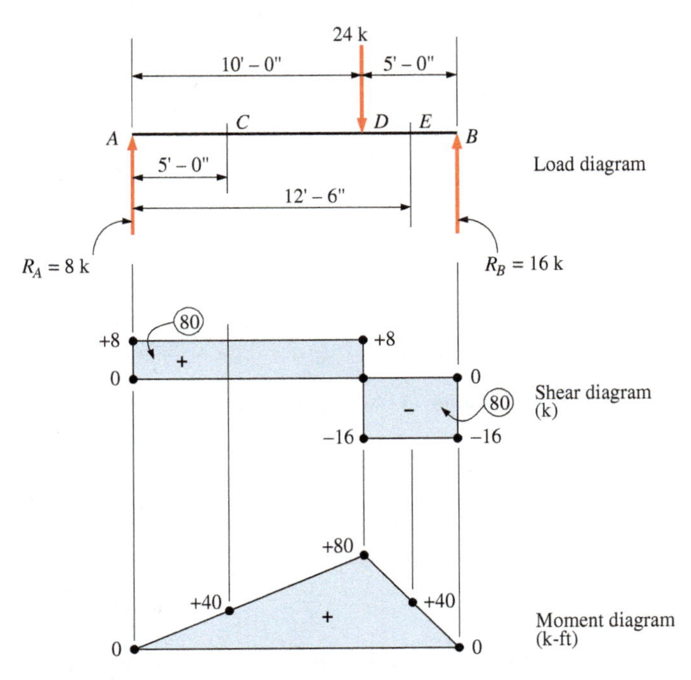

FIGURE 13.33 Load, shear, and moment diagrams.

Solution The load diagram and shear diagram are given. The procedure for computing the bending moment at any point along the length of a beam is discussed in Section 13.4. A section is cut at the point in question and a free body of the beam segment to the left of the section is drawn. The bending moment at the section can then be calculated.

For the beam of this example, the bending moment at each simple support is zero. Cutting sections at points *C*, *D*, and *E* at 5 ft, 10 ft, and 12.5 ft, respectively, from the left end of the beam, bending moments can be calculated as follows:

$$\text{at point } C\text{: } M = +(8\text{ k})(5\text{ ft}) = +40\text{ k-ft}$$

$$\text{at point } D\text{: } M = +(8\text{ k})(10\text{ ft}) = +80\text{ k-ft}$$

$$\text{at point } E\text{: } M = +(8\text{ k})(12.5\text{ ft}) - (24\text{ k})(2.5\text{ ft}) = +40\text{ k-ft}$$

These three values are plotted on the moment diagram of Figure 13.33, resulting in two straight sloping lines.

With reference to Figure 13.33, several interesting relationships between shear and moment diagrams may be observed:

1. The maximum moment occurs at the point where the shear goes through zero.
2. The change in magnitude of the moment between any two points is equal to the area of the shear diagram between those two points. This means that the moment at any point can be determined by the summation of the shear area up to that point, always working from left to right.
3. The slope of the moment diagram at any point is equal in sign and magnitude to the shear at that point. The slope of the moment diagram to the left of the 24-kip load is +8 kips, while to the right of the 24-kip load the slope is −16 kips. (This concept is explained further in Example 13.12.)

EXAMPLE 13.12 Draw the bending moment diagram for the beam in Figure 13.34. (This is the beam that was analyzed in Example 13.7.)

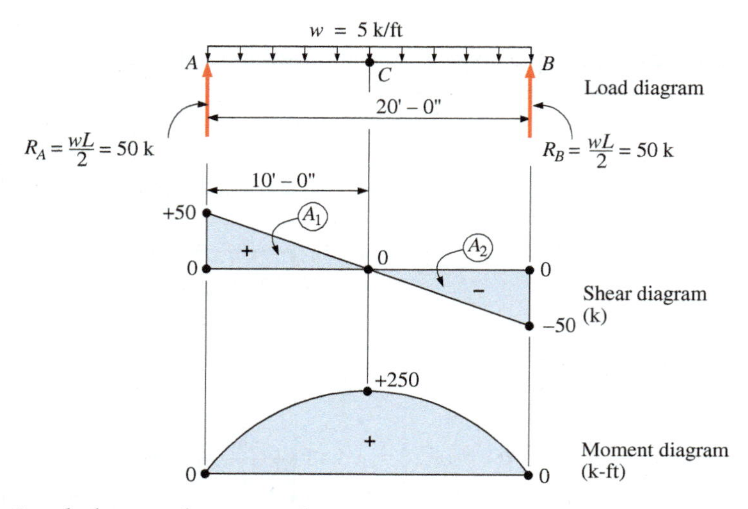

FIGURE 13.34 Load, shear, and moment diagrams.

Solution For the simply supported, single-span beam, the moments at the supports are zero. The only other moment needed is that at *C*, midspan, where the shear is zero. Determine the moment at *C* by noting that the change in moment between *A* and *C* is equal to the area of shear diagram between *A* and *C*. This triangular area, designated A_1 on the shear diagram, is calculated from

$$A_1 = \tfrac{1}{2}(10\text{ ft})(+50\text{ k}) = +250\text{ k-ft}$$

The moment at *A* is zero. Since the change in moment between *A* and *C* is A_1, the moment at *C* is calculated from:

$$M = 0 + A_1 = 0 + 250\text{ k-ft} = 250\text{ k-ft}$$

Note that since A_2 is equal to A_1 in magnitude, but opposite in sign, the summation of shear areas from point *A* to point *B* is zero. Therefore, the moment at point *B* is zero.

Sketch in the actual curve of the moment diagram. Note that proceeding to the right from point *A* to point *C* the shear is positive and decreasing, which means that the slope of a line tangent to the moment diagram is positive, but decreasingly positive as point *C* is approached. Also, from point *C* to point *B*, the shear is negative and is increasingly negative as point *B* is approached, which means that the slope of a line tangent to the moment diagram is increasingly negative proceeding from *C* to *B*. (Signs of slopes are reviewed in Figure 13.35.) At midspan, the shear is zero. Therefore, the slope of a line tangent to the moment diagram at midspan is zero (the tangent to the curve is horizontal). The uniform load results in a moment diagram that is concave downward. This shape could also be verified by computing moments at arbitrary intermediate points.

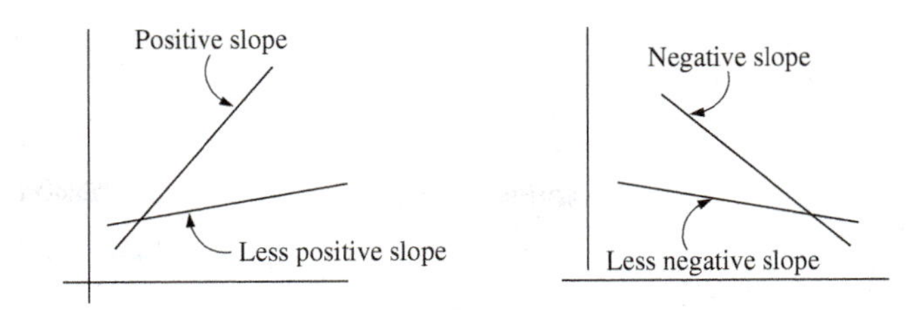

FIGURE 13.35 Signs of slopes.

Note that the shape of the moment diagram in areas of uniformly distributed load (assuming downward, or gravity, load) will be a concave downward curve. Further, a moment diagram will contain a vertical line when a pure moment is applied as a beam load (and also at the fixed end of a cantilever beam).

EXAMPLE 13.13

Draw the bending moment diagram for the cantilever beam for which the load and shear diagrams are given in Figure 13.36. (This is the beam of Example 13.9.)

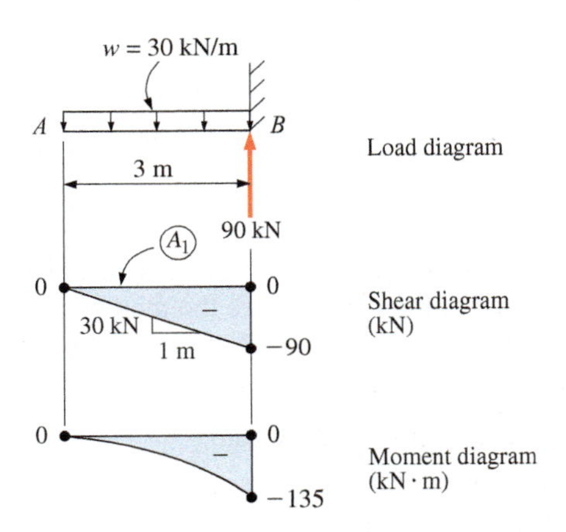

FIGURE 13.36 Load, shear, and moment diagrams.

Solution The load and shear diagrams are given. Since the moment at the free end is zero and the maximum moment will always occur at the fixed end, the only calculation necessary is that for the moment at the fixed end. Calculate the total area of the shear diagram A_1:

$$A_1 = \tfrac{1}{2}(3 \text{ m})(-90 \text{ kN}) = -135.0 \text{ kN} \cdot \text{m}$$

This value is the change in moment from *A* to *B* and, therefore, the moment at *B*. Note that the shear, starting at point *A*, is zero and then becomes increasingly negative. The curve of the moment diagram is sketched in accordingly using relative slopes from Figure 13.35 as a guide.

EXAMPLE 13.14

Draw the moment diagram for the beam for which load and shear diagrams are given in Figure 13.37. (This is the beam of Example 13.8.)

Solution The load and shear diagrams are given. For convenience, intermediate points of interest between the endpoints *A* and *B* have been labeled *C* through *F*. Calculate the various areas of the shear diagram in units of k-ft and obtain, for the triangular areas,

$$A_1 = +169.3 \text{ k-ft}$$
$$A_2 = -1.3 \text{ k-ft}$$

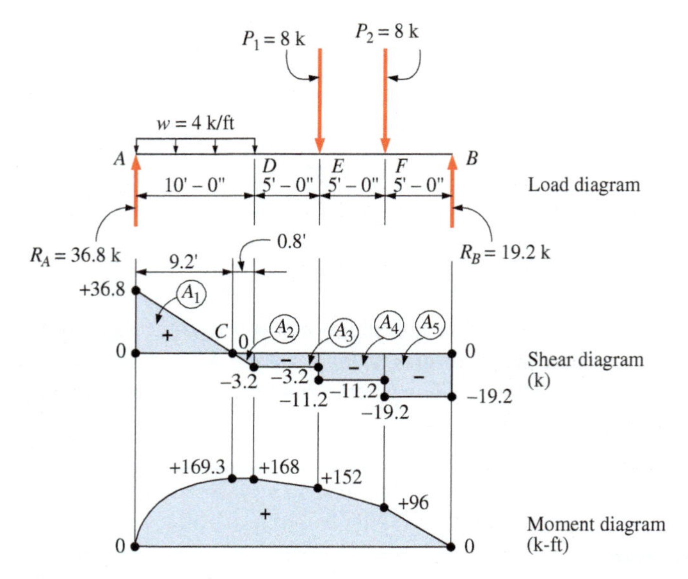

FIGURE 13.37 Load, shear, and moment diagrams.

and, for the rectangular areas,

$$A_3 = -16.0 \text{ k-ft}$$
$$A_4 = -56.0 \text{ k-ft}$$
$$A_5 = -96.0 \text{ k-ft}$$

The moment values can now be calculated using the shear diagram areas. Working from the left end of the beam where the moment is equal to zero, the moments at all of the points of interest can be computed by adding algebraically the successive areas of the shear diagram. The steps are as follows:

1. The moment at A is zero.
2. The moment at C, the point of zero shear, is equal to the shear area between A and C:

$$M = +169.3 \text{ k-ft}$$

3. The moment at D, the end of the distributed load, is equal to the moment at C plus the shear area between C and D:

$$M = 169.3 \text{ k-ft} - 1.3 \text{ k-ft} = +168.0 \text{ k-ft}$$

4. The moment at E, under the load P_1, is equal to the moment at D plus the shear area between D and E:

$$M = 168.0 \text{ k-ft} - 16.0 \text{ k-ft} = +152.0 \text{ k-ft}$$

5. The moment at F, under the load P_2, is equal to the moment at E plus the shear area between E and F:

$$M = 152.0 \text{ k-ft} - 56.0 \text{ k-ft} = +96.0 \text{ k-ft}$$

6. The moment at the right support should be zero and can be checked by adding the shear area between F and B to the moment at F:

$$M = 96.0 \text{ k-ft} - 96.0 \text{ k-ft} = 0 \qquad \textbf{(O.K.)}$$

With all necessary moment values computed, the shape of the moment diagram between the points may be determined by inspection of the shear diagram. The moment diagram will be parabolic from the left end to the end of the distributed load and will be composed of sloping straight lines otherwise. Note that the moment is always positive. This situation is typical for single-span, simply supported beams subjected to vertical downward loads.

13.6.1 Summary of Procedure: Moment Diagram Construction

In summary, the procedure for the construction of the moment diagram, once the shear diagram has been completely defined, may be stated as follows:

1. Draw a horizontal baseline representing zero moment directly below the shear diagram. The baseline has the same length as the beam and shear diagrams.
2. Extend guidelines used for the shear diagram vertically downward to the moment diagram baseline.
3. Calculate the shear areas between key points and note these on the shear diagram. Compute the moments by adding algebraically the successive shear areas starting at the left end of the beam. Alternatively, starting from the left end of the beam, use free-body diagrams to compute moments at the key points (and points of zero shear).
4. Plot the moment values. (Although the moment diagram need not be to scale, using approximately correct proportions is recommended for clarity.)
5. Sketch the correct shape of the moment diagram between the plotted points by referring to the shear diagram.

13.7 SECTIONS OF MAXIMUM MOMENT

In the case of moment, as in the case of shear, the largest numerical value, regardless of sign, is called the maximum. The positive sign merely indicates that the beam, or beam segment, is concave upward, whereas the negative sign indicates that it is concave downward.

As previously stated, except for the cantilever beam the maximum bending moment occurs at some location on the beam where the shear is equal to zero or the shear diagram goes through zero. The location of maximum bending moment is sometimes called the *critical section*. This term is used because the bending moment will develop a bending stress, which is generally the critical or controlling stress with respect to beam design and analysis.

In the case of a beam subjected to distributed load and concentrated loads, the shear diagram may or may not go through zero under a concentrated load. The exact location of zero shear must be calculated. In overhanging beams subject to various types of loads, the shear may equal zero and/or go through zero more than once. Hence, there may exist two or more locations of a possible maximum moment. When this occurs, the moment must be computed at each location, with the largest numerical value representing the maximum moment. In the design process, however, it is sometimes necessary that both a maximum positive and a maximum negative moment be used. This condition has particular relevance for reinforced concrete design.

In single-span, simply supported beams subjected to vertical downward loads, only one critical location will occur resulting in a maximum positive moment. In cantilever beams, both maximum moment and maximum shear occur at the support or fixed end.

EXAMPLE 13.15 Draw the shear and moment diagrams for the overhanging beam in Figure 13.38. The reactions are given.

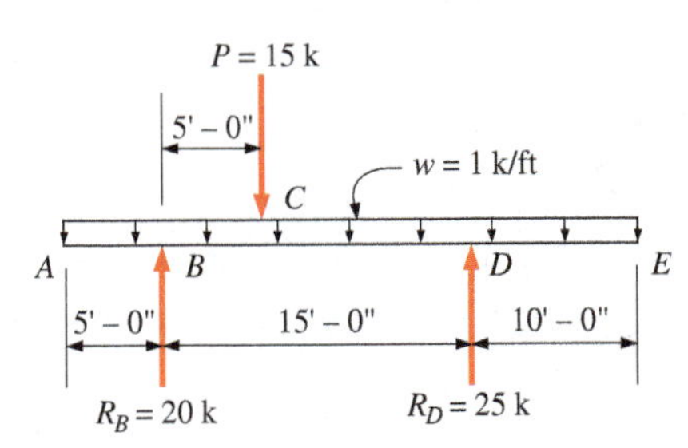

FIGURE 13.38 Load diagram.

Solution **(a)** First, draw the shear diagram by summing the vertical forces beginning at point A and proceeding to the right. The shear at A is zero. Therefore, the shear diagram begins at zero (see Figure 13.39). Between A and B, the shear goes from zero to a value of −5 kips as a result of the uniformly distributed load between these two points. At point B, the upward acting reaction of 20 kips produces an abrupt change in the shear, causing it to go from a value of −5 kips to a value of +15 kips.

This procedure continues across the beam and the shears are computed by successive algebraic addition of each load (or reaction) to the shear already calculated to the left of the load. The shears are plotted as they are computed. Note that the shear must become zero at point E. When the shear diagram is complete, note that the maximum shear value is 15 kips.

(b) To draw the moment diagram, next calculate the shear areas (in units of k-ft). The negative areas are

$$A_1 = \tfrac{1}{2}(5)(-5) = -12.5 \text{ k-ft}$$
$$A_3 = \tfrac{1}{2}(-5 - 15)(10) = -100.00 \text{ k-ft}$$

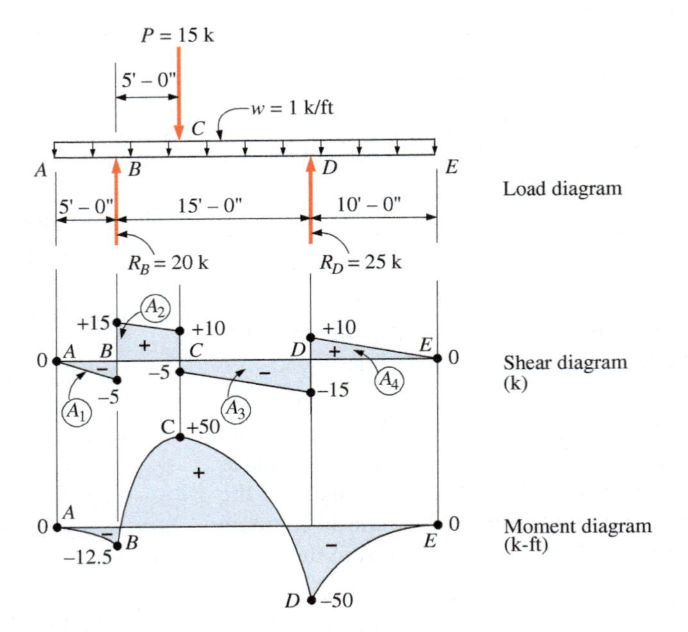

FIGURE 13.39 Load, shear, and moment diagrams.

And for the positive areas,

$$A_2 = \tfrac{1}{2}(15 + 10)(5) = +62.5 \text{ k-ft}$$
$$A_4 = \tfrac{1}{2}(10)(10) = +50 \text{ k-ft}$$

A check on the calculations can be made by ensuring that negative and positive shear areas are equal:

$$50 \text{ k-ft} + 62.5 \text{ k-ft} = 100 \text{ k-ft} + 12.5 \text{ k-ft}$$
$$112.5 \text{ k-ft} = 112.5 \text{ k-ft} \qquad \textbf{(O.K.)}$$

This check indicates that the calculated moment at the right end of the beam will be zero, as it should be in this case.

The moments can now be calculated at those points recommended in Section 13.6. Using the shear diagram areas, the moments are found by adding algebraically the successive areas, starting at the left end:

$$M_A = 0$$
$$M_B = M_A + A_1 = 0 - 12.5 \text{ k-ft} = -12.5 \text{ k-ft}$$
$$M_C = M_B + A_2 = -12.5 \text{ k-ft} + 62.5 \text{ k-ft} = +50.0 \text{ k-ft}$$
$$M_D = M_C + A_3 = +50.0 \text{ k-ft} - 100.0 \text{ k-ft} = -50.0 \text{ k-ft}$$
$$M_E = M_D + A_4 = -50 \text{ k-ft} + 50 \text{ k-ft} = 0$$

The moment diagram is drawn in Figure 13.39. The curvature between all the points may be observed as concave downward based on inspection of the shear diagram:

A to *B*: negative shear, increasing in negative value
B to *C*: positive shear, decreasing
C to *D*: negative shear, increasing in negative value
D to *E*: positive shear, decreasing

Note that at point *C*, a maximum positive moment of 50 k-ft occurs and at point *D* a maximum negative moment of −50 k-ft occurs. Note also that the moment diagram crosses the zero moment baseline at two locations. These points are called *points of inflection* and represent points of zero moment. They can be located by setting up an expression for the bending moment and equating it to zero.

13.8 MOVING LOADS

The structural design of bridge members, as well as that of crane girders in industrial buildings, is based on moving live loads. Thus far, all our discussion has assumed fixed loads.

If a system of concentrated loads with fixed distances between the loads moves across a simple beam, the moment under any given load will increase from zero to a maximum value and then decrease to zero again as the load moves across the beam and approaches the support. If we neglect the weight of the beam and other distributed loads, the maximum moment will always occur directly under a load since the shear goes through zero at a load point. It is evident that the magnitude of both the shear and bending moment will change as the load system moves across the beam. The position of the load system that will develop a maximum bending moment will always be different from the position that will develop a maximum shear.

Taking an arbitrary moving load system as shown in Figure 13.40, the location of load P_2 when the moment under P_2 is a maximum has been established mathematically. It can be shown that when the moment under P_2 is a maximum, the midpoint of the span length bisects the distance between P_2 and the resultant of the load system. This situation is true for any of the loads. Therefore, we state general rule 1 as follows:

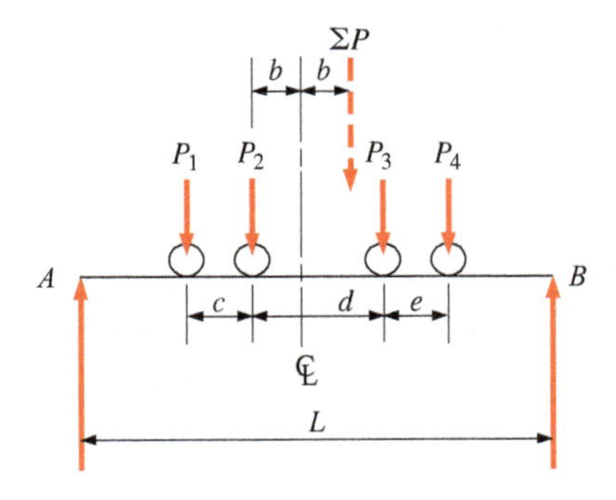

FIGURE 13.40 Moving load system.

1. The moment under any particular chosen load in a moving-load system is maximum when the load system is positioned on the beam so that the centerline of the beam lies midway between the resultant of the system and the chosen load. Therefore, to position the load system for maximum moment under a chosen load:
   - **(a)** Determine the resultant of the load system.
   - **(b)** Determine the distance between the resultant and the chosen load.
   - **(c)** Place the load system on the beam so that the centerline of the beam lies midway between the resultant and the chosen load.

   Each load of the load system may be examined for its maximum moment in accordance with general rule 1. The largest of these moments is the *absolute maximum*, which must be used in the design of the beam. Additional general rules are as follows:
2. It is usually necessary to examine the maximum moment under only the two loads nearest the resultant of the load system to obtain the absolute maximum moment. If the largest load on the span is nearest the resultant, it is necessary to examine under this load only.
3. For a simply-supported beam of span L, with a load system of two equal moving loads having a distance of $0.586L$ or greater between them, the maximum moment will develop at midspan of the beam. Position the load system so that one of the two loads is at midspan and determine the moment in the beam under that load.
4. In most cases, to compute the maximum shear due to the moving-load system, the maximum load is placed over one support with as many of the other loads as possible on the beam span. The maximum shear will be the maximum reaction. In some cases, the absolute maximum shear must be obtained by trial, in which each load is successively placed over the support and the reaction calculated.

The preceding are termed "general" rules, which implies that exceptions may exist under certain conditions.

EXAMPLE 13.16 A two-axle truck weighing a total of 60 kips crosses a bridge that spans 36 ft. The front axle carries 16 kips and is 9 ft from the rear axle. Assuming that this loading is distributed equally to each of two parallel bridge girders, compute the absolute maximum moment and shear. Neglect the weight of the girder.

Solution Using loads to one girder (one-half the total load) as shown in Figure 13.41, compute the magnitude and location of the resultant load. In this parallel force system, the magnitude of the resultant is

$$22\text{ k} + 8\text{ k} = 30\text{ k}$$

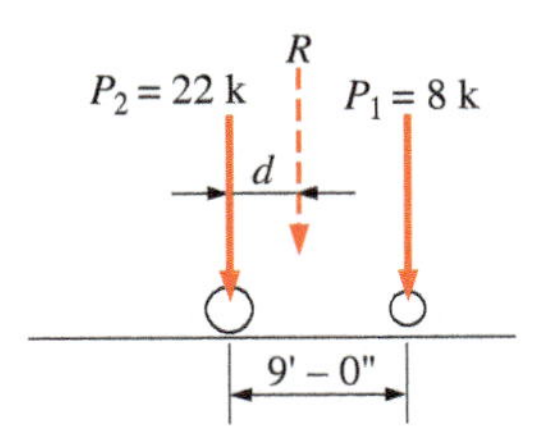

FIGURE 13.41 Resultant of a force system.

The location of the resultant is determined by summing moments with respect to P_2,

$$(8\text{ k})(9.0\text{ ft}) = (30\text{ k})\ d$$
$$d = 2.40\text{ ft}$$

To obtain the absolute maximum moment, place the centerline of the span midway between the resultant force R and the nearest load P_2, as shown in Figure 13.42. The absolute maximum moment will be directly under P_2.

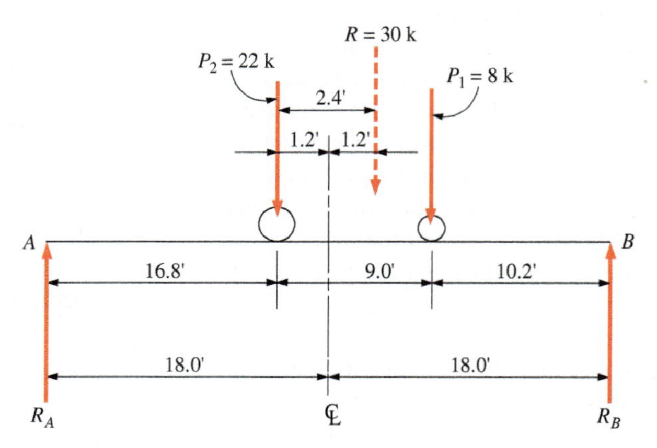

FIGURE 13.42 Load diagram.

Compute R_A:

$$\Sigma M_B = -R_A(36\text{ ft}) + (8\text{ k})(10.2\text{ ft}) + (22\text{ k})(19.2\text{ ft}) = 0$$

from which

$$R_A = 14\text{ k}$$

It is not necessary to compute R_B since only the left beam segment up to P_2 will be taken out as a free body. (Refer to Figure 13.43.)

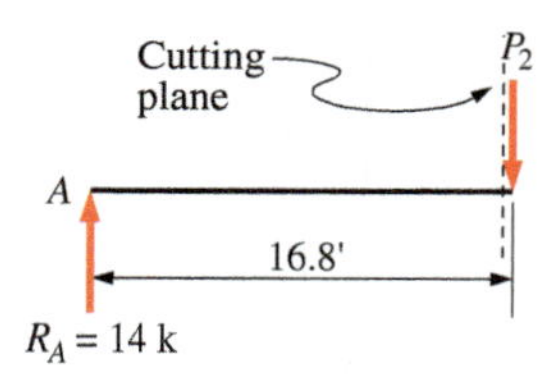

FIGURE 13.43 Free-body diagram.

Calculate the moment at P_2:

$$M_{P_2} = +(14\text{ k})(16.8\text{ ft}) = +235\text{ k-ft}$$

The absolute maximum shear can be obtained by moving the load system as shown in Figure 13.44. The load of 22 kips should be considered to be at an infinitesimal distance to the right of the support at A.

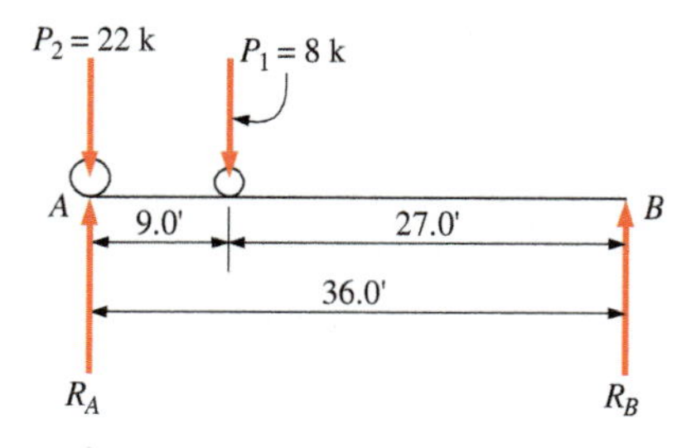

FIGURE 13.44 Load diagram.

Compute R_A:

$$\Sigma M_B = -R_A(36 \text{ ft}) + (22 \text{ k})(36 \text{ ft}) + (8 \text{ k})(27 \text{ ft}) = 0$$

from which

$$R_A = 28 \text{ k}$$

This value represents the absolute maximum shear.

EXAMPLE 13.17

The load system shown in Figure 13.45 is designated as an HS 20–44 truck load and is commonly used in the design of highway bridges. Compute the absolute maximum shear and moment that would occur in a simple span bridge subjected to this load system. The bridge span is 40 ft. Neglect the weight of the bridge itself.

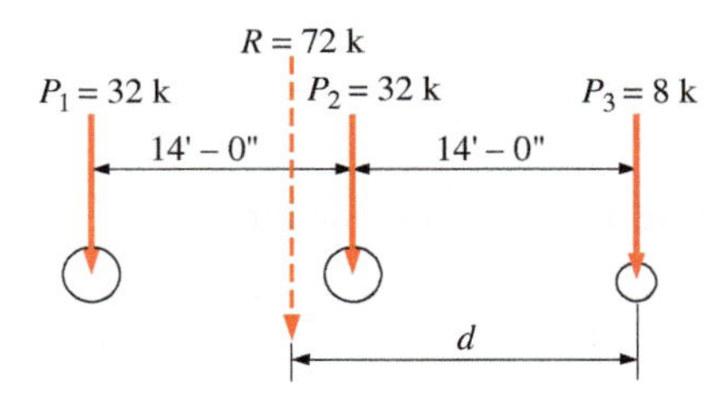

FIGURE 13.45 Standard truck HS 20–44 load.

Solution The magnitude of the resultant load is

$$32 \text{ k} + 32 \text{ k} + 8 \text{ k} = 72 \text{ k}$$

The location of the resultant is determined by summing moments with respect to P_3:

$$32 \text{ k}(28 \text{ ft}) + 32 \text{ k}(14 \text{ ft}) = (72 \text{ k})d$$
$$d = 18.67 \text{ ft}$$

The centerline of the span is placed midway between the resultant force R and the nearest load P_2. (Refer to Figure 13.46.) The absolute maximum moment will be directly under P_2.

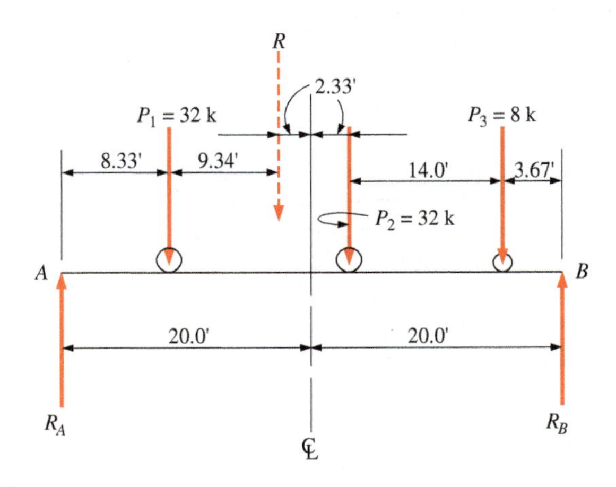

FIGURE 13.46 Load diagram.

Compute R_A:

$$\Sigma M_B = -R_A(40 \text{ ft}) + (8 \text{ k})(3.67 \text{ ft}) + (32 \text{ k})(17.67 \text{ ft}) + (32 \text{ k})(31.67 \text{ ft}) = 0$$

from which

$$R_A = 40.2 \text{ k}$$

Calculate the moment under P_2 using as a free body the beam segment to the left of P_2:

$$M_{P_2} = (40.2\text{ k})(22.33\text{ ft}) - (32\text{ k})(14\text{ ft}) = +450\text{ k-ft}$$

The absolute maximum shear may be obtained by moving the load system as shown in Figure 13.47. Again, compute R_A:

$$\Sigma M_B = -R_A(40\text{ ft}) + (32\text{ k})(40\text{ ft}) + (32\text{ k})(26\text{ ft}) + (8\text{ k})(12\text{ ft}) = 0$$

from which

$$R_A = 55.2\text{ k}$$

This value represents the absolute maximum shear.

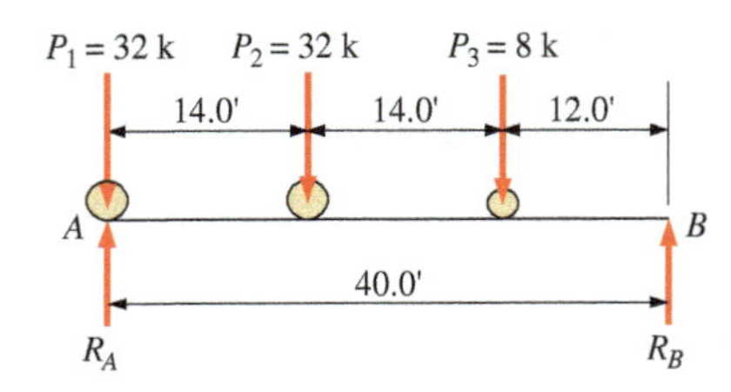

FIGURE 13.47 Load diagram.

SUMMARY BY SECTION NUMBER

13.1 Beams are categorized according to type and/or number of supports. Types of beams in common use are the simple, cantilever, and overhanging beams of which all are statically determinate, since their reactions can be computed using the three laws of equilibrium.

13.2 Loads on beams are either concentrated or distributed. Distributed line loads may be uniform or nonuniform. Concentrated loads act at a single point; distributed line loads are spread out over a length of the beam.

13.3 External beam reactions may be computed by using a free-body diagram of the beam (called a *load diagram*) and applying the three laws of equilibrium. The algebraic sum of all externally applied loads and reactions must equal zero.

13.4 The effect of the externally applied loads and reactions on a beam is to develop internal shear force and bending moment. The shear force is the algebraic sum of the external vertical forces acting on one side of a cutting plane. The bending moment is the algebraic sum of the moments of all the external forces acting on one side of a cutting plane.

13.5 A shear diagram is a graphical representation showing how the shear varies along the length of a beam. It is drawn directly below the load diagram. The shape of the shear diagram is a function of the load diagram.

13.6 A moment diagram is a graphical representation showing how the moment varies along the length of a beam. It is drawn directly below the shear diagram. The shape of the moment diagram is a function of the shear diagram.

13.7 The critical section of a beam occurs where the moment is a maximum. In beams other than the cantilever, the maximum moment will occur at a point of zero shear or where the shear goes through zero.

13.8 Moving-load systems create varying patterns of moment and shear within beams. The absolute maximum moment and the absolute maximum shear are created when the load system is at specific positions on the beam.

PROBLEMS

Section 13.3 Beam Reactions

13.1 through 13.6 Calculate the reactions at points *A* and *B* for the beams shown.

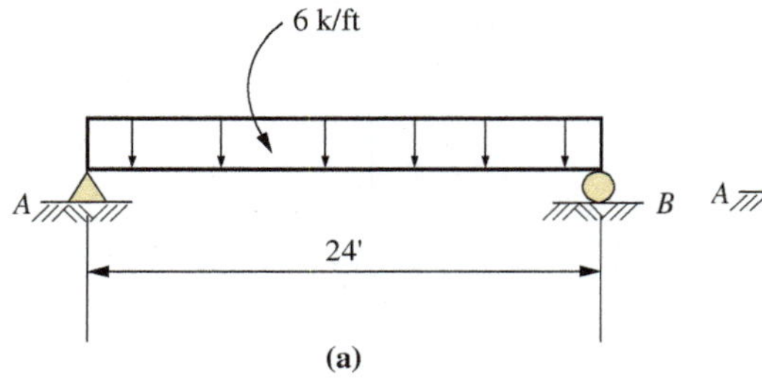

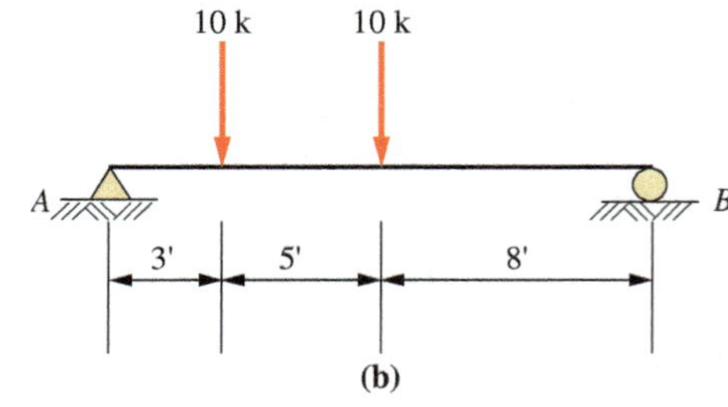

PROBLEM 13.1

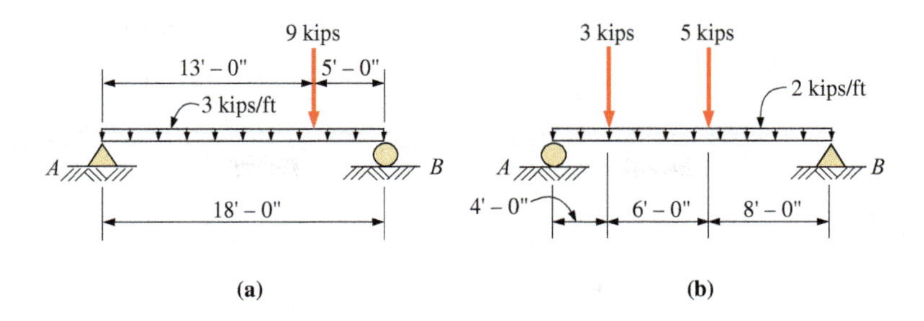

(a) (b)

PROBLEM 13.2

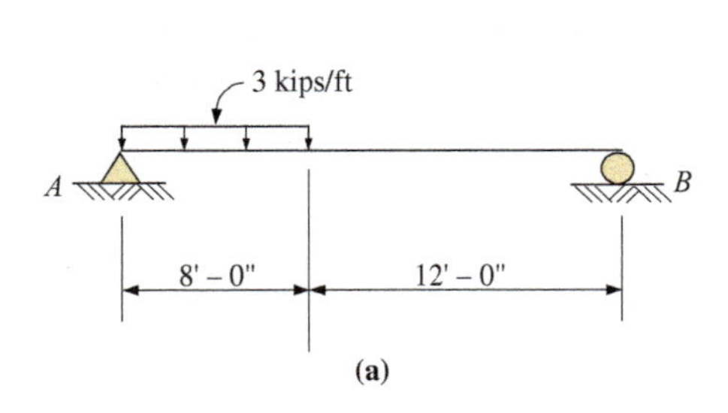

(a)

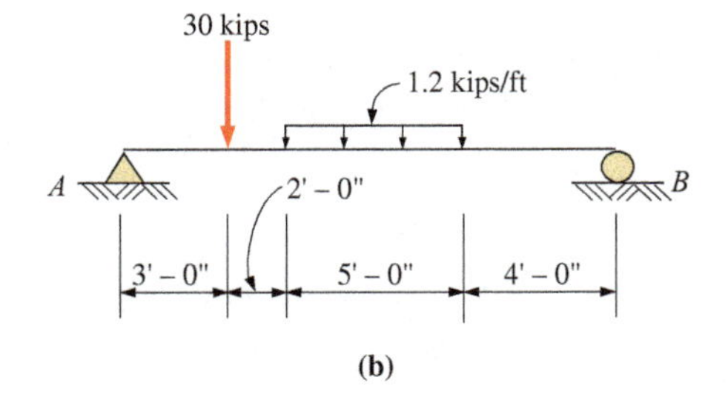

(b)

PROBLEM 13.3

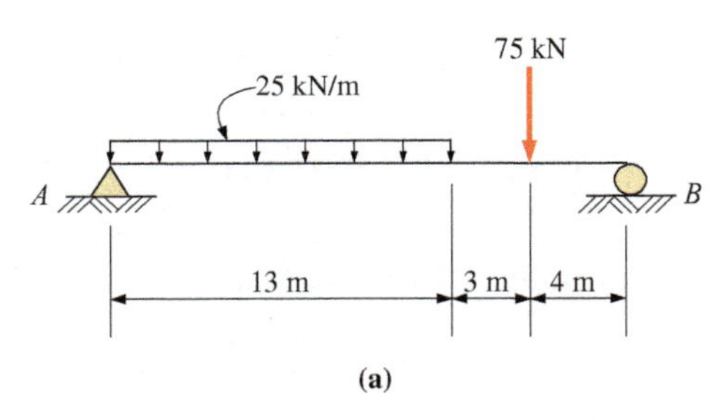

(a)

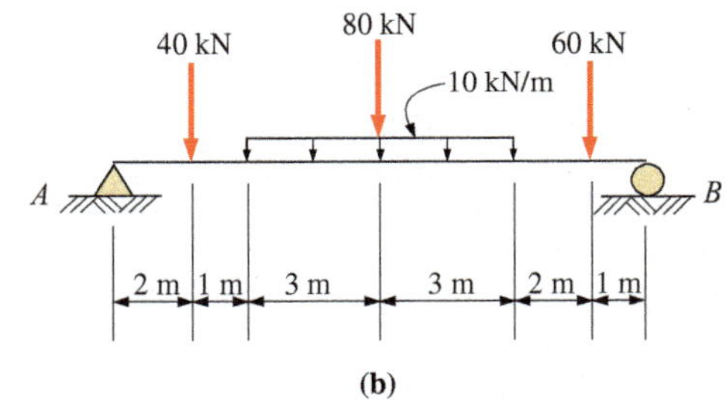

(b)

PROBLEM 13.4

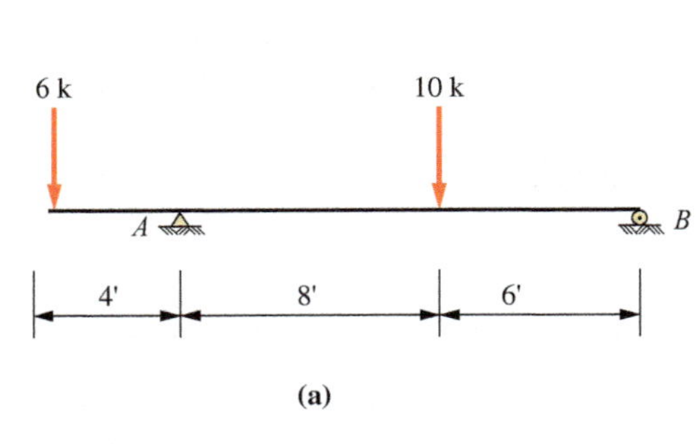

(a)

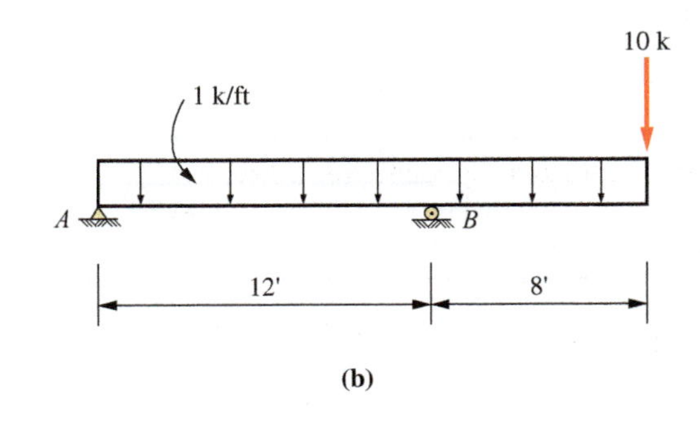

(b)

PROBLEM 13.5

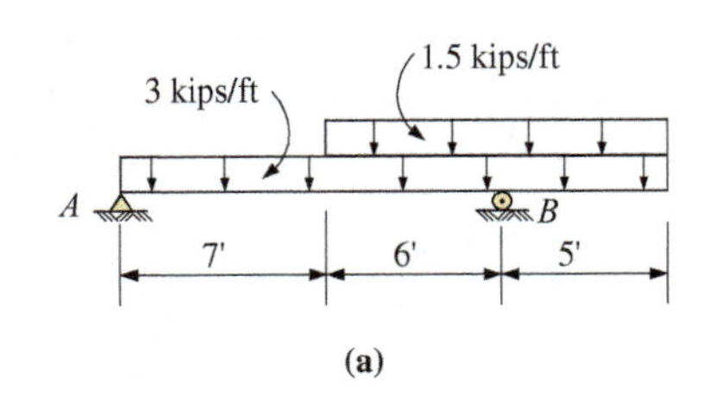

(a)

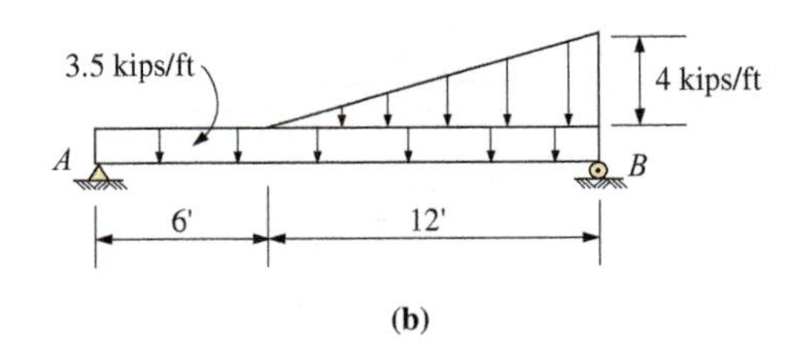

(b)

PROBLEM 13.6

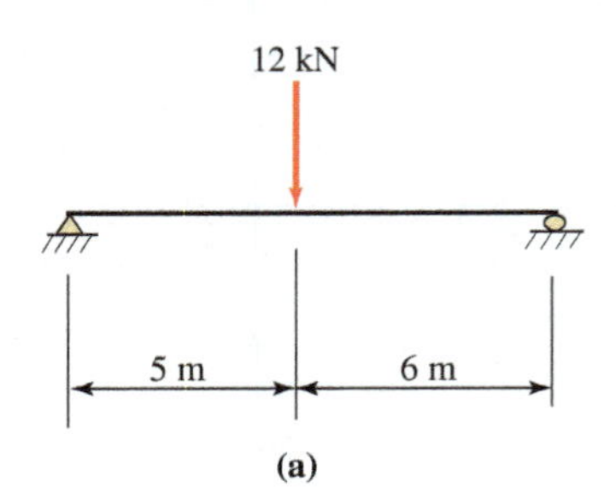

PROBLEM 13.7

(a)

(b)

Section 13.4 Shear Force and Bending Moment

13.7 Calculate the shear and bending moment at 4 m and 7 m from the left end of the beams shown. Show free-body diagrams.

13.8 Calculate the shear and bending moment at 3 ft and at 8 ft from the left for the beams shown. Show free-body diagrams.

13.9 Calculate the shear and bending moment at midspan for the beams shown. Show free-body diagrams.

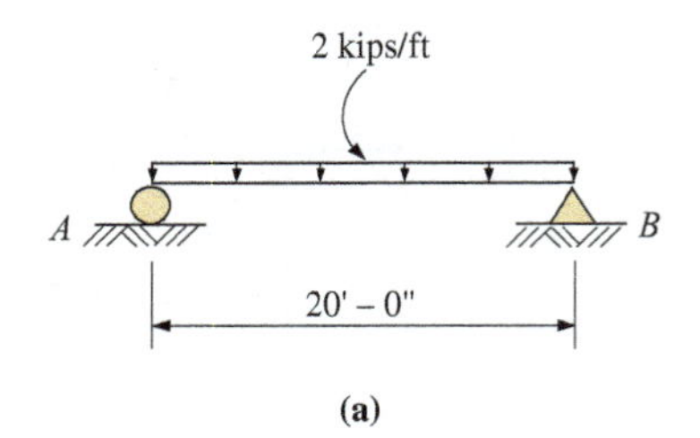

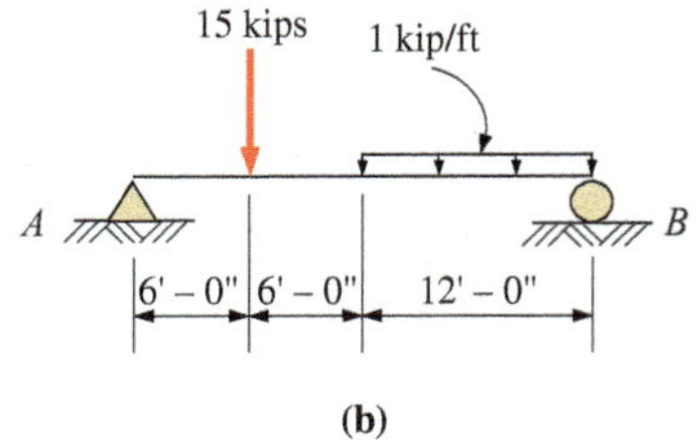

PROBLEM 13.8

(a)

(b)

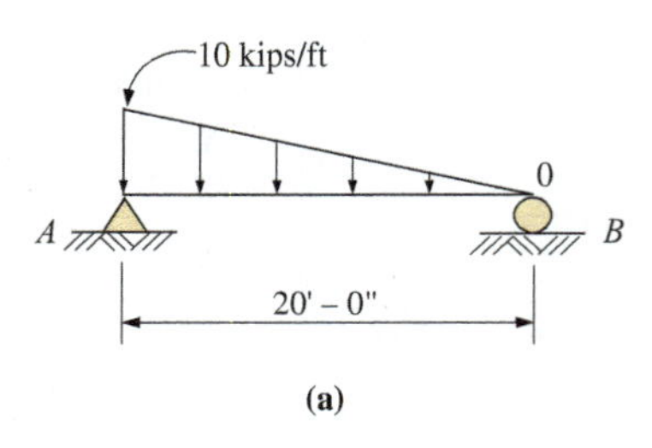

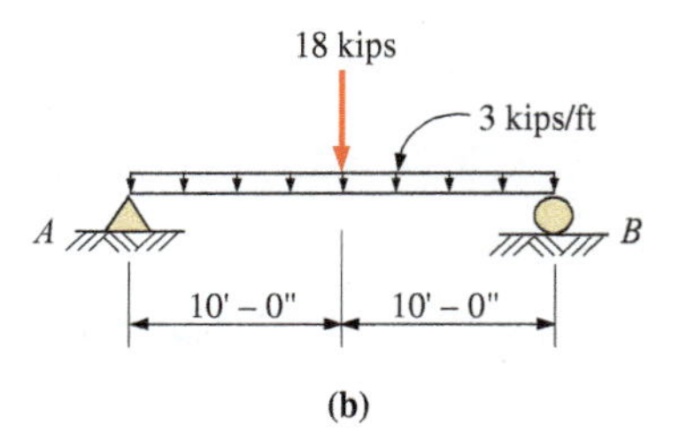

PROBLEM 13.9

(a)

(b)

13.10 Calculate the shear and bending moment at 5 ft and at 15 ft from the left for the beams shown. Show free-body diagrams.

13.11 Calculate the shear and bending moment at 5 m and 10 m from the left end of the beam shown. Show free-body diagrams.

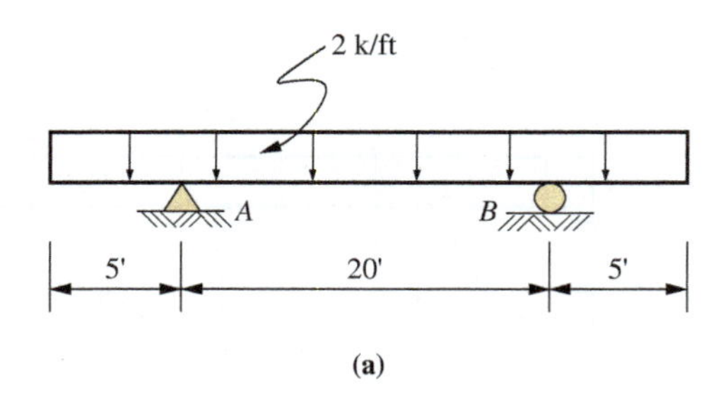

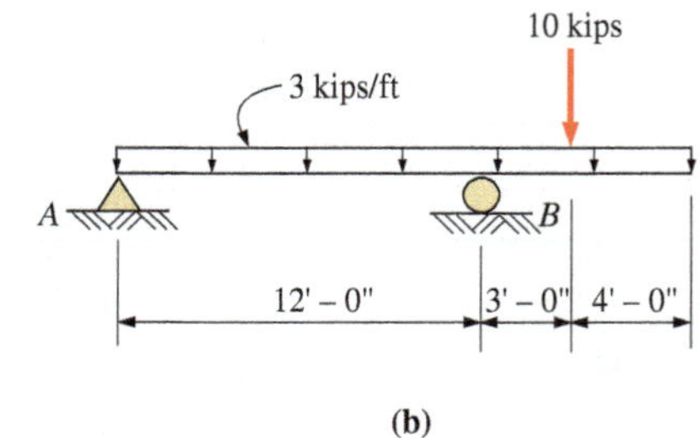

PROBLEM 13.10

(a)

(b)

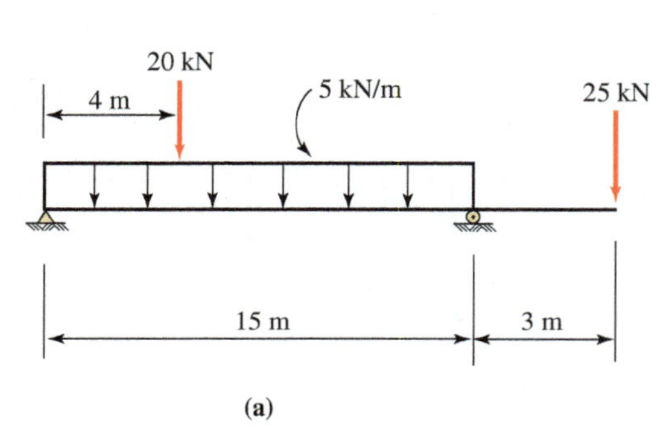

(a)

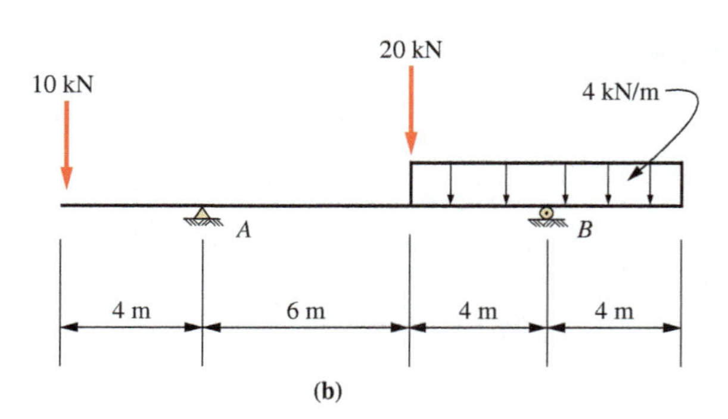

(b)

PROBLEM 13.11

Section 13.5 Shear Diagrams

13.12 through 13.15 For the beams shown, draw complete shear diagrams.

Section 13.6 Moment Diagrams

13.16 through 13.20 For the beams shown (next page), draw complete shear and moment diagrams.

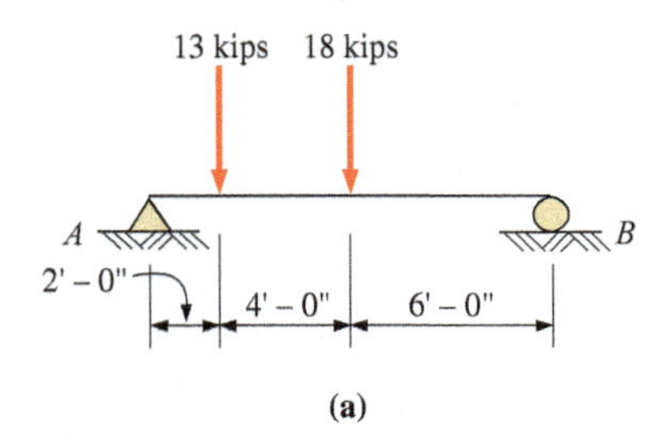

(a)

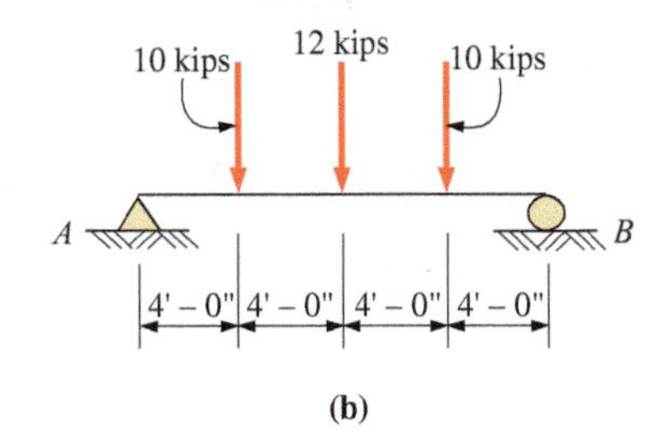

(b)

PROBLEM 13.12

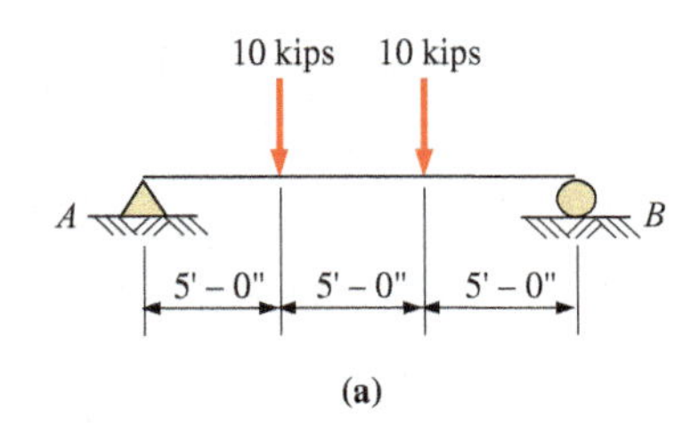

(a)

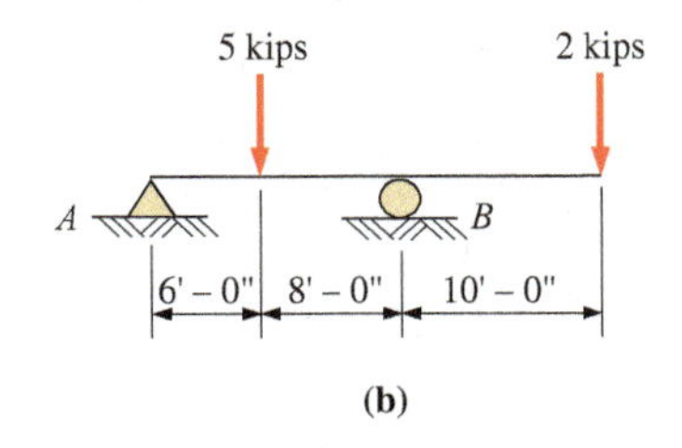

(b)

PROBLEM 13.13

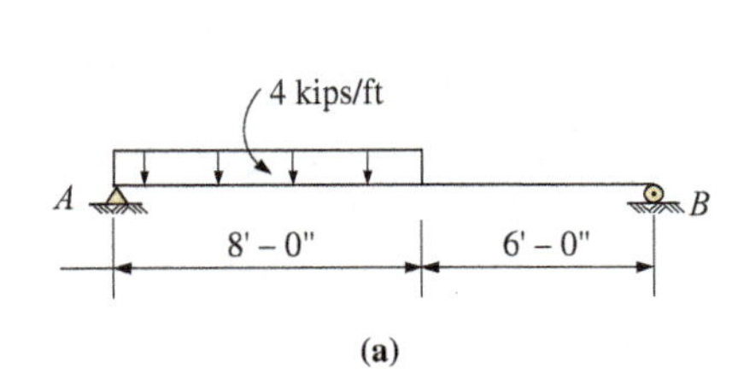

(a)

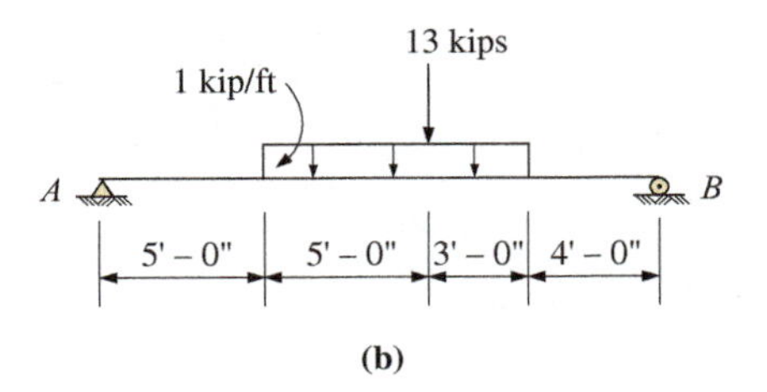

(b)

PROBLEM 13.14

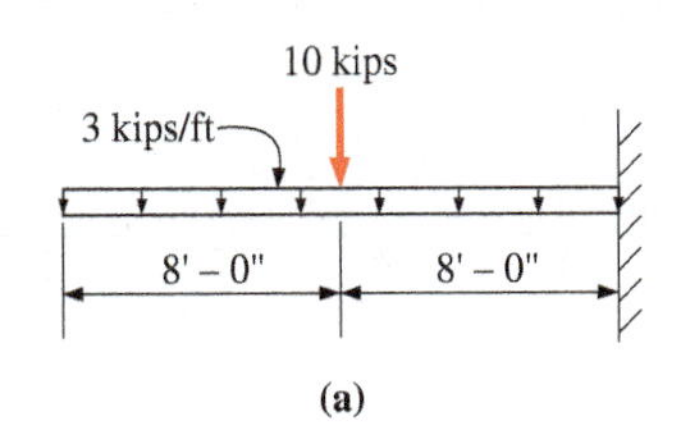

(a)

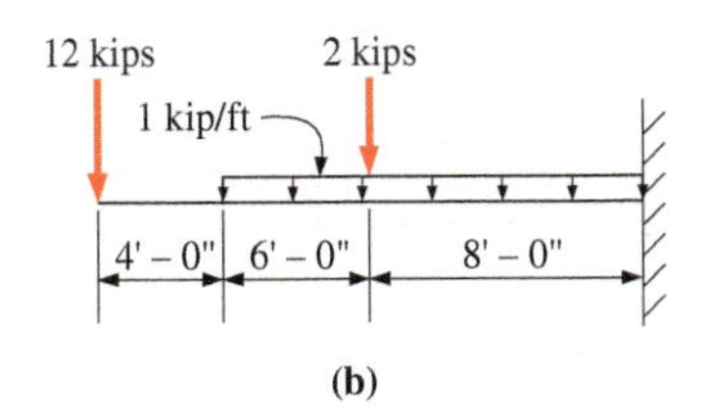

(b)

PROBLEM 13.15

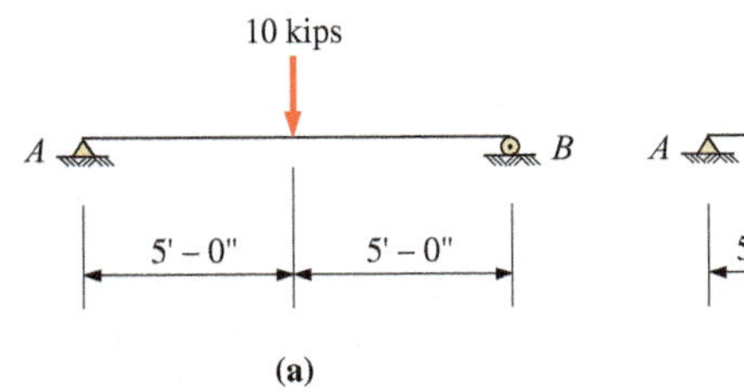

(a)

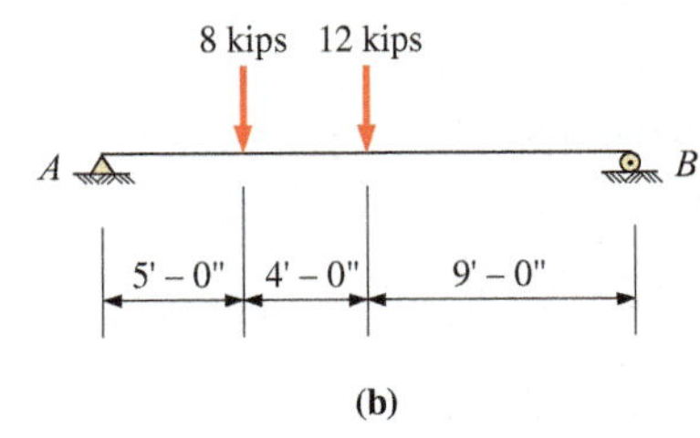

(b)

PROBLEM 13.16

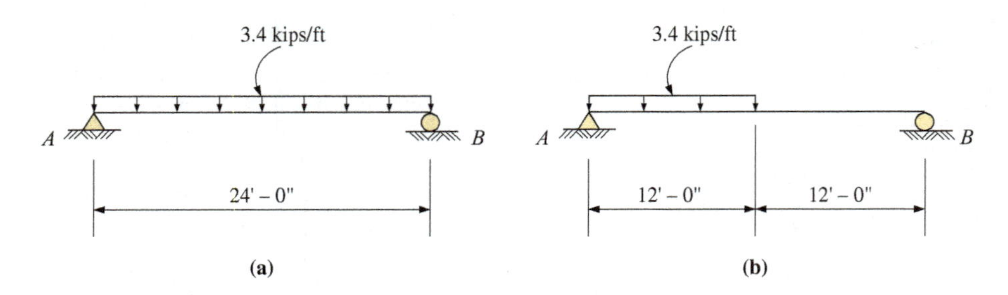

PROBLEM 13.17

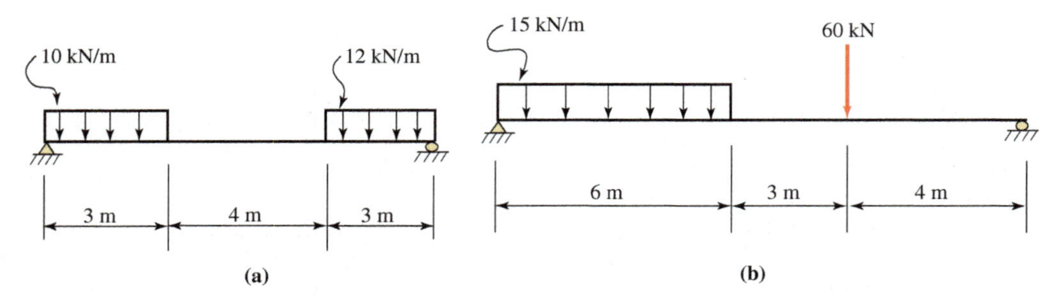

PROBLEM 13.18

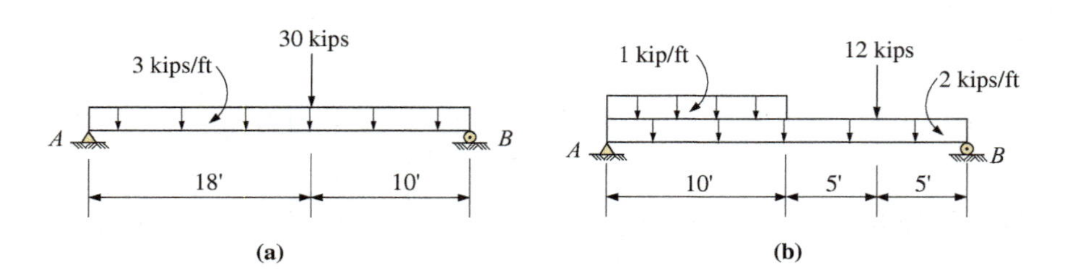

PROBLEM 13.19

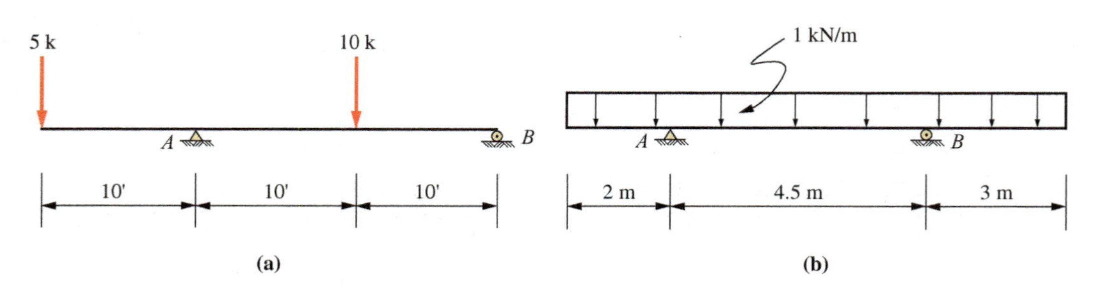

PROBLEM 13.20

Section 13.7 Sections of Maximum Moment

13.21 through 13.23 For the beams shown, draw complete shear and moment diagrams and state the values of the maximum positive and negative moments.

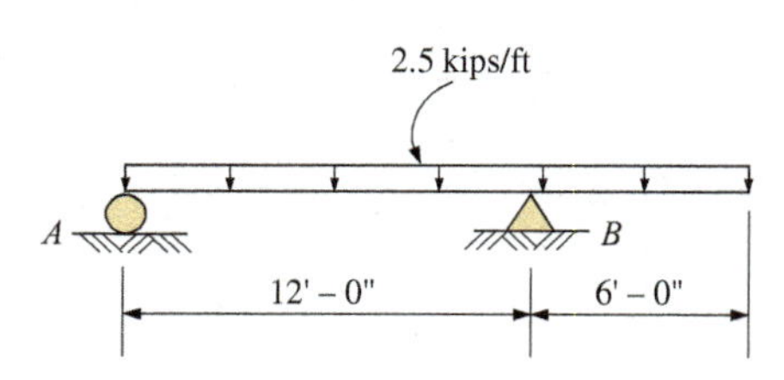

PROBLEM 13.21

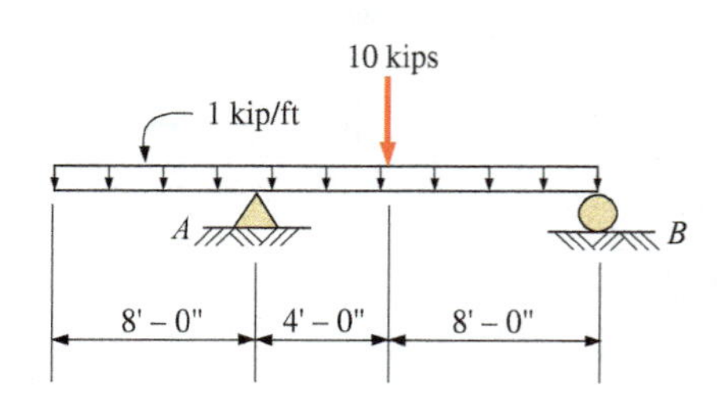

PROBLEM 13.22

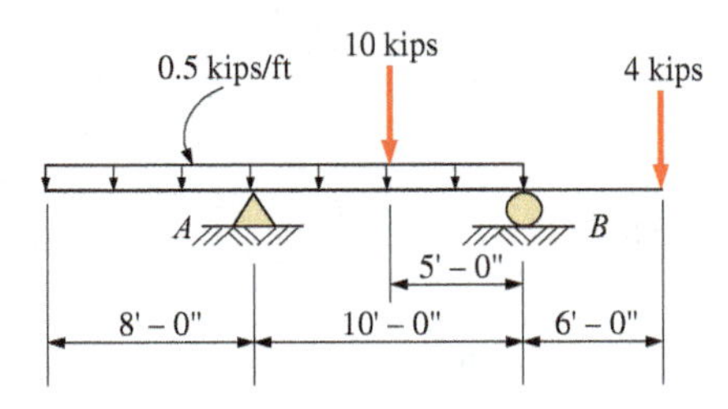

PROBLEM 13.23

Section 13.8 Moving Loads

13.24 A moving-load system is composed of two concentrated loads, each 20 kips, separated by a distance of 10 ft. The loads are to cross a 30-ft simple span. Calculate the absolute maximum shear and bending moment.

13.25 A moving-load system is composed of two concentrated loads separated by 16 ft. One load is 26 kips and the other is 12 kips. The loads are to cross a 40-ft simple span. Calculate the absolute maximum shear and bending moment.

13.26 One of the standard truck loads used in the design of bridges is composed of three concentrated loads, as shown. Calculate the absolute maximum shear and moment produced in a simple bridge span having a length of 26 m.

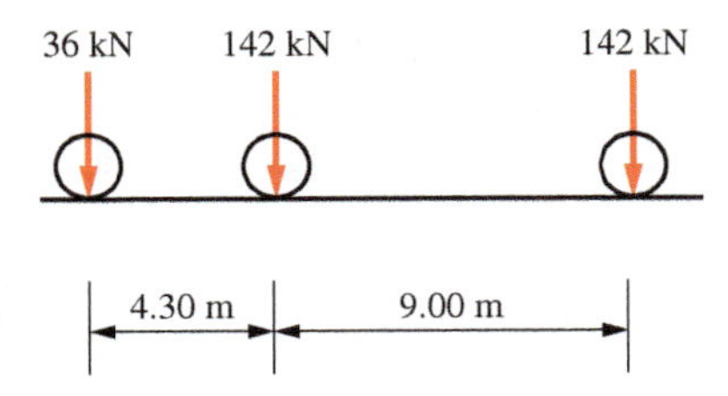

PROBLEM 13.26

Computer Problems

For the following computer problems, any appropriate software may be used. Input prompts should fully explain what is required of the user (the program should be user-friendly). The resulting output should be well labeled and self-explanatory.

13.27 Write a computer program that will calculate the shear and moment at any point along the length of a simply supported beam subjected to a uniformly distributed load. User input is to be beam span length, intensity of distributed load, and the location (with respect to the left support) at which the shear and moment are to be calculated.

13.28 Write a program that will calculate the shear and moment at the tenth points (the increment along the span is to be one-tenth of the overall length) for a simply supported beam subjected to a full-span uniformly distributed load and a concentrated load at midspan. User input is to be the magnitude of the loads and the span length.

13.29 Viking Consultants wishes to generate a table of maximum moments for a range of simply supported beams subjected to uniformly distributed loads. The table is to have beam spans ranging from 10 ft to 50 ft (10-ft increments) on the horizontal axis and loads of 0.5 kips/ft to 5 kips/ft on the vertical axis. Write the program that will generate this table. (*Note*: There is no user input for this program.)

Supplemental Problems

13.30 Calculate the reactions for the simple beams shown.

13.31 Calculate the reactions for the overhanging beams shown.

13.32 and 13.33 Calculate the reactions at points *A* and *B* for the beams shown. Draw the complete load diagram (free-body diagram) in each case.

13.34 For the beams of Problem 13.33, calculate the shear and moment at points 2 m and 8 m from the left end of each beam. Use free-body diagrams.

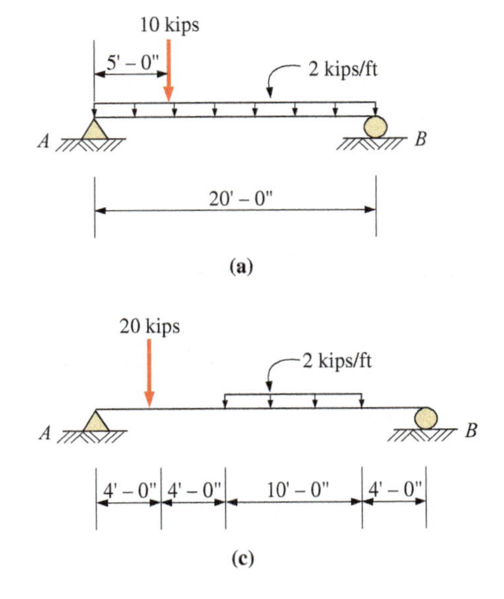

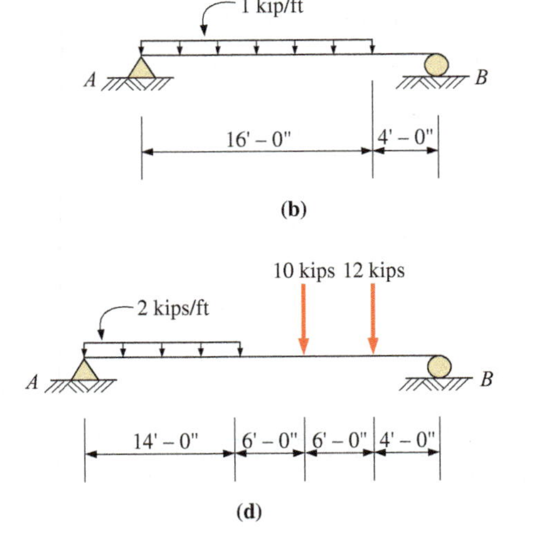

PROBLEM 13.30

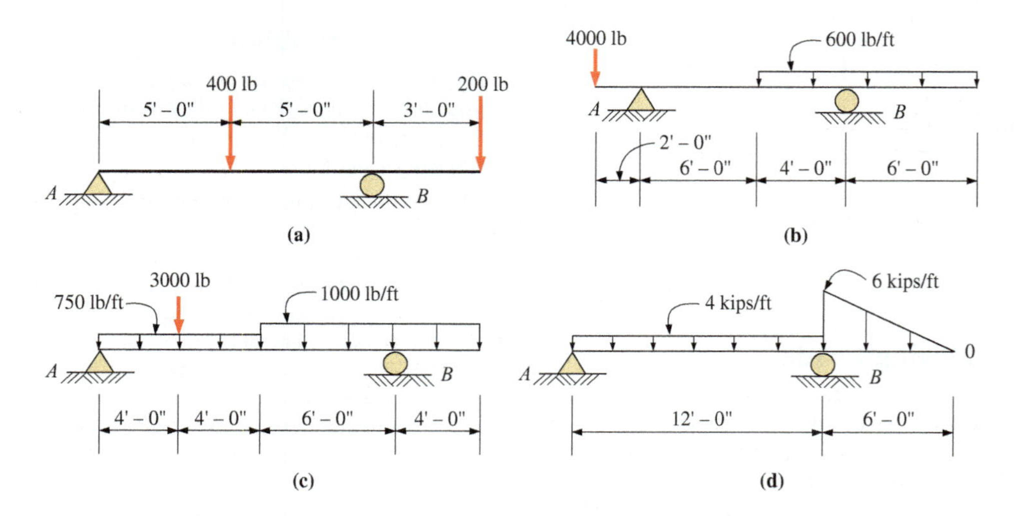

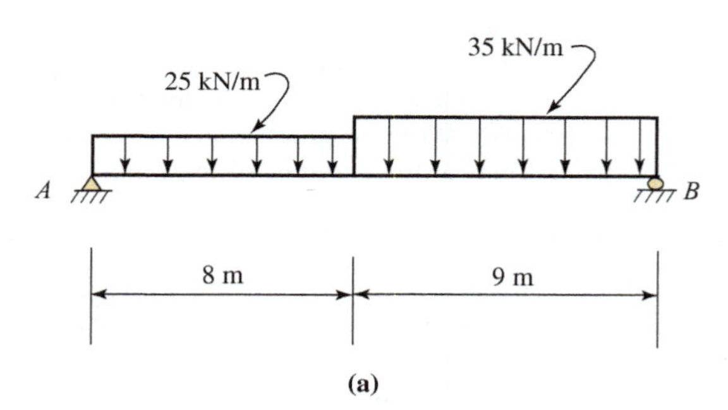

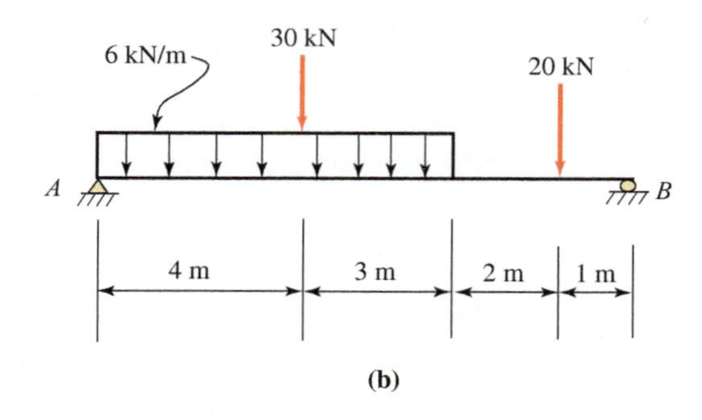

PROBLEM 13.31

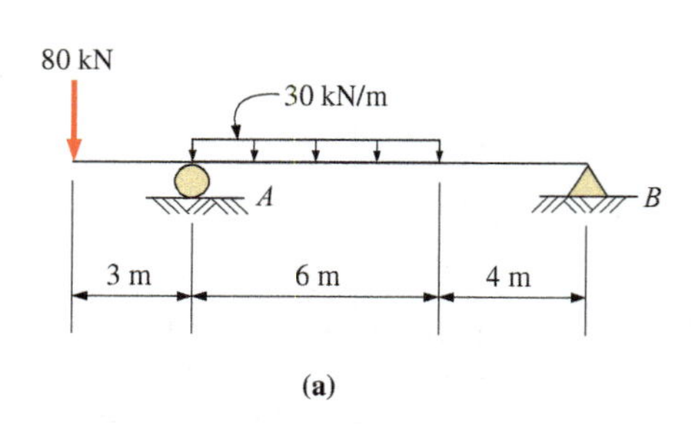

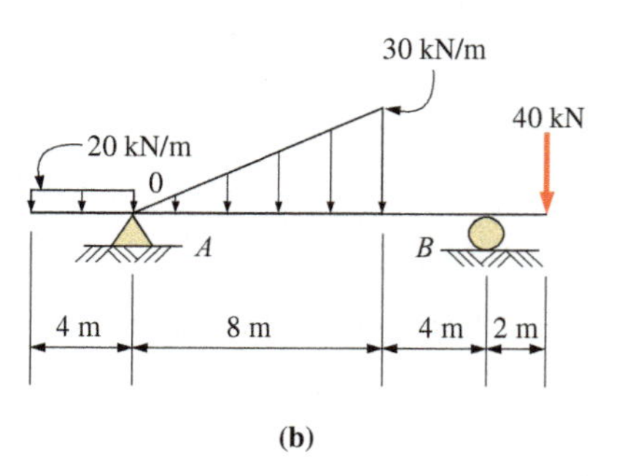

PROBLEM 13.32

80 kN

30 kN/m

A

B

3 m

6 m

4 m

(a)

30 kN/m

20 kN/m

0

40 kN

A

B

4 m

8 m

4 m

2 m

(b)

PROBLEM 13.33

13.35 For the beam shown, calculate the shear and bending moment at points 6 ft and 16 ft from the left end using free-body diagrams.

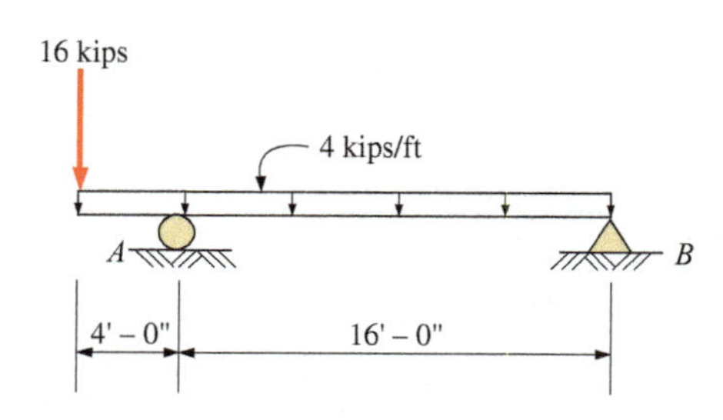

PROBLEM 13.35

13.36 Calculate the shear and bending moment at points 4 ft and 10 ft from the left end for beams (a) and (b) of Problem 13.31 using free-body diagrams.

13.37 Calculate the shear and bending moment at points 2 m and 3.5 m from the left end of the beam shown. Use free-body diagrams.

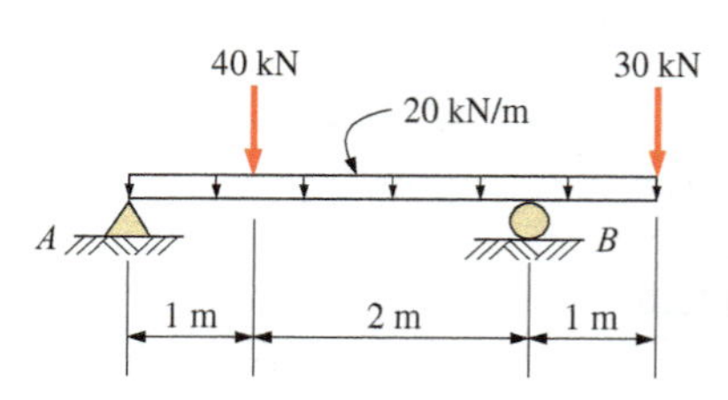

PROBLEM 13.37

13.38 Calculate the shear and bending moment at points 10 m and 16 m from the left end of the beam in Problem 13.4(a). Use free-body diagrams.

13.39 through 13.47 Refer to the beam shown and draw complete shear and bending moment diagrams. Show ordinates at key points and indicate magnitude of shear and moment. Neglect the beam weight.

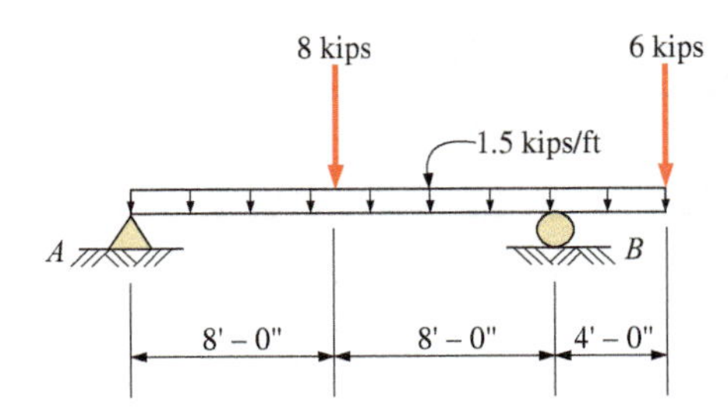

PROBLEM 13.39

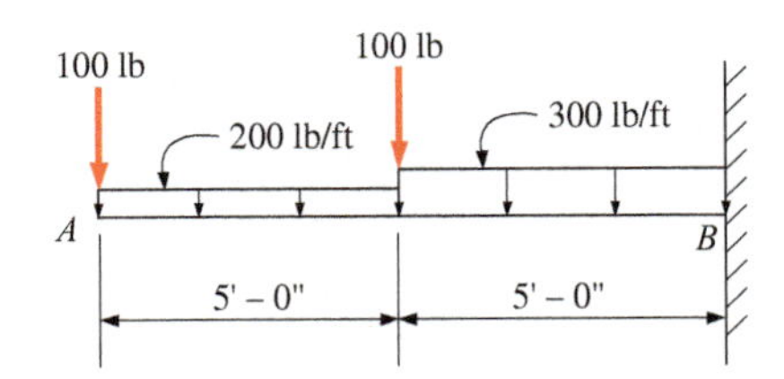

PROBLEM 13.40

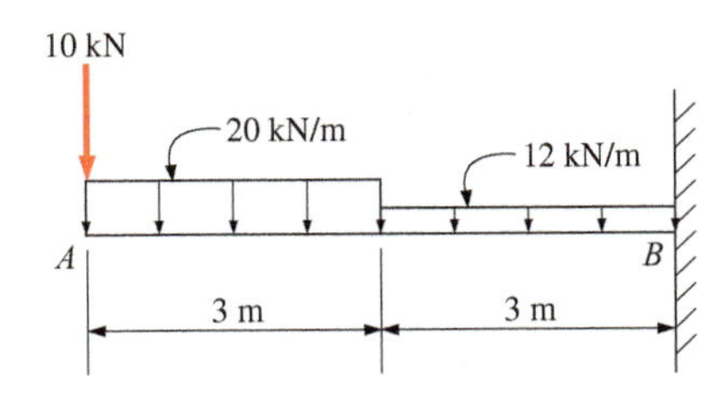

PROBLEM 13.41

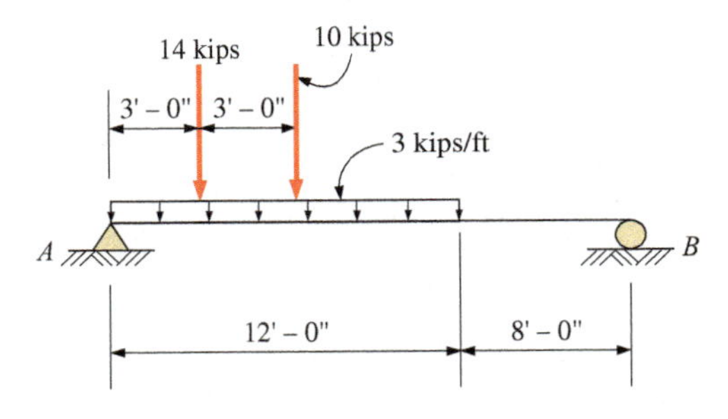

PROBLEM 13.42

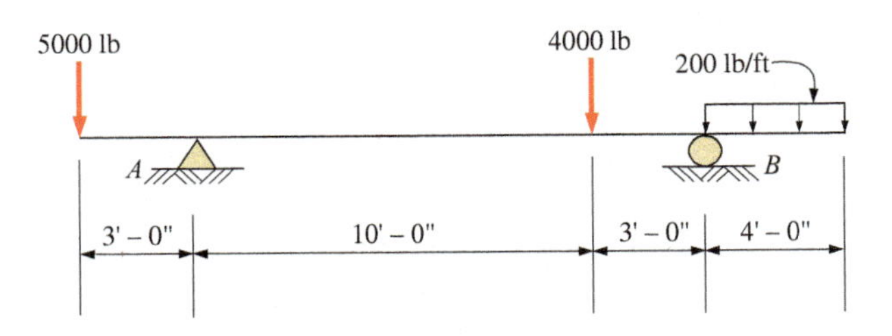

PROBLEM 13.43

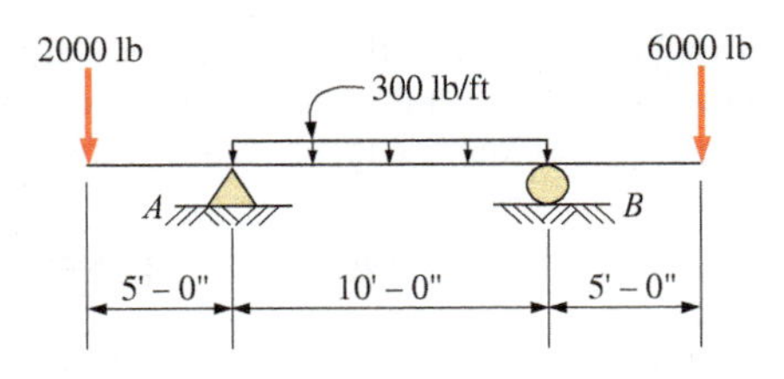

PROBLEM 13.44

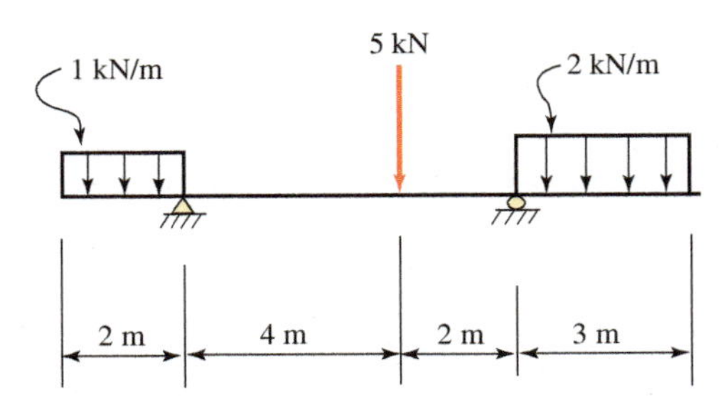

PROBLEM 13.45

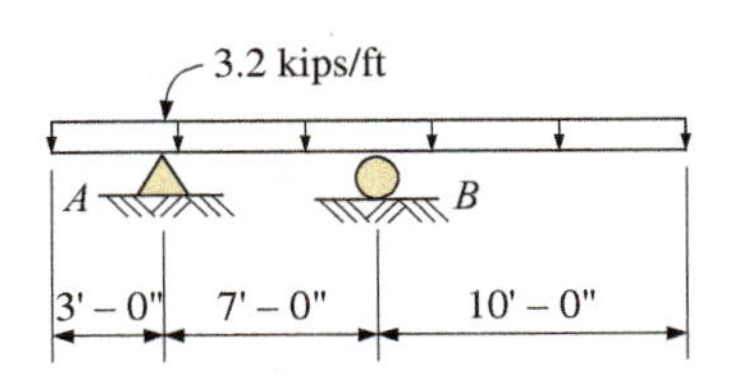

PROBLEM 13.46

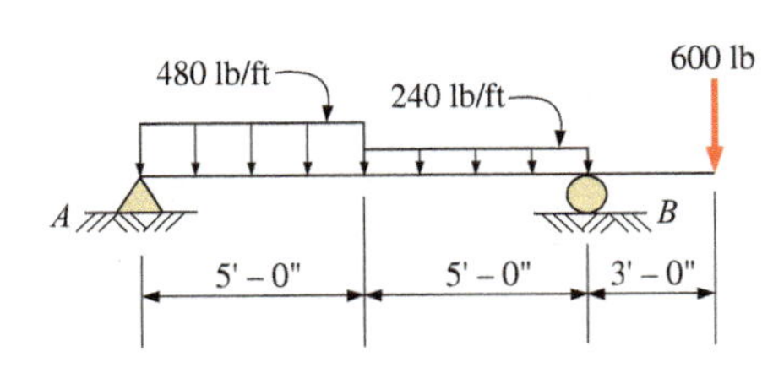

PROBLEM 13.47

13.48 through 13.58 Refer to the indicated problem and draw complete shear and bending moment diagrams. Show ordinates at key points and indicate magnitude of shear and moment. Neglect the beam weight.

13.48 Problem 13.1(b)
13.49 Problem 13.8(a)
13.50 Problem 13.2(a)
13.51 Problem 13.2(b)
13.52 Problem 13.3(a)
13.53 Problem 13.3(b)
13.54 Problem 13.4(b)
13.55 Problem 13.8(b)
13.56 Problem 13.10(a)
13.57 Problem 13.10(b)
13.58 Problem 13.11(a)

13.59 A two-axle roller with axles 5 m apart passes over a 15-m simply supported beam bridge. The load is 200 kN on each axle. Compute the absolute maximum moment and shear. Indicate the position of the wheels and the location of the maximum values.

13.60 A moving load system with wheels at fixed distances apart as shown crosses a 40-ft simply supported beam bridge. Compute the absolute maximum moment and shear.

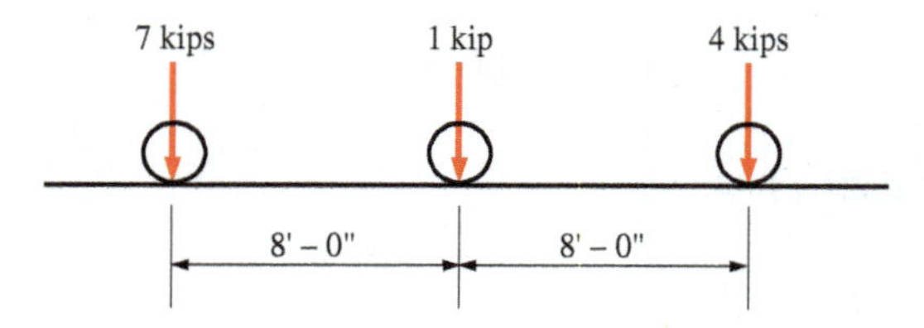

PROBLEM 13.60

13.61 A moving-load system with wheels spaced as shown crosses a 40-ft, single span, simply supported beam bridge. Compute

(a) the maximum moment under wheel *C*

(b) the maximum moment under wheel *D*

(c) the absolute maximum shear

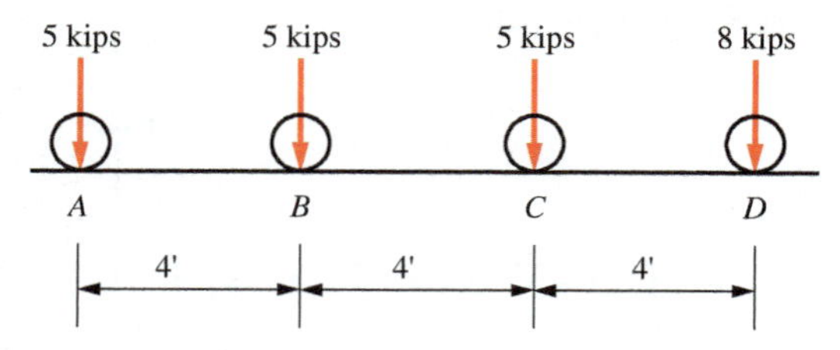

PROBLEM 13.61

CHAPTER FOURTEEN

STRESSES IN BEAMS

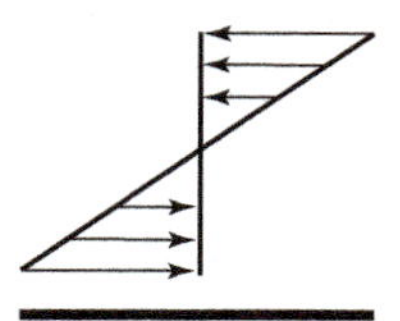

LEARNING OBJECTIVES

Upon completion of this chapter, readers will be able to:

- Discuss the tensile and compressive stresses induced in a beam cross section subjected to bending.
- Compute the section modulus of a beam cross section.
- Calculate the maximum bending stress in the cross section of a beam using the flexure formula.
- Use the bending stress and section modulus to determine the allowable moment that a beam can carry.
- Discuss the shear stresses induced in a beam cross section when subjected to bending.
- Use the general shear equation to calculate the shear stress at various points in a beam cross section.
- Use average shear equations to determine approximate shear stresses in beam cross sections for typical steel shapes, rectangular shapes, and round shapes.
- Analyze beams by comparing the computed stresses to allowable stresses.

14.1 TENSILE AND COMPRESSIVE STRESSES DUE TO BENDING

In Chapter 13 we saw that on every vertical section of a loaded horizontal beam a shear force and/or bending moment will occur, the magnitudes of which can be determined by calculation or from shear and moment diagrams. Therefore, at every vertical section an internal resisting shear and/or moment must be developed for a free body of any segment of the beam to be in equilibrium.

These internal resistances are functions of the shape and area of the cross section of the beam and can be expressed as internal shear stress and bending stress. The stresses may be thought of as representing the effect of the adjacent portion of the beam on the section under consideration.

For the design and analysis of a beam, it is necessary to calculate the induced stresses that occur at specific locations so as to compare these values with some allowable stress for the material used. Since the bending moment is generally the basis for beam design, our discussion at first deals exclusively with the bending stresses developed. It is common practice to use f_b to denote a calculated bending stress and F_b to denote an allowable bending stress. This notation will be used throughout the remainder of this text to be consistent with structural design codes.

Consider the straight horizontal beam of rectangular cross section shown in Figure 14.1, which is subjected to equal vertical loads P. The beam is simply supported and will bend (or deform) as shown by the dashed line. The beam is assumed weightless and the shear and moment diagrams are shown. Assume the beam cross section to be symmetrical with respect to the X–X and Y–Y axes, as shown. The loads are applied in the plane of the Y–Y axis.

The intersection of the two axes represents the centroid of the cross section. Therefore, axis X–X may be termed a centroidal axis. In addition, assume the beam to be homogeneous, of a material that obeys Hooke's law, and with a modulus of elasticity of equal value in both tension and compression.

We now consider the segment of the beam between planes AC and BD as shown in Figure 14.2. This segment lies between the two equal loads P where no shear (and, therefore, no shear stress) exists. This segment of the beam is subjected to uniform bending moment. The straight, unloaded condition and the bent, loaded condition are shown in Figure 14.2. Initially, in the unloaded condition, line segments AB, JF, and CD are of equal lengths.

As the beam deforms (bends) under load, segment AB shortens to $A'B'$ and segment CD lengthens to $C'D'$, as shown in Figure 14.2b. The top of the beam is in compression

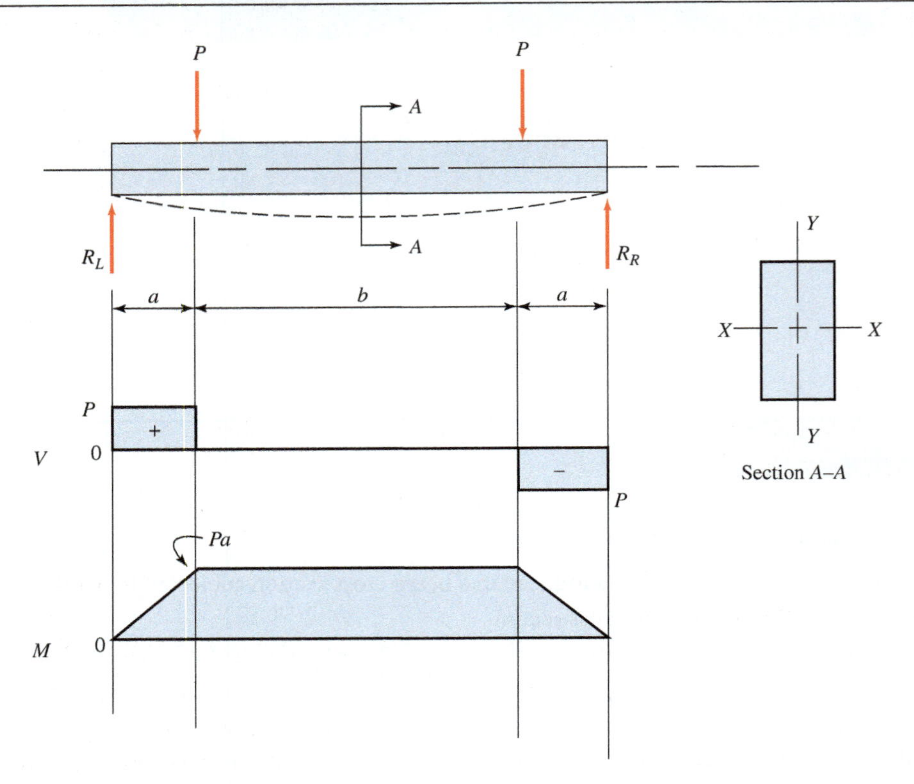

FIGURE 14.1 Load, shear, and moment diagrams.

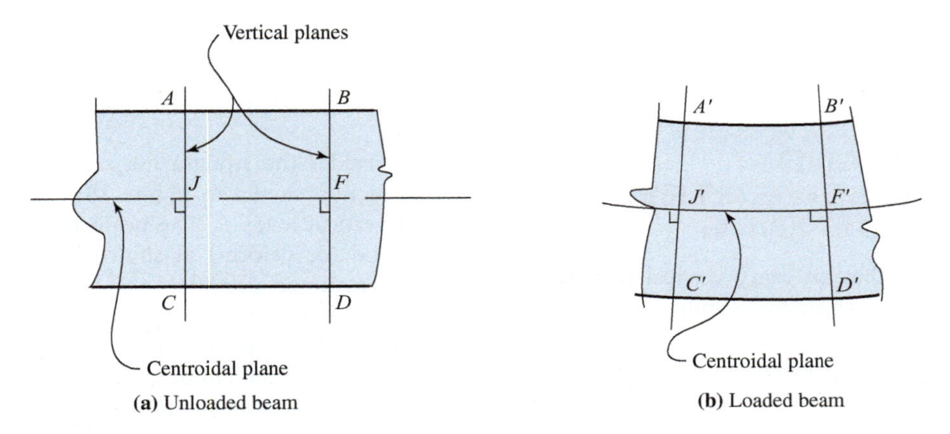

(a) Unloaded beam

(b) Loaded beam

FIGURE 14.2 Beam segment deformation.

and the bottom is in tension. Because of the uniform moment in the central portion of the beam, segments $A'B'$, $J'F'$, and $C'D'$ will be circular arcs. Many experiments have confirmed that the deformation takes place in such a manner that planes AC and BD before bending remain planes after bending. In addition, note in Figure 14.2b that planes $A'C'$ and $B'D'$ remain perpendicular to the neutral plane at their points of intersection.

In Figure 14.2, the centroidal plane remains the same length in both the unloaded and loaded conditions ($JF = J'F'$), indicating no shortening or elongating and, therefore, no compression or tension. The plane in the member on which there is no tension or compression is called the *neutral plane*, and the intersection of the neutral plane with a cross-sectional plane is called the *neutral axis*. In a homogeneous member, the neutral plane passes through the centroid of any cross section and defines a centroidal axis of the cross section.

The dimensional changes at a section in segment AB may be observed in Figure 14.3. The zero change at the neutral axis and the maximum change at the outer fibers in combination with the straight planes create a triangular strain distribution that, in turn, indicates that the change in length of any fiber is proportional to the distance of the fiber from the neutral axis.

Assuming that the stress in any fiber does not exceed the proportional limit of the material, it follows from Hooke's law that the stress in any fiber at a given section is proportional to the distance from the neutral axis to that fiber. Therefore, the stress distribution, like the strain distribution, is triangular in shape. The stress varies from zero

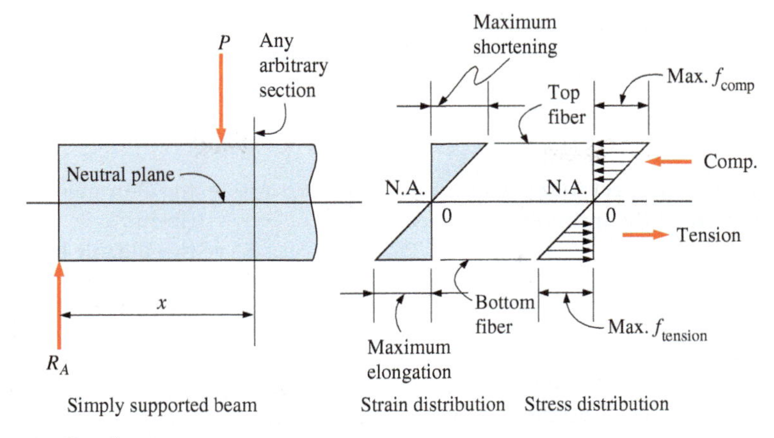

FIGURE 14.3 Stress and strain distribution.

at the neutral axis to a maximum compressive stress at the top outer fiber and a maximum tensile stress at the bottom outer fiber.

14.2 THE FLEXURE FORMULA

For the purpose of analysis and design of beams, it is necessary to work with the relationship between bending stresses, bending moment, and the geometric properties of a cross section. Whether we are dealing with analysis in which, perhaps, a stress is to be determined or with a design in which an allowable stress is used as a factor in the selection of a member, the same basic relationship, called the *flexure formula*, will apply. The derivation and application of the flexure formula requires that we know the location of the neutral (or centroidal) axis. In a symmetrical member, this is easily obtained by inspection. For unsymmetrical members, the location of the centroidal axis must be calculated in the manner described in Chapter 7.

Figure 14.4a shows the side view of a small part of a simply supported beam with a typical stress distribution at some arbitrary location. Figure 14.4b shows the cross section of the beam. The beam is symmetrical with respect to the X–X and Y–Y axes. The section shown is rectangular, but the following discussion is valid for a cross section of any shape having a Y–Y axis of symmetry where the loading is applied in the plane of the Y–Y axis. The line XX is the neutral axis of the cross section. The width of the section is denoted b. The distance from the neutral axis to an infinitesimal area a is denoted y. The distance from the neutral axis to the outer fiber of the cross section is denoted c.

The bending stress (either tension or compression) that develops at the outer fiber (a distance c from the neutral axis) will be referred to, for now, as $f_{b(\max)}$. This maximum bending stress at the outer fiber is the bending stress that is usually of greatest importance. We can also calculate, by proportion, the bending stress that develops at a distance y from the neutral axis. For now, we refer to this lesser stress as f_b. It can be expressed as

$$f_b = \frac{f_{b(\max)}(y)}{c}$$

Recalling that force is equal to the product of stress and area, we write the force developed on infinitesimal area a as

$$\frac{f_{b(\max)}(y)}{c}(a)$$

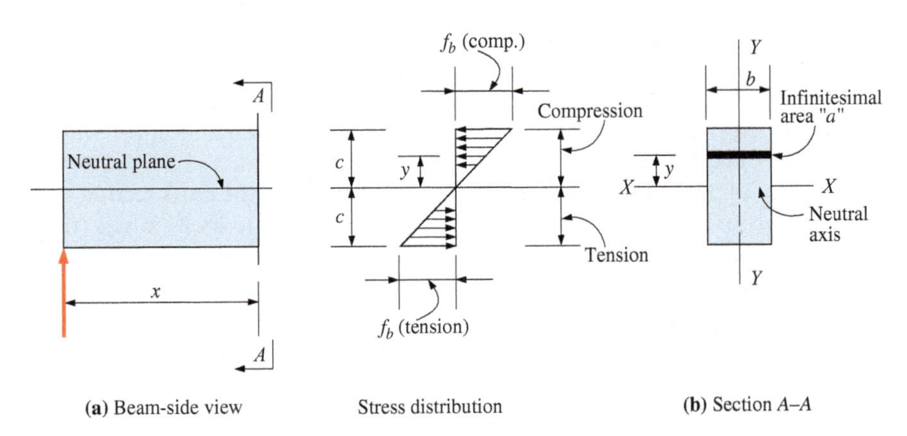

FIGURE 14.4 Flexure formula derivation.

The moment of the preceding force with respect to the X–X axis can be calculated as

$$\frac{f_{b(\max)}\,(y^2)}{c}(a)$$

Finally, the total moment, with respect to the X–X axis, of all the internal forces acting on all the infinitesimal areas can be written as

$$\frac{f_{b(\max)}}{c}\Sigma y^2(a)$$

As shown in Chapter 8, the mathematical quantity $(\Sigma y^2 a)$ is the moment of inertia of a cross section about its X–X axis and is represented by the symbol I. Therefore, the expression for the total moment may be written as

$$\frac{f_{b(\max)}}{c}I$$

Since this total internal moment holds in equilibrium the moment due to the external loads M, it is sometimes called an *internal resisting moment* and can be expressed as

$$M = \frac{f_{b(\max)}\,I}{c} \tag{14.1}$$

where M = the bending moment due to external loads, or the internal resisting moment (lb-in., k-ft) (N • m)
$f_{b(\max)}$ = the bending stress developed at the outer fiber (psi, ksi) (Pa)
I = the moment of inertia about the neutral axis (in.^{4}) (m^4)
c = the distance from the neutral axis to the outer fiber (in.) (m)

Rewriting this expression and solving for the stress gives

$$f_{b(\max)} = \frac{Mc}{I} \tag{14.2}$$

where all terms are as previously defined.

Since stresses are proportional to distance from the neutral axis, we can also write the expression for bending stress developed at any distance y from the neutral axis:

$$f_b = \frac{My}{I} \tag{14.3}$$

where f_b in this case will be less than the maximum bending stress that occurs at the outer fiber. Note that substitution of c for y in Equation (14.3) results in Equation (14.2). Since, for all practical purposes, it is the maximum bending stress that is of importance, we omit the "(max)" from $f_{b\,(\max)}$, with the understanding that it is the maximum bending stress with which we are working (unless otherwise noted).

We can also rewrite Equation (14.1) to find the maximum resisting moment, or allowable moment, for a cross section. To use this expression, the limiting value of bending stress, called the allowable bending stress, must be known:

$$M_R = \frac{F_b I}{c} \tag{14.4}$$

where M_R = the allowable moment or moment strength (lb-in., k-ft) (N • m)
F_b = the alloable bending stress (psi, ksi) (Pa)

and I and c are as previously defined.

In these various forms of the flexure formula, note that the moment of inertia I and the distance c are both functions of the size and shape of the beam cross section. They are both geometric properties of the cross section and do not depend on the material or span length of the beam or on the type of loading on the beam. The quantity I/c, therefore, is also a geometric property. I/c is called the *section modulus* and is generally represented by the symbol S. The section modulus has units of in.^{3} and can be calculated using the moment of inertia from Table 3 (inside the back cover) divided by the distance c.

The flexure formula can thus be rewritten and used in the following forms depending on whether the problem is one of analysis or design. For analysis problems,

$$f_b = \frac{M}{S} \tag{14.5}$$

or

$$M_R = F_b\,S \tag{14.6}$$

For design problems, for some materials, such as timber, the most convenient form is

$$\text{required } S = \frac{M}{F_b} \tag{14.7}$$

The values of the section modulus for standard rolled structural shapes and other standard shapes, both metallic and nonmetallic, can be found in various publications of the American Institute of Steel Construction, the American Forest & Paper Association, and other such organizations. Some of this material is provided in the appendices of this book.

As a physical analogy, the section modulus can be considered a measure of the comparative strength of beams. All other things being equal, if the section modulus of the cross section of beam A is twice as great as that of beam B of the same material, beam A will have double the bending strength.

Note that if the cross section of a beam is not symmetrical with respect to its X–X axis (the axis of bending), it will have two section modulus values of different magnitudes since the c values for the top and bottom fibers will be different. Also, the centroidal axis will not be located at midheight of the cross section.

Since beams are such common members in structures, and since the use of the flexure formula is basic to beam analysis and design, the limitations of the formula should be recognized. Its use is valid only under certain conditions, which may be briefly itemized as follows:

1. The beam must be straight before loading.
2. The beam must be homogeneous, obey Hooke's law, and have equal moduli of elasticity in both tension and compression.
3. The loads and reactions must lie in a vertical plane of symmetry perpendicular to the longitudinal axis of the beam.
4. The beam must be of uniform cross section.
5. The maximum bending stress must not exceed the proportional limit.
6. The beam must have adequate lateral buckling resistance.
7. All component parts of the beam must have adequate localized buckling resistance.
8. The beam must be relatively long in proportion to its depth.
9. The cross section must not be disproportionately wide.
10. The dimensional changes must not be appreciably affected by shear strains that are also present.

Despite the many limiting factors, the flexure formula may be applied quite satisfactorily in the design and analysis of most beams normally encountered.

14.3 COMPUTATION OF BENDING STRESSES

In the analysis of beams, one type of problem involves calculating maximum bending stress. This stress will occur at the outer fibers at the section of the beam where the bending moment is a maximum. In computing the maximum bending stress, the location and the magnitude of the maximum bending moment must be calculated first. The value of the section modulus (or the moment of inertia and the c distance) must then be calculated or obtained from standard tables of properties of sections. By substituting these values in Equation (14.5), or Equation (14.2), we obtain the maximum bending stress.

The following examples illustrate the use of the flexure formula. They are not to be interpreted as complete beam-analysis problems. Various types of more comprehensive problems follow in Section 14.7.

EXAMPLE 14.1 A nominal 8-in.-by-12-in. timber member, shown in Figure 14.5, is used as a beam that is subjected to a vertical loading. For maximum bending strength, the beam is oriented so that the 12-in. dimension is vertical. Calculate the section modulus of the beam with respect to the axis of bending (the X–X axis). Use dressed dimensions.

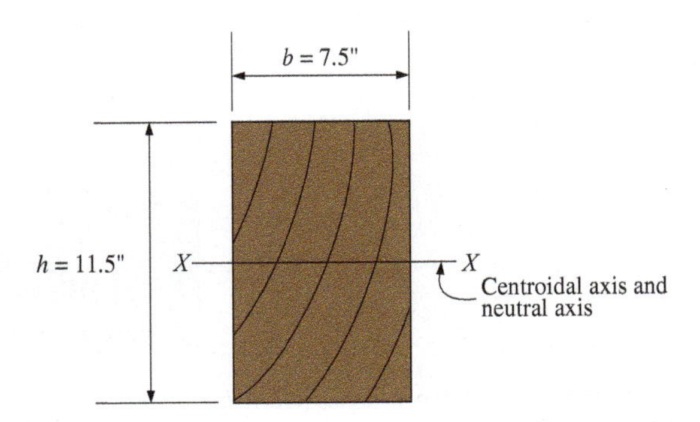

FIGURE 14.5 Beam cross section.

Solution The dressed dimensions for the timber member are $7\frac{1}{2}$ in. by $11\frac{1}{2}$ in. For solid rectangular shapes, an expression for section modulus can be derived as follows:

$$S = \frac{I}{c} = \frac{(bh^3/12)}{h/2} = \frac{bh^2}{6}$$

This convenient expression is used in dealing with solid homogeneous rectangular shapes, such as timber members.

Substitute numerical values:

$$S_x = \frac{bh^2}{6} = \frac{(7.5\text{ in.})(11.5\text{ in.})^2}{6} = 165.3\text{ in.}^3$$

This value may be verified in Appendix E.

EXAMPLE 14.2 The timber member of Example 14.1 is used as a simply supported beam on a 16-ft span carrying a uniformly distributed line load of 400 lb/ft, as shown in Figure 14.6. Calculate the maximum induced bending stress. Neglect the weight of the beam.

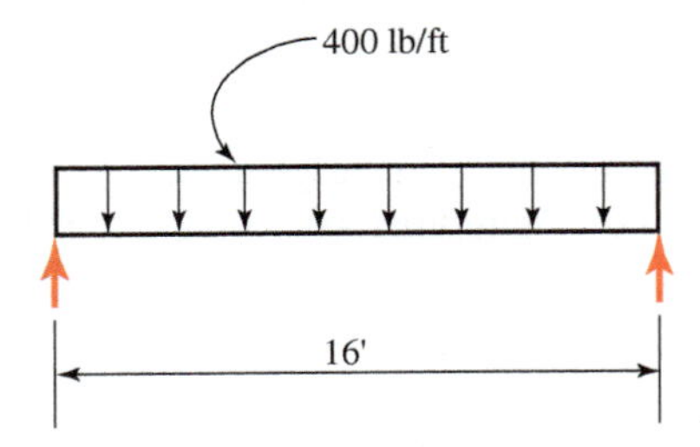

FIGURE 14.6 Load diagram.

Solution The maximum stress can be calculated using Equation (14.5). First, calculate the maximum bending moment. From Appendix H,

$$M = \frac{wL^2}{8} = \frac{(400\ \text{lb/ft})(16\ \text{ft})^2}{8} = 12{,}800\ \text{lb-ft}$$

Using the section modulus for the cross section from the previous example, substitute into Equation (14.5) to find the maximum bending stress:

$$f_b = \frac{M}{S} = \frac{(12{,}800\ \text{lb-ft})(12\text{in./ft})}{165.3\ \text{in.}^3} = 929\ \text{psi}$$

EXAMPLE 14.3 Assume that, due to a construction error, the beam of the previous two examples was placed incorrectly in the field and the small dimension ($7\frac{1}{2}$ in. dressed) was oriented vertically. For the same span length and loading, calculate the new maximum bending stress that would be developed.

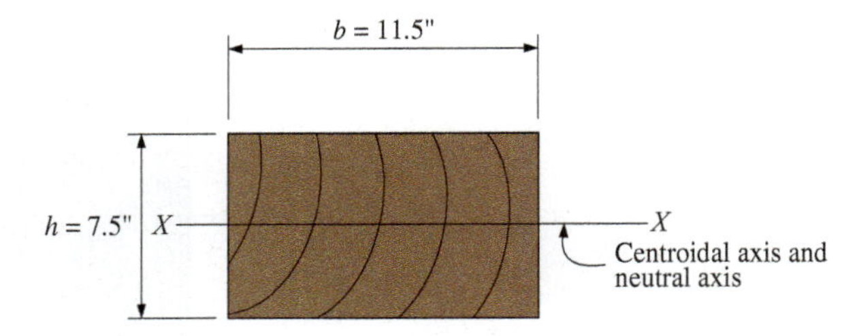

FIGURE 14.7 Beam cross section.

Solution The maximum bending moment remains the same at 12,800 lb-ft. The orientation of the cross section is shown in Figure 14.7. Note that the horizontal axis is designated as axis *X–X*. Calculate the section modulus:

$$S_x = \frac{bh^2}{6} = \frac{(11.5\ \text{in.})(7.5\ \text{in.})^2}{6} = 107.8\ \text{in.}^3$$

Calculate the maximum bending stress:

$$f_b = \frac{M}{S} = \frac{(12{,}800\ \text{lb-ft})(12\ \text{in./ft})}{107.8\ \text{in.}^3} = 1425\ \text{psi}$$

The results of Examples 14.2 and 14.3 indicate that the orientation of a beam cross section is an important strength consideration. For maximum bending resistance, the bending axis (perpendicular to the applied loads) should be the axis about which the section modulus is the largest value.

EXAMPLE 14.4 A W30 × 99 hot-rolled structural steel wide-flange shape is used as a simply supported beam on a span of 32 ft, as shown in Figure 14.8. The beam supports a superimposed uniformly distributed load of 4.0 kips/ft in addition to its own weight. Calculate the maximum bending stress.

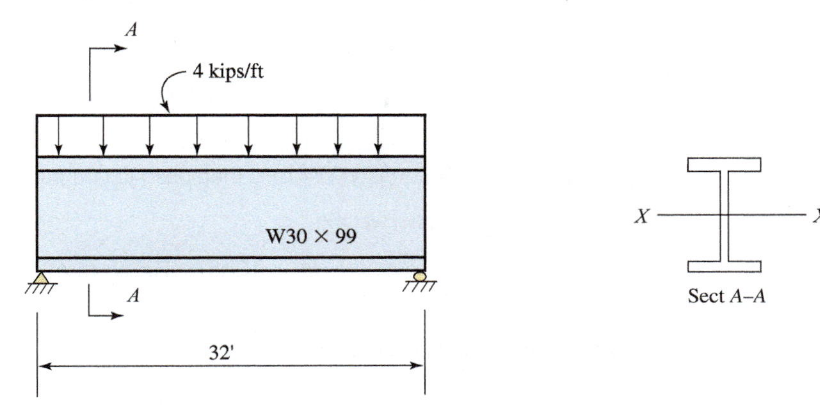

FIGURE 14.8 Beam for Example 14.4.

Solution First, calculate the maximum bending moment. The total load is the sum of the superimposed load (4.0 kips/ft) and the weight of the beam (0.099 kips/ft). From Appendix H,

$$M = \frac{wL^2}{8} = \frac{(4.10\ \text{k/ft})(32\ \text{ft})^2}{8} = 525\ \text{k-ft}$$

The section modulus can be obtained from Appendix A, which is a partial tabulation of the AISC tables of dimensions and properties of structural shapes. Assume that the shape is oriented so that bending is about the strong axis. Hence, $S = 269\ \text{in.}^3$

Calculate the maximum bending stress:

$$f_b = \frac{M}{S} = \frac{(525\ \text{k-ft})(12\ \text{in./ft})}{269\ \text{in.}^3} = 23.4\ \text{ksi}$$

EXAMPLE 14.5 An extra-strong steel pipe having a nominal diameter of 127 mm is to be used as a simple beam with a span length of 7 m, as shown in Figure 14.9. The pipe supports a concentrated load at midspan of 6 kN. Calculate the maximum bending stress due to (a) the weight of the pipe alone and (b) the concentrated load alone.

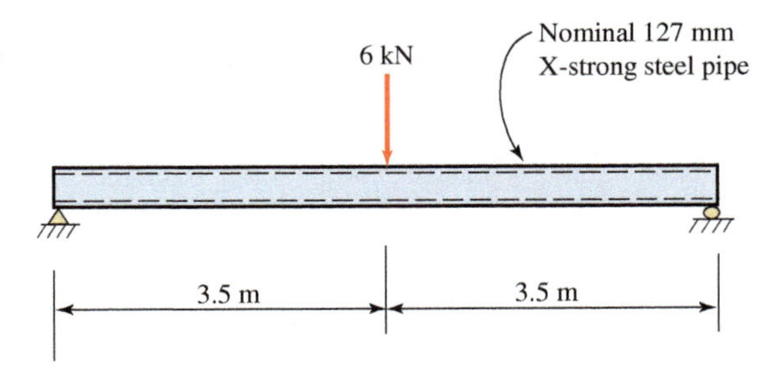

FIGURE 14.9 Beam for Example 14.5.

Solution From Appendix B (SI), the section modulus S is found to be $115 \times 10^3\ \text{mm}^3$. The weight of the pipe is calculated from

$$\begin{aligned} w &= mg \\ &= 30.9\ \text{kg/m}\ (9.81\ \text{m/sec}^2) = 303\ \text{N/m} \end{aligned}$$

For a simply supported beam, the moment due to the beam's own weight is obtained from

$$M = \frac{wL^2}{8} = \frac{(303\text{ N/m})(7\text{ m})^2}{8}$$
$$= 1856\text{ N}\cdot\text{m}$$

The maximum moment due to the applied concentrated load is

$$M = \frac{PL}{4} = \frac{(6\text{ kN})(7\text{ m})}{4} = 10.5\text{ kN}\cdot\text{m}$$

The flexure formula of Equation (14.5) is then used to compute the bending stress, as follows:

(a) Due to the beam's own weight:

$$f_b = \frac{M}{S} = \frac{1856 \times 10^3\text{ N}\cdot\text{mm}}{115 \times 10^3\text{ mm}^3} = 16.14\text{ MPa}$$

(b) Due to the applied concentrated load:

$$f_b = \frac{M}{S} = \frac{10.5 \times 10^6\text{ N}\cdot\text{mm}}{115 \times 10^3\text{ mm}^3} = 91.3\text{ MPa}$$

We could also determine the total stress. Since the calculated stresses occur at the same point in the beam, they are additive:

$$\text{total } f_b = 16.14 + 91.3 = 107.4\text{ MPa}$$

EXAMPLE 14.6 A steel beam is fabricated in the shape of a "T" by welding a 12-in.-by-$\frac{1}{2}$-in. steel plate to a 10-in.-by-$\frac{1}{2}$-in. steel plate, as shown in Figure 14.10. A beam of this type is often used in masonry building construction as a lintel, which is a bending member that spans over wall openings such as for doors and windows.

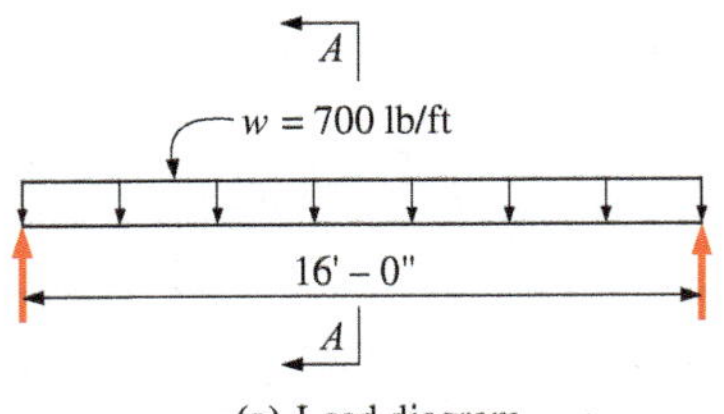

(a) Load diagram

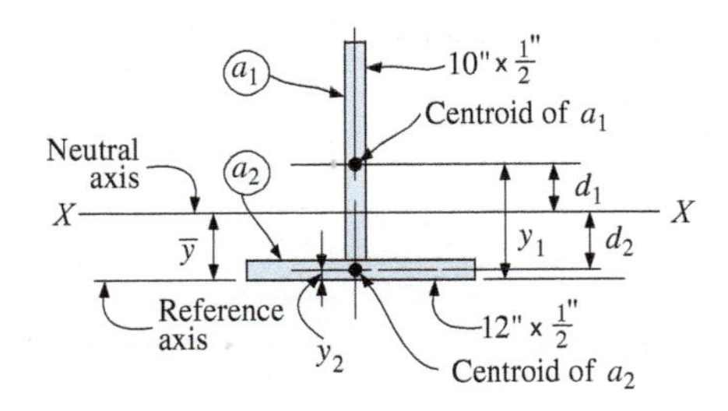

(b) Section A–A

FIGURE 14.10 Lintel beam.

Calculate the maximum bending stresses in tension and compression if the beam supports a wall load of 700 lb/ft (which includes its own weight) and is simply supported with a span length of 16 ft.

Solution This composite member is made up of two simple geometric areas, a_1 and a_2. Since this member is not symmetrical with respect to the X–X bending axis, the neutral axis must first be located. To determine the distance $\bar{y}$, a reference axis is first established at the lower side of the cross section, as shown in

Figure 14.10b. Then, using the principle of moments, the location of the neutral axis will be found using Equation (7.4):

$$\bar{y} = \frac{\Sigma ay}{A}$$

where A is the total area, or the sum of the component areas Σa. From Figure 14.10,

$$a_1 = (10 \text{ in.})(0.5 \text{ in.}) = 5 \text{ in.}^2$$
$$a_2 = (12 \text{ in.})(0.5 \text{ in.}) = 6 \text{ in.}^2$$
$$A = a_1 + a_2 = 11 \text{ in.}^2$$
$$y_1 = (5.0 \text{ in.}) + (0.5 \text{ in.}) = 5.5 \text{ in.}$$
$$y_2 = 0.25 \text{ in.}$$

Substitute into Equation (7.4):

$$\bar{y} = \frac{\Sigma ay}{A} = \frac{a_1y_1 + a_2y_2}{A} = \frac{(5 \text{ in.}^2)(5.5 \text{ in.}) + (6 \text{ in.}^2)(0.25 \text{ in.})}{11.0 \text{ in.}^2} = 2.64 \text{ in.}$$

In the flexure formula, this value represents the c distance to the bottom fiber. The c distance to the top fiber may be calculated from

$$c_{(\text{top})} = 10.5 \text{ in.} - 2.64 \text{ in.} = 7.86 \text{ in.}$$

Next, calculate the moment of inertia about the neutral (centroidal) axis X–X using Equation (8.4):

$$I_x = \Sigma(I_o + ad^2) = (I_o + ad^2)_1 + (I_o + ad^2)_2$$

where

$$d_1 = 7.86 \text{ in.} - 5.0 \text{ in.} = 2.86 \text{ in.}$$
$$d_2 = 2.64 \text{ in.} - 0.25 \text{ in.} = 2.39 \text{ in.}$$

Therefore,

$$I_x = \left[\frac{(0.5 \text{ in.})(10 \text{ in.})^3}{12} + (5 \text{ in.}^2)(2.86 \text{ in.})^2\right] + \left[\frac{(12 \text{ in.})(0.5 \text{ in.})^3}{12} + (6 \text{ in.}^2)(2.39 \text{ in.})^2\right]$$
$$= 82.6 \text{ in.}^4 + 34.4 \text{ in.}^4$$
$$= 117.0 \text{ in.}^4$$

The maximum bending moment from Appendix H is

$$M = \frac{wL^2}{8} = \frac{(700 \text{ lb/ft})(16 \text{ ft})^2}{8} = 22{,}400 \text{ lb-ft}$$

Next, calculate the section moduli values. Since the member is not symmetrical, the section modulus for the top is different from that for the bottom:

$$S_{(\text{top})} = \frac{I}{c_{(\text{top})}} = \frac{117.0 \text{ in.}^4}{7.86 \text{ in.}} = 14.89 \text{ in.}^3$$

$$S_{(\text{bot})} = \frac{I}{c_{(\text{bot})}} = \frac{117.0 \text{ in.}^4}{2.64 \text{ in.}} = 44.3 \text{ in.}^3$$

Now the maximum bending stress can be calculated using Equation (14.5):

$$f_{b(\text{bot})} = \frac{M}{S_{(\text{bot})}} = \frac{(22{,}400 \text{ lb-ft})(12 \text{ in./ft})}{44.3 \text{ in.}^3} = 6070 \text{ psi}$$

$$f_{b(\text{top})} = \frac{M}{S_{(\text{top})}} = \frac{(22{,}400 \text{ lb-ft})(12 \text{ in./ft})}{14.89 \text{ in.}^3} = 18{,}050 \text{ psi}$$

Assuming that this beam has an allowable bending stress of 22,000 psi, note that if the stress in the bottom were to reach the maximum allowable value, the stress in the top would be greater than that allowed and, therefore, would not be acceptable. Thus, the section modulus with respect to the top controls the load-carrying capacity of the member.

EXAMPLE 14.7 A 6 × 8 timber beam of No. 1 Sitka spruce, the cross section of which is shown in Figure 14.11, has an allowable bending stress of 1000 psi. Bending is about the strong axis. Calculate the maximum allowable bending moment (k-ft) to which this timber beam may be subjected.

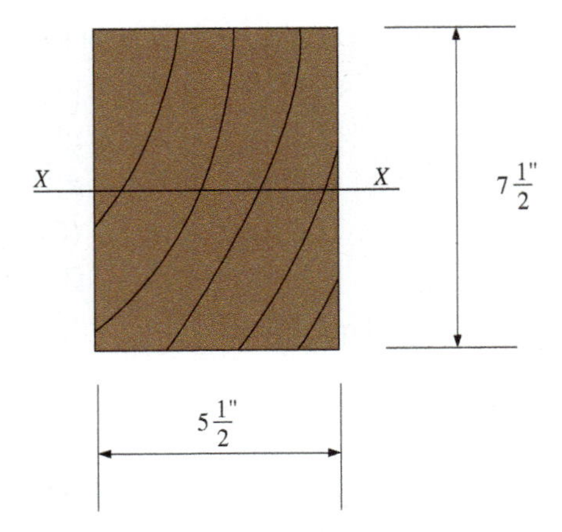

FIGURE 14.11 Beam cross section.

Solution The section modulus S_x can be obtained from Appendix E: $S_x = 51.6$ in.^{3} The allowable bending moment can then be calculated using Equation (14.6):

$$M_{Rx} = F_b S_x = \frac{(1000\text{ psi})(51.6\text{ in.}^3)}{(12\text{ in./ft})(1000\text{ lb/k})} = 4.30\text{ k-ft}$$

Note that this allowable moment (or moment strength) depends on only the cross section and the allowable bending stress.

14.4 SHEAR STRESSES

As discussed in Section 14.1, internal resisting shear stresses are developed on every section of a loaded horizontal beam where the vertical shear force has a numerical value other than zero. The summation of these shear stresses provides the internal resisting shear force that must be equal to the external vertical shear force for any free body of a segment of the beam to be in equilibrium. Chapter 13 dealt with the determination of the vertical shear force at any location along the length of a beam due to the external loads. This section deals with the shear stresses resulting from the vertical shear forces. These shear stresses will be denoted as f_v for a calculated shear stress and F_v for an allowable shear stress to be consistent with design codes.

The distribution of the shear stress developed over the beam cross section is very different from the bending stress distribution. The shear stress is zero at those points on the cross section where the bending stress is a maximum. The location of the point of maximum shear stress is almost always at the neutral axis. It should be noted that this maximum value can occur on horizontal planes other than the neutral axis for odd-shaped, impractical, and uneconomical sections that are seldom encountered.

It was shown in Section 11.6 that if at any point in a stressed member a shear stress exists on a plane, at the same time there must exist a shear stress of equal intensity on a perpendicular plane. Therefore, in a loaded horizontal beam, both vertical and horizontal shear stresses are developed at any given point.

The existence of the horizontal shear stresses is best illustrated by considering several planks of identical cross section. If the planks are stacked together to form a simply supported beam, as shown in Figure 14.12a, and a vertical load is applied, the planks will bend, as shown in Figure 14.12b. Observe that each plank is bending independently. The top fibers of each plank are shortened and the bottom fibers are lengthened. Assuming a frictionless contact surface, the planks tend to slide over each other, as is evident if one notes that the ends of the planks no longer lie in a straight line or plane. The top of one plank will slide inward relative to the bottom of the adjacent plank. If an adhesive is applied between the planks to bond them together, the loaded beam will take the form shown in Figure 14.12c. The sliding that occurred between the planks is now resisted by the adhesive, and a shear stress is developed in the adhesive.

If the beam is a one-piece solid member instead of planks bonded together, the material of which the beam is made will resist the tendency for horizontal sliding of layers. Hence, a horizontal shear stress is developed on any horizontal plane of the beam cross section.

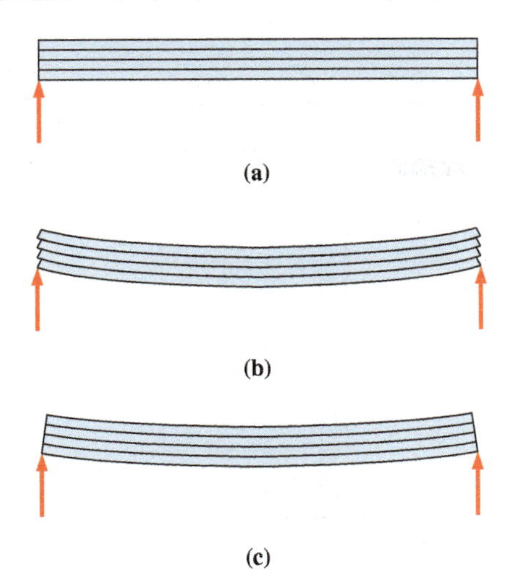

FIGURE 14.12 Horizontal shear in beams.

14.5 THE GENERAL SHEAR FORMULA

Whereas the magnitude of the maximum bending stress can be obtained through the use of the flexure formula, the magnitude of the maximum shear stress can be obtained through the use of the general shear formula. This formula will yield the shear stress on any horizontal plane. As shown in Chapter 11, the horizontal shear stress equals the vertical shear stress at any point in a beam.

For the derivation of the general shear formula, consider the load, shear, and moment diagrams in Figure 14.13a for a loaded simply supported beam. The free-body diagram for an element of this beam bounded by planes D, E, and F is shown in Figure 14.13b. Planes D and E are initially an infinitesimal distance x apart, and plane F lies a distance y_1 above the neutral axis. The object of the derivation is to establish an expression for the horizontal shear stress f_v on the bottom surface of the element.

Let C_D and C_E represent the resultants of the compressive bending stresses on planes D and E. Since the bending moment at E exceeds that at D, C_E is greater than C_D. Since the element must be in equilibrium and the sum of the horizontal forces must equal zero, we must conclude that a horizontal shear force is present on plane F. The horizontal shear force is the product of a horizontal shear stress and the area of the bottom surface of the element and may be expressed as $f_v bx$.

With reference to Figure 14.13c, consider an infinitesimal rectangular area a that lies parallel to the neutral axis on plane E of the element. The centroid of area a lies a distance y from the neutral axis. The compressive bending stress on this area, using the flexure formula, is

$$f_b = M_E \frac{y}{I}$$

and the force acting on area a is

$$f_b(a) = M_E \frac{y}{I}(a)$$

The total force acting on the total area above plane F is then equal to C_E, where

$$C_E = \Sigma \frac{M_E}{I}(ya)$$

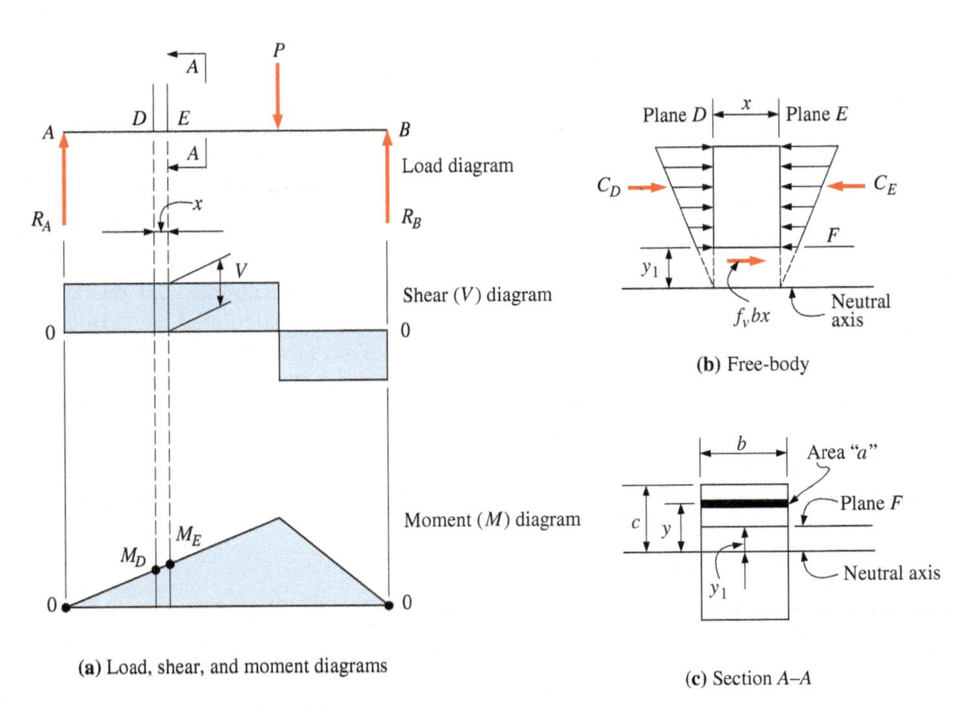

FIGURE 14.13 Derivation of horizontal shear formula.

which can be rewritten as

$$C_E = \frac{M_E}{I} \Sigma ya$$

Note in the preceding that M_E and I are fixed quantities.

The quantity Σya represents the moment of the total area of the element (above plane F) on plane E about the neutral axis. This moment is generally designated by the symbol Q and is termed the *statical moment of the area*. Therefore,

$$C_E = \frac{M_E Q}{I}$$

In the same way, the resultant force C_D acting on plane D of the element is found to be

$$C_D = \frac{M_D Q}{I}$$

The difference between the two forces, then, is

$$C_E - C_D = \frac{M_E Q}{I} - \frac{M_D Q}{I} = (M_E - M_D)\frac{Q}{I}$$

Since the element is in equilibrium, this value must equal the horizontal shear force on the bottom surface of the element (plane F). This force was noted earlier as $f_v bx$. Therefore,

$$f_v bx = (M_E - M_D)\frac{Q}{I}$$

from which

$$f_v = \frac{(M_E - M_D)Q}{Ibx}$$

As we demonstrated in Chapter 13, the change in bending moment between any two points of a beam is equal to the area of the shear diagram between the same two points. With reference to Figure 14.13a, the area of the shear diagram between planes D and E is equal to Vx. Substituting this for $M_E - M_D$ in the previous equation results in an expression commonly called the *general shear formula:*

$$f_v = \frac{VxQ}{Ibx} = \frac{VQ}{Ib} \tag{14.8}$$

where f_v = the horizontal (and vertical) computed shear stress on any given plane of a given cross section of the beam (psi, ksi) (Pa)
V = the computed vertical shear force at the given cross section (lb, kips) (N)
Q = the statical moment about the neutral axis of the cross-sectional area between the horizontal plane where the shear stress is to be calculated and the top (or bottom) of the beam (in.^{3}) (m^3)
I = the moment of inertia of the entire cross section with respect to the neutral axis (the same I used in flexure formula calculations) (in.^{4}) (m^4)
b = the width of the cross section in the horizontal plane where the shear stress is being calculated (in.) (m)

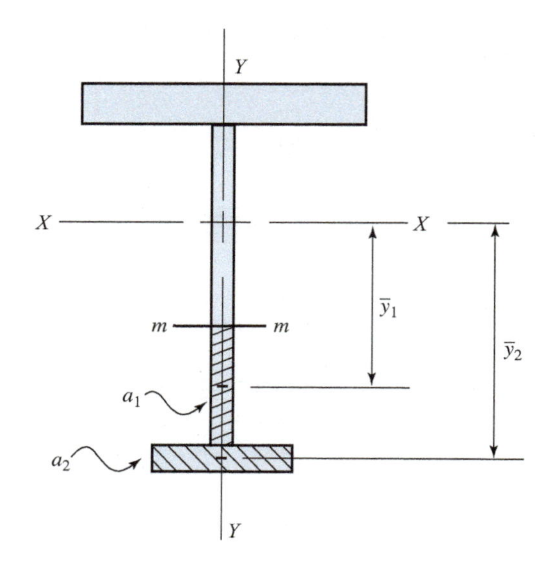

FIGURE 14.14 Determination of Q.

The statical moment of area Q is the product of an area and a distance:

$$Q = A\bar{y}$$

where A is the area above or below the plane where the shear stress is to be found and $\bar{y}$ is the distance from the centroid of that area to the neutral axis of the cross section. For a built-up cross section such as shown in Figure 14.14, which is composed of elements for which areas can be determined and centroids can be located, Q is calculated from

$$Q = \Sigma a\bar{y}$$

where a represents the area of an element of the cross section and $\bar{y}$ is the distance from the centroid of that element to the neutral axis. In Figure 14.14, assume the shear stress is to be calculated on plane m–m located below the centroidal X–X axis. The area below plane m–m is divided into two rectangular areas a_1 and a_2. The centroids of the area are located $\bar{y}_1$ and $\bar{y}_2$ from the neutral axis, and Q is calculated from

$$Q = a_1\bar{y}_1 + a_2\bar{y}_2$$

Several examples that illustrate this procedure follow. Most practical problems dealing with built-up cross sections involve elements (rectangles and triangles) having properties that are readily calculated or standardized shapes whose properties are tabulated.

The general shear formula can be rewritten in a form useful for the calculation of an allowable shear force (or shear capacity) for a bending member. Denoting the shear capacity as V_R, Equation (14.8) yields

$$V_R = \frac{F_v Ib}{Q} \tag{14.9}$$

where V_R = the allowable shear force (or shear capacity) at a given cross section
F_v = the allowable shear stress

and the other terms are as previously defined.

14.6 SHEAR STRESSES IN STRUCTURAL MEMBERS

In the analysis and design of beams in structures, flexure (bending moment) is generally the most critical factor. That is, bending stresses will reach allowable limits first. In some beams, however, due to very heavy loads and/or very short spans, shear may become more critical. The following examples illustrate the application of the general shear formula.

EXAMPLE 14.8 The rough, solid rectangular timber beam in Figure 14.15 is 8 in. wide by 12 in. deep. The beam is subjected to a vertical load, inducing a maximum shear V of 7000 lb. (a) Calculate the maximum shear stress at the neutral axis. (b) Calculate the shear stress at 2 in. above and below the neutral axis. (c) Calculate the shear stress at 4 in. above and below the neutral axis. (d) Plot these stresses showing the distribution of the horizontal shear stress.

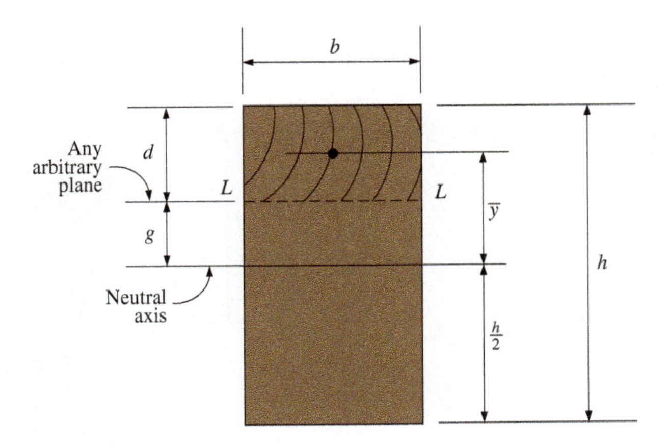

FIGURE 14.15 Beam cross section.

Solution Use the general shear formula, Equation (14.8):

$$f_v = \frac{VQ}{Ib}$$

Two of the terms, V and b, are given. These two terms will remain constant in this example. Next, calculate the moment of inertia with respect to the neutral axis:

$$I = \frac{bh^3}{12} = \frac{(8 \text{ in.})(12 \text{ in.})^3}{12} = 1152 \text{ in.}^4$$

The only other value that must be calculated prior to computing f_v is the statical moment of area, Q.

(a) With reference to Figure 14.15, assume that plane L–L lies at the neutral axis ($g = 0$). Then Q is equal to the statical moment of the area above the neutral axis, with respect to the neutral axis. The area A above the neutral axis is

$$A = bd = (8 \text{ in.})(6 \text{ in.}) = 48 \text{ in.}^2$$

The distance from the centroid of area A to the neutral axis is

$$\bar{y} = \frac{h}{2} - \frac{d}{2} = 6 \text{ in.} - 3 \text{ in.} = 3 \text{ in.}$$

Therefore,

$$Q = A\bar{y} = (48 \text{ in.}^2)(3 \text{ in.}) = 144 \text{ in.}^3$$

and, at the neutral axis,

$$f_v = \frac{VQ}{Ib} = \frac{(7000 \text{ lb})(144 \text{ in.}^3)}{(1152 \text{ in.}^4)(8 \text{ in.})} = 109.4 \text{ psi}$$

(b) With reference to Figure 14.15, assume that plane L–L lies 2 in. above the neutral axis ($g = 2$ in.). The area above the plane is

$$A = bd = (8 \text{ in.})(4 \text{ in.}) = 32 \text{ in.}^2$$

The distance from the centroid of this area to the neutral axis is

$$\bar{y} = \frac{h}{2} - \frac{d}{2} = \frac{12 \text{ in.}}{2} - \frac{4 \text{ in.}}{2} = 4 \text{ in.}$$

Calculate Q:

$$Q = A\bar{y} = (32 \text{ in.}^2)(4 \text{ in.}) = 128 \text{ in.}^3$$

Then

$$f_v = \frac{VQ}{Ib} = \frac{(7000 \text{ lb})(128 \text{ in.}^3)}{(1152 \text{ in.}^4)(8 \text{ in.})} = 97.2 \text{ psi}$$

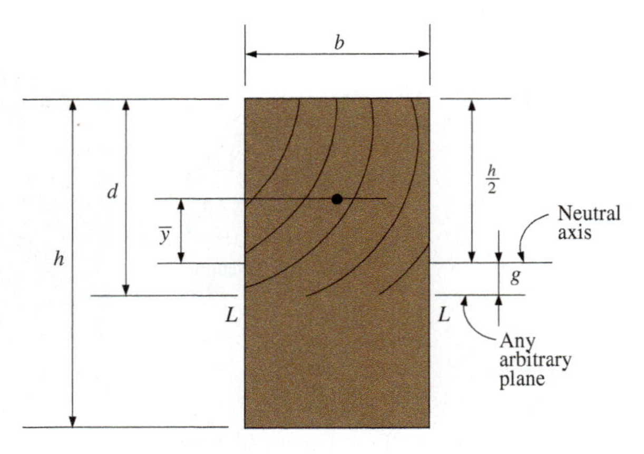

FIGURE 14.16 Beam cross section.

Placing the plane on which to calculate the horizontal shear stress at 2 in. below the neutral axis, see in Figure 14.16 that g remains at 2 in., d is 8 in., and the area above plane L–L is

$$A = bd = (8 \text{ in.})(8 \text{ in.}) = 64 \text{ in.}^2$$

The distance from the centroid of the area to the neutral axis is calculated from

$$\bar{y} = \frac{d}{2} - g = \frac{8 \text{ in.}}{2} - 2 \text{ in.} = 2 \text{ in.}$$

Therefore,

$$Q = A\bar{y} = (64 \text{ in.}^2)(2 \text{ in.}) = 128 \text{ in.}^3$$

and

$$f_v = \frac{VQ}{Ib} = \frac{(7000 \text{ lb})(128 \text{ in.}^3)}{(1152 \text{ in.}^4)(8 \text{ in.})} = 97.2 \text{ psi}$$

As can be observed, the horizontal shear stress is the same whether 2 in. above or 2 in. below the neutral axis, which is due to symmetry with respect to the neutral axis. This situation will exist for all corresponding planes above or below the neutral axis where symmetry with respect to the neutral axis exists.

(c) Placing the plane on which to calculate the horizontal shear stress at 4 in. above (or below) the neutral axis, calculate as follows (using plane L–L above the neutral axis with $g = 4$ in., and referring to Figure 14.15):

$$d = \frac{h}{2} - g = \frac{12 \text{ in.}}{2} - 4 \text{ in.} = 2 \text{ in.}$$

$$A = bd = (8 \text{ in.})(2 \text{ in.}) = 16 \text{ in.}^2$$

$$\bar{y} = \frac{h}{2} - \frac{d}{2} = 6 \text{ in.} - 1 \text{ in.} = 5 \text{ in.}$$

$$Q = A\bar{y} = (16 \text{ in.}^2)(5 \text{ in.}) = 80 \text{ in.}^3$$

$$f_v = \frac{VQ}{Ib} = \frac{(7000 \text{ lb})(80 \text{ in.}^3)}{(1152 \text{ in.}^4)(8 \text{ in.})} = 60.8 \text{ psi}$$

Note that at the outer fibers, Q would be zero. Therefore, the shear stress would also be zero.

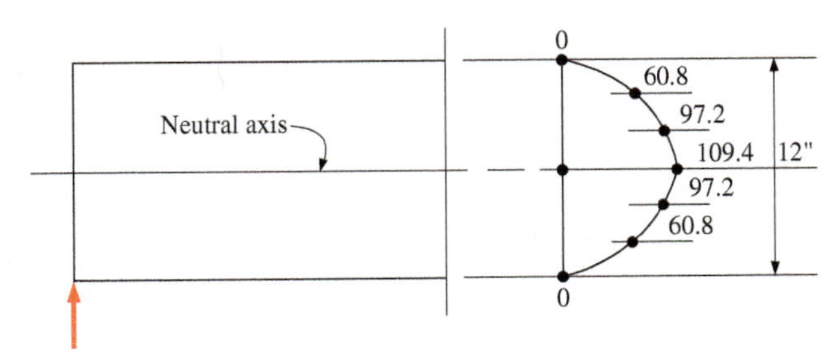

FIGURE 14.17 Shear stress distribution in a beam.

(d) The diagram of the shear stress distribution is shown in Figure 14.17. Note that for a homogeneous solid rectangular cross section, the maximum horizontal shear stress occurs at the neutral axis and the minimum, which is a zero value, occurs at the outer fibers. A closer inspection of how shear stress varies with distance from the neutral axis shows that the curve is parabolic.

EXAMPLE 14.9 The bending member in Figure 14.18 is built up of three steel plates welded together to act as a single unit. It is generally called a *plate girder.* The girder is subjected to a vertical loading, which induces a maximum shear V of 300 kips. (a) Calculate the maximum shear stress in the plane of the neutral axis. (b) Calculate the shear stress at the junction of the flange and the web (plane E–E). (c) Plot the distribution of the shear stress for the entire cross section.

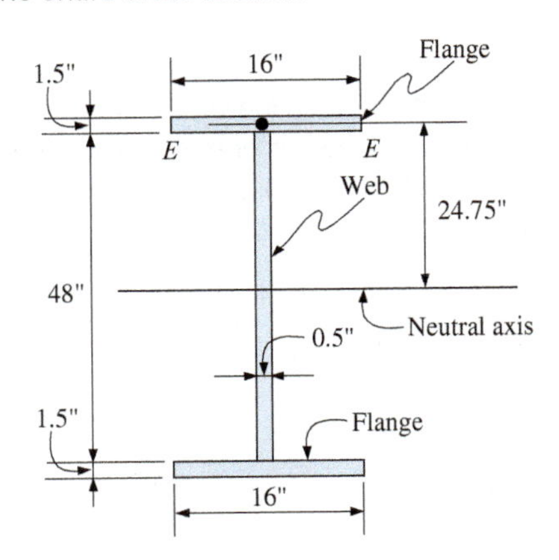

FIGURE 14.18 Girder cross section.

Solution In using the general shear formula, the values of V and I will remain constant when calculating the shear stress on any horizontal plane. V and b are given. The moment of inertia with respect to the neutral axis (I_{NA}) may be computed as follows.

For the bottom or top flange (these are symmetrical with respect to the neutral axis),

$$I_{NA} = I_o + Ad^2 = \frac{bh^3}{12} + Ad^2$$

$$= \frac{(16 \text{ in.})(1.5 \text{ in.})^3}{12} + (16 \text{ in.})(1.5 \text{ in.})(24.75 \text{ in.})^2$$

$$= 14{,}710 \text{ in.}^4$$

For the web,

$$I_{NA} = \frac{bh^3}{12} = \frac{(0.5\text{ in.})(48\text{ in.})^3}{12} = 4610\text{ in.}^4$$

Therefore, the total moment of inertia about the neutral axis is

$$\text{total } I_{NA} = 2(14{,}710\text{ in.}^4) + 4610\text{ in.}^4 = 34{,}000\text{ in.}^4$$

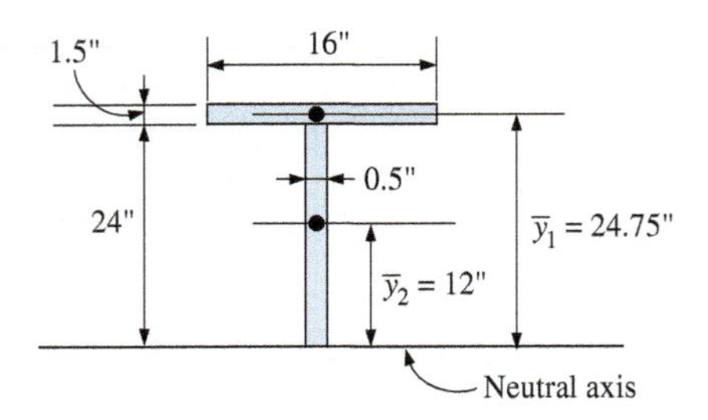

FIGURE 14.19 Beam cross section above the neutral axis.

(a) Calculate the horizontal shear stress at the neutral axis with reference to Figure 14.19 and use subscripts *f* and *w* to designate flange and web, respectively:

$$Q_f = A_f\bar{y}_1 = (16\text{ in.})(1.5\text{ in.})(24.75\text{ in.}) = 594\text{ in.}^3$$
$$Q_w = A_w\bar{y}_2 = (24\text{ in.})(0.5\text{ in.})(12\text{ in.}) = 144\text{ in.}^3$$
$$\text{total } Q = 594\text{ in.}^3 + 144\text{ in.}^3 = 738\text{ in.}^3$$

from which

$$f_v = \frac{VQ}{Ib} = \frac{(300\text{ k})(738\text{ in.}^3)}{(34{,}000\text{ in.}^4)(0.5\text{ in.})} = 13.02\text{ ksi}$$

(b) Calculate the horizontal shear stress at the junction of the flange and the web, with reference to Figure 14.20:

$$\text{total } Q = Q_f = A_f\bar{y}_1 = (16\text{ in.})(1.5\text{ in.})(24.75\text{ in.}) = 594\text{ in.}^3$$

Note that at the junction of the web and the flange the value of b may be either 16 in. or 0.5 in., depending on whether one is considering the stress at an infinitesimal distance above or below the actual junction. The magnitude of the horizontal shear stress at this location, therefore, undergoes an abrupt change. Where b is 16 in.,

$$f_v = \frac{VQ}{Ib} = \frac{(300\text{ k})(594\text{ in.}^3)}{(34{,}000\text{ in.}^4)(16\text{ in.})} = 0.328\text{ ksi}$$

Where b is 0.5 in.,

$$f_v = \frac{VQ}{Ib} = \frac{(300\text{ k})(594\text{ in.}^3)}{(34{,}000\text{ in.}^4)(0.50\text{ in.})} = 10.48\text{ ksi}$$

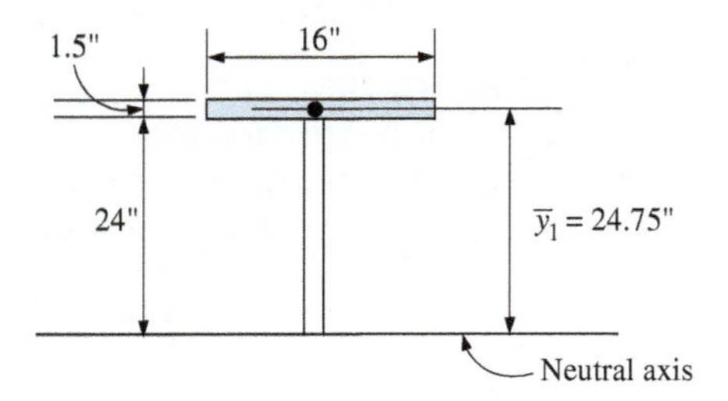

FIGURE 14.20 Beam cross section above the neutral axis.

(c) The diagram of the distribution of the horizontal shear stresses is shown in Figure 14.21.

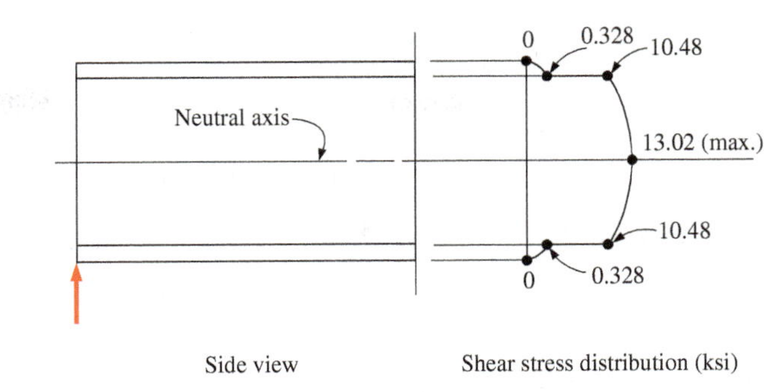

FIGURE 14.21 Shear stress distribution—Example 14.9.

Even though the flange areas of the girder in Example 14.9 are quite large, the low shear stress values in the flange indicate that the flanges resist only a small portion of the total vertical shear force. In plate girders having the I-shape, it is the web that predominantly resists the shear, which is also true for rolled structural steel shapes such as wide-flange beams (W shapes), American standard I-beams (S shapes), and channels (C shapes).

It is estimated that the webs for the types of steel beams just mentioned resist 85% to 90% of the total vertical shear force. For this reason, design specifications allow the use of an "average web shear" approach for the determination of shear stress in rolled and fabricated shapes rather than requiring the use of the general shear formula. The average web shear is calculated from

$$f_v = \frac{V}{dt_w} \tag{14.10}$$

where f_v = the computed maximum shear stress (psi, ksi) (Pa)
V = the vertical shear force at the section under consideration (lb, kips) (N)
d = the *full* depth of the beam (in.) (m)
t_w = the web thickness of the beam (in.) (m)

This method is approximate compared with the theoretically correct general shear formula and assumes that the shear is resisted by the rectangular area of the web extending the full depth of the beam. For the plate girder of Example 14.9,

$$f_v = \frac{V}{dt_w} = \frac{300 \text{ k}}{(51 \text{ in.})(0.5 \text{ in.})} = 11.76 \text{ ksi}$$

This value is almost 10% below the theoretically correct maximum value of 13.02 ksi. Therefore, the average web shear method results in a shear stress that is too low. This situation could be considered unsafe; allowable shear stresses, however, are set intentionally low because the computed average shear stress will always be lower than the actual maximum shear stress.

Using the average web shear approach, we can also write an expression for the allowable shear V_R or the shear capacity of these cross sections. Rewriting Equation (14.10) gives

$$V_R = F_v dt_w \tag{14.11}$$

This approximate approach should be used only for steel beams having symmetry with respect to the neutral axis, such as the types previously mentioned.

EXAMPLE 14.10 A W16 × 100 (steel wide-flange beam) is subjected to a vertical shear of 80 kips. Calculate the maximum horizontal shear stress using (a) the general shear formula and (b) the average web shear approach.

Solution The dimensions and properties of the W16 × 100 can be obtained from Appendix A. The necessary dimensions are shown in Figure 14.22.

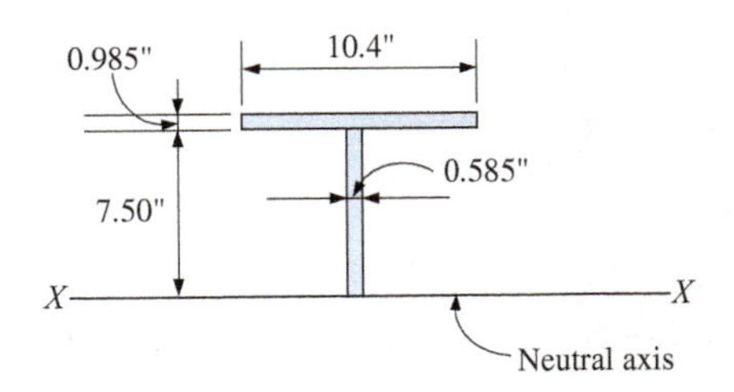

FIGURE 14.22 Beam cross section above the neutral axis.

(a) Calculate Q at the neutral axis, using flange and web elements:

$$Q_f = (10.4\text{ in.})(0.985\text{ in.})\left(7.50\text{ in.} + \frac{0.985\text{ in.}}{2}\right) = 81.9\text{ in.}^3$$

$$Q_w = (7.50\text{ in.})(0.585\text{ in.})\left(\frac{7.50\text{ in.}}{2}\right) = 16.50\text{ in.}^3$$

$$\text{total } Q = 81.9\text{ in.}^3 + 16.5\text{ in.}^3 = 98.4\text{ in.}^3$$

from which

$$f_v = \frac{VQ}{Ib} = \frac{(80\text{ k})(98.6\text{ in.}^3)}{(1490\text{ in.}^4)(0.585\text{ in.})} = 9.05\text{ ksi}$$

(b) Use the average web shear approach:

$$d = 2(7.50\text{ in.} + 0.985\text{ in.}) = 16.97\text{ in.}$$

$$f_v = \frac{V}{dt_w} = \frac{80\text{ k}}{(16.97\text{ in.})(0.585\text{ in.})} = 8.06\text{ ksi}$$

Generally, only the maximum value of the shear stress is of interest since this is the value that must be compared with some allowable shear stress. It is practical, then, to develop equations to apply in special cases. One such case is that of the Example 14.10: a solid, homogeneous rectangular cross section in which the maximum horizontal shear stress is desired. This cross section is very common in timber design. We derive the equation as follows, with reference to Figure 14.23:

$$Q = (b)\left(\frac{h}{2}\right)\left(\frac{h}{4}\right) = \frac{bh^2}{8}$$

$$I = \frac{bh^3}{12}$$

$$f_v = \frac{VQ}{Ib} = \frac{V(bh^2/8)}{(bh^3/12)(b)} = \frac{Vbh^2}{8} \times \frac{12}{b^2h^3}$$

from which

$$f_v = \frac{12V}{8bh} = 1.5\frac{V}{A} \quad \textbf{(14.12)}$$

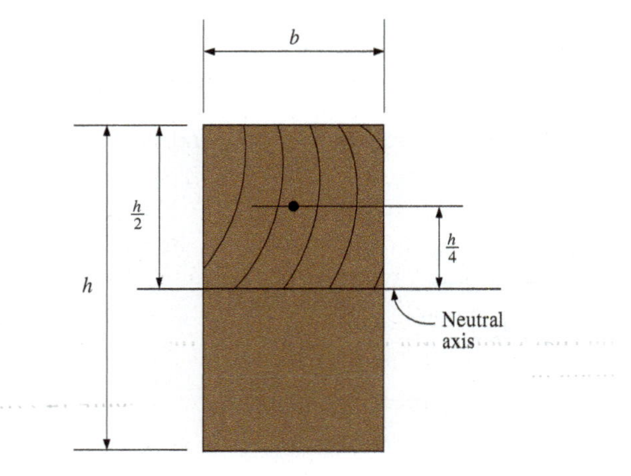

FIGURE 14.23 Beam cross section.

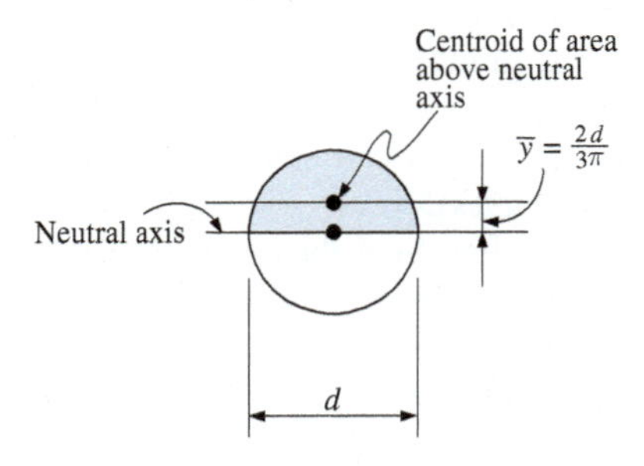

FIGURE 14.24 Circular cross section.

Thus, the maximum horizontal shear stress for a homogeneous solid rectangular beam is 1.5 times the average shear stress V/A, where A represents the *total* cross-sectional area.

For a member of circular cross section, such as a shaft used as a beam, the maximum shear stress also occurs at the neutral axis, even though all other horizontal planes are narrower in width (as may be observed in Figure 14.24). To compute the maximum shear stress, the general shear formula may be used. Also, a simplified expression for the maximum shear stress for this cross section may be developed as follows. Referring to Table 3 for appropriate properties and expressing all values in terms of the diameter d yields

$$b = d$$

$$I_{NA} = \frac{\pi d^4}{64}$$

The area above the neutral axis is calculated from

$$A = \frac{\pi d^2}{8}$$

and the distance from the neutral axis to the centroid of the area above the neutral axis is

$$\bar{y} = \frac{2d}{3\pi}$$

Therefore,

$$Q = A\bar{y} = \frac{\pi d^2}{8}\left(\frac{2d}{3\pi}\right) = \frac{d^3}{12}$$

and

$$f_v = \frac{VQ}{Ib} = \frac{V(d^3/12)}{(\pi d^4/64)(d)} = \frac{16V}{12(\pi d^2/4)}$$

from which

$$f_v = \frac{4V}{3A} \quad \textbf{(14.13)}$$

where A represents the total circular cross-sectional area. Thus, the maximum horizontal shear stress for a solid homogeneous beam of circular cross section is $\frac{4}{3}$ times the average value.

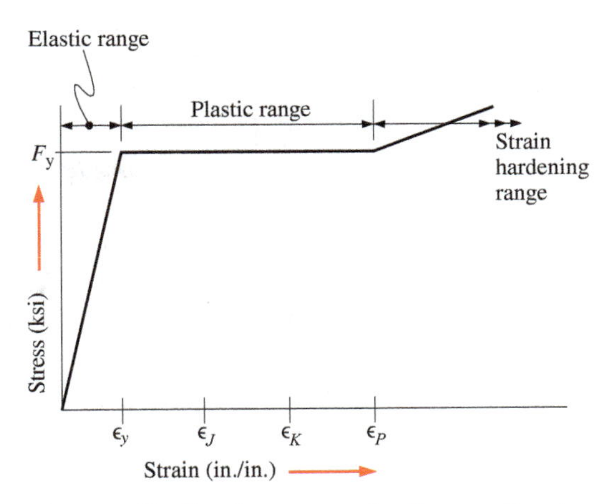

FIGURE 14.25 Idealized stress–strain diagram for structural steel.

14.7 INELASTIC BENDING OF BEAMS

The preceding sections of this chapter included discussion of ductile, homogeneous bending members. A linear bending stress distribution over the depth of the member was assumed, varying from a maximum at the outer fibers to zero at the neutral axis. This stress distribution served as a basis for the design and/or analysis of bending members when using an allowable bending stress. Since the allowable bending stress is always specified as some fraction of a proportional limit stress (or a yield stress), the induced bending stress for a properly designed member will always be in the elastic range. Hooke's law (stress is proportional to strain) was assumed to apply.

Now consider bending members so loaded that they are stressed beyond the proportional limit (and the yield point) into the plastic range. Permanent deformation of the material occurs and stress is no longer proportional to strain. The applied loads considered for this condition are large enough to produce plastic deformations, but not so large that they produce failure of the member or structure. This design approach, as it relates to tension members, was introduced in Section 10.8.

For purposes of discussion, a ductile, homogeneous material such as structural steel is now considered. For such a material, the stress–strain relationship may be reasonably idealized as shown in Figure 14.25. Note that a straight-line relationship between stress and strain is assumed up to the yield point. It is also assumed that the yield point and modulus of elasticity are the same for both tension and compression. Past the yield point, the stress within the plastic range is assumed to be constant even though strain increases. Let us also assume that plane sections before bending remain plane into the plastic range; thus, strains are always proportional to the distance from the neutral axis.

The beam of which the behavior will be examined is assumed to be rectangular in cross section, as shown in Figure 14.26a, and subjected to a gradually increasing bending moment. Further, the cross section is symmetrical about an axis that lies in the plane of the loading. As the moment is increased until the outer fiber strain reaches ϵ_y, the relationship between stress and strain is a linear one, yielding a linear stress distribution, as shown in Figure 14.26b, and the outer fiber stress reaches F_y. The moment at this point is called the *yield moment* M_y. The moment is increased further and the outer fiber strains reach the value ϵ_J, as shown in Figure 14.26c. The corresponding stress in the outer fibers for this strain is still F_y. Those interior fibers that have been strained past ϵ_y will also be stressed to F_y. Other interior fibers, closer to the neutral axis, that have not been strained past ϵ_y, however, will have a linear stress distribution decreasing to zero at the neutral axis. As the moment increases further, more and more of the cross section is stressed to the yield stress F_y, as shown in Figure 14.26d. This process is called *plastification* of the cross section. Finally, almost all the fibers will be strained past ϵ_y and the stress distribution will approach a rectangular shape. This stress distribution is shown idealized in Figure 14.26e (the cross section is fully plastified).

At full plastification of the cross section, it is assumed that no additional moment can be resisted. When this occurs, a plastic hinge is said to have formed and the moment that exists at this point is called the *plastic moment* (M_P), which in effect represents the limiting moment strength of the beam.

To evaluate M_P, consider a rectangular shape that is stressed as shown in Figure 14.27. Assume, for discussion purposes, that this is the cross section of a simply supported beam that is subjected to vertical load only. Since the member must be in equilibrium, the algebraic sum of the internal forces C and T must be equal. Note that these are resultant forces of the bending stresses and are parallel, equal, and opposite in sense, thereby forming an internal couple M_P that must resist the applied bending moment.

Considering the horizontal forces (see Figure 14.27b),

$$C = T$$

$$F_y A_C = F_y A_T$$

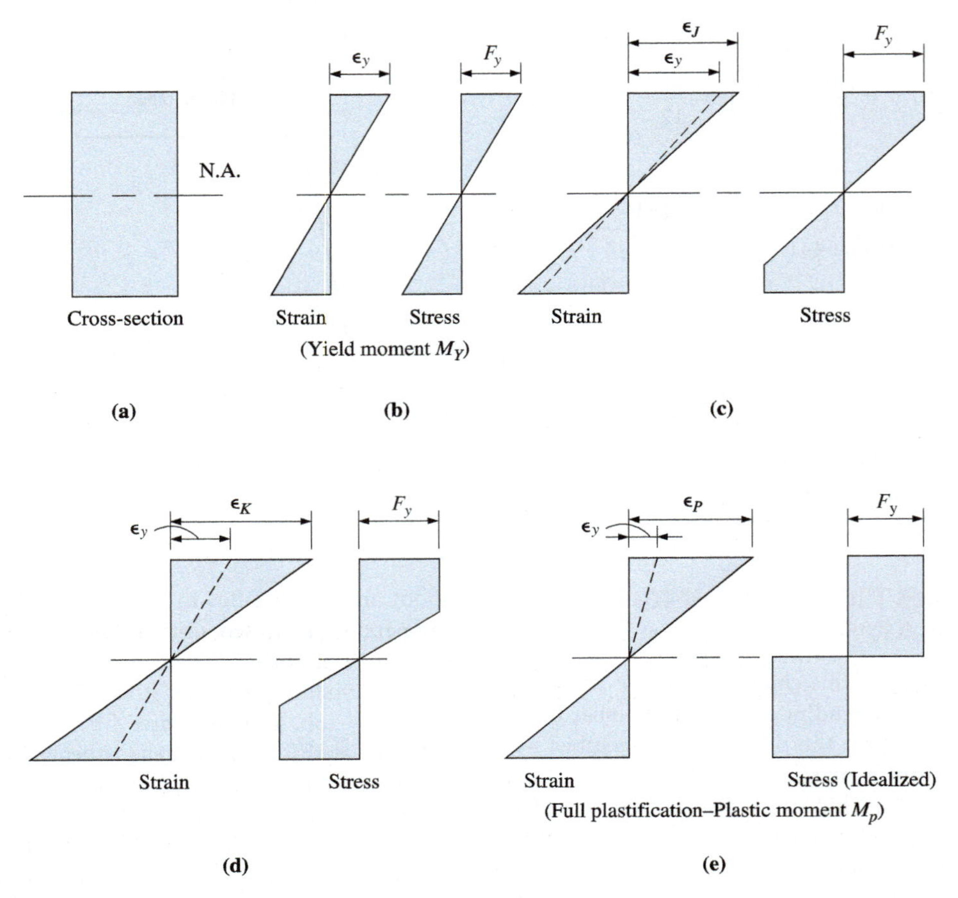

FIGURE 14.26 Stress–strain relationship.

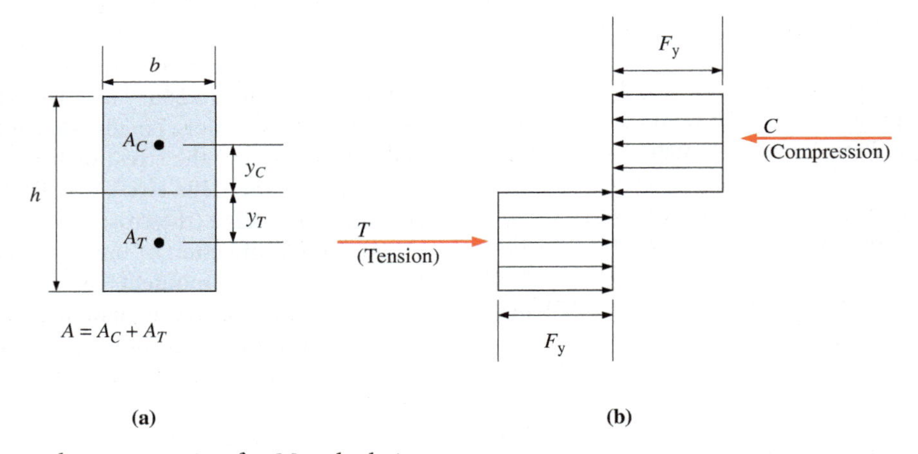

FIGURE 14.27 Rectangular cross section for M_P calculation.

from which

$$A_C = A_T = \frac{A}{2}$$

indicating that the neutral axis of a fully plastified cross section divides the cross section into two parts of equal area. (For cross sections having symmetry about both the X–X and Y–Y axes, the neutral axis is at the centroid of the section for both elastic and plastic loading.)

The internal resistance constitutes a couple, which can be expressed as

$$M_P = C(y_C + y_T) \qquad \text{or} \qquad M_P = T(y_C + y_T)$$
$$= F_y A_C (y_C + y_T) \qquad\qquad = F_y A_T (y_C + y_T)$$

It was previously shown that $A_C = A_T = A/2$. Therefore, by substitution,

$$M_P = F_y\left(\frac{A}{2}\right)(y_C + y_T) \tag{14.14}$$

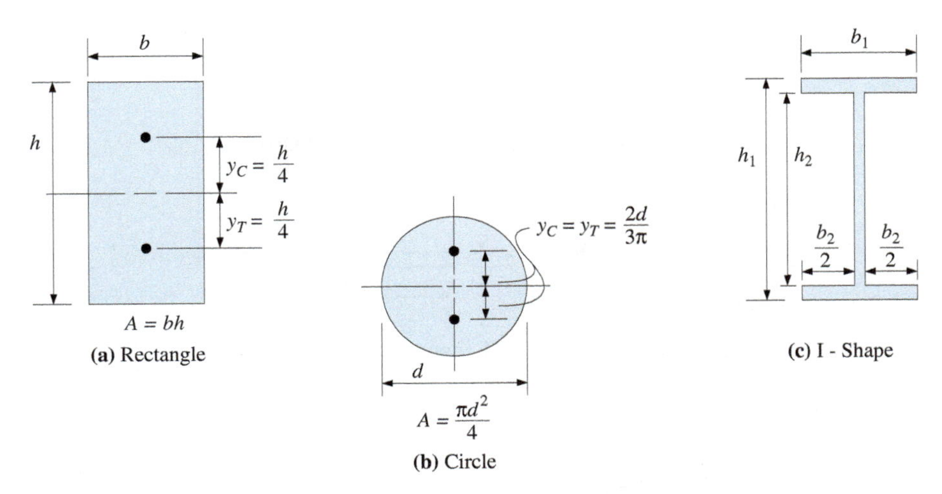

FIGURE 14.28 Shapes for M_P calculations.

Although derived for a rectangular shape, Equation (14.14) can be used to determine M_P for various shapes. M_P can be calculated for the three common shapes shown in Figure 14.28. For a rectangular shape (Figure 14.28a),

$$M_P = F_y\left(\frac{bh}{2}\right)\left(\frac{h}{4} + \frac{h}{4}\right) = F_y\left(\frac{bh^2}{4}\right) \quad \textbf{(14.15)}$$

For a circular shape (Figure 14.28b),

$$M_P = F_y\left(\frac{\pi d^2}{8}\right)\left(\frac{2d}{3\pi} + \frac{2d}{3\pi}\right) = F_y\left(\frac{d^3}{6}\right) \quad \textbf{(14.16)}$$

For an I shape (Figure 14.28c),

$$M_P = F_y\left(\frac{b_1h_1^2}{4} - \frac{b_2h_2^2}{4}\right)$$

$$M_P = \frac{F_y}{4}(b_1h_1^2 - b_2h_2^2) \quad \textbf{(14.17)}$$

The total resistance to bending for first yield at the outer fiber is

$$M_y = F_yS \quad \textbf{(14.18)}$$

where M_y = yield moment
F_y = yield stress
S = elastic section modulus

The total resistance to bending when the plastic moment develops is

$$M_P = F_yZ \quad \textbf{(14.19)}$$

where M_P = plastic moment
Z = plastic section modulus

Note that for a rectangular shape, use Equation (14.15):

$$M_P = F_y\left(\frac{bh^2}{4}\right)$$

Therefore, for a rectangular shape, $bh^2/4$ is the plastic section modulus Z. Plastic section modulus of a cross section can be calculated from the sum of the moments of the areas above and below the neutral axis taken about the neutral axis.

Recall that the elastic section modulus S for a rectangular shape is $bh^2/6$. Therefore, the ratio of the plastic moment and the yield moment is calculated from

$$\frac{M_P}{M_y} = \frac{bh^2/4}{bh^2/6} = \frac{6}{4} = 1.50$$

which indicates that the rectangular beam can carry 50% more moment from the time that it first yields until it reaches its full plastic moment. This ratio is called the *shape factor*.

EXAMPLE 14.11 An A36 structural steel shape is built up using three steel plates, as shown in Figure 14.29. Calculate the yield moment M_y, the plastic moment M_P (both with respect to the X–X axis), and the shape factor. Also calculate the uniformly distributed load that the beam can carry at yield and at full plastification for a simple span length of 40 ft.

Solution (a) *Elastic behavior:*

$$I_x = \Sigma I_O + \Sigma ad^2$$

$$= \frac{(1\text{ in.})(24\text{ in.})^3}{12} + 2\left[\frac{(16\text{ in.})(2\text{ in.})^3}{12}\right] + 2(2\text{ in.})(16\text{ in.})(13\text{ in.})^2 = 11{,}990\text{ in.}^4$$

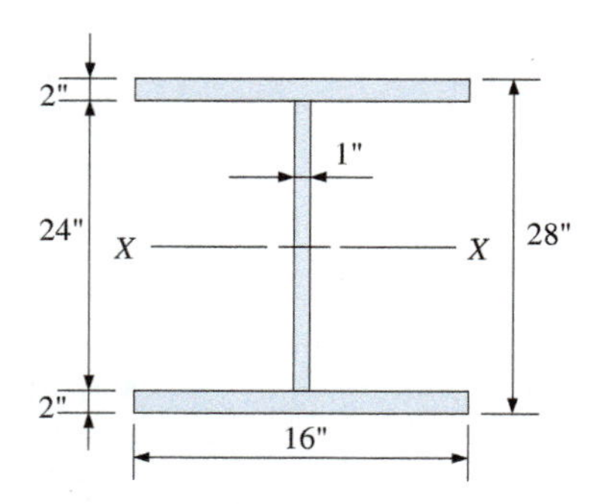

FIGURE 14.29 Built-up structural I shape.

$$S_x = \frac{I_x}{c} = \frac{11{,}990 \text{ in.}^4}{14 \text{ in.}} = 856 \text{ in.}^3$$

$$M_y = F_y S_x = \frac{(36 \text{ ksi})(856 \text{ in.}^3)}{12 \text{ in./ft}} = 2570 \text{ k-ft}$$

From

$$M = \frac{wL^2}{8} \quad \text{or} \quad M_y = \frac{w_y L^2}{8}$$

solve for load at yield w_Y:

$$w_y = \frac{8M_y}{L^2} = \frac{8(2570 \text{ k-ft})}{(40 \text{ ft})^2} = 12.85 \text{ k/ft}$$

(b) *Inelastic (plastic behavior):*

$$Z_x = 2[(1 \text{ in.})(12 \text{ in.})(6 \text{ in.}) + (2 \text{ in.})(16 \text{ in.})(13 \text{ in.})] = 976 \text{ in.}^3$$

$$M_P = F_y Z_x = \frac{(36 \text{ ksi})(976 \text{ in.}^3)}{12 \text{ in./ft}} = 2930 \text{ k-ft}$$

$$\text{shape factor} = \frac{M_P}{M_y} = \frac{2930 \text{ k-ft}}{2570 \text{ k-ft}} = 1.140$$

From

$$M = \frac{wL^2}{8}$$

solve for load at full plastification of the cross section w_P:

$$w_P = \frac{8M_P}{L^2} = \frac{8(2930 \text{ k-ft})}{(40 \text{ ft})^2} = 14.65 \text{ k/ft}$$

In Sections 14.1 through 14.3 we discussed elastic bending stresses in beams and how these stresses could be evaluated. In Chapter 16 we will discuss the design of structural steel beams. It will be seen that under the current AISC design specification, the moment strength is based on the plastic moment M_P divided by a safety factor (see Section 10.7). For structural steel beams, this is then written as:

$$\text{Moment strength } M_R = \frac{M_P}{\text{F.S}} = \frac{F_y Z_x}{\text{F.S.}}$$

For a factor of safety of 1.67 from the AISC specification, this becomes

$$M_R = \frac{F_y}{1.67} Z_x = 0.6\, F_y Z_x \qquad \textbf{(14.20)}$$

This will be illustrated in the following section.

14.8 BEAM ANALYSIS

The analysis problem is generally considered to be the investigation of a beam with a known cross section. Three common types of problems are

1. Determining the calculated induced stresses (bending and shear) for a given beam, loading, and span length.

2. Computing the load-carrying capacity of a given beam.
3. Computing a maximum span length for a given beam and loading.

All these problems are related. They make use of the flexure formula and the general shear formula. In addition, analysis-type problems may involve the connecting of component parts of a built-up member.

EXAMPLE 14.12 A W21 × 73 steel wide-flange beam is to span 40 ft on simple supports, as shown in Figure 14.30. The load shown is a superimposed load, meaning that it does not include the weight of the beam. Assume A992 steel with $F_y = 50$ ksi. Determine whether the beam is adequate by comparing the moment strength (see Equation. (14.20) for moment strength of wide-flange shapes) and the shear strength with the calculated applied moment and shear. Use the average web shear approach and $F_v = 0.4F_y$.

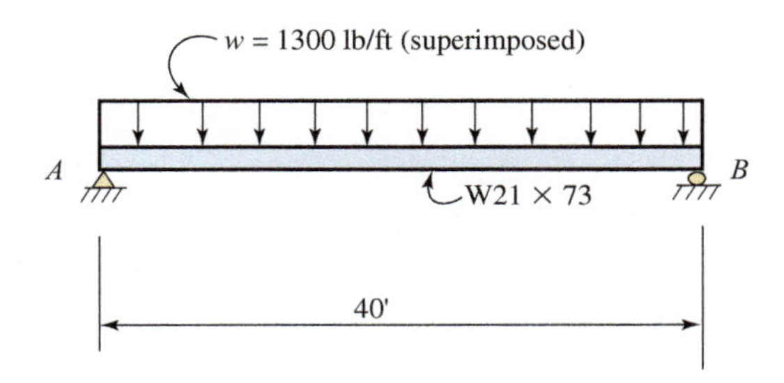

FIGURE 14.30 Beam diagram for Example 14.12.

Solution Calculate the moment strength M_R. The plastic section modulus is obtained from Appendix A.

$$M_R = 0.6F_yZ_x = \frac{0.6(50\text{ ksi})(172\text{ in.}^3)}{12\text{ in./ft}} = 430\text{ k-ft}$$

Calculate the maximum bending moment (see Appendix H). The total uniformly distributed load is 1300 lb/ft + 73 lb/ft = 1373 lb/ft.

$$M = \frac{wL^2}{8} = \frac{(1373\text{ lb/ft})(40\text{ ft})^2}{8} = 275{,}000\text{ lb-ft} = 275\text{ k-ft}$$

$$275\text{ k-ft} < 430\text{ k-ft} \qquad \textbf{(O.K.)}$$

We will determine the shear strength using Equation (14.11) with $F_v = 0.4F_y$. See Appendix A for cross-sectional properties:

$$V_R = F_v d t_w = 0.4(50\text{ ksi})(21.2\text{ in.})(0.455\text{ in.}) = 192.9\text{ k}$$

The maximum applied shear, which is equal to the reaction at either end of the beam, is calculated from

$$V = \frac{wL}{2} = \frac{(1373\text{ lb/ft})(40\text{ ft})}{2} = 27{,}500\text{ lb} = 27.5\text{ k}$$

$$27.5\text{ k} < 192.9\text{ k} \qquad \textbf{(O.K.)}$$

Therefore, the beam is satisfactory with respect to both moment and shear.

EXAMPLE 14.13 A factory building floor is supported by 3-in.-by-16-in. (S4S) Douglas fir joists spaced 24 in. on center and on a span length of 16 ft. Assume the joists to be simply supported beams. (a) Calculate the allowable uniformly distributed load w for each joist in lb/ft. (b) Calculate the allowable uniformly distributed floor load in pounds per square foot (psf).

Solution (a) See Appendices E and F for beam properties and allowable stresses. Properties of the 3 × 16 (S4S) are

$$S_x = 96.9\text{ in.}^3$$
$$A = 38.1\text{ in.}^2$$
$$\text{wt.} = 10.6\text{ lb/ft}$$

Allowable stresses are

$$F_b = 900 \text{ psi}$$
$$F_v = 180 \text{ psi}$$

Considering moment, calculate the allowable bending moment as

$$M_R = F_b S_x = \frac{900 \text{ psi}}{12 \text{ in./ft}}(96.9 \text{ in.}^3) = 7270 \text{ lb-ft}$$

To determine the allowable uniformly distributed load w based on allowable moment, rewrite

$$M = \frac{wL^2}{8}$$

and solve for w:

$$w = \frac{8M}{L^2} = \frac{8(7270 \text{ lb-ft})}{(16 \text{ ft})^2} = 227 \text{ lb/ft}$$

Note that in this case lb/ft means "pounds per lineal foot of joist."

Considering shear, the allowable shear V_R can be calculated by rewriting Equation (14.12) and solving for V, which yields

$$V_R = \frac{F_v A}{1.5} = \frac{(180 \text{ psi})(38.1 \text{ in.}^2)}{1.5} = 4570 \text{ lb}$$

The maximum shear produced in a simply supported uniformly loaded beam is

$$V = \frac{wL}{2}$$

Equate these two shears and solve for w:

$$w = \frac{2V_R}{L} = \frac{2(4570 \text{ lb})}{16 \text{ ft}} = 571 \text{ lb/ft}$$

The controlling value is the smaller of the two allowable loads. Therefore, the allowable total uniformly distributed load is 227 lb/ft as controlled by bending moment. If the allowable superimposed load is desired, the weight of the joist (lb/ft) must be subtracted from the total allowable load. The allowable superimposed load is then

$$227 \text{ lb/ft} - 10.6 \text{ lb/ft} = 216 \text{ lb/ft}$$

(b) The allowable superimposed floor load in pounds per square foot is a function of the joist spacing. In Figure 14.31, note that each joist supports a 24-in. (or 2-ft) width of floor. It has been found that each joist can support a load of 216 lb/ft. The allowable load per foot, divided by the spacing in feet, yields the allowable load in pounds per square foot:

$$\frac{216 \text{ lb/ft}}{2 \text{ ft}} = 108 \text{ psf}$$

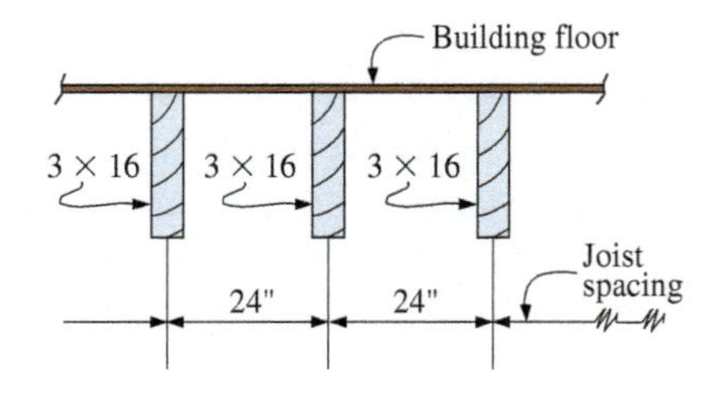

FIGURE 14.31 Floor cross section.

EXAMPLE 14.14 Calculate the allowable superimposed uniformly distributed line load that may be placed on a W610 × 153 hot-rolled structural steel wide-flange section that is simply supported on a 12-m span, as shown in Figure 14.32. The beam is oriented with the strong axis of its cross section horizontal. Assume A992 steel with $F_y = 345$ MPa and $F_v = 0.4F_y$.

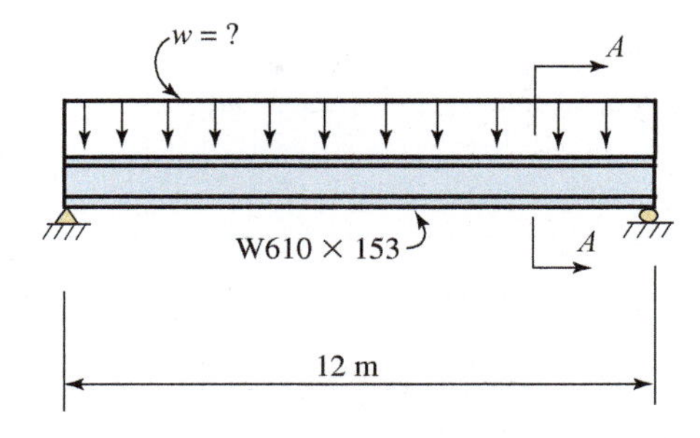

FIGURE 14.32 Beam diagram for Example 14.14.

Solution From Appendix A, for the W610 × 153,

$$Z_x = 4590 \times 10^3 \text{ mm}^3$$
$$d = 622 \text{ mm}$$
$$t_w = 14.00 \text{ mm}$$

The beam weight is calculated from

$$\begin{aligned} w &= mg \\ &= 153 \text{ kg/m } (9.81 \text{ m/sec}^2) = 1.501 \text{ kN/m} \end{aligned}$$

Calculate the moment strengh using Equation (14.20):

$$\begin{aligned} M_{Rx} &= 0.60F_yZ_x \\ &= 0.60(345 \text{ MPa})(4590 \times 10^3 \text{ mm}^3) \\ &= 950 \times 10^6 \text{ N}\cdot\text{mm} = 950 \times 10^3 \text{ N}\cdot\text{m} \end{aligned}$$

From Appendix H, the maximum bending moment for a simply supported beam that supports a uniformly distributed line load is

$$M = \frac{wL^2}{8}$$

Solve for w and substitute:

$$\begin{aligned} w &= \frac{8M}{L^2} = \frac{8(950 \times 10^3 \text{ N}\cdot\text{m})}{(12 \text{ m})^2} = 52{,}700 \text{ N/m} \\ &= 52.7 \text{ kN/m} \end{aligned}$$

Considering shear, using the average web shear approach, the shear strength may be calculated from Equation (14.11), using $F_v = 0.40F_y$.

$$\begin{aligned} V_R &= 0.4F_ydt_w = 0.4(345 \text{ MPa})(622 \text{ mm})(14.00 \text{ mm}) \\ &= 1.202 \times 10^6 \text{ N} \end{aligned}$$

From Appendix H, the maximum shear for a simply supported beam that supports a uniformly distributed line load is

$$V = \frac{wL}{2}$$

Solve for *w* and substitute:

$$w = \frac{2V}{L} = \frac{2(1.202 \times 10^6\ \text{N})}{12\ \text{m}} = 200 \times 10^3\ \text{N/m}$$
$$= 200\ \text{kN/m}$$

$$52.4\ \text{kN/m} < 200\ \text{kN/m}$$

Therefore, moment controls, since it results in the lower value. Subtract the beam weight to determine the allowable superimposed load:

$$52.4\ \text{kN/m} - 1.50\ \text{kN/m} = 50.9\ \text{kN/m}$$

EXAMPLE 14.15 Calculate the maximum allowable span length for a W18 × 50 simply supported steel wide-flange beam subjected to a uniformly distributed load of 2 k/ft (this load includes the weight of the beam). Assume A992 steel with $F_y = 50$ ksi. Use $F_v = 0.4 F_y$.

Solution From Appendix A, the necessary properties for the wide-flange shape are $d = 18.00$ in., $t_w = 0.355$ in., $Z_x = 101$ in.^{3} Considering moment, calculate the moment strength as:

$$M_R = 0.6 F_y Z_x$$
$$= \frac{0.6(50\ \text{ksi})(101\ \text{in.}^3)}{12\ \text{in./ft}} = 253\ \text{k-ft}$$

The allowable span length *L* is then obtained from

$$M = \frac{wL^2}{8}$$

by equating to M_R, rewriting, and solving for *L*:

$$L = \sqrt{\frac{8M_R}{w}} = \sqrt{\frac{8(253\ \text{k-ft})}{2\ \text{k/ft}}} = 31.8\ \text{ft}$$

Considering shear, the shear strength, using Equation (14.11) is:

$$V_R = F_v d t_w = 0.4 F_y d t_w = 0.4(50\ \text{ksi})(18.00\ \text{in.})(0.355\ \text{in.}) = 127.8\ \text{k}$$

The allowable span length *L* is obtained from

$$V = \frac{wL}{2}$$

by equating to V_R, rewriting, and solving for *L*:

$$L = \frac{2V_R}{w} = \frac{2(127.8\ \text{k})}{2\ \text{k/ft}} = 127.8\ \text{ft}$$

The controlling value is the smaller of the two span lengths, indicating that moment controls and that the maximum allowable span length is 31.8 ft.

EXAMPLE 14.16 Three 4-in.-by-6-in. timber members are bolted together so as to act as a single unit (see Figure 14.33). The bolts are spaced 10 in. on centers. The maximum vertical shear *V* is equal to 750 lb. Calculate the maximum force that the cross section of each bolt must resist at each longitudinal joint.

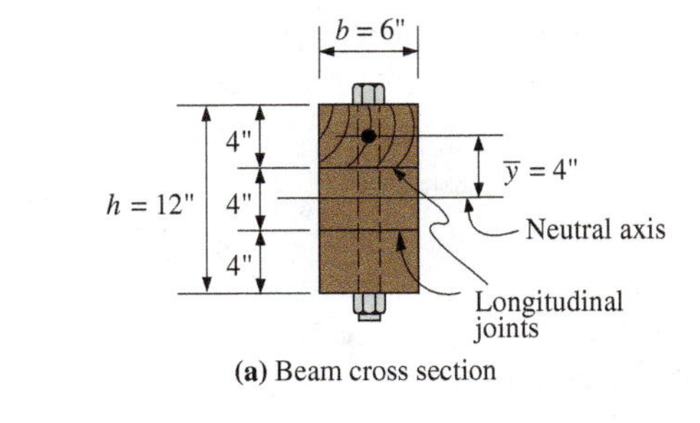

(a) Beam cross section

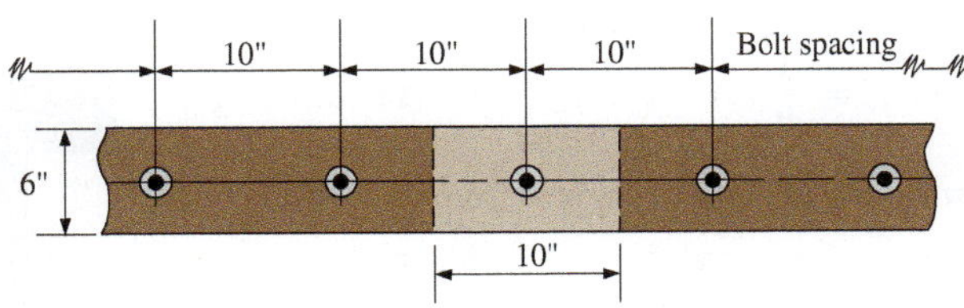

(b) Top view of beam

FIGURE 14.33 Bolted timber beam.

Solution Using the general shear formula, the horizontal shear stress will be computed in the longitudinal joints. Because of symmetry, both joints are subjected to the same stress. Thus,

$$I_{NA} = \frac{bh^3}{12} = \frac{(6\text{ in.})(12\text{ in.})^3}{12} = 864\text{ in.}^4$$

$$Q = A\bar{y} = (4\text{ in.})(6\text{ in.})(4\text{ in.}) = 96\text{ in.}^3$$

$$f_v = \frac{VQ}{Ib} = \frac{(750\text{ lb})(96\text{ in.}^3)}{(864\text{ in.}^4)(6\text{ in.})} = 13.89\text{ psi}$$

In effect, the bolt is compensating for the member not being one solid unit. Therefore, the bolt may be considered to replace a horizontal area of 6 in. by 10 in. that would normally resist the calculated shear stress (see Figure 14.33b). Each bolt, then, must resist a force that can be calculated from

$$P = f_v(A) = (13.89\text{ psi})(6\text{ in.})(10\text{ in.}) = 833\text{ lb}$$

SUMMARY BY SECTION NUMBER

14.1 In beams subjected to loads that produce bending, tensile stresses are developed on the convex side where the fibers have elongated; compressive stresses are developed on the concave side where the fibers have shortened. The plane on which no elongation or shortening of fibers occurs is also the plane of zero bending stress and is called the *neutral plane*. The intersection of the neutral plane with a cross-sectional plane is called a *neutral axis*. The tensile and compressive stresses, called *bending stresses*, vary linearly from zero at the neutral axis to a maximum at the outer fibers.

14.2 The maximum bending stress occurs at the outer fibers (farthest from the neutral axis) and is computed using the flexure formula:

$$f_b = \frac{Mc}{I} = \frac{M}{S} \qquad \textbf{(14.2)} \quad \textbf{(14.5)}$$

where S represents the quantity I/c and is called the *section modulus*.

The maximum resisting moment (called the *allowable moment*) can be computed from

$$M_R = \frac{F_b I}{c} = F_b S \qquad \textbf{(14.4)} \quad \textbf{(14.6)}$$

For design problems, the most convenient form is

$$\text{required } S = \frac{M}{F_b} \qquad \textbf{(14.7)}$$

14.4 Horizontal and vertical shear stresses are developed in beams subjected to vertical loads. The shear stress is zero at the outer fibers and almost always is maximum at the neutral axis.

14.5 The shear stress on any horizontal plane at any cross section of a beam is calculated from the general shear formula:

$$f_v = \frac{VQ}{Ib} \qquad \textbf{(14.8)}$$

The maximum shear stress will occur in the plane of maximum shear. The maximum resisting shear (allowable shear) can be computed from

$$V_R = \frac{F_v Ib}{Q} \quad \textbf{(14.9)}$$

14.6 Shear stresses may become critical in beams that have short spans and/or are heavily loaded. The use of an average web shear approach to simplify shear stress calculations for hot-rolled and fabricated shapes is common:

$$f_v = \frac{V}{dt_w} \quad \textbf{(14.10)}$$

Therefore, the allowable shear can be computed from

$$V_R = F_v dt_w \quad \textbf{(14.11)}$$

The maximum horizontal shear stress occurs at the neutral axis for the following. For a homogeneous solid beam of rectangular cross section,

$$f_v = \frac{1.5V}{A} \quad \textbf{(14.12)}$$

For a homogeneous solid beam of circular cross section,

$$f_v = \frac{4V}{3A} \quad \textbf{(14.13)}$$

14.7 When only the outer fibers of a beam are stressed to the yield point, the beam moment strength is

$$M_y = F_y S \quad \textbf{(14.18)}$$

A considerable amount of strength, however, remains in the interior fibers. When all the fibers are stressed to the yield point, no additional moment can be resisted and the beam moment strength becomes

$$M_P = F_y Z \quad \textbf{(14.19)}$$

where Z is the plastic section modulus and is calculated from the sum of the moments of the areas above and below the neutral axis.

In beam-analysis-type problems, both bending and shear stresses must be considered. Computed stresses must not exceed established allowable stresses. Alternatively, strength may be checked against the applied effect (moment or shear).

Hot-rolled structural steel shapes (such as W shapes) are analyzed and designed for moment using M_P as a basis. Moment strength is calculated from

$$M_R = 0.6F_y Z_x \quad \textbf{(14.20)}$$

PROBLEMS

Unless otherwise noted in the following problems, assume that the beam cross section is oriented so the strong axis is the bending axis.

Section 14.3 Computation of Bending Stresses

14.1 Calculate the section modulus for:

(a) a 6-in.-by-10-in. (S4S) timber cross section, with respect to the strong axis (check with Appendix E)

(b) a pipe cross section having an inner diameter of 100 mm and an outer diameter of 114 mm

(c) the cross section of Example 14.16, with respect to the weak axis

14.2 Calculate the section modulus (with respect to the X–X axis) for the beams shown.

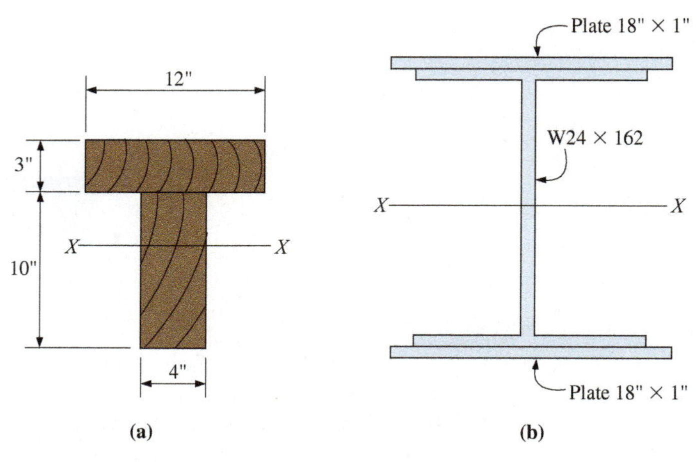

PROBLEM 14.2

14.3 A W18 × 71 steel beam supports a superimposed uniformly distributed load as shown. Calculate the maximum bending stress. Be sure to include the beam weight.

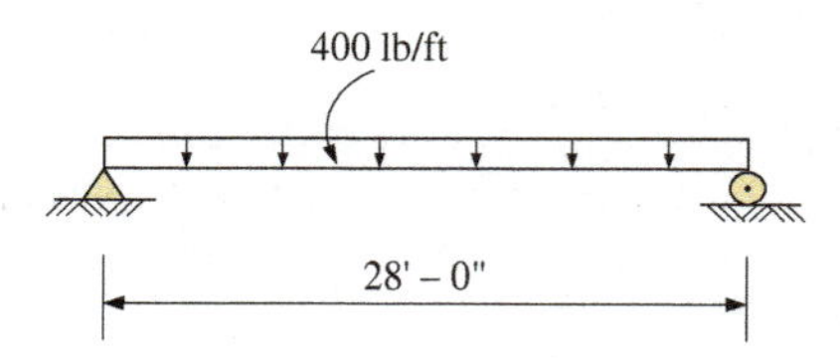

PROBLEM 14.3

14.4 Rework Problem 14.3 changing the orientation of the cross section so the weak axis is the bending axis.

14.5 Assume that the timber member (a) of Problem 14.2 is used as a simply supported beam with a 12-ft span length carrying a uniformly distributed load of 500 lb/ft. The given load includes the weight of the beam. Calculate the maximum induced bending stress.

14.6 The structural steel built-up member (b) of Problem 14.2 is to be used as a simply supported beam on a span of 40 ft. It is to carry a uniformly distributed load of 3.0 kips/ft, which includes its own weight. Calculate the maximum induced bending stress.

14.7 A round steel rod, 25 mm in diameter, is subjected to a bending moment of 160 N·m. Calculate the maximum bending stress.

14.8 A square steel bar, 38 mm on each side, is used as a beam and subjected to a bending moment of 460 N·m. Calculate the maximum bending stress.

14.9 Calculate the moment strength for a W36 × 302 wide-flange section. Assume A992 steel. Consider only strong-axis bending.

14.10 Calculate the allowable bending moment for a solid rectangular 6-in.-by-16-in. timber beam if the allowable bending stress is 1000 psi. Assume that the large dimension is vertical and parallel to the applied loads.

(a) Use nominal dimensions.

(b) Use dressed dimensions.

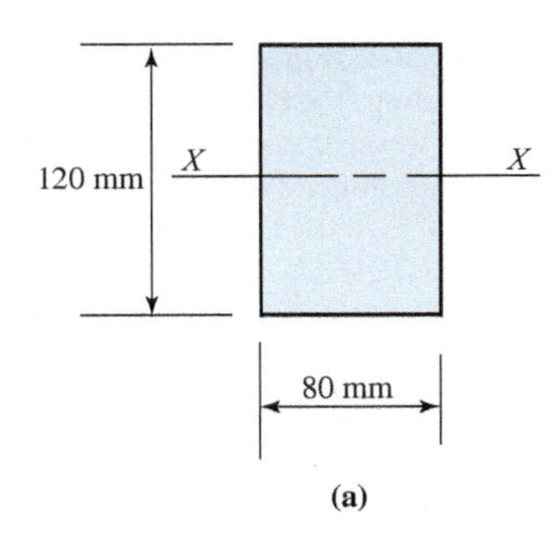

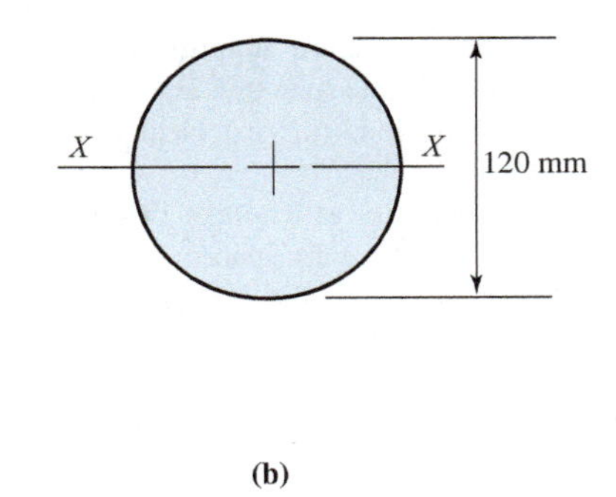

PROBLEM 14.11

Section 14.6 Shear Stresses in Structural Members

14.11 The beams of cross sections shown are subjected to a total shear force of 20 kN. Determine the shear stress at the neutral axis (X–X axis).

14.12 A solid rectangular simply supported timber beam 6 in. wide, 20 in. deep, and 10 ft long carries a concentrated load of 16,000 lb at midspan. Use nominal dimensions.

(a) Compute the maximum horizontal shear stress at the neutral axis.

(b) Compute the shear stress 4 in. and 8 in. above and below the neutral axis. Neglect the weight of the beam.

14.13 A W14 × 30 supports the loads shown. Calculate the maximum shear stress using (a) the general shear formula and (b) the average web shear approach. Be sure to include the weight of the beam.

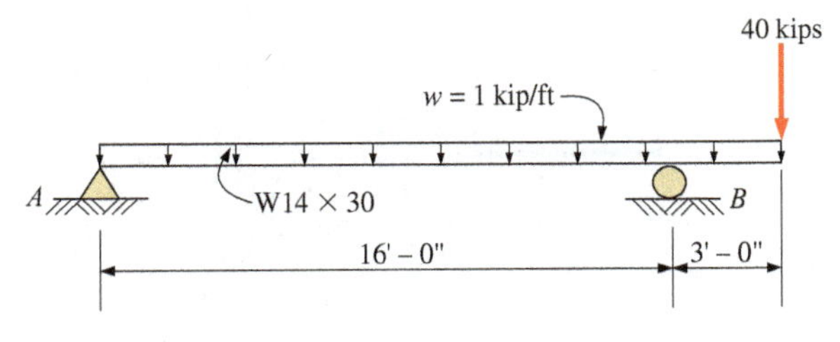

PROBLEM 14.13

14.14 If the allowable shear stress is 100 MPa, calculate the maximum shear force V that a W460 × 60 structural steel wide flange is capable of resisting. Use the average web shear approach.

14.15 A steel pin $1\frac{1}{2}$ in. in diameter is subjected to a shear force of 10,000 lb in the plane of its cross section. Compute the maximum horizontal shear stress.

14.16 A timber power-line pole is 10 in. in diameter at its base where it is solidly embedded in concrete. The pole extends 20 ft vertically upward from its base and is subjected to a horizontal pull of 300 lb at its top. Calculate the maximum bending stress and the maximum shear stress produced in the pole.

Section 14.7 Inelastic Bending of Beams

14.17 Calculate the value of S and Z and the shape factor for the beam cross sections shown.

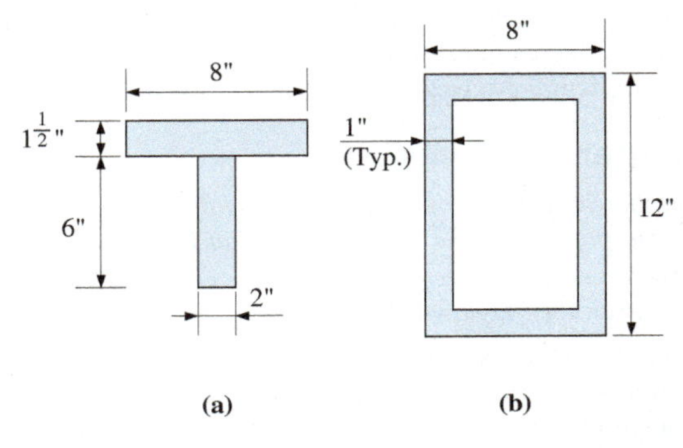

PROBLEM 14.17

14.18 For beams that have cross sections as shown for Problem 14.17, calculate the yield moment M_y, the plastic moment M_P, and the maximum uniformly distributed load that the beams can carry on a simple span of 14 ft. Assume that $F_y = 36$ ksi.

14.19 Calculate the maximum load P that the beam shown can carry based on plastic moment M_P. Assume a yield stress F_y of 40,000 psi. Neglect the weight of the beam.

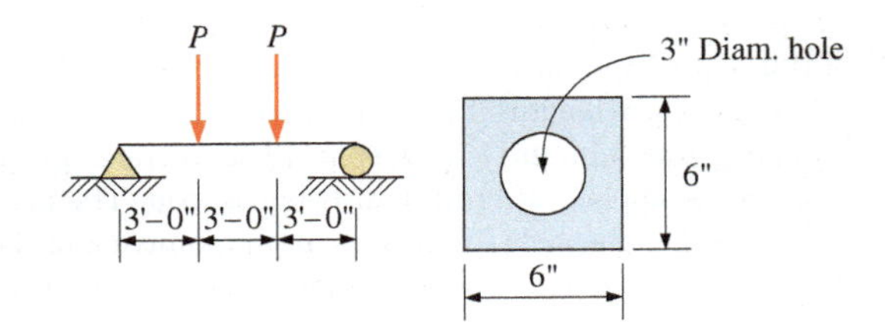

PROBLEM 14.19

Section 14.8 Beam Analysis

14.20 A 4 × 12 (S4S) hem-fir timber beam carries a superimposed uniformly distributed load of 325 lb/ft on a simple span of 12 ft. Determine the maximum bending stress and the maximum shear stress, and compare with the allowable stresses. Be sure to include the beam weight.

14.21 A simply supported W16 × 36 A992 steel beam carries a superimposed uniformly distributed load of 1.25 kips/ft for the full span in addition to a point load of 10 kips at midspan. The span length is 24 ft. Compare moment strength and shear strength with the applied moment and shear. Use the average web shear approach. Be sure to include the beam weight.

14.22 A W250 × 115 steel wide-flange section supports a uniformly distributed line load on a simple span of 5 m. Assume A992 steel. Calculate the allowable load (kN/m). Neglect the weight of the beam.

14.23 Assume that the floor joist dimensions of Example 14.13 are changed to 3 in. by 12 in. and that they are spaced 16 in. on centers. The wood and span length remain the same. Calculate
(a) the allowable superimposed uniformly distributed load for each joist in lb/ft
(b) the allowable superimposed uniformly distributed floor load in psf

14.24 Calculate the allowable superimposed uniformly distributed load that may be placed on a structural steel wide flange W14 × 34 that has a simple span length of 6.5 ft. Assume A992 steel. Consider moment and shear.

14.25 A 3-in.-by-12-in. (S4S) scaffold timber plank is laid flatwise on supports that are 8 ft apart. Calculate the allowable uniformly distributed load that can be carried by the plank. The allowable bending stress is 1600 psi and the allowable shear stress is 100 psi. Neglect the weight of the plank.

Computer Problems

For the following computer problems, any appropriate software may be used. Input prompts should fully explain what is required of the user (the program should be user-friendly). The resulting output should be well labeled and self-explanatory.

14.26 Write a program that may be used to calculate the maximum bending stress and the maximum horizontal shear stress for a simply supported single-span timber beam that carries a uniformly distributed load. User input is to be uniformly distributed load intensity, span, and the nominal dimensions of the cross section. (Use nominal dimensions for the calculations and assume a unit weight for the timber of 35 pcf.)

14.27 Repeat Problem 14.26, except now assume the beam to be a hot-rolled steel wide-flange shape. User input will be span length, uniform load intensity, and the following dimensions for the shape: overall depth, thickness and width of the flange, and the web thickness. Assume the unit weight of the steel to be 490 pcf. Calculate I by approximating the cross section with three rectangles.

14.28 Write a program that will generate a table of maximum allowable span lengths for a group of 2-in.-thick planks (ranging from 4 in. to 12 in. deep in 12-in. increments) as a function of applied uniformly distributed load. The planks are to bend about their strong axes. Both moment and shear should be considered. The load should range from 50 lb/ft to 120 lb/ft in steps of 10 lb/ft. User input is to be allowable bending stress and allowable horizontal shear stress. Assume the wood to have a unit weight of 35 pcf. Use full nominal dimensions.

Supplemental Problems

14.29 Calculate the section modulus with respect to the X–X axis for the sections shown.

14.30 The timber box section (a) of Problem 14.29 is used as a simply supported beam on an 18-ft span length. The beam carries a uniformly distributed load of 500 lb/ft, which includes its own weight. Calculate the maximum induced bending stress.

14.31 A timber beam is subjected to a maximum bending moment of 2600 N · m. The cross section of the beam is 100 mm by 150 mm. The 150 mm side is oriented vertically.
(a) Calculate the maximum bending stress.
(b) Calculate the bending stress 25 mm below the top surface.

14.32 Rework Problem 14.31 assuming that the beam is placed with the 100 mm dimension oriented vertically.

14.33 A $\frac{1}{2}$-in.-diameter steel rod projects 2 ft horizontally from a concrete wall. Calculate the maximum bending stress due to a vertical load of 12 lb at the free end. Neglect the weight of the rod.

14.34 Calculate the maximum bending stress in a W530 × 101 steel beam that spans 13 m on simple supports and supports two equal concentrated loads of 54 kN each. The loads are placed at the third points. Include the weight of the beam.

14.35 A cantilever cast-iron beam is 6 ft long and has a "T" cross section, as shown. Calculate the maximum tensile and compressive bending stresses. The applied load is 450 lb. Neglect the weight of the beam.

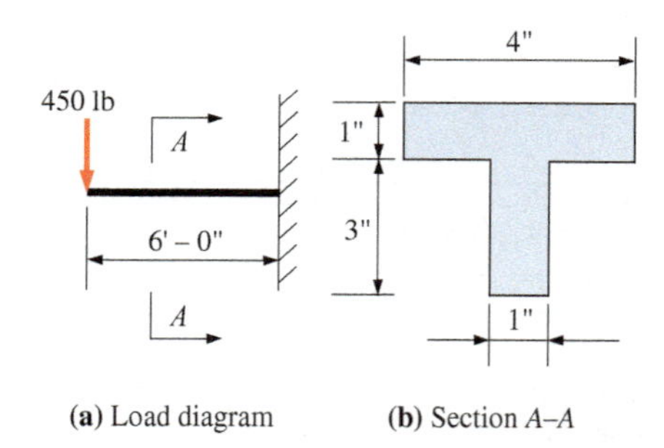

(a) Load diagram **(b)** Section *A–A*

PROBLEM 14.35

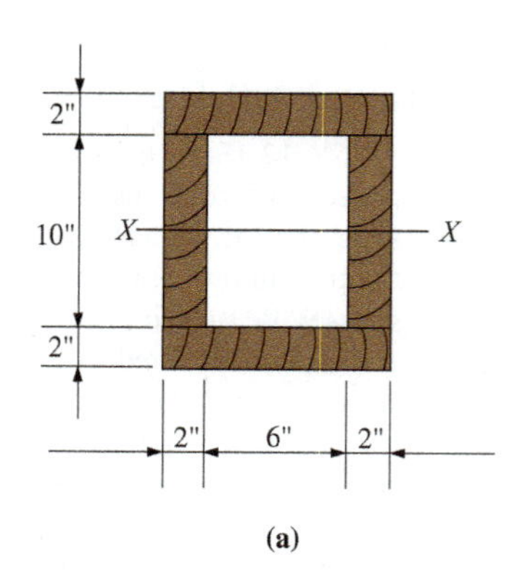

(a)

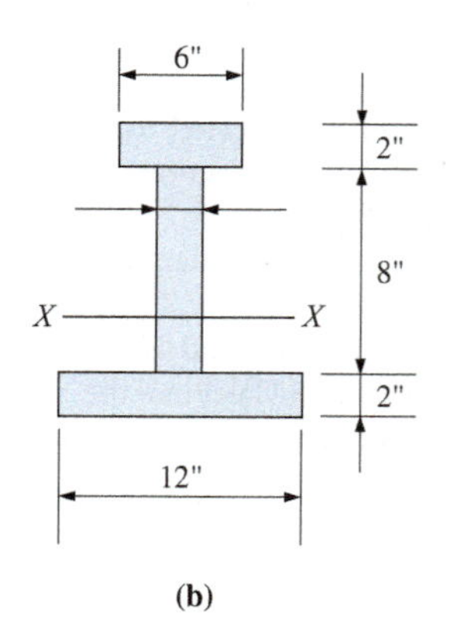

(b)

PROBLEM 14.29

14.36 Calculate the moment strength for a W920 × 390 wide-flange section. Assume A992 steel. Consider only strong axis bending.

14.37 A W8 × 13 steel wide-flange beam on a 20-ft span is supported and loaded as shown. The uniformly distributed load includes the weight of the beam. Calculate the maximum bending stress.

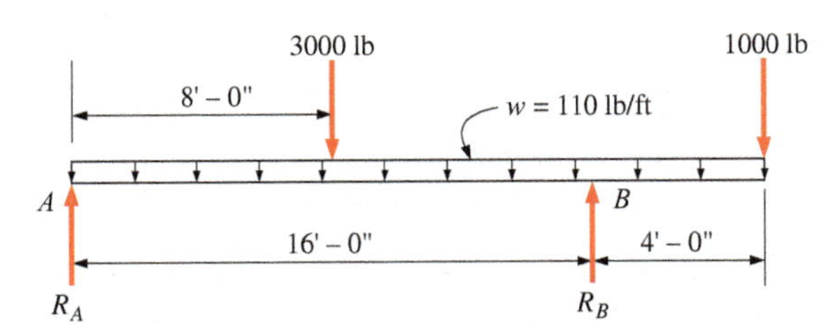

PROBLEM 14.37

14.38 A simply supported beam with a cruciform cross section is loaded as shown. Calculate the maximum horizontal shear stress. Neglect the weight of the beam. (*Hint:* Check two planes in the cross section.)

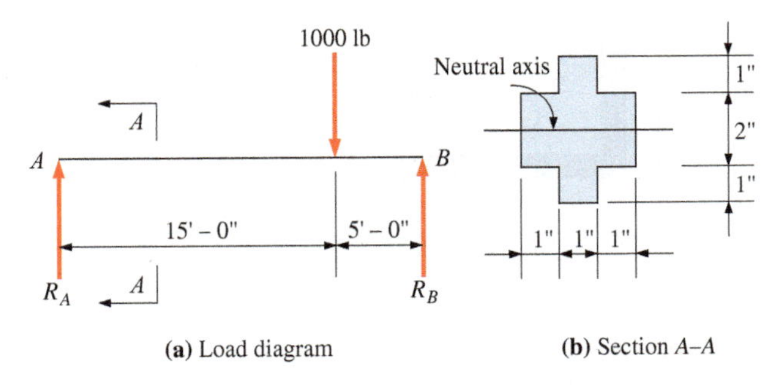

(**a**) Load diagram (**b**) Section *A–A*

PROBLEM 14.38

14.39 A rectangular beam 100 mm in width and 250 mm in depth is oriented with the large dimension placed vertically. Using the general shear formula, calculate the maximum shear stress when the beam is subjected to a maximum shear force of 140 kN.

14.40 The timber box section (a) of Problem 14.29 is made up of four component parts so connected as to act as a single unit. Calculate the maximum horizontal shear stress if the member is subjected to a maximum vertical shear of 4500 lb.

14.41 For the I-shaped timber beam shown, calculate the maximum vertical shear force that will induce a maximum horizontal shear stress of 120 psi. Then calculate the shear stress in the plane of the junction between the web and the flange.

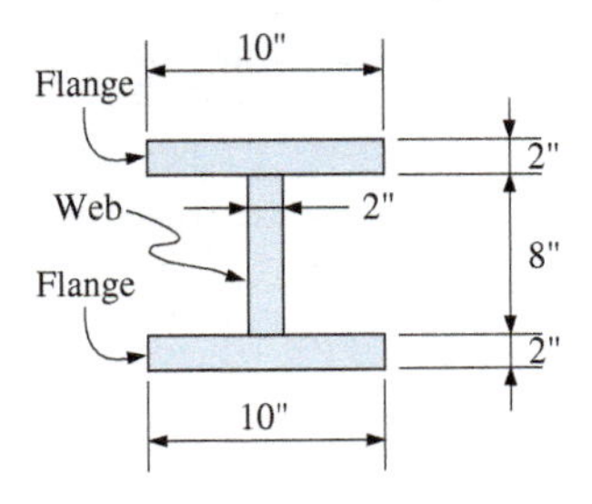

PROBLEM 14.41

14.42 A W36 × 150 steel wide-flange beam is oriented so that bending is about the weak (*Y–Y*) axis. The member is subjected to a maximum shear of 140 kips. Calculate the horizontal shear stress at a sufficient number of points so that a shear stress distribution diagram can be drawn. Draw the diagram.

14.43 A W10 × 45 steel wide-flange beam supports a uniformly distributed load on a simple span of 14 ft.

(**a**) Assume A992 steel and compute the allowable load.

(**b**) Determine the maximum horizontal shear stress using the general shear formula, assuming that the beam carries the load from part (a).

14.44 A W610 × 113 steel wide-flange section is subjected to a vertical shear force *V* of 525 kN. Calculate the maximum horizontal shear stress using

(**a**) the general shear formula

(**b**) the average web shear approach

14.45 A W30 × 108 steel wide-flange beam is simply supported on a span of 10 ft and is subjected to a uniformly distributed load of 42 kips/ft, which includes the weight of the beam. Calculate the maximum bending stress and the maximum horizontal shear stress using the general shear formula.

14.46 A W6 × 12 is strengthened with a $\frac{3}{4}$-in.-by-$\frac{3}{4}$-in. bar welded to its top flange as shown. Because of the weld used, the horizontal shear stress at the connection between the bar and the flange cannot exceed 4.0 ksi. In the web, the shear stress is limited to 20 ksi. Use the general shear formula to determine the allowable vertical shear force on this cross section.

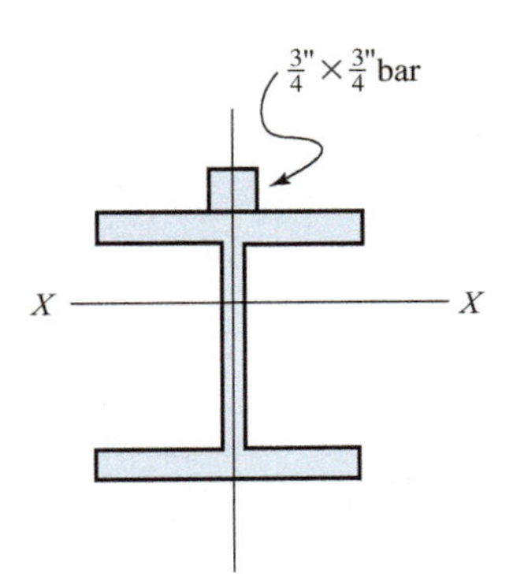

PROBLEM 14.46

14.47 Four wood boards 1 in. by 6 in. in cross section are stacked as shown and used as an 8-ft-long simply supported beam. A 400-lb concentrated load is applied at the center of the span. Neglecting the weight of the beam, calculate the following based on the full nominal (rough) dimensions:

(**a**) the maximum bending stress (boards not glued together)

(**b**) the maximum bending stress if the boards are glued together

(**c**) the maximum shear stress in the middle glued joint

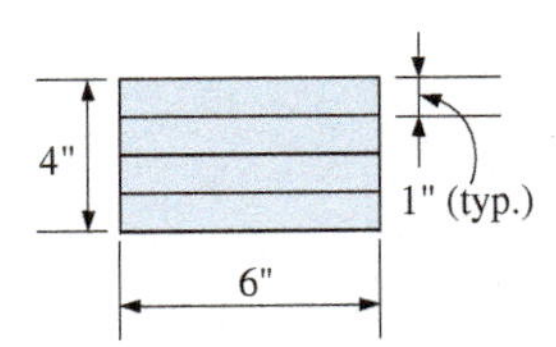

PROBLEM 14.47

14.48 A lintel consists of two 8-in.-by-$\frac{1}{2}$-in. steel plates welded together to form an inverted tee. Calculate the maximum bending stresses in tension and compression when the lintel carries a total uniformly distributed load of 10,000 lb on a simple span of 6 ft. In addition, calculate the maximum shear stress and plot the distribution of the shear stress for the entire cross section. Neglect the weight of the beam.

14.49 A 50-mm-by-300-mm scaffold timber plank, placed flatwise, must carry a uniformly distributed line load of 1.50 kN/m without exceeding an allowable bending stress of 11.0 MPa and an allowable shear stress of 0.83 MPa. Calculate the maximum allowable simple span length. Use full nominal dimensions and neglect the weight of the plank.

14.50 A laminated wood beam is built up by gluing together three 2-in.-by-4-in. members to form a solid beam 4-in.-by-6-in. in cross section, as shown. The allowable shear stress in the glued joints is 60 psi. The beam is a 3-ft-long cantilever with a load P applied at its free end. Neglecting the weight of the beam:

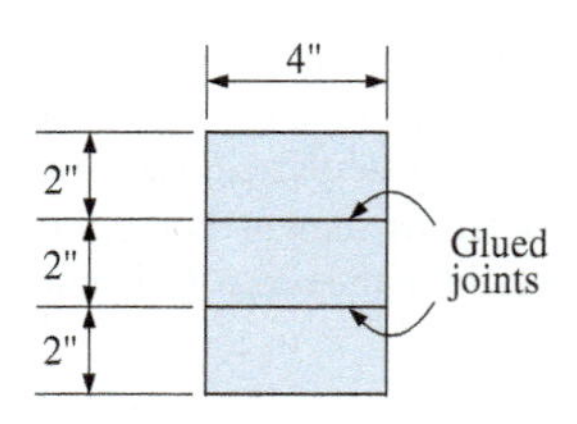

PROBLEM 14.50

(a) Calculate the maximum allowable load P that can be applied at the free end.

(b) Calculate the maximum bending stress in the beam.

14.51 A rectangular hollow shape carries loads as shown. Calculate the maximum value of P if the allowable stresses are 10 MPa for bending and 1.5 MPa for shear. Neglect the weight of the beam.

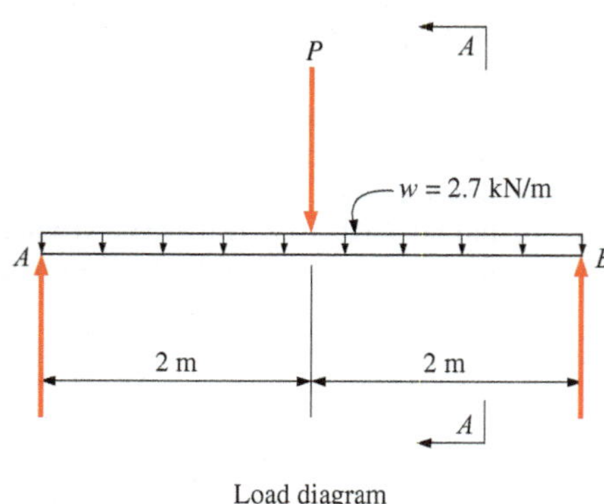

Load diagram

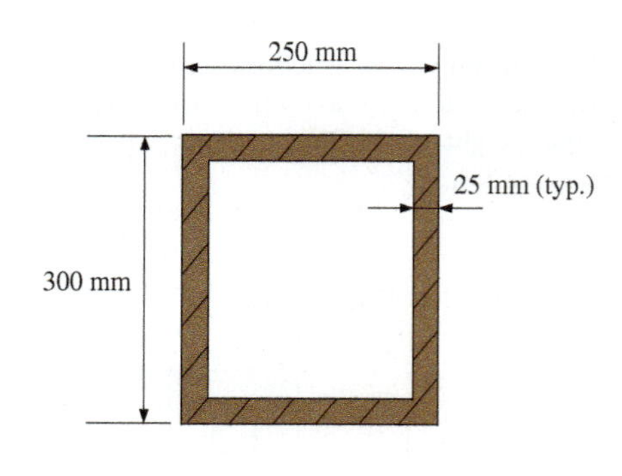

Section A–A

PROBLEM 14.51

14.52 For the beam shown, calculate the maximum tensile and compressive bending stresses and the maximum shear stresses. Neglect the beam weight.

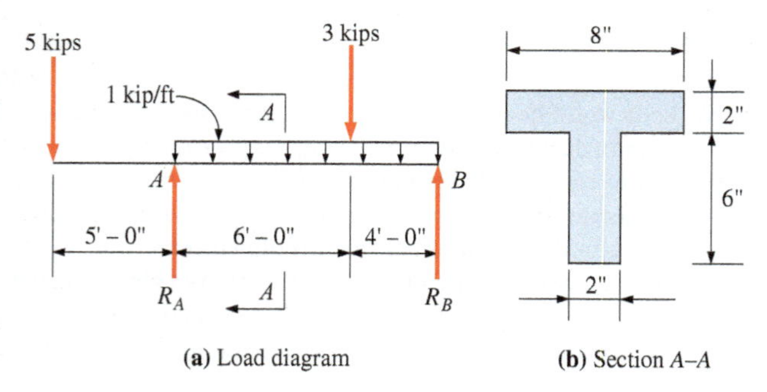

(a) Load diagram

(b) Section A–A

PROBLEM 14.52

14.53 A box beam is built up of four 1-in.-by-6-in. boards as shown. The beam is subjected to a maximum vertical shear force of 920 lb. Calculate the required spacing of wood screws if the allowable shear load on each screw is 250 lb.

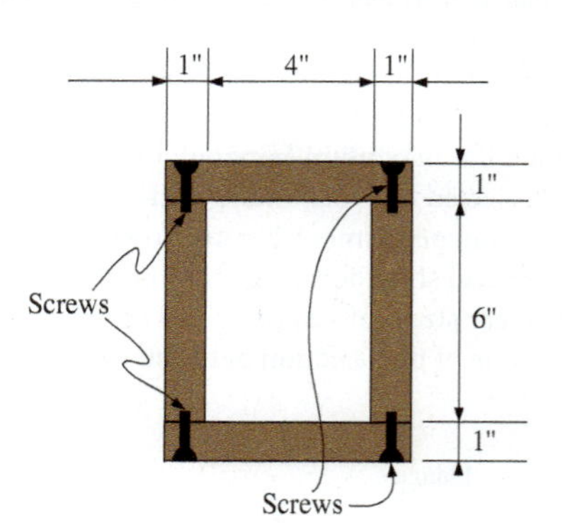

Beam cross section

PROBLEM 14.53

14.54 Find the value of the loads P that can be safely supported by the 6-in.-by-12-in. (S4S) timber beam shown. The allowable bending stress is 1800 psi and the allowable shear stress is 140 psi. Be sure to consider the weight of the beam.

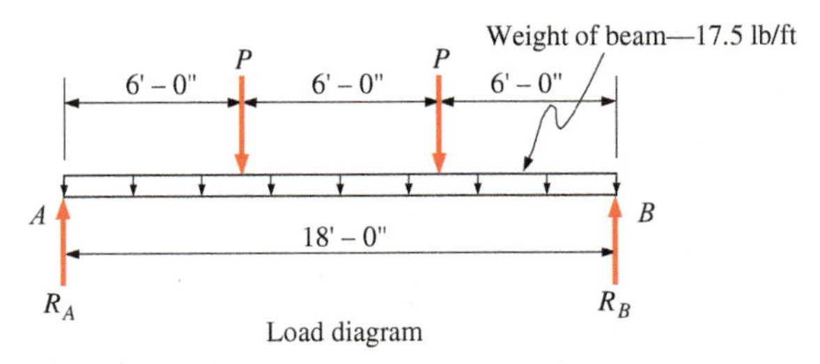

PROBLEM 14.54

14.55 Solve Problem 14.54 assuming that the timber beam is a dressed 6-in.-by-18-in. member.

14.56 Calculate the values of S and Z and the shape factor for the cross section shown.

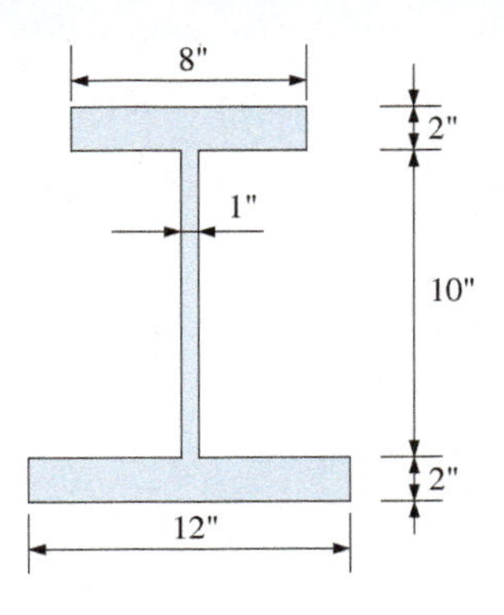

PROBLEM 14.56

14.57 A W18 × 50 is supported on simple supports on a 30-ft span. Calculate the yield moment M_y, the plastic moment M_P, the maximum uniformly distributed load that the beam can carry. Assume that $F_y = 50$ ksi.

CHAPTER FIFTEEN

DEFLECTION OF BEAMS

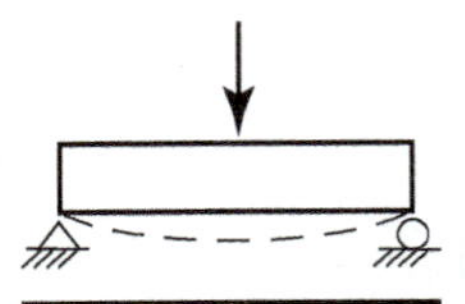

LEARNING OBJECTIVES

Upon completion of this chapter, readers will be able to:

- Discuss deflection of beams and the negative impact of deflection on a building.
- Discuss the relationship between bending stress and deflection as well as bending moment and deflection.
- Calculate the deflection in beams using either formulas or the moment-area method.
- Draw moment diagrams by parts as an alternate method to drawing a moment diagram.

15.1 REASONS FOR CALCULATING BEAM DEFLECTION

When a beam is subjected to a load that creates bending, the beam will sag or deflect, as shown in Figure 15.1. Although a beam may be satisfactory with respect to bending moment and shear, it may be unsatisfactory with respect to deflection. Large deflections are generally indicative of a lack of structural rigidity, despite adequate strength. Therefore, consideration of the deflection of beams is another part of the beam design or analysis process.

Excessive deflections are to be avoided for many reasons. With respect to buildings, excessive deflections can cause substantial cracking in ceilings, floors, and partitions as well as in attached nonstructural elements, such as windows and doors. Other troublesome and potentially serious conditions that could result from beams that lack structural rigidity are

1. Changes in floor and wall alignment
2. Improper functioning of roof drainage systems resulting in the ponding of water on the roof and excessive roof loads
3. Undesirable vibrations from various live loads (e.g., lack of rigidity in a floor system may lead to vibrations caused by pedestrian traffic)
4. Misalignment of precision equipment housed in a building

In addition, visibly sagging beams tend to lessen one's confidence in both the strength of the structure and the skill of the designer.

The proper performance of machine parts is also affected by excessively flexible bending members. Shafts carrying gears must have adequate rigidity so that gear teeth will mesh as designed and synchronization of motion will be maintained. Various parts of metal-shaping equipment, as well as supporting frames of vehicles and machines, must also have sufficient rigidity. Without adequate rigidity, undesirable vibrations, friction, and abrasion can develop.

There are two considerations in evaluating deflection. The first is the computation of the magnitude and direction of the deflection and the second is the establishment of an allowable deflection. This chapter is primarily concerned with the computation of the magnitude of the deflection. Allowable deflections are generally established by design specifications, codes, and/or recommendations based on past acceptable practice and judgment. As an example, the AISC design specification stipulates that for beams and girders supporting plastered ceilings, the maximum live load deflections must not exceed $\frac{1}{360}$ of the span. Additional criteria that, in effect, limit the deflection of a beam are also established by the AISC. Other design specifications furnished by various agencies address the deflection problem

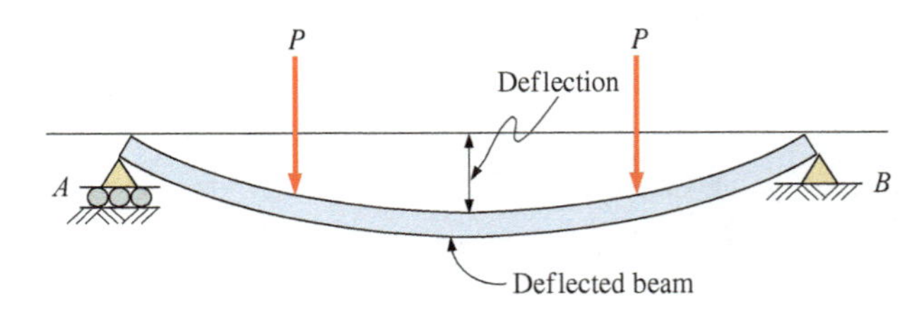

FIGURE 15.1 Beam deflection.

by requiring minimum depth/span ratios. For example, the AASHTO *Standard Specifications for Highway Bridges* stipulates that the ratio of depth to length of span for beams or girders preferably will not be less than $\frac{1}{25}$. This is in addition to upper limits on live load deflections.

15.2 CURVATURE AND BENDING MOMENT

As we described in Section 14.1, when a simple beam is subjected to vertical loads, tensile stresses are developed on one side of the neutral plane and compression stresses on the other side. The fibers subjected to tension are elongated and those subjected to compression are shortened. This situation causes the beam to curve and therefore to deflect from its original unloaded position. When a beam deflects, the curved position assumed by the neutral plane is generally called the *elastic curve.*

The radius of curvature at any point on the elastic curve is equal to the radius of that circle having a circumference that conforms to the shape of the elastic curve at that point. Generally, the elastic curve of a loaded beam is not circular in shape. An infinitesimal segment of it may be considered the arc of a circle, however.

With reference to Figure 14.1, the portion of the beam between the two concentrated loads is a beam segment of zero shear and constant bending moment. As shown in Figure 14.2a, planes A–C and B–D located in that segment are vertical, parallel, and normal to the horizontal unloaded beam. The loaded shape, which exists after the loads have been applied and bending occurs, is shown in Figure 14.2b.

Figure 14.2b is redrawn in Figure 15.2 to show the geometry of the segment of the circle. The planes A'–C' and B'–D' of the loaded beam (which were originally planes A–C and B–D of the unloaded beam) now intersect at point O and the angle between them is designated θ. R is the radius of curvature of the elastic curve and c is the distance from the neutral plane (elastic curve) to the outer fiber.

Line GFH is drawn through point F parallel to plane A'–C' as shown in Figure 15.2. Line segments CH and JF are both equal to the original length of the outer fibers that have increased in length by the amount HD'. Therefore, the strain of the outer fiber is as follows:

$$\epsilon = \frac{\text{change in length}}{\text{original length}} = \frac{HD'}{C'H} = \frac{HD'}{JF}$$

By definition, the modulus of elasticity is

$$E = \frac{\sigma}{\epsilon}$$

Therefore, the stress can be expressed as

$$f_b = E\epsilon = E\left(\frac{HD'}{JF}\right)$$

If the curvature is very slight and the distance JF is very small, $D'FH$ and JOF may be considered to be similar triangles. Therefore,

$$\frac{c}{R} = \frac{HD'}{JF}$$

and, by substitution, the stress equation becomes

$$f_b = \frac{Ec}{R} \quad \textbf{(15.1)}$$

where f_b = the tensile or compressive bending stress at the outer fiber (psi or ksi) (Pa)
E = the modulus of elasticity (psi or ksi) (Pa)
c = the distance from the neutral axis to the outer fibers (in.) (mm)
R = the radius of curvature of the elastic curve (in.) (mm)

From the flexure formula, stress also can be written as

$$f_b = \frac{Mc}{I}$$

Therefore,

$$\frac{Ec}{R} = \frac{Mc}{I}$$

and, solving for R gives

$$R = \frac{EI}{M} \quad \textbf{(15.2)}$$

where all terms have been previously defined.

The equation just derived expresses the relation between the radius of curvature of the beam at any section and the bending moment at that section. Therefore, with E and I constant, R varies inversely with the moment M. If the moment M is zero, the radius of curvature R becomes infinity and the elastic curve is a straight line. As the moment M increases, the radius of curvature R decreases and the smallest value occurs where the moment is a maximum. If the bending moment is constant over part of the length of a beam, the radius of curvature is also constant and the elastic curve is the arc of a circle.

Note also that if the product EI is (theoretically) infinitely large, the radius of curvature will also become infinitely large and the elastic curve will be a straight line. Therefore, EI

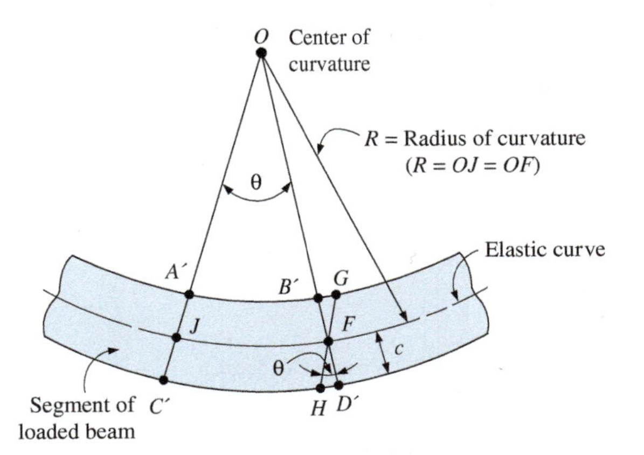

FIGURE 15.2 Loaded beam segment in pure bending.

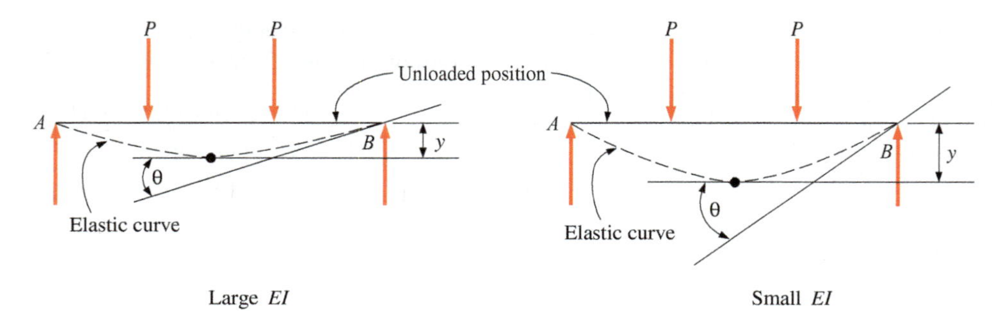

FIGURE 15.3 Relative beam stiffness.

is an important factor in determining the shape of the elastic curve as well as in determining the displacement of points on the curve from the unloaded horizontal position.

The effect of relative magnitudes of EI (other parameters being equal) is shown in Figure 15.3. Note that with a large EI, angle θ (defined as the angle between the tangents to two points on the elastic curve) is small, as is the vertical displacement y of a point on the elastic curve. With a small EI, both θ and y are relatively larger.

EXAMPLE 15.1 The steel blade for a small band saw is 0.025 in. thick and runs on pulleys that are 12 in. in diameter. Compute the maximum bending stress caused by bending the blade around the pulleys. $E = 30{,}000{,}000$ psi for the blade.

Solution As the blade goes over the pulley, it conforms to the radius of the pulley; therefore, $R \approx 6$ in. From Equation (15.1) the maximum stress is written as

$$f_b = \frac{Ec}{R} = \frac{(30{,}000{,}000 \text{ psi})(0.025 \text{ in.}/2)}{6 \text{ in.}} = 62{,}500 \text{ psi}$$

EXAMPLE 15.2 A $\frac{3}{4}$-in.-square steel bar is loaded as shown in Figure 15.4. Calculate the radius of curvature of the beam segment between the supports. Use $E = 30{,}000{,}000$ psi.

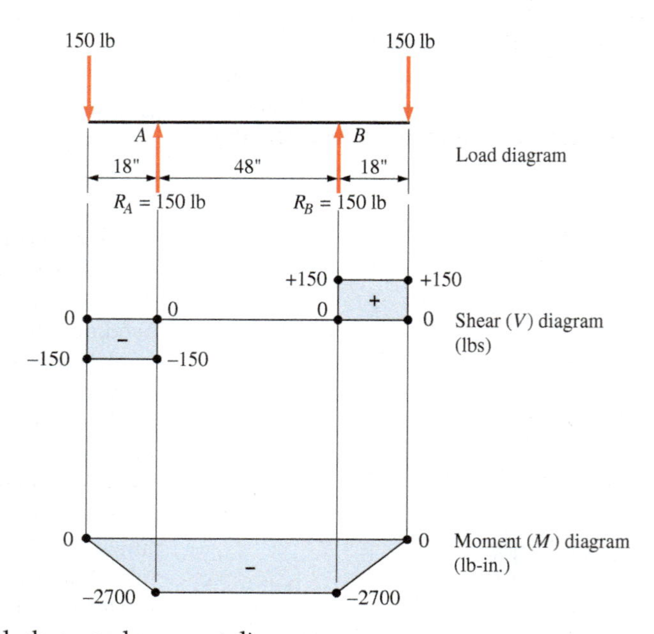

FIGURE 15.4 Load, shear, and moment diagrams.

Solution The shear and moment diagrams are shown. Note that constant moment exists between the supports. For the calculation of moment of inertia, see Table 3, inside the back cover.

$$I = \frac{bh^3}{12} = \frac{(0.75 \text{ in.})(0.75 \text{ in.})^3}{12} = 0.0264 \text{ in.}^4$$

Calculate the radius of curvature from Equation (15.2):

$$R = \frac{EI}{M} = \frac{(30{,}000{,}000 \text{ psi})(0.0264 \text{ in.}^4)}{2700 \text{ lb-in.}} = 293 \text{ in.} = 24.4 \text{ ft}$$

15.3 METHODS OF CALCULATING DEFLECTION

Deflections of bending members can be calculated in a number of ways. Two basic methods presented in this chapter are the *moment-area method* and the *formula method*. The formula method is easier to use and should be applied whenever possible. Numerous derived formulas for beams having various support conditions and subjected to various types and combinations of loadings are given in Appendix H. These formulas have been derived for members with a constant moment of inertia using either the moment-area method or some other theoretical method.

In practice, the most commonly used method of computing deflection is the formula method. When support and loading conditions are such that the deflection formulas cannot be used, however, the moment-area method, which is extremely versatile, may be used. It is particularly useful for bending members with varying moments of inertia. These are frequently encountered in machine design and occasionally in building design.

Regardless of the method used, be aware that deflection calculations (like stress calculations) are based on certain assumptions. These assumptions can be briefly summarized as follows:

1. The maximum bending stress does not exceed the proportional limit.
2. The beam is homogeneous, obeys Hooke's law, and has an equal modulus of elasticity in tension and compression.
3. A plane section through the beam before bending remains a plane after bending takes place.
4. The beam has a vertical plane of symmetry and the loads and reactions act in this plane perpendicular to the longitudinal axis of the beam.
5. Deflections are relatively small and the length of the elastic curve is the same as the length of its horizontal projection.
6. Deflection due to shear is negligible. (Shear deflections are generally much smaller than the deflections due to bending moment.)

15.4 THE FORMULA METHOD

As previously mentioned, the formula method is based on formulas derived by various theoretical means. Typical derivations for specific types of beams and loading conditions are presented in Section 15.7. In general, Appendix H provides derived formulas for calculating (among other quantities) the following:

1. The deflection of any point along the length of the beam
2. The maximum deflection
3. The location of the maximum deflection

In some cases, all three items are not included, which implies that too many complicating factors preclude the derivation of a usable formula; hence, a theoretical method should be used for deflection calculations. Generally, only the location and magnitude of the maximum deflection are of interest since it is the maximum deflection that must be compared with some allowable deflection value.

In the formulas of Appendix H, the Greek uppercase delta and subscript $x(\Delta_x)$ represent the deflection of the beam from its straight unloaded position at any point a distance x from the left end of the beam or from a support. The designation Δ_{max} represents the maximum deflection occurring at a definite location that is a distance x from the left end of the beam or from a support.

Note that the deflection formulas in Appendix H are all shown as indicating a positive $(+)$ deflection; that is, a downward deflection is assumed positive. We follow this practice although, from a practical viewpoint, whether the deflection is upward or downward is less important than the magnitude of the deflection. Generally, the direction of the deflection is evident and may be readily determined from the loading conditions.

Deflection is inversely proportional to E and I. The combined term EI represents the stiffness of a beam and is indicative of the resistance of the beam to deflection. The units of these two quantities are psi (or ksi) and in.^{4}, respectively, in the U.S. Customary System and Pa (or MPa) and m^4 or mm^4 in the SI. Other factors that enter into the problem of deflection determination are the type, magnitude, and location of the applied loads along with the span length and type of supports.

Since deflections are usually numerically small, it is desirable to state the computed deflection in inch units in the U.S. Customary System and in millimeters in the SI. Therefore, units and conversions must be considered carefully since span lengths are usually given in feet in the U.S. Customary System and in meters in the SI and uniformly distributed line loads are in lb/ft or k/ft in the U.S. Customary System and in N/m or kN/m in the SI.

EXAMPLE 15.3

A W14 × 74 structural steel wide-flange section supports a superimposed uniformly distributed line load of 2300 lb/ft on a simply supported span of 24 ft, as shown in Figure 15.5. Calculate the maximum deflection. The allowable deflection is $\frac{1}{360}$ of the span length. Determine whether or not the beam is satisfactory.

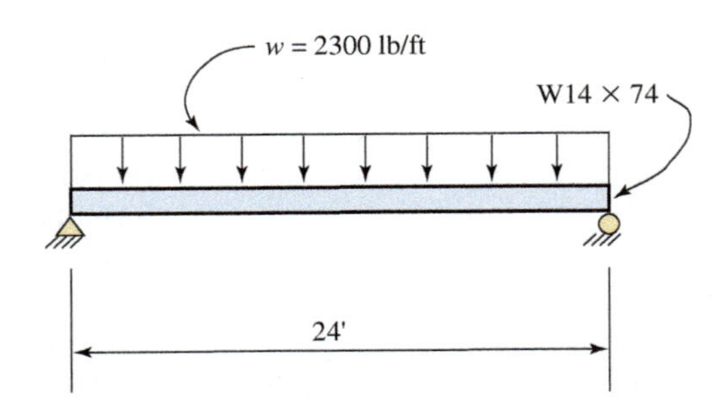

FIGURE 15.5 Beam diagram.

Solution From Appendix H, Case 1, the maximum deflection for a simply supported beam subjected to a uniformly distributed load will occur at midspan ($x = L/2$) and is given by

$$\Delta_{max} = \frac{5wL^4}{384EI}$$

where w = the uniformly distributed load per unit of length (in this case, the sum of the superimposed load and the beam weight)
L = the span length (24 ft)
E = the modulus of elasticity (30,000,000 psi, from Appendix G)
I = the moment of inertia (795 in.^{4} from Appendix A)

Substituting gives

$$\Delta_{max} = \frac{5[(2300 + 74)\text{lb/ft}](24\text{ ft})^4(12\text{ in./ft})^3}{384(30{,}000{,}000\text{ lb/in.}^2)(795\text{ in.}^4)} = 0.74\text{ in.}$$

The allowable deflection is calculated from

$$\Delta_{all} = \frac{\text{span}}{360} = \frac{(24\text{ ft})(12\text{ in./ft})}{360} = 0.80\text{ in.}$$

Since 0.74 in. < 0.80 in., the beam is satisfactory with respect to deflection.

EXAMPLE 15.4

Determine the maximum deflection for the 8-in.-by-18-in. (S4S) Douglas fir beam shown in Figure 15.6. The allowable deflection is $\frac{1}{360}$ of the span length.

Solution From Appendix H, Case 1, the maximum deflection is given by

$$\Delta_{max} = \frac{5wL^4}{384EI}$$

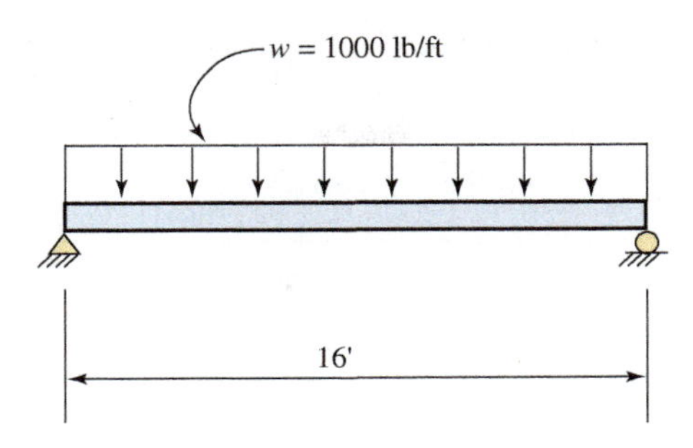

FIGURE 15.6 Beam diagram.

From Appendix E, the moment of inertia for this timber beam is 3350 in.^{4} and the weight is 36.4 lb/ft. From Appendix F, the modulus of elasticity is 1,700,000 psi. Substituting gives

$$\Delta_{max} = \frac{5[(1000 + 36.4)\ \text{lb/ft}](16\ \text{ft})^4(1728\ \text{in.}^3/\text{ft}^3)}{384(1{,}700{,}000\ \text{psi})(3350\ \text{in.}^4)} = 0.268\ \text{in.}$$

$$\Delta_{all} = \frac{\text{span}}{360} = \frac{(16\ \text{ft})(12\ \text{in./ft})}{360} = 0.533\ \text{in.}$$

Since 0.268 in. $<$ 0.533 in., the beam is satisfactory with respect to deflection.

EXAMPLE 15.5 A solid, round, simply supported steel shaft, having a diameter of 50 mm, supports a concentrated load, as shown in Figure 15.7. Compute the deflection under the load and at point *C*. Neglect the weight of the shaft. Use the formula method.

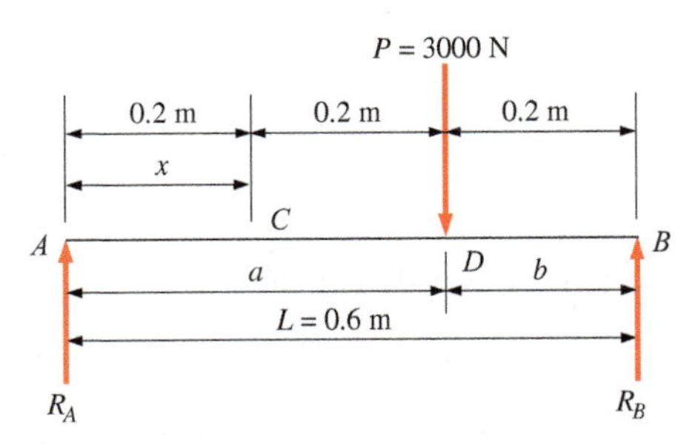

FIGURE 15.7 Load diagram.

Solution From Appendix H, Case 6, the deflection at the point of load (point *D*) is given by

$$\Delta = \frac{Pa^2b^2}{3EIL}$$

From Table 3, obtain an expression for moment of inertia:

$$I = \frac{\pi d^4}{64} = \frac{\pi(50\ \text{mm})^4}{64} = 307 \times 10^3\ \text{mm}^4$$

Substitute using units of newtons and millimeters and noting the modulus of elasticity for steel is 207×10^3 MPa:

$$\Delta_D = \frac{(3000\ \text{N})(400\ \text{mm})^2(200\ \text{mm})^2}{3(207 \times 10^3\ \text{N/mm}^2)(307 \times 10^3\ \text{mm}^4)(600\ \text{mm})}$$
$$= 0.168\ \text{mm}$$

From Appendix H, the deflection at point C is given by

$$\Delta_C = \frac{Pbx}{6EIL}(L^2 - b^2 - x^2)$$

$$= \frac{(3000\text{ N})(200\text{ mm})(200\text{ mm})[(600\text{ mm})^2 - (200\text{ mm})^2 - (200\text{ mm})^2]}{6(207 \times 10^3\text{ N/mm}^2)(307 \times 10^3\text{ mm}^4)(600\text{ mm})}$$

$$= 0.147\text{ mm}$$

15.4.1 The Principle of Superposition

The applicability of the formula method may be expanded. The *principle of superposition* is used in the common situation where a beam is subjected to more than one load and/or different types of loads. The deflection at any given point is determined by computing the deflection (using the formulas) at the point under consideration due to each individual load and then adding the results.

Under some combined loadings the maximum deflection for each of the different loadings may not occur at the same point. As a result, it may be difficult to determine the location of the maximum deflection or its magnitude. In such a situation, the principle of superposition may not be applicable and an alternate solution using the moment-area method may be necessary.

EXAMPLE 15.6 A 4-in. nominal diameter standard-weight steel pipe is used as a cantilever beam projecting 12 ft from the fixed end. Compute the deflection at the free end due to the two loads shown in Figure 15.8. Neglect the weight of the beam.

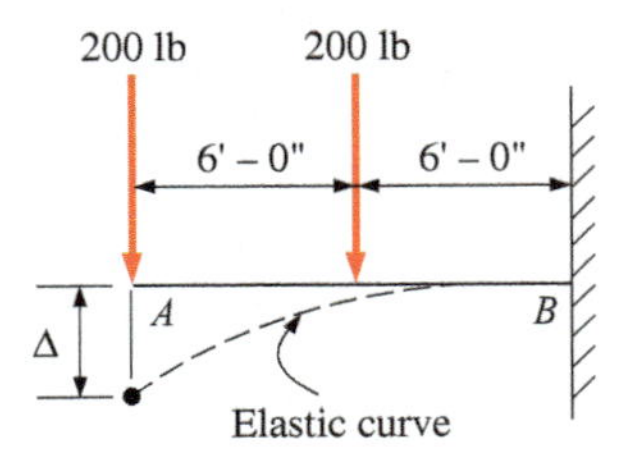

FIGURE 15.8 Load diagram.

Solution The principle of superposition will be used. The loading can be separated into two different cases as shown. The case designations (14 and 13) are from Appendix H and are shown in Figure 15.9. Thus,

$$\Delta_{14} = \frac{PL^3}{3EI} \quad \text{and} \quad \Delta_{13} = \frac{Pb^2}{6EI}(3L - b)$$

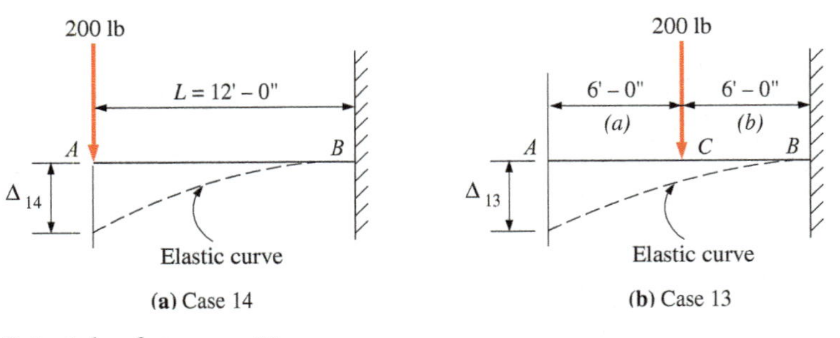

FIGURE 15.9 Principle of superposition.

Use $E = 30{,}000{,}000$ psi (from Appendix G) and $I = 7.23$ in.^{4} (from Appendix B) and substitute:

$$\Delta_{14} = \frac{(200 \text{ lb})(12 \text{ ft})^3(1728 \text{ in.}^3/\text{ft}^3)}{3(30{,}000{,}000 \text{ psi})(7.23 \text{ in.}^4)} = 0.918 \text{ in.}$$

$$\Delta_{13} = \frac{(200 \text{ lb})(6 \text{ ft})^2(12 \text{ in./ft})^2}{6(30{,}000{,}000 \text{ psi})(7.23 \text{ in.}^4)}[3(12 \text{ ft}) - 6 \text{ ft}](12 \text{ in./ft}) = 0.287 \text{ in.}$$

The total deflection at the free end is calculated from

$$\Delta = \Delta_{14} + \Delta_{13} = 0.918 \text{ in.} + 0.287 \text{ in.} = 1.205 \text{ in.}$$

15.5 THE MOMENT-AREA METHOD

As previously mentioned, in the case of beams subjected to combined and/or unsymmetrical loadings, deflection formulas may not be available. Or, if such formulas do exist, they may be too cumbersome for quick and easy use. When this occurs, the moment-area method proves a very valuable noncalculus alternate method for computing deflections. This method is based on the relationships established in Section 15.2 and is generally presented in the form of two theorems. These theorems, as stated here, apply to beams with constant values of modulus of elasticity and moment of inertia (which is the usual situation). The application to beams with varying moments of inertia is discussed in Section 15.7.

The simply supported beam shown in Figure 15.10 is subjected to a uniformly distributed line load. Its shear and moment diagrams are shown in Figure 15.10b and c. The deflected shape of the beam is indicated by the elastic curve that originally (before loading) was straight and horizontal.

We remove an arbitrary portion of the beam between points A and B and show it in exaggerated form in Figure 15.11. The first moment-area theorem can be stated as follows:

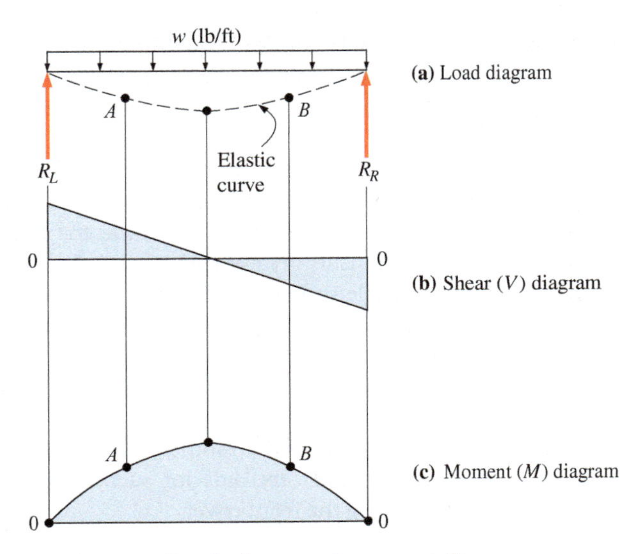

FIGURE 15.10 Load, shear, and moment diagrams.

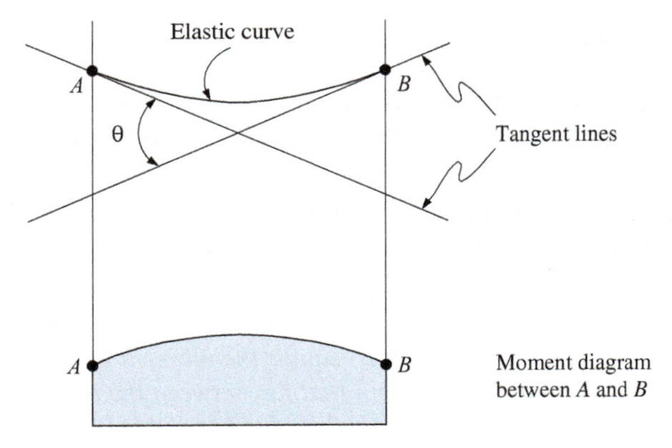

FIGURE 15.11 First moment-area theorem.

> *The angle θ between the tangents to any two points (A and B) on the elastic curve equals the area of the moment diagram that lies between those two points divided by* EI.

Expressing this as a formula yields

$$\theta = \frac{A_M}{EI} \quad \textbf{(15.3)}$$

where θ = the angle between any two tangents to the elastic curve. Its units are *radians* (recall that 1 radian equals 57.3°).

A_M = the area of the moment diagram lying between two vertical planes that pass through points A and B. The basic units for this quantity are lb-in.^{2} in the U.S. Customary System and N · m^2 in the SI.

E = the modulus of elasticity (psi) (Pa)

I = the moment of inertia with respect to the neutral axis of the beam (in.^{4}) (mm^4)

As in all other formulas, the units must be consistent. In the U.S. Customary System, the length units of moment of inertia and modulus of elasticity are always expressed in inch units. Units in the numerator should also be expressed in inch units. Either lb or kips may be used for force units. The following units are recommended for the calculation of θ (radians):

$$\theta = \frac{A_M}{EI} = \frac{(\text{lb-in.})(\text{in.})}{(\text{lb/in.}^2)(\text{in.}^4)} = \frac{\text{lb-in.}^2}{\text{lb-in.}^2}$$

In the SI, the length units of moment of inertia and modulus of elasticity are generally expressed in millimeter units with either kN or N used for force units. The following units are recommended for the calculation of θ (radians):

$$\theta = \frac{A_M}{EI} = \frac{(\text{N} \cdot \text{mm})(\text{mm})}{(\text{N/mm}^2)(\text{mm}^4)} = \frac{\text{N} \cdot \text{mm}^2}{\text{N} \cdot \text{mm}^2}$$

Recall that 1 MPa = 1 N/mm^2.

The same arbitrary portion of the beam (between points A and B) is again removed. It is shown in exaggerated form in Figure 15.12. The second moment-area theorem can be stated as follows:

> *The vertical displacement* y_{AB} *of a point (point* A*) on the elastic curve from a tangent to the elastic curve at a second point (point* B*) equals the moment of the area of the moment diagram that lies between the two points, taken about the first point (point* A*), divided by* EI.

Expressing this as a formula gives

$$y_{AB} = \frac{A_M \bar{x}}{EI} \tag{15.4}$$

where y_{AB} = the vertical displacement of point A on the elastic curve from a tangent line drawn at point B on the elastic curve

$\bar{x}$ = the distance from the centroid of the diagram area between any two points (A and B) to the point where the vertical displacement is desired

The other terms are as previously defined.

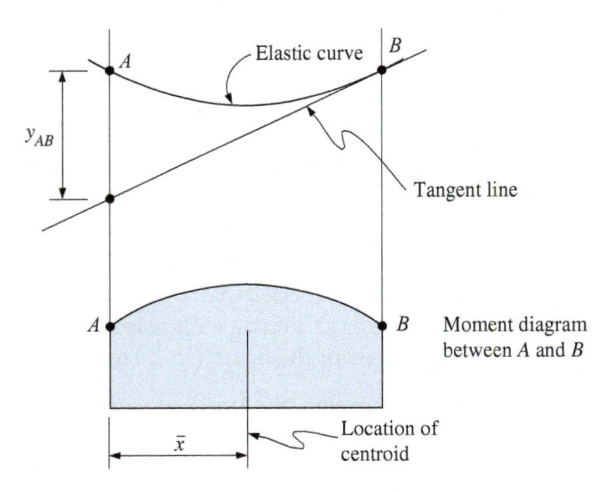

FIGURE 15.12 Second moment-area theorem.

When using Equation (15.4), the units must be consistent. In the U.S. Customary System, the following units are recommended for the calculation of displacement (in inches):

$$y = \frac{A_M \bar{x}}{EI} = \frac{(\text{lb-in.})(\text{in.})(\text{in.})}{(\text{lb/in.}^2)(\text{in.}^4)} = \frac{\text{lb-in.}^3}{\text{lb-in.}^2} = \text{in.}$$

In the SI, the following units are recommended for the calculation of displacement in millimeters (mm):

$$y = \frac{A_M \bar{x}}{EI} = \frac{(\text{N} \cdot \text{mm})(\text{mm})(\text{mm})}{(\text{N/mm}^2)(\text{mm}^4)} = \frac{\text{N} \cdot \text{mm}^3}{\text{N} \cdot \text{mm}^2} = \text{mm}$$

The two preceding theorems, represented by Equations (15.3) and (15.4), are applicable between any two points on the elastic curve of any beam for any type of loading. Only relative rotation of the tangents and relative vertical displacements, however, are obtained directly. The displacement y calculated from Equation (15.4) does not necessarily represent the desired deflection of the loaded beam. The distinction between y and the usually desired deflection Δ of a particular point will become apparent in the examples that follow.

In applying the moment-area method, the following sequence of steps should be followed:

1. Draw a load diagram of the beam. This diagram does not need to be to scale, but approximate proportions should be used.
2. Directly below the load diagram, draw an approximate elastic curve exaggerating the deflections. Indicate all reference tangents to be used as well as all displacements to be calculated.
3. Below the elastic curve diagram, draw a moment diagram, either a combined moment diagram or a moment diagram by parts (discussed in Section 15.6). Draw a shear diagram only if it is necessary for the drawing of the moment diagram.
4. Apply Equation (15.3) and/or Equation (15.4). For Equation (15.4), if the moment diagram involves several parts, a tabular approach may clarify the process.

The use of a horizontal tangent to the elastic curve, assuming its location is known, will generally provide the simplest solution. In cantilever beams, the tangent at the fixed end is always horizontal, thus greatly simplifying the solution of this type of problem. For simply supported beams subjected to a symmetrical loading, the maximum deflection will always occur at midspan and the tangent to the elastic curve will be horizontal at that point, thus simplifying the solution of the problem.

To aid in the application of the moment-area method, areas enclosed by curves and centroids for such areas are furnished in Table 2, inside the front cover.

EXAMPLE 15.7 A cantilever beam *AB* is subjected to a concentrated load of 6000 lb at its free end, as shown in Figure 15.13a. The elastic curve of the deflected beam is shown in Figure 15.13b. Compute the slope of the tangent line at point *A* on the elastic curve. The beam is a W8 × 31 structural steel wide-flange section. Neglect the weight of the beam.

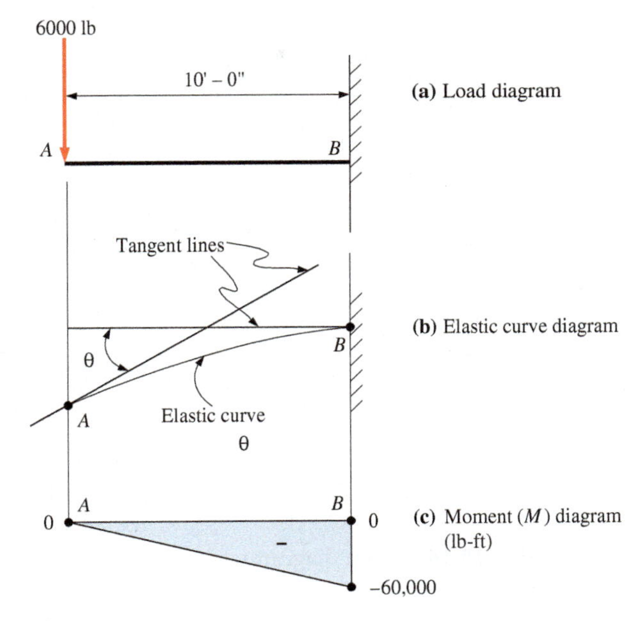

FIGURE 15.13 Beam diagrams.

Solution Note that the tangent line to the elastic curve at the fixed end of a cantilever beam is always horizontal. Therefore, the first moment-area theorem, Equation (15.3), will yield the desired slope at the free end of the beam if the entire moment diagram (between points *A* and *B*) is considered.

Use *E* from Appendix G and *I* from Appendix A:

$$\theta = \frac{A_M}{EI} = \frac{(0.5)(10\text{ ft})(12\text{ in./ft})(60{,}000\text{ lb-ft})(12\text{ in./ft})}{(30{,}000{,}000\text{ psi})(110\text{ in.}^4)} = 0.0131\text{ rad}$$

Convert to degrees:

$$\theta = (0.0131\text{ rad})(57.3°/\text{rad}) = 0.75°$$

EXAMPLE 15.8 Compute the angle between the tangent lines to the elastic curve at midspan and at the left end (points *A* and *B*) of the simply supported beam shown in Figure 15.14. The beam is a W250 × 73 structural steel wide-flange section. Neglect the weight of the beam.

Solution When a simply supported beam is symmetrically loaded, the tangent at midspan is horizontal. The computed angle between this tangent and the tangent to the elastic curve at the left end is designated θ_1 in the elastic curve diagram. The slope of the tangent line at the left end (point *A*) with respect to the original unloaded position of the beam is designated θ_2. This angle is generally termed the *end rotation* of the beam. Since the tangent at *B* is horizontal, $\theta_1 = \theta_2$.

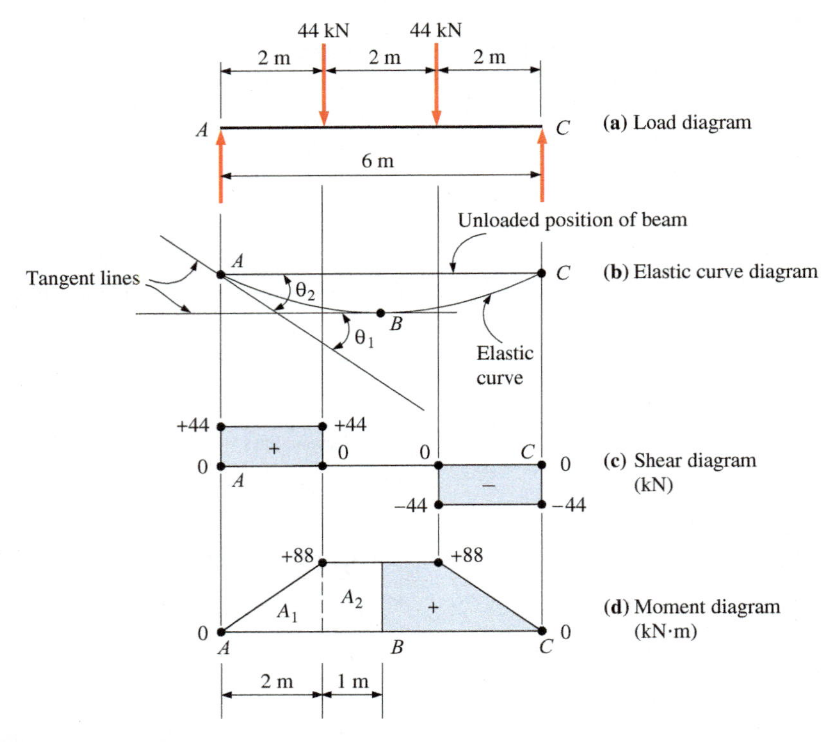

FIGURE 15.14 Beam diagrams.

In this problem, a shear diagram is drawn to simplify the construction of the moment diagram. Using E and I from Appendices Tables G.2 and A.2, respectively, calculate the desired angle using Equation (15.3), considering the moment diagram between points A and B:

$$\theta_1 = \theta_2 = \frac{A_M}{EI} = \frac{A_1 + A_2}{EI}$$

$$= \frac{0.5(2000\text{ mm})(88 \times 10^6\text{ N}\cdot\text{mm}) + (1000\text{ mm})(88 \times 10^6\text{ N}\cdot\text{mm})}{(207 \times 10^3\text{ MPa})(111 \times 10^6\text{ mm}^4)}$$

$$= 7.66 \times 10^{-3}\text{ rad}$$

Convert to degrees:

$$\theta_1 = \theta_2 = (7.66 \times 10^{-3}\text{ rad})(57.3°/\text{rad}) = 0.439°$$

EXAMPLE 15.9 Compute the maximum deflection Δ of the cantilever beam shown in Figure 15.15. The beam is a W8 × 24 structural steel wide-flange section.

Solution The maximum deflection of a cantilever beam subjected to vertical downward loads will always occur at the free end. Since the tangent to the elastic curve at the fixed end is horizontal and parallel to the original unloaded position of the beam, the vertical displacement y_{AB} of point A on the elastic curve from the tangent to the curve at point B is equal to the deflection at the free end. The vertical displacement y_{AB} is equal to the moment, about A, of the moment diagram area between points A and B.

Since the shear diagram (not shown) is a sloping straight line, the moment curve is a second-degree parabola. The area of the moment diagram (see Table 2, inside the front cover) between points A and B is calculated from

$$A_M = 0.333(12\text{ ft})(-36{,}000\text{ lb-ft}) = -144{,}000\text{ lb-ft}^2$$

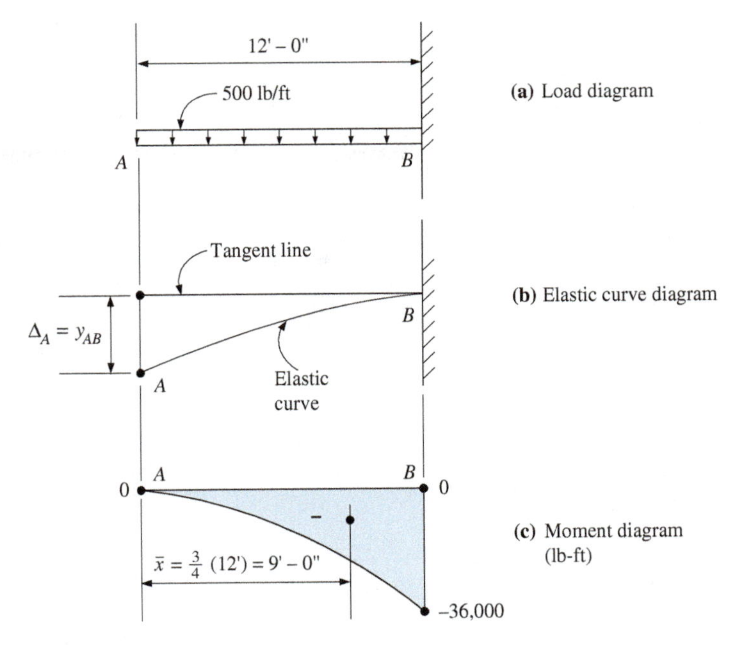

FIGURE 15.15 Beam diagrams.

Note that the area is a negative value. The distance from the centroid of the moment area to point *A* (see Table 2) is

$$\bar{x} = \frac{3}{4}(12\text{ ft}) = 9\text{ ft}$$

Using *E* and *I* from Appendices G and A, respectively,

$$y_{AB} = \frac{A_M\bar{x}}{EI} = \frac{(-144{,}000\text{ lb-ft}^2)(144\text{ in.}^2/\text{ft}^2)(9\text{ ft})(12\text{ in./ft})}{(30{,}000{,}000\text{ psi})(82.8\text{ in.}^4)} = -0.90\text{ in.}$$

The negative sign indicates that the vertical displacement y_{AB} from the tangent line is downward. Or, point *A* will lie below the tangent drawn at point *B*. Similarly, if we drew a tangent line at *A* and calculated y_{BA}, it would be negative, indicating that point *B* was situated below the tangent line drawn at point *A*. Therefore, note that the sign of the vertical displacement from one point to the tangent line drawn at another point will be the same as the sign of the moment diagram between those same two points. Further note that the distance $\bar{x}$ is always considered positive.

Finally, since we consider downward deflection of a beam from the original horizontal position to be positive:

$$\Delta_A = -y_{AB} = 0.90\text{ in.}$$

You will discover that once a point has been located relative to the tangent line using the rationale for signs of the displacements as discussed, the sign of the final deflection can be determined by inspection of the absolute values of the calculated quantities. Therefore, a well-drawn sketch will be invaluable.

EXAMPLE 15.10 A 4-in. nominal diameter standard-weight steel pipe ($I = 7.23\text{ in.}^4$ from Appendix B) is used as a cantilever beam projecting 12 ft from the fixed end, as shown in Figure 15.16. Calculate the maximum deflection at the free end. Neglect the weight of the beam.

Solution This solution is similar to that of Example 15.9. The vertical displacement y_{AB} of point *A* on the elastic curve from the tangent at point *B* is equal to the maximum deflection Δ_A at the free end. From the second moment-area theorem, Equation (15.4), we see that this displacement (and, therefore, the desired

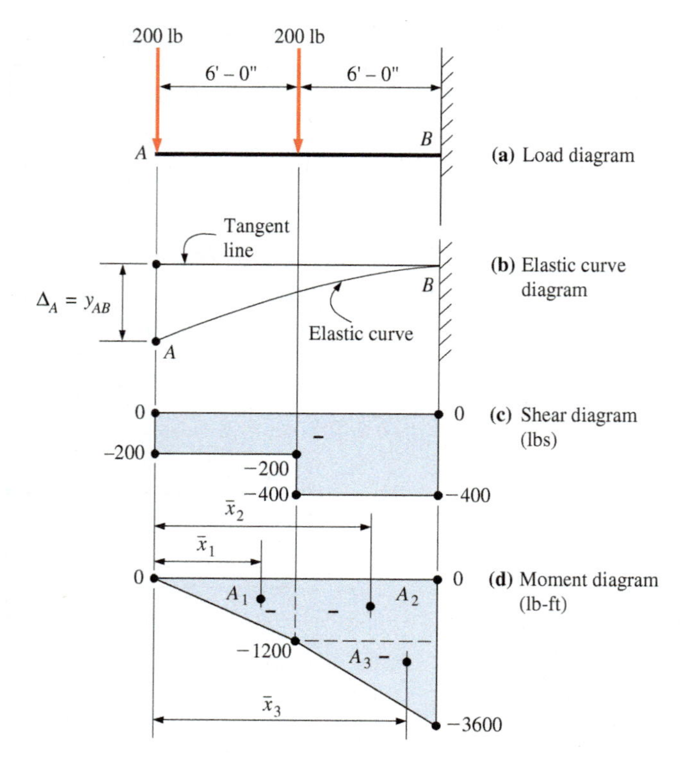

FIGURE 15.16 Beam diagrams.

deflection) is equal to the moment, about A, of the moment diagram area between points A and B. Note that in this case, however, the moment diagram area must be broken up into simple geometric shapes to determine areas and centroids.

The area of the moment diagram of Figure 15.16d is divided into two triangles and one rectangle. The areas are calculated as follows:

$$A_1 = 0.5(6\text{ ft})(-1200\text{ lb-ft}) = -3600\text{ lb-ft}^2$$
$$A_2 = (6\text{ ft})(-1200\text{ lb-ft}) = -7200\text{ lb-ft}^2$$
$$A_3 = 0.5(6\text{ ft})(-2400\text{ lb-ft}) = -7200\text{ lb-ft}^2$$

The distances from the centroid of each area to point A are calculated from

$$\bar{x}_1 = 0.667(6\text{ ft}) = 4\text{ ft}$$
$$\bar{x}_2 = 6\text{ ft} + 0.5(6\text{ ft}) = 9\text{ ft}$$
$$\bar{x}_3 = 6\text{ ft} + 0.667(6\text{ ft}) = 10\text{ ft}$$

Using E and I from Appendices G and B, respectively, the vertical displacement is calculated as

$$y_{AB} = \frac{A_M\bar{x}}{EI} = \frac{(A_1\bar{x}_1 + A_2\bar{x}_2 + A_3\bar{x}_3)12^3}{EI}$$
$$= \frac{-[(3600\text{ lb-ft}^2)(4\text{ ft}) + (7200\text{ lb-ft}^2)(9\text{ ft}) + (7200\text{ lb-ft}^2)(10\text{ ft})]1728\text{ in.}^3/\text{ft}^3}{(30{,}000{,}000\text{ psi})(7.23\text{ in.}^4)}$$
$$= -1.205\text{ in.}$$

The negative sign indicates that the vertical displacement of point A is downward from the tangent line. Since downward deflection is considered positive,

$$\Delta_A = -y_{AB} = 1.205\text{ in.}$$

Note that a tabular format is also convenient to use when there are several individual areas in the calculation. It may be set up as follows:

Element	A_M(lb-ft^2)	$\bar{x}$ (ft)	$A_M\bar{x}$(lb-ft^3)
A_1	0.5(6)(−1200) = −3600	4	−14,400
A_2	6(−1200) = −7200	9	−64,800
A_3	0.5(6)(−2400) = −7200	10	−72,000
Total:			−151,200

from which

$$y_{AB} = \frac{A_M\bar{x}}{EI} = \frac{(-151{,}200 \text{ lb-ft}^3)(12^3 \text{ in.}^3/\text{ft}^3)}{(30{,}000{,}000 \text{ psi})(7.23 \text{ in.}^4)} = -1.205 \text{ in.}$$

EXAMPLE 15.11 Calculate the maximum deflection at midspan for the symmetrically loaded simple beam in Figure 15.17. The beam is a W150 × 24 structural steel wide-flange section. Neglect the weight of the beam.

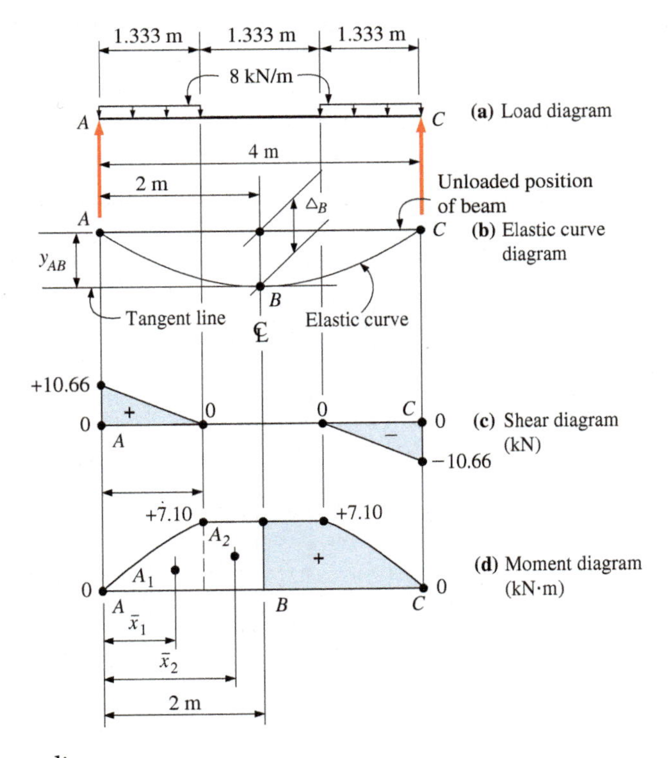

FIGURE 15.17 Beam diagrams.

Solution The maximum deflection occurs at point B. Therefore, the tangent to the elastic curve at B is horizontal. Since the unloaded position of the beam is also horizontal, $y_{AB} = \Delta_B$. Therefore, the determination of y_{AB} using the second moment-area theorem, Equation (15.4), will also yield Δ_B.

The area of the moment diagram between points A and B is broken into two separate geometric shapes: a second-degree parabola and a rectangle, as shown in Figure 15.17d. These two areas are calculated as follows (refer to Table 2):

$$A_1 = \frac{2}{3}bh = \frac{2}{3}(1.333 \text{ m})(7.10 \text{ kN}\cdot\text{m}) = 6.31 \text{ kN}\cdot\text{m}^2 = 6.31 \times 10^9 \text{ N}\cdot\text{mm}^2$$

$$A_2 = bh = (2.00 \text{ m} - 1.333 \text{ m})(7.10 \text{ kN}\cdot\text{m}) = 4.74 \text{ kN}\cdot\text{m}^2 = 4.74 \times 10^9 \text{ N}\cdot\text{mm}^2$$

The distances from the centroid of each area to point *A* are calculated from

$$\bar{x}_1 = \frac{5}{8}(1.333 \text{ m}) = 0.833 \text{ m} = 833 \text{ mm}$$

$$\bar{x}_2 = 1.333 \text{ m} + \frac{1.333 \text{ m}}{4} = 1.666 \text{ m} = 1666 \text{ mm}$$

For convenience, we have changed the moment areas and centroid locations to units of newtons and millimeters. Using *E* and *I* from Appendices G and A, respectively, calculate the vertical displacement as

$$y_{AB} = \frac{A_M\bar{x}}{EI} = \frac{(A_1\bar{x}_1 + A_2\bar{x}_2)}{EI}$$

$$= \frac{(6.31 \times 10^9 \text{ N}\cdot\text{mm}^2)(833 \text{ mm}) + (4.74 \times 10^9 \text{ N}\cdot\text{mm}^2)(1666 \text{ mm})}{(207 \times 10^3 \text{ N/mm}^2)(13.3 \times 10^6 \text{ mm}^4)}$$

$$= 4.78 \text{ mm}$$

Note that the y_{AB} is a positive value, which indicates that the vertical displacement of point *A* is upward from the tangent line at *B*. Therefore, *B* lies below *A*, and since we consider downward deflection positive,

$$\Delta_B = y_{AB} = 4.78 \text{ mm}$$

EXAMPLE 15.12 The beam shown in Figure 15.18 is a 3-in. nominal diameter standard-weight steel pipe ($I = 3.02 \text{in.}^4$). Calculate the deflection at the free end and at midspan between supports. Neglect the weight of the beam.

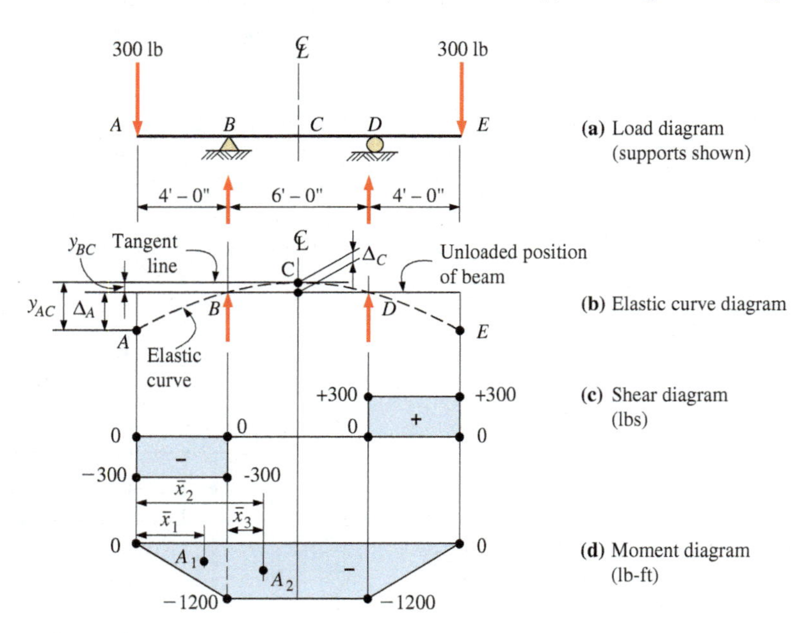

FIGURE 15.18 Beam diagrams.

Solution The shear diagram is drawn to simplify the drawing of the moment diagram. The beam is symmetrically loaded and the tangent to the elastic curve at midspan is horizontal. The desired deflections are Δ_A and Δ_C. Note that these are measured from the original unloaded position of the beam. Compute y_{AC} using the second moment-area theorem, Equation (15.4). Values for *E* and *I* are obtained from Appendices G and B, respectively. Thus,

$$y_{AC} = \frac{A_M\bar{x}}{EI} = \frac{(A_1\bar{x}_1 + A_2\bar{x}_2)(12^3)}{EI}$$

$$= \frac{-[0.5(4 \text{ ft})(1200 \text{ lb-ft})(0.667)(4 \text{ ft}) + (3 \text{ ft})(1200 \text{ lb-ft})(5.5 \text{ ft})]1728 \text{ in.}^3/\text{ft}^3}{(30{,}000{,}000 \text{ psi})(3.02 \text{ in.}^4)}$$

$$= -0.500 \text{ in.}$$

Note that y_{AC} is *not* the deflection at the free end. The deflection at the free end can be calculated from

$$\Delta_A = y_{AC} - y_{BC}$$

Therefore, y_{BC} must first be calculated. Moments of the moment diagram area between points *B* and *C* are taken about point *B:*

$$y_{BC} = \frac{A_M \bar{x}}{EI} = \frac{(A_2 \bar{x}_3)(12^3)}{EI}$$

$$= \frac{(3\text{ ft})(-1200\text{ lb-ft}(1.5\text{ ft})(1728\text{ in.}^3/\text{ft}^3)}{(30{,}000{,}000\text{ psi})(3.02\text{ in.}^4)}$$

$$= -0.103\text{ in.}$$

Neglecting the signs of the displacements y_{AC} and y_{BC}, calculate the deflection of point *A* from

$$\Delta_A = y_{AC} - y_{BC} = 0.500\text{ in.} - (0.103\text{ in.}) = 0.397\text{ in.}$$

Note that Δ_C is the upward deflection of point *C* and that it is equal to y_{BC}. The calculated displacement y_{BC} has a value of −0.103 in., which verifies that point *B* lies below the tangent line at point *C*. (Point *B* is a support point and does not move vertically.) Therefore,

$$\Delta_C = y_{BC} = -0.103\text{ in.}$$

where the negative sign indicates an upward deflection at point *C*.

In this section, the beams used in the examples were loaded to produce moment diagrams consisting of simple geometric shapes or areas that could be broken up into simple geometric shapes. When beams are subjected to combined and/or asymmetrical loadings, however, the moment diagrams can become more difficult to analyze. In such situations, an alternate method of drawing moment diagrams is preferable. This method is presented in Section 15.6. Further applications of the moment-area method using this alternate approach to the drawing of moment diagrams are presented in Section 15.7.

15.6 MOMENT DIAGRAM BY PARTS

The areas encountered in conventional moment diagrams, particularly where asymmetrical and/or combined loadings exist, can often be very cumbersome to work with. These areas and their centroids may be difficult to obtain without some calculus manipulations. To simplify the calculations of areas for the moment-area method, we may draw the moment diagrams in a somewhat different form. Essentially, we draw the various rudimentary parts that comprise the moment diagrams, hence the descriptor "moment diagram by parts."

The moment-diagram-by-parts method offers a practical solution whereby the moment diagram results in common geometric shapes with known properties or properties that can easily be obtained. The technique involves considering each loading separately and then drawing a bending moment diagram for that load alone, as if there were no other loads acting on the member. Actually, all loads are acting simultaneously, and the true value of the moment at any point will be the algebraic sum of the values indicated on the separate diagrams. The moment diagram of each individual load is based on the assumption that the beam is fixed at some location, in effect creating a cantilever beam subject to the one load.

The moment diagram by parts can be started at either end of the beam. At each concentrated load, or reaction, a triangular area will begin. This triangular area will be positive for upward loads or reactions and negative for downward loads or reactions. At locations where a uniformly distributed load exists, a parabolic curve will result. The method is illustrated in Example 15.13.

EXAMPLE 15.13 Draw the bending moment diagram by parts for the beam shown in Figure 15.19. Neglect the weight of the beam.

Solution The beam reactions are computed in the normal way and are shown on the load diagram. In this problem, we work from the left end (point *A*) and assume a fixed end condition at the other end (point *B*).

The moment diagram may be considered to consist of three parts: one part due to the reaction R_A, a second part due to the concentrated load, and a third due to the uniformly distributed load. Beginning at *A*, the moment diagram due to the reaction is designated A_1 in Figure 15.19. As shown in

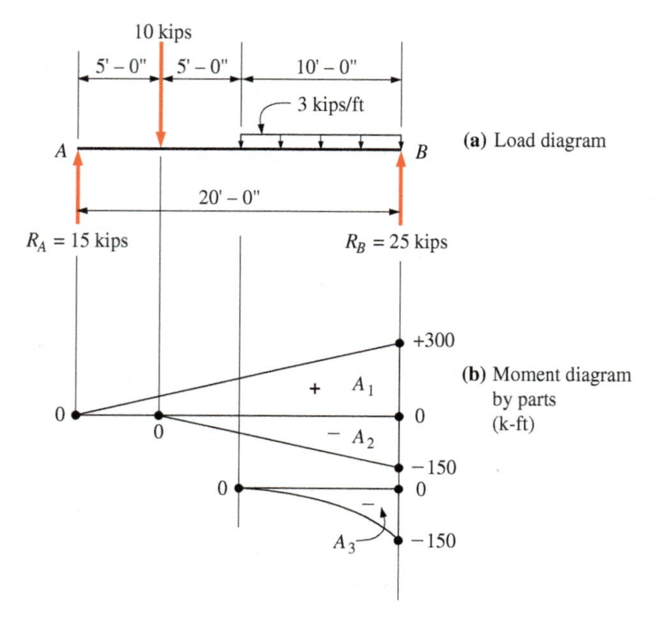

FIGURE 15.19 Beam diagrams.

Figure 15.20, it is the same moment diagram as that for a cantilever beam that is subjected to a concentrated load at its free end. It is positive, since the load is acting upward.

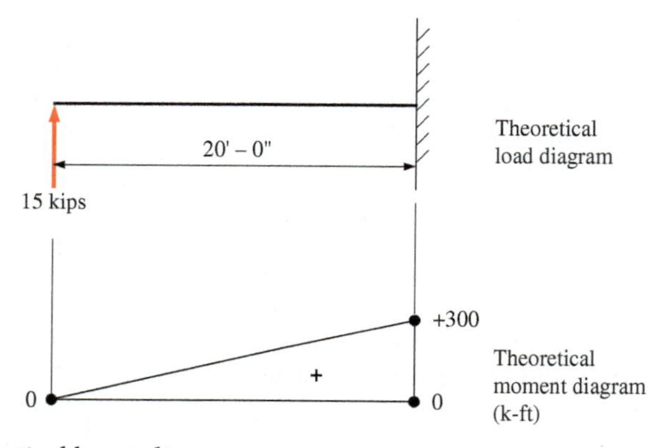

FIGURE 15.20 Theoretical beam diagrams.

The moment diagram for the applied concentrated load of 10 kips is designated A_2 in Figure 15.19. As shown in Figure 15.21, it is seen to be the same moment diagram as that of a cantilever beam with a 15-ft span length subjected to a concentrated load at its free end. It is negative, since the load is acting downward. (Note that in drawing the moment diagrams of Figure 15.19, the same horizontal reference line is used for the two diagrams.)

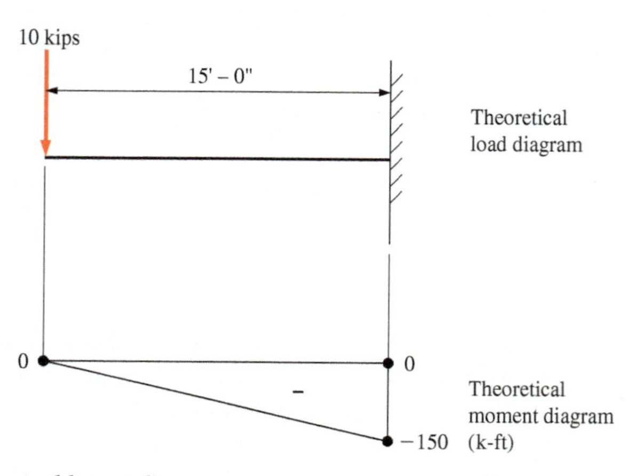

FIGURE 15.21 Theoretical beam diagrams.

The moment diagram for the uniformly distributed load is designated A_3 in Figure 15.19. As shown in Figure 15.22, it is seen to be the same moment diagram as that of a cantilever beam of 10-ft span length subjected to a uniformly distributed load for its full length. It is negative, since the load is acting downward. (Note that this is drawn in Figure 15.19 using a second reference line. The various diagrams may be combined, superimposed, or drawn separately, whichever is preferred.)

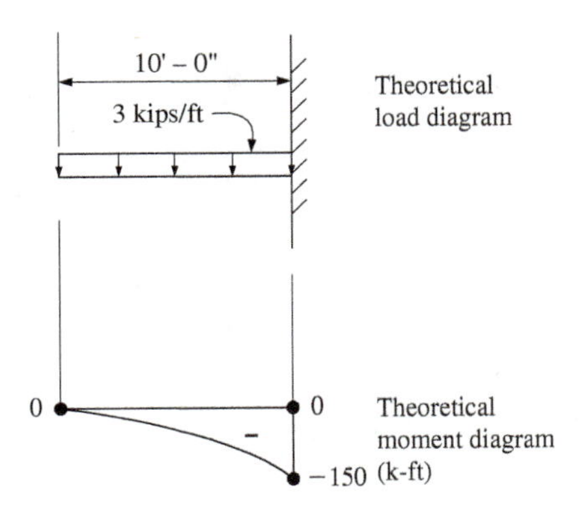

FIGURE 15.22 Theoretical beam diagrams.

If so desired, the moment at any point due to all the loads can be obtained by an algebraic sum of all the ordinates from all the diagrams at any given point. Note, however, that this method does not indicate the location or the magnitude of the maximum moment; therefore, its usage is relatively limited.

Using the same reasoning, the moment diagram by parts can be drawn working from right to left. The fixed condition may be assumed to exist at point *A*. The resulting diagram is shown in Figure 15.23.

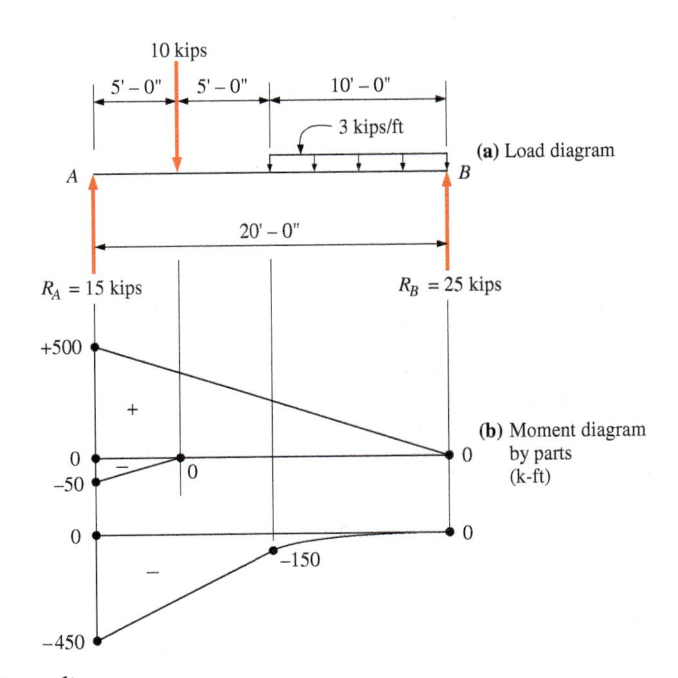

FIGURE 15.23 Beam diagrams.

It is also possible to draw the moment diagram by parts by working both ways (left to right and right to left) to some reference location. Figure 15.24 shows the moment diagram using midspan as the location at which the beam is fixed.

You may wish to verify that each of the three moment diagrams (part b of Figures 15.19, 15.23, and 15.24) yields the same result. For instance, in each case, the midspan moment determined by summing the magnitude of the applicable midspan moment from each part of the diagram is +100 k-ft. (Note that in Figure 15.24 the summation is made just to the left or just to the right of midspan.)

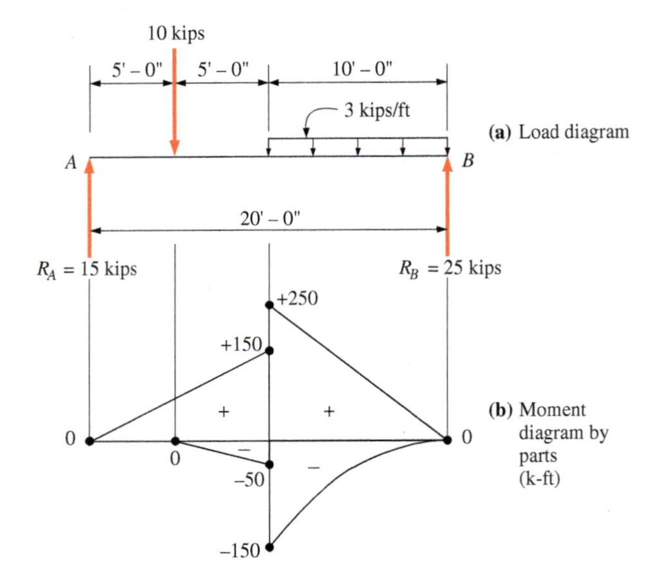

FIGURE 15.24 Beam diagrams.

15.7 APPLICATIONS OF THE MOMENT-AREA METHOD

The more complex deflection problems generally involve beams subjected to some form of unsymmetrical loading. When this occurs, the position of a horizontal tangent to the elastic curve is unknown. Therefore, some other tangent line must be used as a reference tangent. Generally this line will be the tangent at the support farthest from an individual load or from the resultant of the loads. The following two examples illustrate the procedure of calculating the deflection at any given point for a simple beam and an overhanging beam.

EXAMPLE 15.14 Calculate the deflection at midspan for the beam in Figure 15.25. The beam is a 4-in. nominal diameter standard-weight steel pipe. Neglect the weight of the beam.

Solution The beam reactions are computed in the normal manner and are shown on the load diagram. In this example, the shear diagram will not be drawn since the moment diagram by parts will be used. With this method, the resulting moment diagram shapes are sufficiently elementary to make the use of the shear diagram unnecessary.

The bending moment diagram by parts is drawn from left to right and the reference tangent, as shown in the elastic curve diagram, is the tangent at the left support (point *A*). First, calculate y_{BA}, which is the vertical displacement of point *B* from the tangent to the elastic curve at point *A*. The value of y_{BA} is equal to the moment, about *B*, of the moment diagram area between points *A* and *B*.

Use *E* from Appendix G and *I* from Appendix B:

$$y_{BA} = \frac{A_M\bar{x}}{EI} = \frac{\left[\left(\frac{1}{2}\right)(10\text{ ft})(11{,}000\text{ lb-ft})\left(\frac{10\text{ ft}}{3}\right) - \left(\frac{1}{2}\right)(5\text{ ft})(9000\text{ lb-ft})\left(\frac{5\text{ ft}}{3}\right) - \left(\frac{1}{2}\right)(2.5\text{ ft})(2000\text{ lb-ft})\left(\frac{2.5\text{ ft}}{3}\right)\right]\left(1728\,\frac{\text{in.}^3}{\text{ft}^3}\right)}{(30{,}000{,}000\text{ psi})(7.23\text{ in.}^4)}$$

$$= +1.145\text{ in.}$$

The positive sign indicates that point *B* lies above the tangent line. By proportion, y_C can be calculated from

$$y_C = \frac{5}{10}y_{BA} = \frac{+1.145}{2} = +0.572\text{ in.}$$

Note that the quantity *y* with a single subscript denotes the distance from that point on the *undeflected* beam to a line tangent to the elastic curve.

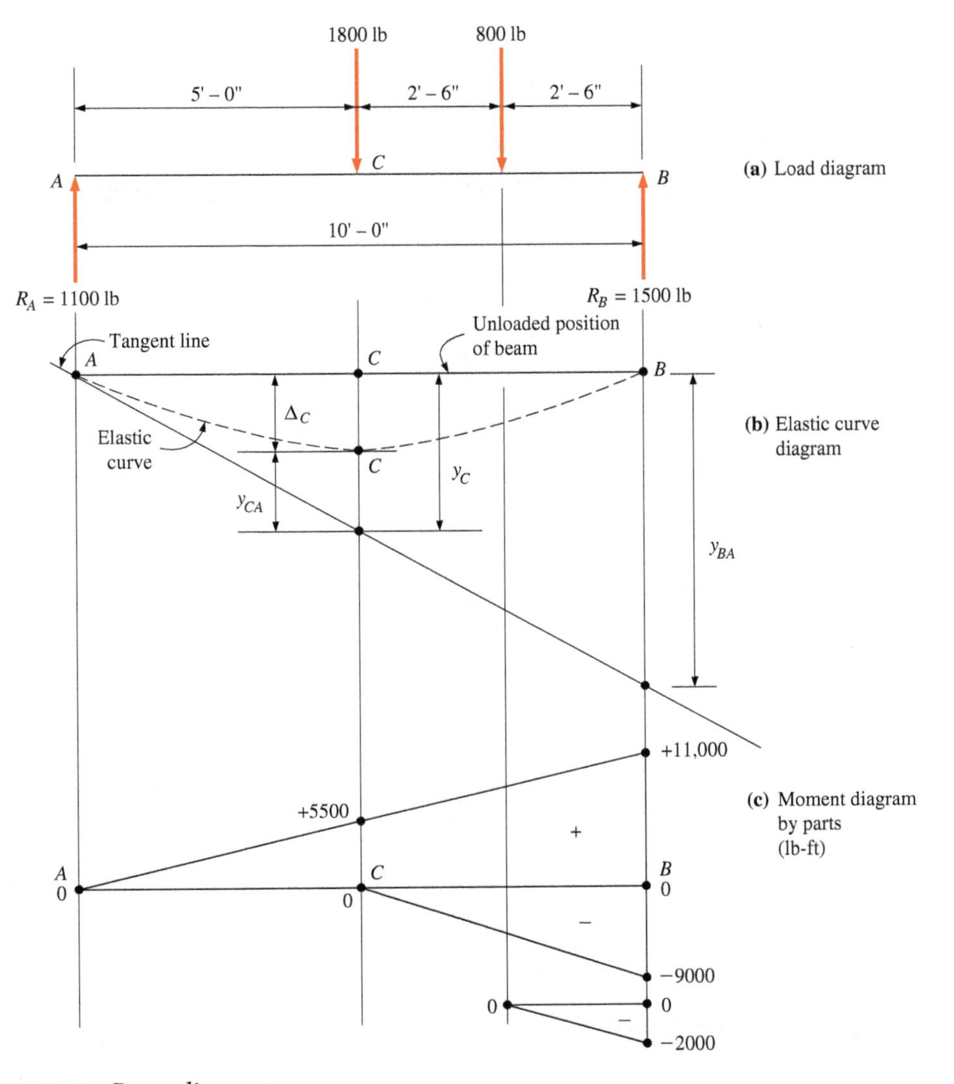

FIGURE 15.25 Beam diagrams.

Now calculate y_{CA}, which is the vertical displacement of point C on the elastic curve from the tangent to the elastic curve at point A. The value of y_{CA} is equal to the moment, about point C, of the moment diagram area between points A and C.

$$y_{CA} = \frac{A_M\bar{x}}{EI} = \frac{\left[\left(\frac{1}{2}\right)(5\text{ ft})(5500\text{ lb-ft})\left(\frac{5\text{ ft}}{3}\right)\right]\left(1728\dfrac{\text{in.}^3}{\text{ft}^3}\right)}{(30{,}000{,}000\text{ psi})(7.23\text{ in.}^4)}$$

$$= +0.183\text{ in.}$$

As shown on the elastic curve diagram,

$$\Delta_C = y_C - y_{CA} = 0.572\text{ in.} - 0.183\text{ in.} = 0.389\text{ in.}$$

which is the deflection at midspan from the unloaded position of the beam.

EXAMPLE 15.15 The beam in Figure 15.26 is a W12 × 50 structural steel wide-flange section. (a) Calculate the deflection at midspan between the two supports A and B. (b) Calculate the deflection of the free end of the beam (point C). Neglect the weight of the beam.

Solution The beam reactions are computed in the normal way and are shown in the load diagram of Figure 15.26. Although the shape of the elastic curve for a simple beam is quite evident, the

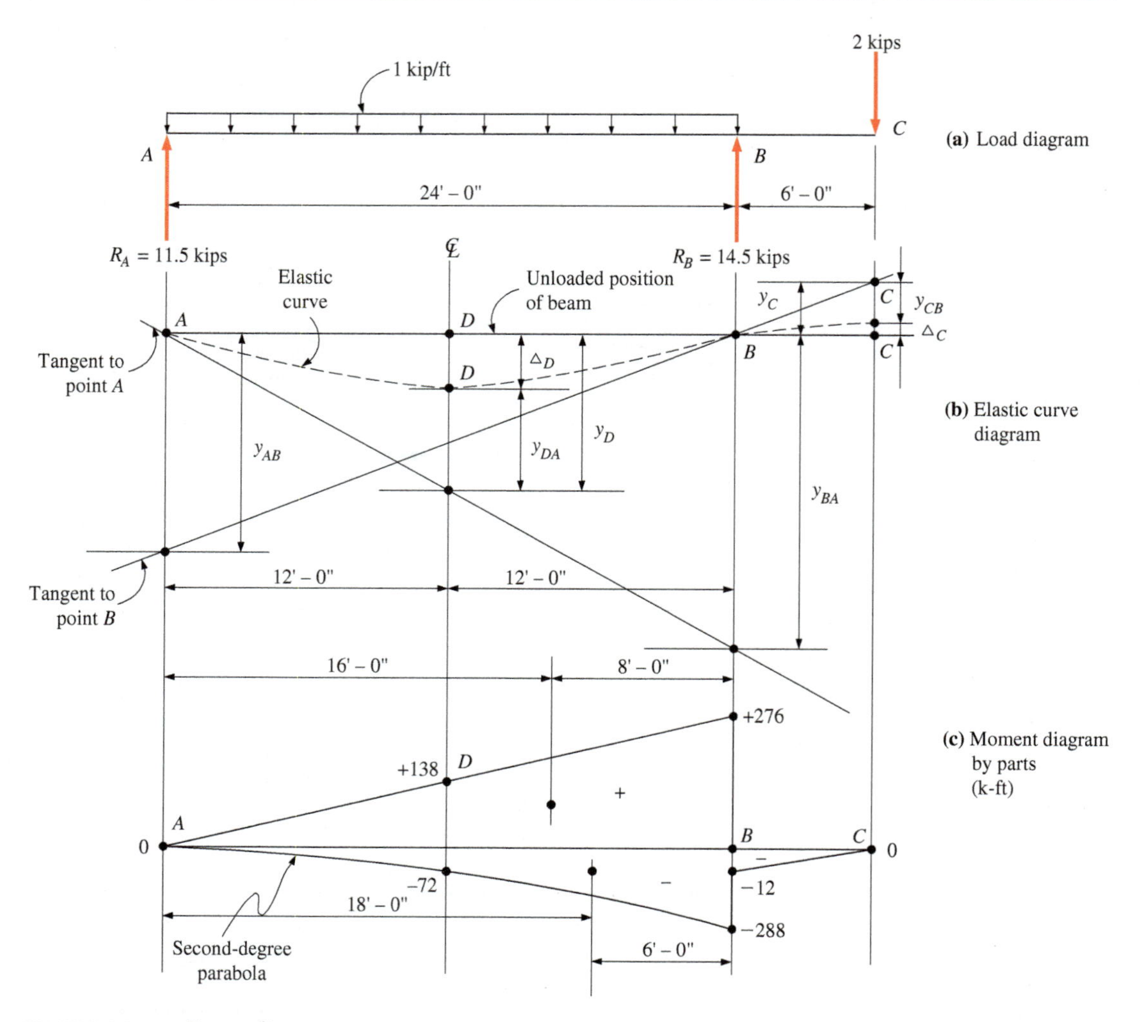

FIGURE 15.26 Beam diagrams.

elastic curve for the overhanging beam will have to be assumed and then verified by calculation. It is difficult to determine by inspection whether the free-end deflection will be upward or downward. The assumed shape is shown in Figure 15.26b.

For this type of beam, draw the moment diagram by parts working from both left to right and right to left, assuming a fixed condition at the right support (point *B*).

(a) To determine the midspan deflection between points *A* and *B*, a reference tangent to the elastic curve at the left support (point *A*) is drawn. The vertical displacement y_{BA} is equal to the moment, about point *B*, of the moment diagram area between points *A* and *B*. From the second moment-area theorem, Equation (15.4), and using *E* and *I* from Appendices G and A, respectively,

$$y_{BA} = \frac{A_M\bar{x}}{EI} = \frac{\left[\left(\frac{1}{2}\right)(24\text{ ft})(276\text{ k-ft})\left(\frac{24\text{ ft}}{3}\right) - \left(\frac{1}{3}\right)(24\text{ ft})(288\text{ k-ft})\left(\frac{24\text{ ft}}{3}\right)\right]\left(1728\,\frac{\text{in.}^3}{\text{ft}^3}\right)}{(30{,}000\text{ ksi})(394\text{ in.}^4)}$$

$$= +1.853\text{ in.}$$

By proportion, the vertical distance from the tangent line at *A* to point *D* on the undeflected beam can be calculated from

$$y_D = \frac{12}{24}(y_{BA}) = \frac{12}{24}(1.853\text{ in.}) = +0.926\text{ in.}$$

Now calculate y_{DA}, which is the vertical displacement (at midspan) of point *D* on the elastic curve from the tangent to the elastic curve at *A*. The value of y_{DA} is equal to the moment, about point *D*, of the moment diagram area between points A and D. From the second moment-area principle, Equation (15.4),

$$y_{DA} = \frac{A_M\bar{x}}{EI} = \frac{\left[\left(\frac{1}{2}\right)(12\text{ ft})(138\text{ k-ft})\left(\frac{12\text{ ft}}{3}\right) - \left(\frac{1}{3}\right)(12\text{ ft})(72\text{ k-ft})\left(\frac{12\text{ ft}}{4}\right)\right]\left(1728\frac{\text{in.}^3}{\text{ft}^3}\right)}{(30{,}000\text{ ksi})(394\text{ in.}^4)}$$

$$= +0.358\text{ in.}$$

As shown in the elastic curve diagram in Figure 15.26b, the deflection at midspan (point D) between the two support points is calculated from

$$\Delta_D = y_D - y_{DA} = 0.926\text{ in.} - 0.358\text{ in.} = 0.568\text{ in.}$$

(b) To obtain the deflection at the free end of the beam (point C), the same tangent line at point A could be used. For purposes of example, however, use a tangent to the elastic curve at point B. First, calculate y_{AB}, the vertical displacement of point A from this tangent line. The value of y_{AB} is equal to the moment, about point A, of the moment diagram area between points A and B. Thus,

$$y_{AB} = \frac{A_M\bar{x}}{EI} = \frac{\left[\left(\frac{1}{2}\right)(24\text{ ft})(276\text{ k-ft})\left(\frac{2}{3}\right)(24\text{ ft}) - \left(\frac{1}{3}\right)(24\text{ ft})(283\text{ k-ft})\left(\frac{3}{4}\right)(24\text{ ft})\right]\left(1728\frac{\text{in.}^3}{\text{ft}^3}\right)}{(30{,}000\text{ ksi})(394\text{ in.}^4)}$$

$$= 0.421\text{ in.} - (0.021\text{ in.})$$

$$= +1.684\text{ in.}$$

Note that y_{AB} is a positive (+) value. If a negative value had resulted, the assumption as to the slope of the tangent line at point B would have been incorrect. A negative value would indicate that point A on the elastic curve was situated below the tangent line, which would imply a clockwise rotation of the line tangent to the beam at point B. But because y_{AB} is a positive value, indicating that point A lies above the tangent line, the assumption as to the direction of rotation of the tangent line at point B is correct.

By proportion, y_C can be calculated from

$$\frac{y_C}{6} = \frac{y_{AB}}{24}$$

$$y_C = \frac{6}{24}(1.684) = 0.421\text{ in.}$$

Note that point C on the undeflected beam lies below the tangent line drawn at point B on the elastic curve.

Now calculate y_{CB}, which is the vertical displacement of point C on the elastic curve from the tangent to the elastic curve at point B. The value of y_{CB} is equal to the moment, about point C, of the moment diagram area between points B and C. From the second moment-area theorem, Equation (15.4),

$$y_{CB} = \frac{A_M\bar{x}}{EI} = \frac{\left[\left(\frac{1}{2}\right)(6\text{ ft})(12\text{ k-ft})\left(\frac{2}{3}\right)(6\text{ ft})\right]\left(1728\frac{\text{in.}^3}{\text{ft}^3}\right)}{(30{,}000\text{ ksi})(394\text{ in.}^4)}$$

$$= -0.021\text{ in.}$$

The negative sign indicates that point C on the elastic curve lies below the tangent line drawn at point B. Therefore, by inspection of Figure 15.26b, and neglecting the sign of y_{CB} (using its absolute value), the free-end deflection at point C is obtained from

$$y_C - y_{CB} = 0.421\text{ in.} - 0.021\text{ in.} = 0.400\text{ in.}$$

Recognizing that the deflection at point C is upward,

$$\Delta_C = -0.400\text{ in.}$$

The previous two examples illustrated the procedure for computing the deflection at any given point. The maximum deflection, however, was not obtained in either case. It is the maximum deflection that is generally the value of interest. Example 15.16 illustrates the procedure for calculating both the location and the magnitude of the maximum deflection.

EXAMPLE 15.16 A 3-m-long aluminum bar is used as a beam and loaded as shown in Figure 15.27. The bar is 50 mm by 50 mm in cross section. Calculate the maximum deflection of the beam. Neglect the weight of the beam. For convenience, lengths are shown in millimeter units.

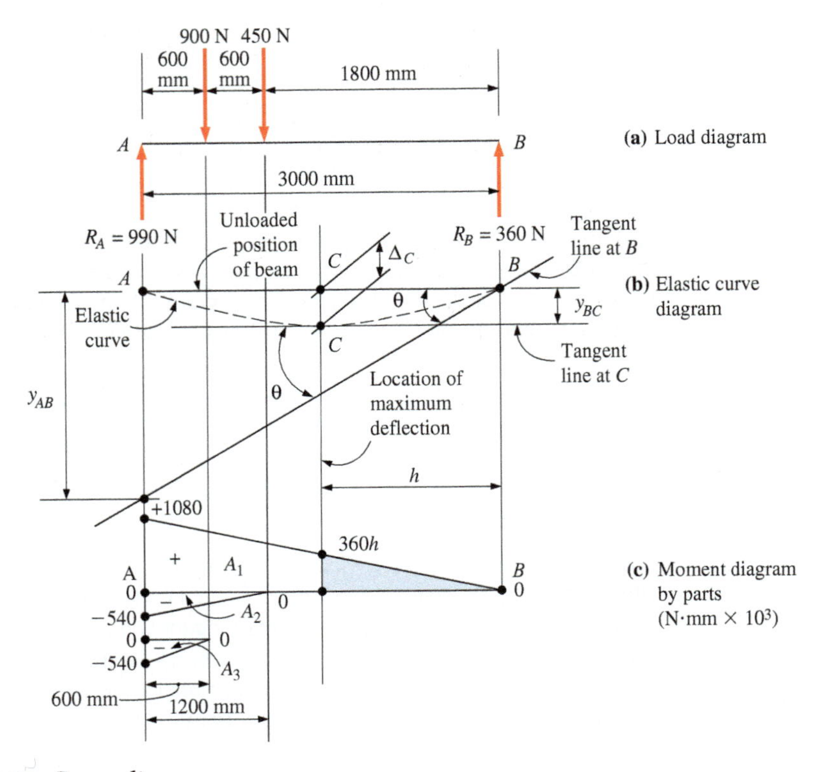

FIGURE 15.27 Beam diagrams.

Solution The beam reactions are computed in the usual way and are shown in the load diagram. The moment diagram by parts will be drawn from right to left. Note that it is the right support that is farthest from the resultant of the loads. The reference tangent to the elastic curve will be the tangent at the right support (point *B*).

The modulus of elasticity for the aluminum bar (70×10^3 MPa or N/mm^2) is obtained from Appendix G. Calculate the moment of inertia (from Table 3 inside the back cover):

$$I = \frac{bh^3}{12} = \frac{(50\text{ mm})(50\text{ mm})^3}{12} = 521 \times 10^3\text{ mm}^4$$

Next, compute y_{AB}, which is equal to the moment, about point *A*, of the moment diagram area between points *A* and *B*. Use the second moment-area theorem, Equation (15.4), units of newtons and millimeters, and a tabular format:

Element	$A_M(\text{N}\cdot\text{mm}^2)$	$\bar{x}$(mm)	$A_M\bar{x}(\text{N}\cdot\text{mm}^3)$
A_1	$0.5(3000)(1080 \times 10^3) = 1.620 \times 10^9$	$\frac{1}{3}(3000) = 1000$	1.620×10^{12}
A_2	$-0.5(1200)(540 \times 10^3) = -324 \times 10^6$	$\frac{1}{3}(1200) = 400$	-129.6×10^9
A_3	$-0.5(600)(540 \times 10^3) = -162.0 \times 10^6$	$\frac{1}{3}(600) = 200$	-32.4×10^9
Total:			1.458×10^{12}

from which

$$y_{AB} = \frac{A_M\bar{x}}{EI} = \frac{1.458 \times 10^{12}\text{ N}\cdot\text{mm}^3}{(70{,}000\text{ MPa})(521{,}000\text{ mm}^4)} = 400\text{ mm}$$

Next, compute the angle between the tangent at point B and the original unloaded position of the beam. Recall that for small angles the tangent of an angle approximately equals the angle itself measured in radians. So,

$$\theta = \tan\theta = \frac{y_{AB}}{L} = \frac{40.0}{3000} = 0.013\,33 \text{ radian}$$

The tangent to the elastic curve at the point of maximum deflection is horizontal. Therefore, it also makes an angle of θ with the line that is tangent to the elastic curve at point B. Using the first moment-area theorem, Equation (15.3), the angle θ between the two tangents is equal to the area of the moment diagram between the same two points (B and C), divided by EI. Therefore, recognize that the slope of the A_1 portion of the moment diagram is

$$\frac{1080 \times 10^3 \text{ N}\cdot\text{mm}}{3000 \text{ mm}} = 360\frac{\text{N}\cdot\text{mm}}{\text{mm}}$$

and arbitrarily locate the point of maximum deflection a distance h (where $h \leq 1800$ mm) from point B:

$$\theta = \frac{A_M}{EI} = \frac{\frac{1}{2}h(\text{slope} \times h)}{EI} = \frac{\frac{1}{2}h\left[(360)\frac{\text{N}\cdot\text{mm}}{\text{mm}}\right]h}{EI}$$

Then solve for h:

$$\frac{\left(360\frac{\text{N}\cdot\text{mm}}{\text{mm}}\right)}{2}h^2 = \theta\, EI = 0.013\,33(70 \times 10^3 \text{ N}\cdot\text{mm}^2)(521 \times 10^3 \text{ mm}^4)$$

from which

$$h = 1643 \text{ mm}$$

Had h been greater than 1800 mm (but not greater than 2400 mm), the solution would have had to be rewritten to include the effect of A_2 of the moment diagram.

Now that the location of the maximum deflection is known, the moment of the moment diagram between points B and C, y_{BC} can be determined. The second moment-area theorem, Equation (15.4):

$$y_{BC} = \frac{A_M\bar{x}}{EI} = \frac{\frac{1}{2}(1643 \text{ mm})(360 \text{ N})(1643 \text{ mm})\left(\frac{2}{3}\right)(1643 \text{ mm})}{(70{,}000 \text{ MPa})(521{,}000 \text{ mm}^4)}$$

$$= 14.59 \text{ mm}$$

Since the tangent at point C is horizontal and the unloaded position of the beam is horizontal, $\Delta_C = y_{BC}$. Therefore, the maximum deflection for the beam is 14.59 mm.

The previous discussion and examples were applicable to beams with constant values for modulus of elasticity, E, and moment of inertia, I. When a beam has a varying moment of inertia, the moment-area method may still be used with only slight modifications to the procedure we have followed thus far. The two moment-area theorems are still valid. Instead of calculating areas of moment diagrams or moments of moment diagram areas and then dividing by EI in the final step, however, each moment area must now be divided by the applicable value of I. This step is usually accomplished by drawing an additional diagram, an M/I diagram, in which the varying moment of inertia is introduced prior to computing any moment diagram areas. Division by E (normally constant) is then accomplished in the final calculation. Example 15.17 illustrates the procedure of calculating the deflection for a beam of varying moment of inertia.

EXAMPLE 15.17 Calculate the maximum deflection for the structural steel cantilever beam shown in Figure 15.28. The beam is stiffened with steel plates in such a manner that for a length of 4 ft at the fixed end, the moment of inertia is 400 in^4. For the remaining 8 ft, the section has no plates and the moment of inertia is 200 in^4. Use a modulus of elasticity value of 30,000,000 psi. Neglect the weight of the beam.

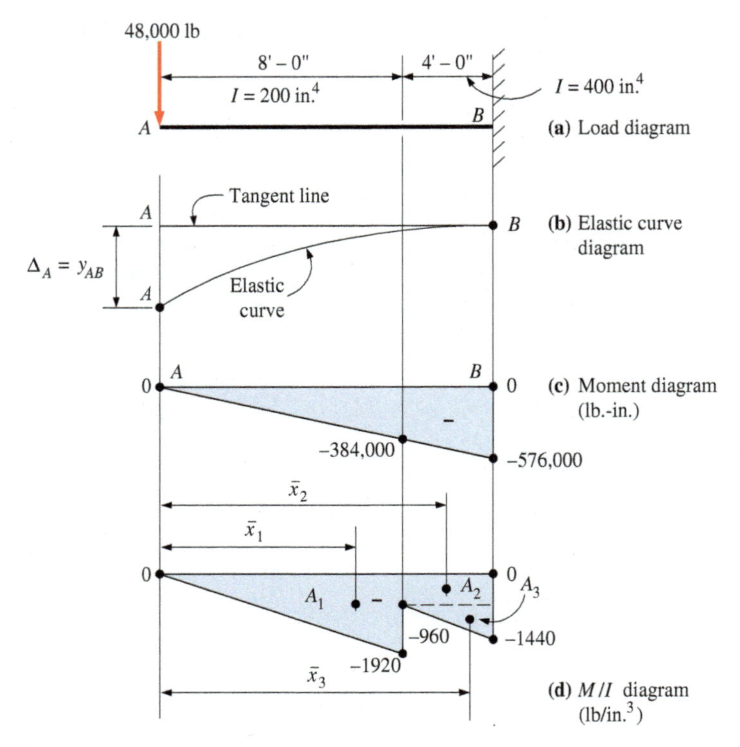

FIGURE 15.28 Beam diagrams.

Solution The solution is similar to that of Example 15.10. The vertical displacement y_{AB} of point A on the elastic curve from the tangent at point B is numerically equal to the maximum deflection Δ_A at the free end. The vertical displacement y_{AB}, from the second moment-area theorem, Equation (15.4), equals the moment, about A, of the M/I diagram between points A and B divided by E.

The M/I diagram ordinates are expressed in units of lb/in^3. Note that the M/I diagram must be broken up into simple geometric shapes for the determination of areas as follows:

$$A_1 = \frac{1}{2}(8\text{ ft})\left(12\frac{\text{in.}}{\text{ft}}\right)\left(-1920\frac{\text{lb}}{\text{in.}^3}\right) = -92{,}160\text{ lb/in.}^2$$

$$A_2 = (4\text{ ft})\left(12\frac{\text{in.}}{\text{ft}}\right)\left(-960\frac{\text{lb}}{\text{in.}^3}\right) = -46{,}080\text{ lb/in.}^2$$

$$A_3 = \frac{1}{2}(4\text{ ft})\left(12\frac{\text{in.}}{\text{ft}}\right)\left(-480\frac{\text{lb}}{\text{in.}^3}\right) = -11{,}520\text{ lb/in.}^2$$

The distances from the centroid of each area to point A can be calculated next:

$$\bar{x}_1 = \frac{2}{3}(8\text{ ft})\left(12\frac{\text{in.}}{\text{ft}}\right) = 64.0\text{ in.}$$

$$\bar{x}_2 = (8\text{ ft})\left(12\frac{\text{in.}}{\text{ft}}\right) + (2\text{ ft})\left(12\frac{\text{in.}}{\text{ft}}\right) = 120.0\text{ in.}$$

$$\bar{x}_3 = (8\text{ ft})\left(12\frac{\text{in.}}{\text{ft}}\right) + \frac{2}{3}(4\text{ ft})\left(12\frac{\text{in.}}{\text{ft}}\right) = 128.0\text{ in.}$$

Use Equation (15.4):

$$y_{AB} = \frac{A_{(M/I)}\bar{x}}{E} - = \frac{A_1\bar{x}_1 + A_2\bar{x}_2 + A_3\bar{x}_3}{E}$$

$$= \frac{-[(92{,}160)(64.0) + (46{,}080)(120.0) + (11{,}520)(128.0)]}{30{,}000{,}000}$$

$$= -0.430\text{ in.}$$

The negative sign indicates that point A lies below the tangent line. Since downward deflection is assumed positive,

$$\Delta_A = 0.430 \text{ in.}$$

The previous examples in this section illustrated some of the applications of the moment-area method. In addition, the method may be used to derive formulas, such as those used in Section 15.4. The following examples illustrate the procedure for the derivation of such formulas.

EXAMPLE 15.18 Derive expressions for the maximum deflection of a cantilever beam subjected to (a) a concentrated load at its free end and (b) a uniformly distributed load. In each case, neglect the weight of the beam.

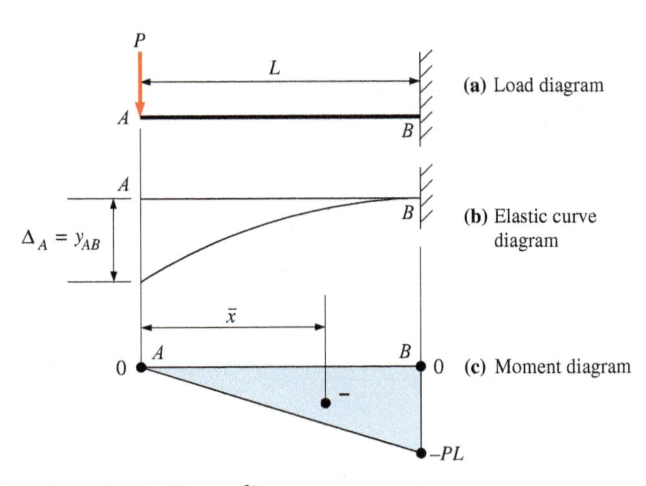

FIGURE 15.29 Beam diagrams.

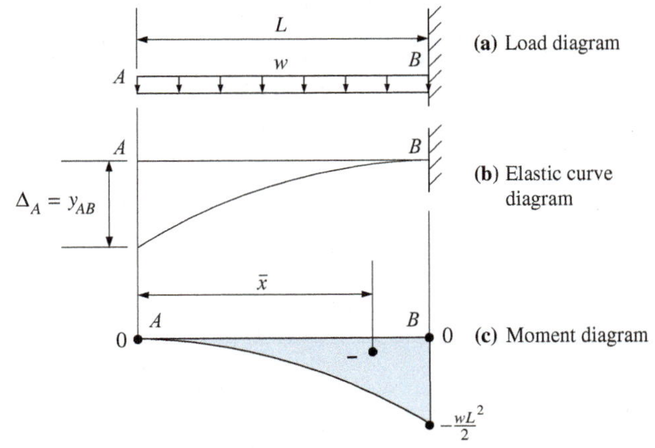

FIGURE 15.30 Beam diagrams.

Solution Figures 15.29 and 15.30 show the load diagrams, elastic curve diagrams, and the moment diagrams for the beams under consideration. The maximum deflection in each case is the vertical displacement y_{AB} of the free end (point A) from the tangent to the elastic curve at B. (Refer to Table 2 for expressions for areas and centroid locations.)

(a)

$$y_{AB} = \frac{A_M\bar{x}}{EI} = \frac{-\left(\frac{1}{2}\right)(PL^2)\left(\frac{2}{3}L\right)}{EI} = \frac{-PL^3}{3EI}$$

Since downward deflection is considered positive,

$$\Delta_A = \frac{PL^3}{3EI}$$

(b)

$$y_{AB} = \frac{A_M\bar{x}}{EI} = \frac{-\left(\frac{1}{3}\right)(L)(wL^2/2)\left(\frac{3}{4}L\right)}{EI} = \frac{-wL^4}{8EI}$$

Since downward deflection is considered as positive,

$$\Delta_A = \frac{wL^4}{8EI}$$

EXAMPLE 15.19 Derive expressions for the maximum deflection of a simply supported beam subjected to (a) a concentrated load at midspan and (b) a uniformly distributed load over the entire span length. Neglect the weight of the beam in each case.

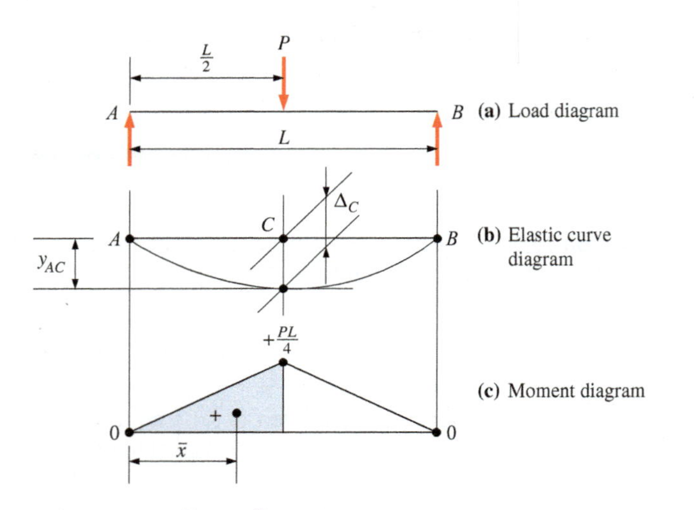

FIGURE 15.31 Beam diagrams.

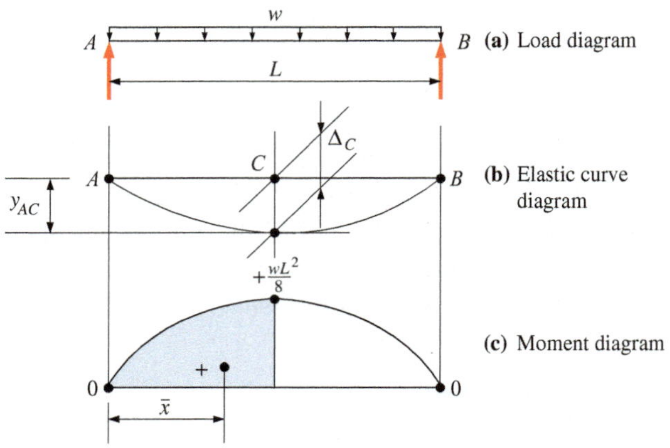

FIGURE 15.32 Beam diagrams.

Solution Figures 15.31 and 15.32 show the load diagrams, the elastic curve diagrams, and the moment diagrams. Using the horizontal tangent at midspan, the vertical displacement y_{AC} at the support equals the maximum deflection at midspan.

(a)
$$\Delta_C = y_{AC} = \frac{A_M\bar{x}}{EI} = \frac{(\frac{1}{2})(L/2)(PL/4)(L/3)}{EI} = \frac{PL^3}{48EI}$$

(b)
$$\Delta_C = y_{AC} = \frac{(\frac{2}{3})(L/2)(wL^2/8)(\frac{5}{16}L)}{EI} = \frac{5wL^4}{384EI}$$

SUMMARY BY SECTION NUMBER

15.2 In a beam subject to pure bending, the relationship between stress in the outer fibers and the radius of curvature of the elastic curve is

$$f_b = \frac{Ec}{R} \quad (15.1)$$

The relationship between the radius of curvature of the beam at any section and the bending moment at that section is

$$R = \frac{EI}{M} \quad (15.2)$$

15.3 Two basic methods for calculating deflections of bending members are the formula method and the moment-area method. The most commonly used method is the formula method.

15.4 The formula method is based on the usage of standard formulas that are readily available. Appendix H provides numerous deflection formulas for different types of beams and loading conditions. Using the principle of superposition, the formula method may also be used to compute the deflection at a point due to many and/or different types of loads.

15.5 Since deflection formulas may not be available for specific cases of combined and/or unsymmetrical loadings, the moment-area method offers a practical, noncalculus-based alternate solution. It involves the use of the bending moment diagram. The two basic moment-area theorems are

$$\theta = \frac{A_M}{EI} \quad (15.3)$$

$$y_{AB} = \frac{A_M\bar{x}}{EI} \quad (15.4)$$

where θ represents an angular rotation and y_{AB} represents a vertical displacement. This method may be used for beams with variable or constant EI values.

15.6 Drawing a moment diagram by parts is an alternate method of drawing a moment diagram. The technique involves creating a hypothetical cantilever beam, loading it with each load separately, and drawing a moment diagram for that load alone. The individual moment diagrams are then superimposed to represent the sum. This technique facilitates the subsequent moment-area calculations.

PROBLEMS

1. *For steel members unless otherwise noted, use a modulus of elasticity of 30,000,000 psi (207×10^3 MPa). For other materials, refer to the appropriate appendix table.*
2. *Unless otherwise noted, the given loads are superimposed loads; neglect the beam weight; neglect shear and moment considerations.*
3. *Unless otherwise noted, use any appropriate method to calculate deflection.*

Section 15.2 Curvature and Bending Moment

15.1 A $\frac{1}{4}$-in.-diameter aluminum rod is bent into a circular ring having a mean diameter of 125 in. Compute the maximum bending stress in the rod.

15.2 Calculate the maximum bending stress produced in a $\frac{1}{32}$-in.-diameter steel wire when it passes around a 20-in.-diameter pulley.

15.3 A 500-mm-long steel bar having a cross section of 1 mm by 20 mm is bent to a circular arc that subtends an angle of 60° as shown. Calculate the maximum bending stress.

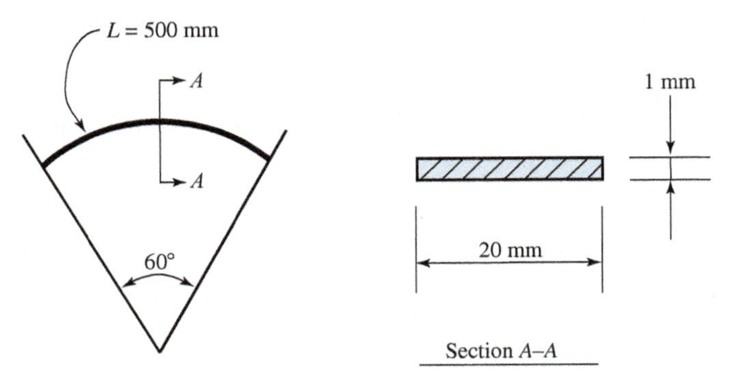

PROBLEM 15.3

15.4 An aluminum wire has a diameter of d in. Determine the minimum diameter D of the coil in which the wire can be wound without exceeding a bending stress of 30,000 psi.

15.5 A 3-in.-wide by $\frac{1}{2}$-in.-thick board is bent to a radius of curvature of 62 in. by a bending moment of 700 lb-in. Calculate the modulus of elasticity.

15.6 A Douglas fir beam is 6 in. wide and 14 in. deep. Compute the radius of curvature of the beam if it is subjected to a constant bending moment of 200,000 lb-in.

Section 15.4 The Formula Method

For Problems 15.7 through 15.14, use the formula method.

15.7 A solid, round simply supported steel shaft has a diameter of 38 mm and a span length of 800 mm. The shaft supports a concentrated load of 3 kN at midspan. Calculate the maximum deflection of the shaft.

15.8 Rework Problem 15.7 changing the diameter of the shaft to 25 mm and the span length to 500 mm.

15.9 Compute the maximum deflection of a 10-in.-by-14-in. simply supported solid rectangular redwood (S4S) beam. The span length is 15 ft and the beam is subjected to a uniformly distributed load of 1000 lb/ft. The allowable deflection is $\frac{1}{300}$ of the span length. Is the beam satisfactory?

15.10 Compute the maximum deflection for the beam of Problem 15.9 if the loading consists of one concentrated load of 5000 lb at midspan instead of the uniformly distributed load. The allowable deflection is $\frac{1}{300}$ of the span length. Is the beam satisfactory?

15.11 A W16 × 45 structural steel beam is simply supported on a span length of 24 ft. It is subjected to two concentrated loads of 12 kips each applied at the third point $(a = 8\text{ ft})$. Compute the maximum deflection.

15.12 A 4-in.-diameter, 12-ft-long solid steel shaft having a circular cross section is used as a simply supported beam. A concentrated load P is to be applied at midspan. Calculate the maximum allowable load P. Consider deflection. The allowable deflection is 0.50 in. Consider the beam weight.

15.13 Assume that a company's design criterion specifies that the deflection at the center of a simply supported solid (circular cross section) steel shaft due to its own weight must not exceed 0.010 in. per foot of span. Calculate the maximum permissible span length for a 3-in.-diameter shaft.

15.14 A steel wide-flange section is used as a cantilever beam having a span of 12 ft. The beam supports a distributed load that varies uniformly from 5 kips/ft at the support to zero at the free end. Determine the required moment of inertia of the beam if the deflection is not to exceed $\frac{1}{240}$ times the span.

Section 15.5 The Moment-Area Method

For Problems 15.15 through 15.26, use the moment-area method.

15.15 A W10 × 22 structural steel wide-flange beam on a simple span of 16 ft supports a concentrated load of 10,000 lb at midspan. Calculate the end rotation of the beam.

15.16 A W24 × 84 structural steel wide-flange beam on a simple span of 30 ft supports a uniformly distributed load of 3.0 kips/ft. Calculate the end rotation of the beam.

15.17 The cantilever beam shown is composed of an aluminum pipe that has an outside diameter of 76.0 mm and an inside diameter of 62.7 mm. A moment of 2.1 kN · m is applied at the free end. Find the rotation at the free end of the pipe. Neglect the weight of the pipe.

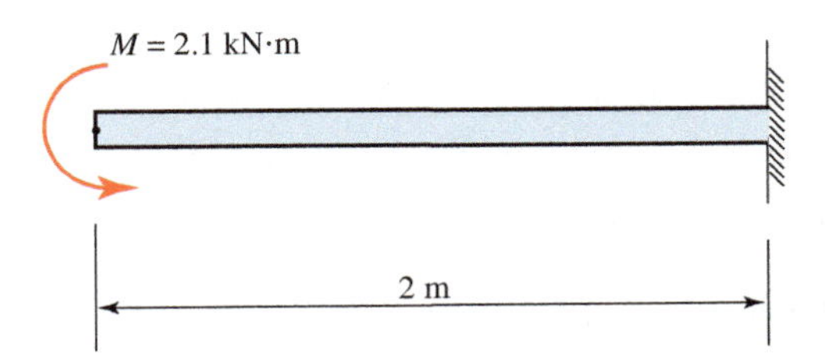

PROBLEM 15.17

15.18 A steel bar 1 in. thick and $1\frac{1}{2}$ in. wide is subjected to the loads shown. Calculate the angle between the tangent lines to the elastic curve at points B and C.

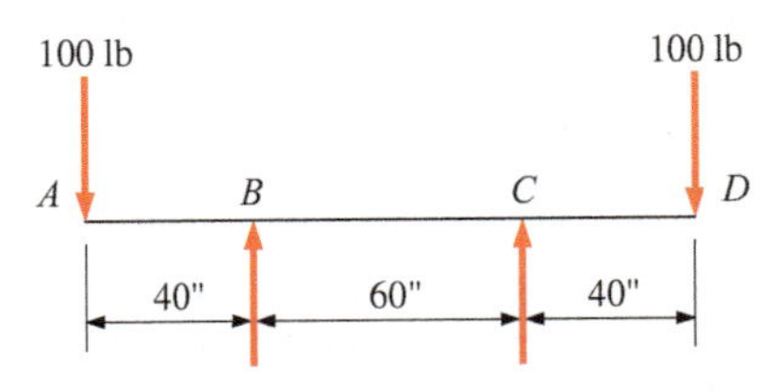

PROBLEM 15.18

15.19 For the beam of Problem 15.18, calculate the slope of the tangent line to the elastic curve at point A.

15.20 Using the moment-area method, check the deflection obtained in Problem 15.9.

15.21 Using the moment-area method, check the deflection obtained in Problem 15.10.

15.22 An 8-in.-by-12-in. Douglas fir (S4S) is used as a 10-ft-long cantilever beam. Compute the concentrated load at the free end that will cause a $\frac{3}{4}$-in. maximum deflection.

15.23 Compute the load required at the midspan of the beam of Problem 15.22 to cause a deflection of $\frac{3}{4}$ in. at the free end.

15.24 Using the moment-area method, check the deflection obtained in Problem 15.11.

15.25 A W12 × 30 structural steel beam is simply supported on a span of 12 ft. It is subjected to a uniformly distributed load of 675 lb/ft and a concentrated load of 14,000 lb at midspan. The allowable deflection is $\frac{1}{360}$ of the span. Is the beam satisfactory?

15.26 Compute the maximum deflection for the beam of Problem 15.16.

Section 15.6 Moment Diagram by Parts

15.27 Draw the moment diagram by parts for the beam in Figure 15.14. (Draw the diagram from left to right.) Compare the value of the maximum moment.

15.28 Draw the moment diagram by parts for the beam in Figure 15.4. (Draw the diagram from left to right.) Compare the value of the maximum moment.

15.29 Draw the moment diagram by parts for the beam in Figure 15.17. (Draw the diagram from left to right.) Compare the value of the maximum moment.

15.30 For the beam shown, draw the conventional moment diagram and left-to-right moment diagram by parts. Compare the maximum moments.

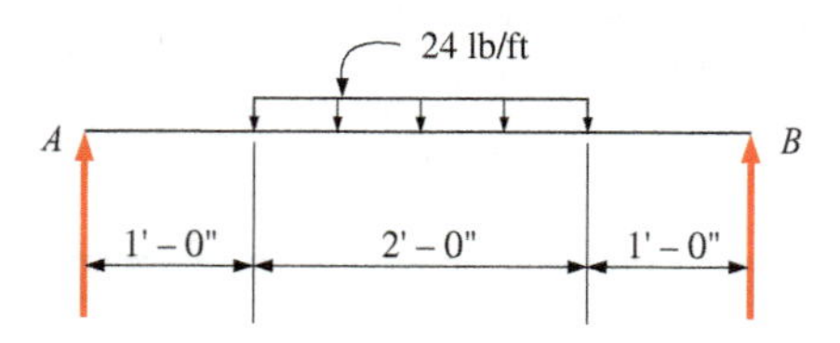

PROBLEM 15.30

Section 15.7 Applications of the Moment-Area Method

For Problems 15.31 through 15.43, use the moment-area method.

15.31 A steel bar is 3 in. wide and 1 in. thick. It is used as a 10-ft-long simply supported beam. Bending is about the weak axis. The beam supports two concentrated loads of 120 lb each spaced 4 ft apart and placed 3 ft from each end. Calculate the maximum deflection and the deflection at one of the loads.

15.32 Verify the deflection at point *C* that was calculated in part (b) of Example 15.15 by using the tangent to the elastic curve at point *A*.

15.33 A wood test beam 1.72 in. wide by 1.75 in. deep spans 28 in. between simple supports. The beam is deflected 0.074 in. under a 200-lb load applied at midspan. Compute the modulus of elasticity *E*.

15.34 A W530 × 138 structural steel wide-flange section is used as a 12-m-long simply supported beam. It is subjected to a uniformly distributed load for its full length. The load induces a maximum bending stress of 152 MPa. Compute the magnitude of the superimposed load (kN/m) and determine the maximum deflection. Include the effect of the beam weight for this problem.

15.35 A structural steel wide-flange section is loaded as shown. Calculate the maximum deflection between the supports and the deflection of the free end. Assume that $I = 228$ in^4.

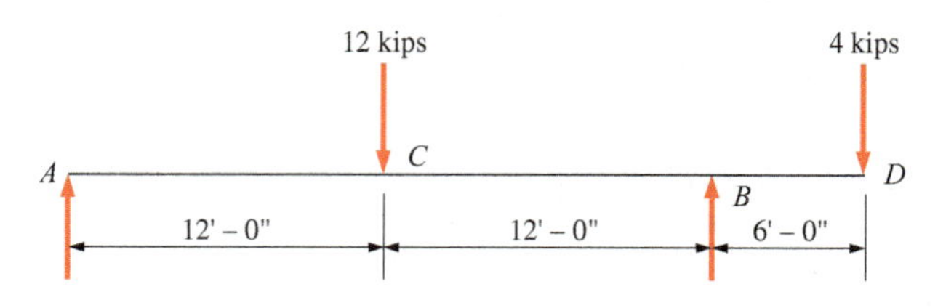

PROBLEM 15.35

15.36 An aluminum beam with a moment of inertia of 200 in.^{4} is used as a cantilever beam and loaded as shown. Calculate the maximum deflection.

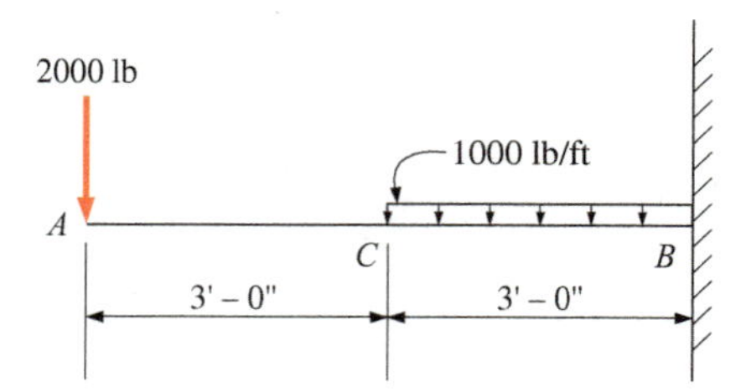

PROBLEM 15.36

15.37 For the W27 × 114 structural steel wide-flange beam shown, calculate the maximum deflection.

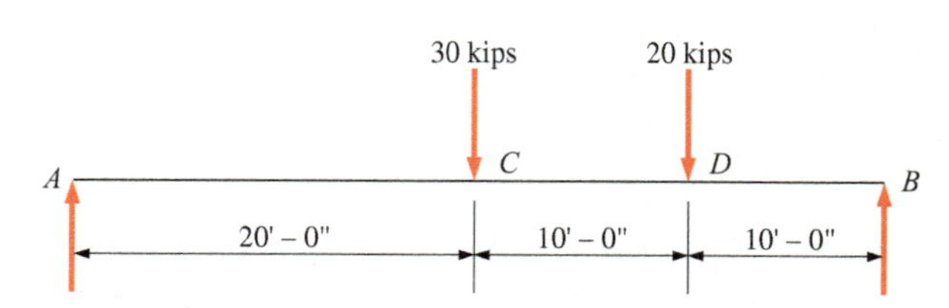

PROBLEM 15.37

15.38 Calculate the minimum required diameter for a solid circular steel cantilever beam 10 ft long. The maximum deflection due to its own weight is not to exceed 0.1 in.

15.39 For the steel beam shown, calculate the slope at the free end and the maximum deflection. Note the varying moment of inertia.

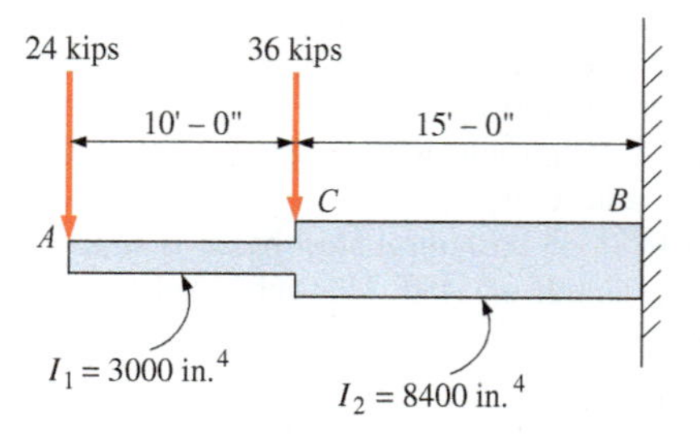

PROBLEM 15.39

15.40 Calculate the maximum deflection for the simply supported steel beam shown. Neglect the weight of the beam.

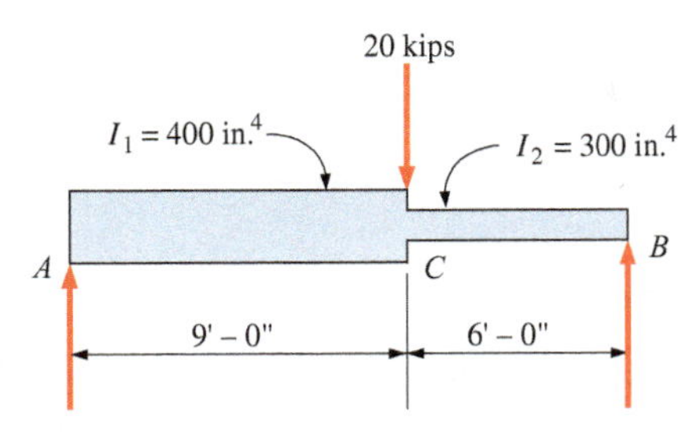

PROBLEM 15.40

15.41 A sign post is composed of standard-weight steel pipe. The top portion is 4-in. nominal diameter and the bottom portion is 5-in. nominal diameter, as shown. The weight of the sign is 100 lb and its center of gravity is located as shown. Assume simple supports between sign and post. Calculate the horizontal deflection at the top of the post due to the weight of the sign.

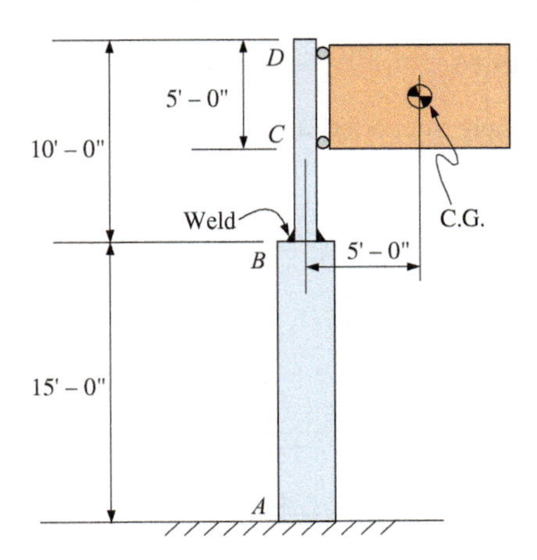

PROBLEM 15.41

15.42 Derive an expression for the maximum deflection of the cantilever beam shown.

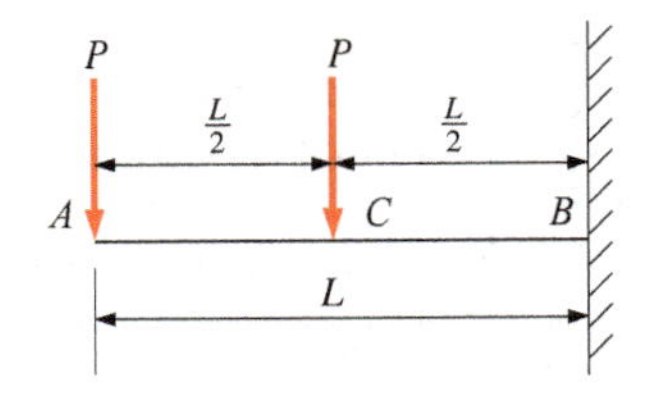

PROBLEM 15.42

15.43 Derive an expression for the maximum deflection of the simply supported beam shown.

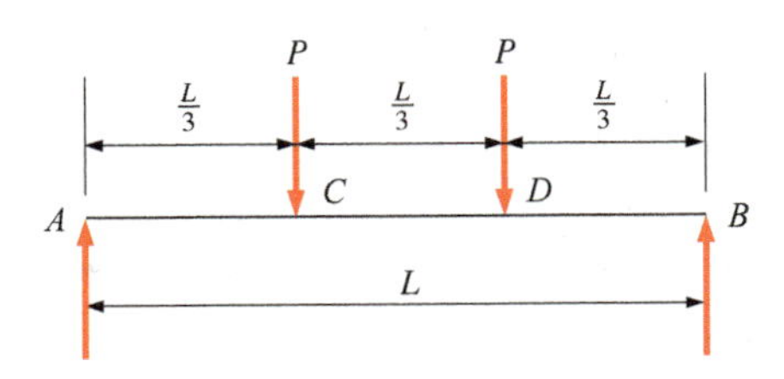

PROBLEM 15.43

Computer Problems

For the following computer problems, any appropriate software may be used. Input prompts should fully explain what is required of the user (the program should be user-friendly). The resulting output should be well labeled and self-explanatory.

15.44 Write a program that will calculate the deflection at any point on a simply supported beam subjected to a uniformly distributed load for its full length and a concentrated load at midspan. User input is to be L, E, I, w, P, and the location of the desired deflection referenced to the left support.

15.45 Rework Problem 15.44, but have the program produce a list of deflections at the tenth points of the span. Input is to be the same, except that no location of the desired deflection will be given.

15.46 Write a program that will calculate the slope of the elastic curve at the tenth points of the span for a cantilever beam subjected to a concentrated load at its free end. User input is to be E, I, L, and P.

15.47 Rework Problem 15.46, but add a uniformly distributed load w over the full length of the beam.

15.48 Develop a spreadsheet application that will calculate the maximum deflection of a simply supported beam subjected to a full-span uniformly distributed load w and a point load P at midspan. The allowable deflection is to be Span/x where x will be user input. Additionally, user input will be w, P, modulus of elasticity E, and moment of inertia I. Calculated deflection and allowable deflection are to be to 0.001 in. span lengths are to range from 10 ft to 30 ft in increments of 1 ft. Show a chart of the results that will display the relationship between calculated and allowable deflection at each value of span length. Use the spreadsheet to verify the solutions for Problems 15.9 and 15.25.

Supplemental Problems

15.49 If the elastic limit of a steel wire is 60,000 psi, compute the diameter of the smallest circle into which a No. 14 (0.080-in.-diameter) wire may be coiled without undergoing a permanent set.

15.50 Calculate the bending moment required to produce a radius of curvature of 1250 ft for a $1\frac{1}{2}$-in. square steel bar.

15.51 A 6-ft-long cantilever beam is subjected to a concentrated load of 5 kips acting at its free end. Calculate the maximum deflection at the free end if the beam is **(a)** a solid rectangular Douglas fir (S4S), 8-in.-by-12-in. beam; **(b)** a W10 × 22 structural steel wide-flange shape; **(c)** an 8-in. nominal diameter extrastrong steel pipe. Use the formula method.

15.52 A W16 × 36 structural steel wide-flange section is simply supported on a span length of 20 ft. It is subjected to a uniformly distributed load of 1.0 kips/ft and a concentrated load of 10 kips at midspan.

(a) Compute the maximum deflection using the formula method.

(b) Compute the maximum bending stress.

15.53 A simply supported W410 × 100 structural steel wide-flange beam spans a length of 9 m. It is subjected to a uniformly distributed load of 4 kN/m and two concentrated loads of 40 kN each applied at the third points of the span. Compute the maximum deflection using the formula method.

15.54 A W10 × 22 structural steel wide-flange shape is simply supported on a span of 20 ft. A superimposed uniformly distributed load of 1000 lb/ft is applied to the beam. As the beam deflects under load, it comes into contact with a third support at midspan that is $\frac{1}{2}$ in. below the level of the two outer supports. Calculate the reaction on the center support. Use the formula method.

15.55 A solid, round simply supported steel shaft is used as a beam with a span length of 700 mm. The shaft supports two concentrated loads of 3 kN each applied at the third points of the span. Calculate the required shaft diameter if its deflection must not exceed 0.20 mm. Using the computed diameter, compute the maximum bending stress and shear stress and compare with allowable stresses of 165 MPa in bending and 100 MPa in shear. Use the formula method.

15.56 Using the moment-area method, check the deflections obtained in Problem 15.51.

15.57 A 1-in.-diameter steel bar is 25 ft long and balanced at the middle in a horizontal position. Calculate the distance that the ends deflect below the middle of the bar due to the weight of the bar.

15.58 A 102-mm nominal diameter standard-weight steel pipe is used as a simple beam with a span length of 4 m. The member is loaded as shown. Calculate the deflection at midspan.

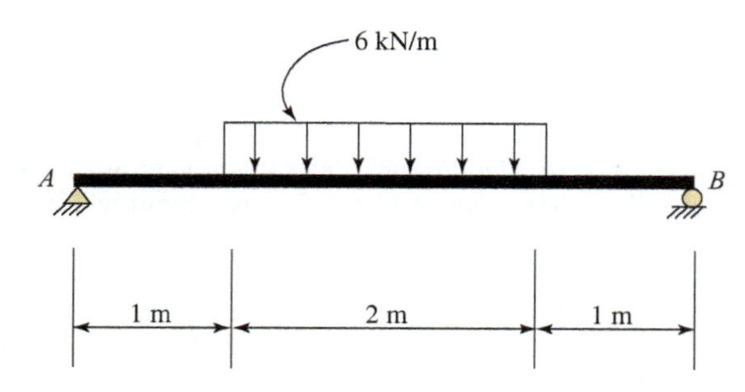

PROBLEM 15.58

15.59 Compute the maximum deflection for the aluminum cantilever beam of Problem 15.17.

15.60 An 8-in.-wide by 12-in.-deep redwood timber beam (S4S) is used as a 20-ft-long simply supported beam. Compute the concentrated load at midspan that will cause a bending stress of 1350 psi. Compute the deflection at midspan and at the quarter points.

15.61 A solid steel shaft 3 in. in diameter and 20 ft long is used as a simply supported beam subjected to a 500 lb load at midspan. Calculate the maximum deflection.

15.62 For the beam shown, draw the conventional moment diagram and the moment diagram by parts (right to left and left to right).

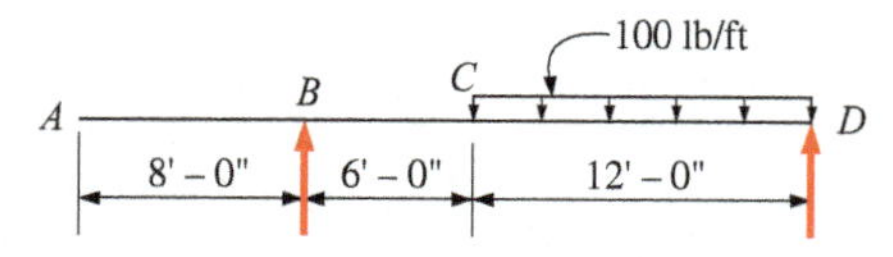

PROBLEM 15.62

15.63 Rework Problem 15.62 with concentrated loads of 5 kips added at points *A* and *C*. In addition, draw the moment diagram by parts right to left and left to right to point *C*.

15.64 A solid steel shaft 3 in. in diameter and 20 ft long is used as a simply supported beam. A concentrated load of 500 lb is located 6 ft from one end. Calculate the maximum deflection, the deflection at midspan, and the deflection under the load.

15.65 A W27 × 114 structural steel wide-flange section is loaded as shown. Calculate the slope at the free end, the deflection at the free end, and the deflection at point *C*.

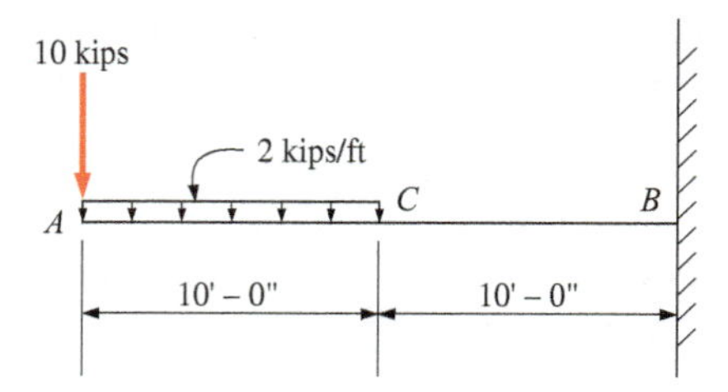

PROBLEM 15.65

15.66 A 6-in.-by-10-in. hem-fir timber beam (S4S) is loaded as shown. Calculate the deflection under the concentrated load and at midspan between supports.

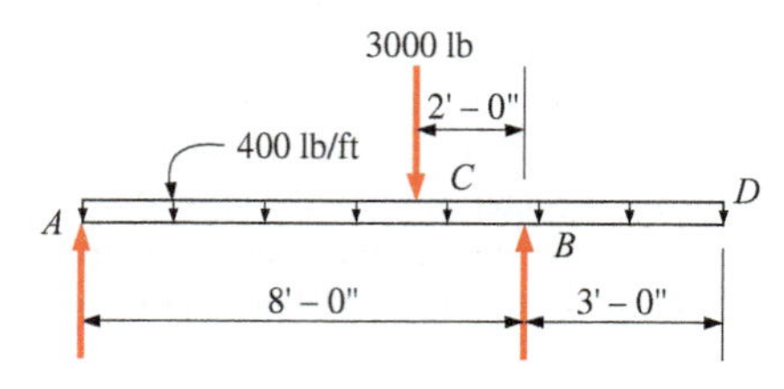

PROBLEM 15.66

15.67 A simply supported structural steel wide-flange section is used as a beam and supports loads as shown. Compute the deflection at the center of the span and at the point of the concentrated load. Use the moment-area method. Use $I = 861 \times 10^6\ \text{mm}^4$.

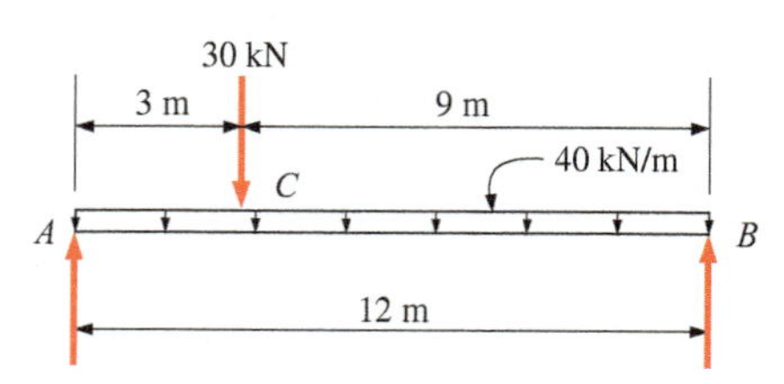

PROBLEM 15.67

15.68 Calculate the maximum permissible span length for a 3-in.-diameter solid steel shaft used as a cantilever beam. The maximum deflection due to the weight of the shaft may not exceed 0.2 in.

15.69 A W10 × 33 structural steel wide-flange section 10 ft long is used as a cantilever beam supporting a uniformly distributed load of 800 lb/ft.

(a) Calculate the deflection at the free end.

(b) What superimposed uniformly distributed load can be placed on the beam if the allowable deflection is 0.25 in.?

15.70 A W10 × 68 structural steel wide-flange section supports loads as shown. What are the maximum allowable loads *P* that the beam can support if the upward deflection under the 10-kip load is limited to 0.05 in.?

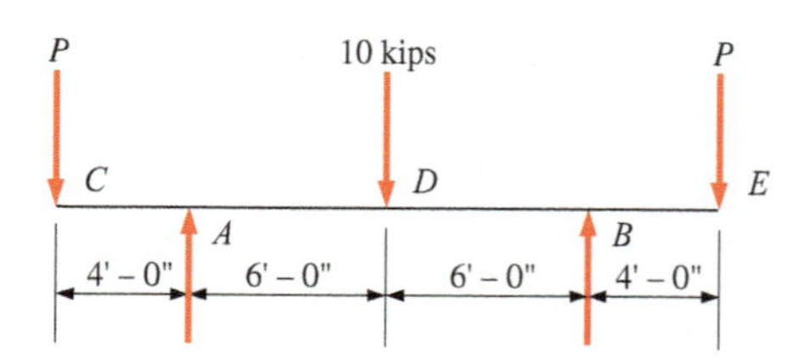

PROBLEM 15.70

15.71 Determine the deflection at point C and midway between the supports for the W36 × 194 structural steel beam shown.

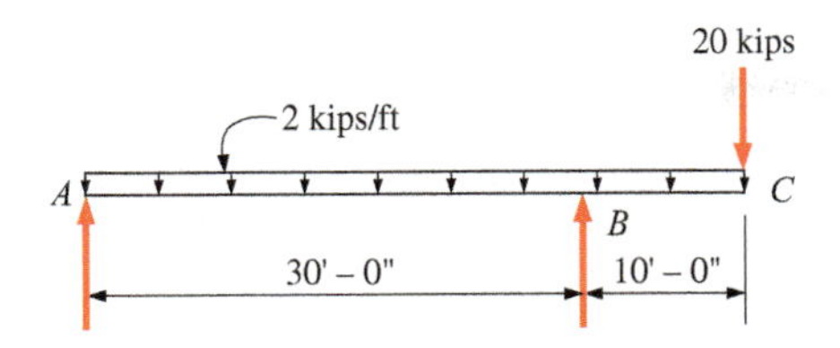

PROBLEM 15.71

15.72 Calculate the deflection midway between the reactions and at the free end for the steel beam shown.

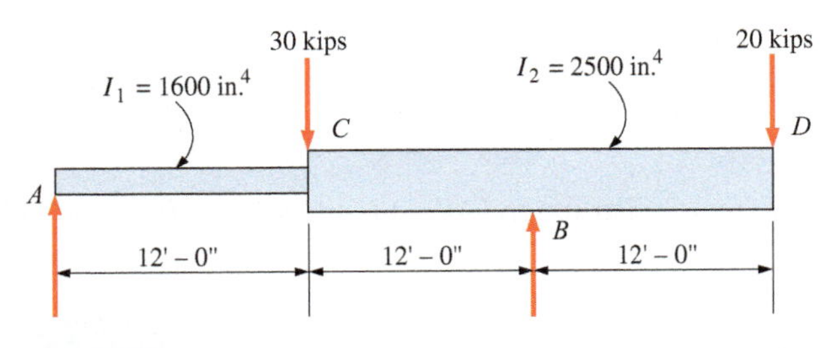

PROBLEM 15.72

15.73 Derive an expression for the maximum deflection of the beam shown.

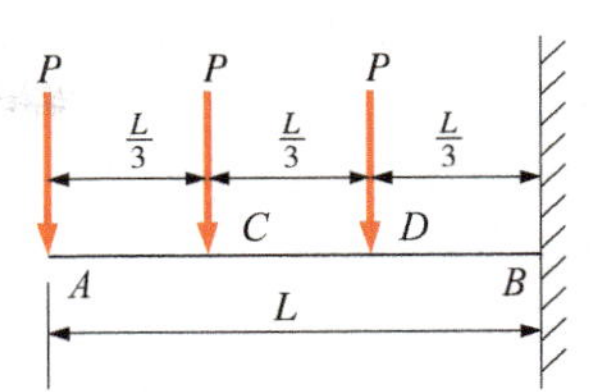

PROBLEM 15.73

15.74 Derive an expression for the maximum deflection of the beam shown.

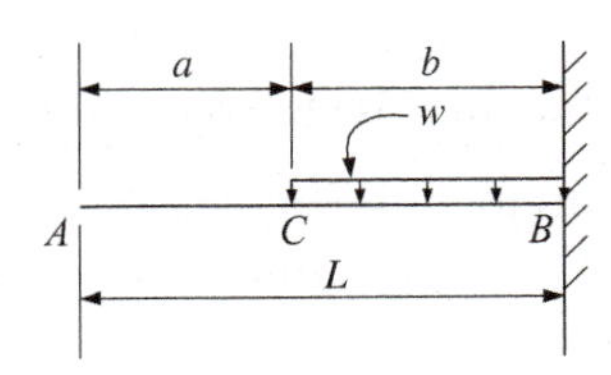

PROBLEM 15.74

CHAPTER SIXTEEN

DESIGN OF BEAMS

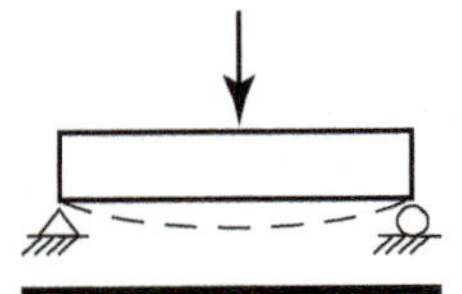

LEARNING OBJECTIVES

Upon completion of this chapter, readers will be able to:

- Design beams to resist bending and shear stresses.
- Design steel beams in accordance with the American Institute of Steel Construction (AISC) *Manual of Steel Construction*.
- Design solid rectangular timber beams based on the member properties in the appendices.

16.1 THE DESIGN PROCESS

All structural design is governed by the basic rule that the strength required of the member, element, or structure to resist applied loads or forces must be less than or equal to the strength furnished. This may be written simply as

$$\text{strength required} \leq \text{strength furnished}$$

For beams, the size and shape of the cross section must be selected to provide, principally, sufficient moment and shear strength and sufficient stiffness to limit deflection.

It is the function of building codes and standards to provide guidelines for the designer as to how strengths should be determined, what the minimum safety factors should be, and as to the types and magnitudes of loads and forces that should be considered. Structural standards currently include both allowable stress design (ASD), also called allowable strength design, and the more recent load and resistance factor design (LRFD) approaches. There are ASD and LRFD standards written for both structural steel design and timber (wood) design. Both methods are currently acceptable. Both result in adequate structural strength. For the individual materials, both methods result in approximately the same end safety factors. The philosophies of the two approaches have some differences and some similarities. The examples presented in this chapter reflect the ASD approach.

Allowable stresses and safety factors have been previously discussed (see Sections 10.6 and 10.7). Recall that allowable stress was defined as

$$\text{allowable stress} = \frac{\text{failure stress}}{\text{safety factor}}$$

For beam design, in which we are selecting size and type of cross section, we may then use this allowable stress as an upper limit for the stress that will develop when we consider the span of the beam, the applied loads, and the support conditions.

We could also write the previous relationship in a slightly different way to reflect strength rather than stress:

$$\text{allowable strength} = \frac{\text{failure strength}}{\text{safety factor}}$$

For bending moment and shear considerations, ASD design for timber uses the allowable stress approach while ASD design for structural steel uses the allowable strength approach. The treatment of deflection is the same for both materials and follows the principles discussed in the previous chapter.

Otherwise, our discussion of beam design involves no new principles. It focuses on the normally used design techniques for structural steel and timber beams. It will be seen that beam design is sometimes a trial-and-error process. The following general procedure is suggested as a step-by-step approach:

1. Establish the location, magnitude, and type of superimposed loads along with the span length and support conditions (both vertical and horizontal) of the beam. In addition, establish all design constraints such as deflection limits, allowable strengths or allowable stresses, and any physical limitations of the beam due to the material of which it is made and/or its dimensions.
2. Draw a load diagram and calculate the reactions.
3. Determine the maximum shear and the maximum bending moment. Use the beam diagrams and formulas

of Appendix H when possible. Otherwise, draw complete shear and moment diagrams. (Use Appendix H with caution. Compare beam types, loading conditions, and types of support very carefully.)

4. Select an economical (least weight, usually) trial cross section that will have sufficient strength to resist the maximum absolute value of applied moment and sufficient stiffness to limit deflection to a specified value, if any.
5. Using the selected section, add the beam weight to the applied loads and revise steps 2 and 3. Then check to ensure that the beam is still satisfactory for moment and deflection.
6. Check shear. Shear is rarely critical for steel beams but may be critical in timber beams. Revise the selection if necessary. The final selection (with its weight included) must provide sufficient strength to resist moment and shear and provide sufficient stiffness to limit deflection.

This procedure is illustrated with a flowchart in Figure 16.1.

This design procedure can be simplified if an estimated beam weight is included in step 1. Various rules of thumb exist as guidelines for timber and steel beam weight estimates. Suggested values range from 3% to 10% of the total load. The use of such rules of thumb should be based on judgment developed through practical experience. In general, assuming a beam loaded to its bending capacity, the longer the span, the greater the proportion of the total load that is due to the weight of the beam. At this point, however, we suggest that the recommended steps in the preceding design procedure be followed.

Another consideration influencing the design process is lateral support of a beam. If a vertically loaded beam does not have adequate lateral support, its compression flange (or compression side, in the case of a rectangular cross section) will have a tendency to buckle or deflect laterally where it is free to do so. If we consider a simply supported beam subjected to vertically downward loads, the bending stresses developed are compressive above the neutral axis and tensile below the neutral axis. The tensile stresses tend

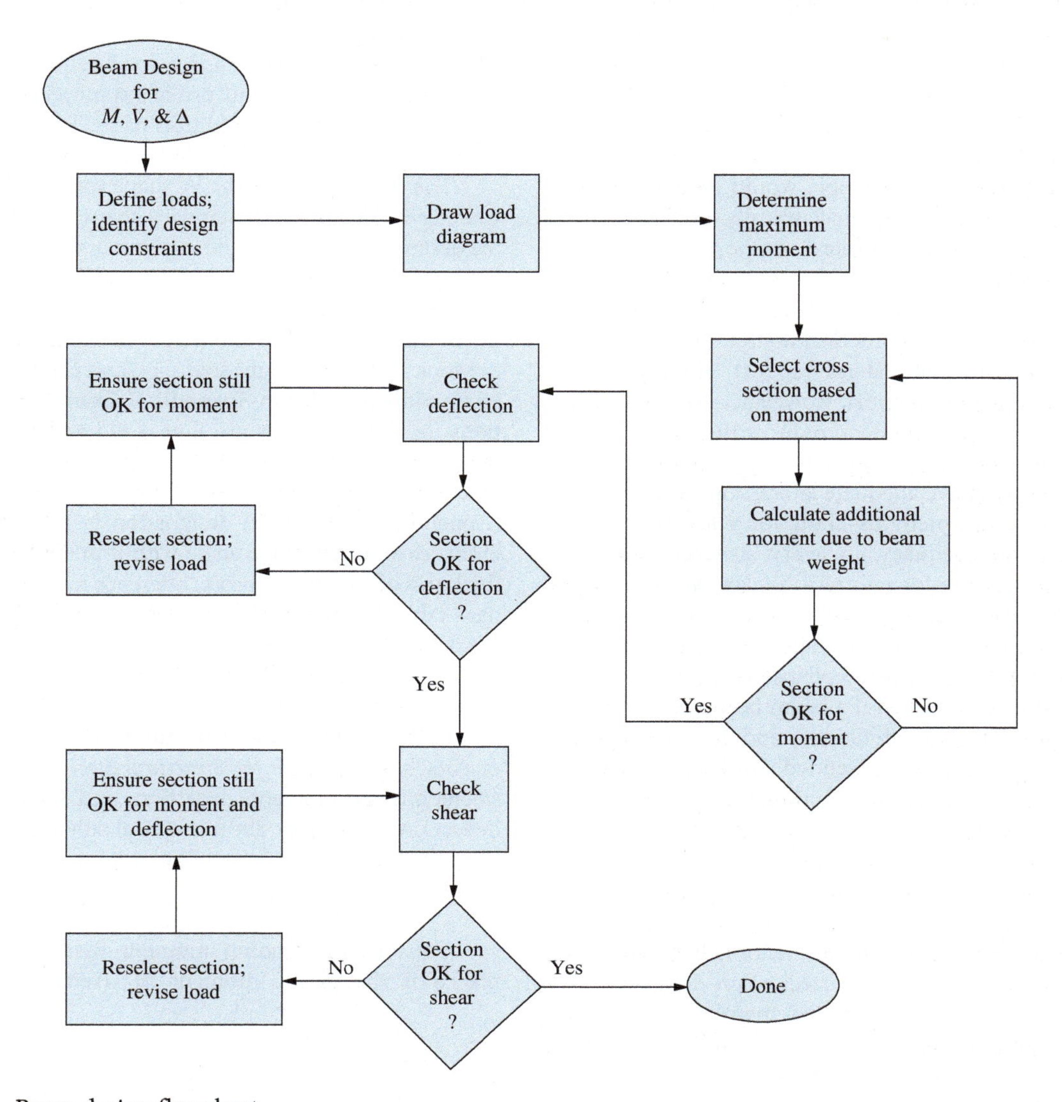

FIGURE 16.1 Beam design flowchart.

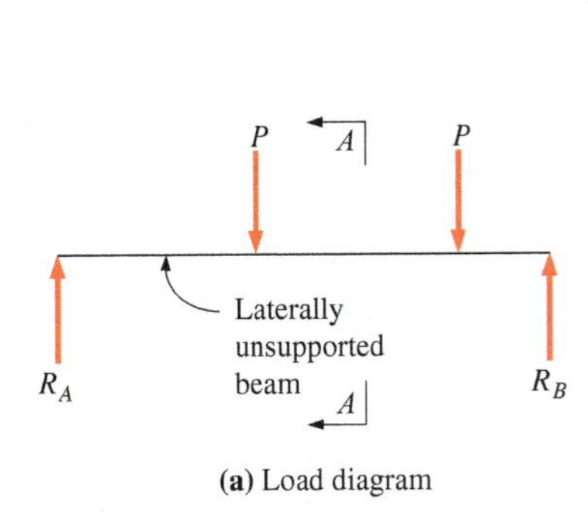

(a) Load diagram

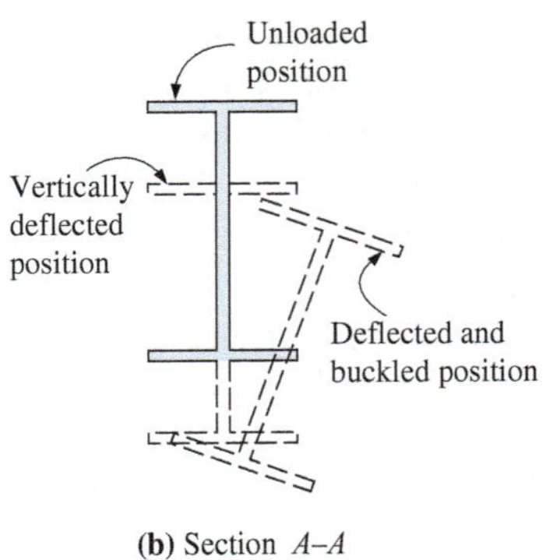

(b) Section A–A

FIGURE 16.2 Beam deflection and lateral buckling.

to hold the beam in a straight line between the supports, whereas the compressive stresses tend to deflect the beam in a lateral direction. This behavior is similar to that of a slender column subjected to axial compressive loads. The opposing tendencies of the top and bottom portions of the beam result in a torsional buckling, which occurs concurrently with a vertical deflection resulting from the applied loads. Figure 16.2 shows a laterally unsupported beam that has deflected vertically and buckled laterally.

The lateral buckling of the compression flange can be prevented by proper beam design or by providing physical lateral support for the compression flange or the compression side of the beam. An example would be a steel beam with its compression flange mechanically anchored to a reinforced concrete floor slab that it supports. Another example is a wood floor joist that has its compression edge laterally supported throughout its length by the floor deck to which it is fastened. As a practical matter, most beams are constructed with adequate lateral support. Should a beam have inadequate lateral support, however, a special design would be required that, in effect, reduces the load-carrying capacity of the beam. For purposes of our discussion, we assume that all beams have adequate lateral support.

Although design methods allow for reasonable assurance of structural adequacy, in every structural member there is a finite chance for failure to occur. An unforeseen overload could lead to a beam failure. An undetected material deficiency could lead to a collapse. Note that failure is not to be interpreted only as a collapse or rupture. Failure may be said to have occurred if a beam becomes unserviceable due to some structural deficiency and does not perform the function for which it was intended. For example, a beam or series of beams may not be functional if excessive vibrations result from some form of applied load.

16.2 DESIGN OF STEEL BEAMS

Since the material included in this section is of an introductory nature, the treatment of steel beam design focuses on conventional standard rolled steel members and simple geometric shapes that may be used as bending members. For a more detailed treatment of beam design, textbooks on structural design should be consulted.

Many varieties of standard hot-rolled structural steel shapes are available for use as beams. Some of the commonly available shapes shown in Appendices A through D may be used individually as beams, or they may be combined with other shapes and/or plates as built-up beams. The most commonly used shape for beams is the W shape. The maximum nominal depth commonly available is approximately 36 in., although deeper sections (up to about 44 in.) do exist. When the span length and/or load requirements become excessive for the economic use of a standard W shape, it may be strengthened by the addition of steel plates, or a section built up entirely of plates (called a *plate girder*) can be used.

The designation for the W shape was briefly discussed in Chapter 9. For American Standard channels (see Appendix C), the designation is interpreted in a similar manner. For instance, in the designation C15 × 50, the initial letter (C) indicates that the shape is a channel, 15 is the actual depth of the channel in inches, and 50 is the weight of the member in pounds per foot. Standard-weight steel pipe and extrastrong steel pipe sections are included in Appendix B. Examples of the designations for these sections are Pipe 4 Std and Pipe 4 X-Strong, which designate a standard-weight and an extrastrong-weight pipe section of 4 in. nominal outside diameter. For angles (Appendix D), a typical designation is L8 × 6 × 1, which indicates an unequal leg angle with 8-in. and 6-in. legs and a thickness of 1 in. In each case, reference must be made to standard tables to obtain most properties. Appendices A through D contain only a few of the many shapes and sizes available. For a more complete listing, refer to manufacturers' literature and/or the AISC *Manual of Steel Construction.*[1]

In the selection of a structural steel W shape we make use of AISC's *Specification for Structural Steel Buildings.* The considerations are moment, deflection, and shear. Moment and deflection are primary. Shear is critical only when loads are very large and spans are very short. We make use of the relationship

$$\text{strength required} \leq \text{strength furnished}$$

Considering bending moment first, the strength furnished is the plastic moment M_p (refer to Section 14.8)

[1]American Institute of Steel Construction, Inc., *Manual of Steel Construction*, 14th ed. (Chicago: AISC, 2011).

divided by a safety factor. The safety factor is denoted by Ω_b and is equal to 1.67. The strength required is the applied bending moment calculated either using formulas or using shear and moment diagrams. We will denote this moment as M_a. As a limit, for design, we may then write

$$M_a = \frac{M_p}{\Omega_b} = \frac{F_y Z_x}{1.67}$$

where Z_x is the plastic section modulus with respect to the x centroidal (strong) axis of the W-shape cross section. The plastic section modulus is a tabulated quantity. Therefore, the moment design equation for minimum required Z_x is

$$\text{required } Z_x = \frac{1.67\, M_a}{F_y}$$

Considering deflection, the relationship is

calculated deflection $\leq$ deflection limit

or

$$\Delta \leq \Delta_{\text{limit}}$$

We may either check the deflection, as in the expression just shown, or determine a required minimum moment of inertia I. The deflection limit is commonly a fraction of the span length, such as span/400. For example, for a single-span, simply supported beam supporting a uniformly distributed load the midspan deflection Δ is (refer to Case 1 in Appendix H):

$$\Delta = \frac{5wL^4}{384EI}$$

For design, as a limit, $\Delta = \Delta_{\text{limit}}$, then solving for a minimum required I we have:

$$\text{required } I = \frac{5wL^4}{384E\Delta_{\text{limit}}}$$

Although shear rarely governs, it should be checked. The average web shear approach (see Section 14.6) is used and for almost all steel W-shape beams the strength relationship is written

$$V_a \leq \frac{0.6F_y A_w}{\Omega_v}$$

where V_a = the applied shear force
F_y = the yield stress of the steel
A_w = the full-depth web area $(d \times t_w)$
Ω_v = the shear safety factor and is equal to 1.50

The preceding equation can be simplified to

$$V_a \leq 0.4F_y d t_w$$

EXAMPLE 16.1 Select the most economical (lightest) W shape to support a superimposed uniformly distributed load of 4 kips/ft on a simply supported span of 25 ft, as shown in Figure 16.3. Assume the yield stress of the steel to be 50 ksi. The deflection limit for total load is span/360.

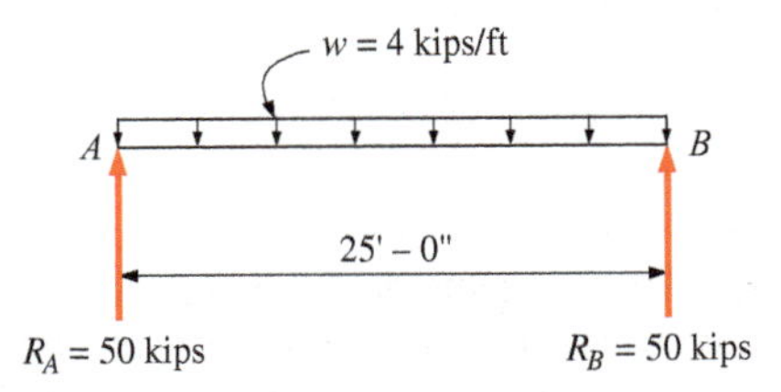

FIGURE 16.3 Load diagram and reactions.

Solution The load diagram with the reactions is shown in Figure 16.3. Since the beam is simply supported and carries a uniformly distributed load, maximum shear and moment can be computed using the shear and moment equations of Appendix H. The maximum moment is calculated from

$$M_a = \frac{wL^2}{8} = \frac{(4\text{ k/ft})(25\text{ ft})^2}{8} = 313\text{ k-ft}$$

This value represents the maximum moment due to the superimposed loads and does not include the weight of the beam. The required strong-axis plastic section modulus is then obtained from

$$\text{required } Z_x = \frac{1.67\, M_a}{F_y} = \frac{1.67(313\text{ k-ft})(12\text{ in./ft})}{50\text{ ksi}} = 125.5\text{ in.}^3$$

Appendix I, Table I.1, is set up to aid in the selection of the lightest W shape that provides a given Z_x. Shapes are listed in the order of decreasing Z_x values. Some of the shapes are grouped and headed with a shape shown in boldface type. The shape at the top of each group is the lightest shape in the group. We will therefore try a W21 × 62 (Z_x = 144 in.^{3}).

Next, add the effect of the beam weight (0.062 k/ft):

$$\text{additional } M_a = \frac{wL^2}{8} = \frac{(0.062\text{ k/ft})(25\text{ ft})^2}{8} = 4.84\text{ k-ft}$$

$$\text{total } M_a = 313\text{ k-ft} + 4.84\text{ k-ft} = 318\text{ k-ft}$$

$$\text{required } Z_x = \frac{1.67\ M_a}{F_y} = \frac{1.67(318\text{ k-ft})(12\text{ in./ft})}{50\text{ ksi}} = 127.5\text{ in.}^3$$

$$127.5\text{ in.}^3 < 144\text{ in.}^3 \qquad \textbf{(O.K.)}$$

Next we will calculate the required moment of inertia based on the maximum allowable deflection:

$$\Delta_{\text{limit}} = \frac{\text{span}}{360} = \frac{25\text{ ft}(12\text{ in./ft})}{360} = 0.833\text{ in.}$$

The total load is 4 k/ft + 0.062 k/ft = 4.06 k/ft:

$$\text{required } I = \frac{5wL^4}{384\ E\Delta_{\text{limit}}} = \frac{5(4.06\text{ k/ft})(25\text{ ft})^4(1728\text{ in.}^3/\text{ft}^3)}{384(30{,}000\text{ ksi})(0.833\text{ in.})} = 1428\text{ in.}^4$$

I_x for the W21 × 62 is 1330 in.^{4} Since 1330 in.^{4} < 1428 in.^{4}, the shape is N.G. for deflection.

Use Appendix J, Table J.1. This table lists W shapes in order of decreasing I_x. Entering the table with a required I_x of 1428 in.^{4}, we select a W21 × 73 (I_x = 1600 in.^{4}). Note also that Z_x for the W21 × 73 is 172 in.^{3} and exceeds the required Z_x of 127.5 in.^{3} The increase of 11 lb/ft in the beam weight is insignificant.

Last, check shear. For the W21 × 73, d = 21.20 in. and t_w = 0.455 in. The maximum applied shear force, including the beam weight, is

$$V_a = \frac{wL}{2} = \frac{(4.07\text{ k/ft})(25\text{ ft})}{2} = 50.9\text{ k}$$

The shear capacity is

$$0.4F_y d t_w = 0.4(50\text{ ksi})(21.20\text{ in.})(0.455\text{ in.}) = 192.9\text{ k}$$

$$50.9\text{ k} < 192.9\text{ k} \qquad \textbf{(O.K.)}$$

The W21 × 73 is satisfactory for moment, deflection, and shear.

Use a W21 × 73.

EXAMPLE 16.2 A 6-in.-thick concrete floor is supported on steel beams and girders. The layout for a typical building interior bay is shown in the plan view of Figure 16.4. The floor is to be designed for a live load of 120 psf. The yield strength of the steel is 50 ksi. The live load deflection limit for each member is span/360. Select the lightest W shapes for the beams and girders.

Solution This structural steel beam-and-girder floor system is typical for a commercial or industrial building. Although some beams are supported by the girders and some by the columns, the span, load, and support conditions are the same. Therefore, the design will be the same for all beams.

Design of Beams (B1) As shown in Figure 16.4, the span length of the beam is 40 ft. In a floor system of this type, it is common practice to assume that the beam is simply supported. The end connections of the beam determine whether it is simply supported or not. It is assumed at this point that the end connections will be so designed that the loaded beam will have a freedom of rotation at each end, which, in effect, creates a simple beam.

The loading on beam B1 includes the live load and the dead load. The dead load consists of the weight of the concrete floor and the weight of the beam itself. The live load is given as 150 psf. The weight of the reinforced concrete may be taken as 150 pcf. Therefore, the weight of a 6-in.-thick slab per square foot of floor can be computed as

$$(150\text{ pcf})\left(\frac{6\text{ in.}}{12\text{ in./ft}}\right) = 75\text{ psf}$$

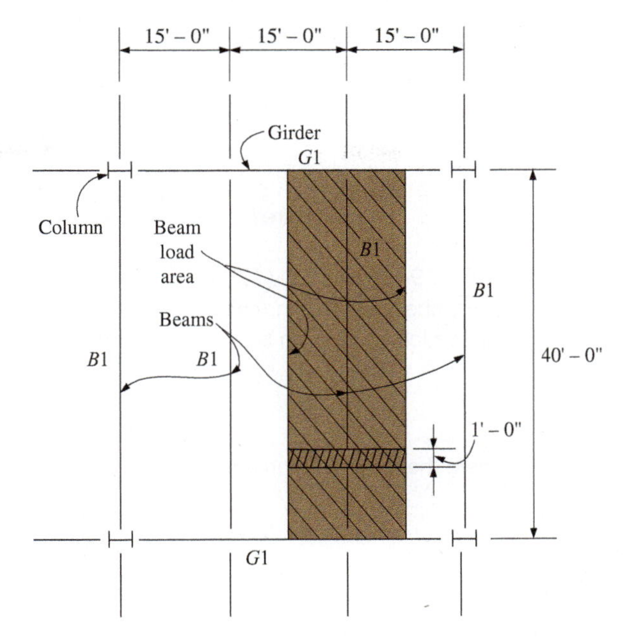

FIGURE 16.4 Framing plan.

The weight of the beam will be neglected for now, since the beam section is unknown. It will be considered, however, after an initial beam selection has been made.

Since the beams are spaced 15 ft on centers, each beam will support a 15-ft width of floor. This load area is cross-hatched in Figure 16.4. Calculate the design load per linear foot of beam (double cross-hatched area):

$$\text{live load} = (120 \text{ psf})(15 \text{ ft}) = 1800 \text{ lb/ft}$$

$$\text{slab dead load} = (75 \text{ psf})(15 \text{ ft}) = 1125 \text{ lb/ft}$$

from which

$$\text{total design load} = 1800 + 1125 = 2930 \text{ lb/ft}$$
$$= 2.93 \text{ k/ft}$$

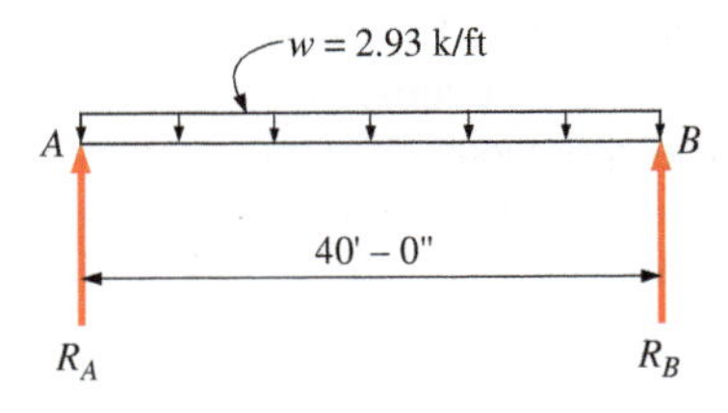

FIGURE 16.5 Beam load diagram.

The beam is isolated as a free body and shown in Figure 16.5. Since the beam is simply supported and supports a uniformly distributed load, maximum shear and moment can be computed using the shear and moment equations of Appendix H. The maximum moment is computed from

$$M_a = \frac{wL^2}{8} = \frac{(2.93 \text{ k/ft})(40 \text{ ft})^2}{8} = 586 \text{ k-ft}$$

from which

$$\text{required } Z_x = \frac{1.67\ M_a}{F_y} = \frac{1.67(586 \text{ k-ft})(12 \text{ in./ft})}{50 \text{ ksi}} = 235 \text{ in.}^3$$

From Appendix I, Table I.1, try a W27 × 94 ($Z_x = 278$ in.^{3}). Add the effect of beam weight (0.094 k/ft):

$$\text{additional } M_a = \frac{wL^2}{8} = \frac{(0.094\text{ k/ft})(40\text{ ft})^2}{8} = 18.80\text{ k-ft}$$

$$\text{total } M_a = 586\text{ k-ft} + 18.80\text{ k-ft} = 605\text{ k-ft}$$

$$\text{new required } Z_x = \frac{1.67\ M_a}{F_y} = \frac{1.67(605\text{ k-ft})(12\text{ in./ft})}{50\text{ ksi}} = 242\text{ in.}^3$$

The W27 × 94 is still O.K. (278 in.^{3} > 242 in.^{3}).

Next, check deflection (calculate the required I_x). The live load has been calculated as 1.8 kips/ft and the live load deflection limit is calculated from

$$\Delta_{\text{limit}} = \frac{\text{span}}{360} = \frac{40\text{ ft}(12\text{ in./ft})}{360} = 1.33\text{ in.}$$

From here the required moment of inertia may be calculated:

$$\text{required } I = \frac{5w_{LL}L^4}{384\ E\Delta_{\text{limit}}} = \frac{5(1.8\text{ k/ft})(40\text{ ft})^4(1728\text{ in.}^3/\text{ft}^3)}{384(30{,}000\text{ ksi})(1.33\text{ in.})} = 2600\text{ in.}^4$$

I_x for the W27 × 94 is 3270 in.^{4} Since 3270 in.^{4} > 2600 in.^{4}, the W27 × 94 is O.K. for deflection as well as for moment.

Last, check shear: For the W27 × 94, from Appendix A, $d = 26.90$ in. and $t_w = 0.490$ in. The maximum shear is the reaction (due to total load):

$$V_a = \frac{wL}{2} = \frac{(2.93\text{ k/ft} + 0.094\text{ k/ft})(40\text{ ft})}{2} = 60.5\text{ k}$$

The shear capacity is

$$0.4F_y dt_w = 0.4(50\text{ ksi})(26.90\text{ in.})(0.490\text{ in.}) = 264\text{ k}$$

$$60.5\text{ k} < 264\text{ k} \qquad \textbf{(O.K.)}$$

Use a W27 × 94.

Design of Girders (G1)

The span length of the girder shown in Figure 16.4 is taken as the center-to-center distance of the supporting members that, in this example, is 45 ft. The loading to G1 is symmetrical, consisting of two concentrated loads and a uniformly distributed load due to the weight of the member. The small portion of slab supported directly on the girder can be neglected, since the entire floor load was taken by the beams and is therefore transferred to the girder as a beam reaction. Since this is an interior bay, the girder will support beams framing into it from each side. The reaction from B1 is 60.5 kips. Therefore, each concentrated load supported by the girder is equal to 2(60.5) = 121.0 kips.

The girder load diagram is shown in Figure 16.6. Only the superimposed loads are included. Note that the beam on the column line does not enter into the problem since it frames directly into the column, transferring its load to the column without physically touching the girder.

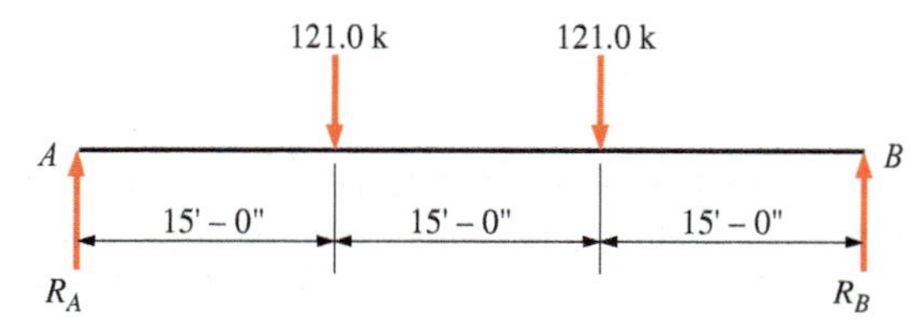

FIGURE 16.6 Girder load diagram.

Since the beam is simply supported and symmetrically loaded, maximum shear and moment can be computed using the equations in Appendix H. The maximum moment can be calculated from

$$M = Pa = (121.0\text{ k})(15\text{ ft}) = 1815\text{ k-ft}$$

from which

$$\text{required } Z_x = \frac{1.67\ M_a}{F_y} = \frac{1.67(1815\text{ k-ft})(12\text{ in./ft})}{50\text{ ksi}} = 727\text{ in.}^3$$

From Appendix I, Table I.1, try a W36 × 194 ($Z_x = 767\text{ in.}^3$). Add the effect of the weight of the girder (0.194 k/ft):

$$\text{additional } M_a = \frac{wL^2}{8} = \frac{0.194\text{ k/ft}(45\text{ ft})^2}{8} = 49.1\text{ k-ft}$$

$$\text{total } M_a = 1815\text{ k-ft} + 49.1\text{ k-ft} = 1864\text{ k-ft}$$

$$\text{new required } Z_x = \frac{1.67\ M_a}{F_y} = \frac{1.67(1864\text{ k-ft})(12\text{ in./ft})}{50\text{ ksi}} = 747\text{ in.}^3$$

The W36 × 194 is O.K. for moment ($767\text{ in.}^3 > 747\text{ in.}^3$). Next, calculate the required I_x. The deflection limit specified is due to the live load:

$$\Delta_{\text{limit}} = \frac{\text{span}}{360} = \frac{45\text{ ft}(12\text{ in./ft})}{360} = 1.50\text{ in.}$$

Determine the live load portion of the applied loads on the girder (this will be twice the live load reaction of beam B1):

$$2\left(\frac{w_{LL}L}{2}\right) = 2\left[\frac{(1.80\ \text{k/ft})(40\text{ ft})}{2}\right] = 72\text{ k}$$

This loading is shown in Figure 16.7. Using Case 7 from Appendix H, the maximum deflection at the center of the girder is

$$\Delta = \frac{Pa}{24EI}(3L^2 - 4a^2)$$

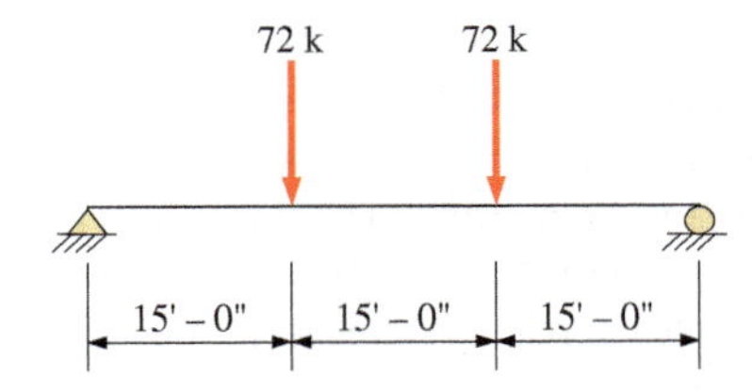

FIGURE 16.7 Load diagram.

Therefore,

$$\text{required } I = \frac{Pa}{24\ E\Delta_{\text{limit}}}(3L^2 - 4a^2)$$

$$= \frac{(72\text{ k})(15\text{ ft})}{24(30{,}000\text{ ksi})(1.50\text{ in.})}\left[3(45\text{ ft})^2 - 4(15\text{ ft})^2\right](1728\text{ in.}^3/\text{ft}^3) = 8940\text{ in.}^4$$

I_x for a W36 × 194 is 12,100 in.^{4} Since 12,000 in.^{4} > 8940 in.^{4}, the W36 × 194 is O.K. for deflection as well as moment.

Last, check shear: For the W36 × 194, $d = 36.50$ in. and $t_w = 0.765$ in. from Appendix A. The maximum shear is the reaction due to total applied load plus the weight of the girder (see Figure 16.5):

$$V_a = \frac{2P}{2} + \frac{wL}{2} = \frac{2(121.0\text{ k})}{2} + \frac{(0.194\text{ k/ft})(45\text{ ft})}{2} = 125.4\text{ k}$$

The shear capacity is

$$0.4F_y dt_w = 0.4(50\text{ ksi})(36.50\text{ in.})(0.765\text{ in.}) = 558\text{ k}$$

$$125.4\text{ k} < 558\text{ k}$$ **(O.K.)**

Use a W36 × 194.

16.3 DESIGN OF TIMBER BEAMS

Timber beam applications can be found predominantly in the residential and small commercial building design and construction field and to a lesser extent in the bridge (particularly temporary bridge) design and construction field. The design procedure, as itemized in Section 16.1, is applicable to timber beams as well as to steel beams.

A timber beam, like a steel beam, is designed for moment, shear, and deflection. Several other elements must be considered in a complete design for timber beams. One of the important considerations is compression perpendicular to the grain. As shown in Figure 16.8, a load (force) applied perpendicular to the grain, such as that which occurs under the ends of a beam or under a column supported by a beam, induces a localized compressive or bearing stress.

As shown in Appendix F, the allowable compressive stress perpendicular to the grain is generally considerably less than the allowable compressive stress parallel to the grain because wood, being an aggregate of cells (or fibers) running primarily in one direction, is not an isotropic material. That is, the strength properties of a piece of wood are not the same in all directions. Wood is strongest when loaded to induce a stress parallel to the grain, either in tension or compression.

Although the solid rectangular member shown in Figure 16.9a is the most common section used when designing in wood, various shapes of built-up timber members, such as those shown in Figure 16.9b through d, may be used. All these shapes (except the solid rectangular shape) are built up of sawn timber elements connected with various types of fasteners. Another type of built-up shape is the structural glued laminated timber member (commonly called *glulam member*). Glulam members are engineered products composed of assemblies of suitably selected and prepared wood laminations bonded together with adhesives, as shown in Figure 16.10.

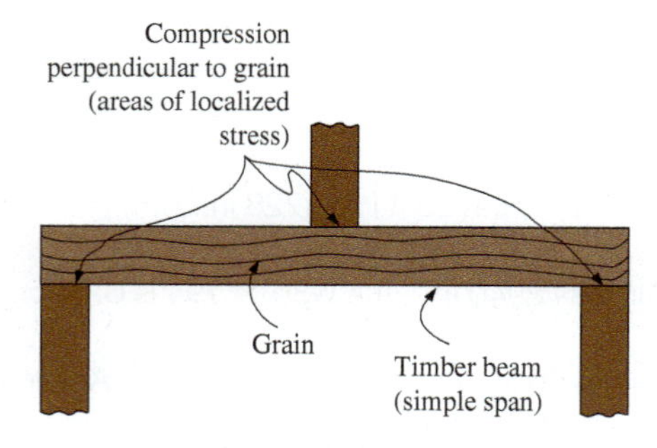

FIGURE 16.8 Compression perpendicular to grain.

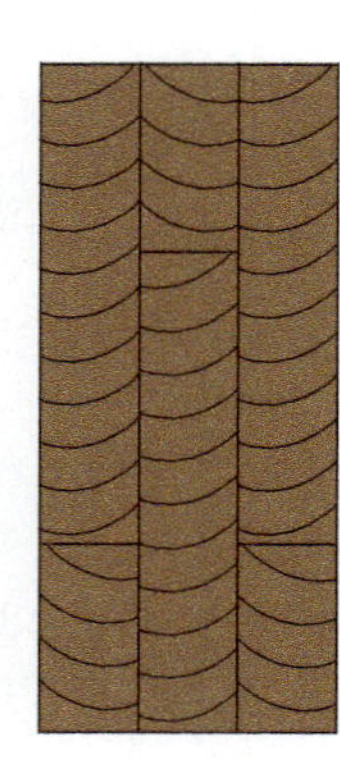

FIGURE 16.10 Glued laminated beams.

Other types of engineered wood products make use of long, thin strands of wood that are structurally bonded together in rectangular cross sections, and some of these products are then used to form other composite products such as I-joists. Since only selected materials are used in the engineering wood products, the structural properties (allowable stresses, modulus of elasticity, etc.) are significantly higher than for the original wood species.

The selection of timber beams (ASD approach) makes use of allowable stresses for bending, shear, and other effects. These values will vary with the multitude of different wood species, grades, sizes, and with the conditions of use. Our discussion is based on the *National Design Specification for Wood Construction* of the American Forest and Paper Association (AF&PA).[2]

In the design of timber members, we again make use of the relationship from Section 16.1:

$$\text{strength required} \leq \text{strength furnished}$$

For moment, shear, and bearing, the left side of the equation will be treated as before, a load diagram along with shear and moment diagrams (or formulas for simple cases) will allow the determination of the appropriate forces and moments (required strengths).

Moment strength furnished, the right side of the equation, will be determined using the flexure formula (see Section 14.2). Equation (14.7) was previously developed:

$$\text{moment strength} = F_b S_x$$

therefore, as a limit,

$$F_v S_x = M_a$$

[2]American Forest & Paper Association, *National Design Specification for Wood Construction* (ANSI/AF&PA—2005) (Washington, DC: AF&PA, 2005).

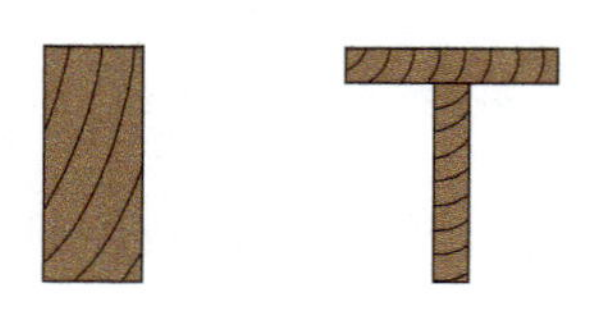

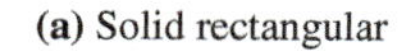

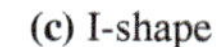

(a) Solid rectangular **(b)** Tee shape **(c)** I-shape **(d)** Box beam

FIGURE 16.9 Timber shapes.

and the design expression for moment is:

$$\text{required } S_x = \frac{M_a}{F_b}$$

where S_x = section modulus referenced to the strong (x) axis.
M_a = the applied moment
F_b = allowable bending stress

The deflection consideration will be handled just as it was with steel beams. Either a required moment of inertia can be calculated using the deflection limit, or a beam can first be selected and the deflection checked against the limit.

Shear stress in structural members that was discussed in Section 14.6 and Equation (14.12), which applied to rectangular sections, may be rewritten for solving a required area:

$$\text{required } A = \frac{1.5V}{F_v}$$

where A = rectangular cross-sectional area
V = the applied shear force
F_v = allowable shear stress

Alternatively, the shear stress may be calculated and compared with the allowable shear stress.

Bearing strength perpendicular to the grain is a direct stress application and is calculated from $f_{cp(\text{all})}A_b$, where $f_{cp(\text{all})}$ is the allowable bearing stress perpendicular to the grain and A_b is the bearing area. Alternatively, the bearing stress may be calculated and compared with the allowable stress.

Appendix F contains some representative values of design stresses and modulus of elasticity values for timber. It is intended solely as a resource to accompany the examples and problems of this text. The tabulated values do not reflect the effect of cross-sectional size. Various adjustments to the allowable stresses due to conditions of use such as inadequate lateral support, load duration, and wet conditions are applicable to these tabular values, but are beyond the scope of this text. Those who require more detailed information are referred to the AF&PA Specification as well as other publications that treat timber design in great detail.[3, 4]

Since the material included in this section is intended solely as an introduction, our treatment of timber beam design is limited to solid rectangular timber members. The design of built-up timber sections is largely a matter of trial and error. This process is generally accomplished using rules of thumb and approximations to create a desired cross section and then performing an analysis to check that allowable stresses will not be exceeded.

The design of timber members should be based on sizes that are readily available. These will usually be the dressed sizes, but they may be the nominal sizes. As shown in Appendix E, the sections are customarily specified in terms of nominal sizes, which are the approximate sizes of rough, green, undressed pieces. The dressed size is the actual dimension of the finished product after surfacing. The dimensions and other properties of the dressed timber (S4S) are also provided in Appendix E. The S4S designation indicates that the member has been surfaced, or planed (for the purpose of obtaining smoothness of surface and uniformity of size), on all four sides. This table may be used for both analysis and design problems. It will be used for the timber design examples that follow.

Due to the many sizes of solid rectangular sections that would satisfy a design, it is sometimes difficult to select the most economical structurally adequate member. To facilitate the selection process, rules of thumb relative to beam depth–width ratio should be considered. Generally, a timber beam depth–width ratio should be between 2.0 and 3.0, with 1.33 as an absolute lower limit. (This ratio does not apply to joists that are closely spaced; for example, 16-in. or 24-in. spacings. Depth–width ratios for joists may range from 4.0 to 6.0.) Another rule of thumb indicates that under normal loading conditions, the depth of a timber beam should be approximately 1 inch per foot of span where a beam is uniformly loaded and approximately $1\frac{1}{4}$ inch per foot of span where the loads are concentrated or combined.

EXAMPLE 16.3 The most common floor-framing arrangement in timber construction consists of closely spaced joists supported by either bearing walls or beams, and supporting a floor deck that is securely fastened to the tops of the joists, as shown in Figure 16.11.

Select the hem-fir (S4S) joist required to support the floor shown. The live load is 40 psf. There will be a superimposed dead load of 15 psf, which includes the dead weight of the floor and the ceiling but not the weight of the joists. The joists are spaced 16 in. on center and span 16 ft. Assume the joists to be simply supported. Deflection due to live load must not exceed span/360. Bearing will be 3 in. minimum at each end (concrete foundation wall or steel beam). Consider moment, shear, deflection, and bearing.

[3]Western Wood Products Association, *Western Woods Use Book–Structural Data and Design Tables*, 4th ed. (Portland, OR: WWPA, 2005).

[4]American Institute of Timber Construction, *Timber Construction Manual*, 5th ed. (New York: Wiley, 2004).

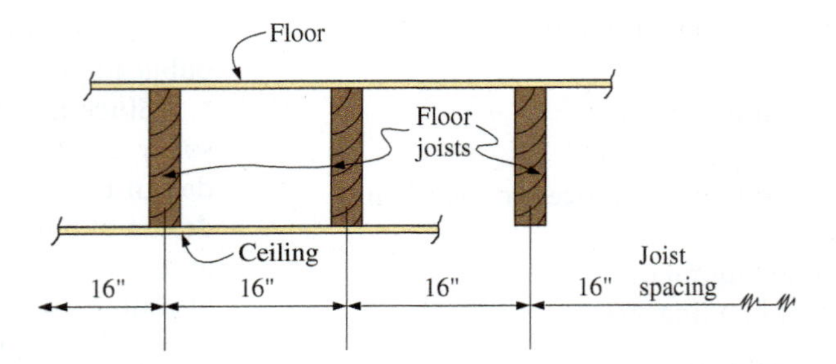

FIGURE 16.11 Floor cross section.

Solution From Appendix F, we obtain the following:

Bending: $F_b = 1000$ psi

Shear: $F_v = 150$ psi

Bearing (perpendicular): $F_{cp} = 400$ psi

Modulus of elasticity: $E = 1400$ ksi (1,400,000 psi)

The floor loading per square foot must be converted to a loading per linear foot of joist. Since each joist supports a 16-in.-wide strip of floor, the loading per foot is calculated as follows:

$$\text{live load} = 40\frac{\text{lb}}{\text{ft}^2}\left(\frac{16\text{ in.}}{12\text{ in./ft}}\right) = 53.3\text{ lb/ft}$$

$$\text{dead load} = 15\left(\frac{16}{12}\right) = 20.0\text{ lb/ft}$$

Therefore, the total load is

$$53.3 + 20.0 = 73.3\text{ lb/ft}$$

The load diagram is shown in Figure 16.12. From Appendix H, the maximum moment is calculated from

$$M = \frac{wL^2}{8} = \frac{(73.3\text{ lb/ft})(16\text{ ft})^2}{8} = 2350\text{ lb-ft}$$

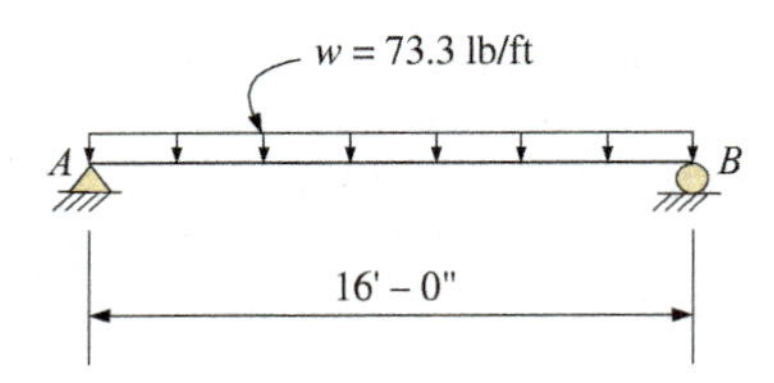

FIGURE 16.12 Joist load diagram.

from which

$$\text{required } S_x = \frac{M}{F_b} = \frac{(2350\text{ lb-ft})(12\text{ in./ft})}{1000\text{ psi}} = 28.2\text{ in.}^3$$

From Appendix E, Table E.1, try a 2 × 12, which has properties of

$$S_x = 31.6\text{ in.}^3$$
$$I_x = 178\text{ in.}^4$$
$$A = 16.9\text{ in.}^2$$
$$\text{wt.} = 4.68\text{ lb/ft}$$

The additional moment due to the weight of the joist is calculated from

$$\text{additional } M = \frac{wL^2}{8} = \frac{(4.68\text{ lb/ft})(16\text{ ft})^2}{8} = 150\text{ lb-ft}$$

The total design moment is then

$$\text{new } M = 2350 + 150 = 2500 \text{ lb-ft}$$

from which

$$\text{required } S_x = \frac{M}{F_b} = \frac{(2500 \text{ lb-ft})(12 \text{ in./ft})}{1000 \text{ psi}} = 30.0 \text{ in.}^3$$

$$30.0 \text{ in.}^3 < 31.6 \text{ in.}^3 \qquad \textbf{(O.K.)}$$

The 2 × 12 joist is satisfactory with respect to bending moment.

Next we'll calculate a required moment of inertia. The live load deflection limit is calculated from

$$\Delta_{\text{limit}} = \frac{\text{span}}{360} = \frac{16 \text{ ft}(12 \text{ in./ft})}{360} = 0.533 \text{ in.}$$

and the live load has been determined to be 53.3 lb/ft. Therefore,

$$\text{required } I_x = \frac{5w_{LL}L^4}{384E\Delta_{\text{limit}}} = \frac{5(53.3 \text{ lb/ft})(16 \text{ ft})^4(1728 \text{ in.}^3/\text{ft}^3)}{384(1{,}400{,}000 \text{ psi})(0.533 \text{ in.})} = 105.3 \text{ in.}^4$$

Since 105.3 in.^{4} < 178 in.^{4}, the 2 × 12 is O.K. for deflection as well as moment.

The maximum shear is equal to the reaction at either support and includes the effects of the superimposed load and the weight of the joist.

$$V = R_A = R_B = \frac{wL}{2} = \frac{[(73.3 + 4.68)\text{lb/ft}](16 \text{ ft})}{2} = 624 \text{ lb}$$

The required area based on shear is calculated from:

$$\text{required } A = \frac{1.5V}{F_v} = \frac{1.5(624 \text{ lb})}{150 \text{ psi}} = 6.24 \text{ in.}^2$$

$$6.24 \text{ in.}^2 < 16.9 \text{ in.}^2 \qquad \textbf{(O.K.)}$$

Therefore, the 2 × 12 is adequate for shear as well as for moment and deflection.

Last, we will check bearing at the supports. The allowable bearing stress perpendicular to the grain is 400 psi. Figure 16.13 shows the bearing length of 3 in. (minimum) and the bearing width which is equal to the width of the (S4S) joist.

$$F_{cp} = \frac{\text{reaction}}{\text{bearing area}} = \frac{624 \text{ lb}}{3 \text{ in.}(1.5 \text{ in.})} = 138.7 \text{ psi}$$

$$138.7 \text{ psi} < 400 \text{ psi} \qquad \textbf{(O.K.)}$$

Use a 2 × 12 (S4S) joist.

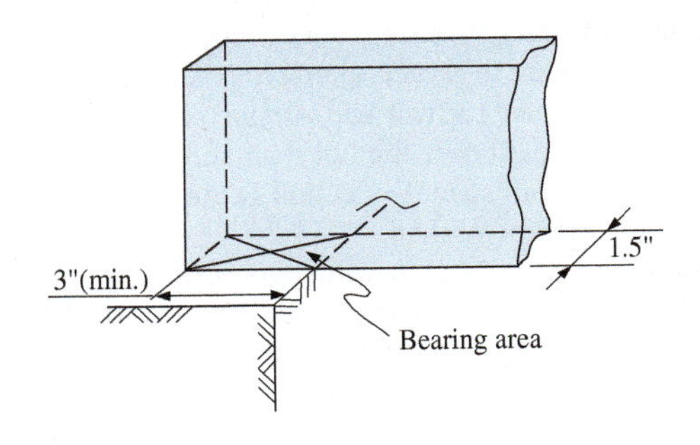

FIGURE 16.13 End bearing.

SUMMARY BY SECTION NUMBER

16.1 Bending stress is the type of stress that usually limits the allowable load on a beam. Therefore, in normal design process, a beam should be selected on the basis of bending moment and then checked for shear. (See the detailed step-by-step design procedure included in this section.)

16.2 Our discussion of the design of steel beams is limited to standard hot-rolled structural steel shapes. For the design of the structural steel shapes, reference must be made to standard tables of properties included in the AISC *Manual of Steel Construction*. Portions of these tables appear in Appendices A through D of this text. In addition, Appendices H, I, and J serve as references to expedite the design process.

16.3 Our discussion of the design of timber beams is limited to solid rectangular shapes since these are, by far, the most commonly used sections in timber design. The necessary member properties and suggested design values needed for design are given in Appendices E and F. Although the design of timber members is generally based on dressed dimensions, nominal dimensions may also be used.

PROBLEMS

In the following problems, the given loads are superimposed service loads; that is, they do not include the weights of the beams (unless noted otherwise). For structural steel beams (unless otherwise noted), assume a yield stress of 50 ksi (345 MPa). For timber beams, all beams are solid, rectangular shapes and Appendices E and F are applicable. Consider only moment and shear (unless otherwise noted).

Section 16.2 Design of Steel Beams

16.1 Select the lightest W shape to support a uniformly distributed load of 2.1 kips/ft on a simple span of 24 ft.

16.2 A simply supported beam is to support a uniformly distributed load of 10 kips/ft. Select the lightest W shape. The span length is 32 ft. Deflection is not to exceed span/240. Neglect beam weight.

16.3 Rework Problem 16.1 given a load of 1.0 kip/ft and a span length of 30 ft.

16.4 Select the lightest W shape to support a concentrated load of 100 kN placed at midspan. The beam is on a simple span of 10 m. Deflection is not to exceed span/240. Neglect beam weight.

16.5 Select the lightest W shape to support a superimposed uniformly distributed line load of 435 kN/m on a simply supported span of 5 m. Deflection is not to exceed span/240.

16.6 A simply supported beam is to span 15 ft. It will support a uniformly distributed load of 2 kips/ft over the full span and a concentrated load of 60 kips at midspan. Deflection is not to exceed span/240. Select the lightest W shape.

16.7 A simply supported beam is to span 24 ft. It will support a uniformly distributed load of 1.2 kips/ft over the full span and a concentrated load of 8 kips located 7 ft from the left support. Select the lightest W shape.

Section 16.3 Design of Timber Beams

16.8 Design a timber beam of hem-fir (S4S) to support a uniformly distributed line load of 600 lb/ft on a simply supported span of 15 ft. Deflection is not to exceed span/360.

16.9 Select simply supported timber beams (S4S) for the following uniformly distributed line loads and spans. Deflection is not to exceed span/240.

(a) Douglas fir, 400 lb/ft, 22-ft span

(b) hem-fir, 400 lb/ft, 18-ft span

16.10 Select a southern pine (S4S) timber beam for the conditions shown.

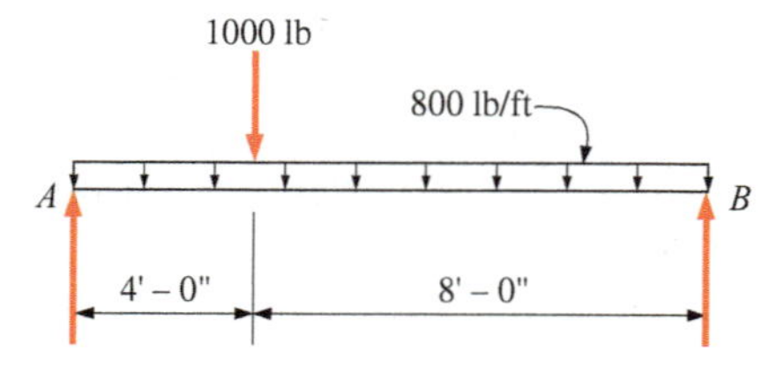

PROBLEM 16.10

16.11 Select simply supported hem-fir (S4S) joists to support a floor live load of 40 psf on a 12-ft simple span. The superimposed dead load will be 15 psf (not including the joist weight). The joists are to be spaced 24 in. on center. Deflection is not to exceed span/360.

16.12 Design simply supported timber beams (S4S) for the following uniformly distributed line loads and spans:

(a) eastern white pine, 5.0 kN/m, 7-m span length

(b) southern pine, 6.3 kN/m, 8-m span length

Computer Problems

For the following computer problems, any appropriate software may be used. Input prompts should fully explain what is required of the user (the program should be user-friendly). The resulting output should be well labeled and self-explanatory.

16.13 Write a program that will select the required depth (2-in. increments) of a 2-in.-thick timber joist that will be spaced 16 in. on center and support a floor carrying a uniformly distributed load. Use nominal dimensions and properties. Consider moment and shear. Limits on the nominal depth are to be 6 in. minimum and 12 in. maximum. User input is to be live load (psf), dead load (psf), and span (ft). Neglect the weight of the joist.

16.14 Write a program that will help you select the lightest W shape from a limited group of W shapes. The beam is to be simply supported and will carry a uniformly distributed line load. Consider moment only. You will need to create a short table of W-shape properties from which the program can make the selection. (Real-world software contains more comprehensive tables of shape properties.) Choose a few shapes (perhaps six) from Appendix I and include nominal depth, weight, and section modulus in the table. User input for this program is to be uniformly distributed line load intensity (kips/ft), span length (ft), and yield stress (ksi).

16.15 Rework Problem 16.14 including shear and deflection.

Supplemental Problems

16.16 Select the lightest W shape to support a uniformly distributed line load of 1600 lb/ft on a simple span of 48 ft. Deflection is not to exceed span/240.

16.17 Select the lightest W shape for the beam shown.

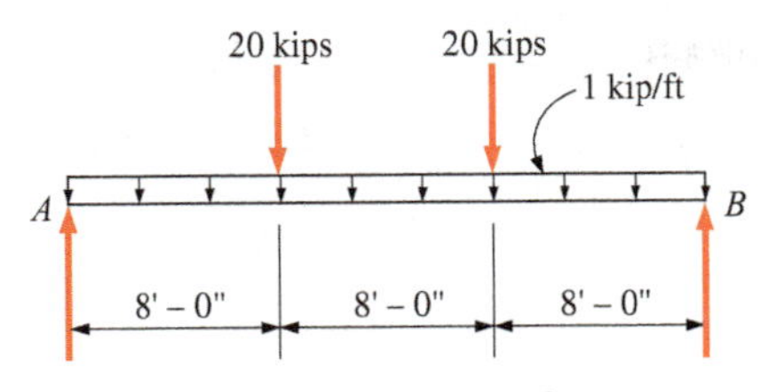

PROBLEM 16.17

16.18 Select the lightest W shape for the cantilever beam shown.

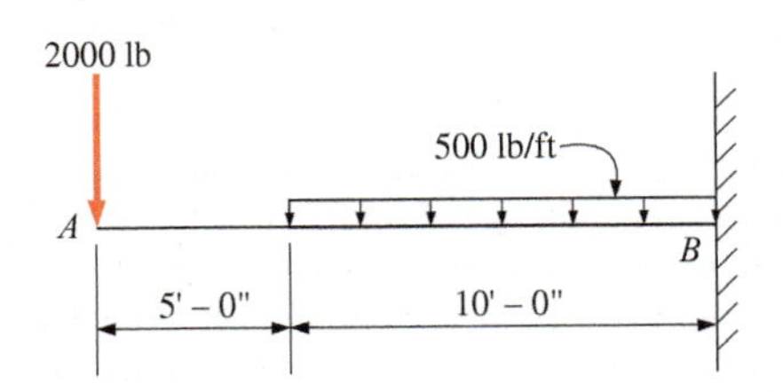

PROBLEM 16.18

16.19 Select the lightest W shape to support a uniformly distributed line load of 30 kN/m and concentrated loads of 50 kN located as shown. Deflection is not to exceed span/360.

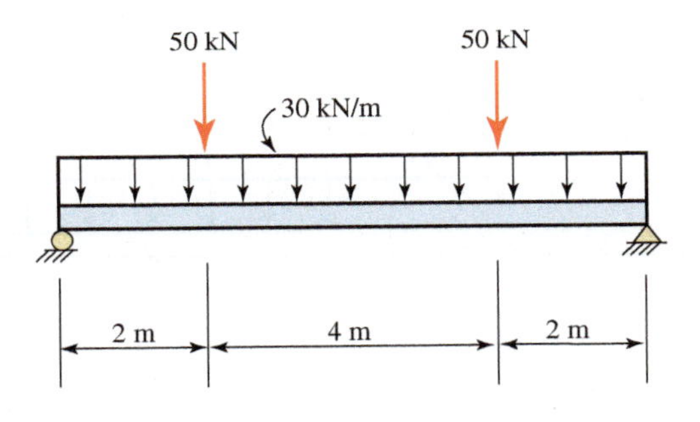

PROBLEM 16.19

16.20 Select the lightest W shape for the beams shown.

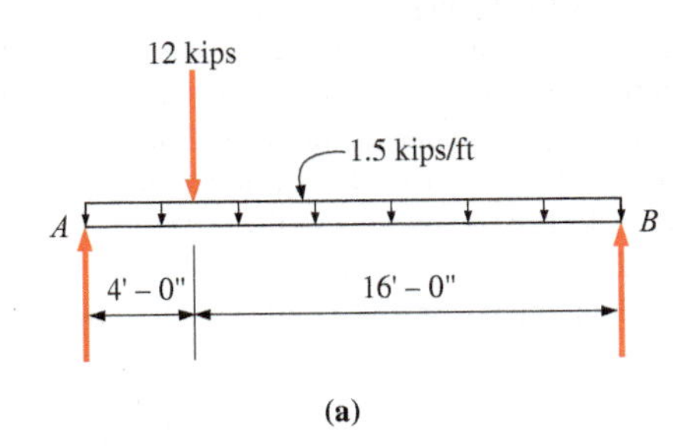

(a)

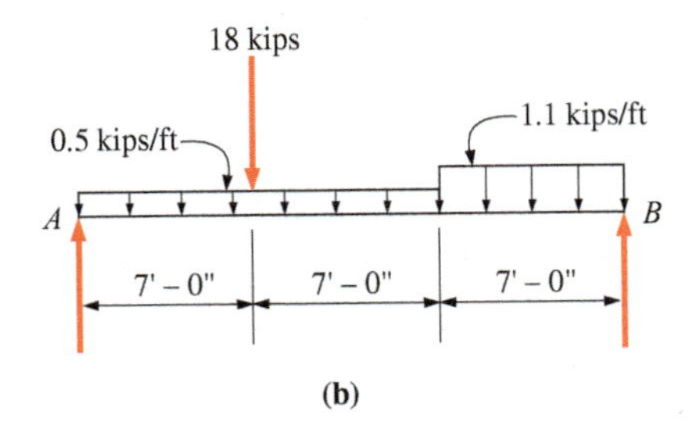

(b)

PROBLEM 16.20

16.21 The structural steel floor system shown is to support a 5-in.-thick reinforced concrete slab, a live load of 100 psf, and an additional future dead load of 15 psf. Select the lightest W shapes for the beams and girders.

16.22 The structural steel framing plan shown represents a floor system for a narrow commercial building. The floor is to support a 6-in.-thick reinforced concrete slab, a live load of 150 psf, and an additional future load of 20 psf. Select the lightest W shapes for B1, B2, and G2.

16.23 Select the lightest steel wide-flange section (W shape) for the beam shown.

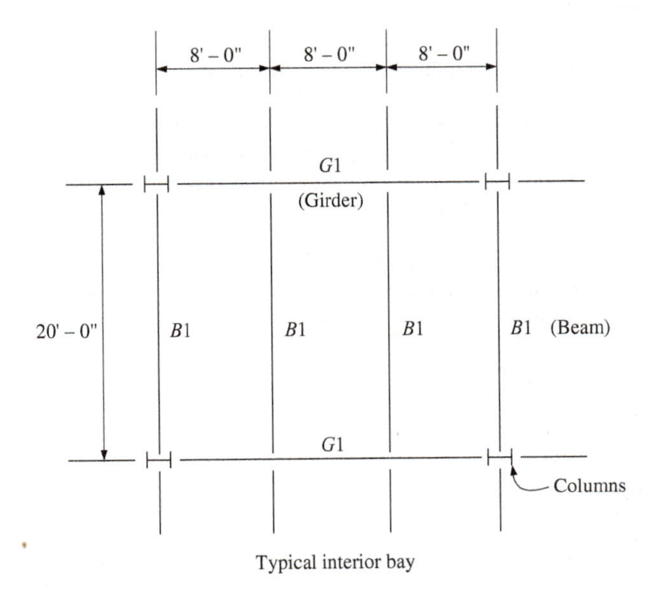

PROBLEM 16.21

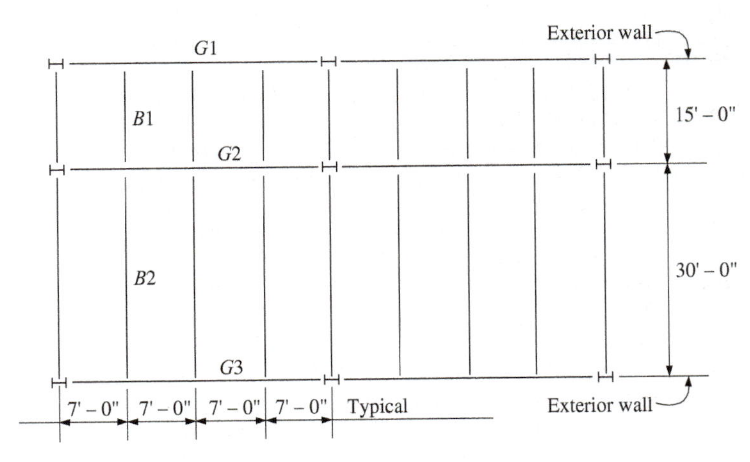

PROBLEM 16.22

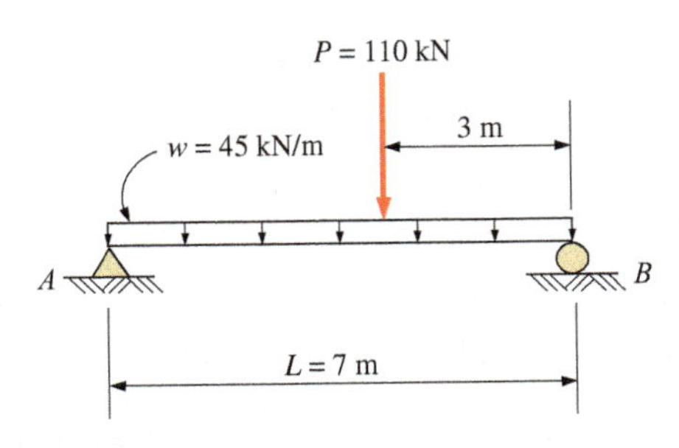

PROBLEM 16.23

16.24 Select the lightest steel wide-flange section (W shape) for the beam shown.

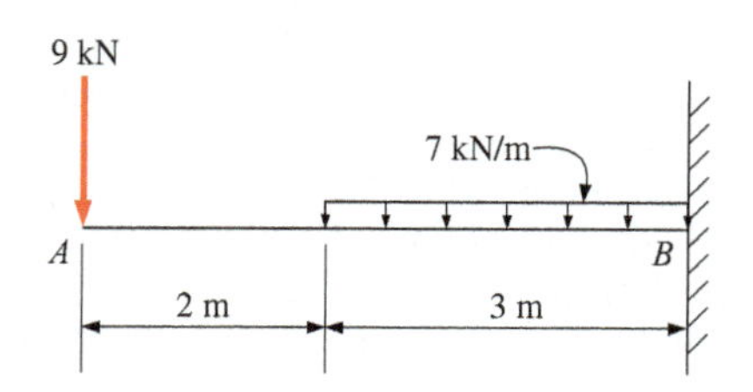

PROBLEM 16.24

16.25 Design the lightest W-shape beams to support a 150-mm-thick reinforced concrete slab as shown. The beams are simply supported, are 2.20 m on center, and span 12 m. The service live load on the slab is 6.0 kPa. (Assume the unit weight of reinforced concrete to be 23.5 kN/m^3.)

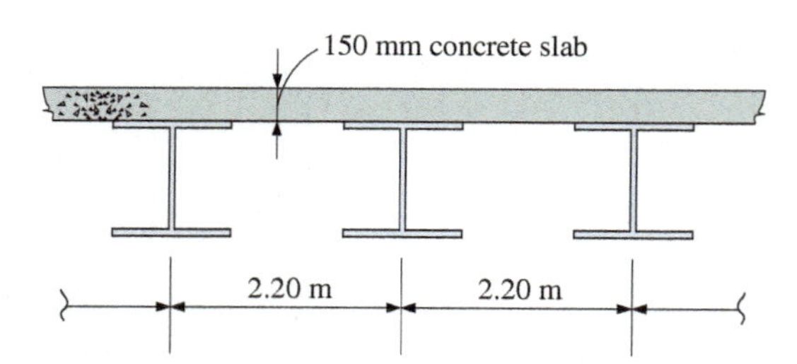

PROBLEM 16.25

16.26 In Problem 16.18, assume that the 500 lb/ft load extends fully across the span. The total load deflection limit at the free end is span/200. Select the lightest W shape.

16.27 Select a southern pine (S4S) simply supported beam to support a uniformly distributed load of 400 lb/ft on a span of 14 ft.

16.28 A redwood beam is to support a uniformly distributed load of 250 lb/ft and a concentrated load of 4.0 kips at midspan. The simply supported beam spans 20 ft. Select a 12-in.-wide (S4S) beam. Deflection is not to exceed span/360.

16.29 A partial plan view for a residential floor is shown. Select the joists. Use 2-in.-thick (S4S) hem-fir joists. Assume 16-in. spacing. Live load = 40 psf. Dead load = 12 psf (excluding the joist weight).

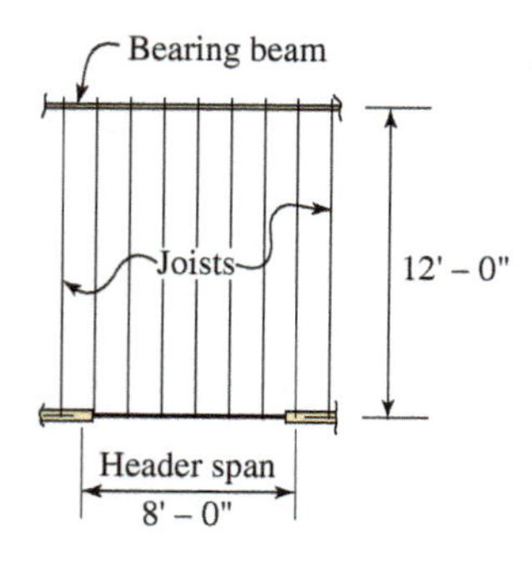

PROBLEM 16.29

16.30 For the floor framing of Problem 16.29, select 2-in.-thick (S4S) redwood planks for the header. Multiple planks may be used. In addition to the floor load, the header must support a wall weight of 50 lb/ft and a total load from the roof of 300 lb/ft. Use a design span of 8′–0″.

16.31 Select a Douglas fir (S4S) beam for the conditions shown.

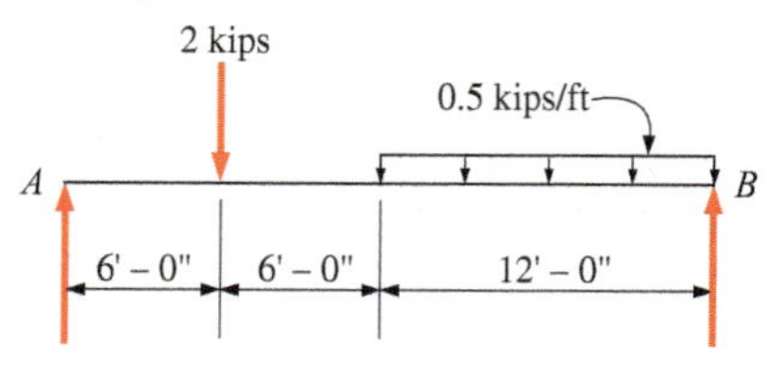

PROBLEM 16.31

16.32 Select southern pine (S4S) simply supported joists to carry a floor live load of 50 psf and a superimposed dead load (flooring and ceiling) of 20 psf. The joists are to be spaced 16 in. on center and the span length is to be 18 ft.

16.33 Rework Problem 16.32 using joists spaced 12 in. on center.

16.34 Select Douglas fir (S4S) simply supported joists to carry a floor live load of 2.5 kPa and a superimposed dead load (floors and ceiling) of 1.0 kPa. The joists are to be spaced 0.4 m on center and the span length is 6 m.

16.35 Select southern pine (S4S) simply supported joists to carry a floor live load of 120 psf and a superimposed dead load of 20 psf. The joist spacing will be 16 in. and the span will be 14 ft.

16.36 A 15-ft-span simply supported hem-fir (S4S) beam supports the contributory load of a width of 10 ft of floor. The live load is 50 psf and the superimposed dead load is 50 psf. Select a beam with a depth of not more than 18 in.

16.37 Select a timber beam (S4S) of Southern pine for the conditions shown.

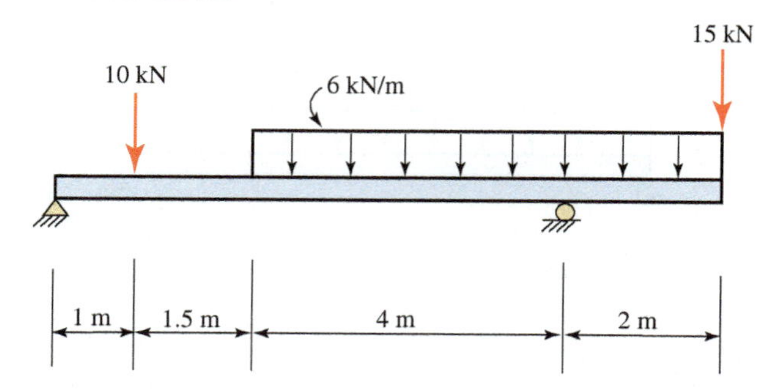

PROBLEM 16.37

16.38 A series of 14-ft-long Douglas fir (S4S) joists carry a floor live load of 40 psf and a superimposed dead load of 20 psf. In addition, there is a concentrated load from a partition running perpendicular to the joists 5 ft from one end. The partition load is 250 lb/ft. The joists are spaced 16 in. on center. Select the required size for the joists.

16.39 A cantilever beam 3 m long is to be made from several 50 × 250 Douglas fir (S4S) planks bolted together (the long dimensions of the planks are placed vertically). A concentrated load of 2 kN is to be carried at the free end. Determine the number of planks necessary to support the load.

16.40 Select an eastern white pine (S4S) beam for the conditions shown.

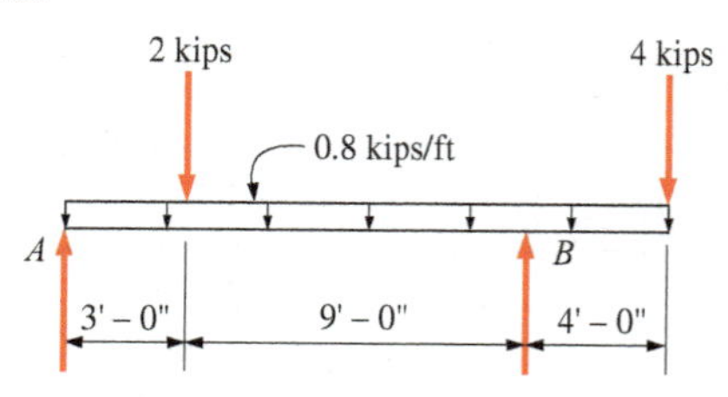

PROBLEM 16.40

CHAPTER SEVENTEEN

COMBINED STRESSES

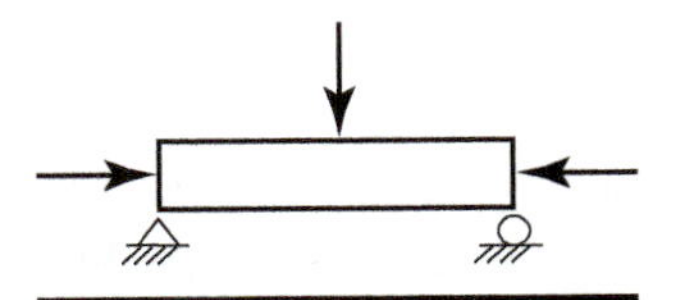

LEARNING OBJECTIVES

Upon completion of this chapter, readers will be able to:

- Discuss biaxial bending in structural members.
- Calculate stresses in members subject to both axial loads and transverse loads.
- Calculate the stresses in short compression members subject to eccentric loads.
- Calculate the maximum allowable eccentricity in a rectangular compression member.
- Learn how to calculate principle stresses in members subjected to both normal and shear stress by either formulas or Mohr's Circle.

17.1 INTRODUCTION

Thus far in our study of strength of materials, we have considered individual fundamental stresses (normal stresses and shear stresses) acting alone, without taking into account any interaction. When stresses acting at a point are collinear and of the same type, they can be added algebraically. For instance, tensile and compressive normal stresses due to bending and axial load can be added algebraically when they are collinear (acting along the same line). When stresses are not collinear, or when they are of different types (e.g., normal stresses and shear stresses), they must be added vectorially.

Although some machine and structural parts can be adequately designed by considering the stresses individually, in some situations combinations of stresses may produce critical conditions that should be investigated. This case is particularly true when the stresses at actual failure of a member are being studied.

17.2 BIAXIAL BENDING

When a load is applied on a beam and the load does not lie in the plane of the weak axis (the usual case) or the strong axis, bending takes place about both axes and is termed *biaxial bending*. The simplest case is shown in Figure 17.1, where the horizontal beam is tilted, subjected to a gravity load P, and the line of action of the load passes through the center of gravity of the cross section. The load can be resolved into components P_x and P_y parallel to the X–X and Y–Y axes, respectively. The component forces cause bending moment M_x (about the X–X axis, due to P_y) and M_y (about the Y–Y axis, due to P_x). Since both moments result in longitudinal stresses (or stresses that are normal to the cross section), the stresses can be added algebraically. This process of algebraic addition of the same kind of stresses to obtain the resultant or combined stress is called *superposition*. The principle of superposition is based on the premise that the resultant stress due to several forces in any system is the algebraic sum of

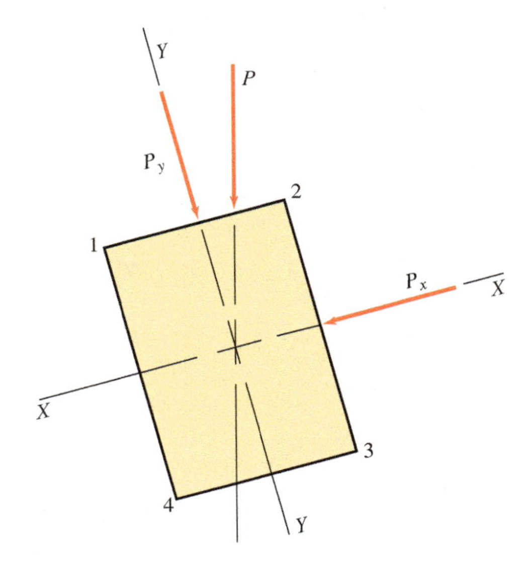

FIGURE 17.1 Cross section of beam subjected to biaxial bending.

the effects caused by the individual forces. Superposition of stresses is permissible if the maximum stress remains within the elastic limit and if deformations are very small.

With reference to Figure 17.1, note that the moment M_x (due to force P_y) would create compressive stress at points 1 and 2 and tensile stress at points 3 and 4. Moment M_y (due to force P_x) would cause compressive stress at points 2 and 3 and tensile stress at points 1 and 4. Using the flexure formula, as developed in Chapter 14, a general expression for the stress at any of the four points can be written as

$$f_{\max} = \pm \frac{M_x c_x}{I_x} \pm \frac{M_y c_y}{I_y}$$

or

$$f_{\max} = \pm \frac{M_x}{S_x} \pm \frac{M_y}{S_y} \qquad \textbf{(17.1)}$$

It is customary to use a sign convention for the stresses. We consider tensile stress to be indicated by a plus sign and compressive stress by a negative sign.

EXAMPLE 17.1 A simple beam is supported on a 15° slope and is loaded as shown in Figure 17.2. The line of action of the 16-kN load passes through the centroid of the cross section. Determine the bending stresses at the flange tips (the four outer corners) of the cross section. Neglect the weight of the beam.

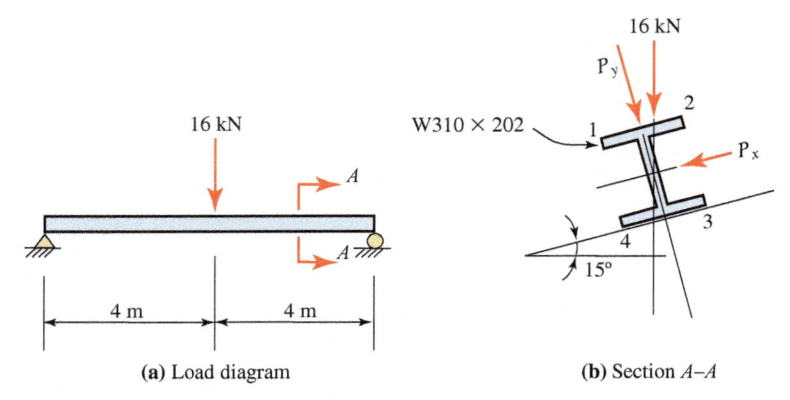

FIGURE 17.2 Sketch for Example 17.1.

Solution For the W310 × 202,

$$S_x = 3030 \times 10^3 \text{ mm}^3$$
$$S_y = 1050 \times 10^3 \text{ mm}^3$$

Resolve the 16-kN load into its components:

$$P_x = 16 \sin 15° = 4.14 \text{ kN}$$
$$P_y = 16 \cos 15° = 15.45 \text{ kN}$$

The moments are calculated from

$$M_x = \frac{P_y L}{4} = \frac{(15.45 \text{ kN})(8 \text{ m})}{4} = 30.9 \text{ kN} \cdot \text{m} = 30.9 \times 10^6 \text{ N} \cdot \text{mm}$$

$$M_y = \frac{P_x L}{4} = \frac{(4.14 \text{ kN})(8 \text{ m})}{4} = 8.28 \text{ kN} \cdot \text{m} = 8.28 \times 10^6 \text{ N} \cdot \text{mm}$$

The stresses will be calculated using Equation (17.1):

$$\begin{aligned} f_{\max} &= \pm \frac{M_x}{S_x} \pm \frac{M_y}{S_y} \\ &= \frac{30.9 \times 10^6 \text{ N} \cdot \text{mm}}{3030 \times 10^3 \text{ mm}^3} \pm \frac{8.28 \times 10^6 \text{ N} \cdot \text{mm}}{1050 \times 10^3 \text{ mm}^3} \\ &= \pm 10.20 \text{ MPa} \pm 7.89 \text{ MPa} \end{aligned}$$

Use a tabular format to summarize the stresses, with reference to section *A–A* in Figure 17.2 and recall that tensile stress is indicated with a positive sign:

Point	$\frac{M_x}{S_x}$ (MPa)	$\frac{M_y}{S_y}$ (MPa)	Resulting Stress (MPa)
1	−10.20	+7.89	−2.31
2	−10.20	−7.89	−18.09
3	+10.20	−7.89	+2.31
4	+10.20	+7.89	+18.09

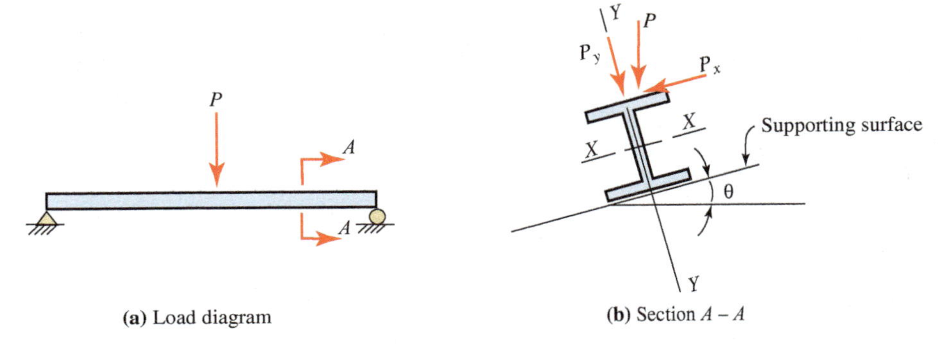

FIGURE 17.3 Beam subjected to load applied at center of top flange.

Although Example 17.1 illustrates how stresses can be algebraically summed for the simple biaxial bending case, the practical situation involves loads that may not pass through the centroid of the cross section. For tilted W shapes, the load is usually applied at the center of the top flange, as shown in Figure 17.3. The force component P_x will induce a twist, or a torque, in the cross section in addition to the bending moment. One way to handle this case is to assume that the top flange provides all the resistance to the bending effect of force component P_x while the full strong-axis strength of the cross section resists the bending effect of the force component P_y. The two flanges are equal in size, and very little of the section modulus S_y (or the moment of inertia I_y) is contributed by the web. Therefore, the section modulus of one flange is approximately $S_y/2$, and the formula for stress in the top flange can be written as

$$f_{\max} = -\frac{M_x}{S_x} \pm \frac{M_y}{S_y/2}$$

17.3 COMBINED AXIAL AND BENDING STRESSES

Many structural and machine members are subjected to both transverse loads and direct axial loads simultaneously, thereby developing a combination of bending stresses and direct axial stresses. One such example, as shown in Figure 17.4a, is a simply supported beam subjected to a transverse uniform loading w in combination with a direct axial load P. The transverse loading will produce a bending moment and a subsequent triangular bending stress distribution over the cross section of the beam, as shown in Figure 17.4c. Although the cross section could be arbitrarily taken at any point, we consider the cross section at midspan (where moment will be maximum) and designate this plane as plane *X–X*. The bending stress will vary from a maximum at the outer fibers to zero at the neutral axis. Note that since the beam in Figure 17.4a is simply supported and subjected to downward-acting transverse loads, the bending stress will be tensile below the neutral axis and compressive above the neutral axis. We use the flexure formula to determine the magnitude of the maximum bending stress:

$$f_b = \frac{Mc}{I} = \frac{M}{S}$$

Since the beam is simultaneously subjected to a direct axial load parallel to the longitudinal axis of the member (this load could be either tensile or compressive), a direct axial stress will be developed as described in Chapter 9. Recall that the direct axial stress distribution is assumed to be uniform over the cross section of the member, as shown in Figure 17.4b, and that the magnitude of the direct axial stress can be obtained from the expression

$$\sigma_t = \frac{P}{A}$$

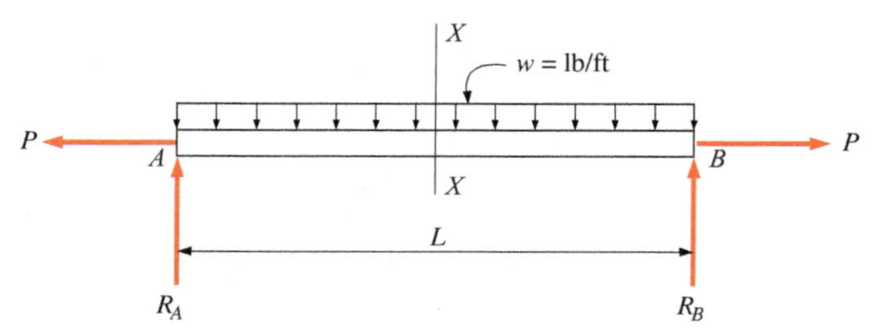

(a) Simply supported beam

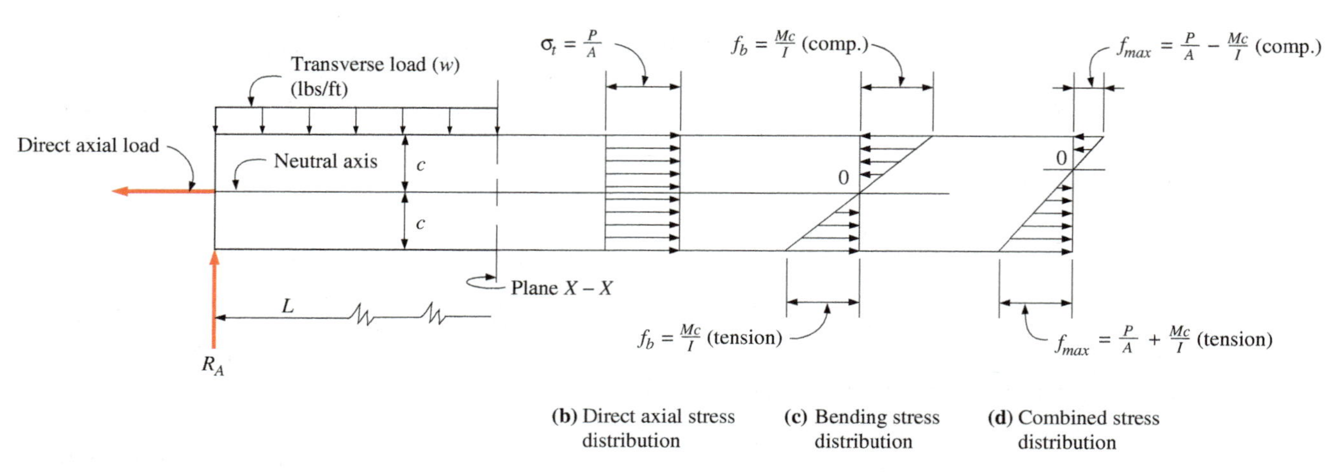

(b) Direct axial stress distribution (c) Bending stress distribution (d) Combined stress distribution

FIGURE 17.4 Combined axial and bending stresses.

All terms were defined in Chapter 9 along with the limitations of usage.

Note that the bending results in tensile and compressive stresses that are normal to the plane of the cross section (plane *X–X*). Similarly, the direct axial tension results in tensile stresses normal to this plane. Since the same kind of stresses (normal) are developed, a resultant combined stress at any location can be determined by a simple algebraic sum of the individual direct axial and bending stresses produced at the same locations, as shown in Figure 17.4d. Using the principle of superposition, this stress is expressed as

$$f_{\max} = \pm\frac{P}{A} \pm \frac{Mc}{I} \tag{17.2}$$

It is customary to use a sign convention for the stresses. Recall that we use the plus sign to designate tensile stress and the minus sign to designate compressive stress. Note that if the axial stress is tensile, it will add to the tensile bending stress; if it is compressive, it will add to the compressive bending stress. In either case, the total combined stress at any point will always equal the algebraic sum of the axial and bending stresses at that point. Therefore, for the beam of Figure 17.4, the resultant combined stress is

$$f_{\max} = +\frac{P}{A} \pm \frac{Mc}{I}$$

Note that this approach is slightly inexact, particularly for slender members or members with relatively low *EI* values where comparatively large deflections may be produced by the transverse loads. With reference to Figure 17.5, the load *P* multiplied by the deflection Δ will induce an additional bending moment (sometimes called the $P\Delta$ *moment*). For the axial compressive load *P* shown, this additional moment would be added to the moment caused by the transverse loads P_1 and P_2. If *P* were tensile, the $P\Delta$ moment would be subtracted from the transverse load moment. Unless members are long and slender, the $P\Delta$ moment is usually neglected since the deflections are very small.

In this chapter, only members of relatively short span lengths (hence, small deflections) are considered. Therefore, the bending stresses caused by the axial loads may be neglected without appreciable error.

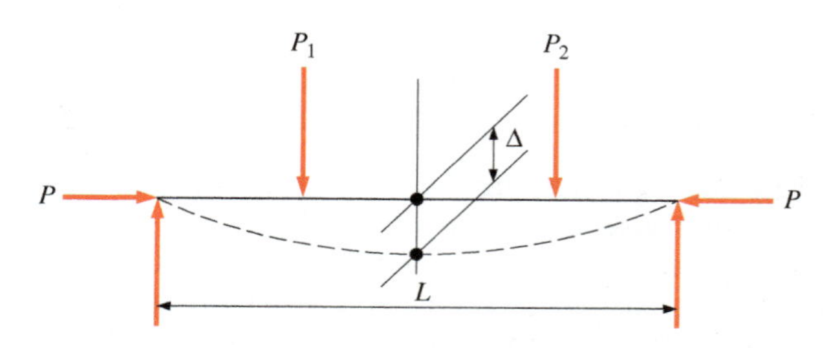

FIGURE 17.5 Axial compression and bending.

EXAMPLE 17.2 A W14 × 61 structural steel wide-flange section is used as a simply supported beam with a span length of 10 ft. It is subjected to a uniformly distributed transverse load of 6 kips/ft, including its own weight, and an axial tensile force of 100 kips. Compute the maximum combined tensile and compressive stresses.

Solution Appendix A contains the necessary properties of the W14 × 61:

$$A = 17.9 \text{ in.}^2$$
$$I_x = 640 \text{ in.}^4$$
$$S_x = 92.1 \text{ in.}^3$$

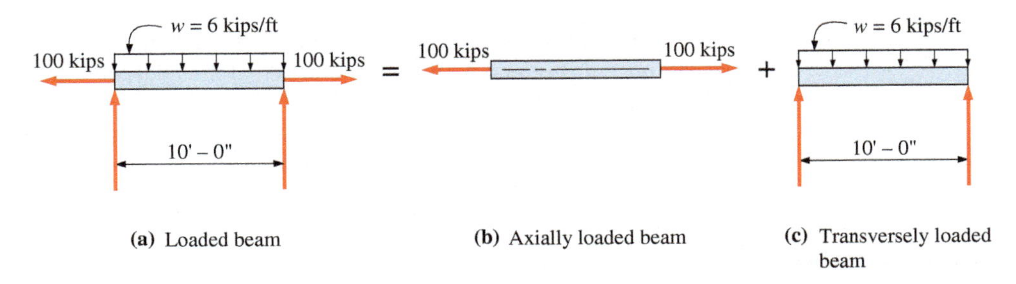

FIGURE 17.6 Method of superposition.

To illustrate the method of superposition, this problem is solved by dividing it into two parts. The loaded beam is shown in Figure 17.6a. It is shown subjected to only the axial load in Figure 17.6b, whereas in Figure 17.6c it is shown subjected to only the transverse load.

For the axial load, the normal direct stress throughout the length of the beam is

$$\sigma_t = \frac{P}{A} = \frac{100 \text{ k}}{17.9 \text{ in.}^2} = +5.59 \text{ ksi (tension)}$$

and is shown in Figure 17.7a.

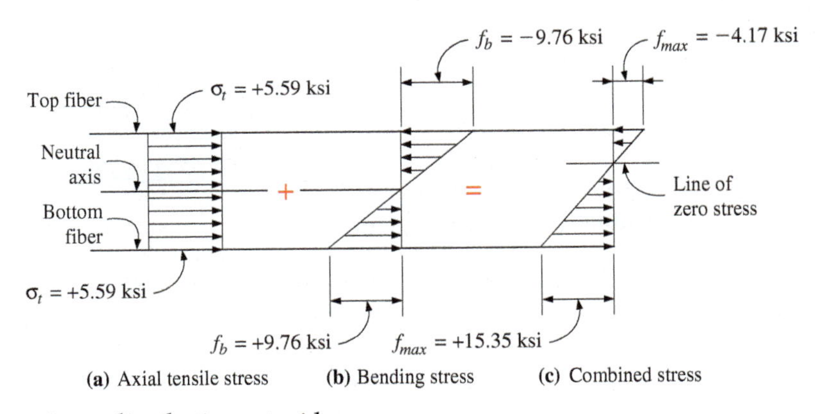

FIGURE 17.7 Stress distributions at midspan.

The normal stress due to the transverse load depends on the magnitude of the maximum bending moment, which in this case occurs at midspan. The maximum bending moment (see Appendix H) is

$$M = \frac{wL^2}{8} = \frac{(6 \text{ k/ft})(10 \text{ ft})^2}{8} = 75 \text{ k-ft}$$

From the flexure formula, the maximum stresses at the outer fibers caused by this moment are

$$f_b = \frac{Mc}{I} = \frac{M}{S} = \frac{(75 \text{ k-ft})(12 \text{ in./ft})}{92.1 \text{ in.}^3} = \pm 9.77 \text{ ksi}$$

These stresses will be normal to the cross section of the beam at midspan and decrease linearly toward the neutral axis, as is shown in Figure 17.7b.

To obtain the resultant or combined stress, the bending stress must be added algebraically to the direct axial tensile stress. Expressing this mathematically gives

$$f_{max} = +\frac{P}{A} \pm \frac{Mc}{I}$$

from which the bottom-fiber stress is

$$f_{max} = +5.59 + 9.77 = +15.36 \text{ ksi (tension)}$$

and the top-fiber stress is

$$f_{max} = +5.59 - 9.77 = -4.18 \text{ ksi (compression)}$$

Thus, as can be seen from Figure 17.7c, the resultant normal stress is 15.36 ksi tension at the midspan bottom fiber and 4.18 ksi compression at the midspan top fiber. Since the bending moment varies along the length of the beam, the bending stresses and therefore the combined stresses will be different at different sections. Note that in Figure 17.7c, the line of zero stress, which lies on the neutral axis if only flexural stresses exist, moves upward.

EXAMPLE 17.3 An 8-in.-diameter steel pipe extends vertically out of a solid concrete anchorage, as shown in Figure 17.8. The pipe is subjected to horizontal and vertical loading as shown. The vertical load includes the weight of the pipe. Compute the maximum combined tensile and compressive stresses at the top of the concrete anchorage. Use $A = 8.40$ in.^{2}, $S = 16.8$ in.^{3}, and $I = 72.5$ in.^{4}

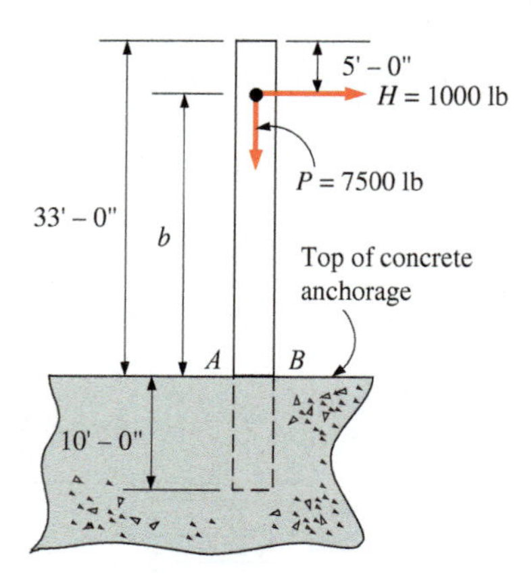

FIGURE 17.8 Load diagram.

Solution The section to be investigated is in the plane of the top of the concrete anchorage (plane *A–B*). The direct axial load *P* is 7500 lb. The direct stress on plane *A–B* caused by the axial load is

$$\sigma_c = \frac{P}{A} = \frac{7500 \text{ lb}}{8.40 \text{ in.}^2} = -893 \text{ psi (compression)}$$

which is shown in Figure 17.9a. Note that this is a projected view of the compressive stress. The actual compressive stress block would be ring-shaped, with a hole in the middle.

The bending stress in plane *A–B* caused by the horizontal load *H* depends on the magnitude of the bending moment. The maximum bending moment (see Appendix H) is

$$M = Hb = (1000 \text{ lb})(28 \text{ ft}) = 28{,}000 \text{ lb-ft}$$

Using the flexure formula, the maximum stresses at the outer fibers caused by this moment are

$$f_b = \frac{Mc}{I} = \frac{M}{S} = \frac{(28{,}000 \text{ lb-ft})(12 \text{ in./ft})}{16.8 \text{ in.}^3} = \pm 20{,}000 \text{ psi}$$

which is shown in Figure 17.9b.

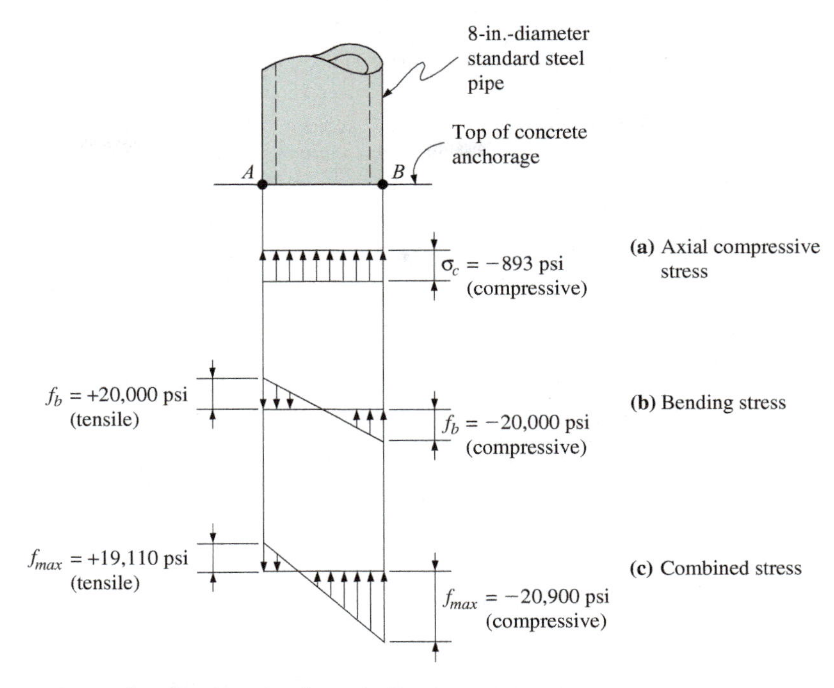

FIGURE 17.9 Stress distributions in plane *A–B*.

The resultant or combined stress is the algebraic sum of the two stresses and is shown in Figure 17.9c:

$$f_{\text{max}} = -\frac{P}{A} \pm \frac{Mc}{I}$$

Therefore, the combined fiber stress at point *A* is

$$f_{\text{max}} = -893 + 20{,}000 = +19{,}110 \text{ psi (tension)}$$

and the combined fiber stress at point *B* is

$$f_{\text{max}} = -893 - 20{,}000 = -20{,}900 \text{ psi (compression)}$$

17.4 ECCENTRICALLY LOADED MEMBERS

A *short* compression member may be defined as a member that will fail by crushing or yielding, as opposed to buckling, when subjected to an increasing axial compressive load. When such a member is subjected to a vertical direct axial load (a load coincident with a longitudinal axis through the centroid of the cross section), the compressive stress developed is assumed to be uniformly distributed over the cross section of the member. If the vertical load is not axial but is parallel to a longitudinal axis through the centroid, as shown in Figure 17.10a, the stress developed on any cross section will not be uniformly distributed. The stress at any point will then be a combined stress similar to the combined stresses developed in a member subjected to both transverse and direct axial loads, as discussed in Section 17.3.

When the vertical load is not coincident with the longitudinal axis through the centroid, it is said to be *eccentric*. The eccentricity *e* of the load is the distance measured along one of the centroidal axes from the centroid *O* to the line of action of the force *P*. Note that in this section the discussion is limited to the case where the load *P* lies on one of the centroidal axes and is eccentric to the other. Also note that since the member is short, its lateral deflection is assumed to be so small that it can be neglected in comparison with the initial eccentricity *e*. Therefore, the method of superposition may be used.

The analysis of an eccentrically loaded member involves what is commonly termed the *load-moment relations*. In Figure 17.10a we see a short compression member subjected to a vertical load *P* applied with an eccentricity *e* on one of the two centroidal axes. In Figure 17.10b, two equal and opposite forces *P* are applied at the centroid of the cross section, each equal in magnitude to the eccentric load *P*. The addition of these two forces does not change the

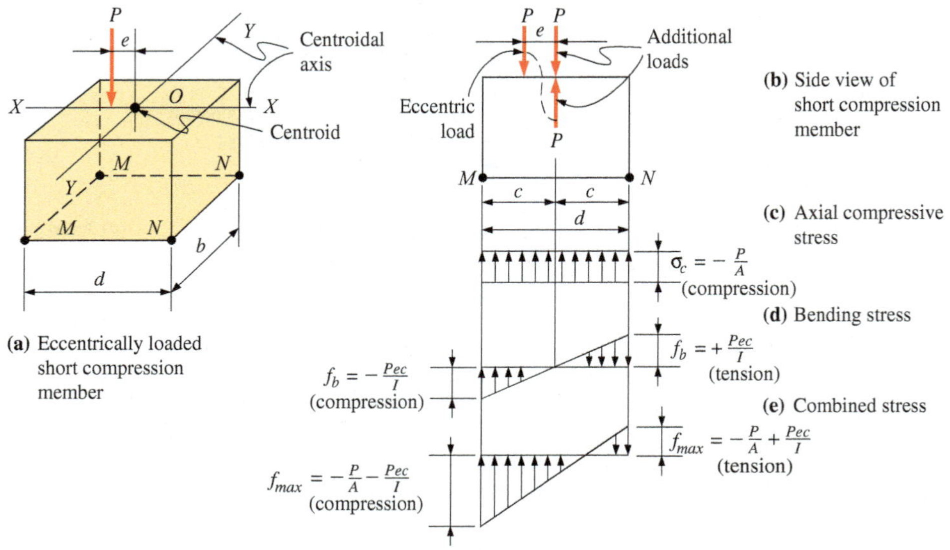

FIGURE 17.10 Combined stresses caused by eccentric load.

problem, since the algebraic sum of the two collinear and equal additional forces is equal to zero.

In effect, we have replaced the eccentric load system with a downward axial compressive force P acting at the centroid and a couple having a magnitude of Pe. As shown in Figure 17.10c and d the compressive force P, at the centroid, produces direct axial compressive stresses of

$$\sigma_c = \frac{P}{A}$$

and the couple Pe, acting about the Y–Y axis, produces bending stresses of

$$f_b = \pm\frac{Mc}{I} = \pm\frac{Pec}{I}$$

The total combined stress can be expressed as follows, using the same sign convention as that used in Sections 17.2 and 17.3:

$$f_{\max} = -\frac{P}{A} \pm \frac{Pec}{I} \qquad \textbf{(17.3)}$$

The diagram of the combined stress is shown in Figure 17.10e. It is assumed that the maximum bending stress is larger than the direct axial stress; therefore, there is a zone of tensile stress on the right and compressive stress on the left. If the maximum bending stress is smaller in magnitude than the direct axial stress, then there will be compressive stress over the entire cross section of the member and no tensile stress will be developed.

EXAMPLE 17.4 A full-size rectangular timber member is used as a short compression post similar to that shown in Figure 17.10a. The cross section of the post has dimensions of $b = 250$ mm and $d = 400$ mm. The post is subjected to an eccentric load P of 175 kN applied 125 mm from the Y–Y axis. Refer to Figure 17.10a and calculate the combined stresses in the outer fibers at edges MM and NN.

Solution The axial compressive stress due to the load P is

$$\sigma_c = -\frac{P}{A} = -\frac{175 \times 10^3\text{ N}}{250\text{ mm }(400\text{ mm})} = -1.750\text{ MPa}$$

The moment caused by the eccentric load is

$$M = Pe = (175 \times 10^3\text{ N})(125\text{ mm}) = 21.9 \times 10^6\text{ N}\cdot\text{mm}$$

The bending stress developed by the moment is

$$f_b = \frac{Mc}{I} = \frac{Pec}{I}$$

where $c = d/2 = 200$ mm and I, with respect to the Y–Y bending axis, is calculated from

$$I_y = \frac{bd^3}{12} = \frac{(250\text{ mm})(400\text{ mm})^3}{12} = 1.333 \times 10^9\text{ mm}^4$$

Therefore,

$$f_b = \frac{Mc}{I} = \frac{(21.9 \times 10^6 \text{ N}\cdot\text{mm})(200 \text{ mm})}{1.333 \times 10^9 \text{ mm}^4} = \pm 3.29 \text{ MPa}$$

The combined stress is the algebraic sum of the two stresses and is shown in Figure 17.10e:

$$f_{max} = -\frac{P}{A} \pm \frac{Pec}{I}$$

The combined fiber stress at edge *MM* is

$$f_{max} = -1.750 - 3.29 = -5.04 \text{ MPa (compression)}$$

The combined fiber stress at edge *NN* is

$$f_{max} = -1.750 + 3.29 = +1.540 \text{ MPa (tension)}$$

EXAMPLE 17.5

A press frame is shown in Figure 17.11. When the press operates, a maximum force *P* of 30,000 lb is exerted between the upper and lower jaws, subjecting the arm of the frame to combined bending and axial stresses. Calculate the maximum tensile and compressive stresses at points *o* and *i* on plane *A–A*, where these points represent the outside and inside faces of the arm of the frame, respectively.

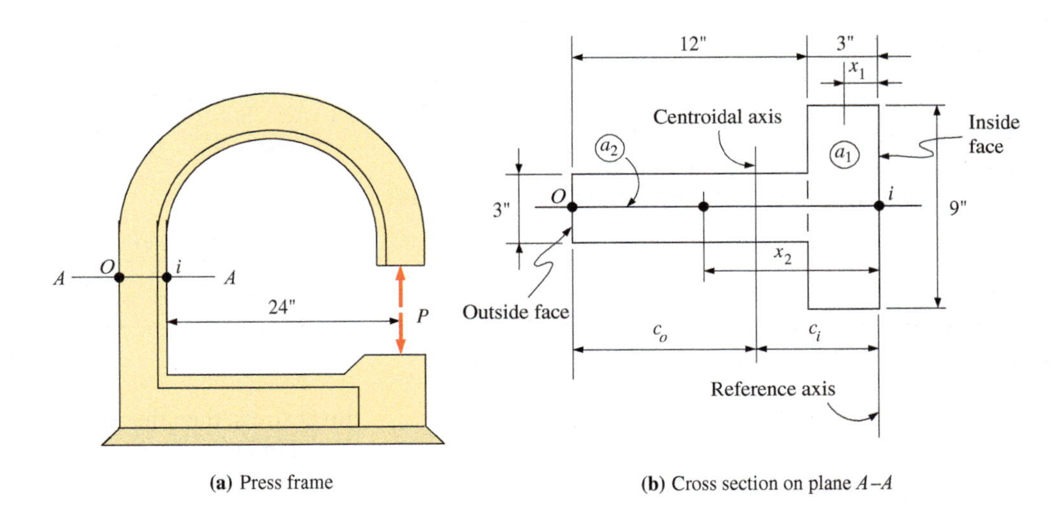

FIGURE 17.11 Eccentrically loaded press frame.

Solution The eccentrically applied load *P* results in a force reaction *P* and a moment reaction *Pe* at plane *A–A*, as shown on the free-body diagram of the frame (Figure 17.12).

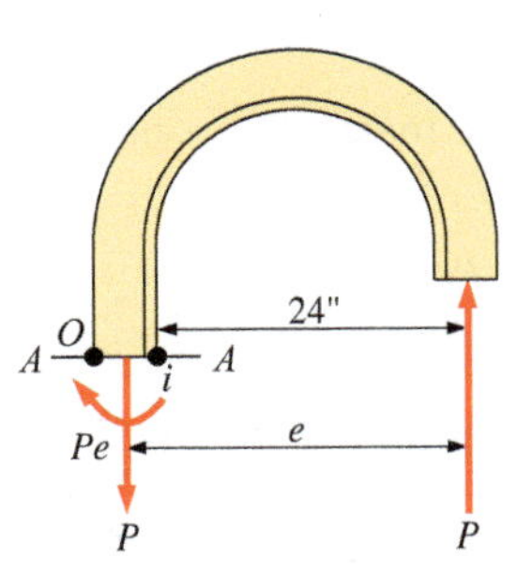

FIGURE 17.12 Press frame free-body diagram.

First, calculate the properties of the cross section of the frame (refer to Figure 17.11b). The cross-sectional area is

$$A = a_1 + a_2 = 27 + 36 = 63 \text{ in.}^2$$

The centroidal axis is located from the designated reference axis. Note that x is the distance from the centroid of each area a to the reference axis.

$$c_i = \frac{\Sigma ax}{\Sigma a} = \frac{a_1x_1 + a_2x_2}{a_1 + a_2} = \frac{(27\text{ in.}^2)(1.5\text{ in.}) + (36\text{ in.}^2)(9\text{ in.})}{27\text{ in.}^2 + 36\text{ in.}^2} = 5.79\text{ in.}$$

from which

$$c_o = 15 - 5.79 = 9.21\text{ in.}$$

The moment of inertia is calculated with respect to the centroidal axis:

$$I = \Sigma(I_o + ad^2)$$

$$= \left[\frac{9(3)^3}{12} + 9(3)(5.79 - 1.5)^2\right] + \left[\frac{3(12)^3}{12} + 3(12)(9.21 - 6)^2\right]$$

$$= 1320\text{ in.}^4$$

The stresses developed on plane A–A can be determined using Equation (17.3). Note that the direct tensile force reaction P develops a tensile stress equal to P/A that is uniformly distributed over the cross section. The P/A term has a positive sign because it is in tension. The bending moment Pe will develop a tensile stress on the inside face and a compressive stress on the outer face. Thus,

$$f_{\max} = +\frac{P}{A} \pm \frac{Pec}{I}$$

From the free-body diagram, the eccentricity $e = c_i + 24 = 29.79$ in. Therefore, the stress at the inside face, point i, is

$$f_i = +\frac{30{,}000\text{ lb}}{63\text{ in.}^2} + \frac{(30{,}000\text{ lb})(29.79\text{ in.})(5.79\text{ in.})}{1320\text{ in.}^4} = +4400\text{ psi (tension)}$$

and at the outside face, point o, the stress is

$$f_o = +\frac{30{,}000\text{ lb}}{63\text{ in.}^2} - \frac{(30{,}000\text{ lb})(29.79\text{ in.})(9.21\text{ in.})}{1320\text{ in.}^4} = -5760\text{ psi (compression)}$$

17.5 MAXIMUM ECCENTRICITY FOR ZERO TENSILE STRESS

When a short compression member is subjected to an eccentric load as discussed in Section 17.4, the maximum combined stresses, either tensile or compressive, will develop at the outer edges of the cross section. In Figure 17.13, these stresses occur at edges AB and CD, since the bending occurs with respect to the Y–Y plane.

As discussed in Section 17.3, the maximum combined stress will be

$$f_{\max} = -\frac{P}{A} \pm \frac{Pec}{I}$$

Note that the stress at edge AB will always be compressive since the eccentric load is closer to AB than it is to CD. The stress at CD could be compressive, tensile, or zero, depending on the eccentricity of the load. If the bending stress is greater than the direct axial stress, outer edge CD will develop a tensile stress, as shown in Figure 17.13e. On the other hand, if the bending stress is less than the direct axial stress, outer edge CD will develop a compressive stress, as shown in Figure 17.13f. If the bending stress is equal to the direct stress, then the stress at outer edge CD will be zero, as shown in Figure 17.13g.

Since several materials (such as concrete) have insignificant tensile strength, it is important to know the largest load eccentricity for which a tensile stress will not develop for given conditions. Although this discussion focuses on the zero tensile stress situation, the same theory applies to the case of zero compressive stress, albeit less important and less practical.

The limiting stress distribution for zero tensile stress is shown in Figure 17.13g, where zero stress occurs along edge CD. This situation will occur when the load eccentricity reaches and does not exceed a specific value. This maximum load eccentricity for the case of zero stress can be calculated, since at zero stress

$$\frac{Pec}{I} = \frac{P}{A}$$

If, in Figure 17.13, $AB = b$, $AC = d$, $c = d/2$, $I = bd^3/12$, and the cross-sectional area $= bd$, the preceding equation becomes

$$\frac{Pe(d/2)}{(bd^3/12)} = \frac{P}{bd}$$

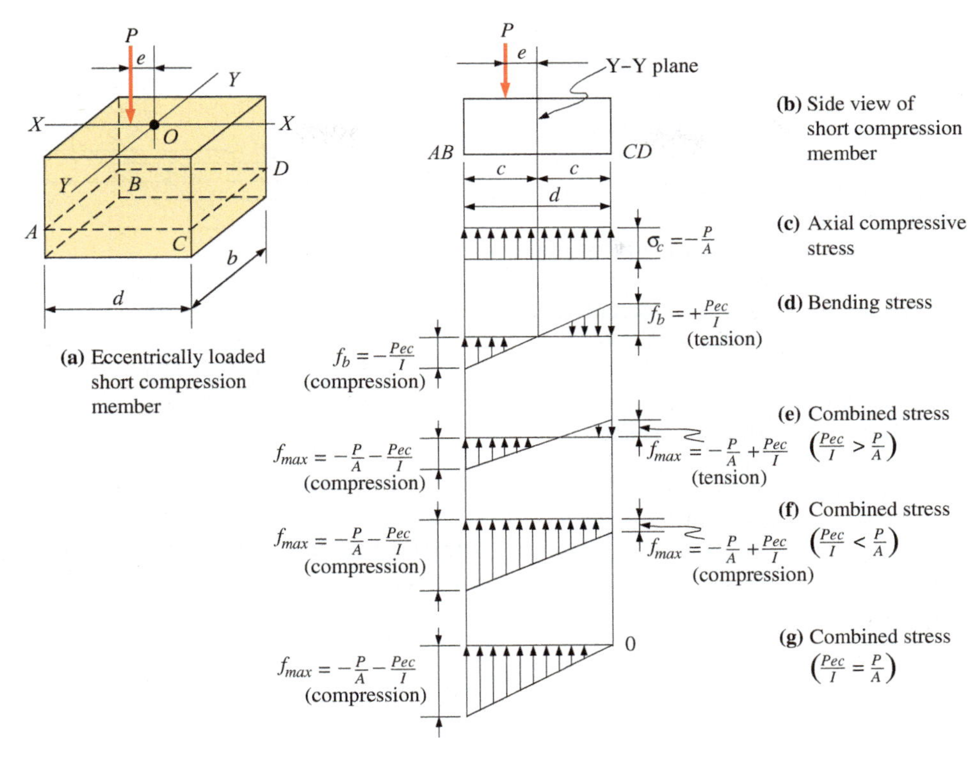

FIGURE 17.13 Combined stresses caused by eccentric load.

Solving for e yields

$$e = \frac{d}{6} \tag{17.4}$$

This equation represents the largest load eccentricity for which a tensile stress will not develop.

If the load eccentricity e becomes greater than $d/6$, then the bending stress becomes greater than the direct axial stress and a tensile stress will develop at the outer edge CD, as shown in Figure 17.13e. If the load eccentricity is less than or equal to $d/6$, then only compressive stresses will develop over the cross section of the member, as shown in Figure 17.13f and g.

Should the eccentricity occur in the other direction (on the opposite side of the Y–Y axis of Figure 17.13a), then the largest eccentricity for zero tensile stress will still be $d/6$, but it will be measured to the right of the Y–Y axis. Therefore, if the eccentric load is applied within the middle one-third $(d/6) + (d/6)$ of the length d, tensile stresses will not be developed.

The same principles would hold if the eccentric load were applied along the Y–Y axis instead of the X–X axis, as shown in Figure 17.13a. The maximum eccentricity for the limiting zero tensile stress would be $b/6$.

17.6 ECCENTRIC LOAD NOT ON CENTROIDAL AXIS

In this section, we consider the case of an eccentric load that does not lie on either centroidal axis, which results in a situation sometimes termed *double eccentricity*, illustrated in Figure 17.14. The combined stress at any point is the algebraic sum of

1. The direct axial stress due to the load P at the centroid O
2. The bending stress due to the moment Pe_1 with respect to the bending axis Y–Y
3. The bending stress due to the moment Pe_2 with respect to the bending axis X–X

For points around the outside of the member, this sum can be expressed in equation form:

$$f_{\max} = -\frac{P}{A} \pm \frac{Pe_1c_1}{I_y} \pm \frac{Pe_2c_2}{I_x} \tag{17.5}$$

where c_1 and c_2 represent the distance from the centroidal axes to the outer fibers of the member where the combined

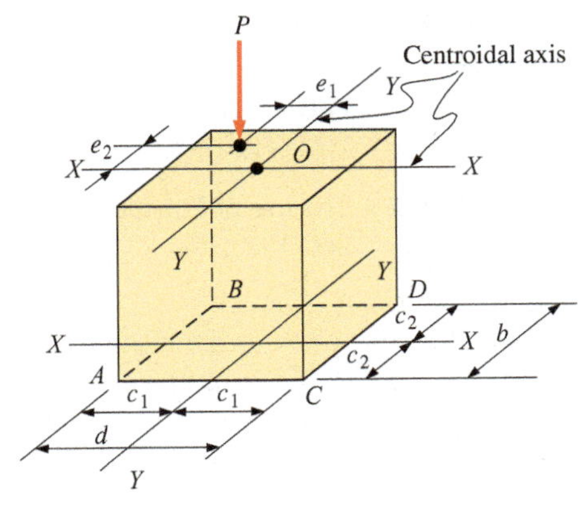

FIGURE 17.14 Short compression member with double eccentricity.

stress is being computed. Although the preceding is written for any point around the outside of the member, where the maximum and minimum stresses will occur, the stress at *any* point within the member can also be found. We simply substitute y_1 and y_2 for the respective c dimensions, where the y dimensions locate the point with respect to the appropriate centroidal axes; refer to Equation (14.3).

The limits of the position of the eccentric load P so that no tensile stress will occur at any location is similar to that established in Section 17.5. The largest load eccentricity along the Y–Y axis is $b/6$ and along the X–X axis, $d/6$. Since the eccentric load does not lie on either axis, however, it must fall within an area that is formed by connecting those designated limiting points in order that no tensile stress develop. As shown in Figure 17.15, if the eccentric load on a rectangular short compression member falls within the diamond-shaped area, no tensile stress will develop anywhere over the cross section of the member. This area is commonly called the *kern* of the cross section, and its shape will depend on the shape of the cross section.

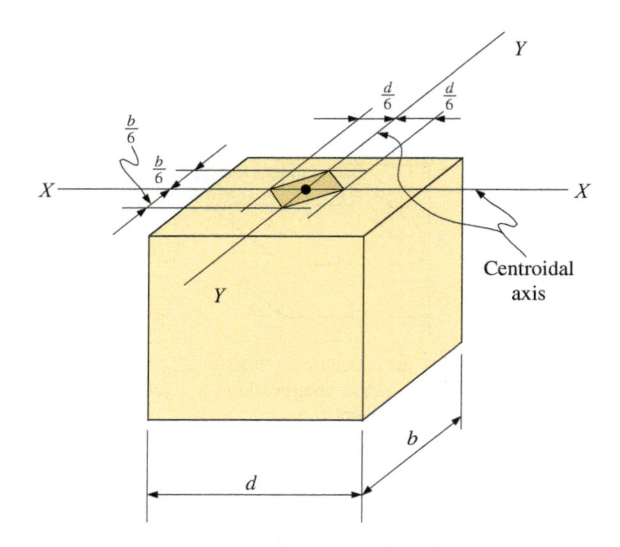

FIGURE 17.15 Kern area of rectangular short compression member.

EXAMPLE 17.6 A short compression member is subjected to a compressive load of 100,000 lb. There is a double eccentricity, as shown in Figure 17.16. The member is 12 in. by 14 in. in cross section. Calculate the combined stress at each of the four corners *A, B, C,* and *D*. Locate the line of zero stress.

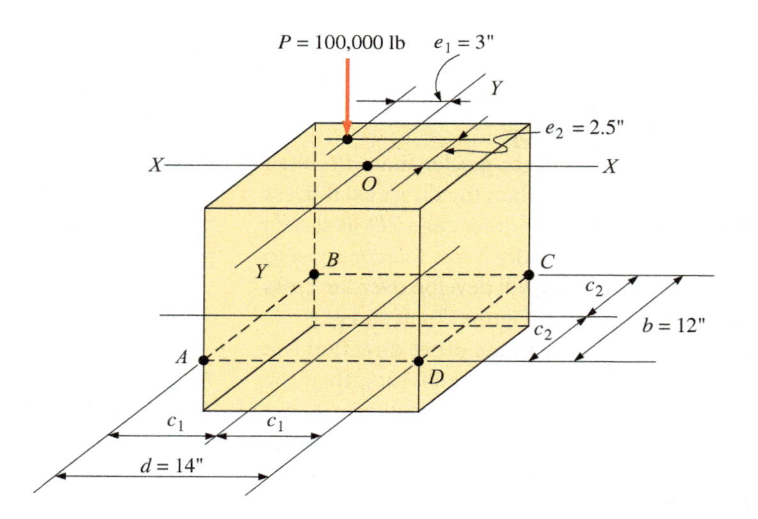

FIGURE 17.16 Rectangular short compression member with double eccentricity.

Solution The use of Equation (17.5) will require the calculation of the moments of inertia with respect to the X–X and Y–Y axes:

$$I_x = \frac{db^3}{12} = \frac{(14\text{ in.})(12\text{ in.})^3}{12} = 2016\text{ in.}^4$$

$$I_y = \frac{bd^3}{12} = \frac{(12\text{ in.})(14\text{ in.})^3}{12} = 2744\text{ in.}^4$$

The various terms of the equation are evaluated as follows:

$$\frac{P}{A} = \frac{100{,}000\ \text{lb}}{(12\ \text{in.})(14\ \text{in.})} = 595\ \text{psi}$$

$$\frac{Pe_1c_1}{I_y} = \frac{(100{,}000\ \text{lb})(3\ \text{in.})(7\ \text{in.})}{2744\ \text{in.}^4} = 765\ \text{psi}$$

$$\frac{Pe_2c_2}{I_x} = \frac{(100{,}000\ \text{lb})(2.5\ \text{in.})(6\ \text{in.})}{2016\ \text{in.}^4} = 744\ \text{psi}$$

The stresses at the four points can now be calculated using Equation (17.5):

$$f_{\text{max}} = -\frac{P}{A} \pm \frac{Pe_1c_1}{I_y} \pm \frac{Pe_2c_2}{I_x}$$

At point *A*,

$$f_{\text{max}} = -595 - 765 + 744 = -616\ \text{psi (compression)}$$

At point *B*,

$$f_{\text{max}} = -595 - 765 - 744 = -2104\ \text{psi (compression)}$$

At point *C*,

$$f_{\text{max}} = -595 + 765 - 744 = -574\ \text{psi (compression)}$$

At point *D*,

$$f_{\text{max}} = -595 + 765 + 744 = +914\ \text{psi (tension)}$$

The line of zero stress is located as shown in Figure 17.17.

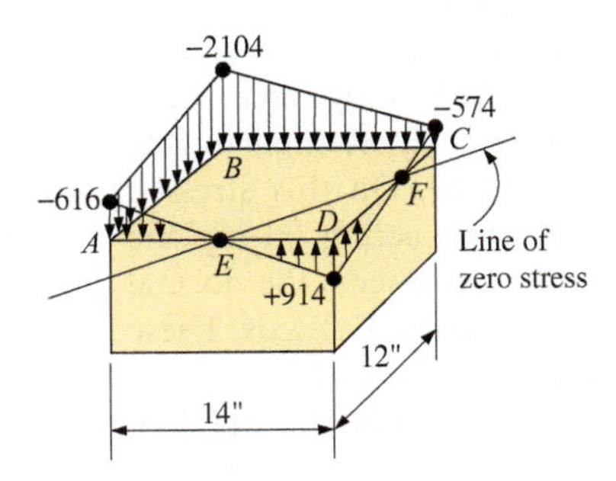

FIGURE 17.17 Line of zero stress.

Calculating *DE* from similar triangles,

$$\frac{\text{DE}}{914} = \frac{14 - \text{DE}}{616}$$

from which $DE = 8.36$ in. Similarly, $DF = 7.37$ in.

17.7 COMBINED NORMAL AND SHEAR STRESSES

It was established in Chapter 14 that in homogeneous elastic beams, where stresses are proportional to strains, two kinds of stresses are developed simultaneously. These are the bending stresses, consisting of tensile and compressive stresses, which are normal to the surfaces on which they act, and shear stresses, which are parallel to the surfaces on which they act. Bending stresses are called *normal stresses;* shear stresses are called *tangential stresses.* As shown in Chapter 14, they can be calculated using the following expressions: From the flexure formula,

$$f_b = \frac{Mc}{I}$$

and from the general shear formula,

$$f_v = \frac{VQ}{Ib}$$

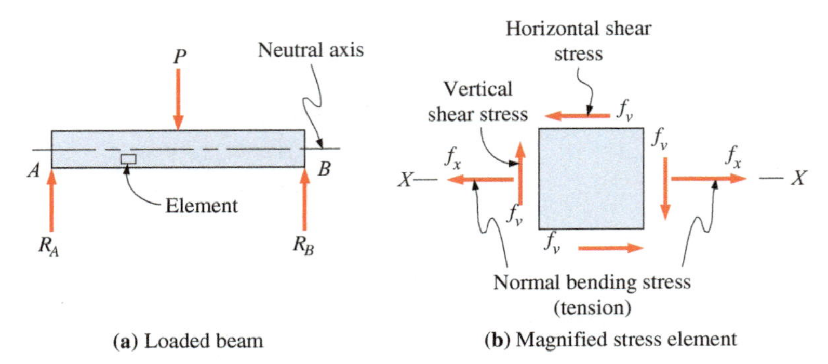

FIGURE 17.18 Combined normal and shear stress.

Consider an infinitesimal element of a simply supported beam, as shown in Figure 17.18. The element is taken at some point where the shear and the bending moment are not equal to zero and at some location other than the extreme fibers or the neutral axis; thus, the element is subjected to both bending and shear stresses. These stresses are shown acting on the faces of the element in Figure 17.18b. The normal stress is represented by f_x and the shear stress is represented by f_v. The combination of these stresses will result in maximum and minimum normal and shear stresses being developed on planes that are inclined with respect to the axis of the beam.

For the development of the various relationships that exist in this situation, we consider a more general case of an element from a member that is acted on by forces from many directions. This general case is represented by the stressed element shown in Figure 17.19. On this stressed element, f_x represents the normal stress acting on a plane perpendicular to the X–X axis and f_y represents the normal stress acting on a plane perpendicular to the Y–Y axis. These two stresses may be either tension or compression (tension is shown), and they may result either from the bending moment or direct load. The shear stress at the same point acting on all the faces of the element is represented by f_v. Recall that in Section 11.6 the shear stresses on mutually perpendicular planes were shown to be equal. The shear stress may be the result of an externally applied torsional moment (Chapter 12) or beam shear (Chapter 14).

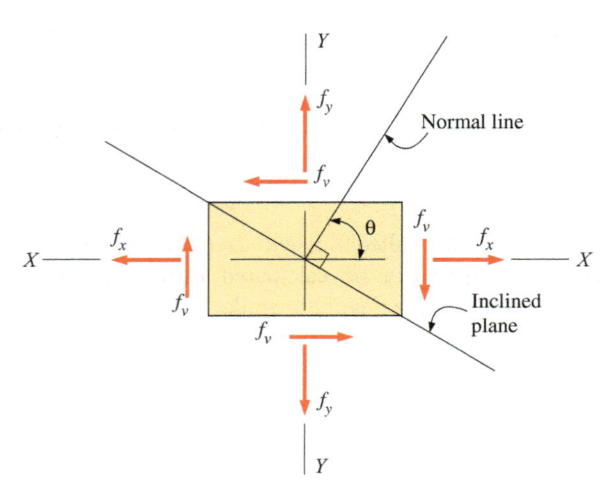

FIGURE 17.19 Stressed element.

By combining the stresses shown in Figure 17.19, the normal and shear stresses on any inclined plane can be determined. We define the direction of a plane by specifying the angle between the X–X axis and a line that is *normal* to the plane. This convention is the same as that used in Chapter 11. If we cut the rectangular element along the inclined plane indicated in Figure 17.19, we obtain the triangular element shown in Figure 17.20 as a free body.

Note in Figure 17.20 that since the top portion of the rectangular element has been removed, it must be replaced by the forces that it exerted upon the inclined plane of the lower portion. These forces consist of both normal and shear forces acting along the inclined plane. These unknown forces (and the resulting stresses) are the ones we evaluate. Note that in Figure 17.19 stresses are shown acting on the various faces of the element, rather than forces. Each of these stresses is assumed to be uniformly distributed over the area on which it acts. Prior to applying equations of equilibrium to the free body, these stresses must first be converted to forces (recall that force equals stress multiplied by area). Since the element is shown as being two-dimensional, the thickness of the element perpendicular to the plane of the paper is taken as unity (a unit thickness).

All stresses are converted to forces in the free body of Figure 17.20 so that equations of equilibrium can now be applied. In this figure, f_n represents the normal stress (either tension or compression) on the inclined plane and f'_v represents the tangential stress (shear stress) on the inclined plane. Taking an algebraic summation of forces normal to the inclined plane with the element faces assigned length dimensions of h, w, and d as shown gives

$$f_n(d)(1) = f_x(h)(1)\cos\theta + f_y(w)(1)\sin\theta - f_v(h)(1)\sin\theta - f_v(w)(1)\cos\theta$$

Dividing through by $(d)(1)$ gives

$$f_n = f_x\left(\frac{h}{d}\right)\cos\theta + f_y\left(\frac{w}{d}\right)\sin\theta - f_v\left(\frac{h}{d}\right)\sin\theta - f_v\left(\frac{w}{d}\right)\cos\theta$$

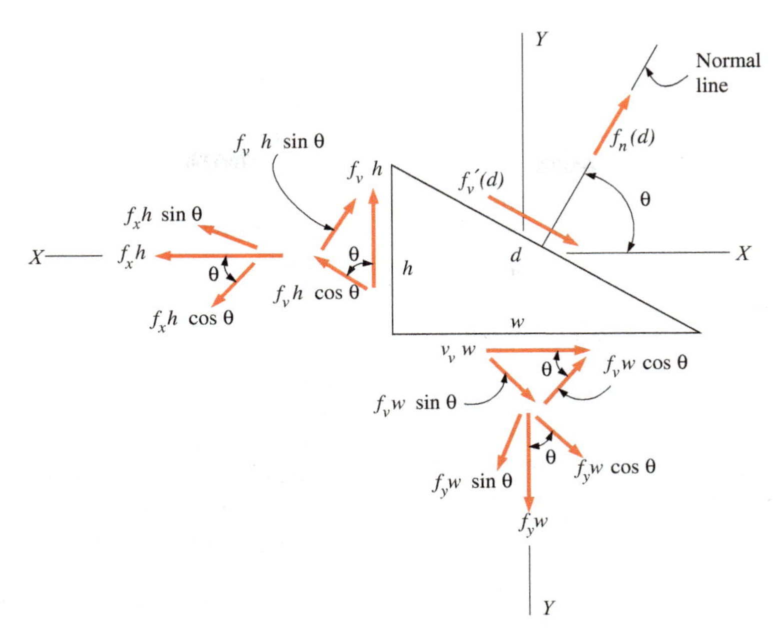

FIGURE 17.20 Free-body diagram.

Since $\sin\theta = w/d$ and $\cos\theta = h/d$,

$$f_n = f_x \cos^2\theta + f_y \sin^2\theta - 2f_v \sin\theta\cos\theta \qquad \textbf{(17.6)}$$

This equation represents the normal stress on any plane, which has a normal inclined at an angle θ with the X–X axis in terms of stresses on the planes perpendicular to the X–X and Y–Y axes.

Next, taking an algebraic summation of forces parallel to the inclined plane gives

$$f_v'(d)(1) = f_v(h)(1)\cos\theta + f_x(h)(1)\sin\theta - f_y(w)(1)\cos\theta - f_v(w)(1)\sin\theta$$

Dividing throughout by $(d)(1)$ gives

$$f_v' = f_v\left(\frac{h}{d}\right)\cos\theta + f_x\left(\frac{h}{d}\right)\sin\theta - f_y\left(\frac{w}{d}\right)\cos\theta - f_v\left(\frac{w}{d}\right)\sin\theta$$

Again, since $\sin\theta = w/d$ and $\cos\theta = h/d$,

$$f_v' = (f_x - f_y)\sin\theta\cos\theta + f_v(\cos^2\theta - \sin^2\theta) \qquad \textbf{(17.7)}$$

This equation represents the shear stress on any plane, which has a normal inclined at an angle θ with the X–X axis in terms of the stresses on planes perpendicular to the X–X and Y–Y axes.

If the inclined plane is rotated by varying the angle θ, it will reach a position for which the normal stress acting on it is a maximum or minimum. The position of the plane for maximum normal stress is perpendicular to the position of the plane for a minimum normal stress. On these planes there are no shear stresses. The planes on which the normal stresses become maximum or minimum are called the *principal planes*, and the corresponding normal stresses acting on these planes are called the *principal stresses*.

To determine the value of the angle θ_P that would locate the principal planes, we set the value of f_v', given by Equation (17.7), equal to zero and solve for the angle in terms of the stresses on the planes perpendicular to the X–X and Y–Y axes:

$$0 = (f_x - f_y)\sin\theta_P\cos\theta_P + f_v(\cos^2\theta_P - \sin^2\theta_P)$$

With the use of some trigonometric identities, the solution for θ_P can be shown to be

$$\tan 2\theta_P = -\frac{2f_v}{f_x - f_y} \qquad \textbf{(17.8)}$$

Since $\tan 2\theta = \tan(2\theta + 180°)$, we see that the use of Equation (17.8) will result in two possible values of θ_P, differing by 90°. On one plane thus described, the *maximum* normal stress will exist, whereas on the other plane the *minimum* normal stress will exist.

Having solved for $\tan 2\theta_P$, the two values of θ_P can be determined. By substituting one of the values of θ_P, from Equation (17.8), into Equation (17.6), it can be determined which of the two principal stresses is acting on that plane. The other principal stress acts on the other plane. We define the principal stresses as f_1 and f_2, where f_1 is the algebraically larger stress. Equation (17.6) may be modified to obtain the principal stresses directly. It will be shown later (in Section 17.9) that the principal stresses can be obtained from

$$f_{1,2} = \left(\frac{f_x + f_y}{2}\right) \pm \sqrt{\frac{(f_x - f_y)^2}{4} + f_v^2} \qquad \textbf{(17.9)}$$

By substituting the value of θ_P as determined from Equation (17.8) into Equation (17.7) to obtain f_v', it may be shown that the shear stresses on the principal planes are equal to zero.

The orientation (θ_s) of the planes of maximum shear stress can be developed mathematically from Equation (17.7):

$$\tan 2\theta_s = \frac{f_x - f_y}{2f_v} \quad \textbf{(17.10)}$$

Substitution of one of the values of θ_s from Equation (17.10) into Equation (17.7) (for f'_v) will yield the maximum shear stress, which can be expressed by the direct equation

$$f'_{v(\max)} = \pm\sqrt{\frac{(f_x - f_y)^2}{4} + f_v^2} \quad \textbf{(17.11)}$$

This equation represents the shear stress developed on the inclined plane, which has a normal inclined at an angle of θ_s with the original X–X axis of the stressed element, as defined by Equation (17.10). Both positive and negative values of shear stress result from the calculation. They could be considered algebraic maximum and minimum values; they are, however, equal in absolute magnitude.

Note that the maximum shear stress expression is the same as the radical that occurs in Equation (17.9). Hence, the principal stresses can be calculated from

$$f_{1,2} = \frac{f_x + f_y}{2} \pm f'_{v(\max)}$$

Note that if θ_s, which defines the position of the maximum shear stress, Equation (17.10), is substituted into Equation (17.6), it will be found that the planes of maximum shear stress are not always free of normal stresses. This situation is unlike that on the principal planes, where no shear stresses exist.

The expression for the normal stress acting on the plane of maximum shear stress is

$$f_n = \frac{f_x + f_y}{2} \quad \textbf{(17.12)}$$

Equations (17.6) through (17.11) are valid for the directions of stresses shown in Figure 17.19. If the stresses are opposite in direction, the signs for the various terms should be changed. For normal stresses, tensile stresses are considered positive; compressive stresses are considered negative. For shear stresses, the positive sense is indicated as shown in Figure 17.19. Also, as shown in Figure 17.19, θ is positive when measured counterclockwise from the horizontal and negative when measured clockwise from the horizontal.

EXAMPLE 17.7 An element of a machine member subjected to biaxial loading is stressed as shown in Figure 17.21. (a) Compute the normal and shear stress intensities on a plane that has a normal rotated 60° counterclockwise from the X–X axis. (b) Compute the principal stresses and the orientation of the principal planes. (c) Compute the maximum shear stress and locate the plane on which it acts.

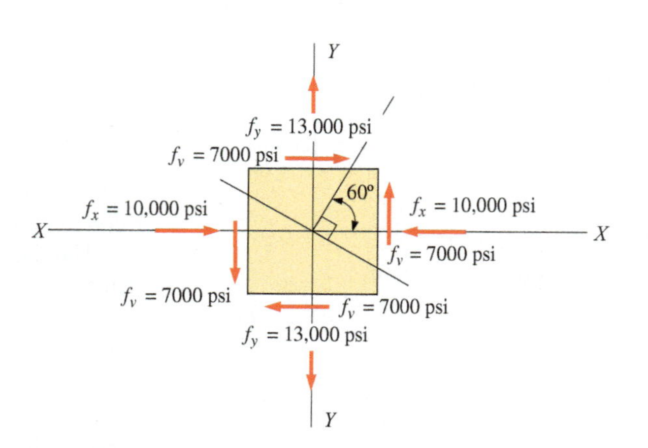

FIGURE 17.21 Original stressed element.

Solution (a) To compute the normal and shear stress on any plane, Equations (17.6) and (17.7) can be used. In accordance with our sign convention for tensile and compressive stresses, and with reference to Figure 17.19 for the shear stress signs, $f_x = -10{,}000$ psi, $f_y = +13{,}000$ psi, $f_v = -7000$ psi, and $\theta = +60°$. From Equation (17.6), the normal stress is

$$\begin{aligned} f_n &= f_x\cos^2\theta + f_y\sin^2\theta - 2f_v\sin\theta\cos\theta \\ &= -10{,}000(0.25) + 13{,}000(0.75) - 2(-7000)(0.8660)(0.500) \\ &= +13{,}310 \text{ psi (tension)} \end{aligned}$$

From Equation (17.7), the shear stress is

$$f'_v = (f_x - f_y)\sin\theta\cos\theta + f_v(\cos^2\theta - \sin^2\theta)$$
$$= (-10{,}000 - 13{,}000)(0.4333) + (-7000)(-0.500)$$
$$= -6470 \text{ psi}$$

Recall that Equations (17.6) and (17.7) were based on shear, as shown in Figure 17.20. Therefore, the minus sign for the shear stress indicates that the direction of the computed stress on the inclined plane is that shown in Figure 17.22.

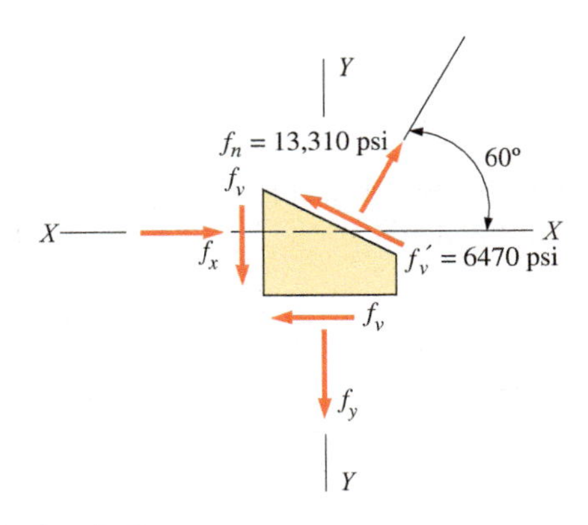

FIGURE 17.22 Stresses on inclined plane.

(b) The magnitudes of the principal stresses can be determined using Equation (17.9). The principal stresses are

$$f_{1,2} = \frac{f_x + f_y}{2} \pm \sqrt{\frac{(f_x - f_y)^2}{4} + f_v^2}$$
$$= \frac{-10{,}000 + 13{,}000}{2} \pm \sqrt{\frac{(-10{,}000 - 13{,}000)^2}{4} + (-7000)^2}$$
$$= +1500 \pm 13{,}460$$

from which

$$f_1 = +14{,}960 \text{ psi (tension)}$$
$$f_2 = -11{,}960 \text{ psi (compression)}$$

The planes on which the principal stresses act can be found using Equation (17.8):

$$\tan 2\theta_P = -\frac{2f_v}{f_x - f_y} = -\frac{2(-7000)}{-10{,}000 - 13{,}000} = -0.609$$

For convenience in understanding how θ_P is determined, Figure 17.23 shows a plot of the tangent function. If $\tan 2\theta_P$ is negative, then $2\theta_P$ must lie in the second or fourth quadrants. Also observe in Figure 17.23 that

$$\tan 2\theta = \tan(180° + 2\theta)$$

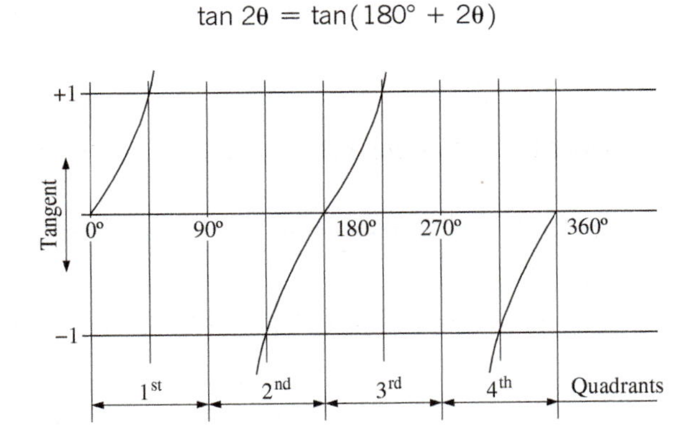

FIGURE 17.23 Tangent function (0° to 360°).

Further, relate angles in the first and second quadrants:

$$\tan 2\theta = -\tan(180° - 2\theta)$$

Relate angles in the first and fourth quadrants:

$$\tan 2\theta = -\tan(360° - 2\theta)$$

For $\tan 2\theta_P = -0.609$, if $2\theta_P$ were in the first quadrant (if the tangent value were positive) it would have the value of 31.34°. In the second quadrant, however,

$$2\theta_P = 180° - 31.34° = 148.66°$$

and in the fourth quadrant,

$$2\theta_P = 360° - 31.34° = 328.66°$$

Therefore, the principal planes are defined by two planes located by θ_P measured counterclockwise from the *X–X* axis:

$$\theta_P = \frac{148.66}{2} = 74.33°$$

$$\theta_P = \frac{328.66}{2} = 164.33°$$

To determine on which plane the maximum principal stress acts, substitute $\theta_P = 74.33°$ into Equation (17.6) along with the original given stresses as shown in Figure 17.21:

$$\begin{aligned} f_n &= f_x\cos^2\theta_P + f_y\sin^2\theta_P - 2f_v\sin\theta_P\cos\theta_P \\ &= -10{,}000(0.0730) + 13{,}000(0.9270) - 2(-7000)(0.2601) \\ &= +14{,}960 \text{ psi} \end{aligned}$$

Thus, the maximum principal stress of +14,960 psi occurs on the principal plane that has a normal oriented at 74.33° counterclockwise from the *X–X* axis, and the minimum principal stress of −11,960 psi occurs on the principal plane that has a normal oriented at 164.33° counterclockwise from the *X–X* axis. These results are commonly shown by rotating the element so that the *X–Y* axes of the original stressed element align with the principal planes and noting the principal stresses, as shown in Figure 17.24. The shear stresses on the principal planes are zero.

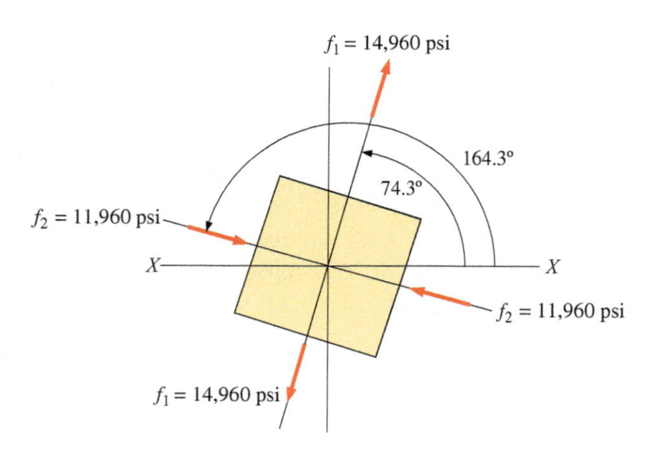

FIGURE 17.24 Principal stress element.

(c) The maximum shear stress can be determined from Equation (17.11):

$$\begin{aligned} f'_v &= \pm\sqrt{\frac{(f_x - f_y)^2}{4} + f_y^2} \\ &= \pm\sqrt{\frac{(-10{,}000 - 13{,}000)^2}{4} + (-7000)^2} \\ &= \pm 13{,}460 \text{ psi} \end{aligned}$$

The angles of inclination of the normals to the planes on which the maximum shear stresses occur can be obtained using Equation (17.10):

$$\tan 2\theta_s = \frac{f_x - f_y}{2f_v} = \frac{-10{,}000 - 13{,}000}{2(-7000)} = +1.643$$

The angle $2\theta_s$ is in the first and third quadrants and has values of 58.67° and 238.67°. Therefore,

$$\theta_s = \frac{58.67}{2} = 29.34°$$

and

$$\theta_s = \frac{238.67}{2} = 119.34°$$

Note that the maximum combined shear stress occurs on planes inclined at 45° to the principal planes. This is always true.

To determine the direction of the shear stress, substitute $\theta_s = 29.34°$ into Equation (17.7):

$$\begin{aligned} f'_v &= (f_x - f_y)\sin\theta_s\cos\theta_s + f_v(\cos^2\theta_s - \sin^2\theta_s) \\ &= (-10{,}000 - 13{,}000)(0.4900)(0.8717) + (-7000)(0.8717^2 - 0.4900^2) \\ &= -13{,}460 \text{ psi} \end{aligned}$$

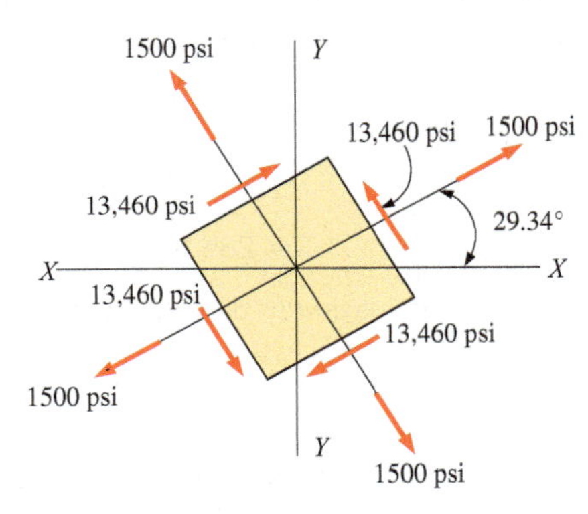

FIGURE 17.25 Maximum shear stress element.

With reference to Figure 17.19, the negative sign indicates that the shear stress on the plane, the normal of which is oriented at 29.34° with the original *X–X* axis, is as shown in Figure 17.25. The element is rotated so that the *X–Y* axes of the original stressed element are parallel with the planes of maximum shear stress and the shear stresses are noted.

The normal stresses acting on the planes of maximum shear stress can be determined from Equation (17.12):

$$f_n = \frac{f_x + f_y}{2} = \frac{(-10{,}000 + 13{,}000)}{2} = +1500 \text{ psi (tension)}$$

EXAMPLE 17.8 A simply supported short-span built-up steel beam is shown in Figure 17.26. At point *A*, the juncture of the web with the bottom flange, (a) calculate the principal stresses, (b) locate the principal planes, and (c) calculate the maximum shear stress. Neglect the weight of the beam and the effects of stress concentrations. The loading is assumed to be a point loading.

Solution (a) Prior to determining the principal stresses, the normal and shear stresses (f_x, f_y, and f_v) must be determined. The bending moment at midspan is

$$M = \frac{PL}{4} = \frac{(60 \text{ k})(2 \text{ ft})}{4} = 30 \text{ k-ft}$$

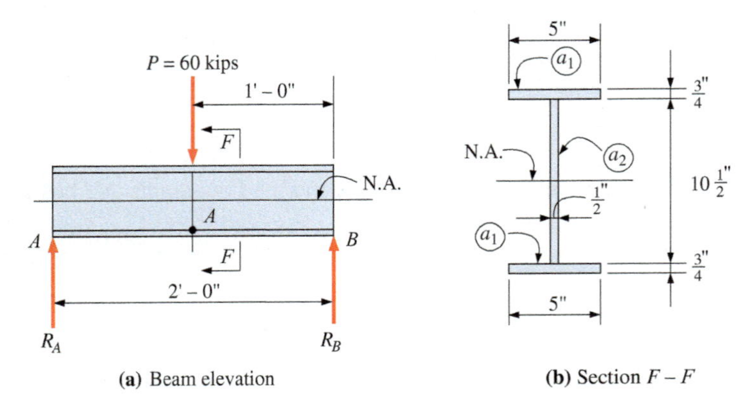

FIGURE 17.26 Built-up steel beam.

The shear at midspan (actually an infinitesimal distance to the left of midspan) is

$$V = \frac{P}{2} = \frac{60\text{ k}}{2} = 30\text{ k}$$

Calculating the bending stress at midspan at the outer fibers, the moment of inertia about the strong axis of the cross section, neglecting I_o for the flanges, is

$$\begin{aligned} I &= \Sigma I_{o2} + \Sigma a_1 d^2 \\ &= \frac{(0.5\text{ in.})(10.5\text{ in.})^3}{12} + 2(5\text{ in.})(0.75\text{ in.})(5.625\text{ in.})^2 \\ &= 285.5\text{ in.}^4 \end{aligned}$$

from which the maximum bending stress (at the outer fibers) is

$$f_b = \pm\frac{Mc}{I} = \pm\frac{(30\text{ k-ft})(12\text{ in./ft})(6\text{ in.})}{285.5\text{ in.}^4} = \pm 7.57\text{ ksi}$$

The bending stress at point A will be denoted as s_x and is calculated by ratio:

$$f_x = \frac{5.25\text{ in.}}{6\text{ in.}}(7.57\text{ ksi}) = +6.62\text{ ksi(tension)}$$

Calculating the shear stress at point A (just within the web), the first statical moment of area Q is determined:

$$Q = (5\text{ in.})(0.75\text{ in.})(5.625\text{ in.}) = 21.1\text{ in.}^3$$

from which

$$f_v = \frac{VQ}{Ib} = \frac{(30\text{ k})(21.1\text{ in.}^3)}{(285.5\text{ in.}^4)(0.5\text{ in.})} = 4.43\text{ ksi}$$

The stresses at point A are shown in Figure 17.27a acting on an infinitesimal element. Note that there is no normal stress on the plane perpendicular to the Y–Y axis ($f_y = 0$).

The principal stresses can now be determined. From Equation (17.9),

$$\begin{aligned} f_{1,2} &= \frac{f_x + f_y}{2} \pm \sqrt{\frac{(f_x - f_y)^2}{4} + f_v^2} \\ &= \frac{6.62}{2} \pm \sqrt{\frac{(+6.62)^2}{4} + 4.43^2} \\ &= 3.31 \pm 5.53 \end{aligned}$$

from which

$$\begin{aligned} f_1 &= +8.84\text{ ksi (tension)} \\ f_2 &= -2.22\text{ ksi (compression)} \end{aligned}$$

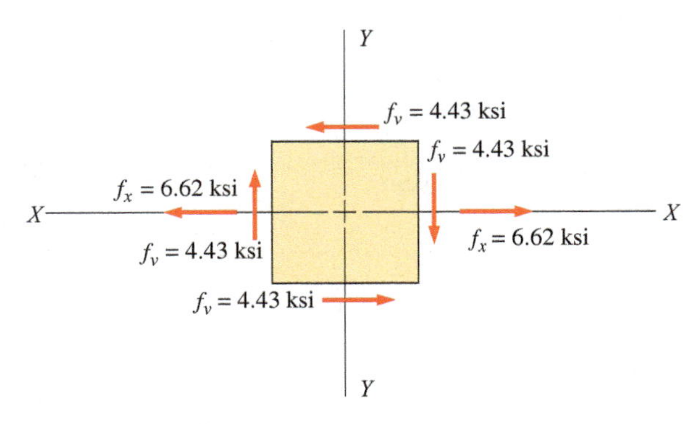

(a) Original stressed element

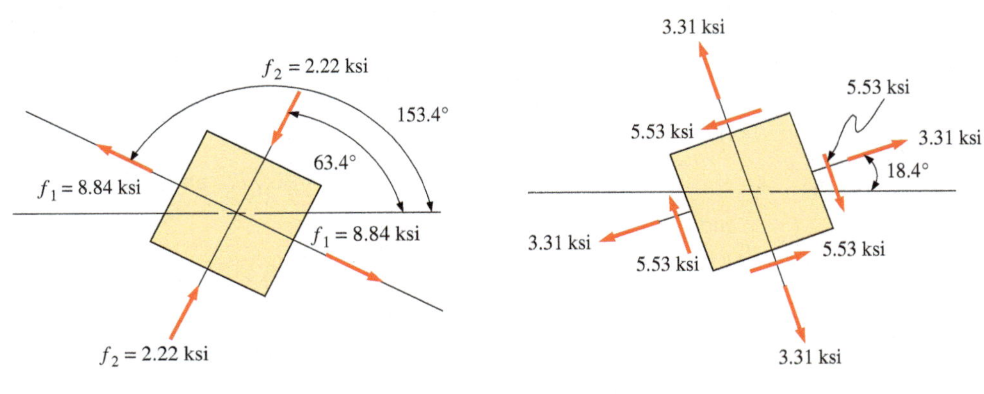

(b) Principal stress element

(c) Maximum shear stress element

FIGURE 17.27 Stress elements.

Note that the maximum principal stress f_1 is larger than the maximum bending stress at the outer fibers.

(b) The principal planes can be located using Equation (17.8):

$$\tan 2\theta_p = -\frac{2f_v}{f_x - f_y} = -\frac{2(4.43)}{6.62} = -1.338$$

Therefore, $2\theta_P$ lies in the second and fourth quadrant (refer to Figure 17.23). If it were in the first quadrant, $2\theta_P$ would be 53.23°. See, then, that in the second quadrant,

$$2\theta_p = 180° - 52.23° = 126.77°$$

and in the fourth quadrant,

$$2\theta_p = 360° - 53.23° = 306.77°$$

Therefore, the principal planes are defined by two planes that have normals located by θ_P measured counterclockwise from the X–X axis, where

$$\theta_P = \frac{126.77}{2} = 63.4°$$

and

$$\theta_P = \frac{306.77}{2} = 153.4°$$

To determine on which plane the maximum principal stress acts, substitute 63.4° for θ_P in Equation (17.6) along with the original computed normal and shear stress, as shown in Figure 17.27a:

$$\begin{aligned} f_n &= f_x\cos^2\theta_P + f_y\sin^2\theta_P - 2f_v\sin\theta_P\cos\theta_P \\ &= 6.62(0.2005) - 2(4.43)(0.4004) \\ &= -2.22 \text{ ksi} \end{aligned}$$

This value is the minimum principal stress. Thus, the maximum principal stress of +8.84 ksi occurs on the principal plane that has a normal oriented 153.4° counterclockwise from the original *X–X* axis of the element. The principal stress element is shown in Figure 17.27b.

(c) The maximum shear stress can be determined from Equation (17.11):

$$f'_v = \pm\sqrt{\frac{(f_x - f_y)^2}{4} + f_v^2}$$

$$= \pm\sqrt{\frac{(-6.62)^2}{4} + 4.43^2}$$

$$= \pm 5.53 \text{ ksi}$$

The angle of inclination of the normal to the plane on which the maximum shear stress occurs, measured counterclockwise from the *X–X* axis of the element, can be determined from Equation (17.10):

$$\tan 2\theta_s = \frac{f_x - f_y}{2f_v} = \frac{6.62}{2(4.43)} = +0.7472$$

Therefore,

$$2\theta_s = 36.77° \quad \text{and} \quad 216.77°$$

and

$$\theta_s = 18.4° \quad \text{and} \quad 108.4°$$

As expected, the principal planes and the planes of maximum shear stress are inclined at 45° to each other. You may wish to verify that the principal planes are free of shear stress by substituting θ values of 63.4° (and/or 153.4°) into Equation (17.7).

To determine the direction of the shear stress, substitute 18.4° for θ in Equation (17.7):

$$f'_v = (f_x - f_y)\sin\theta_s\cos\theta_s + f_v(\cos^2\theta_s - \sin^2\theta_s)$$

$$= 6.62(0.3156)(0.9489) + 4.43(0.9004 - 0.0996)$$

$$= +1.983 + 3.548$$

$$= +5.53 \text{ ksi}$$

The positive sign indicates that the shear stress on the plane, which has a normal at 18.4° with the original *X–X* axis, is oriented as shown in Figure 17.27c.

The normal stress f_n acting on the planes of maximum shear stress is determined from Equation (17.12):

$$f_n = \frac{f_x + f_y}{2} = \frac{+6.62}{2} = +3.31 \text{ ksi (tension)}$$

17.8 MOHR'S CIRCLE

An ingenious method for the solution of the state of stress at a point in a stressed body was presented by Otto Mohr, a German engineer, in 1882. The method, called *Mohr's circle*, is graphical (or can be partially graphical) and can be used to evaluate the variation of shear and normal stresses on all inclined planes at a point in a body.

In this section, we discuss a simple uniaxial application to determine the normal and shear stresses developed on an inclined plane of an axially loaded prismatic member. More complex applications are discussed in Section 17.9.

With reference to Figure 17.28a, the member being considered is a bar of cross-sectional area *A*, loaded in tension with an axial load *P*. Our task is to find the normal and shear stresses on a plane that has a normal inclined at an angle θ measured counterclockwise from the *X–X* axis, as shown in Figure 17.28b. The sequence of steps in the construction and interpretation of the Mohr circle for this application is as follows:

1. Establish a rectangular coordinate system in which normal stresses will be plotted as abscissas (horizontally) and shear stresses will be plotted as ordinates (vertically). (See Figure 17.28c.)
2. To some convenient scale, draw a circle with its center on the horizontal axis at point *M* and tangent to the vertical axis. For this application, the member is subjected

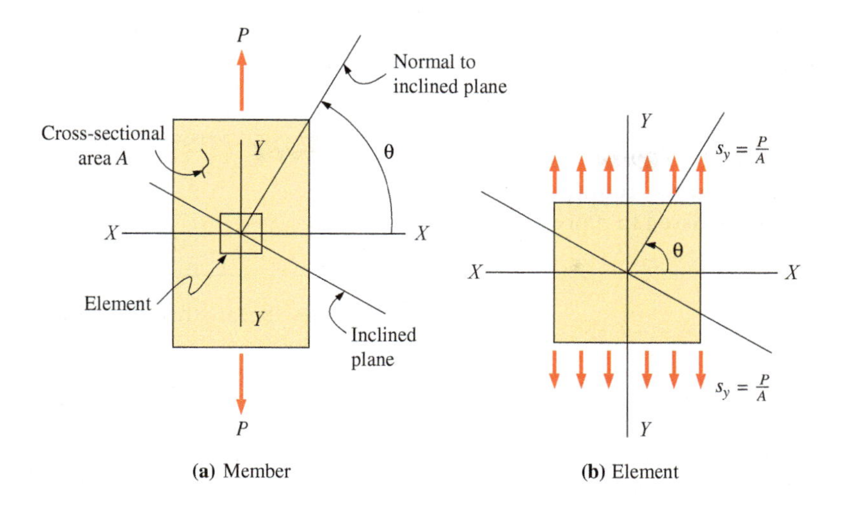

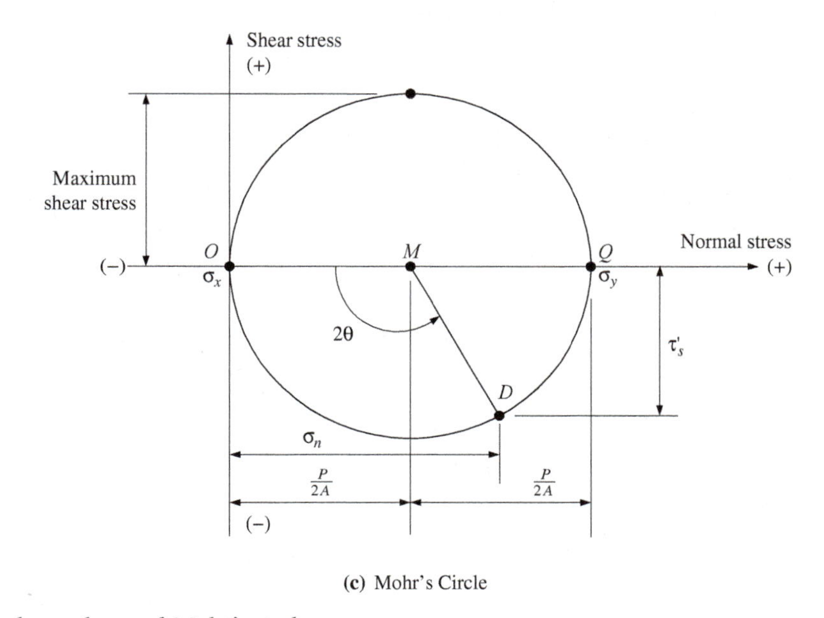

FIGURE 17.28 Axially loaded member and Mohr's circle.

to a tensile load. The circle is drawn to the right of the vertical axis in the area that represents positive (tensile) normal stress. (Compressive stresses are considered negative.) Point Q represents the maximum normal stress σ_y. Therefore, diameter OQ equals P/A and the radius equals $P/(2A)$, as shown.

3. The origin, point O, represents σ_x, which in this example is zero. MO represents the X–X axis. MQ represents the Y–Y axis. From the radius MO (the X–X axis), turn a counterclockwise angle of 2θ, draw the radius from M to intersect the circle, and designate the intersection as point D. Angle θ is as defined in Figure 17.28b. The counterclockwise direction, as shown on the element, is maintained when turning off the angle on the circle.
4. The abscissa of point D represents the tensile normal stress (σ_n) developed on the inclined plane under consideration. The ordinate of D represents the shear stress (τ') developed on the inclined plane. Further, note that the maximum ordinate on the circle represents the maximum shear stress.

As may be observed in Figure 17.28, as θ varies, the shear stress and normal stress on the inclined plane also vary. The circle displays all the combinations of normal and shear stresses.

Although Mohr's circle can be a graphical solution, one may also use trigonometric computations in conjunction with the graphical construction. This procedure permits the use of an approximate sketch, not necessarily to scale, while still maintaining acceptable numerical accuracy in the results.

For this brief introduction to Mohr's circle using a uniaxial stress example, signs of the stresses, as shown in Figure 17.28c, are based on the usual sign convention, positive stresses being plotted upward and to the right from the origin at O. The ordinate of point D indicates that a negative shear stress was determined on the indicated plane. With reference to Figure 17.29, we see that this shear stress tends to turn the element counterclockwise. This rule for Mohr's circle is discussed in more detail in Section 17.9.

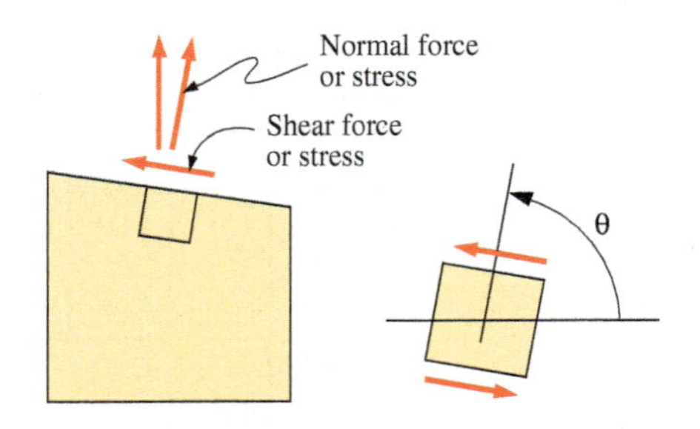

FIGURE 17.29 Shear stress direction on inclined plane.

EXAMPLE 17.9 A 10-in.-long metal block, 4 in. by 4 in. in cross section, is subjected to a compressive load of 32,000 lb. Using Mohr's circle, (a) determine the normal (tensile or compressive) stress and shear stress on a plane that has a normal inclined 60° counterclockwise from the X–X axis, as shown in Figure 17.30, and (b) determine the magnitude of the maximum shear stress and locate the plane on which it acts.

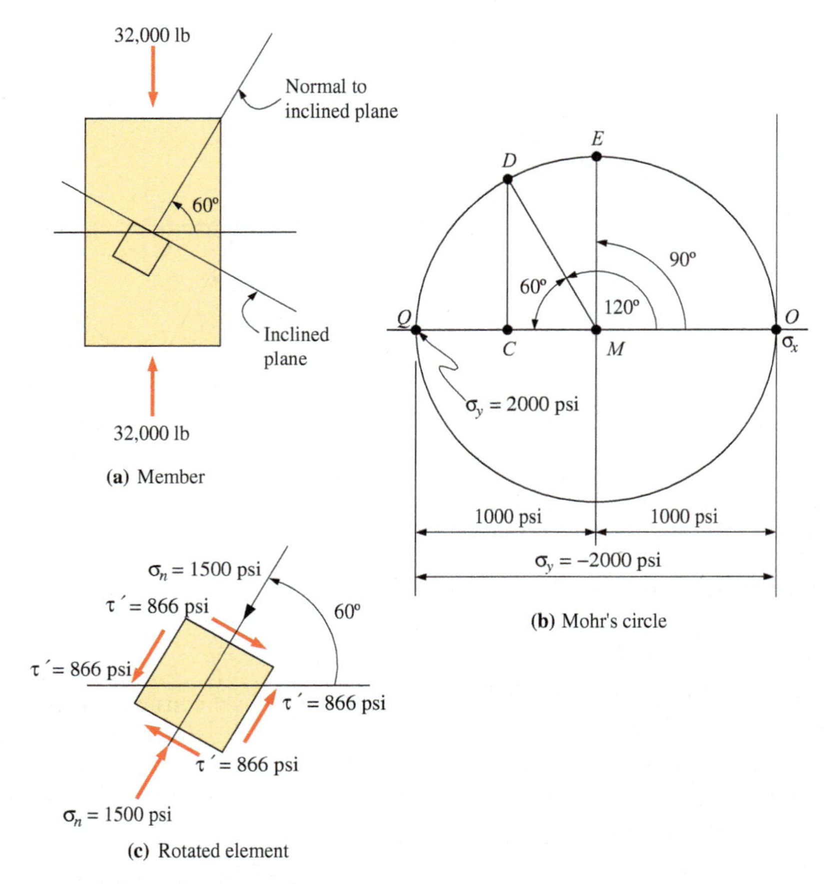

FIGURE 17.30 Mohr's circle: uniaxial stress.

Solution (a) To construct Mohr's circle, compute the maximum compressive stress on a plane perpendicular to the longitudinal axis of the member:

$$\sigma_y = -\frac{P}{A} = -\frac{32{,}000\ \text{lb}}{16\ \text{in.}^2} = -2000\ \text{psi (compression)}$$

The radius of Mohr's circle, then, is $P/(2A)$ or 1000 psi.

Next, draw the circle tangent to the vertical axis using a radius of 1000 psi and having its center on the horizontal axis at M, as shown in Figure 17.30. Note that the circle is drawn on the left side of the origin since σ_y is negative. Point Q represents σ_y, point O represents σ_x, and MO represents the X–X axis.

From radius *MO*, turn off a counterclockwise angle of 120° (this is 2θ), draw the radius from *M* to intersect the circle, and designate this point as *D*. The values of the compressive stress and the shear stress are represented by point *D* and can be scaled graphically or can be computed from the triangular relationships shown in the diagram. (Again, the accuracy of the graphical solution will depend on the accuracy with which the circle was constructed.)

Note that the coordinates of point *D* on Mohr's circle show that the normal stress is a negative value, indicating a compressive stress. Also note that the shear stress is a positive value that, in effect, indicates its direction on the plane under consideration.

Calculate these values using Mohr's circle as the reference and neglect signs (since the resulting signs are obvious by inspection):

$$\begin{aligned}\sigma_n &= OC = 1000 + 1000\cos 60^\circ \\ &= 1500 \text{ psi (compression) (negative)} \\ \tau' &= DC = 1000 \sin 60^\circ \\ &= 866 \text{ psi (positive)}\end{aligned}$$

The two determined stresses are shown on an element rotated so that the normal to the plane being considered forms an angle of 60° (counterclockwise) with the *X–X* axis (see Figure 17.30c). Note that the direction of the positive shear stress on the inclined plane tends to rotate the element clockwise, which is in agreement with our previous discussion and with Figure 17.29.

(b) The maximum shear stress for the uniaxially loaded member always occurs on a plane at 45° with the plane perpendicular to the longitudinal axis of the member. If an angle of 90° (this is $2\theta_s$) is turned off on the circle—refer to the manner in which the 120° angle was turned in part (a)—we obtain line *ME*. By inspection, the ordinate to point *E*, which represents the maximum shear stress, is equal to the radius of the circle and is equal to +1000 psi. Note that the compressive stress represented by *E* is equal to −1000 psi.

17.9 MOHR'S CIRCLE: THE GENERAL STATE OF STRESS

The use of Mohr's circle as an alternate method of determining normal and shear stresses on inclined planes of an axially loaded prismatic member was discussed in Section 17.8. Mohr's circle may also be used in more complex applications. In this section, we study the use of Mohr's circle for the determination of the magnitude and direction of the principal stresses as well as for the magnitude and direction of the maximum shear stress. From a practical design standpoint, these are the values that are usually of concern, rather than the relatively unimportant combined stresses on arbitrary planes.

If we consider an element taken from a machine member subjected to biaxial loading, stresses can be developed as shown in Figure 17.31a. The sequence of steps in the construction and interpretation of Mohr's circle for this application is as follows:

1. Make a sketch of the element indicating the known normal and shear stresses with their directions, as shown in Figure 17.31a.
2. Establish a rectangular coordinate system with normal stresses to be plotted as abscissas (horizontally) and shear stresses as ordinates (vertically), as shown in Figure 17.31b.
3. With respect to a sign convention, tensile normal stresses are positive and compressive normal stresses are negative. Shear stresses are positive if they tend to rotate the element clockwise and negative if they tend to rotate the element counterclockwise. This special rule is for shear stress when using Mohr's circle. For this example, the shear stresses on the vertical face are positive and are accompanied by a positive tensile stress σ_x and the shear stresses on the horizontal face are negative and are accompanied by a positive tensile stress σ_y, as shown in Figure 17.31a. Assume, for purposes of this discussion, that $\sigma_x \geq \sigma_y$ algebraically.
4. With reference to Figure 17.31b, plot σ_x on the horizontal axis as *OG*. This stress is the algebraically larger normal stress. Next, plot σ_y on the horizontal axis as *OH*. This stress is the algebraically smaller normal stress. Both points are plotted to the right of the vertical axis, since both are positive values. (*Note:* A positive (+) stress is algebraically larger than a negative (−) stress; e.g., $+100 \text{ psi} \geq -500$ psi; also, $-200 \text{ psi} \geq -400$ psi.)
5. The shear stress associated with the vertical plane on which σ_x acts is $+\tau$. Lay off τ vertically upward (in a positive direction) from *G*. Call this point *X*. Since the shear stress associated with the horizontal plane on which σ_y acts is $-\tau$, lay off τ vertically downward (in a negative direction) from point *H*. Call this point *Y*.
6. Draw the *XY* line. This line is the diameter of Mohr's circle. Line *XY* intersects the normal stress (horizontal) axis at point *M*. Draw a circle with its center at point *M*

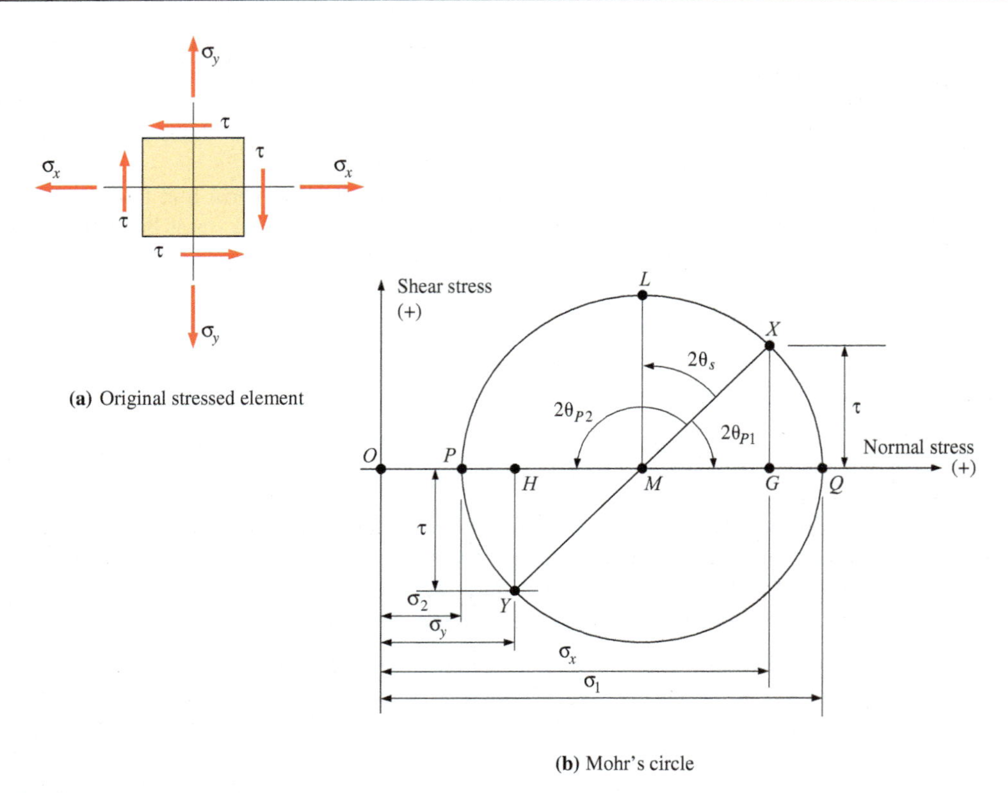

FIGURE 17.31 Mohr's circle example.

and passing through points X and Y, which completes the construction of Mohr's circle. Line XY is the reference line that correlates the circle with the particular problem under investigation. It represents the X–Y axes of the original stressed element shown in Figure 17.31a: MX represents the X–X axis; MY represents the Y–Y axis. All angles in the circle are doubled; hence, the X–X and Y–Y axes appear to be 180° apart when in reality they are 90° apart.

7. The principal stresses can be determined at points Q and P (dimensions σ_1 and σ_2) on the circle. Note that these are both positive and therefore tensile stresses and that at these points the shear stresses are zero. From the circle, the equation for the principal stresses can be written as follows:

$$\begin{aligned}\sigma_{1,2} &= OM \pm \text{circle radius}\\ &= OM \pm MX\\ &= \frac{\sigma_x + \sigma_y}{2} \pm \sqrt{(MG)^2 + (GX)^2}\\ &= \frac{\sigma_x + \sigma_y}{2} \pm \sqrt{\frac{(\sigma_x - \sigma_y)^2}{4} + \tau^2}\end{aligned}$$

This equation was previously presented as Equation (17.9).

8. The orientation of the normals to the principal planes is determined by angle XMQ (and by angle XMP). These angles are measured from the reference line MX clockwise to line MQ and counterclockwise to line MP. Note that angle XMQ is labeled $2\theta_{P1}$ on the circle, which would indicate that the inclination of the maximum principal stress on the stressed element is measured by a clockwise rotation of θ_{P1} from the X–X axis on the stressed element. Similarly, the orientation of the direction of the minimum principal stress is obtained from angle XMP, which is a counterclockwise angle from line MX labeled $2\theta_{P2}$.

9. The orientation of the plane of maximum shear stress can also be determined. Point L represents maximum shear stress $(+\tau')$. It lies on a plane inclined at θ_s with the X–X axis on the stressed element of Figure 17.31a. Angle $2\theta_s$ is indicated on the circle as the counterclockwise angle from line MX to line ML. Note that on the circle it is oriented 90° from the principal planes. Therefore, it is oriented 45° from the principal plane on the stressed element.

10. The results of the analysis are usually displayed graphically on two elements, one rotated to the inclination of the principal planes and the other rotated to the inclination of the planes of maximum shear stress. Appropriate stresses are shown on each element. (See Example 17.10.)

17.9.1 Brief Summary: Mohr's Circle for General Stress

1. Sketch the stressed element showing all applied stresses.
2. Set up the rectangular coordinate system for the circle.

3. Follow the appropriate sign convention: tension—positive; compression—negative; shear—clockwise rotation positive, counterclockwise rotation negative.
4. Plot σ_x and σ_y on the normal stress (horizontal) axis.
5. Plot the shear stress $\pm\tau$ associated with the normal stress for both the vertical face and the horizontal face. Label points *X* and *Y*.
6. Draw the *XY* line and Mohr's circle.
7. Determine the principal stresses and the maximum shear stress.
8. Determine $2\theta_{P1}$ and/or $2\theta_{P2}$ measuring from the *X* end of the reference line.
9. Determine $2\theta_s$.
10. Sketch two rotated elements, one for principal stresses and one for maximum shear stresses.

EXAMPLE 17.10 An element of a machine member is subjected to the stresses shown in Figure 17.32. Using Mohr's circle, (a) determine the magnitudes of the principal stresses and the maximum shear stresses, (b) determine the orientations of the principal planes and the planes on which the maximum shear stresses occur, and (c) show the results for the previous two parts on two rotated stressed elements.

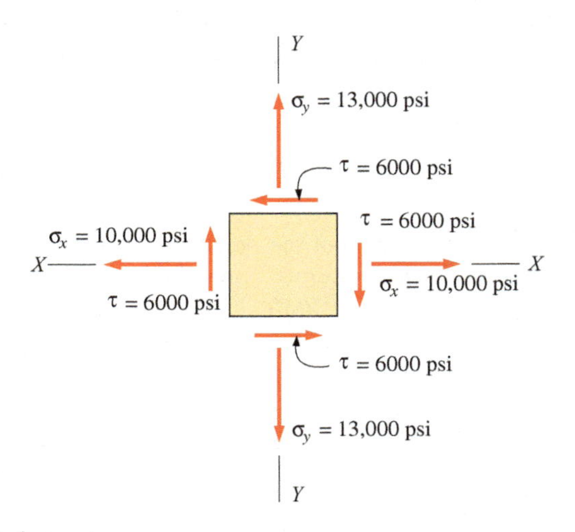

FIGURE 17.32 Stressed element.

Solution (a) In accordance with our sign convention, $\sigma_x = +10{,}000$ psi, $\sigma_y = +13{,}000$ psi, τ on the vertical face $= +6000$ psi, and τ on the horizontal face $= -6000$ psi. Mohr's circle is shown in Figure 17.33. Since it is not drawn to scale, mathematical relationships will be used along with the graphical construction.

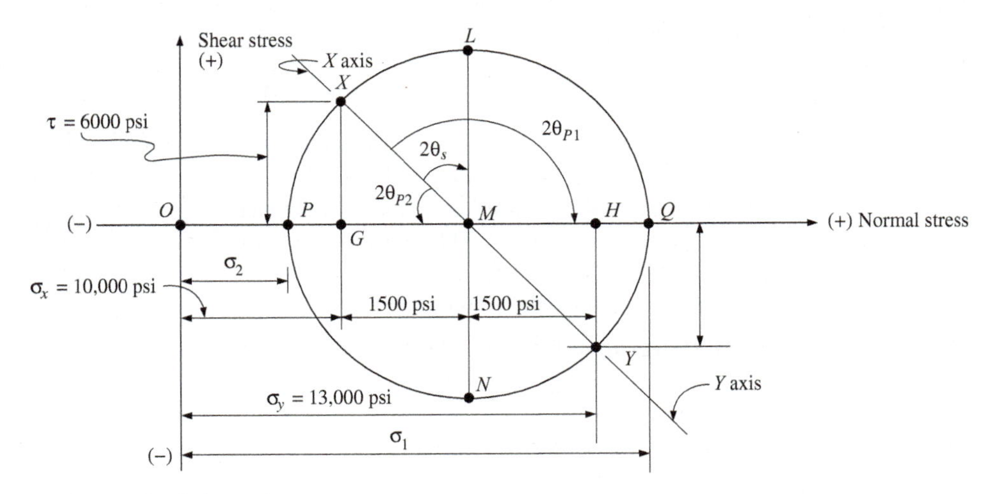

FIGURE 17.33 Mohr's circle.

First, plot $OG = +10{,}000$ psi (this is σ_x) and $OH = +13{,}000$ psi (this is σ_y) on the horizontal axis. Since the shear stress τ associated with σ_x is a positive value, draw a line perpendicular to the horizontal axis (upward) at G and plot point X so that $XG = +6000$ psi. Since τ associated with σ_y is a negative value, draw a line perpendicular to the horizontal axis (downward) at H and plot point Y so that $HY = -6000$ psi.

Then, draw the diameter XY and locate the point at which it crosses the horizontal axis M, which is the center of Mohr's circle. Compute the radius of the circle using triangle MHY:

$$GM = MH = \frac{\sigma_y - \sigma_x}{2} = \frac{13{,}000 - 10{,}000}{2} = 1500 \text{ psi}$$

$$MY = \sqrt{(MH)^2 + (HY)^2} = \sqrt{1500^2 + 6000^2} = 6185 \text{ psi}$$

Distances OQ and OP along the horizontal axis represent the maximum and minimum principal stresses σ_1 and σ_2:

$$\sigma_1 = OM + MQ = (10{,}000 + 1500) + 6185 = +17{,}685 \text{ psi (tension)}$$

$$\sigma_2 = OM - MP = (10{,}000 + 1500) - 6185 = +5315 \text{ psi (tension)}$$

The maximum shear stress τ' is represented by point L (or point N). The magnitude of the maximum shear stress is equal to the radius of the circle, 6185 psi. Note also that the normal stresses acting on the planes of maximum shear stress can be determined at points L and N. Also, compare with Equation (17.12). Thus,

$$\sigma_n = \frac{\sigma_x + \sigma_y}{2} = \frac{10{,}000 + 13{,}000}{2} = +11{,}500 \text{ psi (tension)}$$

(b) The inclination of the principal stresses is determined by measuring from the X end of reference axis X–Y to the maximum principal stress at point Q. The clockwise angle of $2\theta_{P1}$ is labeled on the circle. From triangle XMG,

$$2\theta_{P2} = \angle XMG = \tan^{-1}\frac{XG}{GM} = \tan^{-1}\frac{6000}{1500} = 4.0 = 75.96^\circ$$

$$2\theta_{P1} = 180^\circ - 75.96^\circ = 104.04^\circ$$

$$\theta_{P1} = 52.0^\circ \text{ (clockwise)}$$

Therefore, the maximum principal stress f_1 is inclined at a clockwise angle of 52.0° with the X–X axis of the original stressed element.

The inclination to the plane of maximum positive shear stress is represented by θ_s. Point L on the circle represents this stress and lies clockwise $2\theta_s$ from the X end of the X–Y reference line. The simplest way to obtain this angle is to subtract 90° from $2\theta_{P1}$:

$$2\theta_s = 2\theta_{P1} - 90^\circ$$

from which

$$\theta_s = \theta_{P1} - 45^\circ = 7.0^\circ \text{(clockwise)}$$

Since angle $2\theta_s$ is shown as clockwise to the maximum positive shear stress (radius ML), the element is shown rotated clockwise 7° in Figure 17.34b. Since the shear stress at point L is positive, it is shown acting downward on the face of the element that corresponds to the right vertical face of the original stressed element. This action tends to rotate the element clockwise and is consistent with the previously discussed sign convention. Note that on the stressed element the planes of maximum shear stress are oriented at 45° to the principal planes.

(c) Results from parts (a) and (b) are shown graphically in Figure 17.34.

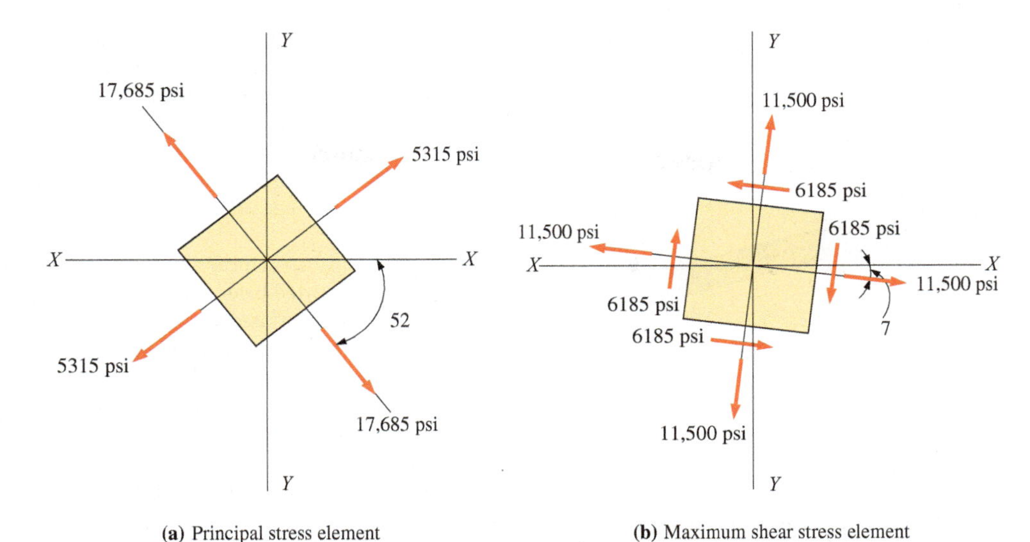

FIGURE 17.34 Results for Example 17.10.

SUMMARY BY SECTION NUMBER

17.2 Biaxial bending occurs where a beam is subjected to a load or loads that produce bending about both the strong and the weak axes. For the simple case where no torsion is produced, the combined stresses of interest can be determined from

$$f = \pm\frac{M_x}{S_x} \pm \frac{M_y}{S_y} \tag{17.1}$$

17.3 When a member of relatively short span length is subjected to both transverse loads and direct axial loads, bending stresses and axial stresses are developed. The combined stress at any location is a simple algebraic sum of the two types of stress and is expressed as

$$f = \pm\frac{P}{A} \pm \frac{Mc}{I} \tag{17.2}$$

17.4 When a short compression member is subjected to a load that lies on only one of the centroidal axes, bending stresses and axial stresses are developed. If the moment is expressed in terms of the load and eccentricity $(M = Pe)$, the combined stress expression is

$$f = -\frac{P}{A} \pm \frac{Pec}{I} \tag{17.3}$$

17.5 When a short compression member is subjected to an eccentric load, combined stresses (either tension or compression) will develop at the outer edges of the cross section. If the bending stress is equal to the axial stress, the stress at one outer edge will be zero. The maximum load eccentricity for which this zero stress will develop in a rectangular shape is

$$e = \frac{d}{6} \tag{17.4}$$

where d represents the dimension parallel to the eccentricity.

17.6 If a load applied to a short compression member is eccentric to both centroidal axes, the member is subjected to a double eccentricity, and the combined stress expression is

$$f = -\frac{P}{A} \pm \frac{Pe_1c_1}{I_y} \pm \frac{Pe_2c_2}{I_x} \tag{17.5}$$

17.7 The combination of normal and shear stress results in maximum and minimum normal and shear stresses being developed on planes that are inclined with respect to the planes of stress being combined. The planes on which the normal stresses become maximum or minimum are called *principal planes*. The normal stresses acting on the planes are called *principal stresses* and can be obtained from

$$f_{1,2} = \left(\frac{f_x + f_y}{2}\right) \pm \sqrt{\frac{(f_x - f_y)^2}{4} + f_{xy}^2} \tag{17.9}$$

Shear stresses on the principal planes are equal to zero. The maximum shear stress developed on an inclined plane is

$$f'_{s\,(\max)} = \pm\sqrt{\frac{(f_x - f_y)^2}{4} + f_{xy}^2} \tag{17.11}$$

17.8 and **17.9** Mohr's circle is a graphical (or semigraphical) alternate method of calculating normal stresses and shear stresses acting on any inclined plane at any point in a stressed body. It can be used both for uniaxial and for biaxial loading conditions.

PROBLEMS

Unless noted otherwise, neglect the weights of the members in these problems.

Section 17.2 Biaxial Bending

17.1 A 100-mm-by-100-mm (S4S) timber beam on a simple span of 4 m is subjected to vertical and lateral point loads placed at midspan, as shown. The lines of action of the loads pass through the centroid of the cross section. Determine the stresses at the corners of the cross section at the point of maximum moment.

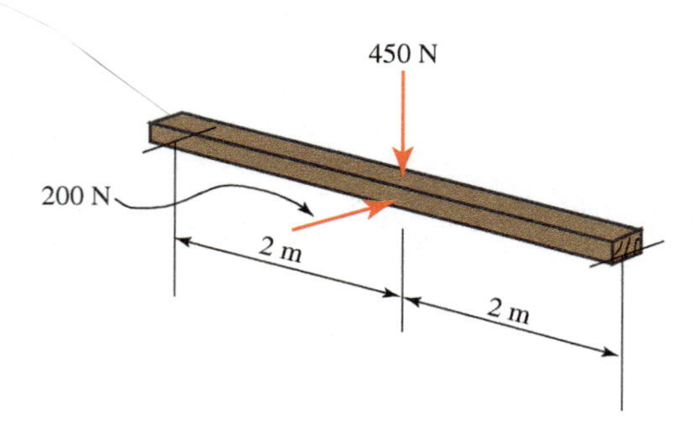

PROBLEM 17.1

17.2 A horizontal 30-ft simple span beam is supported on a 20° slope and is loaded as shown. The uniformly distributed line load covers the full span. Determine the maximum tensile and compressive stresses due to bending. The beam weight is included in the given load.

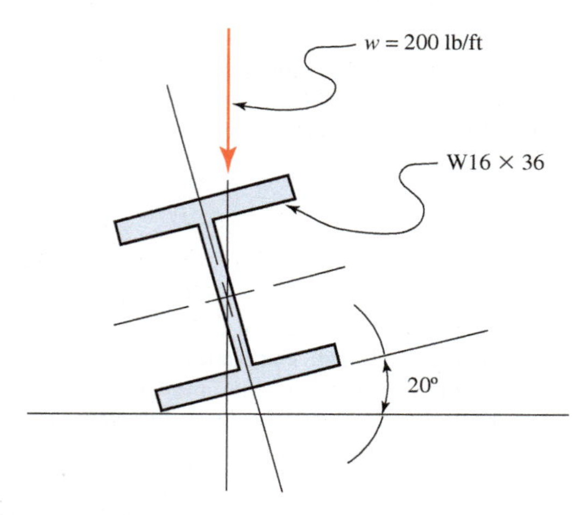

PROBLEM 17.2

Section 17.3 Combined Axial and Bending Stresses

17.3 A 1-in.-by-4-in. steel bar is subjected to the loads shown. Calculate the combined stresses at points *A* and *B*.

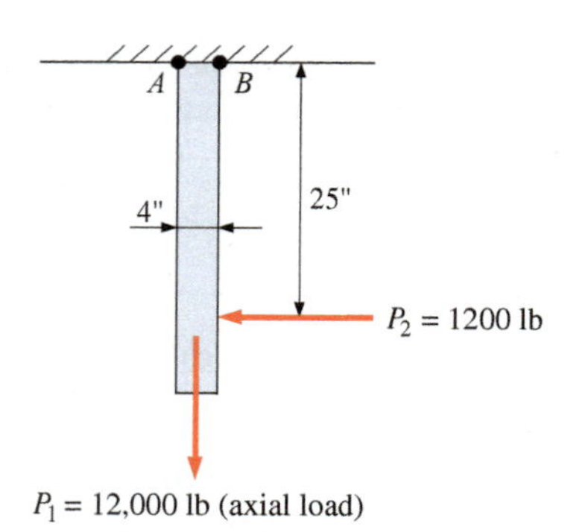

PROBLEM 17.3

17.4 A W410 × 100 structural steel wide-flange section is used as a horizontal lower chord member of a bridge truss. It is subjected to a tensile load of 1200 kN. The orientation of the member is such that the *Y*–*Y* axis (the weak axis) is horizontal. Calculate the maximum and minimum combined stresses due to the direct axial load and the member's own weight. The panel length (length of the member) is 8 m.

17.5 A W12 × 72 structural steel wide-flange section is used as a simply supported beam with a span length of 12 ft. It is subjected to concentrated loads of 20,000 lb at each quarter point and an axial tensile load of 50,000 lb. Calculate the maximum combined tensile and compressive stresses.

17.6 A solid steel shaft 3 in. in diameter and 4 ft long is fixed at one end and free at the other, as shown. The shaft is subjected to an axial tensile load of 50,000 lb. Compute the magnitude of the force *P* at the free end necessary to reduce the stress at point *B* to zero. Then compute the maximum combined stress at point *A*.

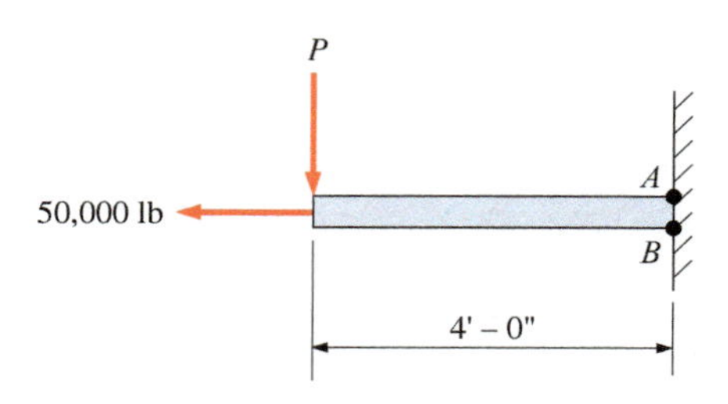

PROBLEM 17.6

Section 17.4 Eccentrically Loaded Members

17.7 A short compression member is subjected to a compressive load of 300,000 lb at an eccentricity of 1.5 in., as shown. The member is 12 in. by 12 in. in cross section. Calculate the combined stresses at the outer edges *AA* and *BB*.

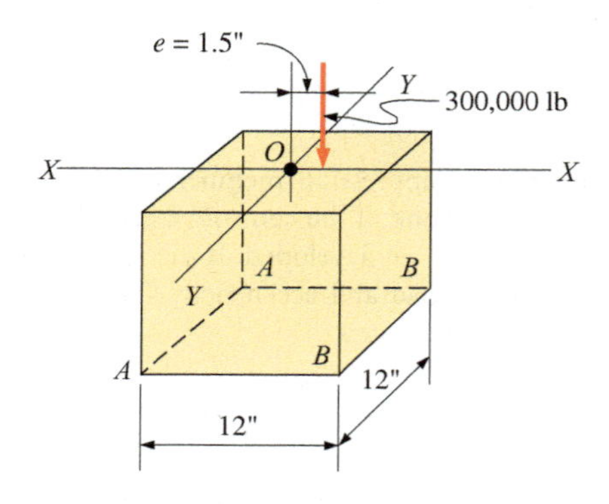

PROBLEM 17.7

17.8 With reference to Problem 17.7, calculate the eccentricity that must exist if the combined stress at outer edge *AA* is to be zero.

17.9 A section of a 51-mm-diameter standard-weight steel pipe is bent into the form shown and rigidly embedded in a concrete footing. Calculate the stresses at points A and B on the pipe. P is 900 N and e is 600 mm.

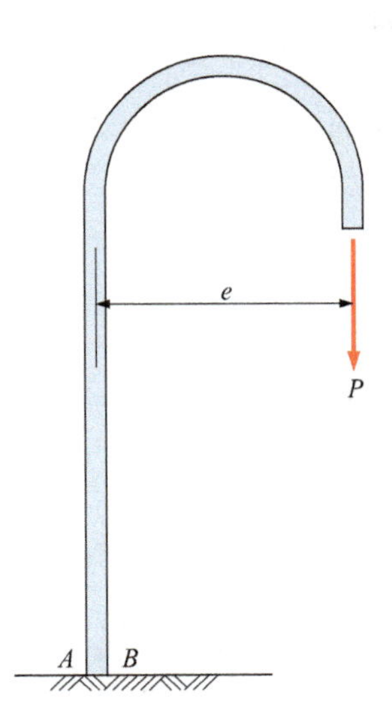

PROBLEM 17.9

17.10 For the pipe of Problem 17.9, compute the maximum load P that may be applied if the allowable compressive stress at point B is 80 MPa.

Section 17.5 Maximum Eccentricity for Zero Tensile Stress

17.11 A concrete pedestal is in the shape of a cube and is 6 ft on each side. The pedestal supports a superimposed load of 40,000 lb located on the Y–Y axis, as shown. Calculate the maximum eccentricity e for zero tension at the base of the pedestal. Include the weight of the pedestal using a concrete unit weight of 150 pcf.

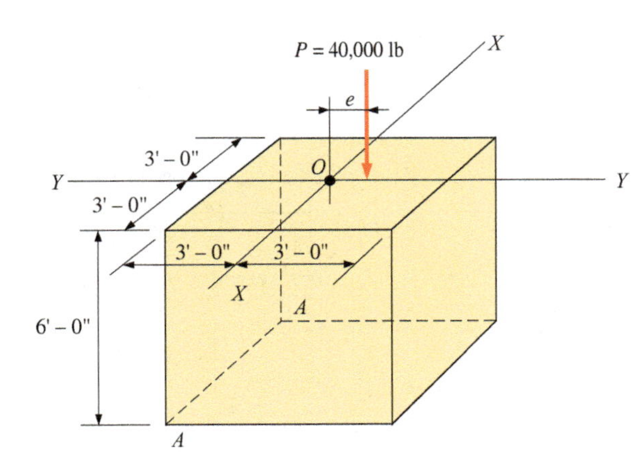

PROBLEM 17.11

17.12 For the pedestal of Problem 17.11, assume that the load is placed at the computed eccentricity. Calculate the maximum tensile and compressive stresses developed on a horizontal plane 3 ft below the top of the pedestal.

17.13 Rework Problem 17.11, but assume that the applied load is 150,000 lb instead of 40,000 lb.

17.14 A 12-in.-square concrete pedestal is subjected to a 100-kip vertical eccentric load applied on the Y–Y axis, as shown. Determine the required magnitude of the horizontal load P (also in the vertical plane through the Y–Y axis) for a combined stress of zero at point A.

Section 17.6 Eccentric Load Not on Centroidal Axis

17.15 A short compression member is subjected to a compressive load of 5000 lb at a double eccentricity, as shown.

(a) Calculate the combined stress at each of the four corners, A, B, C, and D.

(b) Locate the line of zero stress.

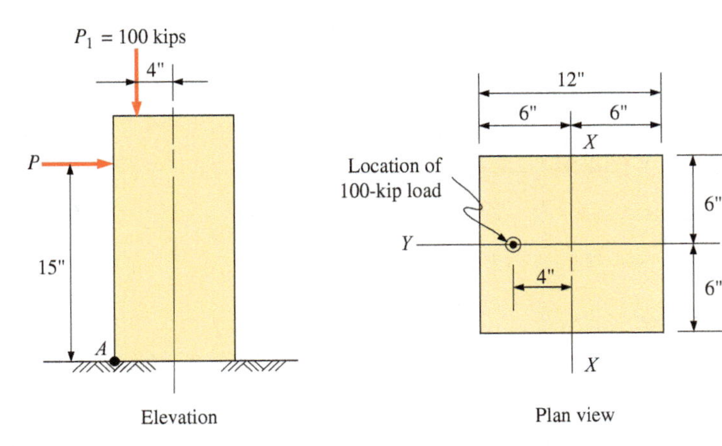

PROBLEM 17.14

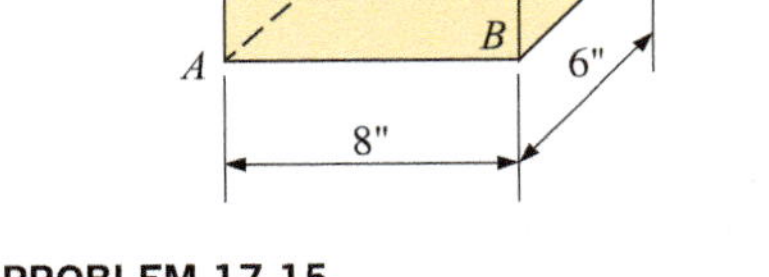

PROBLEM 17.15

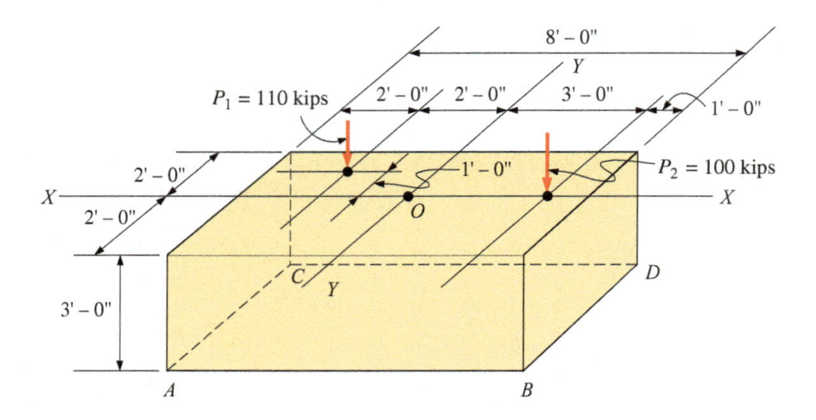

PROBLEM 17.16

17.16 A rectangular concrete footing, 4 ft by 8 ft in plan, is subjected to two column loads, as shown. Calculate the stress (base pressure) at each corner of the footing. Use a concrete unit weight of 150 pcf. Express the answers in pounds per square foot (psf).

Section 17.7 Combined Normal and Shear Stresses

17.17 The bending and shear stresses developed at a point in a machine part are shown acting on infinitesimal elements. For each element, calculate

(a) the principal stresses and the orientation of the principal planes

(b) the maximum shear stress

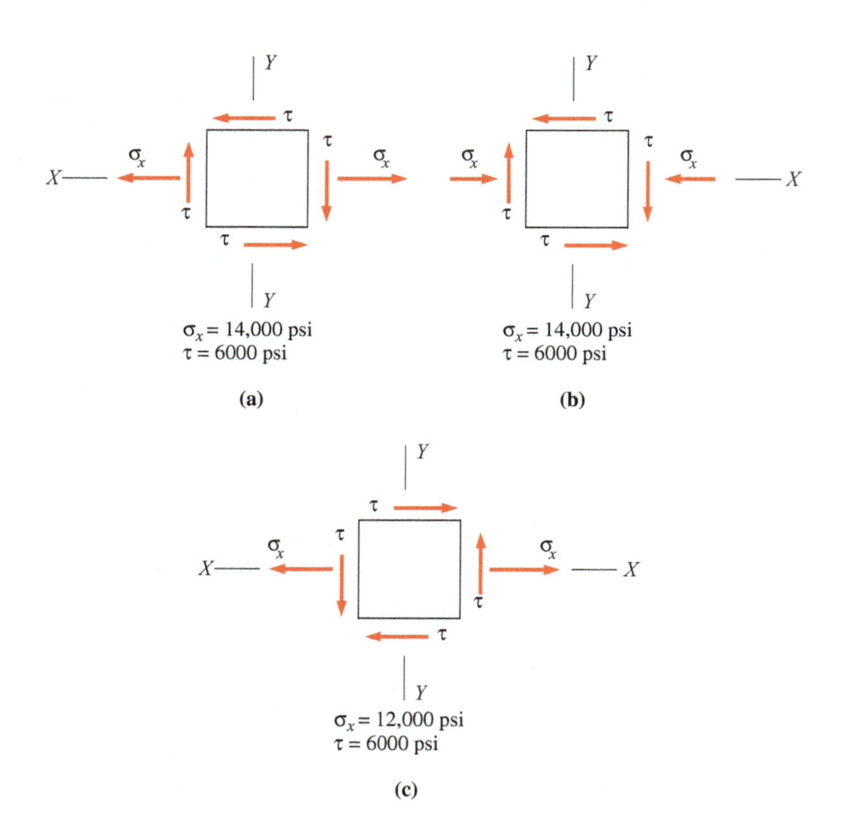

PROBLEM 17.17

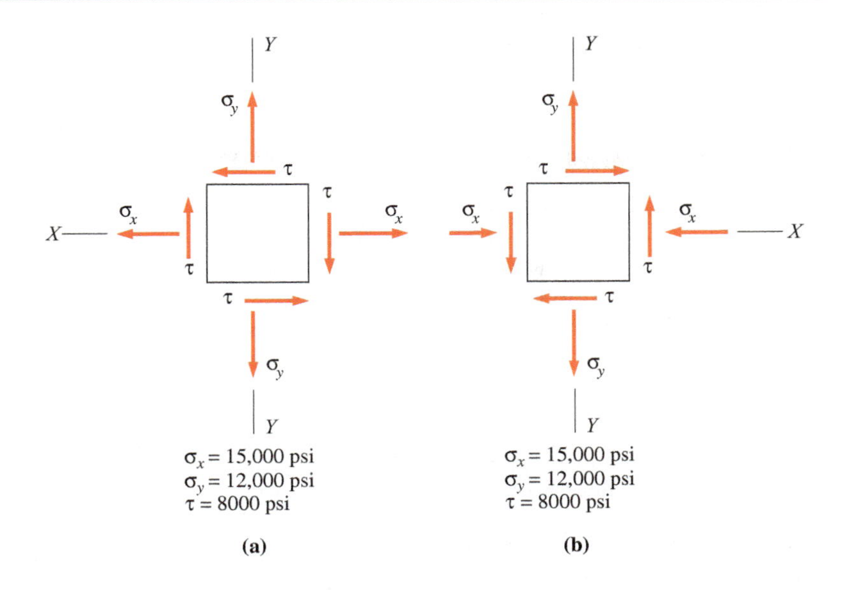

PROBLEM 17.18

17.18 Stresses developed at a point in a machine part are shown acting on infinitesimal elements. For each element, calculate

(a) the normal and shear stress intensities on a plane, the normal of which is inclined at an angle of 60° counterclockwise from the X–X axis

(b) the principal stresses and the orientation of the principal planes

(c) the maximum shear stress

17.19 Calculate the principal stresses at points A and B for the bracket shown. $P = 9000$ lb and $\theta = 30°$. The bracket is 1 in. thick.

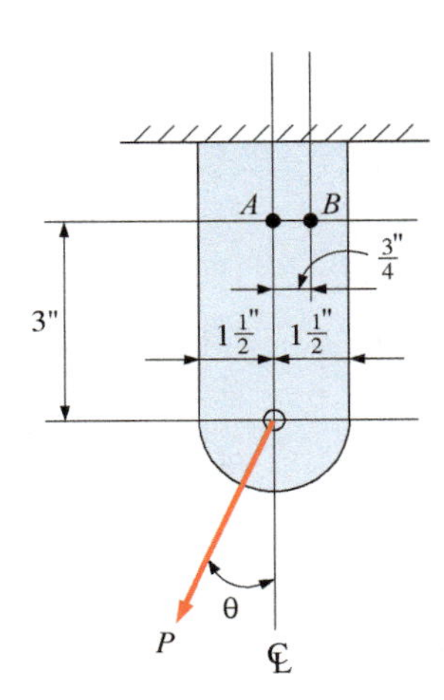

PROBLEM 17.19

17.20 Rework Problem 17.19 using $P = 8000$ lb and $\theta = 45°$.

Section 17.8 Mohr's Circle

17.21 A 1-in.-square steel bar is subjected to an axial tensile load of 10,000 lb. Use Mohr's circle to determine the shear stress and normal stress on a plane that has a normal inclined at 60° counterclockwise from a plane perpendicular to the longitudinal axis of the member.

17.22 A bar having a cross-sectional area of 6 in.^{2} is subjected to an axial tensile load of 39,000 lb. Using Mohr's circle:

(a) Find the normal and shear stresses on planes with normals inclined at 75°, 65°, and 50° counterclockwise from a plane perpendicular to the longitudinal axis.

(b) Find the maximum shear stress.

17.23 Rework Problem 17.22, changing the load to a compressive load of 72,000 lb. Also find the normal and shear stresses on a plane that has a normal inclined at 30° counterclockwise from a plane perpendicular to the longitudinal axis of the bar.

Section 17.9 Mohr's Circle: The General State of Stress

17.24 Solve Problem 17.17 using Mohr's circle.

17.25 For the elements shown in Problem 17.18, use Mohr's circle to determine

(a) the principal stresses and the orientation of the principal planes

(b) the maximum shear stress

17.26 Solve Problem 17.19 using Mohr's circle.

17.27 In Problem 17.19, change the load to 8000 lb and the angle to 45° and solve the problem using Mohr's circle.

Computer Problems

For the following computer problems, any appropriate software may be used. Input prompts should fully explain what is required of the user (the program should be user-friendly). The resulting output should be well labeled and self-explanatory.

17.28 Write a computer program that will allow the user to solve Problem 17.39. Assume constant dimensions for the wall, but a varying water level. User input will therefore be the height of the water behind the wall.

17.29 Write a computer program that will solve for the normal stress and shear stress on any inclined plane in a uniaxially loaded member, such as the metal block of Example 17.9. User input is to be the cross-sectional area of the member, the axial load, and the inclination θ of a normal to the plane measured counterclockwise from the X–X axis. The input prompts should enable the user to understand the required input fully.

17.30 Rework Problem 17.29, but have the program generate a table for normal and shear stresses for values of θ from 0° to 45° in increments of 5°. User input for this program is to be the cross-sectional area of the member and the axial load.

17.31 Write a program that will allow a user to solve for the principal stresses, the orientation of the principal planes, and the maximum shear stress in a stressed element. Input will be the normal stresses and the associated shear stresses. Data from Problem 17.17 or Problem 17.18 may be used to test the program.

Supplemental Problems

17.32 A 4-in.-by-8-in. (S4S) Douglas fir timber beam is supported on a 16-ft simple span and is subjected to the loads shown. The uniformly distributed line load does not include the weight of the beam (the beam weight, a gravity load, should be considered). Find the maximum tensile and compressive stresses.

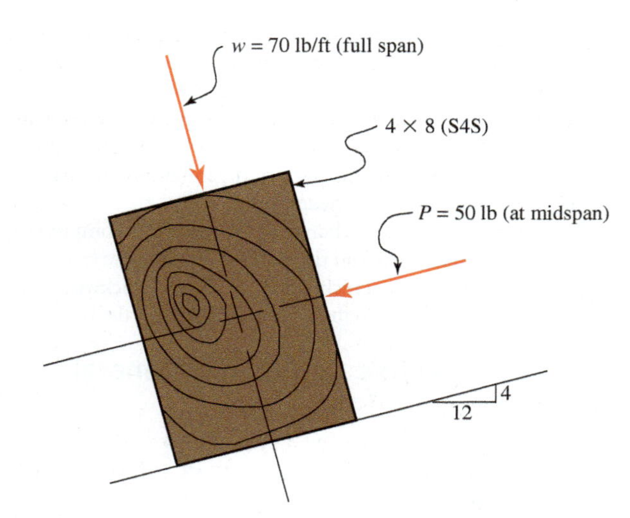

PROBLEM 17.32

17.33 A horizontal flexural member (a girt) in the wall of a prefabricated metal building is composed of a C130 × 13 channel. The girt is to span 4 m between simple supports and is subjected to full-span uniformly distributed line loads as shown. The girt is oriented with the Y–Y (weak) axis in the horizontal plane. The gravity load is 400 N/m and includes the weight of the girt. The wind load is 600 N/m. Find the maximum tensile and compressive stresses.

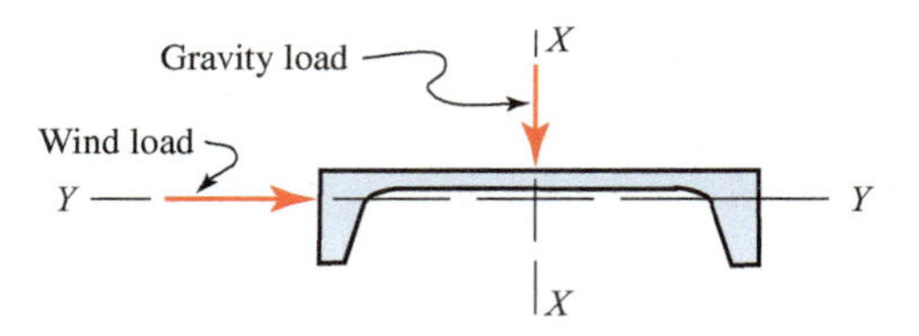

PROBLEM 17.33

17.34 A simply supported W18 × 50 structural steel wide-flange beam has a span length of 15 ft. It is subjected to a concentrated load of 25,000 lb at midspan and a uniformly distributed line load of 350 lb per linear foot. The uniform load includes the weight of the beam. The allowable tensile stress for the beam is 22,000 psi. Calculate the maximum axial tensile force that may be applied to this beam at its ends.

17.35 A steel link in a machine is designed to avoid interference with other moving parts. The link is 100 mm thick. The cross-sectional area of the link is reduced by one-half at section A–A as shown. Compute the maximum tensile stress developed across section A–A. Neglect any stress concentrations.

17.36 An 8-in.-square (S4S) vertical timber post is 8 ft long and fixed at its lower end. It supports a vertical axial compressive load of 7000 lb on top and a horizontal load of 1000 lb at a point 3 ft below the top. Compute the maximum and minimum combined stresses at the base.

17.37 A short 3-in.-square steel bar with a 1-in.-diameter axial hole is fixed at its base and loaded at the top as shown. Calculate the value of the force P for which the maximum combined normal stress at the fixed end will not exceed 22 ksi.

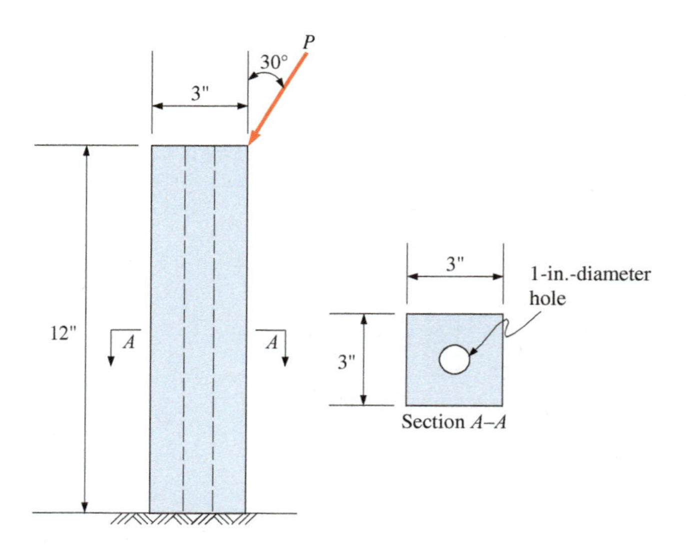

PROBLEM 17.37

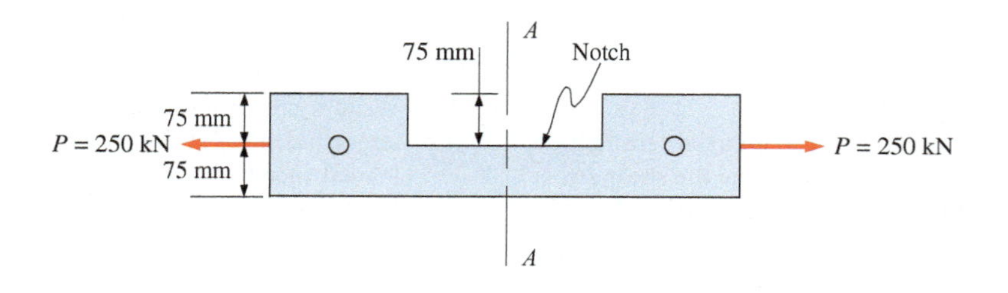

PROBLEM 17.35

17.38 A timber member 150 mm by 250 mm (S4S) is loaded as shown. Calculate the maximum combined stress acting normal to the plane A–A. Neglect the weight of the member.

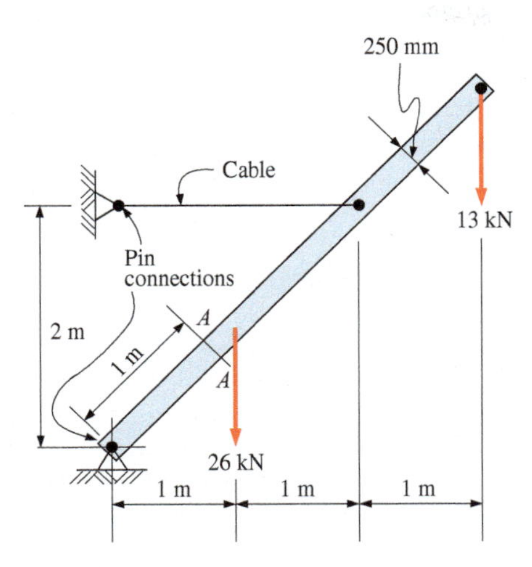

PROBLEM 17.38

17.39 A concrete wall 8 ft high and 3 ft thick is monolithic with a concrete base. The wall has water behind one face to a height of 6 ft, as shown. Calculate the combined normal stresses at points A and B at the bottom of the wall in pounds per square inch and pounds per square foot. Consider a 1-ft length of wall. Assume the unit weight of concrete to be 150 pcf and the unit weight of water to be 62.4 pcf.

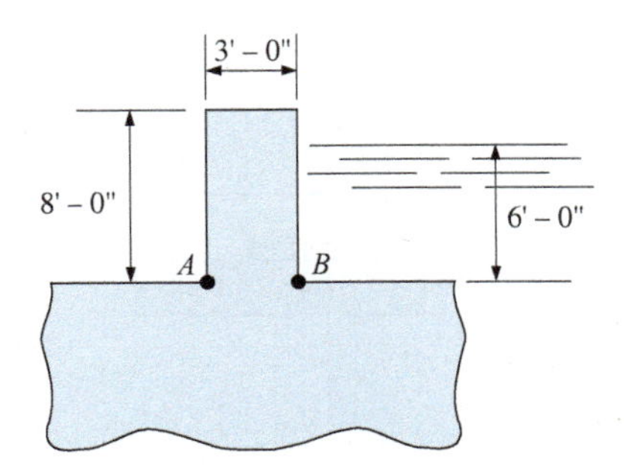

PROBLEM 17.39

17.40 A short compression member is subjected to a compressive load of 25,000 lb at an eccentricity of 1.5 in., as shown. The member is 6 in. by 8 in. in cross section. Calculate the combined stresses at the outer edges AA and BB.

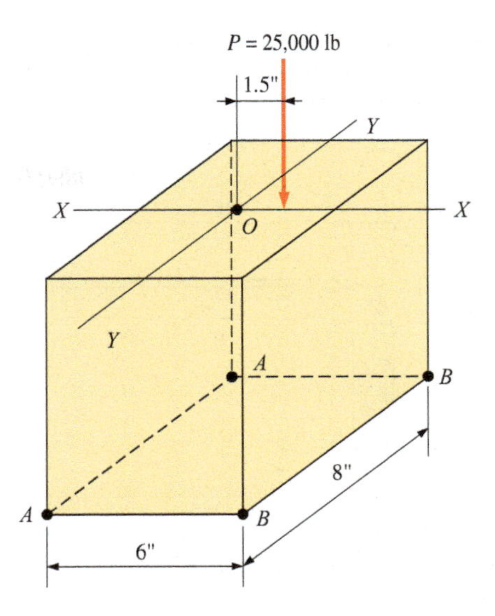

PROBLEM 17.40

17.41 Calculate the maximum eccentric load that can be applied in Problem 17.40 if the maximum allowable compressive stress is 1000 psi.

17.42 A short compression member is subjected to two compressive loads, as shown. The member is 6 in. by 12 in. in cross section. Calculate the combined stress at outer edges AA and BB.

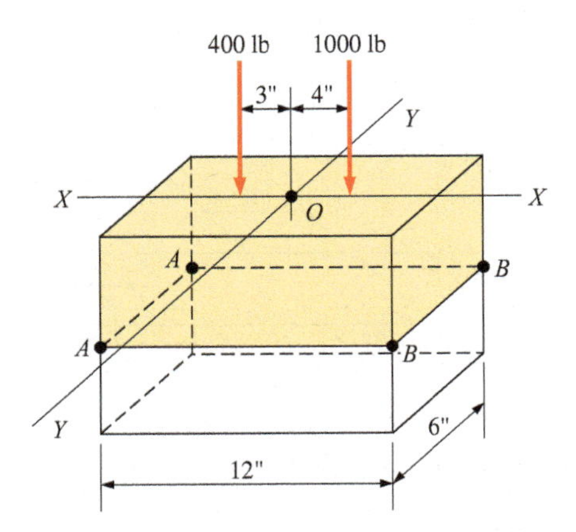

PROBLEM 17.42

17.43 Calculate the force P that may be applied to the punch press frame of Example 17.5 if the allowable stresses for cast iron are 6000 psi in tension and 10,000 psi in compression.

17.44 A load of 1000 lb is supported on a cast-iron bracket with a T cross section, as shown. Calculate the combined stresses at points A and B due to the load.

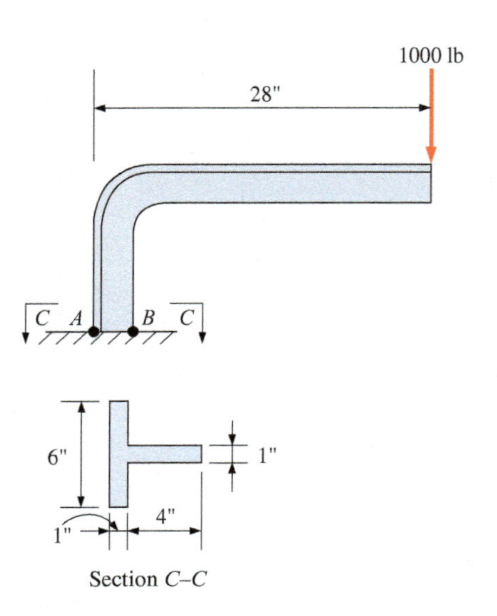

PROBLEM 17.44

17.45 A short compression member is subjected to an eccentric load, as shown. The member is 6 in. by 12 in. in cross section. Calculate the combined stress at each of the four corners (A, B, C, and D) and locate the line of zero stress.

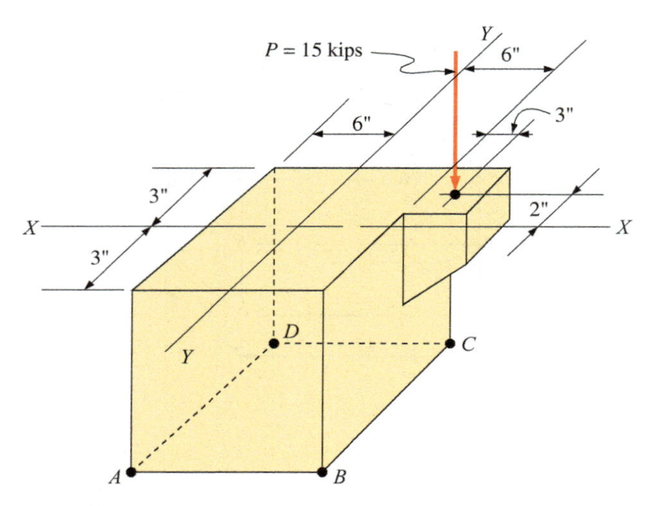

PROBLEM 17.45

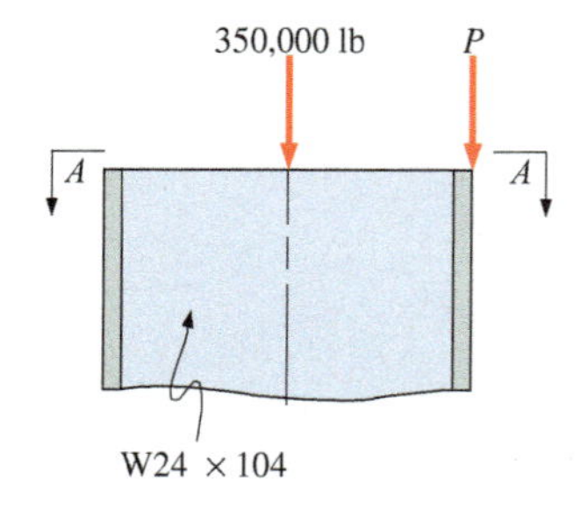

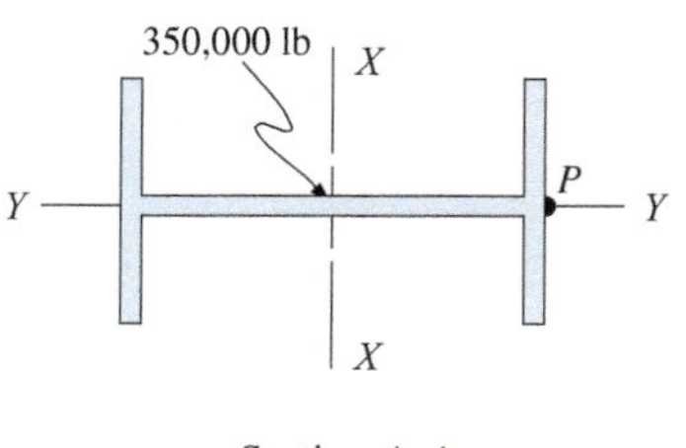

PROBLEM 17.46

17.46 A W24 × 104 structural steel wide-flange section is used as a short compression strut and subjected to a vertical axial load of 350,000 lb, as shown. Calculate the maximum load P that may be applied on the Y–Y axis at the outside face of the flange without exceeding the maximum compressive stress of 20,000 psi.

17.47 A cast-iron frame for a piece of industrial equipment has the dimensions shown. Calculate the force P that may be applied to this frame if the allowable stresses at section A–A are 6000 psi tension and 10,000 psi compression.

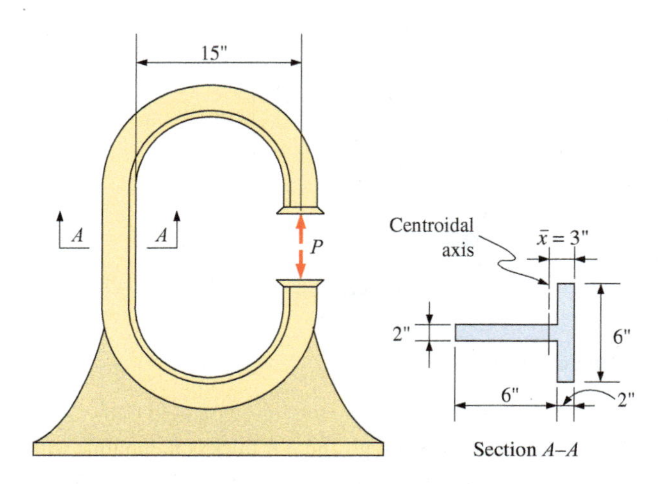

PROBLEM 17.47

17.48 The assembly shown is used in a machine. It supports an axial load P_1 of 8000 lb and an eccentric load P_2 of 8000 lb. The vertical member is a hollow cylinder having a 3-in. outside diameter and a 2-in. inside diameter. Calculate the maximum eccentricity e if the allowable stresses are 8000 psi in tension and 18,000 psi in compression.

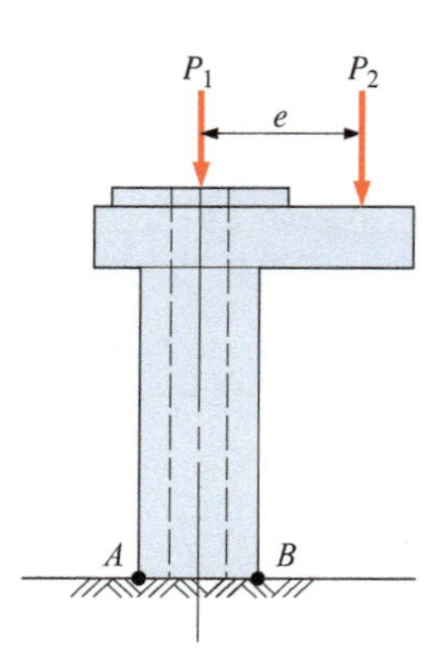

PROBLEM 17.48

17.49 A 50-mm-diameter solid steel shaft is subjected to an axial compressive load of 225 kN and a torque (twisting moment) of 3400 N · m. Calculate the principal stresses and the maximum shear stress on an element on the surface of the shaft.

17.50 An element of a machine member is subjected to the stresses shown. Calculate the magnitude and direction of the principal stresses and the maximum shear stress. Use the analytical approach.

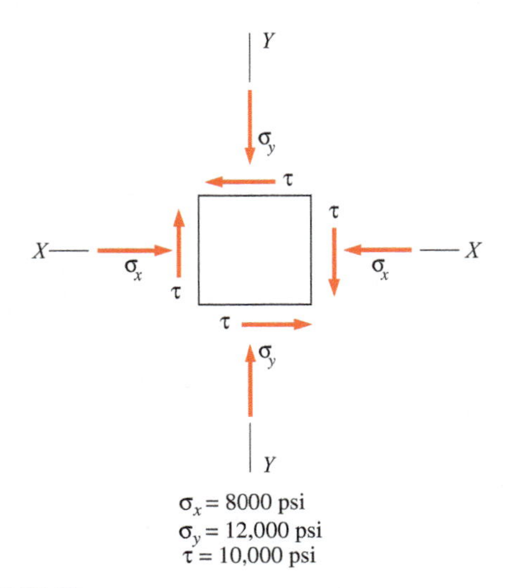

PROBLEM 17.50

17.51 A short-span cantilever built-up beam has the dimensions shown. Calculate the principal stresses, the inclination of the principal planes, and the maximum shear stress at points *A*, *B*, and *C*. Neglect the weight of the beam and the effect of stress concentrations. Use the analytical approach.

17.52 Solve Problem 17.50 using Mohr's circle.

17.53 A cantilever beam is subjected to an inclined load *P*, as shown. The beam is 50 mm by 150 mm in cross section. Calculate the principal stresses, the inclination of the principal planes, and the maximum shear stress at point *A*. Neglect the weight of the beam.

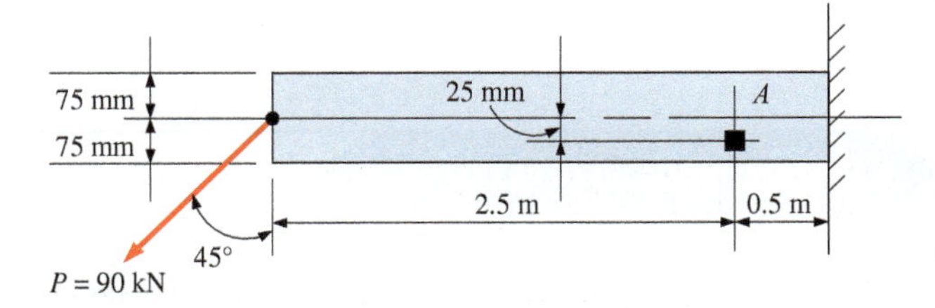

PROBLEM 17.53

17.54 A 6-in.-diameter solid shaft is subjected to a torque of 40 ft-kips. Using Mohr's circle:

(a) Determine the magnitude and direction of the principal stresses on the surface of the shaft.

(b) Determine the stresses acting on the faces of an element rotated 20° clockwise.

17.55 Rework parts (b) and (c) of Example 17.7 using Mohr's circle.

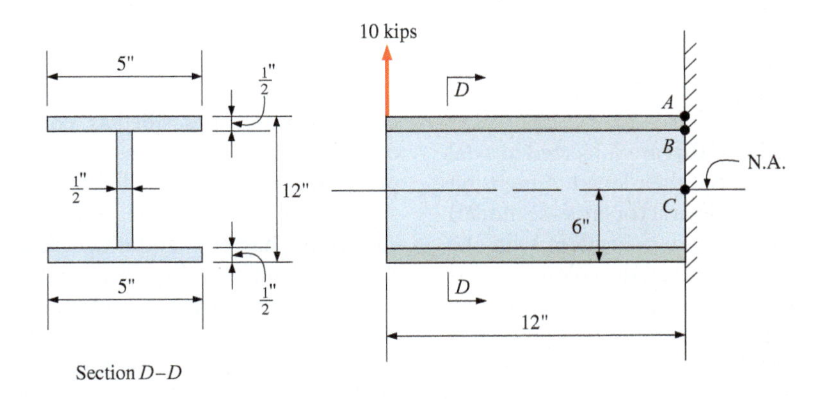

PROBLEM 17.51

CHAPTER EIGHTEEN

COLUMNS

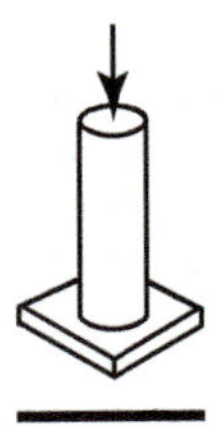

LEARNING OBJECTIVES

Upon completion of this chapter, readers will be able to:

- Identify columns and discuss their typical failure mechanisms.
- Use Euler's equation to calculate a failure load or buckling stress for ideal columns.
- Define Slenderness ratio as it pertains to columns.
- Identify and discuss end restraints on columns and their effect on column analysis.
- Calculate failure loads in columns that do not conform to Euler's equation and discuss how to apply factors of safety for safe design.
- Apply recognized standards for analysis and design of columns in machine parts.
- Design timber columns in conformance with the National Design Specification for Wood Construction.

18.1 INTRODUCTION

Members of structures and machines that are subjected to axial compressive loads are called *columns* if their length dimension is significantly larger than their least lateral (or cross-sectional) dimension. When the applied loads are coincident with the longitudinal centroidal axis of the member, the column is said to be axially loaded. This situation is a special case and one that exists rarely, if at all. Despite this generally accepted fact, a column may still be designed as though it were axially loaded, since an appropriate factor of safety will usually compensate for a minor unintended eccentricity.

When the line of action of the applied load does *not* coincide with the longitudinal centroidal axis, the load is said to be eccentric and the column is said to be eccentrically loaded.

In this chapter, we use the terms *column* and *compression member* interchangeably. Although all axially loaded columns are compression members, not all compression members are called columns. Members that carry compressive loads are commonly given names descriptive of their functions, such as pillars, pedestals, shores, props, supports, masts, and piers, to name a few. Truss members in compression, either chords or web members, are other compression members that are not categorized as columns, even though they are compression members and exhibit a behavior usually described as "column action." Various types of machinery can also have component parts acting as compression members but called something other than columns.

Axially loaded short compression members were discussed in Chapter 9, and eccentrically loaded short compression members were discussed in Chapter 17. In both cases, the use of the direct stress formula ($s = P/A$) was applicable; in addition, in the latter case, the flexure formula ($s = Mc/I$) was employed due to the bending action. Using these generally accepted methods, reasonably reliable solutions were provided for the short compression member. As the length of a compression member increases (maintaining a constant cross section), however, additional factors enter into the problem.

It is convenient to classify columns into three broad categories according to their modes of failure. The three types—short columns, intermediate columns, and long columns—are shown in Figure 18.1.

The long column (also referred to as a *slender column*), shown in Figure 18.1c, will fail by elastic buckling, which occurs at compressive stresses within the elastic range (recall that the upper limit of elastic range is the proportional limit or the elastic limit). Buckling of a column can be described as a bending action developed under compressive loads. Failure is said to occur through a lack of stiffness.

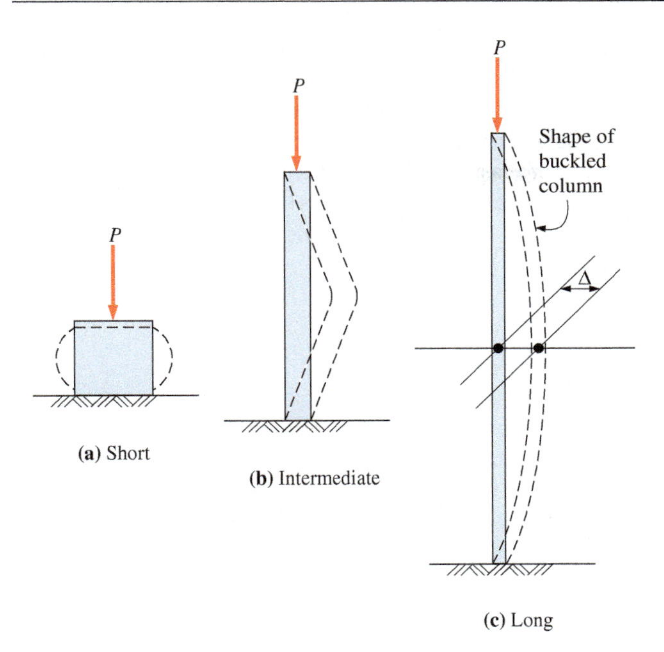

FIGURE 18.1 Types of columns.

A very short and stocky column, as shown in Figure 18.1a, will obviously not fail by elastic buckling. It will crush if made of a brittle material (for example, concrete) or yield if made of a ductile material (structural steel) at compressive stresses in the inelastic range. Since crushing and yielding are material failures, the maximum axial load a short column can support is determined solely by the strength of the material. If yielding is the failure criterion, the failure load may be determined as the product of the yield stress of the material and the cross-sectional area of the member.

The intermediate column lies between the two extremes, as shown in Figure 18.1b. Most columns or compression members are in the intermediate category. These columns will fail by inelastic buckling initiated when a localized yielding occurs at some point of weakness or crookedness. This failure mode represents a combination of buckling and material failure.

The design and analysis of intermediate columns is much more complex than that of the other columns. Intermediate columns are usually designed and analyzed using empirical formulas based on extensive test results, experience, and judgment. Failure of the intermediate column cannot be predicted using either the elastic buckling criterion of the slender column or the yielding criterion of the short column.

The behavior of compression members can best be understood by beginning with the slender column. This type of column was used originally in the development of the theory of axially loaded elastic column behavior.

18.2 IDEAL COLUMNS

The development of the theory of elastic column behavior is attributed to Leonhard Euler (1707–1783), a Swiss mathematician. Euler derived a theoretical equation for the load that would buckle a slender column. The equation is commonly referred to as the *elastic column buckling formula* or, simply, *Euler's formula*. Euler's theory was presented in 1744 and still is the basis for the analysis and design of slender columns.

The buckling of a slender column can be demonstrated by loading an ordinary wooden yardstick in compression. The yardstick will buckle (constituting failure), as shown in Figure 18.1c. If the load that first produced this buckling (or bending) remains constant, the lateral deflection Δ will remain constant and the column will support the load. If, however, the load is increased, the column will deflect further and finally collapse. A decrease in load would permit the column to straighten itself. The load that is just sufficient to hold the column in a bent condition is called the *critical load* for the column. The critical load can also be defined as the greatest load that the column will support. Euler's formula gives the buckling load P_e (also called the *critical load* or the *Euler buckling load*) for a theoretically perfect column (also called an *ideal column* or the *Euler column*). The ideal column can be briefly described as a pin-ended homogeneous slender column, initially straight, of an elastic material that is concentrically loaded. The pin-ended column is a column with ends free to rotate but restrained against lateral movement. The Euler buckling load is expressed as

$$P_e = \frac{\pi^2 EI}{L^2} \tag{18.1}$$

where P_e = the critical load; the concentric load that will cause initial buckling (lb) (N)
π = a mathematical constant (3.1416)
E = the modulus of elasticity of the material (psi) (MPa)
I = the least moment of inertia of the cross section (in.^{4}) (mm^4)
L = the length of the column from pin end to pin end (in.) (mm)

Note that the units must be consistent. For instance, it is sometimes convenient to substitute E in terms of ksi, in which case the units of critical load will be kips.

Tests have verified that Euler's formula accurately predicts the buckling load if conditions are such that the buckling stress is less than the proportional limit of the material and adherence to the basic assumptions is maintained. Since the buckling stress must be compared with the proportional limit, Euler's formula is commonly written in terms of stress, which can be derived from the preceding buckling load formula using the relationship $r = \sqrt{I/A}$ or $I = Ar^2$ (see Section 8.5):

$$f_e = \frac{\pi^2 E}{(L/r)^2} \tag{18.2}$$

where f_e = the critical stress; the uniform compressive stress at which initial buckling occurs (same units as E)
r = the least radius of gyration of the cross section (same units as L)

The unitless ratio L/r is called the *slenderness ratio* of the column. Once the critical stress has been determined, the critical load may be found from $P_e = f_e A$ or from Equation (18.1).

Note that the ideal, or perfect, column is theoretical and does not exist in reality. Even in laboratory conditions, it is impossible to obtain a perfectly straight column, frictionless pinned ends, or a perfectly axial load. Consequently, Euler's formula cannot be used directly as a practical approach to column analysis and design and must be modified.

Note also that the only material property represented in the Euler critical load and critical stress formulas is the modulus of elasticity E. Recall that E is a measure of the stiffness of a material. Material strength, such as the yield stress of steel, does not affect the magnitude of the Euler buckling load.

EXAMPLE 18.1 A 25-mm-diameter steel rod, shown in Figure 18.2, is used as a pin-connected compression member. Calculate the critical load using Euler's formula. The proportional limit for the steel is 235 MPa and the modulus of elasticity is 207×10^3 MPa. Assume the length of the rod to be (a) 1 m and (b) 2 m.

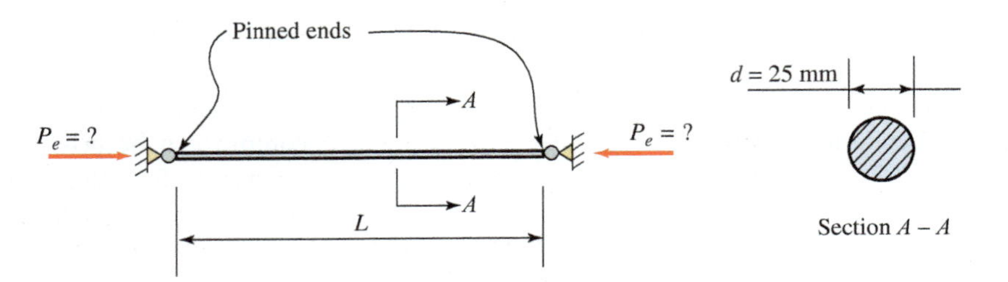

FIGURE 18.2 Pin-connected compression member.

Solution Refer to Table 3 inside the back cover for properties of areas:

$$A = 0.7854d^2 = 0.7854(25\text{ mm})^2 = 491\text{ mm}^2$$

$$r = \frac{d}{4} = \frac{25\text{ mm}}{4} = 6.25\text{ mm}$$

(a) For a 1-m length of rod,

$$f_e = \frac{\pi^2 E}{(L/r)^2} = \frac{\pi^2 (207 \times 10^3\text{ MPa})}{[(1000\text{ mm})/(6.25\text{ mm})]^2}$$

$$= 79.8\text{ MPa} < 235\text{ MPa} \qquad \textbf{(O.K.)}$$

The critical load can then be calculated from

$$P_e = f_e A = (79.8\text{ MPa})(491\text{ mm}^2) = 39.2 \times 10^3\text{ N} = 39.2\text{ kN}$$

(b) For a 2-m length of rod,

$$f_e = \frac{\pi^2 E}{(L/r)^2} = \frac{\pi^2 (207 \times 10^3\text{ MPa})}{[(2000\text{ mm})/(6.25\text{ mm})]^2}$$

$$= 19.95\text{ MPa} < 235\text{ MPa} \qquad \textbf{(O.K.)}$$

$$P_e = f_e A = (19.95\text{ MPa})(491\text{ mm}^2) = 9.80 \times 10^3\text{ N} = 9.8\text{ kN}$$

EXAMPLE 18.2 A 1-in.-diameter steel shaft is used as an axially loaded pin-connected compression member. The high-strength steel has a proportional limit of 50,000 psi and a modulus of elasticity of 30,000,000 psi. Calculate (a) the shortest length L for which Euler's formula is applicable and (b) the critical load if the length of the shaft is 48 in.

Solution From Table 3,

$$A = 0.7854d^2 = 0.7854(1\text{ in.})^2 = 0.7854\text{ in.}^2$$

$$r = \frac{d}{4} = \frac{1.0\text{ in.}}{4} = 0.25\text{ in.}$$

$$I = \frac{\pi d^4}{64} = \frac{\pi(1\text{ in.})^4}{64} = 0.0491\text{ in.}^4$$

(a) The shortest length L for which Euler's formula applies will be that length for which the critical stress is equal to the proportional limit:

$$f_e = \frac{\pi^2 E}{(L/r)^2}$$

Solve for L:

$$L^2 = \frac{\pi^2 E r^2}{f_e} = \frac{\pi^2(30{,}000{,}000\text{ psi})(0.25\text{ in.})^2}{50{,}000\text{ psi}} = 370.1\text{ in.}^2$$

from which

$$L = 19.24\text{ in.}$$

(b) If $L = 48$ in., Euler's formula is valid (48 in. > 19.24 in.). Therefore, the critical load can be calculated from Equation (18.1):

$$P_e = \frac{\pi^2 EI}{L^2} = \frac{\pi^2(30{,}000{,}000\text{ psi})(0.0491\text{ in.}^4)}{(48\text{ in.})^2} = 6310\text{ lb}$$

Equation (18.2) established the relationship between modulus of elasticity, slenderness ratio, and critical stress for slender columns. If E is known and if various values of the slenderness ratio are assumed, the corresponding values of f_e can be computed. This value, the Euler critical stress, can then be plotted as a function of the slenderness ratio. The resulting curve is shown in Figure 18.3. This curve is often referred to as the *Euler curve*, and it clearly depicts the way in which the critical stress at initial buckling decreases, along with the load-carrying capacity of the column, as the slenderness ratio increases.

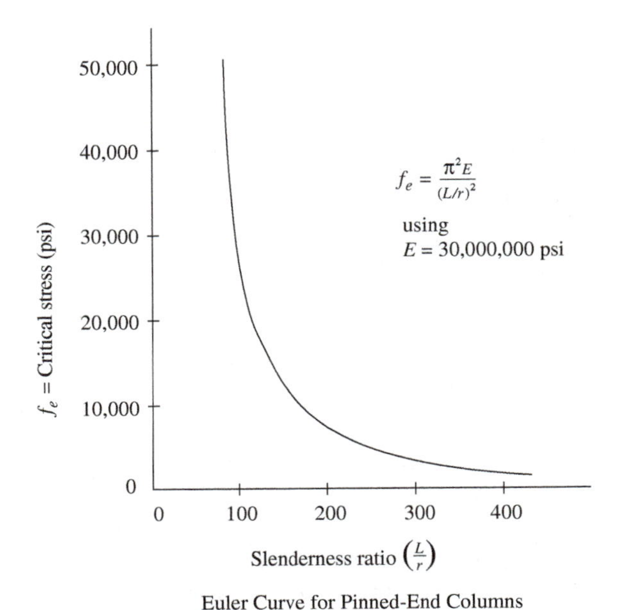

FIGURE 18.3 Euler curve for pinned-end columns.

We see, then, that the slenderness ratio, which is a function of the length and radius of gyration of the column, is a significant factor affecting the load-carrying capacity of the column. The tendency of a member to buckle is a function of the slenderness ratio. A column will always buckle about the axis having the largest slenderness ratio. In other words, if the unbraced length of the column is the same in all directions, it will buckle about the axis with the least radius of gyration. The concept of unbraced length is introduced because some columns are braced at various points. At such a braced point, column buckling (or lateral deflection) is prevented. Further, the bracing may prevent the buckling in a specific direction only. The buckled column must be visualized so as to associate the correct values of radius of gyration with values of length between supports (or braces).

Since Euler's formula is valid only if the resulting critical stress is below the proportional limit, Figure 18.3 shows that the range in which Euler's formula is applicable will vary with material. As an illustration, assume that a material (steel) has a proportional limit of 34,000 psi. If the 34,000 psi line were shown in Figure 18.3, it would intersect the Euler curve at a slenderness ratio value of 93.3. Any value of slenderness ratio less than 93.3 would result in a critical stress in excess of 34,000 psi. Therefore, Euler's formula is not applicable for this material if the slenderness ratio is less than 93.3.

18.3 EFFECTIVE LENGTH

Euler's formula (Equation [18.1]) gives the buckling load for an ideal column with pinned ends. A slender column, however, in addition to being less than perfect in other respects, may have end conditions that provide restraint of some magnitude so that the column ends are prevented from rotating freely. In such a case, the use of Euler's formula may be extended to columns having other than pinned ends through the use of an *effective length* in place of the actual unbraced length. The effective length is the distance between points of inflection (contraflexure) on the deflected shape of the column. These are points of zero bending moment and may be considered analogous to pinned ends, since pinned ends are also points of zero bending moment. For the ideal column, these points occur at the actual column ends. The effect of different end conditions on the deflected shape of a column is shown in Figure 18.4. As may be seen, the effective length of a column can be quite different for various end conditions, even though the actual length of the column does not vary.

End conditions can be accounted for through the use of an effective length factor K, a dimensionless number, that, when multiplied by the actual length of the column, yields the effective length KL. Therefore, the slenderness ratio is more correctly defined as the ratio of the effective length to the radius of gyration (KL/r). Euler's formulas can then be rewritten with the inclusion of the effective length:

$$P_e = \frac{\pi^2 EI}{(KL)^2} \quad \textbf{(18.3)}$$

$$f_e = \frac{\pi^2 E}{(KL/r)^2} \quad \textbf{(18.4)}$$

For example, if the ends of a column are fixed against rotation and translation (lateral movement), as shown in Figure 18.4b, and the column is loaded axially until it buckles, points of inflection will develop at the quarter points of the column length. If the middle portion of this fixed-end column is considered separately, it behaves as a pinned-end column having a length $L/2$ or $0.5L$. The length $0.5L$ is the effective length (KL) of the fixed-end column, where the K factor is 0.5. Substituting this into the critical load formula, Equation (18.3), gives

$$P_e = \frac{\pi^2 EI}{(KL)^2} = \frac{\pi^2 EI}{(0.5L)^2} = \frac{4\pi^2 EI}{L^2}$$

This equation indicates that a slender column with fixed ends that buckles elastically will be four times stronger than the same column with pinned ends.

Note that for the "flagpole" case (Figure 18.4d), one end is fixed and the other end is free with respect to both rotation and translation. In this case, there is an inflection point at the top of the column as well as an imaginary point of inflection at a distance L below the column base. The theoretical effective length factor K is 2.0. Note that the curvature shown is similar to the upper half of the pinned-end column of Figure 18.4a. The effective length for this "fixed/free" column is $2.0L$. Such a column has only one-quarter the strength of the same column with pinned ends.

Table 18.1 summarizes the K values that are shown in Figure 18.4. The theoretical K values are for ideal end conditions. In addition, since the idealized end conditions rarely exist, recommended values of K for design and analysis are provided. The latter K values should be used when the idealized end conditions are approximated.

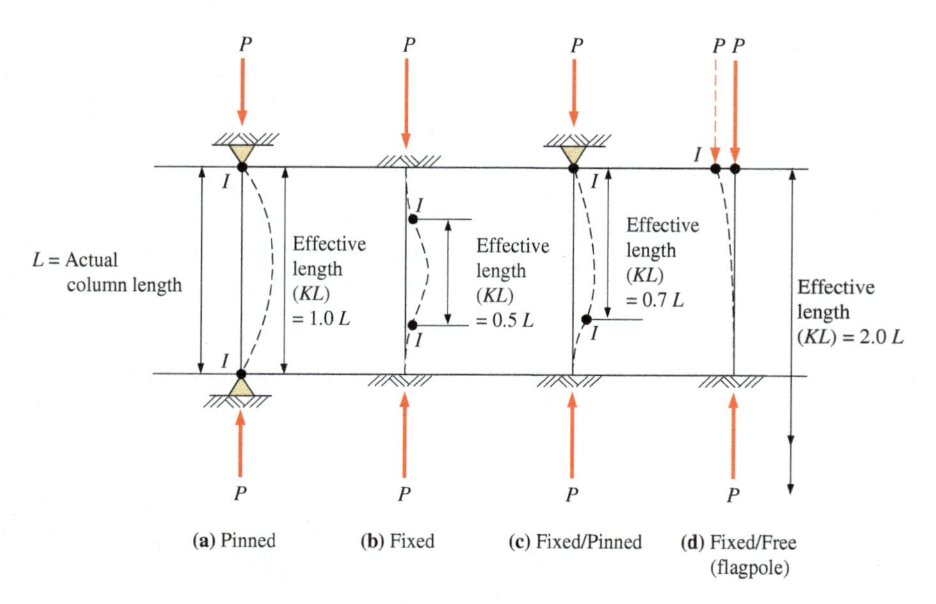

FIGURE 18.4 Effect of end conditions.

TABLE 18.1 Effective length factors

Idealized End Conditions	Theoretical K Value	Recommended K Value for Design and Analysis	Number of Times Stronger Than Pinned-End Column[a]
Pinned	1.0	1.0	1
Fixed	0.5	0.65	4
Fixed/pinned	0.7	0.80	2
Fixed/free	2.0	2.10	$\frac{1}{4}$

[a]Using the theoretical K value.

EXAMPLE 18.3 A column is composed of a 3-in.-diameter standard-weight steel pipe (Pipe 3 Std.) and is solidly embedded in a concrete foundation at its base and pinned at the top. The column is 12 ft long, as shown in Figure 18.5. Assuming the proportional limit to be 34 ksi, find the buckling load using the Euler theory.

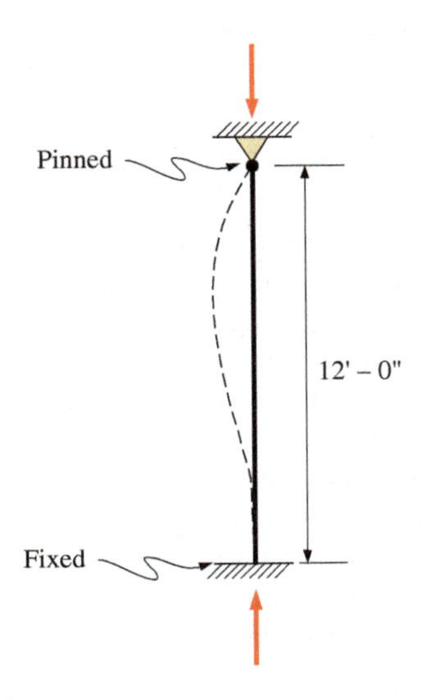

FIGURE 18.5 Column for Example 18.3.

Solution This is a fixed-pinned column. From Appendix B, the cross-sectional properties for this pipe are $A = 2.23 \text{ in.}^2$ and $r = 1.16$ in.

From Table 18.1, select $K = 0.8$ (recommended K for design/analysis). Calculate f_e using Equation (18.4):

$$f_e = \frac{\pi^2 E}{\left(\frac{KL}{r}\right)^2} = \frac{\pi^2(30{,}000 \text{ k/in.}^2)}{\left(\frac{0.8(144 \text{ in.})}{1.16 \text{ in.}}\right)^2} = 30.0 \text{ k/in.}^2$$

Note that 30 ksi < 34 ksi, therefore the Euler theory is applicable.

$$P_e = f_e A = 30.0 \text{ k/in.}^2\ (2.23 \text{ in.}^2) = 66.9 \text{ k}$$

18.4 REAL COLUMNS

In previous sections, we discussed the load-carrying capacity of columns based on the Euler theory which was, in turn, based on a theoretically perfect column. Real, or practical, columns have end conditions that are somewhere between the categories shown in Table 18.1: Hot-rolled steel columns have residual (initial) stresses in the cross section, all are composed of less-than-perfect materials, none are absolutely straight, and some have other imperfections that will affect column strength.

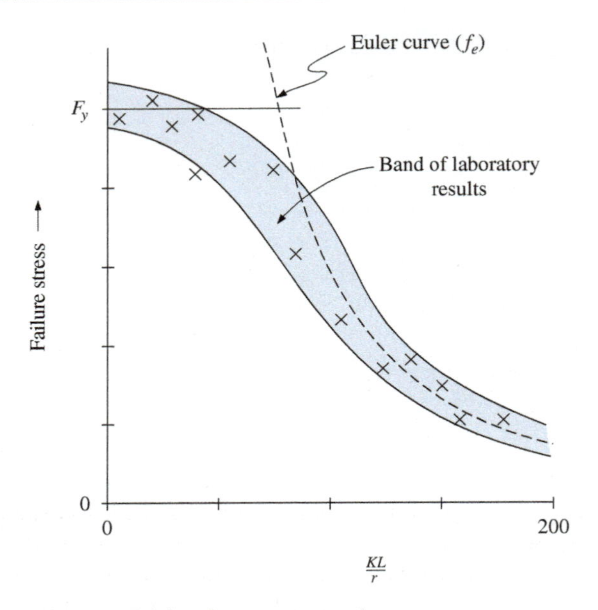

FIGURE 18.6 Real-column test results.

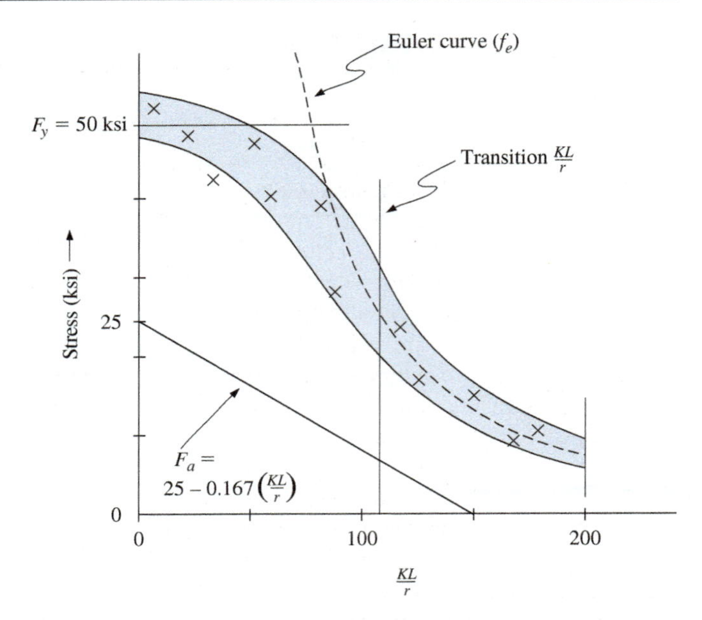

FIGURE 18.7 Axial compressive stress v. KL/r.

If a group of columns composed of a single material (say, ductile steel) and having a range of KL/r values is tested to compression load failure in a laboratory, a plot of the results can be drawn that will appear as shown in Figure 18.6. A column of a given KL/r value (on the horizontal axis) will be expected to fail at a stress within the band of results shown. The upper limit of KL/r has arbitrarily been set at 200. For low values of KL/r, the upper limit of failure stress is seen to be around the yield stress F_y. At low values of KL/r, there is no elastic buckling involved, but rather the material deforms or "squashes." At the higher values of KL/r, the upper limit of failure stress is seen to coincide with the elastic buckling stress predicted by the Euler theory. The deviation of the practical results from the two limits (F_y and the Euler curve) is seen to be the greatest at KL/r values a bit less than the value of KL/r where the Euler theory predicts a buckling stress of about F_y.

18.5 ALLOWABLE STRESSES FOR COLUMNS

The approach to practical column design has always made use of test results such as shown in Figure 18.6. The object is to set allowable stresses to reflect the failure curve of the practical column with a factor of safety applied. One rather simplistic way of doing this is to set a straight line somewhere below the failure curve to represent allowable axial compressive stress as a function of KL/r, as shown in Figure 18.7.

Figure 18.7 is a copy of Figure 18.6 showing an assumed F_y of 50 ksi. The straight line shown has an equation of

$$F_a = 25 - 0.167\left(\frac{KL}{r}\right)$$

where F_a is the allowable axial compressive stress (ksi). It is seen that this equation will result in an allowable stress giving a factor of safety of 2 when $KL/r = 0$, somewhat greater at higher KL/r values, and limiting KL/r to an upper value of about 150. While this is straightforward, it does not approximate the practical column test results very well.

It is common for specifications to more closely reflect the sweep of the practical column curve with a more complex equation or a set of equations. Note in Figure 18.7 that the band of practical results begins to deviate from the Euler curve at a transition KL/r value associated with a stress value around $F_y/2$. The value of the transition KL/r could be calculated from

$$\frac{\pi^2 E}{\left(\text{transition}\dfrac{KL}{r}\right)^2} = \frac{F_y}{2}$$

from which

$$\text{transition}\frac{KL}{r} = \sqrt{\frac{2\pi^2 E}{F_y}} \qquad \textbf{(18.5)}$$

For $F_y = 50$ ksi, this value becomes

$$\sqrt{\frac{2\pi^2(30{,}000\ \text{k/in.}^2)}{50\ \text{k/in.}^2}} = 109$$

Since practical columns have their strengths accurately predicted by the Euler theory for the higher KL/r values, the Euler critical stress can be divided by a factor of safety to give an allowable stress. This is valid for columns with KL/r values greater than the transition value ($KL/r = 109$ in our example). For the lower KL/r values (<109), a curve with a concave downward sweep such as a properly configured parabola can be employed. Factors of safety will vary with different materials.

18.6 AXIALLY LOADED STRUCTURAL STEEL COLUMNS (AISC)

Structural steel columns are primarily composed of wide-flange shapes, although sometimes built-up shapes and other shapes such as tubes are used. The accepted standard for analysis and design of these columns is the *Specification for Structural Steel Buildings*, which is available in the *AISC Manual of Steel Construction*.[1] The method of evaluating column strength set forth in the noted specification is complex and detailed. The procedure outlined here applies to most W-shape columns in simple applications and is presented for introductory purposes only. Notation has been modified slightly to be in accordance with that used elsewhere in the text.

For these columns, slenderness ratios in excess of 200 are felt to reflect columns so slender as to be overly sensitive to uncontrollable factors such as initial straightness of the column. Therefore, slenderness ratios (KL/r) should not exceed 200.

Two formulas are employed to evaluate column strength $P_{a(\text{all})}$. The transition KL/r is calculated from

$$4.71\sqrt{\frac{E}{F_y}} \qquad \textbf{(18.6)}$$

For $KL/r > 4.71\sqrt{E/F_y}$, the allowable column load is calculated from

$$P_{a(\text{all})} = \frac{0.877 f_e A}{1.67} = 0.525 f_e A \qquad \textbf{(18.7)}$$

where f_e is the Euler critical buckling stress for an axially loaded column and is calculated from Equation (18.4), and A is the gross cross-sectional area of the column. For $KL/r \leq 4.71\sqrt{E/F_y}$, the allowable column load is calculated from

$$P_{a(\text{all})} = \frac{\left[0.658^{\left(\frac{F_y}{f_e}\right)}\right] F_y A}{1.67} \qquad \textbf{(18.8)}$$

In Equations (18.7) and (18.8), one could consider the equation to have the form $P_{a(\text{all})} = F_a A$. Figure 18.8 is a plot of the allowable stress F_a as a function of KL/r. The yield stress F_y has been assumed to be 50 ksi. Note how the curve reflects the shape of Figure 18.6.

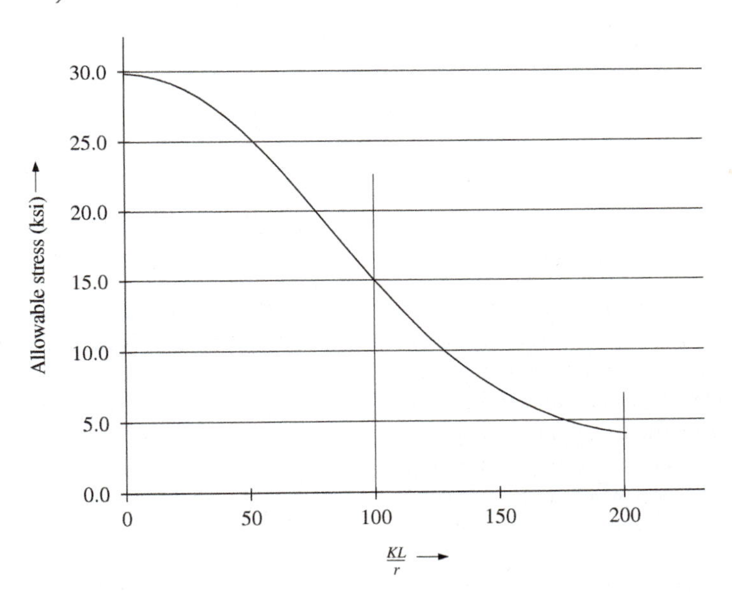

FIGURE 18.8 Steel F_a v. KL/r ($F_y = 50$ ksi).

EXAMPLE 18.4 Calculate the allowable axial compressive load for a W12 × 50 structural steel column using the AISC approach. The column ends are pinned. Assume $F_y = 50$ ksi, and use an unbraced length of (a) 15 ft and (b) 30 ft.

Solution From Appendix A, for the W12 × 50, $A = 14.6$ in.^{2} and $r_y = 1.96$ in. Note that it is only the least radius of gyration that is of interest, since it is this value that will result in the largest slenderness ratio. From Table 18.1, obtain a K value of 1.0 (pinned ends).

[1] American Institute of Steel Construction, Inc., *Manual of Steel Construction*, 13th ed. (Chicago: AISC, 2005).

Next, the transition KL/r value will be used to determine whether Equation (18.7) or Equation (18.8) applies:

$$\text{transition}\frac{KL}{r} = 4.71\sqrt{\frac{E}{F_y}} = 4.71\sqrt{\frac{30{,}000\text{ ksi}}{50\text{ ksi}}} = 115.4$$

(a) For $L = 15$ ft,

$$\frac{KL}{r} = \frac{1.0(15\text{ ft})(12\text{in./ft})}{1.96\text{ in.}} = 91.8$$

This is less than the transition KL/r value of 115.4, therefore Equation (18.8) applies. First we will calculate the value of the Euler critical buckling stress, followed by the $P_{a(all)}$ calculation:

$$f_e = \frac{\pi^2 E}{\left(\frac{KL}{r}\right)^2} = \frac{\pi^2(30{,}000\text{ ksi})}{91.8^2} = 35.1\text{ ksi}$$

$$P_{a(all)} = \frac{\left[0.658^{\left(\frac{F_y}{f_e}\right)}\right]F_y A}{1.67} = \frac{\left[0.658^{\left(\frac{50\text{ ksi}}{35.1\text{ ksi}}\right)}\right](50\text{ ksi})(14.6\text{ in.}^2)}{1.67} = 240\text{ k}$$

(b) For $L = 30$ ft,

$$\frac{KL}{r} = \frac{1.0(30\text{ ft})(12\text{ in./ft})}{1.96\text{ in.}} = 183.7$$

This is greater than the transition value of 115.4, therefore Equation (18.7) applies. Calculate the Euler critical buckling stress, followed by $P_{a(all)}$:

$$f_e = \frac{\pi^2 E}{\left(\frac{KL}{r}\right)^2} = \frac{\pi^2(30{,}000\text{ ksi})}{183.7^2} = 8.77\text{ ksi}$$

$$P_{a(all)} = 0.525 f_e A = 0.525(8.77\text{ ksi})(14.6\text{ in.}^2) = 67.2\text{ k}$$

In the design of axially loaded structural steel columns, since the allowable axial load is a function of the slenderness ratio, there is no direct solution for a required column area. If an equation such as Equation (18.1) governs (and at the outset, there is no assurance that it does), then a required moment of inertia can be determined and a section selected. Otherwise, a trial-and-error procedure can be used whereby a trial section is assumed, then analyzed to determine its allowable load, after which another trial section (either larger or smaller) is selected and the process repeated. One may assume an allowable stress of perhaps $\frac{1}{4}$ to $\frac{1}{3}$ times the yield stress in order to aid in the initial selection of a trial section. The trial-and-error approach will be illustrated in Section 18.7. Usually, the selection of columns is aided through the use of allowable-load tables where allowable loads are tabulated for various shapes and for various effective lengths. The selection of an appropriate shape is then straightforward.

18.7 AXIALLY LOADED STEEL MACHINE PARTS

In the design and analysis of machine parts subjected to axial compressive loads, the AISC column formulas are not strictly applicable, even though they result in solutions that are generally satisfactory. The AISC design specification was developed for the design of steel-framed buildings; it is recognized nationally in that it is incorporated by reference into most state and municipal building codes, thereby making it legal and enforceable. In the area of machine design, however, there is no one nationally accepted or legally adopted design specification that encompasses all machine design applications.

For the design and/or analysis of steel compression members that are a part of some machine, some acceptable and appropriate column equations can be used. They are as follows:

1. The member is categorized as slender if

$$\frac{KL}{r} \geq \sqrt{\frac{2\pi^2 E}{F_y}}$$

The allowable axial compressive stress can then be computed from

$$F_a = \frac{\pi^2 E}{(KL/r)^2(\text{F.S.})} \qquad \textbf{(18.9)}$$

Note that the limiting KL/r value (or the transition KL/r value) that determines whether a member is to be treated as a slender (long) or intermediate column is the same as the value defined by Equation (18.5). Also note that Equation (18.9) is a modified Euler critical stress formula—see Equation (18.2)—except for the factor of safety, which may be varied.

2. The member is categorized as intermediate if

$$\frac{KL}{r} < \sqrt{\frac{2\pi^2 E}{F_y}}$$

and the allowable axial compressive stress can be determined using the empirical J. B. Johnson formula:

$$F_a = \frac{\left[1 - \dfrac{F_y(KL/r)^2}{4\pi^2 E}\right]F_y}{\text{F.S.}} \quad \textbf{(18.10)}$$

where all terms are as were defined previously. This formula is generally considered valid for *KL/r* values ranging down to zero.

The minimum factor of safety used in machine part applications is usually 2.0. Larger values are used under conditions of greater uncertainty. These conditions include all of the possible uncertainties discussed in Section 10.6.

Figure 18.9 is a plot of allowable axial compressive stress determined using Equations (18.10), (18.11), the J. B. Johnson formula (Equation [18.10]), $F_y = 50$ ksi, and a factor of safety of 2.0.

Note again how the curve follows the pattern of the laboratory test results of Figure 18.6.

The analysis and design procedure for steel compression member machine parts is similar to that used for steel wide-flange columns in the previous section. Numerous tables and design aids are available from which to determine the allowable axial compressive stress and to simplify analysis. Similarly, column load tables simplify the design process and eliminate the trial-and-error procedure. In the absence of such design aids, we recommend the analysis and design procedures described in the following examples.

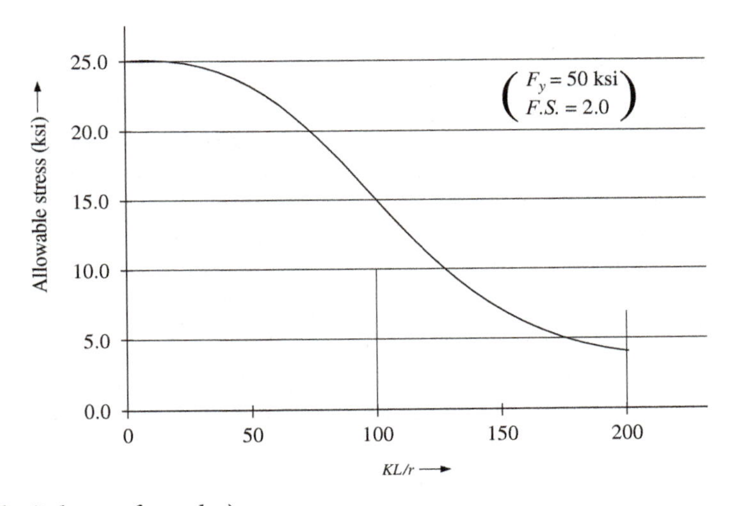

FIGURE 18.9 F_a v. *KL/r* (Euler/Johnson formulas).

EXAMPLE 18.5 A square machine member measuring $\frac{1}{2}$ in. by $\frac{1}{2}$ in. has a length of 12 in. The member is to be loaded in axial compression. End conditions are fixed/pinned. The member is made of AISI 1040 hot-rolled steel. Compute the allowable load using a factor of safety of 3.0. The modulus of elasticity *E* is 30,000,000 psi.

Solution For the $\frac{1}{2}$-in.-square cross section (from Table 3),

$$A = (0.5 \text{ in.})^2 = 0.25 \text{ in.}^2 \quad \text{and} \quad r = \frac{h}{\sqrt{12}} = 0.144 \text{ in.}$$

From Table 18.1, the effective length factor is 0.80. Therefore,

$$\frac{KL}{r} = \frac{0.80(12 \text{ in.})}{0.144 \text{ in.}} = 66.7$$

Next, compute the transition *KL/r* value. The yield stress F_y is obtained from Appendix G:

$$\sqrt{\frac{2\pi^2 E}{F_y}} = \sqrt{\frac{2\pi^2(30{,}000{,}000 \text{ psi})}{42{,}000 \text{ psi}}} = 118.7$$

Since 66.7 < 118.7, the column is intermediate (or short) and the J. B. Johnson formula should be used:

$$F_a = \frac{\left[1 - \dfrac{F_y(KL/r)^2}{4\pi^2 E}\right] F_y}{\text{F.S.}}$$

$$= \frac{\left[1 - \dfrac{(42{,}000 \text{ psi})(66.7)^2}{4\pi^2(30{,}000{,}000 \text{ psi})}\right] 42{,}000 \text{ psi}}{3.0} = 11{,}790 \text{ psi}$$

from which the allowable axial compressive load is

$$P_{a(\text{all})} = F_a A = (11{,}790 \text{ psi})(0.25 \text{ in.}^2) = 2950 \text{ lb}$$

EXAMPLE 18.6 Determine the required diameter of a round linkage bar of AISI 1040 hot-rolled steel with a length of 15 in. The bar is subjected to an axial compressive load of 3000 lb and is pin-connected. A factor of safety of 3.0 is to be used.

Solution For a round bar, the following properties apply (from Table 3):

$$A = 0.7854d^2 \quad \text{and} \quad r = 0.25d$$

From Appendix G, for 1040 hot-rolled steel,

$$F_y = 42{,}000 \text{ psi} \quad \text{and} \quad E = 30{,}000{,}000 \text{ psi}$$

Assume that the member is in the category of an intermediate column. Therefore, an allowable axial compressive stress can be computed in terms of the bar diameter using the J. B. Johnson formula. The validity of the assumption will have to be checked before the solution is finalized. Thus, using units of lb and in.:

$$F_a = \frac{\left[1 - \dfrac{F_y(KL/r)^2}{4\pi^2 E}\right] F_y}{\text{F.S.}}$$

$$= \frac{\left[1 - \dfrac{42{,}000[1(15)/(0.25d)]^2}{4\pi^2(30{,}000{,}000)}\right] 42{,}000}{3.0}$$

$$= \frac{\left[1 - \dfrac{0.128}{d^2}\right] 42{,}000}{3.0}$$

$$= 14{,}000 - \frac{1792}{d^2}$$

Since an allowable axial load of 3000 lb is desired, substitute $P_{a(\text{all})}/A$ for F_a and the expression becomes

$$\frac{P_{a(\text{all})}}{A} = 14{,}000 - \frac{1792}{d^2}$$

Substituting gives

$$\frac{3000}{0.7854d^2} = 14{,}000 - \frac{1792}{d^2}$$

$$\frac{3000}{0.7854} = 14{,}000d^2 - 1792$$

$$14{,}000d^2 = 3820 + 1792$$

from which

$$d = 0.633 \text{ in.}$$

Try a $\frac{3}{4}$-in.-diameter round bar.

Now check the validity of the J. B. Johnson formula for the bar chosen:

$$\frac{KL}{r} = \frac{1(15 \text{ in.})}{0.25(0.75 \text{ in.})} = 80$$

Compute the value of the slenderness ratio, which is the lower limit for long columns:

$$\sqrt{\frac{2\pi^2 E}{F_y}} = \sqrt{\frac{2\pi^2(30{,}000{,}000 \text{ psi})}{42{,}000 \text{ psi}}} = 118.7$$

Since 80 < 118.7, the use of the J. B. Johnson formula is valid.

Use a $\frac{3}{4}$-in.-diameter round bar.

18.8 AXIALLY LOADED TIMBER COLUMNS

The most frequently used wood columns are simple solid pieces of square or rectangular cross section. Simple columns may consist of a single piece of wood or of two or more pieces properly fastened together to form a single member. The column of round cross section is a type of simple solid column that is less frequently encountered. There are also other types of timber columns, such as spaced columns and built-up columns, the complexity of which is beyond the scope of this text. Our discussion is limited to the simple solid wood column of square or rectangular cross section. Generally, axially loaded columns of rectangular cross section have lateral side dimensions that do not differ by more than 2 in.

The most widely used timber-column design approach is that furnished in the National Design Specification for Wood Construction (NDS) issued by the American Forest & Paper Association (AF&PA).[2]

Prior timber specifications classified timber columns (based on slenderness ratio) as short, intermediate, or long, and the allowable stress was then determined using an appropriate formula. The current specification provides a single, uniform method for determining the allowable stress for any column in which the slenderness ratio does not exceed 50. For timber columns, slenderness ratio is the ratio of the unbraced length L of a column to its least lateral dimension d:

$$\text{slenderness ratio} = \frac{L}{d} \leq 50$$

The slenderness ratio is further modified and expressed as L_e/d, where L_e represents the effective unbraced length and is defined as

$$L_e = KL$$

where K is the effective length factor as shown in Table 18.1.

The allowable axial compressive stress for a timber column is determined by multiplying the allowable stress for compression parallel to the grain F_c (from Appendix F) by adjustment factors for load duration, wet service, temperature, size, and column stability. The last of these reflects the effective slenderness ratio of the column and is the predominant adjustment factor for ordinary columns. An in-depth study of the NDS is beyond the scope of this text. The approach, however, is illustrated using the limited properties contained in Appendix F and by restricting discussion to columns of solid, square, or rectangular sections of sawn, visually graded timber used at normal temperature and moisture conditions and subjected to permanent loading. The NDS should be consulted for those cases not included within this narrow range.

For our assumptions, all adjustment factors except column stability are taken conservatively as 1.0. Therefore, the allowable axial compressive stress is calculated from

$$F_a = F_c C_P \tag{18.11}$$

where the column stability adjustment factor C_P is calculated from Equation 3.7-1 of the NDS. For convenience and use of the limited properties of Appendix F, this equation is slightly modified and rewritten as

$$C_P = \frac{1+\alpha}{1.6} - \sqrt{\left(\frac{1+\alpha}{1.6}\right)^2 - \frac{\alpha}{0.8}} \tag{18.12}$$

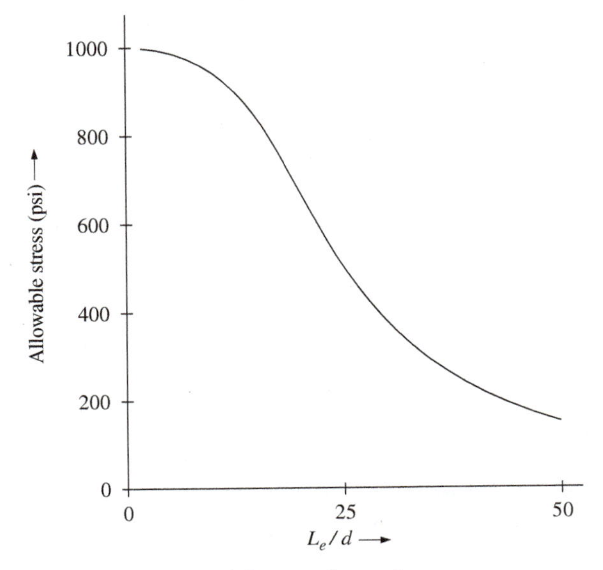

FIGURE 18.10 F_a example—timber columns.

[2]American Forest & Paper Association, *National Design Specification for Wood Construction* (Washington, DC: AF&PA, 2005).

The variable α is determined from

$$\alpha = \frac{0.3E}{(L_e/d)^2 F_c} \quad \textbf{(18.13)}$$

where E is the modulus of elasticity (from Appendix F) and all other terms have been previously defined.

Assuming values of $F_c = 1000$ psi and $E = 1{,}300{,}000$ psi, the three preceding formulas (Equations [18.11] through [18.13]) can be used to plot F_a for a range of L_e/d values from 0 to 50. This plot is shown in Figure 18.10.

As with Figures 18.8 and 18.9, this curve is seen to closely resemble the shape of the laboratory results of Figure 18.6.

EXAMPLE 18.7 Find the allowable axial compressive load for a pin-connected 8-in.-by-10-in. (S4S) Douglas fir column having an unbraced length of 10 ft. Assume normal moisture conditions and duration of loading. See Figure 18.11.

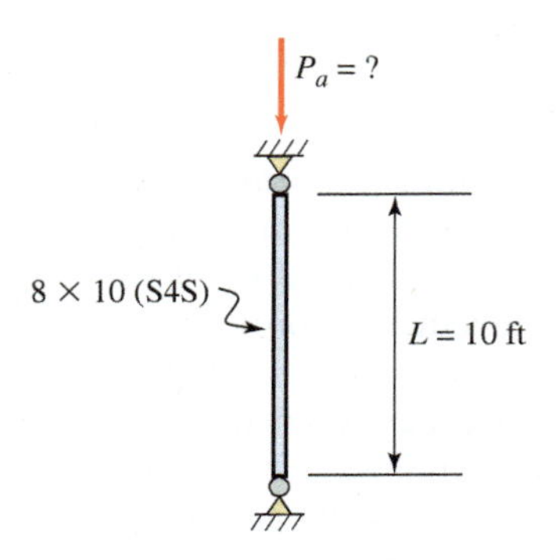

FIGURE 18.11 Pin-connected timber column.

Solution From Table F.1, for Douglas fir,

$$F_c = 1050 \text{ psi} \quad \text{and} \quad E = 1{,}700{,}000 \text{ psi}$$

From Table 18.1, the effective length factor K_e is 1.0; thus,

$$\frac{L_e}{d} = \frac{KL}{d} = \frac{1.0(10 \text{ ft})(12 \text{ in./ft})}{7.5} = 16 < 50 \qquad \textbf{(O.K.)}$$

Calculate α from Equation (18.13):

$$\alpha = \frac{0.3E}{(L_e/d)^2 F_c} = \frac{0.3(1{,}700{,}000 \text{ psi})}{16^2 (1050 \text{ psi})} = 1.897$$

Next, calculate the column stability adjustment factor:

$$C_P = \frac{1+\alpha}{1.6} - \sqrt{\left(\frac{1+\alpha}{1.6}\right)^2 - \frac{\alpha}{0.8}}$$

$$= \frac{1+1.897}{1.6} - \sqrt{\left(\frac{1+1.897}{1.6}\right)^2 - \frac{1.897}{0.8}} = 0.858$$

from which

$$F_a = F_c C_P = 1050 \text{ psi}(0.858) = 901 \text{ psi}$$

The allowable axial compressive load is then

$$P_{a(\text{all})} = F_a A = 901 \text{ psi}(7.5 \text{ in.})(9.5 \text{ in.}) = 64{,}200 \text{ lb}$$

EXAMPLE 18.8 Calculate the allowable axial compressive load for a 200-mm-by-200-mm (S4S) hem-fir column having an unbraced length of 7 m. Assume the column is pinned at the top and fixed at the bottom. Assume normal conditions of temperature, moisture, and duration of loading.

Solution For hem-fir, from Table F.2,

$$F_c = 8.96 \text{ MPa} \quad \text{and} \quad E = 9.7 \times 10^3 \text{ MPa}$$

From Table E.2, for the 200-by-200 (S4S) timber,

$$d = 191 \text{ mm} \quad \text{and} \quad A = 36.5 \times 10^3 \text{ mm}^2$$

From Table 18.1, the effective length factor K is 0.8; thus,

$$\frac{L_e}{d} = \frac{KL}{d} = \frac{0.8(7000 \text{ mm})}{191 \text{ mm}} = 29.3 < 50 \qquad \textbf{(O.K.)}$$

Calculate α:

$$\alpha = \frac{0.3E}{(L_e/d)^2 F_c} = \frac{0.3(9.7 \times 10^3 \text{ MPa})}{29.3^2(8.96 \text{ MPa})} = 0.378$$

Calculate the column stability adjustment factor:

$$C_P = \frac{1+\alpha}{1.6} - \sqrt{\left(\frac{1+\alpha}{1.6}\right)^2 - \frac{\alpha}{0.8}}$$

$$= \frac{1+0.378}{1.6} - \sqrt{\left(\frac{1+0.378}{1.6}\right)^2 - \frac{0.378}{0.8}} = 0.342$$

from which

$$F_a = F_c C_P = (8.96 \text{ MPa})(0.342) = 3.06 \text{ MPa}$$

and

$$P_{a(\text{all})} = F_a A = (3.06 \text{ MPa})\,(36.5 \times 10^3 \text{ mm}^2)$$
$$= 111.7 \times 10^3 \text{ N}$$
$$= 111.7 \text{ kN}$$

The design of axially loaded timber columns is similar to that of steel columns. An abundance of allowable axial load tables for square and rectangular solid timber columns are available. Such tables are usually employed in timber-column design. In their absence, the design process becomes one of trial and error. The following design example will illustrate the trial-and-error procedure.

EXAMPLE 18.9 Design a square pin-connected Douglas fir column (S4S) to support an axial compressive load of 45,000 lb. The unbraced length of the column is 14 ft. Assume normal moisture conditions and duration of load.

Solution For Douglas fir, from Appendix F,

$$F_c = 1050 \text{ psi} \quad \text{and} \quad E = 1{,}700{,}000 \text{ psi}$$

From Table 18.1, the effective length factor is 1.0. Assuming $F_a = F_c$, one can estimate an approximate required area:

$$\text{required } A = \frac{P}{F_a} = \frac{45{,}000 \text{ lb}}{1050 \text{ psi}} = 42.9 \text{ in.}^2$$

Since the assumed allowable stress is likely to be high, the required area quite possibly will be greater than 42.9 in^2. Therefore, try a column with somewhat more area, a nominal 8-in.-by-8-in. (S4S) column with an area of 56.25 in^2:

$$\frac{L_e}{d} = \frac{1.0(14 \text{ ft})(12\text{in./ft})}{7.5 \text{ in.}} = 22.4 < 50 \qquad \textbf{(O.K.)}$$

Calculate α:

$$\alpha = \frac{0.3E}{(L_e/d)^2 F_c} = \frac{0.3(1{,}700{,}000 \text{ psi})}{22.4^2(1050 \text{ psi})} = 0.968$$

Calculate the column stability adjustment factor:

$$C_p = \frac{1+\alpha}{1.6} - \sqrt{\left(\frac{1+\alpha}{1.6}\right)^2 - \frac{\alpha}{0.8}}$$

$$= \frac{1+0.968}{1.6} - \sqrt{\left(\frac{1+0.968}{1.6}\right)^2 - \frac{0.968}{0.8}} = 0.680$$

from which

$$F_a = F_c C_p = (1050 \text{ psi})(0.680) = 714 \text{ psi}$$

and

$$P_{a(\text{all})} = F_a A = (714 \text{ psi})(56.25 \text{ in}^2) = 40{,}200 \text{ lb} < 45{,}000 \text{ lb}$$

Therefore, a larger column cross section must be chosen. A similar analysis for a 10-by-10 (S4S) will show an allowable axial load of 77,600 lb. Therefore, use a 10-by-10 (S4S) column.

SUMMARY BY SECTION NUMBER

18.1 Columns are usually categorized as short, intermediate, or long (slender). Most real-world columns are in the intermediate category where the failure mode is a combination of buckling of the member and yielding (or crushing) of the material. Buckling is an elastic bending behavior due to an axial compressive load.

18.2 The classic Euler formula for obtaining buckling load (or critical load) for a theoretically perfect (ideal) column is expressed as

$$P_e = \frac{\pi^2 EI}{L^2} \quad \textbf{(18.1)}$$

It can also be expressed in terms of a critical stress:

$$f_e = \frac{\pi^2 E}{(L/r)^2} \quad \textbf{(18.2)}$$

L/r is called the *slenderness ratio*. Euler's formula is valid only if the resulting critical stress is less than the proportional limit of the material.

18.3 The effective length (KL) of a column is a function of the unbraced length of the column and its end conditions. Values for the effective length factor K are given in Table 18.1.

18.4 Real columns conform to Euler theory only if they are slender (higher KL/r values). If KL/r is low, column strength is governed by the yielding strength or crushing strength of the material. Real columns are also affected by lack of initial straightness, residual stresses (hot-rolled steel columns), and other imperfections. Figure 18.6, a plot of failure stress as a function of KL/r, shows how real failure stresses in columns vary from the Euler theory.

18.5 Allowable stresses for columns reflect the real column behavior curve over the KL/r range (Figure 18.6). Formulas for allowable stress utilize real column strengths and introduce factors of safety appropriate for the material.

18.6 In the field of steel frame-building design, AISC Equations (18.7) and (18.8) are used for the determination of allowable loads for axially loaded columns in simple applications.

18.7 A combination of the empirical J. B. Johnson formula and a modified Euler formula is frequently used for the analysis and design of steel machine parts subjected to axial compressive loads. The Euler expression is used for the slender columns and the J. B. Johnson formula for the other (short and intermediate) columns.

18.8 The analysis and design of axially loaded timber columns is accomplished using a procedure recommended in the National Design Specification for Wood Construction of the AF&PA. It is similar to that used for steel columns except that the allowable stress formulas are those recommended in the National Design Specification of the AF&PA; see Equations (18.11) through (18.13).

PROBLEMS

In the following problems, unless otherwise noted, assume for steel that the modulus of elasticity is 30,000 ksi $(207 \times 10^3 \text{ MPa})$, the proportional limit is 48 ksi (331 MPa), and the yield stress is 50 ksi (345 MPa).

Section 18.2 Ideal Columns

18.1 Calculate the Euler buckling load for an axially loaded, pin-connected W14 × 22 structural steel wide-flange column. The column length is 12 ft, as shown.

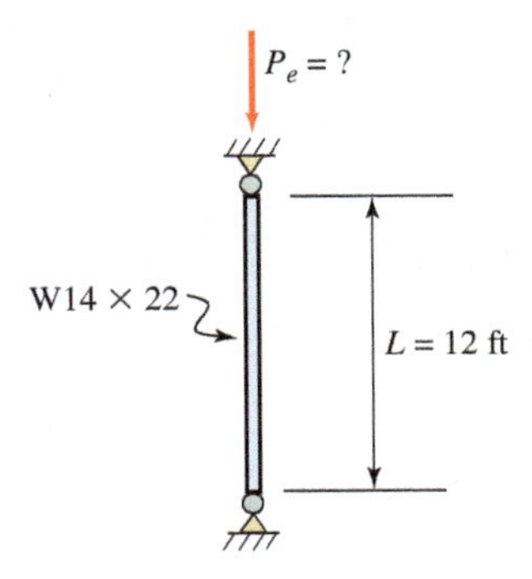

PROBLEM 18.1

18.2 Calculate the Euler buckling load for a pin-connected steel pipe column. Use $A = 8.4 \text{ in.}^2$, $r = 2.94$ in., and the following lengths: **(a)** 50 ft, **(b)** 35 ft, **(c)** 20 ft, and **(d)** 15 ft. The proportional limit is 34,000 psi.

18.3 A pin-connected axially loaded compression member is made of an aluminum alloy. The proportional limit of the material is 220 MPa and the modulus of elasticity is 69.0×10^3 MPa. Calculate the lowest value of slenderness ratio for which Euler's formula is applicable.

18.4 A pin-connected axially loaded steel bar is used as a compression member. The bar has a rectangular cross section 1 in. by 2 in. The proportional limit is 30,000 psi. Calculate the minimum length L for Euler's formula to be applicable. In addition, calculate the critical stress and the critical load if the length of the bar is 5 ft.

18.5 Plot a curve showing the relationship of f_e versus L/r for steel columns. Draw the curve for slenderness ratios that vary from 0 to 200. Assume the proportional limit = 235 MPa.

Section 18.4 Real Columns

18.6 A W12 × 22 structural steel wide-flange section is used as an axially loaded column. Determine the allowable axial compressive load using Euler's formula and a factor of safety of 2.0. The column is pin-connected at both ends and is 14 ft long.

18.7 Steel columns are to have end conditions as shown in Figure 18.4a through d. Calculate, for each case, the minimum slenderness ratio for the application of Euler's formula. The proportional limit of the material is 40 ksi.

18.8 Calculate the allowable axial compressive load for a pin-connected hollow tube of aluminum using Euler's formula and a factor of safety of 2.0. The tube is rectangular, 1 in. by 2 in. in cross section, with a wall thickness of 0.1 in. and a length of 48 in. The proportional limit of the material is 30,000 psi.

18.9 A W200 × 59 structural steel wide-flange section is used as a pin-ended column 7 m long. Find the allowable axial compressive load using Euler's formula and a factor of safety of 2.5.

18.10 Use Euler's formula and a factor of safety of 2.5 to design a W14 structural steel wide-flange column to support an axial load of 350 kips. The length of the column is 34 ft and its ends are pin-connected.

Section 18.6 Axially Loaded Structural Steel Columns (AISC)

For Problems 18.11 through 18.17, unless otherwise noted, yield stress of the steel is 50 ksi (345 MPa) (ASTM A992).

18.11 Calculate the allowable axial compressive load for a W14 × 90 structural steel column having an unbraced length of 16 ft. The ends are pin-connected.

18.12 Calculate the allowable axial compressive load for the column of Problem 18.11:
(a) if the ends are fixed
(b) if they are fixed/pinned

18.13 A W250 × 115 structural steel column is to carry an axial load of 1300 kN. The ends are pinned. Determine whether or not the column is adequate if:
(a) the length is 6 m.
(b) the length is 9 m.

18.14 Calculate the allowable axial compressive load for a steel pipe column having an unbraced length of 20 ft. The ends are pin-connected. Use $A = 11.9 \text{ in.}^2$ and $r = 3.67$ in. $F_y = 35$ ksi.

18.15 Calculate the allowable axial compressive load for a W12 × 50 structural steel column having a 12-in.-by-$\frac{1}{2}$-in. plate bolted to each flange. The unbraced length is 32 ft. The ends are pin-connected. Use $F_y = 36$ ksi.

18.16 Select the lightest standard-weight steel pipe section to support an axial compressive load of 100 kN. The column is pin-connected at each end and has an unbraced length of 5.50 m. Use $F_y = 240$ MPa.

18.17 Select the lightest W14 shape for a column subjected to an axial compressive load of 360 kips. The unbraced length of the column is 24 ft and the ends are pin-connected.

Section 18.7 Axially Loaded Steel Machine Parts

For Problems 18.18 through 18.21, use the Euler–Johnson formulas, Equations (18.9) and (18.10), and a modulus of elasticity of 30,000,000 psi.

18.18 Calculate the allowable axial compressive load for a bar of rectangular cross section that measures 1 in. by 2 in. and is 20 in. long. The member is made of AISI 1040 hot-rolled steel. Use a factor of safety of 3.5 and assume the member is pin-connected.

18.19 Compute the required diameter of a piston rod of AISI 1040 hot-rolled steel with a length of 20 in. The rod is subjected to an axial compressive load of 6000 lb. Use a factor of safety of 2.5. Assume that the rod is pin-connected.

18.20 Compute the required diameter of an air cylinder piston rod of AISI 1040 hot-rolled steel. The rod has a length of 54 in. and is subjected to an axial compressive load of 1900 lb. Assume pinned ends. Use a factor of safety of 3.5.

18.21 Compute the required diameter of a steel rod of AISI 1050 cold-drawn steel with a yield stress of 85,000 psi and a length of 8 ft. The rod is subjected to an axial compressive load of 18,000 lb. Assume pinned ends. Use a factor of safety of 3.0.

Section 18.8 Axially Loaded Timber Columns

For Problems 18.22 through 18.26, assume normal conditions of moisture, temperature, and duration of loading. Refer to Appendix F for values of E and F_c.

18.22 Find the allowable axial compressive load for an 8-in.-by-8-in. (S4S) Douglas fir column. The ends are pin-connected. Use an unbraced length of
(a) 10 ft
(b) 17 ft

18.23 Solve Problem 18.22 if the column is a 250 mm by 250 mm (S4S) eastern white pine column with fixed/pinned ends. Use lengths of
(a) 3.05 m
(b) 5.18 m

18.24 Design a square pin-connected Douglas fir column (S4S) to support an axial compressive load of 90,000 lb. The unbraced length of the column is 18 ft.

18.25 Design a square or rectangular (2-in. nominal increments) pin-connected southern pine column (S4S). The unbraced length of the column is 16 ft. The column is to support an axial compressive load of
(a) 48,000 lb
(b) 60,000 lb

18.26 Solve Problem 18.25 if the end conditions are fixed/ pinned.

Computer Problems

For the following computer problems, any appropriate software may be used. Input prompts should fully explain what is required of the user (the program should be user-friendly). The resulting output should be well labeled and self-explanatory.

18.27 Write a program that will calculate the Euler buckling load for a given column and check whether Euler's formula is applicable. User input is to be area and least radius of gyration for the cross section, the proportional limit and modulus of elasticity for the material, and the length of the column. Assume pinned ends.

18.28 Write a program that will calculate the allowable axial compressive load for an S4S square timber column (4 in. by 4 in. minimum) for lengths ranging from 0 to the maximum allowed (in 2-ft increments). Ends are pinned. User input is to be the nominal column size and the values of F_c and E. Output should consist of allowable axial compressive load versus length in a tabular format.

18.29 Use a spreadsheet to calculate allowable axial load for a square pin-ended timber column. Vary the effective slenderness ratio from 0 to 50 in steps of 2.0. Set up the spreadsheet so that a user can input the column size and material properties. Include a plot of allowable load versus effective slenderness ratio.

18.30 Use a spreadsheet to calculate the AISC allowable loads for W-shape columns. In the left column of the spreadsheet, place the effective length (ft) varying by 1-ft increments. Across the top of the spreadsheet, note the cross section from Appendix A. The body of the spreadsheet will be the allowable loads.
(a) W8 shapes
(b) W10 shapes
(c) W12 shapes
(d) W14 shapes

Supplemental Problems

18.31 Calculate the Euler buckling load for an axially loaded, pin-connected aluminum alloy rod having a diameter of $1\frac{1}{2}$ in. Use a modulus of elasticity of 10,000,000 psi and a proportional limit of 32,000 psi. Calculate the buckling load for the following lengths:
(a) 24 in.
(b) 60 in.
(c) 96 in.

18.32 Calculate the Euler buckling load for an axially loaded, pin-connected steel rod $\frac{1}{2}$ in. in diameter and 5 ft long.

18.33 A W12 $\times$ 40 structural steel shape of ASTM A992 steel is used as an axially loaded column. Calculate the minimum length of the column for which Euler's formula may be used. The column is pin-connected at each end.

18.34 Calculate the Euler buckling load for a pin-connected axially loaded steel rod having a diameter of 25 mm. The length of the rod is
(a) 1 m
(b) 2 m
(c) 4 m

18.35 Rework Problem 18.34 assuming that the material is aluminum with a modulus of elasticity of 70×10^3 MPa and a proportional limit of 220 MPa.

18.36 A built-up steel column is made by welding a 16-in.-by-1-in. plate to each flange of a W14 $\times$ 48 wide-flange section. The column is 22 ft long and the ends are pin-connected. Calculate the maximum slenderness ratio.

18.37 A 2-in.-diameter standard-weight steel pipe is used as an axially loaded column. Calculate the allowable axial compressive load using Euler's formula and a factor of safety of 4. The pipe column is 84 in. long and is fixed at one end and pin-connected at the other end. Assume the proportional limit = 34 ksi.

18.38 A structural steel column is 30 ft long and must support an axial compressive load of 20 kips. Using Euler's formula and a factor of safety of 2.0, select the lightest wide-flange section. Assume that the column is pin-connected at each end. Check the applicability of Euler's formula.

18.39 Compute the allowable axial compressive load for a W250 $\times$ 73 structural steel wide-flange section of ASTM A992 steel. Use the AISC column formulas. End conditions and lengths are as follows:
(a) pinned, 5 m
(b) pinned, 10 m
(c) pinned/fixed, 10 m

18.40 Determine the allowable axial compressive load for a structural steel W14 $\times$ 132 column. The ends are pin-connected and the length is 20 ft.

18.41 Using the AISC column approach, compute the allowable axial compressive load for the pin-connected built-up ASTM A36 steel column shown. The length of the column is 19 ft.

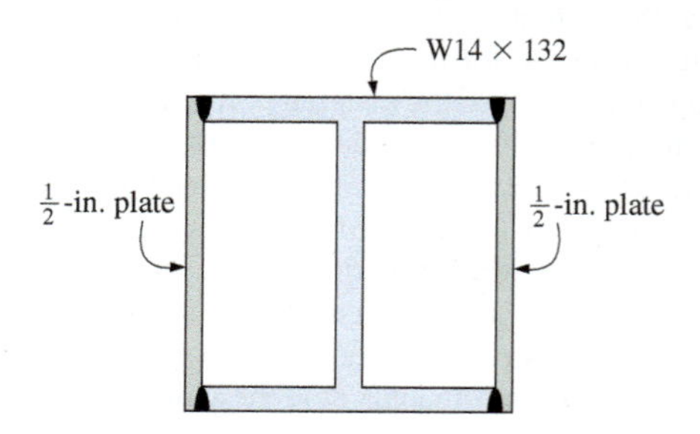

PROBLEM 18.41

18.42 Using the AISC column equations, select the lightest W10 shape for a column subjected to an axial compressive load of 280 kips. The unbraced length of the column is 18 ft and the ends are pinned. The steel is ASTM A992.

18.43 Select the lightest extrastrong steel pipe section to support an axial compressive load of 90 kips. The column is pin-connected and has an unbraced length of 16 ft. Use the AISC column approach. Use $F_y = 35$ ksi.

18.44 Compute the required diameter of a steel push-rod subjected to an axial compressive load of 10 kips. The rod is to be made of AISI 1020 cold-drawn steel (yield stress = 50 ksi). The length is 24 in. and the ends are pinned. Use the Euler–Johnson formulas with a factor of safety of 3.0.

18.45 A 19-mm-diameter steel rod is 350 mm in length and is used as a pin-connected compression member. The rod is AISI 1040 hot-rolled steel. Using the Euler–Johnson formulas with a factor of safety of 2.5, calculate the allowable axial compressive load that the rod can carry.

18.46 A pin-connected linkage bar is 16 in. long and is subjected to an axial compressive load of 4600 lb. The bar is made of a high-strength steel with a yield stress of 110,000 psi. Compute the required dimensions of the bar if its cross section is to be rectangular with the width being twice the depth. Use the Euler–Johnson formulas with a factor of safety of 3.0.

18.47 Using the NDS provisions, calculate the allowable axial compressive load for a 10-in.-by-12-in. (S4S) Douglas fir column. The ends are fixed/pinned and the length is 16 ft.

18.48 Using the NDS provisions, calculate the allowable axial compressive load for a southern pine column having a full nominal size of 12 in. by 12 in. The ends are pin-connected and the length is 20 ft.

18.49 Design a square (S4S) pin-connected Douglas fir column to support an axial load of 62 kips. The unbraced length of the column is 18 ft. Use the NDS provisions.

18.50 A bin weighing 100 tons when full is to be supported by four timber columns, each 20 ft long. Select a square (S4S) timber cross section for this application. The material is to be southern pine. The load is assumed to be equally distributed to each column and the column ends are pin-connected. Use the NDS provisions.

18.51 Design a pin-connected southern pine square column (S4S), using the NDS provisions. The column is to support an axial compressive load of 450 kN. The unbraced length is 6 m.

CHAPTER NINETEEN

CONNECTIONS

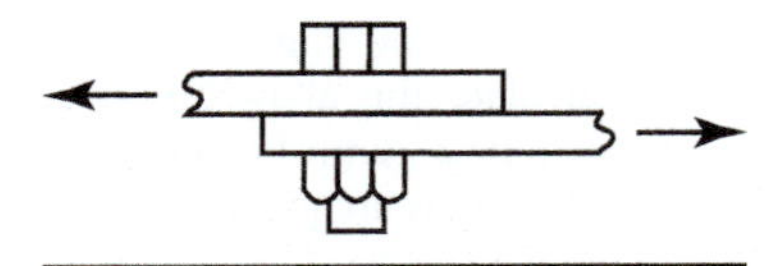

LEARNING OBJECTIVES

Upon completion of this chapter, readers will be able to:

- Define bolted and welded connections.
- Identify the failure modes associated with bolted connections.
- Calculate the strength of bolted connections for each failure mode.
- Identify and analyze multiple types of welded connections.

19.1 INTRODUCTION

In the preceding chapters, we considered bending, tension, compression, and torsion and their effects on structural members. Generally, we considered the members to be isolated entities, which served our purpose in studying the behavior of the members. In the real world, however, rather than being isolated, practically all members act in combination with other members to which they are attached. Most machines and structures are actually assemblies composed of members connected together. Each member must be connected in a way that will enable it to transmit its applied loads to another part of the assembly or to a foundation.

In this chapter, we introduce the analysis and design of connections. Since connections serve primarily to transmit load, the design of connections must be based on the theoretical principles discussed in the previous chapters.

Although our emphasis is on connections for steel, be aware that the same basic principles apply to other metals. For a treatment of connections used in timber construction, refer to publications of the American Institute of Timber Construction (AITC).[1]

Rivets were used for centuries for joining pieces of metal; in the recent past, however, because of their many disadvantages, such as high installation costs and their tendency to loosen under cyclic loads, rivets have become almost obsolete. They have been replaced by bolts and welding in almost all structural steel applications. For our discussion of bolted and welded connections, we introduce material from the applicable design specifications of the AISC[2] and the American Welding Society (AWS).[3]

19.2 BOLTS AND BOLTED CONNECTIONS (AISC)

Several types of bolts can be used for connecting structural steel members. The two types generally used in structural applications are *unfinished bolts* and *high-strength bolts*, which are shown in Figure 19.1. Proprietary bolts incorporating ribbed shanks, end splines, and slotted ends are also available, but all these may be considered modifications of the high-strength bolt.

Unfinished bolts are also known as *machine bolts, common bolts, ordinary bolts*, or *rough bolts*. They are designated as ASTM A307 and are threaded bolts of low-carbon steel with rough, unfinished shanks. These bolts range in diameter from $\frac{5}{8}$ in. to $1\frac{1}{2}$ in. inclusive, by $\frac{1}{8}$-in. increments. They are generally installed in standard holes (circular holes) having a diameter $\frac{1}{16}$ in. larger than the shank of the bolt. Unfinished bolts are relatively inexpensive and can be tightened with a hand wrench. Since permissible loads on these bolts are significantly less than those for high-strength bolts of similar size, however, their application is limited.

[1]American Institute of Timber Construction, *Timber Construction Manual*, 4th ed. (New York: Wiley, 1994).

[2]American Institute of Steel Construction, Inc., *Specification for Structural Steel Buildings* (Chicago: AISC, 2005). (Also included in the AISC *Manual of Steel Construction*, 13th ed. [Chicago, AISC, 2005].)

[3]American Welding Society, *Welding Handbook* (Miami: AWS, latest edition). (A multivolume series.)

FIGURE 19.1 Structural bolts. Left: $\frac{7}{8}$-in.-diameter, 6-in.-long, A325 high-strength bolt. Center: $\frac{3}{4}$-in.-diameter A325 high-strength load-indicator bolt, incorporating a splined end to facilitate installation. Right: $\frac{3}{4}$-in.-diameter unfinished bolt.

High-strength bolts are undoubtedly the most commonly used mechanical fastener for structural steel. In fact, bolting with high-strength bolts has become the primary means of connecting steel members in the shop as well as in the field. These fasteners offer advantages in speed of installation, strength, simplicity, and safety.

The two basic types of high-strength bolts are the ASTM A325 high-strength carbon steel bolt and the ASTM A490 heat-treated high-strength bolt.[4,5] The A490 bolt has the higher material strength. Both are threaded structural bolts used with heavy hex nuts. In general, the A325 and A490 bolts are available in diameters ranging from $\frac{1}{2}$ in. to $1\frac{1}{2}$ in. inclusive, by $\frac{1}{8}$-in. increments, with $\frac{3}{4}$-in. and $\frac{7}{8}$-in. diameters being the most commonly used sizes in structural applications. They are inserted in holes having a diameter $\frac{1}{16}$ in. larger than the diameter of the shank of the bolt.

The performance of a high-strength bolt depends on the proper installation of the bolt. Until recently, the installation of all high-strength bolts required that the bolts be tightened in such a way that the tension induced into the bolt be equal to, or greater than, 70% of the specified minimum tensile strength for that steel, as prescribed by the AISC Specification for Structural Joints.[6] This requirement now applies only to slip-resistant connections that, as discussed in Section 19.4, are designated as *slip-critical* connections. Other types of bolted connections, known as *bearing-type* connections, may be installed by tightening to a "snug tight" condition. *Snug tight* is defined as the tightness that exists when all plies in a joint are in firm contact. The tension induced into the bolt based on the snug-tight concept is somewhat less than the tension induced based on the 70% concept which, in effect, permits some slippage in the joint. The consideration of slip-critical connections requires more involved calculations dealing with the condition of the interface of each joint and is beyond the scope of this text.

When the bolt is tightened in accordance with the AISC requirements, a significant clamping force is developed between the connected parts. As a result of this clamping force, the contact surfaces in the slip-critical connection are capable of transmitting loads entirely by friction. In the bearing-type connection, frictional resistance between the connected parts is neglected and the loads are assumed to be transmitted into and out of the bolts by direct bearing.

Bolted connections serve primarily to transmit loads between members or elements of members; hence, the design of connections must be based on the same structural

[4]American Society for Testing and Materials, *Standard Specification for High-Strength Bolts for Structural Steel Joints* (West Conshohocken, PA: ASTM, latest edition).

[5]American Society for Testing and Materials, *Standard Specification for Heat Treated Steel Structural Bolts* (West Conshohocken, PA: ASTM, latest edition).

[6]American Institute of Steel Construction, Inc., *Specification for Structural Joints Using ASTM A325 or A490 Bolts* (Chicago: AISC, 2004).

principles used in the design of the members. This process involves creating a connection that is structurally adequate, economical, and practical.

The simplest form of bolted connection is the lap joint shown in Figure 19.2a. Some joints in structures are of this general type, but it is not a commonly used detail since, as shown in Figure 19.2b, the connected members have a tendency to bend as a result of the applied tensile loads not being collinear. The joint will deform so that the loads *will* be collinear. A more desirable connection that eliminates this tendency of the connected members to bend is the butt joint shown in Figure 19.3. The connected members lie in the same plane and splice plates are used. The splice plates, in effect, replace the member at the point where it is cut. A few other commonly used bolted connections are illustrated in Figure 19.4:

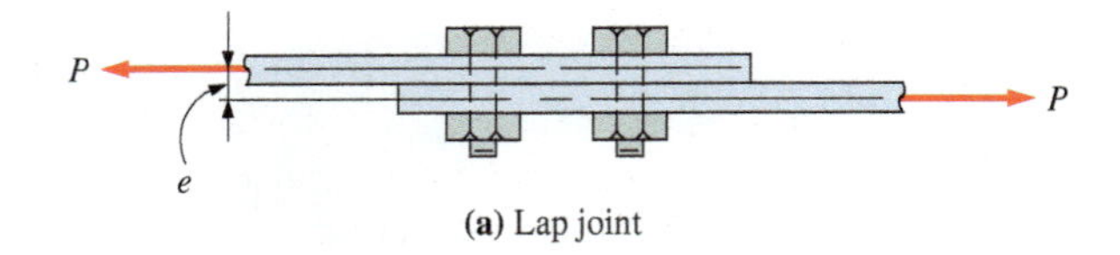

(a) Lap joint

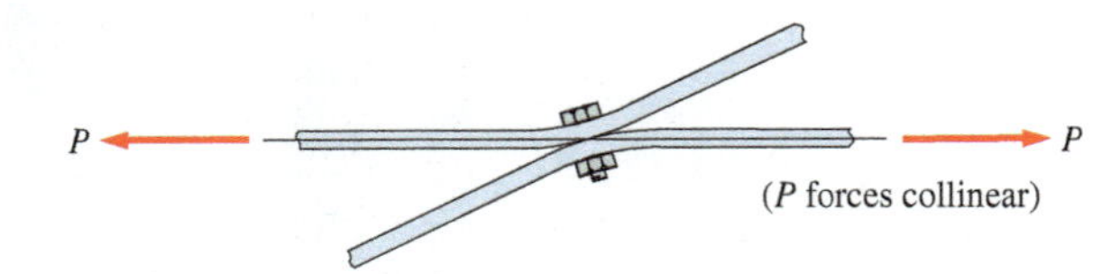

(b) Bending in lap joint

FIGURE 19.2 Typical lap joint.

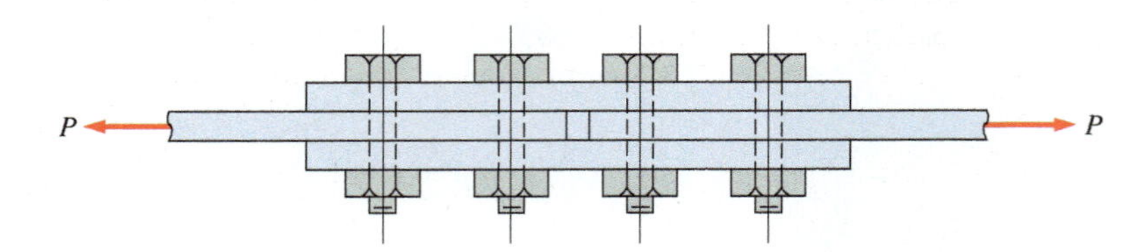

FIGURE 19.3 Butt joint.

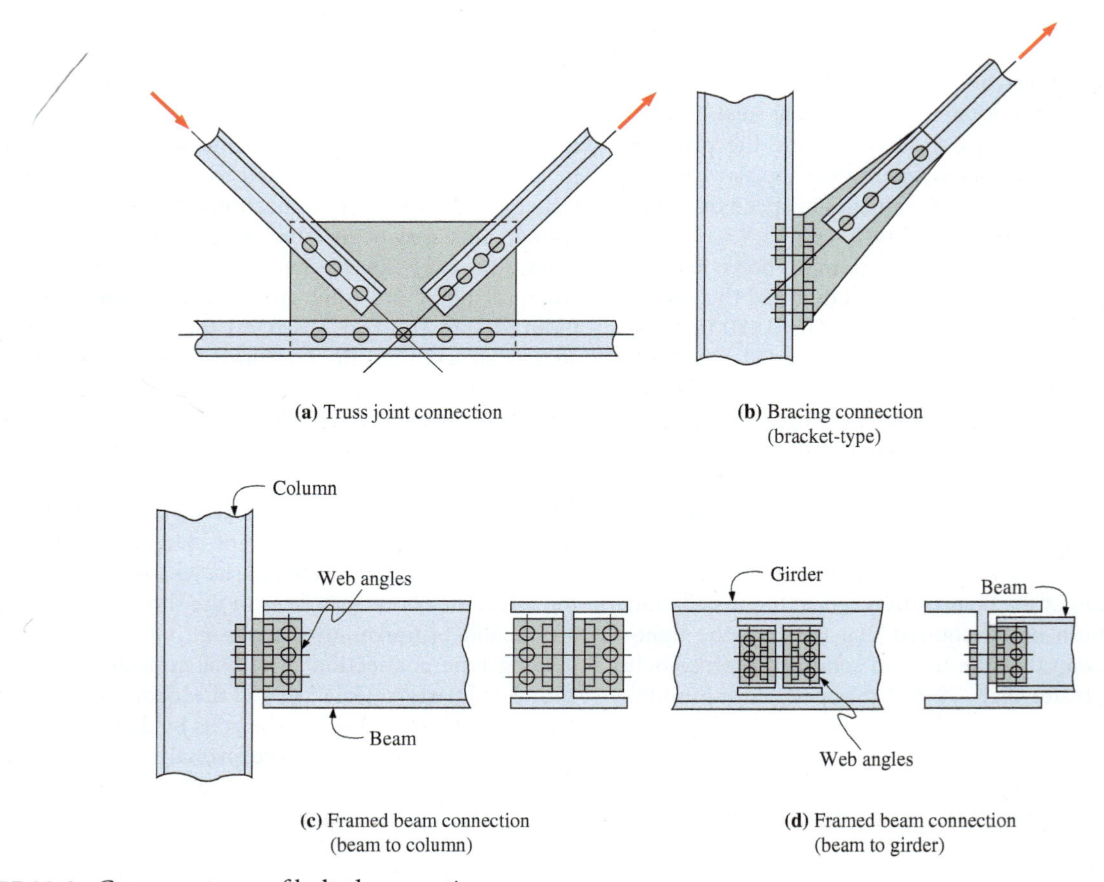

(a) Truss joint connection

(b) Bracing connection (bracket-type)

(c) Framed beam connection (beam to column)

(d) Framed beam connection (beam to girder)

FIGURE 19.4 Common types of bolted connections.

19.3 MODES OF FAILURE OF A BOLTED CONNECTION

A simple bolted lap joint subjected to an axial tensile load is shown in Figure 19.5. The stresses developed in such a joint are quite complex. For analysis and design purposes, the following failure modes are usually considered: (a) shear in the bolts, (b) bearing of the bolts on the connected material, (c) tension in the connected members, (d) end tear-out, and (e) block shear.

(a) *Failure in shear* is a failure in the bolts. It is usually assumed that each bolt will carry its proportional share of the total load on the joint. In the lap joint shown in Figure 19.5, the bolts have a tendency to shear off along the single contact plane of the two plates. Since the bolt is resisting the tendency of the plates to slide past one another along the contact surface and is being sheared on a single plane, the bolt is said to be in *single shear*. It is a single cross-sectional area in each bolt that resists the applied load on that bolt.

In a butt joint, such as that shown in Figure 19.6, there are two contact planes. Therefore, each bolt has a tendency to shear off along two contact planes and is said to be in *double shear*. Hence, two cross-sectional areas in each bolt resist the applied load on that bolt.

(b) *Failure in bearing* is a failure in the connected material. Bolts may be adequate in shear, but the connection may fail if the material joined is not capable of transmitting the load into the bolts. This failure is a crushing or compressive-type behavior whereby the bolts and connected plates bear against each other, as shown in Figure 19.7. The bolt may crush the material of the plate against which it bears, resulting in an elongation of the hole and an associated deformation of the member. In addition, the bolt itself may be deformed by the plate acting on it.

(c) *Failure in tension* is a failure in the connected material, not a tension failure of the bolts. The connected members of a bolted connection, such as the lap joint shown

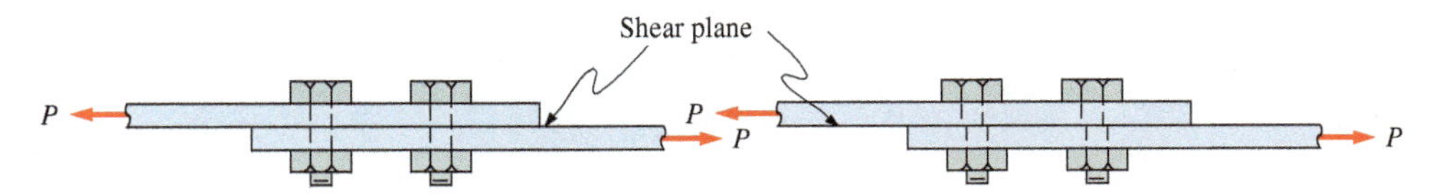

FIGURE 19.5 Lap joint, bolts in single shear.

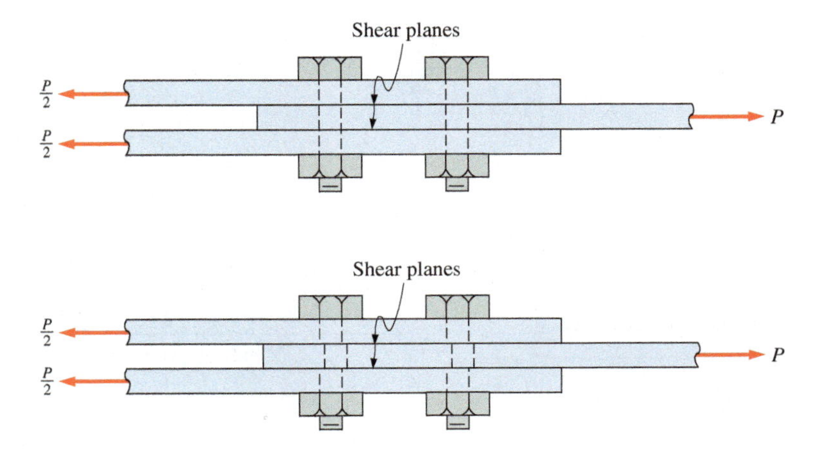

FIGURE 19.6 Butt joint, bolts in double shear.

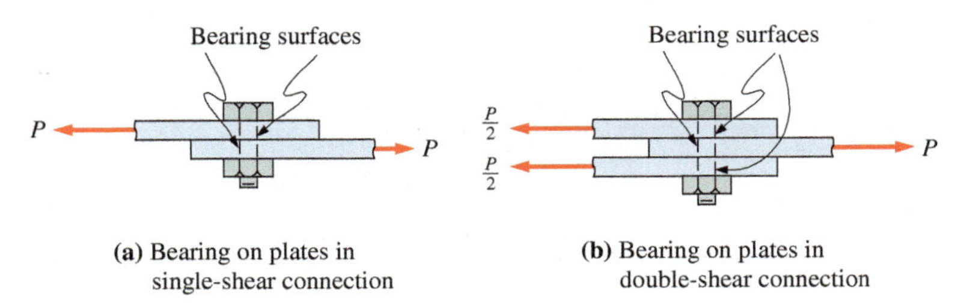

FIGURE 19.7 Bearing stresses.

in Figure 19.8, may fail in tension on the transverse plane through the bolt holes. This is the plane having the least resisting area, a likely area for rupture to occur. This area may be termed the *net area.* A tensile failure could also occur elsewhere in the member where no holes exist; this type of failure, however, is one of excessive deformation due to yielding as opposed to one in which rupture occurs. Both tensile failure modes are recognized by the AISC Specification and are discussed further in Section 19.4.

(d) *End tear-out* is shown in Figure 19.9. It could occur where the distance parallel to the applied load, between the edge of the plate and the bolt hole, is insufficient. This failure mode is easily avoided by providing sufficient edge distance, as recommended in the AISC Specification; hence, it is not investigated in this text.

(e) *Block shear* is another type of failure of structural connections that must be considered for both tension member connections and beam connections. It is depicted in Figure 19.10.

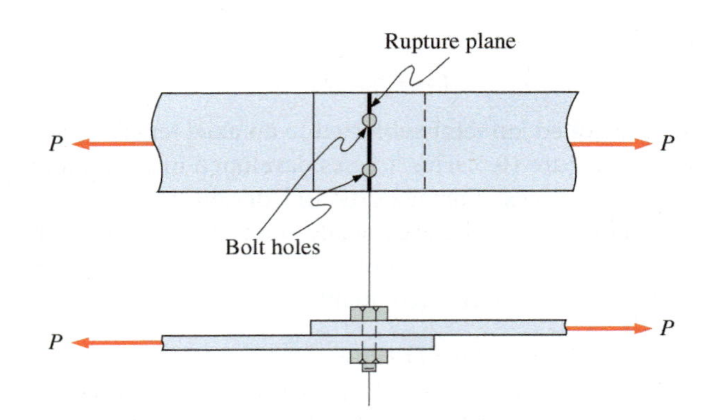

FIGURE 19.8 Lap joint, failure in tension.

Block shear is characterized by a combination of shear failure along a plane through the bolt holes and a simultaneous tension failure along a perpendicular plane. Block shear may be avoided by providing proper connection geometry using adequate bolt spacings and edge distances. Therefore, it is not considered in connection analysis and design in this text.

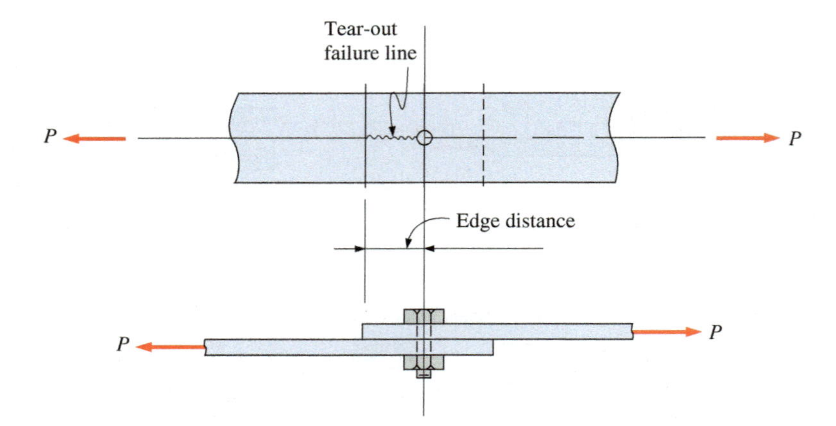

FIGURE 19.9 Lap joint, end tear-out.

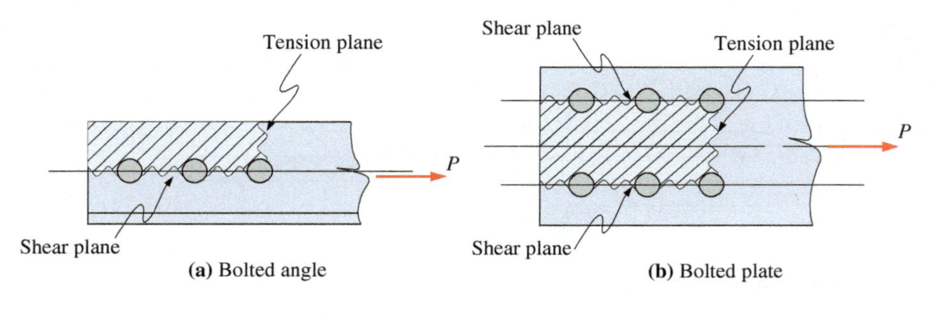

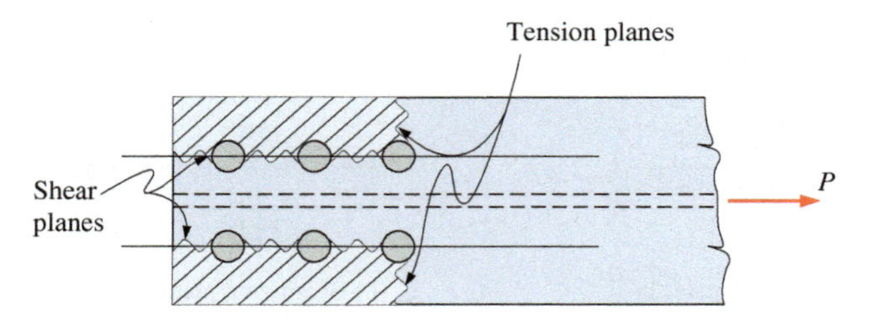

FIGURE 19.10 Block shear in end connections of tension members.

19.4 HIGH-STRENGTH BOLTED CONNECTIONS

The strength of connections composed of high-strength bolts is analyzed by considering shear, bearing, and tension separately. The allowable tensile load, or tensile capacity, for the connection will then be the smallest of the computed values.

19.4.1 Shear Strength

The shear strength for the connection, based on bolt shear, is the product of the cross-sectional area of the shank, its allowable shear stress, and the number of bolts in the connection. The allowable load is determined from

$$P_s = A_B F_v N \quad \textbf{(19.1)}$$

where P_s = the allowable load for the connection, based on bolt shear (lb, kips) (N)
A_B = the circular cross-sectional area of one bolt (in.^{2}) (mm^2)
F_v = the allowable shear stress in the bolt material (psi, ksi) (MPa)
N = the number of bolts contained in the connection being considered

In a lap joint, such as shown in Figure 19.5, each bolt is in single shear. When a bolt is subjected to more than one plane of shear, such as the double shear in the butt joint of Figure 19.6, the shear strength of each bolt must be multiplied by 2, the number of shear planes (or shear areas) in the bolt. Thus, Equation (19.1) can be rewritten as

$$P_s = nA_B F_v N \quad \textbf{(19.2)}$$

where n is the number of shear planes per bolt.

The allowable shear stress is a function of the type of high-strength bolt (A325 or A490), the location of the threads with respect to the joint, and the type of connection (discussed later in this section). The allowable shear stress is determined by dividing the nominal (or failure) stress of the fastener by a safety factor. The type of bolt hole in the connected material also affects the allowable shear stress in the bolt. For purposes of this text, the bolt holes are assumed to be of standard size and of circular shape (as opposed to oversized or slotted holes). Allowable shear stress values for bearing-type connections are presented in Table 19.1. These are based on values from the AISC Specification.

19.4.2 Bearing Strength

The bearing strength of a connection is a function of the bearing (crushing) strength of the connected material and the resisting contact area. The true distribution of the bearing stress on the material around the perimeter of a hole is unknown. Satisfactory results, however, have been obtained by assuming a uniform bearing stress acting on the projection of the contact area. This projected area, a rectangular area, is obtained as the product of the nominal diameter of the bolt and the thickness of the connected material. We obtain the strength of the connection, based on bearing on the connected material, as the product of the resisting contact area, an allowable bearing stress, and the number of bolts in the connection:

$$P_p = dtF_p N \quad \textbf{(19.3)}$$

where P_p = the allowable load for the connection, based on bearing on the connected material (lb, kips) (N)
d = the nominal bolt diameter (in.) (mm)
t = the thickness of the connected part (in.) (mm)
F_p = the allowable bearing stress on the connected material (psi, ksi) (MPa)
N = the number of bolts contained in the connection being considered

The allowable bearing stress is calculated from a multiple of the lowest ultimate tensile strength F_u of the connected material, divided by a factor of safety. It also depends on spacing between fasteners, distance to the nearest edge, and whether deformation at the bolt holes is a consideration. As a maximum, the allowable bearing stress on connected material may be taken as $1.5 \times F_u$. Some values for F_p and F_u for selected steels are given in Table 19.2. The values of F_p are based on the assumption that minimum bolt spacings and edge distances satisfy the requirements of the AISC Specification.

TABLE 19.1 Allowable shear stress on steel fasteners (Bearing-type connection)

Description of Fastener	Allowable Shear Stress (F_v) [ksi (MPa)]
A307 low-carbon bolts	13.5 (94)
A325 bolts—threads in shear plane	27.0 (186)
A325 bolts—threads excluded from shear plane	34.0 (228.5)
A490 bolts—threads in shear plane	34.0 (228.5)
A490 bolts—threads excluded from shear plane	42.0 (289.5)

Note: U.S. Customary System values are based on the AISC Specification. SI values are converted from U.S. Customary System values.

TABLE 19.2 Allowable stresses in ksi (MPa)

		Allowable Bearing Stress		Allowable Tensile Stress	
				(Gross)	(Net)
Structural Steels	F_u	$F_p = (1.5)F_u$	F_y	$F_t = (0.60)F_y$	$F_t = (0.50)F_u$
A36 carbon	58[a] (400)	87[a] (600)	36 (250)	21.6 (150)	29[a] (200)
A992 high-strength low-alloy	65 (448)	97.5 (672)	50 (345)	30 (207)	32.5 (224)

Note: U.S. Customary System values are from the AISC Specification. SI values are converted from U.S. Customary System values.
[a]Minimum values.

19.4.3 Tensile Strength

The tensile strength, or the allowable tensile load, of a connection subjected to an axial tensile load is a function of a resisting area and an allowable tensile stress F_t. In calculating the allowable tensile load, two conditions will be checked. These two conditions involve either a net area A_n or a gross area A_g as the resisting area. Net area and gross area are depicted in Figure 19.11. When the gross area is used, the allowable tensile load is calculated from

$$P_g = A_g F_t = A_g(0.60)F_y \quad \textbf{(19.4)}$$

where P_g = the allowable tensile load on the gross area of the member (lb, kips) (N)
A_g = the gross cross-sectional area (in.^{2}) (mm^2)
F_y = the yield stress of the material (psi, ksi) (MPa)

For the plate shown in Figure 19.11, the gross area is the product of b and t. Note in Equation (19.4) that the allowable tensile stress is $0.60 \times F_y$. Also note that Equation (19.4) ignores the reduction in cross-sectional area due to the bolt holes and is based on a yielding mode of failure.

When the net area is used, it can be calculated as follows:

$$A_n = A_g - A_H \quad \textbf{(19.5)}$$

which can be rewritten as

$$A_n = bt - N_F d_H t$$

The allowable tensile load is calculated from

$$P_n = A_n F_t = A_n(0.50)F_u \quad \textbf{(19.6)}$$

where A_n = the net area of the cross section (in.^{2}) (mm^2)
A_H = the projection $(d_H \times t)$ of the area of the holes (in.^{2}) (mm^2)
b = the width of the plate (in.) (mm)
t = the thickness of the plate (in.) (mm)
N_F = the number of holes in the fracture plane
d_H = the diameter of the holes (in.) (mm)
P_n = the allowable tensile load on the net area of the member (lb) (kips) (N)

and the other terms are as previously defined. Note in Equation (19.6) that the allowable tensile stress is equal to $0.50 \times F_u$. Note also that the hole diameter, for analysis and design purposes, is taken as the bolt diameter plus $\frac{1}{8}$ in., according to the AISC Specification. This value provides for both the clearance around the bolt and an allowance for misshape of the holes. (For selected values for ultimate tensile stress F_u, yield stress F_y, and allowable tensile stress F_t, see Table 19.2.)

The smaller of the two calculated allowable tensile loads (P_n or P_g) is then used as the controlling allowable tensile load when determining the strength of the connection.

As mentioned in Section 19.2, high-strength bolted connections may be of two types, either *slip-critical* or

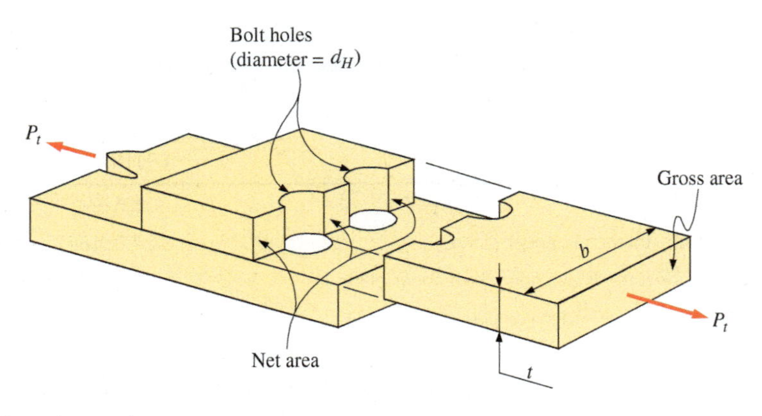

FIGURE 19.11 Lap joint, plates in tension.

bearing-type. In the slip-critical connection, it is assumed that the applied load is transmitted from one connected part to another entirely by the friction that results from the high tension induced in the bolt upon installation.

In the bearing-type connection, it is assumed that the connected parts do slip and come into bearing contact with the bolts and that the load is transmitted by the shear resistance of the bolt as well as by bearing of the connected parts on the bolt. The frictional resistance between the connected parts is neglected and is of no concern.

Bearing-type connections can be designed with the bolt threads in the shear plane or out of the shear plane. The allowable shear stress will reflect the particular condition. As a convenience, the following bolt designations may be used for bolts in bearing-type connections:

Bolt Designation	Type of Application
A325-N, A490-N	Bearing-type connection with threads in the shear plane
A325-X, A490-X	Bearing-type connection with threads excluded from the shear plane

The difference between the slip-critical and bearing-type connections lies in the allowable stresses used in the analysis or design. A slip-critical connection will generally contain more bolts than a bearing-type connection. For both types of connections, the contact surfaces must be brought into solid contact. No loose mill scale, burrs, dirt, or other foreign material is permitted.

EXAMPLE 19.1

Compute the allowable tensile load P for the single-shear lap joint shown in Figure 19.12. The plates are ASTM A36 steel and the high-strength bolts are $\frac{3}{4}$-in.-diameter A325-X in standard holes.

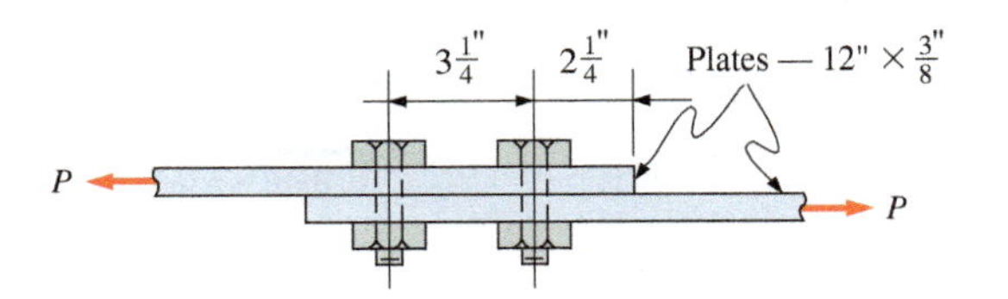

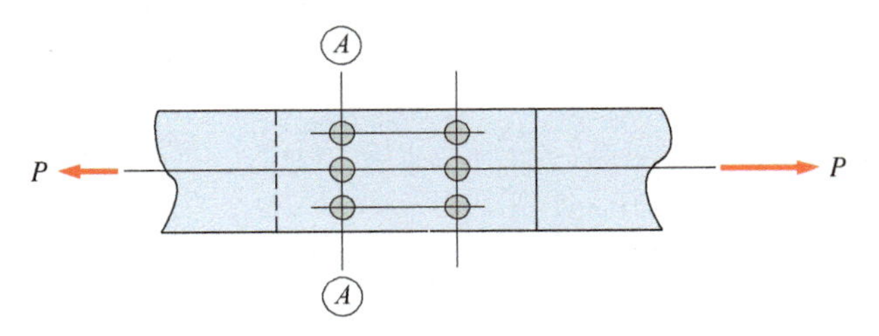

FIGURE 19.12 Single-shear lap joint.

Solution First, check bolt shear. From Table 19.1, the allowable shear stress F_v is 30.0 ksi. The allowable load for the connection, based on bolt shear, is calculated from Equation (19.2):

$$P_s = nA_BF_vN$$

The cross-sectional area of the bolt is calculated from

$$A_B = \pi d^2/4 = 0.7854d^2 = 0.7854(0.75\text{ in.})^2 = 0.442\text{ in.}^2$$

Then

$$\begin{aligned} P_s &= nA_BF_vN \\ &= 1.0(0.442\text{ in.}^2)(34\text{ ksi})(6) = 90.1\text{ k} \end{aligned}$$

Next, check bearing on the $\frac{3}{8}$-in.-thick plate. From Table 19.2, the allowable bearing stress F_p is 87 ksi. The allowable load for the connection, based on bearing on the connected material, is calculated from Equation (19.3):

$$\begin{aligned} P_p &= dtF_pN \\ &= (0.75\text{ in.})(0.375\text{ in.})(87\text{ ksi})(6) \\ &= 147\text{ k} \end{aligned}$$

Last, check the tensile capacity of the plates. Using the allowable tensile stresses from Table 19.2, the allowable tensile load, based on gross area, is calculated from Equation (19.4):

$$P_g = A_g F_t = (12 \text{ in.})(0.375 \text{ in.})(21.6 \text{ ksi}) = 97.2 \text{ k}$$

and, based on net area, from Eqs. (19.5) and (19.6),

$$A_n = bt - N_F d_H t = (12 \text{ in.})(0.375 \text{ in.}) - 3(0.875 \text{ in.})(0.375 \text{ in.}) = 3.52 \text{ in.}^2$$

$$P_n = A_n F_t = (3.52 \text{ in.}^2)(29 \text{ ksi}) = 102 \text{ k}$$

Therefore, bolt shear governs the strength of the lap joint. The joint (or connection) has an allowable tensile load of 79.6 k.

EXAMPLE 19.2

Compute the allowable tensile load P for the double-shear butt joint in Figure 19.13. The plates are ASTM A36 steel and the fasteners are 20-mm-diameter A325 high-strength bolts in standard holes. Assume that the connection is bearing-type, with the threads in the shear plane (A325-N).

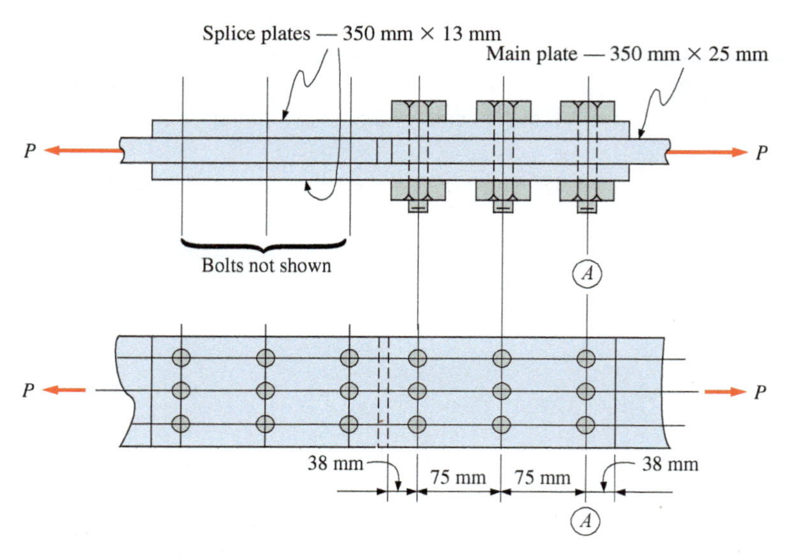

FIGURE 19.13 Double-shear butt joint.

Solution Note that the A325-N bolts are in double shear and that the allowable shear stress F_v from Table 19.1 is 165 MPa. The allowable load, based on bolt shear, is calculated from Equation (19.2):

$$\begin{aligned} P_s &= nA_B F_v N \\ &= 2(0.7854)(20 \text{ mm})^2(186 \text{ MPa})(9) \\ &= 1{,}051{,}000 \text{ N} \\ &= 1051 \text{ kN} \end{aligned}$$

Next, check bearing on the 25-mm-thick main plate using Equation (19.3). (Note that bearing on the splice plates need not be checked since the total splice plate thickness exceeds that of the main plate.) From Table 19.2, the allowable bearing stress F_p is 600 MPa:

$$\begin{aligned} P_p &= dtF_p N \\ &= (20 \text{ mm})(25 \text{ mm})(600 \text{ MPa})(9) \\ &= 2{,}700{,}000 \text{ N} \\ &= 2700 \text{ kN} \end{aligned}$$

Tension in the 25-mm main plate is considered next. Tension on both the gross area and the net area must be considered. The full load acts on bolt line AA in the main plate. The net area is calculated at that location. In the calculation of net area, 3 mm is added to the bolt diameter to determine the hole diameter for analysis purposes. For the gross area,

$$A_g = (350 \text{ mm})(25 \text{ mm}) = 8750 \text{ mm}^2$$

From Equation (19.5), the net area is

$$\begin{aligned} A_n &= A_g - A_H \\ &= 8750 \text{ mm}^2 - 3(20 \text{ mm} + 3 \text{ mm})(25 \text{ mm}) = 7025 \text{ mm}^2 \end{aligned}$$

Using allowable tensile stresses from Table 19.2, tensile capacities based on gross area and net area can be calculated. From Equation (19.4),

$$\begin{aligned} P_g &= A_g F_t = A_g(0.60 F_y) \\ &= (8750 \text{ mm}^2)(150 \text{ MPa}) \\ &= 1{,}313{,}000 \text{ N} \\ &= 1313 \text{ kN} \end{aligned}$$

From Equation (19.6),

$$\begin{aligned} P_n &= A_n F_t = A_n(0.50)(F_u) \\ &= (7025 \text{ mm}^2)(200 \text{ MPa}) \\ &= 1{,}405{,}000 \text{ N} \\ &= 1405 \text{ kN} \end{aligned}$$

Therefore, bolt shear governs the strength of the butt joint. The allowable tensile load for the joint is 933 kN.

EXAMPLE 19.3

A butt splice is to be designed for the plates shown in Figure 19.14. Use $\frac{3}{4}$-in.-diameter A325-N high-strength bolts in standard holes. The plates are ASTM A36 steel and the axial tensile load is 120.5 k. Compute the number of bolts required on each side of the splice and indicate the bolt arrangement by means of a sketch (AISC Specification).

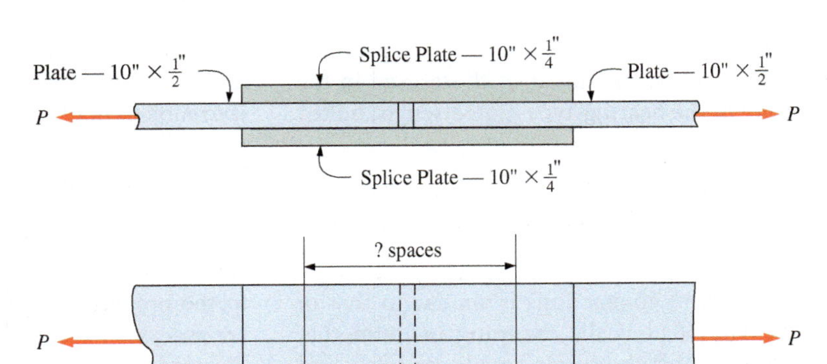

FIGURE 19.14 Double-shear butt splice.

Solution Two $\frac{1}{4}$-in.-thick splice plates are selected, since the sum of the splice-plate thickness should be at least equal to the thickness of the plates being spliced.

Considering bolt shear first, the allowable shear stress F_v from Table 19.1 is 24.0 ksi. Then, solve Equation (19.2) for the required number of bolts N. Note that the bolts are in double shear. Thus,

$$\text{required } N = \frac{P}{nA_B F_v} = \frac{120.5 \text{ k}}{2(0.7854)(0.75 \text{ in.})^2(27 \text{ ksi})} = 5.05 \text{ bolts}$$

Next, consider bearing on the connected material. Since there are two $\frac{1}{4}$-in.-thick splice plates and one $\frac{1}{2}$-in.-thick main plate, the splice plates and the main plate are equally critical, so base the design on a $\frac{1}{2}$-in. material thickness. From Table 19.2, obtain the allowable bearing stress F_p of 87 ksi. Solve Equation (19.3) for the required number of bolts:

$$\text{required } N = \frac{P}{dtF_p} = \frac{107\text{ k}}{(0.75\text{ in.})(0.5\text{ in.})(87\text{ ksi})} = 3.28\text{ bolts}$$

Of the two preceding considerations, shear in the bolts is the more critical. Select a 6-bolt pattern for each side of the splice, as shown in Figure 19.15. Then, check the allowable tensile load based on gross area and net area of the main plate, using allowable stresses from Table 19.2. From Equation (19.4),

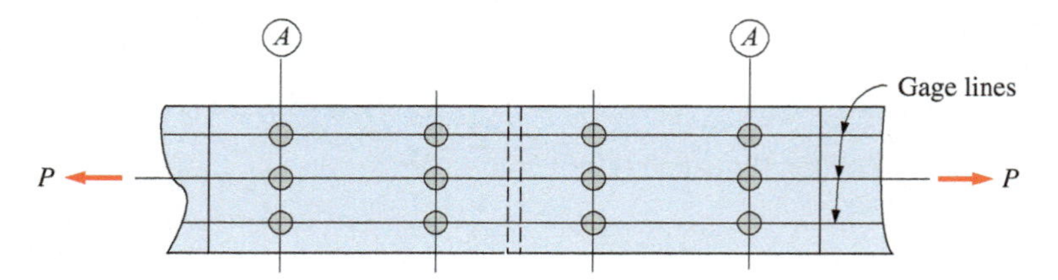

FIGURE 19.15 Bolt pattern.

$$P_g = A_gF_t = (10\text{ in.})(0.5\text{ in.})(21.6\text{ ksi}) = 108\text{ k} > 107\text{ k} \qquad \textbf{(O.K.)}$$

From Equations (19.5) and (19.6):

$$A_n = bt - N_Fd_Ht$$
$$= (10\text{ in.})(0.5\text{ in.}) - 3(0.875\text{ in.})(0.5\text{ in.}) = 3.69\text{ in.}^2$$
$$P_n = (3.69\text{ in.}^2)(29\text{ ksi}) = 107\text{ k} \qquad \textbf{(O.K.)}$$

Therefore, the bolt arrangement shown in Figure 19.15 is satisfactory.

The analysis and design of connections using unfinished bolts involve procedures similar to those used in the analysis and design of the bearing-type high-strength bolted connections.

19.5 INTRODUCTION TO WELDING

The function of welds in connections is similar to that of bolts and rivets. The weld is the fastening medium that provides the path by which the applied loads are safely and predictably transferred from one element to another.

Welding may be defined as a process in which two or more pieces of metal are fused together by heat to form a joint. In most welding processes, this is accomplished through the addition of filler metal from an electrode. The surfaces to be connected or fused are subjected to heat by means of an electric arc, a gas flame, or a combination of electric resistance and pressure. The welding processes used in the design of steel structures, pressure vessels, boilers, tanks, and machines are almost exclusively all the electric-arc type.

In the arc-welding process, an electric arc is formed between the end of a metal electrode and the steel components to be welded. The heat from the arc (approximately 6500°F) raises the temperature of the electrode and the base metal (the material to be connected) in the immediate area of the arc to the point where they melt together to form a localized molten pool of steel on the surface of the member. The electrode, with its electric arc, is moved along the joint to be welded, with the molten pool rapidly solidifying as the temperature of the pool behind it drops below the melting point.

The chemical and mechanical properties of the added weld metal from the electrode should be as similar as possible to the properties of the base metal. A variety of electrodes are available to satisfy the requirements of the various steels. Electrodes are designated by a numbering system with an *E* prefix (indicating electrode). In structural welding, the prefix is followed by two or three digits (e.g., E60, E70, E100). The number denotes the minimum ultimate tensile strength (in kips per square inch) of the weld metal in the electrode. Thus, an E70 electrode would have an ultimate tensile strength of 70,000 psi. Two other digits are added to the electrode designation to denote special factors relative to the welding process other than strength properties. Proper electrode–base metal combinations have been established by the AISC that result in weld metal that is always stronger than the base metal.

The two types of load-carrying welds most commonly used are the *fillet weld* and the *groove weld*. Other load-carrying welds are the *plug weld* and the *slot weld*, which generally are used only under special circumstances (e.g., where space is limited or where the fillet weld lacks adequate load-carrying capacity). These four weld types are shown in Figure 19.16. Our discussion will be limited to the commonly used load-carrying fillet and groove welds.

In structural applications, the fillet weld finds most frequent use, whereas in pressure vessels, boilers, tanks, and ship-building applications, the groove weld predominates. Groove welds generally require extensive edge preparation as well as special fabrication. As a result, they are more costly than fillet welds.

In any given joint, the adjoining members to be connected may be oriented with respect to each other in several ways. The joints are usually categorized as *butt, tee, lap, corner*, and *edge*, as shown in Figure 19.17.

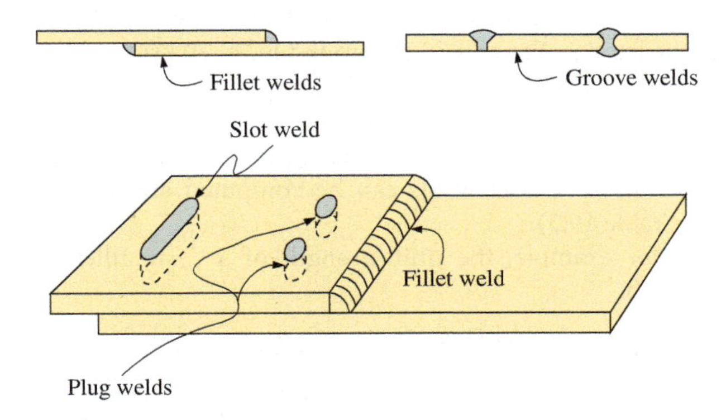

FIGURE 19.16 Weld types.

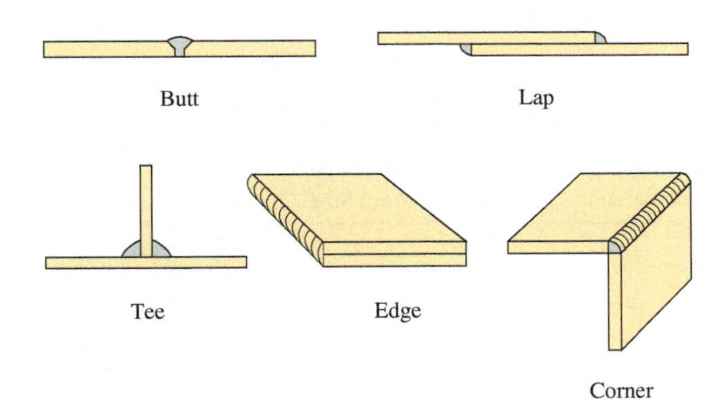

FIGURE 19.17 Joint types.

19.6 STRENGTH AND BEHAVIOR OF WELDED CONNECTIONS (AISC)

Fillet welds are welds of theoretically triangular cross section joining two surfaces that are at approximate right angles to each other, such as in a lap joint or a tee joint. The cross section of a typical fillet weld is a right triangle with equal legs, as shown in Figure 19.18. The size of the weld is designated by the leg size. The *root* is the vertex of the triangle or the point at which the legs intersect. The *face* of the weld is the hypotenuse of the weld triangle and is a theoretical plane since in reality weld faces will form as either convex or concave, as shown in Figure 19.19. The convex fillet weld is the more desirable of the two since it has less of a tendency to crack as a result of shrinkage while cooling. The distance from the theoretical face of a weld to the root is called the *throat*.

Groove welds are welds made in a groove between adjacent ends or surfaces of two parts to be joined, as in a butt, corner, or tee joint. (A tee joint may be fillet welded or groove welded.) The edge preparation for a welded butt joint can be made in any one of a wide variety of configurations, a few of which are shown in Figure 19.20. With the exception of the square-groove weld, some edge preparation is required for either one or both of the members to be connected. Because of their high cost compared with fillet welds, groove welds should be used only when specifically required.

In the groove-weld connections, there is no need to calculate the stresses in the weld or attempt to determine

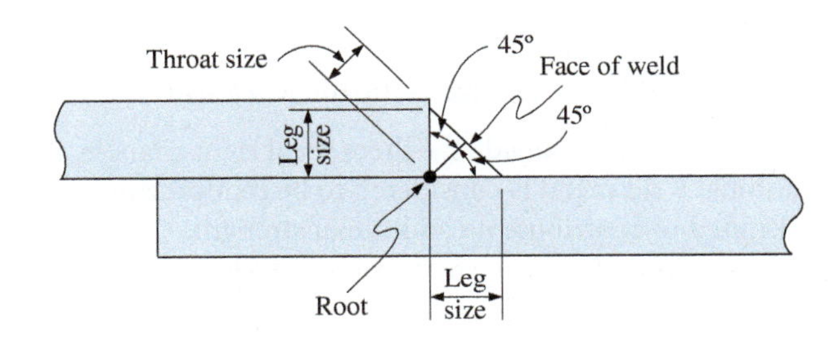

FIGURE 19.18 Typical fillet weld.

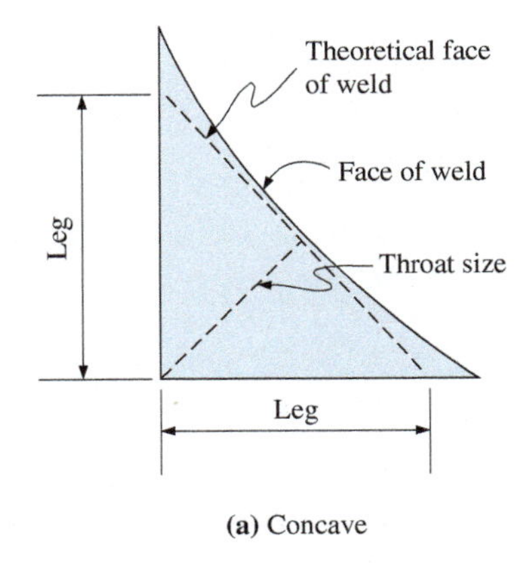

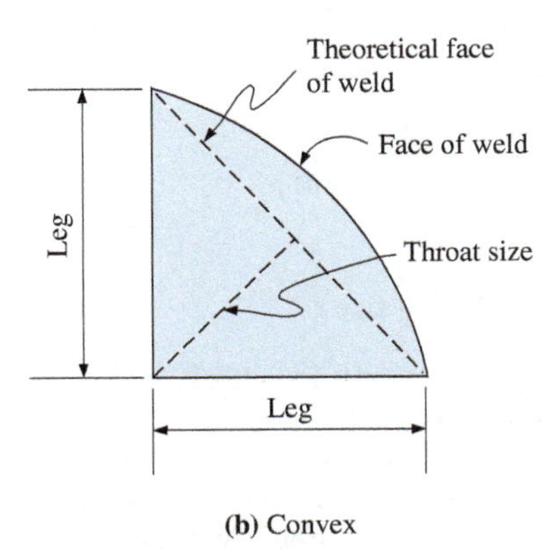

FIGURE 19.19 Fillet welds.

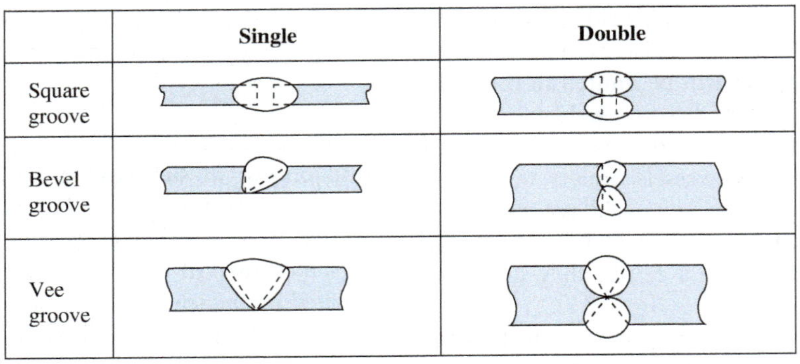

	Single	Double
Square groove		
Bevel groove		
Vee groove		

FIGURE 19.20 Groove welds.

its size, since all groove welds are full strength; that is, their strength is equal to or greater than that of the steel members being joined. Assuming that the proper electrode is used with the base metals, allowable stresses in the weld may be taken as the same as those for the base materials.

With respect to the design of fillet-welded connections, it is necessary to determine the size and length of the fillets to avoid overwelding or underwelding. Tests have shown that fillet welds will fail in shear. Therefore, the strength of a fillet weld is based on the shear strength of the effective throat area of the weld. If we consider a 1-in. length of weld, the effective throat area is the product of the throat dimension and the 1-in. length of the weld, and the strength is the product of the area and an allowable shear stress. Hence, a value for shear strength is usually expressed in units of kips per inch or pounds per inch.

In a fillet with equal leg sizes and where a cross-sectional shape of the weld is theoretically a 45° right triangle, the effective throat size is

$$\sin 45° \times \text{leg size} = 0.707 \times \text{leg size}$$

If weld metal exists outside the theoretical right triangle, this additional weld metal is considered to be reinforcement and is assumed to contribute no additional strength.

The allowable shear stress for the weld metal can be taken as

$$F_v = 0.3F_u \quad \textbf{(19.7)}$$

where F_u is the specified minimum ultimate tensile strength of the electrode weld metal. Therefore, the shear strength of a fillet weld per lineal inch length of weld is

$$\begin{aligned} P_s &= F_v(0.707)(\text{leg size}) \\ &= 0.3F_u(0.707)(\text{leg size}) \end{aligned}$$

from which

$$P_s = 0.212F_u(\text{leg size}) \quad \textbf{(19.8)}$$

Using Equation (19.8), the strength per linear inch (or unit strength) for a fillet weld having a leg size of $\frac{1}{16}$ in. can be calculated. This value is hypothetical since, according to the AISC, the minimum allowable weld size in structural welding is $\frac{1}{8}$ in. This value will be a convenient reference value for calculations, however. The unit strength of other size fillet welds can be obtained by multiplying by the number of sixteenths in the leg size. For an E70 electrode ($F_u = 70$ ksi),

$$P_s = 0.212(70)\left(\tfrac{1}{16}\right) = 0.928 \text{ k/in.}$$

Using 0.928 k/in. as a basic value, the unit strength of the other sizes of fillet welds can be computed and tabulated (see Table 19.3).

For example, the unit strength of a $\frac{3}{16}$-in. fillet weld would be

$$(0.928)(3) = 2.78 \text{ k/in.}$$

A similar approach could be used for other electrodes, such as the E60 electrode, where $F_u = 60$ ksi. This would result in a $\frac{1}{16}$-in. fillet weld having a strength of 0.795 k/in.

TABLE 19.3 Fillet weld sizes

Thickness of Material	Minimum Weld Size[a]	Maximum Weld Size
$\frac{1}{8}$	$\frac{1}{8}$	$\frac{1}{8}$
$\frac{3}{16}$	$\frac{1}{8}$	$\frac{3}{16}$
$\frac{1}{4}$	$\frac{1}{8}$	$\frac{3}{16}$
$\frac{5}{16}$	$\frac{3}{16}$	$\frac{1}{4}$
$\frac{3}{8}$	$\frac{3}{16}$	$\frac{5}{16}$
$\frac{7}{16}$	$\frac{3}{16}$	$\frac{3}{8}$
$\frac{1}{2}$	$\frac{3}{16}$	$\frac{7}{16}$
$\frac{9}{16}$	$\frac{1}{4}$	$\frac{1}{2}$
$\frac{5}{8}$	$\frac{1}{4}$	$\frac{9}{16}$
$\frac{11}{16}$	$\frac{1}{4}$	$\frac{5}{8}$
$\frac{3}{4}$	$\frac{1}{4}$	$\frac{11}{16}$
$\frac{13}{16}$	$\frac{5}{16}$	$\frac{3}{4}$
$\frac{7}{8}$	$\frac{5}{16}$	$\frac{13}{16}$

[a]Weld size is determined by the thicker of the two parts joined, except that the weld size need not exceed the thickness of the thinner part joined.

The strength of a fillet weld depends on the direction of the applied load, which may be parallel or perpendicular to the direction of the weld. In both cases, the weld fails in shear, but the plane of rupture is not the same. If the load is parallel to the weld, failure will occur along the 45° plane of the throat area. If the load is perpendicular to the weld, failure will occur on a plane oriented at approximately 67.5° from one side of the triangular cross section.

Tests have shown that a fillet weld loaded in a direction perpendicular to the weld is approximately one-third stronger than one loaded in a parallel direction (see Figure 19.21). The AISC Specification, however, does not allow for this consideration in weld design. The strengths of all fillets are based on the values calculated for loads applied in a parallel direction.

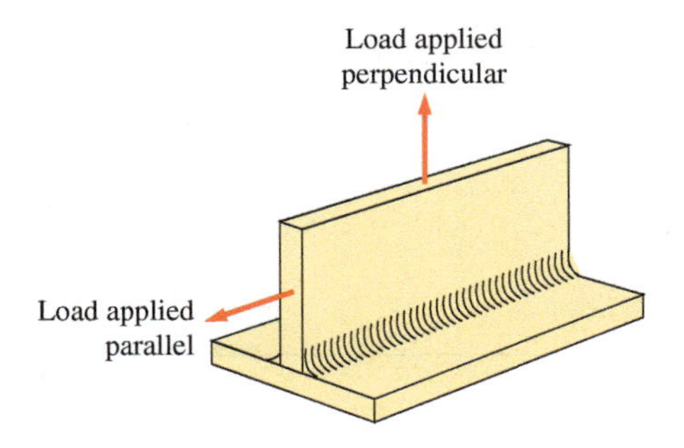

FIGURE 19.21 Direction of applied load.

In addition to the strength criteria, the AISC Specification includes design requirements with respect to minimum and maximum sizes of fillet welds. These sizes are summarized in Table 19.3.

Additional AISC provisions for fillet-welded connections are summarized as follows:

1. The minimum effective length of a fillet weld must not be less than four times the nominal size of the weld. If a weld is intermittent, the minimum length of each weld is $1\frac{1}{2}$ in.
2. Side or end fillet welds may be returned continuously around the corners for a distance not greater than four times the weld size nor half the width of the part. This is called an *end return*. Its use is optional.
3. If fillet welds parallel to the applied load are used alone (without perpendicular welds other than end returns) in end connections of flat-plate tension members, the length of each fillet weld cannot be less than the perpendicular distance between them.
4. Where lap joints are used, the minimum amount of lap should be five times the thickness of the thinner part joined, but not less than 1 in.

EXAMPLE 19.4

Calculate the allowable axial tensile load that can be applied to the connection in Figure 19.22. The plates are ASTM A36 steel. The weld is a $\frac{7}{16}$-in. fillet weld and is made using an E70 electrode.

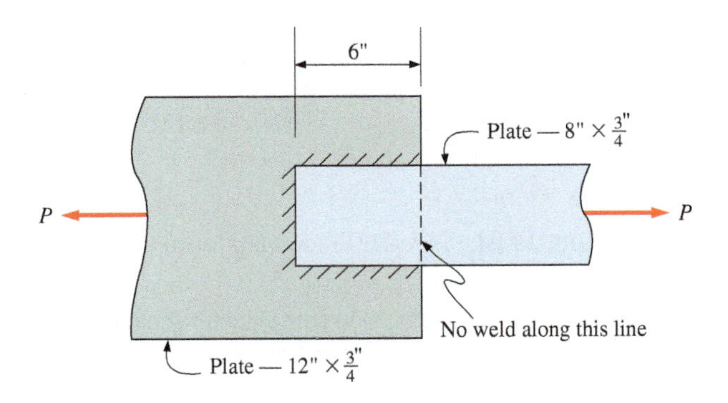

FIGURE 19.22 Welded lap joint.

Solution The total length of the weld is 20 in. The unit strength of a $\frac{7}{16}$-in. weld is 7(0.928 k/in.) = 6.50 k/in. Therefore,

$$\text{weld capacity} = 6.50(20) = 130.0 \text{ k}$$

From Table 19.2, the allowable axial tensile stress is 21.6 ksi. The tensile capacity of the smaller plate is then calculated from

$$P = A_g F_t = (8 \text{ in.})(0.75 \text{ in.})(21.6 \text{ ksi}) = 129.6 \text{ k}$$

Therefore, the allowable axial tensile load is 129.6 kips.

EXAMPLE 19.5 Calculate the allowable axial tensile load that can be applied to the welded connection in Figure 19.23. The plates are ASTM A36 steel. The weld is a $\frac{5}{16}$-in. fillet weld made using an E70 electrode.

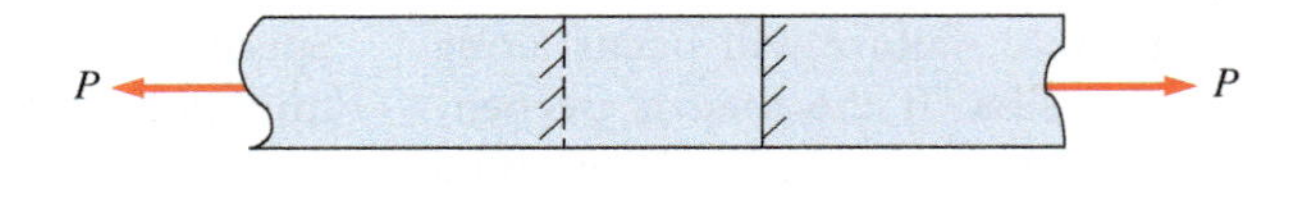

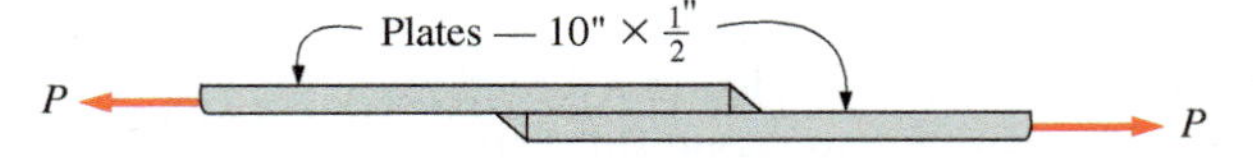

FIGURE 19.23 Transverse-welded lap joint.

Solution The total length of the weld is 20 in. The unit strength of a $\frac{5}{16}$-in. fillet weld is 5(0.928 k/in.) = 4.64 k/in. Therefore,

$$\text{weld capacity} = 4.64(20) = 92.8 \text{ k}$$

From Table 19.2, the allowable axial tensile stress is 21.6 ksi. The tensile capacity of the plate is then calculated from

$$P = A_g F_t = (10 \text{ in.})(0.5 \text{ in.})(21.6 \text{ ksi}) = 108 \text{ k}$$

Therefore, the allowable axial tensile load is 92.8 kips.

EXAMPLE 19.6 Design the fillet welds parallel to the applied load to develop the full tensile capacity of the 8-in.-by-$\frac{1}{2}$-in. plate in Figure 19.24. The plates are ASTM A36 steel and the electrode is an E70. Provide end returns.

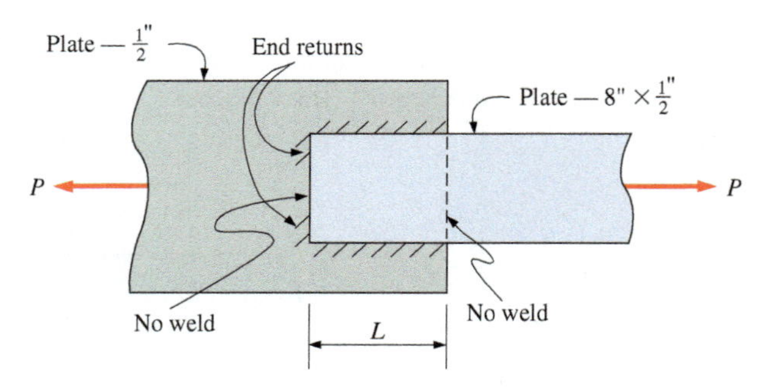

FIGURE 19.24 Parallel-loaded lap joint.

Solution From Table 19.2, the allowable axial tensile stress is 21.6 ksi. The tensile capacity for the 8-in.-by-$\frac{1}{2}$-in. steel plate is then calculated from

$$P = A_g F_t = (8 \text{ in.})(0.5 \text{ in.})(21.6 \text{ ksi}) = 86.4 \text{ k}$$

The maximum weld size, from Table 19.3, is a $\frac{7}{16}$-in. weld. For economy reasons, however, use a $\frac{5}{16}$-in. weld, which is the largest weld that can be made manually in a single pass. The unit strength of a $\frac{5}{16}$-in. weld is 5(0.928 k/in.) = 4.64 k/in. The length of the weld required is then

$$\text{required length} = \frac{86.4 \text{ k}}{4.64 \text{ k/in.}} = 18.62 \text{ in.}$$

The maximum length of an end return is four times the weld size:

$$4 \times \frac{5}{16} \text{ in.} = 1.250 \text{ in.}$$

Use 1-in. end returns.

The required length for each side weld is then

$$\text{required } L = \frac{18.62 \text{ in.} - 2(1 \text{ in.})}{2} = 8.31 \text{ in. (use 8.5 in.)}$$

The minimum length of the side fillet welds must not be less than the perpendicular distance between them. Therefore, the minimum length required is 8 in. Since 8.5 in. > 8 in., the design is O.K.

SUMMARY BY SECTION NUMBER

19.1 The connection of structural steel members or elements of members is accomplished almost exclusively through the use of bolts or welding.

19.2 The most frequently used bolts in structural steel connections are unfinished bolts (ASTM A307) and high-strength bolts (ASTM A325 or A490).

19.3 The possible failure modes for bolted joints are (a) shear in the bolts, (b) bearing of the bolts on the connected material, (c) tension in the connected parts, (d) end tear-out, and (e) block shear.

19.4 The allowable load for a high-strength bolted connection is determined by considering shear, bearing, and tension separately. The shear strength is obtained from

$$P_s = nA_B F_v N \tag{19.2}$$

The bearing strength is obtained from

$$P_p = dtF_p N \tag{19.3}$$

The tensile strength is obtained from

$$P_g = A_g F_t = A_g(0.60)F_y \tag{19.4}$$

or from

$$P_n = A_n F_t = A_n(0.50)F_u \tag{19.6}$$

Allowable yield and ultimate tensile stresses are listed in Tables 19.1 and 19.2. High-strength bolted connections are categorized as either *slip-critical* or *bearing-type* connections. Various allowable stresses apply.

19.5 Welding is a process in which two or more pieces of metal are fused together by heat to form a joint. The most common welding process is electric-arc welding. Fillet welds and groove welds are the two types of load-carrying welds most commonly used.

19.6 The cross section of a typical fillet weld is a right triangle with equal leg sizes. A fillet weld will fail in shear, with its shear strength (per linear inch of weld) equal to

$$P_s = 0.212F_u(\text{leg size}) \tag{19.8}$$

Groove welds are full-strength welds requiring no strength calculations if the proper electrode is used.

PROBLEMS

Where applicable, refer to Tables 19.1 and 19.2 for allowable stresses. Assume standard holes for all bolted connections.

Section 19.4 High-Strength Bolted Connections

19.1 Compute the allowable tensile load for the single-shear lap joint shown. The plates are ASTM A36 steel and the high-strength bolts are $\frac{7}{8}$-in.-diameter A325 bolts in standard holes. Assume a bearing-type connection with threads excluded from the shear plane (A325-X).

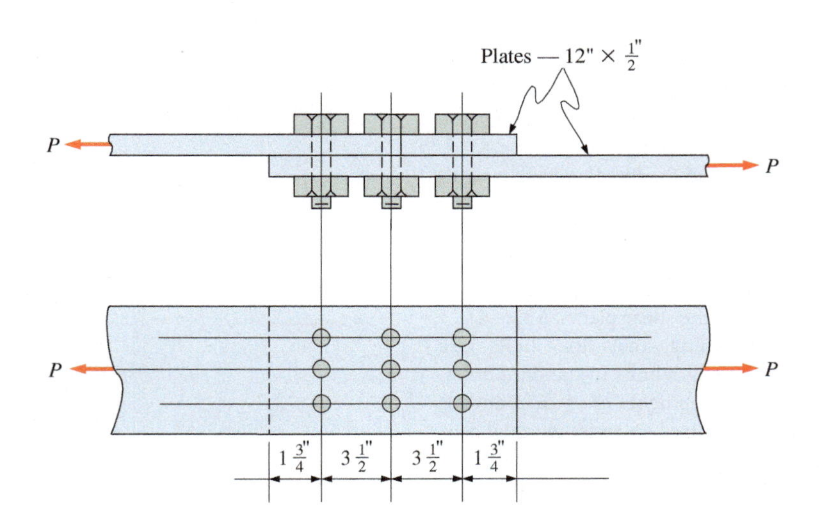

PROBLEM 19.1

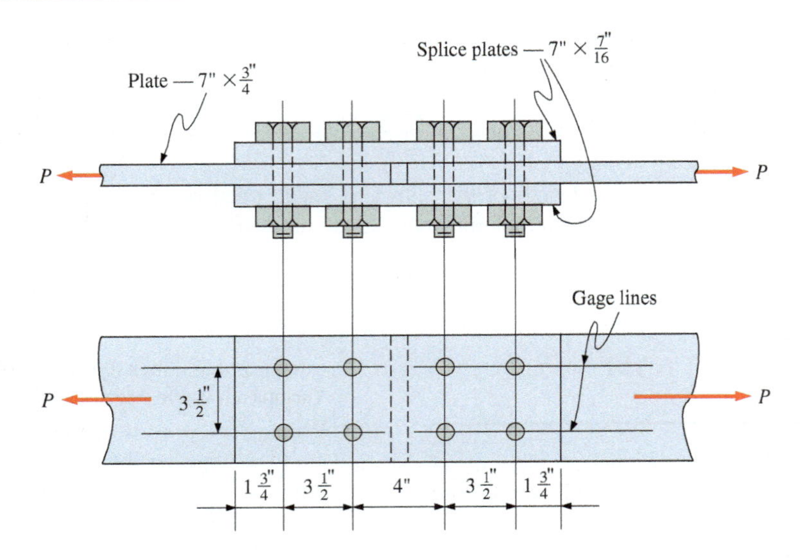

PROBLEM 19.4

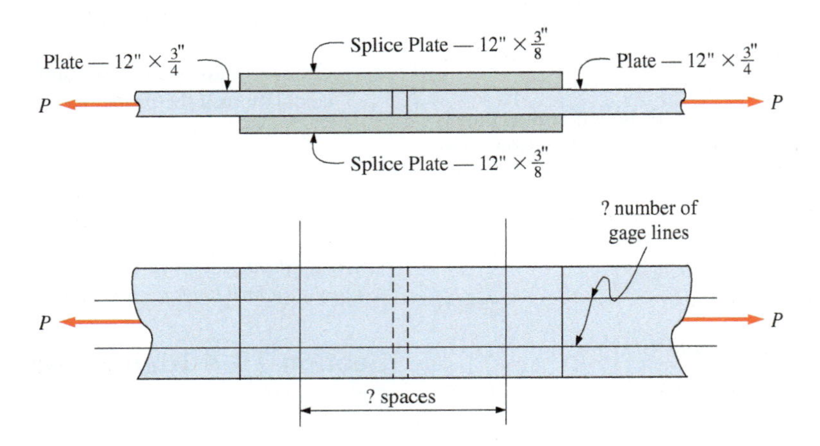

PROBLEM 19.7

19.2 Rework Problem 19.1 assuming a bearing-type connection with bolt threads in the shear plane (A325-N).

19.3 Rework Problem 19.1 assuming a bearing-type connection with bolt threads in the shear plane (A325-N). Assume the plate dimensions to be 300 mm × 13 mm. Bolt pitch (spacing) is 90 mm, the edge distance is 45 mm, and the bolts are 22 mm in diameter.

19.4 Compute the allowable tensile load for the double-shear butt joint shown. The plates are equivalent to A992 steel. The bolts are $\frac{7}{8}$-in.-diameter A325-N high-strength bolts in standard holes.

19.5 Rework Problem 19.4 assuming a bearing-type connection with bolt threads excluded from the shear plane (A325-X).

19.6 Rework Problem 19.4 assuming that the bolts are $\frac{3}{4}$-in.-diameter A490-N high-strength bolts in standard holes.

19.7 Select the number and arrangement of $\frac{3}{4}$-in.-diameter A325-N high-strength bolts required to resist an axial tensile load of 150 kips for the butt joint shown. The plates are all ASTM A36 steel. Indicate the bolt arrangement with a sketch.

Section 19.6 Strength and Behavior of Welded Connections (AISC)

19.8 Calculate the allowable tensile load for the connection shown. The plates are ASTM A36 steel and the weld is a $\frac{3}{8}$ in. fillet weld, which is made using an E70 electrode.

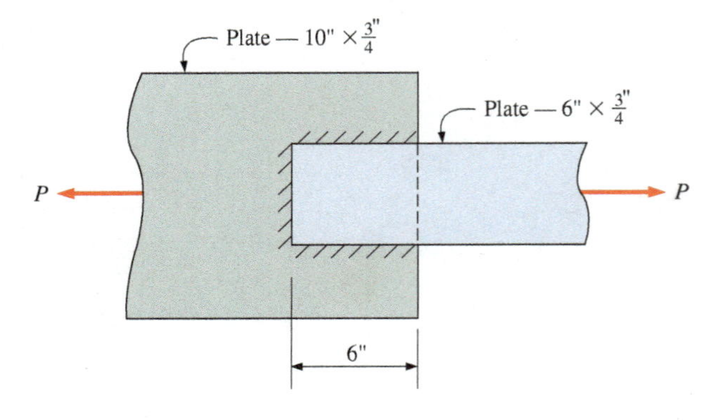

PROBLEM 19.8

19.9 In the connection shown, $\frac{1}{4}$-in. side and end fillet welds are used to connect the 3-in.-by-1 in. tension member to the plate. The applied load is 60,000 lb. Find the required dimension L. The steel is ASTM A36 and the electrode used is an E70.

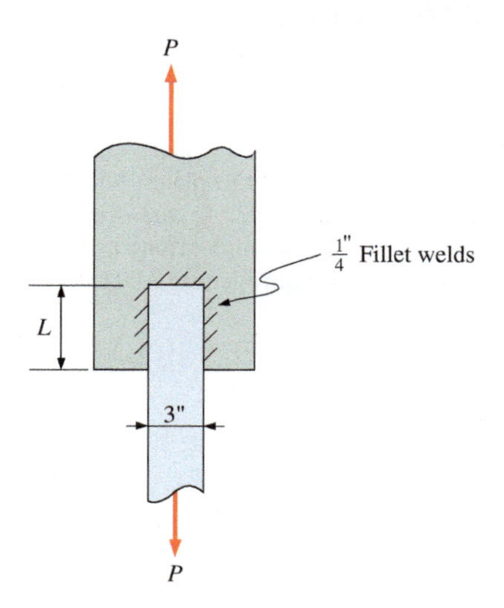

PROBLEM 19.9

19.10 Design the fillet welds parallel to the applied load to develop the full allowable tensile load of the 6-in.-by-$\frac{3}{8}$-in. ASTM A36 steel plate shown. The electrode is an E70.

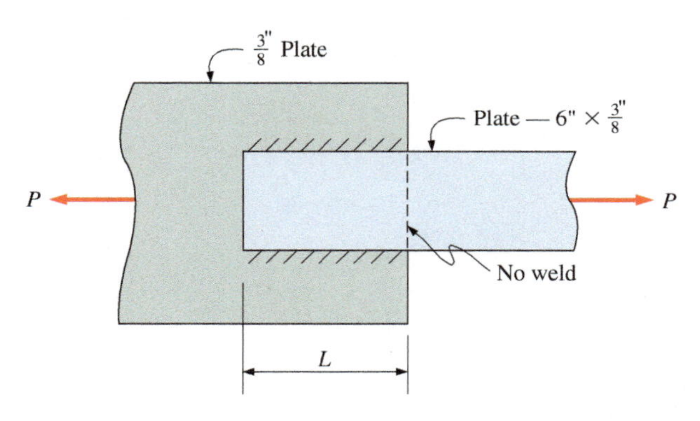

PROBLEM 19.10

19.11 A fillet weld between two steel plates intersecting at right angles was made with one $\frac{1}{4}$-in. leg and one $\frac{9}{16}$-in. leg using an E70 electrode. Determine the strength of this weld in kips per inch.

19.12 Design an end connection using longitudinal welds and an end transverse weld to develop the full tensile capacity of the angle shown. Use A36 steel and E70 electrodes. *Note:* A reduction coefficient for the computation of the net area for a tension member that does not have all its cross-sectional elements connected to the supporting member is commonly used. This factor is neglected in this text.

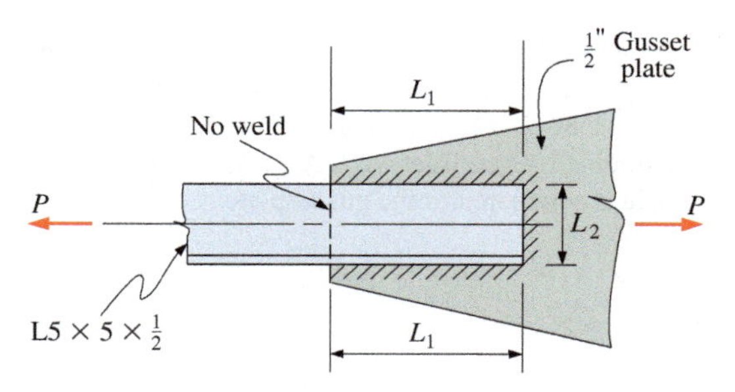

PROBLEM 19.12

Computer Problems

For the following computer problems, any appropriate software may be used. Input prompts should fully explain what is required of the user (the program should be user-friendly). The resulting output should be well labeled and self-explanatory.

19.13 The load that can be transmitted by a single fastener in shear is sometimes called the *shear value*. The shear value is the product of the cross-sectional area of the shank of the fastener and an allowable shear stress. Write a program that will generate a table of shear values for fasteners having the allowable stresses given in Table 19.1. Fastener diameters are to range from $\frac{5}{8}$ in. to $1\frac{1}{2}$ in. (in $\frac{1}{8}$-in. increments).

19.14 A single-shear lap joint is to have fasteners arranged on three gage lines, similar to the joint shown in Figure 19.12. The steel plates are to be of ASTM A36 steel. Write a program that will calculate the allowable tensile load for such a joint assuming $\frac{3}{4}$-in.-diameter A325-X high-strength bolts. Input is to be the number of bolts, which must be in multiples of three, and the cross-sectional dimensions (width and thickness) of the main plates. Assume standard holes.

Supplemental Problems

19.15 Calculate the allowable tensile load for the butt joint shown. The fasteners are $\frac{7}{8}$-in.-diameter A325 high-strength bolts in a slip-critical connection (A325-N). The plates are of ASTM A36 steel.

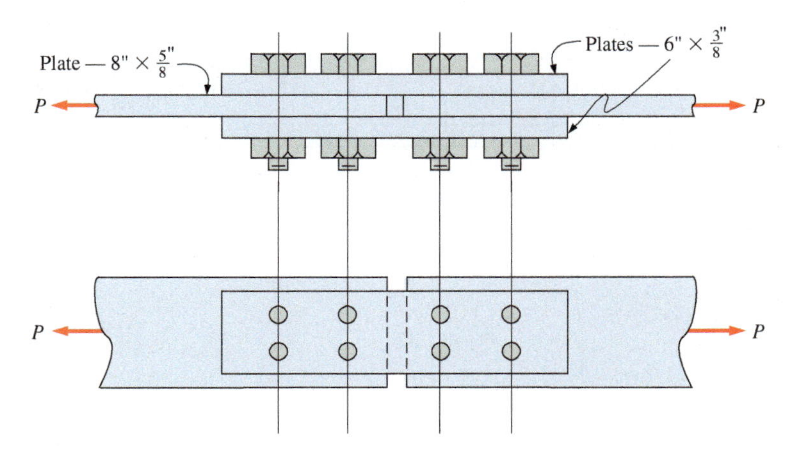

PROBLEM 19.15

19.16 Calculate the allowable tensile load for the lap joint shown. The fasteners are 25-mm-diameter A325-X high-strength bolts. The plates are ASTM A36 steel.

19.17 Calculate the allowable tensile load for the butt joint shown. The fasteners are 20-mm-diameter A325-N high-strength bolts and the plates are ASTM A36 steel.

19.18 Rework Problem 19.10 assuming that both plates are 12 mm thick and that the smaller plate is 150 mm wide. Use a 10-mm fillet weld.

19.19 Rework Problem 19.12 assuming that the angle is an L127 × 89 × 12.7 and that the gusset plate is 13 mm thick. Assume a 10-mm fillet weld. Assume that the long leg of the angle is connected to the gusset plate.

19.20 Two ASTM A36 steel plates, each 12 in. by $\frac{1}{2}$ in., are connected by a lap joint. The fasteners are $\frac{3}{4}$-in.-diameter A325-N high-strength bolts. The connection is subjected to an axial tensile load of 100 kips. Calculate the number of bolts required for the connection and sketch the bolt layout pattern.

19.21 Rework Problem 19.20 changing the fasteners to $\frac{3}{4}$-in.-diameter A490-X bolts.

19.22 Calculate the minimum main plate thickness for the joint of Problem 19.4 so that the plates, which are 7 in. wide, will have a capacity equal to the shear capacity of the $\frac{7}{8}$-in.-diameter A325-N high-strength bolts in standard holes. The plates are equivalent to A992 steel.

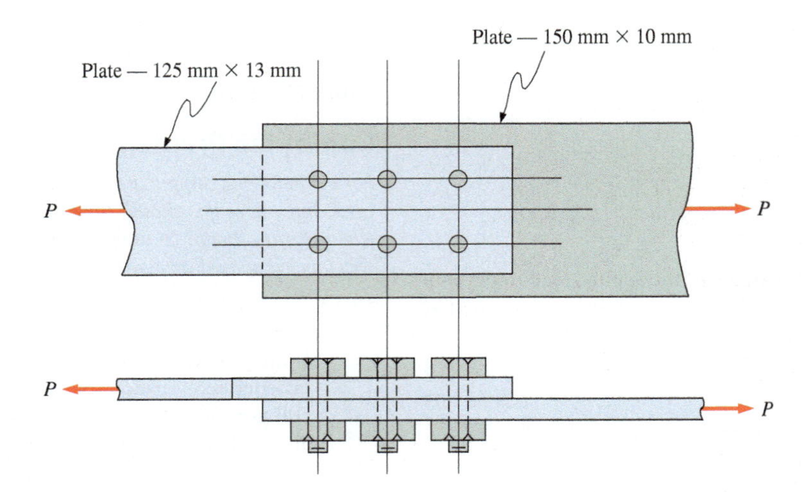

PROBLEM 19.16

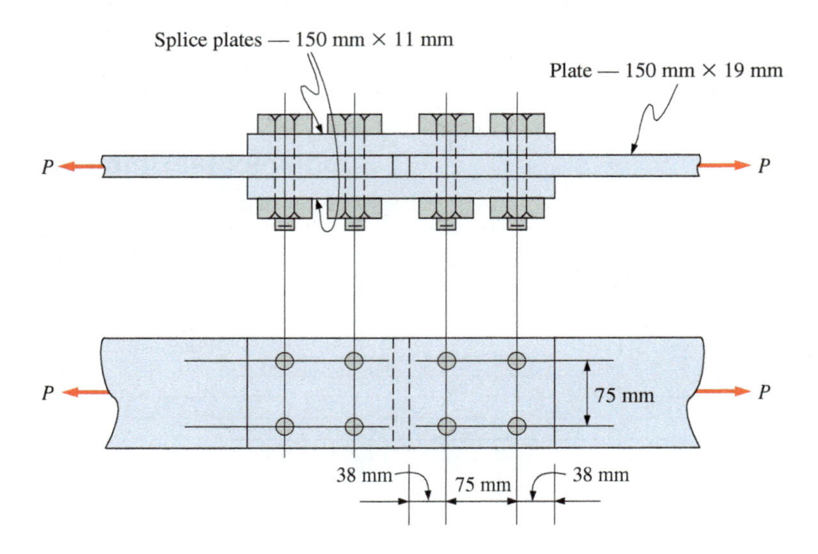

PROBLEM 19.17

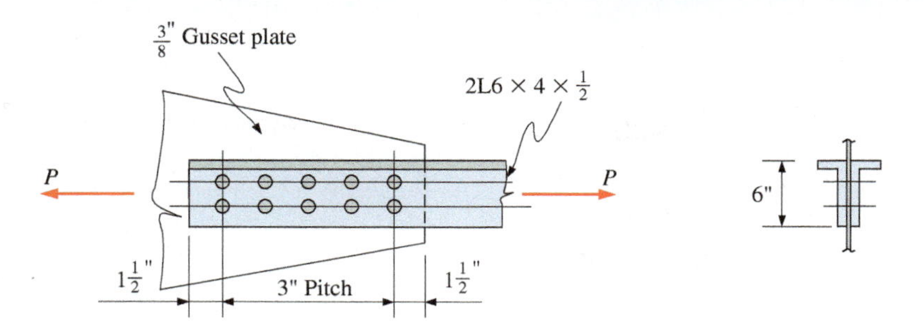

PROBLEM 19.23

19.23 A roof truss tension member is made up of 2L6 × 4 × $\frac{1}{2}$ (total $A_g = 9.50$ in.^{2}) of A36 steel as shown. The bolts are $\frac{3}{4}$-in.-diameter. A325-N high-strength bolts in standard holes. Compute the tensile capacity P for the connection shown. The special note for angle tension connections of Problem 19.12 applies.

19.24 Rework Problem 19.23 changing the fasteners to six $\frac{3}{4}$-in.-diameter A490-X high-strength bolts.

19.25 Determine the allowable tensile load that can be applied to the connection shown. The plates are of ASTM A36 steel and the weld is made using an E70 electrode.

19.26 The welded connection shown is subjected to an axial tensile load P of 100,000 lb. Calculate the minimum size fillet weld required if the length L is 7 in. Assume an E70 electrode.

19.27 In Problem 19.26, use a $\frac{3}{8}$-in. fillet weld, change L to 9 in., and find the allowable tensile load as governed by the welds.

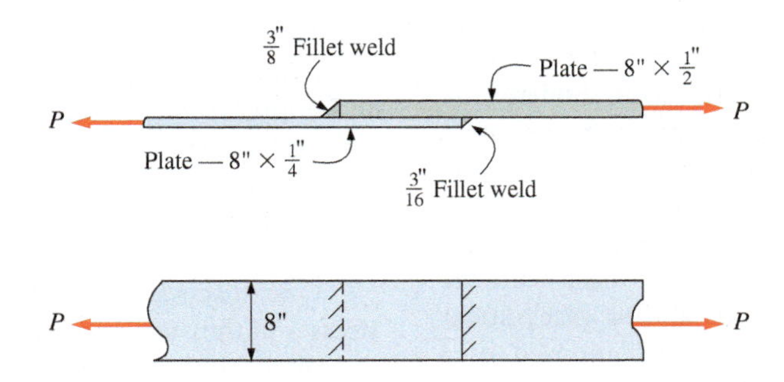

PROBLEM 19.25

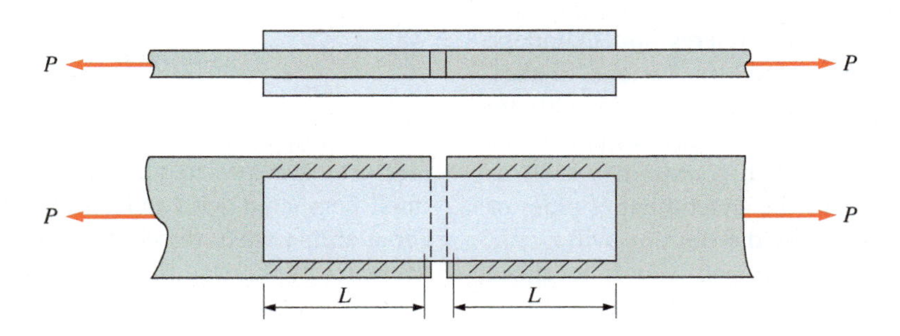

PROBLEM 19.26

CHAPTER TWENTY

PRESSURE VESSELS

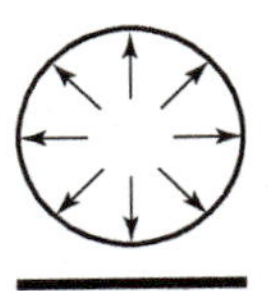

LEARNING OBJECTIVES

Upon completion of this chapter, readers will be able to:

- Identify pressure vessels.
- Calculate the circumferential and longitudinal tensile stresses in the shell of a thin walled pressure vessel.
- Identify considerations critical to the design of joints in pressure vessels.
- Recognize considerations that are necessary in the design and fabrication of thin walled pressure vessels.

20.1 INTRODUCTION

Pressure vessels may be described as leakproof containers. They are found in various shapes and sizes. Boilers, fire extinguishers, shaving cream cans, and pipes are common examples. The sophisticated vessels used in the space, nuclear, and chemical industries are also examples of pressure vessels.

Pressure vessels commonly have the form of spheres, cylinders, ellipsoids, or some composite of these with the primary purpose of containing liquids and/or gases under pressure. In practice, vessels are usually composed of a complete pressure-containing shell together with flange rings and fastening devices for connecting and securing mating parts.

In this chapter, we are primarily concerned with the stresses developed in the walls of spheres and cylinders. Stress analysis can be performed by various means. Our limited discussion assumes a simple vessel with no consideration of shape discontinuities. If the vessel exhibits abrupt cross-sectional changes or discontinuities such as penetrations of the shell wall for the attachment of pipes or nozzles, irregularities in the stress distribution will occur. These changes, in general, are developed in only localized portions of the vessel and are called *localized stresses* or *stress concentrations*. In these cases, the problem becomes too complex for an analytical solution. A computer solution or an experimental solution will then be required. Our discussion is limited to the analytical approach.

The stress analysis is only one phase of a total design. Significant considerations in design are the selection of the material to be used and its relationship to the environment to which it will be subjected. Based on the safety demands of nuclear reactors, space vehicles, and deep-diving submersibles, considerable research has been directed toward new and improved materials and their behavior in specific environments. No one perfect pressure vessel material is suitable for all situations. Material selection must be consistent with the application and the environment. This is particularly true for nuclear plant pressure vessels.

Our discussion in this chapter considers only the stresses in thin-walled pressure vessels at sections other than the discontinuity locations. Thin-walled pressure vessels are defined as having a wall thickness t not in excess of one-tenth (0.1) of the internal radius r_i of the vessel. Mathematically, this relationship can be stated as

$$t \leq 0.1 r_i \quad \text{or} \quad \frac{t}{r_i} \leq 0.1$$

In most cylindrical and spherical vessels, the internal pressures are low so that the wall thickness is relatively small compared with the other dimensions of the vessel. For a thin-walled vessel, we will assume that the tensile stress across the wall thickness is constant. The magnitude of the stress can be obtained using the basic equations of equilibrium, which result in an average stress with an error less than 4% or 5%. For thick-walled vessels $t/r_i > 0.1$, as in the cases of gun barrels or high-pressure hydraulic presses, the variation in the stress from the inner surface to the outer surface becomes appreciable, with a higher stress existing at the inner surface. This problem is more complex and the thin-walled formulas cannot be used.

20.2 STRESSES IN THIN-WALLED PRESSURE VESSELS

A *thin-walled pressure vessel* has been defined as one with a wall thickness not in excess of one-tenth (0.1) of the internal radius of the vessel. It is usually made up of curved plates fastened together by continuous joints or seams. In a cylindrical pressure vessel, the plates are joined using two types of joints: a longitudinal, or lengthwise, joint and a circumferential joint. Both types are shown in Figure 20.1. A spherical pressure vessel, by virtue of its complete symmetry, has only the circumferential-type joint.

Cylindrical and spherical thin-walled pressure vessels are generally subjected to some magnitude of internal gas and/or liquid pressure. As a result of the internal pressure, tensile stresses are developed in the pressure vessel walls. As with other structural shapes, the induced tensile stresses may not exceed specified allowable tensile stresses. The internal pressure tends to rupture the pressure vessel along either a longitudinal or a circumferential joint. The walls of a thin-walled pressure vessel are assumed to act as a membrane in that no bending of the walls takes place.

We begin our analysis of pressure vessels by considering a cylindrical pressure vessel. Figure 20.2a shows section *A–A*, obtained by passing plane *A–A* through the pressure vessel of Figure 20.1. This is a typical cross section of a cylindrical thin-walled pressure vessel subjected to an internal pressure p.

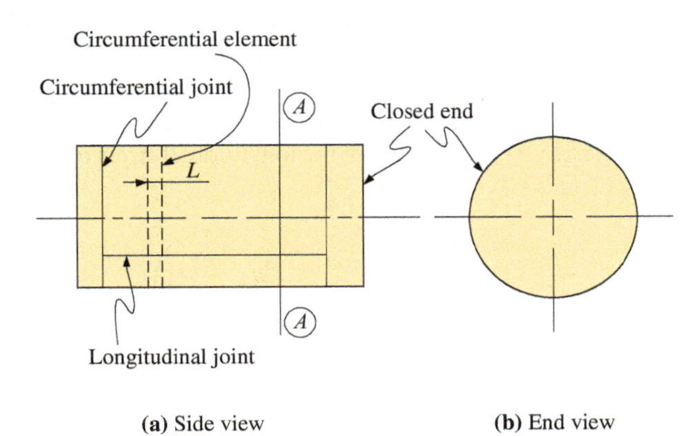

FIGURE 20.1 Cylindrical pressure vessel.

The internal pressure at any point acts equally in all directions and is always perpendicular to any surface on which it acts. Therefore, in Figure 20.2a, the pressure is seen to act radially outward.

The radially acting internal pressure tends to rupture the longitudinal joints. To resist this tendency, tensile stresses are developed in the walls of the pressure vessel. Those stresses are called *circumferential stresses* (σ_{t_c}), and they act on the longitudinal joint. To establish an expression for the circumferential stress, let us consider a circumferential element of length L taken from the cylindrical pressure vessel of Figure 20.1. The free-body diagram of half of this element is shown in Figure 20.2b, which also shows the forces and pressures acting on the free body. Note, as shown in Figure 20.2a, that the internal pressure p acts radially and is uniformly distributed over the curved surface. For the free body shown in Figure 20.2b, the horizontal components p_H of the radial pressure act in opposite directions, thus cancelling each other by virtue of symmetry about the vertical centerline. The vertical components p_V, however, act in the same direction and must be resisted for a state of equilibrium to exist. The total upward acting force on the curved surface is the sum of all the vertical components of the forces due to the internal pressure and can be represented by a single resultant force P. It can be shown that this resultant vertical force P is obtained as the product of the pressure p and the area of the curved wall surface projected on a horizontal plane. Since the projected area is equal to LD_i, the resultant vertical force can be calculated from

$$P = pLD_i$$

The resistance to this force is furnished by the walls of the pressure vessel. A resisting tensile force T_c is developed in each cut wall, as shown in the free body of Figure 20.2b. A summation of vertical forces shows that $P = 2T_c$, from which $T_c = P/2$. Solving for the circumferential tensile stress σ_{t_c} (also called the *hoop stress*) in the wall of the pressure vessel yields

$$\sigma_{t_c} = \frac{T_c}{tL} = \frac{P/2}{tL} = \frac{pLD_i}{2tL} = \frac{pD_i}{2t} \tag{20.1}$$

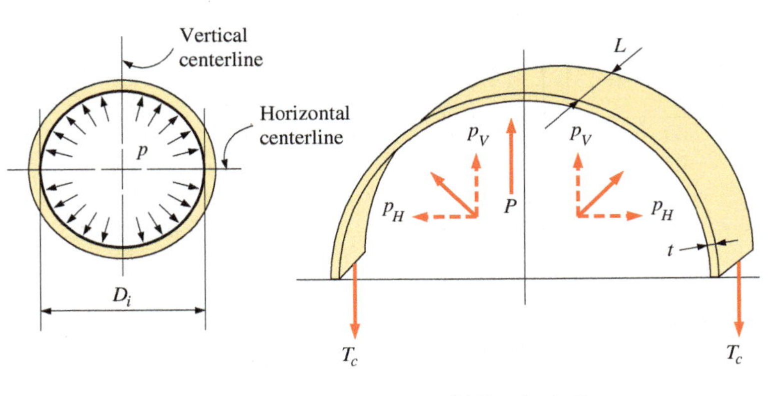

FIGURE 20.2 Cylindrical pressure vessel.

where σ_{t_c} = the circumferential tensile stress in the wall of the pressure vessel (psi, ksi) (Pa)
p = the internal pressure (psi, ksi) (Pa)
D_i = the inside diameter (in.) (mm)
t = the wall thickness of the pressure vessel (in.) (mm)

Equation (20.1) can also be used for design purposes to calculate a required wall thickness to resist a given internal pressure while not exceeding an allowable tensile stress $\sigma_{t_{c(\text{all})}}$. The equation can be rewritten as follows:

$$\text{required } t = \frac{pD_i}{2\sigma_{t_{c(\text{all})}}} \tag{20.2}$$

In addition to the tendency of a closed-end cylindrical pressure vessel to rupture along a longitudinal joint, the pressure tends to push out the ends and pull the vessel apart on a circumferential joint simultaneously. To resist this tendency, tensile stresses are developed in the walls of the pressure vessel that act on a circumferential joint. These *longitudinal stresses* are perpendicular to the circumferential stresses.

To establish an expression for the longitudinal stresses, let us consider a free body of the closed-end portion of a cylindrical pressure vessel subjected to an internal pressure, as shown in Figure 20.3.

Since the end of the pressure vessel is closed, the end area on which the pressure acts is circular in shape. The total resultant force tending to push out the end of the vessel can be calculated as the product of pressure and area:

$$P = p\left(\frac{\pi D_i^2}{4}\right)$$

For a state of equilibrium to exist, force P must equal the total resistance T_L offered by the walls of the vessel around the entire circumferential joint. Assuming the longitudinal stress σ_{t_L} in the walls of the vessel to be uniform throughout the wall thickness, we can calculate the stress as

$$\sigma_{t_L} = \frac{T_L}{\pi D_c t} = \frac{P}{\pi D_c t}$$

where D_c is the vessel diameter center-to-center of walls. Recognizing that $D_i \approx D_c$ (very close), we make this substitution as well as the substitution for P from the preceding expression:

$$\sigma_{t_L} = \frac{p(\pi D_i^2/4)}{\pi D_i t} = \frac{pD_i}{4t} \tag{20.3}$$

where σ_{t_L} is the longitudinal stress in the walls of the pressure vessel (psi, ksi) (Pa) and the other terms are as previously defined for Equation (20.1).

Note that the longitudinal stress developed is one-half the circumferential stress. Thus, if a fluid in a closed vessel (a tank or pipe) freezes, the vessel will rupture along a longitudinal joint. Note that the two expressions for stresses are not accurate in the immediate vicinity of a closed end.

One often encounters a thin-walled spherical pressure vessel subjected to an internal pressure. Although the spherical shape is more efficient than the cylindrical shape, it is not as commonly used and is generally more expensive to manufacture. Whereas the cylindrical shape is subjected to two different magnitudes of stress, as previously discussed, every point in the spherical shape is subjected to the same maximum tensile stress. This tensile stress is the same as that for the longitudinal stress in a thin-walled cylindrical pressure vessel and can be evaluated using Equation (20.3).

Note that our discussion in this section, including the equations developed, applies only to thin-walled pressure vessels and the case of internal pressure. Since we are neglecting all discontinuities and localized stresses, the state of stress for an element of a thin-walled cylindrical-shaped pressure vessel may, for all practical purposes, be considered biaxial, as shown in Figure 20.4. Recall from previous discussion that the circumferential stress σ_{t_c} is twice that of the longitudinal stress σ_{t_L}.

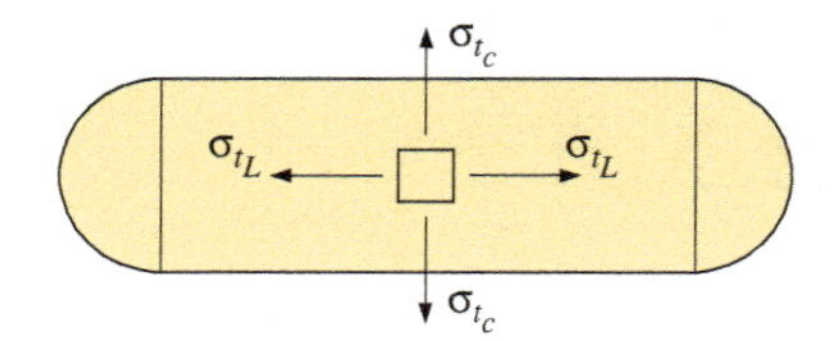

FIGURE 20.4 State of biaxial stress.

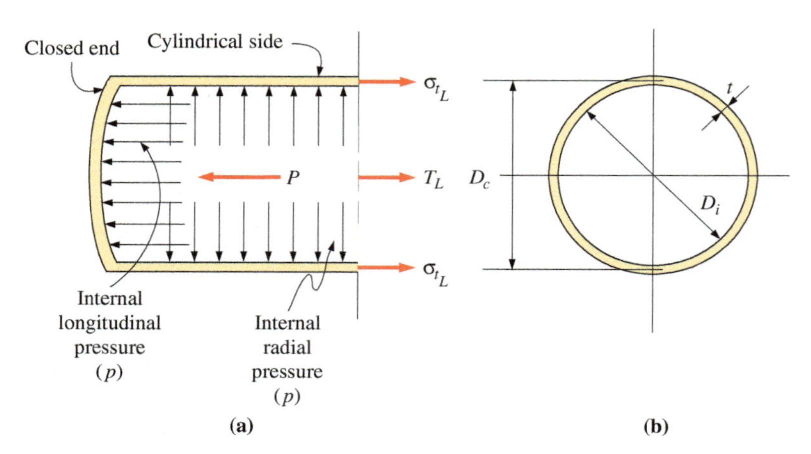

FIGURE 20.3 Longitudinal stresses in cylindrical pressure vessel.

If an additional tensile force P is applied to the closed ends as shown in Figure 20.5, the longitudinal stress σ_{t_L} is increased by the amount of

$$\sigma_{t_P} = \frac{P}{A} = \frac{P}{\pi D_i t}$$

Therefore, the total longitudinal stress can be written as

$$\text{total } \sigma_{t_L} = \sigma_{t_L} + \sigma_{t_P} = \frac{pD_i}{4t} + \frac{P}{\pi D_i t} \quad \textbf{(20.4)}$$

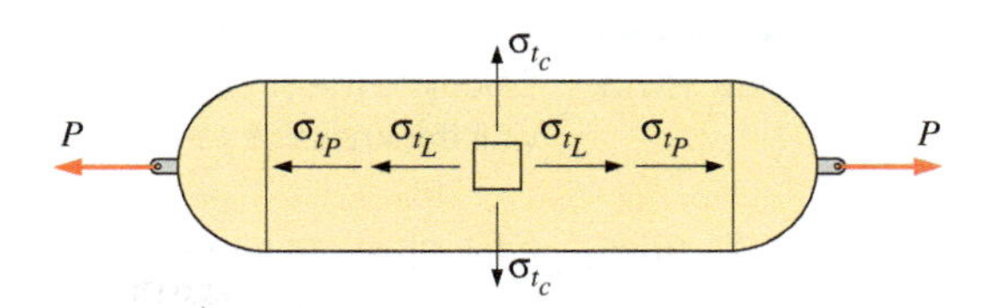

FIGURE 20.5 Modified state of biaxial stress.

EXAMPLE 20.1 Calculate the circumferential and longitudinal stresses developed in the walls of a cylindrical gas storage tank. The tank is of steel, has a 48-in. inside diameter, and has a wall thickness of $\frac{3}{8}$ in. The internal gage pressure is 150 psi. Show that the thin-wall theory is applicable to this problem.

Solution The internal pressure is gage pressure, which means that it is an internal pressure in excess of an external pressure. Usually, the external pressure is atmospheric. The thin-wall theory (along with its equations) is applicable if $t \leq 0.1 \times$ (radius):

$$0.1\left(\frac{48}{2}\right) = 2.4 \text{ in.}$$

$$0.375 \text{ in.} < 2.4 \text{ in.} \qquad \textbf{(O.K.)}$$

The circumferential stress developed in the tank walls can be found from Equation (20.1):

$$\sigma_{t_c} = \frac{pD_i}{2t} = \frac{150(48)}{2(0.375)} = 9600 \text{ psi}$$

The longitudinal stress developed in the tank walls can be found from Equation (20.3):

$$\sigma_{t_L} = \frac{pD_i}{4t} = \frac{150(48)}{4(0.375)} = 4800 \text{ psi}$$

EXAMPLE 20.2 Calculate the maximum allowable internal gage pressure that can be contained by a closed cylindrical steel tank having an internal diameter of 1.00 m and a wall thickness of 7 mm. The steel has an allowable tensile stress of 100 MPa.

Solution First, check the applicability of the thin-wall theory:

$$0.1\left(\frac{1000}{2}\right) = 50 \text{ mm}$$

$$7 \text{ mm} < 50 \text{ mm} \qquad \textbf{(O.K.)}$$

Since the circumferential stress developed will be twice the longitudinal stress developed, the allowable internal pressure must be based on Equation (20.1), where

$$\sigma_{t_c} = \frac{pD_i}{2t}$$

Rewrite and solve for p:

$$p = \frac{2t\sigma_{t(\text{all})}}{D_i} = \frac{2(7)(100)}{1000} = 1.40 \text{ MPa}$$

EXAMPLE 20.3 Calculate the stress developed in the walls of a spherical steel tank having a 12-ft inside diameter and a wall thickness of $\frac{3}{4}$ in. The tank is subjected to an internal gage pressure of 300 psi.

Solution First, check the applicability of the thin-wall theory:

$$0.1(144/2) = 7.2 \text{ in.}$$

$$0.75 \text{ in.} < 7.2 \text{ in.} \qquad \textbf{(O.K.)}$$

The tensile stress developed in the walls of a spherical pressure vessel can be found from Equation (20.3):

$$\sigma_{t_L} = \frac{pD_i}{4t} = \frac{300(12)(12)}{4(0.75)} = 14{,}400 \text{ psi}$$

EXAMPLE 20.4 A steel cylinder with a 300 mm diameter contains a gas at a gage pressure of 1.5 MPa. Using a steel having a yield stress of 350 MPa and a factor of safety (based on yielding) of 3.0, compute the required wall thickness of the cylinder. Check the applicability of the thin-wall theory.

Solution Since the circumferential stress developed will be twice the longitudinal stress developed, the required wall thickness will be determined using Equation (20.2):

$$\text{required } t = \frac{pD_i}{2\sigma_{t_{c(\text{all})}}} = \frac{(1.5 \text{ MPa})(300 \text{ mm})}{2(350/3 \text{ MPa})} = 1.93 \text{ mm}$$

Next, check the applicability of the thin-wall theory:

$$0.1\left(\frac{300}{2}\right) = 15 \text{ mm}$$

$$1.93 \text{ mm} < 15 \text{ mm} \qquad \textbf{(O.K.)}$$

EXAMPLE 20.5 A cylindrical pressure vessel made of steel has a 24-in. inside diameter and a wall thickness of $\frac{3}{8}$ in. The vessel is subjected to an internal gage pressure of 200 psi. The vessel is simultaneously subjected to an axial tensile force of 90,000 lb. Calculate the circumferential and longitudinal stresses developed in the walls.

Solution First, check the applicability of the thin-wall theory:

$$0.1\left(\frac{24}{2}\right) = 1.2 \text{ in.}$$

$$0.375 \text{ in.} < 1.2 \text{ in.} \qquad \textbf{(O.K.)}$$

The circumferential stress developed in the walls of the vessel is due to the internal pressure only; use Equation (20.1):

$$\sigma_{t_c} = \frac{pD_i}{2t} = \frac{200(24)}{2(0.375)} = 6400 \text{ psi}$$

The longitudinal stress developed in the walls is the sum of the stresses due to the internal pressure and the axial tensile load; use Equation (20.4):

$$\sigma_{t_L} = \frac{pD_i}{4t} + \frac{P}{\pi D_i t}$$

$$= \frac{200(24)}{4(0.375)} + \frac{90{,}000}{\pi(24)(0.375)} = 6380 \text{ psi}$$

20.3 JOINTS IN THIN-WALLED PRESSURE VESSELS

The design of joints (frequently called *boiler joints*) in thin-walled pressure vessels, as well as the design of the pressure vessels themselves, is standardized by the American Society of Mechanical Engineers (ASME) Boiler and Pressure Vessel Code.[1] The ASME Code constitutes an international standard providing recommended design and manufacturing criteria for pressure-containing structures.

As stated previously, thin-walled pressure vessels are generally manufactured from curved plates fastened together by designed joints that must remain sealed under internal pressure. The most common type of joint for the pressure vessel is the continuous-welded butt joint for which the plates are groove welded using some form of edge preparation as discussed in Section 19.6. For all practical purposes, the welded boiler joints have replaced the once commonly used riveted boiler joints.

The welds are generally a full-penetration-type weld. The thickness of the weld is equal to the thickness of the plates. The allowable tensile stress for the connected plates, based on the ASME Boiler Code, is generally adopted as either one-fourth of the ultimate tensile strength of the steel or two-thirds of the yield strength of the steel, whichever is less. The allowable tensile stress for the weld may be assumed to be equal to, or some percentage of, the allowable tensile stress of the connected plates (75%, for example). The percentage is often called a *percentage factor*. The factor may vary greatly and has the effect of reducing the internal pressure capacity of a vessel. The percentage factor is also referred to as the *efficiency of the joint*. The maximum efficiency of a welded butt joint may be taken as 100%, which implies a full-strength weld as well as proper and comprehensive inspection during fabrication. Using a radiographic inspection process, a 100% joint efficiency may be assumed if the full continuous welded butt joint is X-rayed. With a partial (or spot) X ray, a joint efficiency of 85% should be used. If the joint is not X-rayed, a joint efficiency of 70% is recommended.

EXAMPLE 20.6 A cylindrical steel gas storage tank has an inside diameter of 18 in. and a wall thickness of $\frac{3}{8}$ in. All joints are full-strength groove-welded butt joints. Assume a joint efficiency (percentage factor) of 70%. Calculate the safe internal gage pressure if the allowable tensile stress for the steel is 12,000 psi. Check the applicability of the thin-wall theory.

Solution First, check the applicability of the thin-wall theory:

$$0.1(18/2) = 0.9 \text{ in.}$$

$$0.375 \text{ in.} < 0.9 \text{ in.} \qquad \textbf{(O.K.)}$$

Since the circumferential stress developed will be twice the longitudinal stress developed, the safe (allowable) internal pressure must be based on Equation (20.1) modified by the joint efficiency (J.E.), where

$$\sigma_{t_c} = \frac{pD_i}{2t(\text{J.E.})}$$

Rewrite and solve for p:

$$p = \frac{2t(\text{J.E.})\sigma_{t_{c(\text{all})}}}{D_i} = \frac{2(0.375)(0.70)(12{,}000)}{18} = 350 \text{ psi}$$

Note that, in effect, the joint efficiency reduces the allowable tensile stress and therefore the allowable internal gage pressure.

EXAMPLE 20.7 A spherical tank used for gas storage is 12 m in diameter and is subjected to an internal gage pressure of 0.400 MPa. Determine and select the required thickness of the tank wall if it is constructed of steel plate with an allowable tensile stress of 138 MPa. Assume butt-welded joints with 85% efficiency. Check the applicability of the thin-wall theory.

[1] American Society of Mechanical Engineers, *Boiler and Pressure Vessel Code* (New York: ASME, 2001).

Solution The tensile stress developed in the welded joints of a spherical pressure vessel can be found from Equation (20.3), where

$$\sigma_{t_L} = \frac{pD_i}{4t}$$

Rewrite, introducing the 85% joint efficiency, and solve for t:

$$\text{required } t = \frac{pD_i}{4\sigma_{t_{L(\text{all})}}(\text{J.E.})} = \frac{0.400(12 \times 10^3)}{4(138)(0.85)} = 10.23 \text{ mm}$$

Use an 11 mm-thick plate.

The thin-wall theory applies if $t < 0.1 \times (\text{radius})$:

$$0.1\left(\frac{12 \times 10^3}{2}\right) = 600 \text{ mm}$$

$$11 \text{ mm} < 600 \text{ mm}$$ **(O.K.)**

In liquid-filled tanks and pipes that are open to the air at the top, the internal pressure at any point is equal to the weight of a column of the liquid of the same height as the vertical distance from the point to the surface of the liquid. This vertical distance is called the *head*. If the head, in feet, is designated h, the pressure in pounds per square inch for water, which weighs 62.4 pounds per cubic foot (9800 N/m^3), is equal to

$$p = \frac{62.4(h)}{144} = 0.433(h)$$

Note that h is in feet and p is in pounds per square inch.

EXAMPLE 20.8 Calculate and select the required wall thickness of the bottom section of a 15-ft-diameter cylindrical steel standpipe subjected to an 80-ft head of water. Assume that the top of the standpipe is open. The allowable tensile stress for the steel plate wall of the standpipe is 12,600 psi. The plates are connected by a full-strength butt weld of 100% efficiency. Add approximately $\frac{1}{8}$ in. to the computed wall thickness to compensate for corrosion. Localized stresses at the junction of the vertical walls and the bottom may be neglected.

Solution Since the contained liquid is water, the pressure p in any direction at the base of the standpipe is calculated as

$$p = 0.433(h) = 0.433(80) = 34.6 \text{ psi}$$

The pressure is equal in all directions and it acts radially against the vertical standpipe wall, as shown in Figure 20.6. It varies from zero at the top of the standpipe to a maximum at the bottom. Since the top of the standpipe is open, there is no longitudinal stress developed. The required wall thickness is based on an allowable circumferential tensile stress and is obtained from Equation (20.2):

$$\text{required } t = \frac{pD_i}{2\sigma_{t_{c(\text{all})}}}$$

The point that must be considered for design is the bottom of the standpipe where the radial pressure is at a maximum. Now substitute:

$$\text{required } t = \frac{pD_i}{2\sigma_{t_{c(\text{all})}}} = \frac{34.6(15)(12)}{2(12{,}600)} = 0.247 \text{ in.}$$

Adding the corrosion allowance of $\frac{1}{8}$ in. yields

$$\text{required } t = 0.247 + 0.125 = 0.372 \text{ in.}$$

Use a $\frac{3}{8}$-in.-thick plate.

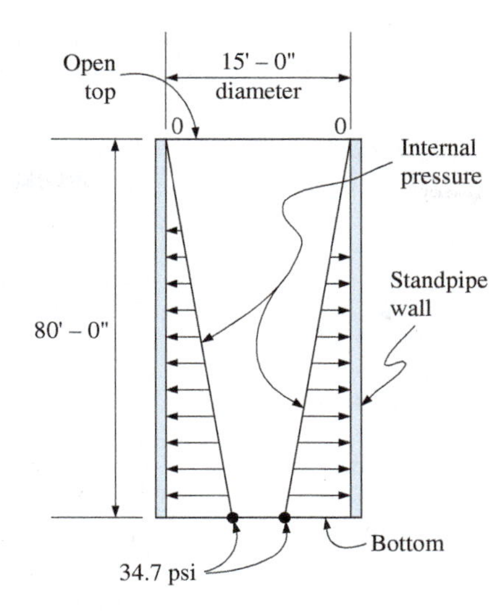

FIGURE 20.6 Section through standpipe.

20.4 DESIGN AND FABRICATION CONSIDERATIONS

The basic expressions developed in Section 20.2 are based on the assumptions that there is continuous elastic behavior throughout the pressure vessel and that the tensile stresses developed are uniformly distributed over the wall cross section. Discontinuities that cause stress concentrations may invalidate these assumptions.

The stress concentrations may be of little significance if the vessel is subjected to a steady internal pressure and is fabricated using a ductile material such as mild steel. The steel should be capable of yielding at highly stressed locations, thereby allowing a transfer of stress from overstressed areas to adjacent understressed ones. When the internal pressure is repetitive, localized stresses become significant because of fatigue, even though the material is ductile and has a large measure of static reserve strength. For this condition, the average stress formulas of Section 20.2 must be modified by a stress concentration factor to obtain a realistic maximum stress value.

Stress concentrations cannot be avoided in pressure vessels. All vessels must have supports and openings for attachments as necessary operating features. These stress concentrations invariably result in geometric discontinuities and cause localized stresses that must be considered in vessel design.

The majority of pressure vessels are made from various parts that have been fabricated, by welding, into shapes such as cylinders and hemispheres to form the base vessel. To this base vessel are also attached, by welding, the necessary appurtenances such as pipes, access openings, and support attachments. Only those closures that must be frequently removed for service or maintenance are attached by bolts, studs, or other mechanical closure devices. This practice contributes to a leakproof vessel since the number of mechanical joints is kept to a minimum. In addition, the welding process, assuming acceptable quality control for the welding procedures, will contribute toward eliminating or controlling the effect of stress concentrations. Actually, almost all failures of pressure vessels occur as a result of fatigue in the areas of high-stress concentrations. Therefore, the elimination or reduction of these localized stresses provides a partial solution to safe pressure vessel construction.

SUMMARY BY SECTION NUMBER

20.1 Pressure vessels (in this text) are leakproof containers with no consideration of shape discontinuities or structural discontinuities.

20.2 Thin-walled pressure vessels are generally cylindrical or spherical. They are made up of curved plates fastened together by continuous longitudinal and/or circumferential welded joints. Pressure vessels generally are subjected to an internal gas or liquid pressure, thereby developing tensile stresses in the walls of the vessel. The circumferential tensile stress (developed on a longitudinal joint) is

$$\sigma_{t_c} = \frac{pD_i}{2t} \qquad \textbf{(20.1)}$$

The longitudinal tensile stress (developed on a circumferential joint) is

$$\sigma_{t_L} = \frac{pD_i}{4t} \qquad \textbf{(20.3)}$$

20.3 The most common type of joint used for pressure vessels is a continuous, full-penetration, welded butt joint. In this type of a joint, the edges of the plates are carefully shaped prior to being groove welded.

20.4 Localized stresses cannot be eliminated in pressure vessels. Almost all failures occur at such locations. Vessel design (including material selection) and fabrication must be such as to minimize the effect of stress concentrations.

PROBLEMS

For each of the following problems, check the applicability of the thin-wall theory. Unless otherwise noted, the stated diameter represents an inside diameter.

Section 20.2 Stresses in Thin-Walled Pressure Vessels

20.1 A welded water pipe has a diameter of 9 ft and a wall of steel plate $\frac{13}{16}$ in. thick. After fabrication, this pipe was tested under an internal pressure of 225 psi. Calculate the circumferential stress developed in the walls of the pipe.

20.2 Calculate the tensile stresses developed (circumferential and longitudinal) in the walls of a cylindrical boiler 2 m in diameter with a wall thickness of 13 mm. The boiler is subjected to an internal gage pressure of 1 MPa.

20.3 Calculate the internal water pressure that will burst a 16-in.-diameter cast-iron water pipe if the wall thickness is $\frac{5}{8}$ in. Use an ultimate tensile strength of 60,000 psi for the pipe.

20.4 Calculate the wall thickness required for a 1.25-m-diameter cylindrical steel tank containing gas at an internal gage pressure of 3.5 MPa. The allowable tensile stress for the steel is 103 MPa.

20.5 A spherical gas container, 52 ft in diameter, is to hold gas at a pressure of 50 psi. Calculate the thickness of the steel wall required. The allowable tensile stress is 20,000 psi.

20.6 Calculate the tensile stresses (circumferential and longitudinal) developed in the walls of a cylindrical pressure vessel. The inside diameter is 18 in. The wall thickness is $\frac{1}{4}$ in. The vessel is subjected to an internal gage pressure of 300 psi and a simultaneous external axial tensile load of 50,000 lb.

Section 20.3 Joints in Thin-Walled Pressure Vessels

20.7 A gas is to be stored in a spherical steel tank 6 m in diameter. The wall thickness is 23 mm. Assume all joints to be groove-welded butt joints with a 90% joint efficiency. Calculate the maximum safe internal gage pressure if the allowable tensile stress for the steel is 100 MPa.

20.8 The longitudinal joint in an 8-ft-diameter closed cylindrical steel boiler is butt welded with a joint efficiency of 85%. The boiler plate is $\frac{3}{4}$ in. thick with an allowable tensile stress of 11,000 psi. Calculate the maximum safe internal pressure.

20.9 A spherical steel tank, 15 m in diameter, is to hold gas at a pressure of 275 kPa. The joints are butt welded with an efficiency of 75%. The allowable tensile stress for the steel plate is 125 MPa. Calculate the required thickness of the walls.

20.10 A closed cylindrical tank for an air compressor is 24 in. in diameter and is subjected to an internal pressure of 450 psi. The tank walls are of steel and have an allowable tensile stress of 11,000 psi. Joint efficiency is 100%. Calculate the required wall thickness.

20.11 Calculate the wall thickness required for a 36-in.-diameter steel penstock with an allowable tensile stress of 11,000 psi. The penstock, which is a pipe for conveying water to a hydroelectric turbine, operates under a head of 400 ft. Assume welded joints of 70% efficiency.

Computer Problems

For the following computer problems, any appropriate software may be used. Input prompts should fully explain what is required of the user (the program should be user-friendly). The resulting output should be well labeled and should be self-explanatory.

20.12 Write a program that will compute the tensile stresses in the wall of a thin-walled pressure vessel. User input should be type of vessel (cylindrical or spherical), internal diameter, wall thickness, and internal pressure. The program should also verify that the pressure vessel is thin walled. The program can be used to check some of the preceding problems.

20.13 Rework the program of Problem 20.12 giving the user the option of specifying the application of an external axial tensile load.

20.14 Viking Vessel Company manufacturers cylindrical gas storage tanks. The company foresees a market for nominal 100-ft^3 tanks of varying proportions. The tanks are to be rated at 60 psi. Viking uses steel with an allowable tensile stress of 15,000 psi. Welded joints are fully inspected, and a 100% joint efficiency may be assumed. Write a program that will generate a table of required wall thicknesses for tanks that have length-to-diameter ratios ranging (approximately) from 1.0 to 4.0. Diameters are to be in 2-in. increments. Lengths should be rounded up to the next 2-in. increment so that the volume of the tank is 100 ft^3, minimum. The output should be a list showing diameter, length, volume, and required wall thickness. (The thickness may be rounded up to 0.01 in.)

Supplemental Problems

20.15 Calculate the circumferential and longitudinal tensile stresses developed in the walls of a cylindrical steel plate boiler. The diameter of the boiler is 8 ft and the wall thickness is 1 in. The boiler is subjected to an internal gage pressure of 230 psi.

20.16 A welded aluminum alloy cylindrical pressure vessel has a diameter of 8 in. and a wall thickness of $\frac{3}{16}$ in. The ultimate tensile strength of the alloy is 65,000 psi. Calculate the internal pressure that will cause rupture.

20.17 Compute the stress developed in the walls of a spherical steel tank having an outside diameter of 500 mm and an inside diameter of 490 mm. The tank is subjected to an internal gage pressure of 14.2 MPa.

20.18 A spherical tank is to hold 400 ft^3 of gas at a pressure of 150 psi. The steel to be used has an ultimate tensile strength of 65 ksi and a yield strength of 42 ksi. Assume a 70% joint efficiency. Determine the required wall thickness.

20.19 A closed cylindrical air-compressor tank is 600 mm in diameter and is subjected to an internal pressure of 5 MPa. The tank walls are of steel, which has an allowable tensile stress of 110 MPa. Calculate the required wall thickness.

20.20 A cast-iron pipe, 3 ft in diameter, is subjected to an internal gage pressure of 100 psi. Calculate the wall thickness required if the allowable tensile stress is 11,000 psi.

20.21 A steel pipe located in a hydroelectric power plant is 24 in. in diameter and has a wall thickness of $\frac{3}{8}$ in. The system operates under a maximum head of 750 ft. Assuming an ultimate tensile strength of 55,000 psi, calculate the factor of safety for the pipe.

20.22 Calculate the tensile stresses (circumferential and longitudinal) developed in the walls of a cylindrical pressure vessel. The inside diameter is 0.50 m. The wall thickness is 7 mm. The vessel is subjected to an internal gage pressure of 2.50 MPa and a simultaneous external axial tensile load of 200 kN.

20.23 A 5-ft-diameter steel pipe is made of $\frac{3}{8}$-in. plate with groove-welded longitudinal joints. Assuming an allowable tensile strength of 13,000 psi and a joint efficiency of 100%, compute the safe head of water on the pipe.

20.24 Rework Problem 20.23, changing the joint efficiency to 80%.

20.25 A piece of 10-in.-diameter pipe of 0.10-in. wall thickness was closed off at the ends. This assembly was placed in a testing machine and subjected simultaneously to an axial tensile load P and an internal gage pressure of 240 psi. If the longitudinal tensile stress in the pipe walls is not to exceed 12,000 psi, calculate the maximum value of the applied load P.

20.26 A vertical cylindrical steel tank, open at the top, is 4 m in diameter and 20 m high. The tank is filled with water. Determine the required wall thickness at the bottom section of the tank if the allowable tensile stress is 125 MPa. The plates are connected by butt welds of 70% efficiency. Add 3 mm to the required wall thickness to compensate for corrosion over the intended life of the structure.

CHAPTER TWENTY ONE

STATICALLY INDETERMINATE BEAMS

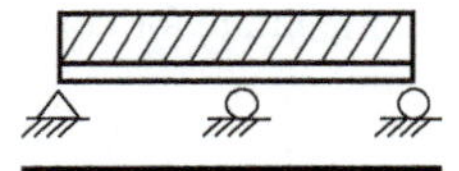

LEARNING OBJECTIVES

Upon completion of this chapter, readers will be able to:

- Identify statically indeterminate beams in structural systems.
- Identify restrained beams such as a propped cantilever beam or fixed beam.
- Use the method of superposition to analyze retrained beams.
- Identify continuous beams in a structural system.
- Use the theorem of three moments to analyze continuous beams.

21.1 INTRODUCTION

The beams considered thus far have been statically determinate; that is, the equations of static equilibrium were sufficient to compute the external reactions. This permitted us to analyze and design the beams based on stresses and deflections. The types of beams introduced in Section 13.1 and categorized as statically determinate were the simple, cantilever, and overhanging beams (see Figure 13.2). Their reactions can be determined by using the three basic laws of equilibrium: $\Sigma F_v = 0$, $\Sigma F_H = 0$, and $\Sigma M = 0$.

In statically indeterminate beams, there are more unknown reactions than there are equations of equilibrium. In this chapter, three types of statically indeterminate beams are considered. The common designations for these types of beams are *fixed beam, propped cantilever beam*, and *continuous beam*. These beams are shown in Figure 13.2. Since their reactions cannot be determined by the three laws of equilibrium alone, additional equations based on deflection or end rotation of the beam must be introduced.

Although statically determinate beams are more common than statically indeterminate beams, statically indeterminate beams frequently provide advantages in economy, function, or aesthetics.

In this chapter, only horizontal beams are considered. In addition, for any given beam, cross-sectional properties (area and moment of inertia) and material properties (modulus of elasticity) will be assumed to be constant.

21.2 RESTRAINED BEAMS

Restrained beams include the propped cantilever beam and the fixed beam. The propped cantilever is fixed at one end and simply supported at either the other end or at a point near the other end. The fixed beam, as the name implies, is fixed at both ends. Both types are shown in Figure 21.1.

The upper diagram for each beam shows the conventional representation of the beam including the shape of the elastic curve. The lower diagram for each beam shows the beam as a free body with applied loads and resisting shears and moments that, in effect, hold the free body in equilibrium.

For purposes of analysis, the important fact about a fixed end is that the tangent to the elastic curve at that point is horizontal (the slope of the tangent line is zero). The fixed end, or point of restraint, is considered to be at the face of the wall. This fact assumes that the restraint is furnished by anchoring or embedding the end of the beam in a wall (other means of restraint may be furnished). Therefore, the beam span extends to the face of the wall (or walls, if the beam is fixed at both ends).

With respect to the propped cantilever, the free-body diagram shows that in addition to the vertical forces R_A and V_B the beam is acted on by a moment of unknown magnitude at B. Note that this constitutes three unknowns. Only two equations of static equilibrium, however, exist for the determination of the forces acting on the beam (there are no horizontal forces). The two equations are not sufficient

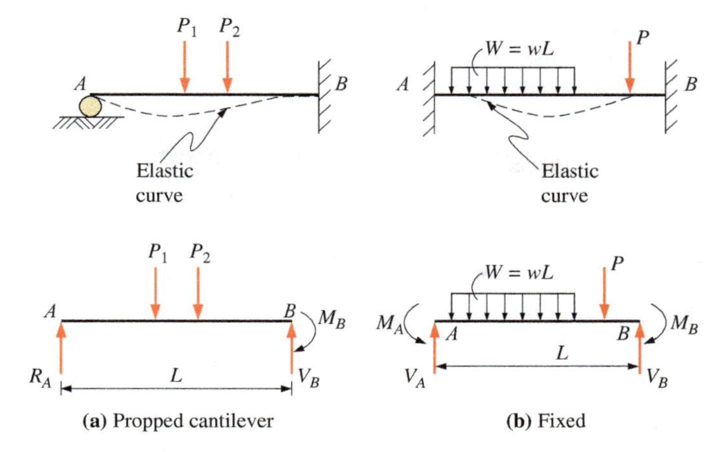

FIGURE 21.1 Restrained beams.

to determine the three unknowns. In fact, there are many combinations of values of M_B, R_A, and V_B that will satisfy the conditions of equilibrium. Therefore, some condition in addition to the two furnished is required to establish which combination of values is the correct one. Usually, for the propped cantilever, this additional condition is that the deflection at point A is zero.

When a beam is fixed at both ends, as shown in Figure 21.1b, there are four unknown reactive elements. As with the propped cantilever, the two equations of static equilibrium are not sufficient to determine the four unknowns. Therefore, two conditions in addition to the two furnished by the equilibrium equations must be introduced. The two conditions are that the slopes of the ends of the beam must remain unchanged. Consider the restrained fixed beam to be equivalent to a simple beam acted on by the given loading as well as end moments M_A and M_B. The end moments must be sufficient to rotate the ends of the beam until the slopes of the tangent at the ends correspond to the slopes at the ends of the restrained beam.

21.3 PROPPED CANTILEVER BEAMS

There are several methods of analysis for the propped cantilever beam. The method of superposition offers the simplest solution by using deflection equations for cantilever beams as furnished in Appendix H.

The method of superposition, as used in this application, is based on the principle that the deflection at any point in the propped cantilever beam is the algebraic sum of two deflections:

1. The deflection at that point due to the given loads with the reaction R_A from the simple support removed (this, in effect, is the deflection of the cantilever)
2. The deflection at that point in the same cantilever beam (with the applied loads removed) caused by the reaction R_A that was previously removed

Generally, the desired final deflection at the simple support of the loaded propped cantilever beam is zero. Thus, each support will be on the same level and the reaction R_A has such a value that it produces an upward deflection at the support equal to the downward deflection at the same point due to the given loads.

This relationship can be expressed mathematically with the solution yielding a value for R_A. Figure 21.2a depicts a propped cantilever beam with both supports on the same level. If the support at A is removed, the beam becomes a cantilever with a maximum downward deflection at the free end, as shown in Figure 21.2b.

From Appendix H, the maximum deflection at the free end is calculated from

$$\Delta_1 = \frac{wL^4}{8EI} = \frac{WL^3}{8EI}$$

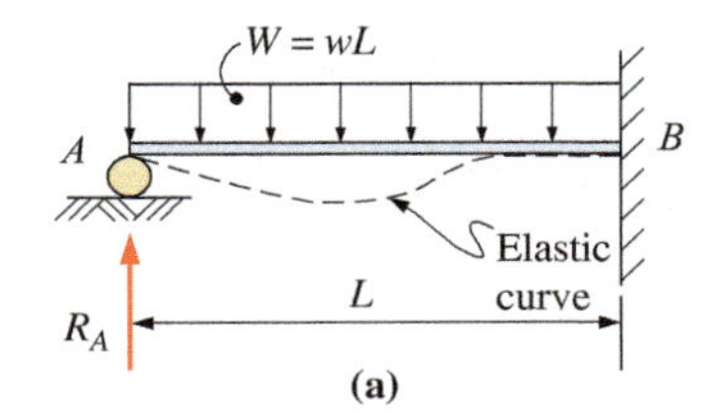

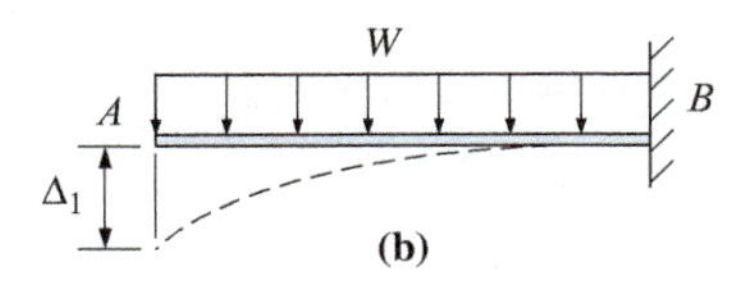

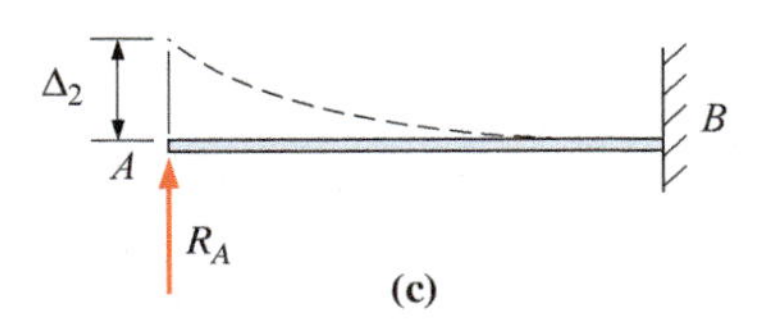

FIGURE 21.2 Propped cantilever beam.

Since it is generally the case that zero deflection is desired at the propped end, the reaction R_A must be of such a magnitude so as to eliminate Δ_1. Since R_A occurs at the end of the beam, as shown in Figure 21.2c, and the uniformly distributed load is removed, the upward deflection due to R_A is calculated from

$$\Delta_2 = \frac{R_A L^3}{3EI}$$

Since the final deflection equals zero, Δ_1 must equal Δ_2. Substitution then yields

$$\frac{WL^3}{8EI} = \frac{R_A L^3}{3EI}$$

from which

$$R_A = \frac{3W}{8}$$

Note that no sign convention has been used.

Since one unknown has now been computed, the other two unknowns can be determined using the two applicable laws of static equilibrium ($\Sigma M = 0$ and $\Sigma F_y = 0$).

EXAMPLE 21.1 Select the most economical (lightest) structural steel wide-flange section (W shape) for the propped cantilever beam that supports the loads shown in Figure 21.3. Consider moment and shear only. Assume A992 steel ($F_y = 50$ ksi). Neglect the weight of the beam. The vertical deflection at support A is zero.

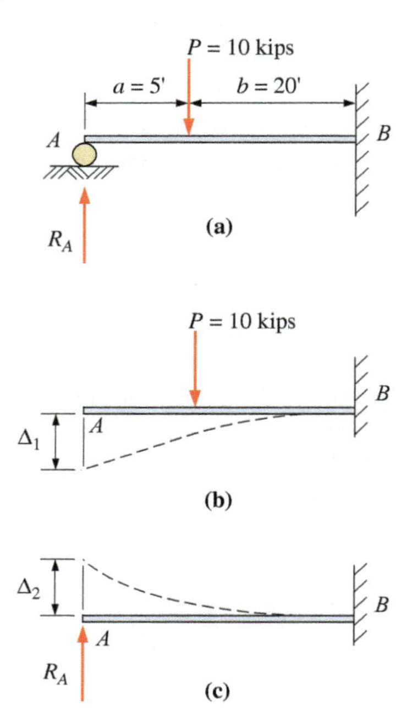

FIGURE 21.3 Propped cantilever beam for Example 21.1.

Solution Removing the support at A as shown in Figure 21.3b results in a cantilever beam loaded with a concentrated load P. From Appendix H,

$$\Delta_1 = \frac{Pb^2}{6EI}(3L - b) = \frac{(10\text{ k})(20\text{ ft})^2(12\text{ in./ft})^2}{6EI}[3(25\text{ ft}) - (20\text{ ft})](12\text{ in./ft})$$

$$= \frac{63{,}360{,}000\text{ k-in.}^3}{EI}$$

With the loading removed and reaction R_A applied as shown in Figure 21.3c,

$$\Delta_2 = \frac{R_A L^3}{3EI} = \frac{R_A(25\text{ ft})^3(12\text{ in./ft})^3}{3EI} = \frac{R_A(9{,}000{,}000\text{ in.}^3)}{EI}$$

Since the final deflection must equal zero, Δ_1 must equal Δ_2:

$$\Delta_1 = \Delta_2$$

$$\frac{63{,}360{,}000}{EI} = \frac{R_A(9{,}000{,}000)}{EI}$$

$$R_A = 7.04 \text{ k}$$

Solve for the shear V_B at the fixed support and refer to Figure 21.4a:

$$\Sigma F_y = 0$$

$$R_A + V_B - 10 = 0$$

$$7.04 + V_B - 10 = 0$$

from which

$$V_B = 2.96 \text{ k}$$

The complete shear and moment diagrams can then be drawn (see Figure 21.4).

Next, select the shape based on the required plastic section modulus:

$$\text{required } Z_x = \frac{1.67\ M}{F_y} = \frac{1.67(35.2 \text{ k-ft})(12 \text{ in./ft})}{50 \text{ ksi}} = 14.11 \text{ in.}^3$$

From Table I-1, select a W12 × 16 ($Z_x = 20.1$ in.^{3}). Next, check shear. From Appendix A, $d = 11.99$ in. and $t_w = 0.220$ in. The shear capacity is calculated from

$$0.4F_y d t_w = 0.4(50 \text{ ksi})(11.99 \text{ in.})(0.220 \text{ in.}) = 52.8 \text{ k}$$

$$7.04 \text{ k} < 52.8 \text{ k}$$ **(O.K.)**

The W12 × 16 is satisfactory for both moment and shear.

Use a W12 × 16.

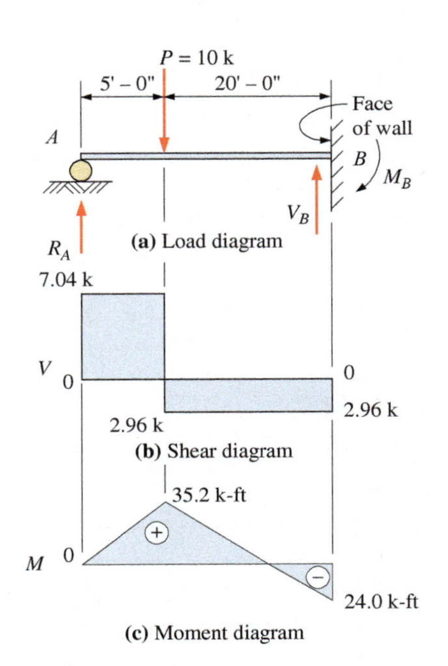

FIGURE 21.4 Shear and moment diagrams.

Note that both supports of the beam in Example 21.1 are on the same level. If the simple support (left end) is raised above the level of the tangent to the elastic curve at the fixed end, the reaction R_A will increase. If the simple support settles below the level of the tangent line at the fixed end, the reaction will decrease. The reaction becomes zero when the simple support is lowered until the beam carries the load as a cantilever (if it is able to do so). The solution of such problems is similar to that when both supports are on the same level.

For instance, in Example 21.1, assume that the support at A settles 1.0 in. The equation then used to determine R_A would be written as

$$\Delta_1 - \Delta_2 = 1.0 \text{ in.}$$

where Δ_1 = downward deflection due to load
Δ_2 = upward deflection due to the reaction R_A

21.4 FIXED BEAMS

Fixed beams are generally designated *fixed-end beams*. It is also generally assumed that the ends have complete fixity, which implies that the tangent to the elastic curve has a slope of zero at each support. It is sometimes convenient to think of a fixed beam as a beam simply supported at each end and acted on not only by a load, or a system of loads, but also by moments applied to the beam at its ends. If these moments are of the proper sense and magnitude, they will rotate the ends of the beam until the tangents to the elastic curve are horizontal at those points. The beam then meets all the requirements of a beam "fixed at both ends."

The method of superposition can also be used for this type beam, but rather than use deflection equations to supplement the static equilibrium equations, slope equations for the elastic curve will be used to evaluate two of the four unknown reactive elements. Tables 21.1 and 21.2

TABLE 21.1 Slopes of beams on two supports

Case	Simple Beam and Loading	Slope at Supports
1	P; $L/2$, $L/2$; A, B; θ_A, θ_B	$\theta_A = \theta_B = \dfrac{PL^2}{16EI}$
2	P; a, b; A, B; θ_A, θ_B; L	$\theta_A = \dfrac{Pb\,(L^2 - b^2)}{6LEI}$ $\theta_B = \dfrac{Pab\,(2L - b)}{6LEI}$
3	w; A, B; θ_A, θ_B; L	$\theta_A = \theta_B = \dfrac{wL^3}{24EI}$
4	M_A; A, B; θ_{A_A}, θ_{B_A}; L	$\theta_{A_A} = \dfrac{M_AL}{3EI}$ $\theta_{B_A} = \dfrac{M_AL}{6EI}$
5	M_B; A, B; θ_{A_B}, θ_{B_B}; L	$\theta_{A_B} = \dfrac{M_BL}{6EI}$ $\theta_{B_B} = \dfrac{M_BL}{3EI}$
6	W; A, B; θ_A, θ_B; L	$\theta_A = \theta_B = \dfrac{5WL^2}{96EI}$

TABLE 21.2 Slopes of cantilever beams

Case	Cantilever Beam and Loading	Slope at Free End
1	P, L, θ	$\theta = \frac{PL^2}{2EI}$
2	P, a, L, θ	$\theta = \frac{Pa^2}{2EI}$
3	w, L, θ	$\theta = \frac{wL^3}{6EI}$
4	a, L, θ	$\theta = \frac{wa^3}{6EI}$
5	M, L, θ	$\theta = \frac{ML}{EI}$
6	W, L, θ	$\theta = \frac{WL^2}{12EI}$

furnish derived equations for the slope of the elastic curve at designated locations for simple and cantilever beams under various types of loads.

Figure 21.5a depicts a horizontal beam fixed at both ends with a concentrated load at any point. The weight of the beam has been neglected. Note that at fixed ends A and B there are four unknown reactive elements: resisting moments M_A and M_B and the shears V_A and V_B. Therefore, two equations, in addition to the two equations of static equilibrium, are required for determining the reactions.

Two equations can be obtained by considering the beam in Figure 21.5a to be made up of the beams in Figure 21.5b through d. Note that for the slopes at A and B to be equal to zero, it is necessary that

$$\theta_A' = \theta_{A_A} + \theta_{A_B}$$

and

$$\theta_B' = \theta_{B_A} + \theta_{B_B}$$

where θ_{A_A} is the slope at A due to a moment at A and θ_{A_B} is the slope at A due to the moment at B.

In other words, the beam in Figure 21.5a is equivalent to the superposition of the beams in Figure 21.5b through d. Using Table 21.1, the equations for the slopes can be substituted

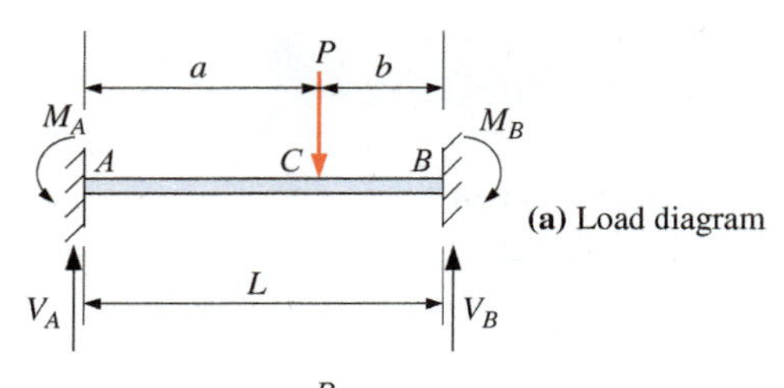

(a) Load diagram

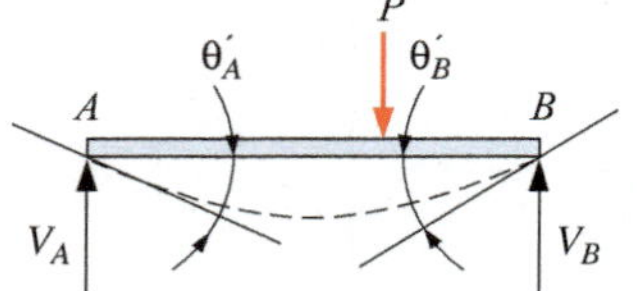

(b) Deflection diagram based on load P — simple beam

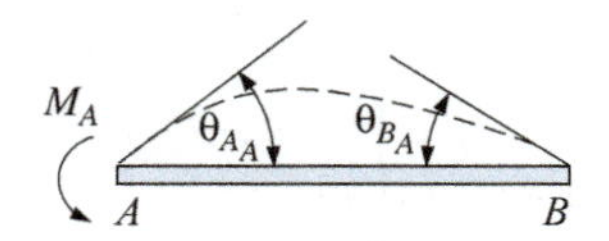

(c) Deflection diagram based on moment at A

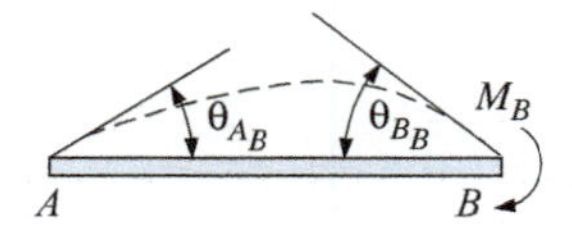

(d) Deflection diagram based on moment at B

FIGURE 21.5 Fixed beam.

in the above relationships and equations for M_A and M_B can be established as follows:

$$\theta'_A = \theta_{A_A} + \theta_{A_B}$$

$$\frac{Pb(L^2 - b^2)}{6LEI} = \frac{M_A L}{3EI} + \frac{M_B L}{6EI} \qquad (1)$$

Also,

$$\theta'_B = \theta_{B_A} + \theta_{B_B}$$

$$\frac{Pab(2L - b)}{6LEI} = \frac{M_A L}{6EI} + \frac{M_B L}{3EI} \qquad (2)$$

Solving Equation (1) and Equation (2) simultaneously yields

$$M_A = \frac{Pab^2}{L^2}$$

$$M_B = \frac{Pa^2 b}{L^2}$$

Since two unknowns are now determined, the other two unknowns can be calculated using the two applicable laws of static equilibrium ($\Sigma M = 0$ and $\Sigma F_y = 0$).

Note that no sign convention has been used with respect to upward or downward rotation.

EXAMPLE 21.2 In the construction of walls, it is necessary to provide beams over openings such as doors or windows. This case is especially true where the walls are constructed of brick or masonry units. The beams are called *lintels*, and they must support the weight of the wall and any other loads above the opening. Due to an arching action that generally develops above the opening, the loading on the lintel will be of a triangular shape. Assuming complete fixity at each end, design a W-shape structural steel lintel beam for the span and loading shown in Figure 21.6. Consider moment and shear. Assume A992 steel ($F_y = 50$ ksi). Neglect the weight of the beam. Due to the wall thickness, the flange width must be between 8 and 12 inches.

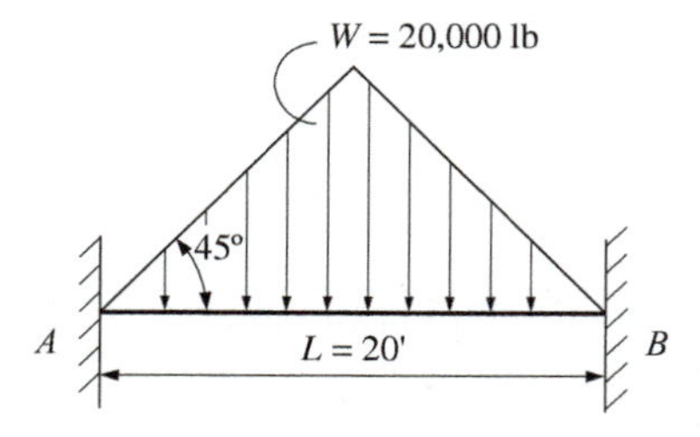

FIGURE 21.6 Fixed lintel beam.

Solution Using the method of superposition, the beam shown in Figure 21.7a is equivalent to the superposition of the beams in Figure 21.7b through d.

For the slopes at A and B to be equal to zero, it is necessary that

$$\theta'_A = \theta_{A_A} + \theta_{A_B}$$

and

$$\theta'_B = \theta_{B_A} + \theta_{B_B}$$

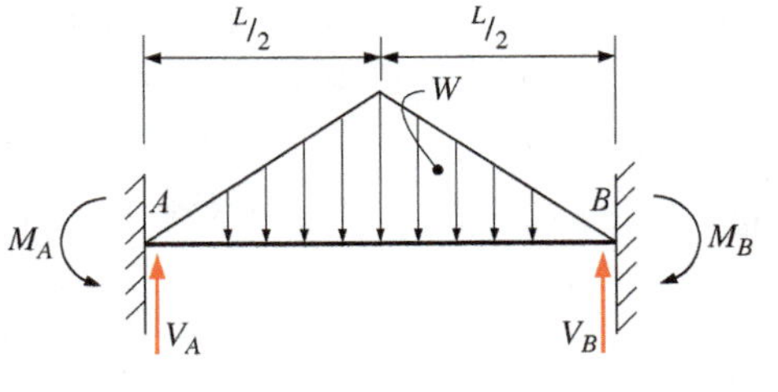

(a) Load diagram

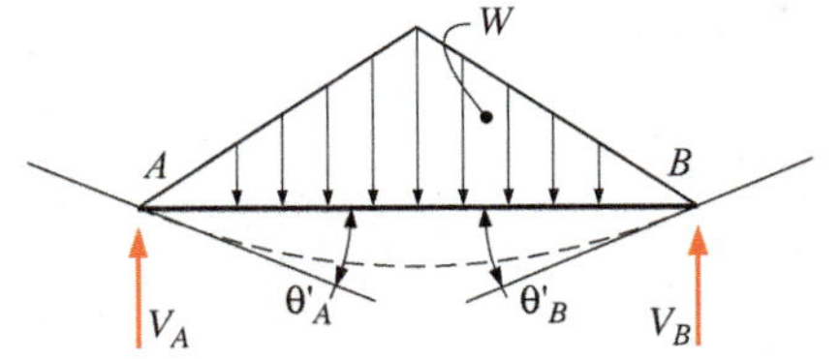

(b) Deflection diagram based on load W—simple beam

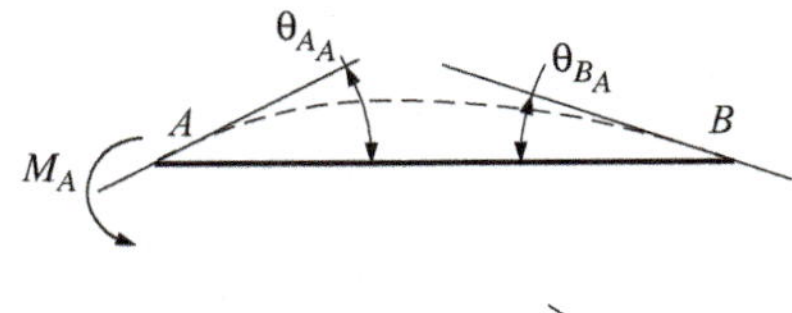

(c) Deflection diagram for end-moment at A

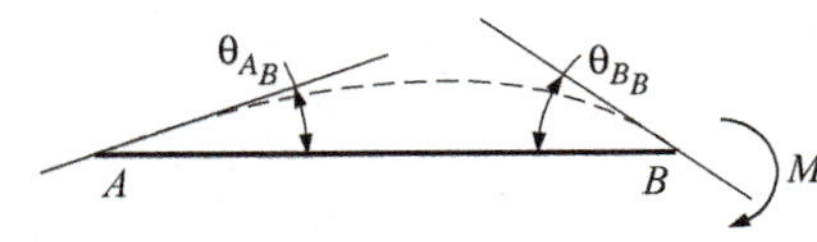

(d) Deflection diagram for end-moment at B

FIGURE 21.7 Fixed lintel beam.

Substitute in both relationships from Table 21.1:

$$\theta'_A = \theta_{A_A} + \theta_{A_B}$$

$$\frac{5WL^2}{96EI} = \frac{M_A L}{3EI} + \frac{M_B L}{6EI}$$

which can be simplified to

$$M_B = \frac{5}{16}WL - 2M_A \tag{1}$$

Also,

$$\theta'_B = \theta_{B_A} + \theta_{B_B}$$

$$\frac{5WL^2}{96EI} = \frac{M_A L}{6EI} + \frac{M_B L}{3EI}$$

which can be simplified to

$$\frac{5WL}{16} = M_A + 2M_B \tag{2}$$

Substitute Equation (1) into Equation (2):

$$\frac{5WL}{16} = M_A + 2\left(\frac{5}{16}WL - 2M_A\right)$$

$$M_A = \frac{5WL}{48}$$

Since the loading is symmetrical,

$$M_A = M_B = \frac{5WL}{48} = \frac{5(20{,}000\text{ lb})(20\text{ ft})}{48} = 41{,}670\text{ lb-ft}$$

Use the laws of static equilibrium:

$$\Sigma F_y = W - V_A - V_B = 0$$

Since the loading is symmetrical $(V_A = V_B)$,

$$W = 2V_A$$

$$V_A = \frac{20{,}000 \text{ lb}}{2} = 10{,}000 \text{ lb} = V_B$$

Find the moment at midspan using laws of static equilibrium and a free-body diagram as shown in Figure 21.8. Take moments at midspan (point O) and assume moments causing compression in the top of the beam to be positive:

$$\Sigma M_O = +V_A\left(\frac{L}{2}\right) - M_A - \frac{W}{2}\left(\frac{1}{3}\right)\left(\frac{L}{2}\right)$$

$$= +10{,}000 \text{ lb}(10 \text{ ft}) - 41{,}670 \text{ lb-ft} - \frac{20{,}000 \text{ lb}}{2}\left(\frac{1}{3}\right)\left(\frac{20 \text{ ft}}{2}\right)$$

$$= 25{,}000 \text{ lb-ft}$$

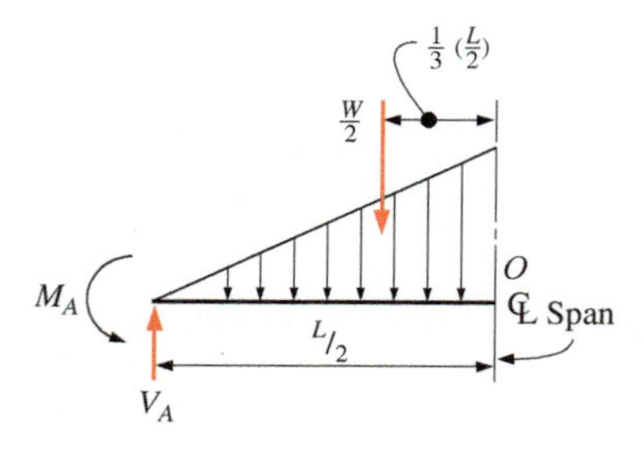

FIGURE 21.8 Free-body diagram.

Therefore, moment is maximum (41,670 lb-ft) at the fixed ends and the required plastic section modulus can be calculated:

$$\text{required } Z_x = \frac{1.67M}{F_y} = \frac{1.67(41{,}670 \text{ lb-ft})(12 \text{ in./ft})}{50{,}000 \text{ ksi}} = 16.70 \text{ in.}^3$$

From Table I.1, select a W8 × 31 (Z_x = 30.4 in.^{3}). There are lighter W shapes that will satisfy the Z_x requirements, but the desired minimum flange width is 8 in. and b_f for the W8 × 31 is 7.995 in. (say O.K.). Next, check shear. From Appendix A, d = 9.13 in. and t_w = 0.285 in. The shear capacity is calculated from

$$0.4F_y d t_w = 0.4(50{,}000 \text{ psi})(9.13 \text{ in.})(0.285 \text{ in.}) = 52{,}000 \text{ lb}$$

$$10{,}000 \text{ lb} < 52{,}000 \text{ lb} \qquad \textbf{(O.K.)}$$

The W8 × 31 is satisfactory for both moment and shear.

Use a W8 × 31.

21.5 CONTINUOUS BEAMS: SUPERPOSITION

A continuous beam is one that rests on more than two supports, as shown in Figure 21.9. This type of member is frequently used in modern structures and generally offers economy (in weight of steel required) when compared with a series of simple beams over the same spans. The effect of the beam continuity is to reduce the maximum bending moment from that of the simple beam, thereby reducing the required beam size. Continuous beams are statically indeterminate; therefore, the external reactions cannot be found by the equations of static equilibrium alone.

In general, the method of superposition used for the one-span indeterminate beams cannot be conveniently used

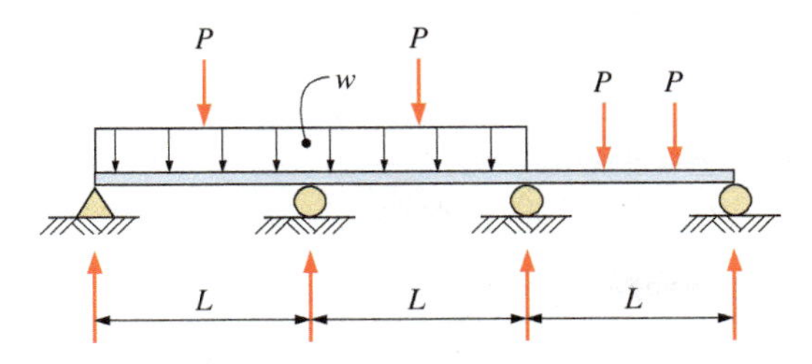

FIGURE 21.9 Continuous beam.

for continuous beams. An exception is a two-span continuous beam such as shown in Figure 21.10. The solution is similar to those of the one-span indeterminate beams.

With reference to Figure 21.10, if the support at C is removed, beam AB becomes a simple beam with a span length equal to the sum of the two spans and will deflect downward a distance Δ_1 due to the applied uniformly distributed load w.

A support does exist at C, however, and the task becomes one of determining the magnitude of the support reaction R_C that will deflect the beam AB upward the same distance that it deflected downward under the load w. Note in Figure 21.10c that the uniformly distributed load is removed when applying R_C. Since the final deflection must be zero,

$$\Delta_1 = \Delta_2$$

$$\frac{5wL^4}{384EI} = \frac{R_C L^3}{48EI}$$

$$R_C = \frac{5}{8}wL$$

With R_C computed, the other reactions R_A and R_B can be found and subsequently shears and moments can be computed using the equations of static equilibrium. Note that the reaction R_C and the beam moments will change if the interior support is not at the same level as the two exterior supports.

This method of solution is not typical for continuous beams, but it is applicable for the case discussed.

21.6 THE THEOREM OF THREE MOMENTS

A very convenient method of finding bending moments in continuous beams is by means of a relationship that exists between the bending moments at the three supports of any two adjacent spans. This relationship is expressed as an equation and is commonly called the *theorem of three moments*. After the moments at the supports have been computed, shears, reactions, and moments at various points can be determined using the equations of static equilibrium.

In Section 21.4, we showed that a loaded beam fixed at each end is equivalent to a simple beam having the same span and load, with moments applied at each end so as to make the tangents to the elastic curve at the ends of the span horizontal. In a continuous beam, the tangents at the end of a span are generally not horizontal. Any span of a continuous beam, however, can be considered equivalent to a simple beam having the same span and load and acted on by end moments of sufficient magnitude to give the tangents at the ends of the beam the slope of the elastic curve of the continuous beam at the supports in question.

Figure 21.11 depicts a continuous beam subjected to uniformly distributed loads w_1 and w_2. An equation will be derived that will relate the given loads and spans to the

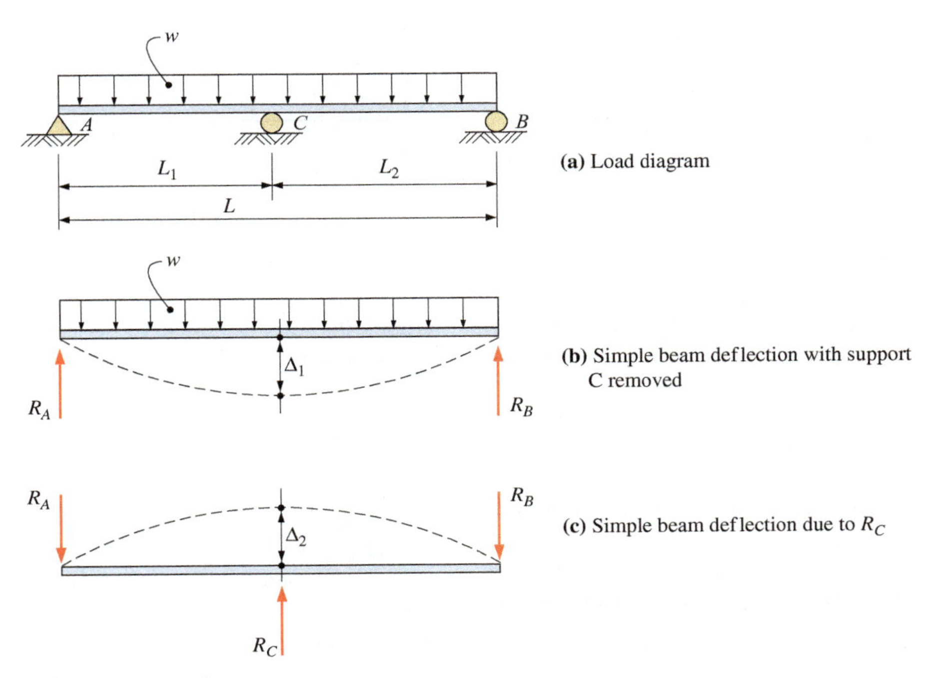

FIGURE 21.10 Two-span continuous beam.

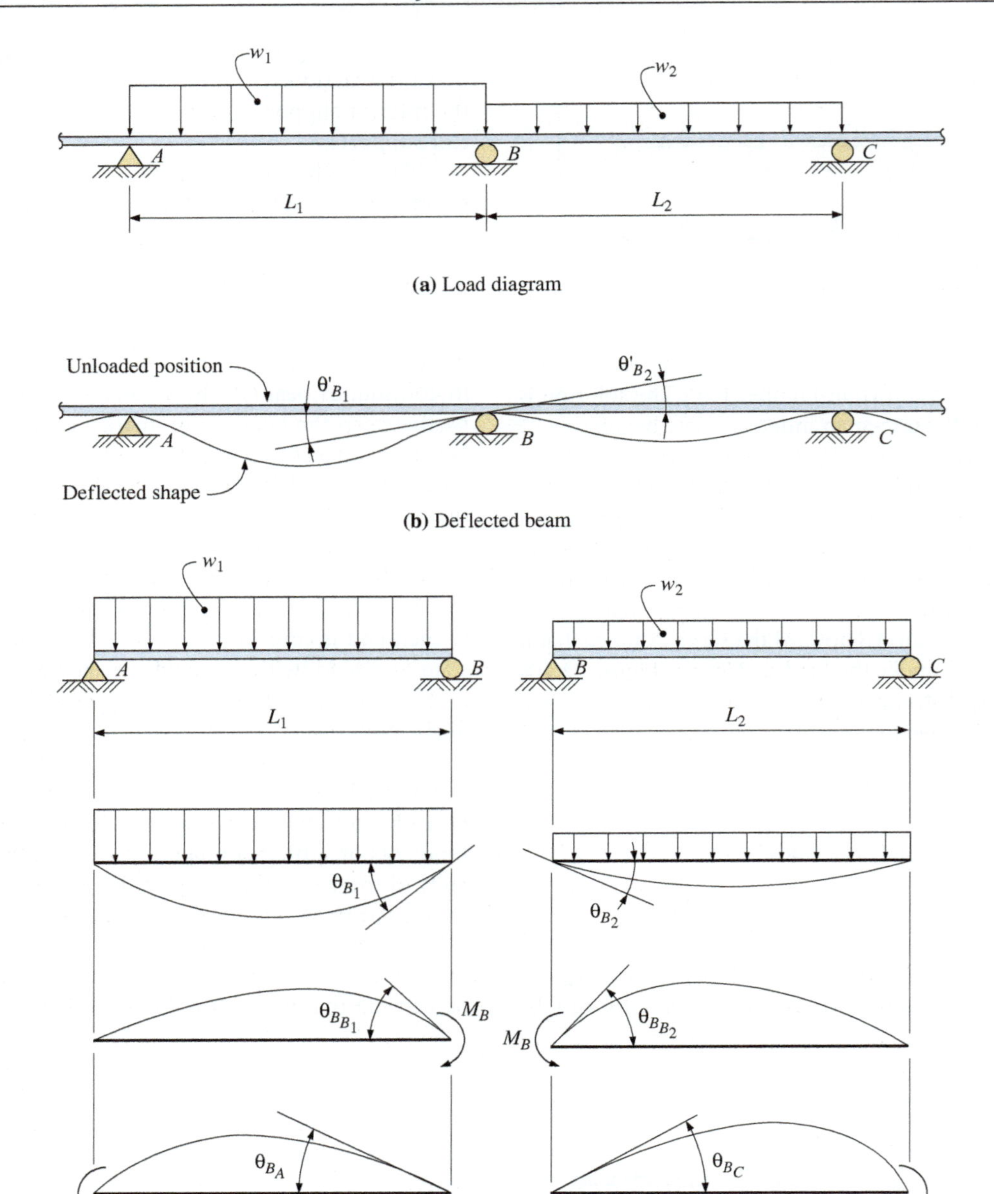

FIGURE 21.11 Continuous beam.

bending moments at points A, B, and C. The key to the derivation is that $\theta'_{B_1} = \theta'_{B_2}$. That is, the beam is continuous so there is one and only one tangent at point B.

The slope angle in each span is composed of three parts in Figure 21.11c, one part produced by the applied load and the other two parts produced by the (unknown) end moments. The slope angle in each span will be computed and then the two will be equated.

Negative moments will be assumed (tension in the top of the beam) and will produce slope components as shown. Therefore,

$$\theta'_{B_1} = \theta_{B_1} + \theta_{BB_1} + \theta_{B_A}$$
$$\theta'_{B_2} = \theta_{B_2} + \theta_{BB_2} + \theta_{B_C}$$

Slopes θ'_{B_1} and θ'_{B_2} must be of equal magnitude, but of opposite sign, so

$$\theta'_{B_1} = -\theta'_{B_2}$$

Substituting yields

$$\theta_{B_1} + \theta_{BB_1} + \theta_{B_A} = -(\theta_{B_2} + \theta_{BB_2} + \theta_{B_C})$$

$$\frac{w_1L_1^3}{24EI} + \frac{M_BL_1}{3EI} + \frac{M_AL_1}{6EI} = -\frac{w_2L_2^3}{24EI} - \frac{M_BL_2}{3EI} - \frac{M_CL_2}{6EI}$$

from which

$$M_AL_1 + 2M_B(L_1 + L_2) + M_CL_2 = -\frac{w_1L_1^3}{4} - \frac{w_2L_2^3}{4} \quad \textbf{(21.1)}$$

TABLE 21.3 Substitution terms for theorem of three moments

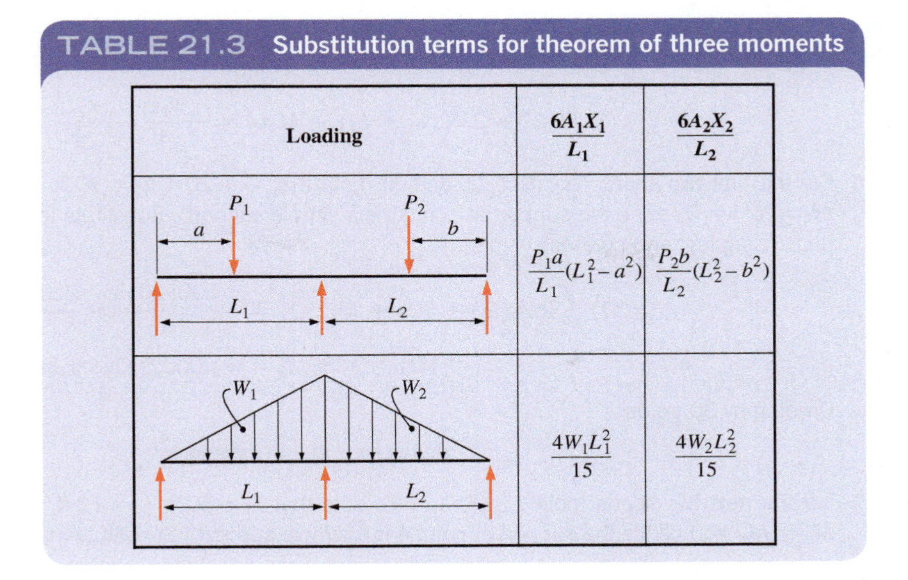

Loading	$\frac{6A_1X_1}{L_1}$	$\frac{6A_2X_2}{L_2}$
P_1, P_2, a, b, L_1, L_2	$\frac{P_1a}{L_1}(L_1^2 - a^2)$	$\frac{P_2b}{L_2}(L_2^2 - b^2)$
W_1, W_2, L_1, L_2	$\frac{4W_1L_1^2}{15}$	$\frac{4W_2L_2^2}{15}$

This equation is a special form of the theorem of three moments. A general form, which can be derived using the moment-area principles of Chapter 16, can be written as

$$M_AL_1 + 2M_B(L_1 + L_2) + M_CL_2 = -\frac{6A_1x_1}{L_1} - \frac{6A_2x_2}{L_2} \tag{21.2}$$

where the A and x terms on the right side result from the area and centroid locations of the moment diagrams. These terms vary with the type of loading. Table 21.3 shows two other types of loading and the appropriate terms for substitution in the right side of the equation. For the case of equal spans and equal uniformly distributed load on each span, the equation becomes

$$M_A + 4M_B + M_C = -\frac{wL^2}{2} \tag{21.3}$$

For a combination of uniformly distributed loads and concentrated loads, the right side of the equation becomes a combination of the terms shown. This case is illustrated in Example 21.4.

The three-moment equation is applicable to continuous beams with any number of spans since the equations give the relationship between the moments at any three successive supports. An equation is written for each two successive spans. Assuming that the ends of the continuous beam are not fixed, there will then be two fewer equations than the number of unknown moments. If the support is at the end of the beam and not fixed, the moment at the support is zero. If the beam overhangs the end support, the moment at the support is computed by static conditions, the overhanging part being used as a free body. Note that if the beam is convex upward at a support, as is the usual case, the moment is negative and should be used as such in the equations.

After the moments are computed, the first span at either end, including the overhang, if any, may be taken as a free body to determine the end reactions. The first and second span together may next be taken as a free body to determine the second reaction, and so on across the span, until finally the entire beam may be taken as a free body.

After the reactions are computed, the complete shear and moment diagrams can be drawn. The entire process is illustrated in the next three examples.

EXAMPLE 21.3 A three-span continuous beam carries the uniformly distributed loads shown in Figure 21.12. The loads include an assumed beam weight. (a) Compute the moments at each support. (b) Compute the four reactions.

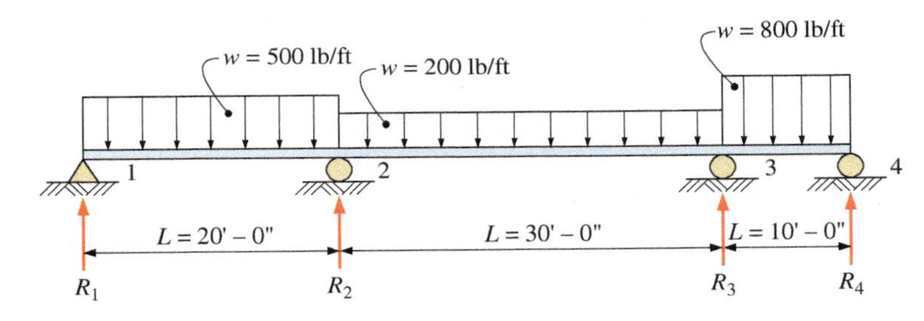

FIGURE 21.12 Beam for Example 21.3.

Solution (a) As indicated previously, the special form of the three-moment equation, Equation (21.1), for a continuous beam supporting uniformly distributed loads is

$$M_A L_1 + 2M_B(L_1 + L_2) + M_C L_2 = -\frac{w_1 L_1^3}{4} - \frac{w_2 L_2^3}{4}$$

For the first two spans (points 1, 2, and 3), note that $L_1 = 20$ ft, $L_2 = 30$ ft, $M_B = M_2$, $M_C = M_3$, and $M_A = M_1 = 0$ (since the support at point 1 is a simple support). Substitute into Equation (21.1) (with units being feet and pounds):

$$0(20) + 2M_2(20 + 30) + M_3(30) = -\frac{500(20)^3}{4} - \frac{200(30)^3}{4}$$

$$100M_2 + 30M_3 = -1{,}000{,}000 - 1{,}350{,}000$$

Dividing by 30 yields

$$3.333M_2 + M_3 = -78{,}333 \qquad \textbf{(1)}$$

For the next two spans (points 2, 3, and 4), note that $L_1 = 30$ ft, $L_2 = 10$ ft, $M_A = M_2$, $M_B = M_3$, and $M_c = M_4 = 0$ (since the support at point 4 is a simple support). Substitute into Equation (21.1):

$$M_2(30) + 2M_3(30 + 10) + M_4(140) = -\frac{200(30)^3}{4} - \frac{800(10)^3}{4}$$

$$30M_2 + 80M_3 + 0(10) = -1{,}350{,}000 - 200{,}000$$

Dividing by 80 yields

$$0.375M_2 + M_3 = -19{,}375 \qquad \textbf{(2)}$$

Solving Equation (1) and Equation (2) simultaneously for M_2 yields

$$M_2 = -19{,}930 \text{ lb-ft}$$

Substituting the value of M_2 into either Equation (1) and Equation (2) yields

$$M_3 = -11{,}900 \text{ lb-ft}$$

(b) Compute the four reactions. Note that the support moments are negative, indicating that at the support the top fibers of the beam are in tension and the bottom fibers are in compression. Also note that in the following calculations for ΣM, counterclockwise is assumed as positive.

For reaction R_1, take $\Sigma M = 0$ at point 2. Refer to Figure 21.13. So,

$$-R_1(20) + \frac{500(20)^2}{2} - 19{,}930 = 0$$

$$R_1 = 4003 \text{ lb}$$

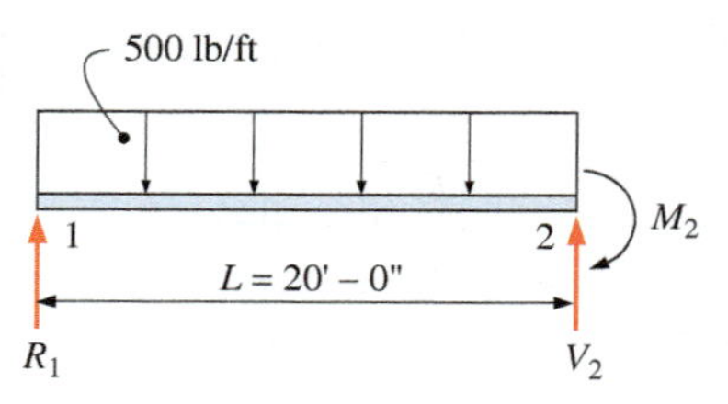

FIGURE 21.13 Free body, span 1–2.

For reaction R_2, take $\Sigma M = 0$ at point 3. Refer to Figure 21.14. So,

$$-4003(50) - R_2(30) - 11{,}900 + 500(20)(40) + 200(30)(15) = 0$$

$$R_2 = 9265 \text{ lb}$$

For reaction R_3, take $\Sigma M = 0$ at point 4. Refer to Figure 21.12. So,

$$-4003(60) - 9265(40) - R_3(10) + 500(20)(50) + 200(30)(25) + 800(10)(5) = 0$$

$$R_3 = 7922 \text{ lb}$$

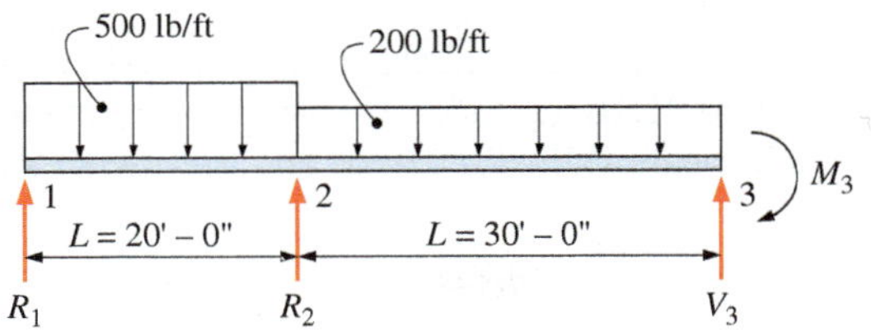

FIGURE 21.14 Free-body diagram, spans 1–2 and 2–3.

For reaction R_4, consider the entire beam as a free body and take a summation of vertical forces ($\Sigma F_y = 0$). Up is taken as positive. So,

$$+4003 + 9265 + 7922 + R_4 - 500(20) - 200(30) - 800(10) = 0$$

$$R_4 = 2810 \text{ lb}$$

EXAMPLE 21.4

For the continuous beam shown in Figure 21.15, calculate the moments at the supports. Neglect the weight of the beam.

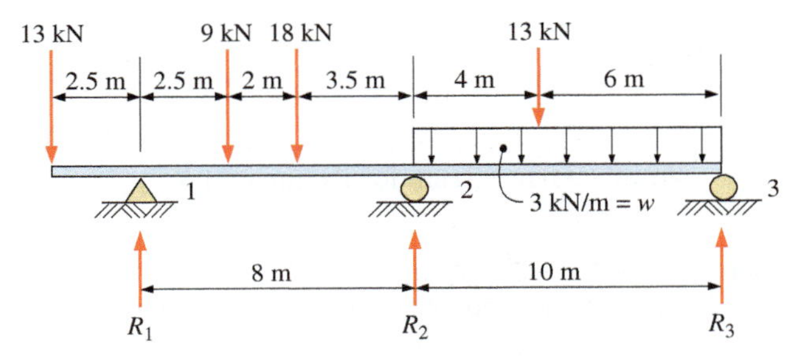

FIGURE 21.15 Beam for Example 21.4.

Solution Due to the overhang at the left end, the moment at support 1 can be computed using a statics equation:

$$M_1 = -13(2.5) = -32.5 \text{ kN}\cdot\text{m}$$

The end support at point 3 is a simple support; therefore, $M_3 = 0$. The only unknown moment is at support 2 (M_2).

A combined form of the three-moment equation is

$$M_A L_1 + 2M_B(L_1 + L_2) + M_C L_2 = -\Sigma\left[\frac{P_1 a}{L_1}(L_1^2 - a^2) - \frac{w_1 L_1^3}{4}\right] - \Sigma\left[\frac{P_2 b}{L_2}(L_2^2 - b^2) - \frac{w_2 L_2^3}{4}\right]$$

There are only two spans to consider, span 1–2 and span 2–3. Note that $L_1 = 8$ m, $L_2 = 10$ m, $M_A = M_1$, $M_B = M_2$, and $M_C = M_3 = 0$. Distances *a* and *b* are defined in Table 21.3.

Substitute in the combined form of the three-moment equation (units are kN and m):

$$-32.5(8) + 2M_2(8 + 10) + 0(10)$$
$$= -\frac{9(2.5)}{8}(8^2 - 2.5^2) - \frac{18(4.5)}{8}(8^2 - 4.5^2) - \frac{13(6)}{10}(10^2 - 6^2) - \frac{3(10)^3}{4}$$

Solving for M_2 yields

$$M_2 = -44.3 \text{ kN}\cdot\text{m}$$

If a continuous beam is fixed at both ends, the foregoing approach must be modified. The moments at the supports can be computed if at each fixed end an auxiliary span, the length of which is zero, is assumed. With this modification, the three-moment equation will furnish enough equations to compute the moments at the fixed ends and at the interior supports. The approach is illustrated in Example 21.5.

EXAMPLE 21.5 A three-span continuous beam is fixed at both ends and is supported and loaded as shown in Figure 21.16. (a) Calculate the moments at the fixed ends and at the interior supports. (b) Calculate the shears at the fixed ends and the interior support reactions.

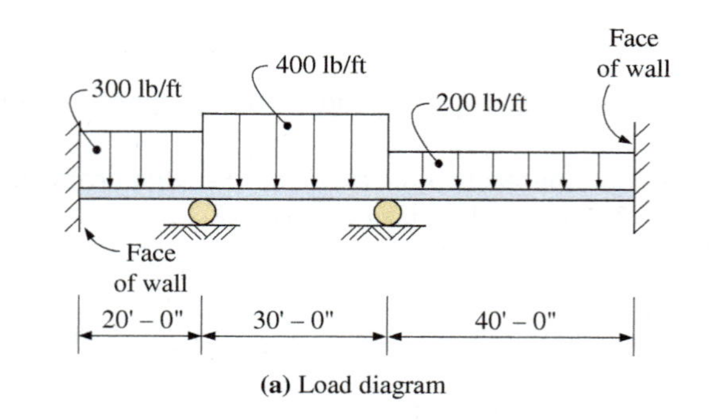

(a) Load diagram

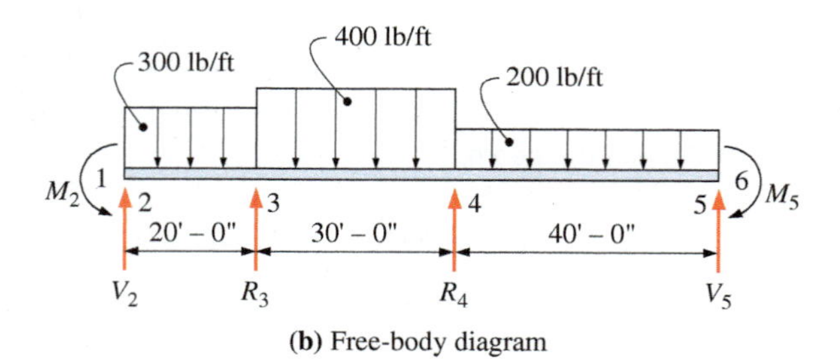

(b) Free-body diagram

FIGURE 21.16 Beam for Example 21.5.

Solution (a) At the left of point 2 and at the right of point 5, auxiliary spans of length zero are assumed. The three-moment equation can then be written for points 1, 2, and 3 and then for points 2, 3, and 4, and so on. As indicated previously, the special form, Equation (21.1), of the equation for a continuous beam supporting uniformly distributed loads is

$$M_A L_1 + 2M_B(L_1 + L_2) + M_C L_2 = -\frac{w_1 L_1^3}{4} - \frac{w_2 L_2^3}{4}$$

For points 1, 2, and 3, where $w_1 = 0$ and $L_1 = 0$,

$$M_1(0) + 2M_2(0 + 20) + M_3(20) = -0 - \frac{300(20)^3}{4}$$

$$40M_2 + 20M_3 = -600{,}000 \text{ lb-ft} \quad \textbf{(1)}$$

For points 2, 3, and 4,

$$M_2(20) + 2M_3(20 + 30) + M_4(30) = -\frac{300(20)^3}{4} - \frac{400(30)^3}{4}$$

$$20M_2 + 100M_3 + 30M_4 = -3{,}300{,}000 \text{ lb-ft} \quad \textbf{(2)}$$

For points 3, 4, and 5,

$$M_3(30) + 2M_4(30 + 40) + M_5(40) = -\frac{400(30)^3}{4} - \frac{200(40)^3}{4}$$

$$30M_3 + 140M_4 + 40M_5 = -5{,}900{,}000 \text{ lb-ft} \quad \textbf{(3)}$$

For points 4, 5, and 6,

$$M_4(40) + 2M_5(40 + 0) + M_6(0) = -\frac{200(40)^3}{4}$$

$$40M_4 + 80M_5 + 0 = -3{,}200{,}000 \text{ lb-ft} \quad \textbf{(4)}$$

Solving Equations 1, 2, 3, and 4 simultaneously yields

$$M_2 = -3333 \text{ lb-ft}$$
$$M_3 = -23{,}333 \text{ lb-ft}$$
$$M_4 = -30{,}000 \text{ lb-ft}$$
$$M_5 = -25{,}000 \text{ lb-ft}$$

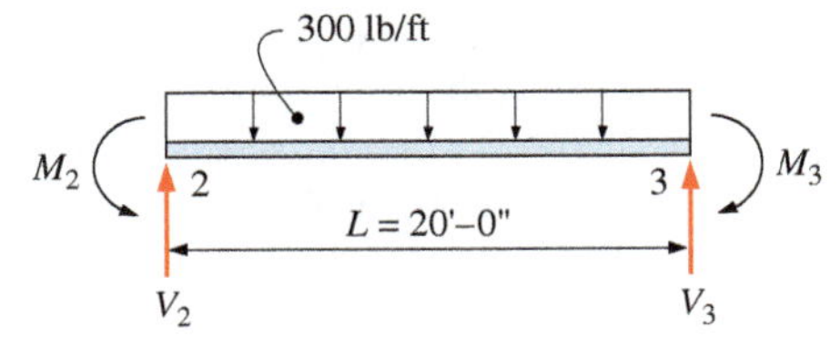

FIGURE 21.17 Free-body diagram, span 2–3.

(b) Solve for shear at points 2 and 5 and reactions at points 3 and 4. With reference to Figure 21.17, take $\Sigma M = 0$ at point 3 to determine V_2 (note that counterclockwise is positive):

$$-V_2(20) + 3333 - 23{,}333 + 300(20)(10) = 0$$

$$V_2 = 2000 \text{ lb}$$

Next, determine R_3 by taking $\Sigma M = 0$ at point 4 (Figure 21.18):

$$-2000(50) - 30R_3 - 30{,}000 + 3333 + 300(20)(40) + 400(30)(15) = 0$$

$$R_3 = 9778 \text{ lb}$$

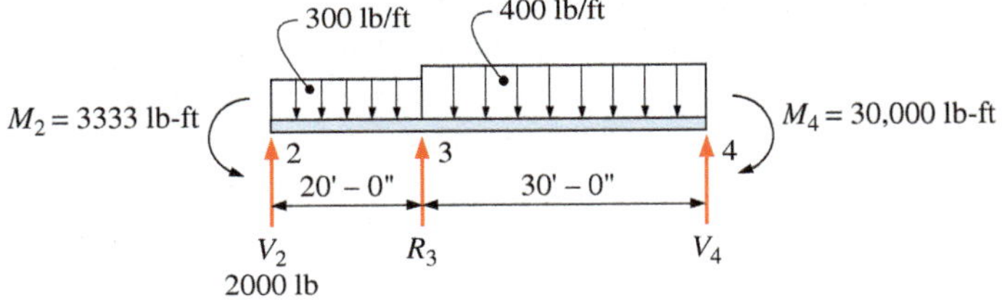

FIGURE 21.18 Free-body diagram, spans 2–3 and 3–4.

Then, using the free-body diagram shown in Figure 21.16b, taking $\Sigma M = 0$ at point 5, solve for R_4:

$$-2000(90) - 9778(70) - R_4(40) + 3333 - 25{,}000 + 300(20)(80) + 400(30)(55) + 200(40)(20) = 0$$

$$R_4 = 10{,}347 \text{ lb}$$

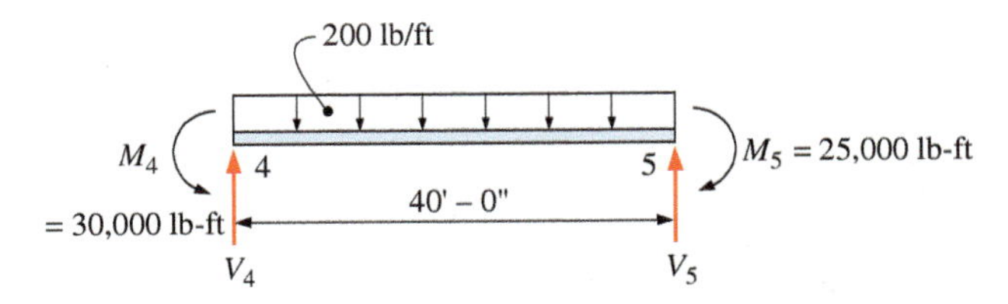

FIGURE 21.19 Free-body diagram, span 4–5.

Last, using the free-body diagram shown in Figure 21.19, taking $\Sigma M = 0$ at point 4, solve for V_5:

$$+40V_5 + 30{,}000 - 25{,}000 - 200(40)(20) = 0$$

$$V_5 = 3875 \text{ lb}$$

Note that the shears at points 2 and 5 are equivalent to vertical reactions at those points. As a check, the sum of all the vertical forces on the beam must equal zero. Taking ΣF_y (up is positive) yields

$$+2000 + 9778 + 10{,}347 + 3875 - 300(20) - 400(30) - 200(40) = 0 \quad \textbf{(O.K.)}$$

SUMMARY BY SECTION NUMBER

21.1 A statically indeterminate beam is one that has more unknown reactions than there are equations of equilibrium. To determine the unknown reactions, additional equations based on deflection or end rotation must be used.

21.2 Restrained beams include the propped cantilever and the fixed beam. A fixed support keeps the end of the beam from rotating (and translating).

21.3 A propped cantilever beam can be conveniently analyzed using the method of superposition. The important fact for this analysis is that the supports are on the same level and the deflection at the simply supported end is to be zero. The reaction at the simply supported end must cause a deflection equal in magnitude and opposite in sense to that caused by the load.

21.4 A fixed beam or a fixed-end beam can be analyzed using the method of superposition. In this case, it is convenient to consider slopes at the supports. Since the slope of the elastic curve at the supports is zero, the unknown moments must cause slopes equal in magnitude and opposite in sense to those caused by the loads. Tables of derived slope equations are furnished in this section.

21.5 A continuous beam is one that rests on more than two supports. In general, the method of superposition used for the one-span indeterminate beams cannot be conveniently used for continuous beams. An exception is a two-span symmetrically loaded continuous beam.

21.6 The theorem of three moments is a convenient method of finding bending moments in continuous beams. The analysis uses the relationship that exists between the bending moments at the three supports of any two adjacent spans. Equations (21.1) and (21.2) allow the determination of moments at the supports, after which reactions and shears can be found. Complete shear and moment diagrams can then be drawn.

PROBLEMS

All beams in these problems are assumed to have constant E *and* I *values. Unless otherwise noted, neglect the beam weight.*

Section 21.3 Propped Cantilever Beams

21.1 Rework Example 21.1 changing dimension a to 10 ft, dimension b to 15 ft, and the point load to 15 kips.

21.2 Use the method of superposition to determine the reactions for the propped cantilever beam shown in Figure 21.2. Span length is 20 ft and the uniformly distributed load w is 2.5 kips/ft.

21.3 Use the method of superposition to determine the reactions for the propped cantilever beams shown. Draw complete shear and moment diagrams.

Section 21.4 Fixed Beams

21.4 through 21.6 Draw complete shear and moment diagrams for the fixed beams shown. Select the lightest structural steel W shape. Assume A992 steel. Consider moment and shear.

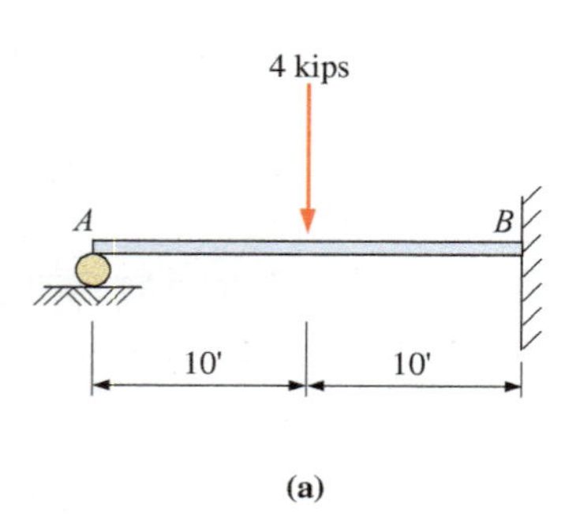

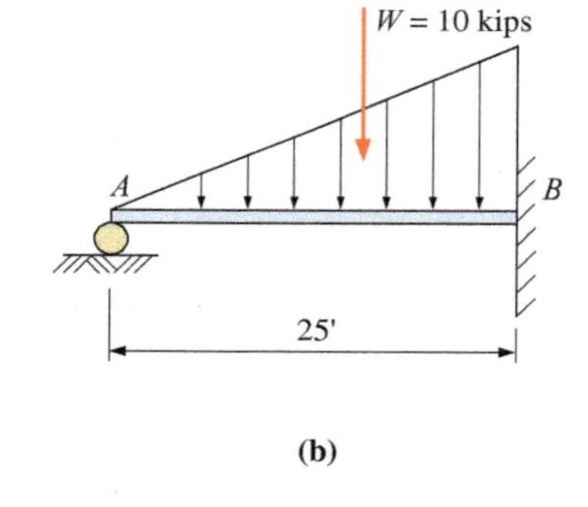

PROBLEM 21.3

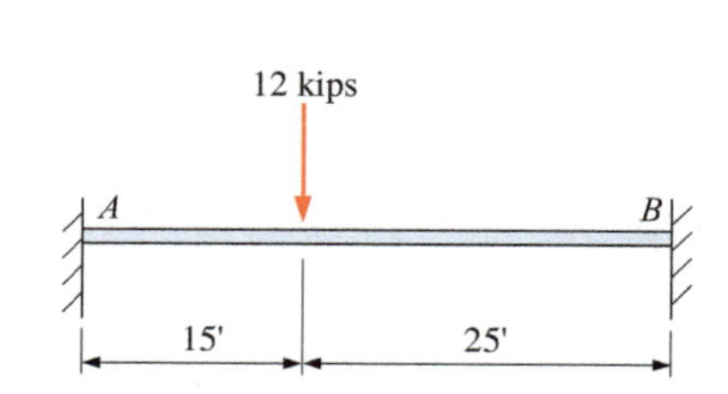

PROBLEM 21.5

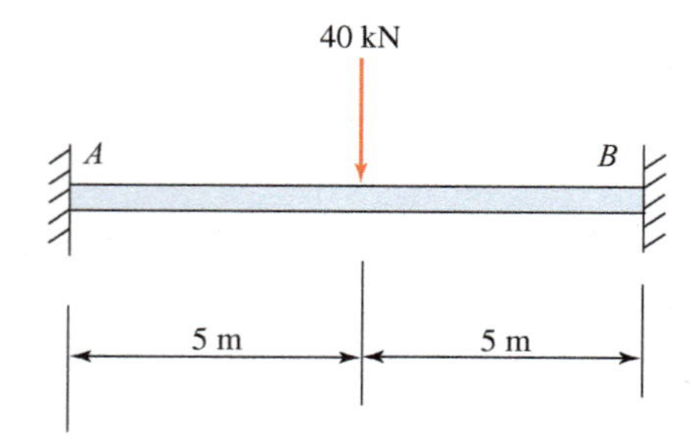

PROBLEM 21.4

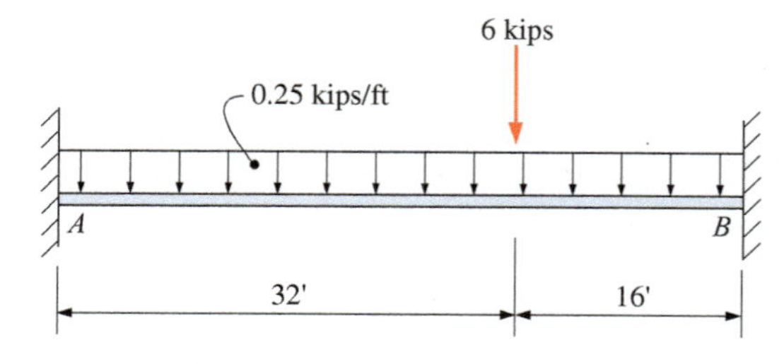

PROBLEM 21.6

Section 21.5 Continuous Beams: Superposition

21.7 Determine the reactions for the beam shown.

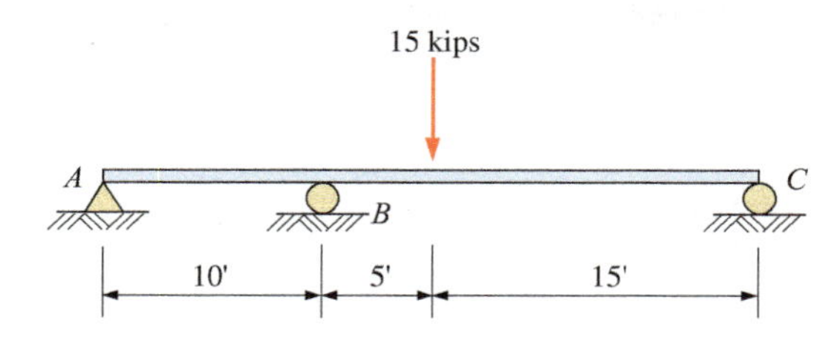

PROBLEM 21.7

21.8 Determine the reactions for the beam shown.

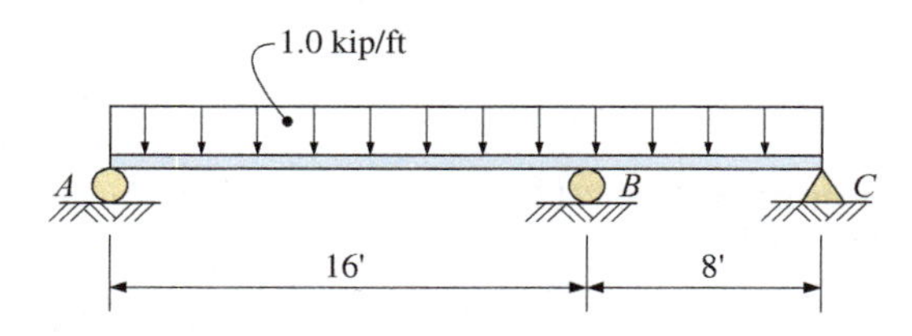

PROBLEM 21.8

21.9 Select a southern pine timber beam (S4S) for the application shown. Consider shear and moment. See Appendix F for design values.

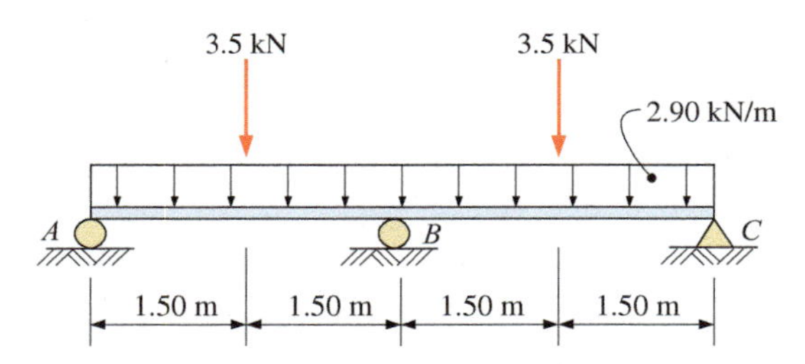

PROBLEM 21.9

Section 21.6 The Theorem of Three Moments

21.10 through 21.12 For the continuous beams shown, find moments at the supports and the reactions. Draw complete shear and moment diagrams.

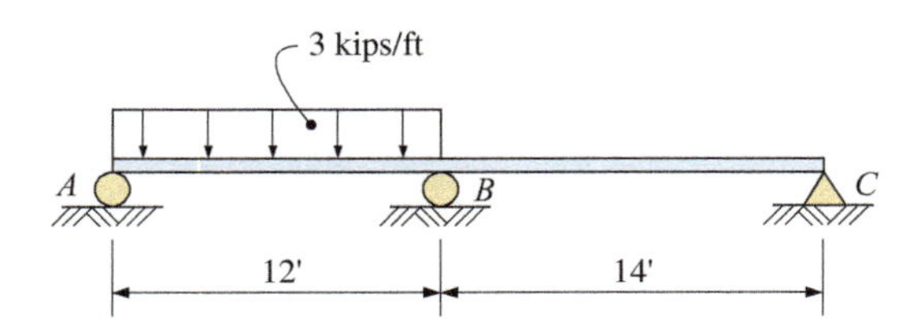

PROBLEM 21.10

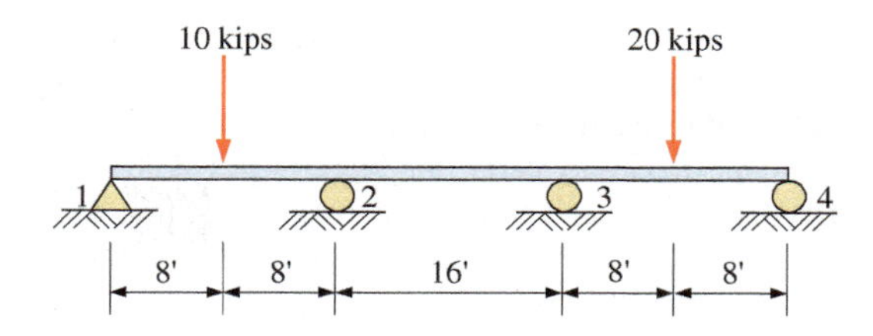

PROBLEM 21.11

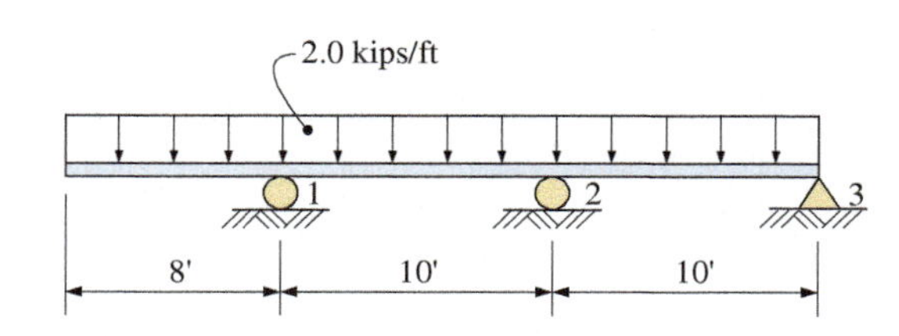

PROBLEM 21.12

Supplemental Problems

For these problems, use any appropriate method of analysis. Unless noted otherwise, neglect the beam weight in all problems.

21.13 A structural steel W12 × 30 has a span length of 24 ft. The beam is fixed at one end and simply supported at the other. Both supports are on the same level. The beam is subjected to a uniformly distributed load of 800 lb/ft, which includes its own weight. Draw the shear and moment diagrams and calculate the maximum bending stress. Label the simple support *A* and the fixed support *B*.

21.14 Rework Problem 21.13 assuming that the simple support settles 0.6 in.

21.15 Rework Problem 21.13 assuming that the simple support is raised 0.6 in. above the level of the fixed end.

21.16 A structural steel W12 × 40 has a span length of 20 ft. The beam is fixed at one end and simply supported at the other end. Both supports are on the same level. The beam is loaded with a concentrated load of 20,000 lb at a point 8 ft from the simply supported end. Compute the maximum bending stress in the beam.

21.17 Rework Problem 21.16 assuming that the simply supported end is raised one inch above the level of the fixed end.

21.18 Draw complete shear and moment diagrams for the beam shown.

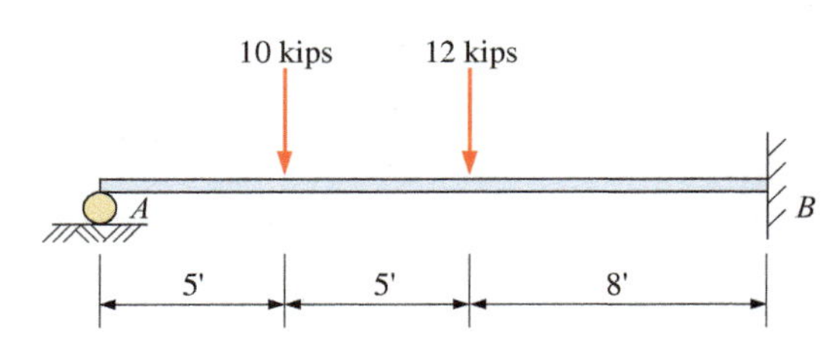

PROBLEM 21.18

21.19 Rework Problem 21.18 adding a uniformly distributed load of 1 kip/ft for the full span.

21.20 Select the most economical (lightest) structural steel W-shape beam to carry the superimposed loads shown. Draw the shear and moment diagrams. Assume A992 steel ($F_y = 50$ ksi) and complete fixity at each end with supports on the same level.

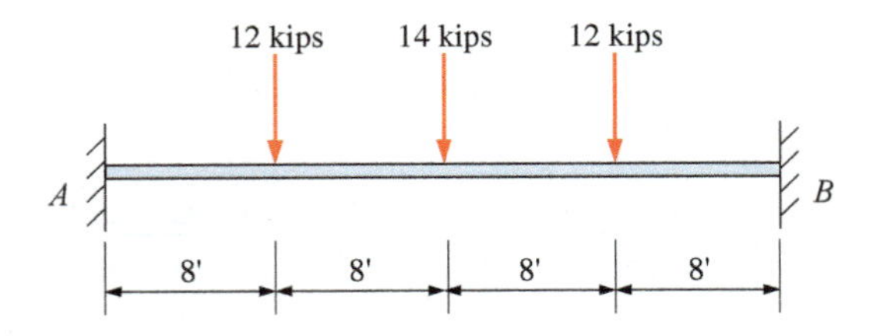

PROBLEM 21.20

21.21 through 21.23 Draw complete shear and moment diagrams for the beam shown.

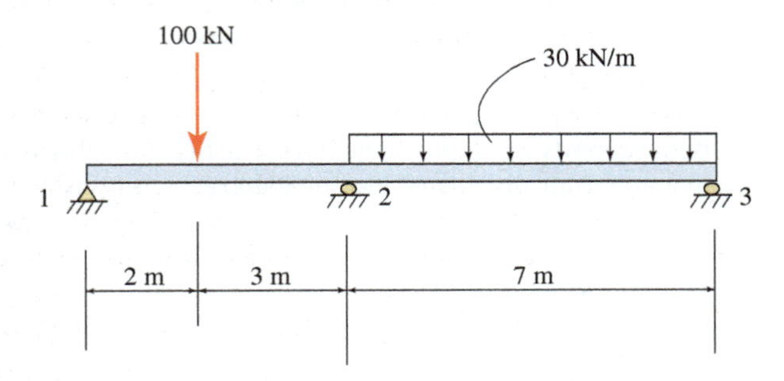

PROBLEM 21.21

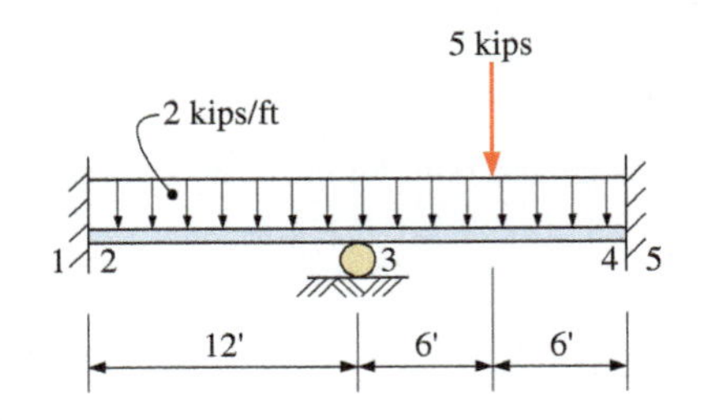

PROBLEM 21.22

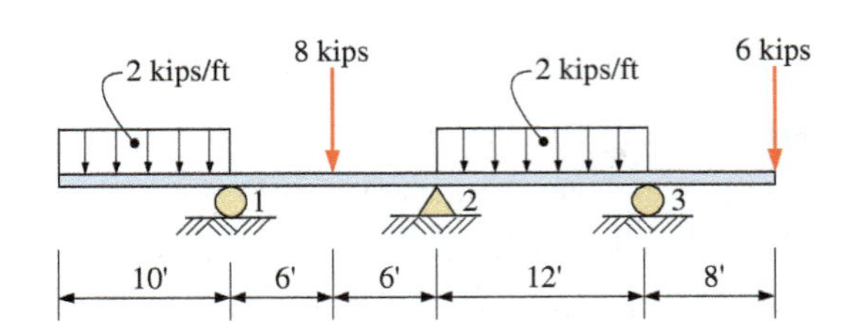

PROBLEM 21.23

21.24 A structural steel W18 × 50 supports loads as shown.

(a) Find the reactions and draw complete shear and moment diagrams.

(b) Assume that the center support settles 0.3 in. Find the reactions and draw complete shear and moment diagrams.

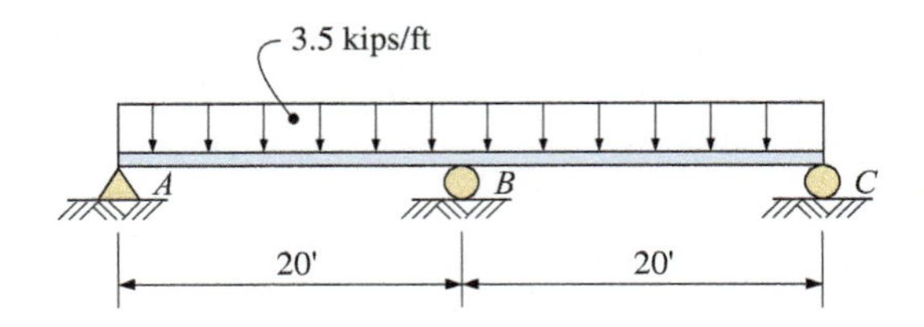

PROBLEM 21.24

APPENDICES

APPENDICES NOTES

The following appendices contain helpful data in both U.S. Customary units and SI. The data are intended only for use in conjunction with the examples and problems in this text.

1. **Appendices A through D.** These appendices furnish dimensions and properties for some standard structural shapes and are presented in both U.S. Customary and SI measurements. The data are from the American Institute of Steel Construction (AISC) *Manual of Steel Construction*, 14th ed., and are reproduced here with permission of AISC.
2. **Appendix E.** For the SI portion of this table, U.S. Customary dressed sizes were converted to millimeter sizes, and these were used to calculate the other properties. Nominal sizes have been rounded to the nearest 10 mm; dressed sizes and properties are shown to three significant digits.
3. **Appendices F and G.** These data on material properties and design values are average values and could vary significantly with specific materials. The reader is again cautioned that these data are intended only for use with this text. Refer to publications from the proper technical organizations such as the Aluminum Association, the American Society for Testing and Materials, the American Forest & Paper Association, and so forth to verify data for specific situations. The SI portions of these appendices represent a "soft" conversion from U.S. Customary units (using the conversion factors printed inside the front cover of this book).
4. **Appendix H.** The material in this appendix is from the *Manual of Steel Construction*, 14th ed., and was reproduced with the permission of the American Institute of Steel Construction.

APPENDIX A

SELECTED W SHAPES: DIMENSIONS AND PROPERTIES

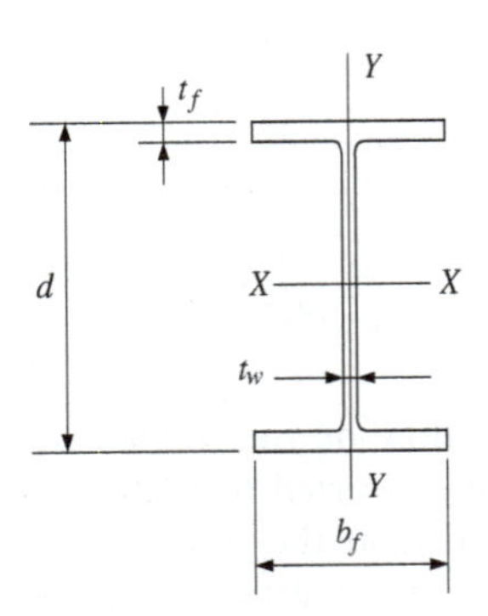

TABLE A.1 U.S. Customary units

Designation	A in.^{2}	d in.	t_w in.	b_f in.	t_f in.	I_x in.^{4}	S_x in.^{3}	r_x in.	Z_x in.^{3}	I_y in.^{4}	S_y in.^{3}	r_y in.
W40 × 277	81.5	39.7	0.830	15.8	1.58	21,900	1100	16.4	1250	1040	132	3.58
W36 × 302	89.0	37.3	0.945	16.7	1.68	21,100	1130	15.4	1280	1300	156	3.82
W36 × 194	57.0	36.5	0.765	12.1	1.26	12,100	664	14.6	767	375	61.9	2.56
W36 × 150	44.3	35.9	0.625	12.0	0.94	9040	504	14.3	581	270	45.1	2.47
W33 × 118	34.7	32.9	0.550	11.5	0.74	5900	359	13.0	415	187	32.6	2.32
W30 × 108	31.7	29.8	0.545	10.5	0.76	4470	299	11.9	346	146	27.9	2.15
W30 × 99	29.0	29.7	0.520	10.5	0.67	3990	269	11.7	312	128	24.5	2.1
W27 × 114	33.6	27.3	0.570	10.1	0.93	4080	299	11.0	343	159	31.5	2.18
W27 × 94	27.6	26.9	0.490	10.0	0.745	3270	243	10.9	278	124	24.8	2.12
W24 × 162	47.8	25.0	0.705	13.0	1.22	5170	414	10.4	468	443	68.4	3.05
W24 × 104	30.7	24.1	0.500	12.8	0.75	3100	258	10.1	289	259	40.7	2.91
W24 × 84	24.7	24.1	0.470	9.02	0.77	2370	196	9.79	224	94.4	20.9	1.95
W24 × 76	22.4	23.9	0.440	8.99	0.68	2100	176	9.69	200	82.5	18.4	1.92
W21 × 83	24.4	21.4	0.515	8.36	0.835	1830	171	8.67	196	81.4	19.5	1.83
W21 × 73	21.5	21.2	0.455	8.3	0.74	1600	151	8.64	172	70.6	17.0	1.81
W21 × 62	18.3	21.0	0.400	8.24	0.615	1330	127	8.54	144	57.5	14.0	1.77
W21 × 50	14.7	20.8	0.380	6.53	0.535	984	94.5	8.18	110	24.9	7.64	1.3
W18 × 71	20.9	18.5	0.495	7.64	0.81	1170	127	7.5	146	60.3	15.8	1.7
W18 × 60	17.6	18.2	0.415	7.56	0.695	984	108	7.47	123	50.1	13.3	1.68
W18 × 50	14.7	18.0	0.355	7.5	0.57	800	88.9	7.38	101	40.1	10.7	1.65
W18 × 40	11.8	17.9	0.315	6.02	0.525	612	68.4	7.21	78.4	19.1	6.35	1.27
W16 × 100	29.4	17.0	0.585	10.4	0.985	1490	175	7.1	198	186	35.7	2.51
W16 × 45	13.3	16.1	0.345	7.04	0.565	586	72.7	6.65	82.3	32.8	9.34	1.57
W16 × 36	10.6	15.9	0.295	6.99	0.43	448	56.5	6.51	64.0	24.5	7.0	1.52
W16 × 26	7.68	15.7	0.250	5.5	0.345	301	38.4	6.26	44.2	9.59	3.49	1.12

TABLE A.1 U.S. Customary units (*Continued*)

Designation	A in.^{2}	d in.	t_w in.	b_f in.	t_f in.	I_x in.^{4}	S_x in.^{3}	r_x in.	Z_x in.^{3}	I_y in.^{4}	S_y in.^{3}	r_y in.
W14 × 132	38.8	14.7	0.645	14.7	1.03	1530	209	6.28	234	548	74.5	3.76
W14 × 109	32.0	14.3	0.525	14.6	0.86	1240	173	6.22	192	447	61.2	3.73
W14 × 90	26.5	14.0	0.440	14.5	0.71	999	143	6.14	157	362	49.9	3.7
W14 × 74	21.8	14.2	0.450	10.1	0.785	795	112	6.04	126	134	26.6	2.48
W14 × 61	17.9	13.9	0.375	10.0	0.645	640	92.1	5.98	102	107	21.5	2.45
W14 × 48	14.1	13.8	0.340	8.03	0.595	484	70.2	5.85	78.4	51.4	12.8	1.91
W14 × 34	10.0	14.0	0.285	6.75	0.455	340	48.6	5.83	54.6	23.3	6.91	1.53
W14 × 30	8.85	13.8	0.270	6.73	0.385	291	42.0	5.73	47.3	19.6	5.82	1.49
W14 × 22	6.49	13.7	0.230	5.0	0.335	199	29.0	5.54	33.2	7.0	2.8	1.04
W12 × 96	28.2	12.7	0.550	12.2	0.9	833	131	5.44	147	270	44.4	3.09
W12 × 72	21.1	12.3	0.430	12.0	0.67	597	97.4	5.31	108	195	32.4	3.04
W12 × 58	17.0	12.2	0.360	10.0	0.64	475	78.0	5.28	86.4	107	21.4	2.51
W12 × 50	14.6	12.2	0.370	8.08	0.64	391	64.2	5.18	71.9	56.3	13.9	1.96
W12 × 40	11.7	11.9	0.295	8.01	0.515	307	51.5	5.13	57.0	44.1	11.0	1.94
W12 × 30	8.79	12.3	0.260	6.52	0.44	238	38.6	5.21	43.1	20.3	6.24	1.52
W12 × 22	6.48	12.3	0.260	4.03	0.425	156	25.4	4.91	29.3	4.66	2.31	0.848
W12 × 16	4.71	12.0	0.220	3.99	0.265	103	17.1	4.67	20.1	2.82	1.41	0.773
W10 × 112	32.9	11.4	0.755	10.4	1.25	716	126	4.66	147	236	45.3	2.68
W10 × 68	19.9	10.4	0.470	10.1	0.77	394	75.7	4.44	85.3	134	26.4	2.59
W10 × 54	15.8	10.1	0.370	10.0	0.615	303	60.0	4.37	66.6	103	20.6	2.56
W10 × 45	13.3	10.1	0.350	8.02	0.62	248	49.1	4.32	54.9	53.4	13.3	2.01
W10 × 33	9.71	9.73	0.290	7.96	0.435	171	35.0	4.19	38.8	36.6	9.2	1.94
W10 × 22	6.49	10.2	0.240	5.75	0.36	118	23.2	4.27	26	11.4	3.97	1.33
W10 × 12	3.54	9.87	0.190	3.96	0.21	53.8	10.9	3.9	12.6	2.18	1.1	0.785
W8 × 40	11.7	8.25	0.360	8.07	0.56	146	35.5	3.53	39.8	49.1	12.2	2.04
W8 × 31	9.13	8.00	0.285	8.0	0.435	110	27.5	3.47	30.4	37.1	9.27	2.02
W8 × 24	7.08	7.93	0.245	6.5	0.4	82.7	20.9	3.42	23.1	18.3	5.63	1.61
W8 × 13	3.84	7.99	0.230	4.0	0.255	39.6	9.91	3.21	11.4	2.73	1.37	0.843
W6 × 25	7.34	6.38	0.320	6.08	0.455	53.4	16.7	2.7	18.9	17.1	5.61	1.52
W6 × 12	3.55	6.03	0.230	4.0	0.28	22.1	7.31	2.49	8.3	2.99	1.50	0.918
W4 × 13	3.83	4.16	0.280	4.06	0.345	11.3	5.46	1.72	6.28	3.86	1.90	1.00

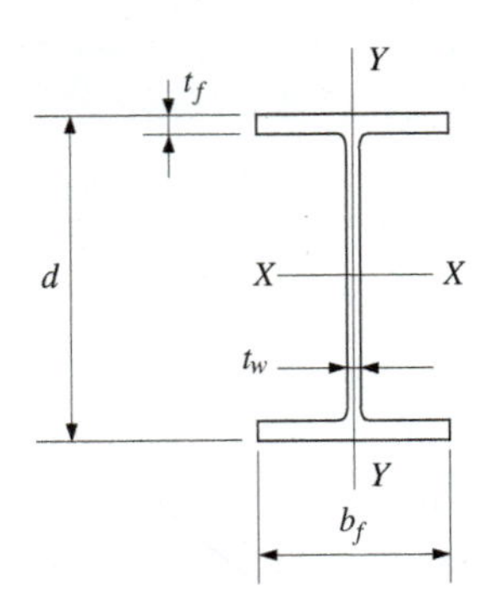

TABLE A.2 SI units

Designation mm × kg/m	A mm²	d mm	t_w mm	b_f mm[a]	t_f mm[a]	I_x 10^6 mm⁴	S_x 10^3 mm³	r_x mm	$Z_x \times 10^3$ mm³	I_y 10^6 mm⁴	S_y 10^3 mm³	r_y mm
W1000 × 222	28,300	970	16.0	298	21.1	4080	8410	381	9800	95.3	636	58.2
W920 × 390	49,800	937	21.30	419	36.6	7450	15,900	389	18,000	454	2160	95.5
W840 × 329	42,100	861	19.7	400	32.5	5370	12,400	358	14,000	350	1740	91.2
W610 × 153	19,500	622	14.0	229	24.9	1250	4010	254	4590	49.5	434	50.5
W610 × 113	14,500	607	11.2	229	17.3	874	2880	246	3280	34.3	302	48.8
W530 × 138	17,600	549	14.7	213	23.6	862	3150	221	3620	38.7	362	46.7
W530 × 101	12,900	536	10.9	210	17.4	616	2290	218	2620	26.9	257	45.7
W530 × 72	9100	523	8.89	206	10.9	399	1520	209	1750	16.1	156	42.2
W460 × 60	7610	455	8.00	152	13.3	255	1120	183	1280	7.95	104	32.3
W410 × 100	12,600	414	10.0	260	16.9	397	1920	177	2130	49.5	380	62.5
W360 × 110	14,100	361	11.4	257	19.9	331	1840	153	2060	55.8	436	63
W310 × 202	25,700	340	20.1	314	31.8	516	3050	142	3510	166	1050	80.3
W310 × 107	13,600	312	10.9	305	17.0	248	1600	135	1770	81.2	531	77.2
W310 × 97	12,300	307	9.91	305	15.4	222	1440	134	1590	72.4	477	76.7
W310 × 38.7	4940	310	5.84	165	9.65	84.9	547	131	610	7.2	87.5	38.4
W250 × 115	14,600	269	13.5	260	22.1	189	1410	114	1600	64.1	493	66
W250 × 73	9290	254	8.64	254	14.2	113	895	110	990	38.9	306	64.5
W200 × 59	7550	210	9.14	206	14.2	60.8	582	89.7	652	20.4	200	51.8
W150 × 24	3060	160	6.60	102	10.3	13.4	167	66	192	1.84	36.1	24.6

[a] Flange and web thicknesses may vary due to mill rolling practices.

APPENDIX B

SELECTED PIPES: DIMENSIONS AND PROPERTIES

TABLE B.1 U.S. Customary units

Nominal Diameter in.	Outside Diameter in.	Inside Diameter in.	Wall Thickness in.	Weight lb/ft	A in.^{2}	I in.^{4}	S in.^{3}	r in.
				Standard Weight				
2	2.38	2.07	0.154	3.66	1.02	0.627	0.528	0.791
3	3.50	3.07	0.216	7.58	2.07	2.85	1.63	1.17
3.5	4.00	3.55	0.226	9.12	2.50	4.52	2.26	1.34
4	4.50	4.03	0.237	10.80	2.96	6.82	3.03	1.51
5	5.56	5.05	0.258	14.60	4.01	14.3	5.14	1.88
6	6.63	6.07	0.28	19.0	5.2	26.5	7.99	2.25
				Extra Strong				
3.5	4.00	3.36	0.318	12.5	3.43	5.94	2.97	1.31
4	4.50	3.83	0.337	15.0	4.14	9.12	4.05	1.48
5	5.56	4.81	0.375	20.8	5.73	19.5	7.02	1.85
6	6.63	5.76	0.432	28.6	7.83	38.3	11.6	2.20
8	8.63	7.63	0.500	43.4	11.90	100	23.1	2.89

TABLE B.2 SI units

Nominal Diameter mm	Outside Diameter mm	Inside Diameter mm	Wall Thickness mm	Mass kg/m	A mm^2	I 10^6 mm^4	S 10^3 mm^3	r mm
				Standard Weight				
51	60.50	52.5	3.91	5.44	658	0.261	8.65	20.1
102	114.00	102	6.02	16.10	1910	2.84	49.7	38.4
152	168.00	154	7.11	28.30	3350	11.0	131	57.2
				Extra Strong				
76	88.90	73.7	7.62	15.30	1830	1.54	34.6	29.0
102	114.00	97.3	8.56	22.30	2670	3.8	66.4	37.6
127	141.00	122	9.53	30.90	3700	8.12	115	47.0

APPENDIX C

SELECTED CHANNELS: DIMENSIONS AND PROPERTIES

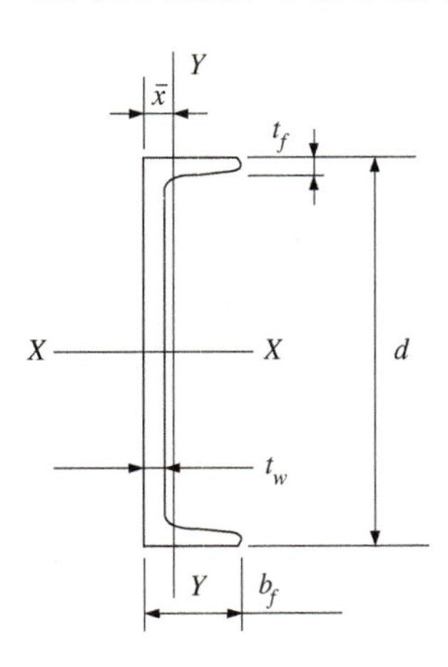

TABLE C.1 U.S. Customary units

Designation	A in.^{2}	d in.	b_f in.	Avg. t_f in.	t_w in.	$\bar{x}$ in.	I_x in.^{4}	S_x in.^{3}	r_x in.	I_y in.^{4}	S_y in.^{3}	r_y in.
C15 × 50	14.7	15.0	3.72	0.65	0.716	0.799	404	53.8	5.24	11	3.77	0.865
C15 × 40	11.8	15.0	3.52	0.65	0.52	0.778	348	46.5	5.43	9.17	3.34	0.883
C12 × 30	8.81	12.0	3.17	0.501	0.51	0.674	162	27	4.29	5.12	2.05	0.762
C12 × 25	7.34	12.0	3.05	0.501	0.387	0.674	144	24	4.43	4.45	1.87	0.779
C12 × 20.7	6.08	12.0	2.94	0.501	0.282	0.698	129	21.5	4.61	3.86	1.72	0.797
C10 × 30	8.81	10.0	3.03	0.436	0.673	0.649	103	20.7	3.43	3.93	1.65	0.668
C10 × 20	5.87	10.0	2.74	0.436	0.379	0.606	78.9	15.8	3.67	2.8	1.31	0.69
C10 × 15.3	4.48	10.0	2.6	0.436	0.24	0.634	67.3	13.5	3.88	2.27	1.15	0.711
C6 × 10.5	3.07	6.0	2.03	0.343	0.314	0.5	15.1	5.04	2.22	0.86	0.561	0.529

TABLE C.2 SI units

Designation mm × kg/m	A mm^2	d mm	b_f mm	Avg. t_f mm[a]	t_w mm[a]	$\bar{x}$ mm	I_x 10^6 mm^4	S_x 10^3 mm^3	r_x mm	I_y 10^6 mm^4	S_y 10^3 mm^3	r_y mm
C310 × 30.8	3920	305	74.7	12.7	7.16	17.7	53.7	352	117	1.61	28.2	20.2
C200 × 17.1	2170	203	57.4	9.91	5.59	14.5	13.5	133	79	0.545	12.7	15.8
C130 × 13	1700	127	48	8.13	8.26	12.1	3.7	58.3	46.7	0.26	7.28	12.3

[a]Flange and web thicknesses may vary due to mill rolling practices.

SELECTED ANGLES: PROPERTIES FOR DESIGNING

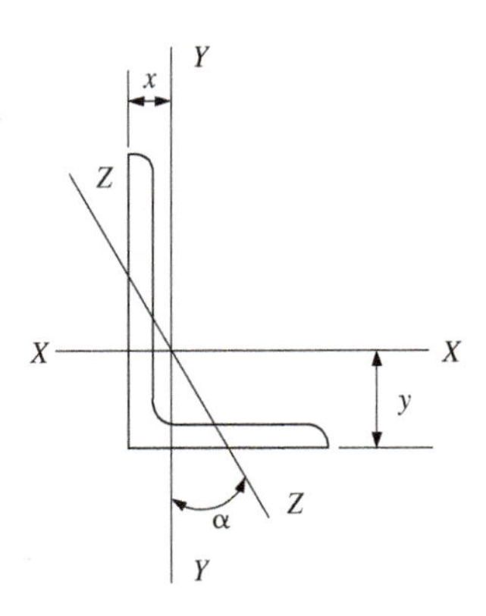

TABLE D.1 U.S. Customary units

Designation	Wt. lb/ft	A in.^{2}	I_x in.^{4}	S_x in.^{3}	r_x in.	y in.	I_y in.^{4}	S_y in.^{3}	r_y in.	x in.	r_z in.	Tan α
L6 × 6 × 1	37.4	11.0	35.4	8.55	1.79	1.86	35.4	8.55	1.79	1.86	1.17	1.00
L6 × 4 × 3/4	23.6	6.9	24.5	6.23	1.88	2.07	8.63	2.95	1.12	1.07	0.856	0.428
L6 × 4 × 3/8	12.3	3.6	13.4	3.3	1.93	1.93	4.86	1.58	1.16	0.933	0.87	0.446
L5 × 5 × 1/2	16.2	4.8	11.3	3.15	1.53	1.42	11.3	3.15	1.53	1.42	0.98	1.00
L4 × 4 × 3/8	9.80	2.9	4.32	1.5	1.23	1.13	4.32	1.5	1.23	1.13	0.779	1.00
L4 × 3 × 1/4	5.80	1.69	2.75	0.988	1.27	1.22	1.33	0.585	0.887	0.725	0.639	0.558
L$2\frac{1}{2}$ × $2\frac{1}{2}$ × 1/4	4.10	1.19	0.692	0.387	0.764	0.711	0.692	0.387	0.764	0.711	0.482	1.00

TABLE D.2 SI units

Designation mm × mm × mm	Mass kg/m	A mm^2	I_x 10^6 mm^4	S_x 10^3 mm^3	r_x mm	y mm	I_y 10^6 mm^4	S_y 10^3 mm^3	r_y mm	x mm
L127 × 89 × 12.7	20.2	2580	4.16	48.7	40.1	41.9	1.67	25.4	25.4	22.9
L102 × 76 × 6.4	8.6	1090	1.14	16.2	32.3	31.0	0.554	9.59	22.5	18.4
L64 × 64 × 6.4	6.1	768	0.288	6.34	19.4	18.1	0.288	6.34	19.4	18.1

APPENDIX
E

PROPERTIES OF STRUCTURAL TIMBER

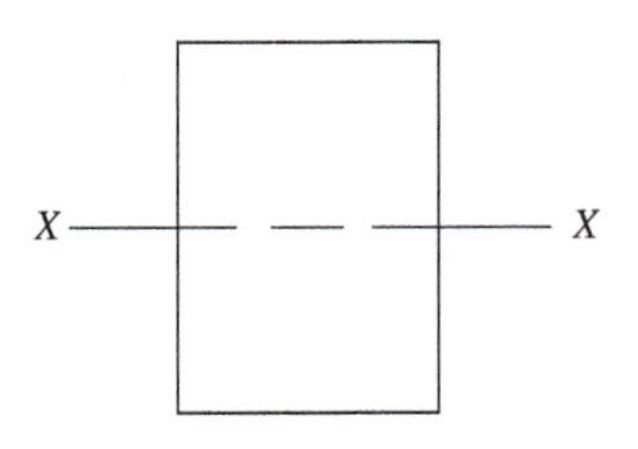

TABLE E.1 U.S. Customary units

Nominal Size in.	Dressed Size in.	A in.^{2}	Wt. lb/ft	I_x in.^{4}	S_x in.^{3}
2 × 4	$1\frac{1}{2} \times 3\frac{1}{2}$	5.25	1.46	5.36	3.06
× 6	$\times 5\frac{1}{2}$	8.25	2.29	20.8	7.56
× 8	$\times 7\frac{1}{4}$	10.9	3.02	47.6	13.1
× 10	$\times 9\frac{1}{4}$	13.9	3.85	98.9	21.4
× 12	$\times 11\frac{1}{4}$	16.9	4.68	178	31.6
× 14	$\times 13\frac{1}{4}$	19.9	5.52	291	43.9
× 16	$\times 15\frac{1}{4}$	22.9	6.35	443	58.2
× 18	$\times 17\frac{1}{4}$	25.9	7.17	642	74.5
3 × 4	$2\frac{1}{2} \times 3\frac{1}{2}$	8.75	2.42	8.93	5.10
× 6	$\times 5\frac{1}{2}$	13.8	3.82	34.7	12.6
× 8	$\times 7\frac{1}{4}$	18.1	5.04	79.4	21.9
× 10	$\times 9\frac{1}{4}$	23.1	6.42	165	35.6
× 12	$\times 11\frac{1}{4}$	28.1	7.81	297	52.7
× 14	$\times 13\frac{1}{4}$	33.1	9.20	485	73.2
× 16	$\times 15\frac{1}{4}$	38.1	10.6	739	96.9
× 18	$\times 17\frac{1}{4}$	43.1	12.0	1070	124
4 × 4	$3\frac{1}{2} \times 3\frac{1}{2}$	12.3	3.40	12.5	7.15
× 6	$\times 5\frac{1}{2}$	19.3	5.35	48.5	17.6
× 8	$\times 7\frac{1}{4}$	25.4	7.05	111	30.7
× 10	$\times 9\frac{1}{4}$	32.4	8.93	231	49.9
× 12	$\times 11\frac{1}{4}$	39.4	10.9	415	73.8
× 14	$\times 13\frac{1}{4}$	46.4	12.9	678	102
× 16	$\times 15\frac{1}{4}$	53.4	14.9	1030	136
× 18	$\times 17\frac{1}{4}$	60.4	16.8	1500	174

Notes: Properties and weights are for dressed sizes. Assumed unit weight of timber is 40 pcf. Moment of inertia and section modulus are about the strong axis.

TABLE E.1 U.S. Customary units (*Continued*)

Nominal Size in.	Dressed Size in.	A in.^{2}	Wt. lb/ft	I_x in.^{4}	S_x in.^{3}
6 × 6	$5\frac{1}{2} \times 5\frac{1}{2}$	30.3	8.40	76.3	27.7
× 8	$\times 7\frac{1}{2}$	41.3	11.4	193	51.6
× 10	$\times 9\frac{1}{2}$	52.3	14.5	393	82.7
× 12	$\times 11\frac{1}{2}$	63.3	17.5	697	121
× 14	$\times 13\frac{1}{2}$	74.3	20.6	1130	167
× 16	$\times 15\frac{1}{2}$	85.3	23.6	1710	220
× 18	$\times 17\frac{1}{2}$	96.3	26.7	2460	281
× 20	$\times 19\frac{1}{2}$	108	29.8	3400	349
8 × 8	$7\frac{1}{2} \times 7\frac{1}{2}$	56.3	15.6	264	70.3
× 10	$\times 9\frac{1}{2}$	71.3	19.8	536	113
× 12	$\times 11\frac{1}{2}$	86.3	23.9	951	165
× 14	$\times 13\frac{1}{2}$	101	28.0	1540	228
× 16	$\times 15\frac{1}{2}$	116	32.0	2330	300
× 18	$\times 17\frac{1}{2}$	131	36.4	3350	383
× 20	$\times 19\frac{1}{2}$	146	40.6	4630	475
× 22	$\times 21\frac{1}{2}$	161	44.8	6210	578
10 × 10	$9\frac{1}{2} \times 9\frac{1}{2}$	90.3	25.0	679	143
× 12	$\times 11\frac{1}{2}$	109	30.3	1200	209
× 14	$\times 13\frac{1}{2}$	128	35.6	1950	289
× 16	$\times 15\frac{1}{2}$	147	40.9	2950	380
× 18	$\times 17\frac{1}{2}$	166	46.1	4240	485
× 20	$\times 19\frac{1}{2}$	185	51.4	5870	602
× 22	$\times 21\frac{1}{2}$	204	56.7	7870	732
× 24	$\times 23\frac{1}{2}$	223	62.0	10,300	874
12 × 12	$11\frac{1}{2} \times 11\frac{1}{2}$	132	36.7	1460	253
× 14	$\times 13\frac{1}{2}$	155	43.1	2360	349
× 16	$\times 15\frac{1}{2}$	178	49.5	3570	460
× 18	$\times 17\frac{1}{2}$	201	55.9	5140	587
× 20	$\times 19\frac{1}{2}$	224	62.3	7110	729
× 22	$\times 21\frac{1}{2}$	247	68.7	9520	886
× 24	$\times 23\frac{1}{2}$	270	75.0	12,400	1060

TABLE E.2 SI units

Nominal Size mm	Dressed Size mm	A $mm^2 \times 10^3$	Mass kg/m	I_x $mm^4 \times 10^6$	S_x $mm^3 \times 10^6$
50 × 100	38.1 × 88.9	3.39	2.17	2.23	0.0502
× 150	× 140	5.33	3.42	8.71	0.124
× 200	× 184	7.01	4.49	19.8	0.215
× 250	× 235	8.95	5.74	41.2	0.351
× 300	× 286	10.9	6.98	74.3	0.519
× 360	× 337	12.8	8.23	122	0.721
× 410	× 387	14.7	9.45	184	0.951
× 460	× 438	16.7	10.69	267	1.22
80 × 100	63.5 × 88.9	5.65	3.62	3.72	0.0836
× 150	× 140	8.89	5.70	14.5	0.207
× 200	× 184	11.7	7.49	33.0	0.358
× 250	× 235	14.9	9.56	68.7	0.584
× 300	× 286	18.2	11.64	124	0.866
× 360	× 337	21.4	13.71	203	1.20
× 410	× 388	24.6	15.79	309	1.59
× 460	× 438	27.8	17.82	445	2.03
100 × 100	88.9 × 88.9	7.90	5.06	5.21	0.117
× 150	× 140	12.4	7.97	20.3	0.290
× 200	× 184	16.4	10.48	46.2	0.502
× 250	× 235	20.9	13.39	96.1	0.818
× 300	× 286	25.4	16.29	173	1.21
× 360	× 337	30.0	19.20	284	1.68
× 410	× 388	34.5	22.1	433	2.23
× 460	× 438	38.9	24.9	623	2.84
150 × 150	140 × 140	19.6	12.56	32.0	0.457
× 200	× 191	26.7	17.13	81.3	0.851
× 250	× 241	33.7	21.6	163	1.36
× 300	× 292	40.9	26.2	290	1.99
× 360	× 343	48.0	30.8	471	2.75
× 410	× 394	55.2	35.3	714	3.63
× 460	× 445	62.3	39.9	1030	4.62
× 510	× 496	69.4	44.5	1420	5.74

Notes: Properties are for dressed sizes. Assumed unit mass of timber is 641 kilograms per cubic meter. Moment of inertia and section modulus are about the strong axis.

TABLE E.2 SI units (*Continued*)

Nominal Size mm	Dressed Size mm	A $mm^2 \times 10^3$	Mass kg/m	I_x $mm^4 \times 10^6$	S_x $mm^3 \times 10^6$
200 × 200	191 × 191	36.5	23.4	111	1.16
× 250	× 241	46.0	29.5	223	1.85
× 300	× 292	55.8	35.7	396	2.71
× 360	× 343	65.5	42.0	642	3.75
× 410	× 394	75.3	48.2	974	4.94
× 460	× 445	85.0	54.5	1400	6.30
× 510	× 495	94.5	60.6	1930	7.80
× 560	× 546	104	66.8	2590	9.49
250 × 250	241 × 241	58.1	37.2	281	2.33
× 300	× 292	70.4	45.1	500	3.42
× 360	× 343	82.7	53.0	810	4.73
× 410	× 394	95.0	60.8	1230	6.24
× 460	× 445	107	68.7	1770	7.95
× 510	× 495	119	76.4	2440	9.84
× 560	× 546	132	84.3	3270	12.0
× 610	× 597	144	92.2	4270	14.3
300 × 300	292 × 292	85.3	54.6	606	4.15
× 360	× 343	100	64.2	982	5.73
× 410	× 394	115	73.7	1490	7.55
× 460	× 445	130	83.3	2140	9.64
× 510	× 495	145	92.6	2950	11.9
× 560	× 546	159	102.2	3960	14.5
× 610	× 597	174	111.7	5180	17.3

APPENDIX F

DESIGN VALUES FOR TIMBER CONSTRUCTION

Note: Values shown are selected from a range and are provided solely for use in solving examples and problems in this text.

TABLE F.1 U.S. Customary units

Species	Allowable Stress[a] (psi)					Modulus of Elasticity E (ksi)
	F_c	F_{c_p}	F_t	F_b	F_v	
Douglas fir	1050	550	625	900	180	1700
Southern pine	1250	410	825	1400	175	1700
Hem-fir	1300	400	500	1000	150	1400
Eastern white pine	725	350	275	600	135	1100
Redwood	1050	650	600	1000	160	1300

[a]Allowable stresses:
F_c: compression parallel to grain
F_{c_p}: compression perpendicular to grain
F_t: tension parallel to grain
F_b: bending
F_v: horizontal shear

TABLE F.2 SI units

Species	Allowable Stress[a] (MPa)					Modulus of Elasticity E (MPa × 10^3)
	F_c	F_{c_p}	F_t	F_b	F_v	
Douglas fir	7.24	3.79	4.31	6.21	1.24	12
Southern pine	8.62	2.83	5.69	9.65	1.21	12
Hem-fir	8.96	2.76	3.45	6.89	1.03	9.7
Eastern white pine	5.00	2.41	1.90	4.14	0.93	7.6
Redwood	7.24	4.48	4.14	6.89	1.10	9.0

[a]Allowable stresses:
F_c: compression parallel to grain
F_{c_p}: compression perpendicular to grain
F_t: tension parallel to grain
F_b: bending
F_v: horizontal shear

TYPICAL AVERAGE PROPERTIES OF SOME COMMON MATERIALS

Note: Values shown are approximate and are provided solely for use in solving examples and problems in this text.

TABLE G.1 U.S. Customary units

Material	Density pcf	Mod. of Elasticity E ksi	Mod. of Rigidity G ksi	Tensile Yield Strength F_y ksi	Ultimate Strength ksi			Coeff. of Thermal Expansion α in./in./F° $\times 10^{-6}$	Poisson's Ratio μ
					Tens.	Comp.	Shear		
Steel ASTM A36 or A501	490	30,000	12,000	36	58			6.5	0.25
Steel (high-strength low-alloy) ASTM A992	490	30,000	12,000	50	65			6.5	0.25
Steel AISI 1020 hot-rolled	490	30,000	11,500	30	55			6.5	0.25
Steel AISI 1040 hot-rolled	490	30,000	11,500	42	76			6.5	0.25
Stainless steel (annealed)	490	29,000	11,600	40	85		60	6.5	0.25
Cast iron (gray)	490	15,000	6000		20	80	32	5.9	0.26
Aluminum alloy 6061-T6	165	10,000	4000	35	38		27	13.1	0.33
Titanium alloy	275	16,500	6500	130	140		100	5.4	
Magnesium alloy	112	6500	2400	20	40		20	14.5	0.34
Brass (rolled)	535	14,000	6000	50	60		50	10.4	0.34
Bronze (cast)	535	12,000	5000	25	33	56		10.1	0.35
Copper (hard drawn)	550	15,000	6000	40	55		38	9.3	0.35
Concrete	150	3120				3.0		5.5	0.20
Concrete	150	3605				4.0		5.5	0.20

TABLE G.2 SI units

Material	Density kg/m^3 × 10^3	Mod. of Elasticity E MPa × 10^3	Mod. of Rigidity G MPa × 10^3	Tensile Yield Strength F_y MPa	Ultimate Strength MPa			Coeff. of Thermal Expansion α m/m/C° × 10^{-6}	Poisson's Ratio μ
					Tens.	Comp.	Shear		
Steel ASTM A36 or A501	7.85	207	83	250	400			11.7	0.25
Steel (high-strength low-alloy) ASTM A992	7.85	207	83	345	450			11.7	0.25
Steel AISI 1020 hot-rolled	7.85	207	79.3	210	380			11.7	0.25
Steel AISI 1040 hot-rolled	7.85	207	79.3	290	520			11.7	0.25
Stainless steel (annealed)	7.85	200	80	280	580		410	11.7	0.25
Cast iron (gray)	7.21	103	41		140	550	220	10.6	0.26
Aluminum alloy 6061-T6	2.64	70	28	240	260		190	23.6	0.33
Magnesium alloy	1.794	45	17	210	280		140	26.1	0.34
Titanium alloy	4.40	114	45	900	970		690	9.7	
Brass (rolled)	8.56	97	41	340	410		340	18.7	0.34
Bronze (cast)	8.56	83	34	170	230	390		18.2	0.35
Copper (hard drawn)	8.81	103	41	280	380		260	16.7	0.35
Concrete	2.40	21.5				21		9.9	0.20
Concrete	2.40	24.9				28		9.9	0.20

APPENDIX H

BEAM DIAGRAMS AND FORMULAS

1. SIMPLE BEAM: UNIFORMLY DISTRIBUTED LOAD

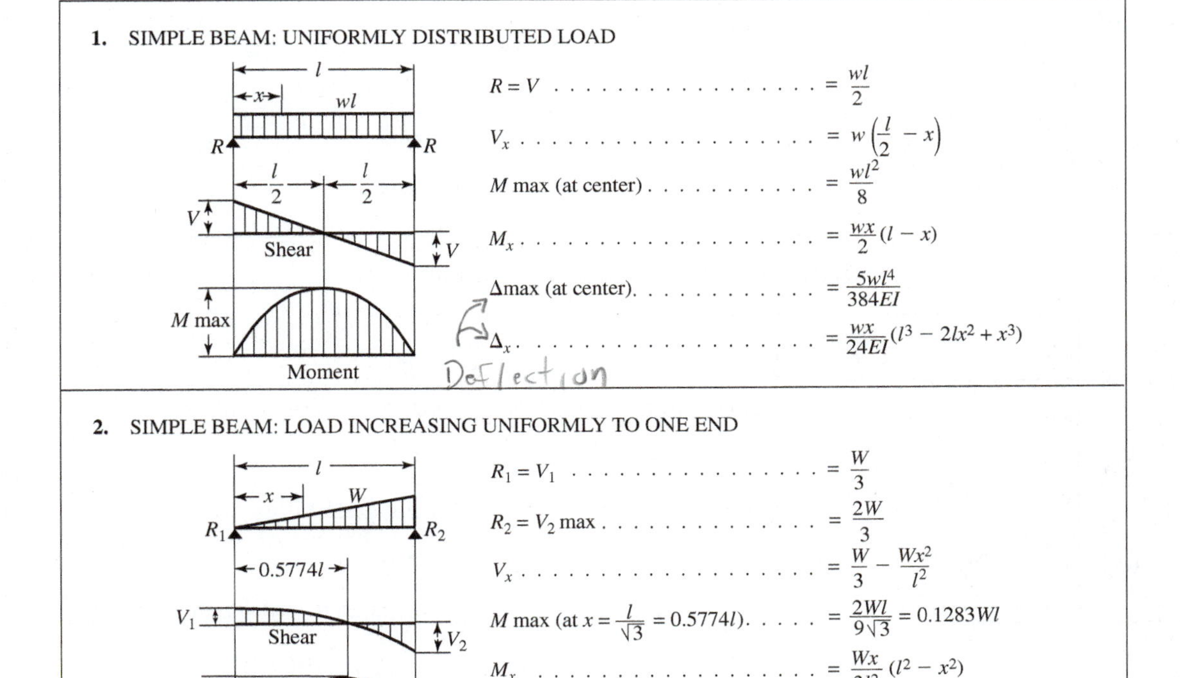

$R = V \dots = \frac{wl}{2}$

$V_x \dots = w\left(\frac{l}{2} - x\right)$

M max (at center) $\dots = \frac{wl^2}{8}$

$M_x \dots = \frac{wx}{2}(l - x)$

Δmax (at center) $\dots = \frac{5wl^4}{384EI}$

$\Delta_x \dots = \frac{wx}{24EI}(l^3 - 2lx^2 + x^3)$

2. SIMPLE BEAM: LOAD INCREASING UNIFORMLY TO ONE END

$R_1 = V_1 \dots = \frac{W}{3}$

$R_2 = V_2$ max $\dots = \frac{2W}{3}$

$V_x \dots = \frac{W}{3} - \frac{Wx^2}{l^2}$

M max $\left(\text{at } x = \frac{l}{\sqrt{3}} = 0.5774l\right) \dots = \frac{2Wl}{9\sqrt{3}} = 0.1283Wl$

$M_x \dots = \frac{Wx}{3l^2}(l^2 - x^2)$

Δmax $\left(\text{at } x = l\sqrt{1 - \sqrt{\frac{8}{15}}} = 0.5193l\right) \dots = 0.01304\frac{Wl^3}{EI}$

$\Delta_x \dots = \frac{Wx}{180EIl^2}(3x^4 - 10l^2x^2 + 7l^4)$

Reproduced courtesy of the American Institute of Steel Construction.

BEAM DIAGRAMS AND FORMULAS
For various static loading conditions

3. SIMPLE BEAM: LOAD INCREASING UNIFORMLY TO CENTER

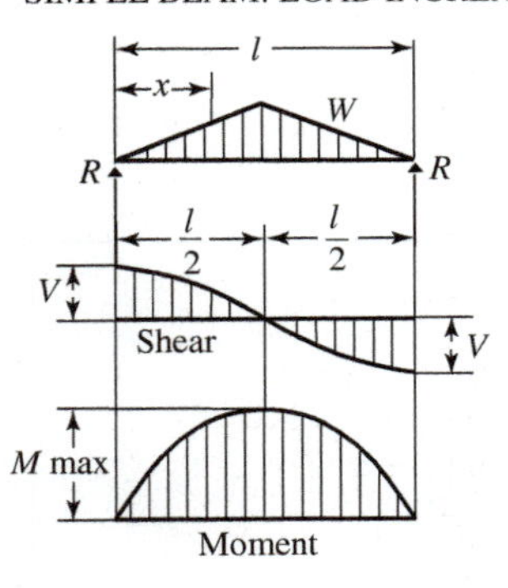

$R = V \quad \dots\dots \quad = \frac{W}{2}$

$V_x \left(\text{when } x < \frac{l}{2}\right) \quad \dots\dots \quad = \frac{W}{2l^2}(l^2 - 4x^2)$

$M \text{ max (at center)} \quad \dots\dots \quad = \frac{Wl}{6}$

$M_x \left(\text{when } x < \frac{l}{2}\right) \quad \dots\dots \quad = Wx\left(\frac{1}{2} - \frac{2x^2}{3l^2}\right)$

$\Delta\text{max (at center)} \quad \dots\dots \quad = \frac{Wl^2}{60EI}$

$\Delta_x \left(\text{when } x < \frac{l}{2}\right) \quad \dots\dots \quad = \frac{Wx}{480EIl^2}(5l^2 - 4x^2)^2$

4. SIMPLE BEAM: UNIFORM LOAD PARTIALLY DISTRIBUTED AT ONE END

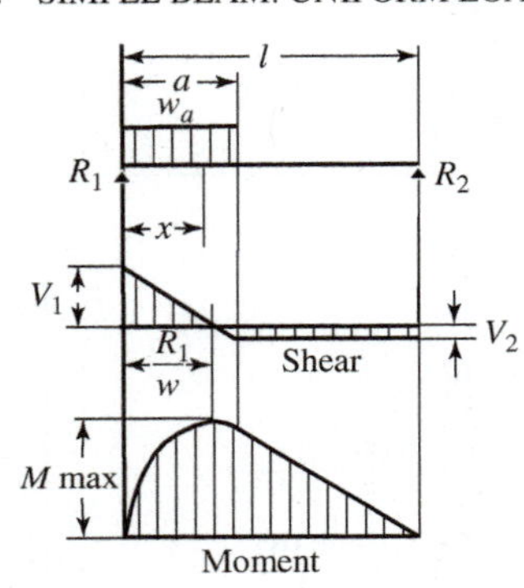

$R_1 = V_1 \text{ max} \quad \dots\dots \quad = \frac{wa}{2l}(2l - a)$

$R_2 = V_2 \quad \dots\dots \quad = \frac{wa^2}{2l}$

$V_x (\text{when } x < a) \quad \dots\dots \quad = R_1 - wx$

$M \text{ max} \left(\text{at } x = \frac{R_1}{w}\right) \quad \dots\dots \quad = \frac{R_1^2}{2w}$

$M_x (\text{when } x < a) \quad \dots\dots \quad = R_1 x - \frac{wx^2}{2}$

$M_x (\text{when } x > a) \quad \dots\dots \quad = R_2(l - x)$

$\Delta_x (\text{when } x < a) \quad \dots\dots \quad = \frac{wx}{24EIl}(a^2(2l - a)^2 - 2ax^2(2l - a) + lx^3)$

$\Delta_x (\text{when } x > a) \quad \dots\dots \quad = \frac{wa^2(l - x)}{24EIl}(4xl - 2x^2 - a^2)$

5. SIMPLE BEAM: CONCENTRATED LOAD AT CENTER

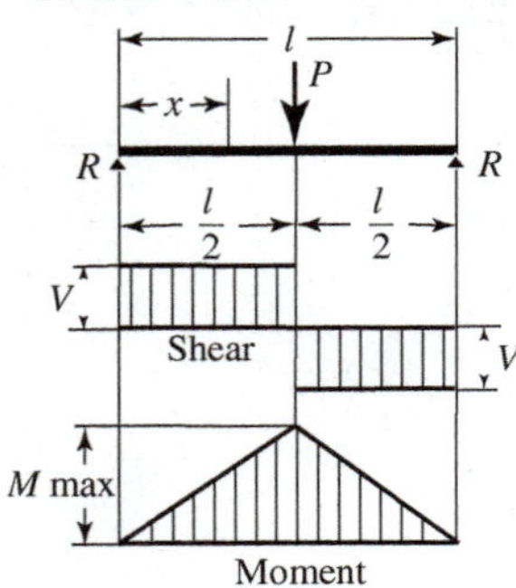

$R = V \quad \dots\dots \quad = \frac{P}{2}$

$M \text{ max (at point of load)} \quad \dots\dots \quad = \frac{Pl}{4}$

$M_x \left(\text{when } x < \frac{l}{2}\right) \quad \dots\dots \quad = \frac{Px}{2}$

$\Delta\text{max (at point of load)} \quad \dots\dots \quad = \frac{Pl^3}{48EI}$

$\Delta_x \left(\text{when } x < \frac{l}{2}\right) \quad \dots\dots \quad = \frac{Px}{48EI}(3l^2 - 4x^2)$

BEAM DIAGRAMS AND FORMULAS
For various static loading conditions

6. SIMPLE BEAM: CONCENTRATED LOAD AT ANY POINT

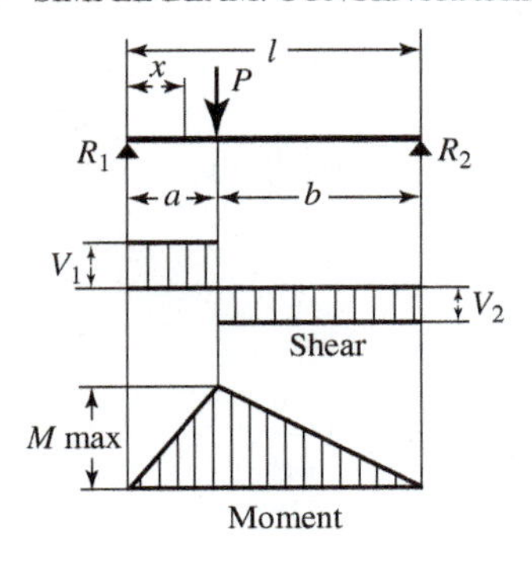

$R_1 = V_1$ (max when $a < b$) . . . . . . $= \dfrac{Pb}{l}$

$R_2 = V_2$ (max when $a > b$) . . . . . . $= \dfrac{Pa}{l}$

M max (at point of load) . . . . $= \dfrac{Pab}{l}$

M_x (when $x < a$) . . . . $= \dfrac{Pbx}{l}$

$\Delta\text{max}\left(\text{at } x = \sqrt{\dfrac{a(a+2b)}{3}} \text{ when } a > b\right) = \dfrac{Pab(a+2b)\sqrt{3a(a+2b)}}{27EIl}$

Δ_a (at point of load) . . . . $= \dfrac{Pa^2b^2}{3EIl}$

Δ_x (when $x < a$) . . . . $= \dfrac{Pbx}{6EIl}(l^2 - b^2 - x^2)$

7. SIMPLE BEAM: TWO EQUAL CONCENTRATED LOADS SYMMETRICALLY PLACED

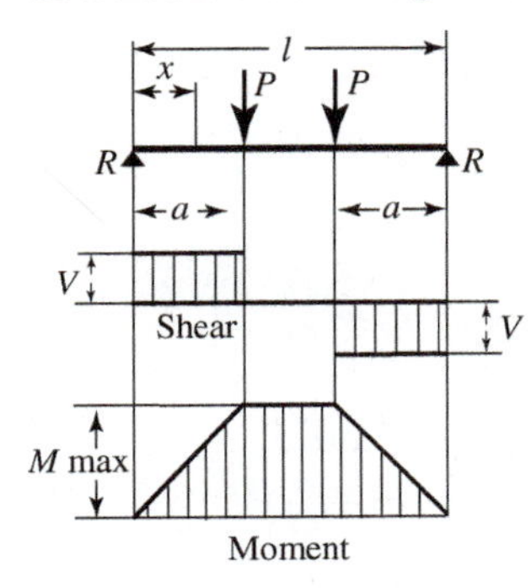

$R = V$ . . . . $= P$

M max (between loads) . . . . $= Pa$

M_x (when $x < a$) . . . . $= Px$

Δmax (at center) . . . . $= \dfrac{Pa}{24EI}(3l^2 - 4a^2)$

Δ_x (when $x < a$) . . . . $= \dfrac{Px}{6EI}(3la - 3a^2 - x^2)$

$\Delta_x\left(\text{when } x > a \text{ and } < (l - a)\right)$ . . . . $= \dfrac{Pa}{6EI}(3lx - 3x^2 - a^2)$

8. BEAM FIXED AT ONE END, SUPPORTED AT OTHER: UNIFORMLY DISTRIBUTED LOAD

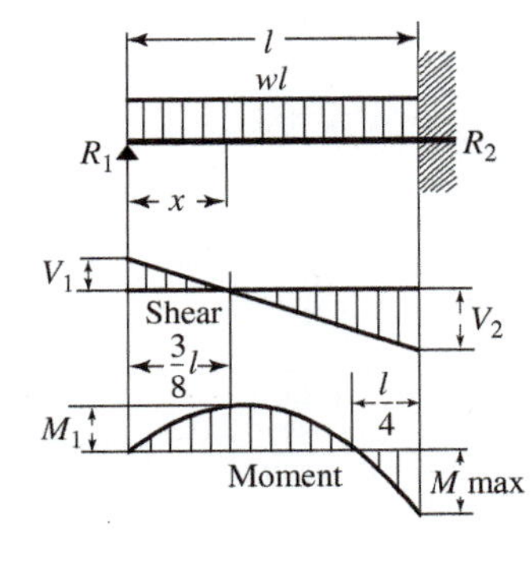

$R_1 = V_1$ . . . . $= \dfrac{3wl}{8}$

$R_2 = V_2$ max . . . . $= \dfrac{5wl}{8}$

V_x . . . . $= R_1 - wx$

M max . . . . $= \dfrac{wl^2}{8}$

$M_1\left(\text{at } x = \dfrac{3}{8}l\right)$ . . . . $= \dfrac{9}{128}wl^2$

M_x . . . . $= R_1x - \dfrac{wx^2}{2}$

$\Delta\text{max}\left(\text{at } x = \dfrac{l}{16}(1 + \sqrt{33}) = 0.4215l\right) = \dfrac{wl^4}{185EI}$

Δ_x . . . . $= \dfrac{wx}{48EI}(l^3 - 3lx^2 + 2x^3)$

BEAM DIAGRAMS AND FORMULAS
For various static loading conditions

9. BEAM FIXED AT BOTH ENDS: UNIFORMLY DISTRIBUTED LOADS

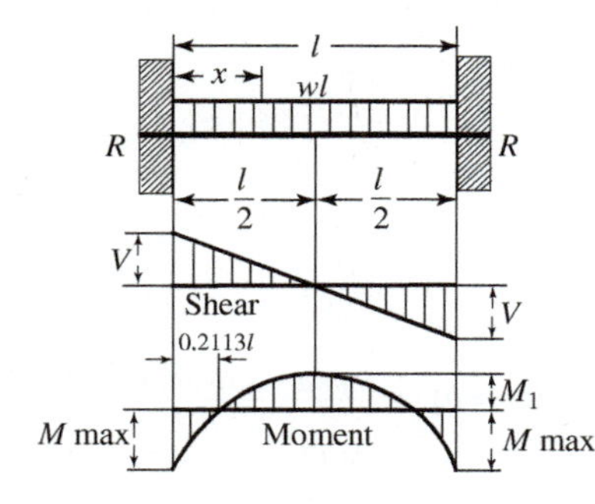

$R = V$ . . . . $= \frac{wl}{2}$

V_x . . . . $= w\left(\frac{l}{2} - x\right)$

M max (at ends) . . . . $= \frac{wl^2}{12}$

M_1 (at center) . . . . $= \frac{wl^2}{24}$

M_x . . . . $= \frac{w}{12}(6lx - l^2 - 6x^2)$

Δmax (at center) . . . . $= \frac{wl^4}{384EI}$

Δ_x . . . . $= \frac{wx^2}{24EI}(l - x)^2$

10. BEAM FIXED AT BOTH ENDS: CONCENTRATED LOAD AT CENTER

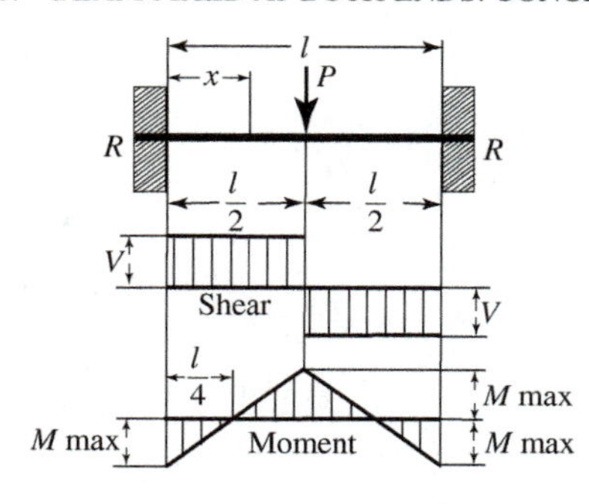

$R = V$ . . . . $= \frac{P}{2}$

M max (at center and ends) . . . . . . $= \frac{Pl}{8}$

$M_x\left(\text{when } x < \frac{l}{2}\right)$ . . . . $= \frac{P}{8}(4x - l)$

Δmax (at center) . . . . $= \frac{Pl^3}{192EI}$

$\Delta_x\left(\text{when } x < \frac{l}{2}\right)$ . . . . $= \frac{Px^2}{48EI}(3l - 4x)$

11. CANTILEVER BEAM: LOAD INCREASING UNIFORMLY TO FIXED END

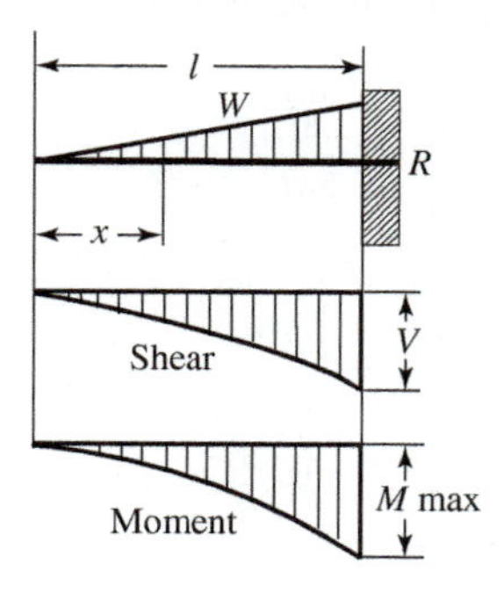

$R = V$ . . . . $= W$

V_x . . . . $= W\frac{x^2}{l^2}$

M max (at fixed end) . . . . $= \frac{Wl}{3}$

M_x . . . . $= \frac{Wx^3}{3l^2}$

Δmax (at free end) . . . . $= \frac{Wl^3}{15EI}$

Δ_x . . . . $= \frac{W}{60EIl^2}(x^5 - 5l^4x + 4l^5)$

BEAM DIAGRAMS AND FORMULAS
For various static loading conditions

12. CANTILEVER BEAM: UNIFORMLY DISTRIBUTED LOAD

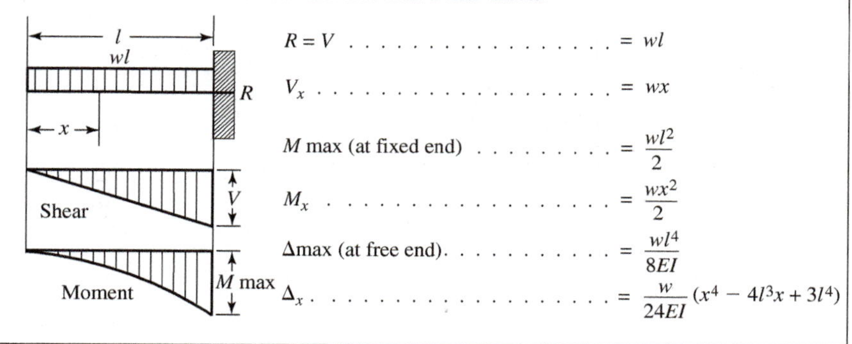

$R = V \ldots\ldots\ldots = wl$

$V_x \ldots\ldots\ldots = wx$

M max (at fixed end) $\ldots\ldots = \dfrac{wl^2}{2}$

$M_x \ldots\ldots\ldots = \dfrac{wx^2}{2}$

Δmax (at free end) $\ldots\ldots = \dfrac{wl^4}{8EI}$

$\Delta_x \ldots\ldots\ldots = \dfrac{w}{24EI}(x^4 - 4l^3x + 3l^4)$

13. CANTILEVER BEAM: CONCENTRATED LOAD AT ANY POINT

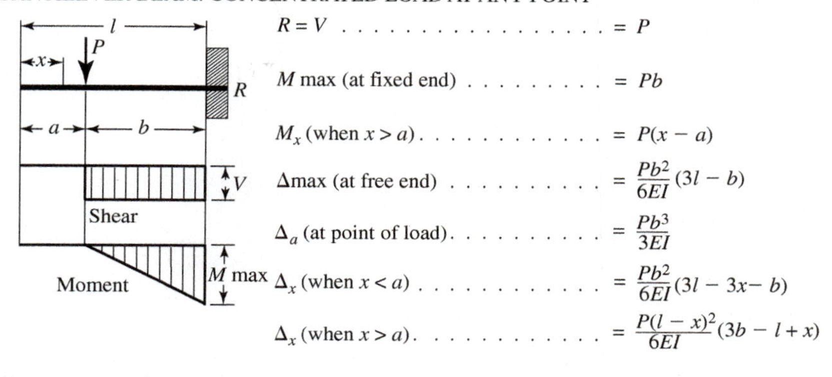

$R = V \ldots\ldots\ldots = P$

M max (at fixed end) $\ldots\ldots = Pb$

M_x (when $x > a$) $\ldots\ldots = P(x - a)$

Δmax (at free end) $\ldots\ldots = \dfrac{Pb^2}{6EI}(3l - b)$

Δ_a (at point of load) $\ldots\ldots = \dfrac{Pb^3}{3EI}$

Δ_x (when $x < a$) $\ldots\ldots = \dfrac{Pb^2}{6EI}(3l - 3x - b)$

Δ_x (when $x > a$) $\ldots\ldots = \dfrac{P(l - x)^2}{6EI}(3b - l + x)$

14. CANTILEVER BEAM: CONCENTRATED LOAD AT FREE END

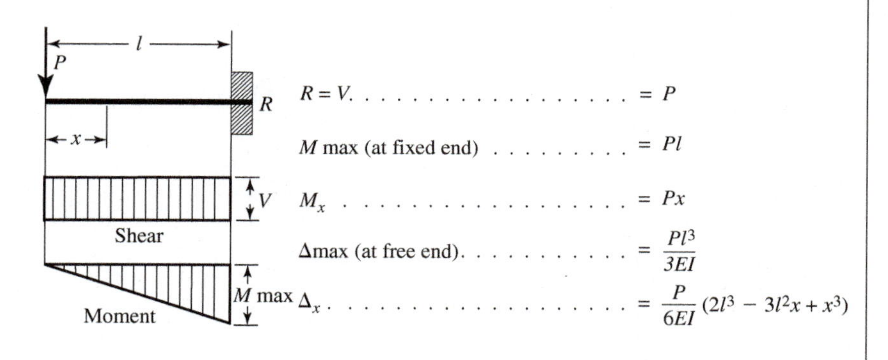

$R = V \ldots\ldots\ldots = P$

M max (at fixed end) $\ldots\ldots = Pl$

$M_x \ldots\ldots\ldots = Px$

Δmax (at free end) $\ldots\ldots = \dfrac{Pl^3}{3EI}$

$\Delta_x \ldots\ldots\ldots = \dfrac{P}{6EI}(2l^3 - 3l^2x + x^3)$

APPENDIX I

STEEL BEAM SELECTION TABLE (Z_X)

TABLE I.1 U.S. Customary units (Listed by Decreasing Z_x)

Shape	Z_x (in.^{3})	I_x (in.^{4})
W36 × 302	**1280**	21,100
W40 × 277	**1250**	21,900
W36 × 194	**767**	12,100
W36 × 150	**581**	9040
W24 × 162	468	5170
W33 × 118	**415**	5900
W30 × 108	**346**	4470
W27 × 114	343	4080
W30 × 99	**312**	3990
W24 × 104	289	3100
W27 × 94	**278**	3270
W14 × 132	234	1530
W24 × 84	**224**	2370
W24 × 76	**200**	2100
W16 × 100	198	1490
W21 × 83	196	1830
W14 × 109	192	1240
W21 × 73	**172**	1600
W14 × 90	157	999
W12 × 96	147	833
W10 × 112	147	716
W18 × 71	**146**	1170
W21 × 62	**144**	1330
W14 × 74	126	795
W18 × 60	**123**	984
W21 × 50	**110**	984
W12 × 72	108	597
W14 × 61	102	640
W18 × 50	101	800
W12 × 58	86.4	475
W10 × 68	85.3	394
W16 × 45	**82.3**	586
W18 × 40	**78.4**	612
W14 × 48	78.4	484
W12 × 50	71.9	391
W10 × 54	66.6	303
W16 × 36	**64**	448
W12 × 40	57	307
W10 × 45	54.9	248
W14 × 34	**54.6**	340
W14 × 30	**47.3**	291
W16 × 26	**44.2**	301
W12 × 30	43.1	238
W8 × 40	39.8	146
W10 × 33	38.8	171
W14 × 22	**33.2**	199
W8 × 31	30.4	110
W12 × 22	29.3	156
W10 × 22	26	118
W8 × 24	23.1	82.7
W12 × 16	**20.1**	103
W6 × 25	18.9	53.4
W10 × 12	**12.6**	53.8
W8 × 13	11.4	39.6
W6 × 12	8.3	22.1
W4 × 13	6.28	11.3

TABLE I.2 SI units (Listed by Decreasing Z_x)

Shape	Z_x 10^3 mm^3	I_x 10^6 mm^4
W920 × 390	**18,000**	7450
W840 × 329	**14,000**	5370
W1000 × 222	**9800**	4080
W610 × 153	**4590**	1250
W530 × 138	**3620**	862
W310 × 202	3510	516
W610 × 113	**3280**	874
W530 × 101	**2620**	616
W410 × 100	**2130**	397
W360 × 110	2060	331
W530 × 72	**1750**	399
W310 × 107	1770	248
W250 × 115	1600	189
W310 × 97	1590	222
W460 × 60	**1280**	255
W250 × 73	990	113
W200 × 59	**652**	60.8
W310 × 39	**610**	84.9
W150 × 24	**192**	13.4

STEEL BEAM SELECTION TABLE (I_X)

TABLE J.1 U.S. Customary units (Listed by Decreasing I_X)

Shape	I_x (in.^{4})	Z_x (in.^{3})
W40 × 277	**21,900**	1250
W36 × 302	21,100	1280
W36 × 194	**12,100**	767
W36 × 150	9040	581
W33 × 118	**5900**	415
W24 × 162	5170	468
W30 × 108	**4470**	346
W27 × 114	4080	343
W30 × 99	**3990**	312
W27 × 94	**3270**	278
W24 × 104	3100	289
W24 × 84	**2370**	224
W24 × 76	**2100**	200
W21 × 83	1830	196
W21 × 73	**1600**	172
W14 × 132	1530	234
W16 × 100	1490	198
W21 × 62	**1330**	144
W14 × 109	1240	192
W18 × 71	1170	146
W14 × 90	999	157
W18 × 60	**984**	123
W21 × 50	**984**	110
W12 × 96	833	147
W18 × 50	800	101
W14 × 74	795	126
W10 × 112	716	147
W14 × 61	640	102
W18 × 40	**612**	78.4
W12 × 72	597	108
W16 × 45	586	82.3
W14 × 48	484	78.4
W12 × 58	475	86.4
W16 × 36	**448**	64
W10 × 68	394	85.3
W12 × 50	391	71.9
W14 × 34	**340**	54.6
W12 × 40	307	57
W10 × 54	303	66.6
W16 × 26	**301**	44.2
W14 × 30	291	47.3
W10 × 45	248	54.9
W12 × 30	238	43.1
W14 × 22	**199**	33.2
W10 × 33	171	38.8
W12 × 22	156	29.3
W8 × 40	146	39.8
W10 × 22	118	26
W8 × 31	110	30.4
W12 × 16	103	20.1
W8 × 24	82.7	23.1
W10 × 12	**53.8**	12.6
W6 × 25	53.4	18.9
W8 × 13	39.6	11.4
W6 × 12	22.1	8.3
W4 × 13	11.3	6.28

TABLE J.2 SI units (Listed by Decreasing I_x)

Shape	I_x 10^6 mm^4	Z_x 10^3 mm^3
W920 × 390	**7450**	18,000
W840 × 329	**5370**	14,000
W1000 × 222	4080	9800
W610 × 153	1250	4590
W610 × 113	**874**	3280
W530 × 138	862	3620
W530 × 101	**616**	2620
W310 × 202	516	3510
W530 × 72	**399**	1750
W410 × 100	397	2130
W360 × 110	331	2060
W460 × 60	**255**	1280
W310 × 107	248	1770
W310 × 97	222	1590
W250 × 115	189	1600
W250 × 73	113	990
W310 × 38.7	**84.9**	610
W200 × 59	60.8	652
W150 × 24	**13.4**	192

CENTROIDS OF AREAS BY INTEGRATION

In Section 7.3, we saw that by applying Varignon's theorem, equations for $\bar{x}$ and $\bar{y}$, which locate the centroid of an area with respect to given reference axes, could be written as

$$\bar{x} = \frac{\Sigma ax}{A} \quad \text{or} \quad \frac{\Sigma ax}{\Sigma a} \tag{7.3}$$

$$\bar{y} = \frac{\Sigma ay}{A} \quad \text{or} \quad \frac{\Sigma ay}{\Sigma a} \tag{7.4}$$

Recall that in using Equations (7.3) and (7.4), the total area in question was divided into component areas of known geometric shapes. For each component area, the centroid location and the area were known.

Alternatively, if an area is bounded by lines and/or curves that can be defined mathematically, the location of its centroid can be obtained using integral calculus. Since integration is the process of summing up infinitesimal quantities, the total area is divided into infinitesimally small component areas (commonly termed *differential areas* or *differential elements*) and the integration process is performed. The equations, with respect to the X and Y axes, then become

$$\bar{x} = \frac{\int x dA}{\int dA} \tag{7.3/K}$$

$$\bar{y} = \frac{\int y dA}{\int dA} \tag{7.4/K}$$

When using integration, note that we must still be able to determine the magnitude of the area of the typical differential element, as well as the location of its centroid. In all cases, limits of integration must be established so that all the differential elements are included in the integration.

EXAMPLE K.1 A triangle is shown in Figure K.1. Using integration, find the vertical distance from the base of the triangle (the reference axis) to its centroid.

Solution A typical differential area dA (where $dA = Ldy$) is shown. The length L is determined by similar triangles:

$$\frac{L}{b} = \frac{h - y}{h}$$

$$L = \frac{b(h - y)}{h}$$

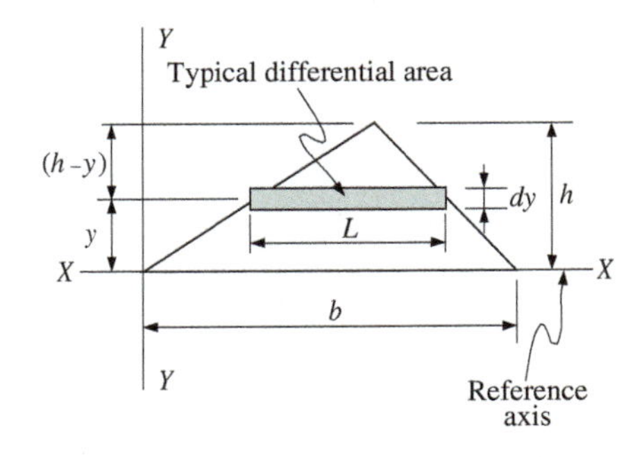

FIGURE K.1 Centroid of a triangle by integration.

From Equation (7.4/K), and integrating from 0 to h:

$$\bar{y} = \frac{\int y dA}{\int dA} = \frac{\int_0^h yLdy}{\int_0^h Ldy} = \frac{\int_0^h y\frac{b(h-y)}{h}dy}{\int_0^h \frac{b(h-y)}{h}dy}$$

$$= \frac{\int_0^h y(h-y)dy}{\int_0^h (h-y)dy} = \frac{\int_0^h hydy - \int_0^h y^2dy}{\int_0^h hdy - \int_0^h ydy}$$

$$= \frac{\left[\frac{hy^2}{2}\right]_0^h - \left[\frac{y^3}{3}\right]_0^h}{[hy]_0^h - \left[\frac{y^2}{2}\right]_0^h} = \frac{\frac{h^3}{2} - \frac{h^3}{3}}{h^2 - \frac{h^2}{2}} = \frac{h}{3}$$

EXAMPLE K.2 A semicircle of radius r is shown in Figure K.2. Using integration, find the distance from the base (reference axis) of the semicircle to its centroid.

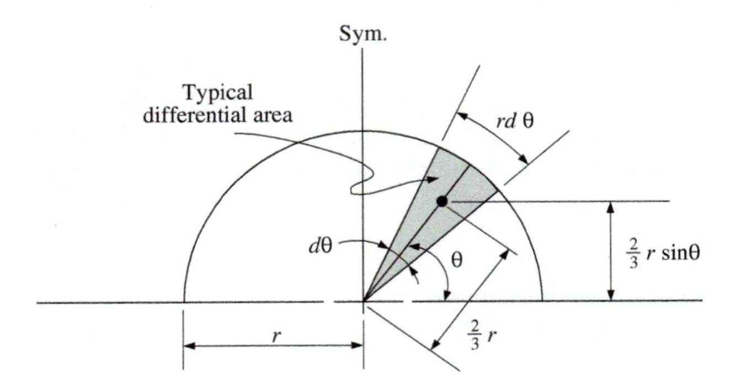

FIGURE K.2 Centroid of a semicircle by integration.

Solution This problem could be solved using rectangular coordinates, but polar coordinates are more convenient. The typical differential area shown is a triangle of height r and base $rd\theta$ (where θ is measured in radians). Therefore, the area of the differential area is

$$dA = \frac{1}{2}(rd\theta)r = \frac{r^2d\theta}{2}$$

From Example K.1, we know that the centroid of a triangle lies two-thirds of its height from the vertex. The distance from the reference axis to the centroid of the differential area is then $\frac{2}{3}r\sin\theta$. From Equation (7.4/K) and integrating from 0 to π:

$$\bar{y} = \frac{\int ydA}{\int dA} = \frac{\int_0^\pi \left(\frac{2}{3}r\sin\theta\right)\left(\frac{r^2d\theta}{2}\right)}{\int_0^\pi \frac{r^2d\theta}{2}}$$

$$= \frac{\frac{r^3}{3}\int_0^\pi \sin\theta\, d\theta}{\frac{r^2}{2}\int_0^\pi d\theta} = \frac{\frac{r^3}{3}[-\cos\theta]_0^\pi}{\frac{r^2}{2}[\theta]_0^\pi}$$

$$= \frac{\frac{r^3}{3}[-(-1)-(-1)]}{\frac{r^2}{2}(\pi - 0)} = \frac{4r}{3\pi}$$

APPENDIX L

AREA MOMENTS OF INERTIA BY INTEGRATION

In Chapter 8, the moment of inertia of a plane area A was defined as

$$I_x = \Sigma ay^2 \quad \textbf{(8.1)}$$

$$I_y = \Sigma ax^2 \quad \textbf{(8.2)}$$

The term a represented a small area, which was a component of area A, and x or y represented the distance from the small component area a to the axis being considered. Equations (8.1) and (8.2) furnished an approximate moment of inertia, the accuracy being a function of the size of the small area chosen.

If the area A can be defined mathematically by lines and/or curves, we may use integral calculus to perform the summation. The integration process is accomplished by dividing the plane area into an infinite number of differential areas (each designated dA) and then summing the moments of inertia of all the differential areas. The result is an exact moment of inertia. With reference to Figure 8.1, the moment of inertia with respect to axes X–X and Y–Y may then be expressed as

$$I_x = \int y^2 dA \quad \textbf{(8.1/L)}$$

$$I_y = \int x^2 dA \quad \textbf{(8.2/L)}$$

The integration process is used to derive the theoretical and exact moment of inertia formulas of the simple geometric shapes presented in Table 8.1. Examples illustrating this process follow.

EXAMPLE L.1 Determine the moment of inertia for the rectangular area shown in Figure L.1 with respect to (a) the centroidal axis parallel to the base and (b) an axis coinciding with the base.

Solution (a) The typical differential area dA (where $dA = bdy$) is shown. Selecting limits of integration as $h/2$ and $-h/2$, the moment of inertia about axis $X_o - X_o$ is calculated from

$$I_{x_o} = \int y^2 dA$$

$$= \int_{-h/2}^{h/2} y^2 bdy = b\int_{-h/2}^{h/2} y^2 dy$$

$$= \frac{b}{3}[y^3]_{-h/2}^{h/2} = \frac{bh^3}{12}$$

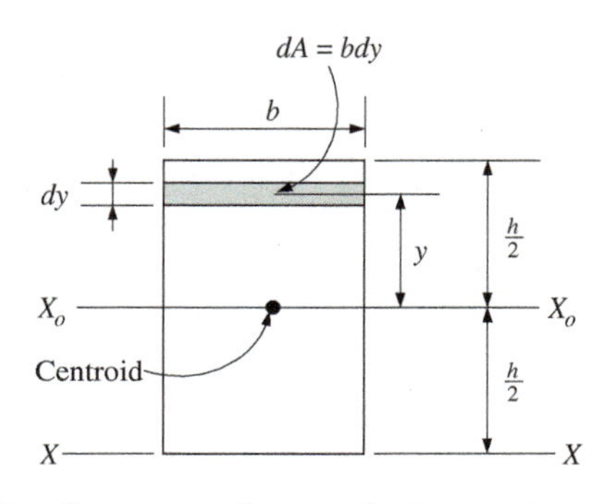

FIGURE L.1 Moment of inertia of a rectangular area by integration.

(b) Selecting limits of integration of 0 and h, the moment of inertia about axis X–X is calculated from

$$I_x = \int_0^h y^2 b dy = b\int_0^h y^2 dy$$

$$= \frac{b}{3}[y^3]_0^h = \frac{bh^3}{3}$$

EXAMPLE L.2 Determine the moment of inertia with respect to the base for the triangular area shown in Figure L.2.

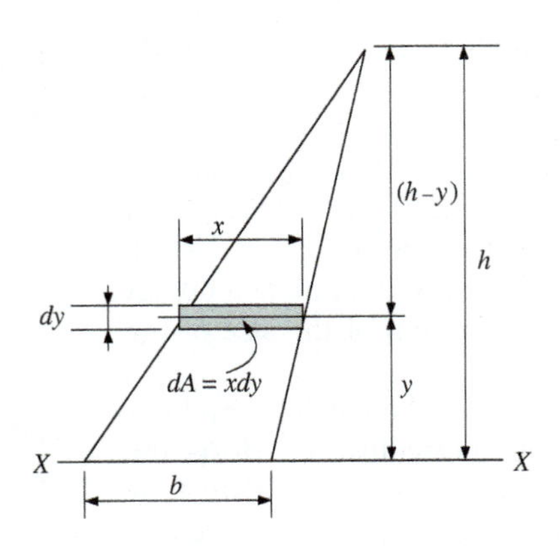

FIGURE L.2 Moment of inertia of a triangular area by integration.

Solution The typical differential area dA (where $dA = xdy$) is shown. The length x must be written in terms of y. From similar triangles,

$$\frac{x}{b} = \frac{h-y}{h}$$

from which

$$x = \frac{b}{h}(h-y)$$

Selecting limits of integration of 0 and h, the moment of inertia with respect to axis X–X is calculated:

$$I_x = \int y^2 dA = \int y^2 x dy$$

$$= \int_0^h y^2 \frac{b}{h}(h-y)dy$$

$$= \frac{b}{h}\left[\int_0^h hy^2 dy - \int_0^h y^3 dy\right]$$

$$= \frac{b}{h}\left[\frac{hy^3}{3} - \frac{y^4}{4}\right]_0^h = \frac{bh^3}{12}$$

NOTATION

Symbols

A = total cross-sectional area, cross-sectional area over which stress develops, required cross-sectional area

A_M = area-of-moment diagram

C = compression, compressive force

C_c = arbitrary value of effective slenderness ratio that separates elastic and inelastic buckling

D = diameter

E = modulus of elasticity (Young's modulus)

F = force, load, frictional force, allowable stress

G = modulus of elasticity in shear, modulus of rigidity

I = area moment of inertia

J = area polar moment of inertia

K = effective length factor

L = length, span length of beam, lead of screw

M = moment of a force

M_P = plastic moment

M_R = moment strength

M_y = yield moment

N = force or reaction acting perpendicular (normal) to a surface, number of bolts

P = force, load, axial load capacity

P_e = critical load for column buckling

Q = force, statical moment of area

R = reaction, resultant force or load, radius of curvature of the elastic curve

S = section modulus

T = tension, tensile force, torque

V = shear force

V_R = shear strength

W = weight of a body, load, total distributed load, total weight

Z = plastic section modulus

a = component area, infinitesimal area

b = width of cross section

c = distance to outer fiber

d = perpendicular distance between moment center and a force or between forces of a couple, diameter or linear dimension of a geometric shape, transfer distance for moment of inertia calculation

e = base of natural logarithms (2.718), eccentricity

k = stress concentration factor

g = acceleration of gravity (9.81 m/sec^2)

h = linear dimension for a geometric shape, height of cross section

m = mass

n = modular ratio, ratio of modulus of elasticity values, number of shear planes per bolt

n_r = number of revolutions per minute

p = pitch of thread, internal pressure

r = radius, mean radius of a screw, radius of gyration, radial distance

f = average computed stress, unit stress

t = thickness

w = uniformly distributed line load intensity, weight of component element

$\bar{x}$ = distance from a resultant force to a reference point or plane, distance to centroidal Y–Y axis from reference line

y = vertical displacement

$\bar{y}$ = distance to centroidal X–X axis from reference line

Abbreviations

AASHTO = American Association of State Highway and Transportation Officials

AF&PA = American Forest & Paper Association

AISI = American Iron and Steel Institute

AITC = American Institute of Timber Construction

ASD = allowable stress design

ASME = American Society of Mechanical Engineers

ASTM = American Society for Testing and Materials

AWS = American Welding Society

C° = Celsius degrees

CG = center of gravity

F° = Fahrenheit degrees

FS = factor of safety

hp = horsepower

MPa = megapascal (= 1 N/mm^2)

N = newton

Pa = pascal (= 1 N/m^2)

W = watt
ft = foot
g = gram
in. = inch
kip = kilopound (1000 lb)
kg = kilogram
kN = kilonewton
kPa = kilopascal
ksi = kips per square inch
kW = kilowatt
lb = pound
m = meter
mm = millimeter
psf = pounds per square foot
psi = pounds per square inch
rpm = revolutions per minute
r/s = radian per second
sym = symmetrical

Greek Letter Symbols

α (lowercase alpha) = angular value, linear coefficient of thermal expansion
β (lowercase beta) = angle of contact for belt friction, angle of wrap
Δ (uppercase delta) = deflection
ΔT = change in temperature
δ (lowercase delta) = total deformation (total change in length)
ε (lowercase epsilon) = strain, unit strain
γ (lowercase gamma) = unit weight of fresh water ($= 62.4\ \text{lb/ft}^3$)
θ (lowercase theta) = angular value, lead angle, angle of twist, angle between two tangents
μ (lowercase mu) = coefficient of friction, Poisson's ratio
ϕ (lowercase phi) = angular value, angle of friction, resistance factor
σ (lowercase sigma) = direct stress
τ (lowercase tau) = Direct shear stress

ANSWERS TO SELECTED PROBLEMS

Chapter 1

1.1. **(a)** 12.21 ft
(b) 12.00 m

1.3. **(a)** $a = 25.59$ ft
(b) $A = 20.8°$, $B = 69.2°$

1.5. $c = 11.66$ ft, $\theta = 31.0°$

1.7. $x = 6.63$ ft, $\theta = 56.4°$

1.9. **(a)** $c = 15.50$ ft, $A = 44.3°$, $B = 55.7°$
(b) $a = 96.0$ ft, $C = 57.4°$, $B = 50.6°$
(c) $A = 16.26°$, $B = 73.7°$, $C = 90°$

1.11. Pads: 3.20 lb; shock: 8.70 lb

1.13. **(a)** 30.0 lb
(b) 480 oz

1.15. 88 ft/sec

1.17. **(a)** 384×10^3 acre-ft
(b) 521×10^6 tons

1.19. 431 gal

1.23. **(a)** 1003 kg
(b) 8.08×10^6 mm^2
(c) 127,300 kPa

1.25. **(a)** 1.103 kN
(b) 16.24 kN
(c) 77.4 kN

1.27. 1571 liters

1.29. **(a)** 13.47 ft
(b) 6.11 ft
(c) 27°

1.31. 45.7 m

1.33. 1350 km

1.35. $h = 500$ ft, slant ht $= 625$ ft, base width $= 750$ ft

1.37. 57.0 m

1.39. 2,240,000 gal/min, 1.071 mi^3/yr

1.41. Bldg: 17,060 ft^2; lot: 72,900 ft^2

Chapter 2

2.1. **(a)** 78.1 lb, $\theta_x = 50.2°$
(b) 284 N, $\theta_x = 50.7°$
(c) 4.27 kips, $\theta_x = 20.6°$

2.3. **(a)** $F_x = +520$ lb, $F_y = -300$ lb
(b) $F_x = -3.54$ k, $F_y = -3.54$ k
(c) $F_x = -600$ lb, $F_y = +1039$ lb

2.5. **(a)** $P_x = +301$ lb, $P_y = -109.4$ lb
(b) $P_x = +554$ lb, $P_y = -320$ lb
(c) $P_x = +245$ lb, $P_y = -206$ lb
(d) $P_x = +11.17$ lb, $P_y = -320$ lb

2.7. $R = 247$ lb, $\theta_x = 45°$

2.11. $P_x = 77.1$ lb, $P_y = 91.9$ lb

2.13. **(a)** $F_y = -611$ lb, $F_x = +222$ lb
(b) $F_x = +277$ lb, $F_y = -115$ lb

2.15. **(a)** $F_y = -693$ lb, $F_x = -400$ lb
(b) $F_x = -90$ lb, $F_y = -120$ lb

2.17. **(a)** $P_y = -75.2$ k, $P_x = -27.4$ k
(b) $P_x = -30.0$ kN, $P_y = +52.0$ kN

2.19. $\theta = 27.6°$

2.21. 1944 lb

2.23. -8250 lb, -4710 lb

2.25. $P_{1x} = +1.351$ k, $P_{1y} = +7.66$ k,
$P_{2x} = +6.43$ k, $P_{2y} = -7.66$ k

2.27. $T_V = 1879$ lb, $T_H = 684$ lb

2.29. 6-lb force: $F_y = 5.65$ lb, $F_x = 2.02$ lb
9-lb force: $P_y = 8.07$ lb, $P_x = 3.99$ lb

Chapter 3

3.1. $R = 64.0$ lb, $\theta_x = 27.4°$

3.3. $R = 65.3$ lb, $\theta_x = 76.6°$

3.5. $R = 13.60$ k, $\theta_x = 68.1°$

3.7. $F = 87.9$ lb, $\theta_x = 3.87°$

3.9. $R = 265$ N vertically downward

3.11. $R = 675$ lb, $\theta_x = 36.3°$

3.13. **(a)** $R = 47.2$ N
(b) $\theta_x = 10.18°$

3.15. $F_1 = 512$ lb, $\theta_x = 55.2°$

3.17. **(a)** (lb-ft) -80, -360, $+750$, -690
(b) -380 lb-ft (clockwise)

3.19. **(a)** -40 lb-ft, $+103.9$ lb-ft, -113.1 lb-ft
(b) -49.2 lb-ft

3.21. **(a)** $M_A = -217 \text{ kN}\cdot\text{m}$
(b) $M_B = +215 \text{ kN}\cdot\text{m}$
3.23. $\Sigma M_o = -1429$ lb-ft
3.25. $\Sigma M_A = 2150$ lb-ft
3.27. $R = +10 \text{ k} \uparrow, \bar{x} = 6.7$ ft
3.29. $R = -3 \text{ k} \downarrow, \bar{x} = 5.67$ ft
3.31. $F_1 = 30 \text{ k} \downarrow, x_1 = 12.67$ ft
3.33. $R = -38 \text{ k} \downarrow, \bar{x} = 13.84$ ft
3.35. **(a)** 1123 psf
(b) 10,100 lb
(c) 6 ft
3.37. $R = 46 \text{ kN}, \bar{x} = 9.99$ m
3.39. -1100 lb-in. (clockwise)
3.41. 64 lb
3.43. 427 lb, $\theta_x = 20.6°, \bar{x} = 22.3$ in.
3.45. 71.6 k, $\theta_x = 77.9°, \bar{x} = 4.43$ ft
3.49. 59.5 N, $\theta_x = 47.8°$
3.51. $F_1 = F_2 = 579$ lb
3.53. $F_1 = 198.7 \text{ lb}, F_2 = 249$ lb
3.55. $R = 47.5 \text{ lb}, \theta_x = 84.9°$
3.57. $F_1 = 161.4 \text{ lb}, F_2 = 240$ lb
3.59. $R = 22{,}200 \text{ lb}, \theta_x = 63°, \bar{x} = 7.06$ ft, Yes
3.61. $F = 2570 \text{ lb}, \theta = 55.3°$
3.63. -76.0 k-ft
3.65. -55.0 k-ft
3.67. $1500 \text{ lb} \downarrow, \bar{x} = 7.20$ ft
3.69. **(a)** 3650 lb, 1.385 ft up
(b) 3276 lb, 1.467 ft up
3.71. **(a)** $M_A = -19{,}100$ lb-ft
(b) $R = 1700 \text{ lb}, \bar{x} = 23.9$ ft up
3.73. **(a)** $M_A = -192.7 \text{ kN}\cdot\text{m}, M_B = +1.825 \text{ MN}\cdot\text{m}$
(b) $R = 126.3 \text{ kN}, \theta_x = 87.2°, \bar{x} = 1.53$ m
3.75. $R = 26.0 \text{ lb}, \theta_x = 56.6°, \bar{x} = 4.82$ ft

Chapter 4

4.9. $F_{AB} = 224 \text{ lb}, \theta = 63.4°$
4.11. 53.2 lb, 68.4 lb
4.13. $F_1 = 70.3 \text{ kN}, F_2 = 55.2$ kN
4.15. $R_A = 12.45 \text{ k} \uparrow, R_B = 12.55 \text{ k} \uparrow$
4.17. 3.25 m
4.19. $R_A = +12.25 \text{ k}, R_B = +16.75$ k
4.21. $R_A = 3.06 \text{ kN} \uparrow, R_B = 18.74 \text{ kN} \uparrow$
4.23. $R_A = 3340 \text{ lb} \uparrow, R_B = 4260 \text{ lb} \uparrow$
4.25. $R_{AV} = 0.5 \text{ k} \uparrow, R_{AH} = 30 \text{ k} \leftarrow, R_B = 19.5 \text{ k} \uparrow$
4.27. $R_{AV} = 640 \text{ lb} \uparrow, R_{AH} = 815 \text{ lb} \rightarrow, R_{BV} = 553 \text{ lb} \uparrow,$ $R_{BH} = 415 \text{ lb} \leftarrow$
4.29. $F_{DB} = 2.54 \text{ kN}; R_A = 3.35 \text{ kN}, \theta_x = 42.9°$
4.31. $R_B = 5226 \text{ lb}; R_{Cx} = 4311 \text{ lb}, R_{Cy} = 4304$ lb
4.39. $CA = 18.07 \text{ kN}, B_H = 12.78 \text{ kN} \leftarrow, B_V = 2.77 \text{ kN} \downarrow$
4.41. $R_A = 3.80 \text{ k} \uparrow, R_B = 12.92 \text{ k} \uparrow$
4.43. $R_A = 12.50 \text{ kN} \uparrow, R_B = 0.50 \text{ kN} \downarrow$
4.45. $L = 10.0$ ft or 8.5 ft
4.47. $T = 7.01 \text{ lb}, \theta = 44.5°$
4.49. $R_A = 4.93 \text{ k} \uparrow, B_V = 7.07 \text{ k} \uparrow, B_H = 4 \text{ k} \rightarrow$
4.51. $R_A = 13.72 \text{ k} \uparrow, B_V = 18.28 \text{ k} \uparrow, B_H = 7 \text{ k} \leftarrow$
4.53. $R_A = 7.98 \text{ k} \uparrow, B_V = 3.18 \text{ k} \downarrow, B_H = 10 \text{ k} \rightarrow$
4.55. $CD = 33.2 \text{ k}, R_{AH} = 25.4 \text{ k} \rightarrow, R_{AV} = 5.33 \text{ k} \downarrow$
4.57. $BD = 15{,}330 \text{ lb}, R_{AH} = 12{,}270 \text{ lb} \leftarrow, R_{AV} = 4100 \text{ lb} \uparrow$
4.59. $BD = 725 \text{ lb}, C_H = 533 \text{ lb} \leftarrow, C_V = 281 \text{ lb} \downarrow$
4.61. Cylinder B: 740 N @ wall, 888 N @ Cyl. A
Cylinder A: 888 N @ Cyl. B, 1227 N @ floor, 740 N @ wall

Chapter 5

5.1. $AB = 7.07$ k (C), $AC = 5.0$ k (T)
5.3. (k) $AB = 20.8$ (C), $BC = 23.3$ (C), $CD = 29.2$ (C), $DE = 23.3$ (T), $EF = 26.7$ (T), $FA = 26.7$ (T), $BF = 15.0$ (T), $CE = 17.5$ (T), $BE = 4.17$ (C)
5.5. (#) $AB = 427$ (C), $BC = 774$ (C), $CD = 670$ (T), $DA = 670$ (T), $BD = 600$ (T)
5.7. (KN) $AB = 10$ (C), $BC = 16.7$ (T), $AC = 100$ (C), $BD = 113.3$ (C), $CD = 30$ (C), $CE = 86.7$ (C), $DE = 50$(T), $DF = 153.3$ (C), $EF = 0$
5.9. (#) $BC = 2700$ (T), $BE = 0$, $FE = 2340$ (C)
5.11. (#) $BC = 8940$ (C), $CI = 0$, $IJ = 8000$ (T)
5.13. (k) $BC = 6$ (T), $BG = 13.4$ (T), $FG = 12$ (C)
5.15. (#) $A = 1167$, $B = 1530$, $C = 769$, $D = 1004$, $E = 833$
5.17. (#) $A = 84.4$, $B = 178.9$, $C = 96.0$, $E = 468$, $F = 517$
5.19. (k) $A = 14.1$, $B = 11.8$, $C = 11.8$
5.21. $A = B = C = 3.40$ k
5.23. $BD = 36.5$ kN (C), $BE = 16.7$ kN (C)
5.25. (k) $AG = 4.0$ (T), $AB = 5.33$ (T), $AE = 6.67$ (C), $GE = 0$, $BC = 5.33$ (T), $BE = 0$, $EC = 6.67$ (T), $EF = 10.67$ (C), $CF = 12$ (C), $CD = 13.33$ (T), $FD = 10.67$ (C)
5.27. (k) $AB = 20$ (C), $AF = 28.3$ (T), $BF = 8$ (C), $BC = 20$ (C), $FC = 17$(C), $FG = 32$ (T), $CD = 24$ (C), $CG = 0$, $GH = 32$ (T), $DE = 24$ (C), $DH = 10$ (C), $HE = 33.9$ (T)
5.29. (k) $AB = 13.0$ (C), $AE = 0.29$ (C), $BC = 13.79$ (C), $BF = 5.47$ (T), $BE = 5.0$ (T), $EF = 0.29$ (C), $CD = 5.08$ (C), $CF = 3.37$ (T), $FD = 3.79$ (T)
5.31. (lb) $AB = 6667$ (C), $AF = 4000$ (T), $BF = 5000$ (T), $BC = 6333$ (C), $BD = 3000$ (T), $BE = 6667$ (C), $EF = 4000$ (T), $CD = 7833$ (T), $DE = 7833$ (T)

5.33. (lb) $BF = 833$ (C), $BC = 1389$ (C), $EF = 1778$ (T)
5.35. (k) $CD = 24$ (T), $BH = 8.94$ (T), $HG = 8$(T)
5.37. (k) $EH = 46$ (C), $CE = 8.49$ (T), CB $= 6$ (T), $AB = 4$ (C)
5.39. (lb) $A = 7767$, $B = 3150$, $C = 6707$, $D = 10{,}742$, $E = 10{,}262$
5.41. (kN) $A = 52.3$, $B = 32.6$, $C = 43.6$
5.43. (lb) $A = 313$, $B = 313$, $C = 280$, $D = 140$, $E = 140$
5.45. On object: 40 lb, at A: 52 lb
5.47. $P = 35.3$ lb

Chapter 6

6.1. 34.66 lb $<$ 39 lb (block will not slide)
6.3. $P = 522$ N
6.5. **(a)** $F = 20$ lb
(b) $F = 0$
(c) $F = 10$ lb
6.7. $P = 96.7$ lb
6.9. $P = 57.5$ lb (sliding)
6.11. $P = 280$ lb
6.13. $P = 325$ lb
6.15. $T_L = 12{,}200$ lb
6.17. $\beta = 132.4°$
6.19. 1.55 turns
6.21. $Q = 51.7$ lb
6.23. $C = 22.2$ kN
6.29. $P = 30$ lb
6.31. **(a)** $P = 700$ lb
(b) $P = 800$ lb
6.33. **(a)** $F = 17.1$ lb
(b) $P = 40.6$ lb
(c) $P = 0$
6.35. Body will remain at rest.
6.37. $P = 280$ lb
6.39. $P = 56$ lb, $h = 57.1$ in.
6.41. $\mu_{SF} = 0.252$
6.43. $P = 46.6$ lb
6.45. $W = 245$ lb
6.47. $W_1 = 244$ lb
6.49. $P = 380$ lb
6.51. $P = 70.8$ lb
6.53. 2.41 turns
6.55. $T_L = 633$ lb
6.57. $T_S = 26.3$ N, $T_L = 296$ N
6.59. **(a)** $W = 12{,}280$ lb
(b) $W = 9834$ lb (decrease $= 2446$ lb)

Chapter 7

7.1. $\bar{x} = 2.68$ ft
7.3. $\bar{x} = 23.9$ in.
7.5. $l = 35.1$ in.
7.7. $\bar{y} = 11.74$ in.
7.9. $\bar{y} = 4.02$ in.
7.11. **(a)** $\bar{x} = 1.94$ in., $\bar{y} = 4.66$ in.
(b) $\bar{x} = 2.80$ in., $\bar{y} = 2.30$ in.
(c) $\bar{x} = 4.92$ in., $\bar{y} = 3.81$ in.
(d) $\bar{x} = 8.55$ in., $\bar{y} = 9.44$ in.
7.13. **(a)** $\bar{x} = 7.77$ in., $\bar{y} = 3.62$ in.
(b) $\bar{x} = 5.5$ in., $\bar{y} = 6.5$ in.
7.19. $\bar{y} = 2.96$ in.
7.21. $\bar{y} = 484.6$ mm
7.23. $\bar{y} = 0.321$ m
7.25. **(a)** $\bar{x} = 18.74$ in., $\bar{y} = 10.25$ in.
(b) $\bar{y} = 1.703$ in.
7.27. **(a)** $\bar{x} = 2.33$ in., $\bar{y} = 4.33$ in.
(b) $\bar{x} = 9.34$ in., $\bar{y} = 3.27$ in.
(c) $\bar{x} = 4.60$ in., $\bar{y} = 6.65$ in.

Chapter 8

8.1. **(a)** $I_x = 13{,}623$ in.^{4}
(b) $I_x = 527$ in.^{4}
(c) $I_x = 7.19 \times 10^6$ mm^4
8.3. $I_x = 1733$ in.^{4}
8.5. **(a)** $I_{xb} = 3402$ in.^{4}
(b) $I_{xb} = 3432$ in.^{4}
8.7. $I_x = 5178 \times 10^6$ mm^4
8.9. **(a)** $I_x = 628$ in.^{4}, $I_y = 2980$ in.^{4}
(b) $I_x = 246$ in.^{4}, $I_y = 61.5$ in.^{4}
(c) $I_x = 596$ in.^{4}, $I_y = 464$ in.^{4}
(d) $I_x = 1528$ in.^{4}, $I_y = 8330$ in.^{4}
8.11. $I_x = 3370$ in.^{4}, $I_{x'} = 13{,}540$ in.^{4}
8.13. $I_x = 1.300 \times 10^9$ mm^4, $I_y = 5.39 \times 10^9$ mm^4
8.15. $s = 7.37$ in.
8.17. $r_x = 3.87$ in., $r_y = 2.09$ in.
8.19. $r_x = 0.692$ in., $r_y = 0.669$ in.
8.21. $r_x = 4.36$ in., $r_y = 4.57$ in.
8.23. $J = 4.12$ in.^{4}
8.29. $I_x = 63.4$ in.^{4}, $I_y = 30.7$ in.^{4}
8.31. $I_x = 2220$ in.^{4}
8.33. $I_x = 153.7$ in.^{4}, $I_y = 151.3$ in.^{4}
8.35. **(a)** $r_y = 88.7$ mm
(b) $r_y = 114.5$ mm

8.37. **(a)** 22.0×10^6 mm^4
(b) 643×10^6 mm^4
(c) 267×10^6 mm^4

8.39. $r_x = 4.20$ in., $r_y = 3.07$ in.

8.41. $r = 1.793$ in., $J = 49.6$ in.^{4}

Chapter 9

9.3. AB: $s = 18.75$ ksi, BC: $s = 31.1$ ksi

9.5. **(a)** 4000 psi
(b) 4180 psi

9.7. $\sigma_t = 17{,}380$ psi

9.9. $\tau_{(\text{ult})} = 373$ MPa

9.11. $P = 66{,}000$ lb

9.15. 0.01875

9.17. 0.0125 in.

9.19. **(a)** 132.4 MPa
(b) 0.00189
(c) 7.56 mm

9.21. $E = 17{,}050$ ksi

9.27. 181.3 mm

9.29. Use square footing: $5' - 0'' \times 5' - 0''$

9.31. 350 kN

9.33. 19.56 ksi

9.35. $1\frac{1}{8}$-in.-diam. rod

9.37. 2.67 ft

9.39. $\tau = 50{,}930$ psi

9.41. 0.24 in.

9.43. **(a)** 710 psi
(b) 0.000418
(c) 0.0426 in.

9.45. **(a)** 9780 psi
(b) 0.000326
(c) 0.00587 in.

9.47. 10.16×10^3 N, 517 MPa

9.49. **(a)** 2.40 in.^{2}
(b) 20,800 psi

9.51. 0.02 in. (CD in compression)

9.53. 0.0192 in. (elongation)

9.55. 0.163 in.

9.57. 1304 psi

9.59. 63.1 mm

Chapter 10

10.1. **(a)** 26,160 psi
(b) 0.000894
(c) 29,300,000 psi

10.3. Point 1: $E = 29{,}200{,}000$ psi;
Point 2: $E = 29{,}100{,}000$ psi;
Point 3: NA (stress $>$ P.L.)

10.5. 0.817 in.

10.7. 27.6 mm

10.9. 3.09 in.

10.11. **(a)** 1.67
(b) 3.94

10.13. **(a)** 66,000 lb
(b) 70,100 lb

10.17. 116.4×10^3 MPa

10.19. 9,700,000 psi

10.21. 21.4%, 26.7%

10.23. 6.01 k

10.25. 9/32 in.

10.27. **(a)** 26.3 mm
(b) 29.9 mm

10.29. **(a)** 10.53 in.^{2}
(b) 8.0 in.^{2}

10.31. **(a)** $\delta_{AB} = 0.0040$ in., $\delta_{CD} = 0.0090$ in.
(b) 0.222

Chapter 11

11.1. **(a)** 0.004
(b) 0.300

11.3. 0.250

11.5. $\delta_x = +0.259$ mm, $\delta_y = -0.0386$ mm, $\delta_z = -1.590 \times 10^{-3}$ mm

11.7. 9.72 ksi, 2.12 ksi

11.9. 0.687 in., 0.313 in.

11.11. 136,300 lb

11.13. $\sigma_{cu} = 15{,}900$ psi; $\sigma_{st} = 31{,}800$ psi

11.15. **(a)** 49.1 kips
(b) 12.28 ksi

11.17. **(a)** 88.8 MPa
(b) 91.2 MPa
(c) 109.3 MPa

11.19. 11,150 lb

11.21. **(a)** 66.7 MPa
(b) 42.9 MPa, 15.60 MPa

11.23. 1.184 ksi

11.25. $\theta = 26.6°$

11.31. Longitudinal: -0.0256 in.; lateral: -0.000266 in.

11.33. 29,500,000 psi, 0.250

11.35. x: 2.9973 in.; y: 12.0162 in.; z: 0.99935 in.

11.37. 61.6° F, 9.44 ksi

11.39. 12.96 MPa

11.41. −1.866 mm

11.43. 155.4 lb

11.45. **(a)** 107.5 F°
(b) 129.9 F°

11.47. Long wires: 183.3 lb; short wires: 4633 lb

11.49. Steel: 22,300 psi; copper: 11,150 psi

11.51. 591 k

11.53. Concrete: 90.6%; steel: 9.4%; $s_C = 0.468$ ksi, $s_{ST} = 3.89$ ksi

11.55. $a = 5$ ft, $s_{ST} = 12{,}000$ psi, $s_{AL} = 8000$ psi

11.57. **(a)** $\sigma_{BR} = 6800$ psi, $\sigma_{ST} = 17{,}010$ psi
(b) 0.136 in.

11.59. $P_B = 95.7$ kN, $P_{ST} = 42.2$ kN

11.61. 75.5 MPa

11.63. 56.7 kN

11.65. 96 kN

11.67. **(a)** $\tau = 70.6$ MPa, $\sigma_n = 122.2$ MPa
(b) $\sigma_{n(\max)} = 163.0$ MPa, $\tau_{(\max)} = 81.5$ MPa

11.69. 60.6 MPa

11.71. **(a)** 9200 psi

Chapter 12

12.1. A: 4 k-in.; B: 13 k-in.; C: 31 k-in.

12.3. 7130 psi

12.5. 245,000 lb-in.

12.7. Use a $5\frac{1}{2}$-in.-diameter shaft.

12.9. 19.8%

12.11. 1.375°

12.13. 1.824°

12.15. 2.407°

12.17. 2.87°

12.19. 1621 lb-in.

12.21. Use a 16-mm-diameter shaft.

12.23. **(a)** 5016 psi
(b) 0.862°

12.25. 4760 psi

12.31. 29.3 kN · m

12.33. 6750 psi

12.35. 42.9 MPa

12.37. 85,900 lb-in., 0.286°

12.39. Use a hollow shaft, $O.D. = 9.5$ in., $I.D. = 6.25$ in., 1.25°.

12.41. 17.35 kW

12.43. 538 hp

12.45. 57.6 hp

12.47. 20.57 mm

12.49. 1.528 in.

12.51. 23.8 hp, 2260 psi

Chapter 13

13.1. **(a)** $R_A = R_B = 72.0$ k ↑
(b) $R_A = 14.13$ k↑, $R_B = 7.88$ k↑

13.3. **(a)** $R_A = 19.2$ k↑, $R_B = 4.80$ k↑
(b) $R_A = 26.4$ k↑, $R_B = 9.64$ k↑

13.5. **(a)** $R_A = 12.00$ k, $R_B = 4.00$ k
(b) $R_A = 3.33$ k↓, $R_B = 33.3$ k↑

13.7. **(a)** 4 m: $V = +6.55$ kN, $M = +26.2$ kN · m; 7 m: $V = -5.45$ kN, $M = +21.9$ kN · m
(b) 4 m: $V = +1.11$ kN, $M = +34.4$ kN · m; 7 m (just left of load): $V = +1.11$ kN, $M = +37.8$ kN · m; 7m (just right of load): $V = -18.89$ kN, $M = +37.8$ kN · m

13.9. **(a)** $V = -8.3$ k, $M = +250$ k-ft
(b) (just left of midspan): $V = +9$ k, $M = +240$ k-ft (just right of midspan): $V = -9$ k, $M = +240$ k-ft

13.11. **(a)** 5 m: $V = +2.2$ kN, $M = +153.5$ kN · m; 10 m: $V = -22.8$ kN, $M = +102$ kN · m
(b) 5 m: $V = +12$ kN, $M = -28$ kN · m; 10 m: $V = +12$ kN, $M = +32$ kN · m

13.13. **(a)** $V_{\max} = \pm 10$ k
(b) $V_{\max} = -3.57$ k

13.15. **(a)** $V_{\max} = -58$ k
(b) $V_{\max} = -28$ k

13.17. **(a)** $V_{\max} = \pm 40.8$ k, $M_{\max} = +245$ k-ft
(b) $V_{\max} = +30.6$ k, $M_{\max} = +137.7$ k-ft

13.19. **(a)** $V_{\max} = -61.3$ k, $M_{\max} = +463$ k-ft
(b) $V_{\max} = -31.5$ k, $M_{\max} = +155.1$ k-ft

13.21. +25.3 k-ft, −45.0 k-ft

13.23. +11.25 k-ft, −24.0 k-ft

13.25. +33.2 kips, +290 k-ft

13.31. **(a)** $R_A = 140$ lb ↑, $R_B = 460$ lb ↑
(b) $R_A = 4200$ lb ↑, $R_B = 5800$ lb ↑
(c) $R_A = 7140$ lb ↑, $R_B = 11{,}860$ lb ↑
(d) $R_A = 21$ kips ↑, $R_B = 45$ kips ↑

13.33. **(a)** $R_A = 230$ kN ↑, $R_B = 30$ kN ↑
(b) $R_A = 153.4$ kN ↑, $R_B = 86.6$ kN ↑

13.35. 6 ft: $V = +30$ kips, $M = -28$ k-ft
16 ft: $V = -10$ kips, $M = +72.0$ k-ft

13.37. 2 m: $V = -36.7$ kN, $M = +6.6$ kN · m
3.5 m: $V = +40$ kN, $M = -17.6$ kN · m

13.39. $V_{\max} = -18.25$ kips, $M_{\max} = +62.0$ k-ft

13.41. $V_{\max} = -106$ kN, $M_{\max} = -384$ kN · m

13.43. $V_{\max} = -5000$ lb, $M_{\max} = -15{,}000$ lb-ft

13.45. $V_{\max} = +6.0$ kN, $M_{\max} = -9$ kN · m

13.47. $V_{\max} = +1920$ lb, $M_{\max} = +3840$ lb-ft

13.49. $V_{\max} = \pm 20$ k, $M_{\max} = +100$ k-ft

13.51. $V_{max} = +22.56$ k, $M_{max} = +107.6$ k-ft

13.53. $V_{max} = +26.36$ k, $M_{max} = +79.1$ k-ft

13.55. $V_{max} = +14.25$ k, $M_{max} = +85.5$ k-ft

13.57. $V_{max} = +31.0$ k, $M_{max} = -103.4$ k-ft

13.59. $V_{max} = +333$ kN, $M_{max} = +1042$ kN·m

13.61. **(a)** 184.2 k-ft
(b) 173.9 k-ft
(c) 20 kips

Chapter 14

14.1. **(a)** 82.7 in.^{3}
(b) 59.3×10^3 mm^3
(c) 170.9 in.^{3}

14.3. 4.36 ksi

14.5. 752 psi

14.7. 104.3 MPa

14.9. 3200 k-ft

14.11. **(a)** 3.13 MPa
(b) 2.36 MPa

14.13. **(a)** 12.72 ksi
(b) 11.57 ksi

14.15. 7550 psi

14.17. **(a)** $S_x = 25.15$ in.^{3}, $Z = 45$ in.^{3}, S.F. = 1.79
(b) $S_x = 108.7$ in.^{3}, $Z = 138$ in.^{3}, S.F. = 1.27

14.19. 55 kips

14.21. $M = 152.6$ k-ft, $V = 20.4$ k,
$M_R = 160$ k-ft, $V_R = 93.8$ k

14.23. **(a)** 115.6 plf
(b) 86.7 psf

14.25. 195.4 plf

14.29. **(a)** $S_x = 255$ in.^{3}
(b) $S_{x(\text{bott})} = 191.5$ in.^{3}, $S_{x(\text{top})} = 129.7$ in.^{3}

14.31. **(a)** 6.93 MPa
(b) 4.62 MPa

14.33. 23,500 psi

14.35. $f_{b(\text{top})} = 4660$ psi, $f_{b(\text{bott})} = 9070$ psi

14.37. 15,840 psi

14.39. 8.40 MPa

14.41. 2270 lb; web $s_s = 103.4$ psi, flange $s_s = 20.7$ psi

14.43. **(a)** 5.56 k/ft
(b) 12.2 ksi

14.45. $f_b = 21.1$ ksi, $f_v = 14.7$ ksi

14.47. **(a)** 2400 psi
(b) 600 psi
(c) 12.5 psi

14.49. 2.71 m

14.51. 14.74 kN

14.53. Max. spacing = 4.76 in.

14.55. 6845 lb

14.57. $M_Y = 370$ k-ft $(w = 3.29$ k/ft$)$
$M_P = 421$ k-ft $(w = 3.74$ k/ft$)$

Chapter 15

15.1. 20.0 ksi

15.3. 217 MPa

15.5. $E = 1{,}389{,}000$ psi

15.7. 1.510 mm

15.9. 0.449 in. < 0.60 in. (O.K.)

15.11. 0.579 in.

15.13. 13.03 ft

15.15. 0.373°

15.17. 3.91°

15.19. 3.06°

15.21. 0.240 in.

15.23. 6.74 kips

15.25. 0.1661 in. < 0.40 in. (O.K.)

15.27. $M_{max} = +88$ kN·m

15.29. $M_{max} = +7.10$ kN·m

15.31. 0.9124 in., 0.747 in.

15.33. 1,609,000 psi

15.35. 0.655 in.↓, 0.292 in. ↑

15.37. 0.825 in.

15.39. 0.45°, 1.511 in.

15.41. 0.680 in. →

15.43. $\Delta = \dfrac{23PL^3}{648EI}$

15.49. 40 in.

15.51. **(a)** 0.385 in.
(b) 0.1757 in.
(c) 0.1956 in.

15.53. 17.77 mm

15.55. Required $d = 65.1$ mm, $f_b = 25.8$ MPa,
$f_v = 1.201$ MPa

15.57. 9.56 in.

15.59. 68.3 mm

15.61. 1.206 in.

15.65. $\theta = 0.291°$, $\Delta_A = 0.859$ in., $\Delta_C = 0.283$ in.

15.67. $\Delta_{\text{midspan}} = 64.8$ mm, $\Delta_C = 46.6$ mm

15.69. **(a)** 0.339 in.
(b) 557 plf

15.71. $\Delta_C = 0.0793$ in.↓; midway between supports
$\Delta = 0.0201$ in.↓

15.73. $\Delta_A = \dfrac{5PL^3}{9EI}$

Chapter 16

16.1. W16 × 36

16.3. W14 × 30

16.5. W1000 × 222

16.7. W14 × 34

16.9. **(a)** 6 × 20 (S4S)
(b) 6 × 16 (S4S)

16.11. 2 × 12 (S4S)

16.17. W21 × 50

16.19. W530 × 101

16.21. Beams: W14 × 22; Girders: W21 × 50

16.23. W530 × 101

16.25. W410 × 100

16.27. 6 × 12 (S4S)

16.29. 2 × 10 (S4S)

16.31. 8 × 18 (S4S)

16.33. 2 × 12 (S4S)

16.35. 2 − 2 × 10 (S4S)

16.37. 200 × 410 (S4S)

Chapter 17

17.1. ±5.55 MPa, ±2.13 MPa

17.3. $\sigma_A = -8250$ psi, $\sigma_B = +14{,}250$ psi

17.5. Bottom: $\sigma_{max} = +17{,}154$ psi;
top: $\sigma_{min} = -12{,}414$ psi

17.7. $\sigma_{AA} = -520$ psi, $\sigma_{BB} = -3646$ psi

17.9. $\sigma_A = +57.6$ MPa, $\sigma_B = -60.3$ MPa

17.11. 21.7 in.

17.13. 14.59 in.

17.15. $\sigma_A = -104.2$ psi, $\sigma_B = -417$ psi, $\sigma_C = +208$ psi,
$\sigma_D = -104.2$ psi

17.17. **(a)** $\sigma_1 = +16{,}220$ psi, $\sigma_2 = -2220$ psi, $\theta_p = 69.7°$,
$\tau = \pm 9220$ psi
(b) $\sigma_1 = +2220$ psi, $\sigma_2 = -16{,}220$ psi, $\theta_p = 20.3°$,
$\tau = \pm 9220$ psi
(c) $\sigma_1 = +13{,}810$ psi, $\sigma_2 = -1810$ psi, $\theta_p = 19.90°$,
$\tau = \pm 7810$ psi

17.19. Point A: $\sigma_1 = +3900$ psi, $\sigma_2 = -1300$ psi
Point B: $\sigma_1 = +7480$ psi, $\sigma_2 = -381$ psi

17.21. $\sigma_n = +7500$ psi, $\tau = -4330$ psi

17.23. For $\theta = 75°$: $\sigma_n = -11{,}200$ psi, $\tau = +3000$ psi
For $\theta = 65°$: $\sigma_n = -9860$ psi, $\tau = +4600$ psi
For $\theta = 50°$: $\sigma_n = -7040$ psi, $\tau = +5910$ psi
For $\theta = 30°$: $\sigma_n = -3000$ psi, $\tau = +5200$ psi
$\tau_{(max)} = +6000$ psi

17.25. **(a)** $\sigma_1 = +21{,}640$ psi, $\sigma_2 = +5360$ psi, $\theta_{p1} =$ 39.69° (clockwise), $\tau_{(max)} = \pm 8140$ psi
(b) $\sigma_1 = +14{,}190$ psi, $\sigma_2 = -17{,}190$ psi, $\theta_{p_2} =$ 15.33° (clockwise), $\tau_{(max)} = \pm 15{,}690$ psi

17.27. Point A: $\sigma_1 = +3925$ psi, $\sigma_2 = -2039$ psi
Point B: $\sigma_1 = +8100$ psi, $\sigma_2 = 556$ psi

17.33. +130.5 MPa, −57.4 MPa

17.35. 133.3 MPa

17.37. 19 kips

17.39. $\sigma_A = -2698$ psf $= -18.74$ psi,
$\sigma_B = +298$ psf $= +2.07$ psi

17.41. 19,200 lb

17.43. 40,940 lb

17.45. $\sigma_A = +1.146$ ksi, $\sigma_B = -0.729$ ksi, $\sigma_C = -1.563$ ksi,
$\sigma_D = +0.313$ ksi

17.47. 13,680 lb

17.49. $\sigma_1 = +92.6$ MPa, $\sigma_2 = -207$ MPa,

17.51. $\tau_{(max)} = \pm 149.9$ MPa
Point A: $\sigma_1 = -3.26$ ksi, $\sigma_2 = 0$,
$\tau_{(max)} = \pm 1.63$ ksi on 45° plane
Point B: $\sigma_1 = +0.48$ ksi, $\sigma_2 = -3.48$ ksi,
$\theta_p = 20.5°$, $\tau_{(max)} = \pm 1.98$ ksi
Point C: $\sigma_1 = \sigma_2 = \pm 1.99$ ksi, $\theta_p = 45°$

17.53. $\sigma_1 = +0.470$ MPa, $\sigma_2 = -274.8$ MPa, $\theta_p = 87.6°$,
$\tau_{(max)} = \pm 137.6$ ksi

17.55. $\sigma_1 = +14{,}960$ psi, $\sigma_2 = -11{,}960$ psi,
$\tau_{(max)} = \pm 13{,}460$ psi, $\theta_{p_2} = 15.67°$ (clockwise),
$\theta_s = 29.3°$ (counterclockwise)

Chapter 18

18.1. 100.2 k

18.3. 55.6

18.7. **(a)** 86.0
(b) 172.1
(c) 122.9
(d) 43.0

18.9. 337 kN

18.11. 656 k

18.13. **(a)** 1300 kN < 1680 kN, O.K.
(b) 1300 kN > 842 kN, N.G.

18.15. 211 k

18.17. W14 × 94

18.19. $\frac{7}{8}$-in.-diam. rod

18.21. $2\frac{1}{2}$-in.-diam. rod

18.23. **(a)** 276 kN
(b) 237 kN

18.25. **(a)** 8 × 12 (S4S)
(b) 10 × 10 (S4S)

18.31. **(a)** 42.6 k
(b) 6.81 k
(c) 2.66 k

18.33. 12.70 ft

18.35. **(a)** 13.25 kN
(b) 3.31 kN
(c) 828 N

18.37. Euler theory not applicable

18.39. **(a)** 1255 kN
(b) 415 kN
(c) 648 kN

18.41. 1042 k

18.43. Pipe 6 x-strong

18.45. 26.6 kN

18.47. 97.8 k

18.49. 10 × 10 (S4S)

18.51. 300 × 300 (S4S)

Chapter 19

19.1. 129.6 k

19.3. 585 kN

19.5. 121.9 k

19.7. 8 bolts each side; 2 gage lines

19.9. $L = 7$ in.

19.11. 4.79 kips/in.

19.15. 87 k

19.17. 395 kN

19.19. Side welds: 126 mm

19.21. 8 bolts on two gage lines

19.23. 205 k

19.25. 43.2 k

19.27. 201 k

Chapter 20

20.1. 14,950 psi

20.3. 4690 psi

20.5. 0.39 in.

20.7. 1.380 MPa

20.9. 11.0 mm

20.11. 0.405 in.

20.15. $\sigma_{t_C} = 11{,}040$ psi, $\sigma_{t_L} = 5520$ psi

20.17. 174 MPa

20.19. 13.64 mm

20.21. F.S. = 5.28

20.23. 374.4 ft

20.25. 18,850 lb

Chapter 21

21.1. W14 × 22

21.3. **(a)** $R_A = 1.25$ kips, $V_B = 2.75$ kips, $M_B = 15$ k-ft
(b) $R_A = 2.0$ kips, $V_B = 8.0$ kips, $M_B = 33.3$ k-ft

21.5. W14 × 22

21.7. $R_A = 3.28$ kips ↓, $R_B = 16.17$ kips ↑, $R_C = 2.11$ kips ↑

21.9. 80 × 250 (S4S)

21.11. $M_1 = 0$, $M_2 = -8$ k-ft, $M_3 = -28$ k-ft, $M_4 = 0$, $R_1 = 4.5$ kips, $R_2 = 4.25$ kips, $R_3 = 13$ kips, $R_4 = 8.25$ kips

21.13. $R_A = 7.2$ kips, $V_B = 12$ kips, $M_B = 57.6$ k-ft, $f_b = 17.91$ ksi

21.15. $R_A = 7.74$ kips, $V_B = 11.46$ kips, $M_B = 44.7$ k-ft, $f_b = 13.90$ ksi

21.17. 19.74 ksi

21.19. $R_A = 15.72$ kips, $V_B = 24.28$ kips, $M_{max} = -105.04$ k-ft

21.21. $R_1 = 31.56$ kN, $R_2 = 193.75$ kN, $R_3 = 84.69$ kN, $M_2 = -142.2$ kN·m

21.23. $R_1 = 33.17$ kips, $R_2 = 1.99$ kips, $R_3 = 22.84$ kips, $M_1 = -100$ k-ft, maximum $+M = 22.86$ k-ft

INDEX

T